Corrosion Mechanisms in Theory and Practice

Third Edition

CORROSION TECHNOLOGY

Editor, Philip A. Schweitzer, P.E.,
Consultant, York, Pennsylvania

Corrosion Protection Handbook, Second Edition, Revised and Expanded, Philip A. Schweitzer
Corrosion Resistant Coatings Technology, Ichiro Suzuki
Corrosion Resistance of Elastomers, Philip A. Schweitzer
Corrosion Resistance Tables: Metals, Nonmetals, Coatings, Mortars, Plastics, Elastomers and Linings, and Fabrics, Third Edition, Revised and Expanded (Parts A and B), Philip A. Schweitzer
Corrosion-Resistant Piping Systems, Philip A. Schweitzer
Corrosion Resistance of Zinc and Zinc Alloys: Fundamentals and Applications, Frank Porter
Corrosion of Ceramics, Ronald A. McCauley
Corrosion Mechanisms in Theory and Practice, P. Marcus and J. Oudar
Corrosion Resistance of Stainless Steels, C. P. Dillon
Corrosion Resistance Tables: Metals, Nonmetals, Coatings, Mortars, Plastics, Elastomers and Linings, and Fabrics, Fourth Edition, Revised and Expanded (Parts A, B, and C), Philip A. Schweitzer
Corrosion Engineering Handbook, Philip A. Schweitzer
Atmospheric Degradation and Corrosion Control, Philip A. Schweitzer
Mechanical and Corrosion-Resistant Properties of Plastics and Elastomers, Philip A. Schweitzer
Environmental Degradation of Metals, U. K. Chatterjee, S. K. Bose, and S. K. Roy
Environmental Effects on Engineered Materials, Russell H. Jones
Corrosion-Resistant Linings and Coatings, Philip A. Schweitzer
Corrosion Mechanisms in Theory and Practice, Second Edition, Revised and Expanded, Philippe Marcus
Electrochemical Techniques in Corrosion Science and Engineering, Robert G. Kelly, John R. Scully, David W. Shoesmith, and Rudolph G. Buchheit
Metallic Materials: Physical, Mechanical, and Corrosion Properties, Philip A. Schweitzer
Corrosion Resistance Tables: Metals, Nonmetals, Coatings, Mortars, Plastics, Elastomers and Linings, and Fabrics, Fifth Edition, Philip A. Schweitzer
Corrosion of Ceramic and Composite Materials, Second Edition, Ronald A. McCauley
Analytical Methods in Corrosion Science and Engineering, Philippe Marcus and Florian Mansfeld
Paint and Coatings: Applications and Corrosion Resistance, Philip A. Schweitzer
Corrosion Control Through Organic Coatings, Amy Forsgren
Corrosion Engineering Handbook, Second Edition - 3 Volume Set, Philip A. Schweitzer
Fundamentals of Corrosion: Mechanisms, Causes, and Preventative Methods, Philip A. Schweitzer
Corrosion Mechanisms in Theory and Practice, Third Edition, Philippe Marcus

Corrosion Mechanisms in Theory and Practice

Third Edition

Edited by Philippe Marcus

CRC Press is an imprint of the
Taylor & Francis Group, an **informa** business

CRC Press
Taylor & Francis Group
6000 Broken Sound Parkway NW, Suite 300
Boca Raton, FL 33487-2742

First issued in paperback 2017

© 2012 by Taylor & Francis Group, LLC
CRC Press is an imprint of Taylor & Francis Group, an Informa business

No claim to original U.S. Government works

Version Date: 20110526

ISBN 13: 978-1-4200-9462-6 (hbk)
ISBN 13: 978-1-138-07363-0 (pbk)

This book contains information obtained from authentic and highly regarded sources. Reasonable efforts have been made to publish reliable data and information, but the author and publisher cannot assume responsibility for the validity of all materials or the consequences of their use. The authors and publishers have attempted to trace the copyright holders of all material reproduced in this publication and apologize to copyright holders if permission to publish in this form has not been obtained. If any copyright material has not been acknowledged please write and let us know so we may rectify in any future reprint.

Except as permitted under U.S. Copyright Law, no part of this book may be reprinted, reproduced, transmitted, or utilized in any form by any electronic, mechanical, or other means, now known or hereafter invented, including photocopying, microfilming, and recording, or in any information storage or retrieval system, without written permission from the publishers.

For permission to photocopy or use material electronically from this work, please access www.copyright.com (http://www.copyright.com/) or contact the Copyright Clearance Center, Inc. (CCC), 222 Rosewood Drive, Danvers, MA 01923, 978-750-8400. CCC is a not-for-profit organization that provides licenses and registration for a variety of users. For organizations that have been granted a photocopy license by the CCC, a separate system of payment has been arranged.

Trademark Notice: Product or corporate names may be trademarks or registered trademarks, and are used only for identification and explanation without intent to infringe.

Library of Congress Cataloging-in-Publication Data

Corrosion mechanisms in theory and practice. -- 3rd ed. / edited by Philippe Marcus.
p. cm. -- (Corrosion technology ; 26)
Summary: "Building on previous editions described as "a useful contribution to the current literature on corrosion science, engineering, and technology" by Corrosion Review, this third installment offers real-world applications and problem-solving techniques to reduce the occurrence of pits, cracks, and deterioration in industrial, automotive, marine, and electronic structures. It also details the electrochemical and chemical properties, adsorption mechanisms, environmental parameters, and microstructural factors that influence corrosion formation. Including contributions from authors around the world, this reorganized third edition covers the most recent advances in the field, adding three new chapters on Fundamentals of Corrosion from a Surface Science and Electrochemical Basis, High-Temperature Oxidation, and Atomistic Modeling of Corrosion"-- Provided by publisher.
Includes bibliographical references and index.
ISBN 978-1-4200-9462-6 (hardback)
1. Corrosion and anti-corrosives. I. Marcus, P. (Philippe), 1953- II. Title. III. Series.

TA462.C65575 2011
620.1'1223--dc22 2010053752

Visit the Taylor & Francis Web site at
http://www.taylorandfrancis.com

and the CRC Press Web site at
http://www.crcpress.com

Contents

Preface vii
Editor ix
Contributors xi

1. **Fundamentals of Corrosion** 1
   Hans-Henning Strehblow and Philippe Marcus

2. **Surface Effects on Hydrogen Entry into Metals** 105
   Elie Protopopoff and Philippe Marcus

3. **Anodic Dissolution** 149
   Michel Keddam

4. **Thin Oxide Film Formation on Metals** 217
   Francis P. Fehlner and Michael J. Graham

5. **Passivity of Metals** 235
   Hans-Henning Strehblow, Vincent Maurice, and Philippe Marcus

6. **Passivity of Austenitic Stainless Steels** 327
   Clive R. Clayton and Ingemar Olefjord

7. **Mechanisms of Pitting Corrosion** 349
   Hans-Henning Strehblow and Philippe Marcus

8. **Sulfur-Assisted Corrosion Mechanisms and the Role of Alloyed Elements** 395
   Philippe Marcus

9. **Further Insights on the Pitting Corrosion of Stainless Steels** 419
   Bernard Baroux

10. **Crevice Corrosion of Metallic Materials** 449
    Pierre Combrade

11. **Stress-Corrosion Cracking Mechanisms** 499
    Roger C. Newman

12. **Corrosion Fatigue Mechanisms in Metallic Materials** 545
    T. Magnin

13. **High-Temperature Corrosion** 573
    Michael Schütze

14. Corrosion Prevention by Adsorbed Organic Monolayers and Ultrathin Plasma Polymer Films 617
Michael Rohwerder, Martin Stratmann, and Guido Grundmeier

15. Atmospheric Corrosion 669
Christofer Leygraf

16. Corrosion of Aluminum Alloys 705
Nick Birbilis, T.H. Muster, and Rudolph G. Buchheit

17. Microbially Influenced Corrosion 737
Dominique Thierry and Wolfgang Sand

18. Corrosion in Nuclear Systems: Environmentally Assisted Cracking in Light Water Reactors 777
F.P. Ford and P.L. Andresen

19. Corrosion of Microelectronic and Magnetic Data-Storage Devices 825
Gerald S. Frankel and Jeffrey W. Braithwaite

20. Organic Coatings 863
J.H.W. de Wit, D.H. van der Weijde, and G. Ferrari

Index 907

Preface

Corrosion is a major issue for the increase of service life and reliability of metallic materials. A detailed understanding of the mechanisms of corrosion is necessary to solve existing corrosion problems and to prevent future problems.

The aim of this book is to review recent advances in the understanding of corrosion and protection mechanisms. A detailed view is provided of the chemical and electrochemical surface reactions that govern corrosion, and of the link between microscopic forces and macroscopic behavior.

In order to cover fundamental as well as practical aspects, the third edition, which has been revised and expanded (four new chapters have been added), contains 20 chapters. Chapters 1 through 13 cover the basic aspects of corrosion: fundamentals of corrosion, entry of hydrogen into metals, anodic dissolution, passivity of metals and alloys, localized corrosion (pitting and crevice corrosion), stress corrosion cracking and corrosion fatigue, and high-temperature corrosion. Chapters 14 through 20 provide the connection between the theoretical aspects of corrosion mechanisms and practical applications in industry: corrosion inhibition, atmospheric corrosion of aluminium alloys, microbially induced corrosion, corrosion in nuclear systems, corrosion of microelectronic and magnetic data-storage devices, and organic coatings.

This book is based on the results of intensive worldwide research efforts in materials science, surface science, and corrosion science over the past few years. The contributors, from leading academic and industrial research institutes, are highly recognized in these disciplines.

Philippe Marcus

Editor

Professor Philippe Marcus is director of research at CNRS (Centre National de la Recherche Scientifique) and director of the Laboratory of Physical Chemistry of Surfaces at Ecole Nationale Supérieure de Chimie de Paris, Chimie ParisTech, France. Dr. Marcus received his PhD (1979) in physical sciences from the University Pierre and Marie Curie, Paris, France.

His field of research is surface chemistry, surface electrochemistry, and corrosion science, with emphasis on the understanding of the structure and properties of metal and alloy surfaces.

His research interests include the growth mechanisms and structure of oxide layers on metals and alloys in gaseous and aqueous environments; adsorption of inorganic, organic, and biomolecules; the mechanisms of corrosion of metals and alloys at the nanoscale; passivity, passivity breakdown, and localized corrosion; high-temperature oxidation; and the applications of advanced surface analytical methods such as x-ray photoelectron spectroscopy, scanning tunneling microscopy, and time-of-flight secondary ion mass spectrometry.

The author or coauthor of over 300 papers in scientific journals, books, and conference proceedings in the areas of corrosion science, surface chemistry and electrochemistry, surface analysis, and materials science, and of two books, *Corrosion Mechanisms in Theory and Practice* (first and second editions) and *Analytical Methods in Corrosion Science and Engineering*, he served or serves on the editorial boards of five major journals in the field of electrochemistry and corrosion: *Electrochimica Acta; Corrosion Science; Materials and Corrosion; Corrosion Engineering, Science and Technology;* and *Corrosion Reviews.*

Professor Marcus has received a number of awards and honors, including the J. Garnier Award (1980) and Rist Award (1986) of the French Metallurgical Society, the 2005 Uhlig Award from the Electrochemical Society, the 2008 Whitney Award from NACE International, the Cavallaro Medal of the European Federation of Corrosion in 2008, and the U.R. Evans Award of the Institute of Corrosion in 2010. He is an elected fellow of the Electrochemical Society (2005) and of the International Society of Electrochemistry (2009). He was the chair of the 2006 Gordon Research Conference on Aqueous Corrosion.

Professor Marcus is currently president of the European Federation of Corrosion, chairman of the EFC Working Party on Surface Science and Mechanisms of Corrosion and Protection, chairman of the International Steering Committee for the European Conferences on Applications of Surface and Interface Analysis, and chairman of the Scientific and Technical Committee of CEFRACOR (Centre Français de l'Anticorrosion). He is the former chairman of the Corrosion, Electrodeposition and Surface Treatment, and of the Electrochemical Materials Science divisions of the International Society of Electrochemistry.

Contributors

P.L. Andresen
General Electric Global Research Center
Niskayuna, New York, U.S.

Bernard Baroux
Grenoble Institute of Technology
Grenoble, France

Nick Birbilis
Department of Materials Engineering
Monash University
Clayton, Victoria, Australia

Jeffrey W. Braithwaite
Sandia National Laboratories
Albuquerque, New Mexico, U.S.

Rudolph G. Buchheit
Department of Materials Science and Engineering
The Ohio State University
Columbus, Ohio, U.S.

Clive R. Clayton
State University of New York
Stony Brook, New York, U.S.

Pierre Combrade
Technical Center
Framatome
Le Creusot, France

Francis P. Fehlner
Corning Incorporated
Corning, New York, U.S.

G. Ferrari
TNO Industries
Den Helder, the Netherlands

F.P. Ford
General Electric Global Research Center
Niskayuna, New York, U.S.

Gerald S. Frankel
The Ohio State University
Columbus, Ohio, U.S.

Michael J. Graham
National Research Council of Canada
Ottawa, Ontario, Canada

Guido Grundmeier
Department for Technical and Macromolecular Chemistry
University of Paderborn
Paderborn, Germany

Michel Keddam
Pierre and Marie Curie University
Paris, France

Christofer Leygraf
Royal Institute of Technology
Stockholm, Sweden

T. Magnin
Ecole Nationale Supérieure des Mines de Saint-Etienne
Saint Etienne, France

Philippe Marcus
Laboratory of Physical Chemical Study of Surfaces
Ecole Nationale Supérieure de Chimie de Paris
and
Centre National de la Recherche Scientifique
Paris, France

Vincent Maurice
Laboratory of Physical Chemical Study of Surfaces
Ecole Nationale Supérieure de Chimie de Paris
and
Centre National de la Recherche Scientifique
Paris, France

T.H. Muster
CSIRO Materials Science and Engineering
Clayton, Victoria, Australia

Roger C. Newman
Department of Chemical Engineering and Applied Chemistry
University of Toronto
Toronto, Ontario, Canada

Ingemar Olefjord
Chalmers University of Technology
Göteborg, Sweden

Elie Protopopoff
Ecole Nationale Supérieure de Chimie de Paris
and
Centre National de la Recherche Scientifique
Paris, France

Michael Rohwerder
Max Planck Institute for Iron Research GmbH
Düsseldorf, Germany

Wolfgang Sand
Biofilm Centre
University of Duisburg-Essen
Essen, Germany

Michael Schütze
Karl Winnacker Institute of DECHEMA
Frankfurt, Germany

Martin Stratmann
Max Planck Institute for Iron Research
Düsseldorf, Germany

Hans-Henning Strehblow
Laboratory of Physical Chemical Study of Surfaces
Ecole Nationale Supérieure de Chimie de Paris
and
Centre National de la Recherche Scientifique
Paris, France

and

Institute for Physical Chemistry and Electrochemistry
Heinrich Heine University Düsseldorf
Düsseldorf, Germany

Dominique Thierry
French Corrosion Institute
Brest, France

D.H. van der Weijde
Corus Research and Development
Ijmuiden, the Netherlands

J.H.W. de Wit
Netherlands Institute for Metals Science
Delft University of Technology
Delft, the Netherlands

and

Corus Research and Development
Ijmuiden, the Netherlands

1

Fundamentals of Corrosion

Hans-Henning Strehblow
Ecole Nationale Supérieure de Chimie de Paris,
Centre National de la Recherche Scientifique
and
Heinrich Heine University Düsseldorf

Philippe Marcus
Ecole Nationale Supérieure de Chimie de Paris
and
Centre National de la Recherche Scientifique

CONTENTS

1.1 Introduction 2
1.2 Electrolytes and Solvents 3
1.3 Conductivity and Transfer Coefficient 5
1.4 The Structure of Electrolytes 6
  1.4.1 XAS and Extended X-Ray Absorption Fine Structure (EXAFS) for Determination of the Short-Range Order 9
  1.4.2 Short-Range Order of Dissolved Species 11
1.5 Debye–Hückel–Onsager Theory of Diluted Strong Electrolytes 13
  1.5.1 Basic Discussion 13
  1.5.2 Equivalent Conductivity of Strongly Dissociated Electrolytes 15
  1.5.3 Equivalent Conductivity of Weakly Dissociated Electrolytes 18
  1.5.4 Activity Coefficient 19
1.6 Some Basic Definitions in Electrochemistry 21
1.7 The Electrochemical Double Layer 23
1.8 Thermodynamics of Chemical Equilibria 30
  1.8.1 Energy U and Enthalpy H 30
  1.8.2 Entropy S, Helmholtz Free Energy F, Gibbs Free Energy or Free Enthalpy G 33
  1.8.3 Electrochemical Equilibrium and Nernst Equation 36
  1.8.4 Potential–pH Diagrams 37
  1.8.5 Reference Electrodes 41
1.9 Liquid Junction Potentials 42
1.10 Electrode Kinetics 43
  1.10.1 Charge Transfer Overvoltage η_{CT} and Butler–Volmer Equation 44
  1.10.2 Diffusion Overvoltage η_D 49
  1.10.3 Ohmic Drop η_Ω and Microelectrodes 52

1.11 Electrochemical Methods 55
1.11.1 Electrochemical Cells 55
1.11.2 Potentiostatic Measurements 57
1.11.3 Galvanostatic Measurements 58
1.11.4 Rotating Disc Electrodes (RDEs) and Rotating Ring Disc Electrodes (RRDEs) 59
1.11.5 Electrochemical Transients 61
1.11.6 Impedance Spectroscopy 62
1.12 Reduction of Redox Systems 66
1.12.1 Hydrogen Evolution 67
1.12.2 Oxygen Reduction 70
1.13 Metal Dissolution 71
1.13.1 Basics of Metal Dissolution 71
1.13.2 High Rates of Metal Dissolution and Salt Precipitation 75
1.13.3 Selective Dissolution of Alloys 76
1.13.4 Metal Dissolution and Complex Formation 79
1.14 Metal Dissolution in Combination with Reduction Reactions 80
1.15 Heterogeneous Metal Surfaces and Local Elements 83
1.16 Protection 85
1.16.1 Cathodic Protection 85
1.16.2 Anodic Protection 87
1.16.3 Stray Currents 87
1.17 Inhibition of Corrosion 87
1.17.1 Inhibition by Adsorption 88
1.17.2 Inhibition by Precipitation of Compounds 91
1.17.3 Inhibition by Passivation 92
1.18 Semiconductor Electrochemistry and Photoelectrochemistry 93
1.18.1 Some Properties of Semiconductors 94
1.18.2 Electron Transfer at Semiconductor Electrodes 95
1.18.3 Photoelectrochemistry 99
1.19 Conclusion 102
1.20 Acknowledgments 102
References 103

1.1 Introduction

Corrosion is the chemical degradation of materials, such as metals, semiconductors, insulators, and even polymers, due to the exposure to environment. The environment may be a gas phase with or without moisture and an aqueous or nonaqueous electrolyte. The corrosion processes occur at the surface and may involve the bulk of the materials and the contacting gas or liquid phase. Traditionally, the community of engineers and scientists distinguishes between corrosion in electrolytes and corrosion in the gas phase at elevated temperatures, i.e., the wet and the hot and dry corrosions. Although the environment and the observed effects are different, both fields often use in basic research similar analytical methods to learn about the leading mechanisms. This book concentrates mainly on electrochemical corrosion phenomena and on atmospheric corrosion in a wet environment.

Corrosion research is a highly interdisciplinary field. It involves surface science, electrochemistry, physics, materials science, metallurgy, engineering, and theoretical calculations.

A broad range of in situ and ex situ methods is applied to study the reaction mechanisms. They will be mentioned in this chapter when discussing the various phenomena and aspects of corrosion. The complexity of the phenomena requires the application of many methods to get a deeper insight. Therefore groups with an expertise in different fields have to work together—theoreticians and experimentalists with specialization in electrochemistry, electrode kinetics, surface science, surface analytical methods like x-ray photoelectron spectroscopy (XPS), Auger electron spectroscopy (AES), and various methods applying synchrotron radiation in the infrared (IR), visible, and x-ray range. This chapter intends to give an introduction to electrochemistry and electrode kinetics with the main reactions involved in corrosion, i.e., anodic metal dissolution and the cathodic reactions, hydrogen evolution and oxygen reduction, the structure of electrolytes and electrode surfaces, surface analytical methods, semiconductor electrochemistry, passivity of metals, and the related breakdown of passivity and localized corrosion.

The corrosion process is determined decisively by the chemical and physical conditions of the interface. The composition and structure of the bulk solid phase are also very important for the corrosion properties of a material. There is seldom a pure single-phase metal, but most often different alloying elements are present with their chemical and electrochemical properties, bi- or multiphase solids, composites or metals with inclusions, and grain boundaries with different composition compared to the grains themselves. This is a consequence of segregation of components at grain boundaries and the material surface with related changes of the chemical and physical properties. In addition, the solids contain defects such as dislocations, voids, stressed areas of the material, etc., each of these factors having influence on the electrochemical properties and possibly leading to localized corrosion phenomena.

The composition of the electrolyte and its changes during the reactions are very important for the corrosion phenomena. The local pH and concentration of corrosion products are very important factors in this sense. The contamination of electrolytes by corrosion products may be another serious problem. All the above-mentioned details demonstrate that it is important to study the basics of electrochemical equilibria and electrode kinetics. One has to apply in situ and ex situ surface analytical methods to understand the chemistry and physics of corrosion phenomena and the problems they induce in engineering, industry processes, and the everyday degradation of materials. Estimates show that corrosion causes a damage of 3.5% of the national gross product in industrialized countries, a large fraction of which might be avoided if the knowledge and the results of corrosion science and engineering were respected.

This chapter on fundamentals of corrosion tries to give an introduction to the basics of physical chemistry, electrochemistry, thermodynamics, surface science, and surface analysis. It addresses to graduate students and informed scientists and tries to call back in mind the basis to understand the physics and chemistry of corrosion phenomena. All these topics may not be described in detail in this chapter, and reference is given to some books for further reading: general physical chemistry [1], electrochemistry and electrode kinetics [2–4], surface analytical methods [5,6], and corrosion [7,8].

1.2 Electrolytes and Solvents

An electrolyte consists in general of a solvent with dissolved species, the solutes, which may dissociate into positively charged cations and negatively charged anions. The most usual solvent in corrosion is water. It dissociates weakly according to the reaction of Equation 1.1.

The dissociation constant is equal to the ionic product $K = a(H_3O^+)\, a(OH^-) = 1.008 \times 10^{-14}$ at 25°C. Therefore, the specific conductivity of pure water is only $\kappa = 6 \times 10^{-8}\ \Omega^{-1}\cdot cm^{-1}$. Due to the large dipole moment of its molecule, water is a good solvent for most solutes forming ions. The conductivity of water is increased by solutes like acids, bases, and dissolved salts. The solutes of Equations 1.2, 1.3, 1.4a, and 1.5 are completely dissociated. Several acids like sulfuric acid show a strong dissociation for the first hydrogen ion and a much weaker for the second one (Equations 1.4a and b). Therefore, the dissolution of sulfuric acid leads to a smaller concentration of hydrogen ions than expected from the assumption of a complete dissociation for the second step. Equations 1.2 and 1.6 represent the dissolution of gases in water, which leads to electrolytes with cations and anions forming acidic and alkaline solutions, respectively.

$$2H_2O \leftrightarrow H_3O^+ + OH^- \quad (1.1)$$

$$HCl + H_2O \leftrightarrow H_3O^+ + Cl^- \quad (1.2)$$

$$HClO_4 + H_2O \leftrightarrow H_3O^+ + ClO_4^- \quad (1.3)$$

$$H_2SO_4 + H_2O \leftrightarrow H_3O^+ + HSO_4^- \quad (1.4a)$$

$$HSO_4^- + H_2O \leftrightarrow H_3O^+ + SO_4^{2-} \quad (1.4b)$$

$$NaOH \leftrightarrow Na^+ + OH^- \quad (1.5)$$

$$NH_3 + H_2O \leftrightarrow NH_4^+ + OH^- \quad (1.6)$$

According to Brönsted, an acid is a proton donor and a base a proton acceptor. Thus, $HClO_4$, HCl, H_2SO_4, and HSO_4^- are acids and NaOH and NH_3 are bases. Water acts as proton acceptor in Equations 1.2 through 1.4a and b and as proton donor in Equation 1.6 (and is both acceptor and donor in Equation 1.1). Salts form in most cases (NaCl, $FeCl_2$, $CuSO_4$, etc.) strong electrolytes with almost 100% dissociation. Many acids and bases are strongly dissociated, as NaOH, $HClO_4$, HCl, and H_2SO_4 (for the first step, Equation 1.4a), others less or only weakly dissociated as HSO_4^- (Equation 1.4b) or acetic acid (Equation 1.7), which have dissociation constants of $K = 1.2 \times 10^{-2}$ and 1.8×10^{-5}, respectively [9].

$$CH_3COOH + H_2O \leftrightarrow H_3O^+ + CH_3COO^- \quad (1.7)$$

The acidity or alkalinity of a solution is characterized by the pH value defined by Equation 1.8.

$$pH = -\log a(H_3O^+) \quad (1.8)$$

where $a(H_3O^+)$ is the activity of the hydrogen ions where the activity is related to the concentration c by the relation $a = \gamma c$, where γ is called the activity coefficient. γ takes into account the ionic forces between the dissolved particles that are described by the Debye–Hückel theory discussed in Section 1.5.

There are many nonaqueous solvents that are involved in corrosion problems too. Methanol or ethanol and pure liquid acetic acid show a weak self-dissociation like water as described by Equations 1.9 and 1.10. As fuels or solvents for chemical processes, they may cause corrosion of container materials. Similarly, liquid HF is an important solvent for many biological macromolecules, such as proteins, enzymes, vitamin B12, etc., which may be recovered without any damage to their composition and properties. Some liquefied gases like liquid NO_2 or molten $HgBr_2$ are solvents with weak self-dissociation (Equations 1.11 and 1.12).

$$2CH_3OH \leftrightarrow CH_3OH_2^+ + OH^- \quad (1.9)$$

$$2CH_3COOH \leftrightarrow CH_3COOH_2^+ + CH_3COO^- \quad (1.10)$$

$$2NO_2 \leftrightarrow NO^+ + NO_3^- \quad (1.11)$$

$$2Hg_2Br_2 \leftrightarrow HgBr^+ + HgBr_3^- \quad (1.12)$$

Another technologically important electrolyte is molten cryolite ($NaAlF_6^{3-}$), which is used for the electrochemical production of aluminum from its oxide (bauxite, Al_2O_3). It forms by dissociation Na^+, ALF_6^{3-}, and F^- ions. NaCl, which melts at $T < 800°C$, is responsible for high-temperature corrosion, which occurs often for metals in contact with an electrolyte of molten salts. Mixtures of salts often have a low melting point. A mixture of 60% $AlCl_3$, 14% KCl, and 26% NaCl melts at 94°C. $N(CH_3)_4SCN$ (tetramethylammoniumthiocyanate) has a melting point of −50°C only. There are many organic solvents that may be used for materials that do not allow water like alkali metals, as for example Li in batteries, or solutes that are weakly soluble in water. Frequently used solvents are acetonitrile (CH_3CN), dimethylsulfoxide, dimethylformamide, ethanol, dioxane, ethylene, and propylene carbonate. These solvents have a much smaller dielectric constant than water and have a high solubility for organic solutes, but not for salts that require dissociation for the dissolution process. However, there exist even for these organic solvents sufficiently soluble salts like alkali perchlorates, $LiAlCl_4$, $LiAlH_4$, and $(NR_4)X$ with organic groups R and halides X. They form ions by dissociation and thus lead to a sufficient conductivity even in these organic solvents.

Another class of electrolytes is solids with high conductivity at elevated temperatures. Only two of them will be mentioned here: $ZrO_2 \cdot Y_2O_3$ and $Na_2O \cdot 11Al_2O_3(\beta)$. β-Alumina reaches the conductivity of sulfuric acid at a relatively low temperature of ca. 500°C.

1.3 Conductivity and Transfer Coefficient

The conductivity of an electrolyte is determined by the concentration c and charge z of the dissolved anions and cations and their mobility. The specific conductivity κ is the conductivity of an electrolyte volume of unit length l and unit surface area A ($l = 1$ cm,

$A = 1\,\text{cm}^2$). The equivalent conductivity Λ equals κ normalized by the concentration and charge of the ions, i.e., their equivalent concentration cz ($\Lambda = \kappa/cz$). The equivalent conductivity $\Lambda_i = \kappa_i/c_iz_i$ of an ion i (Equation 1.14) with its charge z_i and its concentration c_i is related to its mobility u_i by Equation 1.13 with Faraday's constant $F = 96{,}487\ \text{C}\cdot\text{mol}^{-1}$. The total specific conductivity κ of the electrolyte is given by the sum of the contributions of the various ions as given by Equation 1.15. Similarly the total equivalent conductivity is given by Equation 1.16. Although the equivalent conductivity of an ion is a characteristic and specific value, it still depends on the concentration and charge of all anions and cations within the electrolyte due to their electrostatic interaction. The Debye–Hückel theory takes into account these Coulomb forces and it allows calculating Λ_i as given in Section 1.5.1.

$$\Lambda_i = u_i F \tag{1.13}$$

$$\kappa_i = \Lambda_i c_i z_i \tag{1.14}$$

$$\kappa = \sum \kappa_i = \sum_i \Lambda_i c_i z_i = F \sum_i u_i c_i z_i \tag{1.15}$$

$$\Lambda = \sum_i \Lambda_i = F \sum_i u_i \tag{1.16}$$

Another useful quantity is the transfer coefficient t_i given by Equation 1.17 for simple electrolytes with the same value for c_iz_i for cations and anions.

$$t_i = \frac{\kappa_i}{\kappa} = \frac{\Lambda_i}{\Lambda} = \frac{u_i}{\sum_i u_i} \tag{1.17}$$

It describes the amount of charge transported by an ion i due to its mobility. These numbers may be obtained from transfer experiments by measuring the change of the electrolyte composition in the vicinity of a cathode and an anode during an electrolysis experiment. It allows calculating the individual equivalent conductivities Λ_i of the ions from the measured specific conductivity κ of the electrolyte. Transfer coefficients are also important in solid electrolytes, i.e., oxide films on electrodes. Applying marker experiments with isotopes or using implanted markers of noble gas ions into a film allows the determination of the contribution of cations and O^{2-} anions for oxide growth, which is discussed in Chapter 5.

1.4 The Structure of Electrolytes

An electrolyte is a medium that is between the completely disordered gas and the well-ordered crystalline solid. Solutions have no long-range order, but a short-range order exists. This has been investigated by x-ray diffraction of water and of molten metals or by x-ray absorption spectroscopy (XAS). With decreasing temperature, one has an increasing

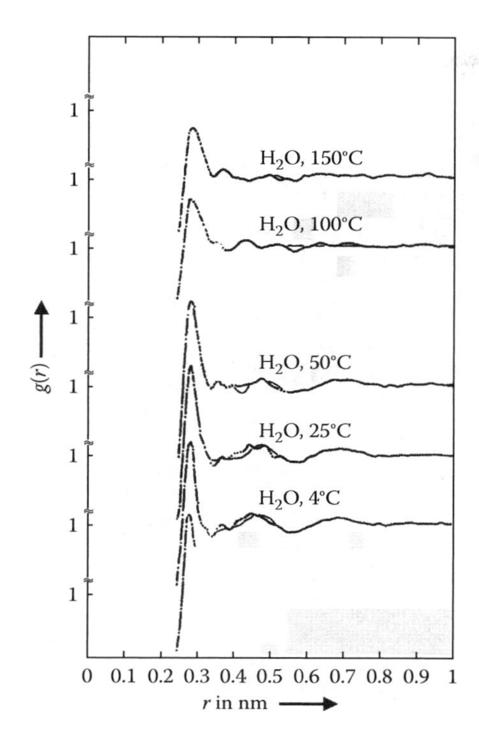

FIGURE 1.1
Distribution function $g(r)$ of water at different temperatures. (From Narten, A.H., *Disc. Faraday Soc.*, 43, 97, 1967.)

order and diffuse x-ray diffraction lines become visible, which allows calculating pair distribution functions $g(r)$ as shown in Figure 1.1 for water at different temperatures [10]. A first coordination shell appears in this presentation at a distance $r = 0.3$ nm of the central water molecule. The data yield a calculated coordination number close to 4. This result corresponds to a tetrahedral coordination of a central water molecule by four others similar to the structure of ice given in Figure 1.2a and b. In this tridymite structure, water molecules are held together by hydrogen bonds with a distance of 0.276 nm. Large voids of this structure cause a low density and a high molar volume. Figure 1.2c shows a cluster model for liquid water with this existing partial order. Figure 1.1 suggests two further weakly developed shells at $r = 0.45$ and 0.7 nm. They get more diffuse with increasing temperature and disappear above 50°C. The smaller peaks of $g(r)$ correspond to less order with increasing distance from the central water molecule. Only a short-range order is realized, which is destroyed with increasing temperature due to the thermal motion of the molecules. As a consequence, one observes a density maximum at 4°C, and a relatively high molar heat C_V, which contains a rest melting of the clusters. Above 4°C when most of the clusters have disappeared, the decrease of density due to a larger motion of the molecules is dominating as for most solids and liquids. The high mobility of H_3O^+ and OH^- ions is also a consequence of the short-range order of water and the clusters held together by hydrogen bonds. The conduction by H_3O^+ and OH^- ions corresponds to a switching of hydrogen bonds of neighboring water molecules as depicted in Figure 1.3. No charged and hydrated ions have to move, which would be much slower than the switching of hydrogen bonds with the same result. Similarly to water, dissolved species attract water molecules forming a short-range order with coordination shells. This may be studied by XAS as described in the next section.

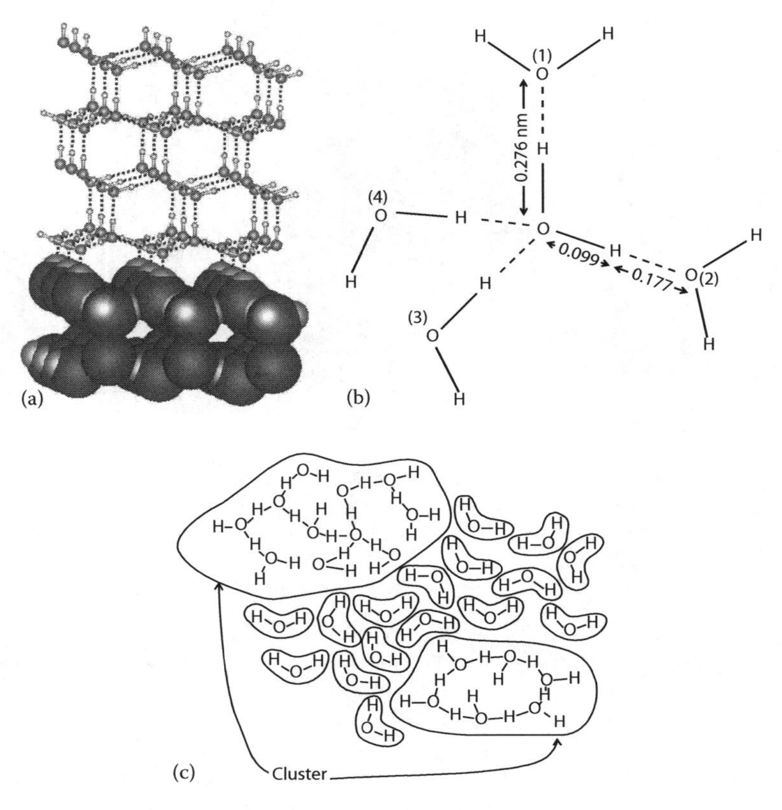

FIGURE 1.2
(a) Ice structure, (b) tetrahedral coordination of water molecules with hydrogen bonds, and (c) short-range order of liquid water.

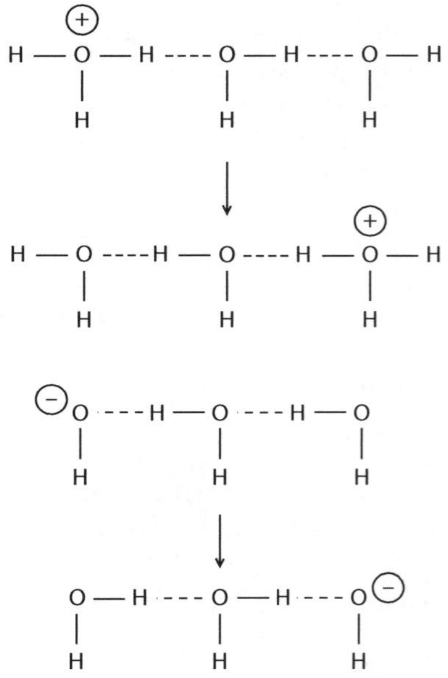

H — O(+) — H ---- O — H ---- O — H (each O bonded to H below)

↓

H — O ---- H — O ---- H — O(+) — H (each O bonded to H below)

(−)O --- H — O --- H — O (each O bonded to H below)

↓

O — H --- O — H --- O(−) (each O bonded to H below)

FIGURE 1.3
High conductivity of H_3O^+ and OH^- ions via rearrangement of hydrogen bonds using the short-range order of water.

1.4.1 XAS and Extended X-Ray Absorption Fine Structure (EXAFS) for Determination of the Short-Range Order

The absorption spectrum of x-rays shows characteristic absorption edges. They are observed when the energy of the incoming photons reaches the value that is necessary to excite electrons from an orbital of an element of the absorber to an empty level or to the continuum corresponding to the ionization of the atoms. Usually several absorption edges are observed with increasing energy corresponding to the different occupied electronic levels of the element. A spectrum taken with an intense and stable x-ray source has characteristic oscillations of the absorption coefficient after the edge, i.e., at higher energy. They contain valuable data on the short-range order in the vicinity of the absorbing atoms of the element under study as shown in Figure 1.4a for nickel metal [11]. All kinds of materials, even liquids, show more or less pronounced oscillations corresponding to a

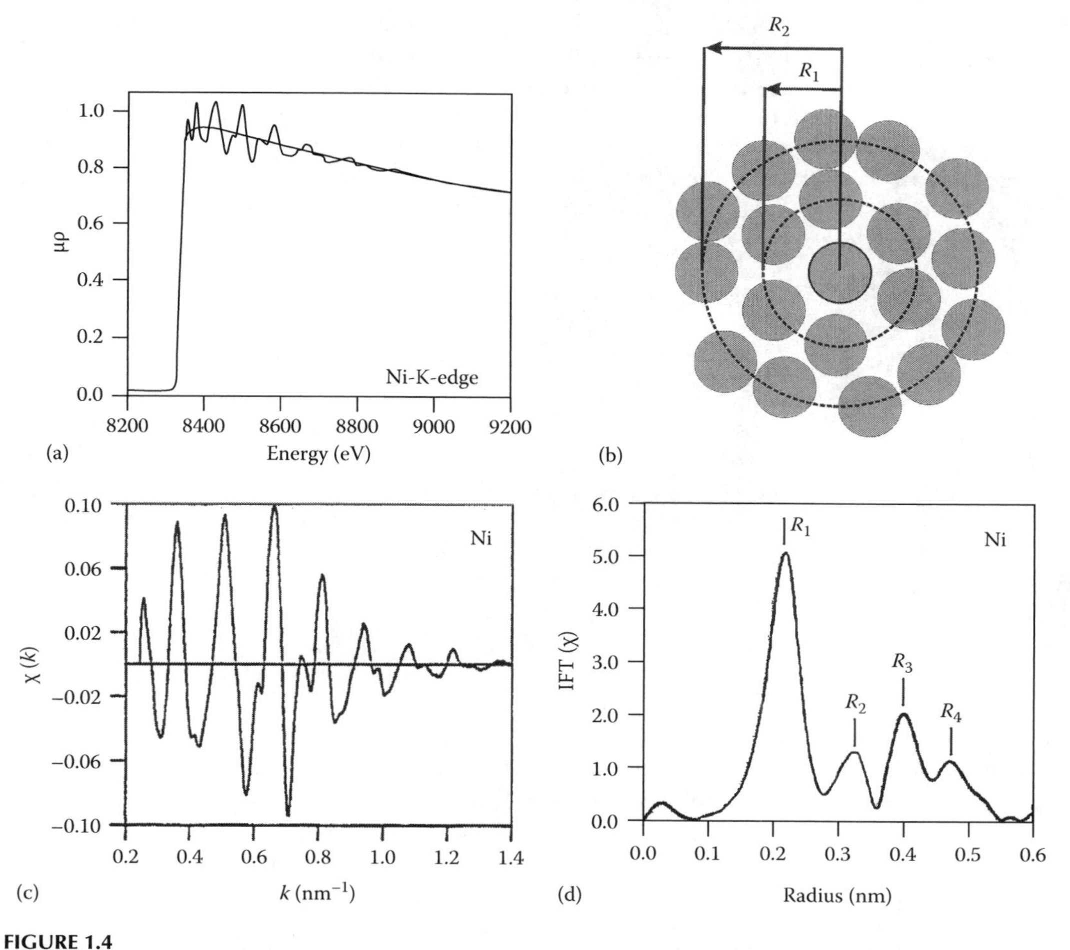

FIGURE 1.4
Explanation of EXAFS and its evaluation for Ni metal, (a) absorption edge of Ni, (b) short-range order and coordination shells, (c) EXAFS function $\chi(k)$, (d) Fourier transform of χ FT(χ) to distance space. (From Rehr, J.J., *Phys. Rev. Lett.*, 69, 3397, 1992.)

well- or less-developed short-range order of crystalline or amorphous materials. They are related to an interference phenomenon of electrons after their excitation by the absorption of x-rays. When an atom absorbs a photon and it gets ionized, the leaving electron will be scattered at the atoms of the surrounding coordination shells (Figure 1.4b). Taking into account the wave nature of electrons characterized by their De Broglie wavelength $\lambda_{DB} = h/(mv)$ with their mass m, velocity v, and Planck constant h, the outgoing electron wave will interfere with the reflected electron wave. Whether this interference will be constructive or destructive depends on several parameters, which are: R_i the distance the electron wave has to overcome, i.e., the radius of the ith coordination shell and the De Broglie wavelength λ_{DB}, which is given by the excess energy of the electron and thus by its velocity v depending on the energy of the absorbed x-ray photon and on the phase shift Φ, which occurs during the reflection at the coordination shell. These factors determine the oscillation of the relative variation of the absorption coefficient, the so-called EXAFS function $\chi(k) = \Delta\mu_a/\mu_{a,0}$. It sums up the contribution of the existing coordination shells with their specific parameters as shown by Equations 1.18 a and b. The energy of the electrons and their wavelength enter by the wave vector k, which is correlated to the energy distance ΔE of the absorption edge by Equation 1.18c. Besides the already-mentioned radius R_j and the phase shift $\Phi_j(k)$, other parameters enter, the amplitude $A_j(k)$, i.e., the coordination number N_j, the exponential expression of the Debye–Waller factor $\exp(-2\sigma_j^2k^2)$ containing the mean square displacement σ_j^2 of the atoms of the coordination shell due to thermal motion and structural disorder. Equation 1.18b contains in addition a decrease with the square of the radius R_j and an exponential attenuation due to inelastic losses of the electrons resulting from their interaction with the matrix, containing their mean free path length $\lambda(k)$.

$$\chi = \frac{\mu - \mu_a^0}{\mu_{a,0}} = \frac{\Delta\mu_a}{\mu_{a,0}} = \sum_j A_j(k)\sin[2kR_j + \Phi_j(k)] \tag{1.18a}$$

$$A_j(k) = N_jS_j(k)F_j(k)\exp[-2\sigma_j^2k^2]\exp\left[-\frac{2R_j}{\lambda(k)}\right]\frac{1}{kR_j^2} \tag{1.18b}$$

$$k = \sqrt{\left(\frac{2m}{\hbar^2}\right)\Delta E} \tag{1.18c}$$

The usual procedure of the evaluation of an EXAFS data of the XA spectrum involves first the background subtraction and the evaluation of χ (Figure 1.4c). Then one has to separate the contribution of the various coordination shells to visualize their data. This is done by a Fourier transform to the distance space. Figure 1.4d clearly shows four well-developed shells of crystalline Ni metal, R_1 to R_4, which are characteristic of a cubic face-centered metal. This result still needs a correction by the phase shift Φ and the evaluation of N_i and σ_i. This occurs for specific specimens by comparison with EXAFS data of well-characterized standards like nickel metal in this case or by calculation of the EXAFS results with a program like FEFF [12] or data analysis packages from the Internet with reasonable assumptions [13]. For this purpose, the χ^- data of each shell are separated and submitted to a Fourier backtransform. Then the parameters are varied to get a closest fit of the experimental and the calculated data or those from a standard. Finally one gets a set of corrected parameters R_i, N_i, σ_i for each coordination shell i as well as the phase shift Φ_i. With this procedure, one may evaluate the EXAFS data of highly disordered systems and amorphous materials as well as dissolved species, which are discussed later in this section. Further information is given in Ref. [11].

1.4.2 Short-Range Order of Dissolved Species

An absorption experiment through a bulk material is rather simple. The investigation of electrode surfaces and surface layers is more complicated. One needs an intense interaction of the beam with the surface structure without picking up too much signal of the substrate. Therefore, these measurements are performed under grazing incidence with very small angles of incidence, which are discussed in a later section. In many cases, the beam has to get through the electrochemical cell without too much absorption. This could cause a problem for aqueous solutions due to the high absorption by water at smaller x-ray energies. Investigations of Ag ions in solution are not difficult due to the high energy of 25.5 keV of the Ag edge. It becomes more difficult for Cu with $E > 9$ keV. The edge energy of an element under study should be larger than 5 keV to get still a reasonable signal.

Figure 1.5 shows an in situ transmission experiment through a solution of 1 M Na_2SO_4 in water with dissolved Ag(I) ions [14]. The x-ray beam with a narrow shape passes close to the Ag electrode at a distance of 200 μm. It will detect and analyze Ag ions, which are formed by dissolution at the electrode surface. The electrochemical conditions of the experiment are controlled with a potentiostat, using a Pt counter electrode (CE) and an Ag/AgCl reference electrode (RE). After Ag dissolution at the electrode surface, Ag^+ ions get to the electrolyte and are analyzed by the incoming beam. The edge height is proportional to the Ag^+ concentration. In the case of Figure 1.5, the Ag^+ concentration has been determined with the edge height to be 8.6×10^{-3} M. At more negative electrode potentials where no Ag^+ ions are released, no Ag edge is found, as expected. The oscillations of the EXAFS function may be evaluated till $k = 8$ A^{-1} as seen in the insertion of Figure 1.5a. For higher k values, the data are too noisy to permit a reliable data evaluation. Their Fourier transform (Figure 1.5b) shows a well-pronounced first coordination shell at 0.17 nm. Higher coordination shells are not seen. Peaks below 0.1 nm are artifacts. The first coordination shell is truncated as shown in Figure 1.5b and then Fourier backtransformed as shown in the inset of Figure 1.5b. Here the Fourier backtransform is phase shift corrected and shows an excellent agreement with the calculated data (dashed curve). The related structure parameters are $R = 0.23$ nm, $N = 2.15$, and $\sigma = 0.0103$. The coordination number fits to $N = 2.0$ of polycrystalline Ag_2O. R and σ are larger than the values for crystalline Ag_2O, $R = 0.196$ nm and $\sigma = 0.0077$ nm, respectively. This is consistent with less order for the dissolved species. The coordination shell suggests a coordination of Ag^+ with two OH^- to form a $[Ag(OH)_2]^-$ complex. This is expected according to electrochemical data [15,16]. The second coordination shell would correspond to the H atoms belonging to OH. The simulation of this second shell yields with $R = 0.23$ nm, $N = 2$, and $\sigma = 0.01$ nm only a very small peak, which is not seen in the EXAFS evaluation of the data.

The short-range order of dissolved cations like Fe^{2+}, Co^{2+}, Ni^{2+}, Mn^{2+}, Cr^{3+}, and Fe^{3+} have been examined with EXAFS but also with x-ray diffraction (XRD) and neutron scattering. The counterions were Cl^-, NO_3^-, and ClO_4^-. Although these cations did not

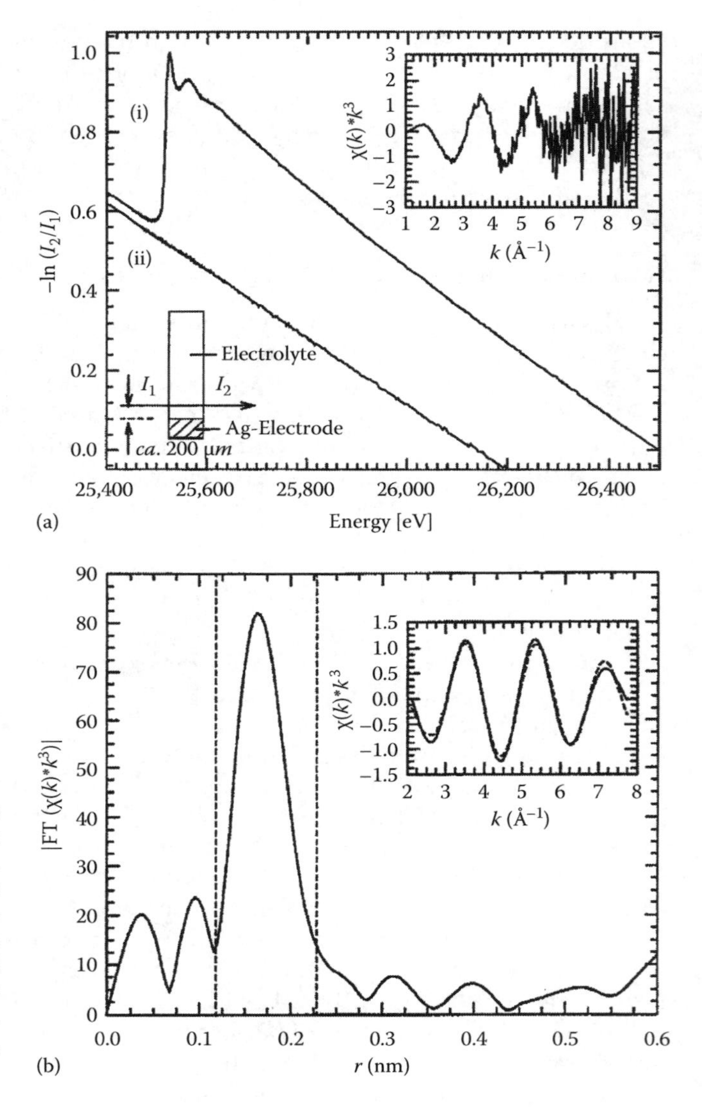

FIGURE 1.5
In situ EXAFS of dissolved Ag^+ ions in front of a dissolving Ag electrode: (a) with and without Ag dissolution, inset: k^3-weighted $\chi(k)$ and (b) FT($\chi(k)k^3$) to distance space showing first coordination shell. Inset: Fourier backtransform of truncated first coordination shell and comparison with simulation (dashed) with $R = 0.23$ nm, $N = 2.15$, $\sigma = 0.0103$. (From Lützenkirchen-Hecht, D. et al., *Corr. Sci.*, 40, 1037, 1998.)

show complexation, one well-developed coordination shell with a radius $R = 0.20–0.21$ and a coordination number of $N = 6$ has been found, which corresponds to the octahedral coordination of the central cations with water molecules [17,18]. This structure is still present up to a concentration of 4 M, which excludes a complexation of the cations by Cl^-. The addition of LiCl to a 2 M $NiCl_2$ solution up to a Cl/Ni ratio of 2–5 yields a first coordination shell with $R = 0.206$ nm and $N = 6$ and a second with $R = 0.238$ and $N = 0.7–1.0$. Apparently the hexagonal water structure is preserved and the approach of Cl^- occurs in a second shell. Thus, for high Cl^- concentrations, the structure of dissolved Ni^{2+} ions approaches that of crystalline $NiCl_2 \cdot 6H_2O$, although the expected coordination number is still not reached.

1.5 Debye–Hückel–Onsager Theory of Diluted Strong Electrolytes

1.5.1 Basic Discussion

Although strong electrolytes are fully dissociated, they still show a change of their equivalent conductivity Λ with concentration. This is a direct consequence of the long-range Coulomb interaction of the cations and anions in solution. Increasing concentration yields a decrease of Λ. Furthermore, the activity coefficient γ_i of ions depends on the concentration of ions and their charge. For diluted solutions, it is also decreasing with increasing concentration. The Debye–Hückel–Onsager theory takes into account these electrostatic interactions of ions and allows a simple calculation of the quantities Λ_i and γ_i of individual ions.

The Coulomb forces of charged ions in solution tend to order the electrolyte. A central cation attracts negatively charged anions and rejects positively charged cations and vice versa. On the other hand, the thermal motion tends to realize a homogeneous distribution of the ions against this order. As a consequence, one has a diffuse preferential accumulation of anions around cations and of cations around anions (Figure 1.6). Thus, a central cation is surrounded by a diffuse negatively charged cloud and a central anion by a diffuse positively charged cloud. This order cannot be observed by XRD or EXAFS or neutron scattering due to the fast thermal motion and its diffuse nature. The preferential attraction of ions with opposite charge and the depletion of ions with the same charge follow the simple Boltzmann relation of Equation 1.19 with the linear approximation valid for small values of the exponent. $N_i(r)$ is the concentration of ions of kind i at a distance r of the central ion, $\langle N_i \rangle$ their average concentration, z_i their charge, $e_0 = 1.6 \times 10^{-19}$ C the elementary charge, $\varphi(r)$ the potential at distance r, k the Boltzmann constant, and T the Kelvin temperature. Taking $\langle N_i \rangle = N_A c_i$, the sum over all ions in solution weighted by their charge $z_i e_0$ yields the density distribution $\rho(r)$ of the excess charge of Equation 1.20 applying the electro neutrality condition of Equation 1.20a.

$$\frac{N_i(r)}{\langle N_i \rangle} = \exp\left\{-\frac{z_i e_0 \varphi(r)}{kT}\right\} \approx 1 - \frac{z_i e_0 \varphi(r)}{kT} \tag{1.19}$$

$$\rho(r) = \sum_i N_i(r) z_i e_0 = \sum_i \langle N_i \rangle z_i e_0 \left[1 - \frac{z_i e_0 \varphi(r)}{kT}\right]$$

$$\rho(r) = \sum_i -\frac{N_A c_i z_i^2 e_0^2}{kT} \varphi(r) = -\frac{N_A e_0^2 2I}{kT} \varphi(r) \tag{1.20}$$

$$\sum_i \langle N_i \rangle z_i e_0 = 0 \tag{1.20a}$$

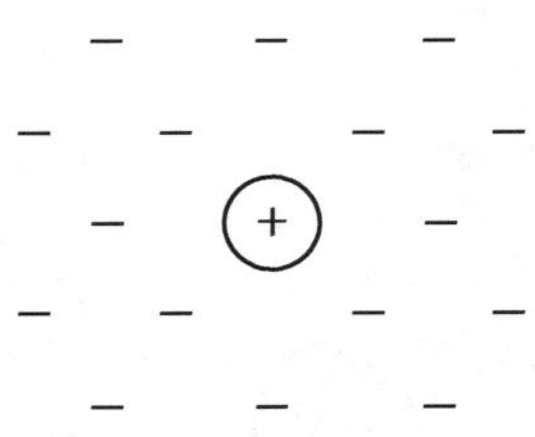

FIGURE 1.6
Ionic cloud around a cation.

where I is the ionic strength, related to the sum of the concentration c_i of all ions with a weight factor z_i^2 for their charge.

$$I = \frac{1}{2}\sum_i c_i z_i^2 \tag{1.20b}$$

Introducing $\rho(r)$ of Equation 1.20 into the Poisson equation yields the differential Equation 1.21 with a solution for $\varphi(r)$ given in Equation 1.22 for the boundary condition $\varphi = 0$ for $r = \infty$ with the parameters defined in Equations 1.22a and b.

$$\frac{1}{r^2}\frac{d}{dr}\left[r^2\frac{d\varphi(r)}{dr}\right] = -\frac{\rho(r)}{\varepsilon\varepsilon_0}\varphi(r) = \frac{N_A e_0^2 2I}{\varepsilon\varepsilon_0 kT}\varphi(r) \tag{1.21}$$

$$\varphi(r) = \frac{Z}{r}\exp\left\{-\frac{r}{\beta}\right\} \tag{1.22}$$

$$Z = \frac{z_i e_0}{4\pi\varepsilon\varepsilon_0}\frac{\exp\{a/\beta\}}{1 + a/\beta} \tag{1.22a}$$

$$\beta = \sqrt{\frac{\varepsilon\varepsilon_0 kT}{2N_A e_0^2 I}} \tag{1.22b}$$

where

$\varphi(r)$ is a shielded Coulomb potential, i.e., the Coulomb potential built up by the central ion Z/r diminished by the shielding effect of the ionic cloud

a is the radius of the central ion

β (Equation 1.22b) is the radius of the ionic cloud also called the Debye length

It describes the distance of the central ion at which the charge distribution $4\pi r^2\rho(r)$ has its maximum. Assuming a point charge for $a \ll \beta$, Equation 1.22a yields the Coulomb potential Z/r of the central ion. With the dielectric constant for the vacuum $\varepsilon_0 = 8.854 \times 10^{-12}\ C^2 \cdot N^{-1} \cdot m^{-2}$, the Boltzmann constant $k = 1.381 \times 10^{-23}\ J\ K^{-1}$ and the Faraday's constant $F = 96{,}487 \cdot C \cdot mol^{-1}$, and the elementary charge $e_0 = 1.602 \times 10^{-19}\ C$, one obtains Equation 1.23:

$$\beta = 6.288 \times 10^{-11}\ \mathrm{mol}^{1/2} \cdot \mathrm{K}^{-1/2} \cdot \mathrm{m}^{-1/2} \cdot \sqrt{\frac{\varepsilon T}{I}} \tag{1.23}$$

With the relative dielectric constant of water $\varepsilon = 78.3$ and $T = 298$ K, Equation 1.24 follows:

$$\beta = 3.04 \times 10^{-8}\frac{1}{\sqrt{I}} \tag{1.24}$$

The Debye length is determined decisively by the ionic strength I. A 10^{-3} M solution of an electrolyte like NaCl ($z = 1$) yields $\beta = 9.6$ nm, which is relatively large compared to the radius of the ions in the range of 0.1–0.2 nm. For a 0.1 M solution, β is 0.96 nm, thus approaching the dimension of ions. For higher concentrations, it gets smaller than the

TABLE 1.1

Debye Length/nm, Radius of Ion Cloud, and Thickness of Diffuse Double Layer for Various Concentrations of Some z,z Valent Electrolytes

Concentration (M)	1,1 Valent	1,2 Valent	2,2 Valent	1,3 Valent
10^{-4}	30.4	17.6	15.2	12.4
10^{-3}	9.62	5.56	4.81	3.92
10^{-2}	3.04	1.76	1.52	1.24
10^{-1}	0.96	0.56	0.48	0.39

ionic radius, which means that the diffuse ionic cloud becomes meaningless. Table 1.1 presents the Debye length for varying concentrations and charges of the ions. The Debye–Hückel theory and especially the Debye length β play also an important role for the structure and dimensions of the double layer at the electrode/electrolyte interface and therefore are important for electrode kinetics. Similarly, the structure of the space charge layer of a semiconductor may be described by this theory especially when it is in contact with the electrolyte and thus is important for semiconductor electrochemistry. Here β is closely related to the size of the space charge layer. These details are discussed later in Section 1.7 (Equation 1.44) and Section 1.18 (Equations 1.179 and 1.180).

The shielded Coulomb potential of Equation 1.21 contains two contributions, the electrostatic part determined by the charge of the central ion $\varphi_z(r)$ given by Equation 1.25 and the part caused by the ionic cloud $\varphi_{Cl}(r)$ given by Equation 1.26. The potential $\varphi_{Cl}(r)$ for $r = a$, i.e., at the surface of the central ion, yields Equation 1.27, which is used to calculate the activity coefficient γ_i as discussed in Section 1.5.4.

$$\varphi_z(r) = \frac{z_i e_0}{4\pi\varepsilon\varepsilon_0 r} \tag{1.25}$$

$$\varphi_{Cl}(r) = \varphi(r) - \varphi_z(r) = \frac{z_i e_0}{4\pi\varepsilon\varepsilon_0 r}\left(\frac{\exp(a/\beta)}{1+a/\beta}\exp\left(-\frac{r}{\beta}\right) - 1\right) \tag{1.26}$$

$$\varphi_{Cl}(r = a) = -\frac{z_i e_0}{4\pi\varepsilon\varepsilon_0}\frac{1}{\beta + a} \tag{1.27}$$

1.5.2 Equivalent Conductivity of Strongly Dissociated Electrolytes

The ionic cloud has consequences on the equivalent conductivity and mobility of ions. Two effects reduce both values with increasing ionic strength I, the relaxation effect and the electrophoretic effect. If an ion is moving under the influence of an electric field, the ionic cloud has to be built in those places the ion is moving in and it has to be destroyed in those the ion has left. These processes occur by diffusion and thus require some time. As a consequence, the ionic cloud becomes asymmetric as shown in Figure 1.7a. In front of the migrating ion, the excess charge is still not fully developed and in the left places it still remains. As a consequence, the center of the negative excess charge surrounding a positively charged cation does not coincide with the center of the ion and thus a retarding effect tries to pull the moving ion backward. As a consequence, the ion moves slower compared to the situation the cation moves in a solution with vanishing ionic strength

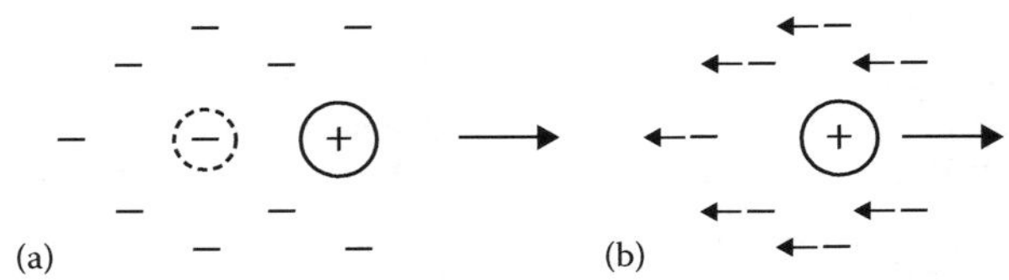

FIGURE 1.7
(a) Separation of the centers of negative and positive charge during migration of a cation causing the relaxation effect; (b) increase of relative velocity during migration of a cation due to the motion of the ionic cloud in the opposite direction leading to the electrophoretic effect.

and thus with $\beta = \infty$, i.e., when no ionic cloud exists. This is the relaxation effect, which is expressed quantitatively by Equations 1.28a and b. Λ_0 is the equivalent conductivity for vanishing electrolyte concentration ($c = 0$), and Λ_{0+} and Λ_{0-} are the related individual equivalent conductivities of the cations and the anions, respectively.

$$\Lambda_{\text{rel}} = \frac{e_0^2}{3kT} \frac{|z_+ z_-| q}{(1+\sqrt{q})} \frac{1}{4\pi\varepsilon\,\varepsilon_0} \Lambda_0 \frac{1}{\beta} \tag{1.28a}$$

$$q = \frac{|z_+ z_-|}{|z_+| + |z_-|} \frac{\Lambda_{0+} + \Lambda_{0-}}{|z_+|\Lambda_{0-} + |z_-|\Lambda_{0+}} \tag{1.28b}$$

In solution, not only the central cation but also the surrounding ionic cloud moves, however, with its opposite excess charge in the opposite direction (Figure 1.7b). Therefore, the central cation feels a larger velocity relative to its local environment and thus is exposed to higher friction forces. In consequence, this ion is slowed down relative to the laboratory system in comparison with a situation with a missing ionic cloud. This is the electrophoretic effect expressed by Equation 1.29, where η_V is the viscosity of the electrolyte.

$$\Lambda_{\text{el}} = \frac{F e_0}{6\pi} \frac{|z_+| + |z_-|}{\eta_V} \frac{1}{\beta} \tag{1.29}$$

Both the relaxation and the electrophoretic effect reduce the equivalent conductivity, and therefore the related terms Λ_{rel} and Λ_{el} have to be subtracted from Λ_0 to get the equivalent conductivity Λ_c at a given concentration c, as shown in Equation 1.30.

$$\Lambda_c = \Lambda_0 - \Lambda_{\text{rel}} - \Lambda_{\text{el}} = \Lambda_0 - (B_1\Lambda_0 + B_2)\sqrt{I} \tag{1.30}$$

B_1 and B_2 are constants to both effects. The correcting terms Λ_{rel} and Λ_{el} given by Equations 1.28 and 1.29 are both proportional to the inverse of β and thus increase with $\sqrt{I}$. Equation 1.31 summarizes both effects by a conductivity coefficient f_Λ, which takes care of the change of Λ_c with the ionic strength I as a factor to Λ_0.

$$\Lambda_c = f_\Lambda \Lambda_0; \qquad f_\Lambda = 1 - \frac{1}{\Lambda_0}\left[B_1\Lambda_0 + B_2\right]\sqrt{I} \tag{1.31}$$

For 1,1-valent electrolytes ($|z_+| = |z_-| = 1$) with $q = 0, 5$ and $I = c$, one obtains Kohlrausch's law (Equation 1.32). For $T = 298\,\text{K}$, $\varepsilon = 78.3$, and the viscosity $\eta_V = 0.890\,\text{kg}\cdot\text{m}^{-1}\cdot\text{s}^{-1}$ for water, one obtains $B_1 = 0.2302\,\text{L}^{1/2}\,\text{mol}^{-1/2}$ and $B_2 = 60.68\,\text{cm}^2\cdot\Omega^{-1}\cdot\text{L}^{1/2}\cdot\text{mol}^{-1/2}$. For a 10^{-2} M NaCl solution with $\Lambda_0 = 126.45\,\text{cm}^2\cdot\Omega^{-1}\cdot\text{mol}^{-1}$, one calculates $\Lambda_c = 117.47\,\text{cm}^2\cdot\Omega^{-1}\cdot\text{mol}^{-1}$, which agrees well with the measured value of $\Lambda_c = 118.51\,\text{cm}^2\cdot\Omega^{-1}\cdot\text{mol}^{-1}$.

$$\Lambda_c = \Lambda_0 - [B_1\Lambda_0 + B_2]\sqrt{c} = \Lambda_0 - \text{const}\sqrt{c} \qquad (1.32)$$

Conductivity measurements are usually performed with two electrodes of an inert metal like platinum. Their surface is covered with electrochemical deposits of high surface platinum (platinum black) in order to reduce the current density for a given current. Usually the surface is increased by a factor of 1000 by the deposit of platinum black. The measurements are performed with alternating current to avoid problems with the overpotential at the electrode surfaces and the necessary minimum voltage for the decomposition of the electrolyte. Figure 1.8a gives a block diagram for the equipment to measure the conductivity. A small alternating voltage in the range of a few mV and a frequency of ca. $f = 10^3\ \text{s}^{-1}$ is applied to the electrochemical cell, which yields an alternating cell current I_C, which is measured after passing a rectifier. I_C is proportional to the conductivity and thus inverse to the ohmic resistance R_Ω of the electrolyte. The cell may be described by the two electrode capacities C_E and the ohmic resistance R_Ω in series (Figure 1.8b). The charge transfer resistance R_{CT} of the electrodes may be neglected due to the small capacitive resistance $1/\omega C_E$ for a sufficiently high frequency $\omega = 2\pi f$ and especially due to the large surface area of the electrodes and a related large capacity C_E. There exists a short circuit of R_{CT} by C_E and its small capacitive resistance. With $C_E = 20 \times 1{,}000 = 20{,}000\,\mu\text{F}\cdot\text{cm}^{-2}$, one obtains $2/\omega C_E = 0.016\ \Omega$ for the capacitive resistance of the two electrodes, which is also negligible relative to typical values of the electrolyte resistance $R_\Omega = 0.1 \times 10^7\ \Omega$ for well and poorly conducting electrolytes. Thus, one measures R_Ω with a good approximation only.

Table 1.2 summarizes some values of the equivalent conductivity Λ of electrolytes with decreasing values for increasing concentration and their largest values Λ_0 for $c = 0\,\text{M}$ [9,19]. The values are large for acids and bases due to the special conduction mechanism for H_3O^+ and OH^- ions using the short-range order of water. Highly charged ions ($CuSO_4$) have small values due to their large hydration shell and the related increase of friction

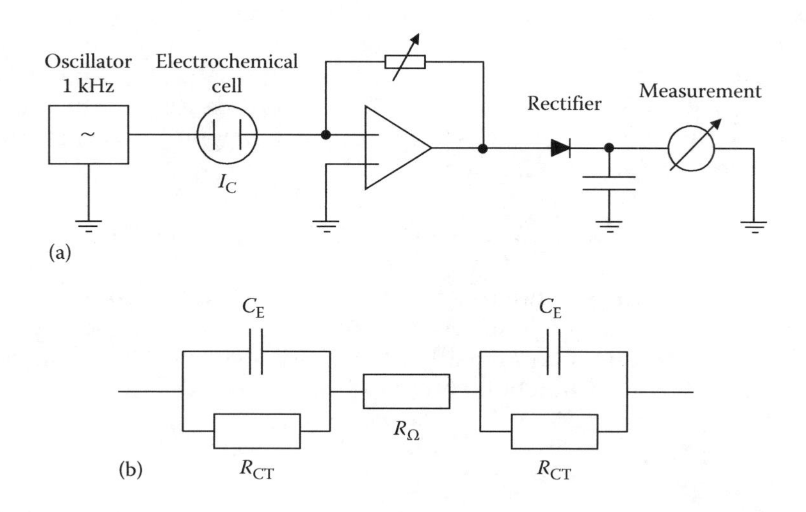

FIGURE 1.8
(a) Block diagram for conductivity measurements with alternating current. (b) Circuit for two-electrode capacities C_E and an ohmic resistance R_Ω in series.

TABLE 1.2

Equivalent Conductivity $\Lambda/\Omega^{-1}\cdot mol^{-1}\cdot cm^2$ of Some Electrolytes for Some Concentrations in Water at 25°C

c/mol L^{-1}	HCl	H_2SO_4	NaOH	KCl	NaCl	$CuSO_4$
0	429.2	429.8	247.8	149.9	126.5	133.6
0.001	412.4		244.7	147.0	123.7	115.3
0.01	412.0		238.0	141.3	118.5	83.1
0.1	391.3	251.2		129.0	106.7	50.6

Sources: *Handbook of Chemistry and Physics*, CRC Press, Cleveland, OH, 1977–1978; D'Ans-Lax, *Taschenbuch für Chemiker und Physiker*, Springer, Berlin, Germany 1960.

with the surrounding water for their migration. NaCl and KCl have similar values and the differences are related to the slightly different hydration shells of the two cations.

1.5.3 Equivalent Conductivity of Weakly Dissociated Electrolytes

Many dissolved acids or bases but also some salts are only weakly dissociated. Equation 1.33 gives an example for an acid HA with the related dissociation constant K of Equation 1.34.

$$HA \leftrightarrow H^+ + A^- \tag{1.33}$$

$$K = \frac{c(H^+)c(A^-)}{c(HA)} = \frac{\alpha^2 c}{1-\alpha} \tag{1.34}$$

The dissociation grade α describes the fraction of dissolved molecules that are dissociated, leading to the concentrations $c(H^+)$ and $c(A^-)$ for the cations and anions. For a solution of pure HA with $c(H^+) = c(A^-) = \alpha c$, one obtains the relation between K and α presented in Equation 1.34. In the case of acetic acid, the dissolved molecules HA are not charged and ions form by dissociation only. This opens the possibility to determine α and K by conductivity measurements. For the very small value of $K = 1.8 \times 10^{-5}$ for acetic acid, Λ_c is mainly determined by the concentration of ions αc and thus by the value of the dissociation grade α and its changes with concentration. The ionic interaction described by the Debye–Hückel theory is of minor influence. This may be seen easily in Equation 1.35, which contains both effects, dissociation, and ionic interaction.

$$\Lambda_c = \alpha[\Lambda_0 - (B_1\Lambda_0 + B_2)\sqrt{\alpha c}]; \quad \Lambda_c = \alpha\,\Lambda_0 \text{ for small } \alpha \tag{1.35}$$

The expression in brackets is multiplied by $\sqrt{\alpha c}$ with small values for α and thus with a negligibly small value subtracted from Λ_0 leading to $\alpha = \Lambda_c/\Lambda_0$ as a good approximation. Insertion of this expression for α in Equation 1.34 yields Ostwald's law (Equation 1.36) for the dissociation constant as a function of the equivalent conductivity Λ_c.

$$K = \frac{(\Lambda_c/\Lambda_0)^2}{(1-\Lambda_c/\Lambda_0)}\,c \tag{1.36}$$

Thus, α and K may be calculated from the measured data for Λ_c. For the example of acetic acid with $c = 0.1\,M$, one obtains with $\Lambda_0 = 390.59\ \Omega^{-1}\cdot cm^2\cdot mol^{-1}$ and $\Lambda_c = 5.20\ \Omega^{-1}\cdot cm^2\cdot mol^{-1}$,

$\alpha = 0.0133$, and $K = 1.79 \times 10^{-2}$ mol L^{-1} = 1.79×10^{-5} mol·cm^{-3}. Experimental data show that Λ_c decreases parallel to α with increasing c, which yields constant values for K as expected.

1.5.4 Activity Coefficient

The real behavior of systems is described by the activity coefficient γ_i. Instead of the concentration c_i of a dissolved species, one uses the activity $a_i = c_i\gamma_i$. In the light of the Debye–Hückel theory, γ_i takes care of the electrostatic interactions of the ions. This is the main interaction for charged species in comparison with the smaller dipole and Van der Waals forces, which may be important in the case of uncharged species, but which are not included in the Debye–Hückel theory. The chemical potential μ depends on the concentration according to Equation 1.37.

$$\mu_{re} = \mu_i^0 + RT\ln(c_i\gamma_i) = \mu_i^0 + RT\ln c_i + RT\ln\gamma_i = \mu_{id} + RT\ln\gamma_i \tag{1.37}$$

The correction term $RT \ln \gamma_i$ describes the work to transfer ions i from an ideal solution with no ionic interaction with a real electrolyte, i.e., the transfer from a solution of vanishing ionic strength and thus of a concentration $c = 0$ of dissolved ions to a solution of a given concentration for which an ionic cloud must be considered around ions. For the Debye–Hückel theory, this equals the electrical work to transfer ions with interaction with water molecules only to the ionic cloud for a solution with a given ionic strength I. This work is obtained from the electrical work for charging of an uncharged particle within its ionic cloud to its final charge z_ie_0, as given by Equation 1.38.

$$w_{el} = \int_0^{z_ie_0} \varphi_z(r)\,d(z_ie_0) + \int_0^{z_ie_0} \varphi_{Cl}(r=a)\,d(z_ie_0) \tag{1.38}$$

This charging contains two parts, the charging of an ion without the influence of the ionic cloud and the additional effect of the ion cloud. The latter involves the potential $\varphi_{Cl}(r = a)$ given by Equation 1.27. The second part of Equation 1.38 takes into account the difference between the real situation ($c > 0$) and the ideal situation ($c = 0$). Only this part is used in the further discussion for the calculation of the activity coefficient. Equation 1.37 refers to 1 mol of ions and thus requires multiplying the charging work in Equation 1.38 with Avogadro's number N_A, which yields Equation 1.39.

$$\mu_{re} - \mu_{id} = RT\ln\gamma_i = N_A \int_0^{z_ie_0} \varphi_{Cl}(r=a)\,d(z_ie_0) \tag{1.39}$$

Introducing Equation 1.27 for $\varphi_{Cl}(r = a)$ yields Equation 1.40 for the work $\mu_{re} - \mu_{id}$ for the transfer of 1 mol of ions from the ideally diluted to the real solution.

$$RT\ln\gamma_i = N_A \int_0^{z_ie_0} -\frac{z_ie_0}{4\pi\varepsilon\varepsilon_0}\frac{1}{\beta + a}\,d(z_ie_0) = -N_A\frac{z_i^2e_0^2}{8\pi\varepsilon\varepsilon_0}\frac{1}{\beta + a} \tag{1.40}$$

$$\ln\gamma_i = -\frac{N_A z_i^2 e_0^2}{8\pi\varepsilon\varepsilon_0 RT}\frac{1}{\beta + a} \tag{1.40a}$$

Neglecting the value $a \ll \beta$ for sufficiently diluted electrolytes and introducing the expression for β according to Equation 1.22b, one obtains Equation 1.41.

$$\ln \gamma_i = -\frac{z_i^2 e_0^2}{8\pi \varepsilon \varepsilon_0 kT} \sqrt{\frac{2N_A e_0^2}{\varepsilon \varepsilon_0 kT}} \sqrt{I} \tag{1.41}$$

This relation is valid for diluted electrolytes only due to the above approximation and all the others made for the Debye–Hückel theory. Introducing the values for Avogadro's constant $N_A = 6.022 \times 10^{23}\,\text{mol}^{-1}$, the charge of the electron $e_0 = 1.602 \times 10^{-19}$ C and the Boltzmann constant $k = R/N_A = 1.3807 \times 10^{23}\,\text{J}\cdot\text{K}^{-1}$, one obtains the factor $1.826 \times 10^6\ (\text{l}\cdot\text{K}^3\cdot\text{mol}^{-1})^{1/2}$ given in Equation 1.41a. Introducing the dielectric constant of water $\varepsilon = 78.56$ and the temperature $T = 298.15$ K yields finally the factor 0.509 also given in Equation 1.41a.

$$\log \gamma_i = -\frac{e_0^3 \sqrt{2N_A}}{2.303\, 8\pi(\varepsilon \varepsilon_0 kT)^{3/2}} z_i^2 \sqrt{I} = -\frac{1.826 \times 10^6}{(\varepsilon T)^{3/2}} z_i^2 \sqrt{I} = -0.509 z_i^2 \sqrt{I} \tag{1.41a}$$

γ_i is the individual activity coefficient of one ionic species. As one has always cations and anions together within an electrolyte, one measures experimentally an average activity coefficient $\gamma_\pm$, which is closely related to the individual activity coefficients γ_i. For a simple electrolyte composition containing only one salt, acid, or base of stoichiometry, $C_x^{z+} A_y^{z-}$, $\gamma_\pm$ is defined by Equation 1.42.

$$\log \gamma_\pm = \frac{1}{x+y}(x \log \gamma_+ + y \log \gamma_-) = -\frac{0.509}{x+y}\sqrt{I}(xz_+^2 + yz_-^2) \tag{1.42}$$

Taking into account the electroneutrality equation $xz_+ + yz_- = 0$, a simple calculation yields Equation 1.43. The numeric value refers to $T = 298$ K and $\varepsilon = 78.54$ for water.

$$\log \gamma_\pm = 0.509 z_+ z_- \frac{\text{dm}^{3/2}}{\text{mol}^{1/2}} \sqrt{I} \quad (I \text{ in mol}\cdot\text{L}^{-1}) \tag{1.43}$$

(NB: z_+ for cations is positive and z_- for anions negative).

This average activity coefficient, derived from the assumptions and approximations of the Debye–Hückel theory, only takes into account the long-range Coulomb forces of solvated ions in diluted solutions and not any other interactions as dipole or Van der Waals forces. Thus, only the charge z_i and concentration c_i of the ions enter and no other special properties as, for example, their chemical properties. Equation 1.43 may be used to compare experimental and calculated values of $\gamma_\pm$. $\gamma_\pm = 1$ refers to an ideal solution (c close to zero). For a 0.001 M K_2SO_4 solution, one obtains $I = 0.5\ (1^2 \times 0.002 + 2^2 \times 0.001) = 0.003$ $(\text{mol}\cdot\text{L}^{-1})$. Equation 1.43 yields $\log \gamma_\pm = -0.0558$ and $\gamma_\pm = 0.879$, which agrees sufficiently with the experimental value of 0.889. Table 1.3 gives a compilation of some activity coefficients [9]. They decrease with increasing concentration and charge of the ions. They show only minor changes within one group with the same concentration, as, for example, 1,1-, 1,2-, or 2,2-valent electrolytes, demonstrating again the validity of the Debye–Hückel theory. One also may see the decrease of $\gamma_\pm$ with increasing charge of the cations and anions, which enters in Equation 1.43 as $z_+ z_-$ and as z_i^2 in the ionic strength I.

For large concentrations of $c > 0.1$ M, the experimental activity coefficients deviate from the results calculated with the Debye–Hückel theory to larger values with a minimum at ca.

TABLE 1.3

Activity Coefficients of Some Acids Bases and Salts

Substance (M)	0.1	0.3	0.5	0.8	1.0
HCl	0.796	0.756	0.757	0.783	0.809
$HClO_4$	0.803	0.768	0.769	0.795	0.823
KOH	0.798	0.742	0.732	0.742	0.756
NaOH	0.766	0.708	0.690	0.679	0.678
$AlCl_3$	0.337	0.302	0.331	0.429	0.539
$CuSO_4$	0.150	0.083	0.062	0.048	0.043
KBr	0.772	0.693	0.657	0.629	0.617
KCl	0.770	0.688	0.649	0.618	0.604
$LiClO_4$	0.812	0.792	0.808	0.852	0.887
NaCl	0.778	0.710	0.681	0.662	0.657
NaI	0.787	0.735	O.723	0.727	0.736
NaAc	0.791	0.744	0.735	0.745	0.757
$NiSO_4$	0.150	0.084	0.063	0.047	0.042

Source: *Handbook of Chemistry and Physics*, CRC Press, Cleveland, OH, 1977–1978.

$c = 1\,M$ and an increase for $c > 1\,M$ up to values even larger than 1. For these conditions, the Debye–Hückel theory is no longer valid. If the amount of water molecules bound to the solvation shell of dissolved ions is getting in the order of the free water molecules, the solvent concentration is less than assumed. As a consequence, the effective electrolyte concentration is getting larger, corresponding to a larger activity coefficient. For further increase of c, the amount of water molecules becomes insufficient for a fully developed solvation shell, which again increases the activity and the activity coefficient. Furthermore, the formation of ion pairs leads to ion association, i.e., ion pairs or even higher associates, which also has influence to the activity coefficient.

The activity coefficients for neutral solutes are not described by the Debye–Hückel theory. Here no long-range Coulomb forces but weak dipole–dipole or van der Waals interactions are effective. They lead only to a minor deviation of the activity coefficient of the value 1. However, the activity coefficients of these solutes are also affected when the amount of bound water molecules within their solvation shell gets close to the concentration of free water, i.e., at very high concentrations of solutes.

1.6 Some Basic Definitions in Electrochemistry

An electrochemical cell consists of two electrodes as shown in Figure 1.9. This set-up separates an overall chemical reaction into two partial electrochemical reactions one at each electrode with an electronic exchange via the outer circuit. Figure 1.9 shows the Fe/Fe^{2+} electrode in combination with the H_2/H^+ electrode as an example. At the iron surface, Fe dissolution as Fe^{2+} is one half reaction, at the platinum electrode hydrogen evolution by H^+ reduction is the other half-reaction. Both partial reactions compensate each other electronically by a transfer of electrons in the outer circuit from the Fe/Fe^{2+} electrode to the H_2/H^+ electrode. Both electrodes need an electrolytic contact to form

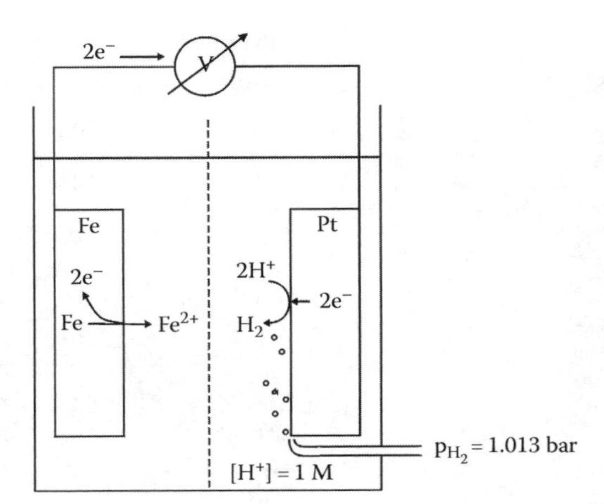

FIGURE 1.9
Electrochemical cell consisting of two electrodes (Fe/Fe^{2+} and H_2/H^+ electrode), separation of a chemical process via a diaphragm in two partial processes with electronic exchange via external circuit.

a cell. A diaphragm between the electrodes prevents the mixing of their electrolytes. At both electrodes, anodic oxidation or cathodic reduction may occur. Depending on the direction of the process, the same electrode may act as an anode or a cathode. An anodic process involves a transfer of positive charge from the metal or semiconductor electrode surface to the adjacent electrolyte. This is equivalent to a transfer of negative charge in the opposite direction. A cathodic process involves a transfer of negative charge from the metal or semiconductor surface to the electrolyte that is equivalent to a transfer of positive charge in the opposite direction. The applied potential determines whether the electrode process occurs in the anodic or cathodic direction. It should be mentioned, however, that iron is dissolving in acidic solutions while having hydrogen evolution at the same surface. Thus, the same electrode may react even as an anode or cathode simultaneously. Any detailed discussion for two different processes occurring at one electrode surface is given in Section 1.14.

One distinguishes metal/metal ion electrodes, which involve the transfer of metal cations across the interface, or redox electrodes with a transfer of electrons across the interface. For electrodes consisting of compounds, as for most semiconductors, cations or anions may be transferred besides electrons thus decomposing the electrode surface. For example, the anodic decomposition of CdS involves the transfer of Cd^{2+} ions to the electrolyte, leaving elemental S at the surface of the CdS electrode.

The potential drop at the electrode/electrolyte interface is a driving force for electrochemical reactions. It determines the rate of the reactions and their directions, for example metal dissolution or metal deposition. This potential drop at one interface may not be measured. Any attempt to measure it directly with a voltmeter involves automatically a second contact with the electrolyte forming an electrochemical cell with two electrodes. Therefore, one can only measure the combination of two potential drops across two interfaces. In order to standardize potential measurements, one uses the standard hydrogen electrode (SHE) as a reference. It is a platinized Pt electrode bubbled with pure hydrogen gas at 1 atm ($=1.013 \times 10^5$ Pa) in a solution of hydrogen ions with an activity $a = 1\,\text{mol L}^{-1}$. Platinizing creates a large surface, thus avoiding any problems of contamination causing deviation of the electrode potentials due to decreased electrochemical reaction rates and related current densities at this surface. Potentials are usually referred to the SHE. One can compare this potential scale to that used in physics with a zero value for the charge-free vacuum. The potential of the SHE has been determined to be −4.5 to −4.6 V (scatter of published data) relative to that of the charge-free vacuum [20–24].

Electrochemical equilibrium is established when the anodic and cathodic partial reactions of one electrochemical process compensate each other, i.e., if no reaction occurs. The equilibrium potential E_{eq} is called the Nernst potential. If the potential is shifted positively with respect to this equilibrium value, the anodic partial process is larger, and an overall anodic reaction will occur. An overall cathodic reaction will occur for a negative deviation of the potential. The deviation of the electrode potential E from its equilibrium value $E - E_{eq} = \eta$ is called overpotential (or overvoltage). η is an expression for the kinetic hindrance of the electrochemical processes. Both partial reactions change exponentially with the electrode potential, which is described by the Butler–Volmer equation (see Section 1.10.1). For large positive overpotentials, the anodic partial reaction becomes predominant and the cathodic partial process may be neglected. For large negative overpotentials, the cathodic process is predominant. For these potential ranges, one obtains a simple exponential i–η relation and a related linear semilogarithmic dependence, the so-called Tafel equation.

The above discussed electrochemical equilibrium and the related Nernst potential E_{eq} with the same processes in opposite direction has to be distinguished from the situation when two different processes compensate each other at the rest or mixed or corrosion potential E_R. Spontaneous iron dissolution with hydrogen evolution in an acidic electrolyte is an example. If no external current is applied, Fe^{2+} dissolution is compensated by H^+ reduction, i.e., the anodic and cathodic processes compensate each other. A deviation from the rest potential $E - E_R = \pi$ is called polarization. For $\pi > 0$, the anodic Fe dissolution is faster and requires an external current. For $\pi < 0$, the hydrogen evolution exceeds iron dissolution with a cathodic overall current.

In electrode kinetics, one distinguishes experiments with controlled potential and experiments with controlled current. For potentiostatic conditions, the potential is held constant by a potentiostat and the current is measured. For galvanostatic conditions, the current is supplied by a constant current source, a galvanostat, and the potential is measured. For potentially controlled experiments, E may be changed with time. For potentiodynamic experiments, E varies linearly with time, for potentiostatic transients E is changed stepwise from one value to another. Similarly, for galvanostatic transients, the current may be changed stepwise. These transient measurements are performed to learn about the kinetics of electrochemical processes and to separate fast and slow reaction steps of an overall process as charge transfer reactions and transport processes (diffusion and migration) as is discussed in Section 1.10.

1.7 The Electrochemical Double Layer

The potential drop at the electrode/electrolyte interface depends on the composition of the electrolyte and the characteristics of the electrode and its interaction with the adjacent solution. A charged metal surface is related to the change of the electrolyte composition in its vicinity and to the potential drop at its surface. A positively charged electrode will attract preferentially negatively charged anions and repel cations as given schematically in Figure 1.10a. A negatively charged surface causes the opposite changes. For specific adsorption of anions, their accumulation may get large, which is compensated by an accumulation of cations at a slightly longer distance from the surface (Figure 1.10b). The situation is more complicated if the electrode is covered by a three-dimensional film, which is discussed in

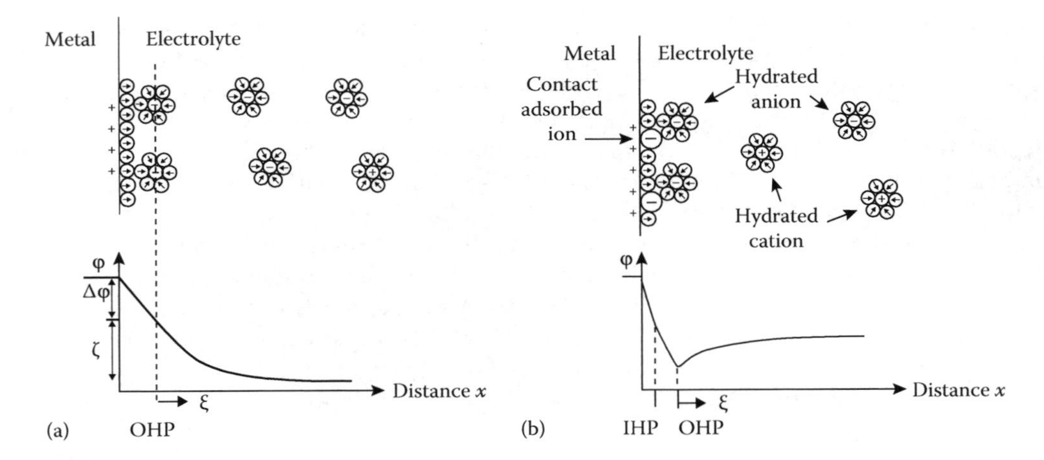

FIGURE 1.10
Model of the electrochemical double layer with potential drop ζ within the diffuse double layer and $\Delta\varphi_H$ within the Helmholtz layer. (a) The centers of the closest hydrated ions define the OHP. (b) The centers of the contact adsorbed ions define the IHP. The hydration layer of the metal surface and the hydration shells of ions are shown.

Chapter 5 (Passivity of Metals). For semiconductor surfaces, the situation gets also more complicated due to the extension of a space charge layer into the electrode itself, which is discussed in Section 1.18 on semiconductor electrochemistry and photo-electrochemistry.

For a charged metal surface, one has to subdivide the total double layer into the Helmholtz layer and the diffuse double layer (Figure 1.10) [25]. The Helmholtz layer is given by the dimensions of the ions, which may approach to the surface only as close as their size allows. The centers of the hydration sphere of ions define the outer Helmholtz plane (OHP) (Figure 1.10a). Specifically or contact adsorbed ions lose their hydration shell and remove the hydrated layer covering the metal surface and thus may approach closer to the surface. Their centers define the inner Helmholtz plane (IHP) (Figure 1.10b). In addition, the thermal motion of the ions will lead to their diffuse distribution causing a diffuse double layer. This situation is similar to that described by the Debye–Hückel theory (Section 1.5). Here one has a simpler one-dimensional distribution of ions and the related change of the potential perpendicular to the electrode surface, which is described by the theory of Gouy Chapman [26–28]. Here again the potential distribution within the diffuse double layer depends decisively on the ionic strength I. With a Boltzmann distribution of the ions perpendicular to the surface and the linear approximation of Equation 1.19, one obtains the charge density distribution of Equation 1.20. Introducing an almost identical relation for the linear change of the charge density $\rho(\xi)$ with potential within the diffuse double layer into Poisson's equation yields the potential change $\Delta\varphi$ with the distance ξ from the OHP (Figure 1.10) as given by Equation 1.44.

$$\Delta\varphi = \varphi(\xi) - \varphi(\xi_\infty) = \zeta \exp\left\{-\frac{\xi}{\beta}\right\} \tag{1.44}$$

The thickness of the diffuse double layer or Debye length β corresponds to the radius of the ionic cloud of the Debye–Hückel theory as given by Equation 1.22b. For the electrode/electrolyte interface, this is the distance ξ at which the potential drops by $1/e$ of its total change ζ within the diffuse double layer. The same relation (Equation 1.44) is important for colloids. The potential drop around these particles keeps them from coagulation due to the repulsive charging of their diffuse double layers. ζ is closely related to the charge density within the diffuse double layer and may be obtained by integration according to Equation 1.45a over the distance $\xi = 0$ to $\xi = \infty$, which yields the total charge Q_d (Equation 1.45b). Due to electro-neutrality, Q_d is opposite to the sum charge Q_M of the metal surface and the Helmholtz layer. Differentiation according to Equation 1.46 yields the capacity of the diffuse double layer. A more precise calculation without the simplifying assumptions of the Debye–Hückel theory, such as linearization of the Boltzmann distribution, yields Equations 1.47 and 1.48 for C_d and Q_d [3].

$$Q_d = -\int_{\xi=0}^{\xi=\infty} \rho\, d\xi = -\int_{\xi=0}^{\xi=\infty} \frac{\varepsilon\varepsilon_0}{\beta^2}\Delta\varphi\, d\xi = -\frac{\varepsilon\varepsilon_0\zeta}{\beta^2}\int_{\xi=0}^{\xi=\infty} \exp\left\{-\frac{\xi}{\beta}\right\} d\xi \tag{1.45a}$$

$$Q_d = -\frac{\varepsilon\varepsilon_0\,\zeta}{\beta} = -Q_M \tag{1.45b}$$

$$C_d = -\left(\frac{dQ_d}{d\zeta}\right) = -\frac{\varepsilon\varepsilon_0}{\beta} = \sqrt{\frac{2\,\varepsilon\varepsilon_0 e_0^2 I}{kT}} = \sqrt{\frac{2\,\varepsilon\varepsilon_0\,F^2 I}{RT}} \tag{1.46}$$

$$C_d = -\left(\frac{dQ_d}{d\zeta}\right) = zF\sqrt{\frac{\varepsilon\varepsilon_0\,2c}{RT}}\cosh\left(\frac{zF\zeta}{2RT}\right) \tag{1.47}$$

$$Q_M = -Q_d = 2\sqrt{\varepsilon\varepsilon_0 RTc}\,\sinh\left(\frac{zF\zeta}{2RT}\right) \tag{1.48}$$

As already discussed in detail for the Debye–Hückel theory for strong electrolytes, the Debye length β varies strongly with the ionic strength I. In the concentration range of 10^{-4} to 1 M of a 1,1 valent electrolyte like a KCl solution, it changes from 30.4 to 0.3 nm. Therefore, ß has ionic dimensions for an excess of conducting electrolyte, so that the potential drop concentrates on the Helmholtz layer only. In general, the double layer consists of the Helmholtz and the outer diffuse double layer. This has the consequence that the total differential capacity C consists of the two partial capacities C_H and C_d in series and is calculated as given by Equation 1.49:

$$\frac{1}{C} = \frac{1}{C_H} + \frac{1}{C_d} \tag{1.49}$$

C_H can be seen as a constant part and C_d as a variable dependent on the ionic strength I. C_H may be obtained from Equation 1.49 with the value for C_d calculated with Equation 1.46 or 1.47. For electrolyte concentrations $c > 1$ M, the contribution of C_d to C is negligible due to its vanishingly small reciprocal value in Equation 1.49.

The composition of the double layer has been investigated directly with XPS. In some cases, electrodes may be emersed from the electrolyte in hydrophobic conditions. This water repelling situation avoids adhering electrolyte but keeps the ions within the charged double layer so that they may be examined quantitatively with XPS as a function of the various parameters such as the potential or the electrolyte composition. The balance of anions and cations yields the charge at the electrolyte side, which equals the charge with opposite sign at the metal surface. Following the potential dependence of this excess charge, one may determine the potential of zero charge (pzc), i.e., the potential at which the interface is not charged. The pzc varies with the orientation of a crystal surface, which is related to the changes of the work function. Figure 1.11 gives an example for an Ag surface emersed from an acidified 1 M solution of $NaClO_4$. The surface concentrations are in the range of some few 0.1 mol · cm^{-2}. The amount of Na^+ decreases and that of ClO_4^- increases with increasing electrode potential. The difference of both is zero at $E_{pzc} = -0.5$ V and has opposite sign for $E < E_{pzc}$ and $E > E_{pzc}$ as expected. This type of investigation has also been performed in various other solutions and with other metals [22,29–32]. From the shift of the binding energy of the XPS peaks with potential, one may even draw conclusions about the location of the ions, i.e., if they are close to the surface within the inner Helmholtz layer or more apart with water layers between the metal surface and the ions, thus experiencing more or less of the

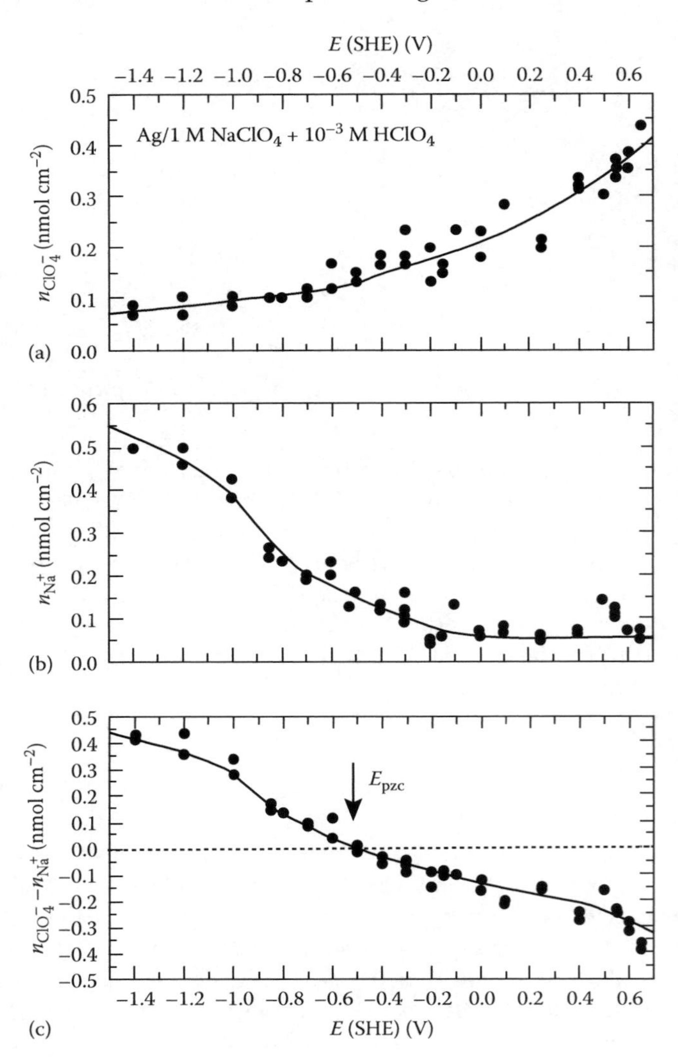

FIGURE 1.11
Composition (nmol cm^{-2}) of the surface of an hydrophobically emersed Ag electrode from 1 M $NaClO_4$ + 10^{-3} M $HClO_4$ solution as a function of the applied electrode potential E, (a) $n(ClO_4^-)$, (b) $n(Na^+)$, (c) $n(ClO_4^-) - n(Na^+)$: charge balance with indication of pzc. (From Hecht, D., *Thesis Heinrich-Heine-Universität*, Düsseldorf, Shaker Verlag, Aachen, Germany, 1997.)

potential drop across the interface. The composition of the double layer and the potential of zero charge E_{pzc} change with the specific adsorption of electrolyte components or during the formation of layers like oxide films. This situation has a pronounced influence on the kinetics of the surface processes including corrosion. These important details are discussed in the following sections of this chapter and Chapter 5 (Passivity of Metals).

A simpler situation exists at the surface of a liquid metal. Liquid metals have a homogeneous surface without different crystal planes, grain boundaries, dislocations, and other defects, which exist on solid metal surfaces. Therefore, many detailed studies on surface properties have been performed on mercury electrodes. They also may be easily cleaned by renewing a hanging droplet from a capillary connected to a mercury reservoir. The most important advantage is that one may measure easily the surface tension or surface energy γ. Figure 1.12 shows one method for measuring γ by a hanging droplet of Hg within electrolyte solutions vs. the applied electrode potential. E is applied *via* a voltage source, the RE being the CE at the same time. This is possible as long as there is no electrochemical reaction, which could polarize the RE. The equilibrium of the forces, i.e., surface tension against gravity, leads to Equation 1.50.

$$2r\pi\gamma\cos\theta = r^2\pi h\rho g; \quad \gamma = \frac{r\rho g h}{2\cos\theta} \tag{1.50}$$

where

- r is the radius of the capillary
- h is the height of the mercury column, which is a measure of the pressure caused by its weight
- ρ is the density of Hg
- $g = 981\ \mathrm{cm \cdot s^{-2}}$ is the gravitational acceleration
- θ is the contact angle of Hg with the glass wall of the capillary

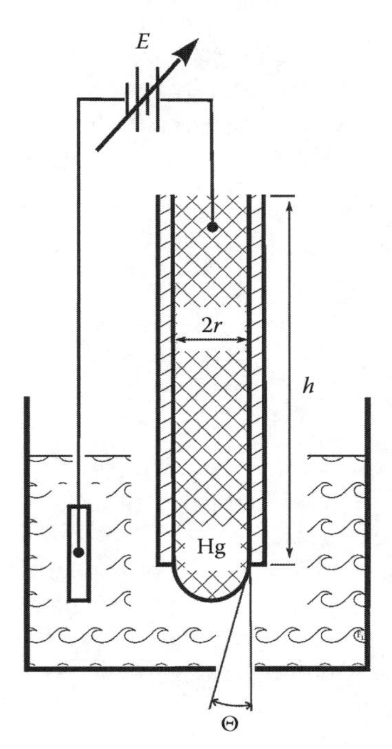

FIGURE 1.12
Measurement of the surface tension as a function of the potential with a hanging mercury drop (electrocapillary curve).

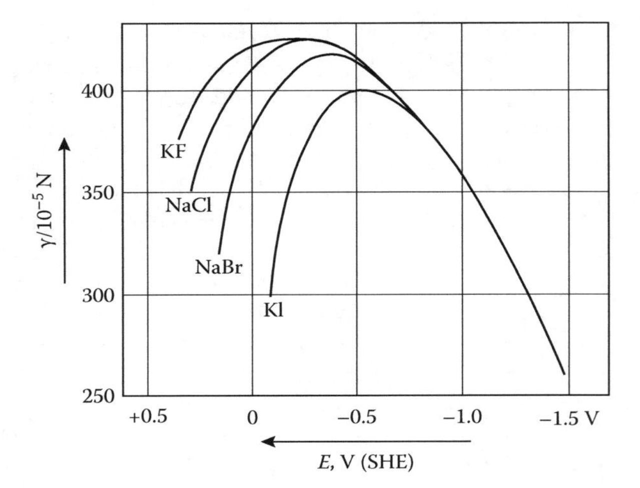

FIGURE 1.13
Electrocapillary curves of Hg in 0.1 M halide solutions. (From Grahame, D.C., *J. Am. Soc.*, 76, 4819, 1954.)

Figure 1.13 shows the electrocapillary curve of a mercury electrode in different halide solutions [33]. The shape of the parabolas gets more asymmetrical going from KF *via* NaCl and NaBr to KI. Simultaneously, its maximum is shifted to more negative potential values. This is related to the increasing specific adsorption of the anions, with I^- being the most strongly specifically adsorbed anion on Hg. F^- does not show any specific adsorption; hence, its electrocapillary curve is symmetrical and its maximum has the most positive potential value. The dependence of γ on E is a consequence of the surface charge on the Hg electrode. The charges lead to repulsive forces and thus reduce γ. These repulsive forces disappear at the maximum when the surface charge is zero. This occurs at pzc. The thermodynamic discussion of this problem leads to Gibbs' adsorption isotherm of Equation 1.51, which describes γ as a function of the electrode potential E and the surface charge Q_M on one hand and the chemical potentials and the surface excesses Γ_i of the different components of the electrolyte on the other. Keeping the electrolyte composition constant and changing the electrode potential only yields the Lippmann equation (Equation 1.52), which gives a quantitative relation between γ and Q_M.

$$d\gamma = -Q_M\, dE - \sum_i \Gamma_i\, d\mu_i \tag{1.51}$$

$$d\gamma = -Q_M\, dE \tag{1.52}$$

The slope of the tangent to the electrocapillary curve, $d\gamma/dE$, gives the surface charge Q_M. In consequence, the maxima of the parabolas of Figure 1.13 correspond to the pzc as already discussed qualitatively above. Apparently, one needs a more negative potential to repel specifically adsorbed anions in order to get the uncharged Hg surface. As F^- is not specifically adsorbed, it serves as a reference for measurements of surface tension and the determination of the charge-free electrode surface. $NaClO_4$ and $KClO_4$ solutions behave similarly without specific adsorption. The first derivative of Lippmann's equation yields Q_M and consequently the second derivative yields the differential capacity C_D (Equation 1.53).

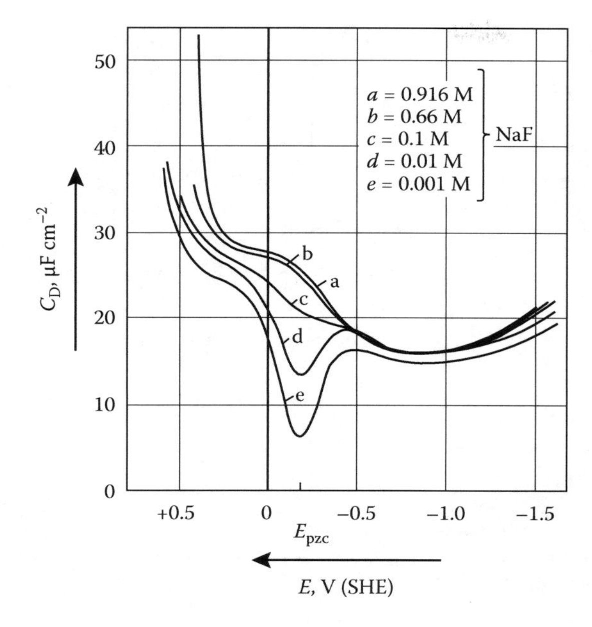

FIGURE 1.14
Differential capacity $C_D = dQ/dE = -d^2\gamma/dE^2$ of Hg in NaF solutions of different concentrations with a peak for $c \leq 0.01$ M showing the influence of the diffuse double layer and the potential of zero charge E_{PZC}. (From Lecoeur, J. and Bellier, J.P., *Electrochim. Acta*, 30, 1027, 1985.)

$$C_D = \frac{dQ_M}{dE} = -\frac{d^2\gamma}{dE^2} \tag{1.53}$$

Without specific adsorption, C_D should be independent of E. This is approximately the case for low concentrations of NaF solutions at negative potentials as shown in Figure 1.14 [34]. However, in the vicinity of pzc, a pronounced minimum is found. This is the influence of the diffuse double layer as expected according to the Gouy–Chapman theory of dilute electrolytes. If the NaF concentration gets larger, this minimum is less pronounced and disappears at ca. 1 M solutions as expected. For this latter situation, the ionic strength is getting large enough to suppress the formation of the diffuse double layer.

For solid electrodes, the determination of γ gets much more difficult. Therefore, C_D is determined directly by electrochemical capacity measurements. The results for a Cu(111) surface in $KClO_4$ solutions without specific adsorption are similar as for Hg as shown in Figure 1.15 [35]. However, the pzc deduced from the minimum of the $C_D - E$ dependence changes appreciably with the crystal orientation as expected. Table 1.4 summarizes the results for Cu, Ag, and Au single crystals in NaF solutions.

The second part of Gibbs' adsorption isotherm of Equation 1.51 describes the change of γ with the chemical potentials μ_i and thus with the electrolyte composition. Here the surface excess Γ_i of a species is the deviation of its amount n_i at the surface from the average $\overline{n_i}$ of the bulk electrolyte divided by the surface area A of the electrode (Equation 1.54). The surface excess of the electrode/electrolyte interface may be determined more directly by quantitative XPS studies on emersed electrodes, which has been described above in this section of the chapter.

$$\Gamma_i = \frac{n_i}{A} - \frac{\overline{n_i}}{A} \tag{1.54}$$

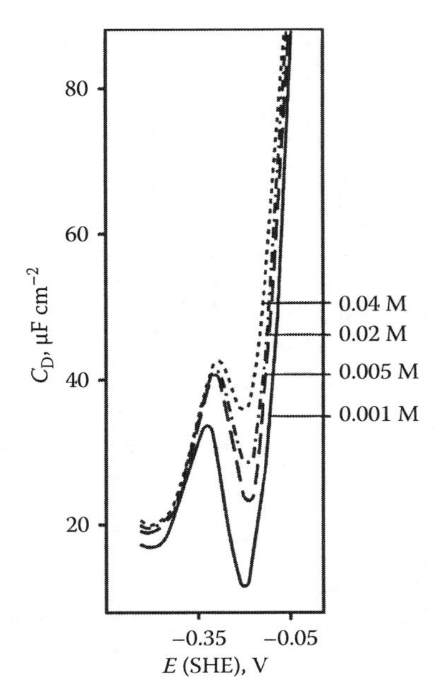

FIGURE 1.15
Differential capacity C_D of Cu(111) in $KClO_4$ solutions of different concentrations as a function of *E*. (From Barin, I., Knacke, O., and Kubaschewski, O., *Thermochemical Properties of Inorganic Substances*, Springer, Berlin, Germany, 1973.)

TABLE 1.4
Potential of Zero Charge of Some Single-Crystal Surfaces in Aqueous NaF Solution

Orientation	Cu	Ag	Au
(111)	−0.200	−0.45	0.643
(100)	−0.540	−0.650	0.350
(210)		−0.790	0.120

Source: Lecoeur, J. and Bellier, J.P., *Electrochim. Acta*, 30, 1027, 1985.

1.8 Thermodynamics of Chemical Equilibria

As in chemistry in general, one may distinguish in electrochemistry between the equilibrium and the kinetics of processes. Both topics will be treated as a basis for corrosion. For the description of electrochemical equilibria, one needs thermodynamics for which a short introduction is given in this section. A more detailed description will be found in textbooks of physical chemistry [1].

1.8.1 Energy *U* and Enthalpy *H*

The energetics of a chemical reaction is described by the internal energy *U* and the enthalpy *H*. The internal energy of a system contains the energy of thermal motion like translation, rotation and oscillation of molecules and atoms, their chemical energy, and for specimens with large surfaces the energy of the surface or interface. In chemistry and corrosion, each compound is characterized by a standard energy of formation ΔU_f defined as the change of internal energy for the reaction of formation of the compound from its elements in their stable forms at 25°C (298.15 K) and 1 atm (1.013×10^5 Pa). From this definition, ΔU_f^{298} has a value zero for elements in their stable form at 25°C and 1 atm. Knowing ΔU_f^{298} and C_V, the molar heat capacity at constant volume, one may calculate ΔU_f at any temperature using Equation 1.55, and adding ΔU_{Ch} for isotherm phase changes like melting, evaporation, etc. The same holds for the enthalpy H, except that the molar heat capacity at constant pressure C_P is used (Equation 1.56). For integration of Equations 1.55 and 1.56, the dependence of C_V and C_P on temperature has been measured and is presented in thermodynamic tables mostly as a polynome in T as given in Equation 1.57 where a, b, c and a', b', c' are empirical constants obtained by fitting of the experimental values.

$$\Delta U_f(T) = \Delta U_f^{298} + \int_{298}^{T} C_V \, dT \quad (1.55)$$

$$\Delta H_f(T) = \Delta H_f^{298} + \int_{298}^{T} C_P \, dT \quad (1.56)$$

$$C_V = a + bT + cT^2 + \cdots; \qquad C_P = a' + b'T + c'T^2 + \cdots \quad (1.57)$$

The standard enthalpy of formation ΔH_f^{298} for a compound is defined as the enthalpy (heat) change for the reaction of formation from the elements under standard conditions, i.e., $T = 25°C$ and $P = 1$ atm. It either can be measured directly by calorimetric measurements on the direct reaction if possible or may be calculated by Hess's law, which is a consequence of the first law of thermodynamics. Energy conservation requires that energy changes are independent of the reaction path. If a reaction may not be measured directly, one may use another reaction path instead, which leads from the same reactants to the desired products. Thus, a proper combination of reactions with known ΔH_r values permits calculating the value for the compound of interest. As an example reaction (Equation 1.58) for burning C to CO does not stop at CO in a calorimetric bomb, but continues up to CO_2. It is more convenient to see Equation 1.58 as a combination of reactions Equations 1.59 and 1.60, where CO_2 is formed by the oxidation of C or CO, respectively. C is burnt to CO_2 according to Equation 1.59 and then changed to CO by the reverse of Equation 1.60. The ΔH values have to be treated according to the reaction equations in order to obtain $\Delta H_f(CO)$ (Equation 1.61). This procedure holds for any other chemical reaction.

$$C + \tfrac{1}{2}O_2 \rightarrow CO \quad (1.58)$$

$$C + O_2 \rightarrow CO_2 \quad (1.59)$$

$$CO + \tfrac{1}{2}O_2 \rightarrow CO_2 \quad (1.60)$$

$$\Delta_f H°(CO) = \Delta_f H°(CO_2) - \Delta_r H°(CO - CO_2) = -393.3 + 283.0 = -110.3 \text{ kJ} \quad (1.61)$$

For a general chemical reaction given by Equation 1.62, one obtains the standard reaction enthalpy by the algebraic stoichiometric sum of the standard enthalpies of formation according to Equation 1.63.

$$\nu_1 S_1 + \nu_2 S_2 + \cdots + \nu_k S_k \rightarrow \nu_1 S_1 + \nu_m S_m + \cdots \quad (1.62)$$

$$\Delta_r H^{\circ 298} = \sum_i \nu_i \Delta H_{f,i}^{298} \quad (1.63)$$

In Equation 1.63, the stoichiometric factors ν_i are positive for products (on the right side of Equation 1.62) and negative for reactants (on the left side). $\Delta_r H^{\circ 298}$ equals $\Delta_r U^{\circ 298}$ for reacting species in the condensed state only. If gases are involved Equation 1.64 must be considered.

$$\Delta H = \Delta U + \Delta(pV) = \Delta U + p\Delta V + V\Delta P \tag{1.64}$$

Applying the ideal gas law, one obtains $\Delta(pV) = \Delta\nu\ RT$. ΔV is the change of moles of gas for the reaction. Burning of 1 mol of CO to CO_2 according to Equation 1.60 yields $\Delta\nu = -0.5$, which results in

$$\Delta\nu\ RT = -0.5 \times 8.314 \times 298 = -1239\ \text{J}\cdot\text{mol}^{-1} = -1.239\,\text{kJ}\cdot\text{mol}^{-1} \quad \text{for } T = 298\,\text{K}.$$

$\Delta_f H^{298}$ for dissolved species includes the standard enthalpy of dissolution of the compound. The formation of 1 mol of HCl in an ideally diluted solution (aq) involves the formation of 1 mol HCl in the gas phase (g) and its dissolution within a large amount of water to a concentration $c \rightarrow 0$ (Equation 1.65a through c).

$$\tfrac{1}{2}\text{H}_2\,(\text{g}) + \tfrac{1}{2}\text{Cl}_2\,(\text{g}) \rightarrow \text{HCl}\,(\text{g}) \quad \Delta_f H^{298}(\text{HCl}) = -92.22\,\text{kJ} \tag{1.65a}$$

$$\text{HCl}\,(\text{g}) \rightarrow \text{H}^+ \times (\text{aq}) + \text{Cl}^- \times (\text{aq}) \quad \Delta_{\text{dis}} H^{298}(\text{HCl}, \text{aq}) = -75.07\,\text{kJ} \tag{1.65b}$$

$$\tfrac{1}{2}\text{H}_2(\text{g}) + \tfrac{1}{2}\text{Cl}_2 \rightarrow \text{H}^+ \times (\text{aq}) + \text{Cl}^-(\text{aq}) \quad \Delta_f H^{298}(\text{HCl}, \text{aq}) = -167.29\,\text{kJ} \tag{1.65c}$$

With the convention that the standard formation enthalpy of H^+ ions in a ideally diluted solution, i.e., for $c \rightarrow 0$ is $\Delta_f H^{298}(\text{H}^+_{\text{aq}}) = 0$, one obtains for $\Delta_f H(\text{Cl}^-, \text{aq})$:

$$\begin{aligned}\Delta_f H^{298}(\text{HCl}, \text{aq}) = \Delta_f H^{298}(\text{Cl}^-, \text{aq}) &= +\Delta_f H^{298}(\text{HCl}) + \Delta_{\text{dis}} H^{298}(\text{HCl}, \text{aq}) \\ &= -92.22 - 75.07 = -167.29\,\text{kJ}\end{aligned} \tag{1.65d}$$

One assumes that the whole enthalpy change refers to the formation of 1 mol Cl^- (aq) ions. On this basis, one may calculate $\Delta_f H^{\circ 298}$ of all ions in solution with the thermodynamical data of the corresponding salt, acid, or base in solution. As an example, the data for KCl in solution allow calculating $\Delta_f H^{\circ 298}$ of K^+ (aq) as follows.

$$\text{K}\,(\text{s}) + \tfrac{1}{2}\text{Cl}_2\,(\text{g}) + \rightarrow \text{KCl}(\text{s}) \quad \Delta_f H^{\circ 298}[\text{KCl}\,(\text{s})] = -438.06\,\text{kJ}\cdot\text{mol}^{-1} \tag{1.66a}$$

$$\text{KCl}\,(\text{s}) \rightarrow \text{K}^+(\text{aq}) + \text{Cl}^-(\text{aq}) \quad \Delta_{\text{sol}} H^{\circ 298}[\text{KCl}(\text{aq})] = +17.18\,\text{kJ}\cdot\text{mol}^{-1} \tag{1.66b}$$

$$\text{K}\,(\text{s}) + \tfrac{1}{2}\text{Cl}_2\,(\text{g}) \rightarrow \text{K}^+(\text{aq}) + \text{Cl}^-(\text{aq}) \tag{1.66c}$$

$$\Delta_f H^{298}(\text{KCl}(\text{aq})) = \Delta_f H^{\circ 298}(\text{KCl}\,(\text{s})) + \Delta_{\text{sol}} H^{\circ 298}(\text{KCl}(\text{aq})) = -420.88\,\text{kJ}\cdot\text{mol}^{-1} \tag{1.66d}$$

$$\Delta_f H^{298}(\text{K}^+(\text{aq})) = \Delta_f H^{\circ 298}(\text{KCl}(\text{aq})) - \Delta_f H^{\circ 298}(\text{Cl}^-(\text{aq})) = -420.87 + 167.29 = -253.58\,\text{kJ}\cdot\text{mol}^{-1} \tag{1.66e}$$

In this way, one may calculate the standard enthalpy of formation of any ionic species. One should keep in mind that all these values refer to the convention that $\Delta_f H^{\circ 298}(H^+(aq)) = 0$. The values of $\Delta_f H^{\circ 298}$ at $p = 1$ atm for any compound in its equilibrium state and for dissolved species including ions are listed in thermochemical tables of references [9,36,37], or frequently in the web [38–40].

The $\Delta_r H^{\circ 298}$ values of any reaction may be calculated from $\Delta_f H^{\circ 298}$ of the reactants. Equation 1.67 gives an example of an ionic reaction (1.67a) and the related enthalpies (1.67b).

$$Ca^{2+}(aq) + CO_2(g) + H_2O(1) \rightarrow CaCO_3(s) + 2H^+(aq) \tag{1.67a}$$

$$\begin{aligned}\Delta_r H^{298} &= 2\Delta_f H^{298}(H^+(aq)) + \Delta_f H^{298}(CaCO_3(s)) - \Delta_f H^{298}(Ca^{2+}(aq)) \\ &\quad - \Delta_f H^{298}(CO_2(g)) - \Delta_f H^{298}(H_2O(1)) \\ &= -0 - 1205.72 + 542.44 + 393.14 + 284.73 = 14.6\,\text{kJ}\cdot\text{mol}^{-1}\end{aligned} \tag{1.67b}$$

Equation 1.68a presents the formation of Fe^{2+} ions from Fe metal by reduction of two H^+ ions to H_2. Equation 1.68b yields $\Delta_f H^{298}(Fe^{2+})$, i.e., the standard formation enthalpy of Fe^{2+} ions ($\Delta_f H^{298} = 0$ for H^+ and H_2).

$$Fe + 2H^+ \rightarrow Fe^{2+} + H_2 \tag{1.68a}$$

$$\Delta_r H^{\circ 298} = \Delta_f H^{\circ 298}(Fe^{2+}) = -89.1\,\text{kJ mol}^{-1} \tag{1.68b}$$

It is often of interest in corrosion to calculate $\Delta_r H^{298}$ for the electrochemical formation of an oxide from its metal in contact with water. As an example, this is given for Fe_3O_4 at $T = 298$ K by Equation 1.69b with zero $\Delta_f H^{\circ 298}$ values for the elements Fe and H_2.

$$3Fe + 4H_2O(1) \rightarrow Fe_3O_4 + 4H_2 \tag{1.69a}$$

$$\Delta_r H^{\circ 298} = \Delta_f H^{\circ 298}(Fe_3O_4) - 4\Delta_f H^{\circ 298}(H_2O,(1)) = -1116.06 - 4(-284.73) = 22.86\,\text{kJ} \tag{1.69b}$$

A reaction is called endothermic when $\Delta_r H^{\circ 298} > 0$ and exothermic if $\Delta_r H^{\circ 298} < 0$. Thus, reaction (1.69a) is endothermic and reaction (1.68b) is exothermic.

1.8.2 Entropy *S*, Helmholtz Free Energy *F*, Gibbs Free Energy or Free Enthalpy *G*

A chemical reaction is driven not only by the energy or enthalpy change $\Delta_r U$ or $\Delta_r H$, but also by the entropy change $\Delta_r S$. $\Delta_r S$ is equal to the stoichiometric sum of the molar entropies S_i of the reactants and products according to Equation 1.70.

$$\Delta_r S = \sum_i \nu_i S_i \tag{1.70}$$

where the stoichiometric factors ν_i are positive for products (on the right side of Equation 1.62) and negative for reactants (on the left side). The same relation holds for reactants and products in their standard state at 25°C (298.15 K):

$$\Delta_r S_0^{298} = \sum_i \nu_i S_i^{298} \tag{1.70a}$$

The standard molar entropies S_i^{298} for elements and compounds can be found in thermodynamic tables [9,36–40]. The entropy is zero at $T = 0\,\mathrm{K}$ for all elements and compounds that are in their energetically stable form (i.e., in an inner equilibrium, without frozen disorder like in disordered or glassy polymers). For a particular phase, the change of the molar entropy between two temperatures T_1 and T_2 may be calculated from C_{pi}, the molar heat capacity at constant pressure by integration of the differential equation 1.71a, which yields Equation 1.71b. For C_{pi}, the polynome in T is used as discussed for the temperature dependence of the enthalpy (Equation 1.57) with the data of thermodynamical tables.

$$\mathrm{d}S_i = C_{pi}\frac{dT}{T} \tag{1.71a}$$

$$S_i(T_2) = S_i(T_1) + \int_{T_1}^{T_2} \frac{C_{pi}}{T}\,dT \tag{1.71b}$$

At each isotherm phase transition like melting, evaporation, etc., the entropy change of Equation 1.72 has to be added where T_{Ch} is the temperature and ΔH_{Ch} the enthalpy change of the transition.

$$\Delta S_{\mathrm{Ch}} = \frac{\Delta H_{\mathrm{Ch}}}{T_{\mathrm{Ch}}} \tag{1.72}$$

Hence, if there is one change from a phase I to a phase II between a temperature T_1 and a temperature T_2, the molar entropy at T_2 may be calculated from the value at T_1 using Equation 1.73.

$$S_i(T_2) = S_i(T_1) + \int_{T_1}^{T_{\mathrm{Ch}}} \frac{C_{pi,\mathrm{I}}}{T}\,dT + \frac{\Delta H_{\mathrm{Ch}}}{T_{\mathrm{Ch}}} + \int_{T\mathrm{Ch}}^{T_2} \frac{C_{pi,\mathrm{II}}}{T}\,dT \tag{1.73}$$

It includes the molar heats $C_{pi,\mathrm{I}}$ and $C_{pi,\mathrm{II}}$ and the related integrals for the temperature increase within both phases and the entropy change for the phase transition. Starting from $T_1 = 0\,\mathrm{K}$, one may calculate all entropies at any temperature T with the given data for $C_{pi}(T)$ starting with the standard value $S_i(T_1) = S_i^0(0) = 0$ for chemicals in internal equilibrium.

The entropy change of any reaction $\Delta_r S$ indicates quantitatively if the reaction may occur or not. $\Delta_r S > 0$ holds for a spontaneous reaction, $\Delta S = 0$ for a reaction in equilibrium, and $\Delta S < 0$ for a nonpossible reaction, i.e., the reverse of a spontaneous reaction. These conditions hold for a closed or isolated system, which includes a heat exchanger to keep the temperature constant for the process without any energy exchange with the environment. Thus, the total entropy change $\Delta S_{\mathrm{Tot}} = \Delta S + \Delta S_{\mathrm{Env}}$ contains the change ΔS_{Env} of the environment, i.e., of the heat exchanger.

For a nonisolated system, i.e., a system that allows exchange of heat and material with the environment, the driving force for a chemical reaction is given by the change of the Helmholtz free energy ΔF, or the change of the Gibbs free energy (or free enthalpy) ΔG, with $G = F + PV$. Both functions are defined by the Gibbs–Helmholtz equation (Equations 1.74 and 1.75). Similar to the relations of ΔU and ΔH, ΔF and ΔG differ for a reaction with the change of the number of moles of gas $\Delta\nu$ as given by Equation 1.75a.

$$\Delta F = \Delta U - T\Delta S \tag{1.74}$$

$$\Delta G = \Delta H - T\Delta S \tag{1.75}$$

$$\Delta G = \Delta F + \Delta(PV) = \Delta F + P\Delta V + V\Delta P = \Delta F + RT\Delta\nu \tag{1.75a}$$

where ΔG is the maximum work, which may be obtained from a chemical reaction at constant pressure. This will be achieved for reversible conditions, which are characterized by the maintenance of equilibrium conditions throughout the process. For $\Delta G < 0$ the reaction is possible, for $\Delta G = 0$ it is in equilibrium, and for $\Delta G > 0$ it will not occur and only the opposite reaction is possible.

This may be easily visualized by an electrochemical cell. The electrochemical reactions at both electrodes lead to a potential difference between them, i.e., the voltage U of the electrochemical cell, which will cause a current flow I in the external circuit. As a consequence, an electrical work $w_{el} = UIt$ will be exchanged for a given time t with a related amount of the chemical process. This work has its maximum value w_{max} if the electrochemical processes at the electrodes occur very slowly with $I \rightarrow 0$ requiring a long time $t \rightarrow \infty$. In this case, the electrode processes stay at equilibrium and the cell voltage keeps its maximum value ($U_{max} = \Delta E_{eq}$). For a given amount of chemical reaction, the electrical work refers to a reversible process, and it has its maximum value, which corresponds to the absolute value of the change of the free enthalpy of the chemical cell reaction $|\Delta_r G| = w_{rev} = w_{max} = U_{max}It$. For a not negligibly small current flow, the process becomes irreversible and the cell voltage decreases due to the kinetic hindrance and the related overpotentials at the electrodes and $w_{el} = UIt < w_{rev} = w_{max} = |\Delta_r G|$.

Two main contributions determine the value of the $\Delta_r G$, the reaction enthalpy $\Delta_r H$, and the reaction entropy $\Delta_r S$ as given by the Gibbs–Helmholtz equation (Equation 1.75). For a large driving force, ΔH should be negative and ΔS positive, i.e., for a negative ΔG the reaction minimizes the energy and maximizes the entropy of the reacting system.

$\Delta_r G$ is given by the stoichiometric sum of the molar free enthalpies usually called chemical potentials μ_i of the reactants and products i according to Equation 1.76.

$$\Delta_r G = \sum_i \mu_i \nu_i \tag{1.76}$$

The standard value of $\Delta_r G$ for a given chemical reaction, $\Delta_r G^0$ ($p = 1$ atm/1.013×10^5 Pa, all $a_i = 1$ M, T) at a given temperature may be calculated from the standard values $\Delta_r H^0$ and $\Delta_r S^0$ according to Equation 1.75. An alternative is the calculation from the standard chemical potentials also called standard Gibbs free energies of formation, according to Equation 1.76.

$$\Delta_r G^0 = \sum_i \nu_i \mu_i^0 \tag{1.76a}$$

$\mu_i^0 = \Delta_f G_i^0$ at a given temperature is the Gibbs free energy of formation of a compound with unit activity at $P = 1$ atm from its elements under their stable form at 25°C (298.15 K). From this definition, its value is zero for elements under their stable form at $T = 298$ K. The values for $\mu_i^0 = \Delta_f G_i^0$ are published in thermodynamic tables, most often at 25°C only [9,36–40].

The dependence of the chemical potentials on the activity can be expressed by the general relation (Equation 1.77).

$$\mu_i = \mu_i^0 + RT \ln a_i \tag{1.77}$$

a_i is the (dimensionless) activity of species i.

For gaseous species, a_i is defined with respect to the partial pressure P_i according to Equation 1.77a:

$$a_i = \frac{\phi_i P_i}{(P_i^\circ = 1)} \tag{1.77a}$$

where ϕ_i is the fugacity coefficient, equal to the ratio of the partial fugacity of the gas to its partial pressure. It represents the deviation of the real gas from the ideal gas behavior, so it tends toward unity when P_i tends to zero, and interactions between i gas atoms or molecules become negligible.

For a species in solution, a_i is defined with respect to the molar concentration c_i (mol L^{-1}) or the molar fraction x_i according to Equations 1.77b and c:

$$a_i = \gamma_i \frac{c_i}{(c_i^\circ = 1)} \tag{1.77b}$$

$$a_i' = \gamma_i' \, x_i \quad \text{for solvents including solid solutions (alloys)} \tag{1.77c}$$

γ_i and γ_i' are dimensionless activity coefficients, representing the deviation from the ideal solution behavior when the concentration varies. The molar concentration scale is the most appropriate for solutes, as γ_i tends toward a constant value γ_i^0 when the concentration tends to zero. The molality scale (mol·kg^{-1}) is more general as it allows treatment at high temperatures, without having to deal with volume variations.

The molar fraction scale where γ_i' tends toward unity when x_i tends to unity is more appropriate for solvents. It is also convenient for solid solutions (alloys).

1.8.3 Electrochemical Equilibrium and Nernst Equation

The electrode potential E_{eq} for a general electrochemical reaction, as in Equation 1.78, is referred to the hydrogen electrode, whose reaction is given by Equation 1.79.

$$\nu_1 S_1 + \nu_2 S_2 + \cdots + ne^- \rightarrow \nu_k S_k + \nu_l S_l + \cdots \tag{1.78}$$

$$\tfrac{1}{2} n\mathrm{H}_2 \rightarrow n\mathrm{H}^+ + ne^- \tag{1.79}$$

where n is the number of electrons, which are consumed for the cathodic reaction of Equation 1.78 and which are produced by the anodic reaction of Equation 1.79. The sum of the two partial reactions (Equations 1.78 and 1.79) gives the overall chemical reaction:

$$\nu_1 S_1 + \nu_2 S_2 + \cdots + \tfrac{1}{2} n\mathrm{H}_2 \rightarrow \nu_k S_k + \nu_l S_l + \cdots + n\mathrm{H}^+ \tag{1.80}$$

E_{eq} is related to the free energy of this reaction by the relation:

$$\Delta_r G = -nFE_{eq} \tag{1.81}$$

Applying Equations 1.76 and 1.77, one obtains the Nernst equation (Equations 1.82 and 1.83).

$$\Delta G = -nFE = \sum_i \mu_i v_i + RT \ln\left[\frac{a_k^{v_k} a_l^{v_l} \cdots a(\mathrm{H}^+)^n}{a_1^{v_1} a_2^{v_2} \cdots P(\mathrm{H}_2)^{n/2}}\right] \tag{1.82}$$

$$E_{eq} = E^0 + \frac{RT}{nF} \ln\left[\frac{a_1^{v_1} a_2^{v_2} \cdots P(\mathrm{H}_2)^{n/2}}{a_k^{v_k} a_l^{v_l} \cdots a(\mathrm{H}^+)^n}\right] = E^0 + \frac{RT}{nF} \ln\left[\frac{a_1^{v_1} a_2^{v_2}}{a_k^{v_k} a_l^{v_l}}\right] \tag{1.83}$$

Equation 1.83 simplifies if the SHE with $p(H_2) = 1$ atm and $a(H^+) = 1$ (molar scale) is used as a reference. E^0 the standard electrode potential is given by Equation 1.84.

$$E^0 = -\frac{1}{nF}\sum_i \mu_i^0 \nu_i^0 = \frac{1}{nF}\left(\nu_1\mu_1^0 + \nu_2\mu_2^0 + \cdots + \frac{n}{2}\mu_{H_2}^0 - n\mu_{H^+}^0 - \nu_k\mu_k^0 - \nu_1\mu_1^0 - \cdots\right) \quad (1.84)$$

If we define the standard chemical potential or Gibbs free energy of a conventional electron by:

$$\mu_{e^-}^0 = \tfrac{1}{2}\mu_{H_2}^0 - \mu_{H^+}^0 \quad (1.85)$$

(which by definition of the standard free energies of formation takes the value zero at 25°C), then the standard electrode potential for the electrochemical equilibrium of Equation 1.78 may be conveniently written as follows:

$$E^0 = \frac{1}{nF}(\nu_1\mu_1^0 + \nu_2\mu_2^0 + \cdots + n\mu_{e^-}^0 - \nu_k\mu_k^0 - \nu_1\mu_1^0 - \cdots) \quad (1.86)$$

Introducing the constants and $T = 298.15$ K to the pre-logarithmic factor of Equation 1.83 and changing to the decadic logarithm, one obtains for a one-electron process $RT \ln 10/F = 8.314 \times 298.15 \times 2.303/96{,}487 = 0.059$ V.

As an example, the equilibrium potentials of the Fe/Fe^{2+} and the Fe^{2+}/Fe^{3+} electrodes, a metal/metal ion and a redox electrode, respectively, can be calculated using Equations 1.83 and 1.86 and are given in Equations 1.87 and 1.88.

$$E_{eq} = -0.440\,\text{V} + 0.030\,\text{lg}\,[a(\text{Fe}^{2+})] \quad (1.87)$$

$$E_{eq} = 0.770 + 0.059\,\text{lg}\left[\frac{a(\text{Fe}^{3+})}{a(\text{Fe}^{2+})}\right] \quad (1.88)$$

The standard potentials indicate the trend of the electrodes to dissolution or oxidation. According to Equation 1.81, the negative value $E^0 = -0.440$ V indicates a positive value for ΔG^0 for reduction of Fe cations by H_2 (Equation 1.80) and thus a spontaneous opposite reaction, i.e., Fe dissolution into Fe^{2+} and reduction of H^+ with H_2 evolution. The positive value $E^0 = +0.770$ V for the Fe^{3+}/Fe^{2+} redox couple indicates that it is a strong oxidant for hydrogen and especially for reactive metals with negative metal/ion standard potentials. Standard potentials for inorganic and organic redox systems and Me/Me^{z+} electrodes are found in Ref. [9]. Table 1.5 summarizes a selection of some systems.

1.8.4 Potential–pH Diagrams

The thermodynamical data are used in corrosion to calculate potential–pH diagrams also called Pourbaix diagrams, which show under which conditions a metal is stable, corroding, or protected by an oxide or hydroxide layer. The data of Equations 1.87 and 1.88 are

TABLE 1.5
Standard Potentials of Redox Systems and Metal/Metal Ion Electrodes

Electrode Reaction	Standard Potential E_0/V
$Li^+ + e^- \rightarrow Li$	−3.045
$K^+ + e^- \rightarrow K$	−2.925
$Na^+ + e^- \rightarrow Na$	−2.714
$Al^{3+} + 3e^- \rightarrow Al$	−1.66
$Zn^{2+} + e^- \rightarrow Zn$	−0.763
$Cr^{3+} + 3e^- \rightarrow Cr$	−0.74
$AsO_4^{3-} + 2H_2O + 2e^- \rightarrow AsO_2^- + 4OH^-$	−0.71
$Cr^{2+} + 2e^- \rightarrow Cr$	−0.557
$Fe^{2+} + 2e^- \rightarrow Fe$	−0.440
$Cr^{3+} + e^- \rightarrow Cr^{2+}$	−0.41
$Cd^{2+} + 2e^{2-} \rightarrow Cd$	−0.403
$Ni^{2+} + 2e^- \rightarrow\rightarrow Ni$	−0.23
$Sn^{2+} + 2e^- \rightarrow Sn$	−0.136
$Pb^{2+} + 2e^- \rightarrow Pb$	−0.126
$Fe^{3+} + 3e^- \rightarrow Fe$	−0.036
$SeO_3^{2-} + 3H_2O + 4e^- \rightarrow Se + 6OH^-$	−0.35
$TeO_3^{2-} + 3H_2O + 4e^- \rightarrow Te + 6OH^-$	−0.02
$2H^+ + 2e^- \rightarrow H_2$	0.000
$Cu^{2+} + e^- \rightarrow Cu^+$	0.158
$AgCl + e^- \rightarrow Ag + Cl$	0.222
$Hg_2Cl_2 + 2e^- \rightarrow 2Hg + 2Cl^-$	0.280
$Cu^{2+} + 2e^- \rightarrow Cu$	0.340
$O_2 + 2H_2O + 4e^- \rightarrow 4OH^-$	0.401
$Fe(CN)_6^{3-} + e^- \rightarrow Fe(CN)_6^{4-}$ (0.01 M NaOH)	0.46
$I_2 + 2e^- \rightarrow 2I^-$	0.536
$Cu^+ + e^- \rightarrow Cu$	0.522
$Fe(CN)_6^{3-} + e^- \rightarrow Fe(CN)_6^{4-}$ (0.5 M H_2SO_4)	0.69
$Fe^{3+} + 3e^- \rightarrow Fe^{2+}$	0.770
$Hg_2^{2+} + 2e^- \rightarrow 2Hg$	0.796
$Ag^+ + e^- \rightarrow Ag$	0.800
$Hg^{2+} + 2e^- \rightarrow Hg$	0.851
$Hg_2{}^{2+} + 2e^- \rightarrow 2Hg$	0.796
$Br_2 + 2e^- \rightarrow 2Br^-$	1.065
$SeO_4^{2-} + 4H^+ + 2e^- \rightarrow H_2SeO_3 + H_2O$	1.15
$O_2 + 4H^+ + 4e^- \rightarrow 2H_2O$	1.229
$Cr_2O_7^{2-} + 14H^+ + 6e^- \rightarrow 2Cr^{3+} + 7H_2O$	1.33
$Cl_2 + 2e^- \rightarrow 2Cl^-$	1.360
$MnO_4^- + 8H^+ + 5e^- \rightarrow Mn^{2+} + 4H_2O$	1.491
$Ce^{4+} + e^- \rightarrow Ce^{3+}$	1.61
$F_2 + 2e^- \rightarrow 2F^-$	2.85

Source: *Handbook of Chemistry and Physics*, CRC Press, Cleveland, OH, 1977–1978.

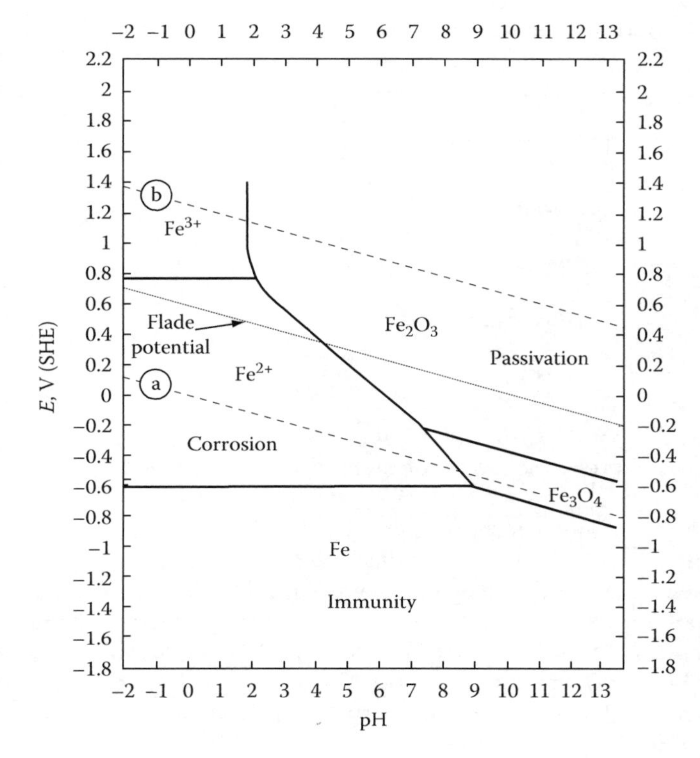

FIGURE 1.16
Potential pH diagram (Pourbaix diagram) for Fe showing the ranges of immunity, corrosion, and passivity. (From Poubaix, M., *Atlas of Electrochemical Equilibria in Aqueous Solution*, Pergamon Press, Oxford, 1996.)

included in the diagram for Fe as presented in Figure 1.16, for an activity $a = 10^{-6}$ of dissolved Fe ions. It especially includes data for the formation of anodic oxides.

As an example, the reduction of Fe_3O_4 into Fe (Equation 1.89) is considered, which illustrates the calculation of ΔG^0 and E^0 from thermodynamic data.

$$Fe_3O_4 + 8H^+ + 8e^- \rightarrow 3Fe + 4H_2O \tag{1.89}$$

$$\begin{aligned}\Delta_r G^0 &= 3\mu^0(\text{Fe}) + 4\mu^0(\text{H}_2\text{O}) - \mu^0(\text{Fe}_3\text{O}_4) - 8\mu^0(\text{H}^+) - 8\mu^0(\text{e}^-) \\ &= 0 + 4(-237.13) - (-1{,}013.23) - 0 - 0 = 64{,}710\,\text{J}\,\text{mol}^{-1}\end{aligned} \tag{1.90}$$

$$E^0 = -\frac{\Delta_r G^0}{8F} = -0.084\,\text{V} \tag{1.91}$$

This electrode process involves H^+ ions and thus is pH dependent. The equilibrium potential E at 25°C depends on pH according to −0.059 V/pH unit (pH = −log $a(H^+)$):

$$E = -0.084 + \left(\frac{0.059}{8}\right)\log a(\text{H}^+)^8 = -0.084 - 0.059\,\text{pH} \tag{1.92}$$

An important redox equilibrium between a dissolved species (Fe^{2+}) and a solid deposit (Fe_2O_3) is considered now:

$$Fe_2O_3 + 6H^+ + 2e^- \rightarrow 2Fe^{2+} + 3H_2O \tag{1.93}$$

The related data are presented in Equations 1.94 through 1.96.

$$\Delta G^0 = 2\mu^0(Fe^{2+}) + 3\mu^0(H_2O) - \mu^0(Fe_2O_3) - 6\mu^0(H^+) - 2\mu^0(e^-)$$
$$= 2(-84.85) + 3(-237.13) - (-740.45) - 0 - 0 = -140,250\ J \quad (1.94)$$

$$E^0 = -\frac{-140,250}{2F} = 0.728\ V \quad (1.95)$$

$$E = 0.728 + \frac{0.059}{2}\log[a(H^+)^6] - \frac{0.059}{2}\log[a(Fe^{2+})^2] = 0.728 - 0.025\log[a(Fe^{2+})] - 0.177\ pH \quad (1.96)$$

Due to six involved hydrogen ions, the equilibrium potential has a strong pH dependence of -0.177 V/pH unit (Equation 1.96). This result leads to the steep line in Figure 1.16 for the Fe^{2+}/Fe_2O_3 equilibrium.

Table 5.1 sums up all equilibria for Fe species represented by the lines in Figure 1.16. The equilibrium potentials vs. pH variations may be calculated with the thermodynamic data as demonstrated above. These lines delimit areas that are the stability domains for Fe species. At sufficiently negative potentials Fe is stable, i.e., it is the domain of immunity. At low pH, Fe should dissolve forming Fe^{2+} for $E > -0.620$ V ($c(Fe^{2+}) = 10^{-6}$ M) and Fe^{3+} for $E > 0.77$ V, which delimits the stability domain of corrosion. The upper right part is the domain of stable oxides, the passivity domain. For these conditions, the oxides should protect the Fe surface against the attack of the electrolyte.

The potential–pH diagrams have been published for most metals. They provide valuable information on conditions for which a metal is immune, passive, or submitted to corrosion by the electrolyte. Figure 1.16 also shows the E–pH lines of the hydrogen (H^+/H_2) and the oxygen (O_2/H_2O) electrodes. These redox systems are important for the attack of metals in many environments. Since the Fe/Fe^{2+} line is below the H^+/H_2 line, it can be predicted that iron is attacked in acidic electrolytes due to oxidation by hydrogen ions. A more detailed discussion is presented in Section 5.3.

However, these diagrams have their limits. They are based on thermodynamic data only, i.e., they refer to equilibria and thus do not contain any information of kinetics, while corrosion is affected by both. Thus, they may be misleading. According to Figure 1.16, Fe should dissolve in strongly acidic electrolytes till to very high potentials. However, it shows passivity, when the potential exceeds the Flade potential, given by Equation 1.97.

$$E_p = 0.58 - 0.059\ pH \quad (1.97)$$

This experimental passivation potential is presented by the central dashed line in Figure 1.16. The standard oxide formation potentials of all known iron oxides are ca. 0.6 V more negative. Thermodynamical data suggest that the Flade potential of Fe should be explained by the oxidation of Fe_3O_4 to γ-Fe_2O_3 according to Equation 1.98.

$$3\gamma - Fe_2O_3 + 2e^- + 2H^+ \rightarrow 2Fe_3O_4 + H_2O \quad (1.98)$$

Apparently Fe_3O_4 and also $Fe(OH)_2$ layers are dissolved quickly in acidic electrolytes so that they are not protective. However, γ-Fe_2O_3 dissolves very slowly, although it is far from its dissolution equilibrium in acidic electrolytes, so that it is protective. Similar discrepancies between the predictions of the potential–pH diagrams and the passive behavior hold for other metals like Ni, Cr, and also for steel. Others follow the predictions of the diagrams like Al or Cu. A more detailed discussion follows in Chapter 5.

1.8.5 Reference Electrodes

Although electrode potentials are referred to the SHE, one uses in most cases electrodes of a second kind. They are very stable and reproducible and simple to use because they avoid the use of hydrogen gas. The potentials of these electrodes are usually determined by the solubility of a weakly soluble reaction product.

The calomel electrode consists of a small container with some Hg and a contacting wire superimposed with weakly soluble Hg_2Cl_2 and a KCl solution of given concentration (0.1 M, 1 M or saturated) (see Figure 1.25a). In the Nernst equation for the Hg/Hg_2^{2+} equilibrium, the very small Hg_2^{2+} concentration is replaced by its expression vs. the known Cl^- activity, using the solubility product of Equation 1.99, which yields Equations 1.100 and 1.101 for the potential of the calomel electrode.

$$K_S = a(Hg_2^{2+})a^2(Cl^-) \tag{1.99}$$

$$E = E^0 + \frac{0.059}{2}\log\left[a\left(Hg_2^{2+}\right)\right] = E^0 + \frac{0.059}{2}\log\left[\frac{K_S}{a(Cl^-)^2}\right] = E^0_{Cal} - 0.059\log[a(Cl^-)] \tag{1.100}$$

$$E^0_{Cal} = E^0 + \frac{0.059}{2}\log\{K_S\} \tag{1.101}$$

The solubility product of Hg_2Cl_2 has been determined with Equation 1.101, $E^0_{Cal} = 0.268$ V and $E^0 = 0.796$ V to be $K_S = 1.27\ 10^{-18}$.

Several other electrodes are used as REs in electrochemical experiments. The Ag/AgCl electrode is an Ag wire anodized in Cl^- solution, thus forming a thin film of insoluble AgCl. This electrode forms in contact with a Cl^-solution a stable RE, which may be easily miniaturized. Table 1.6 presents several REs that are frequently used. In the schematic presentation, the phases are shown with a vertical line depicting a phase boundary. Most REs may be easily produced in the laboratory, and one does not need to buy commercial products. In many cases, their design may be adapted to actual experimental set-ups, even if miniaturized forms are required. Although a diaphragm prevents the mixing of the electrolytes of the RE and the electrochemical cell, one usually uses REs with an

TABLE 1.6

Some Commonly Used REs and Their Standard Potentials E^0/V

Electrode	E^0/V	Nernst Equation
Calomel electrode: Hg | Hg_2Cl_2 | Cl^-	0.268	$E = E^0 - 0.059 \log [Cl^-]$
Hg_2SO_4 electrode: Hg | Hg_2SO_4 | SO_4^{2-}	0.615	$E = E^0 - 0.059 \log [SO_4^{2-}]$
HgO electrode: Hg | HgO | OH^-	0.926	$E = E^0 - 0.059 \log [OH^-]$
AgCl electrode: Ag | AgCl | Cl^-	0.222	$E = E^0 - 0.059 \log [Cl^-]$
$PbSO_4$ electrode: Pb | $PbSO_4$ | SO_4^{2-}	−0.276	$E = E^0 - 0.029 \log [SO_4^{2-}]$

Source: *Handbook of Chemistry and Physics*, CRC Press, Cleveland, OH, 1977–1978.

electrolyte composition close to the one of the main electrolyte of the electrochemical cell. Too large differences in composition and concentration may cause unwanted liquid junction potentials.

1.9 Liquid Junction Potentials

A liquid junction potential or diffusion potential appears at the contact of two electrolytes of different composition. It is a consequence of the concentration gradient. Diffusion of the dissolved species tries to overcome the differences in the composition of the electrolytes. However, their mobility is in general different so that some of the ions are faster than others. A very simple situation is the contact of two solutions of one dissolved salt, acid, or base with a concentration difference, for example, HCl. As shown in Figure 1.17, H^+ and Cl^- ions are transported by diffusion from electrolyte I with activity a_I to electrolyte II with a_{II}. However, the mobility of H^+ is much larger than that of Cl^- (see Section 1.4). As a consequence, H^+ ions move faster and builds up a positive charge at side II, whereas the excess of the slower Cl^- ions at side I forms the negative counter charge. This leads to a potential drop $\Delta\varphi_d = \varphi_{II} - \varphi_I$. This diffusion potential difference accelerates the anions and slows down the cations so that the diffusion of both gets equal. $\Delta\varphi_d$ is superimposed to the cell voltage. Measuring an electrode potential with a RE inevitably introduces the diffusion potential difference between the electrolytes to the measurement. Therefore, one needs methods of calculation of $\Delta\varphi_d$ to subtract it. In the following example, we assume the transfer of 1 F charge across the interface from electrolyte I to II, which involves the electrical work w_{el}:

$$w_{el} = \Delta\varphi_d F \tag{1.102}$$

The amount of charge transported by the cations and anions is given by their charge transfer coefficients t_+ and t_-, respectively. Therefore, t_+/z_+ moles of cations are transferred form I to II and t_-/z_- anions from II to I. This leads to the osmotic work W_{osm+} and W_{osm-}. The total osmotic work is given by the following:

$$W_{osm} = W_{osm+} + W_{osm-} = \frac{t_+}{z_+} RT\left[\frac{a_{II}}{a_I}\right] + \frac{t_-}{|z_-|} RT\ln\left[\frac{a_I}{a_{II}}\right] \tag{1.103}$$

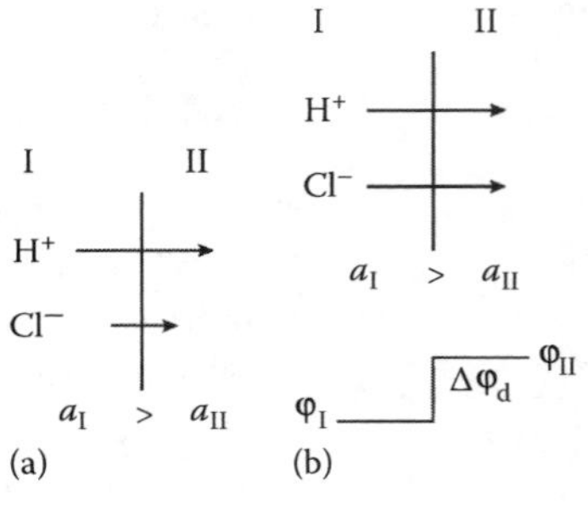

FIGURE 1.17
Liquid junction potential $\Delta\varphi_d$ between two solutions with different HCl concentrations, (a) different mobility of H^+ and Cl^- leading to (b) $\Delta\varphi_d$ and the same diffusion rate for both ions when stationary conditions are achieved.

Assuming that electrical and osmotic work are compensating each other, i.e., $w_{el} + W_{osm} = 0$ and taking into account $z_+ = |z_-| = z$ and $t_+ + t_- = 1$ one obtains:

$$\Delta\varphi_d = \varphi_{II} - \varphi_I = (1-2t_-)\frac{RT}{zF}\ln\left(\frac{a_I}{a_{II}}\right) \tag{1.104}$$

For two contacting HCl solutions with $a_I = 0.1\,M$ and $a_{II} = 0.01\,M$, one obtains with $t_+ = 0.83$ for H^+ and $t_- = 0.17$ for Cl^-: $\Delta\varphi_d = \varphi_{II} - \varphi_I = 0.039\,V$ at 25°C. $\Delta\varphi_d$ may be much larger for larger activity differences. It is also large for differences in OH^- concentrations. It is smaller for most salts due to the smaller difference of the mobilities of cations and anions leading to closer values of the charge transfer coefficients for both.

In general, the composition of the electrolytes is more complicated. For these situations, the equation of Henderson may be applied, which contains only some minor simplifications:

$$\Delta\varphi_d = \varphi_{II} - \varphi_I = \frac{RT}{F}\frac{\sum_i (1/z_i)(a_{iII} - a_{iI})u_i\,|z_i|}{\sum_i (a_{iII} - a_{iI})u_i\,|z_i|}\ln\frac{\sum_i a_{iII}u_i\,|z_i|}{\sum_i a_{iI}u_i\,|z_i|} \tag{1.105}$$

where a_i, z_i, and u_i are the activity, charge, and mobility of the various ions within the solutions I and II. Measured potential differences at the contact of two different electrolytes may be corrected from the values for $\Delta\varphi_d$, calculated by Equations 1.104 or 1.105. Another possibility is the addition of large amounts of KCl to the contacting solutions. Both ions K^+ and Cl^- have the same mobility and do not allow the formation of large $\Delta\varphi_d$ values. KNO_3 is an alternative with similar properties. A preferred choice is the connection via liquid junctions filled with a concentrated solution of KCl, often together with an agar–agar jell. At both contacts, solution I/KCl and KCl/solution II, much smaller diffusion potentials will occur, which even compensate each other partially. For $u_K = 7.6 \times 10^{-4}\,cm^2 \cdot V^{-1} \cdot s^{-1}$, $u_{Cl} = 7.9 \times 10^{-4}\,cm^2 \cdot V^{-1} \cdot s^{-1}$, and $u_H = 36.3 \times 10^{-4}\,cm^2 \cdot V^{-1} \cdot s^{-1}$, one calculates with Equation 1.105 the values $\Delta\varphi_{dI} = 3.9\,mV$ and $\Delta\varphi_{dI} = -1.6\,mV$. The total 2.3 mV is much less than the 39 mV calculated above for the direct contact of the HCl solutions.

1.10 Electrode Kinetics

Electrode processes are submitted to the rules of electrode kinetics. Here one usually has to distinguish reaction steps, which occur sequentially forming the overall electrode reaction. The main charge transfer process with the exchange of charge carriers with the electrode surface is preceded by the transport of reactants to the surface by diffusion and/or migration. The hindrance of the charge transfer or the transport process leads to a charge transfer overvoltage η_{CT} or a diffusion overvoltage η_D, respectively. An intermediate chemical reaction within the electrolyte may cause a reaction overvoltage η_R. Adsorption and desorption are also chemical reaction steps, which may precede or follow the charge transfer. Further chemical reactions and transport of the products to the bulk electrolyte after the charge transfer step may also cause related overvoltages (see Table 1.7).

TABLE 1.7

Sequence of Reaction Steps for Complex Electrochemical Reactions and Related Overpotentials

Transport of reactants	Diffusion overvoltage, η_D
Chemical reaction (homogeneous, heterogeneous)	Reaction overvoltage, η_R
Adsorption on electrode surface	
Charge transfer	Charge transfer overvoltage, η_{CT}
Desorption	
Chemical reaction (homogeneous, heterogeneous)	Reaction overvoltage, η_R
Transport of products	Diffusion overvoltage, η_D

Examples for chemical reactions are the dissociation of the weakly dissociated acetic acid in front of the electrode before the charge transfer to the hydrogen ions may take place. Another chemical reaction step is the combination of adsorbed hydrogen atoms at the electrode surface after their formation by charge transfer to form adsorbed hydrogen molecules, the so-called Tafel reaction, which is discussed in Section 1.12.1 in detail. Any of these reaction steps may be the slowest and thus rate determining depending on the experimental conditions. Diffusion control is usually reached if the reaction rate is high enough so that the transport becomes slower than the charge transfer. It shows up when a concentration gradient has been built up in front of an electrode, which requires some time. Complicated reactions consist in a sequence of reaction steps with different time constants. This property may be used to separate the reaction steps from each other by the application of electrochemical transient measurements in order to examine their individual kinetics. In the following part, the charge transfer and diffusion overpotentials are discussed in detail as they are the main rate-determining reactions in corrosion.

1.10.1 Charge Transfer Overvoltage η_{CT} and Butler–Volmer Equation

For a simple redox reaction, an electron is transferred from the reduced species Red to the electrode leaving the oxidized species Ox within the electrolyte:

$$\text{Red} \rightarrow \text{Ox} + ne^- \tag{1.106}$$

The measured current density i is proportional to the reaction rate according to Faraday's law:

$$i = nF\left[-\frac{dc(\text{Red})}{dt}\right] = nF\left[\frac{dc(\text{Ox})}{dt}\right] = nF(k_+c(\text{Red}) - k_-c(\text{Ox})) = i_+ + i \tag{1.107}$$

i is composed of an anodic i_+ and a cathodic part i_-. They have opposite sign and compensate each other at the Nernst potential, i.e., for electrochemical equilibrium. i is positive for positive overpotential where $i_+ > |i_-|$. n is the number of electrons that are involved in the overall process that has to be distinguished in general from the charge z of the species that is involved in the charge transfer step ($z = 2$ for Fe^{2+} dissolution). $c(\text{Ox})$ and $c(\text{Red})$ are the concentrations of the involved species. k_+ and k_- are the rate constants for the anodic and cathodic partial processes. A simple charge transfer reaction involves one single-electron transfer step. However, there may exist multi-electron transfer reactions, which are a fast sequence of single-electron transfer steps, which one might not be able to resolve. Like any reaction, electrochemical reactions follow the rules of kinetics, with the influence of the applied electrode potential as a decisive factor. For electrochemical reactions, the reaction order is often called electrochemical reaction order. Equation 1.107 shows a reaction order one for both directions of the process of Equation 1.106, i.e., with respect to Ox and Red. However, the rate law may be more complicated and may also involve further species, which do not even show up in the equation for the chemical process. This is the case if a substance reacts as a catalyst or an inhibitor, which is not produced or consumed during the process. A complicated rate equation is usually an indication for a complicated mechanism with various intermediates and a related sequence of reaction steps.

A very important corrosion reaction is the dissolution of a metal according to Equation 1.108. The rate equation is also composed of two parts, the anodic dissolution and the cathodic deposition at the metal surface. In this case, the species that is crossing the electrode/electrolyte interface has a charge of z_+, which equals n ($n = z_+$).

$$\mathrm{Me} \rightarrow \mathrm{Me}^{z+} + n\mathrm{e}^- \quad (z = n) \tag{1.108}$$

$$i = i_+ + i_- = nF(k_+\theta_{\mathrm{Me}} - k_-\, c(\mathrm{Me}^{z+})) \tag{1.109}$$

θ_{Me} is the surface concentration of metal atoms in an energetically favored position, which are involved in the dissolution process. The rate equation may be more complex. The dissolution process may be catalyzed by the formation of adsorbed species like OH, which then enter the rate equation as is shown later for the case of iron dissolution. It has been shown in particular by scanning tunneling microscopy (STM) and synchrotron studies that OH adsorption occurs at much more negative potentials than the equilibrium value for the formation of a three-dimensional anodic oxide layer (see Chapter 5.8.2.1). So dissolution often occurs from a surface covered by an adsorption film.

If the charge transfer step of a redox process (such as Equation 1.106) is rate determining, the Butler–Volmer equation is obtained as follows. A similar equation is obtained for metal dissolution where the concentrations c(Ox) and c(Red) are replaced by $c(Me^{z+})$ and θ_{Me}. As usual in chemical kinetics, the rate constants k contain an exponential term with the ratio of the standard activation free enthalpy $\Delta_r G^\ddagger$ to RT. $\Delta_r G^\ddagger$ is the barrier that the reacting system has to overcome to get to the transition state from which the products are formed. For a simple redox reaction, this involves the change of the coordination shell of solvent molecules and other ligands, i.e., for $(Fe(CN)_6)^{3-,4-}$.

In kinetic studies, the addition of a sufficient concentration of supporting electrolyte (ca. 1 M) allows the whole potential drop to occur immediately at the electrode/electrolyte interface, i.e., at a distance of the size of hydrated ions. Thus, the diffuse double layer becomes negligibly small and the Helmholtz layer takes over the full potential drop, which determines the rate of the charge transfer reaction.

For a potential drop $\Delta\varphi = \varphi_M - \varphi_S$ at the electrode/electrolyte interface, the electrical work for the charge carriers across the interface from the initial state (Red) to the final state (Ox + ne^-) is $w_{el} = -zF\Delta\varphi$. For a positive $\Delta\varphi$, the contribution of this electric work is to decrease the activation free energy for oxidation and increase the activation free energy for reduction. But only a fraction α of this work is kinetically effective up to the maximum of G. Therefore, the standard activation free enthalpy for the oxidation reaction at $\Delta\varphi = 0$, $\Delta_r G^\ddagger_{0+}$, is decreased by $\alpha zF\Delta\varphi$. The standard activation free enthalpy for the reduction reaction at $\Delta\varphi = 0$, $\Delta_r G^\ddagger_{0-}$, is increased by $(1-\alpha)zF\Delta\varphi$. This is shown in Figure 1.18. The charge

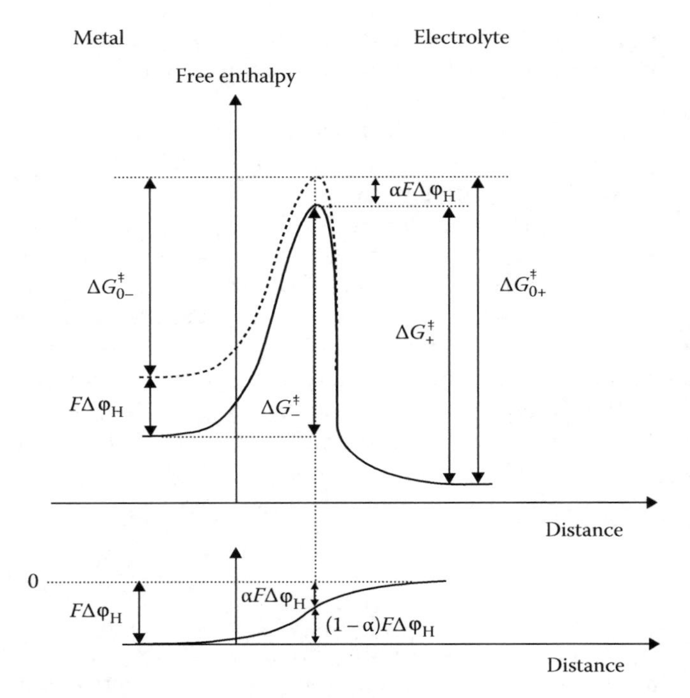

FIGURE 1.18
Diagram of free enthalpy G for charge transfer reactions with variation of free activation enthalpy $\Delta G^{\ddagger}_0$ for the anodic and cathodic processes with potential drop $\Delta\varphi_H$ across the electrode/electrolyte interface (Helmholtz layer) and the influence of the charge transfer coefficient α.

transfer coefficient α is a symmetry factor for the energy barrier whose value is $0 < \alpha < 1$, often close to 0.5. The effect of $\Delta\varphi$ on the rate constants in the oxidation or reduction direction, k_+ and k_-, is shown in Equation 1.110.

$$k_+ = k_0 \exp\left[-\frac{\Delta G^{\ddagger}_{0+} - \alpha z F \Delta\varphi}{RT}\right] = k'_{0+} \exp\left[\frac{\alpha z F \Delta\varphi}{RT}\right] \tag{1.110a}$$

$$k_- = k_0 \exp\left[-\frac{\Delta G^{\ddagger}_{0-} + (1-\alpha) z F \Delta\varphi}{RT}\right] = k'_{0-} \exp\left[\frac{-(1-\alpha) z F \Delta\varphi}{RT}\right] \tag{1.110b}$$

$\Delta\varphi$ cannot be measured or fixed separately, only the electrode potential $E = \Delta\varphi - \Delta\varphi_{SHE}$ may be given or fixed as shown in Figure 1.19 where $\Delta\varphi_{SHE}$ is the potential drop at the SHE. Replacing $\Delta\varphi$ by $(E + \Delta\varphi_{SHE})$ in Equation 1.110 and introducing the rate constants in Equation 1.107 yields Equation 1.111.

$$i = i_+ + i_- = nF\left\{k'_{0+} c(\text{Red}) \exp\left[\frac{\alpha z F \Delta\varphi_{SHE}}{RT}\right] \exp\left[\frac{\alpha z F E}{RT}\right] - k'_{0-} c(\text{Ox}) \exp\left[\frac{-(1-\alpha) z F \Delta\varphi_{SHE}}{RT}\right] \exp\left[\frac{-(1-\alpha) z F E}{RT}\right]\right\} \tag{1.111}$$

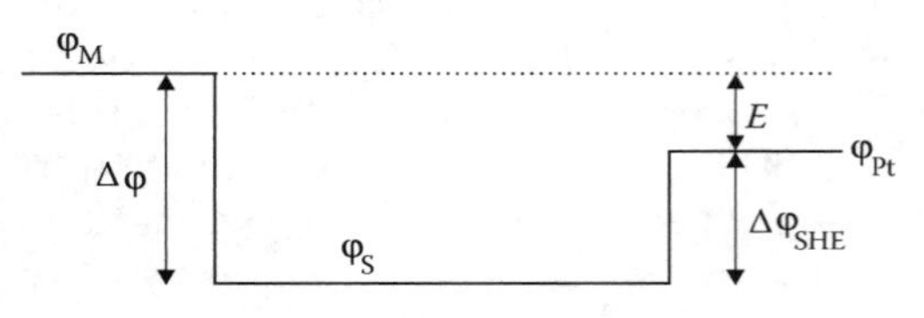

FIGURE 1.19
Potential diagram for the combination of an electrode and the SHE showing the potential drops $\Delta\varphi$ and $\Delta\varphi_{SHE}$ and the electrode potential E referred to SHE.

Considering that at the equilibrium potential E_{eq}, the overall current density $i = 0$, the following relation holds:

$$i_{+}(E_{eq}) = i_{-}(E_{eq}) = i_0 \tag{1.112}$$

i_0 is called the exchange current density and is equal to:

$$i_0 = nFk'_{0+}c(\text{Red})\exp\left[\frac{\alpha zF(E_{eq} + \Delta\varphi_{SHE})}{RT}\right] = nFk'_{0+}c(\text{Ox})\exp\left[\frac{-(1-\alpha)zF(E_{eq} + \Delta\varphi_{SHE})}{RT}\right] \tag{1.113}$$

Defining the charge transfer overvoltage by $\eta_{CT} = E - E_{eq}$ and replacing E by $(\eta_{CT} + E_{eq})$ in Equation 1.111 yields, using Equation 1.113, the Butler–Volmer equation (Equation 1.114).

$$i = i_0\left\{\exp\left[\frac{\alpha zF\eta_{CT}}{RT}\right] - \exp\left[\frac{-(1-\alpha)zF\eta_{CT}}{RT}\right]\right\} = i_{+} + i_{-} \tag{1.114}$$

For large positive overpotentials, i.e., $\eta_{CT} >> RT/F$, the cathodic partial current density i_c may be neglected and the total current density equals i_+ leading to the simple exponential equation $i = i_+ = i_0 \exp(\alpha zF\eta_{CT}/RT)$. For large negative overpotentials, i.e., $\eta_{CT} << -RT/F$, the anodic partial current density i_+ may be neglected and the total current density equals i_- leading to $i = i_- = i_0 \exp[-(1-\alpha)zF\eta_{CT}/RT]$. For these conditions, the logarithm of the measured current density i varies linearly with the overpotential according to the so-called Tafel equation (Equations 1.115a and b).

$$\eta_{CT} = \frac{RT}{\alpha zF}\ln i_0 + \frac{RT}{\alpha zF}\ln i_{+} = a_a + b_a \log i_{+} \tag{1.115a}$$

$$\eta_{CT} = \frac{RT}{(1-\alpha)zF}\ln i_0 - \frac{RT}{(1-\alpha)zF}\ln |i_{-}| = a_c - b_c \log |i_{-}| \tag{1.115b}$$

Figure 1.20 presents schematically the Tafel plot of the Butler–Volmer expression of the current density in a generalized (i/i_0 reduced) form. Deviations from the linear plot are obtained for low overpotentials, when the counter reaction cannot be neglected. For $\alpha = 0.5$, the anodic and cathodic lines are symmetrical and they differ with increasing difference of the charge transfer coefficients α and $1 - \alpha$. The Tafel plot permits the evaluation of important kinetic parameters, i.e., the slope b and the intercept a, which, from Equation 1.115, lead to the charge transfer coefficient α and the exchange current density i_0. i_0 is a measure of the reaction rate. It contains the activation energy and the related Arrhenius factor and is dependent on the concentration of the reacting species. The concentration dependence of i_0 gives information about the electrochemical reaction order of the participating species and thus indirectly on the mechanism of the reaction. In the case of a simple charge transfer process, such as in Equation 1.106, the reaction order is 1 with respect to Ox or Red (Equation 1.113). However, it may depend on other species participating in the reaction with electrochemical reaction orders equal to any number different of 0 including even noninteger numbers. The temperature dependence of i_0 permits conclusions on the standard activation free enthalpy $\Delta G_0^{\ddagger}$.

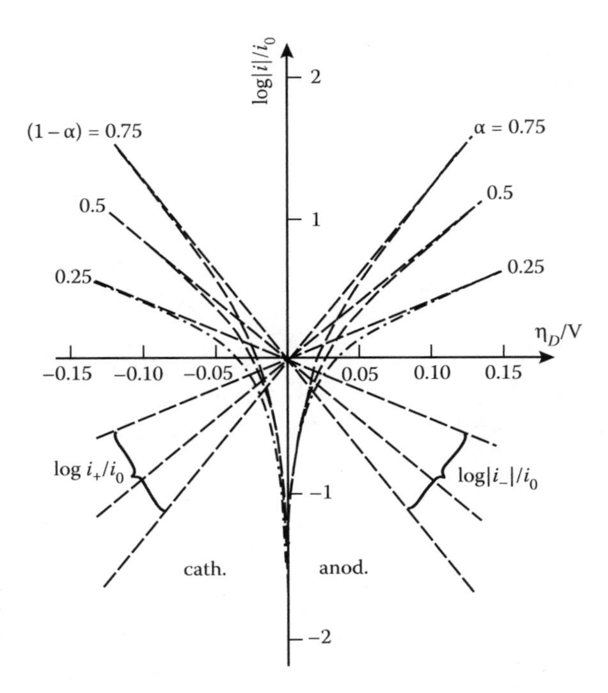

FIGURE 1.20
Tafel plots for a charge transfer reaction with different values of the charge transfer coefficient α. (From Vetter, K.J., *Electrochemical Kinetics*, Academic Press, New York, 1967.)

Close to $\eta = 0$, *V* the Butler–Volmer equation (Equation 1.114) may be linearized with the first term of a McLaurin series:

$$(i)_{\eta_{CT}\to 0} = i_0\left\{1+\frac{\alpha zF\eta_{CT}}{RT}-1-\frac{(1-\alpha)zF\eta_{CT}}{RT}\right\}=\frac{i_0 zF\eta_{CT}}{RT} \tag{1.116}$$

The derivative of Equation 1.116 leads to the slope $(d\eta_T/di)$ for $\eta_T \to 0\,V$, called the charge transfer resistance R_{CT} (Equation 1.117), which allows calculating the exchange current density i_0 without measuring the whole η_{CT} vs. i relationship.

$$R_{CT}=\left(\frac{d\eta_{CT}}{di}\right)_{\eta_{CT}\to 0}=\frac{RT}{i_0 zF} \tag{1.117}$$

A similar relation will be derived for the polarization resistance R_P when two different processes compensate each other like metal dissolution and hydrogen evolution (see Equations 1.162 and 1.163).

Principle studies of the charge transfer have been performed with simple redox systems like Fe^{2+}/Fe^{3+} or $Fe(CN)_6^{4-}/Fe(CN)_6^{3-}$. The central Fe^{2+} ion changes to Fe^{3+} by transfer of one electron to the electrode surface. The coordination sphere of six CN^- ions does not change, only the distance changes slightly. Also the hydration shell of dissolved Fe ions will not change decisively. However, most reactions include much more changes in the environment of the reacting species. For many reactions, the hydration shell is submitted to more changes than for a simple charge transfer process. For metal dissolution, a metal atom will leave the surface and will end within the electrolyte as a hydrated or complexed metal cation with a charge z_+. Here much more changes occur and the electrode reaction gets more complex.

For many reactions, the charge transfer is only one elementary step in a sequence of many others. Some substances break chemical bonds and form new ones. Oxygen reduction is a relatively complicated process with several intermediate species corresponding to a sequence of reaction steps. Nevertheless, a corrosion reaction is often ruled by the Butler–Volmer equation, although reaction steps other than the charge transfer may be rate determining as well.

The details of the charge transfer have been discussed and calculated with several models. The main results are presented in Figure 1.21, showing the density of states of a metal and those of a redox system. Here the energy of the electrons is $|e_0|\ \varphi$ ($e_0 = -1.6 \times 10^{-19}$ C), and the change of the electrode potential appears as a shift of the Fermi level of the metal relative to the one of the redox system. The electronic states of the metal are filled up to the Fermi energy $E_{F,Me}$. At $T = 0$ K, all states below $E_{F,Me}$ are filled and all states above $E_{F,Me}$ are empty. However, at room temperature, some states below $E_{F,Me}$ are free and some states above $E_{F,Me}$ are filled due to thermal activation of electrons. Similarly, the states of the reduced component of the redox system are filled and those of the oxidized components are empty. These states show a distribution on the energy scale due to the variations in the ligand or hydration shell. For Fe^{2+} and Fe^{3+} ions, this is a variation of the distance of water molecules to the central cation and in the case of $Fe(CN)_6^{3-,4-}$, it is that of the CN^- ligands and the water molecules in the second shell. λ is the solvent reorganization energy, taking into account the energy difference of the ligand sphere between the oxidized and the reduced species. The Fermi level $E_{F,Redox}$ is in the middle of the two distributions. Redox reactions require electron transfer from the filled states of the metal to the empty states of the oxidized species for a cathodic reduction and from the filled states of the reduced species to the empty states in the conduction band of the metal for an anodic oxidation. This requires tunnel processes between occupied and empty states at the same energy level. A change of the solvation shell with an energy shift up or down occurs only with the frequency of oscillations of the ligands. The electron transfer occurs in a time range of 10^{-15} s, which is two orders of magnitude faster and thus keeps the energy of the ligand shells constant. Therefore charge transfer occurs at the same level in the energy diagrams of Figure 1.21.

Figure 1.21a depicts the situation of a metal electrode in electrochemical equilibrium with the redox system in the contacting electrolyte, when the overpotential $\eta = \varphi - \varphi_{eq}$ is equal to zero. Therefore, the Fermi energies are at the same level in the energy scale, $(E_{F,Me})_{eq} = E_{F,Redox}$. The overlap between the occupied states of the metal and the unoccupied states of the redox system is equal to the one between the unoccupied states of the metal and the occupied states of the redox system. As a consequence, the partial current densities i_+ and i_- have the same value with opposite sign and compensate to a vanishing total current density $i = i_+ + i_- = 0$. Figure 1.21b depicts the situation for a positive overvoltage ($\varphi > \varphi_{eq}$). The difference between the Fermi levels is $E_{F,Me} - E_{F,Redox} = E_{F,Me} - (E_{F,Me})_{eq} = -|e_0|$ $(\varphi - \varphi_{eq}) = -|e_0|\eta$. Here the overlap between the unoccupied states of the metal and the occupied states of the reduced species is larger and thus leads to $i_+ > |i_-|$ hence to an anodic total current density $i > 0$. Figure 1.21c depicts the situation for a negative overvoltage η with a larger overlap between the occupied states of the metal and the empty states of the oxidized species, which leads to $|i_-| > i_+$ hence to a cathodic total current density $i < 0$. These concepts of charge transfer by tunnel processes are used for the semiconductor/electrolyte interface and are presented in Section 1.18.

1.10.2 Diffusion Overvoltage η_D

In the sequence of elementary reaction steps, the transport of reactants by diffusion may be rate determining. This will cause a concentration gradient in front of an electrode.

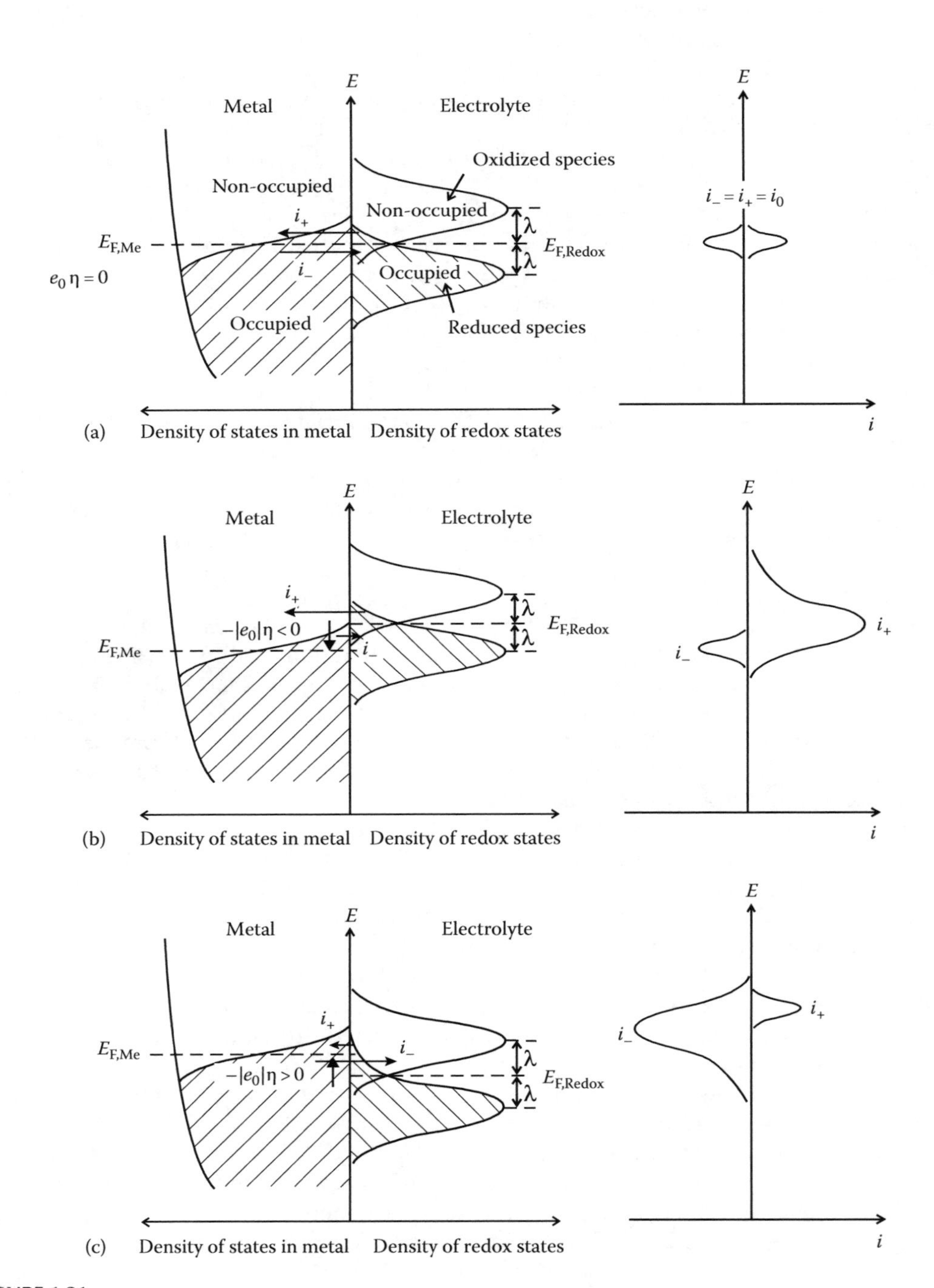

FIGURE 1.21
Band model for charge transfer between a metal electrode and a redox system within the electrolyte and the related current density $i = i_+ + i_-$ for (a) $\eta = 0$, (b) $\eta > 0$, and (c) $\eta < 0$.

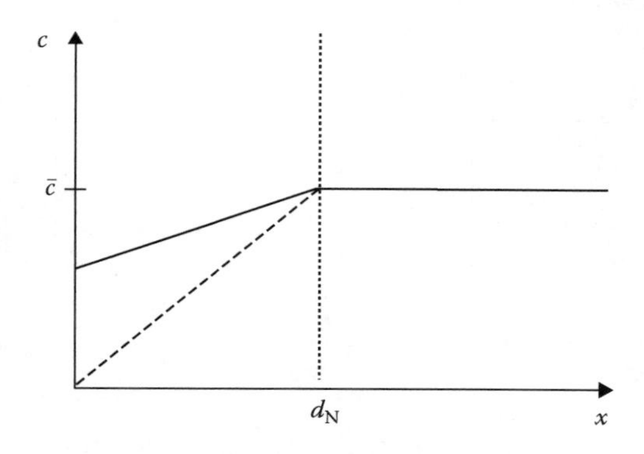

FIGURE 1.22
Concentration profile in front of an electrode within the Nernst diffusion layer d_N leading to diffusion control for the electrode process.

Assuming electrochemical equilibrium for the charge transfer step, the potential drop at the electrode/electrolyte interface will be determined by the concentration of the reactants in its direct vicinity according to the Nernst equation. For the consumption of a redox constituent due to the electrochemical reaction, its concentration in front of an electrode will change in a simple case linearly with the distance. This decrease of the concentration occurs within the Nernst diffusion layer of thickness d_N from c_B in the bulk of the electrolyte to c adjacent to the electrode surface (Figure 1.22). Applying Ficks diffusion law together with Faraday's law leads to Equation 1.118 for the diffusion-limited current density i of a cathodic process at the electrode like $Ox + e^- \rightarrow Red$.

$$i = -\frac{(c_B - c)nFD}{d_N} \tag{1.118}$$

where
D is the diffusion constant
n is the number of exchanged electrons
F is the Faraday constant

The largest concentration gradient $(c_B - c)/d_N$ is obtained for $c = 0$, which causes the maximum diffusion current density i_D of Equation 1.119.

$$i_D = -\frac{nFDc_B}{d_N} \tag{1.119}$$

Introduction of Equation 1.119 in Equation 1.118 yields Equation 1.120.

$$i = i_D + \frac{nFDc}{d_N} \tag{1.120}$$

The cathodic current density for the reduction of an oxidized species Ox gets more negative with decreasing potential and finally levels off to a constant value when i_D is reached (Figure 1.23). Introducing the concentration in front of the electrode in Nernst's equation yields Equation 1.121 for the diffusion overvoltage η_D.

$$E - E_{eq} = \eta_D = \frac{RT}{nF}\{\ln c - \ln c_B\} = \frac{RT}{nF}\ln\frac{c}{c_B} \tag{1.121}$$

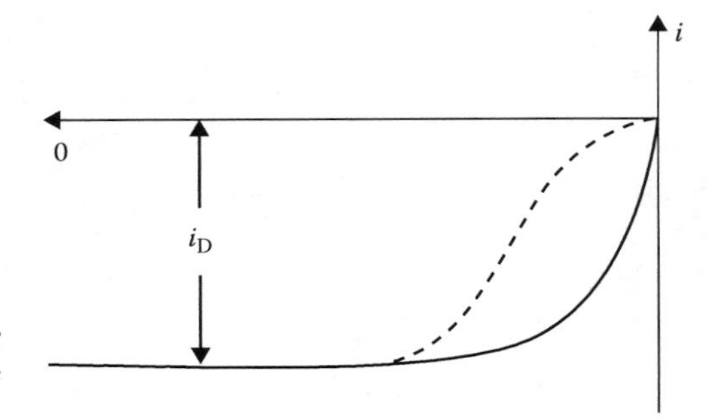

FIGURE 1.23
Current potential diagram for a cathodic reaction under diffusion control, and with superimposed charge transfer control (dashed curve).

The ratio of concentrations c_B and c can be expressed by the current density using Equations 1.118 and 1.119, leading to Equation 1.122.

$$\eta_D = \frac{RT}{nF}\ln\frac{(i_D + i)}{i_D} \tag{1.122}$$

This equation is important for the reduction of oxidants as a counter reaction for metal dissolution for corrosion processes. The same relations will hold for the oxidation of a reduced species Red, however with positive current densities i and i_D and a negative sign in Equations 1.120 through 1.122.

In general, the charge transfer and diffusion overpotential add to each other according to $\eta = \eta_{CT} + \eta_D$, and the measured current density is limited by both effects. At small values, i is determined by the Butler–Volmer equation due to a dominating charge transfer reaction and becomes diffusion limited at larger overpotentials as shown in Figure 1.22 by the dashed line.

Anodic metal dissolution under diffusion control is described by an equivalent relation:

$$i = \frac{(c - c_B)nFD}{d_N} \tag{1.118a}$$

Dissolving metal ions are accumulated at the surface where $c > c_B$. Strong metal dissolution at large overpotentials associated with diffusion-limited transport of cations will cause saturation and precipitation of salt films at the electrode surface. The maximum concentration gradient and the largest diffusion-limited current i_D is reached for $c = c_S$ and a vanishingly small bulk concentration $c_B = 0$, where c_S is the concentration of saturation:

$$i_D = \frac{nFDc_S}{d_N} \tag{1.119a}$$

These effects are important for intense metal dissolution during pitting and electrochemical machining of metals where very high anodic current densities are effective. For anodic potentiostatic transients, the precipitation of a salt film slows down an initially higher dissolution rate and starts diffusion-limited electropolishing of the metal surface.

1.10.3 Ohmic Drop η_Ω and Microelectrodes

Ohmic drops are important for the kinetics of electrode processes and corrosion especially for low-conducting electrolytes or high current densities. Furthermore, the geometry of

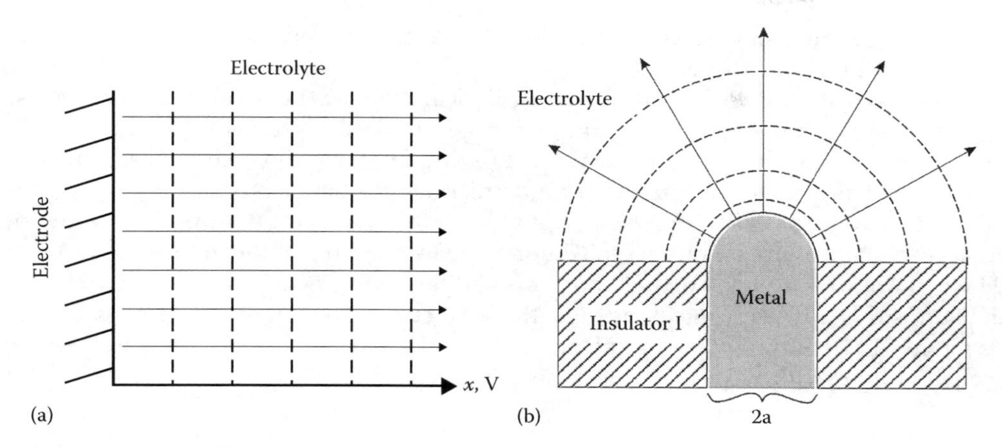

FIGURE 1.24
Current lines and lines of equal concentration (or equal potential) in front of a (a) planar electrode, (b) convex hemispherical electrode.

the electrode surface has a large influence. For a planar electrode, the ohmic resistance within the electrolyte R_Ω changes with the distance x from its surface (Figure 1.24a):

$$R_\Omega = \frac{x}{\kappa} \tag{1.123}$$

where κ is the specific conductivity of the electrolyte.

The related ohmic drop $\Delta U_\Omega = iR_\Omega$ may lead to an appreciable difference between the actual potential and that chosen for the experiment. It can be minimized by an appropriate position of the Haber–Luggin (HL) capillary of the RE, close to the surface of the working electrode (WE). However, it should not be too close in order to avoid the partial blocking of the metal surface and the formation of crevices. As a compromise, the distance should be about three times the diameter of the capillary. The ohmic drop may be also compensated electronically to about 90% of its value by an appropriate circuit built in the potentiostat as shown later in the block diagram of Figure 1.26a. The ohmic drop may be minimized by a small size of the electrode.

The current lines in front of a hemispherical electrode are divergent (Figure 1.24b), and, thus, the local current density and the related ohmic drop are decreasing with increasing distance r. For an electrode of radius a as shown in Figure 1.24b, the resistance and the related ohmic drop are given as a function of the distance r by Equation 1.124, where κ is the specific conductivity.

$$R_\Omega = \frac{ra}{\kappa(r+a)}; \quad R_\Omega = \frac{a}{\kappa} \quad \text{for } r \to \infty \tag{1.124}$$

This situation is used for microelectrodes in low-conducting electrolytes and for high current densities. In this case, a small radius a leads to small ohmic drops ΔU_Ω. A well-conducting electrolyte like 0.5 M H_2SO_4 with a specific conductivity $\kappa = 22\ \Omega^{-1} \cdot cm^{-1}$ yields an ohmic drop $\Delta U_\Omega = 23$ mV for a current density $i = 1\ A \cdot cm^{-1}$ and a distance $r = 0.5$ cm according to Equation 1.123. With a microelectrode of radius $a = 1\ \mu m = 10^{-4}$ cm, one obtains with Equation 1.124 for the otherwise same conditions $\Delta U_\Omega = 4.5 \times 10^{-3}$ mV. ΔU_Ω gets proportionally smaller for electrodes of nanometer dimensions. For weakly conductive electrolytes,

ΔU_Ω gets larger and may be very important for large electrode areas; however, it is still negligible for microelectrodes.

A similar situation holds for small hemispherical pits in the case of localized corrosion. For this situation, the ohmic drop may be large due to a high local dissolution current density; however, it decreases tremendously with the pit radius, i.e., for an early stage of pitting. For this concave situation of a hemispherical corrosion pit, the ohmic drop is increased by a factor of 3 at the pit bottom as has been found by numerical calculations [3,41,42]. The details are discussed in Chapter 7 (Mechanisms of Pitting Corrosion).

The maximum diffusion current density is similarly affected by the size of the electrode. For a hemispherical electrode of radius *a*, the local current density *i*(*r*) decreases according to Equation 1.125 with increasing distance *r* from the electrode surface due to the diverging current lines within the electrolyte (see Figure 1.24b).

$$i(r) = i(a)\frac{a^2}{(a+r)^2} \tag{1.125}$$

where *i*(*a*) is the current density at the electrode surface with $r = 0$.

Applying Fick's diffusion law and Faraday's law to calculate the diffusion-limited current density *i*(*r*) at any distance *r* within the electrolyte yields:

$$i(r) = -nFD\left(\frac{dc}{dr}\right)_r \tag{1.126a}$$

Combining Equations 1.125 and 1.126a yields for the current density *i*(*a*) at the surface of the microelectrode:

$$i(a) = -nFD\frac{(a+r)^2}{a^2}\frac{dc}{dr} \tag{1.126b}$$

Integration according to Equation 1.126c yields Equation 1.127 for the current density *i*(*a*) with c_B and c_S for the concentrations in the bulk and at the surface, respectively.

$$i(a)\int_{r=\infty}^{r=0}\frac{dr}{(a+r)^2} = -\frac{nFD}{a^2}\int_{c_B}^{c_S} dc \tag{1.126c}$$

$$-\frac{i(a)}{a} = -\frac{nFD}{a^2}(c_S - c_B)$$

$$i(a) = -\frac{nFD}{a}(c_B - c_S) \tag{1.127}$$

Assuming $c_S = 0$ at the surface ($r = 0$), one obtains Equation 1.128 for the maximum diffusion current density i_D.

$$i(a) = i_D = -\frac{nFD}{a}c_B \tag{1.128}$$

For a bulk concentration of the redox species Ox, $c_B = 1\,M = 10^{-3}\,mol \cdot cm^{-3}$, $n = 1$, $d_N = 5 \times 10^{-3}\,cm$ for the Nernst diffusion layer, and $D = 5 \times 10^{-6}\,cm^2\,s^{-1}$, one obtains $i_D = -0.1\,A \cdot cm^{-2}$ for the cathodic diffusion-limited current density at a planar electrode according to Equation 1.119. For a microelectrode with a radius $a = 1\,\mu m = 10^{-4}\,cm$, Equation 1.128 yields the much higher value $i_D = -5\,A \cdot cm^{-2}$. Consequently, the change from a charge transfer-controlled to a diffusion-controlled process occurs at higher current densities for microelectrodes. Similarly, the precipitation of a salt layer due to the accumulation of corrosion products up to saturation during anodic metal dissolution requires much higher current densities at the surface of microelectrodes compared to large planar electrodes.

1.11 Electrochemical Methods

This section provides an introduction to electrochemical methods, which are relevant to corrosion studies. More detailed descriptions may be obtained from textbooks of electrochemistry [2].

1.11.1 Electrochemical Cells

The electrochemical cell is usually adapted to the kind of measurements. The cell design for many routine measurements often corresponds to that presented in Figure 1.25a. A glass container of ca. $50\,cm^3$ content is equipped with fittings, holding a CE, the HL capillary, providing connection to the RE and an inlet for nitrogen or argon gas to remove oxygen from the electrolyte. The gas is leaving the closed cell via a water-filled valve. The WE is usually glued with resin to a glass tube with a connecting wire introduced via a central fitting. A flag shape of WE with a thin extension permits a small contact area between resin and metal, which may be critical to many corrosion experiments. A water jacket allows keeping the electrolyte at a chosen temperature with a flow of a heating or cooling liquid from a thermostat.

Special measurements require a special design of the electrochemical cells. For optical in situ measurements, the distance between the electrode and the objective of a microscope should be as small as possible, which may be achieved with a flow cell. Special objectives permit a sufficiently large working distance of a few mm to the specimen surface, although they still have a large optical resolution. An optimized design allows an electrolyte flow with a sufficient convection to increase the transfer of corrosion products or of heat in the case of high dissolution currents. Figure 1.25b presents such a flow cell, which has been applied for in situ microscopic pitting studies. A CE surrounding the WE may minimize ohmic drops, which could be very high in the case of many intensively corroding pits on the metal surface. A glass or quartz window permits optical and spectroscopic studies. The WE is usually embedded in resin or surrounded by a Teflon holder in order to expose a well-defined surface area to the electrolyte. The RE is connected via a channel within the specimen holder filled with electrolyte.

Surface studies using synchrotron radiation work mostly with a grazing incident beam. This is required especially for reflectivity measurements with x-rays to study the chemistry and the structure of the electrode surface. In this case, the surface must be extremely smooth and flat with only a small electrolyte layer on top but still with full control of the electrode potential. Cell designs similar to Figure 1.25b have been used successfully with

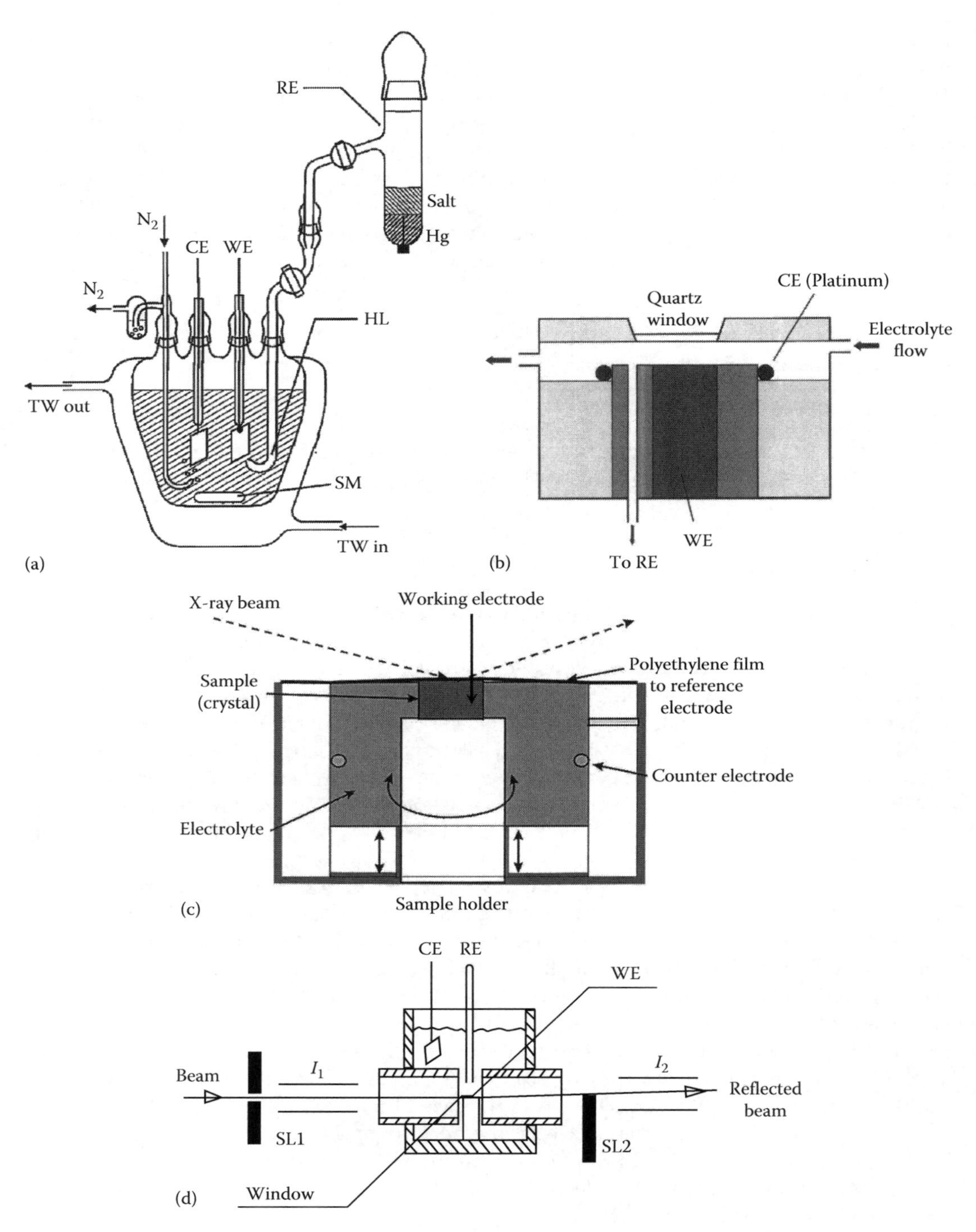

FIGURE 1.25
(a) Standard electrochemical cell with WE, CE, RE, inlet and outlet for protecting gas N_2 (nitrogen, argon), stirring magnet SM, water jacket for thermostat fluid. (b) Electrochemical cell for microscopic (optical) measurements. (c) Cell for in situ reflection and diffraction measurements with flexible membrane. (d) Cell for transmission or reflection measurements, suited for synchrotron measurements, includes ionization chambers I_1 and I_2 for measurements of x-rays.

a thin polymer window on top of the WE in order to reduce the absorption of the x-ray beam (Figure 1.25b). The polyethylene membrane can be blown up with pressed-in electrolyte, to enlarge the electrolyte layer and to reduce ohmic drops during the electrochemical treatment of the specimen WE. It will be taken back close to the specimen surface when the measurement is performed. The cell of Figure 1.25c is suited for studies with reflection or transmission of a light or a x-ray beam. It shows a reflection experiment with x-rays with a grazing incidence of the beam including two ionization chambers to measure the beam intensity. The size of the electrolyte layer and the length of the specimen have to be optimized in order to get enough interaction of the beam with the specimen and to minimize its absorption within the electrolyte. For transmission experiments, the specimen has to be arranged perpendicularly to the beam.

Special miniaturized cell designs are required for in situ studies with STM, scanning force microscopy (SFM), or scanning electrochemical microscopy (SEM) using ultra micro electrodes (UME) as imaging probes, which have to approach very close to the surface. Cell designs and the necessary equipment are described in the appropriate sections of other chapters.

1.11.2 Potentiostatic Measurements

Electrochemical measurements under potential control are performed with a potentiostat. Figure 1.26a shows its block diagram with the most important parts. Most units contain an operational amplifier with the following special electronic properties. It has a very high input resistance of ca. $10^7\ \Omega$ and a very high magnification of 10^5 for signals given to the two inputs, i.e., the inverting (−) and the noninverting one (+). Thus, a small voltage of 1 mV given to the two inputs should yield 100 V at the output, which however exceeds the supply voltage of 15 or 18 V, the maximum at the output. In consequence, the operational amplifier tries to keep both inputs at the same voltage by a sufficient current at its low resistance output. These properties are used for the different parts that are depicted in Figure 1.26a. The inverting input of the operational amplifier of the potentiostat is connected to the RE and the noninverting input to the source that sets the desired voltage between the WE and the RE. The output thus supplies the necessary current I_C to the electrochemical cell *via* the CE to hold the same potential at both inputs, which automatically establishes the electrode potential at WE chosen relatively to RE. The main potentiostatic amplifier is usually followed by booster units (not shown in Figure 1.26a), which improve its properties. The adder sums up all the voltages given to its inputs. It contains the same resistor R_A at all inputs and the connecting resistor $R_S = R_A$ to the output of usually 100 kΩ. Thus, the currents across the input resistors R_A are proportional to the input voltages and sum up to a total current flow across R_S, leading there to a voltage drop, which equals the sum of the input voltages. The various inputs allow adding voltages with different time characteristics like constant values, ramps, and step functions or small sinus waves as used for polarization curves, transients, and impedance measurements, respectively. The adder gives its output voltage to the noninverting input of the operational amplifier of the potentiostat to perform any potentiostatic or potentiodynamic corrosion measurements with sophisticated time characteristics. The cell current I_C is measured usually as a voltage drop across a resistor R_C, which is amplified by a differential amplifier and sent to the recording unit or computer. Similarly, the electrode potential is measured as ΔU(WE–RE) *via* an operational amplifier to avoid any disturbance of the potentiostatic circuit. A compensation unit may give a fraction of the voltage to the adder, which is proportional to the cell current I_C. Thus, one may compensate for the voltage drop between WE and the

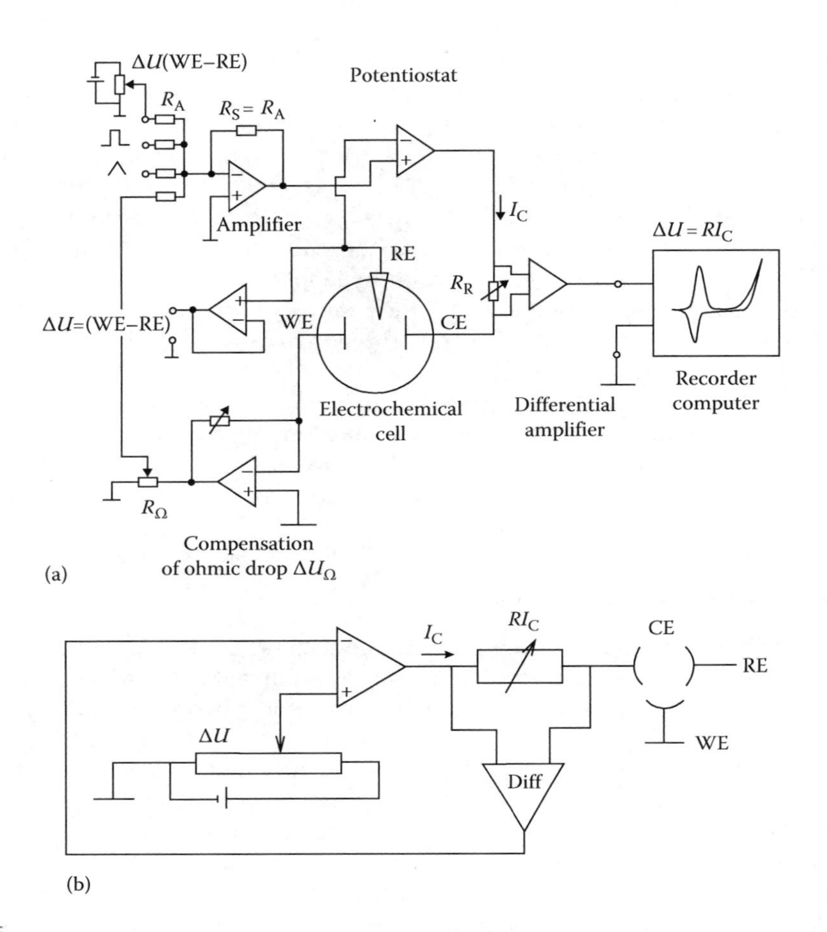

FIGURE 1.26
(a) Block diagram for a potentiostatic circuit including adder with input resistors R_A and coupling resistor $R_S = R_A$, compensation of ohmic drop corresponding to setting at R_Ω and current measurement at R_C via a differential amplifier, WE, CE, and RE. (From Strehblow, H.-H., Phenomenological and electrochemical fundamentals of corrosion, in *Corrosion and Environmental Degradation*, M. Schütze, editor, Vol. 1, Wiley-VCH, Weinheim, Germany, 2000, pp. 1–66.) (b) Galvanostatic circuit with internal supply of ΔU for the setting of cell current I_C measured as RI_C.

tip of the HL capillary of RE within the electrolyte when the setting at resistor R_Ω corresponds to the related ohmic resistance. R_Ω may be determined independently by galvanostatic transients. Potentiostats are usually very fast, so that transient measurements are possible starting ca. 1 μs after the potential change. This time is required to charge the double layer with a capacity of typically $20\,\mu F \cdot cm^{-2}$. With a resistance of $R = 0.1\ \Omega$ for a well-conducting electrolyte, the typical rise time is RC = 2 μs. The necessary output current of the potentiostat is usually 1 A or larger for these very short times and may be easily reached by a well-designed potentiostat.

1.11.3 Galvanostatic Measurements

For galvanostatic measurements, the cell current I_C is set with a galvanostat (Figure 1.26b) and the electrode potential is measured as the voltage of WE against RE as a function of time. I_C is set as a voltage ΔU with a built-in supply. ΔU is given to the noninverting input of an operational amplifier and is compared to the voltage drop RI_C at the output, which is given to the noninverting input of the operational amplifier after passing through a differential amplifier. I_C at the output assumes the appropriate value chosen by ΔU. Usually, galvanostats are built into potentiostats. An input ΔU changing appropriately with time permits galvanodynamic measurements and galvanostatic transients.

1.11.4 Rotating Disc Electrodes (RDEs) and Rotating Ring Disc Electrodes (RRDEs)

RDEs provide well-controlled diffusion conditions (Figure 1.27). The flow of the electrolyte gives access of fresh solution to the surface, whereas the electrochemical reaction at the electrode surface changes the electrolyte composition. Both effects compensate each other, leading to a constant thickness d_N of the Nernst diffusion layer all over the disc with a laminar flow of the solution in vicinity of its surface. The analytical treatment of this convection and diffusion problem leads to Levich's equation:

$$d_N = 1.61\,\omega^{-1/2}\nu^{1/6}D^{1/3} \tag{1.129}$$

where
- $\omega = 2\pi f$, f is the rotational frequency
- $\nu = \nu'/\rho$ with the viscosity of the solution ν' and its density ρ
- D is the diffusion coefficient

For aqueous electrolytes with $\nu = 10^2\,\text{cm}^2\,\text{s}^{-1}$, $\omega = 63\,\text{s}^{-1}$, and $D = 5 \times 10^{-6}\,\text{cm}^2 \cdot \text{s}^{-1}$ as a typical value for ions in water, one obtains $d_N = 1.6 \times 10^{-3}$ cm. With Equation 1.119, one gets for the diffusion-limited current density $i_D = -nFDc_B/d_N = -\,300\,\mu\text{A} \cdot \text{cm}^{-2}$ for a bulk concentration $c_B = 10^{-3}$ M and $n = 1$ for the number of exchanged electrons. Increase of the rotation frequency f will decrease d_N with $\omega^{-1/2}$ according to Equation 1.129 and increase i_D accordingly. Similar results hold for anodic metal dissolution and the maximum current density i_D, which is limited by the precipitation of a salt film.

A RRDE has a concentric ring around a disc electrode, which may be used to analyze the products formed at the disc. The radial flow of the electrolyte parallel to the electrode surface allows a diffusion-limited analytical ring current I_R for an appropriate setting

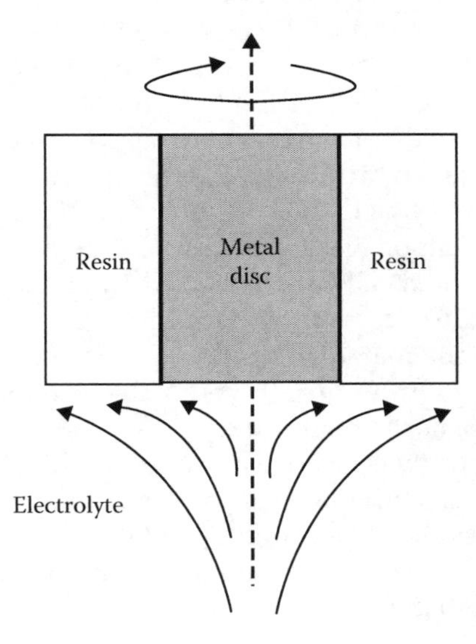

FIGURE 1.27
Cross-section of an RDE and the electrolyte flow in front of its surface. (From Strehblow, H.-H., Phenomenological and electrochemical fundamentals of corrosion, in *Corrosion and Environmental Degradation*, M. Schütze, editor, Vol. 1, Wiley-VCH, Weinheim, Germany, 2000, pp. 1–66.)

of the ring potential. The transfer efficiency (N) for the products of the disc to the ring depends on the radius of the disc and the ring and the ring thickness. It may be calculated or determined experimentally, for example, by the oxidation of dissolved $[Fe(CN)_6]^{4-}$ to $[Fe(CN)_6]^{3-}$ at the disc and its reduction at the ring [43,44]. Transfer efficiencies of $N > 30\%$ may be easily obtained by an appropriate design of the RRD electrode. The disc current I_{Di} and thus the rate of formation of the product may be calculated from the ring current I_{Ri} with Equation 1.130, where n_{Di} and n_{Ri} are the number of electrons for the reactions at the related electrodes.

$$I_{Di} = \frac{n_{Di} I_R}{N n_R} \tag{1.130}$$

The time resolution for the detection of a product at the disk by the ring current is in the range of 0.1 s, which is required for the transport due to the laminar flow of the electrolyte in front of the electrode. Therefore, this method is suited for measurements of potentiostatic dissolution transients in this time frame. One example for corrosion is the separation of currents measured during passivation transients at the disc into a part for metal dissolution and another for oxide formation. The ring current allows calculating the dissolution rate i_C at the disc according to Equation 1.130, whereas the difference to the total disc current $i_{Di} - i_C = i_L$ yields the current density of layer formation i_L. These investigations may be done with the time resolution of the method of ca. 0.1 s, i.e., i_C and i_L may be followed as a function of time.

Using a split RDE, one may analyze two products simultaneously, i.e., the dissolution of Cu^+ and Cu^{2+} or of Fe^{2+} and Fe^{3+}. These measurements require a tripotentiostat, which allows setting the electrode potentials for the disc and the two rings independently whereas an RRD electrode requires a bipotentiostat only. For these measurements, the electronic circuits require a common RE and a grounded CE. Differential amplifiers at the entrance of the three (or two) potentiostats uncouple the WEs (rings and disc) so that the whole circuit is grounded at one point. The related equipment and the procedure to produce electrodes for corrosion studies is described in the literature [45–47].

There have been realized several special developments for the RRDE. One is an electrode with two concentric rings around the central disc. Such an electrode allows, for example, determining the amount of dissolved Cr^{3+} ions at a Cr disc electrode. One may reduce Cr^{3+} ions at an inner GC ring (glassy carbon ring) and reoxidize Cr^{2+} at the outer GC ring. The analytical ring current at the outer ring allows the determination of the Cr^{3+} dissolution at the Cr disc. I_R at the inner ring may not be used directly due to strong hydrogen evolution parallel to the Cr^{3+} reduction at the negative potentials in acidic electrolytes. For this special analytical task, the inner GC ring has been covered by Hg deposits and sensitized by additional Ag deposits in order to reduce hydrogen evolution and to improve Cr^{3+} reduction [47]. A further interesting development is the hydrodynamically modulated RRDE. Here the rotation is modulated to increase the sensitivity of the detection of disc products [48]. If hydrodynamical square wave modulation is applied, one may distinguish between cations dissolved at the disc and those that are already present within the bulk electrolyte [48,49]. The modulation of the ring current shows spikes at the beginning of the changes of rotation, which are exclusively related to the disc products, which has been verified quantitatively by experiments and calculation [48,49]. This method has been applied to the determination of Fe dissolution in 1 M NaOH. The additional application of a Pt split ring allowed the simultaneous determination of soluble Fe^{2+} and Fe^{3+} in dependence of the disc potential when polarization curves or transients are taken [50] (see Chapter 5).

1.11.5 Electrochemical Transients

A very important field in electrochemical corrosion research is the investigation of the kinetics of electrochemical processes with transients. Potentiostatic and galvanostatic transients allow studying the influence of the different reaction steps of an electrochemical reaction in their appropriate time domain. Figure 1.28 presents a potentiostatic transient of a gold RDE separating the different reaction steps after the rise time of the potentiostat: charging of the double layer, the charge transfer controlled range, and finally the diffusion controlled regime. The current densities of Figure 1.28 for long time t and different rotation frequencies of the RDE prove the validity of the $\omega^{-1/2}$ dependence of the Levich equation (Equation 1.129). As already mentioned above, the charging of the double layer occurs for low values of RC within very few μs, i.e., for a well-conducting electrolyte and thus a low resistance ($R = 0.1\ \Omega$) and not too large capacities of typically $C = 20\,\mu\text{F}\cdot\text{cm}^{-2}$. It may be larger for other RC values, i.e., for the example presented in Figure 1.28. At the borders of the different time ranges, the kinetics of two processes are rate determining simultaneously like charge transfer and diffusion. Equation 1.131 describes the current vs. time relation for a redox reaction with a superposition of charge transfer control and diffusion control [51–53].

$$i = i(0)\exp(\lambda^2 t)\, erfc(\lambda\sqrt{t}) \tag{1.131}$$

where

$$\lambda = \frac{i_0}{zF}\left[\frac{1}{c_{B,\text{Red}}\sqrt{D_{\text{Red}}}}\exp\left(\frac{\alpha zF}{RT}\eta\right) + \frac{1}{c_{B,\text{Ox}}\sqrt{D_{\text{Ox}}}}\exp\left(-\frac{(1-\alpha)zF}{RT}\eta\right)\right] \tag{1.131a}$$

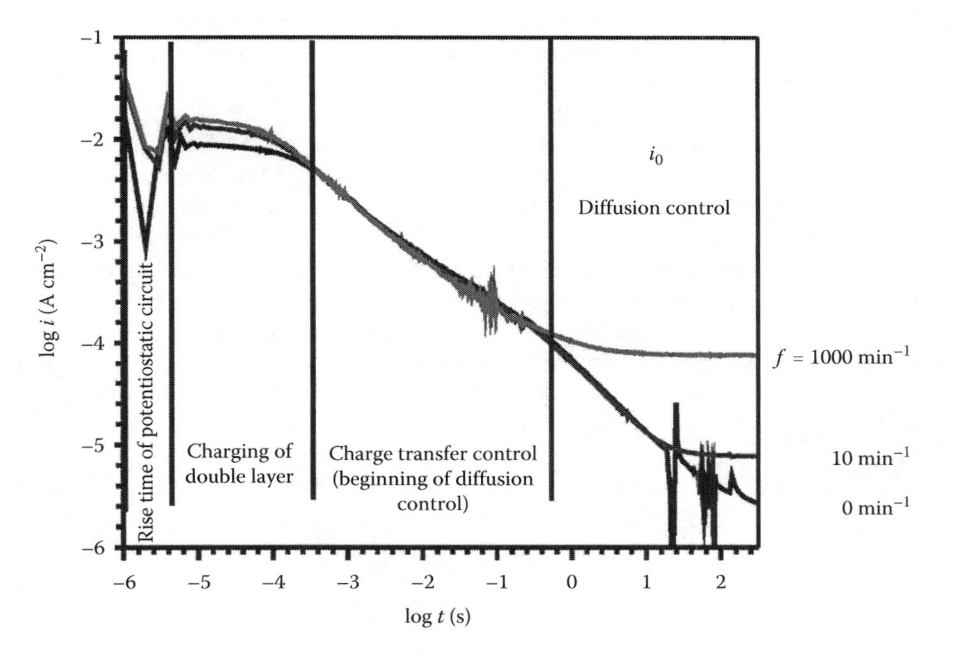

FIGURE 1.28
Potentiostatic transient with time domains for double-layer charging, charge transfer, and diffusion control. Measurement with rotating Au-disc electrode with different rotation frequencies f (min^{-1}) showing Levich behavior for long times. (From Strehblow, H.-H., Phenomenological and electrochemical fundamentals of corrosion, in *Corrosion and Environmental Degradation*, M. Schütze, editor, Vol. 1, Wiley-VCH, Weinheim, Germany, 2000, pp. 1–66.)

and

$$i(0) = i_+ + i_- = i_0 \left\{ \exp\left[\frac{\alpha z F \eta}{RT}\right] - \exp\left[\frac{-(1-\alpha) z F \eta}{RT}\right] \right\} \quad (1.131b)$$

$i(0)$ refers to a vanishingly short time when the current potential dependence is described by the Butler–Volmer equation with no diffusion control. $i(0)$ is obtained by the extrapolation of the $i - \sqrt{t}$ plot to $\sqrt{t} = 0$ according to Equation 1.131c.

$$i = i(0)\left(1 - \frac{2}{\sqrt{\pi}} \lambda \sqrt{t}\right) \quad (\text{for } \lambda \sqrt{t} \ll 1) \quad (1.131c)$$

The values for different potentials E, obtained by extrapolation, may be used to compose the i–E curve without any diffusion control. The approximation in Equation 1.131d holds for long times and permits determining i_D for pure diffusion control by a $i - 1/\sqrt{t}$ plot with extrapolation to $1/\sqrt{t} = 0$.

$$i = i(0) \frac{1}{\sqrt{\pi} \lambda \sqrt{t}} \quad (\text{for } \lambda \sqrt{t} \gg 1) \quad (1.131d)$$

1.11.6 Impedance Spectroscopy

The electrode/electrolyte interface may be seen as a combination of the capacity C of the electrode and a parallel charge transfer resistance R_{CT} or polarization resistance R_P (Figure 1.29). The capacity is of the order of 10–20 μF · cm^{-2} for a free electrode surface. When it is covered by an oxide film, it may be smaller, in the range of some few μF · cm^{-2} only. R_{CT} may change appreciably from ∞ to less than 1 Ω. As an example, a mercury drop is an ideally polarizable electrode for sufficiently negative potentials with a potentially independent small current density leading to $R_{CT} = \Delta E/\Delta i = \infty$. However, it changes to higher values of R_{CT} for more positive potentials when the metal is dissolved ($E > 0.85$ V). In addition, an ohmic resistance R_Ω in series is shown in Figure 1.29 to account for the ohmic drop within the electrolyte between the electrode surface and the HL capillary of the RE. This RC combination determines the time constant for the charging of the double layer for potentiostatic transients. However, this model with passive electronic elements serves only as a first approximation. Many systems require extra elements. This simple description of electrode processes does not contain transport phenomena and chemical reaction steps like diffusion of reactants and adsorption/desorption phenomena in a sequence of the elementary electrochemical reaction step. If one wants to include these complications, one has to introduce further elements into the circuit. Resistors follow Ohm's law with a linear current–voltage characteristic. The charge transfer follows usually the Butler–Volmer equation, which is an exponential relation and thus a simple circuit with a capacitor and

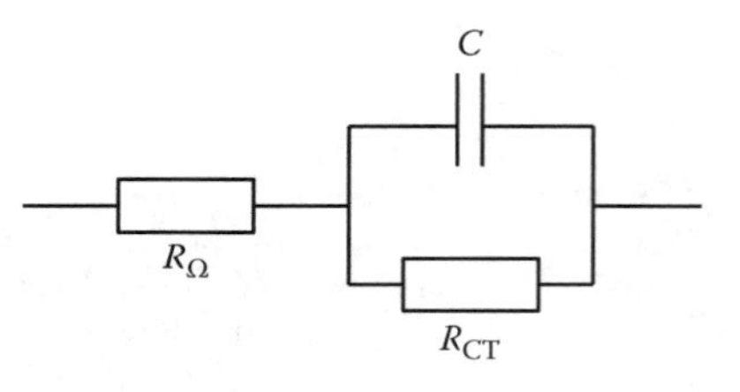

FIGURE 1.29
The electrical properties of an electrode represented by its capacity C and its parallel charge transfer resistance R_{CT} and an ohmic resistance R_Ω in series for the electrolyte between the electrode and the HL capillary of the RE.

resistors seems inadequate. However, in the vicinity of the Nernst equilibrium and the rest potential of corroding systems, the current–voltage characteristic may be linearized if very small potential changes are applied. Therefore, impedance measurements add small alternating voltages of ca. $\Delta E = \pm 10\,\text{mV}$ to the electrode measuring the related current changes. A generator gives a sinusoidal voltage change of varying frequency to the adder of the potentiostat, which holds the electrode usually at its equilibrium or its rest potential. The current and potential changes are separated from their stationary values, and their filtered signals at the chosen frequency are processed. The impedance is calculated from potential and current changes, taking into account the phase shift of both relative to each other. The generator frequency may be varied in a wide range between mHz and several 100 kHz, thus addressing the phenomena with different time constants like the charging of the double layer, the charge transfer reaction, diffusion, chemical reaction steps, etc. The frequency is usually scanned automatically when an impedance spectrum is taken. The upper limits are given by the stray capacities of the experimental set-up, like the capacitance of the cables and the applied measuring instruments. The lower limits are given by the limited long-time stability of the electrode/electrolyte interface. Lock-in technique may help to sort out disturbing signals of other frequency than that chosen by the generator setting. As a large frequency range is measured, the method has been called impedance spectroscopy.

The impedance of a system is the resistance for alternating current (Equation 1.132). Voltage ΔE and current ΔI follow Equations 1.133 with a phase shift Φ between them.

$$Z = \frac{\Delta E}{\Delta I} \tag{1.132}$$

$$\Delta E = \Delta E_0 \cos(\omega t), \quad \Delta I = \Delta I_0 \cos(\omega t + \Phi), \tag{1.133}$$

If the system behaves like an ohmic resistor, i.e., with $Z = R$, the applied voltage and the current are in phase, i.e., with a phase shift $\Phi = 0$. If only a capacitance C is effective, the impedance equals the capacitive resistance $Z = 1/\omega C$ with a phase shift of $\Phi = \pi/2$ with $\omega = 2\pi f$ with the frequency f. An in-series connection of C and R requires the addition of both resistances:

$$R = |Z|\cos\Phi; \quad \frac{j}{\omega C} = j|Z|\sin\Phi \tag{1.134}$$

This is best done in a presentation of complex numbers as in Figure 1.30, with R being the real and $j/\omega C$ the imaginary part ($j = \sqrt{-1}$). The impedance Z is given by a vector addition of the ohmic and capacitive resistance. From Figure 1.30a follows directly the expression for $|Z|$ and $\tan\Phi$ of Equation 1.135.

$$Z = \sqrt{R^2 + \frac{1}{\omega^2 C^2}}; \quad \tan\Phi = \frac{Z_{im}}{Z_{re}} = -\frac{j}{\omega CR} \tag{1.135}$$

For parallel connection of a capacity C and a resistance R_{CT} as the simplest circuit for the electrode/electrolyte interface, one needs the addition of the inverse of both resistances as given in Equation 1.136 and Figure 1.30b. Multiplying the numerator and the denominator

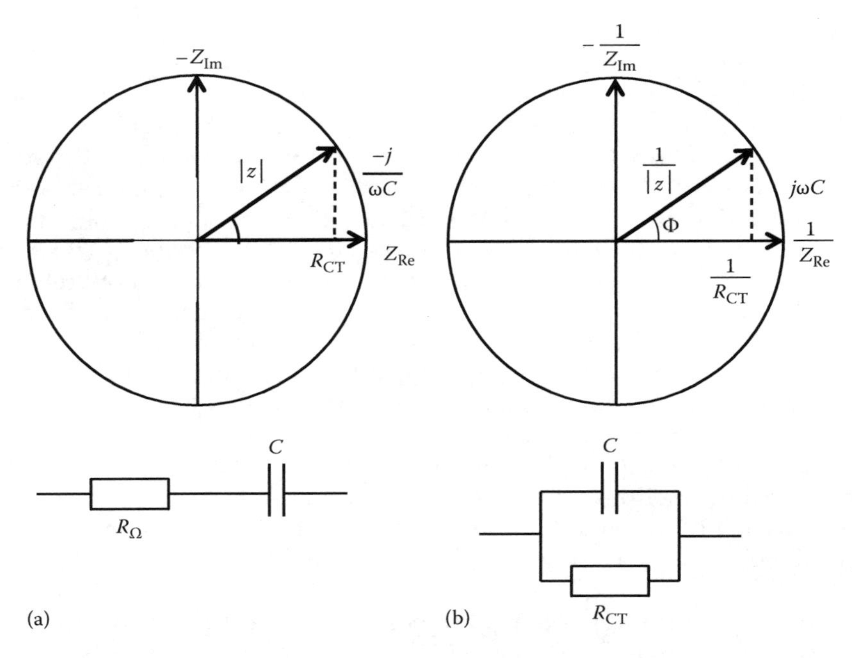

FIGURE 1.30
Impedance Z given as a presentation of complex numbers with the real and the imaginary part, Z_{Re} and Z_{Im}, respectively and the phase shift Φ, (a) for R_{CT} and C in series, (b) for parallel R_{CT} and C (inverse addition of the impedances).

with the expression $(R_{CT} + (j/\omega C))$ and rearranging all terms with the separation of the real and imaginary parts, one obtains Equation 1.137.

$$\frac{1}{Z} = \frac{1}{R_P} - \frac{\omega C}{j} = \frac{j - R_{CT}\omega C}{R_{CT} j} \tag{1.136}$$

$$Z = \frac{R_{CT} j}{j - R_{CT}\omega C} = -\frac{R_{CT}(j/\omega C)}{-(j/\omega C) + R_{CT}} = \frac{R_{CT}(j/\omega C)(R_{CT} + j/\omega C)}{R_{CT}^2 - (j/\omega C)^2}$$
$$Z = -\frac{R_{CT}(j^2/\omega^2 C^2) + R_{CT}^2(j/\omega C)}{R_{CT}^2 + (1/\omega^2 C^2)} = \frac{R_{CT}/\omega^2 C^2}{R_{CT}^2 + (1/\omega^2 C^2)} - \frac{R_{CT}^2/\omega C}{R_{CT}^2 + (1/\omega^2 C^2)} j \tag{1.137}$$

If the circuit contains an additional resistance R_Ω for the electrolyte in series, one obtains the following relation for the simplest equivalent circuit of Figure 1.29 for the electrode/electrolyte interface:

$$Z = R_\Omega + \frac{R_{CT}/\omega^2 C^2}{R_{CT}^2 + (1/\omega^2 C^2)} - \frac{R_{CT}^2/\omega C}{R_{CT}^2 + (1/\omega^2 C^2)} j \tag{1.138}$$

The imaginary and real parts for this simple circuit of an electrode/electrolyte interface with an electrolyte resistance R_Ω in series are depicted in the so-called Nyquist diagram of Figure 1.31. Measurements at the various frequencies yield a semicircle with a distance

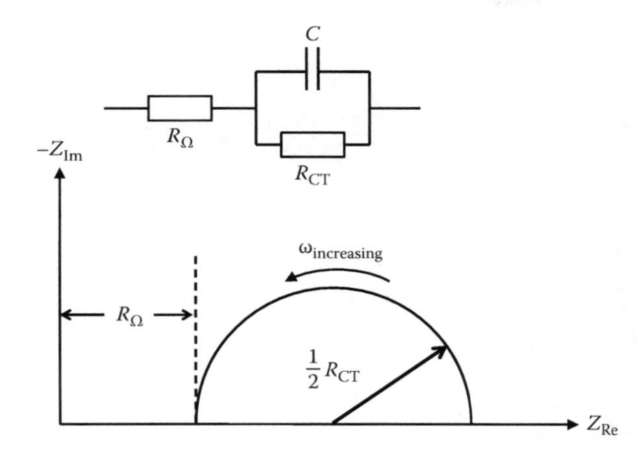

FIGURE 1.31
Nyquist diagram of an electrode with parallel capacitance C and charge transfer resistance R_{CT} and an ohmic resistance R_Ω in series.

R_Ω from the ordinate and with the radius $\frac{1}{2}R_{CT}$. For very high frequencies, the total current may pass via the capacitance C due to the vanishingly small resistance, which yields $Z = R_\Omega$ for the impedance. This situation corresponds to the intersection of the semicircle with the abscissa at the left side. For very low frequencies, the capacitive resistance gets very large and all currents will flow via the two resistors in series with $Z = R_\Omega + R_{CT}$. This corresponds to the intersection of the semicircle with the abscissa at the right side. The presentation of Z in dependence of log ω as given in Figure 1.32 is called the Bode plot. It shows the decrease from $(Z = R_\Omega + R_{CT})$ to $Z = R_\Omega$ with increasing frequency. Simultaneously the phase shift changes from $\Phi = 0$ to a maximum and to $\Phi = 0$ back again. This fits to the result that the resistors are effective only for very low and very high frequencies and with the capacitive resistance being large for low and small for high frequencies. For corroding systems, impedance spectroscopy is performed mostly at the rest potential. For this situation, Equations 1.137 and 1.138 keep their form and the polarization resistance R_P replaces the charge transfer resistance R_{CT}.

The circuit of Figure 1.38 is much to simple to describe the impedance spectra of most real systems including transport phenomena and other reaction steps preceding and following the main charge transfer step. For this purpose, one has to replace the charge transfer resistance R_{CT} by the Faraday impedance Z_F. Z_F contains R_{CT} and a term for the diffusion of reactants called the Warburg impedance Z_W and/or a term for convection called the Nernst impedance Z_N. In the Nyquist diagram, Z_W shows a line, whereas Z_N is a curve like an asymmetric circle. A more realistic situation is described by the equivalent circuit of Randles. It assumes charge transfer in combination with diffusion of a reactant like dissolved metal cations within the electrolyte in front of the electrode. Therefore $Z_F = R_{CT} + Z_W$ is introduced, which changes Equation 1.138 to Equation 1.139.

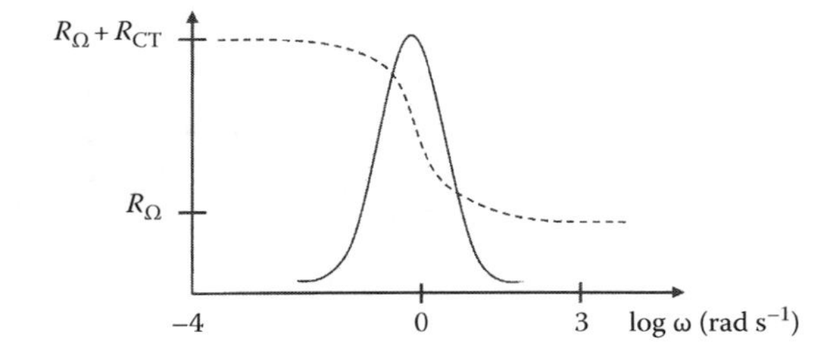

FIGURE 1.32
Bode diagram for an electrode with parallel capacitance C and charge transfer resistance R_{CT} and an ohmic resistance R_Ω in series.

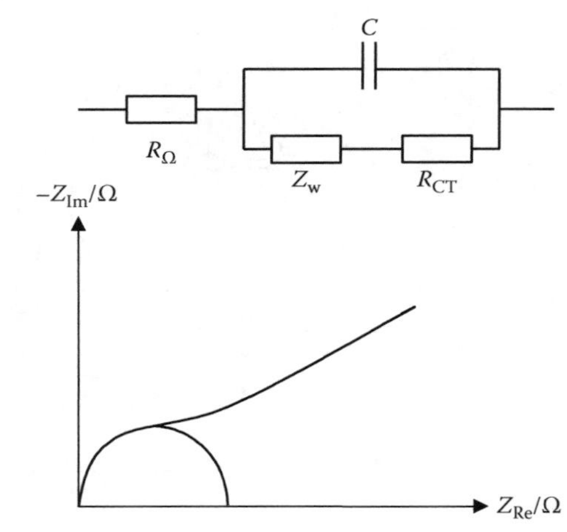

FIGURE 1.33
Randles circle for the electrical circuit of Figure 1.31 with a Warburg impedance Z_W in series to the charge transfer resistance R_{CT} and the related Nyquist diagram.

$$Z = R_\Omega + Z_F \frac{-j/\omega C}{Z_F - j/\omega C} \tag{1.139}$$

$$Z_F = R_{CT} + Z_W \tag{1.139a}$$

Introducing the expression for Z_W in Equations 1.139 and 1.139a yields an expression that describes the experimental findings for these conditions as shown in Figure 1.33. At high frequencies, the system follows the properties of a simple double layer, whereas at low frequencies it is dominated by the Warburg impedance with a −45 degrees line.

Many systems are still far more complicated and cannot be described by Randles equivalent circuit. Besides transport phenomena, the reactions consist of several chemical reactions steps within the electrolyte and at the electrode surface including adsorption and desorption so that the impedance spectra contain several related time constants. Even inductive loops are observed. Therefore in most cases, due to too many parameters, the reaction mechanism may not be deduced from impedance spectra only. Furthermore, the method assumes a homogeneous reaction all over the electrode surface, which is often not the case, especially in corroding systems. However, the method has been successfully used empirically even for complicated systems like metals with coatings. It has been applied in these cases to follow the durability and quality of protecting organic and inorganic films on metals with time.

1.12 Reduction of Redox Systems

Corrosion in open circuit conditions requires a counter reaction to consume the electrons, which are produced by metal dissolution. Important reactions are hydrogen evolution and oxygen reduction. Fe^{3+} reduction is another. $[Fe(CN)_6]^{3-}$ reduction is a simple outer sphere reaction, which may be used for basic corrosion studies. In most cases, one tries for simplicity to study metal dissolution and the redox reactions separately with the use of a potentiostat. In the following, the hydrogen evolution and the oxygen reduction reaction are discussed separately.

1.12.1 Hydrogen Evolution

Hydrogen evolution (HER) is a process with a sequence of several elementary reaction steps (Equations 1.140a through d).

$$H^+ + e^- \rightarrow H_{ad} \quad \text{Volmer reaction} \tag{1.140a}$$

$$H_{ad} + H_{ad} \rightarrow H_{2,ad} \quad \text{Tafel reaction} \tag{1.140b}$$

$$H_{ad} + H^+ + e^- \rightarrow H_{2,ad} \quad \text{Heyrovsky reaction} \tag{1.140c}$$

$$H_{2,ad} \rightarrow H_2 \quad \text{(dissolution or gas bubbles)} \tag{1.140d}$$

Two main mechanisms are accepted, the Volmer–Tafel and the Volmer–Heyrovsky mechanism described by Equations 1.140a and b and 1.140a and c, respectively. The reaction starts with the charge transfer step, i.e., the discharge of hydrogen ions to adsorbed hydrogen atoms (Equation 1.140a), which diffuse on the surface and recombine to adsorbed dihydrogen molecules (Equation 1.140b). As an alternative hydrogen ions pick up an electron in the vicinity of an already present H_{ad} forming $H_{2,ad}$ (Equation 1.140c). Then H_2 molecules desorb from the surface, are dissolved and diffuse to the bulk electrolyte, or coalesce to form gas bubbles on the cathode when the limit of H_2 dissolution is reached at significant overpotentials or current densities (Equation 1.140d).

If the Volmer reaction is the slowest and thus the rate-determining step, the Butler–Volmer equation applies for the HER current.

$$i_H = F k_{H,1,a}\, \theta_H \exp\left\{\frac{\alpha F E}{RT}\right\} - F k_{H,1,c}(1-\theta_H) c(H^+) \exp\left\{-\frac{(1-\alpha)E}{RT}\right\} \tag{1.141a}$$

where

$k_{H,1,a}$ and $k_{H,1,c}$ are the rate constants for the Volmer reaction in anodic or cathodic direction, respectively

θ_H is the surface coverage by adsorbed H atoms

$1 - \theta_H$ is the fraction of free surface accessible to hydrogen adsorption

At the equilibrium potential E_{eq}, the anodic partial current density is the opposite of the cathodic partial current density and its value equals the exchange current density $i_{0,H,1}$.

$$i_{0,H,1} = F k_{H,1,a} \theta_{H,eq} \exp\left\{\frac{\alpha F E_{eq}}{RT}\right\} = F k_{H,1,c}(1-\theta_{H,eq}) c(H^+) \exp\left\{-\frac{(1-\alpha) F E_{eq}}{RT}\right\} \tag{1.141b}$$

where $\theta_{H,eq}$ is the equilibrium surface H coverage.

Introducing the overpotential $\eta = E - E_{eq}$, Equation 1.141a may be rewritten as:

$$i_H = i_{0,H,1}\left[\frac{\theta_H}{\theta_{H,eq}} \exp\left\{\frac{\alpha F \eta}{RT}\right\} - \frac{1-\theta_H}{1-\theta_{H,eq}} \exp\left\{-\frac{(1-\alpha) F \eta}{RT}\right\}\right] \tag{1.141c}$$

Neglecting the anodic part of Equation 1.141c for sufficiently negative overpotentials η, one obtains:

$$i_H = -i_{0,H,1}\left[\frac{1-\theta_H}{1-\theta_{H,eq}}\exp\left\{-\frac{(1-\alpha)F\eta}{RT}\right\}\right] \quad (1.141d)$$

Assuming θ and $\theta_e \ll 1$, and, taking the logarithm, one obtains:

$$\eta = \frac{RT2.303}{(1-\alpha)F}\log i_{0,H,1} - \frac{RT2.303}{(1-\alpha)F}\log|i_H| = a - b\log|i_H| \quad (1.142)$$

Assuming $\alpha = 0.5$ for the charge transfer coefficient, one obtains a cathodic Tafel slope b of 0.120 V, which is found for most electrode/electrolyte combinations as shown in a compilation of literature data shown in Figure 1.34 [54].

Assuming the Tafel reaction or the Heyrovsky reaction as the rate-determining step also leads to semilogarithmic relations similar to Equation 1.142. As an example, if the Tafel reaction (Equation 1.140b) is the slowest, i.e., the rate-determining step, one may write the current for sufficiently negative electrode potentials and vanishing contribution of the anodic partial reaction as follows:

$$i_H = -Fk_{H,2,c}\,\theta^2 \quad (1.143a)$$

with the rate constant $k_{H,2,c}$ for the cathodic direction.

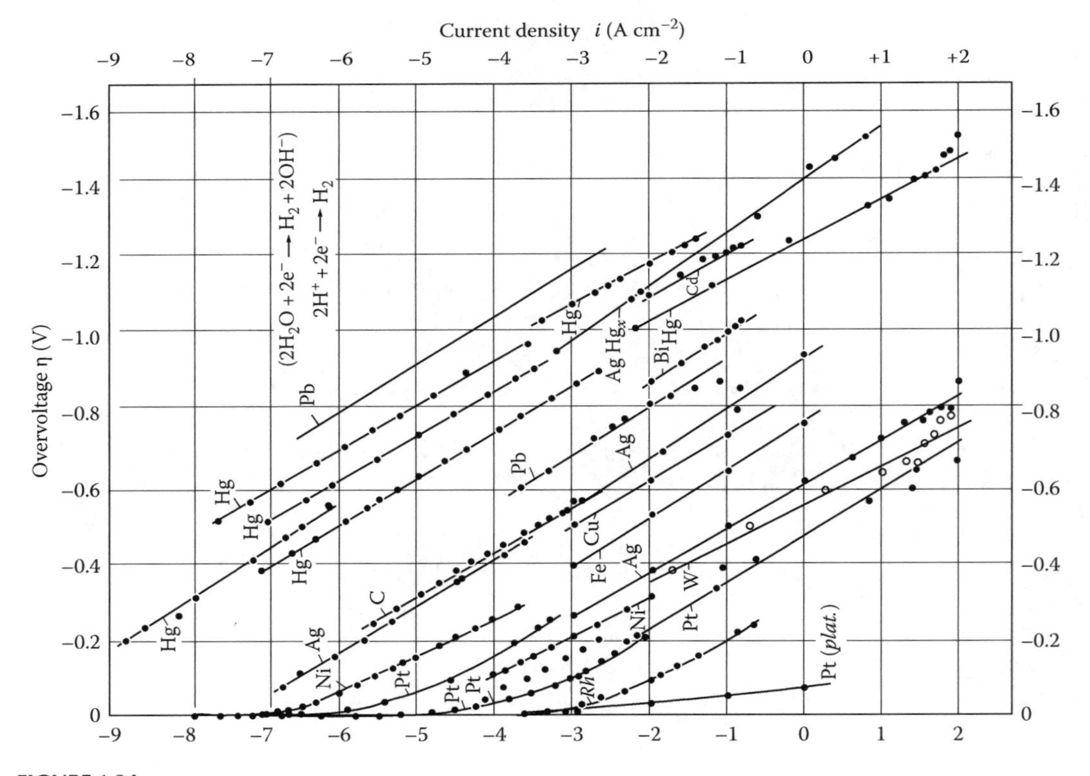

FIGURE 1.34
Tafel plot for hydrogen evolution at different electrodes. (From Vetter, K.J., *Z. Elektrochem., Ber Bunsen Ges. Physik Chem.*, 59, 435, 1955.)

The electrochemical quasi-equilibrium for the preceding fast Volmer reaction leads to:

$$k_{H,1,a}\theta \exp\left\{\frac{\alpha FE}{RT}\right\} = k_{H,1,c}c(H^+)(1-\theta)\exp\left\{-\frac{(1-\alpha)FE}{RT}\right\} \tag{1.143b}$$

$$\frac{\theta}{1-\theta} = K_1 c(H^+)\exp\left\{-\frac{FE}{RT}\right\} \tag{1.143c}$$

with $K_1 = k_{H,1,c}/k_{H,1,a}$.

Assuming $\theta \ll 1$, the coverage can be deduced from Equation 1.143c and introduced in Equation 1.143a, which yields Equation 1.143d.

$$i_H = -Fk_{H,2,c}\theta^2 = -Fk_{H,2,c}K_1^2c^2(H^+)\exp\left\{-\frac{2FE}{RT}\right\} \tag{1.143d}$$

Replacing E by $E_{eq} + \eta$, one obtains the following:

$$i_H = -i_{0,H,2}\exp\left\{-\frac{2F\eta}{RT}\right\} \tag{1.143e}$$

where the exchange current density $i_{0,H,2}$ is given by the following:

$$i_{0,H,2} = Fk_{2,H,c}K_1^2c^2(H^+)\exp\left\{-\frac{2FE_{eq}}{RT}\right\} \tag{1.143f}$$

Taking the logarithm of Equation 1.143e, one obtains the following:

$$\eta = \frac{RT\,2.303}{2F}\log i_{0,H,2} - \frac{RT\,2.303}{2F}\log|i_H| = a - b\log|i_H| \tag{1.144}$$

For these conditions, one expects a slope $b = 0.030$ V for the cathodic Tafel line as found experimentally on platinized Pt (Figure 1.34).

Figure 1.34 shows a large difference between hydrogen evolution overvoltages at a given current for different electrode surfaces, corresponding to different intercepts a in Equations 1.142 or 1.144. This is caused by a large difference of the exchange current densities $i_{H,0}$ and hence of the rate constants $k_{H,c}$ and the related activation energies. Mercury has an extremely large cathodic overvoltage corresponding to a very negative a, thus to a very small $i_{H,0}$ suppressing hydrogen evolution at moderate overpotentials η. Therefore, it may be used as an electrode material in polarography for the quantitative analysis of cations of very reactive metals like Cd by cathodic reduction from aqueous solutions without intense hydrogen evolution. One may even deposit metallic Na into Hg electrodes during electrolysis of aqueous NaCl solutions, the classic industrial process to produce chlorine and NaOH.

In neutral and alkaline solutions, the mechanism of hydrogen evolution is completely different from that in acidic solutions. H^+ ions are not available. Therefore, hydrogen has to be produced by water decomposition. This reaction has larger activation energy and

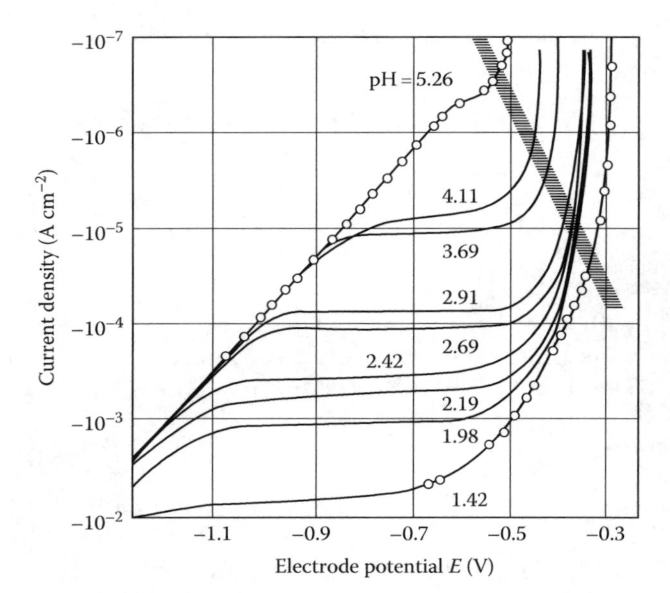

FIGURE 1.35
Polarization curves for hydrogen evolution on Fe surfaces at different pHs of solution. (From Kaesche, H., *Corrosion of Metals*, Springer, Berlin, Germany, 2003; Stern, M., *J. Electrochem. Soc.*, 102, 609, 1955.)

thus is shifted to more cathodic overpotentials. Figure 1.35 shows the log i–E dependence of hydrogen evolution on iron in solutions of different pH [55]. In acidic electrolytes, one observes with increasingly negative potential first a Tafel line with charge transfer control followed by a broad step with diffusion control. This step gets smaller with increasing pH and disappears for pH > 5. For $E < -1.1$ V, the curves for all solutions merge into the same line. Hydrogen evolution is no longer depending on pH due to direct water decomposition. At pH 5.2, only a very small step indicates some diffusion-controlled H^+ reduction merging immediately into line of hydrogen evolution by water decomposition.

1.12.2 Oxygen Reduction

Oxygen is one of the most important oxidants in corrosion reactions. Often only a thin electrolyte film is covering the metal surface, which may easily be saturated with dissolved oxygen. With a bulk concentration $c_B(O_2) = 2 \times 10^{-4}$ M, $D = 10^{-5}$ cm$^2 \cdot$ s^{-1}, $n = 4$, and $d_N = 5 \times 10^{-3}$ cm, the application of Equation 1.119 yields a maximum diffusion current density $i_D(O_2) = -4FD\ c_B(O_2)/d_N = -0.15$ mA cm^{-2}. Thus, oxygen reduction may support metal dissolution current densities up to current densities of this order of magnitude. If diffusion is not rate determining, the current density follows the Butler–Volmer equation.

The mechanism of oxygen reduction is relatively complex. For some electrode surfaces, H_2O_2 has been determined as an intermediate of O_2 reduction, which suggests the following reaction steps for alkaline and acidic solutions.

$$O_2 + 2H_2O + 2e^- \rightarrow H_2O_2 + 2OH^- \quad \text{(alkaline)} \tag{1.145}$$

$$H_2O_2 + 2e^- \rightarrow 2OH^-$$

$$O_2 + 2H^+ + 2e^- \rightarrow H_2O_2 \quad \text{(acidic)} \tag{1.146}$$

$$H_2O_2 + 2H^+ + 2e^- \rightarrow 2H_2O$$

Each reaction may be subdivided into further detailed reaction steps. The overall reaction follows a Butler–Volmer equation. For large cathodic overvoltages, reduction may become diffusion limited and thus might reach the potentially independent maximum current density i_D as described in Equation 1.119 and in Figure 1.23 and as discussed again for metal dissolution with oxygen reduction in Section 1.14.

1.13 Metal Dissolution

1.13.1 Basics of Metal Dissolution

Metals are usually crystalline materials, and their structure has a strong influence on the dissolution reaction, as well as on the metal deposition. Therefore, the model of crystal surfaces of Stranski and Kossel is often applied to explain the results of both reactions [56,57]. A metal may be polycrystalline, exposing different crystal surfaces with varying arrangement and density of metal atoms. Even a well-prepared single-crystal surface used in fundamental corrosion studies has several energetically different sites like a position well within a terrace, the step, kink, and the ad-site on a surface, depicted schematically in Figure 1.36. These structural details have been recently visualized directly with atomic resolution by the application of STM and atomic force microscopy (AFM) on metals like Au, Ag, Cu, Ni, Co, Cr, Fe, and several alloys. Examples are given in Chapter 5 (Section 5.8). Metal evaporation to the gas phase is ruled by the free enthalpy of evaporation, which contains the energy, which is necessary to break the bonds of a surface atom with its neighbors. This process may involve a transfer from any of the mentioned sites, with different activation energies. Ad atoms are in an energetically favored position, so that they might be transferred more easily with a smaller activation energy compared to those from a kink or a step site or even from a position within the terrace (see Figure 1.37). Therefore one discusses mostly a sequence of elementary reactions, with an atom leaving the kink site creating a new kink position and then going to the step and the ad-site, leaving finally the surface. These structural parameters influence also metal dissolution and metal deposition within electrolytes (Figure 1.37). However, these processes are more complicated in comparison with the reactions with the gas phase. Dissolution involves in addition the ionization of the metal atom to form a cation and the hydration of the cation or its solvation

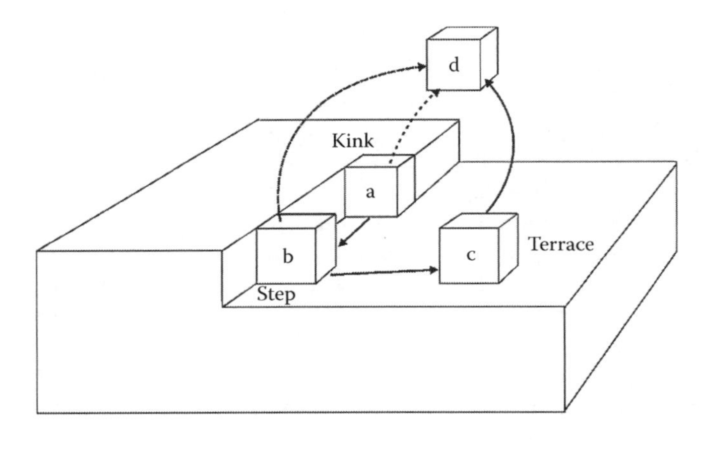

FIGURE 1.36
Model of metal surface with step, kink, and terrace sites after Stranski and Kossel. (From Stranski, I.N., *Z. Physik. Chem.*, 136, 259, 1928; Kossel, W., *Nachr. Ges. Wiss. Göttingen. Math. Physik. K.I.*, 135, 1927.)

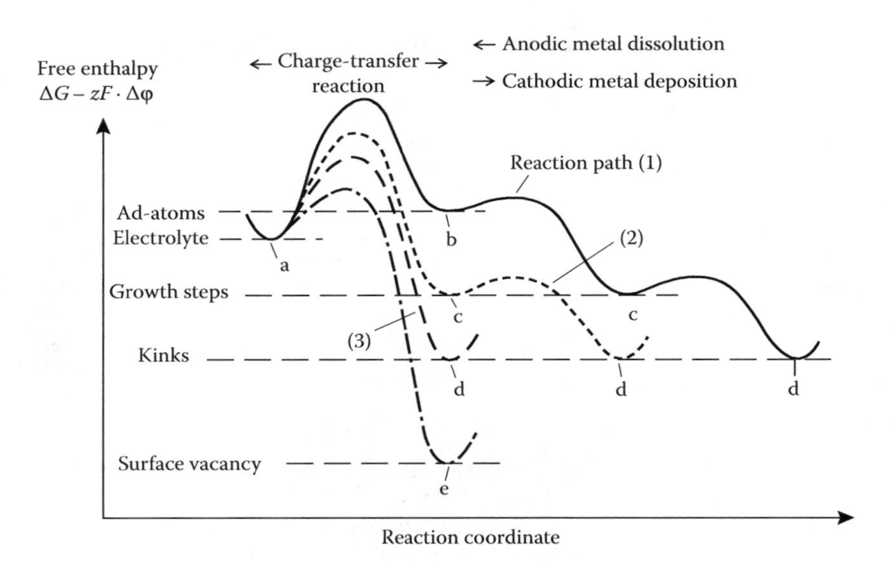

FIGURE 1.37
Free energy diagram for metal dissolution. (From Strehblow, H.-H., Phenomenological and electrochemical fundamentals of corrosion, in *Corrosion and Environmental Degradation*, M. Schütze, editor, Vol. 1, Wiley-VCH, Weinheim, Germany, 2000, pp. 1–66.)

in the case of a nonaqueous electrolyte. Furthermore, cations are often complexed by OH^- and other anions, and the transfer of cations from the metal surface may be preceded by adsorption phenomena. Similar complications occur during metal deposition. Thus, metal dissolution is a much more complex process, which also includes several interactions with the environment with the related exchange of energy.

In situ studies with STM allow investigating these processes on metal single crystals in contact with electrolytes. Below the Nernst equilibrium potential of the metal/metal–ion electrode, one observes in the images so-called frizzy steps on terraces, which are due to a high surface mobility of atoms and dissolution and re-deposition phenomena. They have been studied in detail by STM investigations of Ag(111) surfaces in 0.05 M H_2SO_4 with additions of 5×10^{-3} M $CuSO_4$ [58]. Intense surface diffusion causes atoms to leave the step edge and to come back again and to go into solution and to be re-deposited resulting in its frizzy appearance. The frizzy images get more pronounced, i.e., wider with increasing potential, i.e., when approaching the Nernst potential of the metal/metal–ion electrode. Metal dissolution above the equilibrium potential shows the permanent movement of the step edges of the terraces. The adsorption of anions, including OH^-, has a pronounced influence on the appearance of the step edges in STM images and the dissolution process. Geometrically well-oriented and long well-pronounced step edges are found as well as those with an irregular shape. Metal ion transfer may occur via ad-sites (terrace sites) as well as directly from the kink or step sites depending on the driving force of the reaction, i.e., the electrode potential. Figure 1.37 describes schematically the energy diagram for the transfer of metal cations involving the various sites at the surface. Although the transfer via the kink, step, and ad-site on the terrace seems to be favored due to smaller activation energies (reaction path 1 or 2 in Figure 1.37), more direct transfer might be favored because surface diffusion could be hindered. This has been shown by in situ video STM on Cu(111) surfaces covered with an adsorption layer of Cl^- [59]. Cl^- blocks the surface and most of the step positions due to the strong bonds of the anion with the metal atoms. This may be seen as the formation of a monolayer of weakly soluble CuCl. Those kink sites, which are free and energetically favored, show an intense dissolution reaction, and their movement has been visually followed and measured.

Metal dissolution and deposition are complex processes often involving ionization and hydration of the cations, adsorption of anions, and complexing of dissolving cations by anions. Moreover, the formation of anodic oxide or hydroxide layers influences metal dissolution decisively and may cause drastic changes of the effective mechanisms. Often metal surfaces are covered by anions such as OH^- or halides, and the formation of a hydroxide monolayer may occur at potentials far negative to the equilibrium potential of oxide formation as has been found for Cu(111) and Cu(100) surfaces [60–62]. Thus, metal dissolution starting with a free metal surface seems to be a rare case. The understanding of the mechanistic details of metal dissolution on an atomic basis requires still much more intense studies including the application of modern surface analytical tools. The formation of hydroxide layers and oxide films is described in detail in Chapter 5. Therefore, only some few outlines may be presented here. As an example, the OH^- catalyzed dissolution of iron is discussed now.

Many studies have been performed on the dissolution on iron. This reaction is catalyzed by OH^- ions. Kinetic studies have shown that the electrochemical reaction order for OH^- ions is 1.5. These findings started a discussion about two mechanisms: the Bockris and the Heusler mechanism [63,64]. According to Heusler, it starts with the decomposition of water resulting in the adsorption of OH as there are no OH^- ions, which could adsorb directly in acidic media.

$$H_2O \rightarrow OH^-_{ad} + H^+(aq) \quad \text{fast} \tag{1.147a}$$

$$Fe + OH^-_{ad} \rightarrow FeOH_{ad} + e^- \quad \text{fast} \tag{1.147b}$$

This fast reaction is followed by two single-electron steps to form the complex $FeOH^+$, which is finally transferred into solution where it is decomposed.

$$FeOH_{ad} + OH^-_{ad} + Fe \leftrightarrow FeOH^+(aq) + FeOH_{ad} + 2e^- \quad \text{slow} \tag{1.147c}$$

$$FeOH^+(aq) + H^+(aq) \rightarrow Fe^{2+}(aq) + H_2O \quad \text{fast} \tag{1.147d}$$

Equation 1.147c should be interpreted as dissolution of $FeOH_{ad}$ from a kink site forming $FeOH^+$ in solution with the simultaneous replacement of OH^- to reform $FeOH_{ad}$ at the newly formed kink site, which then is ready for a next reaction step of $FeOH^+$ dissolution. Equation 1.147c is a two-electron reaction step. The first two reactions are assumed to be fast so that they are in quasi-equilibrium. Applying the Butler–Volmer equation to the slow reaction step of Equation 1.147c leads to an electrochemical reaction order of 1 for OH^-. However, if changes of the surface concentration of $(FeOH)_{ad}$ with potential take place, so that it may not be seen as constant, one obtains an electrochemical reaction order of 2. In this case, the surface concentration of $FeOH_{ad}$ should be calculated by writing the equilibrium of the two preceding fast reactions (Equations 1.147a and b).

Assuming Equation 1.147c as the rate-determining process, one gets the following Butler–Volmer equation where the factor 2 of the exponential expression takes into account the two transferred electrons of the step.

$$i_{Fe} = 2Fk_{Fe,a}\theta_{FeOH}\theta_{OH}\theta_{Fe}\exp\left\{\frac{\{2\alpha FE\}}{RT}\right\} = i'_{0,Fe}\exp\left\{\frac{\{2\alpha F\eta\}}{RT}\right\} \tag{1.148a}$$

with

$$i'_{0,Fe} = 2F\,k_{Fe,a}\theta_{FeOH}\theta_{OH}\theta_{Fe}\exp\left\{\frac{2\alpha FE_0}{RT}\right\} \tag{1.148b}$$

The exchange current density of Equation 1.148b shows an electrochemical reaction order 1 for θ_{OH} and therefore also for the bulk concentration $c(OH^-)$ in solution. This is a consequence of the quasi-equilibrium of Equation 1.147a combined with the dissociation equilibrium of water giving the ionic product $K_W = c(H^+)\,c(OH^-)$.

$$\theta_{OH} = \frac{K_1}{c(H^+)} = \frac{K_1\,c(OH^-)}{K_W} \tag{1.148c}$$

Equation 1.148d presents the related Tafel equation.

$$\eta = -\left(\frac{RT2.303}{2\alpha F}\right)\log i'_{0,Fe} + \left(\frac{RT2.303}{2\alpha F}\right)\log i_{Fe} \tag{1.148d}$$

The Tafel slope is $d\eta/d\log i_{Fe} = 0.059$ V for $\alpha = 0.5$. However, this result assumes a constant concentration or surface coverage of $FeOH_{ad}$. The surface coverage of $FeOH_{ad}$ should follow the potentially dependent equilibrium of reaction (Equation 1.147b), which, expressing θ_{OH} as a function of $c(OH^-)$ according to Equation 1.148c, leads to the expression:

$$E = E^0 + \frac{RT}{F}\ln\left\{\frac{\theta_{FeOH}}{\theta_{Fe}\theta_{OH}}\right\} = E^0 + \frac{RT}{F}\ln\left\{\frac{\theta_{FeOH}K_W}{\theta_{Fe}c(OH^-)K_1}\right\} \tag{1.149a}$$

Equation 1.149a gives an expression for θ_{FeOH}, which may be introduced in Equation 1.148a resulting in Equation 1.149b showing a reaction order 2 for OH^-.

$$i_{Fe} = 2FKk_{Fe,a}\theta^2_{Fe}c^2(OH^-)\exp\left\{\frac{\{(1+2)FE}{RT}\right\} \tag{1.149b}$$

K sums up several constants like K_1, K_W, and an exponential expression with E^0 from Equation 1.149a.

Introducing $E = E_{eq} + \eta$, Equation 1.149b may be rewritten as:

$$i_{Fe} = i_{0,Fe}\exp\left\{\frac{(1+2\alpha)F\eta}{RT}\right\} \tag{1.150a}$$

with

$$i_{0,Fe} = 2FKk_{Fe,a}\theta^2_{Fe}\,c^2(OH^-)\exp\left\{\frac{(1+2\alpha)FE_{eq}}{RT}\right\} \tag{1.150b}$$

which yields the semilogarithmic relation:

$$\eta = -\frac{2.303RT}{(1+2\alpha)F}\log i_{0,\text{Fe}} + \frac{2.303RT}{(1+2\alpha)F}\log i_{\text{Fe}} \tag{1.151}$$

Equation 1.151 yields the Tafel slope $d\eta/d\log i_{\text{Fe}} = 0.030\,\text{V}$ for $\alpha = 0.5$. Experimentally, a Tafel slope of 0.03 V and a reaction order 1.5–2.0 for OH^- are found for stationary values of Fe dissolution [64], which supports the assumptions of the proposed mechanism. However, if one uses the potential values at the very beginning of galvanostatic transients, a Tafel plot of the potential vs. current density dependence yields a slope of 0.060 V. Apparently, the reaction of Equation 1.147b needs some seconds to get to equilibrium. The initial potential values refer to the unchanged surface coverage of $FeOH_{ad}$ before the transient, whereas the later values refer to the new value when the equilibrium has been established to the changed potential.

1.13.2 High Rates of Metal Dissolution and Salt Precipitation

For high dissolution rates of metals, one reaches diffusion control after some time when the accumulation of corrosion products at the surface has been achieved. Metal dissolution with extremely high anodic current densities from several 10 A · cm^{-2} up to more than 100 A · cm^{-2} has been studied in solutions with high concentrations of chloride. A concentration ≥1 M Cl^- is necessary to prevent the formation of passive layers at very positive electrode potentials where these extremely large current densities have been measured. Galvanostatic transients have been performed with small electrodes of Fe and Ni of less than 1 mm^2 surface area to ensure the same potential and the same concentration of products all over the surface. Figure 1.38 gives an example for the dissolution of Ni in $NiCl_2$ solution [65]. Studies in several other electrolytes with less $NiCl_2$ or pure HCl have been studied as well as the dissolution of Fe in similar solutions. The observed transients show

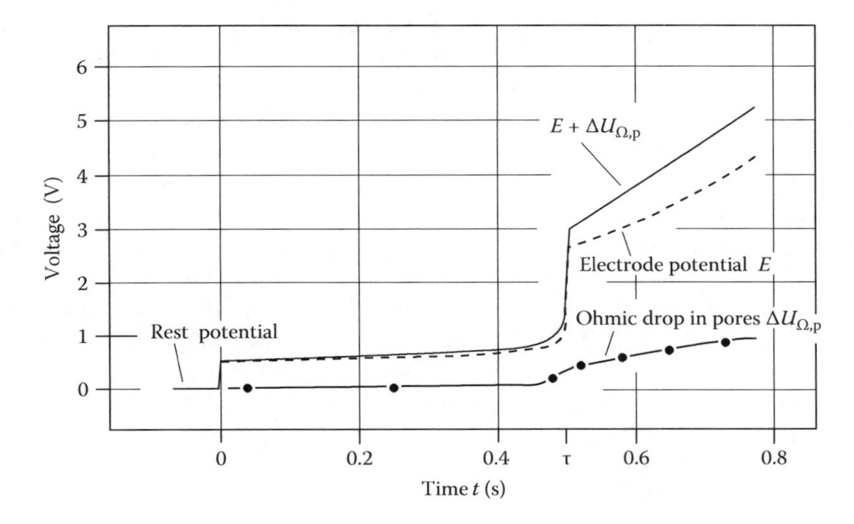

FIGURE 1.38
Galvanostatic transient for Ni dissolution in saturated $NiCl_2$ solution with potential increase after the induction time τ and the ohmic drop in pores measured by superimposed fast short additional galvanostatic transients. (From Strehblow, H.-H. and Wenners, J., *Electrochim. Acta*, 22, 421, 1977.)

a large potential plateau and a final steep increase. The transition time τ up to this steep increase follows Sand's equation [66,67].

$$i\sqrt{\tau} = 0.5\, z_M F \sqrt{\pi D}(c_{M,S} - c_{M,B}) \tag{1.152a}$$

$$i\sqrt{\tau} = 0.5 z_M F \sqrt{\pi D}(c_{M,sat} - c_{M,B}) \tag{1.152b}$$

where

$c_{M,S}$ and $c_{M,B}$ are the metal ion concentrations at the surface and within the bulk electrolyte, respectively

z_M and z_A are the charge of the metal ions and anions (Cl^-), respectively

D' is the effective diffusion coefficient containing a correction for the contribution of migration to ion transport within the electrolyte when only the metal salt and no excess of supporting electrolyte is present ($D' = (1 + |z_M|/|z_A|)\,D$)

The time $t = \tau$ at a given current density i is required to accumulate metal ions to obtain a saturated solution at the surface with $c_{M,S} = c_{M,sat}$ and a final precipitation of a salt film (Equation 1.152b). The product $i\sqrt{\tau}$ is a constant independent of the applied current density i for a given metal and composition of the bulk electrolyte. The data for $FeCl_2$ and $NiCl_2$ with a saturation concentration of $c_{M,sat} = 4.0$ and 4.2 M, respectively, give a good quantitative agreement with the measured results. These studies require a sufficiently large concentration of chloride (≥1 M) in order to prevent passivation at the whole surface. Pure sulfuric, nitric, and perchloric acid lead to passivation of the surface, i.e., to the formation of an oxide layer with different transients. No or a much shorter potential plateau is observed for these solutions. For smaller chloride concentrations, pitting is observed with the formation of a passive layer and corrosion pits and a different current time behavior for transients. These details are discussed in Chapter 7.

Small shorter galvanostatic transients superimposed to the larger transient permit the determination of the ohmic drop $\Delta U_{\Omega,P}$, which is assumed to occur in a porous layer of the precipitated salt film (Figure 1.38). An additional larger potential drop ΔE should be located within the inner poreless salt film. The electrode potential E assumes constant values during the transition time $t < \tau$ and becomes large when a salt film is formed at $t > \tau$ with a steep increase of the voltage $E + \Delta U_{\Omega,P}$. The further linear increase of the voltage is related to the growth of the salt film, which necessarily requires an additional increase of $\Delta U_{\Omega,P}$. On the basis of these measurements, a model with an inner poreless and an outer porous salt layer has been postulated, both taking over a fraction of the increase of the voltage. The increase of the voltage of the inner poreless layer may be attributed to the potential of an electrode covered with a layer similar to a metal with a passive layer.

These conditions of general metal dissolution at high current densities are closely related to electrochemical machining and to the high dissolution currents within corrosion pits at an early growth stage as will be discussed in Chapter 7 (Mechanisms of Pitting Corrosion). The electropolishing of the metal surface under these conditions has been interpreted by oscillation of pores within the salt layer, leading to a smooth and equal metal dissolution all over the attacked surface.

1.13.3 Selective Dissolution of Alloys

For alloys, one has to distinguish single-phase and multiphase alloys. If an alloy is single phase, it contains the two or more metal components in a homogeneously distributed ordered arrangement at atomic level. For systematic studies, single-phase alloys with two chemically very different elements have been investigated in detail, i.e., alloys with one noble and one reactive component like CuZn, CuSn, CuAu, or PdAg with varying quantitative composition. Some of these alloys are technically important (e.g., brass and bronze). In principle, the components may dissolve simultaneously or selectively, which depends on the details of the chemical properties of the elements, the environmental or experimental conditions like the composition of the solution or the electrode potential, and possible transport mechanisms in the solid state like surface and bulk diffusion of metal components. Without any diffusion, the dissolution of one component will only cause the accumulation of the other at the surface. Dissolution will stop when no active atom is in contact with the electrolyte. This may be the case when the reactive component has a low bulk concentration. But diffusion of metal atoms will rearrange the structure and will expose new atoms of the reactive component to the electrolyte and cause their continuous dissolution. Bulk diffusion with the help of imperfections like vacancies or di-vacancies is an effective mechanism to get new atoms of the reactive component to the surface even from deeper sites of the material.

AuCu alloys show a typical behavior with a very low dissolution current at potentials positive of the Cu/Cu^{2+} electrode but negative of the Au/Au^{3+} electrode. Pure copper shows a constant current increase with the potential starting at the Nernst equilibrium potential, whereas AuCu alloys have a current plateau with low current densities of $1\,\mu A \cdot cm^{-2}$ or less as shown in Figure 1.39 [68,69]. Above a critical potential E_{Cr}, the dissolution current is increasing steeply by several orders of magnitude up to more than $10\,mA \cdot cm^{-2}$. E_{Cr} increases with the Au content, whereas the plateau current decreases. This behavior has been explained by a Cu dissolution leaving Au behind, which causes Au enrichment at the metal surface. According to the model of Stranski and Kossel, the dissolution should occur at the kink sites probably via step and ad-sites on the terraces (Figure 1.36). During this dissolution process, the kink sites are blocked by remaining Au atoms, which have to be removed by surface diffusion so that the dissolution reaction may proceed. This surface diffusion is apparently a slow process, which leads to the accumulation of Au clusters at the surface. The fast dissolution above the critical potential is still very selective leaving a spongy Au layer behind. This has been explained by a faster Cu dissolution involving

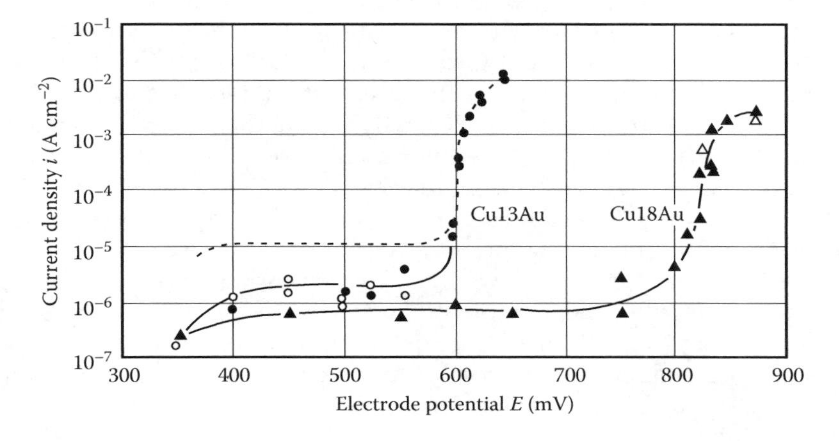

FIGURE 1.39
Log $i - E$ plot for the dissolution of CuAu alloys with 13 and 18 at % Au in showing the current plateau and the steep current increase at the critical potentials of 0.60 and 0.80 V, respectively. (From Kaesche, H., *Corrosion of Metals*, Springer, Berlin, Germany, 2003; Pickering, H.W. and Burne, P.J., *J. Electrochem. Soc.*, 118, 209, 1971.)

a direct transfer of atoms from sites well within the terraces, which needs a larger driving force and thus more positive potentials [69]. The Cu dissolution from inner parts of the metal despite the accumulation of Au has been explained by an enhanced outward diffusion of Cu atoms from the bulk metal to the surface via di-vacancies [70]. The accumulation of the remaining Au atoms leads to the formation of Au dendrites [68]. This porous Au layer is growing linearly in thickness with time. A possible dissolution and precipitation mechanism of Au could be ruled out by RRD experiments showing no dissolved Au atoms even at intermediate stages [70].

Similar studies have been performed with brasses of different composition. Cu-65Zn shows a plateau current of ca. $1\,\mu A \cdot cm^{-2}$ with a steep increase at potential $E > 0.0\,V$ up to $i > 10\,mA \cdot cm^{-2}$. For Cu-30Zn (α-brass), the current plateau reaches up to $E = 0.10\,V$, with a similar increase to $i > 10\,mA \cdot cm^{-2}$. Below these potentials, both metals show a very slow Zn dissolution only, whereas the dissolution of both metal components is reported for $E \geq 0.10\,V$ [71]. The equilibrium potential of the Cu/Cu^{2+} electrode with $E^0 = 0.337\,V$ coincides with the observed current increase for both alloys as its value decreases with the Cu content of the alloy according to the Nernst equation:

$$E = E^0 + \frac{RT}{2F} \ln \frac{a_{Cu^{2+}}}{a_{Cu}} \tag{1.153}$$

This result is different from the one for CuAu alloys. Au cannot dissolve even for potentials $E > E^0 = 1.50\,V$ for the Au/Au^{3+} electrode because its surface is protected by a passive layer even in acidic electrolytes. In the case of CuZn alloys, both metals dissolve at $E > 0.1\,V$ although with different rates. Thus, a Cu-rich phase is formed at the surface due to preferential Zn dissolution. Diffraction studies show the presence of Cu and brasses enriched in Cu. Preferential Zn dissolution requires a supply of Zn atoms by diffusion from the bulk metal via di-vacancies. Cu-86Zn starts high dissolution rates already at $E \geq -0.9\,V$ increasing to more than $1\,mA \cdot cm^{-2}$. The critical potential of Zn dissolution is too negative for this alloy to show a plateau with small dissolution rates. Dissolution of Cu is not possible for these negative potentials similar to the situation of AuCu alloys.

Active dissolution of alloys is a complex process, and a detailed understanding of the leading mechanisms has to be proven by investigation with modern surface analytical tools under well-controlled electrochemical conditions. It also requires the study of single-crystal surfaces with a very careful surface preparation. One has to start fundamental investigations with a chemically and structurally well-defined surface. Methods like XPS and STM might help to get a deeper insight into this complex matter. This also includes the study of alloys with components of similar reactivity. Besides these conditions for dissolving alloys, one has to be well aware that oxidation and dissolution change dramatically when entering the potential range of the formation of passive layers. This might even occur at potentials negative to the potential of oxide formation due to the existence of adsorption layers including chemisorption of OH and the formation of hydroxide monolayers. Oxidation studies of alloys are even more difficult for oxide-covered metals. Here again preferential oxidation of one metal component and its transfer to the oxide film is sometimes observed, while the other remains at the metal surface. However, in many cases, the oxidation of both metal components and their transfer to the oxide is found. In addition, a preferential dissolution of one cation component relative to the other occurs at the oxide/electrolyte interface happens quite often, which is a further complication for the system. These aspects are discussed in detail in Chapter 5.

TABLE 1.8

Dissociation Constants of Some Cation Complexes K_D and the Related Standard Potentials E^0

Reaction	K_D	E^0/V
$AuCl_4^- \leftrightarrow Au^{3+} + 4Cl^-$	2.2×10^{-22}	0.994
$Au(CN)_2^- \leftrightarrow Au^+ + 2CN^-$	$2.3 + 10^{-39}$	−0.6
$Ag(CN)_2^- \leftrightarrow Ag^+ + 2CN^-$	$1.5 + 10^{-19}$	−0.31
$Cu(NH_3)_4^{2+} \leftrightarrow Cu^{2+} + 2NH_3$	$3.5 + 10^{-14}$	−0.05
$Cu(CN)_2^- \leftrightarrow Cu^2 + 2CN^-$	$7.3 + 10^{-17}$	−0.43

Source: *Handbook of Chemistry and Physics*, CRC Press, Cleveland, OH, 1977–1978.

1.13.4 Metal Dissolution and Complex Formation

The formation of complexes has a strong influence on metal dissolution. Some anions form strong complexes with very small values of their dissociation constants. Table 1.8 gives some examples. Au forms very stable complexes according to Equation 1.154a with the related equilibrium constant of Equation 1.154b.

$$Au(CN)_2^- \leftrightarrow Au^+ + 2CN^- \tag{1.154a}$$

$$K_D = \frac{c^2(CN^-)c(Au^+)}{c(Au(CN)_2^-)} \tag{1.154b}$$

The Nernst equation for the oxidation of Au metal to Au^+ gives:

$$E = E^0_{Au} + \frac{RT}{F}\ln c(Au^+) = 1.68 + \frac{RT}{F}\ln c(Au^+) \tag{1.155}$$

Replacing the very small concentration of Au^+ of Equation 1.155 by its expression from Equation 1.154b yields Equation 1.156 with the standard potential E^0_{AuCN} for the $Au/Au(CN)_2^-$ electrode in Equation 1.156a.

$$E = 1.68 + 0.059\log K_D + 0.059\log\frac{c(Au(CN)_2^-)}{c^2(CN^-)} = E^0_{AuCN} + 0.059\log\frac{c(Au(CN)_2^-)}{c^2(CN^-)} \tag{1.156}$$

$$E^0_{AuCN} = 1.68 + 0.059\log K_D \tag{1.156a}$$

Introducing the value $K_D = 2.3 \times 10^{-39}$, one obtains $E^0_{AuCN} = 1.68 + 0.059 \log 2.3 \times 10^{-39} = 1.68 - 2.28 = -0.6\,V$.

The presence of CN^- in the solution shifts the standard potential of the Au/Au^+ electrode by −2.28 V. Therefore Au may be oxidized by oxygen in cyanide solutions, which is used to extract Au from minerals. Table 1.8 summarizes several other examples. Au may be also dissolved by a mixture of hydrochloric and nitric acid. Even the oxidizing power of concentrated nitric acid is not sufficient to oxidize Au to Au^{3+}. The complexing properties of chloride helps to shift the standard potential from $E^0_{Au} = 1.42\,V$ to $E^0_{AuCl_4^-} = 0.994\,V$, which is a consequence of the small value of the dissociation constant $K_D = 2.2 \times 10^{-22}$ of the $AuCl_4^-$ complex. Similar examples refer to the other systems mentioned in Table 1.8, like the dissolution of Ag in cyanide solutions and the complexing of Cu^{2+} cations by ammonia, which prevents the formation of insoluble CuS in the presence of H_2S.

1.14 Metal Dissolution in Combination with Reduction Reactions

In practice, metal dissolution occurs in combination with the reduction of an oxidizing species. Hydrogen evolution and oxygen reduction are frequent counter reactions compensating partially or totally the metal dissolution of such a mixed electrode. Figure 1.40 presents schematically the reactions at an Me/Me^{z+} electrode and the H_2/H^+ electrode. Both electrodes show their Nernst equilibrium potentials E_{eq} and the anodic and cathodic branches of their polarization curves following the exponential relation of the Butler–Volmer equation. The combination of the anodic part of the Me/Me^{z+} polarization curve and the cathodic part of the H_2/H^+ polarization curve (solid curves) yields the measured polarization curve. The current is equal to zero at the corrosion potential or rest potential E_R. As both Nernst potentials $E_{eq,Me}$ and $E_{eq,H}$ are apart from each other, one may neglect in the further discussion the cathodic metal deposition and the anodic hydrogen oxidation. This holds especially for negligible concentrations of metal cations and hydrogen gas within the electrolyte. Writing the Butler–Volmer equations for both reactions in a simplified form with the constants A_M, a_M for the anodic metal dissolution and A_H, a_H for the cathodic hydrogen evolution one obtains:

$$i = i_M + i_H = A_M \exp\left\{\frac{E}{a_M}\right\} - A_H \exp\left\{-\frac{E}{a_H}\right\} \tag{1.157}$$

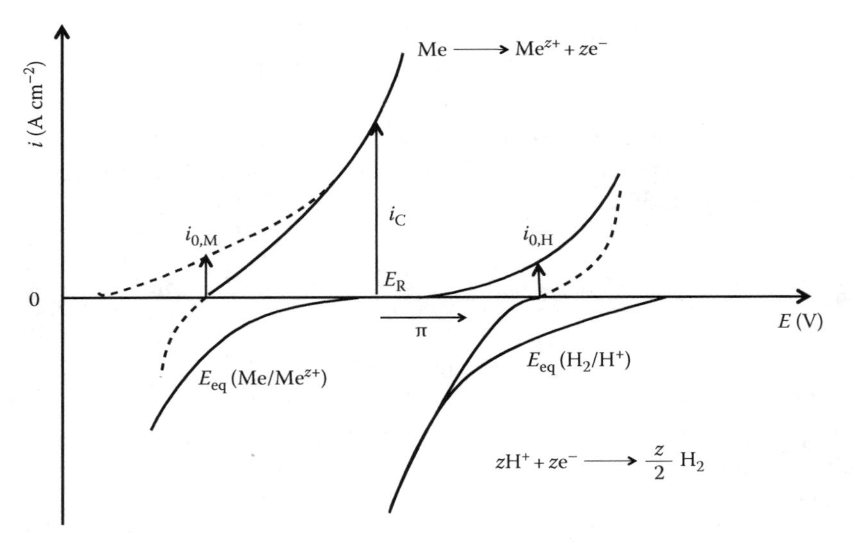

FIGURE 1.40
Superimposed metal dissolution and hydrogen evolution with equilibrium potentials $E_{eq}(Me/Me^{z+})$ and $E_{eq}(H_2/H^+)$, respectively, leading to the rest or corrosion potential E_R for disappearing current in the external circuit and the corrosion current density i_C. Solid curves correspond to the partial current densities of anodic metal dissolution and cathodic hydrogen evolution and to the sum curve of both reactions.

For open circuit conditions, the net current is zero and the mixed electrode is at the corrosion or rest potential E_R. The corrosion current density is:

$$i_C = i_M(E_R) = A_M \exp\left\{\frac{E_R}{a_M}\right\} = -i_H(E_R) = A_H \exp\left\{-\frac{E_R}{a_H}\right\} \tag{1.158}$$

Defining the polarization $\pi = E - E_R$, Equation 1.157 may be rearranged as the Butler–Volmer equation for a mixed electrode:

$$i = i_C\left[\exp\left\{\frac{\pi}{a_M}\right\} - \exp\left\{-\frac{\pi}{a_H}\right\}\right] \tag{1.159}$$

The total current equals the anodic metal dissolution current for high anodic polarizations, and the cathodic hydrogen evolution current for high cathodic polarizations, leading to the semilogarithmic Tafel presentation (Equations 1.160a and b):

$$\log i = \log i_C + \frac{\pi}{b_M} \quad \text{(large anodic } \pi) \quad b_M = 2.303a_M \tag{1.160a}$$

$$\log |i| = \log i_C - \frac{\pi}{b_H} \quad \text{(large cathodic } \pi) \quad b_H = 2.303a_H \tag{1.160b}$$

Figure 1.41 shows the Tafel plot for the Fe/H_2 mixed electrode. The extrapolation of the lines yields i_C and E_C at their intersection. The dashed lines indicate the influence of the counter reaction for small polarizations π.

In the vicinity of the rest potential $E = E_R$ (for $\pi \to 0$), Equation 1.159 may be linearized leading to:

$$i = i_C\left[\frac{\pi}{a_M} + \frac{\pi}{a_H}\right] \tag{1.161}$$

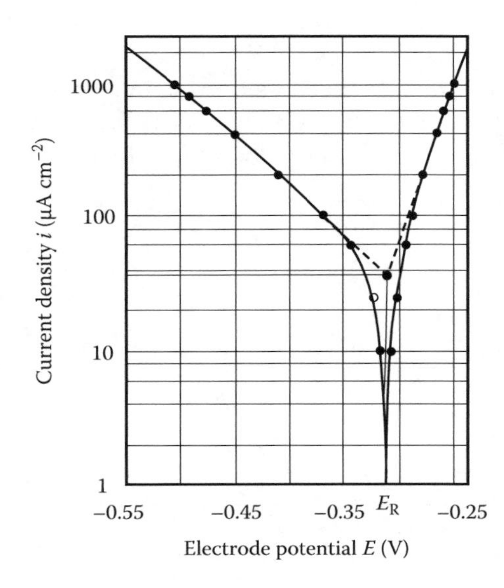

FIGURE 1.41
Tafel plot for dissolution of iron with hydrogen evolution in 0.025 M H_2SO_4 + 0.45 M Na_2SO_4, pH 1.7 and determination of E_R by extrapolation of the Tafel lines. (From Kaesche, H., *Corrosion of Metals*, Springer, Berlin, Germany, 2003.)

The slope of the current vs. polarization variation at $E = E_R$ yields the reciprocal of the polarization resistance R_P (Equation 1.162).

$$\frac{1}{R_P} = \left(\frac{di}{d\pi}\right)_{\pi \to 0} = i_C\left(\frac{1}{a_M} + \frac{1}{a_H}\right) \tag{1.162}$$

Hence the corrosion current i_C can be calculated from the measurement of R_P and the anodic and cathodic Tafel slopes:

$$i_C = \frac{a_M a_H}{(a_M + a_H)R_P} = \frac{b_M b_H}{(b_M + b_H)2.303R_P} \tag{1.163}$$

If the oxygen reduction is the counter reaction of metal dissolution, its kinetics is often determined by diffusion due to the small solubility of O_2 gas and long diffusion distances especially in unstirred solution. In this case, the cathodic reaction gets potential independent at negative potentials with a diffusion-limited cathodic current density $i_{D,O2}$ in the vicinity of the rest potential (Figure 1.42). In this situation, the corrosion current density i_C is equal to the limiting cathodic current density $i_{D,O2}$, and Equation 1.159 simplifies to Equation 1.164, which becomes Equation 1.165 for $\pi \to 0$ ($E \to E_R$). Hence i_C can be calculated from R_P and the anodic Tafel slope (Equation 1.166).

$$i = i_C\left[\exp\left\{\frac{\pi}{a_M}\right\} - 1\right] \tag{1.164}$$

$$i = i_C\frac{\pi}{a_M} \quad (\text{for } \pi \to 0) \tag{1.165}$$

$$i_C = \frac{b_M}{2.303R_P} \tag{1.166}$$

It should be mentioned that a free metal dissolution is not observed very often in practice. Usually the metal is covered by a film of corrosion products like oxide layers, passive films, well-soluble or weakly soluble salt layers, adsorption layers, and organic films, which have

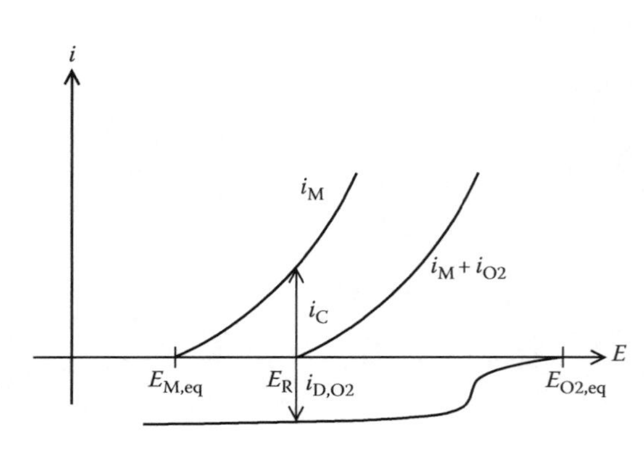

FIGURE 1.42
Superimposed metal dissolution and oxygen reduction with a diffusion-limited potentially independent current density $i_{D,O2} = i_C$.

a decisive influence on the kinetics and mechanism of metal dissolution. Furthermore, the study of the details of fast metal dissolution is not easy to follow with most of the rather slow analytical methods. As a consequence, not many details are known and a lot of work is required to sort out the different factors that have an influence on the kinetics in order to confirm proposed mechanisms.

1.15 Heterogeneous Metal Surfaces and Local Elements

Metal surfaces are heterogeneous. Besides the already mentioned surface atomic structure with step, kink, and ad-sites at a terrace, there are grain boundaries, inclusions, vacancies, dislocations, etc. A very important aspect is the existence of crystallites of a polycrystalline material exposing different orientations of its grains at the surface. The chemical composition may also vary for the different grains. Low solubility of a metal within the matrix of the major metallic component may cause segregation, for example, Cu segregation from Al. Cu-rich parts are good cathodes for the reduction of oxidants, leading to enhanced corrosion. These particles may not be sufficiently passivated because the Al oxide film cannot grow on them. Due to nonappropriate heat treatment, one may get precipitation of Cr carbides at grain boundaries of stainless steel with Cr-depleted zones. Thus, these sites are less protected by the very stable Cr-oxide films. As a consequence, one may have a grain boundary attack, i.e., intergranular corrosion. Another example is the deposition of Cu from Cu^{2+}-containing solutions on Zn surfaces. The Cu sites are good cathodes for hydrogen evolution, whereas the Zn surface is not, even in acidic electrolytes where a passivating film on Zn does not grow.

In conclusion, a surface often does not react homogeneously and preferential dissolution of some sites may occur. Furthermore, the kinetics of cathodic processes may differ from one crystallite to the other for the same reasons. As a consequence, the reactivity of surface sites is different. There are regions with high anodic metal dissolution and others with a pronounced cathodic reduction of redox systems, i.e., local anodes and local cathodes. This does not necessarily mean that local anodes and cathodes show exclusively their related anodic or cathodic reactions. A local anode is a site with preferentially anodic metal dissolution, but some cathodic reduction may occur as well. At a local cathode, cathodic reduction will occur predominantly, but anodic dissolution may occur as well. One has in general preferential anodic and preferential cathodic sites. For open circuit conditions, electroneutrality requires that anodic and cathodic reactions have to compensate each other. This will occur by electronic exchange via the connecting metal substrate. The potential within the substrate is usually the same due to the high electronic conductivity of the metal. In front of the electrode, a potential variation may occur due to a low conductivity of the electrolyte solution. Figure 1.43a shows schematically the polarization curves of two different types of surfaces sites A and B forming the local anodes and cathodes of an electrode surface. If the electrolyte is well conducting, there is no voltage drop within the electrolyte, the excess current at the anodes will be compensated by an excess current at the cathodes and an equivalent charge will be transferred by a flow of electrons within the metal substrate from the local anodes A to the local cathodes B. The common rest potential $E_{R,AB}$ is between the rest potentials $E_{R,A}$ and $E_{R,B}$ of the pure materials A and B. The polarization curve of the composite material measured with a potentiostat will follow a polarization curve with a polarization π referring to $E_{R,AB}$. For $\pi = 0$, i.e., for open circuit conditions, no current flows in an outer circuit. Figure 1.43b describes the situation for an electrolyte

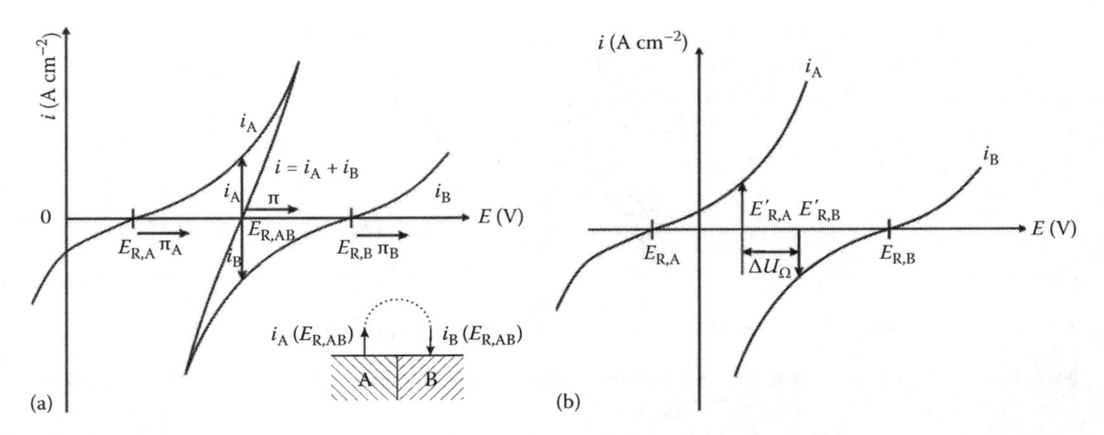

FIGURE 1.43
(a) Polarization curve of a metal with sites A and B of different electrochemical activity. $E_{R,A}$ and $E_{R,B}$ are the rest potentials of the pure sites A and B and $E_{R,AB}$ for the mixed electrode, respectively with polarizations π_A, π_B, and π, respectively. Inset shows current flow $i_A = i_B$ from sites A to B in the external circuit (short circuit). (b) Same as (a) with low-conductivity electrolyte causing ohmic drops ΔU_Ω between sites A and B.

with low conductivity so that an ohmic drop $\Delta U_\Omega = E'_{R,A} - E'_{R,B}$ will occur between the local anodes and local cathodes with the local potentials $E'_{R,A}$ and $E'_{R,B}$. Local elements therefore will show potential differences in front of the electrode following their spatial dimensions (Figure 1.44). If the conductivity is very low, steep changes will occur at the boundaries of the local elements [case (1)] and a constant part in between. The profile will smoothen with increasing conductivity [case(2)] and will disappear for well-conducting electrolytes [case(3)]. The potential profile also smoothens with the distance to the metal surface due to the spreading current lines as shown in Figure 1.44.

These potential profiles may be measured with two REs, one with a fixed position and the other scanning parallel to the surface. With increasing distance to the surface, it may also scan the profile perpendicularly. A variation of this method uses two scanning REs with their HL capillaries close to each other. One may also use two small isolated Pt tips, which allow a smaller set-up. This scanning reference electrode technique (SRET) samples

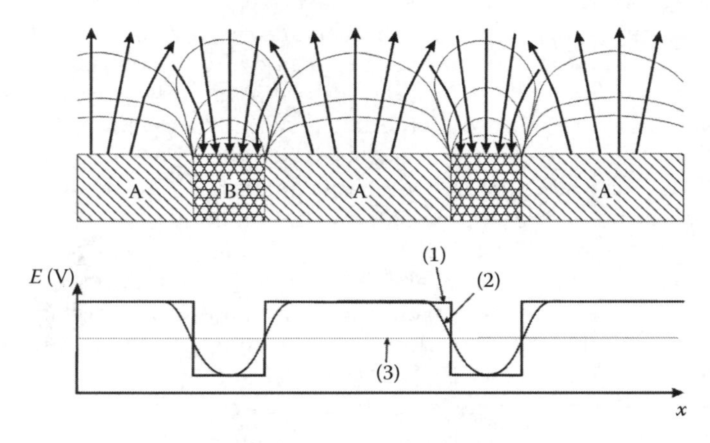

FIGURE 1.44
Current lines (arrows) and lines of equal potential within the electrolyte in front of a corroding metal surface consisting of local elements with metals A and B working as local anodes and cathodes, respectively; potential profile for electrolytes with low (1), medium (2), and high (3) conductivity.

the small changes of the potential drop at the two close tips of the capillaries and thus measures the potential gradient $d\Delta U_\Omega/dx$ in front of the electrode in comparison with the total potential difference ΔU_Ω as described for the first method. A third method uses a vibrating metal tip scanning across the surface and a second RE apart. This scanning vibrating electrode technique (SVET) samples again the potential gradient in front of the electrode with the applied oscillation frequency and amplitude. It may be used in combination with lock-in technique to separate the signal from the background and the disturbing noise of voltages of different frequencies. SVET and SRET are used mostly for imaging sites of different electrochemical activity at an electrode surface. Thus, one may locate inclusions, local elements, and corroding pits and crevices on a passive surface. These methods may be also used to calculate the current density distribution in front of an electrode. For this purpose, one has to take into account possible local changes of the conductivity of the electrolyte, which may occur due to the accumulation of corrosion products.

A technical application of potential measurements is the detection of corrosion damages of underground metal constructions like gas or water pipes. Two electrodes are put to the wet ground, one of them changes place to detect sites of intense dissolution currents and the related ohmic drops. These sites are the locations of corrosion damage.

1.16 Protection

1.16.1 Cathodic Protection

Cathodic protection of metals is achieved by a negative potential, which may be applied by contact to a very reactive metal electrode or by the application of a negative current. In both cases, the potential is shifted to values where the metal does not dissolve or dissolves very slowly. A shift below the Nernst potential of the metal/metal ion electrode will stop dissolution due to thermodynamic reasons. This critical protection potential E_{Pr} may be calculated simply with Nernst's equation with a negligibly small cation concentration, usually 10^{-6} M as used for example in Pourbaix diagrams (see Section 1.8.4). As an example, iron dissolves as Fe^{2+}, and the Nernst potential is given by Equation 1.87. With $a(Fe^{2+}) = 10^{-6}$ M, one obtains $E_{Pr} = -0.614$ V. This potential is more negative than that of the H/H^+ electrode, which causes Fe dissolution for $E > E_{Pr}$ by hydrogen evolution in acidic electrolytes. Therefore the potential should be shifted to $E < -0.614$ V. More noble metals require less negative potentials. Reaction 1.167 yields with Equation 1.168 $E = 0.166$ V for the Cu/Cu^{2+} electrode.

$$Cu^{2+} + 2e^- \leftrightarrow Cu \tag{1.167}$$

$$E = 0.340 - 0.029 \times 6 = 0.166\ \mathrm{V} \tag{1.168}$$

As a consequence, Cu corrosion with hydrogen evolution should not occur. However, Cu dissolution will occur if Cu is exposed to air with the much more positive potential of the O_2/H_2O electrode. For $p(O_2) = 0.2$ bar, one obtains according to Equation 1.169 a pH-dependent potential expressed by Equation 1.170.

$$H_2O \leftrightarrow O_2 + 4H^+ + 4e^- \tag{1.169}$$

$$E = 1.23 - 0.059\ \mathrm{pH} + 0.0148 \log p(O_2) = 1.22 - 0.059\ \mathrm{pH} \tag{1.170}$$

As a consequence, Cu will corrode in solutions in contact with air. This might be prevented by cathodic protection with $E < E_{Pr} = 0.166$ V.

For practical applications, the current of protection of a surface area A, $I_{Pr} = i_{Pr} \times A$, is important because it shifts the potential to the desired value. Anodic dissolution is negligible, so that the total current density is given by the cathodic current of the redox system, i.e., its value for $E \leq E_{Pr}$. For some systems, this is given by the Butler–Volmer equation, for others by the diffusion-limited current density i_D. The hydrogen evolution current in acidic electrolytes is mostly given by the Butler–Volmer equation as discussed in Section 1.12.1. Oxygen reduction often occurs at very large overvoltage and thus is diffusion controlled. With the saturation concentration $c(O_2) = 2 \times 10^{-4}$ M for dissolved oxygen, its diffusion constant $D = 10^{-5}\ cm^2 \cdot s^{-1}$ and a Nernst diffusion layer of $d_N = 5 \times 10^{-3}$ cm, one obtains the maximum diffusion current density $i_D = 0.15\ mA \cdot cm^{-2}$ as given in Section 1.12.2. If the surface is covered by a porous film, diffusion must occur in pores and thus is much smaller, corresponding to the effective metal surface in contact with the electrolyte. A protecting organic layer may further reduce the cathodic current corresponding to a very small effective metal surface area at defects within the film. For a free steel surface of $A = 1\ m^2$, $I = 10$–50 mA is reported in wet soil, 20–150 mA in sea water, whereas for a coverage by 2 mm polyethylene 0.5 μA is a realistic value [74].

As indicated before, there are two kinds of cathodic protection, the contact with a reactive metal that serves as a sacrificial electrode, and the application of a cathodic current from a current supply. If metals with a very negative redox potential like Mg ($E^0 = -1.5$ V), Al ($E^0 = -1.28$ V), Zn ($E^0 = -0.76$ V), and their alloys are in contact with a metal like iron, they dissolve whereas iron is protected. A mixed electrode potential is established as in the case of local elements (see Section 1.15). Due to the very negative standard potentials of these reactive metals, the potential for Fe becomes more negative than E_{Pr}. Mg is often used for the protection of steel structures in soil because of its large potential difference with iron, which may overcome even large ohmic drops in environments of low conductivity. Al and Zn are used for metal constructions in well-conducting sea water, like ships, bridges, offshore drilling stations, etc. The mass of these electrodes must be large enough (several kg) to serve for a current of ca. 0.1 A for 10 years and more. Zn-coated steel is a well-known protected material for cars. Metal dissolution may occur at defects of the Zn layer; however, it is suppressed by the negative mixed potential induced by the reactive Zn electrode.

A comparable effect to the sacrificial electrodes is provided by the direct supply of a cathodic current to the dissolving metal. A proper cathodic current applied to the metal structure sets the potential to a value where corrosion is prevented. A disadvantage is the requirement of a permanent connection to a current supply (or even a potentiostat). However, with this approach, the metal construction may be tuned in for complicated situations. Metallic structures often consist of several metals with different corrosion properties. The environmental conditions may be very difficult. For example, passivation should be maintained but the potential should not become more positive than the critical pitting potential. In these complicated cases, a potentiostat with a CE and a RE is useful. The WE of this circuit is the metal construction. Protection by a current source is applicable to chemical reaction vessels or constructions with permanent location but not always to mobile devices like cars, ships, etc. Therefore both methods are useful, and the choice depends on the specific requirements for the construction in service conditions. In well-conducting electrolytes, one has to take care of the equilibrium potentials of the involved electrodes (metal/metal ion and redox electrode). If the environment has a low conductivity (wet soil), ohmic drops have to be taken into account in order to establish an appropriate protecting potential all over the surface of the construction for the supplied total current. The calculation of ohmic drops may be very complicated for a difficult geometry. Only very simple cases have been treated in Section 1.11.3.

In the presence of aggressive anions like chloride, a cathodic shift to $E < E_{pit}$ is required to avoid localized corrosion. This is achieved by a cathodic current from an external source, which takes over the above-mentioned current density of diffusion-limited oxygen reduction of $i_D = 0.15\,mA \cdot cm^{-2}$ and thereby shifts the potential within the passive range below E_{Pit}. This is another important example for cathodic protection.

1.16.2 Anodic Protection

In some cases, passivity should be maintained. Anodic protection can be used to shift from the range of active dissolution to the passive range of the polarization curve. The necessary anodic current densities to maintain passivity are usually very small, in the range of a few $\mu A \cdot cm^{2-}$. If however pitting is a problem in the presence of halides, the potential has to be maintained below E_{Pit}. The presence of oxygen may cause a positive potential shift of a passive metal within the passive or even the transpassive range.

1.16.3 Stray Currents

Electrical stray currents may cause serious corrosion problems. They occur when large currents enter the ground, like at tracks of tramways or trains or similar situations. These currents may enter into better conductive underground metal structures such as cables or tubes during their way within the less-conductive environment like soil. When these currents leave the metal structure, an electrode reaction causes localized corrosion. Cathodic protection may help to keep the potential sufficiently negative so that these stray currents do not enter the metal but are forced to proceeds within soil. A similar problem is related to welding for ship repair and construction in sea water. If the CE of the welding device has not a good contact to the ship body or even gets only contact to the surrounding sea water, an anodic current enters the metal parts and leaves to the water, causing corrosion of the metal surface. This damage may easily be avoided if the CE is in good contact with the metal parts that are welded.

1.17 Inhibition of Corrosion

The corrosion rate of metals may be reduced by the addition of inorganic or organic compounds, called inhibitors, to their environment. The inhibitor efficiency η_I is defined as the relative reduction of the corrosion current density:

$$\eta_I = \frac{i_{C,0} - i_C}{i_{C,0}} \tag{1.171}$$

where $i_{C,0}$ and i_C are the corrosion current densities at open circuit, i.e., at the rest potential E_R without and with the addition of an inhibitor, respectively. Inhibitors may be added to the aqueous electrolyte, organic solvents, or even the gas phase. Gas phase inhibitors are volatile organic chemicals, which are added to the packaging material of metals and which adsorb at their surface and thus provide temporary protection. Dicyclohexylaminonitrite is one of these compounds. The reduction of the corrosion rate is achieved in several ways. The inhibitor may adsorb at the metal surface and thus reduce its anodic dissolution or the cathodic reaction like hydrogen evolution or oxygen reduction. Both partial reactions may be hindered by one inhibitor. This primary inhibition has to be distinguished from the secondary inhibition, when inhibitors react with the electrode surface forming protective compounds by their reduction (or oxidation). Inhibitors may also form insoluble complexes with dissolving metal ions and precipitate as three-dimensional layers at the surface. Some chemicals are used to reduce traces of oxygen dissolved within the electrolytes of closed systems like the fluids of heat exchangers of cooling and heating systems. They thus avoid corrosion by air oxidation, i.e., they block indirectly the reduction of oxygen as a cathodic reaction. Some inhibitors such as chromates are oxidizing the metal surface, thus facilitating the formation of passivating layers. There are various mechanisms of inhibition. They depend on the chemistry of the systems, i.e., the kind of metal, which is involved, the chemicals added as inhibitors, and further properties of the environment like the pH of the solution. Thus, several mechanisms may be effective, which prevent or reduce corrosion of metals. Furthermore, the composition of added chemicals is often very complicated. In many cases, a mixture of chemicals is required to obtain corrosion protection for systems involving different metals as for example heat exchangers. Each metal may require a different inhibitor for a good corrosion protection.

The detailed mechanism is often not fully understood. A blocking of active sites is often postulated for anodic dissolution or cathodic reduction of redox systems. However, the conclusions were mostly drawn indirectly from electrokinetic data, and a detailed investigation with surface analytical methods is still missing. The chemical reaction steps and their inhibition at surface sites should be further investigated systematically by methods like XPS, AES, STM, SFM, etc. Most of the knowledge is still empiric. However, the results of electrochemical investigations are very valuable for a better corrosion protection of materials in industrial applications. Some few principles of inhibition are discussed in the following sections.

1.17.1 Inhibition by Adsorption

Inhibitors are often chemicals with alkyl chains and an active head group like –CN, –SH, $-NH_2$, –COOH, $-O-PO_3$. The long organic chain protects the surface by its coverage, whereas the head group ensures a strong binding to the atoms of the metal surface. Lonely electron pairs are very important for this binding. Groups that induce a high electron density at the binding head group increase the efficiency of the inhibitor. CH_3 groups in meta or para position of pyridine (C_5H_5N) induce a higher electron density at the nitrogen atom and thus improve its inhibiting properties [72]. The addition of electronegative groups like –Cl has the contrary effect. Similar results are obtained for benzonitriles (C_6H_6-CN) with ligands in the equivalent position at the ring of the aromatic molecule [73]. These adsorbates are mostly active in acidic electrolytes where metal surfaces are often unprotected and dissolve actively, with hydrogen evolution.

As mentioned above, inhibitors may block anodic metal dissolution or the cathodic reduction reaction or both processes simultaneously. If the cathodic reaction is inhibited, the related Tafel line of Figure 1.45a is shifted to negative potentials. As a consequence, the rest potential E_R shifts to a more negative value $E_{R,I}$ and the related corrosion current density from i_C to the smaller value $i_{C,I}$. If the anodic metal dissolution is inhibited, the related

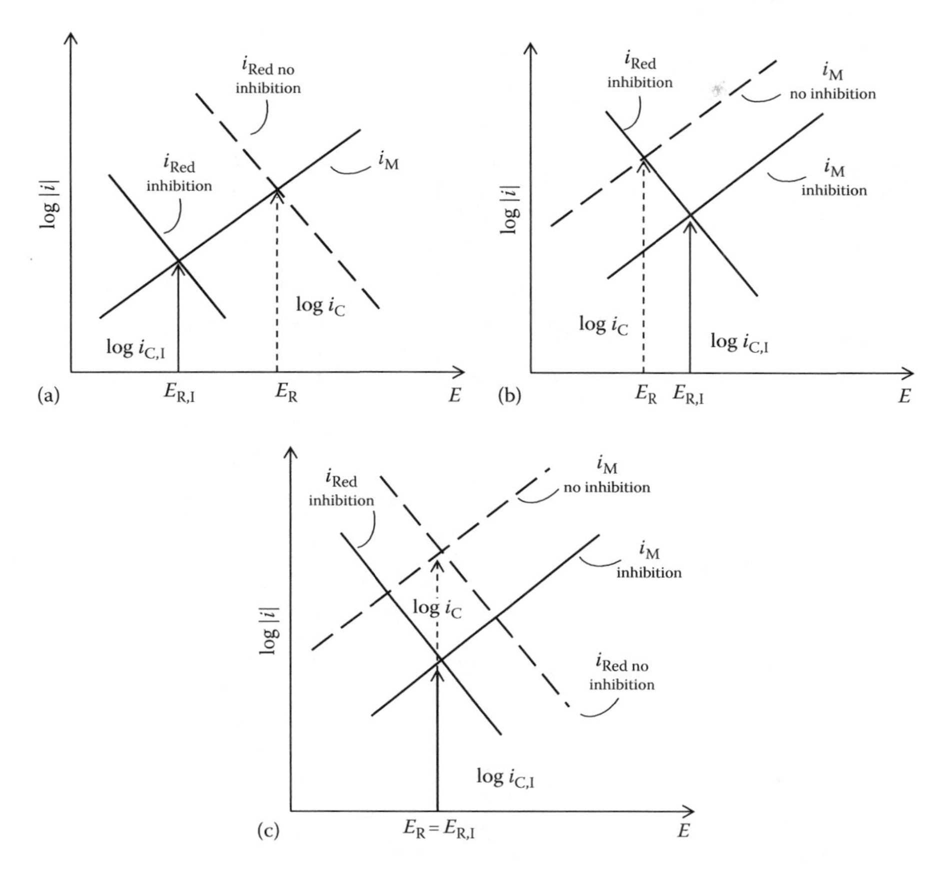

FIGURE 1.45
Tafel plot of a corrosion system with (a) inhibition of the cathodic reaction of a redox system, shift of E_R to negative potentials; (b) inhibition of the anodic metal dissolution, shift of the rest potential E_R to more positive values; and (c) inhibition of both partial reactions, E_R remains at its value.

anodic Tafel line shifts to more positive potentials (Figure 1.45b). E_R becomes more positive ($E_{R,I}$) and the dissolution rate becomes smaller again ($i_{C,I}$). Figure 1.46 gives an example for the inhibition of dissolution of iron by the addition of 2,6 dimethylchinolin in 5% H_2SO_4 [74,75]. The Tafel line for cathodic hydrogen evolution is not affected, whereas that for the iron dissolution is shifted parallel to more positive potentials. As a consequence, the intersections of the anodic Tafel lines with that for hydrogen evolution lead to increasing corrosion potentials E_R and to decreasing corrosion current densities i_C with increasing inhibitor concentration. The parallel anodic Tafel lines suggest that the mechanism of metal dissolution does not change, and the inhibition is only a consequence of the partial coverage of the metal surface. If both reactions are inhibited, E_R remains approximately at the same value, but the corrosion rate is reduced due to the smaller rates of both partial reactions (Figure 1.45c). As an example, iron corrosion in 6N HCl is mentioned with additions of homopiperazin (0.011–0.084 M) [76]. Here the rest potential remains at $E = -0.20$ V, whereas i_C decreases by one order of magnitude.

Due to an increasing surface coverage θ of the adsorbed inhibitor, the inhibition efficiency increases with the inhibitor concentration c. θ changes in the simplest case according to a Langmuir isotherm given by Equation 1.172.

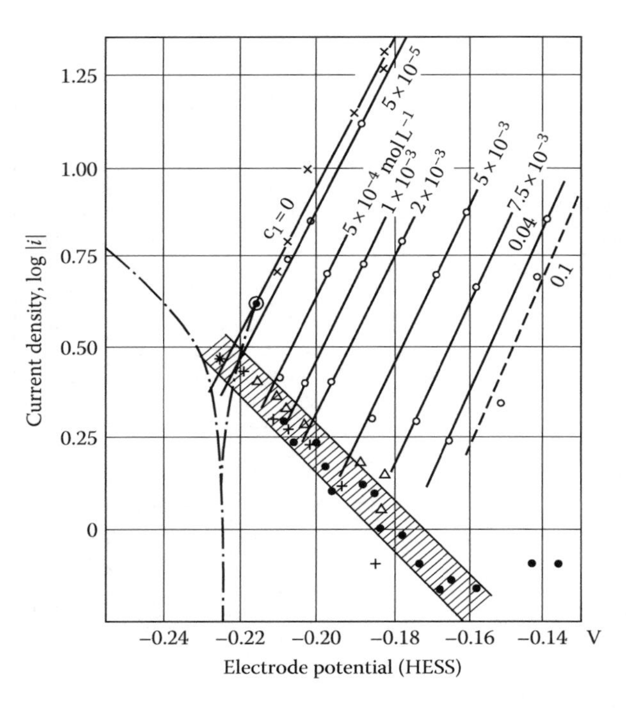

FIGURE 1.46
Inhibition of the anodic dissolution of steel in 5% H_2SO_4 with addition of 2,6 dimethylcholin as indicated. Parallel shift of Tafel lines of anodic metal dissolution and unchanged hydrogen evolution. Increase of E_R and decrease of i_C, metal dissolution × = without inhibitor, o = with inhibitor, • = hydrogen evolution, Δ = corrosion current density i_C, E referring to hydrogen electrode in same solution. (From Kaesche, H., *Corrosion of Metals*, Springer, Berlin, Germany, 2003; Hoar, T. P. and Hollody, R. D., *J. App. Chem.*, 3, 502, 1953.)

$$\frac{\theta}{1-\theta} = ac \tag{1.172}$$

$$a = a_0 \exp\left(-\frac{\Delta G_{ad}}{RT}\right) \tag{1.172a}$$

This isotherm assumes that there are no interactions between the adsorbed molecules and that all surface sites are identical with respect to adsorption, i.e., the free adsorption enthalpy ΔG_{ad} is independent of the surface site and of the surface coverage. Furthermore adsorption leads at maximum to a monolayer of the inhibitor molecules, i.e., there is no formation of a second layer at high inhibitor concentrations. The adsorption constant a and the adsorption enthalpy does not change with θ (Equation 1.172a). The Langmuir isotherm is an ideal approximation of the inhibitor adsorption, which is not very realistic. Adsorbed molecules will have lateral interactions when they get close to each other at high surface coverage. The attractive or repulsive forces cause a change of ΔG_{ad} with θ. Furthermore the surfaces are usually not homogeneous and there are sites of different ΔG_{ad}. Thus, first the sites with strongly negative ΔG_{ad} will be occupied, followed by those with less negative ΔG_{ad}. ΔG_{ad} becomes less negative with increasing θ, leading to more complicated isotherms like the Frumkin isotherm or the Freundlich isotherm shown in Equations 1.173 and 1.174.

$$\frac{\theta}{1-\theta} = ac = a_0 \exp\left(-\frac{\Delta G_{ad}}{RT}\right) = a_0 \exp\left(-\frac{\Delta G_{0,ad}}{RT}\right)\exp\left(-\frac{b\Theta}{RT}\right) \tag{1.173}$$

$$\theta = Kc^m \tag{1.174}$$

The Frumkin isotherm assumes a linear decrease of ΔG_{ad} with Θ, which introduces an exponential term with Θ as shown in Equation 1.173. The Freundlich isotherm uses the empiric parameter m to fit Equation 1.174 to the experimental conditions, thus taking care of the change of ΔG_{ad} with θ.

Assuming the conditions of the Langmuir isotherm, corrosion current densities $i_{C,0}$ and $i_{C,I}$ refer to the surface fractions without and with inhibitor ($\theta = 1$), which add to the total corrosion current density i_C according to Equation 1.175.

$$i_C = \theta\, i_{C,I} + (1-\theta)\, i_{C,0} \tag{1.175}$$

Rearrangement yields Equation 1.176, which allows calculating θ from corrosion current densities.

$$\theta = \frac{i_{C,0} - i_C}{i_{C,0} - i_{C,I}} \tag{1.176}$$

For vanishing $i_{C,I}$, Equation 1.176 equals Equation 1.171. Equation 1.176 allows calculating the surface coverage from $i_{C,0}$ and i_C (and $i_{C,I}$). θ may be calculated similarly from capacity data of the metal surface in contact with an electrolyte with and without an inhibitor as shown in Equations 1.177 and 1.178.

$$C = \theta C_I + (1-\theta) C_0 \tag{1.177}$$

$$\theta = \frac{C_0 - C}{C_0 - C_I} \tag{1.178}$$

where

C_0 is the capacity of a free metal surface with values in the range of $20\,\mu F \cdot cm^{-2}$

C_I is the value for the surface with complete coverage ($\theta = 1$) by the inhibitor, with usually a much smaller value of a few $\mu F \cdot cm^{-2}$ only

The capacity of a partially covered surface C is given by Equation 1.177, which yields after appropriate arrangement Equation 1.178. The equivalent form of Equations 1.176 and 1.178 demonstrates that the reduced corrosion rate in the presence of an inhibitor is a consequence of the partial coverage of the metal surface by inhibitor molecules, which suppress metal dissolution and/or the cathodic counter reaction. Figure 1.47 shows the inhibition efficiency η_I and the surface coverage as a function of the concentration of several alkyl amines for steel in 0.5M H_2SO_4 [77]. The S-shaped form of the dependence suggests a Langmuir isotherm, but it is better described by a Frumkin isotherm of Equation 1.173 with an appropriate fit of the parameters.

1.17.2 Inhibition by Precipitation of Compounds

Precipitates at surfaces block both reactions, metal dissolution and redox processes. In both cases, diffusion is most likely the rate-determining step. Dissolved metal ions have to diffuse through pores from the metal surface to the bulk electrolyte and similarly the redox species from the bulk to the metal surface. Benzotriazole (BTA) is an inhibitor, which forms thick layers with cations, which act as a diffusion barrier. For example, copper forms an insoluble compound with BTA. It has been shown that the dissolution of a

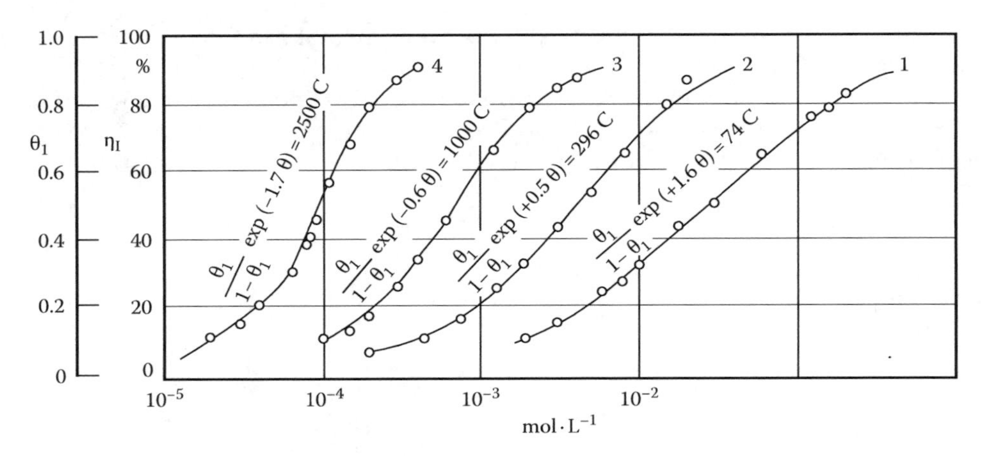

FIGURE 1.47
Inhibition efficiency η_I and surface coverage θ for steel in 0.5M H_2SO_4 with addition of n-hexylamine, n-octylamine, n-decylamine, and n-dodecylamine as inhibitors. (From Szklarska-Smialowska, Z. and Wieczorek, G., in *Proceedings of 3rd European Symposium on Corrosion Inhibitors*, Ferrara, 1970, Annali dell'università di Ferrara, Sezione V, Suppl. N. 5, 1971, p. 453.)

rotating Cu disk is a diffusion-limited process. The formation of a Cu-BTA layer reduces the Cu dissolution current density in 0.5M H_2SO_4 by ca. 1 order of magnitude. It follows the Levich equation (Equation 1.129) with an inverse square root dependence of the rotation frequency of the Cu disc.

Another group of inhibitors for weakly acidic up to weakly alkaline solutions are polyphosphates $(NaPO_3)_n$ and organophosphates, which form layers by precipitation. They essentially inhibit the cathodic reaction. The access of dissolved oxygen to the metal surface has to occur by diffusion within the pores of this layer, which slows down the corrosion process, i.e., metal dissolution controlled by oxygen reduction. Similarly carbonates of calcium and magnesium form diffusion barriers on steel surfaces, especially steel pipes. The formation of carbonate rust layers provides protection in weakly acidic and alkaline solutions. If the concentration of dissolved CO_2 is too high, $CaCO_3$ will dissolve as $CaHCO_3$ and thus is no longer protective. Low pH and high CO_2 content may attack this natural rust layer, whereas it is protective around pH 7. This layer is very important for steel pipes used for fresh water supply. Increasing temperature will stabilize the carbonate layer.

1.17.3 Inhibition by Passivation

Passivity of metals is provided easily in neutral and weakly acidic solutions. Even metals with pronounced active dissolution in strongly acidic electrolytes will be passive if the pH is high enough with a related small active dissolution rate for these conditions. Most oxides films are insoluble at high pH and will give protection against metal dissolution. In consequence, no high reaction rates are required for the cathodic counter reaction to compensate strong active metal dissolution during the passivation process if compared to acidic electrolytes. Thus, the pH is decisive for inhibition by passivation. Buffers are good corrosion inhibitors by maintaining a neutral or weakly alkaline pH, for example, phosphates, borates, carbonates, and silicates. They prevent a low local pH for localized dissolution, which otherwise will occur due to hydrolysis of corrosion products. However, even in strongly acidic electrolytes with very large active dissolution rates, the addition of a high concentration of an oxidizing agent may cause passivity for many metals as

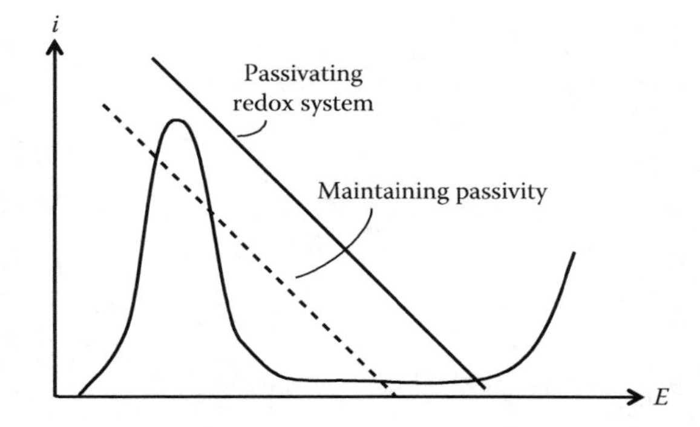

FIGURE 1.48
Schematic polarization curve of a metal with active dissolution and passive range and a passivating redox system creating passivity (solid line) and maintaining passivity (dashed line).

shown schematically in Figure 1.48 by the solid Tafel line. This is the case for concentrated nitric acid causing passivation of iron, steel, and nickel. Its cathodic reduction rate is large enough to overcome the high active dissolution rates of these metals. Nitrites and chromates or dichromates (alkaline and acidic electrolytes) are good inhibitors for metals in a less aggressive environment with moderate active dissolution rates. Due to their toxic properties, they are replaced in practice by molybdates and silicates. Good passivating properties are easily obtained for pure chromium and for stainless steels by the addition of sufficient chromium as an alloying element. The formation of a passivating Cr_2O_3-rich layer allows only very small dissolution rates in the active and passive potential range. For stainless steel and Cr-rich alloys of Fe and Ni, this layer forms even without any oxidizing species, due to water decomposition (see Chapter 5). If the reduction rate of a redox system is smaller than the active metal dissolution, it still may maintain passivity (Figure 1.48, dashed line). However, if a specimen is activated, a potential is established in the active dissolution range of the metal. This redox system may not overcome the maximum dissolution current density and thus may not passivate the metal.

In summary, there are two kinds of inhibition by passivation:

1. If the cathodic reaction is fast enough to compensate active metal dissolution, the oxidizing agent creates passivity. Thus, the redox system allows passing the peak of active metal dissolution, and a potential well within the passive range is established. After passivation, the cathodic process has to compensate only the very small passive corrosion density (Figure 1.48).
2. Other redox systems may only maintain existing passivity. They compensate the small passive dissolution rates when passivity has been already achieved. These redox systems are too slow to overcome the active dissolution peak. Once the metal has been activated, the potential drops to a value in the active dissolution range with the related increased corrosion rate (Figure 1.48).

1.18 Semiconductor Electrochemistry and Photoelectrochemistry

The degradation of materials does not only occur with metals but also with semiconductors or insulators. Furthermore, many metals are covered with oxide and hydroxide films, which have semiconductor properties. Therefore concepts have to be developed to treat semiconductor electrochemistry.

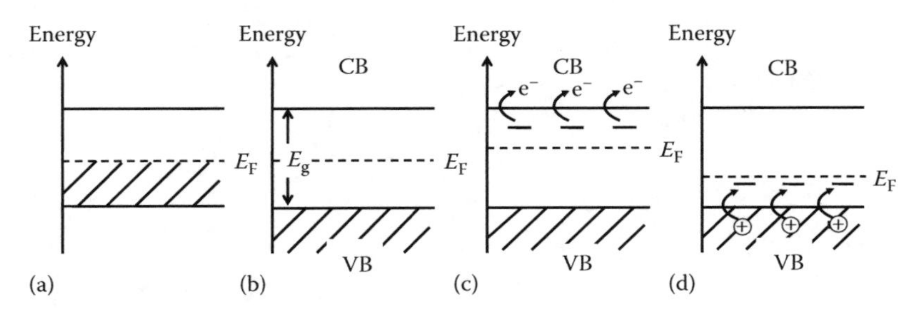

FIGURE 1.49
Band model of (a) a metal, (b) intrinsic, (c) n-type, and (d) p-type semiconductor with Fermi energy E_F, valence band VB, conduction band CB, and donor and acceptor levels.

1.18.1 Some Properties of Semiconductors

Metals have partially filled bands with empty states directly above and filed states below the Fermi level E_F at $T = 0\,K$ (Figure 1.49a). At room temperature, there exists a thermal activation of electrons so that levels slightly above E_F are partially occupied and slightly below are partially empty. With an applied electric field, electrons will move, which leads to electron conduction. They pick up small energies and thus get to levels above the Fermi edge. This is the reason for the high conductivity of metals. In contrast, semiconductors have a completely filled valence band VB, and the transfer of the band gap energy E_G is required to excite electrons from the valence band edge to empty states of the conduction band CB (Figure 1.49b). This is the case for intrinsic semiconductors with no dopings. As a consequence, the conductivity is very low for these intrinsic semiconductors. An insulator is a material with a very large band gap of $E_G > 3\,eV$ and thus has no or an extremely small conductivity. For an intrinsic semiconductor, E_F is in the center of the band gap. Dopings are impurities, which create electronic levels within the band gap. n-dopings are impurities with levels close to the conduction band edge. A small energy is required to excite electrons from these doping levels to the conduction band with remaining positively charged species. These electrons cause a pronounced increase of the conductivity of the semiconductor (Figure 1.49c). For Si, this may be achieved by phosphorous atoms, which will be integrated within the lattice forming four bonds like Si and giving their fifth electron to CB. p-doping is caused by impurities with levels close to the valence band edge, accepting an electron and leaving a positive hole, which also will lead to an increased conductivity by electron hopping within VB (Figure 1.49d). A typical dopant for Si is B, which will accept a fourth electron to be integrated in the bindings of the Si lattice and thus leaves an electron defect at an Si atom, the positive hole.

The contact of an n-type and a p-type semiconductor leads to electronic equilibrium with an equal energy for both Fermi levels. The small electron transfer from the n- to the p-side when a contact is made causes a band bending at the interface with a space charge layer extending into both semiconductors (Figure 1.50a). This n–p junction has diode characteristics. A positive voltage at n relative to p increases the bad bending at the interface, preventing electrons to flow from n to p due to the increased potential barrier they have to overcome. The same will happen to positive holes going from p to n. The diode is blocking electronic flow for these conditions. Applying a negative voltage to n lifts the energy level of n and thus decreases the voltage drop. As a consequence, electron flow from n to p will be facilitated similarly to the flow of positive holes in the opposite direction. Figure 1.50b shows the related diode characteristics with an increasing current when a negative bias $E_n - E_p$ is applied, whereas a positive bias will block the current.

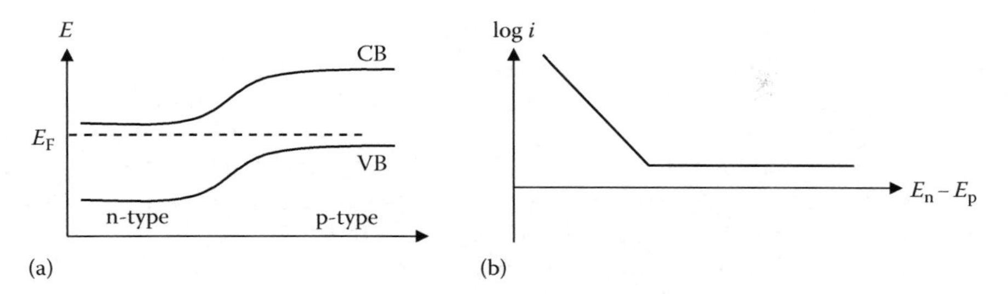

FIGURE 1.50
(a) p–n junction of a p- and n-doped semiconductor forming a space charge layer at the contact (b) related current potential curve with diode characteristics showing vanishing current for increasing voltage $E_n - E_p$ (n-type more positive) and transmission for decreasing voltage.

1.18.2 Electron Transfer at Semiconductor Electrodes

A similar situation is obtained for a semiconductor in contact with an electrolyte. Here the Fermi level of the semiconductor becomes equal to that of the redox system within the electrolyte during contact at open circuit conditions. The states of the redox system show a distribution as given in Figure 1.51, which is related to the changes of the solvation shell of the redox species. The lower states are occupied and the higher states are empty referring to the reduced and the oxidized component of the redox system. The Fermi level is between the distribution of these states. The density of charge carriers within the semiconductor

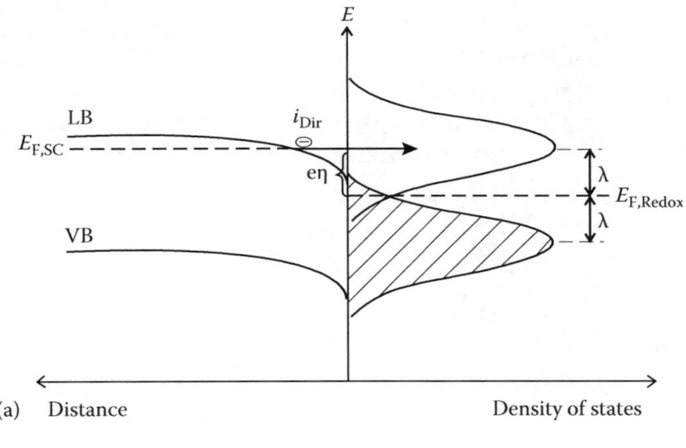

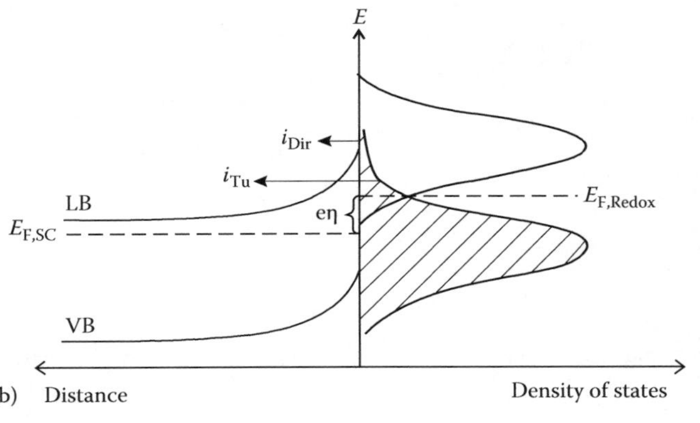

FIGURE 1.51
Electron transfer at an n-type semiconductor electrode surface, band model with occupied (reduced species) and empty states (oxidized species) with direct electron transfer i_{dir}, and tunnel transfer i_T for (a) negative overvoltage η and (b) positive overvoltage.

is much smaller than the one within the electrolyte, i.e., the concentration of ions within the solution. A 1 M solution contains ca. $2N_A \times 10^{-3} = 1.2 \times 10^{21}$ ions cm^{-3} ($N_A = 6.02 \times 10^{23}$ mol^{-1}), whereas a semiconductor has 10^{15} up to 10^{19} dopings · cm^{-3}. As a consequence, the space charge layer extends well into the semiconductor and not into the electrolyte. Applied potential changes to a semiconductor/electrolyte interface will be located within the semiconductor. This is opposite to the situation of a diffuse double layer of a metal/electrolyte interface with small ionic strength I. For the metal/electrolyte interface, the potential drop is located within the electrolyte due to the much higher electron concentration within the metal. A metal has approximately a density of electrons of $N_A/V_M = 10^{23}$ cm^{-3} ($V_M = 7\,cm^3 \cdot mol^{-1}$ = approximate molar volume of many metals). Small electrolyte concentration $c < 1$ M leads to a diffuse double layer as discussed in Section 1.7 in addition to the Helmholtz layer.

Thus, the situation of a semiconductor is similar to the diffuse double layer described by the Debye–Hückel theory. In the case of the semiconductor, the charge carriers are electrons for an n-type semiconductor and positive holes for a p-type semiconductor. The width of the space charge layer is given by the Debye length β. For the semiconductor, the expression $2N_AI$ of Equation 1.22b has to be replaced by the concentration of charge carriers within the semiconductor, i.e., for an n-type semiconductor by the concentration of electrons, which equals approximately that of the donors N_D as shown in Equation 1.179.

$$\beta = \sqrt{\frac{\varepsilon\varepsilon_0 kT}{e_0^2 N_D}} \tag{1.179}$$

In the case of a p-type semiconductor, one has to introduce the concentration of positive holes, which equals that of the acceptors N_{AC}. The width of the space charge layer d_{SC} is potential dependent. As presented in Equation 1.180, it depends on the deviation of the applied electrode potential E from the flat band value E_{FB}. The band bending changes with the applied electrode potential as shown schematically in Figure 1.52 of a semiconductor electrode and disappears at $E = E_{FB}$ with a flat course of the bands (Figure 1.52b).

$$d_{SC} = \beta\sqrt{\frac{2e_0(E - E_{FB})}{kT}} \tag{1.180}$$

Applying a positive potential to an n-type semiconductor with $E > E_{FB}$, one increases the upward band bending as shown in Figure 1.52a, which leads to a shift of electrons within CB into the inner parts of the semiconductor following the potential drop. This causes

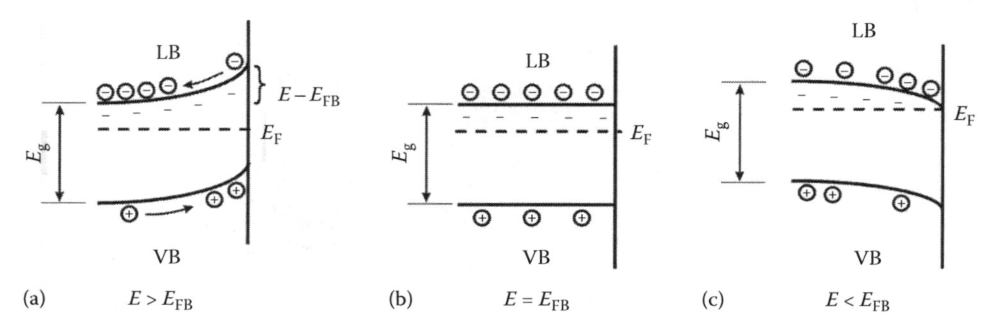

FIGURE 1.52
n-type semiconductor/electrolyte contact and space charge layer with decreasing electrode potential, (a) upward bending, accumulation of positive holes, (b) flat band situation, and (c) downward bending, accumulation of electrons at the interface.

a depletion layer with a decrease of the electron concentration at the interface, which are the majority carriers for n-type material. (NB: A more positive electrode potential E corresponds to a low energy for the electrons in the band model.) The minority carriers, i.e., the positive holes, are accumulated at the interface. At the flat band potential $E = E_{FB}$, no potential drop occurs with no band bending (Figure 1.52b). For $E < E_{FB}$, a downward bending is obtained with the accumulation of the majority carriers, i.e., the electrons at the interface and a shift of positive holes into the inner parts of the semiconductor. In this case, one has an accumulation layer (Figure 1.52c). The situation is different for a p-type semiconductor. The donors and the Fermi level are close to the valence band. For $E > E_{FB}$, an upward bending causes an accumulation of the majority carriers, i.e., of the positive holes and the electrons are depleted. For $E < E_{FB}$ with a downward bending, one obtains a depletion layer for the majority carriers, i.e., the positive holes and the electrons are accumulated.

The capacity of the semiconductor electrode contains two parts, C_{SC} of the space charge layer of the semiconductor and C_H of the Helmholtz layer when the diffuse double layer within the electrolyte may be neglected due to a high electrolyte concentration ($I \geq 1\,M$). According to Equation 1.181, their inverse values have to be added because they are connected in series.

$$\frac{1}{C} = \frac{1}{C_H} + \frac{1}{C_{SC}} \approx \frac{1}{C_{SC}} \tag{1.181}$$

C_H is within 10–20 μF cm^{-2} and much larger than C_{SC} so that its inverse may usually be neglected as shown in Equation 1.181. This result is closely related to the location of potential changes within the semiconductor as shown in Figure 1.52. One may apply the Schottky–Mott equation (Equation 1.182) to the semiconductor electrolyte combination with the Boltzmann constant k and the charge of the electron e_0.

$$\frac{1}{C^2} = \frac{2}{e_0 N_D}\left[E - E_{FB} - \frac{kT}{e_0}\right] \tag{1.182}$$

The $1/C^2 - E$ plot yields for an n-type material a line increasing with E. Its slope gives a value for N_D and its intersection with the abscissa yields E_{FB} if one may neglect the expression kT/e_0. Figure 1.53 gives the results for SnO_2 layers with different doping levels, which have been obtained by thermal decomposition of $SnCl_4$ and organic $Sn(But)_4$ doped differently by the addition of $SbCl_5$ [78]. The extrapolation of all lines gives $E_{FB} = 0.4\,V$ (SCE), which corresponds to $E = 0.64\,V$ (SHE) (saturated calomel electrode SCE: $E = 0.240\,V$). The doping concentrations have been obtained from the slopes of the lines. It varies over two orders of magnitude as indicated. The Schottky–Mott plot of a p-type semiconductor yields a decreasing line with increasing potential and allows the determination of N_{AC} and E_{FB} correspondingly.

For the discussion of electrochemical redox reactions at a semiconductor surface, one has to involve its special characteristics as shown in Figure 1.51 for an n-type material. For potentials $E < E_{FB}$, one has an accumulation of electrons at the surface. The empty states of the oxidized species of the redox system are close to the conduction band. Due to the good overlap of the states, a direct transfer of electrons from the conduction band to the oxidized species within the electrolyte may occur. The more negative the potential, the better the overlap of the involved electronic states with a steep increase of the cathodic branch of the polarization curve as shown in the log $I - E$ diagram of Figure 1.54, which is similar to the behavior of a metal electrode. However, the anodic branch is quite different with much smaller current densities and a steeply increasing line for very high doping concentrations only. The upward bending for $E > E_{FB}$ causes an overlap of only very few filled states of

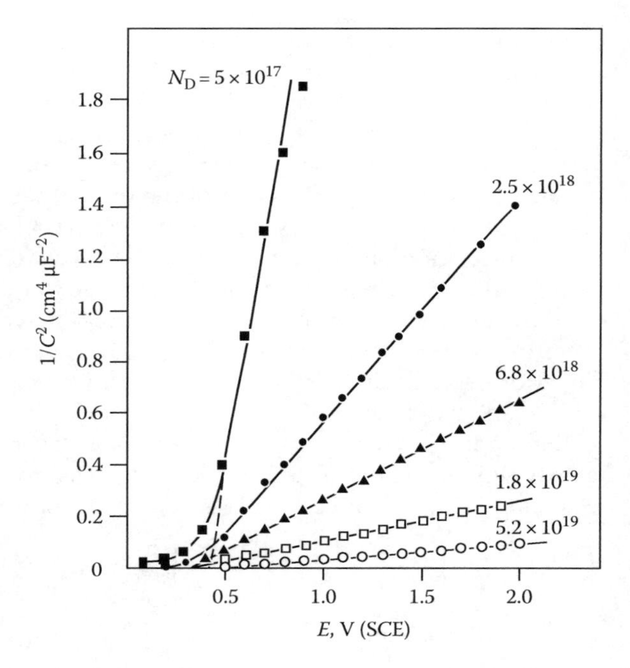

FIGURE 1.53
Schottky–Mott plot for differently doped n-type SnO_2 layers formed by chemical vapor deposition. Extrapolation leading to E_{FB} and slopes to concentration of dopings. (From Memming, R. and Möllers, F., *Ber. Bunsen Ges. Physik. Chem.*, 76, 475, 1972.)

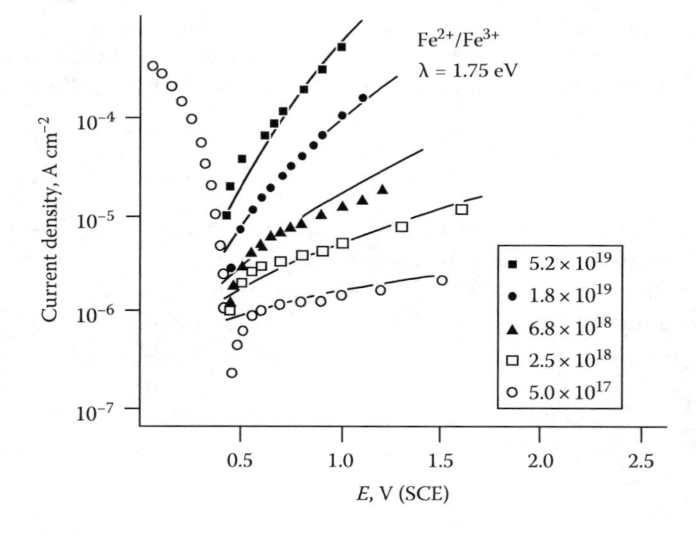

FIGURE 1.54
Tafel plot of current density as a function of the electrode potential for different dopings of SnO_2 layers. (From Memming, R. and Möllers, F., *Ber. Bunsen Ges. Physik. Chem.*, 76, 475, 1972.)

the redox system with those of the conduction band causing a small direct anodic current i_{Dir}. However, a larger part of states may be involved in the anodic reaction with a tunnel transfer from the redox system through the space charge layer to the conduction band causing a tunnel contribution i_{Tu} (Figure 1.51). With increasing potential E, the band bending gets larger and more states will overlap. Furthermore, the width of the space charge layer gets smaller so that the tunnel distance for the electrons decreases. Both effects lead to the increase of i_{Tu}, whereas the small direct contribution i_{Dir} will remain the same. The result is the observed increase of the anodic branch of the current density with potential of Figure 1.54. As the tunnel distance gets smaller with the doping, the anodic branch gets steeper approaching almost the properties of a metallic electrode for $N_D = 5 \times 10^{19}$ cm^{-3} [78].

The n-type semiconductor contacting an electrolyte thus shows a diode behavior with large cathodic currents and a blocking of the anodic part with much smaller current densities. The calculation of the discussed example of the SnO_2/Fe^{2+}, Fe^{3+} system leads to a solvent reorganization energy of $\lambda = 1.2\,\text{eV}$, which is close to 1.5 eV found for a Pt electrode [78]. λ is the energy that is needed to reorganize the shell of solvent molecules of a redox species after exchange of an electron with the metal as indicated in Figure 1.51a.

1.18.3 Photoelectrochemistry

If light is absorbed by a semiconductor surface with photon energies $h\nu > E_G$, electrons may be excited from the valence band to the conduction band. A positive hole is left in the valence band and thus an electron–hole pair is generated. The strong electrical field within the space charge layer separates these charge carriers spatially, so that they cannot recombine. In an n-type semiconductor, the photoelectrons will follow the electrical field and migrate to the bulk. The photogenerated positive holes will be drawn to the surface and may oxidize redox systems within the electrolyte as shown in Figure 1.55. These processes cause an anodic photocurrent $i_{Ph,SC} > 0$, which will be measured during illumination of the semiconductor surface in contact to the electrolyte. The opposite effect will be observed for a p-type semiconductor with a downward bending of the bands. Here the photogenerated electrons will accumulate at the surface and will reduce a redox system, whereas the positive holes will be drawn to the bulk of the semiconductor, which will cause a cathodic photocurrent $i_{Ph,SC} < 0$.

On the basis of this mechanism, W. Gärtner deduced a quantitative expression of the photocurrent i_{Ph} vs. the energy $h\nu$ of the incident light and the applied electrode potential E [79], which will be discussed here for an n-type semiconductor with an anodic photocurrent. Besides some simplifying approximations, his model assumes that all light is absorbed within the semiconductor and that no recombination of charge carriers is possible within the space charge layer. It also assumes that no interband states exist, which may act as recombination centers for electron–hole pairs. Adsorbed impurities and other defects at the surface of semiconductors are often effective recombination centers. Besides the photogeneration of charge carriers within the space charge layer leading to the photocurrent $i_{Ph,SC}$, light absorption may also generate charge carriers in inner parts of the

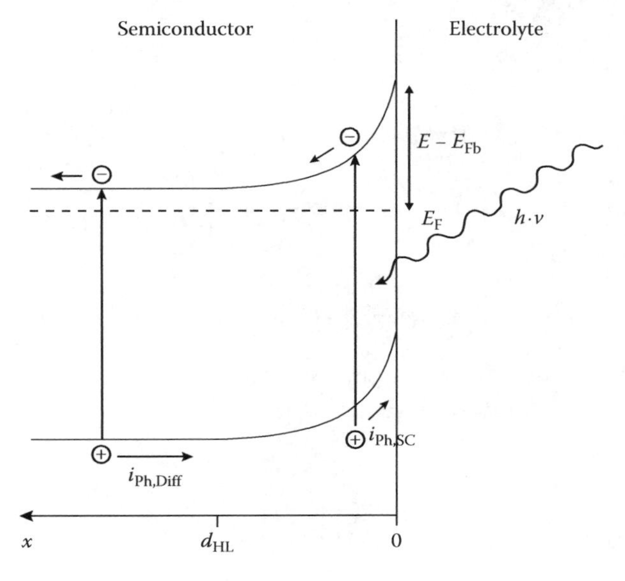

FIGURE 1.55
Mechanism of photocurrent at an n-type semiconductor/electrolyte interface.

semiconductor where no electric field exists. A fraction of the photogenerated electrons diffuse to the inner bulk of the n-type semiconductor, whereas the related positive holes diffuse to the space charge layer where they are drawn effectively by the strong electrical field to the semiconductor/electrolyte interface where they oxidized a redox system. This leads to the contribution $i_{Ph,Diff}$. Both contributions sum up to the total measured photocurrent i_{Ph} of Equation 1.183.

$$i_{Ph} = i_{Ph,SC} + i_{Ph,Diff} \tag{1.183}$$

The absorption of light with an incoming photon flux Φ_0 follows the Lambert–Beer law with the absorption coefficient α and the depth coordinate x (Equation 1.184).

$$\Phi = \Phi_0 \exp(-\alpha x) \tag{1.184}$$

Assuming that each photon absorbed in the space charge layer will generate an electron–hole pair, $\alpha e_0 \Phi\, dx$ is the photocurrent contribution of a layer at depth x of thickness dx. Integration of this expression within the limits $x = 0$ and $x = d_{SC}$ yields together with Equation 184 the contribution $i_{Ph,SC}$ of the space charge layer given by Equation 1.185.

$$i_{Ph,Sc} = -e_0\Phi_0 \int_{x=0}^{d_{SC}} \alpha[\exp\{-\alpha x\}]\,dx = -e_0\Phi_0[\exp\{-\alpha d_{SC}\} - 1] \tag{1.185}$$

The contribution $i_{Ph,Diff}$ caused by diffusion of electrons from $x > d_{SC}$ to $x = d_{SC}$ is determined by a differential equation describing the balance of the electron–hole pairs. It contains three terms, one for the generation of electron–hole pairs by light absorption, a second for the losses by recombination, and a third for the diffusion of positive holes to the space charge layer, where they are transferred effectively to the interface by the high electrical field strength. A detailed discussion yields Equation 1.186 with the diffusion length $L_h = \sqrt{D_h \tau}$ for the positive holes given by their diffusion coefficient D_h and their life time τ.

$$i_{Ph,Diff} = e_0\Phi_0 \frac{\alpha L_h}{1 + \alpha L_h} \exp\{-\alpha\, d_{SC}\} \tag{1.186}$$

The total photocurrent is given by Equation 1.187:

$$i_{Ph} = i_{Ph,SC} + i_{Ph,Diff} = -e_0\Phi_0[\exp\{-\alpha\, d_{SC}\} - 1] + e_0\Phi_0 \exp\{-\alpha d_{SC}\} \frac{\alpha L_h}{1 + \alpha L_h}$$

$$i_{Ph} = e_0\Phi_0\left[1 - \exp\{-\alpha d_{SC}\}\left(1 - \frac{\alpha L_h}{1 + \alpha L_h}\right)\right] = e_0\Phi_0\left[1 - \frac{\exp\{-\alpha d_{SC}\}}{1 + \alpha L_h}\right] \tag{1.187}$$

A similar equation is obtained for p-type semiconductors when D_n and L_n. are replaced by the parameters of the positive holes D_h and L_h.

Assuming that the optical penetration depth $1/\alpha$ is large relative to d_{SC}, i.e., $\alpha d_{SC} \ll 1$ and $\alpha L_h \ll 1$ and introducing Equation 1.180 for d_{SC} in Equation 1.182, one obtains the following approximation (Equation 1.188):

$$i_{Ph} = e_0\Phi_0\left[1 - \exp\left\{-\alpha\beta\sqrt{\frac{2e_0(E - E_{FB})}{kT}}\right\}\right] \tag{1.188}$$

Linearization of the exponential expression via a McLaurin series leads to Equation 1.189:

$$i_{Ph} = \alpha e_0 \Phi_0 \beta \sqrt{\frac{2e_0}{kT}} \sqrt{E - E_{FB}} \quad (1.189)$$

Equation 1.189 allows determining E_{FB} by a plot of i_{Ph} as a function of the square root of the electrode potential E. Its extrapolation to $i_{Ph} = 0$ yields the flat band potential E_{FB}. For vanishing band bending, i.e., $E - E_{Fb} = 0\,V$, the photogenerated electron–hole pairs are no longer separated and recombine. The absorption coefficient α for photon energies in the vicinity of the band gap energy E_G is given by Equation 1.190. Its introduction in Equation 1.189 yields Equation 1.191, which may be used to determine the band gap.

$$\alpha = \text{const}\frac{(h\nu - E_G)^{n/2}}{h\nu} \quad (1.190)$$

$$\frac{h\nu i_{Ph}}{e\Phi_0} = \eta_{eff}\,h\nu = \text{const}\sqrt{E - E_{FB}}\,d_{HI}\,(h\nu - E_G)^{n/2} \quad (1.191)$$

$\eta_{eff} = i_{Ph}/e\,\Phi_0$ is the efficiency for the formation of electron–hole pairs from the absorbed photons. The exponent $n = 1$ has to be introduced for a direct electronic transition whereas $n = 4$ is used for an indirect transition. The resulting equations (1.192) and (1.193) may be used to obtain E_G from the extrapolation of the appropriate plots to $h\nu = 0$.

$$(\eta_{eff}\,h\nu)^2 = \text{const}(h\nu - E_G) \quad \text{for } n = 1, \text{direct transition} \quad (1.192)$$

$$\sqrt{\eta_{eff}\,h\nu} = \text{const}(h\nu - E_G) \quad \text{for } n = 4, \text{ indirect transition} \quad (1.193)$$

In Figure 1.56, the results for electrochemically deposited Cu_2O films of different thicknesses are presented. The plots yield for a direct transition $E_G = 3.0\,eV$, whereas for an indirect transition $E_G = 2.3\,eV$ is obtained.

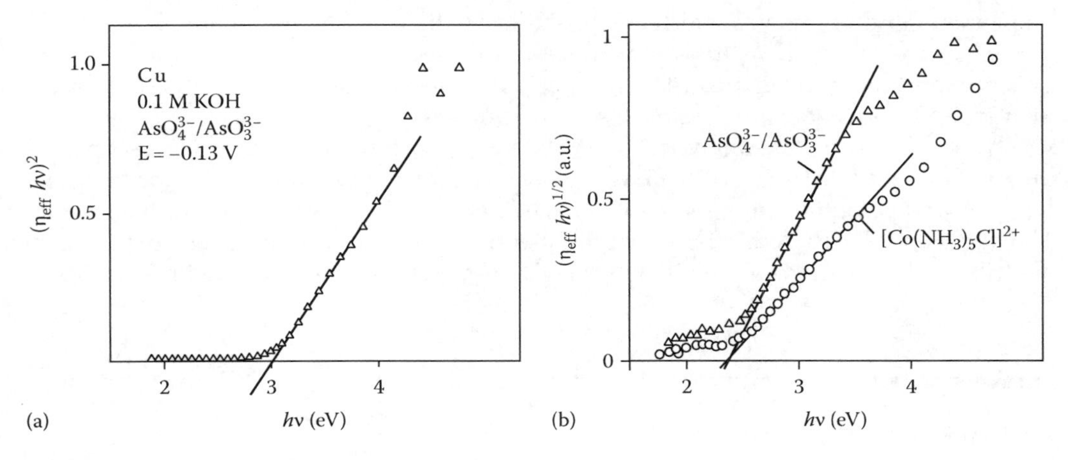

FIGURE 1.56
Efficiency of the photocurrent η_{eff} of a Cu/Cu_2O electrode in alkaline solution with a redox system within the electrolyte (AsO_4^{3-}/AsO_3^{3-} and $[Co(NH_3)_5Cl]^{2+}$) as a function of the energy $h\nu$ of the incident light beam with direct and indirect band gap, (a) $(\eta_{eff}\,h\nu)^2 - h\nu$ plot for the determination of band gap of direct transition, (b) $(\eta_{eff}\,h\nu)^{1/2} - h\nu$ plot for the determination of band gap of indirect transition. (From Collisi, U. and Strehblow, H.-H., *J. Electroanal. Chem.*, 219, 213, 1986.)

It has to be mentioned here that passive films are often crystalline, but with small crystals in the nm range and with a lot of defects. In an early stage, they might be even amorphous and crystallize with time often with regions of amorphous structures in between. Therefore, their photoelectrochemical properties have been explained on the basis of amorphous materials with high concentrations of dopings within the band gap especially in the vicinity of the band edges. These details are discussed in Chapter 5.

Photoelectrochemical measurements are usually performed in an electrochemical cell as described in Section 1.11.1, Figure 1.25a. A quartz window permits shining light on the electrode that is held at a fixed potential by a potentiostat. The light source is mostly a high-intensity Xe high-pressure lamp. The light passes a water filter to absorb any heat radiation and passes then a monochromator. A mechanical chopper produces an alternating light beam, causing a chopped photocurrent. This permits the application of lock-in technique separating the oscillation of i_{Ph} from constant currents and disturbing current oscillations with different frequency as that set at the chopper. i_{Ph} may be rather small in case of passive layers due to their small thickness, which permits the absorption of only a small fraction of the light.

If the electrode potential is not fixed by a potentiostat, i.e., if the electrode is at open circuit conditions, its illumination will cause photo-potentials. In this case, the electrons and holes from the photogenerated pairs will be separated in the space charge layer and will accumulate at the surface and the inner part of the semiconductor, respectively. This situation decreases the potential drop in the space charge layer and thus causes a potential change, i.e., the photo-potential.

1.19 Conclusion

This chapter on the fundamentals of corrosion is a short introduction in those parts of thermodynamics and electrochemistry, which are required for an understanding of corrosion phenomena and the related mechanisms. To keep the chapter small enough, only a condensed overview could be given and it should be seen as a recapitulation of the basics. Other important topics for corrosion, especially methods for corrosion research, are not mentioned here. Modern corrosion research applies various in situ and ex situ methods, spectroscopic and surface analytical tools like XPS, AES, Raman and IR-spectroscopy; scanning techniques like STM, AFM, SEM, and electron microprobe analysis; impedance spectroscopy and potential scanning methods like SRET and SVET; and theoretical calculations. The application of these methods will be mentioned in the different chapters. Literature describes these methods in detail, which is recommended to the interested reader [5,6].

1.20 Acknowledgments

The authors acknowledge the careful reading and the improvements of the text by E. Protopopoff. The computer design of various figures by A. Seyeux and the coworkers of the Laboratoire de Physico-Chimie des Surfaces of the Ecole Nationale Supérieure de Chimie de Paris are gratefully acknowledged.

References

1. P. W. Atkins, *Physical Chemistry*, 5th edition, Oxford University Press, Oxford, U.K., 1994.
2. C. H. Hamann, A. Hamnett, W. Vielstich, *Electrochemistry*, Wiley-VCH, Weinheim, Germany, 2007.
3. K. J. Vetter, *Electrochemical Kinetics*, Academic Press, New York, 1967, p. 780.
4. E. Gileadi, *Electrochemical Kinetics for Chemists*, Chemical Engineers and Materials Scientists, VCH, New York, 1993.
5. D. Briggs, M. P. Seah, *Practical Surface Analysis*, Wiley, Chichester, U.K., 1990.
6. P. Marcus, F. Mansfeld, *Methods in Corrosion Research*, CRC Press, Taylor & Francis, Boca Raton, FL, 2006.
7. D. Landolt, *Corrosion and Chemistry of Metals*, CRC Press, Taylor & Francis, Boca Raton, FL, 2007.
8. H. Kaesche, *Corrosion of Metals*, Springer, Berlin, Germany, 2003.
9. *Handbook of Chemistry and Physics*, CRC Press, Cleveland, OH, 1977–1978.
10. A. H. Narten, *Disc. Faraday Soc.* 43, 97, 1967.
11. D. Lützenkirchen-Hecht, H.-H. Strehblow, Synchrotron methods for corrosion research, in *Methods in Corrosion Science and Engineering*, P. Marcus, F. Mansfeld, editors, CRC Press, Boca Raton, FL, 2005, pp. 169–235.
12. J. J. Rehr, *Phys. Rev. Lett.* 69, 3397, 1992.
13. IXAS XAFS database, International X-ray Absorption Society, http//ixs.csrri.iit.edu/database/
14. D. Lützenkirchen-Hecht, C. U. Waligura, H.-H. Strehblow, *Corr. Sci.* 40, 1037, 1998.
15. B. Miller, *J. Electrochem. Soc.* 117, 491, 1970.
16. R. D. Gilles, J. A. Harrison, *J. Electroanal. Chem.* 27, 161, 1970.
17. D. R. Sandstrom, B. R. Stults, R. B. Greegor, Structural evidence for solutions from EXAFS measurements, in *EXAFS Spectroscopy, Techniques and Application*, Plenum Press, New York, 1981.
18. D. R. Sandstrom, F. W. Lytle, *Ann. Rev. Phys. Chem.* 30, 215, 1979.
19. D'Ans-Lax, *Taschenbuch für Chemiker und Physiker*, Springer, Berlin, Germany, 1960.
20. S. Trasatti, *J. Electroanal. Chem.* 52, 313, 1974.
21. R. Gomer, S. Tryson, *J. Chem. Phys.* 66, 4413, 1977.
22. D. M. Kolb, *Z. Phys. Chem. NF* 154, 179, 1987.
23. E. R. Kötz, H. Neff, K. J. Müller, *J. Electroanal. Chem.* 215, 331, 1986.
24. D. Lützenkrichen Hecht, H.-H Strehblow, *Electrochim. Acta* 43, 2957, 1998.
25. O. Stern, *Z. Elektrochem.* 30, 508, 1924.
26. A. Gouy, *J. Phys.* 4, 457, 1910.
27. A. Gouy, *Ann. Phys.* 7, 129, 1917.
28. C. L. Chapman, *Phil. Mag.* 6, 25, 475, 1913.
29. D. Hecht, *Thesis Heinrich-Heine-Universität*, Düsseldorf, Shaker Verlag, Aachen, Germany, 1997.
30. W. N. Hansen, D. M. Kolb, *J. Electroanal. Chem.* 100, 493, 1979.
31. S. Haupt, U. Collisi, H. D. Speckmann, H.-H. Strehblow, *J. Electroanal. Chem.* 194, 179–190, 1985.
32. D. Hecht, H.-H. Strehblow, *Electrochim. Acta* 43, 19–20, 1998.
33. D. C. Grahame, *Chem. Rev.* 41, 441, 1947.
34. D. C. Grahame, *J. Am. Soc.* 76, 4819, 1954.
35. J. Lecoeur, J. P. Bellier, *Electrochim. Acta* 30, 1027–1033, 1985.
36. I. Barin, O. Knacke, O. Kubaschewski, *Thermochemical Properties of Inorganic Substances*, Springer, Berlin, Germany, 1973.
37. M. Pourbaix, *Atlas of Electrochemical Equilibria in Aqueous Solution*, Pergamon Press, Oxford, U.K., 1996.
38. Chemistry Webbok, NIST Standard Reference Database No69, D. J. Linstrom, W. G. Mallard, editors, http://Webbook.nist.gov
39. *Thermodynamic Properties of Substances and Ions at 25°C*, data from D. D. Ebbing, Houghton-Mifflin-Harcourt, *General Chemistry*, 3rd edition, Appendix C, 1990, http://hyperphysics.phy-astr.gsu.edu/hbase/tables/therprop2.html

40. *Standard Thermodynamic Properties of Substances*, http://www.update.uu.se/~jolkkonen/pdf/CRC-TD.pdf
41. K. J. Vetter, H.-H. Strehblow, *Ber. Bunsen Ges. Physik. Chem.* 74, 1025, 1970.
42. J. Newman, D. N. Hansen, K. J. Vetter, *Electrochim. Acta* 22, 829, 1974.
43. H.-H. Strehblow, Phenomenological and electrochemical fundamentals of corrosion, in *Corrosion and Environmental Degradation*, M. Schütze, editor, Vol. 1, Wiley-VCH, Weinheim, Germany, 2000, pp. 1–66.
44. W. J. Albery, M. L. Hitchman, *Ring Disc Electrodes*, Clarendon, Oxford, U.K., 1971.
45. B. P. Löchel, H.-H. Strehblow, *Werkst. Korr.* 31, 353, 1980.
46. H.-H. Strehblow, H. D. Speckmann, *Werkst. Korr.* 35, 512, 1984.
47. S. Haupt, H.-H. Strehblow, *J. Electroanal. Chem.* 216, 229, 1987.
48. G Engelhardt, D. Schaepers, H.-H. Strehblow, *J. Electrochem. Soc.* 139, 2170, 1992.
49. G. Engelhardt, T. Jabs, H.-H. Strehblow, *J. Electrochem. Soc.* 139, 2176, 1992.
50. S. Haupt, H.-H. Strehblow, *Langmuir* 3, 873, 1987.
51. K. J. Vetter, *Electrochemical Kinetics*, Academic Press, New York, 1967, p. 365.
52. F. G. Cottrell, *Z. Physik. Chem.* 42, 385, 1903.
53. H. Gerischer, W. Vielstich, *Z. Physik. Chem. NF* 3, 16, 1955.
54. K. J. Vetter, *Z. Elektrochem., Ber Bunsen Ges. Physik Chem.* 59, 435, 1955.
55. M. Stern, *J. Electrochem. Soc.* 102, 609, 1955.
56. I. N. Stranski, *Z. Physik. Chem.* 136, 259, 1928.
57. W. Kossel, *Nachr. Ges. Wiss. Göttingen. Math. Physik. K.I.* 135, 1927.
58. M. Ditterle, T. Will, D. M. Kolb, *Surf. Sci.* 327, L495–L500, 1995.
59. W. Poleuska, R. J. Bohm, and O. M. Magnussen, *Electrochem. Acta* 48, 2915, 2003.
60. J. Kunze, V. Maurice, H.-H. Strehblow, P. Marcus, *Electrochim. Acta* 48, 1157, 2003.
61. J. Kunze, V. Maurice, L. H. Klein, H.-H. Strehblow, P. Marcus, *Corr. Sci.* 46, 245, 2004.
62. J. Kunze, V. Maurice, L. H. Klein, H.-H. Strehblow, P. Marcus, *J. Electroanal. Chem.* 113, 554–555, 2003.
63. J. O. M. Bockris, D. Deazic, A. R. Despic, *Electrochim. Acta* 4, 325, 1961.
64. K. E. Heusler, *Z. Elektrochem. Ber. Bunsen Ges. Physik. Chem.* 62, 582, 1958.
65. H.-H. Strehblow, J. Wenners, *Electrochim. Acta* 22, 421, 1977.
66. H. J. S. Sand, *Phil. Mag.* 1, 45, 1900.
67. H. J. S. Sand, *Z. Phys. Chem.* 35, 541, 1900.
68. H. W. Pickering, P. J. Burne, *J. Electrochem. Soc.* 118, 209, 1971.
69. H. Gerischer, Schöppel H. Rickert, *Z. Metallkunde.* 46, 681, 1955.
70. H. W. Pickering, C. Wagner, *J. Electrochem. Soc.* 11, 698, 1967.
71. H. W. Pickering, P. J. Burne, *J. Electrochem. Soc.* 116, 1492, 1969.
72. D. Landolt, *Corrosion and Chemistry of Metals*, CRC Press, Taylor & Francis, Boca Raton, FL, 2007, p. 501.
73. E. McCafferty, *Corrosion Control by Coatings*, H. Leidheiser, editor, Science Press, Princeton, NJ, 1979, p. 279.
74. G. Trabanelli, *Corrosion Mechanisms*, F. Mansfeld, editor, Marcel Dekker, New York, 1987, 119 pp.
75. T. P. Hoar, R. D. Holliday, *J. App. Chem.* 3, 502, 1953.
76. N. Hackerman, D. D. Justice, E. McCafferty, *Corrosion*, Nace 31, 240, 1975.
77. Z. Szklarska-Smialowska, G. Wieczorek, in *Proceedings of 3rd European Symposium on Corrosion Inhibitors*, Ferrara, 1970, Annali dell'università di Ferrara, Sezione V, Suppl. N. 5, 1971, p. 453.
78. R. Memming, F. Möllers, *Ber. Bunsen Ges. Physik. Chem.* 76, 475, 1972.
79. W. W. Gärtner, *Phys. Rev.* 116, 84, 1959.
80. U. Collisi, H.-H. Strehblow, *J. Electroanal. Chem.* 210, 213, 1986.

2

Surface Effects on Hydrogen Entry into Metals

Elie Protopopoff
Ecole Nationale Supérieure de Chimie de Paris
and
Centre National de la Recherche Scientifique

Philippe Marcus
Ecole Nationale Supérieure de Chimie de Paris
and
Centre National de la Recherche Scientifique

CONTENTS

2.1 Introduction 106
2.2 Hydrogen Surface Reactions 107
2.2.1 H_2 Dissociative Adsorption and Absorption Reactions in the Gas Phase 107
2.2.1.1 H_2 Dissociative Adsorption 108
2.2.1.2 H Surface-Bulk Transfer 108
2.2.2 Hydrogen Surface Reactions in Aqueous Electrolyte 108
2.2.2.1 Elementary Surface Reactions 109
2.2.2.2 Overall Electrode Reactions 111
2.3 Thermodynamics of Metal–Hydrogen Systems 113
2.3.1 Thermodynamics of H Adsorption and Absorption in the Gas Phase 113
2.3.1.1 Isotherms of the Metal–Hydrogen Equilibria 113
2.3.1.2 Thermochemical Data 114
2.3.2 Thermodynamics of the Metal–Hydrogen Equilibria in Aqueous Electrolyte 117
2.3.2.1 Isotherms of the Metal–Hydrogen Equilibria 117
2.3.2.2 H_2 Pressure/Potential Correspondence for Adsorption/Absorption Equilibria 119
2.3.2.3 Gibbs Free Energy of Adsorption for UPD H and OPD H 119
2.3.2.4 Thermochemical Data for H Underpotential Electroadsorption on Pt in Aqueous Electrolyte 120
2.3.2.5 Relation between the Energetic and Structural Aspects of H Adsorption 120
2.4 Mechanisms of the H Surface-Bulk Transfer 121
2.4.1 Structural Aspects 121
2.4.1.1 H Adsorption Sites 121
2.4.1.2 H Bulk Interstitial Sites 121
2.4.1.3 Role of the Surface Structure in H Absorption: Entry Sites 122

2.4.2 Energetic Aspects ... 123
2.4.2.1 Gas Phase ... 123
2.4.2.2 Aqueous Electrolyte ... 124
2.5 Kinetics of HER and HAR ... 126
2.5.1 Expressions of the Absolute Rates of the Elementary Surface Steps ... 126
2.5.1.1 Electroadsorption and Electrodesorption ... 127
2.5.1.2 Chemical Combination and Dissociative Adsorption (Equation 2.3) ... 127
2.5.1.3 Electrocombination and Electrodissociation ... 128
2.5.1.4 Surface-Bulk Transfer ... 128
2.5.2 Steady-State Equations for the H Cathodic Reactions ... 129
2.5.3 Dependence of the HER and HAR Rates on the Potential and the Adsorption Gibbs Free Energy ... 131
2.5.3.1 Analysis of the Dependence on the Potential ... 131
2.5.3.2 Analysis of the Dependence on the Gibbs Free Energy of Adsorption ... 131
2.6 Influence of the Surface Modifiers on H Entry ... 135
2.6.1 Effects of Metal Oxide Films ... 135
2.6.2 Effects of Electronegative Species ... 135
2.7 Mechanisms of Action of H Absorption Promoters ... 135
2.7.1 Conditions for H Absorption Promotion ... 135
2.7.2 Proposed Mechanisms of Action of Promoters ... 137
2.7.3 Description of the Adsorption Blocking Effect ... 139
2.7.4 Modeling of the Effects of Adsorption Site Blockers on the HER and HAR ... 140
2.7.4.1 Bases of the Model ... 140
2.7.4.2 Predictions of the Effects of Adsorption Site Blockers on H Entry ... 141
2.7.5 Additional Suggestions for Explaining Promoting Effects ... 144
References ... 145

2.1 Introduction

It is well known that H entry (absorption) into the bulk may lead to embrittlement of metals and alloys. This process, known as H embrittlement or H-induced cracking, often combined on nonnoble metals with corrosion cracking, is particularly detrimental to the resistance of metallic materials [1]. High H concentrations are produced in metals by thermal charging (H_2 dissociation) or electrochemical charging (proton or water reduction); H can accumulate and combine at internal defects such as microcracks present in most commercial metals or alloys; high H_2 pressures can build up within these microcavities which then grow and coalesce, leading to loss of ductility [1]. In aqueous solution, the reaction of H absorption (HAR) into an electrode proceeds in parallel to the reaction of H_2 evolution (HER) at the surface at cathodic overpotentials and, at the corrosion potential (on nonnoble metals), simultaneously with the anodic dissolution or oxidation reactions in the mixed process controlling the rate of corrosion. The H entry into transition metals increases drastically in the presence of dissolved compounds of some electronegative elements, which are poisons of H adsorption. Although many explanations have been proposed to account for these effects, the detailed mechanisms of the action of these H entry promoters are not yet fully understood.

The HAR is usually studied from charging or permeation experiments [2–7]. With the development of techniques using hydrogen storage in solid materials, there has been a renewal of interest in the H absorption reaction and a need to monitor the conditions for obtaining high concentrations of H in the bulk. Accordingly, the established equilibrium and kinetic equations describing the H reactions must not be restricted to low H concentrations and low overpotentials, as was done previously for describing permeation into iron from aqueous solution [8–11].

Surface reaction steps involving adsorbed (chemisorbed, to be accurate) hydrogen, such as H adsorption, desorption, H–H combination, and surface–bulk transfer, play a determining role in the H cathodic reactions. The HAR and the HER most often share a common step of H electroadsorption from protons or water, and only a study of the overall mechanism of these reactions makes it possible to predict the conditions in which the H uptake under the surface can increase. The problem of analyzing all the data on H entry rate is that this rate, even in a pure metal, depends on many variables: the nature of the metal, its thermal–mechanical history, the surface conditions (especially on iron, surface states are not easily reproducible due to the difficulty of removing oxide films on the electrodes), composition of the electrolyte, cathodic current density or electrode potential, temperature, etc. The determining factors in the kinetics of the H cathodic reactions on bare metal surfaces are the cathodic overpotential and the surface parameters, which are the density of sites for H adsorption and the free energy of adsorbed H, both dependent on the structure and the chemical composition of the surface.

The aim of this chapter is to analyze how these surface parameters control the H surface reactions and hence the energetics and the mechanisms of the HER and HAR at electrodes of transition metals in aqueous electrolytes. A prerequisite to a comprehensive understanding of these phenomena is to define precisely the characteristics of H adsorption and absorption common to the gas and liquid phases. First we present the elementary surface steps occurring in the two phases. Then we report the thermodynamic and structural data existing for the two phases. Then we derive the kinetic relations for all the elementary surface reactions in aqueous solution, making clear the surface parameters. This allows a unified treatment of the H reactions at the metal–gas interface and at the metal–liquid electrolyte interface. The role of the structure of H adsorption sites and of the H adsorption energy in these sites is examined in detail. H absorption promoting effects by certain species are detailed; then a model of the effects of the species blocking the H adsorption sites on the rates and the mechanisms of the HER and HAR is presented that makes it possible to predict the surface conditions for H entry promotion.

2.2 Hydrogen Surface Reactions

2.2.1 H_2 Dissociative Adsorption and Absorption Reactions in the Gas Phase

Entry of H into the bulk of a transition metal from a gaseous dihydrogen phase is commonly considered to occur from atomic H adsorbed (more exactly chemisorbed) on the surface and thus involves the following steps [12,13]:

Transport of H_2 molecules to the surface

Molecular adsorption (physisorption)

Dissociation of H_2 and formation of a chemisorption bond between atomic hydrogen and a surface site consisting of a certain number of substrate atoms. H_2 dissociative adsorption is nonactivated on most transition metals

Surface-bulk transfer from adsorbed H to H dissolved beneath the surface (also called absorption step)

Diffusion into the bulk

Possibly hydride formation near the surface when the H concentration reaches a critical value

2.2.1.1 H_2 Dissociative Adsorption

Dissociative adsorption of hydrogen from the gas phase is described in detail in Ref. [13]. It proceeds as follows:

$$H_{2(g)} + 2(M) = 2H_{ads}(M) \tag{2.1}$$

where

(M) denotes an empty metal surface site

$H_{ads}(M)$ a H atom adsorbed at a surface site

The dissociative adsorption process may either occur directly (i.e., the rate is limited by the probability of finding two adjacent empty sites for dissociation) or involve a physisorbed H_2 precursor state mobile on the surface that may migrate until it finds empty sites on which it may dissociate.

2.2.1.2 H Surface-Bulk Transfer

The surface-bulk transfer may be written as

$$H_{ads}(M) + [M] = H_{diss}[M] + (M) \tag{2.2}$$

where

[M] represents an empty metal bulk interstitial site beneath the surface

$H_{diss}[M]$ is a dissolved (absorbed) H atom bonded to a bulk site

Because both the coordination of the H atoms and the energy of the H state change drastically between surface and bulk, it is realistic to consider that the absorption step is not an elementary step but involves a H subsurface state (sorbed H), intermediate between the surface adsorbed state and the bulk absorbed state, and located in the interstitial sites in the first metal layers beneath the surface [14]. This state has been characterized by ultra-high-vacuum (UHV) techniques on Pd, Ni, Pt, and Cu [13,15–22], and is believed to play a role in ordering of the H surface phases [19]. The subsurface sites become accessible by relaxation or reconstruction of the top layer of the metal substrate [13]. Figure 2.1 shows the different H sites at the vicinity of a metal surface [23].

An analysis of isotope effects on H thermal desorption spectra on Pd(111) has also shown a pathway of direct H entry into the bulk without equilibration with the chemisorption state [24]. However, this mechanism observed after H_2/D_2 exposure at 115 K may be related to a quantum tunneling effect not operating at room temperature. Alternatively, it might be due to the presence of two types of surface entry sites on this particular face, one being more stable than the other for chemisorption (see Section 2.2.2.2.3).

2.2.2 Hydrogen Surface Reactions in Aqueous Electrolyte

The H entry into a metal from an aqueous electrolyte is believed to involve the same surface-bulk transfer step as in the gas phase, but the preliminary adsorption step is a more

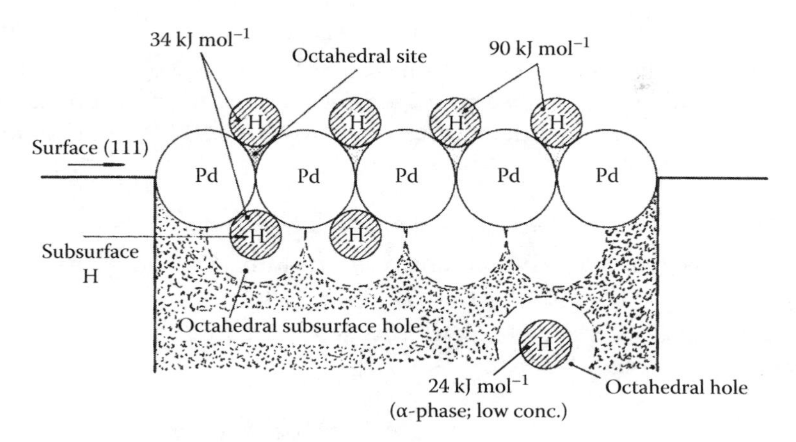

FIGURE 2.1
Structural model of a Pd(111) surface showing the surface, subsurface, and bulk sites. (From Konvalinka, J.A. and Scholten, J.J.F., *J. Catal.*, 48, 374, 1977, Fig. 9.)

complex process because more H sources are involved in aqueous solution, allowing more possible H surface reactions, and also because of the specificity of the electrolyte–metal interface. Whereas H adsorption in the gas phase occurs by dissociative adsorption of gaseous H_2, possibly (if under UHV) on the free sites of a metallic surface, H adsorption in aqueous solution may occur either chemically by dissociation of dissolved H_2 or electrochemically from solvated (hydrated) protons or water molecules; it takes place on a hydrated surface and thus implies the displacement of adsorbed water molecules or specifically adsorbed ions and a local reorganization of the double layer [25]; in some regions of potential, competition with the adsorption of oxygen species formed from the dissociation of water occurs [26–28]. The adsorbed H layer itself is also in interaction with surrounding water molecules, i.e., it is hydrated [13c,29,30].

2.2.2.1 Elementary Surface Reactions

2.2.2.1.1 Chemical H_2 Dissociative Adsorption and H–H Combination (Tafel Reaction)

Under conditions in which there is no specific adsorption of ions or adsorption of oxygen species, the reaction of adsorption from $H_{2(g)}$ in aqueous electrolyte occurs through the following steps:

Dissolution of H_2 gas into water, in equilibrium: $H_{2(g)} \leftrightarrows H_{2(aq)}$
Transport of dissolved H_2 molecules to the surface
Direct dissociative adsorption:

$$H_{2(aq)} + 2(M)_{aq} \rightleftarrows 2H_{ads}(M)_{aq} \tag{2.3}$$

where
$H_{2(aq)}$ is a dissolved H_2 molecule
$2(M)_{aq}$ represents a pair of adjacent hydrated sites available for adsorption of hydrogen
$H_{ads}(M)_{aq}$ represents a H atom adsorbed in a site (M) and hydrated

The reverse reaction is the combinative desorption of H atoms, or chemical combination, also known as the Tafel reaction.

2.2.2.1.2 H Electroadsorption/Electrodesorption (Volmer Reaction)

In aqueous electrolyte, because the metal substrate is an electrode with the electric potential as an additional variable, H adsorption may occur electrochemically (i.e., assisted by the potential) by reduction of hydrated protons or water molecules, depending upon the pH [31].

1. *Proton discharge in acid medium*. It involves the following:
   a. Fast transport of hydrated protons from the bulk solution to the double layer between the electrolyte and the cathode surface
   b. H electroadsorption by electron transfer from the electrode to the hydrated protons in the outer Helmotz plane [31]:

$$H^{+}_{(aq)} + e^{-}_{(M)} + (M)_{aq} = H_{ads}(M)_{aq} \quad (2.4)$$

   where
   $H^{+}_{(aq)}$ represents a hydrated proton
   $e^{-}_{(M)}$ an electron supplied by the metal M

2. *Reduction of water in neutral or basic medium*:

$$H_2O_{(1)} + e^{-}_{(M)} + (M)_{aq} = H_{ads}(M)_{aq} + OH^{-}_{(aq)} \quad (2.5)$$

Although there is no limitation to the supply of water molecules at the electrode in not too concentrated electrolytes (the water activity at the surface is equal to unity), this reaction has a lower rate constant than proton discharge for a given potential hence it is predominant only when the proton concentration is very low.

2.2.2.1.3 Electrodissociation/Electrocombination (Heyrovsky Reaction)

The H atoms can be desorbed electrochemically by combination with hydrated protons or water molecules, depending upon the pH.

1. *Proton plus H atom reaction in acid medium*:

$$H_{ads}(M)_{aq} + H^{+}_{(aq)} + e^{-}_{(M)} = H_{2(aq)} + (M)_{aq} \quad (2.6)$$

2. *Water plus H atom reaction in neutral or basic medium*:

$$H_{ads}(M)_{aq} + H_2O_{(1)} + e^{-}_{(M)} = H_{2(aq)} + OH^{-}_{(aq)} + (M)_{aq} \quad (2.7)$$

2.2.2.1.4 H Surface-Bulk Transfer

The transfer step from the adsorbed to the absorbed state is equivalent to the reaction described for the gas phase (Equation 2.2) except that it occurs from adsorbed H atoms surrounded by water molecules or anions and it liberates hydrated surface sites, as follows:

$$H_{ads}(M)_{aq} + [M] = H_{diss}[M] + (M)_{aq} \quad (2.8)$$

2.2.2.2 Overall Electrode Reactions

The H sources in aqueous solution, that is, hydrated protons and water molecules, can be transformed through the preceding elementary adsorption and desorption steps into $H_{2(g)}$ (HER) or H_{abs} (hydrogen electroabsorption reaction, HEAR).

2.2.2.2.1 Hydrogen Evolution Reaction

This overall reaction involves the following two-step pathways proceeding alone or in parallel:

1. *Volmer–Tafel pathway*: electroadsorption followed by chemical combination
2. *Volmer–Heyrovsky pathway*: electroadsorption followed by electrocombination

These steps are followed by transport of dissolved H_2 molecules away from the electrode via diffusion or by gas evolution (coalescence of H_2 molecules and phase separation).

On a noble metal, equilibrium of the HER (reversible H^+/H_2 electrode) may be attained, provided that the H_2 partial pressure in solution is high enough, because the kinetics are fast; from thermodynamic arguments, the HER occurs with a net rate if the potential is cathodic with respect to the equilibrium potential at the given H_2 pressure and pH. Conversely, if the overpotential is anodic, the reverse reaction, the H_2 oxidation reaction (HOR), prevails [32].

On a corroding metal surface, a mixed process occurs because the main anodic reaction is metal dissolution or oxidation (passivation); the net current is a function of the overpotential with respect to the mixed (corrosion) potential (see Chapter 1). The anodic metal reaction and the cathodic H reaction proceed simultaneously close to the corrosion potential, which implies that the HER may occur on a surface covered by an oxide film (passivated) and that the detrimental effects of anodic dissolution and H entry may be combined in the embrittlement process.

The mechanisms of the HER on the metals currently studied in H entry experiments, (Pd, Ni, Fe) are relatively well known and depend on the overpotential. At high overpotentials, a common mechanism prevails on most metals, because the chemical combination rate and the reverse steps become negligible hence electroadsorption becomes coupled to electrocombination.

On electrodes of noble metals of the platinum group, among which the strong H absorber palladium, the mechanism of the HER involves at low overpotentials electroadsorption in quasiequilibrium, followed by a rate-determining step n.d.s of chemical combination [11,31].

On nickel, the reported mechanism is electroadsorption in quasiequilibrium followed by rate-determining electrocombination [31,33].

On iron, early experiments indicated a coupled mechanism, electroadsorption followed by chemical combination at low overpotential (low H coverage) or electrocombination at high overpotential (high H coverage) [34,35]. However, more recent studies reported a parallel pathways mechanism where the chemical and electrochemical combination steps occur simultaneously, the former being predominant in 0.1 N H_2SO_4 up to high cathodic current densities, whereas in 0.1 N NaOH or neutral solutions the latter (involving water molecules) already predominates at low cathodic current densities [36,37]. The exact mechanism seems to depend on the structure and purity of the iron electrodes and their surface composition.

2.2.2.2.2 Underpotential and Overpotential H Electroadsorption

The process of reversible hydrogen electroadsorption occurs at the surface of metal electrodes of the platinum group at potentials anodic with respect to the reversible hydrogen electrode at the given pH and at a pressure of 1 atm (half-standard reversible H^+/H_2 electrode at 1 atm, denoted RHE1 here): H is electroadsorbed at equilibrium at $E > 0\,V_{(RHE1)}$ up to a quasifull monolayer. Because this process occurs, going in the cathodic direction, before the equilibrium potential of the RHE1, by analogy with underpotential metallic deposition, it is often called H underpotential deposition (UPD) and the adsorbed H the UPD H [38]. Contrary to the HER, which is a steady faradic reaction consuming the H intermediate electroadsorbed at overpotential, for which a stationary state is reached when the net rate of electroadsorption of reactants is equal to twice the net rate of H_2 formation, H UPD is a pseudocapacitive faradic process; that is, at a given potential the adsorption current goes to zero once the equilibrium H coverage is reached, corresponding to equality between the rate of adsorption and the rate of desorption of H atoms. It can be characterized only by transient techniques such as cyclic linear sweep voltammetry, giving for cathodic and anodic potential sweeps the well-known H adsorption/desorption pseudocapacitance peaks [38–42], symmetrical if the sweep rate is not too high [38]. By integration of a voltammogram, the electroadsorption/desorption charge flowing through the interface per unit area during a potential scan is obtained, which allows determination of the H surface density at a given potential. If the metal atomic density is known (e.g., on a well-ordered single crystal surface), the coverage θ_H (number of adsorbed H atoms per substrate atom) is obtained as a function of the potential, giving electroadsorption isotherms [40,41,43–45].

Detection of UPD H is limited cathodically by the onset of the HER, producing a steady current that is rapidly predominant as the overpotential increases; underpotential (UP) H adsorption requires a negative standard free energy of adsorption from $H_{2(g)}$, which is the case for transition metals. On the noble metals (Ag, Au) for which the electronic d-band is filled, no UP H adsorption occurs because adsorption from $H_{2(g)}$ is endothermic [13]. UPD is prevented on electrodes covered with electronegative adsorbates (see Section 2.6), and even on a clean electrode UP H adsorption is often hindered by competitive anodic processes such as UP adsorption of O or OH and electrode oxidation [28,33,46,47]. This explains why UPD H may be detected only on the noble catalytic metals where the anodic preoxidation processes occur at relatively high anodic potentials so that there is a potential range where H atoms compete only with water molecules forming the metal hydration layer and with specifically adsorbed anions [28,38,42,47,48]. On Pd, significant H absorption occurs at underpotential in bulk samples [49,50], so UP H adsorption may be characterized by cyclic voltammetry only on Pd thin films in which the low number of bulk H sites limits the absorption current [51–54]. On corroding or passivating metal electrodes, UPD H detection is impeded by the metal dissolution and oxidation processes and the voltammograms are not easily interpreted [46,55]; although some workers claimed that H electroadsorption occurs in aqueous electrolytes above 0 V (RHE1) on Ni [56], Fe [57], or W [58], there is no strong experimental evidence for this.

Conversely, the electroadsorption step involved in the HER at cathodic overpotentials on all electrodes has been called H overpotential deposition (OPD) and the H_{ads} intermediate involved in the HER called OPD H [38]. Because H OPD does not occur alone but is only a step in the steady HER process, it is much more difficult to characterize than H UPD. A technique of measurement of potential relaxation transients (potential decay) has been developed, which allowed determining pseudocapacitance versus potential curves and obtaining the OPD H coverage by integration [33], but this is not straightforward. The OPD H fractional coverage is usually estimated from analysis of the kinetics of the HER and HAR. This H intermediate is likely to also be the intermediate of the HOR on Pt, in an anodic potential range overlapping that of UPD H [32]. Electroadsorption of this H species occurs under true equilibrium only at the H^+/H_2 equilibrium potential, at overpotential the H coverage versus potential variation depends upon the HER mechanism. It seems likely that the OPD H atoms reacting in the HER on Pt are adsorbed on top of Pt atoms, while the UPD H atoms are adsorbed in high coordination sites [59,60] (see later). There are no similar data for other metals.

2.2.2.2.3 Hydrogen Electroabsorption Reaction

In aqueous solution, the H absorption reaction occurs mainly by electrochemical reduction of protons or water molecules. The overall H electroabsorption reaction from protons is

$$H^+_{(aq)} + e^-_{(M)} + [M] = H_{diss}[M] \quad (2.9)$$

It was demonstrated for iron, from the analysis of the relation between the stationary cathodic and anodic currents measured on each side of a permeation membrane, that the HER and the HEAR share the first step of electroadsorption (see Equations 2.4 and 2.5) and a common adsorbed H intermediate (H_{ads}) [8]; then H enters the metal lattice by the surface-bulk transfer step (Equation 2.8), assumed to be in quasiequilibrium because in most cases the permeation rate is found to be limited only by H diffusion through the bulk. This mechanism has been verified to be valid for permeation through iron and nickel and their alloys and it is the generally accepted one [11].

However, it has been suggested by Russian workers [61,62] for explaining the effects of adsorbed I^- ions on H permeation into Pd membranes that the process of H electroadsorption in the HER and the process of electroabsorption "should be regarded as occurring simultaneously and to a certain extent independently of each other" [62]. Thus on Pd the HEAR would not occur by the pathway described before, but by what has been interpreted since as direct H entry from protons [8,63]; this mechanism was invoked for explaining Pd membrane permeation data showing an anomalous relationship between steady-state cathodic and permeation currents [64]. Actually, because a direct entry would involve a quantum tunneling effect not likely at room temperature (see Section 2.2.1.2), a classical activation mechanism of H entry through a surface intermediate state different from the adsorbed state involved in the HER (OPD H) is more rational and has to be considered as a possible alternative to the preceding mechanism for Pd. The observation that significant H electroabsorption into bulk Pd already occurs at positive potentials versus RHE1, before the HER takes place, and hence is likely to involve UPD H [48,50,65] could be consistent with the so-called direct entry mechanism as UPD H is a surface species different from the HER intermediate (see Section 2.3.2.5).

2.3 Thermodynamics of Metal–Hydrogen Systems

2.3.1 Thermodynamics of H Adsorption and Absorption in the Gas Phase

2.3.1.1 Isotherms of the Metal–Hydrogen Equilibria

2.3.1.1.1 Isotherm of H_2 Dissociative Adsorption

The equation of the isotherm of equilibrium adsorption from $H_{2(g)}$ (see Equation 2.1) is

$$\frac{\theta_H}{1-\theta_H} = \left[\frac{-\Delta G_{Hads}(\theta_H)}{RT}\right] \quad (2.10a)$$

where

θ_H is the H surface fractional coverage in the sites (M)

f_{H_2} is the fugacity of hydrogen gas expressed in atm

ΔG_{Hads} is the half-standard free energy of H adsorption from $\frac{1}{2}H_{2(g)}$ ($\Delta G_{Hads}(\theta_H) = \mu_{Hads}(\theta_H) - \frac{1}{2}\mu_{H_2}$)

ΔG_{Hads} is in the general case a function of θ_H [66]. θ_H is equal to the ratio Γ_H/Γ_{Hsat} of the H surface density to the H surface density at saturation. Γ_{Hsat} may be lower than the surface density of H sites on the given face due to H–H repulsive interactions [13].

2.3.1.1.2 Isotherm of H_2 Dissociative Absorption

The equation of the isotherm of equilibrium absorption (dissolution) from $H_{2(g)}$ (reaction obtained by combination of Equations 2.1 and 2.2) is the following:

$$\frac{X_H}{1-X_H} = f_{H_2}^{1/2} \exp\left[\frac{-\Delta G_{Hdiss}(X_H)}{RT}\right] \tag{2.11}$$

where

X_H is the H bulk fractional concentration in the sites [M]

ΔG_{Hdiss} is the half-standard free energy of H dissolution from $\frac{1}{2}H_{2(g)}$ · ΔG_{Hdiss} is in the general case a function of X_H

X_H, the fractional concentration of H in the metal or fraction of the interstitial sites occupied by H, may be expressed as the ratio c_H/c_{HSat} of the H volumic concentration to the concentration at saturation, or as r_H/r_{Hsat} where r_H is the hydrogen/metal (H/M) atomic ratio and r_{Hsat} is the ratio at saturation. In a particular lattice, only one kind of interstitial sites is populated by H at moderate pressures. In the ideal case where there were no H–H interactions, r_{Hsat} would be equal to the number of those interstitial sites per metal atom (1 for octahedral sites in face-centered cubic [fcc] metals, 2 for tetrahedral sites in hexagonal close-packed [hcp] metals, 6 for tetrahedral sites in body-centered cubic [bcc] metals) [6].

2.3.1.1.3 Isotherm of H Surface-Bulk Transfer

The equation of the isotherm for the surface-bulk transfer step at equilibrium (Equation 2.2) is [60,67,68]

$$\frac{X_H}{1-X_H} = \frac{\theta_H}{1-\theta_H} \exp\frac{\Delta G_{Hads}(\theta_H) - \Delta G_{Hdiss}(X_H)}{RT} \tag{2.12a}$$

2.3.1.2 Thermochemical Data

2.3.1.2.1 H Adsorption

All transition metals from column 3 to 10, plus Cu, are exothermic H adsorbers ($\Delta H_{Hads} < 0$; θ_H decreases when the temperature increases). Experimental heats of H adsorption (per mole of H_{2g}) obtained on polycrystalline transition metal surfaces are shown in Figure 2.2a [69].

The M–H_{ads} bond energy may be obtained from the relation:

$$E_{M-Hads} = \frac{1}{2}E_{dH2} - \Delta H_{Hads} \tag{2.13}$$

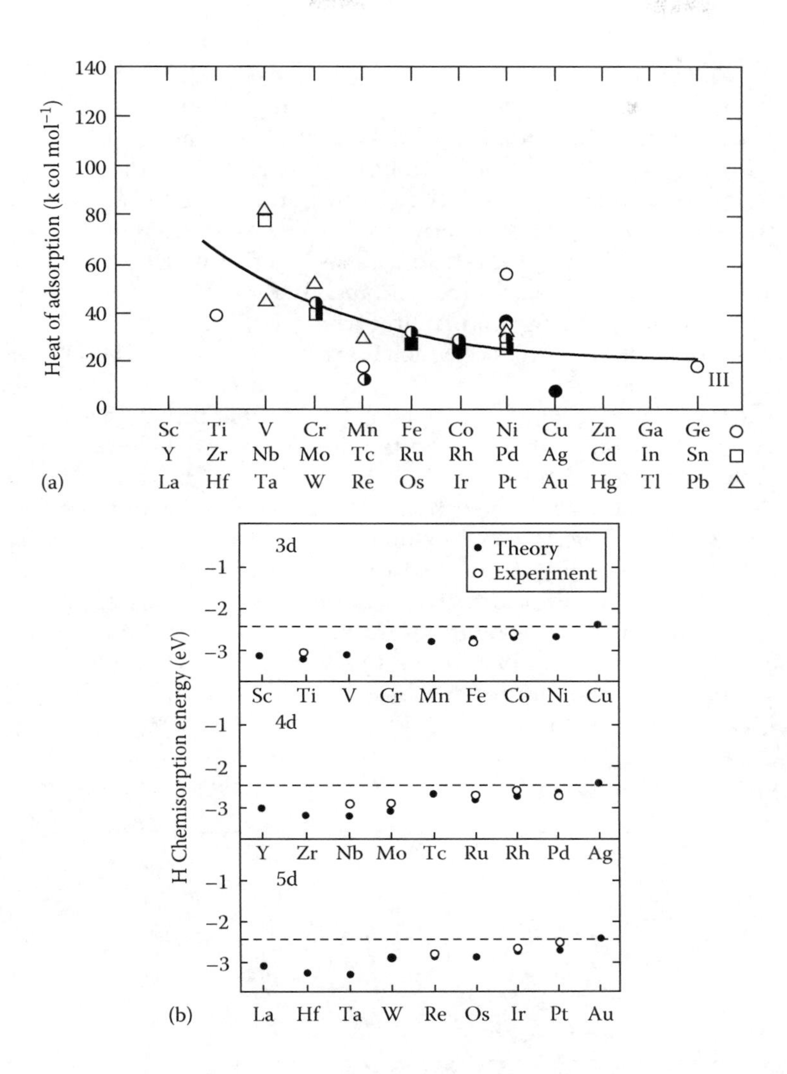

FIGURE 2.2
(a) Experimental heats of adsorption of hydrogen on polycrystalline transition metal surfaces. (From Toyoshima, I. and Somorjai, G.A., *Catal. Rev. Sci. Eng.*, 19,105, 1979, Fig. 2a.) (b) Calculated and experimental chemisorption bond energies for *H* on the most close-packed surfaces of transitions metals (1 eV ≡ 96.5 kJ mol^{-1}). (From Nordlander, P., Holloway, S., and Norskov, J.K., *Surf. Sci.*, 136, 59, 1984, Fig. 2.)

where

E_{dH2} is the dissociation energy of H_2, equal to 436 kJ mol^{-1} at 25°C

ΔH_{Hads} is the enthalpy of adsorption of H per mole of H: $\Delta H_{Hads}(\theta_H) = H_{Hads}(\theta_H) - \frac{1}{2}H_{H_2}$

On single-crystal faces of the transition metals, the values of the initial H adsorption heat [$-\Delta H_{Hads}(\theta_H = 0)$] are between 20 and 70 kJ mol(H)$^{-1}$ [13]. The values of the M–H_{ads} bond energy at zero coverage are in the range 240–290 kJ mol^{-1}, that is, 2.5–3.0 eV [13c]. These values depend slightly on the surface orientation [13].

Calculated and experimental adsorption bond energies for H on the most close-packed surfaces of the transition metals are given in Figure 2.2b [70]. A diminution of the chemisorption bond strength is observed from center left to the right in each metal series (3d, 4d, and 5d).

2.3.1.2.2 H Absorption

The solubility of H in metals varies over a great range, from 10^{-10} to 10^{-1} mol cm^{-3} [11]. According to the sign of the enthalpy of dissolution from $H_{2(g)}$, metals are exothermic H absorbers ($\Delta H_{Hdiss} < 0$; the H concentration decreases with increasing temperature) or endothermic absorbers ($\Delta H_{Hdiss} > 0$; the H concentration increases with temperature). Among the transition metals, the exothermic absorbers most often undergo a transition to a hydride phase when the bulk H/M ratio increases. These are metals of columns 3 to 5 [71]. Most other transition metals are endothermic absorbers and have very low H solubility at 25°C and 1 atm, especially Cu, Ag, and Au in column 11 [72].

The well-known exception is Pd (fcc), which is an exothermic absorber. When the H_2 pressure (or cathodic overpotential) increases, Pd forms with H at 25°C a solid solution (α) up to the terminal solubility 0.03 H/Pd, above which a hydride phase (β) with 0.55 H/Pd precipitates; then the H-saturated phase α and the hydride phase β coexist until all the metal is converted into hydride and the ratio H/Pd = 0.55 is passed; for very high pressures or overpotentials, the hydride may be enriched in H up to the stoichiometric compound PdH. Another special case is Ni (fcc), which exhibits the highest H absorption capacity of any endothermic absorber (solubility at 25°C and 1 atm: ~10^{-6} mol cm^{-3} [11], 1–5 × 10^{-5} H/Ni [2]); it may form a hydride out of equilibrium, that is, under very high pressures or overpotentials, up to NiH [73]; Cr (bcc) also forms hydrides in these conditions. Both Mn and Cr exhibit a minimum solubility as a function of temperature, which means that they change from exothermic absorbers at low temperatures to endothermic at high temperatures [73]. Fe (bcc) has low H solubility (about 3 × 10^{-9} mol cm^{-3}, 2–3 × 10^{-8} H/Fe [2] at 25°C and 1 atm).

Figure 2.3 shows ΔH_{Hdiss} at infinite dilution plotted for most metals [74]. Figure 2.4 shows plots of $-\Delta S_{Hdiss}$ versus ΔH_{Hdiss} [75]. The fcc and hcp metals give one correlation line and

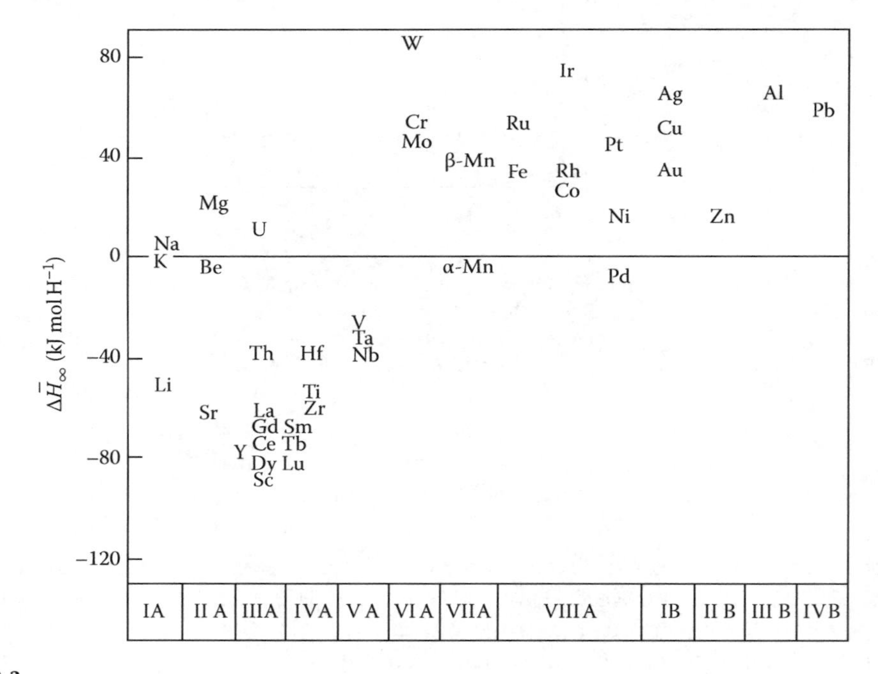

FIGURE 2.3
Enthalpies of solution of *H* at infinite dilution in metals; $\Delta\bar{H}_\infty \equiv \Delta H_{Hdiss}$ in our conventions. (From Griessen, R. and Riesterer, T., in *Hydrogen in Intermetallic Compounds*, Part I, L. Schlapbach, ed., Springer-Verlag, New York, 1988, p. 230, Fig. 6.3.)

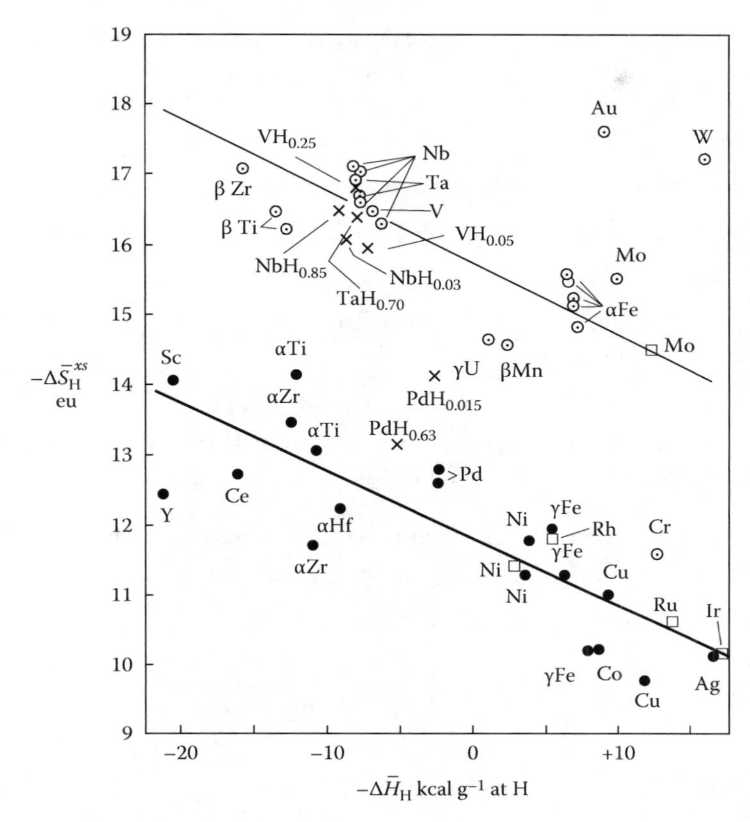

FIGURE 2.4
Correlation between excess entropy (cal mol^{-1} K^{-1}) and enthalpy (kcal mol^{-1}) of solution of H in metals. $\Delta\bar{S}_H^{xs} \equiv \Delta S_{Hdiss} = S_{Hdiss} + R\ln\frac{X_H}{1-X_H} - \frac{1}{2}S^0_{H_2}$. (From McLellan, R.B. and Oates, W.A., *Acta Metall.*, 21, 181, 1973, Fig. 3.)

the bcc metals another one [75]. The approximate linear correlation ($-\Delta S_{Hdiss}$ increases with $-\Delta H_{Hdiss}$) is rather classic: the higher the value of $-\Delta H_{Hdiss}$, the more tightly bound the interstitial H and the lower its vibration frequency, hence the higher the value of $-\Delta S_{Hdiss}$, the reduction in entropy from the gaseous state to the dissolved state.

2.3.2 Thermodynamics of the Metal–Hydrogen Equilibria in Aqueous Electrolyte

2.3.2.1 Isotherms of the Metal–Hydrogen Equilibria

2.3.2.1.1 Isotherm of H_2 Dissociative Adsorption/H–H Combination

Under conditions in which there is no competitive specific adsorption of ions or adsorption of oxygen species in the H_{ads} potential range, equilibrium of the dissociative adsorption from $H_{2(g)}$ in aqueous solution (or the reverse reaction, the chemical combination) (see Equation 2.3) leads to the same adsorption isotherm as in Equation 2.10a:

$$\frac{\theta_H}{1-\theta_H} = f_{H_2}^{1/2} \exp\left[\frac{-\Delta G_{Hads(aq)}}{RT}\right] \tag{2.10b}$$

where $\Delta G_{Hads(aq)}$ is the half-standard free energy of H adsorption from $\frac{1}{2}H_{2(g)}$ in aqueous solution, which is a priori different from ΔG_{Hads} in the gas phase because adsorption in aqueous solution implies displacement of adsorbed water molecules and reorganization of the double layer [25,29].

2.3.2.1.2 Isotherm of H Electroadsorption/Electrodesorption

The equation of the equilibrium H electroadsorption isotherm (see Equations 2.4 and 2.5) is [48,60,65,67,76]:

$$\frac{\theta_H}{1-\theta_H} = a_{H^+} \exp\left[\frac{-FE_{(SHE)}}{RT}\right] \exp\left[\frac{-\Delta G_{Hads(aq)}}{RT}\right] \tag{2.14}$$

where

$E_{(SHE)}$ is the electrode potential referred to the standard reversible hydrogen electrode

$\Delta G_{Hads(aq)}$ is the same as in Equation 2.10b

If the equation is rearranged to involve the potential referred to the half-standard reversible hydrogen electrode in the same electrolyte (RHE1, for $p_{H_2} = 1$ atm, a_{H^+}), $E_{(RHE1)} = E_{(SHE)} - \frac{RT}{F}$? $\ln a_{H^+}$, a simpler expression is obtained [43,44,60,65,76]:

$$\frac{\theta_H}{1-\theta_H} = \exp\left[\frac{-FE_{(RHE1)}}{RT}\right] \exp\left[\frac{-\Delta G_{Hads(aq)}}{RT}\right] \tag{2.15}$$

2.3.2.1.3 Isotherm of Electrodissociation/Electrocombination

Similarly, the isotherm of equilibrium electrodissociation of $H_{2(g)}$ (see Equations 2.6 and 2.7) can be expressed as [67]

$$\frac{\theta_H}{1-\theta_H} = \frac{f_{H_2}}{a_{H^+}} \exp\left[\frac{FE_{(SHE)}}{RT}\right] \exp\left[\frac{-\Delta G_{Hads(aq)}}{RT}\right] \tag{2.16}$$

or

$$\frac{\theta_H}{1-\theta_H} = f_{H_2} \exp\left[\frac{FE_{(RHE1)}}{RT}\right] \exp\left[\frac{-\Delta G_{Hads(aq)}}{RT}\right] \tag{2.17}$$

2.3.2.1.4 Isotherm of H Electroabsorption (HEAR)

The equation of the equilibrium H electroabsorption isotherm (Equation 2.9) is [60,67,68]:

$$\frac{X_H}{1-X_H} = a_{H^+} \exp\left[\frac{-FE_{(SHE)}}{RT}\right] \exp\left[\frac{-\Delta G_{Hdiss}}{RT}\right] \tag{2.18}$$

or

$$\frac{X_H}{1-X_H} = \exp\left[\frac{-FE_{(RHE1)}}{RT}\right] \exp\left[\frac{-\Delta G_{Hdiss}}{RT}\right] \tag{2.19}$$

with ΔG_{Hdiss} as defined before for the gas-phase reaction (Equation 2.11).

2.3.2.1.5 *Isotherm of H Surface-Bulk Transfer*

The equation of the isotherm for the surface-bulk transfer step at equilibrium (Equation 2.8) is identical to that in Equation 2.12a except that ΔG_{Hads} is replaced by $\Delta G_{\mathrm{Hads(aq)}}$.

2.3.2.2 H_2 Pressure/Potential Correspondence for Adsorption/Absorption Equilibria

A comparison of Equations 2.10a and 2.11 with Equations 2.15 and 2.19 shows that, for isotherms in aqueous solution, the term $\exp[-FE_{(\mathrm{RHE1})}/RT]$ plays the role of the fugacity term $f_{\mathrm{H_2}}^{1/2}$ in isotherms of chemical adsorption/absorption from $\frac{1}{2}H_{2(g)}$. So in solution, at equilibrium, a given potential versus RHE1 (reversible hydrogen electrode at 1 atm) is equivalent to the fugacity of H_2:

$$(f_{\mathrm{H_2}})_{\mathrm{eqv}} = \exp\left[\frac{-2FE_{(\mathrm{RHE1})}}{RT}\right] \quad (2.20)$$

This equation provides the correspondence between adsorption/absorption equilibria in the gas phase and electroadsorption/electroabsorption equilibria in solution [11,77]. The H coverage or bulk concentration at 0 V/RHE1 is equal to the coverage or concentration at an H_2 pressure of 1 atm. It is noteworthy that Equation 2.20 applies only to electrosorption reactions in equilibrium or at least in quasiequilibrium and is not valid at large overpotentials where coupled HER mechanisms prevail (see Section 2.2.2.2.1).

2.3.2.3 Gibbs Free Energy of Adsorption for UPD H and OPD H

For gas-phase adsorption, from Equation 2.10a, it is deduced that the fugacity $f_{\mathrm{H_2}}$ corresponding to $\theta_{\mathrm{H}} = \frac{1}{2}$ is equal to $\exp[2\Delta G_{\mathrm{Hads}}(\frac{1}{2})/RT]$. Hence, adsorption sites with $\Delta G_{\mathrm{Hads}}(\frac{1}{2})$ negative, that is, in which the Gibbs partial molar free energy (chemical potential) of the adsorbed H species at half-saturation is lower than $\frac{1}{2}\, G^{\circ}_{\mathrm{H2(g)}}$, attain half-saturation for a H_2 pressure lower than 1 atm, whereas sites with $\Delta G_{\mathrm{Hads}}(\frac{1}{2})$ positive attain half-saturation only for a pressure higher than 1 atm. Similarly, analysis of Equation 2.15 shows that H electroadsorption may occur significantly at $E_{(\mathrm{RHE1})} > 0$ V (at cathodic underpotential) in sites with $\Delta G_{\mathrm{Hads}}(\frac{1}{2})$ negative, whereas the sites with $\Delta G_{\mathrm{Hads}}(\frac{1}{2})$ positive reach half-saturation only for $E_{(\mathrm{RHE1})} < 0$ V (at cathodic overpotential) [60,78]. On Pt, the value of the HER Tafel slope at low cathodic overpotentials (~30 mV) indicates that the HER intermediate is electroadsorbed under quasiequilibrium at very low coverage close to 0 V/RHE1 and reaches significant coverage only at significant cathodic overpotentials; it is said to be overpotentially deposited, and is often denoted OPD H. Hence it can be inferred that it is a weakly bonded H species that has a positive $\Delta G_{\mathrm{Hads}}(\frac{1}{2})$ [60] and is adsorbed on or among a quasifull monolayer of strongly bonded UPD H atoms with negative $\Delta G_{\mathrm{Hads}}(\frac{1}{2})$ [33,38,60,79]. This simple analysis shows that, at least on Pt and neighboring metals, H underpotential electroadsorption can be correlated with low H_2 pressure adsorption in the gas phase [40,43,80] and overpotential electroadsorption with high H_2 pressure adsorption [60].

Similarly, the fact that H electroabsorption in Pd already occurs at E(RHE1) > 0 V [50,65] can be correlated with the exothermic character of H absorption from H_2 in Pd ($\Delta H_{\mathrm{Hdiss}} < 0$), giving a value of $\Delta G_{\mathrm{Hdiss}}$ only slightly positive at room temperature (calculated from the values in Figure 2.4).

2.3.2.4 Thermochemical Data for H Underpotential Electroadsorption on Pt in Aqueous Electrolyte

As mentioned earlier, thermodynamic measurements of H adsorption in aqueous electrolytes are possible only on the UPD H electroadsorbed reversibly above 0 V/RHE1 on some noble metals of the Pt group [32,40,41,48,76] for which $\Delta G_{ads}(\frac{1}{2})$ is negative. The analysis of the data obtained in aqueous electrolytes is complicated by the fact that the potential region of UPD H adsorption/desorption often overlaps with the potential region of the specific adsorption/desorption of anions [42,48], in which case the two processes are competitive and the overall charge density obtained by integration of the voltammograms corresponds to a replacement reaction. The two processes occur in separated potential ranges on Pt(111) single-crystal electrodes in $HClO_4$ or in diluted H_2SO_4 electrolytes [42,48]. Thermodynamic measurements on Pt(111) show that the values of the heat of adsorption of UPD H at zero H coverage and the Pt-UPD H bond energy at zero coverage [≈260 kJ mol^{-1} for Pt(111)] [32,48] are close to those measured in low-pressure gas-phase experiments ($E_{Pt\text{–}Hads} \approx 255$ kJ mol^{-1}) [13].

2.3.2.5 Relation between the Energetic and Structural Aspects of H Adsorption

These thermodynamic measurements indicate that the M–H_{ads} bonds involved at the metal–electrolyte and metal–gas interfaces at low pressure are of the same nature, hence that UPD H atoms probably occupy the same adsorption sites as those characterized by UHV techniques [29,43,60,48], that is, the highly coordinated or hollow sites [13].

With the use of a ^{35}S radiotracer allowing accurate measurement of the sulfur coverage, the present authors studied quantitatively and compared the blocking by chemisorbed sulfur of the H underpotential (UP) adsorption and the HER on Pt single-crystal surfaces. It was concluded that OPD H on Pt is not adsorbed in the same sites as UPD H [60,79]. More direct evidence was obtained by in situ infrared (IR) spectroscopy on a Pt(111) electrode at overpotential that the OPD H atoms reacting in the HER are adsorbed on top of substrate atoms [59], while UPD H atoms are adsorbed in hollow sites.

Thus, a joint analysis of experiments in the gas phase and in aqueous electrolyte gives evidence of the existence of two main kinds of H adsorption sites on a Pt surface: high coordination sites (hollow sites) with a strong M–H_{ads} bond ($\Delta G_{Hads} < 0$) and low coordination sites (on-top sites) with a weak M–H_{ads} bond ($\Delta G_{Hads} > 0$); UPD H is adsorbed in the hollow sites as H adsorbed from H_2 in low-pressure gas-phase adsorption, whereas OPD H is adsorbed at the on-top sites, as the labile and highly reactive H involved at atmospheric or high pressure in the catalytic reactions [60,78].

On nonnoble transition metals, the existence of two H_{ads} species with different bond energies and site coordinations can only be assumed. H UPD is not detected by cyclic voltammetry on oxidizable metals, although these metals readily adsorb H in the gas phase (ΔG_{Hads} for the hollow sites on these metals is more negative than on Pt), because the UPD oxygen species (O_{ads} or OH_{ads}) are more strongly chemisorbed than H in the same sites. A thermodynamic calculation based on gas-phase data shows that on Ni and Fe, H electroadsorption is in direct competition with electroadsorption of O species that also occurs near 0 V/RHE1; hence H can electroadsorb in the hollow sites only at potentials where most of the oxygen layer is desorbed, that is, slightly under or over 0V/RHE1, depending on the binding strength of the O species and their possible inhibiting effect on the M–H_{ads} bond [28]. Hence it is likely that on nonnoble transition metals, there is coexistence of a stable monolayer of H species electroadsorbed in the hollow sites around 0 V/RHE1 by replacement of the O species and a mobile one electroadsorbed at cathodic overpotential at on-top sites and reacting in the HER [28].

2.4 Mechanisms of the H Surface-Bulk Transfer

2.4.1 Structural Aspects

2.4.1.1 H Adsorption Sites

Figure 2.5 shows the atomic arrangement of the theoretical (not reconstructed) low-index faces of face-centered cubic (fcc), hexagonal close-packed (hcp), and body-centered cubic (bcc) metals, with the high-symmetry sites of H adsorption. There is general agreement from UHV structural and vibrational measurements that the H adsorption sites at low hydrogen pressure are the sites of highest coordination, hollow or bridge sites [13,81]. On both the (111) face of fcc and the (0001) face of hcp, there are two kinds of threefold hollow sites: sites over one second-layer metal atom, called hcp-type in surface science conventions, and sites over one interstice between three second-layer atoms, called fcc-type sites. It has been reported that H atoms adsorbed at low pressure tend to occupy the hcp sites on Pt(111) [82] and also that D atoms occupy preferentially the fcc sites on Ni(111) [83].

2.4.1.2 H Bulk Interstitial Sites

Figure 2.6 shows the different kinds of bulk interstitial sites in perfect fcc, hcp, and bcc metal lattice structures. The sites populated by H atoms in the concentration range of solid solution are known to be the octahedral sites for the fcc metals, whereas they are most probably the tetrahedral sites for the bcc and hcp metals [72]. In fcc and hcp metals, there are one octahedral site and two tetrahedral sites per metal atom, whereas in bcc metals there are three octahedral and six tetrahedral sites per metal atom.

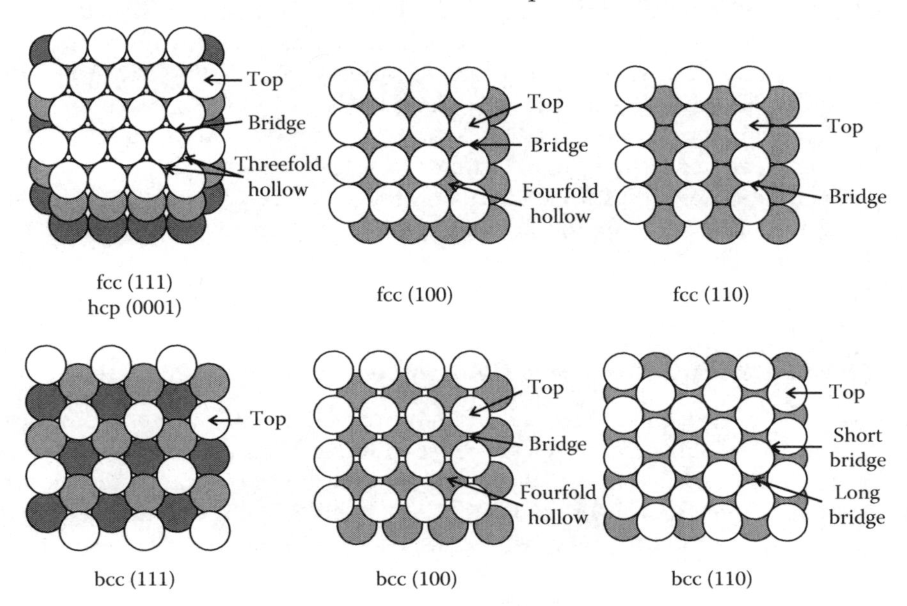

FIGURE 2.5
Top views of the atomic arrangement in the first two or three layers of the low-Miller-index faces of fcc, hcp, and bcc crystals. High-symmetry adsorption sites are indicated. (Adapted from Mate, C.M., Bent B.E. and Somorjai, G.A., in *Hydrogen Effects in Catalysis*, Z. Paal and P. G. Menon, eds., Marcel Dekker, New York, 1988, Chapter 2, pp. 57–81, Fig. 1.)

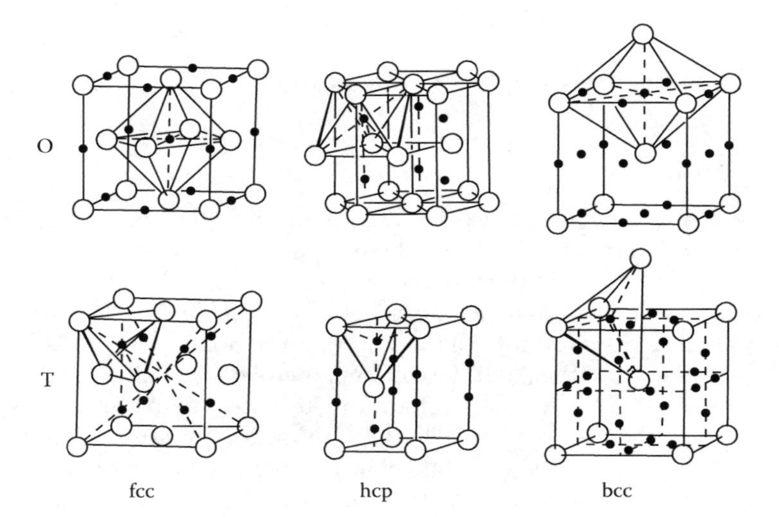

FIGURE 2.6
Interstitial sites: octahedral (O) and tetrahedral (T) sites in fcc, hcp, and bcc lattices. (From Fukai, Y., *The Metal–Hydrogen System*, Springer-Verlag, New York, 1993, Fig. 1.14.)

For diffusion in fcc metals, because the close-packed atoms rows prevent the direct jumps from one octahedral site to another in the ⟨110⟩ directions, the H migration paths involve intermediate jumps in the <111> directions to nearest neighbor tetrahedral sites, then from tetrahedral to octahedral sites. An octahedral site is surrounded by 8 nearest neighbor tetrahedral sites in the <111> directions and 12 octahedral sites in the ⟨110⟩ directions (see Figure 2.5). A tetrahedral site is surrounded by four octahedral sites.

For the bcc metals, the possibilities of jumps between tetrahedral sites are direct from one tetrahedral site to four nearest neighbor tetrahedral sites along the ⟨110⟩ directions or across the octahedrals sites to two next neighbors in the ⟨100⟩ directions. The distances between nearest neighbor tetrahedral sites in bcc metals are smaller by a factor of 2 than those between nearest neighbor octahedral sites in fcc metals [72], which leads to much lower activation energies for H diffusion than in fcc metals [84,85]. This explains the relatively high H diffusion coefficient in bcc iron (between 10^{-5} and $10^{-4}\,cm^2\,s^{-1}$), whereas in fcc metals it is about $10^{-6}\,cm^2\,s^{-1}$ for Pd and between 10^{-10} and $10^{-9}\,cm^2\,s^{-1}$ for Ni and Pt [72].

2.4.1.3 Role of the Surface Structure in H Absorption: Entry Sites

For the low-index faces of fcc crystals, the possible directions of H jumps from the surface adsorption sites to the subsurface sites are represented in Figure 2.7.

The lack of bonding above the surface leads to adsorbed H atoms being closer to the underlying metal atoms than absorbed H atoms in the bulk, so the distances of jumps from the adsorption sites to the nearest subsurface sites are shorter than the nearest neighbor distances in the bulk.

On the (111) face, each hcp surface site is located just above one subsurface tetrahedral site T_{12}^{+} (located between the first and the second metal layers, below three adjacent surface atoms and above one second-layer atom), from which three octahedral subsurface sites O_{12} (below three surface atoms and above three second-layer atoms) may be reached. Each fcc surface site is just above one such subsurface O_{12} site, from which, via three T_{12}^{+} sites, three

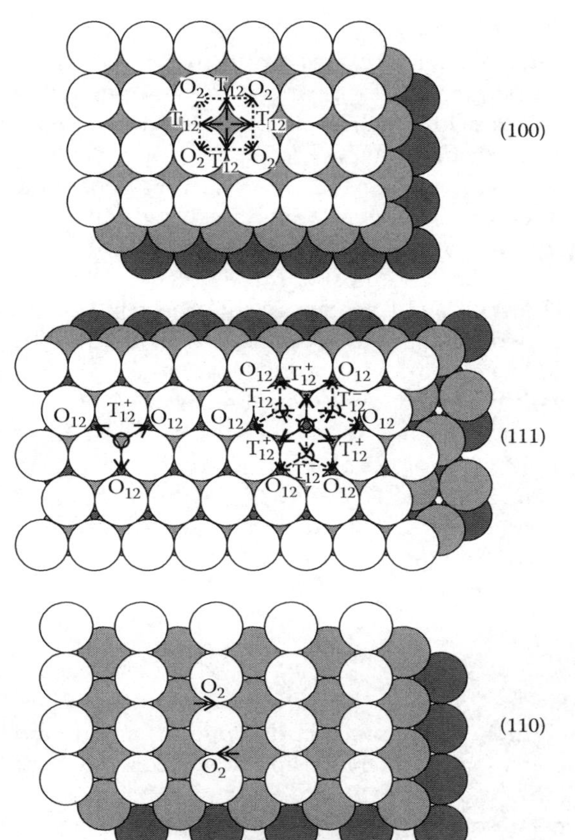

FIGURE 2.7
Possible paths of surface-bulk transfer on the low-index faces of fcc metals. The full line arrows indicate the directions of first H atom jumps from surface sites; the broken line arrows indicate second or third jumps from subsurface sites. On the (111) face, first jumps occur vertically, either from an hcp surface site to a tetrahedral site T_{12}^+ (left) or from a fcc site to an octahedral site O_{12} (right).

T_{12}^- sites (below one surface atom), or one T_{23}^+ site (below the O_{12} site), six O_{12} sites and three O_{23} sites (below the T_{12}^- sites) may be reached.

On the (100) face (if it is not reconstructed), one H atom adsorbed in a fourfold hollow site may jump to four tetrahedral subsurface sites T_{12} (below two adjacent surface atoms), from each of which two octahedral sites O_2 (at the second layer level, below surface atoms) may be reached.

On the (110) face (the more open face of the three low-index faces for fcc metals) the adsorbed H atoms are alternatively adsorbed on each side of the close-packed [110] rows, in the pseudo-threefold fcc sites close to the bridge sites [86]. They may jump to one subsurface site O_2 (at the second layer level, under the bridge sites). On the (110) faces of Ni and Pd, above a coverage of one, hydrogen induces a reconstruction of the surface into a (110)-(1 × 2) structure, probably of the pairing-row type [13,15,82]. This opens the surface even more and allows accommodation of H on the second metal layer, leading to a coverage of 1.5 H per first layer metal atom, and makes it easier to reach subsurface sites [8,87,88].

2.4.2 Energetic Aspects

2.4.2.1 Gas Phase

Energetic analysis using temperature-programmed desorption (TPD), also called thermal desorption spectroscopy (TDS), on a well-defined surface makes it possible to determine the heights of the energy barriers of the surface processes. It was shown that the role of the substrate surface orientation in the population of subsurface sites is crucial. On the reconstructed (110) faces of Ni and Pd or on more open faces, subsurface sites are populated at temperatures as low as 100 K and low H_2 pressures (10^{-6} Pa) [13,87,88], whereas higher temperatures and pressures are necessary for the more densely packed planes [13].

The overall energetics of the hydrogen-metal reactions are illustrated in the schematic drawing of Figure 2.8a showing the one-dimensional potential energy versus distance curves of the various H states at the metal–gas interface. Actually, the energy levels indicate the chemical potentials (partial molar Gibbs free energies) of these states, referred to the standard chemical potential of $\frac{1}{2}H_{2(g)}$. The standard free energy differences between the various states are indicated. Two H adsorbed states (corresponding to the hollow and on-top sites) are represented. The diagram shows that at 1 atm nonactivated adsorption occurs spontaneously in the deep energy wells of the hollow sites, whereas adsorption is activated in the shallow wells of the on-top sites, which need a higher pressure to be filled. The diagram shows a case of slightly endothermic H absorption, with a high activation energy barrier for the transition between strongly bonded H_{ads} and H_{diss}. The depths of the wells and the heights of the energy barriers depend on the nature of the metal, the surface structure, and the H coverage. The barrier separating the chemisorbed and the subsurface H states may be lowered by a high coverage or a surface reconstruction [13].

2.4.2.2 Aqueous Electrolyte

An equivalent energy diagram is shown in Figure 2.8b for the metal–aqueous solution interface [78]. It illustrates a case such as Pt(111) in $HClO_4$, where specific adsorption of anions and electroadsorption of oxygen species occur at high anodic potentials and do not compete with H adsorption [28]. The free energy of adsorption of water from the liquid state can be neglected for transition metals. The specificity of the metal–solution interface, apart from the competition of H adsorption with adsorption of water, anions, and oxygen species, is that the potential versus RHE1 is an external variable that controls the energy level (more exactly the standard electrochemical potential) of the initial state $\left(H^+_{(aq)} + e^-\right)$, with respect to the reference level of $\frac{1}{2}H_{2(g)}$, and hence the standard activation free energy of electroadsorption $\left(\Delta G^{\neq}_{ea}\right)$ in the two kinds of sites. Thus two main situations occur, according to the potential versus RHE1:

1. The potential $E_{(RHE1)}$ is positive (anodic).
   This is the range of underpotential (UP) H adsorption on metals of the Pt group. The energy levels of the initial state $\left(H^+_{(aq)} + e^-\right)$ and the final state (strongly bonded H_{ads}, with $\Delta G_{ads} < 0$) are close and far below the energy levels of $\frac{1}{2}H_{2(g)}$ and the weakly bonded H_{ads}; hence at these potentials UP H electroadsorption in the strongly bonded sites is in equilibrium and there is negligible H electroadsorption in the weakly bonded sites and no H_2 evolution. In the case of palladium, the energy level of H_{diss} is located near the level of $\frac{1}{2}H_{2(g)}$ ($\Delta G_{diss} \approx 0$), and a proximity of the subsurface state (H_{ss}) to the strongly bonded H_{ads} state might explain why H electroabsorption already occurs at positive $E_{(RHE1)}$ [48–50,65] directly from UPD H, without H_2 evolution. The path is likely to be the following:

$$H^+_{(aq)} + e^-_{(M)} \leftrightarrows H_{ads}(\text{M-hollow}) \rightarrow H_{ss}[M] \rightarrow H_{diss}[M] \quad (2.21)$$

   This possibility of absorption from UPD H in Pd electrodes is consistent with what was called the "direct entry" mechanism, as explained in Section 2.2.2.2.3.

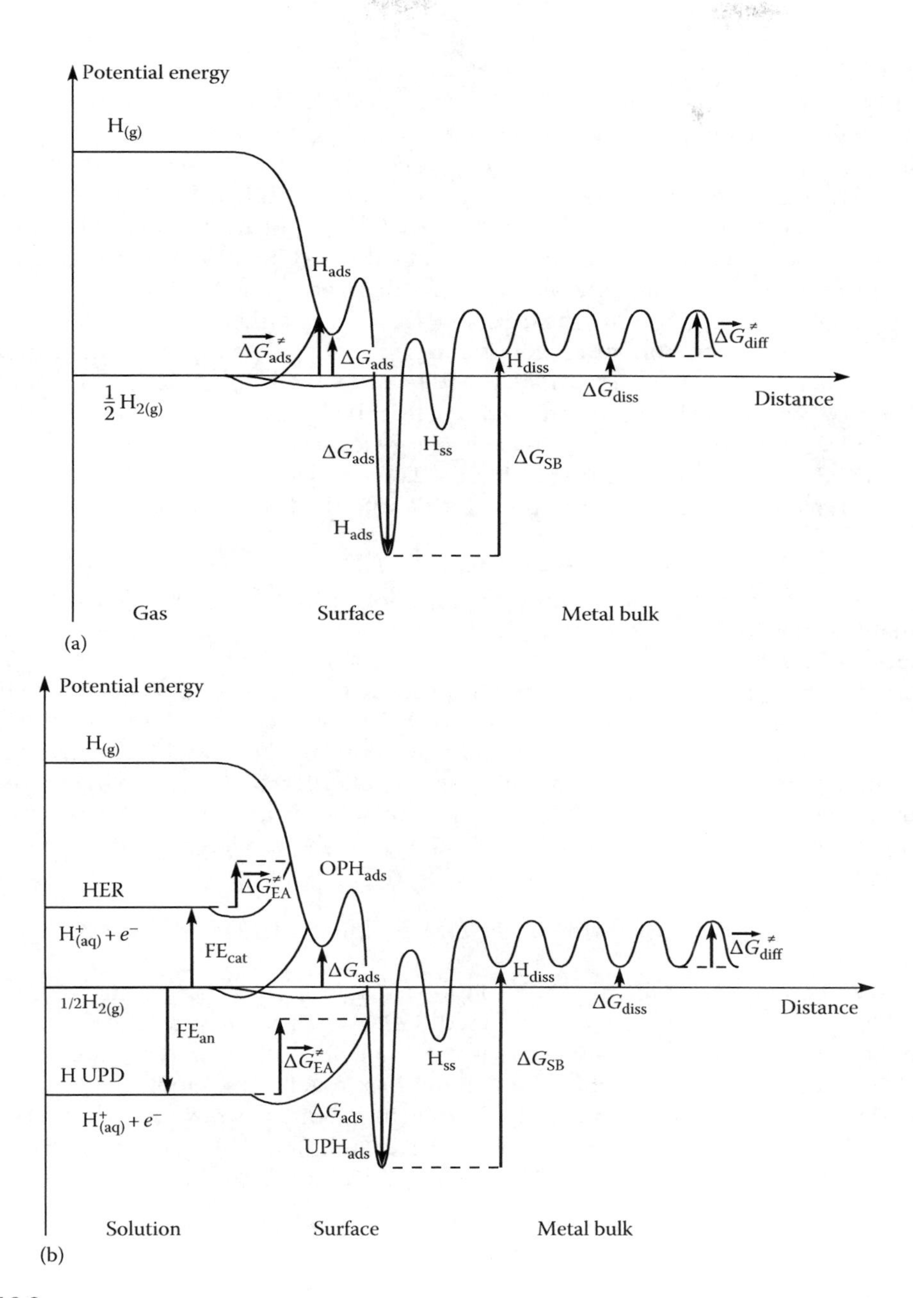

FIGURE 2.8
(a) Schematic diagram of the potential energy versus distance curves for the various H states at the metal–gas interface, namely the two H adsorption states (weakly bonded, corresponding to on-top sites, strongly bonded, corresponding to hollow sites), the subsurface state (H_{ss}), and the bulk dissolved (absorbed) state (H_{diss}). The free energy changes and activation free energies for the different H processes are indicated (SB denotes surface-bulk transfer). (b) Same diagram as (a) for the metal–aqueous solution interface in an electrolyte where competition with adsorption of other species is negligible. The states are the same as for (a) except the specific state ($H^+_{(aq)} + e^-$), initial state of electroadsorption (EA), whose energy level with respect to the $H_{2(g)}$ level depends on the potential referred to RHE1. If E is positive (anodic), electroadsorption occurs in the strongly bonded state, hence denoted underpotential (UP) H_{ads}. If E is negative (cathodic), electroadsorption occurs in the weakly bonded state, hence denoted overpotential (OP) H_{ads}, and H_2 evolution (HER) occurs from this intermediate state.

2. The potential $E_{(RHE1)}$ is negative (cathodic).
This is the range of overpotential (OP) H adsorption. H electroadsorption occurs irreversibly in the strongly bonded sites with negative ΔG_{ads}, so these sites are saturated, and OP H electroadsorption occurs in the weakly bonded sites with positive ΔG_{ads}. The energy level of OP H_{ads} is close or higher than the energy level of $\frac{1}{2}H_{2(g)}$, so H_2 evolution occurs. The energy level of the initial state $\left(H^+_{(aq)} + e^-\right)$ is raised when the potential increases cathodically, so OP H electroadsorption may be in quasiequilibrium only at low cathodic overpotentials and then becomes irreversible, giving rise to the coupled HER mechanisms. Only H_2 evolution by the chemical combination step is described by this simplified diagram, but the principle of the analysis may easily be extended to the electrocombination (Heyrovsky) step. H absorption at overpotential probably also occurs via the saturated strongly bonded H sites, whose energy levels are raised compared to those at low H coverage represented in Figure 2.8b; consequently at overpotential, the activation energy barrier for H penetration into the bulk is lower than shown in Figure 2.8b.

As discussed earlier, such mechanisms involving two adsorbed H species with different bond energies and sites probably apply to all transition metals. It is likely that the sites for the HER adsorbed intermediate (OP H_{ads}) are the weakly bonded on-top sites. But on oxidizable metals, H electroadsorption in the hollow strongly bonded sites followed by absorption is possible only at potentials below the limit of desorption of the oxygen species, calculated for Ni and Fe to be close to $0\,V_{(RHE1)}$ [28]. It is admitted that on iron and ferrous alloys the HER and the HAR share a common H_{ads} intermediate and that the surface-bulk transfer step is in quasiequilibrium [8] (see Section 2.2.2.2.3). The mechanism of H absorption into the bulk of transition metals at cathodic overpotential must then involve a sequence of two consecutive adsorption steps:

$$H^+_{(aq)} + e^-_{(M)} \rightarrow H_{ads}(\text{M-on top}) \leftrightarrows H_{ads}(\text{M-hollow}) \leftrightarrows H_{ss}[M] \leftrightarrows H_{diss}[M] \quad (2.22)$$

The weakly bonded (OP) H_{ads} in the on-top sites simultaneously participates in the HER and the HAR, the strongly bonded (UP) H_{ads} in the hollow sites is involved only in H absorption. Only one adsorption step was considered previously [8,11]. The mechanism proposed here has no implications for the kinetics if the overall surface-bulk transfer step is in quasiequilibrium, but if it is not—a situation occurring, for example, for H permeation in thin membranes [60,89]—the treatment of the kinetics of H absorption will become more complex.

2.5 Kinetics of HER and HAR

2.5.1 Expressions of the Absolute Rates of the Elementary Surface Steps

The absolute rates of all the H elementary surface reactions occurring during the HER and the HAR at the surface of an electrode M in aqueous solution may be expressed as functions of the potential and the Gibbs free energy of H adsorption, using the theory of absolute reaction rates [90] and the Brönsted–Polanyi relations between the activation free energies and the free energies of reactions [91,92]. As usual, a model of regular localized solution is taken for adsorbed H and dissolved H (Frumkin–Fowler model), assuming random distribution of atoms among their respective sites and pair interactions between nearest neighbor atoms [93]. Hence, $\Delta G_{ads(aq)}$ varies linearly with θ_H, according to

$$\Delta G_{ads} = \Delta G_{ads}(0) + g_{HH} RT\,\theta_H \tag{2.23}$$

where
$\Delta G_{ads}(0)$ is the initial adsorption free energy or free energy at zero H coverage
g_{HH} is a pair-energy parameter describing the lateral interactions between adsorbed H atoms [93]

Similarly,

$$\Delta G_{diss} = \Delta G_{diss}(0) + h_{HH} RT\,X_H \tag{2.24}$$

where
$\Delta G_{diss}(0)$ is the H dissolution free energy at infinite dilution
h_{HH} is a parameter for interactions between dissolved H atoms [93]

2.5.1.1 Electroadsorption and Electrodesorption

Only adsorption by proton discharge is considered here (Equation 2.4); the expressions of the rates of adsorption from water are similar. The rates for the forward and reverse reactions are

$$\vec{v}_{EA} = k_{EA} \exp \frac{-\beta(\Delta G_{ads(aq)} + FE)}{RT} \frac{\gamma_{H^+}}{\gamma^{\neq}_{EA}} c_{H^+}(1-\theta_H) \tag{2.25}$$

$$\overleftarrow{v}_{EA} = k_{EA} \exp \frac{(1-\beta)(\Delta G_{ads(aq)} + FE)}{RT} \frac{1}{\gamma^{\neq}_{EA}} \theta_H \tag{2.26}$$

with $k_{EA} = k_{EA} \dfrac{kT}{h} \Gamma_s \exp - \dfrac{\Delta G^{\neq}_{EA}}{RT}$ (for definition of the terms used, see at the end of this section).

2.5.1.2 Chemical Combination and Dissociative Adsorption (Equation 2.3)

$$\vec{v}_{CC} = k_{CC} \exp \frac{2\gamma\,\Delta G_{ads(aq)}}{RT} \frac{1}{\gamma^{\neq}_{CC}} \theta_H^2 \tag{2.27}$$

$$\overleftarrow{v}_{CC} = k_{CC} \exp \frac{-2(1-\gamma)\Delta G_{ads(aq)}}{RT} \frac{f_{H_2}}{\gamma^{\neq}_{CC}} (1-\theta_H)^2 \tag{2.28}$$

with $k_{CC} = k_{CC} \dfrac{kT}{h} \dfrac{z_1\Gamma_s}{2} \exp - \dfrac{(\Delta G^{\neq}_{CC} + \gamma \Delta G^0_{sol})}{RT}$.

Note: The reactions of chemical and electrochemical combination (Equations 2.3, 2.6, and 2.7) lead to H_2 dissolved in water, so the rate equations for the reverse adsorption reactions involve the free energy of adsorption from $H_{2(aq)}$ and the activity of $H_{2(aq)}$, a_{H_2}. However, as the reaction of dissolution of $H_{2(g)}$ is in equilibrium, $a_{H_2} = f_{H_2} \exp - \Delta G^0_{sol}/RT$, where $\Delta G^0_{sol} = \mu^0_{H_2(aq)} - \mu^0_{H_2(g)}$. So the expressions for the rates may be rearranged to involve the fugacity f_{H_2} and $\Delta G_{ads(aq)}$, the free energy for H adsorption from $\frac{1}{2}H_{2(g)}$.

2.5.1.3 Electrocombination and Electrodissociation

Only proton plus atom reaction is considered here (Equation 2.6):

$$\vec{v}_{EC} = k_{EC} \exp \frac{\delta(\Delta G_{ads(aq)} - FE)}{RT} \frac{\gamma_{H^+}}{\gamma^{\neq}_{EC}} c_{H^+} \theta_H \quad (2.29)$$

$$\overleftarrow{v}_{EC} = k_{EC} \exp \frac{(1-\delta)(-\Delta G_{ads(aq)} + FE)}{RT} \frac{f_{H2}}{\gamma^{\neq}_{EC}} (1-\theta_H) \quad (2.30)$$

with $k_{EC} = \kappa_{EC} \frac{kT}{h} \Gamma_s \exp - \frac{(\Delta G^{\neq}_{EC} + \delta \Delta G^0_{sol})}{RT}$.

2.5.1.4 Surface-Bulk Transfer

In the absence of detailed energetic characterization of the subsurface state and knowledge of the transition between the two adsorbed H states (see Section 2.4.2.2.), the overall surface-bulk transfer reaction is considered here as an elementary step from one single adsorption state, as is usually done.

$$\vec{v}_{SB} = \vec{k}_{SB} \theta_H (1 - X_H) \quad (2.31)$$

$$\overleftarrow{v}_{SB} = \overleftarrow{k}_{SB} (1 - \theta_H) X_H \quad (2.32)$$

with

$$\vec{k}_{SB} = k_{SB} \exp \frac{\upsilon(\Delta G_{ads(aq)} - \Delta G_{diss})}{RT}$$

$$\overleftarrow{k}_{SB} = k_{SB} \exp \frac{(1-\upsilon)(\Delta G_{diss} - \Delta G_{ads(aq)})}{RT}$$

where

$$k_{SB} = \kappa_{SB} \frac{kT}{h} \Gamma_s \exp - \frac{\Delta G^{\neq}_{SB}}{RT}$$

Definition of the terms used in the preceding relations: κ_{EA}, κ_{CC}, κ_{EC}, κ_{SB} are the transmission factors for the different reactions; Γ_s is the surface density of H adsorption sites (number per unit area); z_1 (in k_{CC}, Equations 2.27 and 2.28) is the number of adsorption sites adjacent to a given site (lateral or surface coordination), so that the surface density of dual sites is equal to $\frac{1}{2} z_1 \Gamma_s$ [90]. The dimension of the rates is given by the product $kT/h\ \Gamma_s$ ($cm^{-2}\ s^{-1}$). Also, $\gamma^{\neq}_{EA}$, $\gamma^{\neq}_{CC}$, $\gamma^{\neq}_{EC}$ are the activity coefficients of the activated complexes for the first three reactions, dependent on c_{H^+} or/and p_{H2}; $\Delta G^{\neq}_{EA}$, $\Delta G^{\neq}_{CC}$, $\Delta G^{\neq}_{EC}$ and $\Delta G^{\neq}_{SB}$ are the intrinsic activation free energies of the respective reactions (activation free energies for $\Delta G = 0$) [91,92]; β, γ, δ, and ν are Brönsted coefficients or energy-barrier symmetry factors ($0 < \beta, \gamma, \delta, \nu < 1$).

The $\Delta G^{\neq}$'s and the symmetry factors are independent of ΔG (i.e., of the potential and the nature of the substrate) in a range of moderate variation of ΔG [92]. Hence k_{EA}, k_{CC}, k_{EC}, and k_{SB} (intrinsic rate constants) are almost constant for a given reaction, provided that Γ_s and z_l, that is, the surface structure, stay unchanged.

2.5.2 Steady-State Equations for the H Cathodic Reactions

The relation between the H atomic fraction in the bulk sites of the first planes beneath the surface and the potential and surface parameters depends on the mechanism of the global cathodic reaction, which is determined by the steady-state conditions. The following analysis is restricted to the current case of overpotential adsorption. In the steady state, the OP H_{ads} surface coverage is constant with time so the net rate of the electroadsorption step is equal to the sum of the net rates of the consecutive parallel steps consuming adsorbed H:

$$\Gamma_s \frac{d\theta_H}{dt} = (\vec{v}_{EA} - \overleftarrow{v}_{EA}) - (\vec{v}_{EC} - \overleftarrow{v}_{EC}) - 2(\vec{v}_{CC} - \overleftarrow{v}_{CC}) - (\vec{v}_{SB} - \overleftarrow{v}_{SB}) = 0 \tag{2.33}$$

The steady overall cathodic current density, i_c, is given by

$$\frac{i_c}{F} = \vec{v}_{EA} - \overleftarrow{v}_{EA} + \vec{v}_{EC} - \overleftarrow{v}_{EC} = 2(\vec{v}_{EC} - \overleftarrow{v}_{EC} + \vec{v}_{CC} - \overleftarrow{v}_{CC}) + (\vec{v}_{SB} - \overleftarrow{v}_{SB}) \tag{2.34}$$

According to the conditions at the surface of the sample, two main cases are distinguished in electrochemical experiments: charging or permeation.

1. In charging experiments, the H concentration under the surface is imposed by galvanostatic or potentiostatic conditions. H penetrates into the bulk until this concentration is attained in the whole sample and the rate of entry in the bulk is zero, i.e., $\vec{v}_{SB} = \overleftarrow{v}_{SB}$. This condition applied to the rate equations (2.34), (2.31–2.32) leads to $i_c = Fv_{HER}$ and to the isotherm expressing the equilibrium of the surface-bulk transfer (Equation 2.12a).
2. In permeation experiments, a concentration gradient in the bulk is imposed by different conditions at the two faces of a membrane. The stationary diffusion (permeation) rate is attained when the H concentration varies linearly through the membrane (see Figure 2.9). The H concentration under the entry surface is imposed by galvanostatic or potentiostatic conditions, and the concentration under the exit surface is usually fixed to zero by potentiostatic anodic conditions [11,94], such that H diffusing out of the membrane is electrochemically oxidized. In this case, the anodic current density i_p provides a measure of the H permeation rate J_d. In the steady state, J_d, proportional to the concentration gradient through the membrane according to the first Fick law, is equal to the net absorption rate at the entry side. We obtain:

$$v_{HAR} = \vec{v}_{SB} - \overleftarrow{v}_{SB} = J_d = \frac{i_p}{F} = \frac{Dc_H}{l} = \frac{Dc_{Hsat}X_H}{l} \tag{2.35}$$

where
 D is the H diffusion coefficient in the bulk
 l is the thickness of the membrane
 c_H is the volumic concentration of H under the entry face
 c_{Hsat} is the same quantity at saturation

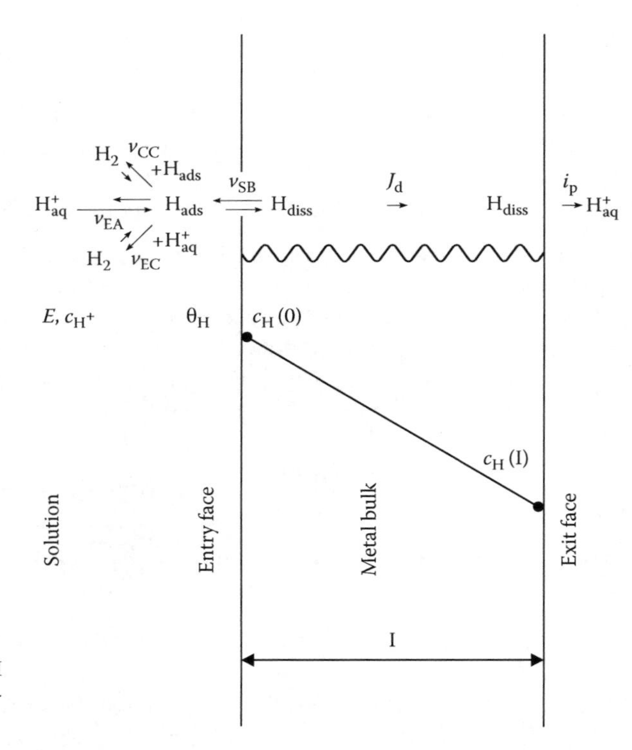

FIGURE 2.9
Balance of reactions during permeation of H through a metal membrane in aqueous solution, and concentration profile in the bulk.

From the steady-state equations (2.33) and (2.35), the variations of the surface and bulk atomic fractions θ_H and X_H are interrelated and depend on the mechanism of the whole cathodic reaction, that is, both the HER and HAR.

Equation 2.35, expressing $\vec{v}_{SB}$ and $\overleftarrow{v}_{SB}$ from Equations 2.31 and 2.32, yields the following expression for X_H versus θ_H:

$$\frac{X_H}{1-X_H} = \frac{\vec{k}_{SB}\theta_H}{[\overleftarrow{k}_{SB}(1-\theta_H)+Dc_{Hsat}/l]} \tag{2.36}$$

In the limiting case where $Dc_{Hsat}/l \ll \overleftarrow{k}_{SB}(1-\theta_H)$ (this corresponds to conditions where D is low, the membrane is thick, θ_H is not close to unity, and $\overleftarrow{k}_{SB}$ is high), the permeation rate is limited by the bulk diffusion rate in the membrane, and the surface-bulk transfer step is in quasiequilibrium ($\vec{v}_{SB} \approx \overleftarrow{v}_{SB}$). It gives, as an approximation, the same expression for X_H versus θ_H as in the charging conditions (Equation 2.12a) except that this equation applies only to the region of the bulk close to the entry surface:

$$\frac{X_H}{1-X_H} \approx \frac{\theta_H}{1-\theta_H}\exp\frac{\Delta G_{ads}-\Delta G_{diss}}{RT} \tag{2.12b}$$

As in this case the permeation rate is negligible with respect to the H_2 evolution rate, as for charging, $(\vec{v}_{SB} - \overleftarrow{v}_{SB})$ may be neglected in Equation 2.33 and then the variation of θ_H with the potential depends only on the HER mechanism and so does the variation of X_H via Equation 2.12b. This is a good approximation for most experiments of permeation through iron, nickel, and their alloys, with relatively thick membranes, where $i_p/i_c < 2/100$

(low permeation rates). In this case, X_H, determined by Equation 2.12b, is thickness independent so, from the last equality of Equation 2.35, i_p is proportional to $1/l$ [11]. It must be noted that this limiting case no longer applies when θ_H approaches unity. As θ_H increases, the complete Equation 2.36 must be considered.

2.5.3 Dependence of the HER and HAR Rates on the Potential and the Adsorption Gibbs Free Energy

Only the usual conditions, charging or permeation limiting case described before are considered here. Application of the steady-state equations (2.33) and (2.12a) or (2.12b), using the rate expressions (2.25) through (2.30), with Equations 2.23 and 2.24 allows one to derive expressions of θ_H and X_H versus the potential and $\Delta G_{ads}(0)$ (adsorption free energy at zero coverage) for each HER mechanism [60,67,68].

2.5.3.1 Analysis of the Dependence on the Potential

Differentiation of the preceding equations yields the partial derivatives of the logarithms of θ_H, i_c the HER current, and X_H (proportional to the stationary permeation current i_p) with respect to the potential for given $\Delta G_{ads}(0)$ and (given metal and surface composition). These derivatives, usually called diagnostic parameters, may be compared with the experimental variations for the particular cases of low H coverage ($\theta_H \ll 1$), midcoverage ($\theta_H = \frac{1}{2}$), or saturation coverage ($\theta_H \approx 1$). The parameters calculated for the six limiting HER mechanisms are reported in Table 2.1 and may be compared with previous calculations [9–11,63] made for particular values of the symmetry factors ($\beta = \delta = \frac{1}{2}$), assuming Langmuir-type adsorption for low coverage and Temkin-type adsorption for medium coverage, with $\gamma = \frac{1}{2}$ (activated H_2 adsorption) or 1 (nonactivated H_2 adsorption), and $X_H \ll 1$. In the simplified form of the Temkin adsorption model used previously, the variations of θ_H, $1 - \theta_H$, $\theta_H/(1 - \theta_H)$ at $\theta_H \approx \frac{1}{2}$ were neglected, whereas the Fowler–Frumkin adsorption model used here is more general. In Table 2.1, it can be checked that at $\theta = \frac{1}{2}$, only for large positive values of g ($g \gg 4$), corresponding to strong repulsive interactions between adsorbed H atoms or high surface heterogeneity, the Tafel slopes are independent of the coverage and hence of the potential and the transfer coefficient α takes the values calculated previously [9–11,63] [$\alpha = -RT/F(\partial \ln i_c/\partial E)$ is related to the Tafel slope b for the HER by the relation $b = -\partial E/\partial \log i_c = RT \ln 10/\alpha F$]. The coefficient $(\partial \ln X_H)/(\partial \ln i_c)$, also reported in Table 2.1, gives the power of i_c to which X_H or the permeation current is proportional. Rigorously, the last two derivatives of Table 2.1 are to be multiplied by the factor $(1 - X_H)/[1 + hX_H(1 - X_H)]$, which represents the limitation of the bulk population due to possible interactions between dissolved H atoms and saturation of the bulk sites. This factor cannot be neglected when θ_H (then X_H) reaches unity, as it drops to zero. However, for weakly absorbing metals, it can be considered as practically equal to unity for moderate coverages ($\theta_H \leq \frac{1}{2}$), where $X_H \ll 1$.

2.5.3.2 Analysis of the Dependence on the Gibbs Free Energy of Adsorption

Similarly, the partial derivatives of the logarithms of θ_H, i_c, and X_H with respect to the free energy of H adsorption at zero coverage, for fixed potential (potentiostatic control) are given in Table 2.2.

From Table 2.2, the variation of the logarithm of the HER current i_c at a given potential with $\Delta G_{ads}(0)$ is a volcano curve for each mechanism, increasing linearly for values of

TABLE 2.1

Dependence of θ_H, i_c, X_H, i_p on the Potential or the Cathodic Current for Various HER Mechanisms

	$-RT/F(\partial \ln \theta_H/\partial E)$		$\alpha = -RT/F(\partial \ln i_c/\partial E)$			$-RT/F(\partial \ln X_H/\partial E) = -RT/F(\partial \ln i_p/\partial E)$		$\partial \ln X_H/\partial \ln i_c = \partial \ln i_p/\partial \ln i_c$	
Mechanism of HER	$\theta_H \ll 1$	$\theta_H = \frac{1}{2}$	$\theta_H \ll 1$	$\theta_H = \frac{1}{2}$	$\theta_H \approx 1$	$\theta_H \ll 1$	$\theta_H = \frac{1}{2}$	$\theta_H \ll 1$	$\theta_H = \frac{1}{2}$
Volmer q.e. Heyrovsky r.d.s.	1	$\frac{2}{4+g}$	$1+\delta$	$\frac{2(1+2\delta+\delta g)}{(4+g)}$	δ	1	1	$\frac{1}{1+\delta}$	$\frac{4+g}{2(1+2\delta+g\delta)}$
Volmer q.e. Tafel r.d.s.	1	$\frac{2}{4+g}$	2	$\frac{2(2+\gamma g)}{(4+g)}$	0	1	1	$\frac{1}{2}$	$\frac{4+g}{2(2+g\gamma)}$
Volmer r.d.s. Heyrovsky q.e.	−1	$\frac{-2}{4+g}$	β	$\frac{2(1+2\beta+\beta g)}{(4+g)}$	$1+\beta$	−1	−1	$\frac{-1}{\beta}$	$\frac{-(4+g)}{2(1+2\beta+g\beta)}$
Volmer r.d.s. Tafel q.e.	0	0	β	β	β	0	0	0	0
Coupled Volmer–Heyrovsky	$\beta-\delta$	$\frac{2(\beta-\delta)}{4+(\beta+\delta)g}$	β	$\frac{2(\beta+\delta+\beta\delta g)}{4+(\beta+\delta)g}$	δ	$\beta-\delta$	$\frac{(\beta-\delta)(4+g)}{4+(\beta+\delta)g}$	$\frac{\beta-\delta}{\beta}$	$\frac{(\beta-\delta)(4+g)}{2(\beta+\delta+g\beta\delta)}$
Coupled Volmer–Tafel	$\frac{\beta}{2}$	$\frac{2\beta}{6+(\beta+2\gamma)g}$	β	$\frac{2\beta(2+\gamma g)}{6+(\beta+2\gamma)g}$	0	$\frac{\beta}{2}$	$\frac{\beta(4+g)}{6+(\beta+2\gamma)g}$	$\frac{1}{2}$	$\frac{4+g}{2(2+g\gamma)}$

Notes: Derivatives expressing the dependency of the stationary H coverage, HER current and H bulk fractional concentration beneath the surface/permeation current, on the potential or on the HER current, at given $\Delta G_{ads}(0)$ and g, for the limiting HER mechanisms, and particular values of the surface coverage q.e. is for quasiequilibrium, r.d.s. for rate-determining step. The Tafel slope for the HER is related to the transfer coefficient α by the relation: $b = -\partial E/\partial \log i_c = RT \ln 10/\alpha F$. Rigorously, the derivatives of $\ln X_H$ are to be multiplied by the factor $(1 - X_H)/[1 + hX_H(1 - X_H)]$.

TABLE 2.2

Dependence of θ_H, i_c, X_H, i_p on the Adsorption Free Energy (at Zero Coverage) for Various HER Mechanisms, at Fixed Potential or Fixed Current

	$-RT(\partial \ln \theta_H/\partial \Delta G_{ads})_E$		$-RT(\partial \ln i_c/\partial \Delta G_{ads})_E$			$-RT(\partial \ln X_H/\partial \Delta G_{ads})_E = -RT(\partial \ln i_p/\partial \Delta G_{ads})_E$		$-RT(\partial \ln X_H/\partial \Delta G_{ads})i_c = -RT(\partial \ln i_p/\partial \Delta G_{ads})i_c$	
Mechanism of HER	$\theta_H \ll 1$	$\theta_H = \frac{1}{2}$	$\theta_H \ll 1$	$\theta_H = \frac{1}{2}$	$\theta_H \approx 1$	$\theta_H \ll 1$	$\theta_H = \frac{1}{2}$	$\theta_H \ll 1$	$\theta_H = \frac{1}{2}$
Volmer q.e. Heyrovsky r.d.s.	1	$\frac{2}{4+g}$	$1-\delta$	$\frac{2(1-2\delta)}{4+g}$	$-\delta$	0	0	$\frac{\delta-1}{1+\delta}$	$\frac{2\delta-1}{1+2\delta+g\delta}$
Volmer q.e. Tafel r.d.s.	1	$\frac{2}{4+g}$	$2(1-\gamma)$	$\frac{4(1-2\gamma)}{4+g}$	-2γ	0	0	$\gamma-1$	$\frac{2(2\gamma-1)}{2+\gamma g}$
Volmer r.d.s. Heyrovsky q.e.	1	$\frac{2}{4+g}$	β	$\frac{2(2\beta-1)}{4+g}$	$-(1-\beta)$	0	0	1	$\frac{1-2\beta}{1+2\beta+g\beta}$
Volmer r.d.s. Tafel q.e.	1	$\frac{2}{4+g}$	β	$\frac{2(2\beta-1)}{4+g}$	$-(1-\beta)$	0	0	0	0
Coupled Volmer–Heyrovsky	$\beta+\delta$	$\frac{2(\beta+\delta)}{4+(\beta+\delta)g}$	β	$\frac{2(\beta-\delta)}{4+(\beta+\delta)g}$	$-\delta$	$\beta+\delta-1$	$\frac{4(\beta+\delta-1)}{4+(\beta+\delta)g}$	$2\delta-1$	$\frac{4\beta\delta-\beta-\delta}{\beta+\delta+g\beta\delta}$
Coupled Volmer–Tafel	$\frac{\beta}{2}+\gamma$	$\frac{2(\beta+2\gamma)}{6+(\beta+2\gamma)g}$	β	$\frac{4(\beta-\gamma)}{6+(\beta+2\gamma)g}$	-2γ	$\frac{\beta}{2}+\gamma-1$	$\frac{4(\beta+2\gamma)-6}{6+(\beta+2\gamma)g}$	$\gamma-1$	$\frac{2(2\gamma-1)}{2+\gamma g}$

$\Delta G_{ads}(0)$ such that $\theta_H \ll 1$, decreasing linearly for values of $\Delta G_{ads}(0)$ such that $\theta_H \approx 1$. In each case, the curve has a maximum for a value of θ_H depending on the values of β, δ, and γ. This result is similar to a classic result obtained for the exchange current densities of the three HER steps [95–98]. From Table 2.2, the maximum of the HER current occurs at $\theta_H = \frac{1}{2}$ if the symmetry factors are equal to $\frac{1}{2}$ or for the two coupled mechanisms if $\beta = \delta$ or $\beta = \gamma$. For high values of g (so-called Temkin case), the current is nearly independent of $\Delta G_{ads}(0)$ in the midcoverage range [96,97].

The steady-state equations for each HER mechanism where one step is quasiequilibrium lead to the adsorption isotherms given in Equations 2.10b and 2.14 through 2.17 hence, since the surface-bulk transfer step is also considered in quasiequilibrium, to a quasiequilibrium between H species in solution and H dissolved in the bulk (Equations 2.18 and 2.19). In this case, there is no dependence of the H bulk concentration, at fixed potential, on the Gibbs energy of adsorption as seen in Table 2.2.

The steady-state conditions for the two HER mechanisms where the electroadsorption step is coupled to one of the subsequent steps lead to pseudoequilibrium adsorption isotherms [11,31,67]. It is seen in Table 2.2 that: for a coupled Volmer–Heyrovsky mechanism, the dependence of X_H on $\Delta G_{ads}(0)$ is slight if $\beta + \delta \approx 1$; only for the coupled Volmer–Tafel mechanism may X_H depend significantly on $\Delta G_{ads}(0)$, unless $\gamma + \beta/2 \approx 1$ for $\theta_H \ll 1$ or $\beta + 2\gamma \approx 3/2$ for $\theta_H \approx \frac{1}{2}$. If $\beta = \gamma = \frac{1}{2}$, X_H is minimum for $\theta_H = \frac{1}{2}$ and maximum for extreme values of the coverage ($\theta_H \ll 1$ or $\theta_H \approx 1$), that is, of the adsorption free energy ($\Delta G_{ads}(0) \gg RT$ or $\Delta G_{ads}(0) \ll -RT$).

Figure 2.10a and b show the theoretical variations of log i_c (here $i_c = i_{HER}$), θ_H, and X_H versus $-\Delta G_{ads}/RT$ for a HER mechanism with the Volmer step in quasiequilibrium and for the coupled Volmer–Tafel mechanism. The symmetry factors are taken equal to $\frac{1}{2}$, and a Langmuir adsorption model is taken, where ΔG_{ads} is not coverage dependent.

In order to analyze experiments in which instead of the potential, the cathodic current is the controlled variable (galvanostatic control), the partial derivatives of ln X_H with respect to $\Delta G_{ads}(0)$ at fixed cathodic current density are also given in Table 2.2. At fixed current density i_c, X_H depends on $\Delta G_{ads}(0)$ for most mechanisms, which may explain some surface effects reported for H permeation in galvanostatic conditions (see Section 2.7.2).

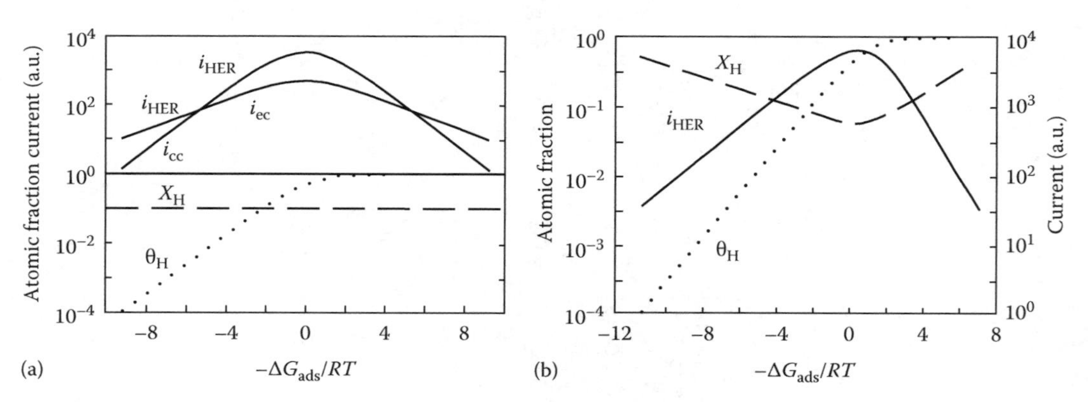

FIGURE 2.10
Theoretical variations at a fixed potential of the HER current, the surface coverage θ_H, and the bulk fractional concentration X_H with the H adsorption free energy ΔG_{ads} for two HER mechanisms: (a) electroadsorption step in quasiequilibrium (a case where the rate-determining step changes from electrocombination to chemical combination); (b) electroadsorption coupled with chemical combination. The symmetry factors are taken equal to $\frac{1}{2}$ and H adsorption is Langmuir-type.

2.6 Influence of the Surface Modifiers on H Entry

2.6.1 Effects of Metal Oxide Films

Perfect oxide surfaces are essentially inert to H_2 dissociative adsorption, although atomic H may be adsorbed on and diffuse relatively fast in stoichiometric oxides [99]. Oxide layers on metal surfaces impede or prevent the H_2 adsorption and H absorption reactions at temperatures below 400°C [100]. Monolayer amounts of oxygen on transition metals reduce the H_2 adsorption and desorption rate by orders of magnitude [99,100]. The effect of oxygen precoverage on transition metals increases with the degree of oxide formation. Thin suboxide layers (in which cations are not in their highest valence state) formed during the initial stages of oxidation or by partial oxide reduction in H_2 impede the H absorption much less than layers of the maximal valence oxides. Transition metal overlayers on the oxide layer restore the original uptake rate of the substrate, which confirms that the effect of oxygen is essentially to impede the H_2 dissociation step [99,100]. Similarly, the effect of passive films formed anodically in solution is to impede H absorption [101,102].

Catalytic effects of low valence Fe–O species have been suggested for explaining H entry enhancement after prolonged cathodic or slightly anodic polarizations in near-neutral solutions [103] or alkaline solutions [104]. Another possible explanation is the formation of protons and acidification at the metal surface associated with oxide reactions, resulting in faster H entry kinetics [105].

2.6.2 Effects of Electronegative Species

As the H cathodic reactions are controlled by surface adsorption processes, they are drastically affected by electronegative species known to poison hydrogenation reactions on transition metals [106]. Electronegative elements are strongly chemisorbed (far more than H) on transition metal surfaces and, even at low concentrations, inhibit coadsorption of reactive species in the gaseous phase [107] as in aqueous medium [60,48,108–110]; they hinder H_2 dissociation in the gas phase [107,111] and the HER in electrolytes, increasing the cathodic overpotential [60,79,106,108–110], even if opposite effects have also been observed, particularly on Fe [112]. Compounds of elements of columns 15 and 16 of the periodic table, P, As, Sb, S, Se, and Te, added to the electrolyte, while being strong inhibitors of H underpotential adsorption, are known to promote H penetration into the bulk of transition metals in aqueous medium [106], particularly in iron and steel [8,36,37,113–121]; palladium [77]; and platinum [122]. Additions of CN^-, I^-, and naphthalene to the electrolyte also increase the permeation rate into iron, although less efficiently, but additions of organic nitriles decrease it; however, all these additives increase the HER overpotential at fixed current density [8]. The surface step has clearly a determining role in these processes. The mechanisms of action of promoters are investigated in detail now.

2.7 Mechanisms of Action of H Absorption Promoters

2.7.1 Conditions for H Absorption Promotion

It has been noted [113] that all the promoters of H entry into iron (and steel) formed from elements of columns 15 and 16 form gaseous hydrides, AsH_3, H_2Se, H_2Te, H_2S, PH_3, and SbH_3 and that the promotion is effective only in the cathodic overpotential region where the hydrides form [113,114,118], and in the range of pH in which they are stable [117]. For example, after addition of a compound of arsenic in the +III oxidation state, H entry inhibition is observed at low overpotentials, attributed to deposition by reduction of elemental arsenic, whereas at higher overpotentials, H uptake increases drastically [37,115,118] and AsH_3 gas is detected [115]. An additional proof is that a significant acceleration of H permeation results if PH_3, H_2S, or AsH_3 produced outside is bubbled into the electrolyte [119].

The paramount role of hydride in promoting H entry into iron and steel has been further confirmed by investigations conducted with a rotating-disk electrode, which also permitted a comparison of the effects of different compounds over a wide range of pH and electrode potentials [63,119]. For a given compound, the critical cathodic overpotential at which the H entry efficiency begins to increase was related to the equilibrium potential of hydride formation from elements (for H_2S the latter is positive versus NHE, which explains why the permeation efficiency increases from the free corrosion potential of steel while for AsH_3, where it is more negative, the permeation efficiency increases at higher overpotentials). Moreover, for S, Se, and Te compounds in moderately acid electrolyte, the permeation rate increases with the cathodic overpotential, reaches a maximum, and then decreases before attainment of the limiting cathodic current density for proton discharge; this was explained by the fact that at high current densities, due to limited proton transport, the activity of protons in the vicinity of the cathode falls to a value giving a local pH equal to the critical value from which the molecular hydrides begin to dissociate into anionic forms. For As and Sb, because the molecular hydrides AsH_3 and SbH_3 are stable in the whole pH range, the permeation rate may increase until the limiting cathodic current density for which the local pH is equal to 7 is reached, after which it decreases because the HER from water molecules is slower [63,119]. These results proved that only the molecular forms of the hydrides are active in promoting H entry [63,117,119].

The elements involved in the promoter species have been classified by various investigators according to their effectiveness in promoting H entry; it seems to depend upon the pH, the conditions of charging, and the method used for measuring the H concentration, which explains why the order in the series varied according to the authors [11,63,116,117]. It has been suggested that the promoter efficiency is related to the strength of the element-hydrogen bond in the hydride [113].

Usually the H concentration for a given polarization increases with the concentration of promoter up to a maximum, for relatively low concentrations, beyond which it decreases [63,117,119]. Figure 2.11 shows the H permeation rate at a given cathodic current density as a function of the concentration of various compounds of the elements discussed here. All the curves show maxima, except the one for Na_2S, where a continuous increase is observed. It is believed that the decrease for concentrations higher than the optimal value is due to the formation of deposits of elemental As, Sb, Se, or Te that block the metal surface for H entry and that there is no such effect with Na_2S solutions, probably because the thermodynamic conditions for S formation are restrictive and even deposited S does not form a blocking film [63,119].

The embrittlement of steels (stress corrosion cracking) in H_2S-containing media is of great industrial importance, so the effect of dissolved H_2S deserves a special analysis. According to most authors [63,120,121], H_2S enhances the HER on iron. This is to be compared with similar effects observed on Fe single crystals precovered by a fraction of a monolayer of adsorbed sulfur [60,112]. A classic explanation is that in acid medium H_2S molecules adsorbed on the cathode catalyze the proton discharge step by being proton transfer centers [63,121,123]. The HAR seems to be promoted only in the low pH region

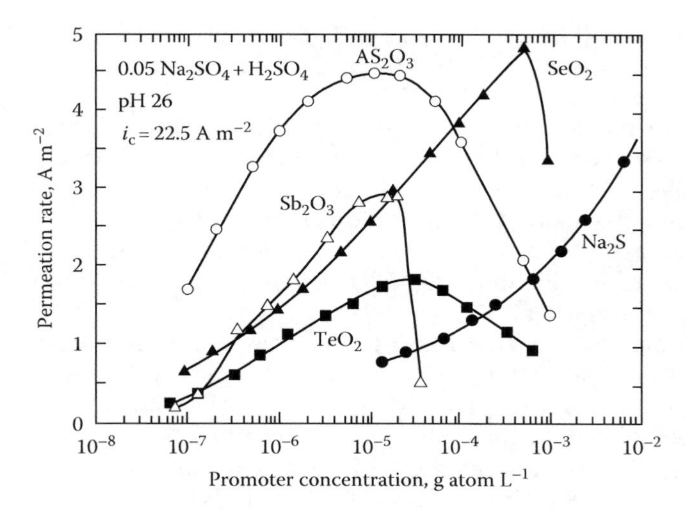

FIGURE 2.11
Effect of promoter concentration on hydrogen permeation rate through steel membranes. (From Zakroczymski, T., in *Hydrogen Degradation of Ferrous Alloys*, R.A. Oriani, J.P. Hirth, and M. Smialowski, eds., Noyes Publications, Park Ridge, NJ, 1985, chapter 11, pp. 215–250, Figure 11-11.)

where H_2S is dissolved molecularly; the rate of H permeation through iron or steel increases with H_2S concentration, beginning from very low values [63].

The effect of dissolved H_2S on the HER and hydrogen permeation on/in passivated and anodically depassivated surfaces of steel was analyzed [120], assuming that the HER occurs by the Volmer–Tafel path only. It was concluded that H_2S plays a multiple role: it increases the rates of iron dissolution and proton discharge, probably by roughening the metal and removing the passivation layer. It then poisons the chemical combination reaction on the depassivated surface, thereby permitting a large fraction of the H atoms to enter the metal [120].

2.7.2 Proposed Mechanisms of Action of Promoters

From the many suggestions made in the past to explain the mechanism of action of promoters of H entry into ferrous metals, the more realistic ones are summarized here. It was often considered that the promoter is adsorbed on the surface, although for elements of columns 15 and 16 it was not clearly stated whether it is in hydride or elemental form [117,124].

1. The promoter increases the strength of the M–H_{ads} bond, thus decreasing the rate of H–H combination and increasing the surface H coverage, which thereby increases the permeation rate [125].
2. The promoter lowers the M–H_{ads} bond energy (weakens the M–H_{ads} bond). This decreases the height of the energy barrier for H surface-bulk transfer [8,61,117].
3. The promoter inhibits the recombination step, so that it increases the H entry rate [36,77,116,120].
4. The promoter is adsorbed in the sites of H_{ads}, lowering the probability of finding a pair of H adsorbed at contiguous sites, necessary for H_2 formation; it thus poisons the H chemical combination step, so a high concentration of atomic H builds up on the surface, which increases the rate of H entry into the metal [121,124,126].

Explanation 1 is valid only for a Vomer–Tafel path with θ_H close to unity (see Figure 2.10b). Furthermore, the promoters are known to be poisons for adsorption of H in the gas phase [107], adsorption of UPD H on Pt [44,45,108–110], and also adsorption of the HER intermediate (OPD H) [79,109,110]. They lower the H adsorption capacity of the surface and are likely to decrease the M–H_{ads} bond strength [44,45,108].

Explanation 2 was proposed in Ref. [8] for a coupled Volmer–Tafel path and is only valid at fixed current density, assumed to be due mainly to the combination rate (see Table 2.2). However, a lowering of the rate constant for surface-bulk transfer cannot by itself explain why H permeation is increased under potentiostatic conditions [122]. At fixed potential, the weakening of the M–H_{ads} bond, associated with a strong reduction of the number of surface adsorption sites, should lead to a reduction of the H coverage, which would probably be the determining factor, decreasing H entry.

Explanation 3, rather phenomenological, is often given for an increase of permeation current under galvanostatic conditions, for a HER mechanism assumed to occur only by the Volmer–Tafel path. In this case, the constant cathodic current is proportional to the electroadsorption rate, which is equal to the sum of the chemical combination rate plus the permeation rate (see Equation 2.34). If the promoter inhibits the chemical combination step, the permeation rate is increased [36,120]. This trivial explanation fails to explain promoter effects occurring under potentiostatic conditions.

Explanation 4 is valid if the reaction consuming H_{ads} is mainly the chemical combination: in addition to assuming the Volmer–Tafel path for the HER, the permeation rate must be negligible to allow an increase of the H coverage (see Equation 2.33). This classic explanation is apparently not self-consistent because it proposes that a high H coverage may appear whereas the H sites are poisoned. However, a modeling of the promoter effects (presented in Section 2.7.4.2) shows that the relative coverage in the sites left active may well increase while, due to the blocking effect, the overall coverage decreases [60,67,68].

There are two recent significant contributions to the understanding of the mechanism of action of promoters on H permeation into iron:

A research group [121], using the rather rough assumption that the mechanism of HER on iron is only coupled Volmer–Tafel but taking the permeation current into account in the establishment of the steady-state equations (so-called IPZ model), analyzed its data on iron in acid solutions containing H_2S and concluded that H_2S increases the proton discharge rate constant and lowers the combination rate constant, leading to a coupled Volmer–Tafel mechanism at low overpotentials, a decreased overvoltage, an increased H coverage, and an increased permeation rate [121]. These conclusions are essentially similar to explanation 4. The model has been refined later by introducing the poison coverage beside the H coverage (so-called IPZA analysis) [127].

The IPZ model was questioned in a subsequent permeation study [37], in which the effects of dissolved species of As on iron and mild-steel were analyzed by accurate fitting of the experimental current–potential curves with the complete steady-state equations and taking into account the reduction of adsorbed As into AsH_3 at cathodic potentials (yet a simplifying choice of parameters could not be avoided). It was demonstrated that the mechanism of HER cannot be or become coupled Volmer–Tafel, except in a very short range of current (and potential). It was observed that the H entry was inhibited a low overpotentials, which was attributed to deposition of bulk As, and that promoting effects of As occurred only at relatively high overpotentials, where As is likely to desorb into AsH_3. From the fit, the authors deduced that the H absorption promotion is associated with an increase of the H coverage. They suggested that the strongly adsorbed poison disturbs the bonding between substrate atoms so that the desorption of As into

AsH_3 provides new active sites for H adsorption, leading to larger H coverage and hence enhanced H entry [37]. In a following paper, the effects of dissolved HS^- were also analyzed, considering the reduction of adsorbed S into HS^- [128]. This reduction occurred at lower overpotentials than with As species and also lead to a permeation enhancement, but less marked and associated with a H coverage lower than on the clean metal. For the two poison species, the common factor for the promotion of H absorption was analyzed as being a substantial reduction of the rate of H bulk-surface transfer (called desorption in the paper) by the chemisorbed poison [128].

2.7.3 Description of the Adsorption Blocking Effect

The mechanisms of action of the promoters of H absorption may be rationalized in terms of their ability to poison adsorption reactions. In the gas phase, the use of UHV surface techniques, correlated with theoretical treatments, has allowed a detailed understanding of the mechanisms of poisoning of catalytic reactions by adsorbed electronegative elements of columns 14–17, called surface modifiers [107]. In aqueous solution, the effects of adsorbed sulfur on the HER on single-crystal surfaces of Ni, Fe, Ag, and Pt [60 and references therein] and the H underpotential adsorption on single-crystal Pt surfaces [44,45] were investigated, as well as the effects of the addition of compounds of S and As to the electrolyte on the HER or HAR [37,109,110,122]. Correlating the results of these studies with the knowledge gained in the gas phase, the following behavior of the surface modifiers in solution can be summarized [60,68]: these species are strongly chemisorbed in the atomic form on surfaces of transition metals in the domains of H UPD and HER at low overpotential (i.e., from 1.0 V to at least −0.1 V/RHE1 for S adsorbed on Pt); in this range their coverage is almost independent of potential, that is, they are irreversibly adsorbed, as compared with H (for example, the $M–S_{ads}$ bond energy is 464 and 414 kJ mol^{-1} for Ni and Fe, respectively, whereas the $M–H_{ads}$ bond energy is only 265 and 270 kJ mol^{-1} [13]. At high cathodic overpotentials, they are reduced to molecular hydrides [26,27,37,109,110]. They have two main effects on H adsorption, arising from the electronic properties of the strong metal–modifier bond and the hydrogen–modifier lateral interactions [107]. The first is a short-range blocking effect: an adsorbed modifier atom induces a weakening of the $M–H_{ads}$ bond into neighboring sites, so drastic that adsorption of H atoms is prevented in these sites; the range of this effect depends on the modifier and on the nature and surface orientation of the metal substrate; this effect diminishes the underpotential H adsorption capacity of a Pt surface, leading to total inhibition for a surface covered by a complete modifier monolayer [44,45]. The second effect is a longer range effect on the energy of the $M–H_{ads}$ bond in the sites left active (not blocked), evidenced by a slight change of the Gibbs energy of adsorption [44,45]. It was demonstrated that for sulfur chemisorbed on Pt, the latter effect may be a weakening of the metal–UPD H bond and a strengthening of the metal–OPD H bond [60].

These data show that the electronegative elements and their derived compounds adsorbed on a surface are characterized by irreversible chemisorption in a large domain of potential and a strong blocking of H adsorption, while they may impede or promote the HER and the HAR. Consequently, because they may be poisons for one reaction and promoters for another, instead of calling such modifier elements poisons or promoters, it is more logical to denominate these species referring to their common property, hence to call them adsorption site blockers (previously referred to as site blocking elements [60,68,108b] or site blocking species [48]).

2.7.4 Modeling of the Effects of Adsorption Site Blockers on the HER and HAR

2.7.4.1 Bases of the Model

The effects of adsorption site blockers on the rates of the surface reactions involving adsorbed hydrogen may be modeled simply by taking into account simultaneously but separately the blocking effect that reduces the number of sites for H adsorption and the effect on the M–H_{ads} bond in the sites not blocked, without any a priori assumption on the direction of the latter effect.

On a surface irreversibly covered by an adsorption site blocker (ASB) on a significant range of potential, the H coverage at a given potential E, θ_H, defined as the ratio of the density of H sites occupied by adsorbed H atoms to the total density of H sites existing on the clean surface (Γ_s), can be expressed as

$$\theta_H(E) = \theta_{Hsat}\, \tau_H(E) \tag{2.37}$$

where

θ_{Hsat} is the saturation coverage of H coadsorbed with ASB, equal to the fraction of H adsorption sites left active (not blocked by ASB), which depends only on the ASB coverage

$\tau_H(E)$ is the "local" coverage in the active sites, equal to the fraction of these sites that are occupied by H atoms, which depends on the potential and on the energetics of H adsorption in these sites

The fraction of sites available for H adsorption, $(1 - \theta_H)$ on a clean surface, is replaced on a surface covered by coadsorbed ASB by $\theta_{Hsat}(1 - \tau_H)$. $\theta_{Hsat}\,\tau_H$ and $\theta_{Hsat}(1-\tau_H)$ replace the coverage terms in the expressions for the rates of the reactions requiring one H surface site, that is, H electroadsorption, electrocombination, surface-bulk transfer and the reverse reactions (Equations 2.25, 2.26, 2.29 through 2.32). So the separation of the global coverage θ_H into two variables allows clearly distinguishing the parameter that is sensitive to the ASB blocking effect on the neighboring sites (θ_{Hsat}) from the parameter sensitive to the ASB effect on the energetics of H adsorption in the sites remaining active (τ_H).

The combination and dissociative adsorption steps require two H surface sites in nearest neighbor position (dual site), so the rates of these reactions are proportional to the density of pairs of nearest neighbor H sites occupied by H atoms, $\frac{1}{2}z_l\Gamma_s\theta_H^2$, or unoccupied, $\frac{1}{2}z_l\Gamma_s(1-\theta_H^2)$ (Equations 2.27 and 2.28). The fraction of pairs occupied by H atoms, equal to the probability of finding two nearest neighbor sites occupied by H atoms, θ_H^2 on a clean surface, is replaced for a surface covered with coadsorbed ASB by $p\tau_H^2$, where p is the fraction of pairs of nearest neighbor sites left active (not blocked by ASB). The fraction of pairs available for H_2 adsorption, equal to the probability of finding two nearest neighbor sites available for H, $(1 - \theta_H)^2$ on a clean surface, is replaced for a surface covered with coadsorbed ASB by $p(1-\tau_H)^2$. The term p is the equivalent of θ_{Hsat} for a dual site reaction and depends only on the ASB coverage.

The variations of θ_{Hsat} and p with the ASB coverage, which reflect the strength of the site blocking effect, are interrelated [60]. They depend on (a) the structure and the orientation of the substrate plane, (b) the nature of the adsorption sites for H and ASB, (c) the blocking range of the ASB, that is, the number of neighboring H sites blocked by one isolated ASB adatom, and (d) the degree of order in the ASB overlayer. These variations have been modeled for low-index single-crystal faces of fcc metals [60]. In the particular case in which one ASB adatom blocks only one H site (geometric blocking effect), $\theta_{Hsat} = (1 - \theta_{ASB})$ and $p = (1 - \theta_{ASB})^2$ for random adsorption of ASB or $(1 - 2\theta_{ASB})$ if the ASB layer is ordered [60].

The confusion between the overall coverage θ_H and the local coverage τ_H leads to frequent misunderstanding of ASB effects and to apparent discrepancies between authors, some explaining H entry promotion by an increase of the coverage of the HER intermediate OPD H [121,124,125], whereas others report a decrease of the H coverage [108–110]. Whereas on the clean surface θ_H is equal to τH, when ASB is adsorbed these two quantities become different. Some techniques of characterization of the OPD H coverage yield the overall coverage $\theta_H = \theta_{Hsat}\,\tau_H$, which decreases when θ_{ASB} increases because of the preponderant effect of decrease of θ_{Hsat}; other techniques yield the local coverage τ_H, which may a priori decrease or increase [60,68]. Integration of H adsorption pseudocapacity versus potential curves for different ASB coverages, measured at underpotential using cyclic voltammetry [44,45] or at overpotential using transient techniques [109,110,129], yields θ_H, which corresponds to the capacity of the surface to adsorb H for a given θ_{ASB}. The local coverage τ_H can then be obtained by normalizing θ_H with respect to the saturation coverage θ_{Hsat} [44,45]. Alternatively, analysis of the dependence of the HER and permeation currents on the potential (via the Tafel $\partial E/\partial \log i_c$ and $\partial E/\partial \log i_p$ slopes) for a cathode irreversibly covered by ASB gives information on the potential-dependent part of the OPD H coverage, which is τ_H [60,68]. In some cases, τ_H has been shown to increase [45,60], indicating a strengthening of the M–H_{ads} bond in the sites not blocked; in such cases, the ASB blocking effect and energetic effect in the sites left active were antagonistic. The blocking effect was preponderant, so the resulting effect was a decrease of the overall coverage $\theta_H = \theta_{Hsat}\,\tau_H$ [45,60,68]. In consequence, the suggestion made by some authors for explaining the action of H entry promoters (see Section 2.7.2) of an increase of the (in fact local) H coverage caused by so-called poisons [121,124,125] is not inconsistent with measurements of diminution of the (overall) H coverage, contrary to what was claimed elsewhere [108].

2.7.4.2 Predictions of the Effects of Adsorption Site Blockers on H Entry

On the basis of this model, it is possible to predict the ASB effects on the bulk H concentration [60,67,68], assuming that there is a common H_{ads} intermediate for the HER and the HAR and that in the steady state the surface-bulk transfer step is in equilibrium (H charging) or quasiequilibrium (H permeation limited by bulk diffusion).

The rates of all the steps in the forward and reverse directions may be expressed as functions of θ_{Hsat}, p, τ_H, X_H, ΔG_{ads}, E, and the steady-state equations (Equations 2.12a,b, 2.33, and 2.34) applied to these modified rate equations. The H bulk fractional concentration beneath the surface is expressed by the following quasiequilibrium equation, analogous to Equations 2.12a and b, where the local coverage τ_H replaces the overall coverage θ_H:

$$\frac{X_H}{1-X_H} \approx \frac{\tau_H}{1-\tau_H} \exp\frac{-\Delta G_{diss}(X_H)}{RT} \exp\frac{\Delta G_{ads}(\tau_H)}{RT} \tag{2.38}$$

This equation does not involve θ_{Hsat}, which means that the reduction of the number of active H adsorption sites has not direct influence on X_H and that X_H is sensitive to surface effects only via τ_H the local coverage and ΔG_{ads} the adsorption free energy in the sites not blocked. If the H_{ads} intermediate involved in the HAR is the HER intermediate, that is, OPD H (which should in general be the case), the expression of the ratio $\tau_H/(1 - \tau_H)$ and hence the ASB effects on the bulk H concentration X_H depend upon the mechanism of the HER [60,67,68]. Let us consider the three main cases:

1. One of the steps of the HER is in quasiequilibrium.
   This situation is the aqueous solution equivalent of gas-phase charging from H_2. For example, let us consider the case where the Volmer step is in quasiequilibrium, that is, $\vec{\upsilon}_{EA} \approx \overleftarrow{\upsilon}_{EA}$ (this mechanism is operating on noble metals at low overpotentials). Equating modified Equations 2.25 and 2.26, the following expression is obtained:

$$\frac{\tau_H}{1-\tau_H} \approx \exp-\frac{\Delta G_{ads}}{RT}\exp-\frac{FE}{RT} \tag{2.39}$$

   Combining Equations 2.38 and 2.39 leads to the quasiequilibrium H electroabsorption isotherm, in which there is no surface parameter (same as in Equation 2.19):

$$\frac{X_H}{1-X_H} \approx \exp-\frac{\Delta G_{diss}}{RT}\exp-\frac{FE}{RT} \tag{2.40}$$

   Because both the H electroadsorption and surface-bulk transfer steps are in quasiequilibrium, the HAR is also in quasiequilibrium. X_H is dependent only on the potential and the free energy of H dissolution in the bulk from $\frac{1}{2}H_{2(g)}$. In this case, no H entry promoting or inhibiting effects induced by ASB surface effects can exist in the steady state [60,67,68,122]. This conclusion can be extended to all the HER mechanisms where one step is in quasiequilibrium. This analysis, which can also be performed by a thermodynamic treatment in expressing equalities of the chemical potentials for hydrogen in the different phases, disagrees with the conclusions of a thermodynamic treatment of poison effects proposed by other authors, suggesting that a surface site blocking effect may induce H absorption promotion in this case [108a].
2. The Volmer and the Heyrovsky steps are coupled (coupled electroadsorption–electrocombination).
   This mechanism is operating on most metals at high overpotentials (see Section 2.2.2.2.1). Here $\vec{\upsilon}_{EA} \approx \vec{\upsilon}_{EC}$. Equating modified Equations 2.25 and 2.29, a pseudoequilibrium adsorption isotherm is obtained where θ_{Hsat} is not involved:

$$\frac{\tau_H}{1-\tau_H} \approx \frac{k_{EA}}{k_{EC}}\frac{\gamma_{EC}^{\ddagger}}{\gamma_{EA}^{\ddagger}}\exp-\frac{(\beta+\delta)\Delta G_{ads}}{RT}\exp-\frac{(\beta-\delta)FE}{RT} \tag{2.41}$$

   Combining Equations 2.38 and 2.41 leads to the following expression:

$$\frac{X_H}{1-X_H} \propto \exp-\frac{\Delta G_{diss}}{RT}\exp(1-\beta-\delta)\frac{\Delta G_{ads}}{RT}\exp-\frac{(\beta-\delta)FE}{RT} \tag{2.42}$$

   X_H is not dependent on θ_{Hsat} and, as long as $\beta + \delta$ is close to unity, there should be only a slight dependence of X_H on ΔG_{ads}. Hence weak surface effects are expected.

3. The Volmer and the Tafel steps are coupled (coupled electroadsorption–chemical combination).

   This mechanism has been reported for iron at low cathodic overpotentials [8,34,35]. Here $\vec{v}_{EA} \approx \vec{v}_{CC}$. Equating modified Equations 2.25 and 2.27 leads to a pseudoequilibrium adsorption isotherm:

$$\frac{\tau_H^2}{1-\tau_H} \approx \frac{k_{EA}}{k_{CC}} a_{H^+} \frac{\gamma_{CC}^{\neq}}{\gamma_{EA}^{\neq}} \frac{\theta_{Hsat}}{p} \exp - \frac{(\beta + 2\gamma)\Delta G_{ads}}{RT} \exp - \frac{\beta FE}{RT} \tag{2.43}$$

   Combining Equations 2.38 and 2.43 leads to the following expression for $\tau_H \ll 1$:

$$\frac{X_H}{1-X_H} \propto a_{H^+}^{1/2} \left(\frac{\theta_{Hsat}}{p}\right)^{1/2} \exp - \frac{\Delta G_{diss}}{RT} \exp\left(1 - \frac{\beta}{2} - \gamma\right) \frac{\Delta G_{ads}}{RT} \exp - \frac{\beta FE}{2RT} \tag{2.44}$$

   This equation shows that surface effects have an influence on X_H both via ΔG_{ads} and θ_{Hsat}/p.

The curve in Figure 2.10b, where θ_H is to be replaced by τ_H, shows that for this mechanism, for values of β and γ close to $\frac{1}{2}$, X_H decreases or increases with the M–H_{ads} bond energy, depending on the value of the local coverage τ_H in the sites not blocked; for low values of τ_H (at low overpotentials), an H entry increase results from a weakening of the M–H_{ads} bond in the sites not blocked, as considered previously [8,61,117]. In contrast, in the particular case of nonactivated adsorption from H_2 ($\gamma \approx 1$), the H concentration increases with the M–H_{ads} bond energy at any coverage (see Table 2.2).

The influence of the adsorption blocking effect on X_H via the parameter θ_{Hsat}/p in Equations 2.43 and 2.44 is considered now. Modeling of the blocking effect shows that this ratio always increases with the ASB coverage [60], the magnitude of the increase depending on the ASB blocking range and the degree of order in the ASB overlayer. Figure 2.12 shows the variation of X_H with θ_{ASB} for the lowest increase of θ_{Hsat}/p, when one ASB atom

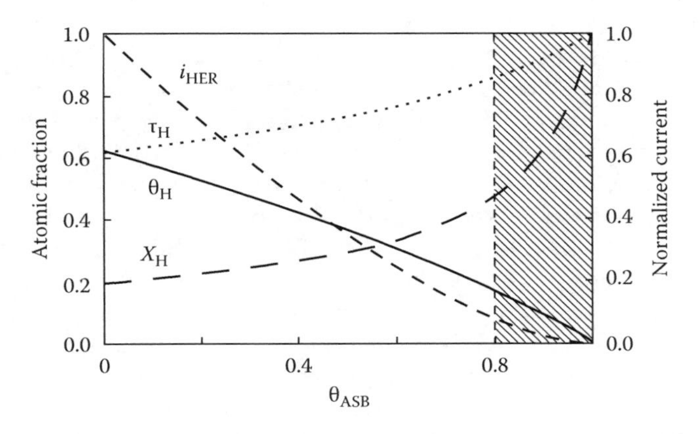

FIGURE 2.12
Theoretical variations, for a coupled electroadsorption–chemical combination HER mechanism, of the overall H coverage θ_H (full line), the local H coverage τ_H (dotted line), the HER current (dashed line), and the bulk H fraction X_H (dashed line) with the coverage of adsorption site blocker (ASB). The represented case is for Langmuir type H adsorption and a "geometric" blocking effect, where one ASB atom blocks one H site, with random ASB adsorption: $\theta_H = (1-\theta_{ASB})\tau_H$; $i_{HER} = k_{CC}(1-\theta_{ASB})^2\tau_H^2$. Above a certain value of θ_{ASB} (arbitrarily taken here), a three-dimensional metal–ASB compound forms.

blocks only one H_{ads} site and the ASB overlayer is disordered, so $\theta_{Hsat}/p = (1 - \theta_{ASB})^{-1}$ [60,68]. More drastic variations are expected in other conditions. It is remarkable that the ASB effect, which reduces the number of H adsorption sites and lowers the H–H combination current, leads to an increase of the local coverage τ_H in the sites left active and from Equation 2.38 to an increase of the bulk concentration X_H, while the overall coverage θ_H is decreased [60,67,68,122]. In this case, H entry promotion is induced by the reduction of the number of active H adsorption sites, as postulated previously [124].

As a conclusion, the promoting effects of adsorption site blockers on H entry can be understood not by considering the effective diminution of θ_H the overall coverage of OPD H (determined by integration of adsorption pseudocapacitance versus potential curves [109,110,129]), but by considering the variation of τ_H the local OPD H coverage in the sites not blocked (obtained by normalization of θ_H or from the Tafel and permeation slopes; see Table 2.1). If the surface-bulk transfer step is in equilibrium or quasiequilibrium, a significant increase of the bulk H concentration beneath the surface can be induced by ASB surface effects only for the HER mechanism where the steps of electroadsorption and chemical combination are coupled. This analysis provides a quantitative explanation of the effects of promotion of H absorption into iron and ferrous alloys.

2.7.5 Additional Suggestions for Explaining Promoting Effects

The preceding analysis provides no explanation for ASB promoting effects on H absorption observed in cases where the HER is not controlled by a coupled Volmer–Tafel mechanism, for example, on noble metals such as Pt and Pd [77,78,108,122]. On these metals, the electroadsorption step in the HER is in quasiequilibrium at low cathodic overpotentials (chemical combination being the rate-determining step); when the overpotential increases, it becomes coupled to the electrocombination step [129]. So H absorption promotion by ASB on these metals can be explained only if the surface-bulk transfer step is not in equilibrium, that is, Equations 2.12a and 2.38 do not apply, at least on the clean surface, and the ASB increases the kinetics of this step [60,68].

A suggestion made to explain ASB effects on Pt was the following [78]: the H_{ads} state strongly bonded in the hollow sites (characterized by a deep energy well in Figure 2.8b), responsible for underpotential adsorption, prevents the surface-bulk transfer through these sites on clean Pt ("UPD H blocks H absorption"); the adsorbed ASB drastically weakens the H adsorption bond in all the hollow sites in the vicinity of ASB atoms, that is, raises the energy level of the corresponding H state in such a way that adsorption in those sites cannot occur at underpotential while the activation energy barrier for the surface-bulk transfer from these sites is lowered. So the ASB would block these H sites for UPD but unblock them for H entry [78]. This model describing the ASB action as the lowering of the activation energy barrier for H entry provides an explanation for detecting H permeation through Pt membranes when the entry side is poisoned for UP H adsorption [78,122].

A similar explanation for the ASB effects involves the subsurface sites. Whereas surface adsorption is inhibited, subsurface adsorption could still occur and even be promoted [130a]. However, no direct experimental evidence exists.

An alternative explanation for the promoter effects exists for nonnoble metals, such as Fe or Ni. The HER rarely proceeds on a bare surface of these metals because, except at high overpotentials, they are covered by corrosion products or passive oxide films [55,102]. Adsorption site blockers such as sulfur block sites for O or OH adsorption and thus impede or inhibit surface oxidation. In the gas phase, it has been shown that at low oxygen pressure and room temperature, Ni oxidation is inhibited when the surface is covered by a complete sulfur monolayer [131]; in aqueous solution, sulfur inhibits the formation of the passive oxide film on Ni and alloys [132]. This may explain why on iron a fraction of a monolayer of sulfur may promote the HER in acid solution [112]. Similarly, by preventing the formation of a passivating oxide layer, which is a barrier to H penetration [101,102], sulfur adsorbed on the surface might promote hydrogen penetration into the metal compared with the oxide-covered metal [130b]. Another possibility consistent with the observations that promoting effects are related to promoter gaseous hydride formation could be that the hydride (H_2S) reduces the metal oxide film, allowing H entry into the depassivated metal [120].

There is still a lack of experiments on the hydrogen cathodic reactions with well-controlled surface coverages by hydrogen, oxygen, and adsorption blockers, characterizing particularly the influence of the blockers on the occupancy of the subsurface sites. Such studies are needed to fully understand the surface processes involved in H-induced cracking of metals and alloys.

References

1. R. A. Oriani, J. P. Hirth, and M. Smialowski, eds., *Hydrogen Degradation of Ferrous Alloys*, Noyes Publications, Park Ridge, NJ, 1985.
2. G. C. Smith, *Hydrogen in Metals*, American Society of Metals, Materials Park, OH, 1963.
3. W. M. Mueller, J. P. Blackledge, and G. Libowitz, eds., *Metal Hydrides*, Academic Press, New York, 1968.
4. G. Alefeld and J. Volkl, eds., *Hydrogen in Metals*, Parts I and II, Springer-Verlag, New York, 1978.
5. L. Schlapbach, ed., *Hydrogen in Intermetallic Compounds*, (a) Part I, Springer-Verlag, New York, 1988; (b) Part II, Springer-Verlag, New York, 1992.
6. Y. Fukai, *The Metal-Hydrogen System*, Springer-Verlag, New York, 1993.
7. H. Wipf, ed., *Hydrogen in Metals*, Part III, Springer-Verlag, Berlin, Germany, 1997.
8. J. O'M. Bockris, J. McBreen, and L. Nanis, *J. Electrochem. Soc. 112*:1025 (1965).
9. J. McBreen and M. A. Genshaw, *Proceedings of the First International Symposium on Corrosion*, Ohio State University, Columbus, OH, 1967.
10. J. McBreen and M. A. Genshaw, *Proceedings of the First Conference on Fundamental Aspects of Stress Corrosion Cracking* (R. W. Staehle, A. J. Forty, and D. van Rooyen, eds.), NACE, Houston, TX, 1969, p. 51.
11. P. K. Subramanyan, *Comprehensive Treatise on Electrochemistry* (J. O'M. Bockris, B. E. Conway, E. Yeager, and R. E. White, eds.), Vol. 4, Plenum, New York, 1981, Chapter 8, pp. 411–462.
12. C. Wagner, *Z. Phys. 193*:386 (1944); *193*:407 (1944).
13. K. Christmann: (a) *Hydrogen Effects in Catalysis* (Z. Paal and P. G. Menon, eds.), Marcel Dekker, New York, 1988, Chapter 1, pp. 3–55, (b) *Surf. Sci Rep. 9*: 1988; (c) *Electrocatalysis* (J. Lipkowski and P. N. Ross, eds.), Wiley-VCH, New York, 1998, Chapter 4, pp. 1–41.
14. J. F. Lynch and T. B. Flanagan, *J. Phys. Chem. 77*:2628 (1973).
15. W. Eberhardt, F. Greuter, and E. W. Plummer, *Phys. Rev. Lett. 46*:1085 (1981).
16. G. Comsa and R. David, *Surf. Sci. 117*:11 (1982).
17. C. T. Chan and S. G. Louie, *Solid State Commun. 48*:417–420 (1983).
18. R. J. Behm, V. Penka, M. G. Cattania, K. Christmann, and G. Ertl, *J. Chem. Phys. 78*:7486–7490 (1983).
19. T. E. Felter, S. M. Foiles, M. S. Daw, and R. H. Stulen, *Surf. Sci. 171*:L379–386 (1986).
20. J. W. He, D. A. Harrington, K. Griffiths, and P. R. Norton, *Surf. Sci. 198*:413–430 (1988).
21. E. Kirsten, G. Parschau, W. Stocker, and K. H. Rieder, *Surf. Sci. Lett. 231*:L183–188 (1990).
22. D. Hennig, S. Wilke, R. Lober, and M. Methfessel, *Surf. Sci. 287/288*:89–93 (1993).
23. J. A. Konvalinka and J. J. F. Scholten, *J. Catal. 48*:374–385 (1977).
24. G. E. Gdowski, R. H. Stulen, and T. E. Felter, *J. Vac. Sci. Technol. A 5*:1103–1104 (1987).

25. E. Gileadi, *Electrosorption* (E. Gileadi, ed.), Plenum, New York, 1967, Chapter 1.
26. P. Marcus and E. Protopopoff, *J. Electrochem. Soc. 137*:2710 (1990); *J. Electrochem. Soc. 140*:1511 (1993).
27. P. Marcus and E. Protopopoff, *J. Electrochem. Soc. 144*:1586 (1997); *Corros. Sci. 39*:1741 (1997).
28. P. Marcus and E. Protopopoff, *Electrochemical Surface Science of Hydrogen Adsorption and Absorption* (G. Jerkiewicz and P. Marcus, eds.), PV 97-16, The Electrochemical Society, Pennington, NJ, 1997, pp. 211–224.
29. P. N. Ross Jr., *Chemistry and Physics of Solid Surfaces IV* (R. Vanselow and R. Howe, eds.), Springer Series in Chemical Physics, Vol. 20, Springer-Verlag, New York, 1982, Chapter 8.
30. F. Wagner, *Structure of Electrified Interfaces* (J. Lipkowski and P. N. Ross, eds.), VCH, New York, 1993, Chapter 9, pp. 309–400.
31. J. O'M. Bockris and A. K. N. Reddy, *Modern Electrochemistry*, Vol. 2, Plenum, New York, 1970.
32. N. M. Markovic, S. T. Sarraf, H. A. Gasteiger, and P. N. Ross Jr., *J. Chem. Soc. Faraday Trans. 92*:3719–3725 (1996); N. M. Markovic, B. N. Grgur, and P. N. Ross Jr., *J. Phys. Chem. B. 101*:5405 (1997).
33. B. E. Conway and L. Bai, *J. Chem. Soc. Faraday Trans. 181*:1841 (1985).
34. M. A. V. Devanathan and Z. Stachurski, *J. Electrochem. Soc. 111*:619 (1964).
35. W. Beck and P. Fischer, *Corros. Sci. 15*:757 (1975).
36. E. G. Dafft, K. Bohnenkamp, and H. J. Engell, *Corros. Sci. 19*:591–612 (1979).
37. S. Y. Quian, B. E. Conway, and G. Jerkiewicz, *J. Chem. Soc. Faraday Trans. 94*:2945–2954 (1998).
38. B. E. Conway, H. Angerstein-Kozlowska, and F. Ho, *J. Vac. Sci. Technol. 14*:351 (1977); B. E. Conway, H. Angerstein-Kozlowska, and W. B. A. Sharp, *J. Chem. Soc. Faraday Trans. 174*:1373 (1978).
39. A. Eucken and B. Weblus, *Z. Elektrochem. 55*:114 (1951).
40. M. Breiter, *Electrochim. Acta 7*:25 (1962); *Ann. N.Y. Acad. Sci. 101*:709 (1963).
41. F. Will, *J. Electrochem. Soc. 112*:451 (1965).
42. J. Clavilier, *J. Electroanal. Chem. 107*:211 (1980).
43. P. N. Ross Jr., *Surf. Sci. 102*:463 (1981).
44. P. Marcus and E. Protopopoff, *Surf. Sci. 161*:533 (1985).
45. E. Protopopoff and P. Marcus, *Surf. Sci. 169*:L.237 (1986).
46. J. L. Weininger and M. W. Breiter, *J. Electrochem. Soc. 110*:484 (1963); *J. Electrochem. Soc. 111*:707 (1964).
47. E. Protopopoff and P. Marcus, *Electrochim. Acta 51*:408–417 (2005).
48. G. Jerkiewicz, *Prog. Surf. Sci. 57*:137–186 (1998).
49. J. Horkans, *J. Electroanal. Chem. 209*:371 (1986).
50. J. McBreen, *J. Electroanal. Chem. 287*:219 (1990).
51. G. A. Attard and A. Bannister, *J. Electroanal. Chem. 300*:467 (1991).
52. M. Baldauf and D. M. Kolb, *Electrochim. Acta 38*:2145–2153 (1993).
53. C. Gabrielli, P. P. Grand, A. Lasia and H. Perrot, *J. Electrochem. Soc.* 151: A1925–1936; A1937–1942; A1943–1949 (2004).
54. H. Duncan and A. Lasia, *Electrochim. Acta 53*:6845–6850 (2008).
55. D. D. Macdonald, in *Hydrogen Degradation of Ferrous Alloys*, Noyes Publications, Park Ridge, NJ, 1985, p. 97, Chapter 4.
56. M. A. V. Devanathan and M. Selvaratnam, *Trans. Faraday Soc. 56*:1820 (1960).
57. A. Caprani and P. Morel, *J. Appl. Electrochem. 7*:65 (1977).
58. L. I. Krishtalik and B. B. Kuz'menko, *Elektrokhimiya 9*:664 (1973).
59. R. J. Nichols and A. Bewick, *J. Electroanal. Chem. 243*:445 (1988); R. J. Nichols, *Adsorption of Molecules at Metal Electrodes* (P. N. Ross and J. Lipkowski, eds.), VCH, New York, 1992, Chapter 7, pp. 347–389.
60. E. Protopopoff and P. Marcus, *J. Chim. Phys. 88*:1423 (1991).
61. I. A. Bagotskaya, *Zh. Fiz. Khim. 36*:2667 (1962).
62. A. N. Frumkin, *Advances in Electrochemistry and Electrochemical Engineering* (P. Delahay, ed.), Vol. 3, Interscience, New York, 1963, p. 379.
63. T. Zakroczymski, in *Hydrogen Degradation of Ferrous Alloys*, Noyes Publications, Park Ridge, NJ, 1985, pp. 215–250, Chapter 11.
64. G. Zheng, B. N. Popov, R. E. White, *J. Electrochem. Soc. 142*:154–156 (1995).

65. G. Jerkiewicz and A. Zolfaghari, *J. Electrochem. Soc. 143*:1240 (1996).
66. E. Gileadi and B. E. Conway, *Modern Aspects of Electrochemistry* (J. O'M. Bockris and B. E. Conway, eds.), Vol. 3, Butterworths, London, 1964, Chapter 5, p. 384.
67. E. Protopopoff and P. Marcus, *C.R. Acad. Sci. Paris 308 II*:1127–1133 (1989).
68. E. Protopopoff and P. Marcus, *Electrochemistry and Materials Science of Cathodic Hydrogen Absorption and Adsorption* (B. E. Conway and G. Jerkiewicz, eds.), PV 94-21, The Electrochemical Society, Pennington, NJ, 1995, pp. 374–386.
69. I. Toyoshima and G. A. Somorjai, *Catal. Rev. Sci. Eng. 19*:105 (1979).
70. P. Nordlander, S. Holloway, and J. K. Norskov, *Surf. Sci. 136*:59 (1984).
71. F. E. Wagner and G. Wortmann, in *Hydrogen in Metals*, Parts I and II, Springer–Verlag, New York, 1978, p. 131.
72. H. Wipf, in *Hydrogen in Metals*, Part III, Springer–Verlag, Berlin, Germany, 1997, pp. 51–91.
73. B. Siegels and G. G. Libowitz, in *Metal Hydrides*, Academic Press, New York, 1968, p. 631, and reference therein.
74. R. Griessen and T. Riesterer, in *Hydrogen in Intermetallic Compounds*, Part I, Springer–Verlag, New York, 1988, p. 230.
75. R. B. McLellan and W. A. Oates, *Acta Metall. 21*:181 (1973).
76. G. Jerkiewicz and A. Zolfaghari, *J. Phys. Chem. 100*:8454 (1996).
77. M. Enyo, *Comprehensive Treatise of Electrochemistry* (B. E. Conway, J. O'M. Bockris, E. Yeager, S. U. M. Khan, and R. E. White, eds.), Vol. 7, Plenum, New York, 1983, pp. 241–300.
78. E. Protopopoff and P. Marcus, *Electrochemical Surface Science of Hydrogen Adsorption and Absorption* (G. Jerkiewicz and P. Marcus, eds.), PV 97-16, The Electrochemical Society, Pennington, NJ, 1997, pp. 159–179.
79. E. Protopopoff and P. Marcus, *J. Vac. Sci. Technol. A 5*:944 (1987); *J. Electrochem. Soc. 135*:3073 (1988).
80. P. Stonehart and P. N. Ross Jr., *Catal. Rev. Sci. Eng. 12*:1 (1975); D. Ferrier, K. Kinoshita, J. McHardy, and P. Stonehart, *J. Electroanal. Chem. 61*:233 (1975).
81. C. M. Mate, B. E. Bent, and G. A. Somorjai, *Hydrogen Effects in Catalysis* (Z. Paal and P. G. Menon, eds.), Marcel Dekker, New York, 1988, Chapter 2, pp. 57–81.
82. A. M. Baro, H. Ibach, and H. D. Bruchmann, *Surf. Sci. 88*:384 (1979).
83. K. Mortensen, F. Besenbacher, I. Stensgaard, and W. R. Wampler, *Surf. Sci. 205*:433–446 (1988).
84. J. Volkl and G. Alefeld, in *Hydrogen in Metals*, Part I and II, Springer–Verlag, New York, 1978, pp. 321–348.
85. M. I. Baskes, C. F. Melius, and W. D. Wilson, *Hydrogen Effects in Metals* (I. M. Bernstein and A. W. Thompson, eds.), The Metallurgical Society of AIME, Warrendale, PA, 1981; R. P. Messmer and C. L. Briant, in *Hydrogen Degradation of Ferrous Alloys*, Noyes Publications, Park Ridge, NJ, 1985, p. 162, Chapter 7.
86. T. Engel and K. H. Rieder, *Surf. Sci. 109*:140–166 (1981); K. H. Rieder and W. Stocker, *Surf. Sci. 164*:55–84 (1985).
87. M. G. Cattania, V. Penka, R. J. Behm, K. Christmann, and G. Ertl, *Surf. Sci. 126*:382–391 (1983).
88. M. Baumberger, K. H. Rieder, and W. Stocker, *Appl. Phys. A 41*:151 (1986).
89. R. N. Iyer, H. W. Pickering, and M. Zamanzadeh, *J. Electrochem. Soc. 136*:2463–2470 (1989); *Scr. Metall. 22*:911–916 (1988).
90. S. Glasstone, K. J. Laidler, and H. Eyring, *Theory of Rate Processes*, McGraw-Hill, New York, 1941.
91. J. Horiuti and M. Polanyi, *Acta Phys. Union des républiques socialistes soviétiques 2*:505 (1935).
92. L. I. Krishtalik, *Comprehensive Treatise of Electrochemistry*, (B. E. Conway, J. O'M. Bockris, E. Yeager, S. U. M. Khan, and R. E. White, eds.), Vol. 7, Plenum, New York, 1983, pp. 87–172.
93. R. Fowler and E. A. Guggenheim, *Statistical Thermodynamics*, Cambridge University Press, Cambridge, UK, 1952.
94. M. A. V. Devanathan and Z. Stachurski, *Proc. R. Soc. Lond. 270A*:90 (1962).
95. H. Gerischer, *Z. Phys. Chem. N.F. 8*:137 (1956); *Bull. Soc. Chim. Belge 67*:506 (1958).
96. R. Parsons, *Trans. Faraday Soc. 54*:1053–1063 (1958).
97. L. I. Krishtalik, *Advances in Electrochemistry and Electrochemical Engineering* (P. Delahay, ed.), Vol. 7, Interscience, New York, 1970, pp. 283–339.

98. A. J. Appleby, *Comprehensive Treatise of Electrochemistry* (B. E. Conway, J. O'M. Bockris, E. Yeager, S. U. M. Khan, and R. E. White, eds.), Vol. 7, Plenum, New York, 1983, pp. 173–239.
99. L. Schlapbach, in *Hydrogen in Intermetallic Compounds*, Part II, Springer–Verlag, New York, 1992, pp. 14–95 and references therein.
100. E. Fromm, *Z. Phys. Chem. N. F. 147*:61–75 (1986).
101. A. M. Brass, *Ann. Chim. Fr. 14*:273–300 (1989).
102. A. M. Brass and J. Collet-Lacoste, *Hydrogen Transport and Cracking in Metals* (A. Turnbull, ed.) *J. Inst. Mater*. 142 (1995); A. M. Brass, J. Collet-Lacoste, M. Garet, and J. Gonzalez, in: *Rev. Metall. CIT Sci. Gen. Mater. 1998*:197–207; A. M. Brass and J. R. Collet-Lacoste, *Acta Mater. 46*:869–879 (1998).
103. J. Flis and T. Zakroczymski, *Corrosion 48*:530 (1992).
104. I. Flis-Kabulska, T. Zakroczymski, J. Flis, *Electrochim. Acta, 52*:2966–2977 (2007); I. Flis-Kabulska, J. Flis, T. Zakroczymski, *Electrochim. Acta, 52*:7158–7165 (2007).
105. I. Flis-Kabulska, J. Flis, T. Zakroczymski, *Electrochim. Acta 53*:3094–3101 (2008); I. Flis-Kabulska, *Electrochem. Commun. 11*:54 (2009).
106. B. J. Berkowitz, J. J. Burton, C. R. Helms, and R. S. Polizzotti, *Scr. Metall. 10*:871–873 (1976).
107. M. P. Kiskinova, *Surf. Sci. Rep. 8* (1988).
108. (a) B. E. Conway and G. Jerkiewicz, *J. Electroanal. Chem. 357*:47–66 (1993); (b) G. Jerkiewicz, J. J. Borodzinski, W. Chrzanowski, and B. E. Conway, *J. Electrochem. Soc. 142*:3755–3763 (1995).
109. L. Gao and B. E. Conway, *Electrochim. Acta 39*:1681–1693 (1994); L. Gao and B. E. Conway, *J. Electroanal. Chem. 395*:261–271 (1995).
110. J. H. Barber and B. E. Conway, *J. Chem. Soc. Faraday Trans. 92*:3709–3717 (1996).
111. C. H. F. Peden, B. C. Kay, and D. W. Goodman, *Surf. Sci. 175*:215 (1986).
112. P. Marcus, S. Montes, and J. Oudar, *Atomistics of Fracture* (R. M. Latanision and J. Pickens, eds.), Plenum, New York, 1983, p. 909.
113. M. Smialowski, *Hydrogen in Steel*, Addison-Wesley, Reading, MA, 1962.
114. B. Baranowski and M. Smialowski, *Bull. Acad. Pol. Ser. Sci. Chim. 7*:663 (1959); B. Baranowski, *Bull. Acad. Pol. Ser. Sci. Chim. 7*: 887, 891, 897, 907 (1959); B. Baranowski and Z. Szklarska-Smialowska, *Electrochim. Acta 9*:1497 (1964); H. Jarmolowicz and M. Smialowski, *J. Catal. 1*:165 (1962).
115. H. Angerstein-Kozlowska, *Bull. Acad. Pol. Ser. Sci. Chim. 6*:739 (1958); *7*:881 (1959); *8*:49 (1960).
116. T. P. Radhakrishnan and L. L. Shreir, *Electrochim. Acta 11*:1007–1021 (1966).
117. J. F. Newman and L. L. Shreir, *Corros. Sci. 9*:631–641 (1969).
118. M. G. Fontana and R. W. Staehle, *Stress Corrosion Cracking of Metallic Materials*, Part III, *Hydrogen Entry and Embrittlement in Steel*, NTIS AD/A-010 265, National Technical Information Service, Springfield, VA, 1975, p. 5285.
119. T. Zackroczymski, Z. Szklarska-Smialowska, and M. Smialowski, *Werkst. Korros. 26*:617 (1975); E. Lunarska, Z. Szklarska-Smixalowska, and M. Smialowski, *Werkst. Korros. 26*:624 (1975); T. Zackroczymski, Z. Szklarska-Smialowska, and M. Smialowski, *Werkst. Korros. 27*:625 (1976).
120. B. J. Berkowitz and H. H. Horowitz, *J. Electrochem. Soc. 129*:468–474 (1982).
121. R. N. Iyer, I. Takeuchi, M. Zamanzadeh, and H. W. Pickering, *Corrosion 46*:460 (1990).
122. P. Marcus and E. Protopopoff, *Proceedings 4th International Conference on Hydrogen and Materials* (P. Azou and N. Chen, eds.) Beijing, China, May 9–13, 1988, p. 168.
123. A. Kawashima, K. Hashimoto, and S. Shimodaira, *Corrosion 32*:321 (1976).
124. U. R. Evans, *The Corrosion and Oxidation of Metals*, Edward Arnold, London, U.K., 1961, p. 397.
125. K. E. Shuler and K. J. Laidler, *J. Chem. Phys. 17*:212 (1949).
126. N. Amokrane, C. Gabrielli, G. Maurin, and L. Mirkova, *Electrochim. Acta 53*:1962–1971 (2007).
127. F. M. Al-Faqeer, K. G. Weil, and H. W. Pickering, *J. Electrochem. Soc. 150* (5): B211–216 (2003); *Electrochim. Acta 48*:3565–3572 (2003).
128. S. Y. Quian, B. E. Conway, and G. Jerkiewicz, *Phys. Chem. Chem. Phys*. 1:2805–2813 (1999).
129. B. E. Conway and L. Bai, *J. Electroanal. Chem. 198*:149–175 (1986); *Electrochim. Acta 31*:1013 (1986).
130. P. Marcus and J. Oudar, in *Hydrogen Degradation of Ferrous Alloys*, Noyes Publications, Park Ridge, NJ, 1985, (a) p. 57; (b) p. 59, Chapter 3.
131. I. Olefjord and P. Marcus, *Surf. Interface Anal. 4*:23–28 (1982).
132. J. Oudar and P. Marcus, *Appl. Surf. Sci. 3*:48 (1979).

3

Anodic Dissolution

Michel Keddam
Pierre and Marie Curie University

CONTENTS

3.1 Introduction: Anodic Dissolution, General Considerations, and Specific Aspects 150
3.1.1 Anodic Dissolution and Corrosion Phenomena 150
3.1.1.1 Anodic Dissolution as Part of Corrosion 150
3.1.1.2 Circumstances of Anodic Dissolution with Respect to Corrosion 150
3.1.2 Kinetics of Anodic Dissolution 151
3.1.3 The Nature of the Intermediate Surface Species in Anodic Dissolution 151
3.1.3.1 Activated State and Anodic Dissolution 151
3.1.3.2 Surface Species and Kinks: A Dual, Kinetic and Atomistic, Description of Anodic Dissolution 152
3.1.4 Electrochemical Techniques in the Study of Anodic Dissolution 154
3.1.4.1 Classification of Electrochemical Techniques 154
3.1.4.2 Nonsteady-State versus Steady-State Techniques 155
3.1.4.3 Background of Time-/Frequency-Resolved Measurements with an Upstream (Emitter)–Downstream (Collector) Electrode Setup 159
3.1.4.4 Anodic Dissolution of Metals and Inductive Behaviors 161
3.2 Anodic Dissolution of Pure Metals 165
3.2.1 Mechanism of Heterogeneous Reaction: Dissolution in the Active State 165
3.2.1.1 Active Dissolution of Iron in Acidic Solutions 165
3.2.1.2 Role of Anions in the Anodic Dissolution of Iron 173
3.2.1.3 Active Dissolution of Other Metals 174
3.2.1.4 Recent Advances in the Mechanism of Passivation by Downstream Collector Electrode Techniques 175
3.2.2 Dissolution in the Passive State 178
3.2.2.1 Film Relaxation and Dissolution in the Passive State 178
3.2.3 Anodic Dissolution under Mass Transport Control 188
3.2.3.1 Mass Transport Control and Corrosion 188
3.2.3.2 Active Dissolution of Iron under Mass Transport Control 189
3.2.3.3 Anodic Dissolution Controlled by Transport in the Presence of Solid Layers 190
3.3 Anodic Behavior of Alloys 192
3.3.1 The Basic Concepts: A Survey 192
3.3.1.1 Thermodynamics and Rate Constant Approaches 192
3.3.1.2 Simultaneous Dissolution and Associated Formalism 193
3.3.1.3 Selective Dissolution 194

3.3.2 Simultaneous Dissolution: Fe–Cr Alloys 194
3.3.2.1 Reaction Mechanism of the Anodic Dissolution of Fe–Cr........ 195
3.3.2.2 Analytical and Mass Balance Approaches to Dissolution and Passivation of Fe–Cr Alloys........ 197
3.3.3 Selective Dissolution 201
3.3.3.1 Anodic Dissolution of Binary Alloys Studied by Electrochemistry, Solution, and Surface Analysis Techniques........ 201
3.3.3.2 Atomistic Modeling of Selective Dissolution and Related Passivation by Percolation Theory........ 203
References........ 207

3.1 Introduction: Anodic Dissolution, General Considerations, and Specific Aspects

The anodic dissolution of a metal is an electrochemical oxidation of its surface atoms resulting finally in the liberation of cations into the electrolyte. Anodic dissolution is of basic importance not only in corrosion but also in several technologies: negative electrodes of primary and secondary batteries, electromachining, electropolishing, and electrowinning of metals, anodes for electrodeposition and cathodic protection, and so forth.

3.1.1 Anodic Dissolution and Corrosion Phenomena

3.1.1.1 Anodic Dissolution as Part of Corrosion

As far as electrochemically assisted corrosion is concerned, anodic dissolution is only one half of the overall phenomenon because the oxidation current is being exactly counterbalanced by the cathodic component so as to maintain the condition of zero net current at the corroding specimen. In fact, the anodic dissolution of the material plays a central role in corrosion research. The kinetics and, to a greater extent, the mechanism of the cathodic process (oxygen or water or proton reduction in most cases) are taken into consideration far less. The outstanding importance given to the anodic dissolution is easily understood if one considers that it is the direct cause of the material decay. This explains why the potential dependence of the anodic dissolution, known from electrochemical experiments at controlled potential, is considered able to restitute the whole set of corrosion circumstances (rate and morphology) encountered by a material in a wide range of free corroding conditions.

Dissolution mechanisms of iron and iron group metals (Co, Ni) have been covered by some review articles [1–3]. This chapter is aimed primarily at introducing the reader to the present state of knowledge of the electrochemistry of corrosion [4–9]. Emphasis will be put on the decisive contribution of modern electrochemical techniques in the elucidation of these complex reaction paths. The examples are selected among the systems most relevant to the corrosion of widely used materials such as iron (mild steels), iron base alloys, and stainless steels. The anodic dissolution of some other nonferrous metals is also dealt with briefly.

3.1.1.2 Circumstances of Anodic Dissolution with Respect to Corrosion

It is usual to consider that various classes of anodic mechanisms exist depending on the range of potential with respect to the passivity domain. Active dissolution taking place at potentials preceding the passivation on a film-free surface is of major importance for the homogeneous corrosion in weakly oxidizing media such as acidic solutions of stable anions (e.g., sulfuric, perchloric, phosphoric, hydrochloric). Localized corrosion in pits, crevices, cracks, etc. is also assumed to proceed through active dissolution stabilized at passive potentials by ohmic drops and/or local chemistry.

The mechanism of active dissolution will be illustrated by the cases of iron, the metal most extensively investigated in this domain, and iron–chromium alloys. The heterogeneous kinetics side of the mechanism of dissolution is first dealt with. Its coupling with mass transfer is then emphasized because it plays a determining role in the characteristic behavior of this metal during localized corrosion.

Mechanisms of dissolution in the full passive state are dealt with for introducing models currently invoked in passivity breakdown and localized corrosion phenomena. Dissolution at transpassive potentials apparently requires some solubilization of the passive layer often associated with higher oxidation states of the metal (Fe^{3+}, Cr^{6+}, Mo^{6+}, Ni^{3+}, etc.). Transpassive dissolution is essentially relevant to corrosion in strongly oxidizing media mainly for alloys such as austenitic stainless steels. This class of dissolution is also observed in the electrochemical processes initiated by the growth of an oxide layer at the metal surface, hence contributing to smooth dissolution (electropolishing and electromachining).

3.1.2 Kinetics of Anodic Dissolution

Energetics and thermodynamics aspects of corrosion phenomena are dealt with in different chapters of this book, and emphasis of this chapter is on kinetics. The background of anodic dissolution considered as a reversible process can be found in [11–14]. This principle can be questioned even for systems considered as examples of equilibrium in textbooks [15], it allows applying to anodic dissolution the concepts elaborated in the theory of crystallization [16]. The Pourbaix diagrams [17] predict from thermodynamic data the stability domains of dissolved cations (anodic dissolution) and of solid species supposed to be adherent and protective (passivity) in a potential-pH coordinate system. It was early intuited that anodic dissolution proceeds through the formation of chemical bonds between surface atoms and solution species (solvent molecules and ions) and cannot be described by the oversimplified view of vacuum ionization of surface atoms. Most of the works devoted to the mechanism of anodic dissolution are aimed at identifying the species formed at the lattice surface and their kinetic role in the reaction mechanism leading to stable cationic species in the solution as the final products. Advances in the field are largely due to the development and application in corrosion research of steadily improved electrochemical techniques and particularly AC techniques (electrochemical impedance spectroscopy [EIS] and related techniques). Three review articles [5–7] give an updated view of the applications of these techniques in basic and applied corrosion science. Considering their widespread importance in research today, their background is further detailed in the following.

3.1.3 The Nature of the Intermediate Surface Species in Anodic Dissolution

It is now generally accepted that anodic dissolution involves the existence of intermediate surface bonds between the metallic state and the solution species. The nature and kinetic behavior of these entities are inferred from classical transient techniques relating the time or frequency response of the current, or of the potential, to the relaxation of their surface concentrations.

3.1.3.1 Activated State and Anodic Dissolution

Identification of the surface species taking part in anodic dissolution can be tentatively dealt with in the framework of the absolute reaction rate and activated complex theory [18].

A description of the activated state in metal dissolution is central to the understanding of corrosion and passivation. However, the identification of this activated state is difficult. For active metal dissolution, the ionization is a very fast process (characteristic time estimated to be less than 10 μs). Following the chemical relaxation technique introduced by Eigen [19,20] for investigating fast homogeneous reactions, so-called scrape potential measurements were applied for the determination of the initial potential and of its relaxation time on fresh surfaces exposed to aqueous solution [21].

This approach, referring to an initially film-free surface, is likely to be particularly relevant to the types of corrosion involving constant or sequential renewal of the corroding surface: stress corrosion cracking (SCC), abrasion and wear corrosion, etc.

3.1.3.2 Surface Species and Kinks: A Dual, Kinetic and Atomistic, Description of Anodic Dissolution

Mechanisms aimed at explaining the anodic dissolution of metals must account for the following features:

1. Transfer of one or several electrons to the electrode electron sea
2. Dependence of the kinetics on the electrode potential and solution composition
3. Role of the electrode structure, as reflected in the crystallographic orientation, and the nature and surface density of the active sites at the atomistic level

At first glance, a striking similarity exists with heterogeneous kinetics in gas-phase and surface catalysis. But, in fact, the situation is much more intricate for the following reasons:

- Adsorbable species and surface interactions are more numerous in the liquid condensed phase than in a simple gas.
- The surface is continuously modified by the dissolution itself, exposing new atoms to the solution.

This last point must be emphasized. Because of atom removal from the lattice surface, chemical bonds of surface atoms with solution species can be thought either to be translated to a first nearest neighbor atom or to remain attached to the dissolving cation. In fact, each of these two processes is at the origin of one of the two principal mechanisms of Fe dissolution, the so-called catalyzed or catalytic mechanism [18,22] and the two consecutive charge transfer steps mechanism (noncatalyzed or consecutive mechanism) [23].

It is now generally agreed that the preceding conditions 1 and 2 are fulfilled by the formation of surface compounds between the metallic element in various oxidation states and species, essentially anions and solvent molecules, present in the solution. The complex pH and anion dependence of the dissolution rate is assumed to reveal the competitive role of various types of mechanisms [24]. According to Sato [25], the following mechanisms are involved:

1. The aquo-ligand mechanism:

$$M \rightarrow M^{+}_{ads} + e^{-}$$

$$M^{+}_{ads} \rightarrow M^{2+}_{aq} + e^{-}$$

2. The hydroxo-ligand mechanism:

$$M + H_2O \rightarrow MOH_{ads} + H^+ + e^-$$

$$MOH_{ads} \rightarrow MOH^+_{aq} + e^-$$

3. The aniono-ligand mechanism:

$$M + A^- \rightarrow MA_{ads} + e^-$$

$$MA_{ads} \rightarrow MA^+_{aq} + e^-$$

It is also generally accepted that the hydroxo-ligand MOH_{ads} is the precursor in the competitive buildup of the prepassive and passive states. Branching mechanisms such as

$$M + H_2O \rightarrow MOH_{ads} + H^+ + e^-$$

$$MOH_{ads} \rightarrow MOH^+_{aq} + e^- = \text{dissolution step}$$

$$MOH_{ads} + H_2O \rightarrow M(OH)_2 + H^+ + e^- = \text{prepassivation step}$$

are most usually postulated. An exhaustive analysis of the active–passive transition of the iron group metals can be found in Ref. [26].

In fact, even though it is a very common experience that corrosion is deeply influenced by the metallurgical state of the material (e.g., cold working, annealing, grain size and boundaries, twinning), there are only few attempts to account for this in mechanistic studies. The cathodic process is merely dealt with from a crystal growth point of view as a part of the phase transition in the metal–vapor system [27]. In contrast, anodic dissolution is essentially considered in terms of electrochemical kinetics [10]. The contribution to this field is mainly due to Lorenz et al., who claimed in a series of papers [6,28] that the catalytic mechanism of iron dissolution is favored on highly active material, i.e., iron with a high density of crystal imperfections and dislocations produced by cold working. On the contrary, the consecutive mechanism would take place preferentially on relatively inactive, annealed, or single-crystal samples. However, at this time, no clear atomistic view of the structure–mechanism relationship was reached.

An elaborated interpretation of the catalytic role played by FeOH in the catalytic mechanism was more recently proposed by Heusler et al. [1,2,29–31]. According to these authors, FeOH is associated with a kink or half-crystal atom on a dissolving edge of terrace (or ledge). If one atom is removed from such a position, the active site for dissolution is simply transferred onto the next neighboring atom on the edge. In other words, FeOH is self-regenerated and can be regarded as a catalyst, a single entity being able to dissolve a very large numbers of atoms. Figure 3.1 displays, in a schematic way on a two-dimensional (2D) lattice, the shift of the active site for dissolution from one atom to the next one. Annihilation of kinks takes place by collision of two kinks of opposite sign, and generation of new kinks occurs at corner positions, intersections of steps.

In the case of alloy dissolution, where further complexity arises from selective dissolution, computer simulations seem to be a promising approach [32].

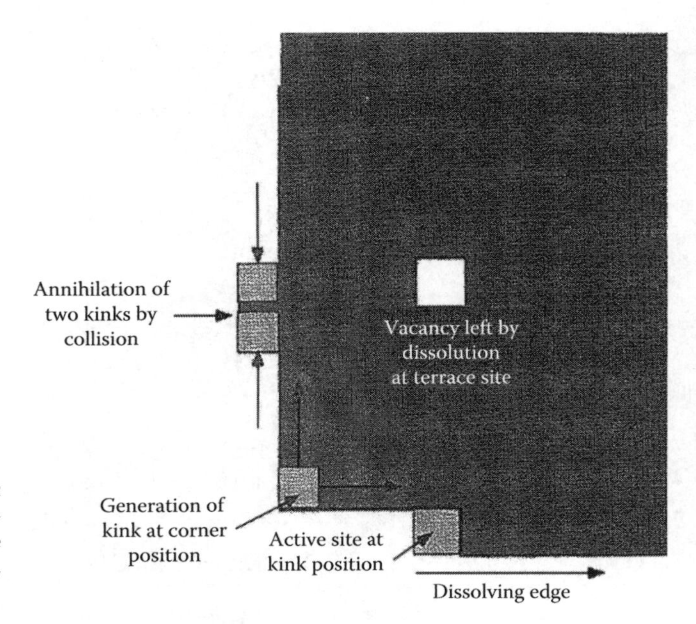

FIGURE 3.1
Definition of atomic crystallographic sites of significance in anodic dissolution. Schematic on a square 2D lattice kinks at steps (edges or ledges), corners, and terraces positions.

3.1.4 Electrochemical Techniques in the Study of Anodic Dissolution

The anodic dissolution of metal has long been and remains an important field of electrochemistry at solid electrodes. Advances were tightly conditioned by the introduction of new experimental techniques.

3.1.4.1 Classification of Electrochemical Techniques

They can be split into two main groups, each of them including steady-state and non-steady-state techniques.

3.1.4.1.1 Current–Voltage Techniques

The electrochemical behavior of the electrode is defined by its current (potential) response to a potential (current) excitation under potentiostatic (galvanostatic) control of the polarization. Some rare kinetics may require specially designed regulating devices [33,34]. Steady-state or voltage sweep current–voltage curves and EIS are the main techniques. To this group also belong related techniques based on a more sophisticated processing of the same raw data (e.g., coulometry, determination of dissolution valencies). The application of Faraday's law to anodic dissolution is of primary significance. In principle, it yields the valence of dissolution, i.e., an integer number. However, there are a number of well-documented examples of metal exhibiting the so-called negative difference effect, i.e., a noninteger electrovalence lower than the oxidation state of the metal in the final cationic species in the solution [35]. Three possible causes can be considered:

1. Chunk effect
2. Dissolution with lower valence and subsequent chemical oxidation of the cation
3. Mixed potential effect where cathodic reduction of the solution coexists with the dissolution

3.1.4.1.2 Nonexclusively Current–Voltage Techniques

In this class of techniques, one handles an extra experimental parameter (either excitation or response) in order to obtain additional information relevant to nonstrictly electrical aspects of the dissolution phenomena.

To this group belong the techniques based on the rotating disk electrode (RDE) [36]. They allow control of the rate of mass transport to and from the dissolving anode to determine the degree of diffusion control in the overall process. A closely related technique of primary interest for anodic dissolution is the rotating ring-disk electrode (RRDE) in which the flux of soluble cations resulting from the dissolution of the RDE is collected through a proper redox reaction on a concentric outer ring electrode [37]. An AC extension is most usefully associated with simultaneous classical impedance measurements of the disk [38]. It is also worth pointing out the promising contribution to anodic dissolution studies of the electrochemical quartz crystal microbalance (EQCM), which yields in situ gravimetric information with a fair time or frequency resolution and a mass sensitivity far better than one monolayer of oxygen (50 ng cm^{-2}) [39].

3.1.4.2 Nonsteady-State versus Steady-State Techniques

3.1.4.2.1 Historical Trend

The development of the electronic potentiostat from the early 1950s was at the origin of a new approach in the investigation of the anodic behavior of metals and alloys, particularly those exhibiting an active–passive transition not amenable to galvanostatic techniques [4]. It soon appeared that reliable mechanistic information cannot be drawn from steady-state techniques alone. Voltage sweep (potentiodynamic plots) was applied, but the sweeping rate was never introduced as a significant experimental parameter because the classical theories of cyclic voltammetry are not valid for multistep heterogeneous reactions with adsorbed intermediates. Even a single one-electron charge transfer coupled to adsorption requires a heavy numerical computation in order to obtain the voltammetric response [40].

In the case of iron dissolution, it was observed early that the time dependence of the faradic current following a potential step is more complex than that expected for simple charge transfer kinetics and may be, in some way, related to the mechanism of reaction [41]. Approximately at the time, similar kinetic situation in the hydrogen evolution reaction was being dealt with by Gerischer and Mehl [42] in terms of faradic impedance. For some time, both transient (time domain) and impedance (frequency domain) techniques have coexisted in anodic dissolution studies. At the end of the 1960s and during the 1970s, decisive advances were realized in the devices for complex impedance measurements down to the millihertz domain and in their interfacing with electrochemical cells. The technique became perfectly suitable for investigating in the frequency domain the long-range time responses of the dissolution-passivation kinetics [43–48].

3.1.4.2.2 Fundamentals of Nonsteady-State Techniques

The basic idea behind these techniques is to extract information about the reaction mechanism and the rate constants from the time dependence of some measurable property of the electrochemical interface in response to a perturbation of its steady state. This relaxation is determined by the nonsteady-state solutions of the kinetic equations. Each of the state-determining kinetic parameters is assumed to contribute by its own time of relaxation.

3.1.4.2.3 Mathematics of Nonsteady State

The time dependence of the system is controlled by a set of nonlinear partial derivative equations, expressing the mass balance of chemical species under the influence of reaction and transport phenomena. The concentrations C_i of bulk species and C_{Si} of surface species obey the kinetic equation

$$\frac{\partial C_i}{\partial t} = -\text{div}\,\psi_i + S_i(C_j) \tag{3.1}$$

where

ψ_i is the local flux of species i under the influence of the gradient of electrochemical potential

S_i is a formal source term (positive or negative) due to homogeneous chemical reactions

Similarly, mass balance due to heterogeneous reactions is given by an ordinary differential equation such as

$$\frac{dC_{Si}}{dt} = S_{Si}(E, C_j, C_{s_j}) \tag{3.2}$$

where

S_{S_i} is also a formal source term due to heterogeneous reactions

E is the applied electrode potential

Boundary conditions at the electrode surface are given by continuity equations between the flux ψ_i and the rate of reaction of species i expressed by surface kinetics laws.

The instantaneous intensity of the faradic current is given by

$$\frac{I}{F} = \Phi(E, C_j, C_{Sj}) \tag{3.3}$$

where C_j and C_{S_j} are relative to those species reacting with electrons.

Integration of the set of Equations 3.1 and 3.2, with the proper initial and boundary conditions and the $E(t)$ function selected as a perturbation, yields the relaxation behavior of the interface in terms of current response.

Most generally, this problem has no analytical solution and must be solved numerically unless the linearized form of these equations is used. A linearization procedure is allowed by using a small-amplitude perturbation $\Delta E(t)$ so as to neglect the nonlinear terms (degree higher than one) in the Taylor expansion of S_i, S_{Si} and ψ_i around the mean steady state of the system. It is known from linear system theory that under these conditions the response $\Delta I(t)$ is proportional to the perturbation $\Delta E(t)$. The dynamic behavior of the electrode at this particular polarization point is completely described by its complex impedance $Z(j\omega) = \Delta E(j\omega)/\Delta I(j\omega)$ in the frequency domain where $\Delta E(j\omega)$ and $I(J\omega)$ are the Fourier transforms of $\Delta E(t)$ and $\Delta I(t)$.

The same approach can be generalized to any other pair of quantities for which equations such as Equations 3.1 through 3.3 can be established. It must be emphasized that all these transmittances are derived from the same model at only the expense of introducing a few specific parameters such as the molecular weight of surface compounds in the case of gravimetry. Therefore, the confrontation of the model with experiments is much more meaningful than applying techniques restricted to the current–voltage relationship.

3.1.4.2.4 A Fast Clue to the Origin and Interpretation of Impedance

The background of impedance derivation is available in several monographs and review articles [7,8,49] and will not be detailed here. Only a shortcut to a qualitative understanding is given here, allowing one in principle to relate the main shapes of the impedance plots to basic features of the electrode processes.

This so-called faradic impedance Z_F basically displays the frequency response of the elementary phenomena participating in the electrochemical transfer of charges across the interface.

The general structure of the frequency dependence of Z_F stems readily from Equation 3.3 in its differentiated form:

$$\frac{\Delta I_F}{F} = \frac{\partial \Phi}{\partial E}\Delta E + \frac{\partial \Phi}{\partial C_{Sj}}\Delta C_{Sj} + \frac{\partial \Phi}{\partial C_j}\Delta C_j \tag{3.4}$$

Each of these partial derivatives gives rise to a specific contribution to the faradaic impedance. They can be split into two groups:

- The partial derivative with respect to E (at constant C_j and C_{Sj}) is assumed to represent the instantaneous (on our timescale) response of the charge transfer process. A pure resistance is therefore associated with this term: the charge transfer resistance denoted by R_t.
- The partial derivative with respect to C_j and C_{Sj} introduced through the delayed responses of C_j and C_{Sj} the frequency dependence of Z_F.

The simplest situation is found when the bulk concentrations are kept constant by an appropriate stirring device (usually RDE), hence $\Delta C_j = 0$. The ΔC_{Sj} are given by the solution of the set of ordinary linear first-order differential equations obtained by linearization of Equation 3.2 under a sine wave potential perturbation $\Delta E = |\Delta E| \exp j\omega t$. Resolution by the Kramers method immediately shows that $\Delta C_j/\Delta E$ is expressed by a rational function of the imaginary angular frequency $j\omega$:

$$\frac{\Delta C_{Sj}}{\Delta E} = \frac{N_j(j\omega)}{D(j\omega)}$$

where the degree of D is equal to the number k of concentrations C_{Sj} involved in the kinetics; the degree of N is $k - 1$. More intricate situations are found when the bulk concentrations C_j also contribute to the relaxation. Integration of the linearized form of Equation 3.1 with proper boundary conditions at the electrode surface, superscript (o), yields the complex ratio $\Delta C_j/\Delta \phi_j$ as a function of the mass transfer parameters.

Diffusion across a Nernst boundary layer (thickness δ, diffusivity D) produces the well-known convective-diffusion term:

$$\frac{\Delta C_j^\circ}{\Delta \phi_j} = -\frac{\delta}{D}\frac{\tanh\sqrt{(j\omega\delta^2/D)}}{\sqrt{(j\omega\delta^2/D)}} \tag{3.5}$$

Finally, Z_F is given by

$$\frac{1}{Z_F} = \frac{1}{R_t} + F\sum_j \frac{\partial \phi}{\partial C_{Sj}} \frac{\Delta C_{Sj}}{\Delta E} + F\sum_j \frac{\partial \phi}{\partial C_j^\circ} \frac{\Delta C_j^\circ}{\Delta \phi_j} \frac{\Delta \phi_j}{\Delta E} \tag{3.6}$$

where the charge transfer resistance R_t is the limit of Z_F for $\omega \Rightarrow \infty$.

As illustrated in this chapter, Equation 3.6, in parallel with the double-layer capacitance $C_{dl,}$ generates identifiable shapes on the impedance curves in the Bode or Nyquist plane making possible to determine the number of chemical entities C_{Sj} and C_j participating in the reaction mechanism and thus providing information on the reaction pattern. In terms of dissolution–passivation processes, capacitive responses and negative resistances are related to inhibition or passivation whereas inductive behaviors arise from catalytic effects or activating intermediates [4–8]. Acquisition and processing of the transient response of electrochemical systems are easily performed by modern laboratory equipments [5,6,49] and do not deserve special attention in this chapter.

3.1.4.2.5 Local Electrochemical Measurements

Practically any real-life solid electrode exhibits, for structural and/or geometric reasons, heterogeneities in surface properties and therefore in reactivity. The characteristic dimension may range between the nanometric scale and the macroscopic size of the electrode. The traditional electrochemical measurements provide surface-average quantities, in both current and potential. The information can be extremely biased, in some cases totally obscured, when sharp differences in reactivity are present. This is often the case for anodic dissolution under the influence of metallurgy and/or composition. In order to overcome the problem, techniques have been introduced for collecting local values of the potential and current densities at short distances above the electrode surface. Potential probes have long been known for attenuating the ohmic drop (Haber–Luggin capillaries). Current probes have been developed intensively in the last 15 years. They are based on the measurements of the ohmic drop along a short current path in the solution of known resistivity. Most of the studies used the so-called scanning vibrating electrode technique (SVET) technology or to a lesser extent twin electrodes. Current mapping thus obtained, with a spatial resolution at best about 15 μm, is essentially applied for imaging galvanic currents associated with local cells in corrosion. The techniques were extended to transient regimes and spatially resolved impedance measurements [local electrochemical impedance spectroscopy (LEIS)] [50–54] are now available with valuable performance. Scanning electrochemical microscopy (SECM) allowing to probe the local electrochemistry and chemistry, in either feedback or generation–collection mode, is also increasingly applied to spatially resolved investigations of dissolution and corrosion. Recent progresses including AC applications of the techniques can be found in Ref. [55].

Undoubtedly, further advances in the interpretation of the kinetics of anodic dissolution are expected in the near future from the growing application of these techniques.

Complex transmittances relevant to corrosion phenomena have been introduced [36,38,39]. Examples are given in several parts of the chapter. Two of them deserve a particular interest. The case of techniques pertaining to the RRDE is dealt with here. Electrogravimetric transmittance, a frequency-resolved technique based on the EQCM, was presented in a paper in 1996 [56]. More recently, a new transmittance, relative to the double layer capacity response, the modulation of interfacial capacitance transfer function (MICTF), has been introduced. It allows gaining information on the chemical conformation of the interface, linked for instance to the nature and concentration of intermediate species, through its coupling with modifications of the double layer [57].

3.1.4.3 Background of Time-/Frequency-Resolved Measurements with an Upstream (Emitter)–Downstream (Collector) Electrode Setup

These techniques are based on the discrimination between the dissolution and the surface film growth components of the working electrode current [58]. The following derivation makes used of the RRDE parameters indexed D (disk) and R (ring). The transposition to an upstream–downstream pair of electrodes in any kind of flow cell (e.g., channel flow double electrode [CFDE]) is straightforward.

3.1.4.3.1 Electrical Charge Balance

In the nonsteady-state regime the electrical charge balance equation at the disk surface is

$$I_{\mathrm{D}} = n_{\mathrm{d}} F \Phi_{\mathrm{d}} + \frac{\mathrm{d}Q_{\mathrm{S}}}{\mathrm{d}t} \tag{3.7}$$

where

I_{D} is the electrical current supplied to the disk (at the exclusion of the double-layer charging)
n_{d} is the dissolution valence of the disk metal
Φ_{d} is the flux of n_{d}-valent cations released in the solution
Q_{S} is the faradic charge stored in surface layers

Equation 3.7 can be extended to the emission of more than one species with different oxidation states: $n_{\mathrm{d}}, m_{\mathrm{d}}, \ldots$

$$I_{\mathrm{D}} = F(n_{\mathrm{d}}\Phi_{\mathrm{d}}^{n} + m_{\mathrm{d}}\Phi_{\mathrm{d}}^{m})\frac{\mathrm{d}Q_{\mathrm{S}}}{\mathrm{d}t} \tag{3.8}$$

3.1.4.3.2 The Collection Efficiencies

A fraction of the flux Φ_{d} is collected and converted into an electrochemical current I_{R} on a downstream electrode owing to the transfer of n_{R} electrons in a convenient redox reaction. At steady state, the dimensionless ratio

$$N = \frac{I_{\mathrm{R}}}{n_{\mathrm{R}}\Phi_{\mathrm{d}}} \tag{3.9}$$

called the *collection efficiency* is determined entirely by the electrode layout. At the nonsteady state, Φ_{d} is obtained:

In the time domain, through a deconvolution because one has

$$I_{\mathrm{R}}(t) = n_{\mathrm{R}} \int \Phi_{\mathrm{d}}(t) N(t-\tau)\mathrm{d}\tau \tag{3.10}$$

where $N(t)$ is the characteristic *transient collection efficiency* of the device experimentally defined for $\Phi_d(t)$ = Dirac function $\delta(t)$. Unlike N, it depends on both the electrode layout and the hydrodynamic parameters (fluid velocity, viscosity, diffusivity).

In the frequency domain, through a simple division by the *complex collection efficiency* $N(\omega)$, which yields

$$\Phi_d(\omega) = \frac{I_R(\omega)}{N(\omega)n_R} \tag{3.11}$$

The collection efficiencies are most easily determined by calibration experiments on redox systems. In practice, the deconvolution of (3.10) is carried out by fast Fourier transform (FFT) using the classical equation $N(\omega)$ = Fourier transform of $N(t)$ and

$$N(0) = \int_{-\infty}^{+\infty} N(t)\mathrm{d}t = N \tag{3.12}$$

3.1.4.3.3 Emission Efficiency and True Capacitance

Exploitation in the time domain of the differential (Equation 3.7) or of the integrated form $Q_d = n_d F P_d + Q_S$, where P_d is the amount of species generated by the disk dissolution, yields as the main information the charge Q_S involved in the surface layer. In the frequency domain, if Δ stands for small-amplitude variables, Equation 3.7 becomes

$$\Delta I_D = n_d F \Delta\Phi_d + j\omega\Delta Q_S \tag{3.13}$$

or

$$1 = \frac{n_d F \Delta\Phi_d}{\Delta I_D} + \frac{j\omega\Delta Q_S}{\Delta I_D} \tag{3.14}$$

where the ratio $N_D = (n_d F \Delta\Phi_d)/(\Delta I_D) = 1-(j\omega\Delta Q_S)/(\Delta I_D)$ is the emission efficiency of the disk, a complex dimensionless quantity representing the fraction of the disk current actually consumed in dissolution. Graphs of N_d in the complex plane provide a direct criterion about the kinetic role of the surface charge Q_S. For instance $(\Delta Q_S)/(\Delta I_D) < 0$; i.e., charge buildup producing a decrease of the dissolution current (e.g., an electrochemically grown passivating layer) is displayed as a positive imaginary part.

In order to get the true electrode capacitance Γ_s, related exclusively to the film growth, Equation 3.14 can be further coprocessed with the faradic impedance of the disk:

$$Z_{D,F} = \frac{\Delta E_D}{\Delta I_D} \quad \text{according to } \Gamma_s = \frac{\Delta Q_S}{\Delta E_D} = \frac{\Delta Q_S}{\Delta I_D}\frac{1}{Z_{D,F}}$$

The application of the technique in the time and frequency domains is illustrated hereafter for the passivation of iron in acidic solutions.

3.1.4.4 Anodic Dissolution of Metals and Inductive Behaviors

It has long been recognized, originally in the time domain (transient measurements) and later in the frequency domain (impedance measurements), that the non-steady-state response of anodically dissolving metals displays in most cases an inductive-like behavior. With the galvanostatic step used in the earlier work, this behavior showed up as a voltage peak (overshoot), its counterpart in potentiostatic mode being a minimum of current density. In the frequency domain, an inductive loop of the Nyquist plot was first identified by Keddam et al. [43–46] in the case of iron in acidic media and interpreted by the so-called consecutive dissolution mechanism (see later).

3.1.4.4.1 Classical Interpretation

It was known from the previous work by Gerischer and Mehl [42] on hydrogen evolution that a reaction path involving two consecutive steps coupled by an adsorbed intermediate species can generate either an inductive or a capacitive reaction impedance depending on the kinetic parameters. The detailed derivation is given in the next section devoted to iron dissolution. It can be shown according to Equation 3.6 that, for a Langmuir type of adsorption and single-electron transfers, Z_F is given by the equation

$$\frac{1}{Z_F} = \frac{1}{R_t} + \frac{1}{\rho}\frac{1}{[F_j\omega\beta/(J_1 + J_2)] + 1} \tag{3.15}$$

where

- R_t and ρ are resistances depending on the kinetic parameters of the reaction steps
- β is the maximum surface concentration of the adsorbed intermediate
- J_1 and J_2 are the current densities of the first and second steps
- F is the Faraday constant
- Z_F is inductive for parameter values such that ρ is positive; this is shown to occur for $J_1 < J_2$

In this case, Z_F can be split immediately into two parallel branches by identification of Equations 3.15 to 3.16:

$$\frac{1}{Z_F} = \frac{1}{R_t} + \frac{1}{jL\omega + \rho} \tag{3.16}$$

the self-inductance L being given by

$$L = \frac{F\beta\rho}{J_1 + J_2} \tag{3.17}$$

The characteristic frequencies associated are in the admittance plane

$$\omega_0^a = \frac{\rho + R_t}{L} = \frac{(J_1 + J_2)}{F\beta} \tag{3.18}$$

and in the impedance plane

$$\omega_0^z = \frac{\rho + R_t}{L} = \frac{[1 + (R_t/\rho)](J_1 + J_2)}{F\beta} \tag{3.19}$$

Both quantities are increasing functions of the DC current density:

$$J = 2\frac{J_1 J_2}{J_1 + J_2} \tag{3.20}$$

The ratio $F\beta$ is about one monolayer of elementary electrical charges per area unit of the interface. Many experiments were found to agree with this model even though several authors mentioned β values definitely exceeding one monolayer and serious deviations from Equations 3.18 through 3.20. Inductive loops at lower frequencies with little or no current dependence are ignored in this discussion.

3.1.4.4.1.1 Weak Points and New Model This problem was revisited in two papers [59,60]. The first one established from a thorough survey of the literature of metal and semiconductor dissolution that the frequency of the inductive impedance, in contradiction of Equation 3.19, is remarkably proportional to the DC current over several orders of magnitude:

$$\omega_0^z = \frac{J}{F\beta} \tag{3.21}$$

The meaning of the deviation between Equations 3.19 and 3.21 can be perceived by noting that according to Equation 3.18, the frequency is determined by the faster of the two reaction steps wherever Equation 3.21 relates it to the slower one. In the second paper, an original interpretation is proposed. It is based as well on a two-step process between the metal atom at the lattice surface and the cation in solution on the outer side of the double layer. A detailed presentation of this approach is far beyond the scope of this chapter. It will be only outlined briefly. Basically, the relaxation is ascribed to the time dependence of the surface concentration of an intermediate state of dissolution. In that sense, the model ingredients are quite similar to those of the consecutive mechanism, namely, the following:

- The ionization proceeds through a two-step path involving an intermediate state (partially solvated).
- The interface is (statistically) shared between the intermediate state (fractional coverage θ) and the ground metallic state $(1 - \theta)$.
- The model is still compatible with monovalent dissolution ($Ag \Rightarrow Ag^+$, for instance).
- The time dependence is due to the difference between the fluxes of metal species crossing the interface on the inner and outer sides of the intermediate state.

The new ideas lie essentially in the adjustment of the potential profiles in the metal side (jellium) and the electrolyte side of the double layer in order to comply with charge balance in conduction and electrostatic properties at the interface. This is sketched in Figure 3.2. According to Ref. [60], the impedance has a structure similar to Equation 3.15. It can be cast in the form

$$\frac{1}{Z_F} = j\omega\left(\frac{C_1 C_2}{C_1 + C_2}\right) + \frac{1}{R_t} + \frac{j\omega[C_2/(C_1 + C_2)] + (1/F\beta)(\partial J/\partial\theta)}{jL\omega + \rho} \tag{3.22}$$

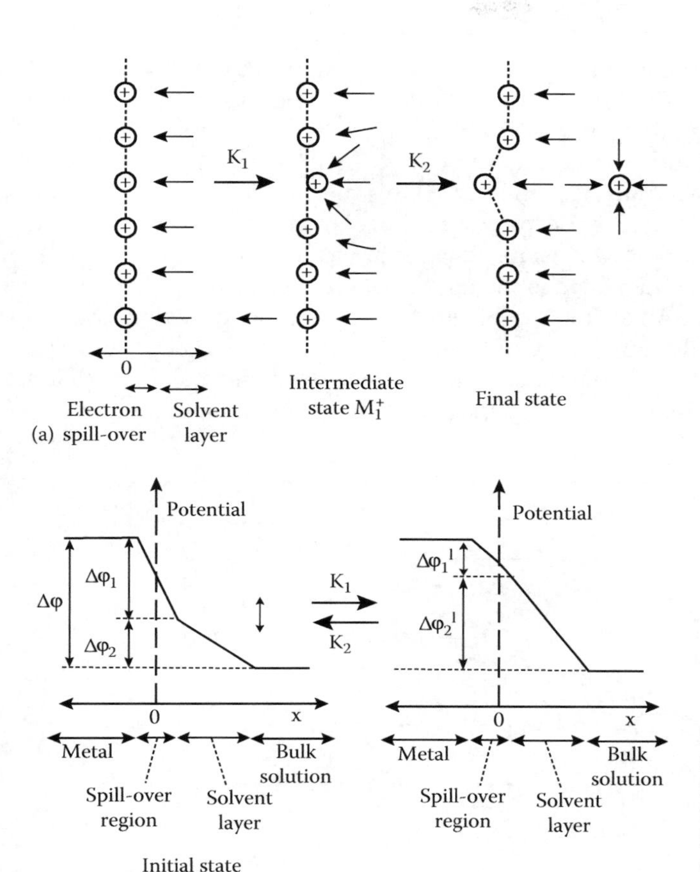

FIGURE 3.2
(a) Sketch of the two consecutive ion-tranfer steps in anodic dissolution. (b) Sketch of the distribution of the interfacial potential drop Δφ across the electrochemical double layer. (From Vanmaekelbergh, D. and Erné, B.H., *J. Electrochem. Soc.*, 1546, 2488, 1999.)

where C_1 and C_2 are the capacitances of the inner and outer parts of the double layer, respectively. With this model, the linear dependence of frequency on DC current can be simulated for acceptable sets of parameters including values of C_1 and C_2 consistent with modern theories of the double layer. However, the ratio of one monolayer is hardly obtained.

Actually, the model introduces a strong coupling between charge transfer and double-layer structure, a feature generally discarded in spite of the high interfacial concentration of cations released by the dissolution. In some way, it may be regarded as a ζ-potential approach to the problem. MICTF data [57] have brought new information in that sense but not validating completely the theory. The topic of the kinetics content of the inductive impedance of anodically dissolving metals is still an open question.

It must be emphasized that the foregoing models, purely kinetic in essence, are unable to account for microscopic aspects of the chemical and geometrical configurations of the interface. In the reaction model of iron dissolution detailed in the next section, a Langmuir isotherm for the reaction intermediate means no interaction between adsorbed species. The surface is also assumed to remain flat in spite of the roughness produced by the random dissolution of the metal lattice. Therefore, the question arose of the contribution of these factors in the time (or frequency) response of the current to a potential change. The problem has been dealt with recently in a series of papers [61–65], based on models of anodic dissolution of flat (2D) and roughened (3D) interfaces using a microscopic approach by cellular automaton and Monte Carlo simulations. Among important features like the prevision of a so-called chunk effect, it is essentially established that in the 3D interface reaction kinetics and morphology are strongly interacting. Taking into account the relaxation of the rough surface generated by the dissolution mechanism leads to a more intricate impedance than the same reaction path operating on a 2D flat surface. It is established that the higher frequency domain reflects, as in the 2D case, the contribution of the surface coverage relaxation by adsorbed intermediates, but now strongly coupled to a fast response of the surface area. The most striking contribution of the surface rearrangement is the emergence of an additional loop at the low frequency end of the diagrams. An example of Nyquist diagram yielded by simulation is shown in Figures 3.3 and 3.4. Finally, it must be emphasized that in the case of electrode processes giving rise to surface roughening, the independent relaxing parameters in the model may be less than the frequency domains shown by the experimental impedance.

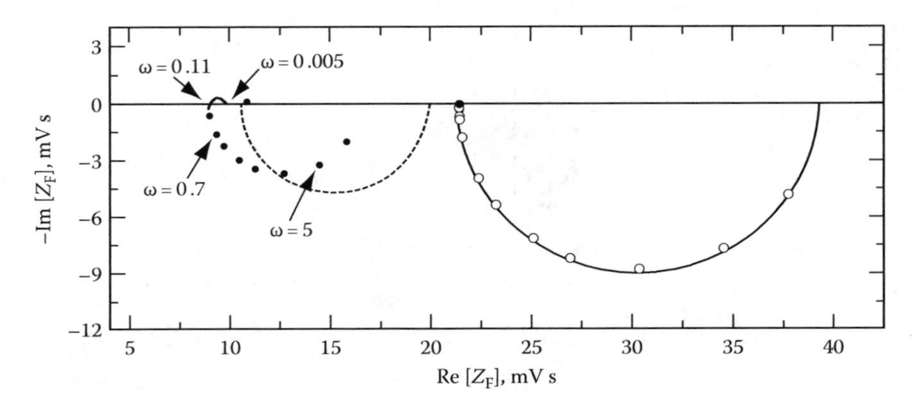

FIGURE 3.3
Nyquist diagrams of the electrochemical impedances corresponding to a two-consecutive step anodic dissolution. Points stand for the result of Monte Carlo simulations without surface relaxation (open circles) and with surface relaxation (solid circles) for *a*. Number of cycles is $n = 20$ and $V = 0.1$ mV. Frequency values range from $\omega = 0.005$ to $\omega = 10\,s^{-1}$. Continuous solid line corresponds to inductive loop behavior of Z_F (ω) predicted by Equation 3.15. Dashed line stands for an analytical approach of the rough case. (From Córdoba-Torres, P., Keddam, M., and Nogueira, R.P., *Electrochim. Acta*, 54, 518, 2008. With permission.)

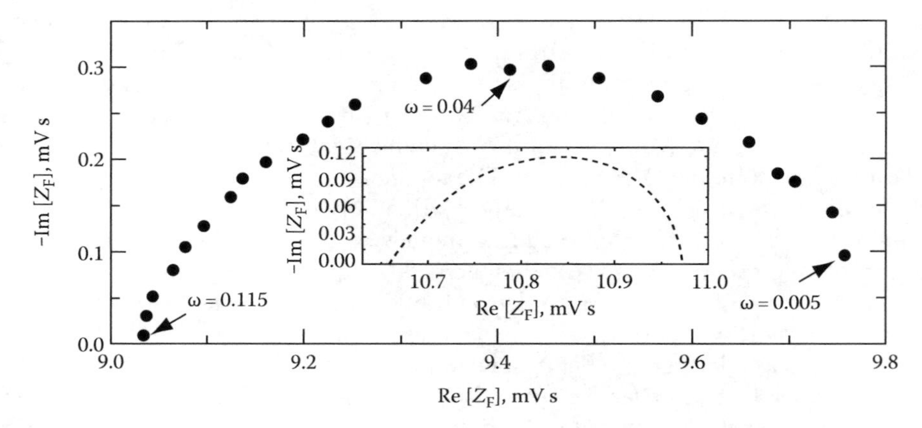

FIGURE 3.4
Details of the low frequency capacitive loop of the rough case displayed in Figure 3.3. Solid points correspond to results from Monte Carlo simulations while dashed line in the inbox stands for its analytical approach.

3.2 Anodic Dissolution of Pure Metals

Even though of little interest for practical corrosion problems, pure metals are extremely basic to any approach to the mechanism of anodic dissolution, which is believed to be much simpler than for alloys. However, the electrochemical behaviors of high-purity metals can be deeply influenced by vanishingly small amounts of impurities able to segregate at the interface during anodic dissolution and also by metallurgical factors. This may explain misleading interpretations and serious discrepancies between materials from different sources with the same purity grade.

3.2.1 Mechanism of Heterogeneous Reaction: Dissolution in the Active State

The case of iron will be taken as an example of heterogeneous reaction in the active range of dissolution. Some other metals are also dealt with briefly.

3.2.1.1 Active Dissolution of Iron in Acidic Solutions

The active dissolution of iron in acidic media has been the subject of a very large number of papers for the last 40 years. All the reaction mechanisms are based on the generally agreed-upon experimental evidence that the dissolution rate increases with the solution pH at potentials well below the onset of passivity processes. The apparently unacceptable participation of hydroxyl ions in the reaction at these low pH values can be related to the strong dissociative power of transition metals with respect to water, an assumption supported for Fe by experimental evidence in ultrahigh vacuum [66]. Both groups of mechanisms stem from a common initial hydrolysis step assumed to be at equilibrium:

$$Fe + H_2O \rightleftarrows (FeOH)_{ads} + H^+ + e^- \quad (3.23)$$

thus accounting for the pH dependence of the subsequent dissolution step. They diverge from each other by the nature of the dissolution steps, leading to the soluble divalent species.

According to the catalytic mechanism [22], $(FeOH)_{ads}$ enters in a catalytic sequence of dissolution at the end of which $(FeOH)_{ads}$ is regenerated and Fe dissolved as $(FeOH^+)$:

$$Fe + (FeOH)_{ads} \rightleftarrows Fe + (FeOH)_{ads} \quad (3.24)$$

$$Fe + (FeOH)_{ads} + OH^- \rightarrow FeOH^+ + (FeOH)_{ads} + 2e^- \quad (3.25)$$

A reluctant criticism of the catalytic mechanism is directed against the quantum mechanically unacceptable transfer of two electrons in a single step. It must be pointed out that on considering even a very crude atomistic content of this two-electron step, it appears less definitely excluded. Interpretation in terms of kink Fe_k shifted to the next neighboring atom, denoted Fe′, in the dissolving edge,

$$Fe'(Fe_kOH)_{ads} + OH^- \rightarrow FeOH^+ + Fe'_k(OH) + 2e^- \quad (3.26)$$

clearly shows that the two electrons are not transferred from the same Fe species.

According to the consecutive mechanism [23], $(FeOH)_{ads}$ enters in a noncatalytic one-electron transfer step by which it is oxidized to $(FeOH^+)$:

$$(FeOH)_{ads} \rightarrow FeOH^+ + e^- \quad (3.27)$$

Finally, $FeOH^+$ is assumed to react with solution protons:

$$FeOH^+ + H^+ \rightarrow Fe^{2+}_{aq} + H_2O \tag{3.28}$$

3.2.1.1.1 Mechanistic Criteria Based on Steady-State and Transient Polarization Data

For a long time, with the exception of the pioneering impedance approach of Epelboin and Keddam [67], the controversy about the validity of these mechanisms remained based on kinetic criteria drawn from true steady-state and fast polarization techniques. Tafel slopes and orders of reaction with respect to OH^- are the two main parameters taken into consideration.

In theory, the transient state is assumed to take place at times short enough to keep the concentration of $(FeOH)_{ads}$ "frozen" at its initial value. Therefore, the initial transition is due only to the change of the reaction rate of step (3.26) or (3.27) under the effect of potential at constant $(FeOH)_{ads}$ concentration. The same discrimination between instantaneous and delayed contributions is at the origin of the frequency dependence of the faradic impedance. Table 3.1 shows the theoretical and experimental values of the steady-state and transient kinetic parameters for both mechanisms, according to Ref. [12].

According to Table 3.1, the delimitation between the two mechanisms relies on a quite clear-cut difference in the set of kinetics criteria and would not involve controversy at all. In fact, a careful analysis of literature data placing the emphasis on experimental conditions leads to the following remarks [68,69].

Tafel slope values are spread over a wide range from less than 30 mV to more than 100 mV depending on ill-identified parameters. The role of metallurgical factors such as dislocations and subgrain boundaries was interpreted as favoring the catalytic mechanism (30 mV), whereas annealed iron with a low density of structural defects would obey the consecutive mechanism (40 mV) [68,69].

In many early studies, iron dissolution was investigated as part of corrosion and inhibition at open-circuit potential under an H_2-saturated atmosphere, no account being taken of the interference of the H_2/H^+ reaction with the Tafel slopes [70,71]. Even under an inert atmosphere, the anodic parameters may be affected by slow hysteresis phenomena resulting in steeper Tafel lines [72–76].

Reaction order measurements are not always very reliable and are sometimes reported with fractional values [12].

Current transients are much too long to be consistent with the minimal exchange current density still compatible with reaction (3.23) being at pseudoequilibrium [77]. Alternative explanations based on local increase of pH have been proposed [72] and further elaborated [77].

TABLE 3.1

Kinetic Data for Iron Dissolution at $T = 298$ K

	Catalytic Mechanism		Consecutive Mechanism	
	Calculated	Experimental	Calculated	Experimental
Steady-state				
Tafel slope, mV	29.6	30 ± 2	39.4	40 ± 2
Reaction order/OH^-	2	2 ± 0.3	1	1 ± 0.1
Transient				
Tafel slope, mV	59.2	60 ± 7	40	60 ± 7
Reaction order/OH^-	1	1 ± 0.1	0.5	00.5 ± 0.1

Source: Hubert, F. et al., *J. Electrochem. Soc.*, 118, 1919, 1971.

From the impedance approach, it was then established that the iron mechanism is far more complex than expected from early experiments and can hardly be investigated correctly by current–voltage plots in spite of continued efforts [2,78]. Thorough analysis over extended ranges of electrode potential, pH, and frequency is presented in the following.

3.2.1.1.2 Atomistic Interpretation of the Catalytic Mechanism

Anodic dissolution of iron is probably the sole case for which an atomistic description was elaborated with a somehow successful issue. An early attempt to explain the transient regime of iron dissolution by relaxation of etched pits by a nucleation and growth model can be found in Ref. [72] and is analyzed in Ref. [67]. Starting in the middle of the 1970s, Heusler et al. have developed in a series of papers [29–31] an experimental and modeling approach aimed at giving a crystallographic basis to the catalytic mechanism. Correlation of kinetic and morphological data was performed on a surface vicinal to {211} (misorientation ≅ 1°) of iron single crystals in acidic solutions. These surfaces are found to develop a relatively simple steady-state morphology made of triangular pyramids limited by nearly perfect {211} planes and ⟨311⟩ edges. A schematic model of steps and kinks projected on a {111} plane is given in Figure 3.5.

Steps are generated at the pyramid apex and move away on the three {211} planes. The structure of the dissolving surface is determined by x_s, the mean distance of steps perpendicular to the ⟨311⟩ direction:

$x_s = bN_P/N_S$ with b = interatomic distance

N_P = probability of generation of six steps at the apex

N_S = probability of kink generation at the intersection of two monatomic steps

and x_k, the mean distance of kinks:

$x_k = bN_k/N_P$

where N_k is the probability of removing an atom from a kink.

The current density j is the product of the current on one kink ze_0N_k by the density of kinks: $j = ze_0N_k/x_sx_k$.

Scanning electron microscopy (SEM) examinations [29] of the macroscopic morphology of dissolution were then substantiated by transmission electron microscopy (TEM) pictures of gold-decorated surfaces [30]. The better resolution, even though not truly atomic, allowed

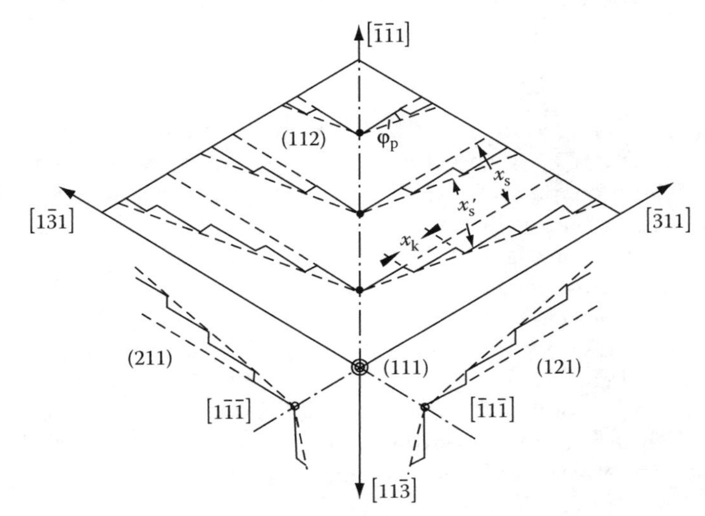

FIGURE 3.5
Schematic model of steps and kinks on crystallographic (211), (112), and (121) planes projected into a (111) plane. (From Allgaier, W. and Heusler, K.E., *J. Appl. Electrochem.*, 9, 155, 1979.)

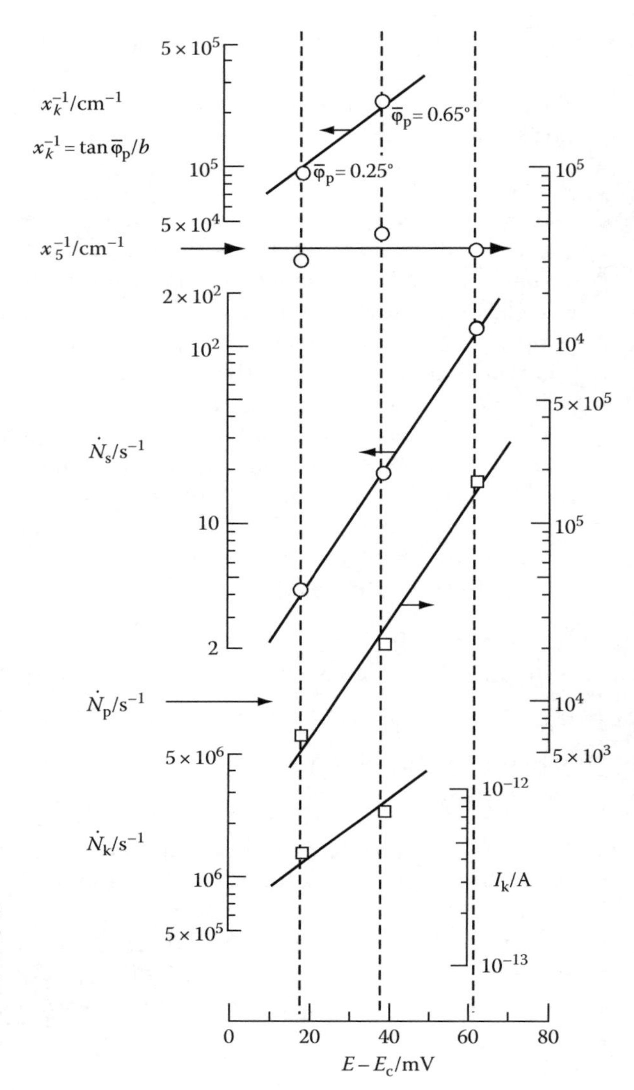

FIGURE 3.6
Potential dependence of structural parameters and mean rates of elementary processes for three different polarization conditions. TEM measurements on replica of gold-decorated surfaces after anodic dissolution. Johnson-Matthey iron. (From Allgaier, W. and Heusler, K.E., *J. Appl. Electrochem.*, 9, 155, 1979.)

an estimate of x_s and x_k and of their dependence on the dissolution rate as shown in Figure 3.6. The surface concentration of kinks $(x_s x_k)^{-1}$ is an exponential function of the potential with an exp(FE/RT) dependence. The potential-dependent parameter is $x_k \cong \exp-(FE/RT)$ while the mean distance of steps, x_s remains constant. Consistent with the catalytic mechanism, at steady state, the dissolution rate of atoms at kinks obeys an exp$(1 + 2\alpha)FE/RT$ Tafel law and, at constant kink density, the elementary rate of atom dissolution is proportional to exp(RE/RT), i.e., $\alpha = 0.5$ and transfer of $n = 2$ electrons in a single step.

A rather convincing justification of the catalytic mechanism is reached in terms of atomistic events, at least on conveniently oriented crystal faces. Now, in turn, the problem remains of understanding why the step nucleation probability depends on pH [31]. Further investigation of iron dissolution by impedance analysis has clearly proved that both catalytic and consecutive mechanisms are operative with a relative contribution depending upon electrode potential and solution pH.

3.2.1.1.3 Advanced Study of Iron Dissolution by Impedance Techniques

Early application of impedance measurement to the anodic dissolution of iron in acidic media demonstrated the inductive character of the faradic impedances. The reaction scheme initially proposed [13,67] is modified according to:

$$\mathrm{Fe} + \mathrm{H_2O} \underset{K_0}{\overset{K_0}{\rightleftarrows}} \mathrm{FeOH^-} + \mathrm{H^+} \tag{3.29}$$

$$\mathrm{FeOH^-} \xrightarrow{K_1} (\mathrm{FeOH})_{\mathrm{ads}} + \mathrm{e^-} \tag{3.30}$$

$$(\mathrm{FeOH})_{\mathrm{ads}} \xrightarrow{K_2} \mathrm{FeOH^+} + \mathrm{e^-} \tag{3.31}$$

The initial fast chemical hydrolysis step is introduced to account for the order of the first reaction with respect to OH^- confirming the results of previous authors [13]. This modification, introduced later [78,79], implies clearly that the origin of chemisorbed hydroxyls is the dissociation of water.

Steady-state current–voltage characteristics and faradaic impedance associated with the mechanism (3.24) through (3.31) were derived and numerically simulated with the following assumptions:

- Step (3.29) in fast equilibrium, i.e., surface concentration of $FeOH^-$ proportional to $[OH^-]$ and time independent
- 2D surface coverage θ by $(FeOH)_{ads}$ governed by a Langmuir adsorption isotherm
- Rate constants K_1 and K_2 obeying Tafel kinetics with symmetry factors ($0 < \alpha_i < 1$) but not constrained to 0.5:

$$K_i = k_i \exp \frac{\alpha_i}{RT} FE$$

With this particular mechanism, Equations 3.2 and 3.3 are written as

$$\frac{\beta \, \mathrm{d}\theta}{\mathrm{d}t} = K_1(1-\theta)[\mathrm{OH^-}] - K_2\theta\beta \tag{3.32}$$

where β is the surface concentration of $(FeOH)_{ads}$ at full coverage ($\theta = 1$).

$$\frac{J}{F} = K_1(1-\theta)[\mathrm{OH^-}] + K_2\theta\beta \tag{3.33}$$

The steady-state coverage θ_S obtained by solving (3.32) for $\mathrm{d}\theta/\mathrm{d}t = 0$, introduced in (3.33), gives a current–voltage relationship:

$$J_S = \frac{2J_1J_2}{J_1 + J_2} \tag{3.34}$$

where J_1 and J_2 are the partial current densities flowing through steps (3.1) and (3.2):

$$J_1 = FK_1[\mathrm{OH^-}] \quad \text{and} \quad J_2 = FK_2\beta$$

Similarly to Equation 3.6, the solution of the linearized form of (3.32) and (3.33) yields the impedance expression

$$\frac{1}{Z_F} = \frac{J_s F}{2RT}\left[\alpha_1 + \alpha_2 + \frac{(\alpha_1 - \alpha_2)(J_2 - J_1)}{Fj\omega\beta + (J_1 + J_2)}\right] \tag{3.35}$$

and an associated

$$R_t J_s = \frac{2RT}{F(\alpha_1 + \alpha_2)} \tag{3.36}$$

A fair model–experiment fit is obtained for both DC and AC responses with the same set of kinetic parameters and a value of β, 3×10^{-9} mol cm^{-2}, in good agreement with one monolayer of adsorbed OH^- ions on a low-index plane of the iron lattice (2×10^{-9} mol cm^{-2} on {100}).

The next advance was achieved owing to the extension of the frequency domain into the millihertz range by digital transfer function analyzer (TFA) [47]. The existence of a second inductive loop was then established at lower frequencies. A branching reaction mechanism was proposed by introducing in parallel with the consecutive mechanism a catalytic dissolution path [82]. This branching reaction path was already prefiguring the main features of the more recent models covering broader ranges of pH and current density.

Later on, more accurate impedance measurements were reported by Lorenz et al. [83] and Keddam et al. [68,69] in 1 M sulfate solutions at $0 \le \text{pH} \le 5$ and $0 \le j \le 150$ mA cm^{-2}. In spite of previous contractictory explanation [84], both groups agreed reasonably about the main impedance shapes and particularly the presence of three relaxation processes showing up at intermediate pH values (around 2). The models elaborated by these groups have in common the interpretation of the frequency dependence in terms of degrees of coverage by adsorbed species. However, very large differences lie in the basic assumptions regarding

- The actual nature of the surface species participating in the impedance response and the injection of an ad hoc "surface relaxation time constant" [75]
- The existence or not of a consecutive mechanism in one of the reaction routes
- The ability to interpret DC and AC data by a unique model in the active, transition, and prepassive range

A detailed discussion of the relative performances and flaws of these approaches can be found in Ref. [69].

A systematic analytic screening of all the possible 40 reaction schemes in which three Fe-containing surface species are involved completed by a numerical simulation finally led to selecting the reaction pattern (3.36) shown below. The three coverages determining the impedance properties in the active and transition domains are related to $Fe(I)_{ads}$, $Fe^*(I)_{ads}$, and $Fe^*(II)_{ads}$. The superscript (*) indicates species involved in catalytic dissolution paths. $Fe(II)_{ads}$ is a precursor of the passive film whose contribution is significant only near the second maxima and beyond.

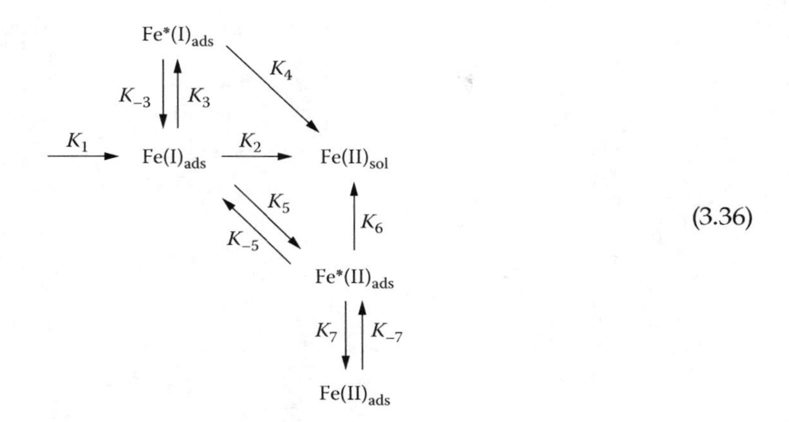

(3.36)

Numerical simulations showed the following:

- The K_1, K_2 path is the consecutive mechanism early established by impedance measurements bounded to the hertz range and covering low pH and low current density [67].
- The K_1, K_3, K_4 path is a catalytic route mainly operative in the transition range, the $Fe^*(I)_{ads}$ coverage being responsible for the decrease of dissolution rate past the first peak shown in Figure 3.7a in this range at higher pH.
- The K_1, K_5, K_6 path is the main dissolution route at higher pH and current density ($j > 50\,mA\ cm^{-2}$ at pH 2 and $j > 10\,mA\ cm^{-2}$ at pH 5) controlling the second peak in Figure 3.7a.

Both catalytic paths introduced in this model were previously proposed by Lorenz et al. [85–89] for interpreting the prepassive dissolution range.

Comparison of Figure 3.7a and b at pH 5 illustrates the outstanding power of impedance measurements associated with modeling for mechanistic analysis.

The pH dependence of the rate constants proves that chemical bonds of Fe atoms with solution species contribute, to a determining extent, to the generation of active sites of dissolution, even in atomistic models where surface lattice features (kinks, steps, terraces, etc.) are generally put forward. Large similarities of glassy metals (amorphous) with crystalline ones [68,90] must also be regarded as arguing in favor of chemical bonds as predominant entities with respect to lattice related sites.

Further support for the existence of a branching reaction pattern was obtained from RRDE investigations [91]. It is found, as shown in Figure 3.8, that the collection efficiency N for the oxidation to the trivalent state of the divalent Fe species released by the dissolution of the iron disk decreases with pH and current density.

This is interpreted by paths K_4 and K_6 producing less reactive Fe(II) species than K_2 does. In agreement with this view, the lower the rotation speed, the larger the collection efficiency [91] because a homogeneous chemical reorganization of the cation into a more reactive one is taking place between disk and ring. Such a chemical step is formally accounted for in most reaction mechanisms by the following step:

$$FeOH^+ + H^+ \rightarrow Fe^{2+} + H_2O$$

Similar results are reported in Ref. [92] from CFDE.

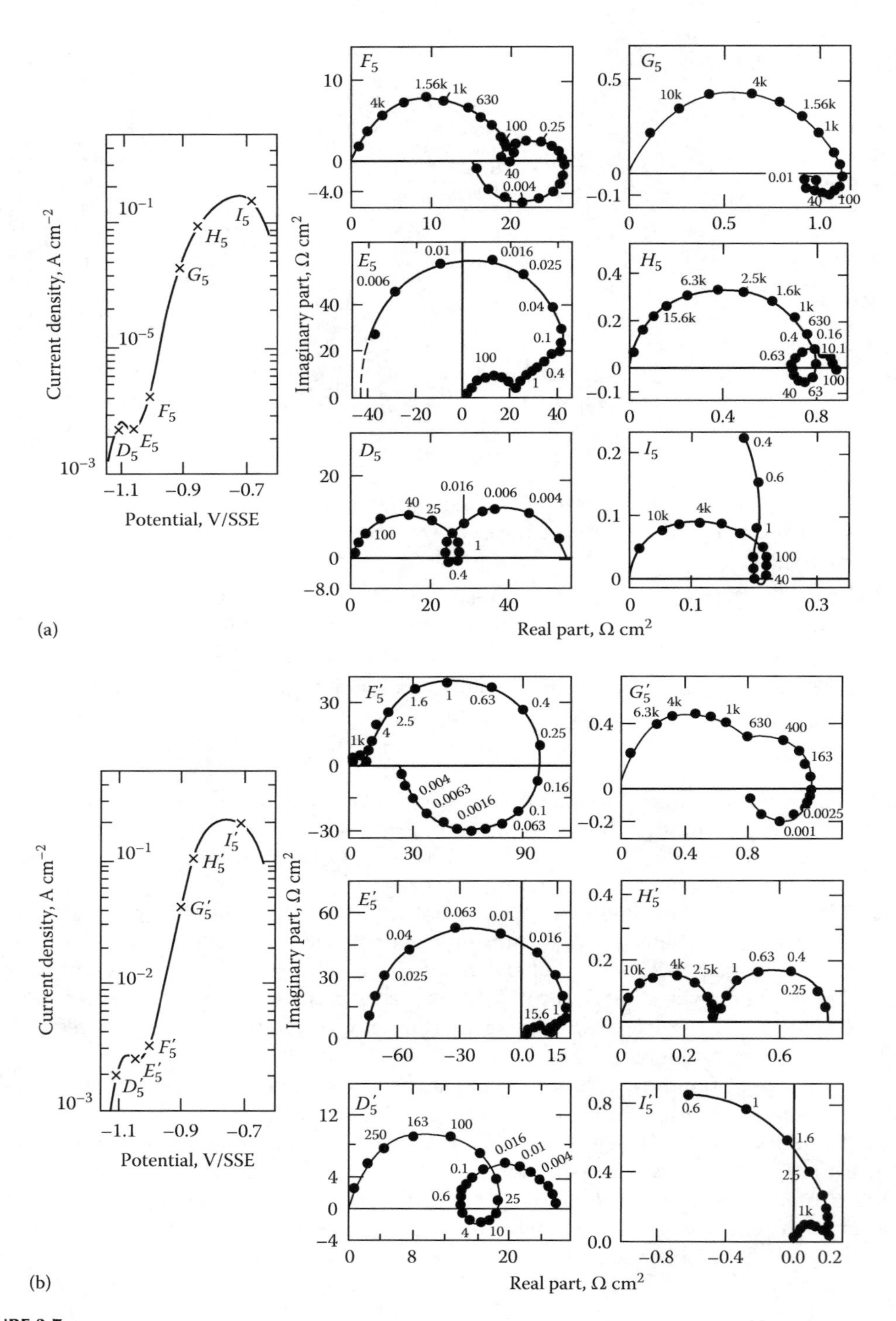

FIGURE 3.7
Experimental (a) and calculated (b) current–potential curves and complex impedance diagrams at the corresponding points labeled on I(E) curves (frequency in Hz) (Johnson-Matthey iron, 1 M sulfate, pH = 5, 25°C). Numerical simulation (b) according to the reaction mechanism (3.36). (From Keddam, M. et al., *J. Electrochem. Soc.*, 128, 257, 1981.)

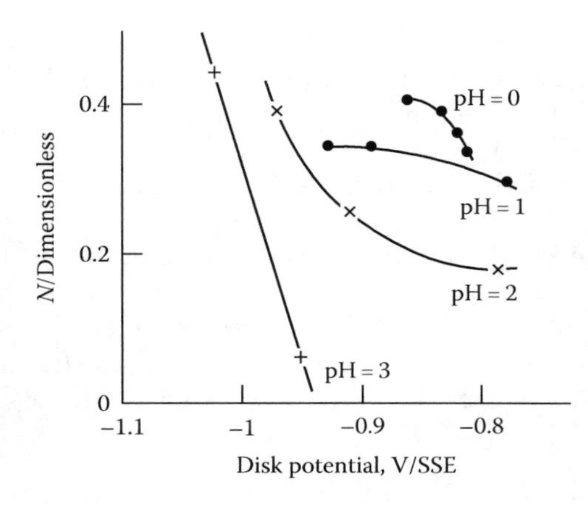

FIGURE 3.8
Potential dependence of the collection efficiency: influence of solution pH (ring reaction $Fe^{2+} \rightarrow Fe^{3+}$) for iron dissolution (Johnson-Matthey iron 1 M sulfate). Theoretical collection efficiency [37]: 0.48. Shift to cathodic potentials at increasing pH is consistent with Figure 3.7. (From Benzekri, N., Contribution au développement de l'électrode disque-anneau en courant alternatif. Applications aux mécanismes de dissolution et passivation anodiques, Thesis, Université P.M. Curie, Paris, 1988.)

3.2.1.2 Role of Anions in the Anodic Dissolution of Iron

The specific role of the electrolyte composition at a given pH is an extremely common experience. The contribution of anions other than hydroxyl is expected through the acid-catalyzed or anion-ligand mechanism [24,25]. It must be emphasized that the role of pH in polyacids, such as H_2SO_4, may be obscured by the correlated changes of electrolyte composition depending on acidity constants [93].

Specific anion dependence is expected to occur in the passive and transpassive domains, and dissolution in the active range can be made to deviate from the hydroxo-ligand mechanism [94] only by anions able to replace OH^-, essentially SH^- [95] and the halide ions. In the case of iron, due to the well-known passivity breakdown and subsequent localized corrosion by halide ions and particularly Cl^-, chloride effects have been investigated extensively. Complexing anions such as acetate have also been considered to a lesser extent.

In the case of acetate, a series of papers [78,93,96] considered the participation of the anion from the initial reaction steps of dissolution and it was also claimed that this anion contributes only to the dissolution of aged solid phases [97]. Later studies [93,96] in acetate solutions at neutral and slightly acid pH reported, a clear negative order with respect to A^- or HA. A model is derived on the basis of the participation of HA and OH^-, which are assumed to play a symmetric role with the formation of $(Fe(OH)_2)_{ads}$, $(Fe(OHA))_{ads}$, and $(FeA_2)_{ads}$. The kinetics of iron dissolution in the active range in the presence of halide ions X^- is largely dominated by the competitive adsorption of X^- [95] with the dissolution activating OH^-. A critical survey of the possible reaction paths in which Cl^- competes with other anion adsorption is given in Ref. [80]. The mechanism is claimed to depend on the pH range. The catalytic step of dissolution in Cl^--free media [70] is considered to become at medium acidities ($pH > 0.6$):

$$FeCl^-_{ads} + FeOH^-_{ads} \rightarrow FeOH^+ + Fe + Cl^- + 2e^- \tag{3.37}$$

In strongly acidic solutions, the formation of the surface complex $FeCl^-H^+$ and dissolution as $FeCl^-$ are proposed to interpret the accelerating effect of H^+ on the dissolution rate [80].

According to other authors [98,99], Cl^- and OH^- would play a symmetric role by forming $(FeClOH)^-$ as an intermediate in a consecutive mechanism followed by the rate-determining step

$$(FeClOH)^-_{ads} \rightarrow FeClOH + e^- \tag{3.38}$$

and in highly acidic solutions [99] the contribution of Cl^- in the dissolution path is depicted by the formation of $(FeCl)_{ads}$ and $FeClH^+$ as intermediate species.

Impedance measurements have been applied to iron dissolution in Cl^--containing media [100–102]. Based on the mechanism for sulfate media and similarly to earlier work [68,69], a branching pattern was proposed in (HCl/NaCl) involving a consecutive and a catalytic dissolution path having $(FeOH)_{ads}$ as a common bifurcation species. A third reaction path of the consecutive type with $(FeClOH)^-$ as an intermediate is in agreement with Ref. [98]. An interpretation of the role of Cl^- was advanced [101,102] in the framework of a progressive modification of the sulfate mechanism [68,69] as a function of the chloride concentration, up to the sulfate-free medium.

Mechanistic changes with pH are interpreted in a semiquantitative way as follows: a gradual increase of Cl^- content initially reduces the contribution of the (K_1, K_5, K_6) dissolution path in favor of a catalytic step similar to (K_4), the $Fe^*(I)_{ads}$ catalyst now being a Cl^--containing species.

3.2.1.3 Active Dissolution of Other Metals

The metals belonging to the so-called iron group (Co, Ni), which exhibit an active–passive transition, have been far less extensively investigated than iron itself. The state of the art can be found in Ref. [2]. Little contribution is due to impedance measurements and the mechanisms must be regarded as much less solidly established than for iron.

3.2.1.3.1 Cobalt

Cobalt seems to behave much similarly to iron and its dissolution is assumed to take place through the formation of $(CoOH)_{ads}$, a catalyst surface species produced in an initial step of dissociative adsorption of water analogous to Equation 3.23. Regarding the structural description of the dissolution mechanism in term of kinks at the lattice surface, cobalt and iron apparently follow the same process in which atomic kinks are assimilated to catalytic sites $(CoOH)_{ads}$. Unfortunately, in the case of Co, microscopic evidence of the changes of surface morphology and consequently estimation of the kink–kink distance are lacking.

3.2.1.3.2 Nickel

The mechanism of active dissolution of nickel remains highly controversial. It seems that there is no agreement so far even on the actual shape of the current–voltage profiles. Two-peak or single-peak curves are reported, depending on the potential sweeping rate, the prepolarization conditions, the structural properties of the metal, etc. This erratic behavior is tentatively ascribed to either the remanence of stable natural oxides on the metal surface or strong interaction with hydrogen. It must be pointed out that experiments performed in situ on a surface continuously refreshed by mechanical abrasion never yielded more reproducible results.

3.2.1.3.3 Chromium

Due essentially to its role in the corrosion resistance of stainless steels, the anodic behavior of Cr has been far more extensively considered. An overview can be found in Ref. [103]. Early studies were based on polarization curves [104] and potential transient experiments [105].

Most of the significant data are established by impedance [43,106,107] and RRDE [108,109] and reveal rather poor reproducibility, probably because of interference with hydrogen overvoltage and its metallurgical dependence. Simultaneous dissolution as Cr^{2+} and Cr^{3+} in the active and active–passive transition ranges [110] and no pH dependence of the dissolution rate led, in contrast to Fe group metals, to more direct dissolution mechanisms being proposed [111], The following oversimplified model [112,113] was considered with the purpose of accounting for the Cr dissolution in the description of Fe–Cr alloys [114].

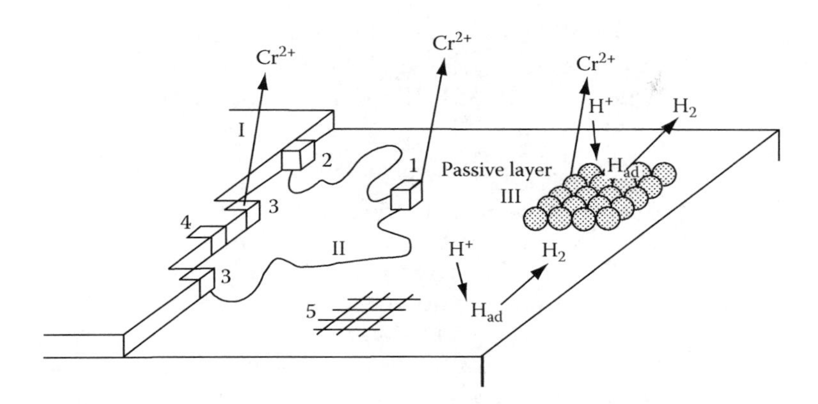

FIGURE 3.9
The corrosion processes occurring at a chromium–sulfuric acid solution interface. (From Dobbelaar, J.A.L., The use of impedance measurements in corrosion research. The corrosion behaviour of chromium and iron–chromium alloys, Thesis, Technical University of Delft, Delft, the Netherlands, 1990.)

By means of CFDE, the existence of adsorbed hydroxylated intermediates CrOH and $CrOH^+$ was proposed [108] in the active dissolution range. From impedance data on polycrystalline and (100) single crystal plane [103,106], simultaneous hydrogen evolution was introduced in the anodic behavior of Cr by assuming a strong dependence of the anodic reaction on the H–Cr bonding. Evidence of diffusion-controlled kinetics in the impedance was tentatively interpreted by surface diffusion of $Cr(I)_{ads}$, and Cr dissolution and H evolution were assumed to be competing on terraces. Figure 3.9 shows the main aspects of this approach, which attempted to elaborate a kinetic description on the basis of an atomistic picture of the metal surface.

One of the most innovative ingredients in this model is the coexistence of H evolution and Cr dissolution on the passive areas, but divalent state of Cr in the passive range appears quite contradictory to available data from x-ray photoelectron spectroscopy (XPS).

3.2.1.3.4 Other Metals

Titanium [115–117] and copper [118] are among the more extensively investigated metals not pertaining to the iron group or being major constituents of stainless steels.

3.2.1.4 Recent Advances in the Mechanism of Passivation by Downstream Collector Electrode Techniques

In spite of the very elaborate approaches based on current–voltage techniques, credible steady-state and transient mechanisms have hardly been derived for the transition from the active to the passive range. An accurate description of the whole set of elementary steps requires additional data on the relative contribution of the oxidation current to the formation and growth of 2D and 3D layers and on the release of cations in solution. Passivation mechanisms can be elaborated on the basis of the active–passive transition in two complementary ways:

- Analysis of passivation transients on an initially active surface either by applying a steep potential jump into the passive range or by creating fresh surfaces at constant applied potential by nonelectrochemical depassivation (chemical: passivity breakdown; mechanical: scratching, ultrasonic waves, etc.; radiative: laser beam impact [119,120]). These techniques have proved to be of outstanding importance for the investigation of the mechanism of localized corrosion associated with passivity breakdown [121,122].

- Analysis of tiny changes of the anode maintained potentiostatically between active and passive states in the negative slope of the dissolution peak. Electrochemical impedance and frequency-resolved RRDE data are used together.

3.2.1.4.1 Dissolution during Transient Passivation of Freshly Generated Iron Surfaces

Pulse depassivation of the upstream iron anode and collection of the Fe(II) released into the solution on a downstream glassy carbon were performed in a CFDE [123]. Transient currents I_D on the Fe anode and collected currents I_R for Fe(II) oxidation are shown in Figure 3.10a and b. After processing for double-layer charging and collection efficiency, the charges involved in dissolution Q_d and Q_s were discriminated from the overall charge Q_D supplied to the disk. The results as a function of pH and disk potential E_D are shown in the Figure 3.10c and d. The whole set of data was interpreted by the reaction model (3.39), a simplified form of the one derived for the faradic impedance of active iron (3.36). The kinetic equations for the time dependence of the two coverages by Fe(II) and Fe(III) were integrated numerically. Considering the wide range of current densities covered by the transient, both noncatalytic and catalytic reactions paths were considered. Only Fe(II) species participate in the dissolution mechanism and passivation is ascribed to Fe(III). These assumptions are justified by the large degree of (super) saturation in ferrous species during short transients.

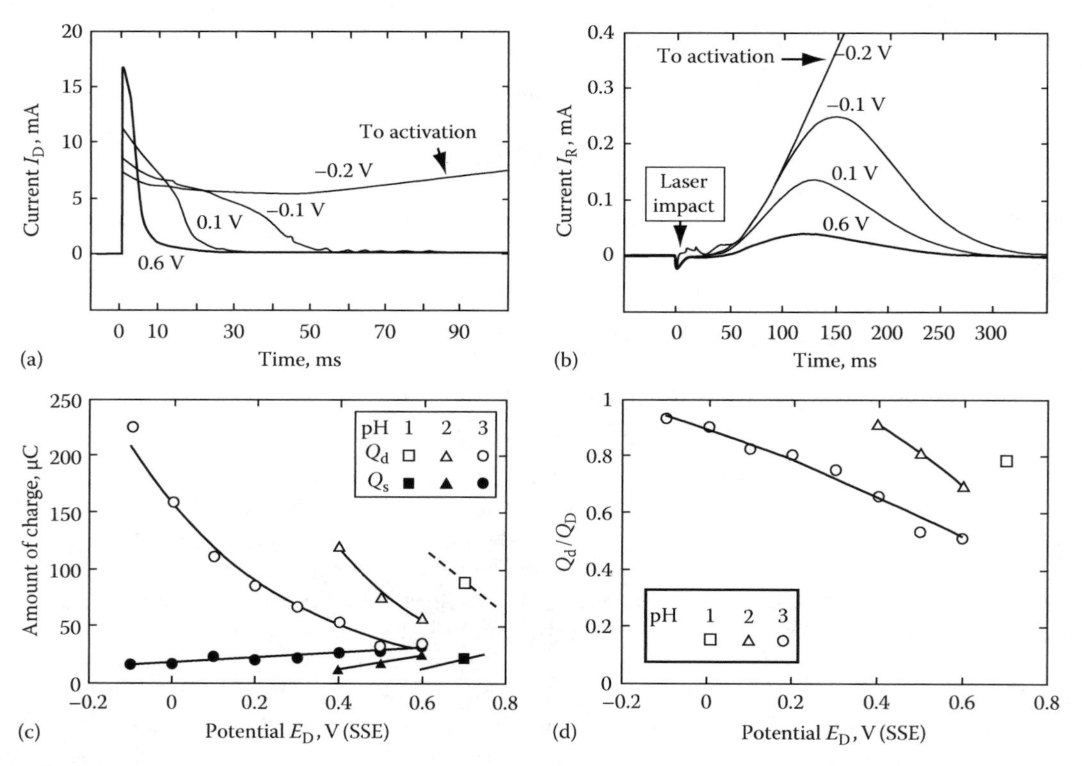

FIGURE 3.10
(a) Transient currents on the Fe working (upstream) electrode. (b) Detection currents on the glassy carbon collector (downstream) electrode 1 M Na_2SO_4 acidified to pH 3. (c) Potential dependence of the charge Q_d for dissolution of ferrous cations and Q_s for passive film formation at various pH. Depassivated surface area 0.08 cm^2. (d) Potential dependence of the charge ratio Q_d/Q_D (integral emission efficiency) at various pH.

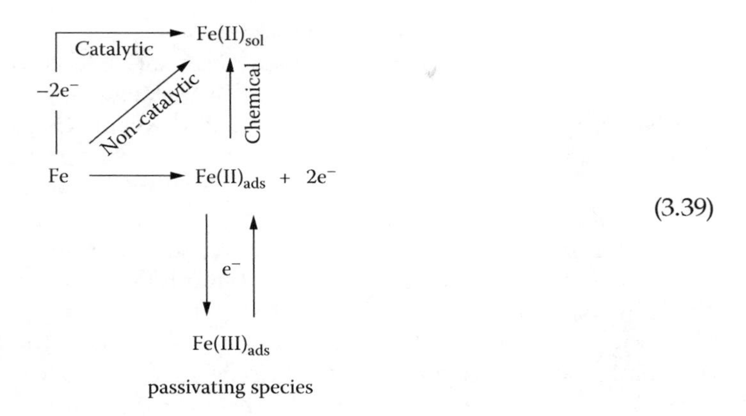

(3.39)

That is supported by the results of simulations indicating in the first milliseconds of the transient a fast buildup of the Fe(II) coverage to a maximum of about 0.5, which then falls off during the growth of the passive film. The higher the pH and the applied potential, the sooner and the steeper the transition from Fe(II) to Fe(III).

3.2.1.4.2 Impedance and AC RRDE Study of the Active–Passive Range of Iron

Impedance and the AC component of the ring current were simultaneously measured at a series of polarization points in the negative slope region of Fe in 1 M sulfuric acid [124]. This region immediately prior to the Flade potential is very meaningful for understanding the imbrication of dissolution and passivation steps. The ring potential E_R was settled at two different potentials for collecting Fe(II) by oxidation (0.80 V/SSE) and Fe(III) by reduction (−0.40 V/SSE).

Typical data obtained for one point of the polarization curve are shown in Figure 3.11. Impedance shows the well-known behavior of a passivation range characterized by the negative zero-frequency limit of the real part consistent with the slope of the steady-state

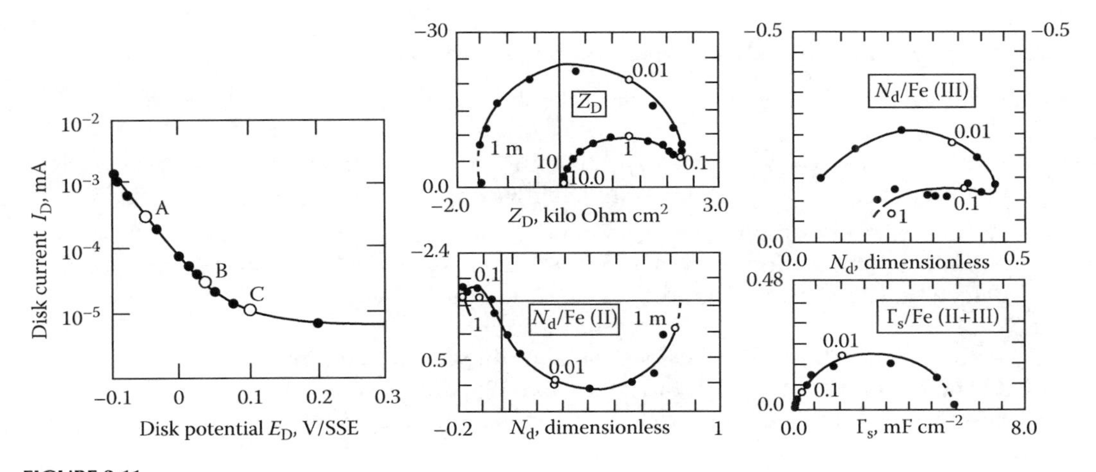

FIGURE 3.11
Impedance and AC-resolved RRDE of the active–passive transition of iron in 1 M H_2SO_4. Rotation speed 900 rpm. Current–voltage curve of the iron disk. Data at polarization B. Z_D: disk impedance, N_d/Fe(II) and N_d/Fe(III): emission efficiencies respectively for Fe(II) and Fe(III) species, ring potential respectively: 0.8 and −0.4 V/SSE, Γ_s/Fe(II + III): faradaic capacitance of the passivating layer.

current–voltage curve. The low-frequency capacitive arc displays the frequency response of the film growth. Emission efficiencies N_d relative to Fe(II) and Fe(III) going to the solution contain two frequency domains, i.e., a surface charge stored in two different species. For Fe(II), the largest changes take place in the same frequency range as for impedance. Its positive imaginary part establishes the passivating role of the faradic surface charge with respect to the active path of iron dissolution (see earlier). A small loop at higher frequencies is poorly visible. It reveals a feature not detected on the impedance spectrum and lies in the half-plane of negative imaginary parts reflecting a surface charge stored in a dissolution intermediate. The emission efficiency for Fe(III) contains the same two domains of frequency but with quite different properties. Both are located in the negative real part region with modulus maximum around the transition frequency (0.1 Hz). Therefore, both surface species participate as sources of dissolution of Fe(III). This is consistent with the generally accepted concept of a passivity current carried by the Fe(III) released by the film lattice. Finally, the capacitance Γ_s associated with the total charge as Fe(II) plus Fe(III) indicates that around 5.5 mC V^{-1} cm^{-2} are needed for passivation. Most of the current decay takes place within 0.2 V; therefore, the charge content of the passive film at the Flade potential can be estimated as 1 mC cm^{-2}. This value is practically equal to the charge equivalent to one monolayer (one cell unit of a ferric oxide layer).

A kinetic model very close to that elaborated more recently for interpreting the transient dissolution of laser beam–depassivated iron was proposed. It made it possible to reproduce the main features of N_d and Γ_s shown in Figure 3.11.

3.2.2 Dissolution in the Passive State

Anodic dissolution in the active state generally results in equivalent corrosion rates unacceptable in practical applications (order of mA cm^2 or more). Self-limitation of the dissolution rate by the buildup of a thin surface layer under sufficiently oxidizing conditions, known as passivation, is the only relevant phenomenon in terms of intrinsic corrosion control. The mechanism of dissolution in the passive state is dealt with in the following. A thorough survey of the field can be found in the proceedings volumes of passivity symposia [126–129].

3.2.2.1 Film Relaxation and Dissolution in the Passive State

For simplicity, the dissolution behavior of passive metals can be split into two groups:

- Passive electrodes with a nearly constant corrosion current over an appreciable potential range. They are generally dealt with in terms of high field migration.
- Passive electrodes with a potential-dependent current, most commonly exhibiting a minimum. Less advanced models are available; they involve the contribution of defects (e.g., point defect model).

Nonsteady-state techniques played an outstanding role in reference papers in the field [4]. Galvanostatic and potentiostatic transients [133–136] are exploited either at short times, for plotting current–voltage characteristics at constant (frozen) film thickness, or at longer times as a function of increasing (or decreasing) film thickness estimated from the total charge $\int I\,dt$. Transient techniques were also extended to rapid changes of solution composition [137]. AC impedance in the very low–frequency range (millihertz and submillihertz) proved quite successful for achieving finely resolved information on corrosion and film growth [4].

3.2.2.1.1 *Passive Iron and the High-Field Migration*

The high-field migration model (also known as hopping mechanism) was applied to interpret the steady-state and nonsteady-state behaviors of iron investigated by many different techniques during the last 40 years. Pioneering works by the German school [138,139] have been revisited in the past decade by more advanced experiments [140] and improved modeling [141].

Salient features accounted for by this model are the following:

- A steady-state corrosion current independent of the electrode potential over an appreciable range.
- The existence of a specific potential, known as the Flade potential E_F, at the cathodic bound of passivity, to which all the film properties are referred in a simple form. It can be regarded as the zero thickness origin.
- A linear dependence of the film thickness (estimated coulometrically but also supported by surface-sensitive techniques such as ellipsometry) on the anodic overvoltage $(E - E_F)$.

In its original form [132], the model was based on a multibarrier thermally activated cation transport giving rise to a current

$$I = I_0 \exp \frac{B(E - E_F)}{l} \tag{3.40}$$

where

$B = zaF/RT$ with a the half-jump distance
z is the number of elementary charges on the mobiles species
l is the film thickness
F, R, and T are the usual physical constants
I_0 is a standard current assumed to depend on the density of charge carriers in the film

At steady state, the potential-independent corrosion current is totally supported by a constant field strength in the film, i.e., a film thickness l proportional to $(E - E_F)$. The conservative flux of cations Fe(III) across the interface requires a potential-independent voltage drop at the film–electrolyte junction.

Most of the nonsteady-state results were obtained by RRDE [142,143] and monitoring of Fe(III) in solution by analytical chemistry [138,139] or radiotracer techniques. Transfer of cations into the solution at the film–electrolyte interface was found to depend on the overvoltage according to redox kinetics and to be roughly proportional to the current forced in the electrode in excess of the steady-state value. Accordingly, as sketched in Figure 3.12, any change of the steady current, or potential, triggers a modification (positive or negative) of the film thickness that finally restitutes the initial electrical field strength and potential drop at the outer film surface, i.e., potential-independent DC current.

The overall problem of the nonsteady-state passive iron has been reexamined with the purpose of reaching a consistent interpretation of the empirical equation firmly established in acidic [135] and neutral [133,134] media and also for nickel in acidic solutions [135] in classical papers. As stated earlier [141], this equation contradicts the high-field mechanism because it predicts a direct logarithmic $Q(t)$ relationship consistent with a place exchange mechanism, whereas the high field should generate an inverse logarithmic dependence [4].

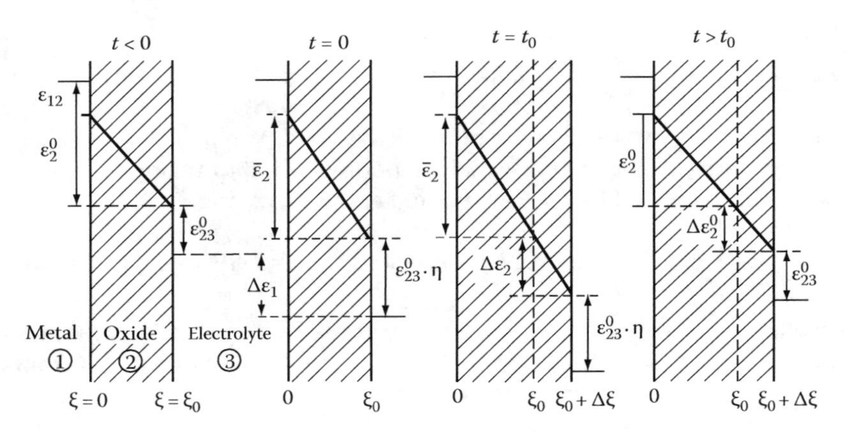

FIGURE 3.12
Positional changes of the potential from the metal (1) through the passive film (2) to the electrolyte (3) for different times of a galvanostatic square pulse. According to Vetter and Gorn [138] and Wagner [130] at steady state, the film solution potential drop ε_{23} is potential independent ($\eta = 0$).

$$I = I_0 \exp\left(\beta E - \frac{Q}{B}\right)x$$

It can be concluded that the key point lies in the potential distribution between the film bulk and the film–electrolyte interface in the transient regime. Subsequently, an attempt was made to incorporate the role of defects (cation vacancies) in the high-field mechanism [144].

3.2.2.1.2 AC Impedance of Passive Iron in Acidic Media

Film resistance to cation migration and capacitance displaying the potential dependence of the charge stored by the film growth are expected. Figure 3.13 shows two complex impedance diagrams in the passive range in H_3PO_4 and H_2SO_4 solutions [140].

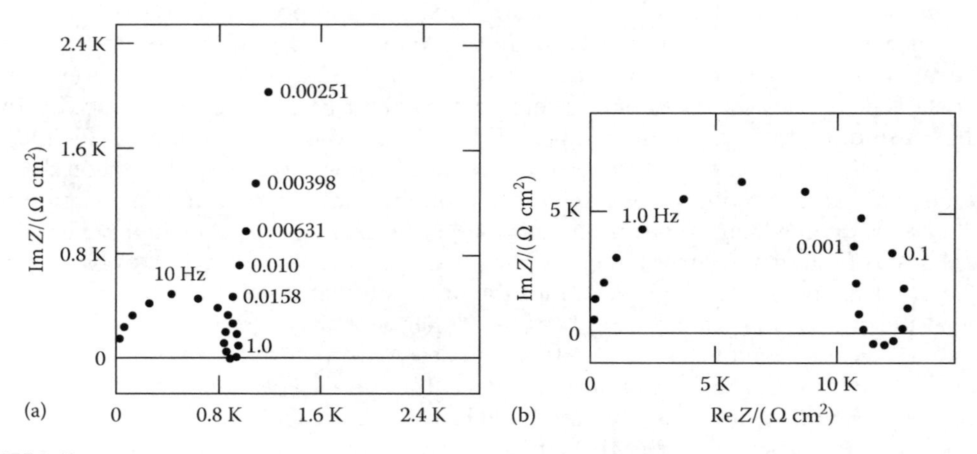

FIGURE 3.13
Impedance diagram of iron in the passive state. Johnson-Matthey iron. (a) 1M H_3PO_4, 37°C, $E = -0.65$ V/SSE. (b) 1M H_2SO_4, 25°C, $E = -0.35$ V/SSE frequency in Hertz. (From Keddam, M. et al., *J. Electrochem. Soc.*, 131, 2016, 1984.)

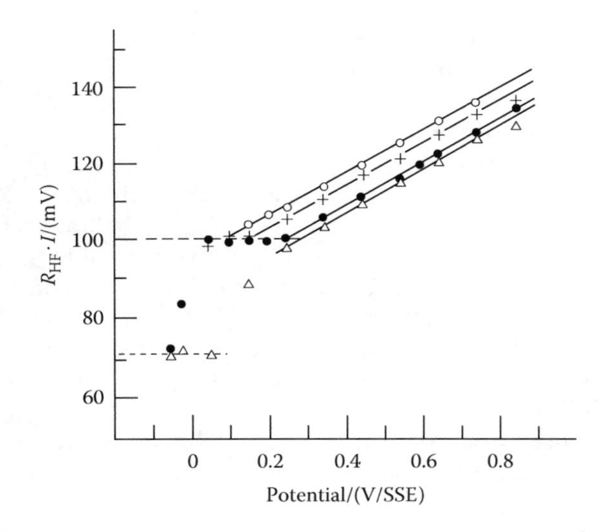

FIGURE 3.14
Potential dependence of the $R_{HF} \cdot I$ product for passive iron in various solutions (Johnson-Matthey iron. 25°C). Δ: 1 M H_2SO_4, pH = 0.03. •: 1 M H_3PO_4, pH = 1 +: 1 M (H_3PO_4 + KH_2PO_4), pH = 2.2. o: 1 M (H_3PO_4 + KH_2PO_4), pH = 2.8. (From Keddam, M. et al., *J. Electrochem. Soc.*, 131, 2016, 1984.)

According to Equation 3.40, the size of the HF loop is attributed to the derivative of the flux due to high-field migration, with respect to E at constant film thickness:

$$R_{HF} = \frac{l}{BI} \tag{3.41}$$

The DC current I being potential independent, a linear dependence of $R_{HF}I$ on E is expected and fairly verified as shown in Figure 3.14 for various solution compositions and pH. The straight lines begin at the Flade potential E_F, where the film is restricted to its inner and outer boundaries and above which the 3D film starts growing.

From the slope, 0.06, of the straight line in Figure 3.14 and assuming a = 0.5 nm, a calculated field strength of 3 × 10^8 V m^{-1} and a potential dependence of the film thickness of 3.5 nm V^{-1} are calculated, in agreement with literature data. The inductive behavior at intermediate frequencies is physically equivalent to the overshooting transient considered in Ref. [144] and may be related to a relaxation of the carrier density. The low-frequency capacitive branch is perfectly consistent with the constant DC current. Its value is about one order of magnitude larger than estimated by Faraday's law applied to the thickness-potential dependence of the film. Little dependence is found with respect to the solution composition, supporting the interpretation of anion effect through the change of I_0 with the density of charge carriers.

According to the high-field migration model, iron dissolution across the film and film growth are only weakly coupled by the transient overvoltage η (see Figure 3.12) at the film–electrolyte interface. In contrast, any model based on defect mobility necessarily implies a dynamic situation with two nonconservative reactions [145] (film growth at the inner interface and film dissolution at the outer one). The whole film is therefore translating outward with a velocity depending on the corrosion current.

3.2.2.1.3 Investigation of the Transient Dissolution of Passive Iron by Impedance and Electrogravimetric Transmittance

One of the recurrent topics in passivity studies is the transient behavior of the passive state. That is, how is the current shared between film growth and metal dissolution during the restoration of a new steady state after a change in the applied potential? As stated before, the predictions of the high-field model in the transient regime are extremely sensitive to the distribution of a potential increment between film bulk and interfaces. No ab initio answer is available and the efforts have been concentrated on the analysis of time or frequency domain experiments. Emphasis was placed on the presence of an overshoot or inductive response, but the interpretation of the inductance remained quite ambiguous without no additional data. Quartz crystal microbalances are very well adapted to this problem because the low corrosion rate provokes no drift of the mean mass of the electrode while for the same reason the emitted flux of cation is below the detection limit of collector electrodes. Impedance when associated with frequency-resolved electrogravimetry is able to discriminate between the fraction of the overall charge Q taken up in film growth (mass gain) and consumed for dissolution (mass loss).

It was shown analytically that the high-field migration model is not able on its own to explain the inductive response shown in Figure 3.15a. Correlatively, the calculated mass transmittances $\Delta m/\Delta E$ and $\Delta m/\Delta Q$ simply reflect Faraday's law [56]. In particular, $\Delta m/\Delta Q$ is a real quantity determined by the antagonistic contributions of transitory film growth and metal dissolution. A new time constant besides that related to the relaxation of the film thickness was introduced [146,147] by deriving the impedance and AC mass response for the model proposed by Kirchheim [144]. It assumed that the density of cationic vacancies (or the density of mobile Fe(III) cations) in the film depends on the excess of current with respect to the corrosion current I_c. Consequently, the film resistance R_{HF} is lowered, producing an inductive-like transient of the migration current. Impedance is foreseen to contain, in addition to the series capacitance at low frequencies standing for the film growth, an inductance generated by the relaxation of the film conductance.

Figure 3.15 displays experimental and computed Nyquist plots of the impedance. Similarly, the electrogravimetric transmittance was derived. It exhibits the mass counterpart of the relaxation of the film conductance. Again, consistent with Faraday's law, the low-frequency limit

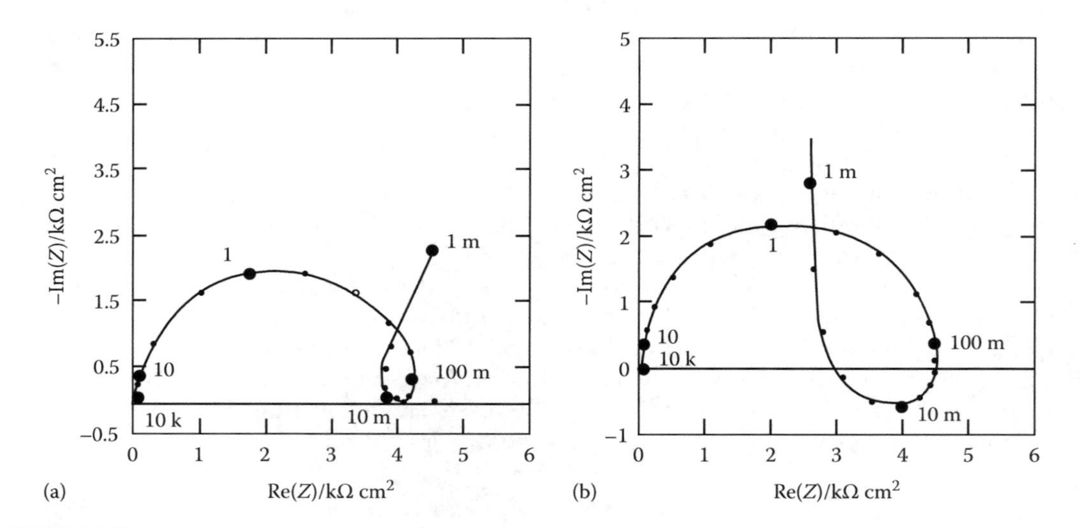

FIGURE 3.15
Impedance of iron in 1M H_2SO_4, potential 0.5V/SSE. Experimental (a) and numerically simulated (b), with a model based on a relaxation of the charge carriers density in the passive film. (From Gabrielli, C. et al., *Mater. Sci. Forum*, 185–188, 631, 1995; Gabrielli, C. et al., *Electrochim. Acta*, 41, 1217, 1996.)

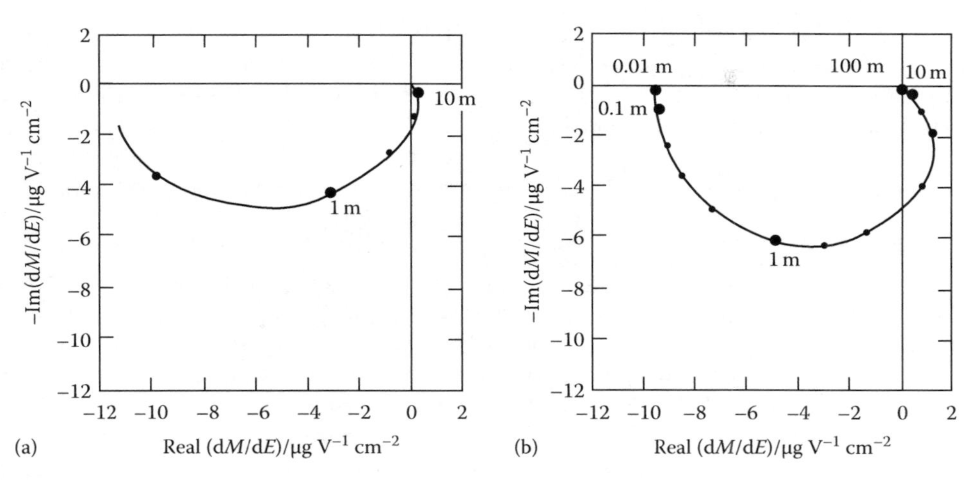

FIGURE 3.16
Electrogravimetric transfer function of iron in 1 M H_2SO_4, potential 0.5 V/SSE. Experimental (a) and numerically simulated (b), with a model based on a relaxation of the charge carriers density in the passive film. Same set of parameters as for Figure 3.15a. (From Gabrielli, C. et al., *Mater. Sci. Forum*, 185–188, 631, 1995; Gabrielli, C. et al., *Electrochim. Acta*, 41, 1217, 1996.)

$$\left(\frac{\Delta m}{\Delta Q}\right)_{\omega \Rightarrow 0} = \frac{1}{\Gamma}\left(\frac{\Delta m}{\Delta E}\right)_{\omega \Rightarrow 0}$$

is fixed by the mass equivalent to the film growth. Experimental and simulated complex electrogravimetric transmittances are shown in Figure 3.16.

The negative real limit at low frequencies reveals that the mass loss by the corrosion current I_c is much larger than the mass gain by the film growth. In agreement with values of the capacitance Γ, this limit, −12 mg V^{-1} cm^{-2} is about 10-fold that of a 2.5 nm V^{-1} thick layer of Fe_2O_3. In contrast, the humped shape in the positive real part region at high frequencies visualizes the mass gain due to film growth. It is noteworthy that this takes place in the same frequency range as that covered by the inductive loop. With this set of parameters, which may be not optimal, both features are larger on simulated than on experimental curves. Paired contribution of these transmittances confirmed that

- Transient is controlled by film interface with respect to bulk
- One intensive property of the film, probably its ionic conductance, is at the origin of the inductive behavior
- The major part of the charge supplied to the film produces corrosion rather than film growth

3.2.2.1.4 Passive Nickel and the Point Defect Model

The passive state of nickel shows much more complicated behavior than for iron in the same solutions. A minimum of dissolution current is most often observed in the middle of the passive range, and increasing dissolution takes place at higher potentials (transpassive range). It is now well established that passive and transpassive dissolutions are so tightly related that they cannot be dealt with separately. According to this behavior, the film is expected to undergo important modifications in its structure and thickness as a function of potential and solution composition.

An advanced form of the point defect model [130,131] has been applied to the passivation of Ni in neutral solutions in a series of papers [145,148,149]. The model involves many more assumptions and parameters than the high-field migration and is able to account for a broad spectrum of electrochemical responses to potential, pH, and cation concentration. Some aspects of the model were investigated by impedance in spite of the extremely low frequencies involved in the processes and control by cation vacancies was concluded [148,149].

3.2.2.1.5 Passive and Transpassive Dissolution of Nickel in Acidic Solutions

The kinetics of nickel dissolution in the passive and transpassive ranges M remained totally unclear until the application of a very low–frequency impedance technique. A general model was proposed on the basis of an extensive study of anion effects [150]. In the passive state, the frequency domain had to be extended far below 1 mHz and long-term stability was obtained only by using single-crystal electrodes [151].

Figure 3.17 shows two current–voltage profiles in molar sulfuric and phosphoric solutions. The behavior contrasts drastically with that of iron; a minimum is reached close to 5 μA cm^{-2}, thus suggesting the contribution of two antagonistic processes. This is substantiated by the impedance data of Figure 3.18 for the phosphoric medium (the lower frequency limit is less than 100 μHz). Diagram Ap has the characteristic shape associated with passivation kinetics (see earlier).

At the minimal current density, Bp, typical diffusion control is observed consistent with the local plateau of the current–voltage curve. A diffusion term is associated with a transport limitation across the passive layer, whereas the inductive loop at the low-frequency end visualizes the contribution of a decaying film protection at increasing potentials. Application of a diffusion impedance equation to a finite-thickness layer ($\delta = 3\,\text{nm}$) given by

$$Z_D = R_D \frac{\tanh\sqrt{(j\omega\delta^2/D)}}{\sqrt{(j\omega\delta^2/D)}} \tag{3.42}$$

led to a diffusivity D of the order of 10^{-16} $cm^2\ s^{-1}$, This value is in agreement with solid-state diffusion at room temperature, and a decrease of the film thickness by a factor of 3 takes place between C_p and E_p.

An early paper [152] claimed on the basis of linear Tafel plots that the dissolution in the transpassive range is controlled by a single charge transfer reaction. It was then suggested

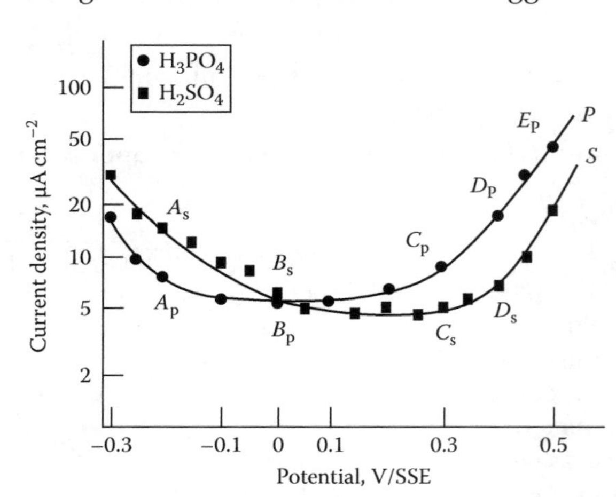

FIGURE 3.17
Steady-state polarization curves for nickel, (111) single crystal, in 1 M H_2SO_4 and 1 M H_3PO_4 solution, 25°C. (From Keddam, M. et al., *Corros. Sci.*, 27, 107, 1987.)

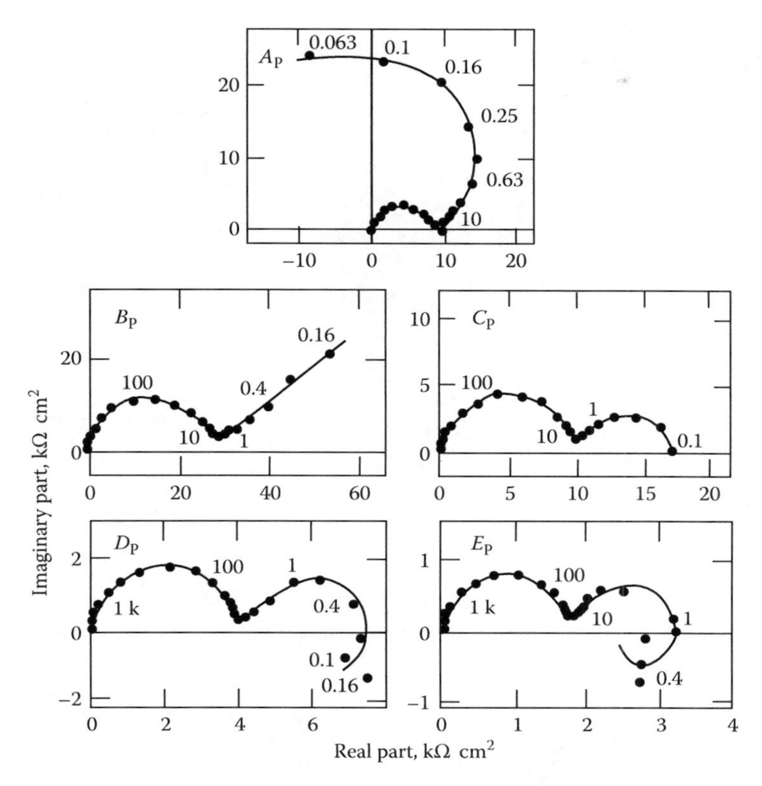

FIGURE 3.18
Impedance diagrams of passive Ni, (111) single crystal, in 1 M H_3PO_4 solution. Potentials labeled in Figure 3.17. (Frequency in millihertz.) (From Keddam, M. et al., *Corros. Sci.*, 27, 107, 1987.)

that the electrochemical impedance behaves as a resonant circuit, i.e., at least two relaxation times [153]. This prediction was directly verified shortly after Ref. [154]. Finally, a thorough investigation by impedance analysis at various pH values and anion concentrations was at the origin of a more elaborate model.

Figure 3.19 shows the current–voltage characteristics measured on three different crystal plan orientations. As reported in Ref. [152], a Tafel law compliance is observed over more than two decades, but the impedance shapes display surprisingly intricate kinetics. A very characteristic impedance feature is observed at point B (but not on the polarization curves), where the impedance behaves like a perfect wave-trap circuit (the impedance becomes infinite at a resonant frequency ≅ 0.4 Hz). Above this transition point, e.g., C, the diagram is totally consistent with the electrochemical resonator described in Ref. [153].

Figure 3.20a shows a detailed evolution of the impedance. The impedance diagram turning counterclockwise when the frequency is decreased was related to the reaction pattern shown below elaborated and compared with experiments by numerical simulations [155].

$$\mathrm{Ni} \underset{K-1}{\overset{K1}{\rightleftharpoons}} \mathrm{Ni(II)_{ads}} \underset{K-3}{\overset{K3}{\rightleftharpoons}} \mathrm{Ni^{*}(II)_{ads}} \underset{K-5}{\overset{K5}{\rightleftharpoons}} \mathrm{Ni(II)_{ads}}$$

$$\mathrm{Ni(II)_{ads}} \xrightarrow{K2} \mathrm{Ni(II)_{sol}} \xleftarrow{K4} \mathrm{Ni^{*}(II)_{ads}}, \qquad \mathrm{Ni(II)_{ads}} \xrightarrow{K_6} \mathrm{O_2} \tag{3.43}$$

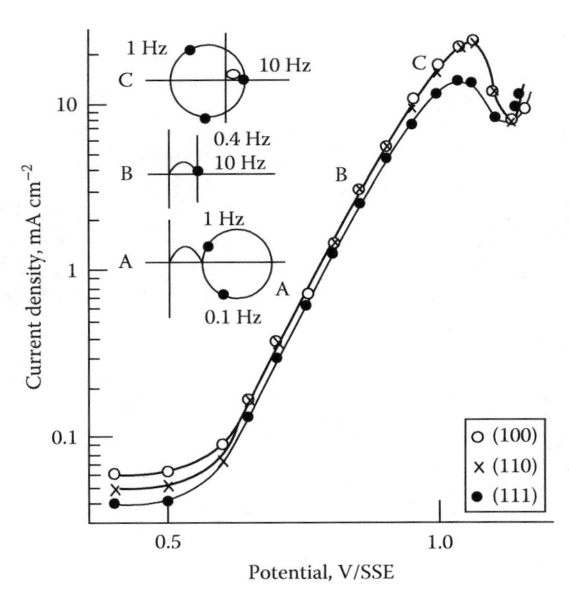

FIGURE 3.19
Current–potential curves and schematic impedance diagrams of transpassive Ni, three different single crystal orientations. 1 M H_2SO_4. (From Keddam, M. et al., *J. Electrochem. Soc.*, 132, 2561, 1985.)

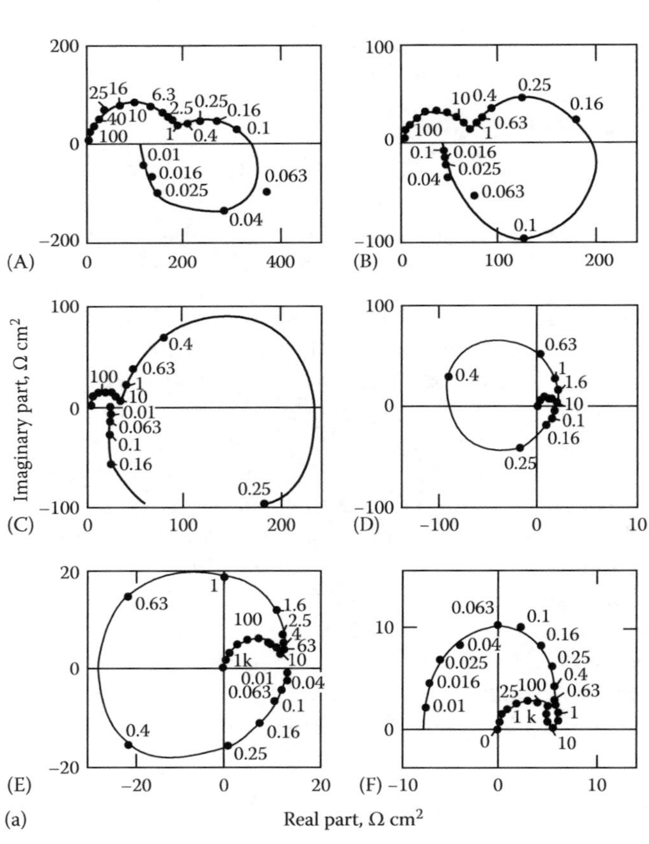

FIGURE 3.20
Experimental and calculated impedance diagrams of transpassive nickel (111) single crystal. 1 M H_2SO_4, 25°C. (a) Detailed evolution; (b) numerical simulations. Measurement points are: (A) 0.75 V/SSE; (B) 0.8 V/SSE; (C) 0.8 V/SSE; (D) 0.9 V/SSE; (E) 0.95 V/SSE; (F) 1.1 V/SSE. (From Keddam, M. et al., *J. Electrochem. Soc.*, 132, 2561, 1985.)

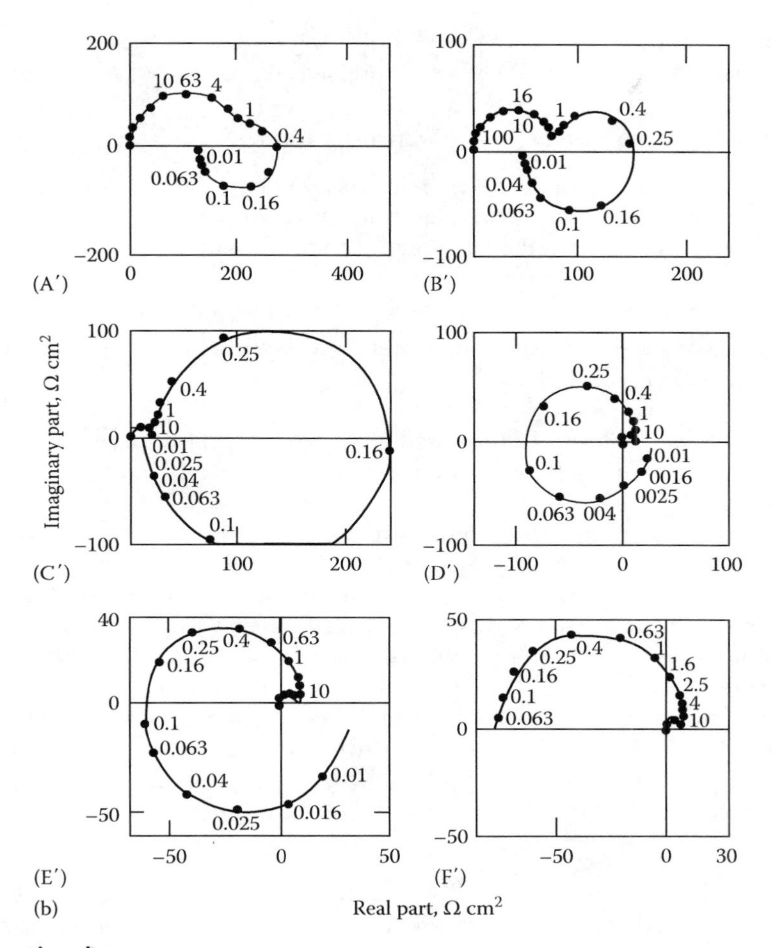

FIGURE 3.20 (continued)

Here Ni(II) stands for Ni cations pertaining to the lattice of the passive oxide film and Ni*(II) is a cation in the film solubilized by chemical bonding with anions.

The likelihood of this species is also supported by the evidence of $Ni{=}SO_4$ band structures in the in situ Raman spectra of the $Ni–H_2SO_4$ interface in this particular potential region. At higher potentials, all the film cations are converted into Ni(III) and Ni enters the secondary passivity region. Numerical simulations are presented in Figure 3.20b and reproduce with a fair accuracy the main features of experimental data, particularly the typical change of impedance trajectory between C′ and D′.

Anodic dissolution of chromium in the transpassive range as well shows a behavior generally attributed to the oxidation of the Cr(III) of the passive film to soluble Cr(V1) oxyanions [43,156,157].

Transpassive dissolution of chromium, molybdenum, and alloys containing these elements has been extensively investigated by Bojinov et al. [158–160] using impedance analysis and steady-state RRDE. Additional in situ information concerning the conductance properties of the films is obtained by contact electric resistance (CER).

Many-parameter models were proposed combining reaction steps coupled by adsorbed species and bulk conduction in 3D layers controlled by the mass balance of vacancies.

3.2.3 Anodic Dissolution under Mass Transport Control

In corrosion practice, the rate of the heterogeneous anodic processes dealt with in the foregoing sections is ideally low enough that mass transfer is never a limiting factor. However, fast dissolution kinetics may occur in the following circumstances of obvious concern to corrosion:

- Active dissolution as a transient regime in the initial stages of metal passivity
- Dissolution of locally depassivated metals following passivity breakdown (e.g., pits, crevices, grain boundaries)

Because of the irreversibility of the dissolution reactions at high overvoltages, the coupling to mass transfer cannot be simply understood in the framework of classical transfer-diffusion theory:

A high concentration of dissolved cations in the Nernst boundary layer is incompatible with the supporting electrolyte approximation.

Limitation of the reaction rate by mass transfer must imply either back diffusion of an acceptor of metal cations (complexing or solvating species) toward the electrode surface or growth of a new phase on the surface with limitation of the reaction rate by ohmic drop or space charge overvoltage.

Little is understood at a mechanistic level because of an almost total lack of knowledge of the local chemistry within the boundary diffusion layer or solid films. The problem is essentially investigated in the framework of pitting, crevices, and SCC [161] by modeling and experiments on occluded cells (artificial pit designing).

3.2.3.1 Mass Transport Control and Corrosion

It is generally accepted that transport effects (diffusion and electromigration) are at the origin of a buildup of metal cation concentration concomitant with a depletion of H^+ and a overconcentration of anions (for electroneutrality reasons). Homogeneous chemical reaction can add their own contribution, e.g., hydrolysis of Cr(III) resulting in a local acidification of neutral media [162].

The most relevant consequence of the change of interfacial chemistry for corrosion is undoubtedly the modification of the active–passive transition and the possible generation of self-sustained large-amplitude oscillations between active and passive states [163,164]. These phenomena were tightly associated with the manifestations of passivity from the earliest research, particularly in the case of iron in acid solutions. Theoretical investigations imply difficult nonlinear mathematics. A significant renewal of the field was observed in relation to the increasing interest in nonlinear phenomena and such concepts as bifurcation theory and chaos in chemistry [165–167].

In spite of this new sophistication, the basic model is still that due to Franck and FitzHugh [163] in which the following reactions are supposed to participate in a cyclic sequence.

An oversimplified form of dissolution and passivation reaction is considered:

$$Fe \rightarrow Fe^{2+} + 2e^- \tag{3.44}$$

$$Fe + nH_2O \Leftrightarrow FeO_n + 2nH^+ + 2ne^- \tag{3.45}$$

Reaction (3.44) at high dissolution rates tends to decrease the pH by H^+ electromigration and to shift the passivation equilibrium reaction (3.45) to the right. At the same constant potential, the iron surface turns into the passive state, and back diffusion of H^+ takes place and restores the active dissolution by displacing the passivation equilibrium to the left. The same sequence restarts, generating a periodic regime. It should be observed that this model ignores the likely participation of salt layers ($FeSO_4$, $7H_2O$) and of ohmic drop and is unable to explain the coexistence at a same electrode potential of side-by-side active and passive areas, i.e., of stable localized corrosion coexisting with a passive sample. The latter self-stabilizing process of dissolution is very generally ascribed to the depassivating role of Cl^-, enriched by electromigration toward the active areas. However, similar behavior can be observed in chloride-free solutions [33,81].

3.2.3.2 Active Dissolution of Iron under Mass Transport Control

The anodic dissolution of iron in sulfuric acid medium exhibits, with well-defined convection on a RDE, a current plateau as shown in Figure 3.21, attributed to the limiting rate of transport between the dissolving surface and the solution bulk [168]. The curves shown in Figure 3.21 suggest that some critical phenomenon is associated with the common branching point from which all the curves merge toward their plateau region, whereas they overlap perfectly below. The nature of the process controlling the heterogeneous steps was investigated by a detailed analysis of the electro-hydro-dynamic (EHD) impedance.

EHD impedances are interpreted in terms of the existence of a viscosity gradient in the boundary layer estimated to be four- to fivefold between solution bulk and electrode surface. On this basis, the profiles of species participating in the mass and charge transport have been computed and are plotted in Figure 3.22.

The interfacial concentration of Fe^{2+} is in good agreement with solubility data. The increase of pH by H^+ migration is reflected in the profile of SO_4^{2-}. It is concluded that no barrier layer is involved in the mass transport control of iron dissolution. The heterogeneous reaction rate is reported as not dependent on the HSO_4^- concentration; therefore, the only acceptor species likely to limit the dissolution rate is water [81]. This is compatible

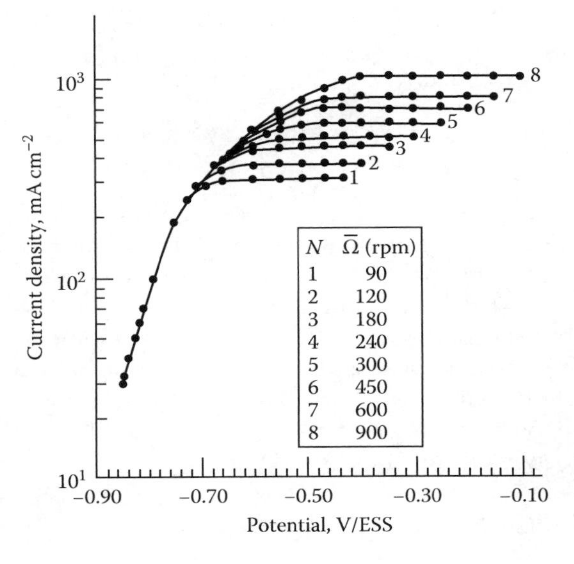

FIGURE 3.21
Current–potential curves for iron at different rotation speeds. Johnson-Matthey iron. 1.8 M H_2SO_4, pH = 0, 25°C. (From Barcia, O.E. et al., *J. Electrochem. Soc.*, 139, 446, 1992.)

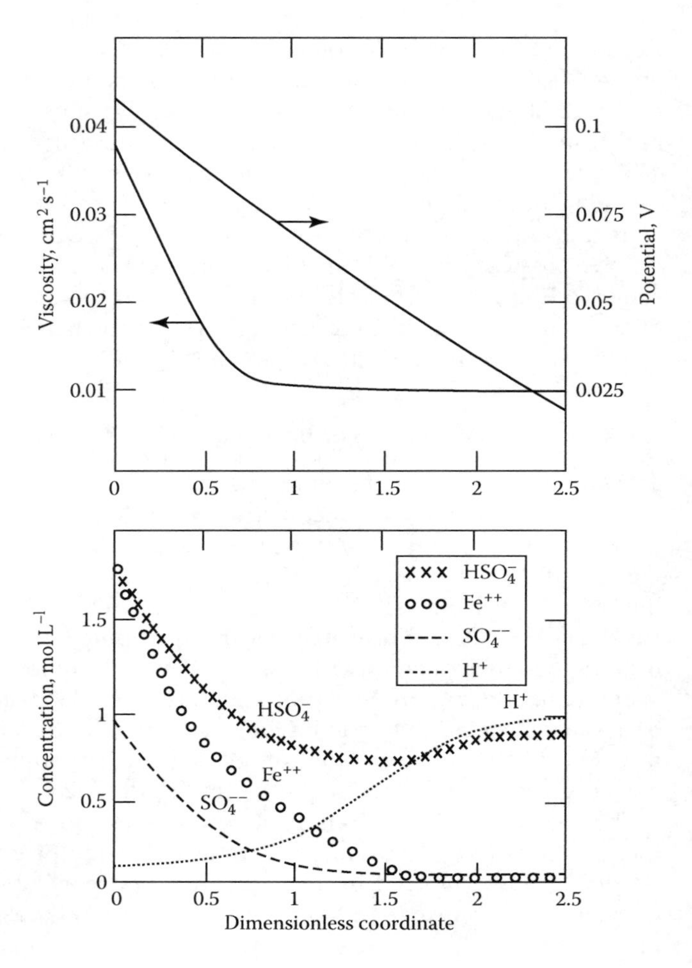

FIGURE 3.22
Concentration field calculated under the conditions of Figure 3.21. Top: viscosity and potential profiles. (From Barcia, O. E. et al., *J. Electrochem. Soc.*, 139, 446, 1992.)

with the modified form (3.29) of the initial step of dissolution in which one water molecule is dissociated and the depletion of free water at the electrode surface by Fe(II) hydration and electromigration of hydrated Fe(II) away from the surface.

3.2.3.3 Anodic Dissolution Controlled by Transport in the Presence of Solid Layers

In many practical cases of corrosion, the anodic dissolution takes place on a surface partially or totally covered by thick (order of micrometers) solid layers grown by several possible mechanisms (heterogeneous nucleation and growth, dissolution–reprecipitation) and usually regarded as porous. Until the past decade, the mechanism by which this layer interferes, with the anodic dissolution, and particularly the transport phenomena, remained poorly understood. The last developments deal with iron and copper in HCl solutions and take advantage of steady-state RDE and RRDE associated with AC impedance (EIS and EHD) techniques. Mass transport and solution chemistry [169,170] let to a dissolution model [171] in which the electrochemical monoelectronic surface step

$$Cu + Cl^- \Leftrightarrow CuCl_{ads} + e \qquad (3.46)$$

is followed by a series of complexation reactions, the more likely being

$$CuCl^-_{ads} + Cl^- \Leftrightarrow CuCl_2^- \tag{3.47}$$

More advanced RDE studies were interpreted by the formation of a porous 3D layer of $CuCl_{ads}$ [163] in a step such as

$$CuCl_{ads} \rightarrow CuCl_{film} \tag{3.48}$$

Because of the reversibility of reactions (3.46) and (3.47), both Cl^- and $CuCl_2^-$ diffusing in opposite directions may control the dissolution rate. In the plateau region, it was shown [171] by RRDE that compact coverage is attained. Therefore, in the model elaborated in Ref. [172] the contribution of the layer to the rate of dissolution is twofold and constitutes the central ingredient of the model: the CuCl film is considered at the same time to dissolve at its outer boundary with the electrolyte

$$CuCl_{film} + Cl^- \rightarrow CuCl_2^-$$

and to limit the rate of Cl^- diffusion to the Cu surface to form the film by reaction (3.46). This form of the model is sketched in Figure 3.23.

It is noteworthy that this model is formally similar to the point defect model (transport of a metal acceptor toward the metal surface, film growth at its inner interface, dissolution at its outer one). EIS and EHD responses associated with this model have been derived [172] taking into account the diffusion of both Cu^- and $CuCl_2^-$ in the solution (diffusion layer δ) and of Cl^- through the film (diffusion layer λ) and allowing for the modulation of the layer thickness by the AC potential, an effect discarded in Ref. [171]. Comparison with the experiments substantiated the model. The mass transport by diffusion and migration and the potential drop at soluble anodes covered by salt films have been dealt with [173,174] by assuming compact (saturated) or porous (supersaturated) coverages (wet) on iron. Even more intricate situations on nickel have been tentatively approached [175].

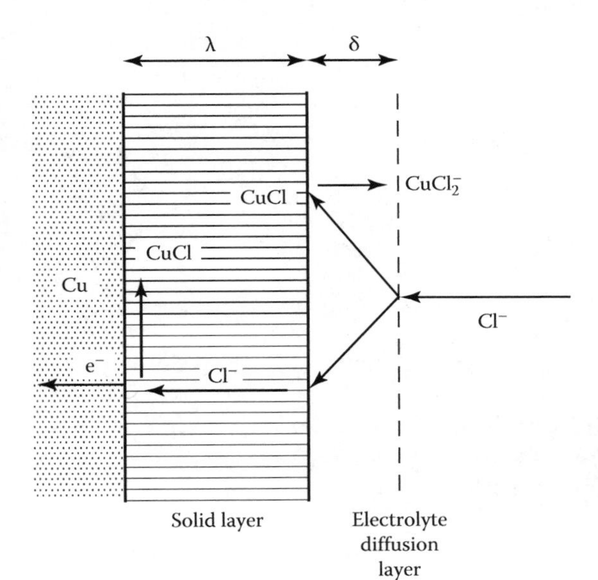

FIGURE 3.23
Model of dissolution of copper through a solid layer growing by a Tafel kinetics at the metal interface and dissolving by a chemical kinetics at the solution interface. (Adapted from Barcia, O.E. et al., *J. Electrochem. Soc.*, 140, 2825, 1993. With permission.)

3.3 Anodic Behavior of Alloys

A detailed understanding of the electrochemistry of alloys is still far from being achieved on the basis of the rather crude theory of pure metals. The complexity is expected to increase more rapidly than the number of components (cross-interactions) and it is doubtful whether a unique mechanism can be at work in view of the extreme variety of compositions. As far as mechanistic information is concerned, only homogeneous (single phase) and binary alloys are considered in this section.

3.3.1 The Basic Concepts: A Survey

3.3.1.1 Thermodynamics and Rate Constant Approaches

It was thought very early that the change of rest potential and of dissolution rate with alloy composition cannot be understood without assuming surface modifications induced by the anodic processes themselves [176]. The nature of these modifications has been the aim of many electrochemical and surface analysis studies but is still not fully elucidated.

In an ideal case, in the absence of any free energy of mixing, the free energy of a metal in an alloy with respect to that of the pure metal is given by

$$\Delta G = RT \ln x$$

where x is the molar fraction of the metal in the alloy, and the corresponding shift of electrode potential is

$$\Delta E = -\frac{RT}{zF} \ln x$$

This equation is obeyed only by a few alloys and dilute amalgams [10]. The individual dissolution rates of Cu and Ni in Ni–Cu were investigated as a function of alloy composition [176–178]. In the framework of the activated state theory, a standard rate constant of the solution of the ith component of an alloy, $k°ia$, is calculated from the exchange current density at the standard potential:

$$(i_0)_i = Z_i F C_i \gamma_i a_{iz^+} (1 - \alpha_{ia}) k^o_{ia} \tag{3.49}$$

where

$\gamma_i C_i$ is the activity of the element in the alloy
a_{iz^+} is the activity of the element cation in the solution
α_{ia} is the transfer coefficient of the metal i-cation equilibrium

The $k°ia$ is found independent of the alloy composition up to 60% Cu, above which a new phase is invoked.

This purely formal kinetic treatment is of very little utility because it is quite unable to dissociate the more elementary contributions to the alloy electrochemistry, namely from metal solid-state physics to surface physical chemistry:

1. Change of elementary rates of charge transfer by electronic interaction in alloy structure (alloying effect)

2. Change of the surface composition of alloy by selective dissolution of the less noble element
3. Growth of 2D or 3D layers resulting from chemical or electrochemical reactions of the alloy components with the solution

Following Rambert and Landolt [179], dissolution of single-phase alloys can be classified in two categories:

1. *Simultaneous dissolution*, where at steady state the alloy elements go into the solution at a rate proportional to their atomic concentration in the alloy. Well-known examples pertain to the iron-based stainless steels and especially ferritic Fe–Cr, on which this type of dissolution is repeatedly found [180,181].
2. *Selective dissolution*, where the less noble element dissolves selectively leaving behind a "porous" metal phase enriched in the more noble constituent, resulting in dealloying. The more common case is the selective dissolution of Zn (dezincification) from α-brass (Cu–Zn). Selective dissolution is in some instances associated with circumstances of SCC [182].

3.3.1.2 Simultaneous Dissolution and Associated Formalism

The basic formulation was elaborated sometime ago [182–184]. A linear relationship of the form

$$I_{AB} = I_A\gamma_0 + I_B(1-\gamma_0) \tag{3.50}$$

was suggested for representing the dissolution current of an alloy AB as a function of the elementary current densities of A and B, γ_0 and $(1 - \gamma_0)$ being the atomic proportions of A and B [commonly denoted $A - (1 - \gamma_0)B$].

It was simultaneously assumed [185] that, as a result of the alloy dissolution, its surface composition, $(\gamma, 1 - \gamma)$, can deviate from the bulk one. Consequently, the proportionality of the atomic fluxes of A and B going into the solution to their alloy content is expressed by

$$\frac{Z_B I_A \gamma_0}{Z_A I_B (1-\gamma_0)} = \frac{\gamma}{1-\gamma} \tag{3.51}$$

where Z_A and Z_B are the dissolution electrovalences.

In principle, Equation 3.51 yields a γ_0 value matching simultaneous dissolution at any potential, I_A and I_B being known from the individual curves of A and B in the same electrolyte. In general, there is no particular reason why this γ_0 would restitute the correct value of I_{AB}. In the case of Fe–Cr alloys, the essential features of the current–voltage characteristics in the transition range (7%–12%) are not satisfactorily reproduced [185] and nonlinear interactions must be introduced [112,113]. Application of EIS to this problem in the case of Fe–Cr alloys in the active and passivation ranges is illustrated in the following. In the full passive domain, compositional changes in the passive film, determined by XPS, play a role similar to the adjustment of the surface composition of the alloy phase to achieve the simultaneous dissolution [186]. According to Ref. [187], percolation phenomena, apparently relevant to selective dissolution, could also explain some features of the dissolution of Fe–Cr in the incompletely passivated state.

3.3.1.3 Selective Dissolution

Different mechanisms have been proposed to explain the selective dissolution of alloys and the formation of a porous dealloyed layer. A dissolution–redeposition mechanism [187] was proposed for α-brass dealloying. A larger group of models requires the description of atomistic processes of restructuring of the more noble atoms A in order to allow the dissolution of the more soluble element B to proceed across an A-enriched porous layer. A critical review of the mechanisms likely to participate in these surface phenomena can be found in Ref. [32]. It was suggested [189] that the rate-determining step is a solid-state diffusion of the less noble atoms via divacancies. Surface diffusion can also be taken into account. Roughening by a mechanism of "negative" aggregation known to generate fractal interfaces [168] constitutes a fruitful approach. In the 1980s, a series of papers [182,187,191] underlined the existence of critical compositions and of a threshold concentration of the less noble constituent below which no dealloying appears. This sharp dealloying threshold is not consistent with any of the diffusion-based models of selective dissolution, which are supposed to produce essentially continuous behaviors. A model based on percolation phenomena was proposed and extensively worked out. Its main background and developments will be exposed.

3.3.2 Simultaneous Dissolution: Fe–Cr Alloys

In contrast to selective dissolution, evenly dissolving alloys can be dealt with, up to a sophisticated level including nonsteady-state responses, by macroscopic, i.e., kinetic descriptions. As shown in Figure 3.24, Olivier [192] pointed out that the steady polarization curves of Fe–Cr alloys in a 0.5 M sulfuric solution display the decay of active and passive currents with increasing Cr content and the emergence of the transpassive dissolution of Cr to the hexavalent state.

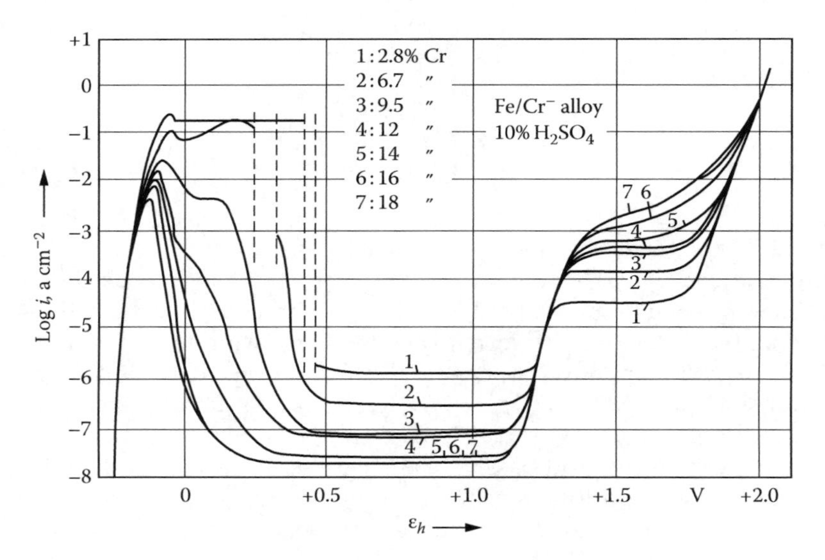

FIGURE 3.24
Current–potential curves of iron–chromium alloys as a function of Cr content. (From Moreau, A. et al., *Electrochim. Acta*, 27, 1281, 1982.)

$$
\begin{array}{ccccc}
 & & Fe^*(I)_{ads} & \longrightarrow & Fe^*(II)_{ads} \\
 & & K_{-3}\downarrow\uparrow K_3 & & \downarrow K_4 \\
Fe(0) & \xrightarrow{K_1} & Fe(I)_{ads} & \xrightarrow{K_2} & Fe(II)_{sol}
\end{array}
$$

$$
Cr \overset{Q1}{\rightarrow} Cr(I)_{ads} \overset{Q2}{\rightarrow} Cr(II)_{sol} \qquad Cr(I)_{ads} \underset{Q-3}{\overset{Q3}{\rightleftharpoons}} Cr(II)_{ads} \tag{3.52}
$$

The studies have focused on two aspects of the behavior of Fe–Cr: the modifications brought about by the addition of chromium to iron in the active dissolution and prepassive ranges on the one hand and in the passive state on the other hand. Electronic interaction by filling up of the *d* level of Fe [193] was put forward. Most of the subsequent contributions concluded in some progressive change from an iron- to a chromium-like behavior without a further detailed mechanism in the absence of a satisfactory model for pure iron on its own. A particular ability of Cr to enhance the passive state of iron, even with low surface coverage, was repeatedly reported [194,195]. Regarding the passive state, advances of in situ surface analysis [Auger electron spectroscopy [AES] and XPS] associated with electrochemical measurements concluded in the formation of Cr-enriched passive films. Mechanistic aspects of the process have been investigated in detail by Kirchheim et al. [185,196,197] and several other groups [108,181,198].

3.3.2.1 Reaction Mechanism of the Anodic Dissolution of Fe–Cr

The anodic behavior of Fe–Cr alloys was investigated by using steady-state polarization curves and sampling by a large number of impedance measurements over the concentration domain (0%–22% Cr) in H_2SO_4–Na_2SO_4 media ($0 < pH < 3$) [112,113]. The OH^- and Cr contents are found to play very similar roles in the reaction mechanism. The more salient characteristics of the role of Cr are visible at 7% Cr in pH 0 solution, as shown in Figure 3.25a, which exhibits a two-peak polarization curve and two perfectly separated passivation loops.

The passivating feature attributed to Cr at point D (capacitive loop of characteristic frequency 4 Hz) is therefore already visible at the beginning of the active domain (point A). Simulations of the current–voltage curves and of the impedance diagrams were performed on the basis of a mechanistic description derived from the reaction pattern previously elaborated for pure iron [68,69]. Therefore, coverages by five surface species were introduced in the derivation. The whole body of data led to taking into consideration, at the same time, the three types of interaction between Fe and Cr listed earlier namely

1. Alloying effects
2. Surface composition of the alloy distinct from the bulk one
3. Interaction of surface species with the neighbor atoms of either of the two alloy components

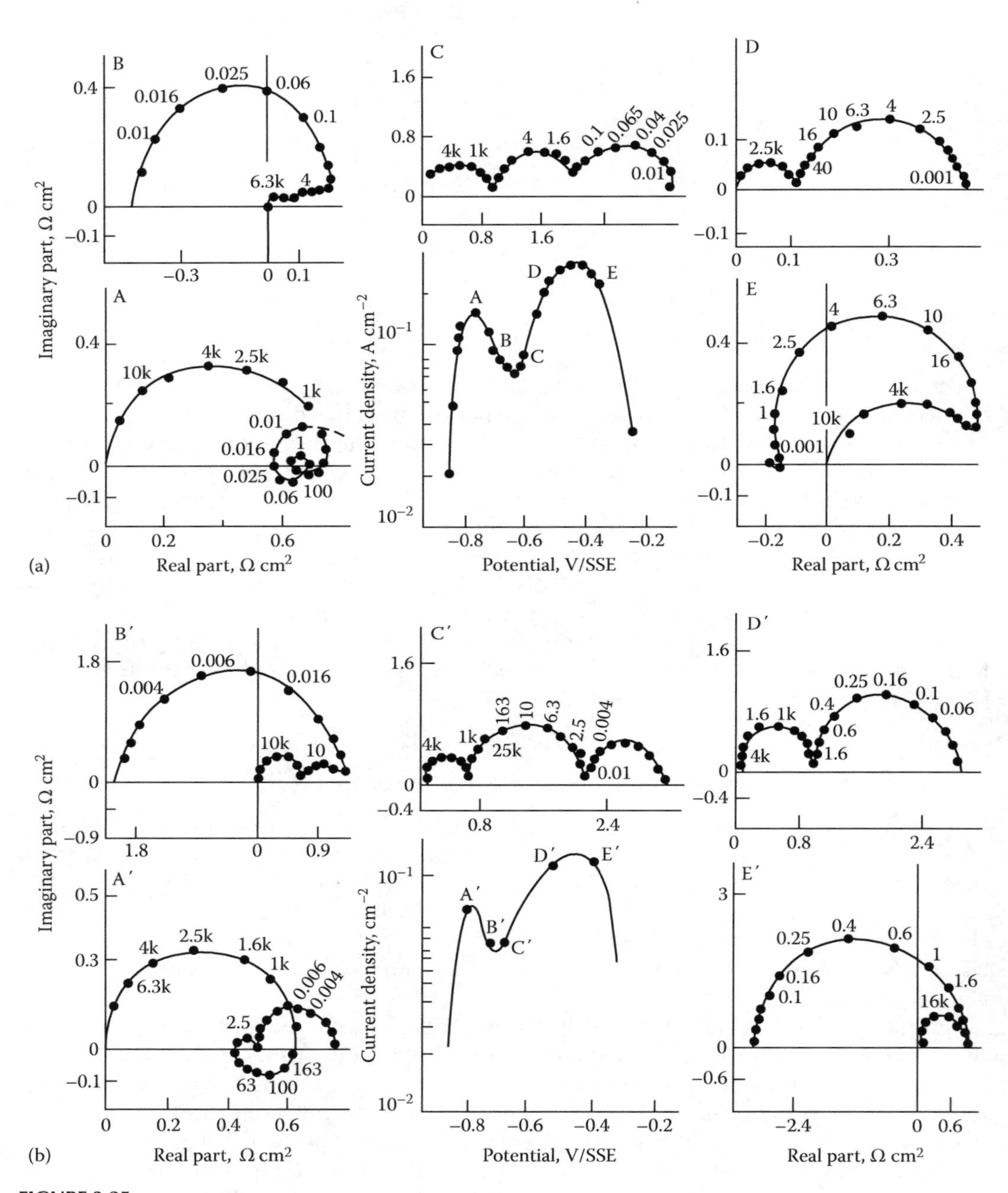

FIGURE 3.25
Experimental (a) and calculated (b) current–potential curves and impedance diagrams for a Fe-7Cr alloy (IRSID, France), 1 M H_2SO_4, 25°C. Impedance diagrams at the corresponding polarization points on the *I*(*E*) curve. Frequency in Hertz. Simulations according to the reaction mechanism (3.52). (From Keddam, M. et al., *Electrochim. Acta*, 31, 1147, 1986.)

The last interaction was found to be the determining one for interpreting the impedance data and the steep onset of passivity. The mechanism is shown below. A specific nonlinear interaction with the chromium passivating species $Cr(II)_{ads}$ was supposed to affect the rate of iron dissolution. Figure 3.25b shows the result of simulations of the polarization curves and impedance diagrams. The sharp decay of the iron dissolution at increasing coverage θ_{Cr}(III) by $Cr(II)_{ads}$ was expressed by introducing the factor $\left(1-\theta^{x}_{Cr(II)}\right)$ with $x \cong 0.5$. A similar interaction function is adopted in the description of poisoning effects by foreign atoms in binary alloy catalysts. As emphasized for the first time in Ref. [113], this power law suggests that the threshold behavior arises from surface percolation of passivated chromium sites.

According to this interpretation, passivation would occur on an incompletely covered surface via long-range interactions between bare iron atoms and passive chromium connected in large clusters. A computer simulation later demonstrated the validity of this idea [191,199]. In the same paper [113], attention was also drawn to the atomistic description of the catalytic mechanism of dissolution (see earlier) in the case of a binary alloy. Therefore, purely macroscopic modeling of the kinetics of dissolution is no longer possible without consideration of the alloy microstructure, distribution of first and second nearest neighbors, and statistical composition at the atomic level. Concepts such as the size of clusters of atoms A and B and percolation phenomena [200] appear as determining and actually have been extensively worked out (see later).

3.3.2.2 Analytical and Mass Balance Approaches to Dissolution and Passivation of Fe–Cr Alloys

An important contribution is due to Japanese authors who developed rotating double-ring and split-ring disk electrode devices [198] and introduced channel flow arrangements [92] and data processing in the transient regime [108]. The respective behavior of Fe and Cr in the presence of Cl^- was investigated [198] for an Fe-30Cr alloy. It was concluded that chloride enters the soluble ferro-complex species while it behaves as an inhibiting species with respect to Cr. By employing time-resolved measurements on CFDE, previously introduced for iron studies [92], it was then shown [108] that in a sulfate medium the amounts of Fe and Cr dissolving at steady state are proportional to their concentrations in the alloy.

3.3.2.2.1 Active and Active–Passive Transition; Impedance and Frequency-Resolved RRDE Measurements

The association of impedance and frequency-resolved RRDE techniques initially introduced for iron [38,124,204] has been extensively applied to the prepassive and passivation ranges of Fe–Cr alloys with and without chloride added [181,201,202]. Of course, the interpretation is more intricate than for pure metals (even with multiple dissolution valences). For kinetic reasons, Cr species are not detectable on the ring. In order to draw unambiguous conclusions, reasonable assumptions had to be made concerning simultaneous alloy dissolution at steady state (see Section 3.3.1.1) and dissolution valences of Cr.

Completely different behaviors were found depending on whether or not chloride is present. The salient features are

Emission efficiency for Fe(II) greater than 1 (for Fe-12Cr), an apparent paradox for a pure metal because that violates the principle of electrical charge conservation

A positive imaginary part of N_d (passivating charge) in the absence of Cl^-

A negative imaginary part of N_d (dissolution intermediate) in the presence of Cl^-

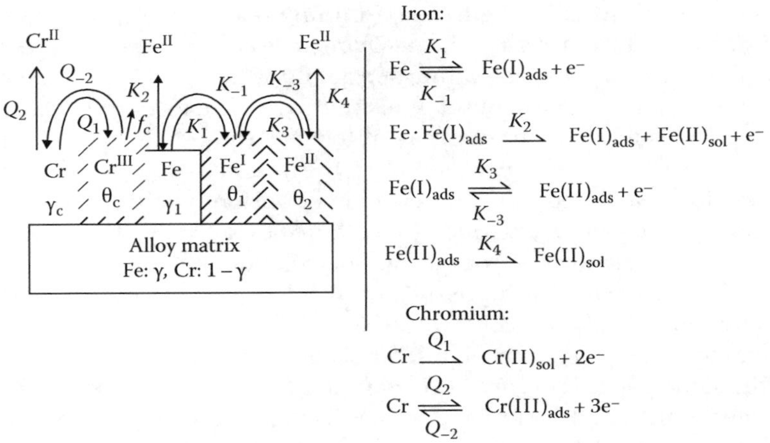

Iron:

$$Fe \underset{K_{-1}}{\overset{K_1}{\rightleftharpoons}} Fe(I)_{ads} + e^-$$

$$Fe \cdot Fe(I)_{ads} \xrightarrow{K_2} Fe(I)_{ads} + Fe(II)_{sol} + e^-$$

$$Fe(I)_{ads} \underset{K_{-3}}{\overset{K_3}{\rightleftharpoons}} Fe(II)_{ads} + e^-$$

$$Fe(II)_{ads} \xrightarrow{K_4} Fe(II)_{sol}$$

Chromium:

$$Cr \xrightarrow{Q_1} Cr(II)_{sol} + 2e^-$$

$$Cr \underset{Q_{-2}}{\overset{Q_2}{\rightleftharpoons}} Cr(III)_{ads} + 3e^-$$

FIGURE 3.26
Schematic representation of the reaction mechanism and of the topographical interaction of reaction step and surface coverages. (From Annergren, I. et al., *Electrochim. Acta*, 41, 1121, 1996.)

A model depicting the surface processes in terms of reaction paths and topographic interaction has been elaborated. It incorporates the key points previously introduced in the model based on impedances [37]. In particular:

Modification of the iron rate constants by chromium

Sharp passivation of the alloy represented by a nonlinear dependence of the blocking on the Cr species coverage

The reaction model and the corresponding surface picture are shown in Figure 3.26. It accounts for the nature and position of the first-order nearest neighbor for making explicit the interaction and mass balance of the surface species. This kind of approach was introduced in pioneering work [114] and then worked out in the framework of percolation models of passivation [187,199]. In view of the process complexity, an accurate fit is hopeless, but the main features could be semiquantitatively simulated, including the amazing $N_d > 1$. A comparison of the experimental and computed N_d relative to Fe(II) is displayed in Figure 3.27.

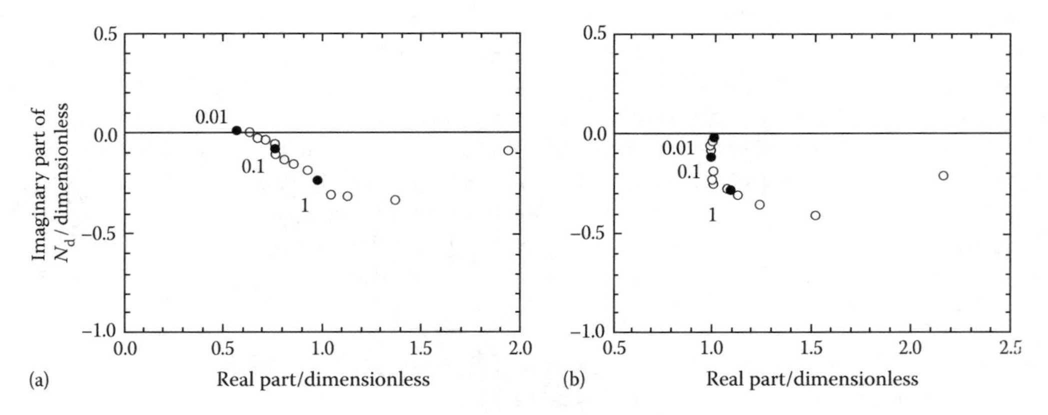

FIGURE 3.27
Emission efficiency of Fe(II) for Fe–12Cr alloy in 0.5 M H_2SO_4. In the active dissolution ([a] E = 0.07 V SSE^{-1}) and passivation ([b] E = 0.18 V SSE^{-1}) ranges. (From Annergren, I. et al., *Electrochim. Acta*, 41, 1121, 1996.)

According to the model, this extra emission of Fe(II) with respect to the electrical current across the electrode can be understood in terms of the contribution of chemical dissolution of an Fe(II) species by a step such as K_4.

Later on, the same model was considered for interpreting the results in chloride-containing media. The main modification was to enhance considerably the rate of dissolution (catalytic) via the $Fe(III)_{ads}$, whereas in Cl^--free media most of the iron dissolves from the $Fe(II)_{ads}$ (step K_2). The resulting change in the sign of N_d arises from the role of the dissolution intermediate of $Fe(II)_{ads}$. These results supported to some extent those of Ref. [198].

3.3.2.2.2 *Dissolution in the Passive State*

In a series of papers by Kirchheim et al. [180,196,197], the dissolution rates of Fe and Cr in Fe–Cr alloys were investigated within their passive range in H_2SO_4 solution. Time-resolved chemical analysis of the solution was performed by atomic absorption spectroscopy of samples of electrolyte. Selective dissolution of iron during the transient passivation stages was exploited in terms of Cr enrichment in the passive layer, and once the steady state was reached, simultaneous dissolution was accurately verified.

Figure 3.28 is an illustration of the time-resolved monitoring of the percentage of Cr in the dissolution products during a phase of film growth triggered by a galvanostatic square pulse [180]. In-depth concentration profiles of Fe and Cr in the film were concomitantly

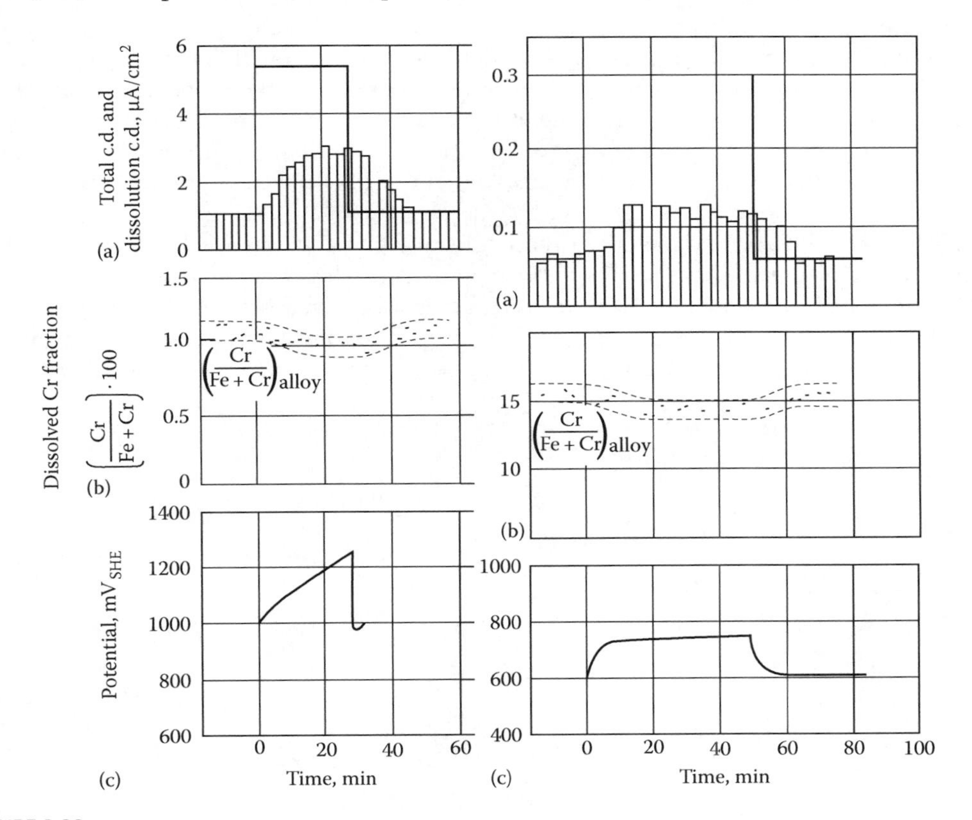

FIGURE 3.28
Galvanostatic transients for the dissolution (bars in [a]), the Cr fraction in the dissolution products (b) and the potential (c) for two different alloys (left: 1 at% Cr, right: 14 at% Cr) after attainment of steady state. (From Heine, B. and Kirchheim, R., *Corros. Sci.*, 31, 533, 1990.)

measured by XPS, and the significant enrichment in Cr for corrosion in the passive state was attributed to the outer first layer of the film. Equation 3.51 expressing simultaneous dissolution is applied to both the alloy phases (a) and the film outer layer (e):

$$i = i_{\mathrm{Fe}} + i_{\mathrm{Cr}}$$

and for the same dissolution valences of both components (3+).

$$i_{\mathrm{Fe}} = i x_{\mathrm{Fe,a}}$$

$$i_{\mathrm{Cr}} = i x_{\mathrm{Cr,a}}$$

where the x's refer to the molar fractions in the alloy. At the outer side of the film, the same partial currents are supported by the molar fractions of cations $x_{\mathrm{Fe,e}}$ and $x_{\mathrm{Cr,e}}$. The ratio of the individual dissolution rates of each of the elements from the metal (a) and the first layer of the passive film (e) is reflected in the dissolution currents of the pure metal components in the same passive conditions, $i_{\mathrm{co,Fe}}$ and $i_{\mathrm{co,Cr}}$, so that

$$\frac{x_{\mathrm{Cr,e}}}{x_{\mathrm{Fe,e}}} = \frac{i_{\mathrm{co,Fe}}}{i_{\mathrm{co,Cr}}} \frac{x_{\mathrm{Cr,a}}}{x_{\mathrm{Fe,a}}} \tag{3.53}$$

Chromium enrichment in the film, according to Equation 3.51, is directly correlated with the lower dissolution rate of pure Cr with respect to pure iron. This is consistent with a lower diffusivity of this element in the film and tentatively explained by the high-field migration model. However, this is done at the expense of introducing, nonclassically, the diffusivity in the preexponential factor of Equation 3.40 [195]. After simple transformation, Equation 3.53, can be written as a reciprocal dependence of the passive current of the alloy with respect to its Cr content.

Figure 3.29 shows an example of the good agreement reported in Ref. [186] and considered as supporting the model. The model validity has been also discussed in the case of Fe–Mo and Fe–Al alloys [196]. Subsequently [197], precipitation from supersatured electrolyte in the initial stage of active dissolution was proposed as a determining step of the passive film growth on alloys with a low Cr content at low pH. This hypothesis can be traced back to the 1930s [11].

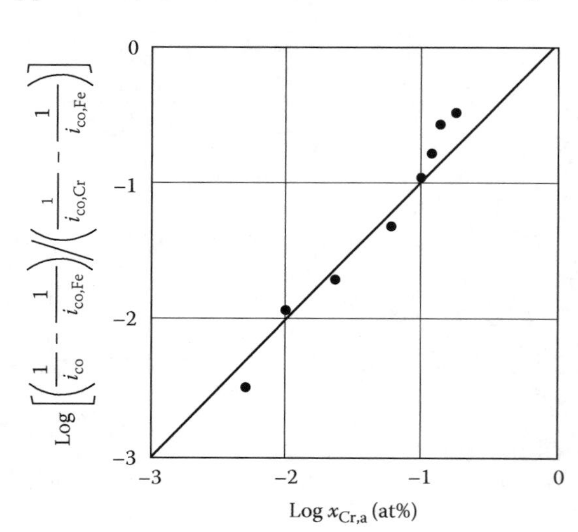

FIGURE 3.29
Relation between the passive current densities, i_{co}, and the concentration of chromium in Fe–Cr alloy, $x_{\mathrm{Cr,a}}$ according to the model relating the dissolution of the alloy to the Cr/Fe proportion at the outer side of the film. (From Kirchheim. R. et al., *Corros. Sci.*, 29, 899, 1989.)

3.3.3 Selective Dissolution

Conditions for simultaneous dissolution generally involve the formation of a 2D or 3D surface layer in which the relative dissolution rates of the elements are equal to their alloy fraction. Even in cases where simultaneous dissolution is obeyed, the process must begin by an initial period of selective dissolution at least at a 2D level [198]. As briefly mentioned earlier, many mechanisms have been invoked to explain how, for a nonpassive alloy, the dissolution of the more reactive element can continue across the dealloyed structure [32,201]. Surface changes have been attributed to surface diffusion and recrystallization, redeposition, short-range atomic rearrangement, and a roughening transition by capillary effects.

Electrochemical measurements coupled with solution analysis were progressively backed up by analysis of the surface and in-depth profiling of the alloy composition in the dealloyed layer [205]. Current–voltage, current–time, and potential–time relationships, as a function of the alloy composition and of the degree of selective dissolution, are the techniques employed extensively. Contributions to the field using this kind of approach can be found in Refs. [180,203,206].

3.3.3.1 Anodic Dissolution of Binary Alloys Studied by Electrochemistry, Solution, and Surface Analysis Techniques

These investigations have dealt with alloys such as AgPd, CuPd, NiPd, and AgAu under conditions of no passivation. Alloy–electrolyte (LiCl) combinations offer the possibility of observing both selective and simultaneous dissolution behaviors depending on the concentration of the noble metal. Two quantities are denned for describing the amount and in-depth property of the dealloyed surface:

e_B = surface excess of the more noble metal B:

$$e_B = \int_0^{\infty} (X_B(z) - X_{B,b})dz$$

where

$X_B(z)$ is the mole fraction of B at distance z from the surface (z)

$X_{B,b}$ is the bulk alloy

S_B is the selectivity coefficient (dimensionless):

$$S_B = 1 - \frac{Q_B}{Q_{B,th}} \tag{3.54}$$

where

Q_B is the charge corresponding to the amount of B dissolved (from solution analysis)

$Q_{B,th}$ its amount calculated for simultaneous dissolution by the same charge ($S_B = 1$ corresponds to complete selective dissolution of A, $S_B = 0$ to simultaneous dissolution)

The value of e_B can be estimated from potential–time transients or AES analysis, and the value of S_B is determined by weight loss and solution analysis. The potential increases with the dissolving charge passed at the interface in the selective dissolution stage, then reaches a critical value E_c and exhibits a plateau where either both components dissolve in stoichiometric proportion or only one keeps dissolving with formation of surface roughness. On Ag–Pd alloys the e_{pd} values at the critical potential, estimated by electrochemistry and

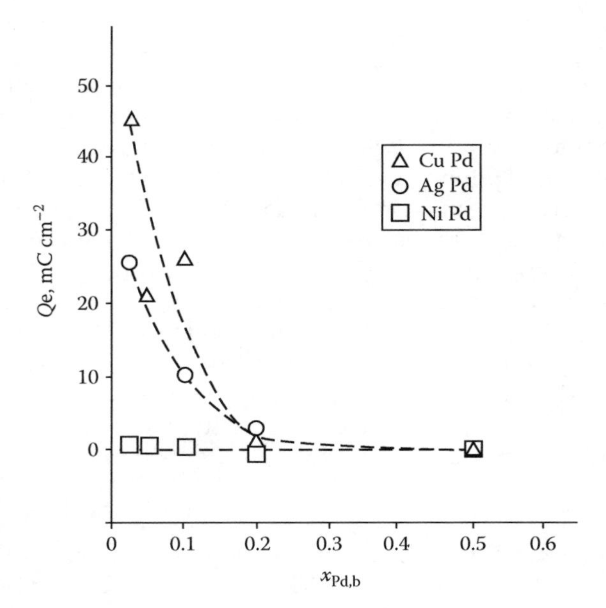

FIGURE 3.30
Measured charge corresponding to Pd enrichment as a function of Pd bulk mole fraction. (From Rambert, S. and Landolt, D., *Electrochim. Acta*, 31, 1433, 1986.)

AES, are in reasonably good agreement; they show a steady decay when the Pd content in the alloy is increased and become negligible above 50%. Figure 3.30 shows an example of this dependence for three alloys [206]. A threshold of Pd concentration is observed around 0.2, beyond which no Pd enrichment takes place. This can be understood in the framework of percolation theory (see the next section).

Some mechanistic data on surface enrichment can be inferred. Surface enrichment is essentially constant after a minimal amount of dissolution (5 mC cm^{-2}), independent of the dissolution rate. Redeposition and surface diffusion are ruled out in favor of short-range rearrangements of surface atoms. A reciprocal shift of binding energy between Ag and Pd atoms is assumed to participate to some degree in simultaneous dissolution of Pd at potentials lower than from the pure metal [163]. Quite identical conclusions are reached for Cu–Pd, Ni–Pd, and Ag–Au alloys [187].

Investigation of the anodic dissolution below the critical potential is likely to provide more relevant information on the mechanism of selective dissolution [185]. The emphasis is generally put on the current–time relationships, which are supposed to reflect the mechanism of transport within the surface layer. Dynamic polarization curves and current–time decays have been investigated for Cu–Au and Ag–Pd in LiCl and acidic sulfate solutions. In all cases, current–voltage profiles display a plateau followed by a steep increase at the same critical potential E_C evidenced on the potential–time transient. The shape of these curves is shown schematically in Figure 3.31.

In agreement with previous work, the current–time transients recorded at various potentials below the critical potential are correctly represented by a reciprocal power law:

$$I \propto t^{-m} \quad \text{with } 0.8 \leq m \leq 1 \tag{3.55}$$

regardless of the alloy composition. The interpretation suggested in Ref. [203] is derived from the model of divacancy diffusion [189]. Diffusivity of the vacancies is supposed to depend on concentration and potential. The resulting nonlinear diffusion is most probably responsible for a t^{-m} current dependence, as indicated by numerical simulations.

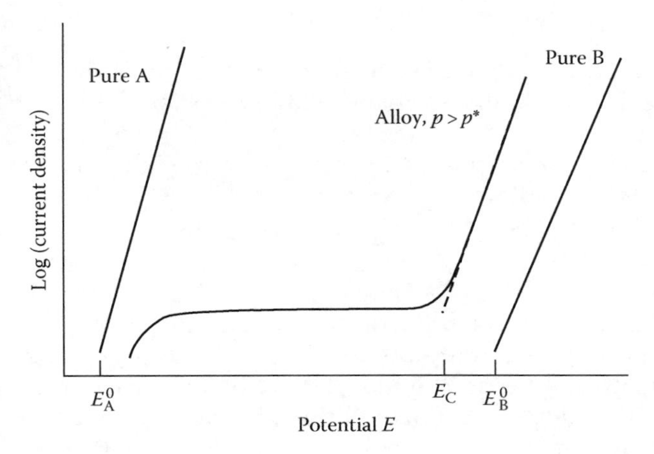

FIGURE 3.31
Schematic illustration of the anodic polarization curves for a binary A–B alloy with respect to the curves for the individual elements A and B. (From Sieradzki, K. et al., *Philos. Mag.*, 59, 713, 1989.)

Actually, even the sophisticated studies combining electrochemistry and surface analysis seem unable to yield any further decisive information on the detailed mechanism of selective dissolution. Atomic arrangements in which dissolution of atoms A can proceed throughout a rough electrode structure enriched in B type have been recognized as amenable to percolation theory [200]. The same theory was also applied to the passivation of binary Fe–Cr alloys in which, as suggested in Refs. [112,113], passivation of the more soluble element, Fe, is enhanced by a connected surface lattice of passive Cr atoms. The main power of percolation theory is to provide diagnostic criteria for computer simulations in terms of concentration thresholds.

3.3.3.2 Atomistic Modeling of Selective Dissolution and Related Passivation by Percolation Theory

The now popular concept of percolation has proved quite successful in many fields in which a macroscopic property depends on the existence of a connected path within a two-phase discrete medium, most often a regular 2D or 3D arrangement of sites (site lattice percolation). Typical features related to percolation are the existence of a critical phenomenon, for instance, a threshold concentration of conducting sites when conduction is considered, and of a power law dependence with respect to the critical quantity in the close vicinity of the critical point. Both the site percolation thresholds and the power exponents have well-established theoretical values for any given lattice geometry and connection rules [200].

In spite of its extensive use in the description of heterogeneous systems, including electrolytic crystal growth, it seems that the percolation concept was considered relatively recently in the field of electrode processes for explaining sharply varying properties of alloys. The characteristic feature of a critical concentration of Zn in α-brass and Al in Al–Cu alloy was correlated with the percolation threshold on an fee lattice [182].

Computer simulations of selective dissolution. An extensive contribution was reported [32] addressing the problems raised by experimental dealloying thresholds p^* at variance with theoretical site percolation thresholds p_c.

Possible interpretations are listed:

1. Dissolution from low-coordination sites (kinks and ledges) faster than from terraces
2. Not randomly distributed atoms in the alloy

3. Surface diffusion of B (noble atoms)
4. Contribution of first-order and higher order nearest neighbors to the rate of dissolution of A (soluble atoms) leading to a high-density percolation problem

Computer simulations with a Monte Carlo algorithm of the selective dissolution of 2D square and 3D cubic lattices modeling binary alloys (A-B) are performed.

3.3.3.2.1 Dealloying Thresholds

Dealloying thresholds p^* are denned by the value of p, the fractional occupancy by A-type atoms, at which vacancy percolation occurs; i.e., continuous dissolution pathways start penetrating all the way through the lattice. Of course, in the absence of any surface diffusion, vacancy percolation maps closely the percolation of A atoms in the initial lattice. This is seen in Table 3.2, where no diffusion or diffusion restricted to atoms B with zero or one first nearest neighbor (lines 1 and 3) generates p^* close to the theoretical percolation threshold on 2D and 3D lattices (0.6 and 0.31, respectively). On the contrary, easy diffusion tends to produce smooth front and layer-by-layer dealloying on 3D lattices as indicated by p^* close to the 2D percolation threshold (lines 5 and 6). This likely explanation of p^* larger than p_c was questioned by in situ scanning tunneling microscopy (STM) demonstrating that Ag atoms dissolve from terraces well below the 3D percolation threshold [207].

3.3.3.2.2 Morphology of Dealloyed Structures on 2D Lattices

The roughness of the dissolution front is characterized by its fractal dimension, d (estimated from the mass theorem [197]). Variation of d as a function of p is shown in Figure 3.32, displaying the classical transition toward the dimension of the infinite cluster at the percolation threshold ($d \cong 1.60$).

TABLE 3.2

Summary of the Dissolution and Diffusion Rules Used in 2D and 3D Computer Simulations

Rule		Dimension	Dealloying Threshold p^*	Percolation Threshold p_c
Dissolution when nearest neighbors ≤3	Diffusion when nearest neighbors ≤1	2	0.595 ± 0.005	0.6
Dissolution when nearest neighbors ≤3	Dissolution when nearest neighbors ≤2	2	0.35 ± 0.05	0.6
Dissolution when nearest neighbors ≤5	No diffusion	3	0.310 ± 0.005	0.31
Dissolution when nearest neighbors ≤5	Dissolution when nearest neighbors ≤3	3	0.225 ± 0.025	0.31
Dissolution when nearest neighbors ≤4	Dissolution when nearest neighbors ≤2	3	0.60 ± 0.05	0.31
Dissolution when nearest neighbors ≤3	Dissolution when nearest neighbors ≤3	3	0.65 ± 0.05	0.31

Source: Sieradzki, K., *Philos. Mag.*, 59, 713, 1989.

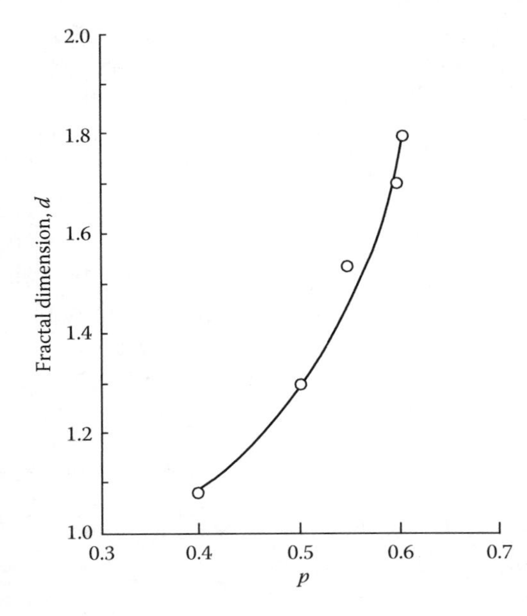

FIGURE 3.32
The fractal dimension d of the dissolution front as a function of p. For the 2D rule (line 1 in Table 3.2). (From Sieradzki, K. et al., *Philos. Mag.*, 59, 713, 1989.)

Easier diffusion of B produces features such as a smoother dissolution front, internal vacancy clusters (polyatomic voids), and islands of A-type atoms hindered from dissolution. Qualitatively similar conclusions are drawn on 3D lattices except for the specific generation of pores with easier diffusion of B atoms as predicted [208] by a nonstochastic approach. This tendency to generate a tunneling attack at the cost of only surface diffusion could be considered as a likely explanation of pit nucleation at the atomic level, with no need for the concept of passive film breakdown.

3.3.3.2.3 Kinetics of Dealloying and Structure of the Dealloyed Layer

The dealloying kinetics are represented by the time dependence of the flux of creation of vacancies, available from the computer data. The results are compared with the prediction by percolation theory, below and above the percolation threshold p^*. Below p^*, near-exponential decay of the current with time is predicted in agreement with experiments. Above p^*, experiments show a decay much steeper than expected, attributed to ohmic and mass transport effects in the porous interface. No mention is made of the negative power law decay reported in Ref. [203] (Equation 3.54).

3.3.3.2.4 The Critical Potential E_C

The nature of the critical potential (see Figure 3.31) was addressed in the framework of this model. No definite conclusions are established. E_c may be the balance point between the smoothing action of surface diffusion and the roughening action of dissolution. Another explanation regards E_c as a threshold potential above which percolation becomes possible within a connected subset of A-type atoms with lowered reactivity due to their atomic surrounding (high-density percolation problem).

The use of a percolation approach clarifies substantially the selective dissolution of single-phase binary alloys. Computer simulations support the existence of a tight relationship between dealloying thresholds and site percolation thresholds. However, a careful survey of the literature over the last 10 years leads to the striking constatation that this important piece of work is totally ignored by other groups active in the field. It should be emphasized that a significant dissolution of the noble metal is repeatedly mentioned, e.g., Pd in Ag–Pd alloys [179], and is not allowed in the percolation model of selective dissolution. Dissolution of both alloy components at different rates is taken into account in the model of passivation of Fe–Cr alloys [191,199], presented in the following.

3.3.3.2.5 *Curvature Effects, Roughening Transition, and Critical Potential*

Following the results of Monte Carlo simulations, the major contribution to the roughening process and macroscopic dealloying comes from a kinetically determined competition between dissolution and diffusion. The origin of roughening during the dissolution of single-phase binary alloys has been revisited [186]. The role of capillary phenomena, i.e., of surface curvature in the free surface energy, has been considered in a quantitative approach to this problem. Dynamic balance between dissolution and surface diffusion generates a modification of the surface profile and is shown to give rise to unstable modes (roughening transition). Equality between the wavelength of instabilities and the cluster size on the surface is thought to occur at the critical potential E_c. Reasonable consistency is claimed with literature data on Ag–Au alloys [183].

3.3.3.2.6 *Percolation Model for Passivation of Binary Alloys, Case of Fe–Cr*

The Fe–Cr alloys are well known to display easier passivation above a critical Cr concentration close to 13 at% [187,193]. A computer approach was developed to demonstrate the main atomistic features of a percolation model for passivation of binary alloys. In a way quite similar to that in Ref. [32], most aspects show up on 2D lattices. Computer simulations [198] were then extended to 3D [199]. Along the same lines as sketched earlier [112,113], the Monte Carlo simulations make use of dissolution probability rules of any atom that depend on its own atomic neighborhood. The Cr atoms having at least two Cr atoms as first or next nearest neighbors are considered as passivated. This allows Cr–O–Cr bridging bonds to block the Fe atoms of the underlying atomic row. Application of these rules and the resulting passivation at the surface of a random 2D lattice of Fe–Cr sites [191] is sketched in (58). A wide set of p and relative dissolution rates were explored. In all cases, larger Cr contents tend to passivate the alloy at the cost of an initial selective dissolution confined to only a few monatomic layers.

Current–voltage trends in the dissolution–passivation range are restituted by assuming Tafel kinetics for iron dissolution and a chemical rate-determining step for chromium. Extension to 3D lattices [199] leads to similar conclusions but highlights a critical concentration $p_c\{1, 2\}$ (the percolation threshold for interactions up to the second nearest neighbor) on a 3D lattice in better agreement with the transition of Fe–Cr to stainless steel behavior.

The atomistic approach to alloy dissolution shows a real possibility of achieving a unified description, incorporating at the same time the kinetic and geometric aspects of these intricate phenomena. Serious difficulties remain for estimating the probabilities of reaction of individual atoms and their dependence on local interaction. These models may involve as many parameters as in the macroscopic models of nonsteady-state (impedance) responses of the anodic dissolution and passivation processes. In that sense, they are subject to the same kind of criticism. Estimation of the ability of the atomistic models at dealing with these nonsteady-state behaviors is certainly one of the next challenges in the field.

References

1. K. E. Heusler, *Encyclopedia of Electrochemistry of the Elements*, Vol. 9A (A. J. Bard, ed.), Marcel Dekker, New York, Vol. 9A, 1982, p. 230.
2. W. J. Lorenz and K. E. Heusler, Anodic dissolution of iron group metals, *Corrosion Mechanisms* (F. Mansfeld, ed.), Marcel Dekker, New York, 1987.
3. J. O'M. Bockris and S. U. M. Khan, *Surface Electrochemistry. A Molecular Level Approach*, Plenum, New York, 1993.
4. I. Epelboin and M. Keddam, Electrochemical techniques for studying passivity and its breakdown, *Passivity of Metals* (R. P. Frankenthal and J. Kruger, eds.), Electrochemical Society, Pennington, NJ, 1978, p. 184.
5. C. Gabrielli, M. Keddam, and H. Takenouti, The use of a.c. techniques in the study of corrosion and passivity, *Treatise on Materials Science and Technology*, Vol. 23, *Corrosion: Aqueous Processes and Passive Films* (J. C. Scully, ed.), Academic Press, New York, 1983, p. 395.
6. F. Mansfeld and W. J. Lorenz, Electrochemical impedance spectroscopy (EIS): Application in corrosion science and technology, *Techniques for Characterization of Electrodes and Electrochemical Processes* (R. Varma and J. R. Selman, eds.), Wiley, New York, 1991, p. 581.
7. D. D. Macdonald, Application of electrochemical impedance spectroscopy in electrochemistry and corrosion science, *Techniques for Characterization of Electrodes and Electrochemical Processes* (R. Varma and J. R. Selman, eds.), Wiley, New York, 1991, p. 515.
8. M. E. Orazem and B. Tribollet, Electrochemical impedance spectroscopy, Wiley-Interscience (2008).
9. D. M. Drazic, *Modern Aspects of Electrochemistry*, Vol. 19 (J. O'M Bockris and B. E. Conway, eds.), Plenum, New York, 1989.
10. A. R. Despic, Deposition and dissolution of metals and alloys, *Comprehensive Treatise of Electrochemistry*, Vol. 7, Part B (B. E. Conway, J. O'M Bockris, E. Yeager, S. U. M. Khan, and R. E. White, eds.), Plenum, New York, 1983, p. 451.
11. K. J. Vetter, *Electrochemical Kinetics: Theoretical and Experimental Aspects*, Academic Press, New York, 1967.
12. F. Hubert, Y. Miyoshi, G. Eichkorn, and W. J. Lorenz, Correlation between the kinetics of electrolytic dissolution and deposition of iron, *J. Electrochem. Soc. 118*:1919–1927 (1971).
13. J. O'M. Bockris and D. Drazic, The kinetics of deposition and dissolution of iron: Effect of alloying impurities, *Electrochim. Acta 7*:293 (1962).
14. T. Hurlen, Reaction models for dissolution and deposition of solid metals, *Electrochim. Acta 11*:1205 (1966).
15. C. Cachet, C. Gabrielli, F. Huet, M. Keddam, and R. Wiart, Growth mechanism for silver electrodeposition-A kinetic analysis by impedance and noise measurements, *Electrochim. Acta. 28*:899–908 (1983).
16. J. N. Stranski, Zur Theorie des Kristallwachstums, *Z. Phys. Chem. 136*:259 (1928).
17. M. Pourbaix, Emploi de la thermodynamique et de la cinétique électrochimique pour l'étude des phénomènes de passivation, *Ber. Bunsenges. Z. Elektrochem. 62*:670 (1958).
18. K. J. Laidler, *Chemical Kinetics*, 2nd ed., McGraw-Hill, New York, 1965, p. 72.
19. M. Eigen, Ionic reactions in aqueous solutions with half-times as short as 10^{-9} second. Application to neutralization and hydrolysis reactions, *Discuss. Faraday Soc. 17*:194 (1954).
20. M. Eigen, Determination of general and specific ionic interactions in solution, *Discuss. Faraday Soc. 24*:251 (1957).
21. R. T. Foley and P. P. Trzaskoma, The identification of the activated state in the dissolution and passivation of metals, *Passivity of Metals*, Corrosion Monographs Series (R. P. Frankenthal and J. Kruger eds.), Electrochem. Society, Pennington, NJ, 1978, p. 337.
22. K. E. Heusler, Der Einflub der Wasserstoffionenkonzentration auf das electrochemische Verhalten des aktiven Eisen in säurer Lösungen. Der Mechanismus der Reaktion: $Fe = Fe^{2+} + 2e^-$, *Z. Elektrochem. 62*:582 (1958).

23. J. O'M. Bockris, D. Drazic, and A. Despic, The electrode kinetics of the deposition and dissolution of iron, *Electrochim. Acta 4*:325 (1961).
24. S. Haruyama, Accelerating factors of anodic dissolution, *Proceedings of Second Japan–USSR Corrosion Seminar: Homogeneous and Heterogeneous Anodic Dissolution of Metals and Their Inhibition*, Japan Society of Corrosion Engineering, Tokyo, 1980, p. 128.
25. N. Sato, Mechanism of homogeneous and heterogeneous anodic dissolution of metals, *Proceedings of Second Japan–USSR Corrosion Seminar: Homogeneous and Heterogeneous Anodic Dissolution of Metals and Their Inhibition*, Japan Society of Corrosion Engineering, Tokyo, 1980, p. 35.
26. J. R. Vilche and A. J. Arvía, Los modelos de reacción relacianados con la transición estado activo-estado pasivo en los metales de Ia familia del hierro, *An. Acad. Nac. Cien. Exactas Fis. Nat. Buenos Aires 33*:33 (1981).
27. E. B. Budevski, Deposition and dissolution of metals and alloys, *Comprehensive Treatise of Electrochemistry*, Vol. 7 (B. E. Conway, J. O'M. Bockris, E. Yeager, S. U. M. Khan, and R. E. White, eds.), Part A, Plenum, New York, 1983, p. 399.
28. G. Eichkorn, W. J. Lorenz, L. Alberg, and H. Fischer, Einflußder Oberflächenaktivität auf die anodischen Auflösungmechanismen von Eisen in sauren Lösungen, *Electrochim. Acta 13*:183 (1968).
29. W. Allgaier and K. E. Heusler, Morphology of (211) surfaces of iron during anodic dissolution, *Z. Phys. Chem. N.F. 98*:161 (1975).
30. W. Allgaier and K. E. Heusler, Steps and kinks on (211) iron surfaces and the kinetics of the iron electrodes, *J. Appl. Electrochem. 9*:155 (1979).
31. B. Folleher and K. E. Heusler, The mechanism of the iron electrode and the atomistic structure of iron surfaces, *J. Electroanal. Chem. 180*:77 (1984).
32. K. Sieradzki, R. R. Corderman, K. Shukla, and R. C. Newman, Computer simulations of corrosion: selective dissolution of binary alloys, *Philos. Mag. 59*:713 (1989).
33. I. Epelboin, C. Gabrielli, M. Keddam, J. C. Lestrade, and H. Takenouti, Passivation of iron in sulfuric acid medium, *J. Electrochem. Soc. 119*:1632 (1972).
34. I. Epelboin, C. Gabrielli, M. Keddam, and H. Takenouti, The study of the passivation process by electrode impedance analysis, *Comprehensive Treatise of Electrochemistry*, Vol. 4, *Electrochemical Materials Science* (J. O'M. Bockris, B. E. Conway, E. Yeager, and R. E. White, eds.), Plenum, New York, 1981, p. 151.
35. M. Garreau, Etude du mécanisme de la formation des ions métalliques à l'interface métal-électrolyte au cours de Ia dissolution anodique des métaux, *Met. Cor. Ind. 541*:3 (1970).
36. C. Deslouis and B. Tribollet, Flow modulated techniques in electrochemistry, *Advances in Electrochemical Science and Engineering*, Vol. 2 (H. Gerischer and C. W. Tobias, eds.), VCH Weiheim, New York, 1991, p. 205.
37. W. J. Albery and M. L. Hitchman, *Ring-Disc Electrodes*, Oxford Science Research Papers, Clarendon Press, Oxford, 1971.
38. N. Benzekri, M. Keddam, and H. Takenouti, A.C. response of a rotating ring-disc electrode: Application to 2-D and 3-D film formation in anodic processes, *Electrochim. Acta 34*:1159 (1989).
39. S. Bourkane, C. Gabrielli, and M. Keddam, Investigation of gold oxidation in sulphuric medium. II. Electrogravimetric transfer function technique, *Electrochim. Acta 38*:1827 (1993).
40. H. Angerstein-Kozlowska, J. Klinger, and B. E. Conway, Computer simulation of the kinetic behavior of surface reactions driven by a linear potential sweep, *J. Electroanal. Chem. 75*:45 (1977).
41. K. F. Bonhoeffer and K. E. Heusler, Bemerkung über die anodische Auflösung von Eisen, *Ber. Bunsenges. Z. Elektrochem. 61*:122 (1957).
42. H. Gerischer and W. Mehl, Zum Mechanismus der kathodischen Wasserstoffabscheidung am Quecksilber, Silber und Küpfer, *Z. Elektrochem. 59*:1049 (1955).
43. I. Epelboin, M. Keddam, and P. Morel, Evidence of multi-step reactions on iron, nickel and chromium electrodes immersed in a sulfuric acid solution, *Proceedings of 3rd Congress Metal Corrosion*, Moscow, 1966, p. 110.
44. M. L. Boyer, I. Epelboin, and M. Keddam, Une nouvelle méthode potentiocinétique d'étude des processus électrochimiques rapides, *Electrochim. Acta 11*:221 (1966).

45. I. Epelboin, M. Keddam, and J. C. Lestrade, Variation de certaines impédances électrochimiques en fonction de la fréquence, *Rev. Gen. Electr. 76*:777 (1967).
46. I. Epelboin, M. Keddam, and J. C. Lestrade, Faradaic impedances and intermediates in electrochemical reactions, *Faraday Discuss. Chem. Soc. 56*:264 (1973).
47. C. Gabrielli and M. Keddam, Progrès récents dans la mesure des impédances électrochimiques en régime sinusoïdal, *Electrochim. Acta 19*:355 (1974).
48. R. D. Armstrong, Electrode impedance for the active–passive transition, *J. Electroanal. Chem. 34*:387 (1972).
49. I. Epelboin, C. Gabrielli, and M. Keddam, Non-steady state techniques, *Comprehensive Treatise of Electrochemistry*, Vol. 9, *Electrodics: Experimental Techniques* (E. Yeager, J. O'M. Bockris, B. E. Conway, and S. Sarangapani, eds.), Plenum, New York, 1984, p. 61.
50. E. Bayet, F. Huet, M. Keddam, K. Ogle, and H. Takenouti, A novel way of measuring local electrochemical impedance using a single vibrating probe, *J. Electrochem. Soc. 144*:L87 (1997).
51. E. Bayet, F. Huet, M. Keddam, K. Ogle, and H. Takenouti, Adaptation of the scanning vibrating electrode technique to AC mode: Local electrochemical impedance measurement, *Mater. Sci. Forum 289–292*:57 (1998).
52. E. Bayet, F. Huet, M. Keddam, K. Ogle, and H. Takenouti, Local electrochemical impedance measurement: Scanning vibrating electrode technique in ac mode, *Electrochim. Acta 44*:4117 (1999).
53. I. Annergren, F. Zou, and D. Thierry, Application of localized electrochemical techniques to study kinetics of initiation and propagation during pit growth, *Electrochim. Acta 44*:4383 (1999).
54. F. Zou and D. Thierry, Diffusion effects in localized electrochemical impedance measurements by probe methods, *J. Electrochem. Soc. 146*:2940 (1999).
55. M. Keddam, N. Portail, D. Trinh, and V. Vivier, Progress in scanning electrochemical microscopy by coupling with EIS and EQCM, *J. Chem. Phys.* Chem, *10*:3175–3182, (2009).
56. C. Gabrielli and M. Keddam, Contribution of impedance spectroscopy to the investigation of the electrochemical kinetics, *Electrochim. Acta 4*:957 (1996).
57. R. Antaño-López, M. Keddam, and H. Takenouti, A new experimental approach to the time-constants of electrochemical impedance: frequency response of the double-layer capacitance, *Electrochim. Acta 46*:3611–3617 (2001).
58. C. Gabrielli, M. Keddam, and H. Takenouti, New Advances in the investigation of passivation mechanisms and passivity by combination of ac relaxation techniques: Impedance, RRDE and quartz electrogravimetry, *Corros. Sci. 31*:129 (1990).
59. B. H. Erné and D. Vanmeakelbergh, The low-frequency impedance of anodically dissolving semiconductor and metal electrodes. A common origin? *J. Electrochem. Soc. 144*:3385 (1997).
60. D. Vanmaekelbergh and B. H. Erné, Coupled partial charge-transfer steps in the anodic dissolution of metals, *J. Electrochem. Soc. 1546*:2488 (1999).
61. P. Córdoba-Torres, R. P. Nogueira, L. de Miranda, L. Brenig, J. Wallenborn, and V. Fairén, Cellular automaton simulation of a simple corrosion mechanism: Mesoscopic heterogeneity versus macroscopic homogeneity, *Electrochim. Acta 46*:2975–2989 (2001).
62. P. Córdoba-Torres, R. P. Nogueira, and V. Fairén, Forecasting interface roughness from kinetic parameters of corrosion mechanisms, *J. Electroanal. Chem. 529*:109–123 (2002).
63. P. Córdoba-Torres, K. Bar-Eli, and V. Fairén, Non-diffusive spatial segregation of surface reactants in corrosion simulations, *J. Electroanal. Chem 571*:189–200 (2004).
64. P. Córdoba-Torres, M. Keddam, and R. P. Nogueira, On the intrinsic electrochemical nature of the inductance in EIS. A Monte Carlo simulation of the two-consecutive-step mechanism. The flat surface 2 D case. *Electrochim. Acta, 54*:518–523 (2008).
65. P. Córdoba-Torres, M. Keddam, and R. P. Nogueira, On the intrinsic electrochemical nature of the inductance in EIS. A Monte Carlo simulation of the two-consecutive steps mechanism: The rough 3D case and the surface relaxation effect, *Electrochim. Acta* 6779–6787 (2009).
66. Y. Ishikawa, T. Yoshimura, and T. Ozaki, Adsorption of water on iron surfaces with reference to corrosion, *Corros. Eng. 40*:643 (1991).
67. I. Epelboin and M. Keddam, Faradaic impedances: Diffusion impedance and reaction impedance, *J. Electrochem. Soc. 117*:1052 (1970).

68. M. Keddam, O. R. Mattos, and H. Takenouti, Reaction model for iron dissolution studied by electrode impedance. I. Experimental results and reaction model, *J. Electrochem. Soc. 128*:257 (1981).
69. M. Keddam, O. R. Mattos, and H. Takenouti, Reaction model for iron dissolution studied by electrode impedance. II. Determination of the reaction model, *J. Electrochem. Soc. 128*:266 (1981).
70. H. Fischer and H. Yamaoka, Zum Mechanismus der Inhibitionswirkung organischer Verbindungen in System Eisen/Säure, *Chem. Ber. 94*:1477 (1961).
71. W. J. Lorenz, H. Yamaoka, and H. Fischer, Zum electrochemishen Verhalten des Ei-sens in salzsauer Lösungen, *Ber. Bunsenges. Phys. Chem. 67*:932 (1963).
72. J. O'M. Bockris and H. Kita, Analysis of galvanostatic transients and application to the iron electrode reaction, *J. Electrochem. Soc. 108*:676 (1961).
73. E. J. Kelly, The active iron electrode. I. Iron dissolution and hydrogen evolution reaction in acidic sulfate solutions, *J. Electrochem. Soc. 112*:124 (1965).
74. W. J. Lorenz and G. Eichkorn, Discussion of "The active iron electrode. I. Iron dissolution and hydrogen evolution reactions in acidic sulfate solutions" by E. J. Kelly, *J. Electrochem. Soc. 112*:1255 (1965).
75. M. Keddam, O. R. Mattos, and H. Takenouti, Discussion of "Impedance measurements on the anodic iron dissolution", by H. Schweickert, W. J. Lorenz, and H. Friedburg, *J. Electrochem. Soc. 128*:1294 (1981).
76. I. Epelboin, P. Morel, and H. Takenouti, Corrosion inhibition and hydrogen adsorption in the case of iron in a sulfuric aqueous medium, *J. Electrochem. Soc. 118*:1283 (1971).
77. G. J. Bignold and M. Fleischmann, Identification of transient phenomena during the anodic polarisation of iron in dilute sulphuric acid, *Electrochim. Acta 19*:363 (1974).
78. G. Bech-Nielsen, The anodic dissolution of iron V. Some observations regarding the influence of cold working and of annealing on the two anodic reactions of the metal, *Electrochim. Acta 19*:821 (1974).
79. G. Bech-Nielsen, The anodic dissolution of iron. XI. A new method to discern between parallel and consecutive reactions, *Electrochim. Acta 27*:1383 (1982).
80. E. McCafferty and N. Hackerman, Kinetics of iron corrosion in concentrated acidic chloride solutions, *J. Electrochem. Soc. 119*:999 (1972).
81. I. Epelboin, C. Gabrielli., M. Keddam, and H. Takenouti, A coupling between charge transfer and mass transport leading to multi-steady states. Application to localized corrosion, *Z. Phys. Chem. N.F. 98*:215 (1975).
82. B. Bechet, I. Epelboin, and M. Keddam, New data from impedance measurements concerning the anodic dissolution of iron in acidic sulphuric media, *J. Electroanal. Chem. 76*:129 (1977).
83. H. Schweickert, W. J. Lorenz, and H. Friedburg, Impedance measurements on the anodic iron dissolution, *J. Electrochem. Soc. 127*:1693 (1980).
84. J. A. Harrison and W. J. Lorenz, A comment on the transient dissolution of pure iron in acid solution, *Electrochim. Acta 22*:205 (1977).
85. D. Geana, A. A. El Miligy, and W. J. Lorenz, Zür anodischen Auflösung von Reineisen im Bereich zwischen aktiven und passiven Verhalten, *Corros. Sci. 13*:505 (1973).
86. D. Geana, A. A. El Miligy, and W. J. Lorenz, The kinetics of iron dissolution and passivation, *Passivity of Metals* (R. P. Frankenthal and J. Kruger, eds.), Electrochemical Society, Pennington, NJ, 1978, p. 607.
87. D. Geana, A. A. El Miligy, and W. J. Lorenz, Galvanostatic and potentiostatic measurements on iron dissolution in the range between active and passive state, *Corros. Sci. 14*:657 (1974).
88. D. Geana, A. A. El Miligy, and W. J. Lorenz, A theoretical treatment of the kinetics of iron dissolution and passivation, *Electrochim. Acta 20*:273 (1975).
89. J. Bessone, L. Karakaya, P. Loorbeer, and W. J. Lorenz, The kinetics of iron dissolution and passivation, *Electrochim. Acta 22*:1147 (1977).
90. K. Juttner, W. J. Lorenz, and G. Kreysa, Dynamic system analysis on metallic glasses, *Corrosion, Electrochemistry and Catalysis of Metallic Glasses* (B. B. Diegle and K. Hashimoto, eds.), Vol. 88-1, Electrochemical Society, Pennington, NJ, 1988, p. 14.

91. N. Benzekri, Contribution au développement de l'électrode disque-anneau en courant alternatif. Applications aux mécanismes de dissolution et passivation anodiques, Thesis, Université P.M. Curie, Paris, 1988.
92. T. Tsuru, T. Hishimura, and S. Haruyama, Anodic dissolution of iron as studied with channel-flow double electrode, *Proceedings of EMCR'2*, Toulouse, France, *Mater. Sci. Forum 8*:429 (1986).
93. K. Takahashi, J. A. Bardwell, B. MacDougall, and M. J. Graham, Mechanism of anodic dissolution of iron. I Behaviour in neutral acetate buffer solutions, *Electrochim. Acta 37*:477 (1992).
94. K. E. Heusler and G. H. Cartledge, The influence of iodide ions and carbon monoxide on the anodic dissolution of active iron, *J. Electrochem. Soc. 108*:732 (1961).
95. Z. A. lofa, V. V. Batrakov, and Cho-Ngok-Ba, Influence of anion adsorption on the action of inhibitors on the acid corrosion of iron and cobalt, *Electrochim. Acta 9*:1645 (1964).
96. K. Takahashi, J. A. Bardwell, B. MacDougall, and M. J. Graham, Mechanism of anodic dissolution of iron. II. Comparison of the behaviour in neutral benzoate and acetate buffer solutions, *Electrochim. Acta 37*:498 (1992).
97. R. Schrebler, L. Basaez, I. Guardiazabal, H. Gomez, R. Cordova, and F. Queirolo, Electrodissolution and passivation of iron electrode in acetic/acetate solution, *Bol. Soc. Quim. 36*:65 (1991).
98. R. J. Chin and K. Nobe, Electrodissolution kinetics of iron in chloride solutions, *J. Electrochem. Soc. 119*:1457 (1972).
99. H. C. Kuo and K. Nobe, Electrodissolution kinetics of iron in chloride solutions. VI. Concentrated acidic solutions, *J. Electrochem. Soc. 125*:853 (1978).
100. D. R. MacFarlane and S. I. Smedley, The dissolution mechanism of iron in chloride solutions, *J. Electrochem. Soc. 133*:2240 (1986).
101. O. E. Barcia and O. R. Mattos, The role of chloride and sulphate anions in the iron dissolution mechanism studied by impedance measurements, *Electrochim. Acta 35*:1003, (1990).
102. O. E. Barcia and O. R. Mattos, Reaction model simulating the role of sulphate and chloride in anodic dissolution of iron, *Electrochim. Acta 35*:1601 (1990).
103. J. A. L. Dobbelaar, The use of impedance measurements in corrosion research. The corrosion behaviour of chromium and iron–chromium alloys, Thesis, Technical University of Delft, Delft, the Netherlands, 1990.
104. Y. M. Kolotyrkin, Electrochemical behaviour and anodic passivity mechanism of certain metals in electrolyte solution, *Z. Elektrochem. 62*:649 (1958).
105. T. Heumann and F. W. Diekötter, Untersuchugen über das electrochemische Verhalten des Chroms in Schwefelsäuren Lösungen in Hinblick auf die Passivitä, *Ber. Bunsenges. Phys. Chem. 67*:671 (1963).
106. J. A. L. Dobbelaar and J. H. W. de Wit, A detailed analysis of impedance measurements in the study of the passivation of chromium, *Corros. Sci. 31*:637 (1990).
107. R. D. Armstrong, M. Henderson, and H. R. Thirsk, The impedance of chromium in the active–passive transition, *J. Electroanal. Chem. 35*:119 (1972).
108. T. Tsuru, Anodic dissolution mechanisms of metals and alloys, *Mater. Sci. Eng. A 146*:1 (1991).
109. S. Haupt and H. H. Strehblow, The analysis of dissolved Cr III with the rotating ring-disc technique and its application to the corrosion of Cr in the passive state, *J. Electroanal. Chem. 216*:229 (1987).
110. M. S. El-Basiouny and S. Haruyama, The polarization behaviour of chromium in acidic sulphate solution, *Corros. Sci. 17*:405 (1977).
111. M. Okuyama, M. Kawakamal, and K. Ito, Anodic dissolution of chromium in acidic sulphate solutions, *Electrochem. Acta 30*:757 (1985).
112. M- Keddam, O. R. Mattos, and H. Takenouti, Mechanism of anodic dissoltion of iron-chromium alloys investigated by electrode impedances. I. Experimental results and reaction model, *Electrochim. Acta 31*:1147 (1986).
113. M. Keddam, O. R. Mattos, and H. Takenouti, Mechanism of anodic dissolution of iron-chromium alloys investigated by electrode impedances. II. Elaboration of the reaction model, *Electrochim. Acta 31*:1159 (1986).

114. I. Epelboin, M. Keddam, O. R. Mattos, and H. Takenouti, The dissolution and passivation of Fe and Fe-Cr alloys in acidified sulphate medium: Influence of pH and Cr content, *Corros. Sci. 19*:1105 (1979).
115. R. D. Armstrong and R. E. Firman, Impedance of titanium in the active–passive transition, *J. Electroanal. Chem. 34*:391 (1972).
116. A. Caprani, I. Epelboin, and P. Morel, Valence de dissolution du titane en milieu sulfurique fluoré, *J. Electroanal. Chem. 43*:App2 (1973).
117. A. Caprani and J. P. Frayret, Behaviour of titanium in concentrated hydrochloric acid: Dissolution–passivation mechanism, *Electrochim. Acta 24*:835 (1979).
118. A. Jardy, A. Legal Lasalle-Molin, M. Keddam, and H. Takenouti, Copper dissolution in acidic sulfate media studied by QCM and R.R.D.E. under A.C. signal, *Electrochim. Acta 37*:2195 (1992).
119. R. Oltra, G. M. Indrianjafy, M. Keddam, and H. Takenouti, Laser depassivation of a channel flow double-electrode: A new technique in repassivation studies, *Corros. Sci. 35*:827 (1993).
120. I. Efimov, M. Itagaki, M. Keddam, R. Oltra, H. Takenouti, and B. Vuillemin, Laser activation of passive electrodes, *Mater. Sci. Forum, 185–188*:937 (1995).
121. F. Huet, M. Keddam, X. R. Novoa, and H. Takenouti, Frequency and time resolved measurements at rotating ring-disc electrodes for studying localized corrosion, *J. Electrochem. Soc. 140*:1955 (1993).
122. F. Huet, M. Keddam, X. R. Nóvoa, and H. Takenouti, Time resolved RRDE applied to pitting of Fe-Cr alloy and 304 stainless steel, *Corros. Sci. 38*:133 (1996).
123. M. Itagaki, R. Oltra, B. Vuillemin, M. Keddam, and H. Takenouti, Quantitative analysis of iron dissolution during repassivation of freshly generated metallic surfaces, *J. Electrochem. Soc. 144*:64 (1997).
124. N. Benzekri, R. Carranza, M. Keddam, and H. Takenouti, a.c. response of R.R.D.E. during the passivation of iron, *Corros. Sci. 31*:627 (1990).
125. R. P. Frankenthal and J. Kruger, eds., *Passivity of Metals*, Corrosion Monographs Series, Electrochemical Society, Pennington, NJ, 1978.
126. M. Froment, ed., *Passivity of Metals and Semiconductors*, Elsevier, Amsterdam, the Netherlands, 1983.
127. German-American Colloquium on Electrochemical Passivation, *Corros. Sci. 29* (2–3) (1989).
128. *Passivation of Metals and Semiconductors. Proceedings of the 7th International Symposium on Passivity*, Clausthal, Germany, 1994, Materials Science Forum (K. E. Heusler, ed.), Trans Tech Publications, Vol. 185–188, 1995.
129. *Passivity-8, Proceedings of the 8th International Symposium on Passivity of Metals and Semiconductors*, Jasper, Canada, 1999, M. B. Ives, J. L. Luo, and J. R. Rodda, eds. Electrochemical Society, Pennington, NJ, 2001.
130. C. Wagner, Beitrag Zur Theory des Anlaufvorgangs, *Z. Phys. Chem. B21*:25 (1933).
131. C. Wagner, Models for lattice defects in oxide layers on passivated iron and nickel, *Ber. Bunsenges. Phys. Chem. 77*:1090 (1973).
132. M. J. Dignam, The kinetics of the growth of oxides, *Comprehensive Treatise of Electrochemistry*, Vol. 4, *Electrochemical Materials Science* (J. O'M Bockris, B. E. Conway, E. Yeager, and R. E. White, eds.), Plenum, New York, 1981, p. 247.
133. N. Sato and M. Cohen, The kinetics of anodic oxidation of iron in neutral solutions. I. Steady state growth region, *J. Electrochem. Soc. 111*:512 (1964).
134. N. Sato and M. Cohen, The kinetics of anodic oxidation of iron in neutral solutions. II. Initial stages, *J. Electrochem. Soc. 111*:519 (1964).
135. I. A. Ammar and S. Darwish, Growth of passive layers on iron and nickel, *Electrochim. Acta 12*:225 (1967).
136. M. Keddam and C. Pallotta, Galvanostatic response of the passive film on iron in acidic media, *Electrochim. Acta 30*:469, (1985).
137. M. Keddam and C. Pallotta, Electrochemical behavior of passive iron in acid medium, *J. Electrochem. Soc. 132*:781 (1984).
138. K. J. Vetter and F. Gorn, Kinetics of layer formation and corrosion processes of passive iron in acidic solutions, *Electrochim. Acta 18*:321 (1973).

139. K. J. Vetter and F. Gorn, Die instationäre Korrosion des passiven Eisens in saurer Lösung, *Werkst. Korros. 21*:703 (1970).
140. M. Keddam, J. F. Lizée, C. Pallotta, and H. Takenouti, Electrochemical behavior of passive iron in acid medium. I. Impedance approach, *J. Electrochem. Soc. 131*:2016 (1984).
141. R. Kirchheim, Growth Kinetics of passive films, *Electrochim. Acta 32*:1619 (1987).
142. K. E. Heusler, Untersuchung der Auflosung des passiven Eisen in Schwefelsäure mit der Rotierenden Scheiben-Ring Elektrode, *Ber. Bunsenges. Phys. Chem. 72*:1197 (1968).
143. T. Tsuru, E. Fujii, and S. Haruyama, Passivation of iron and its cathodic reduction studied with a rotating ring-disc electrode, *Corros. Sci. 31*:655 (1990).
144. R. Kirchheim, The growth kinetics of passive films and the role of defects, *Corros. Sci. 29*:183 (1989).
145. D. D. Macdonald, The point defect model for the passive state, *J. Electrochem. Soc. 139*:3434 (1992).
146. C. Gabrielli, M. Keddam, F. Minouflet, and H. Perrot, Investigation of the anodic behaviour of iron in, sulfuric acid medium by the electrochemical quartz microbalance under ac regime, *Mater. Sci. Forum, 185–188*:631 (1995).
147. C. Gabrielli, M. Keddam, F. Minouflet, and H. Perrot, AC electrogravimetry contribution to investigation of the anodic behaviour of iron in sulfuric medium, *Electrochim. Acta 41:*1217 (1996).
148. C. Y. Chao, L. F. Lin, and D. D. Macdonald, A point defect model for anodic passive film. III. Impedance response, *J. Electrochem. Soc.129*:1874 (1982).
149. D. D. Macdonald and S. I. Smedley, An electrochemical impedance analysis of passive films on nickel (111) in phosphate buffer solutions, *Electrochim. Acta 35*:1949 (1990).
150. A. Jouanneau, M. Keddam, and M. C. Petit, A general model of the anodic behaviour of nickel in acidic media, *Electrochim. Acta 21*:287 (1976).
151. M. Keddam, H. Takenouti, and N. Yu, New data on the kinetics of passive nickel from very low frequency impedance measurements, *Corros. Sci. 27*:107 (1987).
152. K. J. Vetter and K. Arnold, Korrosion und Sauerstoffüberspannung des passiven nickels in Schwefelsaure, *Z. Electrochem. 64*:244 1960).
153. J. Osterwald. Zum Stabilitätverhalten stationären Elektrodenzustand, *Electrochim. Acta 7*:523 (1962).
154. I. Epelboin and M. Keddam, Kinetics of formation of primary and secondary passivity in sulfuric aqueous media, *Electrochim. Ada 17*:177 (1972).
155. M. Keddam, H. Takenouti, and N. Yu, Transpassive dissolution of Ni in acidic sulfate media: a kinetic model, *J. Electrochem. Soc. 132*:2561 (1985).
156. R. D. Armstrong and M. Henderson, The impedance of transpassive chromium, *J. Electroanal. Chem. 40*:121 (1972).
157. R. D. Armstrong and M. Henderson, The transpassive dissolution of chromium, *J. Electroanal. Chem. 32*:1 (1971).
158. M. Bojinov, I. Betova, and R. Raicheff, A model for the transpassivity of molybdenum in acidic sulphate solutions based on ac impedance measurements, *Electrochim. Acta* 1173 (1996).
159. M. Bojinov, I. Betova, R. Raicheff, G. Fabricius, T. Laitinen, and T. Saario, Mechanism of transpassive dissolution and secondary passivation of chromium in sulphuric acid solutions, *Mater. Sci. Forum 289–292*:1019 (1998).
160. M. Bojinov, G. Fabricius, T. Laitinen, and T. Saario, Transpassivity mechanism of iron-chromium-molybdenum alloys studied by AC impedance, DC resistance and RRDE measurements, *Electrochim. Acta 44*:4331 (1999).
161. A. Turnbull, ed., *Corrosion Chemistry within Pits, Crevices and Cracks*, National Physical Laboratory, London, 1987.
162. J. L. Crolet and J. M. Defranoux, Calcul du temps d'incubation de la corrosion caverneuse des aciers inoxydables, *Corros. Sci. 130*:515 (1973).
163. U. F. Franck and R. FitzHugh, Periodische elekroden Prozesse in ihre Beschreibung durch eine mathematisches Modell, *Z. Elektrochem. 65*:156 (1961).
164. U. F. Franck, Instabilität Erscheinungen an passivierbaren Metallen, *Z. Elektrochem. 62*:649 (1958).

165. D. Sazou and M. Pagitsas, Current oscillations associated with pitting corrosion processes induced by iodide ions on the partially passive cobalt surface polarized in sulfuric acid solutions, *Electrochim. Acta 38*:835 (1993).
166. M. Pagitsas and D. Sazou, The improved Franck-FitzHugh model for the electro-dissolution of iron in sulphuric acid solutions: linear stability and bifurcation analysis. Derivation of the kinetic equations for the forced Franck-FitzHugh model, *Electrochim. Acta 36*:1301 (1991).
167. P. Glansdorff and I. Prigogine, *Structures, stabilité et fluctuations*, Masson, Paris, France, 1971.
168. O. E. Barcia, O. R. Mattos, and B. Tribollet, Anodic dissolution of iron in acidic sulfate under mass transport control, *J. Electrochem. Soc. 139*:446 (1992).
169. A. Moreau, Etude du mécanisme d'oxydo-réduction du cuivre dans les solutions chlorurèes acides. II. Systèmes Cu-CuCl-$CuCl_2$ et Cu-$Cu_2(OH)_3Cl$-CuCl$^+$$Cu^{2+}$, *Electrochim. Acta 26*:1609 (1981).
170. A. Moreau, J. P. Frayret, F. Del Rey, and R. Pointeau, Etude des Phénomènes électrochimiques et des transports de matières d'un système mètal-électrolyte: cas d'un disque tournant en cuivre dans des solutions aqueuses d'acide chlorhydrique, *Electrochim. Acta 27*:1281 (1982).
171. F. K. Crundwell, The anodic dissolution of copper in hydrochloric acid solutions, *Electrochim. Acta 37*:2707 (1992).
172. O. E. Barcia, O. R. Mattos, N. Pebere, and B. Tribollet, Mass-transport study for the electrodissolution of copper in 1 M hydrochloric acid solution by impedance, *J. Electrochem. Soc. 140*:2825 (1993).
173. A. C. West, Comparison of modeling approaches for a porous salt film, *J. Electrochem. Soc. 140*:403 (1993).
174. A. C. West, R. D. Grimm, D. Landolt, C. Deslouis, and B. Tribollet, Electro-hydrodynamic impedance study of anodically formed salt films on iron in chloride solutions, *J. Electroanal. Chem. 330*:693 (1992).
175. C. Clerc and D. Landolt, a.c. impedance study of anodic films on nickel in LiCl, *Electrochim. Acta 33*:859 (1988).
176. R. P. Tischer and H. Gerischer, Electrolytische Auflösung von Gold-Silber Liegerungen und die Frage der Resistenzgrenzen, *Z. Elektrochem. 62*:50 (1958).
177. J. O'M Bockris, B. Rubin, A. R. Despic, and B. Lovrecek, The electrodissolution of copper–nickel alloys, *Electrochim. Acta 17*:973 (1972).
178. H. P. Lee and K. Nobe, Rotating ring-disk electrode studies of Cu-Ni alloy electro-dissolution in acidic chloride solutions II.95/5, 90/10 and 70/30 Cu-Ni alloys, *J. Electrochem. Soc. 140*:2483 (1993).
179. S. Rambert and D. Landolt, Anodic dissolution of binary single phase alloys. I. Surface composition changes on Ag-Pd studied by Auger electron spectroscopy, *Electrochim. Acta 31*:1421 (1986).
180. B. Heine and R. Kirchheim. Dissolution rate of Fe and Cr in Fe–Cr alloys in the passive state, *Corros. Sci. 31*:533 (1990).
181. I. Annergren, M. Keddam, H. Takenouti, and D. Thierry, Application of electrochemical impedance spectroscopy and rotating ring-disc measurements on Fe-Cr alloys, *Electrochim. Acta 38*:763 (1993).
182. K. Sieradskl, J. S. Kim, A. T. Cole, and R. C. Newman, The relationship between dealloying and transgranular stress corrosion cracking, *J. Electrochem. Soc. 134*:1635 (1987).
183. M. Stern, Surface area relationships in polarisation and corrosion, *Corros. NACE 14*:329t (1958).
184. W. A. Mueller, Derivation of anodic dissolution curves of alloys from those of metallic components, *Corros. NACE 18*:73t (1962).
185. R. F. Steigerwald and W. D. Greene, The anodic dissolution of binary alloys, *J. Electrochem. Soc.* 11:1026 (1962).
186. R. Kirchheim. B. Heine, H. Fischmeister, S. Hofman, H. Knote, and U. Stolz, The passivity of iron–chromium alloys, *Corros. Sci. 29*:899, (1989).
187. R. C. Newman, F. T. Meng, and K. Sieradski, Validation of a percolation model for passivation of Fe-Cr alloys. I. Current efficiency in the incompletely passivated state, *Corros. Sci. 28*:523 (1988).

188. H. Kaiser, *Corrosion Mechanisms*, Marcel Dekker, New York, 1987.
189. H. W. Pickering and C. Wagner, Electrolytic dissolution of binary alloys containing a noble metal, *J. Electrochem. Soc. 114*:698 (1967).
190. H. E. Stanley and N. Ostrowsky, eds., *On Growth and Form, Fractal and Non-Fractal Patterns in Physics*, Martinus Nijhot, Dordrecht, the Netherlands, 1986.
191. Song Qian, R. C. Newman, R. A. Cottis, and K. Sieradski, Validation of a percolation model for passivation of Fe–Cr alloys: Two-dimensional computer simulations, *J. Electrochem. Soc. 137*:435 (1990).
192. R. Olivier, Passiviteit van ljzer en ljzer-Chroom Legeringen, *6th International Commitee of Electrochemical Thermodynamics and Kinetics (C.I.T.C.E.) (Poitiers)*, Butterworths, London, 1954, p. 314; Thesis, Leiden, the Netherlands, 1955.
193. H. H. Uhlig, Electron configuration in alloys and passivity, *Z. Elektrochem. 62*:700 (1958).
194. R. P. Frankenthal, On the passivity of iron-chromium alloys. I. Reversible primary passivation and secondary film formation, *J. Electrochem. Soc. 114*:542 (1967).
195. R. P. Frankenthal, On the passivity of iron-chromium alloys. II. The activation potential, *J. Electrochem. Soc. 116*:580 (1969).
196. R. Kirchheim, B. Heine, S. Hofman, and H. Hofsass, Compositional changed of passive films due to different transport rates and preferential dissolution, *Corros. Sci. 37*:573 (1990).
197. R. Kirchheim, Kinetics of film formation on Fe-Cr Alloys, *Modification of Passive Films* (P. Marcus, B. Baroux, and M. Keddam, eds.), Institute of Materials, London, 1994.
198. M. Okuyama and S. Kambe, Dissolution of Fe-30 Cr alloy in active and passive states in acidic chloride solutions, *Proceedings of EMCR'3*, Zürich, Switzerland, *Mater. Sci. Forum 44–45*:63 (1989).
199. Song Qian, R. C. Newman, R. A. Cottis, and K. Sieradski, Computer simulation of alloy passivation and activation, *Corros. Sci. 31*:621 (1990).
200. J. W. Essam, Percolation theory, *Rep. Prog. Phys. 43*:833 (1980).
201. I. Annergren, M. Keddam, H. Takenouti, and D. Thierry, Modelling of the passivation mechanism of Fe-Cr binary alloys from ac impedance and frequency resolved RRDE. 2. Behaviour of Fe–Cr alloys in 0.5 M H_2SO_4 with an addition of chloride, *Electrochim. Acta 42*:1595 (1997).
202. I. Annergren, M. Keddam, H. Takenouti, and D. Thierry, Modelling of the passivation mechanism of Fe-Cr binary alloys from ac impedance and frequency resolved RRDE. 1. Behaviour of Fe–Cr aloys in 0.5 M H_2SO_4, *Electrochim. Acta 41*:1121 (1996).
203. J. Laurent and D. Landolt, Anodic dissolution of binary single phase alloys at sub-critical potential, *Electrochim. Acta 36*:49 (1991).
204. N. Benzekri, M. Keddam, and H. Takenouti, a.c. measurements at a rotating ring-disk electrode for corrosion investigations, *Surfaces, Inhibition and Passivation*, Vol. 86–7, Electrochemical Society, Pennington, NJ, 1986, p. 507.
205. J. Gniewek, J. Pezy, B. G. Baker, and J. O'M. Bockris, The effect of noble metals additions upon the corrosion of copper: an Auger spectroscopy study, *J. Electrochem. Soc. 125*:17 (1978).
206. S. Rambert and D. Landolt, Anodic dissolution of binary single phase alloys. II. Behaviour of CuPd, NiPd and AgAu in LiCl, *Electrochim. Acta 31*:1433 (1986).
207. I. C. Oppenheim, D. J. Trevor, C. E. D. Chidsey, P. L. Trevor, and K. Sieradski, In situ scanning tunneling microscopy of corrosion of silver–gold alloys, *Science 254*:687 (1991).
208. A. J. Forty and G. Rowlands, A possible model for corrosion pitting and tunneling in noble metal alloys, *Philos. Mag. A43*:171 (1981).
209. K. Sieradski, Curvature effects in alloys dissolution, *J. Electrochem. Soc. 140* (1993).

4

Thin Oxide Film Formation on Metals

Francis P. Fehlner
Corning Incorporated

Michael J. Graham
National Research Council of Canada

CONTENTS

4.1 Introduction ... 217
4.2 Mechanism of Thin Oxide Growth ... 218
4.3 Controlling Factors ... 220
4.3.1 Metal Structure ... 221
4.3.2 Oxide Structure ... 222
4.3.3 Impurities ... 223
4.4 Techniques for Measuring Thin Film Growth ... 223
4.4.1 Kinetics of Oxide Growth ... 224
4.4.2 Oxide Structure: Chemical ... 224
4.4.3 Oxide Structure: Physical ... 224
4.5 Examples of Metal Oxidation at Low Temperatures ... 224
4.5.1 Silicon ... 225
4.5.2 Iron ... 226
4.5.3 Nickel ... 227
4.5.4 Chromium ... 227
4.5.5 Copper ... 230
4.5.6 Tantalum ... 231
4.6 Conclusion ... 231
References ... 231

4.1 Introduction

A thin oxide film on base metals provides the protective layer required to make metals useful. Without such a layer, these metals would adhere to and react with each other as well as with other materials. The film, invisible to the naked eye, forms at low temperatures. By low, we mean near room temperature. This contrasts with high-temperature oxidation, which occurs at several hundred degrees Celsius and above.

The formation of an oxide film at low temperatures may be illustrated by the machining of an aluminum block. The cutting tool removes the surface of the metal, exposing pristine metal. In less than a millisecond, atmospheric oxygen attacks the exposed metal atoms and a thin oxide layer begins to form. It attains a limiting thickness after a few days. The thickness is in the range of a few nanometers.

The kinetic rate laws of low-temperature oxidation are logarithmic, either direct or inverse. At high temperatures, parabolic kinetics are usually observed. For intermediate temperatures, logarithmic kinetics transform with time into the parabolic type.

The mechanism by which a thin oxide film forms on a metal must explain the transition from a two-dimensional adsorbed oxygen layer to a three-dimensional oxide film. The process at one time appeared to be impossible at room temperature because growth of an oxide requires that ions overcome an energy barrier to move into and through the oxide. The thermal energy available at room temperature is insufficient to overcome this barrier, which is approximately 1 eV. Fortunately, the work of Cabrera and Mott [1] showed how tunneling electrons and an electrochemical mechanism could explain the phenomena.

The process of oxygen adsorption on metals has been discussed in Chapter 2. The transition to an oxide film is a gradual one in the sense that islands of oxide nucleate from the adsorbed oxygen and then grow laterally across the surface. Fehlner and Mott [2] proposed that islands of oxide grow on the metal by a process of place exchange. This requires cooperative movement of both cations and anions. The driving force is postulated to be the image force between an oxygen ion and the metal. Mitchell and Graham [3] suggested that activated processes are unnecessary for island growth when it occurs at a step edge on a metal surface. However, they subsequently reported [4] that misorientation of the (111) plane on a nickel surface did not affect the rate of low-temperature oxidation.

Island growth is the initial stage of three-dimensional oxide growth and has been emphasized for the low-temperature oxidation of Ba and Mg [5a]. During the transition from island growth to a three-dimensional oxide film, the kinetics of the process appear to undergo a change from linear to logarithmic. This is readily observed only at low temperatures.

4.2 Mechanism of Thin Oxide Growth

The model of Cabrera and Mott is discussed in the book on low-temperature oxidation by Fehlner [5b]. The basis of the model is the quantum mechanical concept of electron tunneling. An electron can penetrate an energy barrier without the requirement for thermal activation. As soon as a three-dimensional oxide forms on a metal, electrons tunneling through the oxide are captured by adsorbed oxygen on the oxide surface. The charge separation thus established between the oxide surface and the metal sets up an electric field across the oxide. The proposed mechanism is illustrated in Figure 4.1.

Ion movement into and through an oxide is required if an oxide film is to thicken. Such movement requires that an activation energy of approximately 1 eV be supplied. At high temperatures, thermal energy is sufficient to overcome the energy barrier and the ion can move from site to site within the oxide. This is no longer true at low temperatures. However, the electric field across the oxide lowers the activation energy for ion movement into the oxide, allowing the oxide to continue growing thicker.

The process is self-limiting. The amount of charge is fixed so that the voltage across the film is relatively constant. Thus, the field across the oxide decreases as the oxide thickens. At some critical thickness, the oxide essentially stops growing because the reduction of the activation energy by the field is insufficient to allow continued ion movement. The overall process is quite analogous to anodic oxidation at constant voltage. Thus, the rate of oxide growth is dependent on a preexponential factor and an activation term. The activation

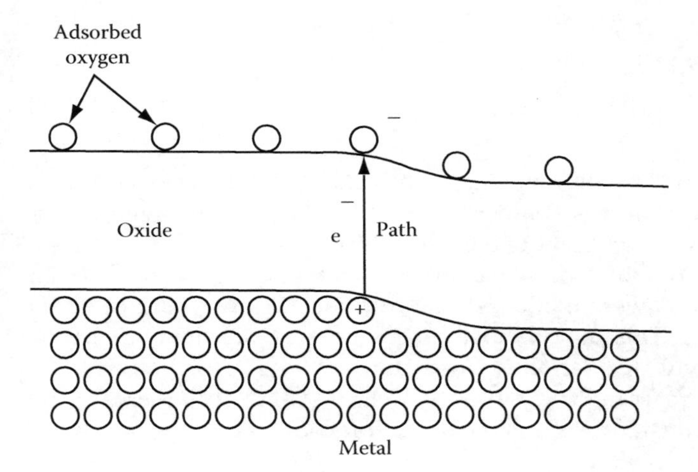

FIGURE 4.1
Proposed mechanism for low-temperature oxidation. (From Cabrera, N. and Mott, N.F., *Rep. Prog. Phys.*, 12, 163, 1948.)

energy in the exponential is reduced by the electric field across the oxide. Corrosion, too, has been discussed in terms of an electrochemical mechanism [6,7] (see Chapter 1).

The rate of oxide growth according to the Cabrera–Mott expression can be written as

$$\frac{dx}{dt} = N\Omega v \exp\left(\frac{W - qaE}{kT}\right) \tag{4.1}$$

where
- x is the oxide thickness
- t is the time
- N is the number of potentially mobile ions
- Ω is the oxide volume per mobile ion
- v is the atomic vibration frequency
- W is the energy barrier to ion movement into the oxide
- q is the ionic charge Ze
- a is half the ion jump distance
- E is the electric field V/x
- k is Boltzmann's constant
- T is the temperature

Once the ion enters the oxide, it is assumed to pass easily to the other side, where reaction takes place.

Ghez [8] has integrated Equation 4.1 to give the inverse logarithmic law:

$$\frac{x_1}{x} = -\ln\left(\frac{t+\tau}{x^2}\right) - \ln(x_1 u) \tag{4.2}$$

where

$$u = N\Omega v \exp\left(-\frac{W}{kT}\right) \tag{4.3}$$

and

$$x_1 = \left| \frac{ZeaV}{kT} \right| \tag{4.4}$$

The constant τ can be neglected if it is small compared with t. The value of u is expressed as nanometers per second and x_1 as nanometers. Fehlner [9] has utilized these equations to fit literature data on low-temperature oxidation. Some of these results will be included in this work.

Fehlner and Mott [2] pointed out that direct logarithmic kinetics could be derived from Equation 4.1 if two conditions were met. First, the film must grow under constant field rather than constant voltage. This may happen when the growing field across the oxide causes ion movement to occur through the oxide before the maximum value of V is attained. Second, the oxide must rearrange with time so that the value of W increases by some factor, say, ξx, where ξ is a constant and x is the oxide thickness. Thus, the activation term in Equation 4.1 becomes

$$\exp\left[\frac{-(W + \xi x - qaE)}{kT}\right] \tag{4.5}$$

which integrates to a logarithmic expression.

The implication then is that the kinetics of low-temperature oxidation depend on the structure of the oxide. This can be considered in terms of network formers and modifiers. A network-forming oxide is one in which covalent bonds connect the atoms in a three-dimensional structure. There is short-range order on the atomic scale but no long-range order. For example, oxygen atoms form tetrahedra around ions such as silicon, triangles around aluminum. These continuous random networks can be broken up by the introduction of modifiers. Oxides such as sodium oxide have ionic bonding. When added to a network-forming oxide, they break the covalent bonds in the network, introducing ionic bonds, which change the properties of the mixed oxide.

It seems likely that direct logarithmic kinetics would be observed for modifying oxides rather than for network-forming oxides because the former have lower single oxide bond strengths and could rearrange more easily (see Table 4.1). The network formers would be expected to follow the inverse logarithmic kinetics of Equation 4.2.

Kingery et al. [10a] have collected the experimental results for cation and anion self-diffusion coefficients in various oxides. For instance, the rate of diffusion of copper in Cu_2O, a modifier, is six orders of magnitude larger than that of aluminum in Al_2O_3, a network former when the aluminum is four coordinate. This result is for a temperature of approximately 1500°C, but the relative relationship is expected to hold at room temperature. This finding also supports the postulate that network modifiers can recrystallize more easily than network formers.

4.3 Controlling Factors

Ion entry into a growing oxide occurs at the metal–oxide interface for cations and at the oxide–gas interface for anions. Davies et al. [11] have shown that one or both species can be mobile in an oxide undergoing anodization. The choice of mobile ion is related to the structure of the oxide, whether it is crystalline or noncrystalline, as discussed below. In the case

TABLE 4.1

Values of Single Oxide Bond Strengths[a]

M in MO_x	Valence	Bond Strength ($kcal\ mol^{-1}$)	Coordination Number
B	3	119	3
Si	4	106	4
Ta	5	104	6
W	6	104	6
Al	3	101–79	4
P	5	111–88	4
Zr	4	81	6
Ti	4	73	6
Zn	2	72	2
Pb	2	73	2
Al	3	67–53	6
Zr	4	61	8
Cr	3	47	6
Sn	4	46	6
Cu	2	44	6
In	3	43	6
Mg	2	37	6
Li	1	36	4
Zn	2	36	4
Ba	2	33	8
Ca	2	32	8
Fe	2	32	6
Ni	2	28	6
Na	1	20	6
K	1	13	9

[a] Glass formers have bond strengths >75 $kcal\ mol^{-1}$, intermediates lie between 75 and ~50 $kcal\ mol^{-1}$, and modifiers lie below 50 $kcal\ mol^{-1}$.

of cation movement, the properties of the metal affect the rate of oxide growth. Such characteristics as crystal orientation, defects, and impurities are important. In the case of anion movement, gas pressure and the presence of moisture are most significant.

4.3.1 Metal Structure

The effect of crystallographic orientation on the rate of low-temperature oxidation is shown dramatically in the case of copper. Rhodin [12] published an early study of the copper–oxygen system at temperatures from –195° to 50°C. The (100) face oxidized approximately twice as fast as the (110) and (111) faces. Young et al. [13] showed that impurities could affect the rate of oxidation, especially on the (110) plane. Their results at 70°C are shown in Figure 4.2. Oxidation rates follow the order (100) > (111) > (110) > (311). Film thickness as a function of time appears to follow a linear relationship, but logarithmic kinetics have also been applied to the data.

The value of N in Equation 4.1 is related to the number of sites where ion entry into the oxide can occur. However, in the case of cation movement, it has not been possible to

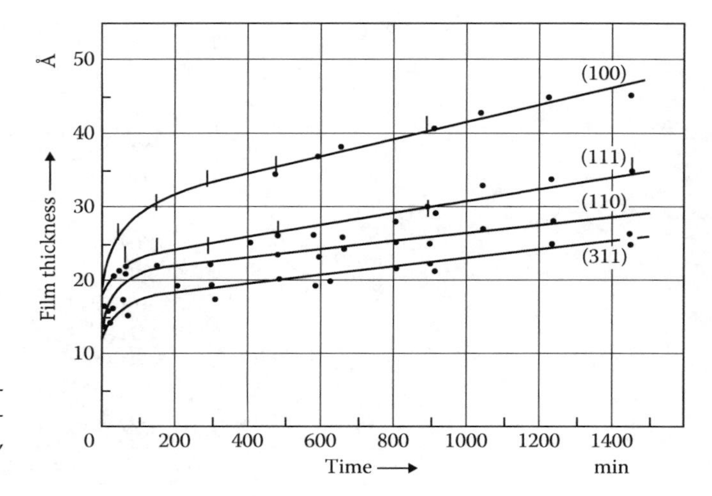

FIGURE 4.2
Copper oxidation. Oxidation of single-crystal copper surfaces in an atmosphere of oxygen at 70°C. (From Young, F.W. Jr. et al., *Acta Metall.*, 4, 145, 1956.)

predict the relative rates of oxidation according to the density of atomic packing on the various crystal faces. Cation loss from a surface occurs preferentially from steps and kink sites, so metal surface roughness and impurities play a major role. All these factors need to be considered when evaluating *N*.

Single-crystal studies as referenced above reveal the basic behavior of oxide growth on metals, but most metals are found as polycrystalline bodies. The gain boundaries that separate randomly oriented grains must accommodate the discontinuities between misoriented lattices. The resulting region of disorder serves as a sink for impurities and often a region of fast oxidation. As a result, a polycrystalline body can develop a very uneven, polycrystalline oxide layer. Metal grains with different crystallographic orientations oxidize at different rates. In addition, grain boundaries serve as points of weakness. The advent of glassy metal alloys has overcome some of these problems, at least at low temperatures [14,15]. The lack of both a crystal lattice and grain boundaries eliminates orientation effects as well as impurity concentrations.

4.3.2 Oxide Structure

The book by Fehlner [5] emphasizes the role that vitreous (glassy or noncrystalline) oxide structure can play in oxide growth on metals. The structure of a vitreous oxide can have perfection equal to that of a single crystal. This fact is not well recognized. At the atomic level, both structures are similar in that the short-range order is the same. For example, the local environments of silicon atoms in cristobalite and fused silica are very much alike. Each silicon atom is coordinated with four oxygen atoms in a tetrahedral arrangement. The difference in the structures is seen in the long-range order present in the crystal but absent in the fused silica glass. As a result, physical properties that depend on the local bonding are similar in the two structures but those dependent on long-range order differ. Both structures are expected to be more protective of a metal surface than a polycrystalline oxide would be. This is because the grain boundaries present in the polycrystalline oxide provide paths for easy ion movement and hence more rapid oxide growth.

The tetrahedral arrangement of atoms in silica lends itself to the formation of a three-dimensional network, which makes up the polymer-like structure of a glass. Glass can be formed from a single oxide or a mixture of oxides. Silicon dioxide is the premier glass-forming oxide either by itself or with other oxides. Additives to silica can either enter into the silica network or modify it. Modifiers such as sodium oxide break up the network, leading to lower softening points (glasses have no sharp melting points), lower chemical durability, and higher ionic conductivity. Some oxides like zinc oxide can act as an intermediate oxide, i.e., either a network former, where it enters the silica tetrahedral network, or a modifier, where it breaks up the network. The choice depends on the other constituents in the glass. A basic review of the subject can be found in Kingery et al. [10b].

Several schemes are used to classify metal oxides as network formers, intermediates, or modifiers. A useful one is based on single oxide bond strengths. Values for the more common metals are collected in Table 4.1 from Sun [16] and Fehlner and Mott [2]. Glass formers tend to have single oxide bond strengths greater than 75 kcal mol^{-1}. The directional covalent bonds interfere with the crystallization of an oxide while it is being formed on the surface of a metal. Intermediates lie between 75 and approximately 50 kcal mol^{-1} and modifiers lie below this value. Their ionic bonds are not directional, so it is easier for the atoms to align in crystalline order.

This division can be related to thin oxide films on metals. The metals that fall into the network-forming or intermediate classes tend to grow protective oxides that support anion or mixed anion–cation movement. The network formers are noncrystalline, while the intermediates tend to be microcrystalline at low temperatures. The Cabrera–Mott inverse logarithmic expression (Equation 4.2) best fits the oxides that are noncrystalline, although a fine-grained oxide may be uniform enough to simulate a noncrystalline one. The metals that are in the modifier class have been observed to grow crystalline oxides by cation transport. These oxides are thicker and less protective. Their kinetics of growth fit direct logarithmic kinetics rather than the inverse type.

4.3.3 Impurities

Water is a major impurity that affects the rate of oxidation at low temperatures. It can, of course, be the sole source of oxygen at higher temperatures where dissociation of water can occur [17], but at low temperatures, it serves to modify the structure of oxides. It has been found, for instance, that the rate of aluminum oxidation is increased [18] and that of copper is decreased [19] by the presence of water in oxygen. The details of the process as it affects low-temperature oxidation have yet to be elucidated. However, it may be postulated that water can act as a modifying oxide when added to network-forming oxides and thus weaken the structure. This would allow oxidation to proceed at a faster rate. On the other hand, water incorporation into modifiers may result in polymeric species [20] that are more stable than the polycrystalline oxides. They would form a stable protective gel layer.

The presence of other impurities such as sodium chloride, sulfur dioxide, or nitrogen oxides can also change the rate of thin-film formation. Examples of this are found in the literature of atmospheric corrosion, which is covered in Chapter 15.

4.4 Techniques for Measuring Thin Film Growth

A primary goal of the analytical work is to combine several methods so that a comprehensive picture of the metal–oxide system results. The features of particular interest include oxide structure and kinetics of the oxidation reaction.

4.4.1 Kinetics of Oxide Growth

Measurement of the rate of oxide growth on a metal surface is critical to sorting out the effects of time, temperature, oxygen pressure, metal structure, and impurities. The early use of manometric techniques has been supplemented by recording microbalances. Either method can give a continuous record of the weight gain or loss of a metal treated in a controlled-atmosphere furnace. Sensitivities in the submonolayer range are possible.

Resistivity measurements can supplement the information obtained from weight gain. Fehlner [5c] has pioneered a method that uses discontinuous metal films. The gaps between islands of metal accentuate the effect of surface oxidation.

4.4.2 Oxide Structure: Chemical

The initial stages of three-dimensional oxide growth can be studied using x-ray photoelectron spectroscopy (XPS) and Auger electron spectroscopy. The valence state of the atoms can be determined from energy shifts of the characteristic peaks. The escape depth of excited electrons from the oxide limits applications to the first 1–2 nm. This is, however, an important stage of growth where valence changes of impurities can be critical.

Examination of the composition of thicker films relies on depth profiles using XPS, Auger, or secondary ion mass spectrometry (SIMS), as well as Rutherford backscattering (RBS) and other nuclear techniques. In SIMS, an ion beam is used to bore a hole through the oxide. Simultaneously, a mass spectrometer records the ions that are generated during the boring. Thus, a continuous record in depth is obtained of the atomic composition. RBS also gives a depth profile of the oxide, but in a nondestructive manner. The energy distribution of backscattered ions such as helium is analyzed to reveal atomic mass as well as depth information. In both techniques, some interferences can be encountered for certain atomic masses.

4.4.3 Oxide Structure: Physical

Again, the methods can be divided into those best used for the initial stages of three-dimensional oxide growth and those useful in the latter stages. Low-energy electron diffraction (LEED) and reflection high-energy electron diffraction (RHEED) fall into the former class. They reveal the periodic structure and changes in that structure as an adsorbed oxygen layer evolves into the three-dimensional oxide. Thicker oxide is best examined using transmission electron microscopy (TEM) to look at both the plane of the film and the cross section. The physical features can be seen directly. Crystallography is studied using transmission electron diffraction. These techniques can be enhanced by applying selective etching to the samples.

Scanning tunneling microscopy (STM) and atomic force microscopy (AFM) are developments that allow the atomic arrangement on a surface to be monitored. Changes in structure with oxidation time can be observed using these techniques.

Additional details on these chemical and physical methods can be found in a recent reference book [21].

4.5 Examples of Metal Oxidation at Low Temperatures

The usefulness of a thin oxide film is often the reason for studying its growth rate and properties. For this reason, each of the examples below will be introduced by an application.

4.5.1 Silicon

Thin silica films of tunneling dimensions (≤5 nm thick) are useful to the electronics industry. Previously, such films interfered with good electrical contact at metal-to-metal junctions. Lateral dimensions on integrated circuits have reached the submicrometer range and film thicknesses have scaled accordingly. Metal oxide semiconductor (MOS) transistors with tunneling oxides as gate dielectrics are fabricated [22]. In the field of solar energy, the use of a thin oxide film between the silicon solar cell and the positive electrode has led to a conversion efficiency of 22% [23]. This equals the semiempirical limit predicted by Bolton [24] and may be compared with the thermodynamic limit of 30%. We see then that thin silicon dioxide films are certainly useful.

Logarithmic growth of oxide on silicon is found at temperatures up to approximately 500°C for pressures of 1 atm. Higher temperatures over 1000°C can be used in reduced oxygen environments provided the total pressure is 1 atm. Kamigaki and Itoh [25] worked with oxygen–nitrogen mixtures, and Ahn et al. [26] used nitrous oxide. In the latter case, some nitrogen was incorporated at the oxide–silicon interface. Other methods for forming thin oxides on silicon include rapid thermal oxidation [27], plasma oxidation [28], and moist oxygen in the presence of ultraviolet light [29]. The detailed mechanism of oxidation involving oxygen transport has become clearer [30–32], as has the growth of ultrathin silica films [33,34].

A noncrystalline oxide is formed on silicon. This is because the high oxide bond strength as given in Table 4.1 results in a very stable material. At the same time, it makes self-diffusion of oxygen or silicon difficult. It has been found that at high temperatures diffusion of oxygen molecules through the silica accounts for continued oxide growth. Oxygen atoms may be involved in the interfacial reaction with silicon. Under high electric fields, i.e., at low temperatures or during anodization, oxygen anions are postulated to be the mobile species. Fehlner [5d] discusses this point further.

The integrated form of the Cabrera–Mott expression (Equation 4.2) has been applied to the dry oxidation of silicon [35–37]. Values of N, W, and V have been calculated and are included in Table 4.2, which lists values for a number of elements. The values of W and V

TABLE 4.2

Values of N, W, and $|V|$ Obtained from Kinetic Data by Fitting the Cabrera–Mott Expression[a]

Periodic Group	Element	N (cm^{-2})	W (eV)	$\|V\|$ (V)
IB	Cu	1×10^{21}	1.9	—
	Cu	2×10^{7}	0.5	86
IIIA	Al	6×10^{8}	1.6	~0.6
	Al	2×10^{3}	~0.9	~3.0
IVA	Si	1×10^{6}	1.5	2.6
	Ge	2×10^{6}	1.1	0.2
	Pb	7×10^{2}	0.4	1.5
VB	Ta	3×10^{12}	1.6	1.5
VIB	Cr	5×10^{13}	1.8	2.8
VIII	Fe	5×10^{8}	0.8	—
	Ni	3×10^{13}	1.6	2.3

Source: Fehlner, F.P., *J. Electrochem. Soc.*, 131, 1645, 1984.

[a] Repeat runs for the same metal indicate two sources of experimental data.

are reasonable but that of N is surprisingly low [9]. It may be related to the fact that anions are the mobile species in silica and the number of entry sites is limited.

4.5.2 Iron

This metal is used in many facets of industry, commerce, and home life. As such, the fact that it rusts away assumes vast economic importance. Scientists have labored to explain the mechanism of corrosion so that the destruction can be controlled. Both the nature of the corrosion film and the kinetics of the reaction have proved useful in understanding and avoiding the losses.

Iron can assume a valence of two or three in oxides. The former acts as a modifier as listed in Table 4.1, and the latter can be a network former, at least in mixed oxide glasses [38].

A logarithmic rate law is followed for the oxidation of polycrystalline iron in the temperature range 24°C–200°C. Results of Graham et al. [39] are shown in Figure 4.3. They were unable to distinguish between a direct and inverse logarithmic expression. Nevertheless, Equation 4.2 has been applied to the data and values of the significant variables are included in Table 4.2. The number of active sites N on the metal surface, assuming cation transport through the oxide, may be related to the grain boundary population. The value of W is believable, but V was unreliable and is not included in the table.

Single-crystal studies show that the rate of oxidation varies with crystal face. The order was found to be polycrystalline > (110) > (112). Observations of the low-temperature oxide by electron diffraction were interpreted as epitaxial Fe_3O_4. XPS has been used to determine the composition of thin oxide films on iron [40–42].

A parabolic oxidation rate was found for temperatures above 200°C. It was controlled by oxide grain size as dictated by the grain size of the prior oxide. This was concluded from a study of the effect of surface pretreatment on the rate of oxidation [43]. Surface preparation is found to play an important role in the oxidation of many metals and alloys.

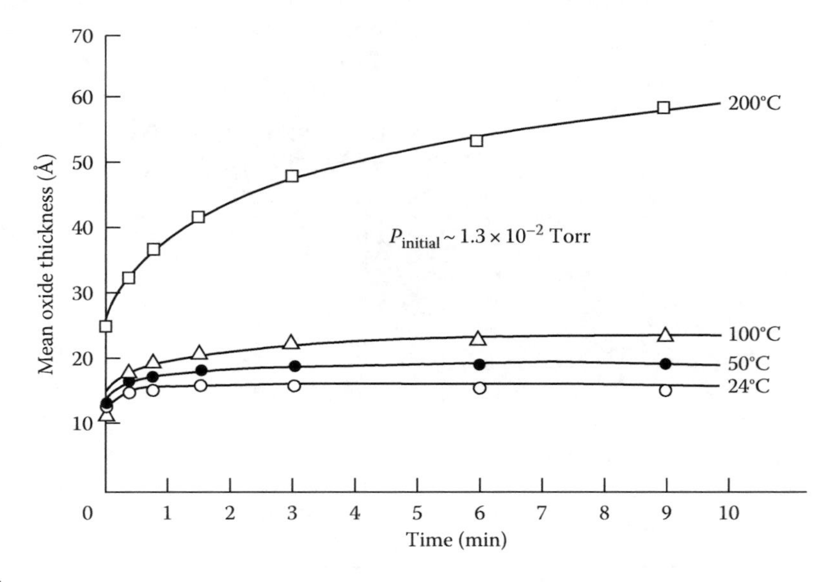

FIGURE 4.3
Iron oxidation. Oxide growth on polycrystalline iron as a function of temperature in an oxygen partial pressure of ~1.3 × 10^{-2} Torr. (From Graham, M.J. et al., *J. Electrochem. Soc.*, 117, 513, 1970.)

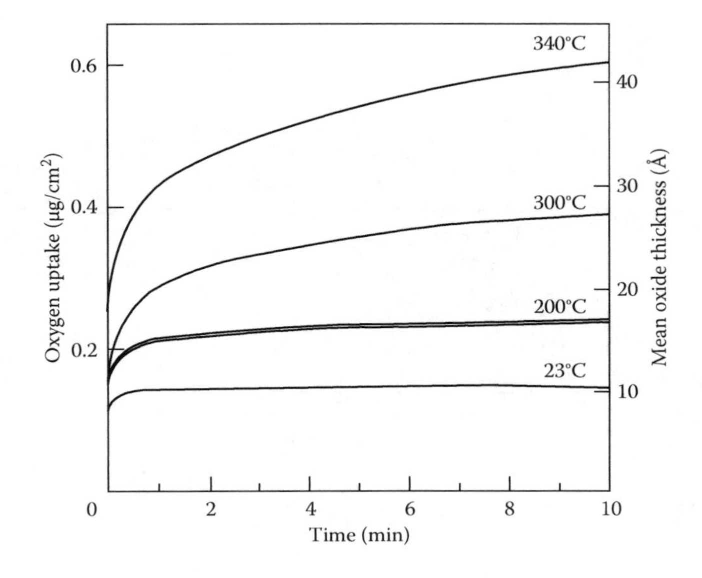

FIGURE 4.4
Nickel oxidation. Oxide growth on zone-refined, polycrystalline nickel sheet at 5×10^{-3} Torr oxygen as a function of temperature. (From Graham, M.J. and Cohen, M., *J. Electrochem. Soc.*, 119, 879, 1972.)

4.5.3 Nickel

This metal is less reactive with oxygen than iron. As such, it is used as a protective coating on base metals and as an additive in iron alloys to improve their corrosion resistance.

Graham and Cohen [44] reported on the low-temperature oxidation of polycrystalline nickel. Their results for the temperature range 23°C–340°C are shown in Figure 4.4. A direct logarithmic law was followed for oxide thicknesses up to 3 nm. Parabolic kinetics were followed for thicker films. The crossover temperature was approximately 300°C.

The valence of nickel in the oxide is 2. As a result, the oxide is a modifier (Table 4.1) and direct logarithmic kinetics are expected. The application of Equation 4.2 to NiO as shown in Table 4.2 gave reasonable values but the applicability of the equation is in question. This may be a case where the factor ξx in Equation 4.5 is insufficient to dominate the electric field so that inverse logarithmic kinetics still apply.

The nucleation and growth of oxide on nickel single crystals have been extensively studied. The results of Mitchell and Graham [3] are shown in Figure 4.5, which illustrates the transformation from oxygen chemisorption to three-dimensional oxide formation. At 40°C (Figure 4.5a), the uptake of oxygen at low pressures shows two distinct plateaus, one corresponding to fractional monolayer adsorption and the other at longer exposures to the formation of an oxide film two or three atomic layers thick. A dramatically higher rate of oxidation has been reported for atomic oxygen-induced oxidation [45]. At 200°C (Figure 4.5b), the second plateau is not observed. Instead, continuous uptake of oxygen occurs, indicative of continuing oxide growth. The kinetics of oxide formation can be explained in terms of oxide nucleation and growth, the density of nuclei depending on the particular crystallographic orientation, temperature, and oxygen pressure. Continuing oxide thickening, as on polycrystalline nickel, follows a logarithmic rate law.

4.5.4 Chromium

This metal by itself is very corrosion resistant. It is therefore used as a coating for base metals to maintain their integrity. An example is decorative trim for appliances. Of even greater importance is the use of chromium in alloys. It is the key component of stainless steel.

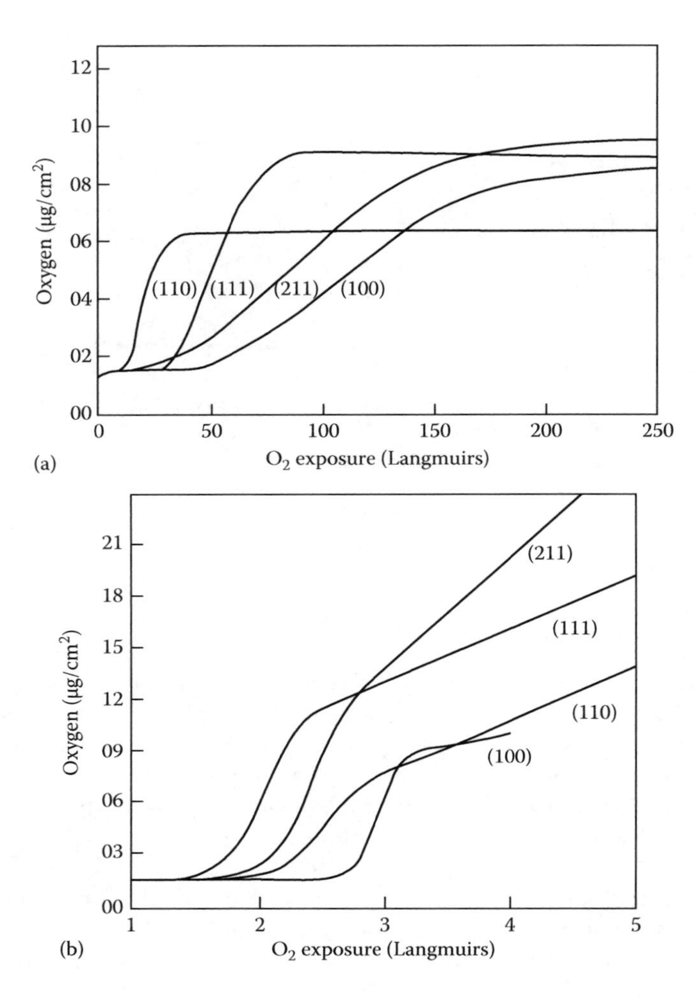

FIGURE 4.5
Nickel oxidation. Oxidation kinetics on various faces of nickel at (a) 40°C and (b) 200°C. One Langmuir equals 10^{-6} Torr s. (From Mitchell, D.F. and Graham, M.J., *Intern. Corrosion Conf. Ser. NACE-6*, p. 18, 1981.)

McBee and Kruger [46] have shown that the addition of chromium to iron in an alloy causes the oxide film to go from polycrystalline to noncrystalline as the amount of chromium increases. In general, oxidation of an alloy causes the more base component to segregate to the surface while the more noble component concentrates in the bulk of the alloy.

Low-temperature oxidation of polycrystalline chromium has been studied by Young and Cohen [47]. They found that a transition from inverse logarithmic to parabolic kinetics occurred at approximately 450°C for the data shown in Figure 4.6. However, the transition was a function of both time and temperature. It also correlated with a change in the oxide structure, from a fine-grained, almost vitreous oxide to a strongly preferred fiber texture. Ionic transport by cations was assumed and values for N, W, and V were calculated as shown in Table 4.2. However, Arlow et al. [48] present evidence that oxygen diffusion is the main mechanism for oxide film growth at temperatures below 300°C. Fortunately, Equation 4.2 applies to both cation and anion movement. Only the value of V would change if the charge on the mobile ion was assumed to be different. At higher temperatures, mixed anion–cation movement takes place. These findings concerning ion motion agree with the fact that Cr_2O_3 can be an intermediate oxide, i.e., having properties between those of network-forming and modifying oxides, according to Table 4.1.

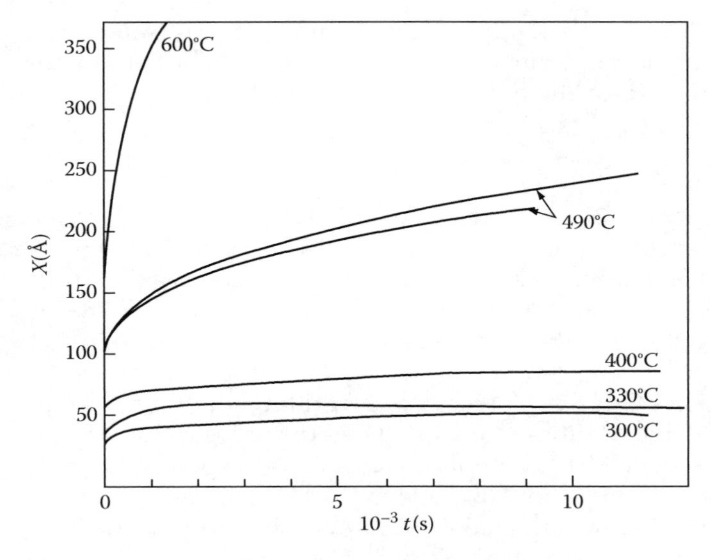

FIGURE 4.6
Chromium oxidation. Oxide growth on polycrystalline chromium at 4×10^{-3} Torr oxygen as a function of temperature. (From Young, D.J. and Cohen, M., *J. Electrochem. Soc.*, 124, 769, 1977.)

Arlow et al. [48] examined in detail the structure of thin oxides formed on a Cr(100) single-crystal surface. RHEED patterns distinguished the evolving atomic arrangements as shown in Figure 4.7. Two face-centered cubic (fcc) phases and a tetragonal phase were observed before the precipitation and growth of α-Cr_2O_3. There is also a transition from initial linear kinetics to a logarithmic reaction rate that approaches zero at some limiting thickness.

Work on Cr(110), under conditions where the kinetics of oxide growth are governed by chromium ion diffusion, shows that below 450°C Cr_2O_3 grows on a layer-by-layer basis whereas at higher temperatures the oxide surface roughens and cavities are formed at the metal–oxide interface [49,50]. In situ STM and Auger studies reveal that the oxidation of

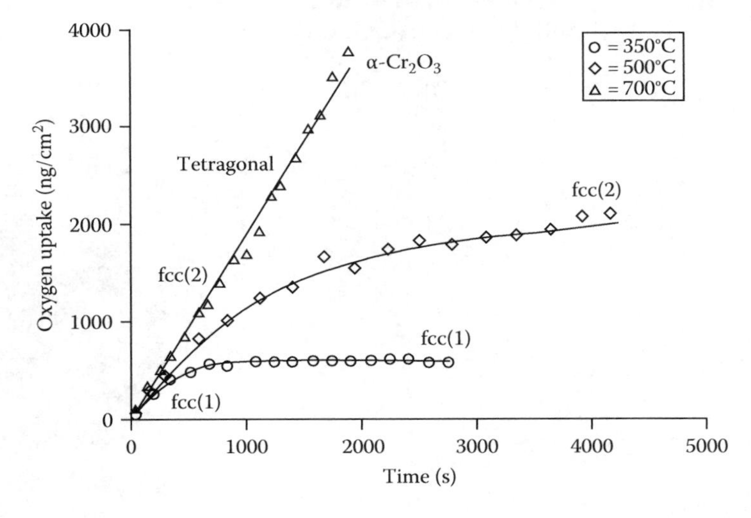

FIGURE 4.7
Chromium oxidation. Oxygen uptake by Cr(100) as a function of time at a pressure of 4.7×10^{-8} Torr oxygen at 350°C, 500°C, and 700°C. Indicated on the kinetic curves are the temperature/mass regions over which the various oxide phases are observed. The two face-centered cubic phases are denoted as fcc(1) and fcc(2). At 700°C, the sticking factor is unity over the range shown. (From Arlow, J.S. et al., *J. Vac. Sci. Technol.*, A5, 572, 1987.)

Cr(110) proceeds in three stages [51]. After an initially formed superlattice is saturated, it rearranges and characteristic oxide stripes form at terraces or above 700°C at steps on the surface. Subsequently, the stripes coalesce and bulk Cr_2O_3 growth continues.

4.5.5 Copper

Usefulness of the red metal is in the same class as that of iron. It is a mainstay of the modern industrial world. Electrical applications dominate, but it is also an important ingredient of alloys used for structural and decorative applications. The formation of a thin oxide on copper does not pose the electrical problem encountered with aluminum or silicon. Current passes quite easily through the oxide. Extensive corrosion of alloys, however, is a continuing problem.

Copper is not expected to follow Cabrera–Mott inverse logarithmic kinetics since its oxide is a modifier according to Table 4.1. In fact, copper follows direct logarithmic kinetics. This was emphasized by results (Table 4.2) from the analysis of experimental data [5e,9] including the results shown in Figure 4.2. No attempt is made here to apply the Fehlner–Mott direct logarithmic expression (Equation 4.5). This is because the evidence for oxide recrystallization with time is very complex. Onay [52] reported on the formation of multiphase, multilayer scales on copper at 300°C. He found that they result from the dissociation of compact cuprous oxide scale that has lost contact with the copper substrate.

In situ TEM studies by Yang et al. [53–56] present evidence that the "passive" oxide film on copper nucleates and grows as oxide islands, not as a uniform layer. Figure 4.8 from Ref. [53] shows the oxidation data of Young et al. [13] for (100) Cu (Figure 4.2) together with a dark-field image from the Cu_2O reflection, where the bright specks are Cu_2O islands. These islands form at both atmospheric [53] and very low oxygen pressures [54,55] and in the latter case can be observed to grow and coalesce with time. Because the oxide island coverage is approximately 30%, the local oxide island thickness in Figure 4.8 is estimated to be ~400 Å, outside the range where an electron tunneling model would apply, and Yang et al. [53–55] consider that oxygen surface diffusion is the dominant mechanism for the transport, nucleation, and initial oxide growth on copper. The surface models have been

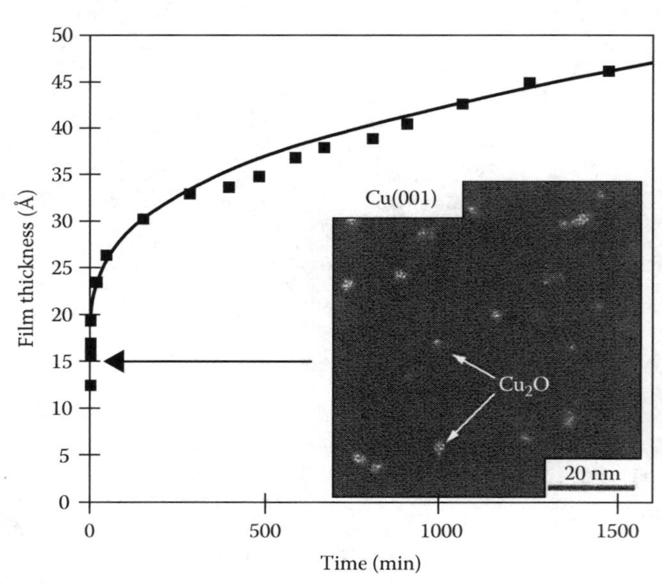

FIGURE 4.8
Copper oxidation. Oxidation of Cu(100) (Fig. 2 from Young, Jr. F.W. et al., *Acta Metall.*, 4, 145, 1956.) and dark-field TEM image from the Cu_2O reflection, where the bright specks are Cu_2O islands. (From Yang, J.C. et al., *Appl. Phys. Lett.*, 73, 2841, 1998.)

extended to quantitatively represent the coalescence behavior of copper oxidation in the framework of the Johnson–Mehl–Avrami–Kolmogorov theory [56]. In contrast to previous speculations, Yang et al. [57] did not observe clear evidence that surface steps are preferential oxide nucleation sites. By adjusting the oxygen partial pressure, it is possible to reversibly grow or shrink oxide islands and accurately determine the equilibrium phase boundary [58].

4.5.6 Tantalum

This example illustrates the similarities between anodization and low-temperature oxidation. Tantalum has formed the basis for a solid-state capacitor that is extensively used in high-performance electronics. The dielectric is formed by anodization, a process studied by Vermilyea [59]. The kinetics are expected to follow inverse logarithmic kinetics because the oxide is a network former, as seen in Table 4.1.

Ghez [8] used Equation 4.2 to fit Vermilyea's results for the temperature range 150°C to 300°C, and the resulting values of N, W, and given V are given in Table 4.2. All three are reasonable. The results of thermal oxidation of tantalum from 25°C to 275°C [60] were also interpreted in terms of Cabrera–Mott theory.

4.6 Conclusion

The growth of thin oxide films on metals and semiconductors can be a useful as well as a destructive phenomenon. The study of the reactions involved leads to a measure of control. Specific oxide thickness and properties can be achieved for electronic applications. Materials that grow network-forming oxides are better for these purposes.

Metals that grow network-modifying oxides more easily undergo degradation by corrosion, a property useful in battery technology. The existence of grain boundaries or other paths of easy ion movement in the oxide allows continued film growth beyond the electron tunneling limit. A partial solution to achieving perfection is to alloy the metal with one that forms a network oxide. The alloying metal tends to oxidize preferentially. It can segregate to the surface as a vitreous oxide film that protects the alloy from further attack.

The need for thin oxides on metals is increasing. Our ability to control the properties of the oxides will depend on our understanding at the atomic level of the processes involved.

References

1. N. Cabrera and N. F. Mott, *Rep. Prog. Phys. 12*:163 (1948–1949).
2. F. P. Fehlner and N. F. Mott, *Oxid. Metals* 2:59 (1970).
3. D. F. Mitchell and M. J. Graham, *Intern. Corrosion Conf. Ser. NACE-6*, p. 18 (1981).
4. D. F. Mitchell and M. J. Graham, *Surf. Sci. 114*:546 (1982).
5. F. P. Fehlner, *Low-Temperature Oxidation: The Role of Vitreous Oxides*, Wiley, New York, 1986, (a) p. 110, (b) pp. 148ff, (c) pp. 94ff, (d) pp. 228ff, (e) p. 199.
6. U. R. Evans, *An Introduction to Metallic Corrosion*, 2nd ed., Edward Arnold, London, 1963.

7. H. H. Uhlig, *Corrosion and Corrosion Control*, 2nd ed., Wiley, New York, 1971.
8. R. Ghez, *J. Chem. Phys. 58*:1838 (1973).
9. F. P. Fehlner, *J. Electrochem. Soc. 131*:1645 (1984).
10. W. D. Kingery, H. K. Bowen, and D. R. Uhlmann, *Introduction to Ceramics*, 2nd edn., Wiley, New York, 1976, (a) p. 240, (b) pp. 91ff.
11. J. A. Davies, B. Domeij, J. P. S. Pringle, and F. Brown, *J. Electrochem. Soc. 112*:675 (1965).
12. T. N. Rhodin, Jr., *J. Am. Chem. Soc. 72*:5102 (1950); *73*:3143 (1951).
13. F. W. Young, Jr., J. V. Cathcart, and A. T. Gwathmey, *Acta Metall. 4*:145 (1956).
14. R. Ramesham, S. DiStefano, D. Fitzgerald, A. P. Thakoor, and S. K. Khanna, *J. Electrochem. Soc. 134*:2133 (1987).
15. D. J. Siconolfi and R. P. Frankenthal, *J. Electrochem. Soc. 136*:2475 (1989).
16. K.-H. Sun, *J. Am. Ceram. Soc. 30*:277 (1947).
17. O. Kubaschewski and B. E. Hopkins, *Oxidation of Metals and Alloys*, 2nd ed., Butterworths, London, 1962, pp. 266ff.
18. J. Grimblot and J. M. Eldridge, *J. Electrochem. Soc. 128*:729 (1981).
19. W. E. Campbell and U. B. Thomas, *Trans. Electrochem. Soc. 91*:263 (1947).
20. C. S. G. Phillips and R. J. P. Williams, *Inorganic Chemistry*, Vol. I, Oxford University Press, New York, 1965, p. 533.
21. P. Marcus and F. Mansfeld, eds., *Analytical Methods in Corrosion Science and Engineering*, CRC/Taylor & Francis Group, Boca Raton, FL, 2006.
22. W. E. Dahlke and S. M. Sze, *Solid-State Electron. 10*:865 (1967).
23. *Solid State Tech.*, August 1992, p. 12.
24. J. R. Bolton, *Solar Energy 31*:483 (1983).
25. Y. Kamigaki and Y. Itoh, *J. Appl. Phys. 48*:2891 (1977).
26. J. Ahn, W. Ting, T. Chu, S. N. Lin, and D. L. Kwong, *J. Electrochem. Soc. 138*:L39 (1991).
27. C. A. Paz de Araujo, R. W. Gallegos, and Y. P. Huang, *J. Electrochem. Soc. 136*:2673 (1989).
28. T. Sugano, *Thin Solid Films 92*:19 (1982).
29. Y. Ishikawa, T. Shibamoto, and I. Nakamichi, *Jpn. J. Appl. Phys. 31*:L750 (1992).
30. G. F. Cerofolini, G. La Bruna, and L. Meda, *Appl. Surf. Sci. 89*:361 (1995).
31. H. C. Lu, T. Gustafsson, E. P. Gusev, and E. Garfunkel, *Appl. Phys. Lett. 67*:1742 (1995).
32. L. Verdi and A. Miotello, *Phys. Rev. B. 51*:5469 (1995).
33. H. Z. Massoud, I. J. R. Baumvol, H. Hirose, and E. H. Poindexter, eds., *Physics and Chemistry of SiO_2 and the Si–SiO_2 Interface-4, ECS Proceedings*, Vol. 2000-2, Electrochemical Society, Pennington, NJ (2000).
34. By H. Z. Massoud, J. H. Stathis, T. Hattori, D. Misra, and I. J. R. Baumvol, eds. *Physics and Chemistry of SiO_2 and the Si–SiO_2 Interface-5, ECS Proceedings* Vol. 2005-1, Electrochemical Society, Pennington, NJ (2005).
35. A. M. Goodman and J. M. Breece, *J. Electrochem. Soc. 117*:982 (1970).
36. F. P. Fehlner, *J. Electrochem. Soc. 119*:1723 (1972).
37. R. J. Archer and G. W. Gobeli, *J. Phys. Chem. Solids 26*:343 (1965).
38. W. A. Weyl, *Coloured Glasses*, Dawson's of Pall Mall, London, 1959, pp. 108ff.
39. M. J. Graham, S. I. Ali, and M. Cohen, *J. Electrochem. Soc. 117*:513 (1970).
40. P. C. J. Graat and M. A. J. Somers, *Appl. Surf. Sci. 100/101*:36 (1996).
41. T.-C. Lin, G. Seshadri, and J. A. Kelber, *Appl. Surf. Sci. 119*:83 (1997).
42. S. J. Roosendaal, B. van Asselen, J. W. Elsenaar, A. M. Vredenberg, and F. H. P. M. Habraken, *Surf. Sci. 442*:329 (1999).
43. M. J. Graham, *Int. Corrosion Conf. Ser. NACE-6*, p. 139 (1981).
44. M. J. Graham and M. Cohen, *J. Electrochem. Soc. 119*:879 (1972).
45. J. A. Slezak, B. D. Zion, and S. J. Sibener, *Surf. Sci. 442*:L983 (1999).
46. C. L. McBee and J. Kruger, *Electrochem. Acta 17*:1337 (1972).
47. D. J. Young and M. Cohen, *J. Electrochem. Soc. 124*:769, 775 (1977).
48. J. S. Arlow, D. F. Mitchell, and M. J. Graham, *J. Vac. Sci. Technol. A5*:572 (1987).
49. A. Stierle, P. Bödeker, and H. Zabel, *Surf. Sci. 327*:9 (1995).

50. A. Stierle and H. Zabel, *Surf. Sci. 385*:167 (1997).
51. M. Müller and H. Oechsner, *Surf. Sci. 387*:269 (1997).
52. B. Onay, *J. Electrochem. Soc. 136*:1578 (1989).
53. J. C. Yang, B. Kolasa, J. M. Gibson, and M. Yeadon, *Appl. Phys. Lett. 73*:2841 (1998).
54. J. C. Yang, M. Yeadon, B. Kolasa, and J. M. Gibson, *Appl. Phys. Lett. 70*:3522 (1997).
55. J. C. Yang, M. Yeadon, B. Kolasa, and J. M. Gibson, *Scr. Mater. 38*:1237 (1998).
56. J. C. Yang, D. Evan, and L. Tropia, *Appl. Phys. Lett. 81*:241(2002).
57. J. C. Yang, M. Yeadon, B. Kolasa, and J. M. Gibson, *Proceedings of Spring ECS Meeting*, Seattle, Washington, DC, May 1999.
58. J. A. Eastman, P. H. Fuoss, L. E. Rehn, P. M. Baldo, G.-W. Zhou, D. D. Fong, and L. J. Thompson, *Appl. Phys. Lett. 87*:051914 (2005).
59. D. A. Vermilyea, *Acta Metall. 6*:166 (1958).
60. P. B. Sewell, D. F. Mitchell, and M. Cohen, *Surf. Sci. 29*:173 (1972).

5

Passivity of Metals

Hans-Henning Strehblow
Ecole Nationale Supérieure de Chimie de Paris,
Centre National de la Recherche Scientifique
and
Heinrich Heine University Düsseldorf

Vincent Maurice
Ecole Nationale Supérieure de Chimie de Paris
and
Centre National de la Recherche Scientifique

Philippe Marcus
Ecole Nationale Supérieure de Chimie de Paris
and
Centre National de la Recherche Scientifique

CONTENTS

5.1 Introduction 236
5.2 Historical Aspects 238
5.3 Thermodynamic Aspects of Passivity 239
5.4 Kinetic Aspects of Passivity 242
5.5 Electrode Kinetics in the Passive State 244
5.5.1 Reactions at the Layer–Electrolyte Interface 246
5.5.2 Ion Transfer Through the Film, High Field Mechanism 249
5.6 Chemical Composition and Chemical Structure of Passive Films 253
5.6.1 Brief Summary of UHV Methods for the Study of Passive Layers 254
5.6.2 Passivity of Metals 255
5.6.2.1 Passivity of Iron 255
5.6.2.2 Passivity of Copper 264
5.6.2.3 Passivity of Silver 266
5.6.2.4 Passivity of Tin 268
5.6.2.5 Passivity of Cobalt 269
5.6.2.6 Passivity of Vanadium 271
5.6.2.7 Passivity of Chromium 272
5.6.2.8 Passivity of Nickel 272
5.6.2.9 Valve Metals 274

5.6.3 Passivity of Binary Alloys....275
5.6.3.1 Passivity of FeCr Alloys....275
5.6.3.2 Passivity of FeNi Alloys....278
5.6.3.3 Passivity of NiCr Alloys....280
5.6.3.4 Passivity of FeAl Alloys....281
5.6.3.5 Passivity of FeSi Alloys....281
5.6.3.6 Passivity of CuNi Alloys....283
5.6.3.7 Passivity of CuSn Alloys....284
5.6.3.8 Passivity of AlCu Alloys....285
5.6.4 Conclusion on the Chemical Composition of Passive Layers....286
5.7 Electronic Properties of Passive Layers....288
5.7.1 Electron Transfer at Passivated Metallic Surfaces....289
5.7.2 Photoelectrochemical Measurements....294
5.7.3 General Electronic Properties of Passive Layers....294
5.7.4 Photoeffects on Passivated Cu....296
5.8 Structure of Passive Films....302
5.8.1 STM Experimental Details....302
5.8.2 Growth and Structure of Passive Films on Cu....303
5.8.2.1 OH Electrosorption on Cu(111) and Cu(001)....303
5.8.2.2 Cu_2O Formation on Cu(111) and Cu(001)....307
5.8.2.3 Duplex Layer on Cu(111) and Cu(001)....310
5.8.3 Growth and Structure of Passive Films on Ag....311
5.8.3.1 STM Studies....311
5.8.3.2 XAS Studies....312
5.8.4 Structure of Passive Films on Fe, Ni, Cr, and Their Alloys....314
5.8.5 Passive Monolayer on Co(0001) and Its Reduction....317
5.8.6 Conclusions on the Structure of Passive Films....320
5.9 Conclusion....321
References....322

5.1 Introduction

Most metals are reactive and are subject to degradation in an oxidizing environment. Nevertheless passivating metals are covered by a thin anodic oxide film (i.e., the passive film) that slows down the corrosion process and thus protects the metal underneath. Passivity is an extremely important surface property of metals that allows using them as structure materials or components.

Passivating metals have a characteristic current density potential curve as shown in Figure 5.1. Starting with a cathodic current in the range of hydrogen evolution, the current density increases at potentials positive to the Nernst equilibrium potential of the Me^{z+}/Me electrode. In solutions of high solubility of the anodic oxides of the metals, the current density of metal dissolution reaches very high values. This is usually the case in strongly acidic electrolytes and in strongly alkaline solutions for amphoteric metals like aluminum. At low potentials, the process is charge transfer controlled but reaches transport control at more positive potentials. This often leads to the precipitation of a salt film, which limits the dissolution kinetics at a high level in the range of some 100 mA cm^{-2}. If the potential gets above a critical value, the passivation potential E_P, the current density drops by many orders of magnitude to the range of some μA

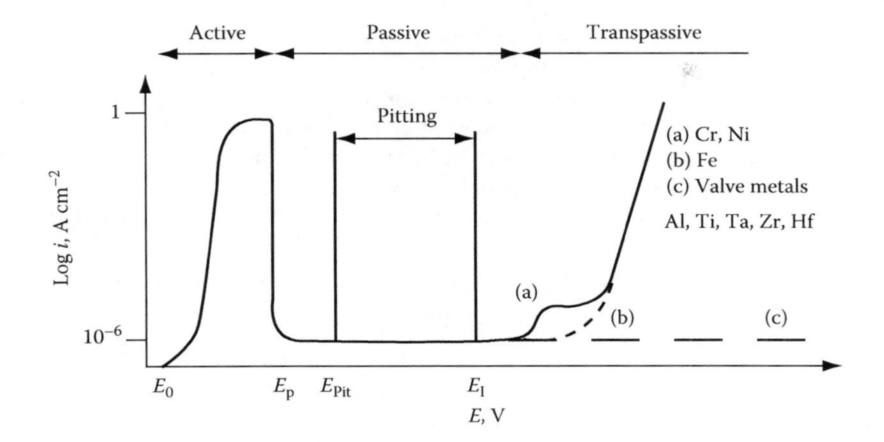

FIGURE 5.1
Polarization curve of metals showing hydrogen evolution, active dissolution above the Nernst potential (E_0) of the Me/Me^{z+} electrode, passivation at the passivation potential (E_P), the passive range, the transpassive range for metals like Cr and Ni (a), oxygen evolution for metals like Fe, Cr, Ni (b), and continuing passivity up to $E > 100$ V for metals with insulating passive layers like Al, Ta, Zr, Hf (c). The potential range of localized corrosion in the presence of aggressive (and inhibiting) anions above the pitting potential (E_{Pit}) and below a possible inhibition potential (E_I) is indicated.

cm^{-2} or less although intense metal dissolution is expected due to the increased driving force for this reaction. The current density stays at this very low level within the passive range and will increase only at very positive potentials when oxygen evolution becomes possible. In acidic solutions, this is the case at ca. $E > 1.6$ V for metals like iron, nickel, or chromium. On metals forming electronically insulating passive films like Al or Ta, no oxygen evolution is observed up to potentials of more than 100 V. Thus, the passive range extends to more than 100 V for these systems. Other metals like Sn show passivity up to potentials of $E = 2.5$ V. The different behavior is related to the dissolution kinetics and the electronic properties of the anodic oxides. Many metals form higher valent cations with higher solubility in the transpassive range so that a related current increase is observed. This is the case for Cr, which forms soluble $Cr_2O_7^{2-}$ or CrO_4^{2-} in acidic or alkaline solutions, respectively. A similar increase of transpassive metal dissolution is observed for Ni in acidic electrolytes due to Ni(III) formation.

Uhlig gave two definitions of passivity in his historical review [1,2]:

- A metal is passive if it substantially resists corrosion in a given environment resulting from marked anodic polarization.
- A metal is passive if it substantially resists corrosion in a given environment despite a marked thermodynamic tendency to react.

The two main reasons behind passivity are as follows:

- The thermodynamics of the metal–electrolyte interface with the formation of a protecting anodic oxide or hydroxide film that cannot dissolve because it is in dissolution equilibrium with the contacting solution.
- The dissolution kinetics of the passivating oxide. The transfer rate of cations from the oxide matrix to the electrolyte is so slow that the dissolution current gets negligible and thus the layer protects the metal because any dissolution has to occur via this surface film. In this case, the passive layer may be far from dissolution equilibrium.

In this chapter, these thermodynamic and kinetics aspects of passivity are presented after a brief historical survey. The following section discusses the electrode kinetics in the passive state. Next the chemical composition and chemical structure of passive films form on pure metals are reviewed with an emphasis on iron. This is followed by a compilation of data for binary alloys. The electronic properties of passive layers are then discussed, and the last section covers the structural aspects of passivity.

Figure 5.1 also shows that most passive metals are subject to intense local corrosive attack if the electrolyte contains aggressive anions. All halides but also other anions like SCN^- or ClO_4^- are effective to trigger localized corrosion at very positive potentials. The details of localized corrosion are presented in Chapter 7.

5.2 Historical Aspects

Passivity has been known for several hundred years. Uhlig mentioned in his review on passivity that Lomonosov was the first in 1738 to detect that iron does not dissolve in concentrated nitric acid [3]. Ostwald described in his history of electrochemistry [4] that Keir observed in 1790 the passivity of iron again in concentrated nitric acid [5]. Similar observations were made in 1782 by Wenzel according to citations of Gmelin [6]. In 1807, Hiesinger and Berzelius found that one could achieve passivity by anodic polarization [6], which was confirmed by Schönbein for iron in diluted nitric, sulfuric, and phosphoric acids [7]. In their correspondence, Schönbein and Faraday discussed the nature of these observations and Faraday came to the conclusion that a thin film provides protection to the metal. He postulated its electronic conductivity based on his own experiments [8].

Since then, many investigations have been performed due to the technical importance of passivity. The progress in understanding was closely linked to the availability of surface analytical methods that brought detailed insight to the systems. In the 1950s and 1960s, many electrochemical studies were performed applying sophisticated electrode techniques. In situ optical methods like ellipsometry and electroreflectivity and reflection spectroscopy were added in the late 1960s [9]. There was a debate whether passivity is caused by a two-dimensional adsorption film or a three-dimensional layer, although the charges detected by potentiostatic and galvanostatic transients strongly suggested a 3D layer, even taking into account the surface roughness. The application of the rotating ring disk (RRD) technique allowed the separation of the total current and charge at the disk electrode into a film formation part and a dissolution part [10,11]. As this could be done with a time resolution of ca. 1 s or less, one could study the kinetics of dissolution during passivation transients and the growth of the passive layer on iron [10]. Similar studies were performed by a quantitative time-resolved analysis of the electrolyte of a flow cell after reaction of dissolved Fe(III) ions with SCN^- forming a red colored complex, which could be analyzed quantitatively by colorimetry [11].

In the 1970s and 1980s, ex situ surface analysis methods like Auger electron spectroscopy (AES), x-ray photoelectron spectroscopy (XPS), ion scattering spectroscopy (ISS), and Rutherford backscattering (RBS) were applied to investigate the chemical composition of passive layers. The development of fast entry locks and flexible specimen handling of ultra-high vaccum (UHV) spectrometers in the mid-1970s made the systematic investigation of corrosion systems and other electrode surfaces possible. There was a strong debate whether one may use these methods that operate in vacuum to study electrode surfaces. In the 1980s, the development of specimen preparation and transfer in a closed system allowed investigation at negative potentials and for short passivation times even in the millisecond range avoiding unwanted oxidation of lower valent species and uncontrolled oxide formation due to air contact [12]. The chemistry of numerous systems has been studied with these methods [13–17]. The effect of exposure to air was discussed using secondary ion mass spectrometry and the ^{18}O isotope marker [18–21]. The intense research with the precautions mentioned here combined to the large number of reliable data obtained from systematic studies, and the fact that only x-ray absorption near edge structure [22–24] is available for in situ chemical analysis of passive film justify the application of ex situ methods to study the chemical aspects of passivity up to now.

Another big debate was on the crystalline or amorphous structure of passive layers. For about 20 years, structure-sensitive in situ methods have been applied to the investigation of passive layers like scanning tunneling microscopy (STM) and atomic force microscopy (AFM) and methods using synchrotron radiation like x-ray diffraction (XRD) and x-ray absorption spectroscopy (XAS). All these methods show that the passive layers are in most cases crystalline [25].

5.3 Thermodynamic Aspects of Passivity

The aqueous corrosion and passivation of metal surfaces involve electrochemical processes at the electrode–electrolyte interface. Like for all chemical reactions, the two aspects of equilibrium and kinetics have to be treated. Thermodynamic data give an important first insight into layer formation. They have been used to compose potential–pH diagrams for all elements and thus for all metals [26,27]. In Chapter 1 of this book on the fundamentals of corrosion, passivity of iron has been mentioned and the calculation of some of the lines based on thermodynamic data has been described. Here a more detailed description of the potential–pH diagrams of iron and copper are presented.

Iron is one of the most studied metals due to its technical importance.

Table 5.1 gives a full compilation of the thermodynamic equilibriums that are used to construct the diagram presented in Figure 5.2. The lines of Figure 5.2 refer to the equilibriums given in Table 5.1. The dissolution to Fe^{2+} shows four lines, which correspond to cation concentrations of 10^{-6}–10^{-1} M. Fe_3O_4 is in equilibrium with Fe metal and is oxidized to Fe_2O_3 at potentials described by Equations 3 and 4, respectively. Both have a −0.059 V/pH dependence. The vertical line describes the potential independent dissolution equilibrium of Fe_2O_3 with four

TABLE 5.1

Electrochemical and Chemical Equilibria for the Potential–pH Diagram of Iron at 298 K

Reaction	Potential
1. $Fe \leftrightarrow Fe^{2+} + 2e^-$	$E = -0.440 + 0.029 \log c(Fe^{2+})$
2. $Fe^{2+} \leftrightarrow Fe^{3+} + e^-$	$E = -0.771 – 0.059 \log c(Fe^{3+})/c(Fe^{2+})$
3. $3Fe + 4H_2O \leftrightarrow Fe_3O_4 + 8H^+ + 8e^-$	$E = -0.085 – 0.059$ pH
4. $2Fe_3O_4 + H_2O \leftrightarrow 3Fe_2O_3 + 2H^+ + 2e^-$	$E = 0.221 – 0.059$ pH
5. $3Fe^{2+} + 4H_2O \leftrightarrow Fe_3O_4 + 8H^+ + 2e^-$	$E = 0.98 – 0.236$ pH $- 0.089 \log c(Fe^{2+})$
6. $2Fe^{2+} + 3H_2O \leftrightarrow Fe_2O_3 + 6H^+ + 2e^-$	$E = 0.728 – 0.177$ pH $- 0.059 \log c(Fe^{2+})$
7. $2Fe^{3+} + 3H_2O \leftrightarrow Fe_2O_3 + 6H^+$	$E = \log c(Fe^{3+}) = -0.72 – 3$ pH
a. $H_2 \leftrightarrow 2H^+ + 2e^-$	$E = 0 – 0.059$ pH
b. $2H_2O \leftrightarrow 4H^+ + 4e^- + O_2$	$E = 1.23 – 0.059$ pH $+ 0.015 \log p(O_2)$

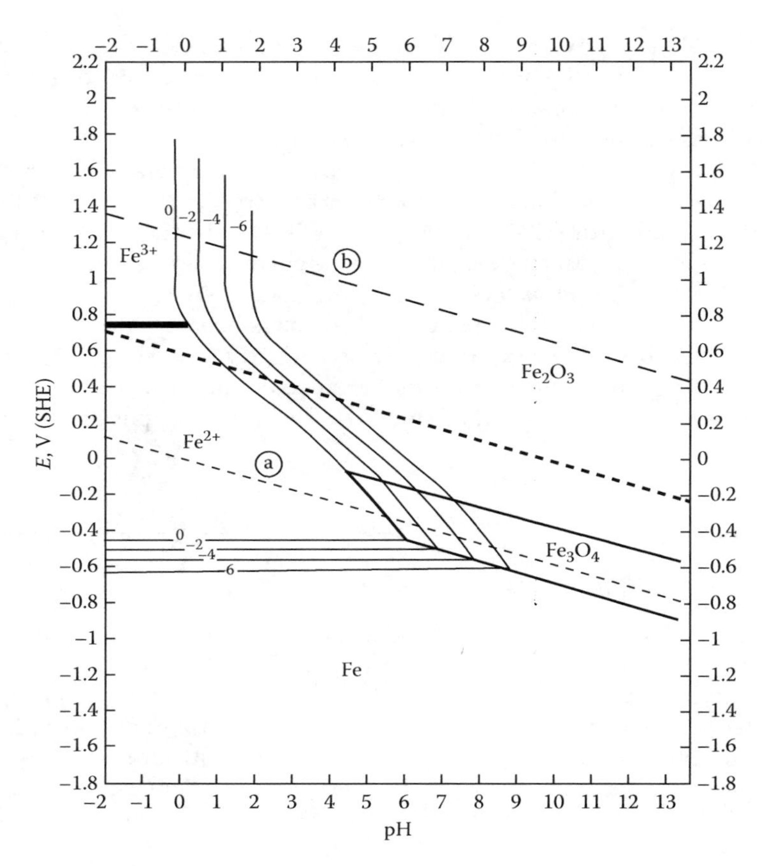

FIGURE 5.2
Potential–pH diagram for Fe showing the lines for electrochemical equilibria and the fields of immunity (stable Fe), passivity (stable oxides Fe_3O_4 and Fe_2O_3), and corrosion (dissolution of Fe^{2+} and Fe^{3+}). The added central dashed line is the Flade potential. (From Pourbaix, M., *Atlas d'Equilibres Electrochimiques*, Guthiers Villars+ Cie, Paris, 1963; Pourbaix, M., *Atlas of the Electrochemical Equilibria in Aqueous Solutions*, Pergamon, Oxford, 1966.)

related Fe^{3+} concentrations. Equation 6 corresponds to the Fe^{2+} oxidation forming deposits of Fe_2O_3 at the metal surface. Equation 5 takes into account the Fe^{2+} oxidation to Fe_3O_4. Reactions 5 and 6 have slopes of −0.236 and −0.177 V/pH depending on the number of H^+ and electrons involved. The four parallel lines refer to the four Fe^{2+} concentrations of 10^{-6}–10^{-1} M.

At very negative potentials, Fe metal is stable and one has the region of immunity. Above the equilibrium characterized by Equation 1, one observes Fe dissolution. Above $E = 0.77$ V, Fe^{2+} is oxidized to Fe^{3+}. Within the field of stability of Fe cations, one observes corrosion. In the fields of stability of Fe_3O_4 and Fe_2O_3, the passive range, the metal is covered by a protecting passive layer. The diagram contains also the potential of the redox systems (a) H^+/H_2 and (b) O_2/H_2O. They are the most important oxidants for corrosion reactions. If their Nernst potential is positive to the values of the reaction under study, they will serve as driving counterreactions. Thus, Fe will be oxidized in acidic electrolytes by hydrogen evolution. It may also be oxidized by oxygen reduction. However, this reaction is often relatively slow due to its large kinetic hindrance and the small concentration of dissolved oxygen and thus has a large overvoltage. However, it is very effective in weakly acidic to alkaline solutions where H^+ reduction is very slow due to its small concentration. Water reduction forming H_2 is a slow process with a large overvoltage by itself (see Section 1.12.1, Chapter 1).

The experimental results for the passivation potential of iron in neutral and alkaline solutions follow the predictions of Fe_3O_4 formation given by Equation 3 in Table 5.1 as discussed below in this chapter. In strongly acidic electrolytes, anodic dissolution as Fe^{2+} and Fe^{3+} is expected at $E > -0.6\,V$ and $0.77\,V$, respectively, with no limit to increasing positive potentials. However, passivation is obtained at $E = 0.58\text{–}0.058$ pH/V experimentally at the so-called Flade potential. This result is marked by the dashed line 8 in Figure 5.2. It is related to the dissolution kinetics of the passivating Fe_2O_3 film and is in apparent contradiction to the solubility of all oxides of iron, i.e., to the thermodynamic values. Similarly Cr forms a very effectively passivating film at potentials positive to 0 V in acidic electrolytes, when the formation of a Cr_2O_3 layer becomes possible. However, this film is not in dissolution equilibrium with a strongly acidic electrolyte, but dissolves extremely slowly, thus forming an excellent passive layer. Similar results are obtained for Ni in acidic solutions.

The potential–pH diagram of Cu is shown as another example in Figure 5.3. The related reactions and electrode potentials are presented in Table 5.2. In acidic solution, Cu dissolution starts at $E = 0.2\,V$ and no passivity is observed. Similarly, Cu should dissolve as CuO_2^{2-} without passivation in strongly alkaline solutions (Equation 3). This is at least partially observed with an increasing active dissolution for high pH. Passivation is expected when Cu_2O formation becomes possible (reaction 4 in Table 5.2). The stability range of Cu_2O is relatively narrow due to its easy oxidation to CuO and $Cu(OH)_2$ at slightly more positive

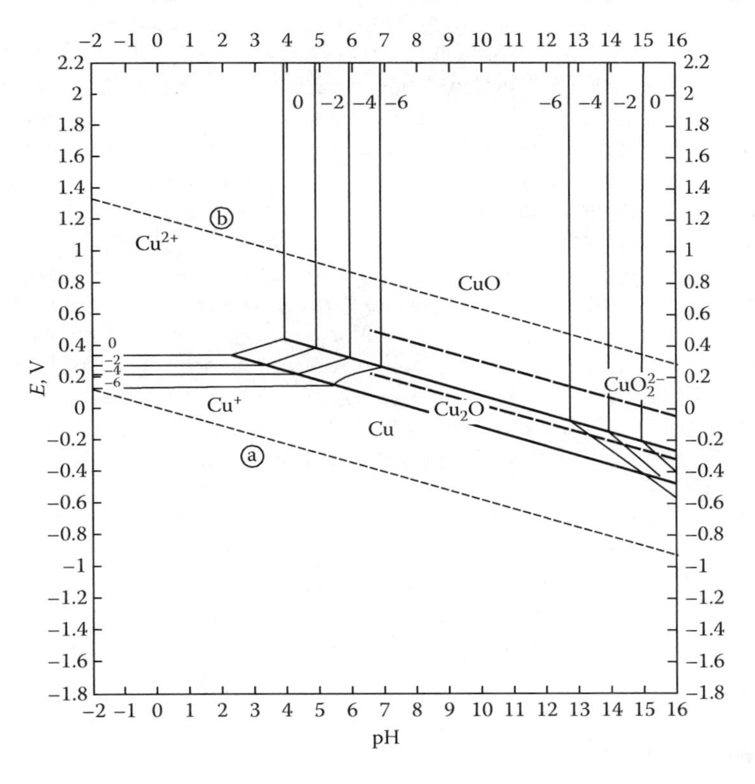

FIGURE 5.3
Potential–pH diagram for Cu showing the lines for electrochemical equilibria and the fields of immunity (stable Cu), passivity (stable oxides Cu_2O and CuO), and corrosion (dissolution of Cu^+ and Cu^{2+}). The two central dashed lines are passivation potentials $E_{P,1}$ and $E_{P,2}$. (From Pourbaix, M., *Atlas d'Equilibres Electrochimiques*, Guthiers Villars+ Cie, Paris, 1963; Pourbaix, M., *Atlas of the Electrochemical Equilibria in Aqueous Solutions*, Pergamon, Oxford, 1966.)

TABLE 5.2

Electrochemical and Chemical Equilibria for the Potential–pH Diagram of Cu at 298 K

Reaction	Potential
1. $Cu \leftrightarrow Cu^{+} + e^{-}$	$E = 0.520 + 0.591 \log c(Cu^{+})$
2. $Cu \leftrightarrow Cu^{2+} + 2e^{-}$	$E = 0.337 + 0.0295 \log c(Cu^{2+})$
3. $Cu + 2H_2O \leftrightarrow CuO_2^{2-} + 4H^{+}$	$E = 1.515 - 0.1182\,pH + 0.0293 \log c\,(CuO_2^{2-})$
4. $2Cu + H_2O \leftrightarrow Cu_2O + 2H^{+} + 2e^{-}$	$E = 0.471 - 0.0591\,pH$
5. $Cu_2O + H_2O \leftrightarrow 2CuO + 2H^{+} + 2e^{-}$	$E = 0.669 - 0.0591\,pH$
6. $Cu_2O + H_2O \leftrightarrow 2Cu^{2+} + H_2O + 2e^{-}$	$E = 0.203 + 0.0591\,pH + 0.0591 \log (Cu^{2+})$
7. $Cu^{2+} + H_2O \leftrightarrow CuO + 2H^{+}$	$E = \log c\,(Cu^{2+}) = 7.89 - 2\,pH$
8. $Cu_2O + 3H_2O \leftrightarrow 2CuO_2^{2-} + 6H^{+} + 2e^{-}$	$E = 2.560 - 0.1773\,pH + 0.059 \log c\,(CuO_2^{2-})$
9. $CuO + H_2O \leftrightarrow CuO_2^{2-} + 2H^{+}$	$E = \log c\,(CuO_2^{2-}) = -31.98 + 2\,pH$
a. $H_2 \leftrightarrow 2H^{+} + 2e^{-}$	$E = 0 - 0.059\,pH$
b. $2H_2O \leftrightarrow 4H^{+} + 4e^{-} + O_2$	$E = 1.23 - 0.059\,pH + 0.015 \log p(O_2)$

potentials (Equation 5). Both Cu oxides are protective in the pH range 4–13. Equations 6 and 8 in Table 5.2 refer to the equilibrium of soluble Cu^{2+} and CuO_2^{2-} with solid Cu_2O, respectively. They have a +0.059 and −0.177 V/pH dependence, respectively. Reduction of soluble CuO_2^{-} species from alkaline solutions allows growing films of Cu_2O with varying thicknesses at the surface. The vertical lines in Figure 5.3 refer to the dissolution equilibrium of CuO with Cu^{2+} and CuO_2^{2-} in acidic and alkaline solutions, respectively.

Thus, the field of immunity is relatively large for this seminoble metal. Dissolution in acidic electrolytes is not limited to positive potentials. Apparently, the oxides of Cu are dissolving with rather large rates in acidic electrolytes so that they cannot be protective for kinetic reasons in contrast to the behavior of Fe, Ni, and Cr. The predictions of the potential–pH diagram are confirmed experimentally and passivity is obtained in neutral to alkaline solutions. The passivation potentials found experimentally are close to the equilibrium data as shown by the dashed lines in Figure 5.3. They correspond to the maximum of the anodic peaks for potentiodynamic polarization curves. The H_2/H^{+} equilibrium (line a) is too negative to allow Cu dissolution by hydrogen evolution. However, dissolved oxygen may oxidize Cu metal in agreement with the thermodynamic data and the results of corrosion studies. Line b in Figure 5.3 is sufficiently positive to allow this process.

Potential–pH diagrams have been calculated for all elements at 298 K but also for higher temperatures and are compiled in Refs. [26,27]. They are very useful for a first understanding of a corrosion system. Any thorough investigation also needs electrokinetic studies and the application of surface analytical methods to get insight into the mechanisms and properties of passivated metal surfaces and their corrosion protection properties.

5.4 Kinetic Aspects of Passivity

Polarization curves are very instructive for examining the combined thermodynamic and kinetic effects of metal passivation. Figure 5.4 presents the examples of Fe, Cr, and Ni in 0.5 M H_2SO_4 [17]. The active dissolution current density of all three metals gets up to

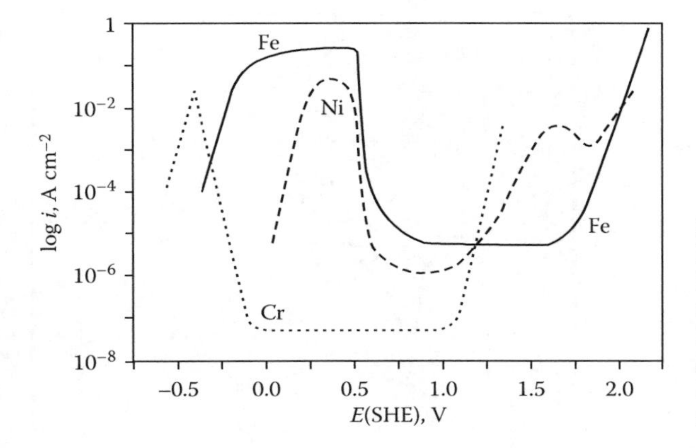

FIGURE 5.4
Polarization curves of Fe, Cr, and Ni in 0.5 M H_2SO_4 showing active dissolution, and passive and transpassive behavior. (From Strehblow, H.-H., in *Passivity of Metals*, R.C. Alkire, D.M. Kolb, eds., *Advances in Electrochemical Science and Engineering*, Vol. 8, Wiley-VCH, Weinheim, Germany, pp. 271–374, 2003.)

100 mA cm^{-2} or more with a steep fall, at the passivation potential E_P, to some few μA cm^{-2} or less. For Fe and Ni, $E_P = 0.6$ V is obtained, whereas for Cr the value is about $E_P = 0$ V only. All three metals show an extended passive potential range although their anodic oxides are not in dissolution equilibrium. Therefore, passivity is a consequence of their slow dissolution kinetics. For Cr, the stationary passive current density is less than 1 μA cm^{-2}, whereas it is in the range of several μA cm^{-2} for the other two metals. Due to the very negative passivation potential and the small passive current density, Cr is a very important alloy component for steels. This is the reason for the excellent corrosion resistance of stainless steel. At more positive potentials, the dissolution of Cr and Ni is increasing due to the formation of higher valent ions like the fast-dissolving $Cr_2O_7^{2-}$ and Ni^{3+}. Fe shows an increased dissolution in the transpassive potential range together with oxygen evolution, the latter being also found for Ni and Cr at $E > 1.6$ V. For Ni, the passive current density changes with the electrode potential. This has been interpreted by an increase of the oxidation state of the cations at the surface of the passive layer, and therefore a change of the composition of the passive film with potential.

Figure 5.5 depicts the polarization curve of Cu in 0.1 M KOH with indication of soluble products and composition of the passive layer indicated. There are two pronounced anodic peaks AI and AII and closely related cathodic peaks CII and CI [28]. Peak AI is relatively small and reaches 0.1 mA cm^{-2} at maximum. Here Cu_2O, which has a very small solubility at pH 13, is formed. The corrosion efficiency, $\eta = i_C/(i_C + i_l)$ with current densities i_C for dissolution and i_l for layer formation, is ca. 40% from passivation transients with RRD studies [29]. η is about 80% at potentials in the range of peak AII with a current density of CuO_2^{2-} formation of several 0.1 mA cm^{-2}. The solubility of CuO_2^{2-} is higher and thus the CuO film is less protective compared to the Cu_2O film. At peak AII, the layer thickness may increase by a dissolution and precipitation mechanism and even grow to visible dimension. The anodic reactions of layer formation and cathodic reactions of reduction marked in Figure 5.5 are discussed more in detail in Section 5.6.2.2 together with findings of surface analysis by XPS and ISS. During the cathodic scan, CuO_2^{2-} dissolution is detected when CuO and $Cu(OH)_2$ are present in the passive layer. Once the Cu(II) part of the film is reduced, no CuO_2^{2-} formation can be detected by RRD studies as expected. Reduction of the Cu(II) layer at peak CII goes along with the release of soluble Cu^+ ions, indicating the partial formation of soluble Cu(I) ions when the passive layer is reduced.

The passivation of many metals and alloys has been studied with polarization curves and potentiostatic transients. In the following, iron is further discussed because it has

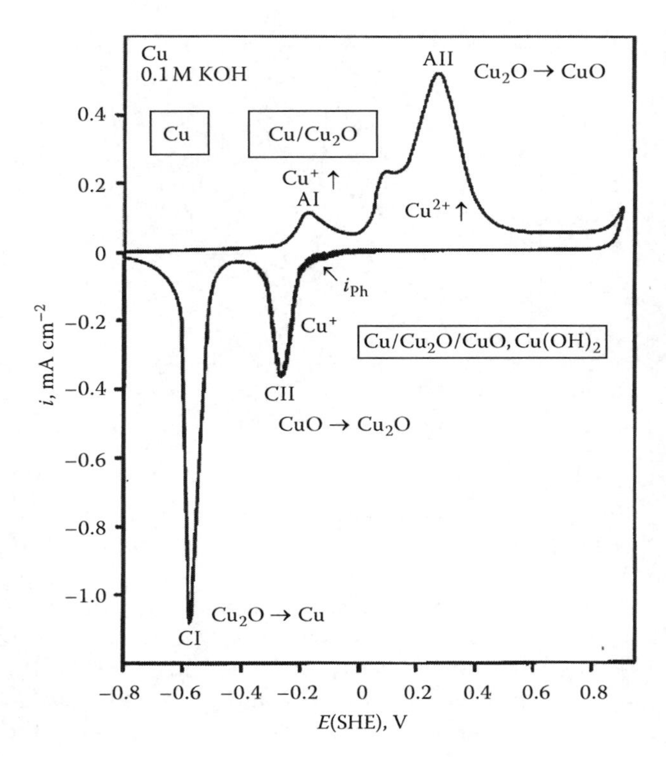

FIGURE 5.5
Polarization curve of Cu in 0.1M KOH with anodic and cathodic current peaks and the related reactions of formation and reduction of oxide layers, the indication of soluble products and the stability range of Cu metal and its oxides. The oscillations at peak CII are due to an oscillating photocurrent density i_{Ph} caused by a chopped light beam. (From Strehblow, H.-H., in *Passivity of Metals*, R.C. Alkire, D.M. Kolb, eds., *Advances in Electrochemical Science and Engineering*, Vol. 8, Wiley-VCH, Weinheim, Germany, pp. 271–374, 2003.)

been investigated in large detail with electrochemical methods. The procedures and the applied methods are typical for studies of passivity and similar investigations have been performed with other metals.

5.5 Electrode Kinetics in the Passive State

Metal dissolution also occurs in the passive state, but at a much smaller rate. The extremely small passive current densities are a consequence of a thin, solid, poreless, and adherent film that covers the metal surface. For iron, the stationary passive current density is potential independent and equals $i_P = 6\,\mu A\ cm^{-2}$ in 0.5 M H_2SO_4. It is smaller in more alkaline solutions and solutions without complexing properties. The film thickness grows linearly with the potential and reaches ca. 5 nm for a ca. 1 V increase above the passivation potential. Any dissolution phenomenon involves the transfer of cations across this layer and thus a transfer through the solid state.

Figure 5.6 presents a simple single layer model of the passive film on a metal surface with the related reactions that have to be discussed. Passive dissolution (1) involves metal oxidation and transfer of the Me^{z+} cations to the layer through the film and across the layer–electrolyte interface. The cation transport through the oxide layer is possible due to a potential drop within the film. The film thickness of a few nanometers generates a high field strength of several $10^6\ V\ cm^{-1}$. Oxide (or hydroxide) growth (2) involves the formation of O^{2-} (or OH^-) ions from water molecules and their transfer into and across the film. New oxide (or hydroxide) may form at the layer surface or at the metal surface by combination

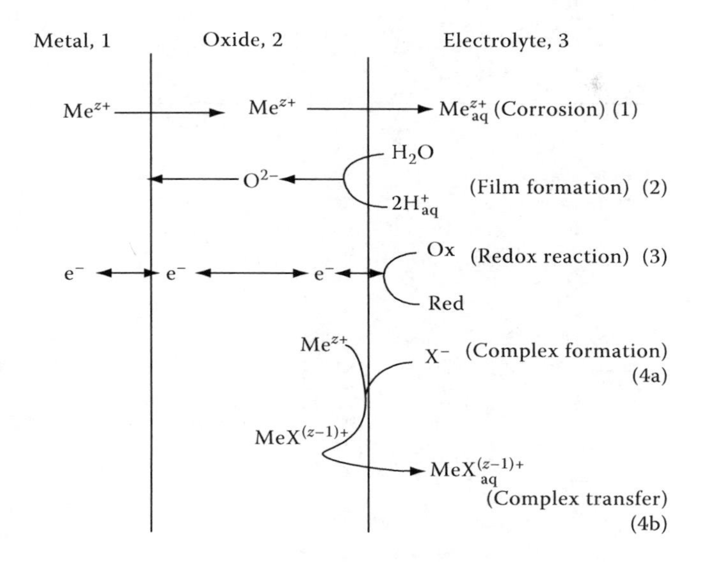

FIGURE 5.6
Model of a passive layer with the related reactions of corrosion, layer formation, and complex formation. (From Strehblow, H.-H., in *Passivity of Metals*, R.C. Alkire, D.M. Kolb, eds., *Advances in Electrochemical Science and Engineering*, Vol. 8, Wiley-VCH, Weinheim, Germany, pp. 271–374, 2003.)

of O^{2-} (or OH^-) ions with metal cations. Where most of the layer growth occurs is a consequence of the mobility and transfer rates of the cations and anions. For iron, experimental studies with markers have shown that both ions have comparable transfer rates and that new oxide grows occurs at both interfaces [30–32]. Electron transfer (3) is another important partial process. It is involved when redox reactions like oxygen evolution or the oxidation/reduction of the Fe^{2+}/Fe^{3+} or $Fe(CN)_6^{3-/4-}$ redox couples occur at passive metal surfaces. Electron transfer is also involved during layer formation or passive corrosion at open circuit conditions, when redox reactions are required as counter reactions for metal oxidation. These processes require conducting or semiconducting properties of the passive layer, which are discussed in detail in Section 5.7. Complexing of cations at the layer surface (4) is another important process. This reaction may enhance the transfer of metal ions from the layer to the electrolyte as has been observed in several cases. If this process occurs locally, an enhanced local dissolution of the film and its subsequent removal may result in the formation of a pit. These reactions are discussed in Chapter 7.

Figure 5.7 shows the potential drops $\Delta\varphi_{1,2}$ and $\Delta\varphi_{2,3}$ at the metal–oxide and the oxide–electrolyte interfaces, respectively, and the potential drop $\Delta\varphi_1$ within the passive layer for stationary conditions (index s) and for two potentials corresponding to two thicknesses. The exact potential drops are not known. Only the change of the sum of all three drops may be measured or determined by the fixed electrode potential. However, the changes of $\Delta\varphi_{2,3}$ may be given with some assumptions as shown below. Usually, $\Delta\varphi_{1,2}$ is assumed as a constant value and $\Delta\varphi_{2,3,s}$ changes with the pH as shown below. Assuming that the potential increase for stationary conditions $\Delta\varphi_1$ is located within the passive layer with constant $\Delta\varphi_{1,2,s}$ and $\Delta\varphi_{2,3,s}$, one obtains a field strength of $1/10^{-7} = 10^7$ V cm^{-1}. This is the order of magnitude for the potential drop at electrode surfaces in the Helmholtz layer. As a consequence, an exponential relationship between current density and potential is found, as discussed for the Butler–Volmer equation of charge transfer controlled processes. Ion transfer in such a high electrical field is treated by the high field mechanism (Section 5.5.2).

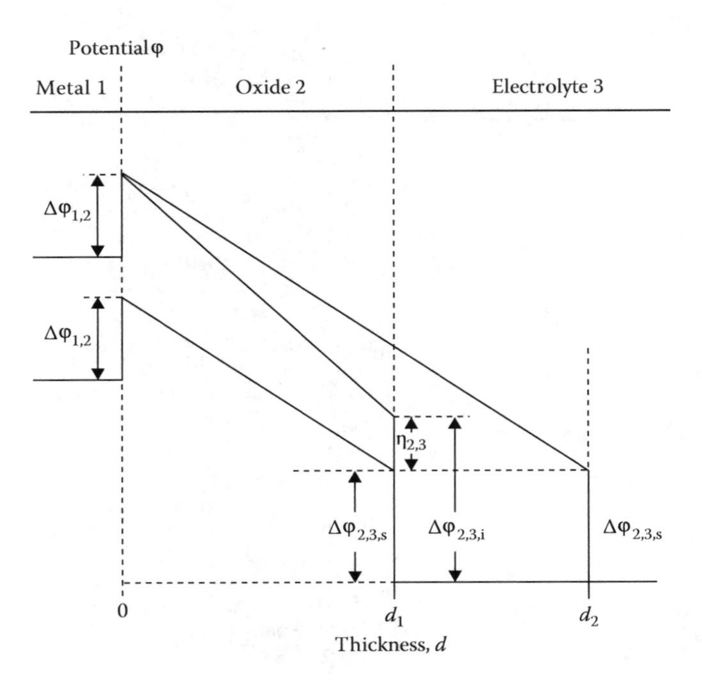

FIGURE 5.7
Potential drops $\Delta\varphi_{1,2}$ and $\Delta\varphi_{2,3}$ across the metal–oxide and oxide–electrolyte interfaces and $\Delta\varphi_l$ across the oxide film for stationary (index s) and instationary (index i) conditions. Instationary conditions after potential increase cause a larger voltage drop $\Delta\varphi_l$ within the layer and an overpotential $\eta_{2,3}$ at the oxide–electrolyte interface where oxide growth and the related O^{2-} production occur. (From Strehblow, H.-H., in *Passivity of Metals*, R.C. Alkire, D.M. Kolb, eds., *Advances in Electrochemical Science and Engineering*, Vol. 8, Wiley-VCH, Weinheim, Germany, pp. 271–374, 2003.)

5.5.1 Reactions at the Layer–Electrolyte Interface

Both partial reactions at the layer surface, passive dissolution, and its related passive corrosion current density i_c (Equation 5.1) and O^{2-} formation and its related layer formation current density i_l (Equation 5.2), depend on the same potential drop $\Delta\varphi_{2,3}$ at the oxide–electrolyte interface. For stationary conditions, no layer growth proceeds ($i_l = 0$), i.e., the reaction of O^{2-} formation (Equation 5.2) is at equilibrium. In these conditions, a Nernst equation can be written (Equation 5.3). The charge transfer of metal cations follows the Butler–Volmer equation with the potential drop $\Delta\varphi_{2,3,S}$, the charge transfer coefficient α, the charge $z = +3$ for dissolving Fe^{3+} ions and the exchange current density $i^0_{C,S}$ (Equation 5.4). Deducing $\Delta\varphi_{2,3,S}$ from Equation 5.3, one obtains the pH dependence of $i_{C,S}$ (Equation 5.4) where $i^0_{C,S}$ (0) is the exchange current density of passive corrosion at pH 0. For iron, $i_{C,S}$ has been measured as a function of the pH, which yields the pH dependence expressed in Equation 5.6. The experimental value of 0.84 yields a charge transfer coefficient $\alpha = 0.28$ for $z = 3$ [34]:

$$Fe^{3+}(ox) \rightarrow Fe^{3+} * aq \tag{5.1}$$

$$H_2O \leftrightarrow O^{2-} + 2H^+ \tag{5.2}$$

$$\Delta\varphi_{2,3,S} = \Delta\varphi_{2,3,S}(0) + \frac{RT}{2F}\ln[a^2(H^+)] = \Delta\varphi_{2,3,S}(0) - 0.059\text{pH} \tag{5.3}$$

$$i_{C,S} = i^0_{C,S} \exp\left[\frac{\alpha zF\Delta\varphi_{2,3,S}}{RT}\right] = i^0_{C,S}(0)\exp[-\alpha z\, 2.303\text{pH}] \tag{5.4}$$

$$i^0_{C,S}(0) = i^0_{C,S} \exp\left[\frac{\alpha zF\Delta\varphi_{2,3,S}(0)}{RT}\right] \tag{5.5}$$

$$\frac{\mathrm{d}\log i_{C,S}}{\mathrm{d}\,\text{pH}} = -\alpha z = -3\alpha = -0.84 \tag{5.6}$$

If the potential is increased, there is a transient increase of the potential drop $\Delta\varphi_l$ and field strength within the layer. This accelerates the cation transfer through the layer, which is described below by the high field mechanism. In addition, an overvoltage, $\eta_{2,3}$, is established at the oxide–electrolyte interface. The layer is no longer in electrochemical equilibrium and there is layer formation according to Equation 5.2. The potential drop at the oxide–electrolyte interface $\Delta\varphi_{2,3} = \Delta\varphi_{2,3,S} + \eta_{2,3}$ enters Equation 5.4. Combining with Equation 5.3, one obtains Equation 5.7 instead of Equation 5.5. Equation 5.7 yields Equation 5.8 for $\eta_{2,3}$. This overvoltage is effective for both reactions at the oxide–electrolyte interface, i.e., for metal corrosion and for layer formation and their related current densities i_C and i_L, respectively:

$$i_C = i^0_{C,S} \exp\left[\frac{\alpha zF\Delta\varphi_{2,3}}{RT}\right] = i_{C,S} \exp\left[\frac{\alpha zF\eta_{2,3}}{RT}\right] = i_{C,S}(0)\exp[-\alpha z\, 2.303\,\text{pH}]\exp\left[\frac{\alpha zF\eta_{2,3}}{RT}\right] \tag{5.7}$$

$$\eta_{2,3} = \frac{RT}{\alpha zF}\ln i_C - \frac{RT}{\alpha zF}\ln i_{C,S} = \frac{0.02}{\alpha}\log i_C - \frac{0.02}{\alpha}\log i_{C,S} \tag{5.8}$$

Thus, one obtains indirectly with Equations 5.5, 5.7, and 5.8 an expression for the potential drop $\Delta\varphi_{2,3}$ for stationary and nonstationary conditions, which requires the measurement of the related corrosion current densities i_C and $i_{C,S}$. This situation is similar to the charging of colloid particles, like a protein or an oxide particle, with a potential drop at their surface, which is established by the pH of the contacting solution with a reaction involving hydrogen ions. For a protein, the dissociation equilibrium of $-NH_3^+/-NH_2$ and $-COOH/-COO^-$ groups is established and, at oxides surfaces, the formation of O^{2-} (or OH^-) is described by Equation 5.2.

The passive corrosion current density i_C can be measured with appropriate analytical methods. For iron, measurements with the RRD electrode allow to measure the dissolution of Fe^{3+} ions by their reduction to Fe^{2+} at the ring [10]. The transfer efficiency between the disk and the ring is about 70% and may be calculated from the geometry of the Fe-disk and Pt-ring electrodes (radius of disk and ring and distance ring–disk) [33]. It can also be calibrated experimentally with a redox couple, e.g., reduction of $Fe(CN)_6^{3-}$ to $Fe(CN)_6^{4-}$ at the disk and reoxidation at the ring. With the measured total current density i measured at an Fe-disk, one can calculate $i_L = i - i_C$. These measurements can be performed with a time resolution of 0.1 s, which is the transfer time for the dissolved cations from the disc to the ring. An alternative method is the determination of dissolved Fe^{3+} ions using their reaction with SCN^- anions to form the red $[Fe(SCN)_2]^-$ complex. The dissolved cations have been collected in a cylindrical flow cell and then the concentration of their colored complex has been analyzed by colorimetry [11].

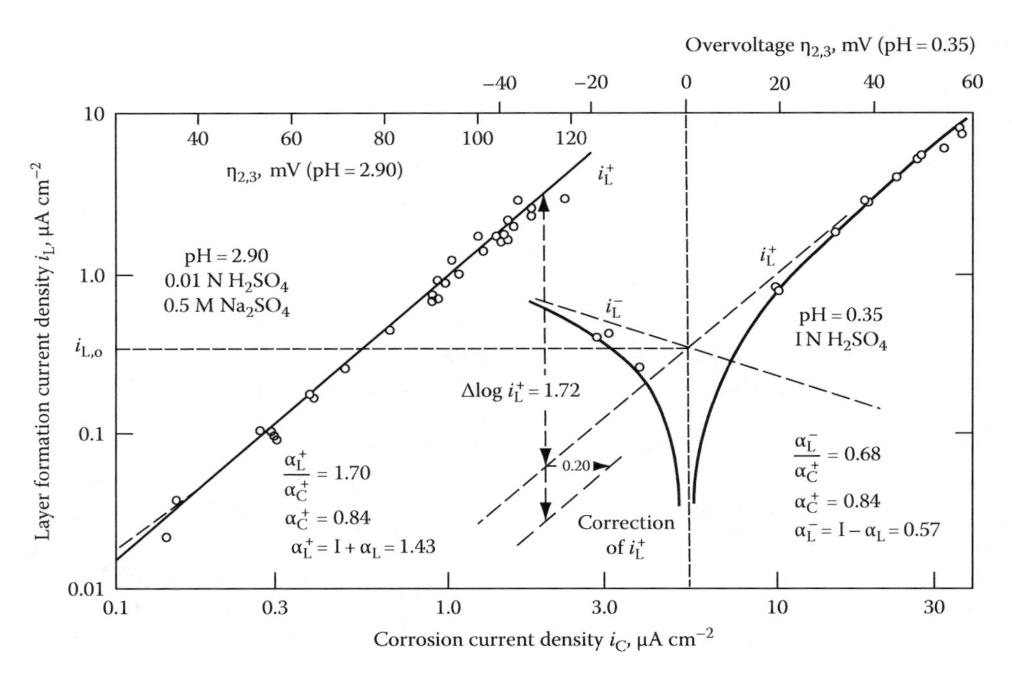

FIGURE 5.8
Tafel plots of the current densities of corrosion (i_C) and layer formation (i_L) of iron in acidic electrolytes as a function of the overvoltage $\eta_{2,3}$ at the oxide–electrolyte interface. (From Vetter, K.J. and Gorn, F., *Electrochim. Acta*, 18, 321, 1973.)

Such potentiostatic transients have been measured for iron and the total current has been separated in its corrosion and layer formation contributions. With the determination of the overvoltage $\eta_{2,3}$ as mentioned above, the i_C, $i_L - \eta_{2,3}$ dependence and their Tafel plots have been determined. Figure 5.8 shows the results of Vetter and Gorn [34] for iron in 0.5 M H_2SO_4 (pH 0.35) and 0.005 M H_2SO_4 + 0.5 M Na_2SO_4 (pH 2.9). The linear relation between $\eta_{2,3}$ and log i_C according to Equation 5.8 has been taken into account. Corrosion is far from any equilibrium and thus the deposition of metal ions as the reverse reaction has not to be taken into account. However, for layer formation (Equation 5.2), the reverse reaction has to be included as shown in Figure 5.8. At $\eta_{2,3} = 0$, both reactions equal each other with $i_C = i_{C,S} = i_{C,+} + i_{C,-} = 0$. Close to $\eta_{2,3} = 0$, the measured total current density i_C deviates from the Tafel line as indicated in Figure 5.8 and as discussed for charge transfer processes in general (see Chapter 1).

Further investigations of Fe dissolution have been performed to determine $i_{C,S}$ in electrolytes with the same pH and ionic strength to keep the same structure of the double layer but with different sulfate concentration in the range 0.0001–0.075 M. They concluded a promoting effect of sulfates on passive dissolution with an electrochemical reaction order $\nu_s = 0.16$. A mechanism has been proposed starting with the formation of a complex Fe^{3+} with SO_4^{2-} ions at the passive surface (Equation 5.9a) followed by transfer to the electrolyte as a rate determining step (Equation 5.9b) and by a fast reaction within the electrolyte (Equation 5.9c) [34]:

$$Fe^{3+} \cdot ox + SO_4^{2-} \cdot aq \rightarrow FeSO_4^{+} \cdot ad \tag{5.9a}$$

$$FeSO_4^+ \cdot ad \rightarrow FeSO_4^+ \cdot aq \quad (5.9b)$$

$$FeSO_4^+ \cdot aq \rightarrow Fe^{3+} \cdot aq + SO_4^{2-} \cdot aq \quad (5.9c)$$

Apparently, metal dissolution at the oxide–electrolyte interface is influenced by the formation of complexes, similar to the OH^- catalytic effect for active Fe dissolution (see Chapter 1). The small order of the electrochemical reaction has been explained with a Temkin adsorption isotherm for SO_4^{2-} anions. The passive current density of iron in 1 M $HClO_4$ is more than one order of magnitude smaller than in 0.5 M H_2SO_4. The complexing effect plays an important role for many anions including especially the aggressive halides. Their presence causes an increased dissolution of the passive layer, which can lead to general or local passivity breakdown and to an intense metal dissolution as in Chapter 7. The investigation of passivated surfaces with modern surface analytical methods like XPS and STM reveals a multilayer structure for most metals with a hydroxide layer on top of the passivating barrier oxide layer. In the light of these results, one presumably has to discuss the exchange of OH^- ions with anions of the electrolyte and the subsequent transfer of these complexes to the solution.

5.5.2 Ion Transfer Through the Film, High Field Mechanism

Depending on the mobility of the cations and anions, new oxide is formed at the layer–electrolyte or the metal–layer interface. This implies that cations as well as anions can migrate through the layer. These aspects have been investigated with SIMS or RBS. The experiments start with a prepassivated metal surface. The site of further oxide growth in water enriched with the isotope O^{18} can be determined with SIMS depth profiles showing where the O^{18} enrichment is located. Thick oxide layers on valve metals may be implanted with noble gas markers like Xe. In this case, further oxide growth does not move the neutral gas atoms, which thus mark the location of fresh oxide formation. RBS profiles allow determining the exact location before and after oxide formation. These studies show that new oxide is formed at both interfaces although with different rates. However, the deduced transfer numbers for cations (t_+) and anions (t_-) are comparative [31,32]. For crystalline oxide films on Zr and Hf, the migration of cations is completely suppressed and the new oxides form at the metal surface [32]. This result has been explained by an effective inward migration of O^{2-} ions at grain boundaries.

All three processes of ion transport across the metal–oxide and oxide–electrolyte interfaces and through the oxide film are governed by the potential drops and the related high field strength of several 10^6 V cm^{-1}. This leads to the same exponential form of the current density potential relation, i.e., a Butler–Volmer equation as described for charge transfer-controlled processes in Chapter 1. In principle, any of the three processes may be rate determining. For electroneutrality reasons, the transfers of charges across the interfaces and through the film have to compensate each other. In the following, ionic transport through the film is discussed within the frame of the high field mechanism [35]. The charge transfer across the interfaces follows the same law.

Passive layers are in most cases nanocrystalline with many defects and in some cases amorphous, so that transport occurs via interstitial sites, vacancies, and other imperfections. The transport of ions, i.e., the outward migration of cations as well the inward migration of anions, is submitted to an exponential dependence on the electrical field strength. The opposite transport, i.e., the inward migration of cations and the outward migration of anions can be neglected due to the extremely high field strength. Figure 5.9 depicts the

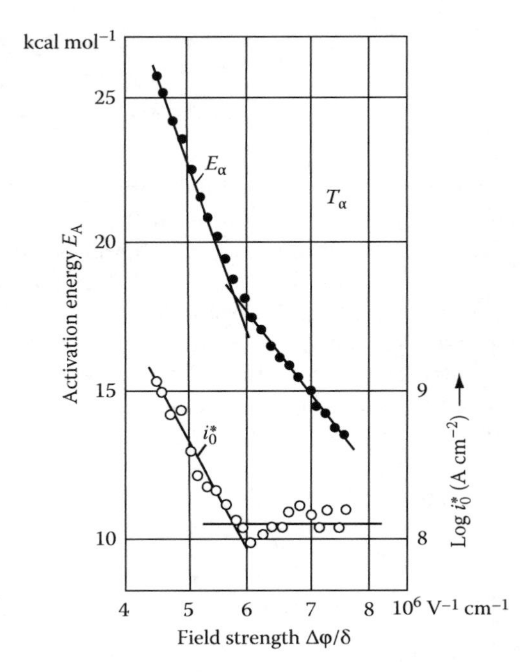

FIGURE 5.9
Activation energy E_A ($=\Delta G^\ddagger$) and preexponential term ν_0 ($=zFN\delta\nu$) of Equation 5.10 as a function of the electrical field strength $\Delta\varphi/\delta$ ($=\Delta\varphi_L/d$). (From Vermilyea, D.A., *J. Electrochem. Soc.*, 102, 655, 1955.)

diagram of the Gibbs free energy in dependence of the distance for the migration of cations [36]. A similar diagram holds for the inward migration of anions. To be transported, a cation has to overcome the maximum of ΔG. This process requires the free enthalpy of activation $\Delta G^\ddagger$. This leads to the exponential expression given by Equation 5.10 for the transfer rate of cations dN/dt with N being the concentration of the mobile cations, i.e., in an interstitial site or close to a defect, δ the jump distance between two maxima of ΔG and ν the oscillation frequency of the mobile ions:

$$\frac{dN}{dt} = N\delta\nu \exp\left[-\frac{\Delta G^\ddagger}{RT}\right] \tag{5.10}$$

$$\Delta G^\ddagger = \Delta G_0^\ddagger - \left(-\frac{\alpha z F \Delta\varphi_L \delta}{d}\right) \tag{5.11}$$

$$i = zF\frac{dN}{dt} = i_0 \exp\left[\frac{\alpha z F \delta \Delta\varphi_L}{RTd}\right] \tag{5.12}$$

$$i = i_0 \exp\left[\frac{\beta(E - E_P)}{d}\right] \tag{5.13}$$

$$\beta = \frac{\alpha z F \delta}{RT} \tag{5.14}$$

$$i_0 = zFN\delta\nu \exp\left[-\frac{\Delta G_0^\ddagger}{RT}\right] \tag{5.15}$$

The activation energy $\Delta G_0^{\ddagger}$ in the absence of the electric field is modified by the electrical work $\alpha zF\Delta\varphi_L\delta/d$ according to Equation 5.11 due to the presence of the potential drop $\Delta\varphi_L$ within the layer of thickness d and thus the electrical field strength $\Delta\varphi_L/d$. The symmetry factor α takes into account the fact that only the fraction of the electrical work up to the maximum of $\Delta G^{\ddagger}$ is kinetically effective. Combining Equations 5.10 and 5.11 and introducing Faraday's law with a cationic charge z leads to Equation 5.12. The potential drop $\Delta\varphi_L$ equals the difference between the applied potential E and the passivation potential E_P assuming constant potential drops $\Delta\varphi_{1,2}$ and $\Delta\varphi_{2,3}$ at the interfaces. Thus the potential increase $E - E_P$ is fully located within the film as described by Equation 5.13 with the Equations 5.14 and 5.15 for the constant β and the exchange current density i_0, respectively.

Equation 5.12 permits the determination of the layer growth. However one needs the information of the current density of layer growth $i_L = i - i_C$. In the simplest case, i_C is negligibly small, i.e., $i = i_L$ or one has to determine i_C as mentioned above and subtract it from the total current density i. With Faraday's law and the molar volume V_M of the anodic oxide, one obtains Equation 5.16 for the increase of the oxide thickness d. There is no analytical solution of this differential equation but a good approximation is given by Equations 5.17 and 5.18, the direct and the inverse logarithmic law of film growth, respectively:

$$\frac{dd}{dt} = \frac{V_M}{zF} i_0 \exp\left[\frac{\beta\Delta\varphi_L}{d}\right] \tag{5.16}$$

$$d = A + B\log[t + t_0] \tag{5.17}$$

$$\frac{1}{d} = A - B\log[t + t_0] \tag{5.18}$$

$$B = \frac{1}{\beta(E - E_P)} \tag{5.19}$$

The growth may be examined by the evaluation of electrochemical currents and charges taking into account the corrosion current density as described above. Other methods are based on the application of in situ ellipsometry or electrochemical quartz crystal microbalance or of surface analytical methods working in UHV like XPS, AES, ISS, and RBS, sometimes in combination with sputter depth profiling. Examples are given in the following section.

A log $i/E - E_P$ plot according to Equation 5.13 allows the determination of the exchange current density i_0 and the factor β, which in turn permits to calculate the jump distance δ according to Equation 5.14. The kinetic parameters i_0 and β are correlated to each other. Small changes of β require large ones of i_0. Therefore, their exact determination from experimental results is difficult. The concentration of mobile ions in the range of $N = 10^{14}$–10^{23} cm^{-3} and the frequency in the range $\nu = 10^{11}$–10^{13} s^{-1} agree with the kinetic data. The jump distance is in the range of $\delta = 0.1$–1 nm. For stationary conditions, no oxide growth is expected with $dd/dt = 0$ according to Equation 5.16. As a consequence, a constant field strength $\Delta\varphi_L/d$ is obtained which refers to $i = i_{C,S}$. This has been verified for iron, which has a potential independent voltage drop $\Delta\varphi_{2,3,s}$ at the oxide–electrolyte interface corresponding to a constant stationary corrosion current density $i_{C,S}$. This picture implies that the oxide thickness increases linearly with the potentials drop $\Delta\varphi_L = E - E_P$ and thus with the electrode potential E for stationary conditions.

After a potential increase, an overvoltage $\eta_{2,3} > 0\,V$ and a larger voltage drop $\Delta\varphi_{L,2}$ within the layer are established. With time, the oxide grows from d_1 to d_2, the overvoltage $\eta_{2,3}$ gets smaller, and disappears finally when the new stationary state is reached. The increase of oxide thickness $d_2 - d_1$ takes over the increased voltage drop $\Delta\varphi_{L,2} - \Delta\varphi_{L,1}$ within the film so that the field strength is the same as before, as depicted in Figure 5.6. The overpotential $\eta_{2,3} > 0\,V$ is necessary to allow the formation of O^{2-} ions following Equation 5.2 during the oxide growth and is the reason for an increased corrosion current density $i_C > i_{C,S}$ at the same interface for these nonstationary conditions.

The exponential dependence of the current density i on the field strength $\Delta\varphi_L/d$ within the film has been studied by potentiostatic transients of iron electrodes in 0.5 M H_2SO_4 starting with stationary conditions at different electrode potentials E and thus different related film thicknesses [37]. A log $i-E$ plot yields lines with slopes that allow calculating the kinetic parameters from the expression $\alpha z F \delta \Delta\varphi_L / RTd$ (Equation 5.12). For galvanostatic experiments on Fe in 0.5 M H_2SO_4 starting from stationary conditions at $E = 0.68\,V$, the charge of layer formation and thus the oxide layer thickness increases linearly with the electrode potential. The slope of the lines increases with the current density and the lines meet at the passivation potential of $E = 0.58\,V$ [38]. This result is consistent with a constant subdivision of the total current density i in a constant layer formation part and a corrosion part with a constant overvoltage $\eta_{2,3}$ for each galvanostatic experiment. The potential increase is linear with time and is located within the oxide film, i.e., it leads to a linear increase of $\Delta\varphi_L$. Similar to iron the passive layers on other metals like aluminum and tantalum grow linearly with the electrode potential. This is followed easily by galvanostatic experiments with a linear growth of the oxide thickness and electrode potential and thus a linear increase of the potential drop $\Delta\varphi_L$ within the film.

Equation 5.15 allows to determine the free activation enthalpy $\Delta G_0^\ddagger$ or the activation energy E_A from the temperature dependence of the passive exchange current density i_0. $\Delta G^\ddagger$ should change linearly with the field strength $\Delta\varphi_L/d$ within the film (Equation 5.11). This has been found for passive Ta, however with a break of the line at $\Delta\varphi_L/d = 6 \times 10^6\ V\ cm^{-1}$ [36]. The preexponential frequency factor in Equation 5.15 should be temperature independent, which apparently is not the case for a field strength below this critical value. This result has been explained by the field dependence of the concentration of mobile cations within the film. Thus an additional free activation enthalpy $\Delta G_{0,N}^\ddagger$ has to be added for a small field strength to activate cations to get to a mobile state (Equation 5.20). The total free activation enthalpy $\Delta G_0^\ddagger$ is thus the sum of this term $\Delta G_{0,N}^\ddagger$ and of the activation term $\Delta G_{0,M}^\ddagger$ for their migration (Equation 5.21). When all ions are in a mobile state a further increase of the field strength cannot increase the number of mobile ions anymore and the activation energy contains the part for the migration only:

$$N = N_0 \exp\left(-\frac{\Delta G_{0,N}^\ddagger}{RT}\right) \tag{5.20}$$

$$\Delta G_0^\ddagger = \Delta G_{0,M}^\ddagger + \Delta G_{0,N}^\ddagger \tag{5.21}$$

Passive layers usually consist of nanocrystals, which can be embedded in amorphous oxides or hydroxides matrices, and are expected to contain many defects like vacancies that allow an effective anion and cation transfer through the layer. On this basis and using the concept of Wagner [39], Macdonald developed the point defect model of passive film growth considering anion transfer via vacancies [40–42]. The outward transport of cations is described as an inward transfer of cation vacancies, and similarly the inward movement of O^{2-} ions is described as an outward movement of anion vacancies. The original work used linear transport equations for diffusion and migration. However, the high electrical field strength, confirmed for most passive layers, requires an exponential current–voltage relation. A linear relationship is only valid for a small electrical field strength. In later development, a small barrier layer was assumed with a high electrical field strength. However, a larger part of the film should still have a small field strength. This assumption is reasonable because passive layers of most metals have a multilayer structure with an inner oxide and an outer hydroxide part. The barrier part of the film seems to be the inner oxide where transport occurs by a high field mechanism following the exponential dependence of Equation 5.12.

5.6 Chemical Composition and Chemical Structure of Passive Films

The model of a single layer film as presented schematically in Figure 5.6 is too simple. It served as a first assumption for electrochemical studies when modern surface analytical methods were still not available. Many investigations have been performed in the last 25 years for the determination of the chemical composition and structure of passive layers. In almost all cases, passive films have at least a bilayer structure. In some cases, it is even more complicated. For the detailed studies of the chemical composition methods working in the UHV have been most effective. A brief summary of the methods mentioned in this chapter and briefly described below is given in Table 5.3. More detailed information is presented in Ref. [43].

This section presents a detailed discussion of the passivity of iron. It is intended as an example of the procedure and strategy in which the results of several electrochemical and surface analytical methods can be combined to get to a clear view and a sound model of the nature of passivity. For other pure metals and alloys, the same methods have been applied. However, the results for the other systems will be presented in a more condensed form to avoid redundancies and to keep the chapter sufficiently short. References to a more detailed description in literature are given.

TABLE 5.3

Characteristics of Some UHV Surface Analytical Methods

Method	In-Depth Resolution	Lateral Resolution	Information
XPS (x-ray photoelectron spectroscopy)	1 nm	25 μm	Composition of surface and surface films, binding and oxidation state
UPS (UV photoelectron spectroscopy)	1 nm	>1 mm	Work function, threshold energies, band structure, binding orbitals
AES, SAES [(scanning) Auger electron spectroscopy]	1 nm	10 nm	Composition of surface and surface films, high lateral information
ISS (ion scattering spectroscopy)	0.3 nm	25 μm	Composition of surface and surface films, modest lateral information
RBS (Rutherford backscattering)	5 nm	0.1 mm	Quantitative quasinondestructive depth profile

5.6.1 Brief Summary of UHV Methods for the Study of Passive Layers

XPS has been used for the study of many passive metals and alloys and the results are discussed in this section. A more detailed description of the method can be found in Ref. [44]. XPS is a soft method with usually a negligible change of the specimen's surface during the measurement. An x-ray beam (MgKα 1253.6 eV or AlKα 1486.6 eV) photoionizes atoms in the material and only electrons originating from the near surface region leave to the vacuum for measurement. The kinetic energy of the photoelectrons is analyzed by passing an electrostatic energy analyzer, and their intensity is measured by a channeltron, a specially designed secondary electron multiplier. The electronic levels of the elements at the surface fingerprint the signal in the XP spectrum. The known energy of the x-rays allows calculating the binding energy (E_B) of the photoelectrons. Slight changes of E_B due to the charge of the atoms allow drawing qualitative and quantitative conclusions on their chemical environment like the oxidation state and the nature of the binding matrix. The integration of the XPS intensity signals provides quantitative results on the amount of the related species and thus gives a measurement of the composition of surface films. The mean free path of the photoelectrons in the solid determines the information depth of the method, which is in the range of a few nanometers. Thus XPS is an excellent method to get information on the composition of surfaces and surface films.

An important improvement is the specimen preparation and transfer within the spectrometer. An electrochemical chamber attached to the spectrometer allows to start with a sputter cleaned surface. Reactive metals may be studied by XPS without artifacts caused by impurities and uncontrolled oxide growth due to air exposure. Thus one may investigate anodic oxide growth at negative potentials and for short times in the millisecond range. One may study lower valent species within the film, i.e., Fe(II) ions, and investigate the reduction of passive films. Any further experimental details are described elsewhere [44].

AES is a good method for chemical analysis with higher lateral resolution than XPS due to the fine focus of the exciting electron beam [45]. Electrons with a primary energy of several keV ionize core levels, which are filled with electrons of higher levels. The energy difference of the involved electronic levels is used to emit a third electron, the Auger electron, which leaves the surface and is analyzed with an electrostatic energy analyzer. The AES signals are therefore characteristic of each element with its electronic levels. They are described by the three involved electronic orbitals, e.g., a KLL AES line starts with the ionization of an electron of the K shell which is filled by an electron from the L shell causing a another L electron to leave the atom. The photoionization process of XPS also leads to x-ray-induced Auger lines. The AES signals also show chemical shift although its interpretation is usually more complicated due to the interaction of three electrons. However, it is sometimes larger and is therefore used when the chemical shift is too small for XPS signals. One example is the distinction of Cu metal atoms and Cu(I) ions with a too small shift of the XPS signals of ca. 0.1 eV. The use of x-ray-induced AES for this problem is illustrated in the following section.

An interesting complementary method to XPS is ISS. A beam of noble gas ions with primary energy E_0 (<2 keV) hits the specimen surface where the ions are backscattered by the atoms of the target surface. This backscattering process is described by an elastic binary collision process between a noble gas atom and a target atom with a well-defined energy loss of the projectile, which leaves the surface with the smaller energy E. This energy ratio E/E_0 depends on the mass of the noble gas ion m and target atom M and on the backscattering angle Θ between the incoming and the reflected beam. For a $\Theta = 90°$, the energy ratio depends on the masses only. Thus, each element at the specimen surface causes an

ISS signal with a characteristic energy ratio. Most of the backscattered ions are neutralized close to the surface after backscattering and only the few percents that remain charged will pass the electrostatic energy analyzer of the spectrometer. Ions that are backscattered from the second or any deeper atomic layer are completely neutralized and thus do not contribute to the ISS signal. Therefore, ISS in combination with a soft sputter process opens a unique possibility for depth profiles with monolayer sensitivity. XPS in combination with noble gas ion sputtering may be also used for depth profiling, however with a depth resolution that corresponds to the mean free path of the photoelectrons of a few nanometers.

RBS uses He^{2+} ions with a much higher primary energy of 2 MeV [46]. In this case, the ions may penetrate up to ca. 1 μm into the target and are backscattered to the vacuum where they are analyzed with a solid state detector. The helium ions are backscattered at different depths with an angle slightly smaller than $\Theta = 180°$ when their nucleus gets close to that of the target atoms. In addition to the energy loss caused by the elastic collision process with a target atom, inelastic losses take place due to excitations of other atoms on the way in and out. Therefore a broad signal is obtained that mirrors qualitatively and quantitatively the depth profile of the near surface of the specimen. If the masses of the target atoms sufficiently differ, their signals do not overlap. A good example is given below for anodic films on AlCu alloys. In this case, the Cu signal is found at much higher energy compared to that of Al due to its high mass. The depth information of the method is large, but the depth resolution is poor and in the range of ca. 5 nm as determined by the energy resolution of the detector. Therefore, the method is applicable for depth profiles of thick films like anodic layers on valve metals.

5.6.2 Passivity of Metals

5.6.2.1 Passivity of Iron

The critical passivation potential of metals E_P is usually described as the potential of oxide formation according to the reaction given by Equation 5.22. If the dissolution equilibrium is established, E_P follows the potential–pH diagram, which has been discussed in Section 5.3. However, experimental data for Fe in strongly acidic electrolytes are in contradiction to this simple interpretation. The thermodynamic values of all three FeO_n oxides, i.e., FeO, Fe_3O_4, and Fe_2O_3 with n = 1, 4/3, and 3/2, respectively, lead to similar values obtained from Equation 5.23. The values are $E_P(0) = -0.060$ V (FeO), $E_P(0) = -0.082$ V (Fe_3O_4), $E_P(0) = -0.040$ V (α-Fe_2O_3), and $E_P(0) = -0.010$ V (γ-Fe_2O_3) [47]. These values are close to the finding for $E_{P,1}$ in alkaline solutions. The experimental value in acidic electrolytes $E_{P,2}$, the so-called Flade potential, follows Equation 5.24. This value was first found by Flade [48] and has been confirmed later by several authors [49–55]. There is a large potential difference of ca. $E_{P,2} - E_{P,1} = 0.6$ V between both experimental values and the calculated result on the basis of Equation 5.22. However, if the Flade potential is related to the oxidation of Fe_3O_4 to γ-Fe_2O_3 according to Equation 5.25, the thermodynamic values get close to the experimental results of $E_{P,2}$. The thermodynamic values are not sufficiently reliable. Estimates with the ΔG_f^0 values of γ-Fe_2O_3 and Fe_3O_4 yield $E_{P,2}(0) = 0.43$ V and 0.63 V, respectively, which are not too far from the experimental value of 0.58 V. The value of $E_{P,2}(0) = 0.221$ V of the Pourbaix diagram (Figure 5.2, Table 5.1) 1 is too small to fit this result. The only oxide that is in thermodynamical equilibrium with Fe metal is thus Fe_3O_4. FeO and Fe_2O_3 will decompose in contact with Fe metal with negative ΔG_{298}^0 values according to Equations 5.26 and 5.27 [56]:

$$\mathrm{Me} + n\mathrm{H_2O} \leftrightarrow \mathrm{MeO}_n + 2n\mathrm{H}^+ + 2n\mathrm{e}^- \tag{5.22}$$

$$E_{P,1} \approx E_{P,1}(0) - 0.059\,\text{pH} = -0.05 - 0.059\,\text{V} \tag{5.23}$$

$$E_{P,2} = E_{P,2}(0) - 0.059\,\text{pH} = 0.58 - 0.059\,\text{pH} \tag{5.24}$$

$$2Fe_3O_4 + H_2O \leftrightarrow 3\gamma\text{-}Fe_2O_3 + 2H^+ + 2e^- \tag{5.25}$$

$$4FeO \leftrightarrow Fe_3O_4 + Fe \quad \Delta G^0_{298} = -17.1\,\text{kJ} \tag{5.26}$$

$$Fe + 4\gamma\text{-}Fe_2O_3 \leftrightarrow 3Fe_3O_4 \quad \Delta G^0_{298} = -96.1\,\text{kJ} \tag{5.27}$$

Thus one has to assume that Fe_3O_4 forms directly on Fe metal and will be oxidized to γ-Fe_2O_3 at the Flade potential, i.e., at $E = 0.58$–0.058 pH. This led to the idea of a bilayer structure with an inner Fe_3O_4 and an outer γ-Fe_2O_3 part [56,57]. Apparently, only γ-Fe_2O_3 forms a protecting layer and Fe_3O_4 is dissolving with too high rates at low pH. As Fe_3O_4 and γ-Fe_2O_3 both have an inverse spinel crystalline structure, they are miscible without any miscibility gap [58]. Thus a continuous change of the anodic oxide from Fe_3O_4 at the metal surface to γ-Fe_2O_3 in contact to the electrolyte is very likely (Figure 5.10 [59]), and the difference between the passivation potential $E_{P,1}$ found by extrapolation from experimental values at high pH (Equation 5.23) and the Flade potential $E_{P,2}$ (Equation 5.24) would relate to a bilayer structure. Any further increase of the electrode potential $E - E_{P,2}$ would be located within the outer layer of γ-Fe_2O_3. Attempts to prove this interpretation by electrokinetic studies of Fe in strongly acidic electrolytes were not successful [60]. Charging curves for anodic layer formation or cathodic reduction studies of the passive layer did not show the existence of an inner Fe_3O_4 part. Fe_3O_4 is not only very soluble at low pH like all other iron oxides but also very fast dissolving so that its presence could not be detected. Only in alkaline solution, a protecting lower valent Fe oxide was found by electrochemical measurements [51,52]. However, the anticipated bilayer structure has been confirmed by XPS studies of specimens passivated in 1 M NaOH as discussed below [61].

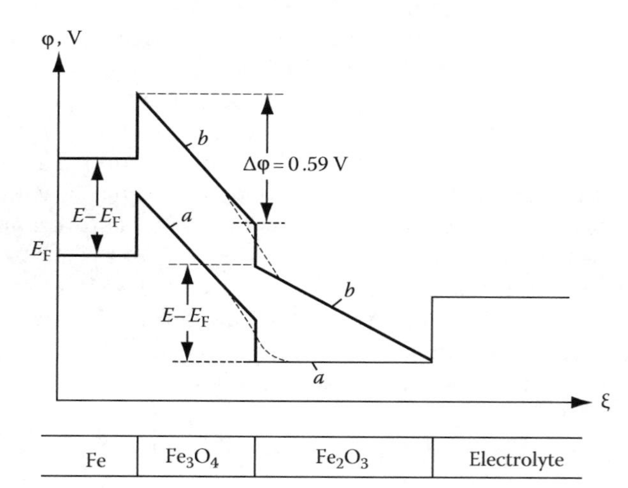

FIGURE 5.10
Schematic diagram of the Fe_3O_4/Fe_2O_3 bilayer structure at $E > E_{P,2}$ of passive iron with the potential profile. (From Vetter, K.J., *Electrochemical Kinetics*, Academic Press, New York, p. 753, 1967.)

Figure 5.11 shows the polarization curve of a rotating Fe disk electrode in 1 M NaOH degassed with purified argon [61]. The polished specimen was preactivated by etching with sulfuric acid to remove the native oxide and introduced at $E = -1.0$ V into electrochemical cell with the surface still wetted with a drop of fresh H_2SO_4. These starting conditions are similar to sputter cleaning of the metal surface for XPS studies under UHV. The polarization curve shows three anodic peaks, which are assigned to the oxidation at $E = -0.9$ V of hydrogen formed at $E \leq -1.0$ V, the formation at $E = -0.7$ V of Fe(II) and some Fe(III) (peak AI) and the oxidation at $E = -0.3$ V of Fe(II) to Fe(III) (peak AII). AI and AII agree with $E_{P,1} = -0.63$ V (Equation 5.23) and with $E_{P,2} = -0.2$ V (Equation 5.26), respectively. The differential capacity of the iron disk (determined with a lock-in technique) decreases during the anodic scan from several 10 μF cm^{-2} to below 10 μF cm^{-2}, which indicates the formation of a dielectric anodic oxide at the metal surface. The reverse scan shows a pronounced cathodic peak C at $E = -1.0$ V with the capacity getting back to large

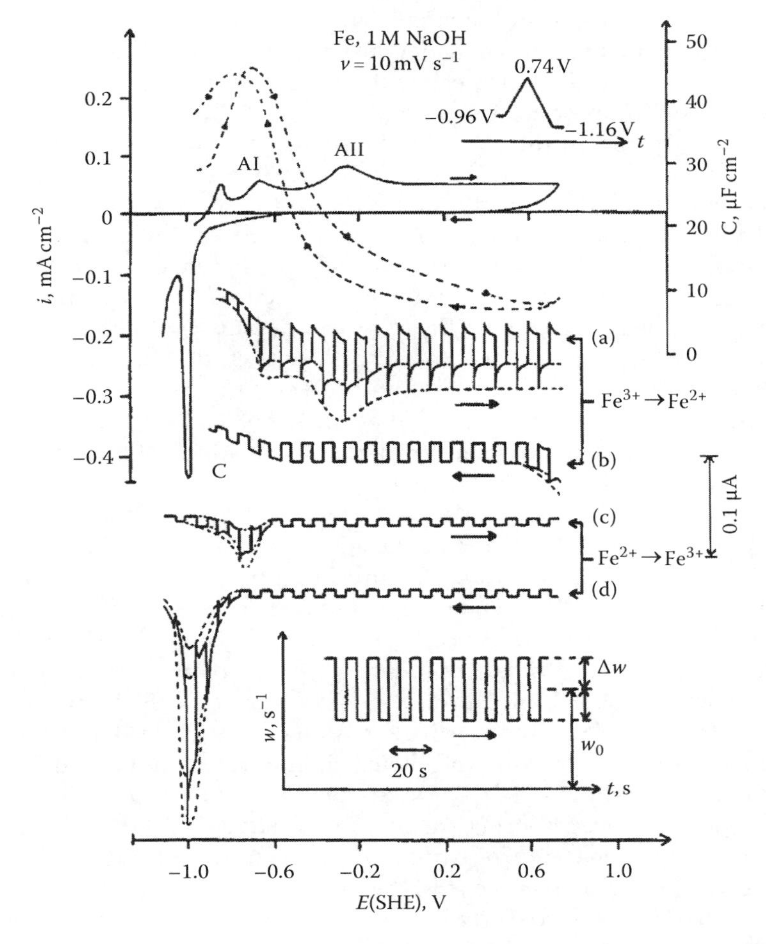

FIGURE 5.11
Polarization curve of the preactivated iron disk of a HMRRD electrode in 1 M NaOH ($dE/dt = 10$ mV s^{-1}) with hydrogen oxidation, Fe(II) and Fe(III) formation at peak AI ($E_{P,1}$), Fe(II) oxidation at AII ($E_{P,2}$), and oxide reduction at peak C. Simultaneous capacity measurements (dashed curves) (a,b) Fe(III) and (b,c) Fe(II) dissolution as detected at the Pt half-ring electrodes. Frequency, $W = 450$ min^{-1} with square wave modulation amplitude, $\Delta w = \pm 250$ min^{-1} and modulation frequency, $f = 0.1$ s^{-1}. (From Haupt, S. and Strehblow, H.-H., *Langmuir*, 3, 873, 1987.)

values indicative of an oxide-free metal surface. The anodic oxide is thus reduced at the potential of C. The anodic polarization curve shows a relatively large background current, which is not observed in usual electrochemical experiments. This can be easily explained by the use of an oxide-free iron specimen, which was contacted to the electrolyte surface at $E = -1.0$ V at the beginning of the experiment. Thus oxide formation starts on a bare metal surface, a condition that is not realized in most electrochemical experiments with specimens having formed already a native oxide layer during their preparation due to exposure to the laboratory atmosphere.

Figure 5.11 also contains the ring currents. One half ring measures the dissolving Fe(III) ions by their reduction to Fe(II) (Figure 5.11a and b) and the other one the Fe(II) ions by their oxidation to Fe(III) (Figure 5.11c and d). The rotation speed has been varied by square wave modulation (hydrodynamically modulated rotating ring disc electrode, HMRRD), which allows the distinction of species of the bulk electrolyte from those which are formed at the disk. It also allows the measurement of small concentrations of dissolving species on a larger background current. The spikes at the ring current immediately after the change of the rotation frequency are related to the Fe ions formed at the Fe disk as verified quantitatively by experimental tests and theoretical calculations [61–64]. Small amounts of Fe(III) are detected starting at $E = -0.7$ (peak A1) for the anodic scan and become negligibly small for the cathodic scan when no current at the Fe disk can be detected. Apparently, Fe dissolution disappears when no Fe oxidation occurs with a vanishing disc current. Fe(II) formation is observed to a very small extent close to peak AI only for the anodic scan. The formation of an outer Fe_2O_3 film prevents any Fe(II) dissolution. It is found again when the reduction of the passive layer occurs at peak C during the cathodic scan. These results support the interpretation of a first Fe(II)-containing film formed at AI, partially oxidized to Fe(III) to form a duplex Fe_3O_4/Fe_2O_3 layer at peak AII and further to grow Fe_2O_3 on top at increasing electrode potentials. In 1 M NaOH, Fe_3O_4 is sufficiently insoluble to form a stable film protecting the metal surface at potentials $E_{P,1} < E < E_{P,2}$. Although the oxides are insoluble in 1 M NaOH, one detects soluble species due to supersaturation phenomena for nonstationary conditions (potentiodynamic scan) and when Fe(II) is formed during reduction.

XPS studies have been performed in order to confirm the composition of the surface film with Fe(II) expected up to peak AI during the anodic scan and at C during reduction, and Fe(III) expected within the passive range. Figure 5.12 shows a $Fe_2p_{3/2}$ spectrum with contributions of Fe(0), Fe(II), and Fe(III). The Fe(II) component or even a reference spectrum can only be obtained by electrochemical preparation at sufficiently negative potentials and if the access of any spurious oxygen is prohibited by a purified argon atmosphere and a clean vacuum. After background subtraction, the contributions to the Fe signal are separated as shown in Figure 5.12. For this procedure, the binding energy and shape of the individual peak components are kept constant and their height is varied to get the closest fit to the measured signal after addition [44]. Angle-resolved measurements (ARXPS) are used to prove the presence of a layered structure. Figure 5.13 shows the ARXPS data [17,61]. With increasing angle Θ between the surface normal and the direction of detection of the photoelectrons, the contribution of the species of the outer part of the film (Fe(III)) increases, and that of the species of the inner part (Fe(II)) decreases. The opposite structure would yield the opposite variation and a homogenously mixed film would show no variation. Figure 5.13 shows a very small Fe(III)/Fe(II) ratio for a very short oxidation time $t_P = 100$ ms with no angle dependence Apparently, only Fe(II) is formed at $E = -0.16$ V in 1 M NaOH during the first 100 ms. With increasing t_P, the ratio increases and bends upward with increasing Θ due to the growth of Fe(III) on top of the inner layer. At $t_P = 100$ s, the film structure is

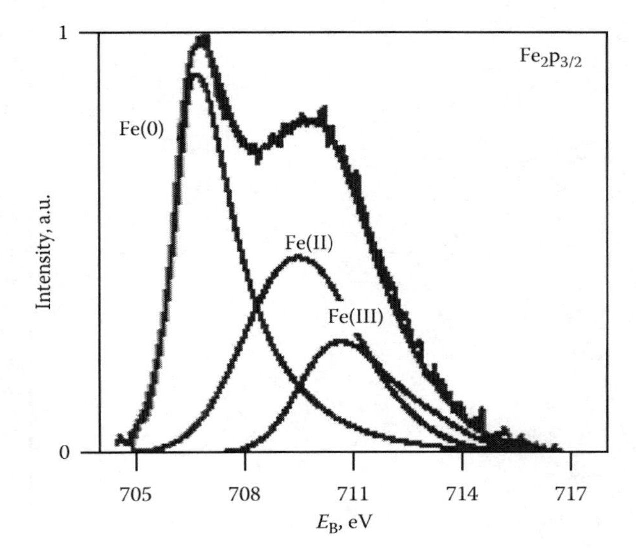

FIGURE 5.12
Deconvolution of an Fe XPS signal to its contributions of Fe(0) substrate, Fe(II) and Fe(III) of the passive layer. (From Strehblow, H.-H., in *Passivity of Metals*, R.C. Alkire, D.M. Kolb, eds., Wiley-VCH, Weinheim, Germany, pp. 271–374, 2003.)

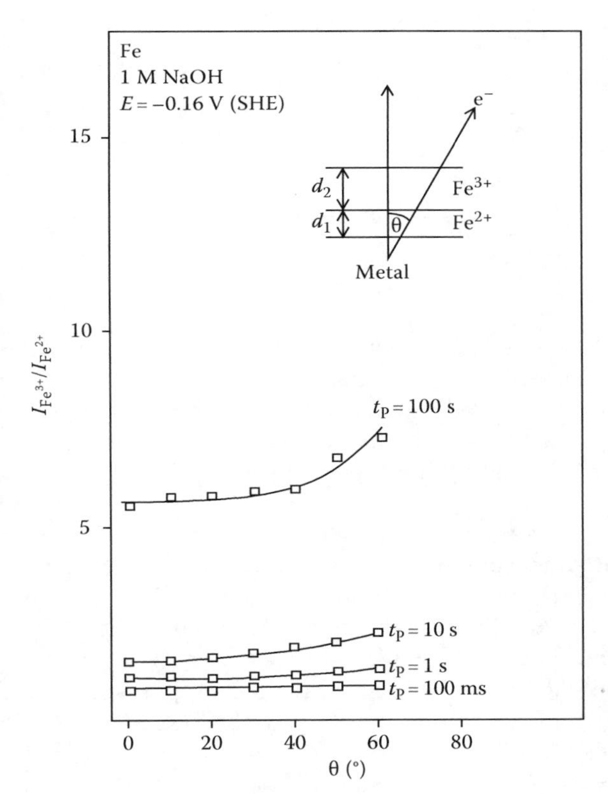

FIGURE 5.13
Angular-dependent Fe(III)/Fe(II) XPS intensity ratio for Fe passivated for different times t_P at $E = -0.16$ V in 1 M NaOH. The insert shows the bilayer model and the take-off angle θ. (From Strehblow, H.-H., in *Passivity of Metals*, R.C. Alkire, D.M. Kolb, eds., Wiley-VCH, Weinheim, pp. 271–374, 2003; Haupt, S. and Strehblow, H.-H., *Langmuir*, 3, 873, 1987.)

finished with an outer major Fe(III) part. These measurements justify the interpretation of a bilayer with an Fe(II)-rich inner and a Fe(III) outer part.

Figure 5.14 shows the partial and total layer thickness ($d = d_1 + d_2$) of the passive film at $t_P = 300$ s as a function of the electrode potentials. At E_{P1}, an Fe(II) film is formed with a small but increasing contribution of Fe(III) for $E_{P1} < E < E_{P2}$. At E_{P2}, Fe(II) is oxidized to

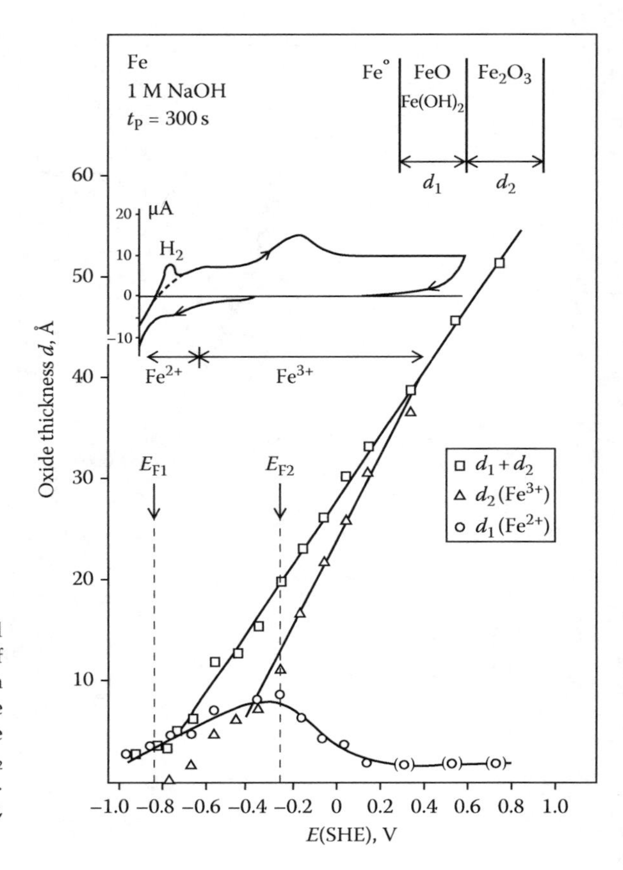

FIGURE 5.14
Chemical structure of the passive layer formed on Fe in 1M NaOH after 300s as a function of potential on the basis of a bilayer model with thicknesses d_1 and d_2 as shown in the insert. The polarization curve to the same potential scale shows the passivation potentials E_{P1} and E_{P2} and the potential range of the soluble products. (From Haupt, S. and Strehblow, H.-H., *Langmuir*, 3, 873, 1987.)

Fe(III). The polarization curve (insert) shows the clear relation of its features to the XPS results and the calculated values for the passivation potentials. For $E > 0.2$ V, the Fe(II) photoelectrons may be fully attenuated by the Fe(III) outer part and the presence of Fe(II) cannot be proved. The evaluation of the O1s signal indicates the presence of 45% of hydroxide for $E = -0.9$ V, decreasing continuously with increasing potential to less than 10% at $E = 0.8$ V.

Figure 5.15 shows the variation of the cation fraction calculated assuming an homogeneous distribution in the passive film after completion at 300s as a function of the passivation potential [17,61]. The decreasing Fe(II) and the increasing Fe(III) fractions yield a composition of 66% Fe(III) (i.e., Fe_3O_4) at $E = E_{P2}$. This result is remarkable since the interpretation of thermodynamic and electrochemical data presented above suggests an oxidation of Fe_3O_4 to γ-Fe_2O_3 at exactly this potential. Hence, the XPS results support strongly the interpretation of the passivation potentials E_{P1} and E_{P2} for the passive film formed in 1M NaOH. At high pH, the Fe(II) and Fe_3O_4 oxides are preserved and can be analyzed ex situ with XPS. These results may not be obtained in strongly acidic electrolytes due to the fast dissolution of Fe(II)-containing oxides.

Figure 5.16 compares the inverse capacity $1/C$ [65] and the oxide thicknesses $d = d_1 + d_2$ and d_2 [61] at various potentials and passivation times t_P for iron in 1M NaOH. The total thickness of the passive layer increases linearly with the potential starting with $d = 0.5$ nm at $E = E_{P1}$ up to $d = 5$ nm at $E = 0.8$ V. In the same way, the total inverse capacity ($1/C = 1/C_{OX} + 1/C_H$) increases linearly with the electrode potential from

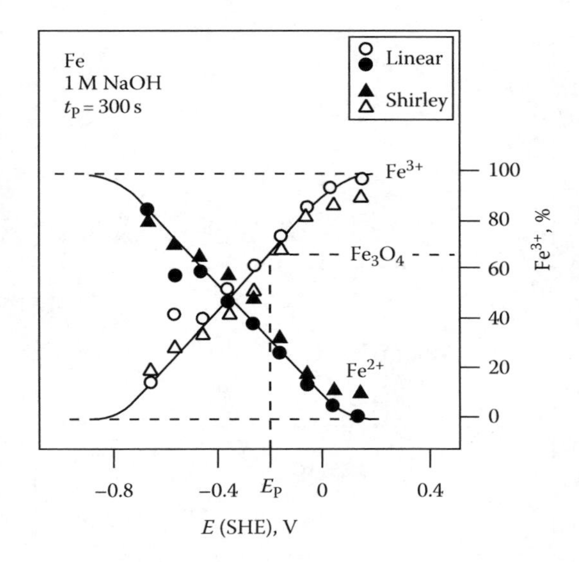

FIGURE 5.15
Fe(II) and Fe(III) cationic fraction in the passive layer formed on Fe 1M NaOH for t_P = 300s versus the passivation potential from XPS data evaluated with two background corrections (linear and Shirley). The potential E_{P2} and the cationic fraction for Fe_3O_4 are marked. (From Strehblow, H.-H., in *Passivity of Metals*, R.C. Alkire, D.M. Kolb, eds., Wiley-VCH, Weinheim, Germany, pp. 271–374, 2003; Haupt, S. and Strehblow, H.-H., *Langmuir*, 3, 873, 1987.)

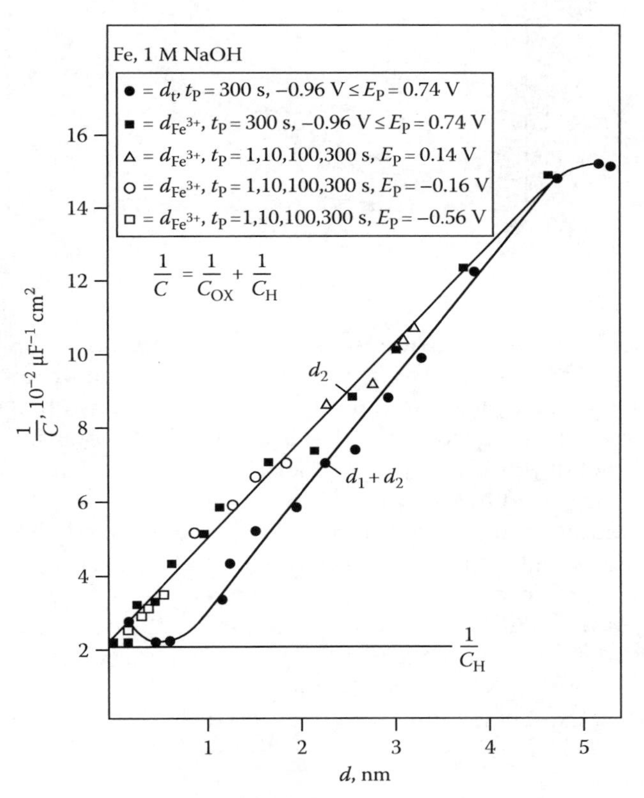

FIGURE 5.16
Inverse capacity $1/C$ of passive iron from Ref. 65 in dependence of the oxide thicknesses $d_1 + d_2$ and d_2 determined by XPS at various potentials and passivation times t_P in 1M NaOH. (From Haupt, S. and Strehblow, H.-H., *Langmuir*, 3, 873, 1987.)

0.02 to 0.15 F^{-1} cm^2. At E_{P2}, the total capacity C equals that of the Helmholtz layer C_H = 25 µF cm^{-2} at the Fe–electrolyte interface when no passive layer is present, if a roughness factor r = 2 is taken into account. $1/C_{OX}$ increases with potential for the passive layer growing in thickness. This is expected for the two capacitors C_H and C_{OX} in series and from $C = r\varepsilon\varepsilon_0/d$ for a condenser with a dielectric constant $\varepsilon_0 = 8.85419 \times 10^{-12}$ J^{-1} C^2 m^{-1}, ε = 18

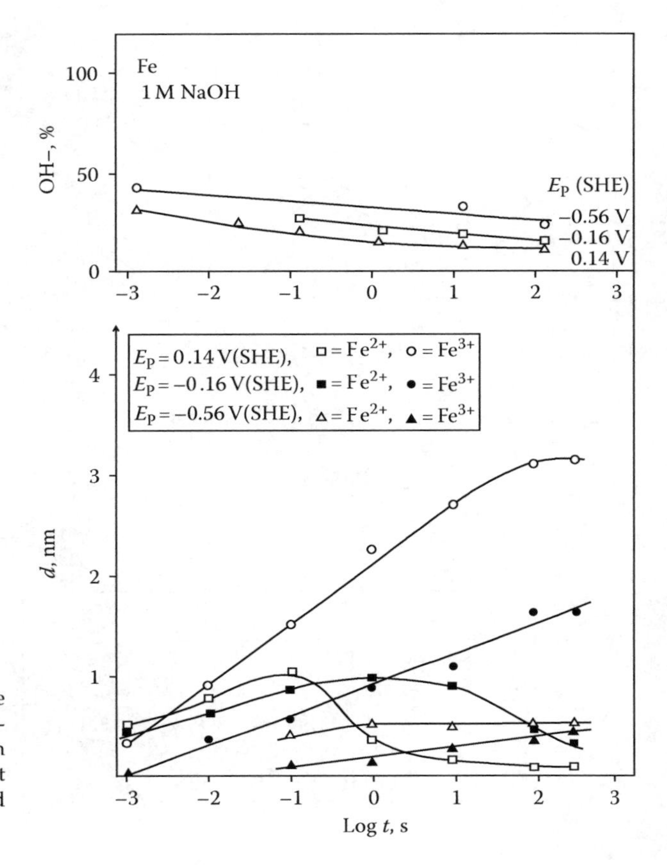

FIGURE 5.17
Change of the composition of the passive layer on Fe (partial and total oxide thickness and OH content in %) in 1 M NaOH in dependence of the passivation time t_P and at the potentials indicated. (From Haupt, S. and Strehblow, H.-H., *Langmuir*, 3, 873, 1987.)

for Fe_2O_3, and the distance d corresponding to the oxide thickness. Thus, both the XPS results and the capacity measurements show qualitatively and quantitatively the linear growth of the layer with the increasing electrode potential. Figure 5.16 also shows that d_2 mirrors the behavior of the capacity data suggesting that the location of the potential decay is mainly within the outer γ-Fe_2O_3 layer. Usually, the inner Fe_3O_4 layer is considered as a better electronic conductor compared to the outer γ-Fe_2O_3 part.

The passivation of Fe in alkaline solution can also be studied with time-resolved XPS measurements as already shown in Figure 5.13. Figure 5.17 shows the formation of Fe(II) at $E = 0.14\,V$ with a maximum at $t_P = 0.1\,s$ and a decrease up to $t_P > 1\,s$ whereas the formation Fe(III) increases linearly with the log of time [61]. Similar results are obtained for $E = -0.16$ and $-0.56\,V$, however with a slower change within more than 100 s. Here again, the O 1s signal shows that the change from Fe(II) to Fe(III) goes along with that from more hydroxide to more oxide. These time-resolved studies show that similar processes occur in the time and in the potential domain. First, the less stable lower valent species is formed followed by its oxidation with time.

XPS studies allow also the measurement of the changes in film composition during reduction of a passive layer. Figure 5.18 depicts the variation of the passive film composition for a prepassivated Fe specimen with the reduction potential [61]. With decreasing potential (potential scale from right to left), Fe(II) is formed at $E \leq -0.6\,V$ with a maximum at $E = -0.9\,V$ and disappears at $E = -1.0\,V$ when the whole layer is completely reduced. Some of the Fe(II) is dissolved as shown by the HMRRD electrode studies at potentials

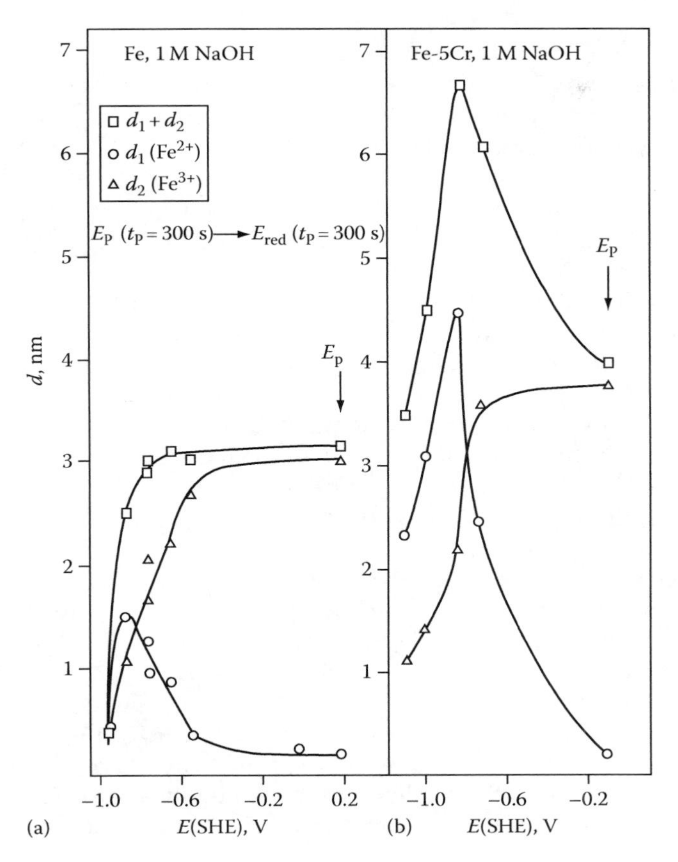

FIGURE 5.18
Stepwise reduction of the passive layer preformed on Fe in 1 M NaOH for 300 s at $E = 0.14$ V with the related layer thicknesses for (a) pure Fe and (b) Fe-5Cr. (From Haupt, S. and Strehblow, H.-H., *Langmuir*, 3, 873, 1987.)

corresponding to the cathodic peak C in Figure 5.11. Figure 5.18 also shows that the Fe(III) content and the total layer thickness decrease continuously and the film changes from oxide to hydroxide. However, some small fraction of the layer seems to remain present at the surface because repetitive potentiodynamic scans show an increasing new anodic peak at $E = -0.45$ V and a related cathodic peak at $E = -0.8$ V, which are related to the oxidation of the Fe(II) remains to Fe(III) and their reduction, respectively. The slightly different result obtained for Fe containing 5% Cr is discussed in Section 5.6.3.1.

These XPS data on the formation and reduction of the passive layer on Fe in a strongly alkaline solution show a chemical structure that can be described by a bilayer model. This does not contradict the idea of a layer with compositional changes from Fe_3O_4 to Fe_2O_3 with depth. XPS shows that the Fe(II) and Fe(III) species are predominantly located in the inner and outer parts of the film, respectively, with a thicker outer Fe(III) part for potentials well above E_{P2}. The average composition at $E = E_{P2}$ is Fe_3O_4, which can be postulated from thermodynamic arguments and agrees well with the nature of the dissolving species found from HMRRD electrode measurements. XPS also shows that at low potentials the passive film is more hydroxylated. An Fe(II)-rich film consists of up to 50% hydroxide. One may even follow the formation and reduction of the passive layer with time-resolved measurements.

5.6.2.2 Passivity of Copper

The polarization curve of Cu has been presented in Figure 5.5. The peak AI is attributed to the formation of Cu_2O and the peak AII to the formation of CuO and $Cu(OH)_2$, the latter being suggested by a large OH component in XP O 1s spectra. The related cathodic peaks CII and CI are attributed to the reverse reactions, i.e., reduction of Cu(II) to Cu(I) oxides and of Cu(I) to Cu metal, respectively. The detection of soluble species by RRD studies provides information about the composition of the outer part of the film. Only the species that are present at the passive film surface may dissolve.

The chemical shift of Cu(I) is too small to allow the detection of a Cu_2O film by XPS. However x-ray induced AES allows a clear distinction. Figure 5.19 gives an example for the decomposition of the measured L3MM signal of a Cu surface covered with a Cu_2O film into its components of Cu metal and Cu(I) [66]. The Cu metal and Cu(I) peak components have been calibrated using standards. The quantitative evaluation yields thickness data, which agree well with those from the charges of the electrochemical reduction of passivated Cu specimens. Anodic oxidation at potentials above peak AII yields a larger chemical shift of the Cu(II) component than for Cu(I). Furthermore, a shake-up satellite indicates the presence of Cu(II) with free *d*-states. Here the photoionization process occurs in combination with the excitation of a second electron into an empty d-level, which leads to smaller kinetic energy and thus to higher binding energy of the shake-up peak.

Based on the clear interpretation of the electrochemical features of the polarization curve of Figure 5.5 provided by the XPS data, the charges for electrochemical reduction can be reevaluated [67]. The cathodic charge of peak CI is always larger than that of CII. This means that at CI two Cu(I) parts are reduced, the fraction which was Cu(II) before being reduced to Cu(I) at CII and the fraction which was present as Cu(I) before reduction. This result suggests the presence of a duplex film at potentials above AII. Confirmation has been obtained by quantitative ISS depth profiling of a Cu specimen passivated for 10 min

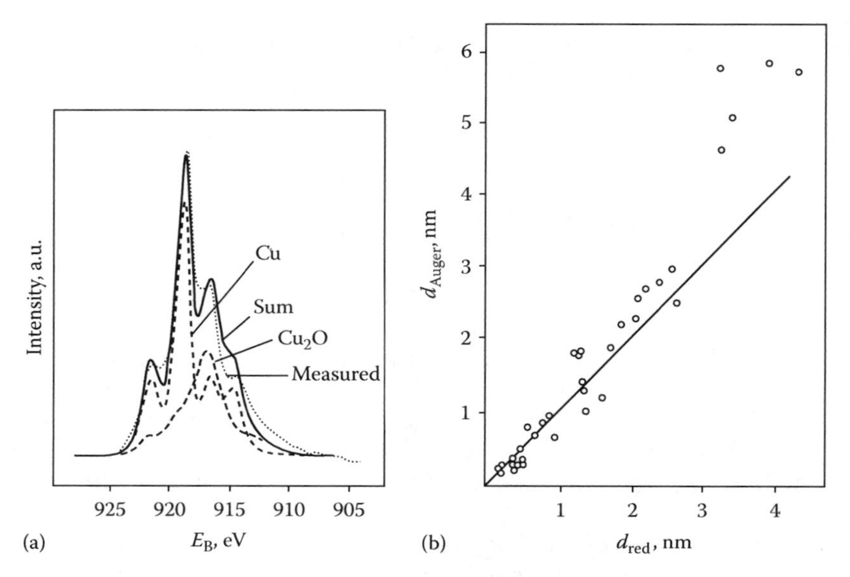

FIGURE 5.19
(a) X-ray-induced AES L_3MM spectrum and its decomposition in Cu(0) and Cu(I) components for Cu passivated at 0.0 V in 0.1 M NaOH; (b) thickness of Cu_2O layer deduced from the AES data (d_{Auger}) and calculated from the charge of electrochemical reduction (d_{red}). (From Speckmann, H.D. et al., *Surf. Interf. Anal.*, 11, 148, 1988.)

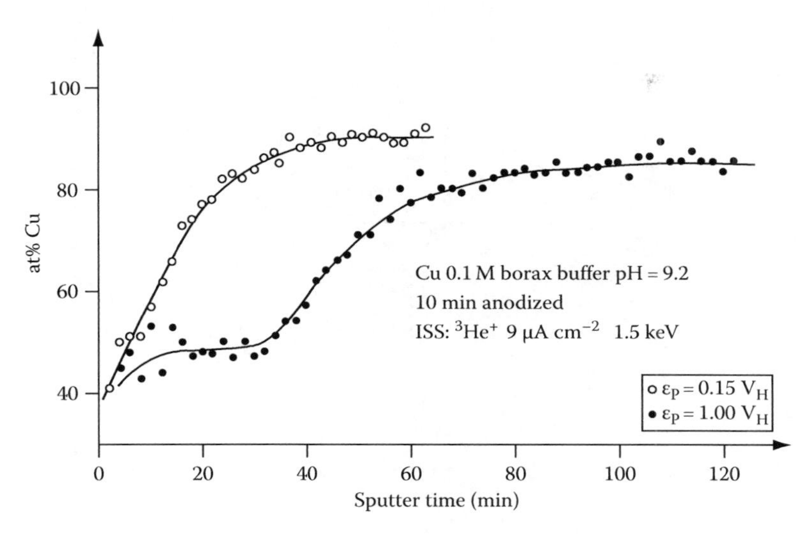

FIGURE 5.20
ISS depth profile of Cu passivated at $E = 0.15$ and 1.00 V in 0.1 M borax buffer pH 9.2 for 10 min, 50% ^{3}He + 50% ^{20}Ne primary ion beam, $E_0 = 1.5$ keV. (From Strehblow, H.-H. and Titze, B., *Electrochim. Acta*, 25, 839, 1980.)

at $E = 1.00$ V in a 0.1 M borax solution (pH 9.2) as shown in Figure 5.20 [67]. The soft sputtering process with intermediate ISS analysis yields a step at 50% Cu, which is related to the outer layer of CuO. After removal of this outer film, the ISS depth profile shows a slow increase of the Cu content like for a specimen that has been passivated at potentials below AII when only Cu_2O is present with a higher stoichiometric Cu content. The combination of electrochemical studies, XPS measurements, and ISS depth profiles clearly indicates the presence of a simple Cu_2O layer at $E(AI) < E < E(AII)$ and a Cu_2O/CuO, $Cu(OH)_2$ duplex layer for $E > E(AII)$. The quantitative evaluation of the data yields an inner Cu_2O film of up to 1.2 nm thickness and an outer up to 2 nm in thickness if CuO or 4 nm if $Cu(OH)_2$ [67].

In strongly alkaline solutions, the anodic oxidation of Cu at $E = E(AII)$ for longer time yields thick visible films. They apparently form by a dissolution–precipitation mechanism. RRD studies at AII in 0.1 M KOH yield an efficiency of up to 90% for CuO_2^{2-} dissolution and of 10% for the layer formation. The Cu(II) dissolution increases with increasing pH due to a larger solubility of Cu(II) species. Less stirring causes more precipitation of Cu(II) species at the surface. At $E > E(AII)$, i.e., within the passive range, a dense film is formed which grows to a thickness of 2 nm only as mentioned above. Apparently, the higher field strength at larger potentials causes a dense protecting film forming by a solid state mechanism. Anodic oxidation studies using potentiostatic transients followed by an immediate reduction suggest the formation of a precursor similar to a Cu(I) oxide within the first milliseconds. This precursor changes from Cu_2O to the duplex film within 10–100 s depending on the applied electrode potential [68]. Structural details of the anodic oxidation of Cu have been examined with in situ electrochemical STM and are presented below.

When soluble CuO_2^{2-} ions are present in strongly alkaline solutions like 1 M NaOH, their reduction at a Cu electrode at $E(CII) < E < E(CI)$ yields a continuously growing film of Cu_2O. By this method, variable Cu_2O layers have been formed up to several nanometer thickness for photoelectrochemical studies [69,70]. The potentiodynamic reduction of these thick films behave differently and yields only one large cathodic peak at more negative potentials ($E = -0.65$ V). The reduction of thin passivating Cu_2O layers suggests a nucleation mechanism since potentiodynamic polarization after a cathodic transient yields the

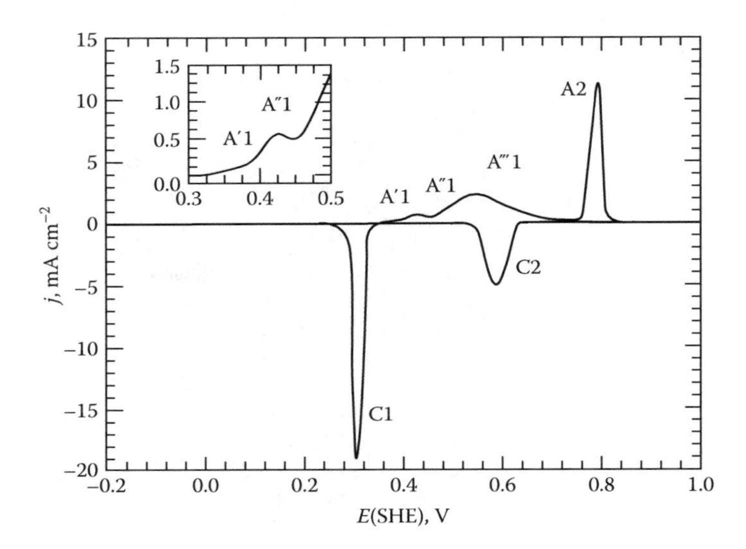

FIGURE 5.21
Polarization curve of Ag in 1 M NaOH with peaks A1 and A2 of formation of Ag_2O and AgO, respectively, and peaks C1 and C2 for their reduction. Insert shows the initiation of oxide formation at potentials negative to peak A‴1. (From Hecht, D. et al., *Surf. Sci.*, 365, 263, 1996.)

polarization curve like for a bare Cu surface, however with a lower current level [71]. This result suggests the repassivation of small metallic Cu islands (i.e., nuclei) formed within a still existing passive layer matrix during the short cathodic reduction transient.

5.6.2.3 Passivity of Silver

Silver is chemically close to copper. Oxide layers can be easily reduced on this seminoble metal and the polarization curve is qualitatively similar. However, there are also many differences for their passive behavior. Figure 5.21 shows the polarization curve in 1 M NaOH with two pronounced anodic and cathodic peaks [72]. Like in the case of Cu, Ag_2O is formed at A1 and it is further oxidized to AgO at A2. A duplex film structure is suggested for oxidation at $E > 0.73$ V (A2) in 1 M NaOH.

Oxide formation on Ag takes place already below the Nernst equilibrium potential at $E_P = 0.357$ for pH 13.8 and $E_P = 0.30$ for pH 12.7 as can be seen by analysis of the XPS Ag 3d and O 1s core levels. The chemical shift of the Ag 3d signals is small (E_B(Ag) = 368.3 eV and $E_B(Ag_2O)$ = 368.0 eV) so that the ratio of the O^{2-} signal to the Ag metal signal was taken in addition for a quantitative evaluation of the layer thickness. It is in the range of 1 nm below $E = 0.36$ V with a steep increase above $E = 0.4$ V as shown in Figure 5.22 [73]. The cathodic current peak for potentiodynamic reduction after a 300 s oxide formation at $E = 0.37$ V is small with a relatively positive potential of $E = 0.34$ V. It shifts to more negative values and develops a second larger and more negative peak with increasing potential of oxide formation. This indication of an inner partially crystalline Ag_2O film covered by a much thicker and better crystallized top AgO layer has been proved by in situ STM [74] and XAS [75], respectively. A 0.5–2 nm thick Ag_2O layer is found below $E = 0.4$ V with a steep increase to >5 nm at higher potential. At $E = 0.40$ V in 1 M NaOH, the Ag_2O layer increases with $t^{0.5}$, which suggests a diffusion-controlled mechanism. Thickness values obtained by XPS and extended absorption fine structure spectroscopy (EXAFS) agree very well with a $t^{0.5}$ increase to 2.5 nm within 30 s [75]. The passive film

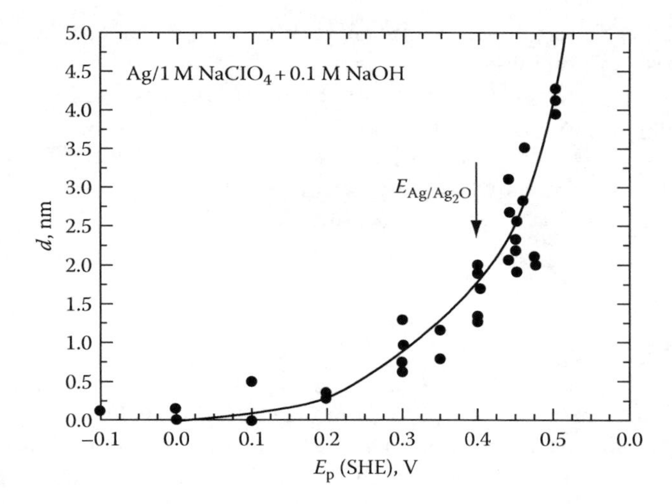

FIGURE 5.22
Thickness of Ag_2O formed on Ag in 1M NaOH after 300s of oxidation at potentials as indicated as deduced from XPS measurements. Arrow indicates the equilibrium potential for pH = 13.8. (From Lützenkirchen-Hecht, D. and Strehblow, H.-H., *Surf. Interf. Anal.*, 38, 686, 2006.)

gets very thick for $E > 0.4$ V, e.g., 70 nm for $E = 0.7$ V. The EXAFS data confirm that the inner Ag_2O layer is disordered and is covered with a thicker crystalline Ag_2O film in which up to seven coordination shells can be detected [72].

The AgO formation occurs by a nucleation and growth mechanism with characteristic current peaks for potentiostatic transients. These peaks are high and sharp for high potentials, i.e., $E = 0.8$ V, and get smaller and broader with decreasing potential. The total charge of AgO formation Q(C2) has been determined by quantitative evaluation of peak C2 in potentiodynamic reduction after potentiostatic layer formation. The specimens were oxidized in two steps with first Ag_2O formation at $E = 0.7$ V for 600 s and a subsequent oxidation for 600 s at potentials $E > 0.7$ V. Q(C2) goes through a sharp maximum of 125 mC cm^{-2} at $E =$ 0.71 V, decreases to a minimum of 0.25 mC cm^{-2} at 0.9 V and increased up to 0.75 mC cm^{-2} for $E > 0.9$ V as shown in Figure 5.23 [75]. Apparently, the nuclei merge together during lateral growth leading to a continuous film that reduces the anodic current for further

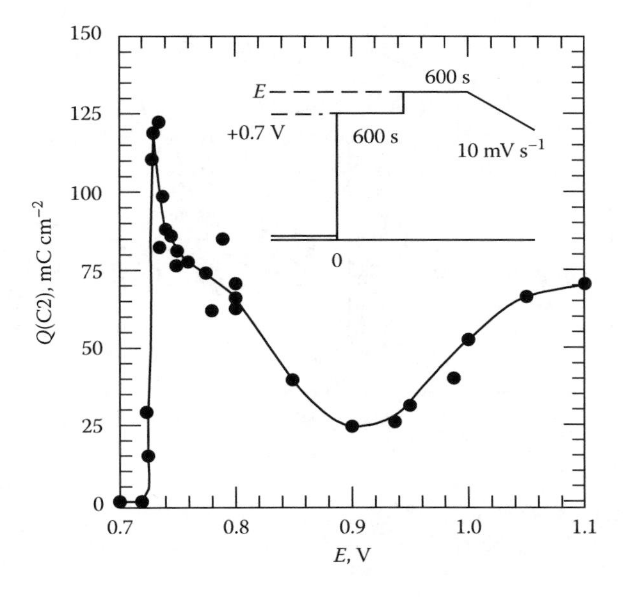

FIGURE 5.23
Charge of potentiodynamic reduction of AgO to Ag_2O formed on Ag in 1M NaOH by two potentiostatic transients, 600s at $E = 0.70$ V + 600s at E as indicated. (From Lützenkirchen-Hecht, D. and Strehblow, H.-H. *Surf. Interf. Anal.*, 41, 820, 2009.)

AgO formation. The number of nuclei is thought to increase with potential that leads to the much faster formation of a continuous layer by lateral growth.

XPS studies show a relatively broad Ag $2p_{3/2}$ XPS signal that can be decomposed by the superposition of two components, one for Ag^+ and one for Ag^{3+} of equal intensity and shape separated by a small chemical shift (ΔE = 0.8 eV) [75]. This agrees well with the description of AgO as an oxide consisting Ag(I) and Ag(III) ions. This interpretation was also used for the evaluation of structural data of anodic AgO films by XAS, which is discussed in Section 5.8.3.

5.6.2.4 Passivity of Tin

Figure 5.24 shows the characteristic polarization curve for Sn in alkaline solution with a pronounced double peak at E(A1) = −0.8 V and E(A2) = −0.5 V followed by a large passive range with current densities of some μA cm^{-2} [76]. The passive range from E = −0.4 to 1.4 V is followed by a secondary passivity (transpassive range) up to 2.2 V, the potential at which oxygen evolution starts. Passive layers formed well within the passive range at E = 1.0 V do not block the redox reactions with $[Fe(CN)_6]^{3-,4-}$. However, the layer loses its electronic conductivity when formed at more positive potentials, i.e., at 2.1 V. The composition of the passive layer changes also in a characteristic way with the potential. According to XPS studies, the Sn surface is covered by a monolayer of SnO and $Sn(OH)_2$ in the range of E = −1.0 to −0.3 V. In the passive range of E = −0.1 to 1.4 V, a film of 0.7–3 nm of SnO_2 with some outer $Sn(OH)_4$ is formed on top of 1 nm SnO at the metal surface [76]. The inner SnO layer has been found by XPS depth profiling. The outer surface of the passive layer contains a mixed monolayer of SnO and $Sn(OH)_2$, which may be seen as a residue of the part of the film that has not been oxidized. The oxide grows linearly with the applied potential. In the secondary passive range, at E = 1.4–2.2 V, the metals surface is covered by 3–12 nm of SnO_2 with some outer $Sn(OH)_2$ and 1 nm of SnO underneath. In borate buffer (pH 9)

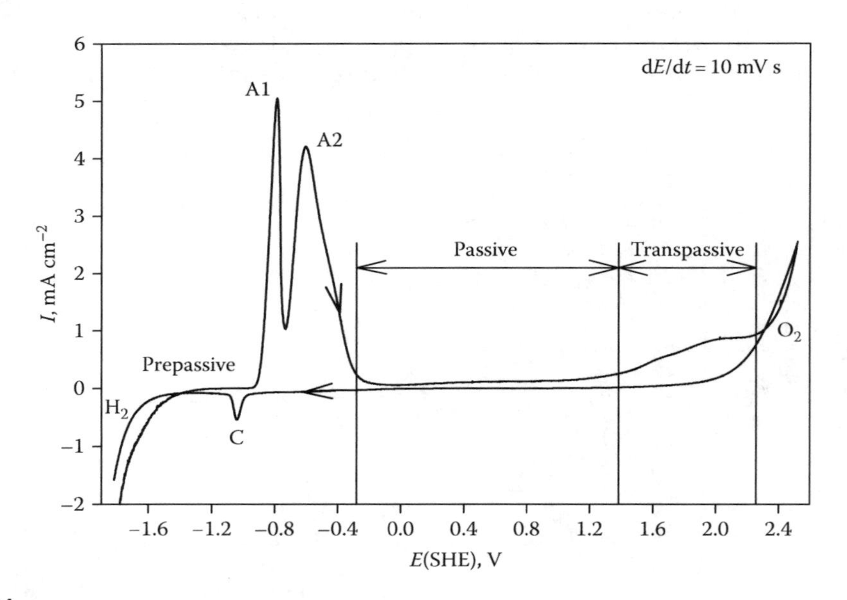

FIGURE 5.24
Polarization curve of Sn in 0.1 M KOH. (From Keller, P., PhD thesis, Elektrochemische und oberflächenanalytische Untersuchungen zur anodischen Deckschichtbildung auf Zinn und Kupfer/Zinn Legierungen, University of Düsseldorf, Dusseldorf, Germany, 2005.)

and phthalate buffer (pH 5), the passive layer has the same structure. The dissolution at the anodic peaks is smaller due to lower solubility of the oxide. The anodic layer may be reduced if formed within the primary passive range. However, a complete reduction is not achieved for passivation in the secondary passivity range.

5.6.2.5 Passivity of Cobalt

Figure 5.25 shows that the polarization curve of Co also presents a primary passive range for $E = -0.4$ to 0.3 V and a secondary passive range from $E = 0.3$–0.7 V at which oxygen evolution starts [77]. Passivity is well developed in alkaline solutions but not in strongly acidic electrolytes where dissolution occurs in the range of some mA cm^{-2} and only a shoulder is seen in the polarization curve before the onset of oxygen evolution [77,78]. In alkaline solutions at the transition between the primary and secondary passive ranges, a sharp peak A2 is found, which is attributed to a fast oxidation of freshly formed $Co(OH)_2$ to CoOOH. This reaction involves the release of H^+ ions only. The $Co(OH)_2$ hydroxide has a layered structure with OH–Co–OH trilayers interspaced by a gap containing water molecules [79]. H^+ ions may easily leave this gap or go into it during oxidation to CoOOH or the reverse reaction [60]. This causes a fast process with a sharp oxidation peak A3. Its charge of 750 μC cm^{-2} corresponds to the oxidation of 4.7 monolayers of $Co(OH)_2$ to CoOOH. After aging the oxidation of hydroxide is no longer that easy and consequently the sharp peak disappears.

In 0.1 M NaOH, the layer has a thickness of up to 15 nm in the primary passive range ($E = -0.4$–0.2 V) and drops to 4 nm with a slight increase to 7 nm in the range of secondary passivity ($E = 0.2$–0.9 V) (Figure 5.26c). In borate buffer (pH 9.2), the thickness is smaller, i.e., 1 and 4 nm in the primary and the secondary passive ranges, respectively. XPS investigations of the $Co2p_{1/2}$ and O1s signal suggest a very thin film of $Co(OH)_2$ of monolayer thickness in the pre-passive and active potential range. In the primary passive range, a thicker film of $Co(OH)_2$ is present, which changes to Co(III) oxide at the transition of primary to secondary passivity. This is observed at $E = 0.2$–0.4 V for 0.1 M NaOH with almost pure Co(III) for $E > 0.4$ V (Figure 5.26a and b). In borate buffer (pH 9.3), the pure

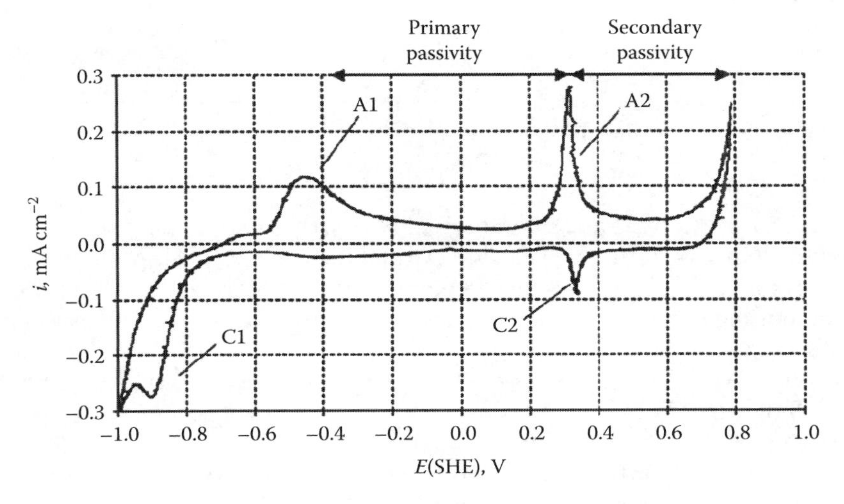

FIGURE 5.25
Polarization curve of Co in 0.1 M NaOH. (From Foelske, A. and Strehblow, H.-H., *Surf. Interf. Anal.*, 34, 125, 2002.)

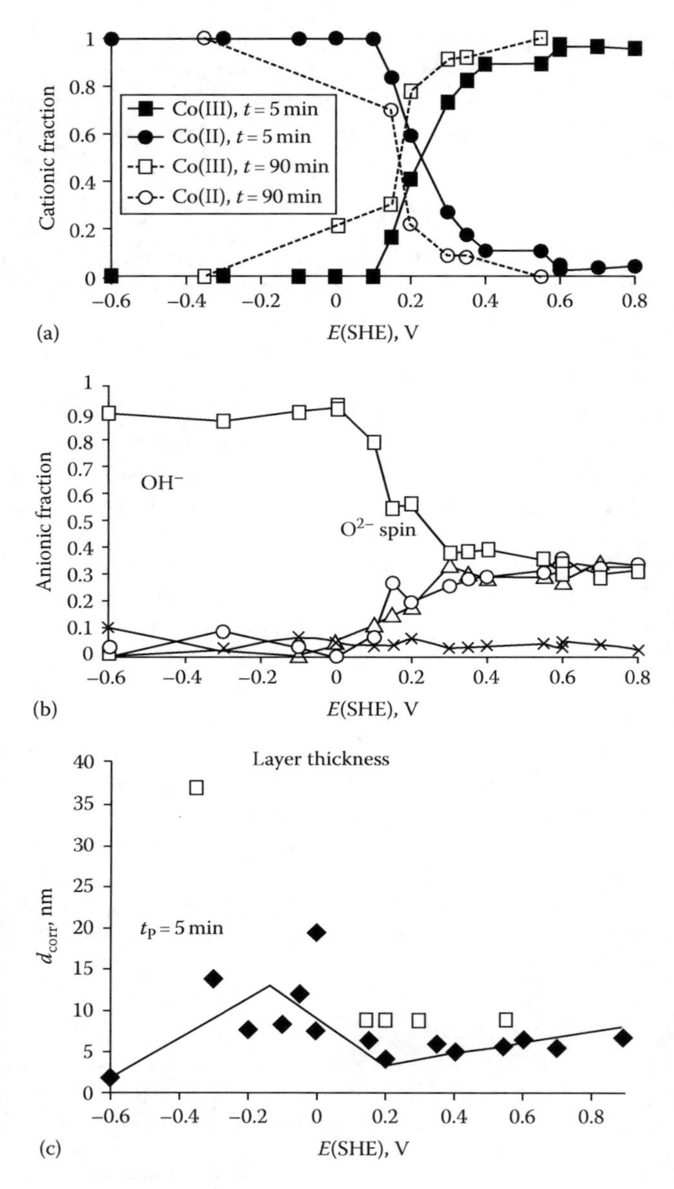

FIGURE 5.26
Passive layer composition and thickness on Co in 0.1 M NaOH pH 13 for passivation time t_P as a function of the electrode potential E as deduced from XPS measurements, (a) cationic fraction for $t_P = 5$ min (solid line) and 90 min (dashed line), (b) anionic fraction for $t_P = 5$ min, (c) layer thickness from XPS sputter depth profiles relative to Ta_2O_5 scale. (From Foelske, A. and Strehblow, H.-H., *Surf. Interf. Anal.*, 34, 125, 2002.)

$Co(OH)_2$ film in the active range contains 20% Co(II) oxide, which increases to ca. 80% Co(III) oxide from $E = 0.7$–1.0 V [77]. The evaluation of the O1s signal suggests for 0.1 M NaOH and borate pH 9.2, the presence of equal amounts of hydroxide, oxide, and a spinel type of oxide in the range of secondary passivity. Figure 5.26b gives the example for 0.1 M NaOH [77].

Time-resolved XPS investigations after potentiostatic transients show first the formation of $Co(OH)_2$ during the first seconds even in the range of secondary passivity, which changes later partially to Co(III) oxide and a spinel-type Co(III) oxide. Here again, one observes the lower valent hydroxide, which changes to the stable chemical structure of the passive layer with time. For $E = 0.25$ V in 1 M NaOH, no reduction is found for the passive film formed at $E = 0.6$. For $E \leq 0.2$ V, the reduction to $Co(OH)_2$ is observed, which remains at the surface even at $E = -0.8\,V$ for 5 min. Thus, the reduction to Co metal requires more negative potentials. In these studies, $E = -1.0\,V$ has been applied to obtain a slow reduction of the hydroxide film. More negative potentials might lead to a faster and complete reduction of $Co(OH)_2$, which, however, will cause unwanted additional hydrogen evolution as can be seen in Figure 5.25. STM studies on a Co(0001) surface in 0.1 M NaOH at $E = -1.0\,V$ are discussed in Section 5.8.5.

5.6.2.6 Passivity of Vanadium

Vanadium shows passivity in alkaline solutions with transpassive dissolution at the potentials of formation of soluble VO_4^{3-} (Figure 5.27) [80]. In the potential range $E = -0.8$ to $-0.6\,V$ in 1 M NaOH, the polarization curve shows a shoulder in the current density of some $10\,\mu A\ cm^{-2}$ with a small current peak A1 at $E = -0.6\,V$ and a related anodic charge of $0.244\,mC\ cm^{-2}$. A related cathodic peak C1 is found during the reverse scan. This corresponds to about half an atomic layer being oxidized to form a monolayer of V(III) hydroxide. At $E > -0.3\,V$, the current increases due to the transpassive dissolution of $VO4^{2-}$ with a current plateau of $10\,mA\ cm^{-2}$ and superimposed current peaks AII and AIII at $E = 0.2$ and 0.4 V, respectively, and a further increase at $E > 0.6$. In borax buffer pH 9.3, similar polarization curves are obtained with a peak A1 of anodic charge $0.163\,mC\ cm^{-2}$ only. Peak AI is shifted as compared to 1 M NaOH by ca. 0.25–0.35 V corresponding to the expected $-0.059\,V/pH$ variation. In acidic electrolytes like 0.5 M H_2SO_4 and phthalate buffer pH 5.5, active dissolution as V^{3+} is observed for $E > 0.4\,V$.

XPS studies including ARXPS show the formation of a bilayer with inner V_2O_3 layer and an outer $V(OH)_3$ part in 1 M NaOH [80]. In the range of $E = -0.8$ to $-0.5\,V$, the film consists of ca. 60% oxide and 40% hydroxide. The layer composition changes linearly to almost pure V_2O_3 film from $E = -0.5$ to $-0.3\,V$. The total layer thickness increases linearly in the range of $E = -0.8$ to $-0.3\,V$ from 0.8 to 1.8 nm. In phthalate buffer, a pure oxide film is

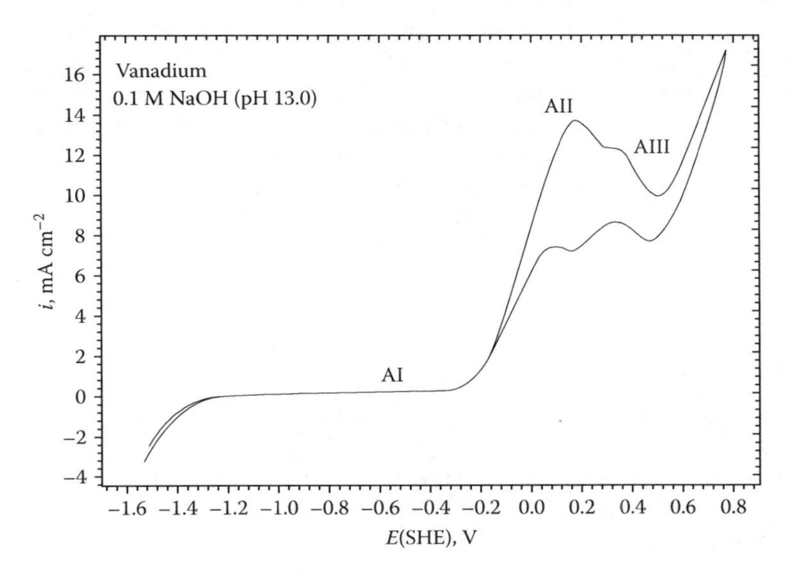

FIGURE 5.27
Polarization curve of *V* in 1 M NaOH. (From Drexle, A. and Strehblow, H.-H., *Proceedings Kurt Schwabe Symposium*, Helsinki, Espoo, Finland, p. 433, 2004.)

found for $E = -0.8$ to 0.1 V and the composition is 40% $V(OH)_3$ and 60% V_2O_3 for $E = 0.2–0.4$ V. The film is only 0.5 nm thick up to $E = 0.1$ V, which corresponds to about two oxide monolayers. For $E > 0.1$ V, it increases linearly up to 2 nm at $E = 0.4$ V.

5.6.2.7 Passivity of Chromium

Chromium is one of the most important alloying components for metals due to its resistance to corrosion. Cr^{2+} dissolution starts in strongly acidic electrolytes at very negative potentials. The current density is exceeding cathodic hydrogen evolution and gets positive at $E = -0.4$ V up to 0 V with a steep fall to values below 0.1 μA cm^{-2} at $E \geq -0.1$ V due the formation of a protecting Cr(III) oxide layer (Figure 5.4). These extremely small current densities have been measured with radioactive tracer methods. A determination with HMRRD electrodes with two concentric rings with reduction of Cr^{3+} to Cr^{2+} and its reoxidation to Cr^{3+} was not successful since the dissolution rate of Cr^{3+} was below the detection limit of the method of some μA cm^{-2} [81]. An indirect determination by comparison of the anodic charge of passivation transients of preactivated specimens and the thickness of the layer determined by XPS yields passive current densities of less than 0.1 μA cm^{-2} [82]. The current density increases again at $E > 1.0$ V due to oxidation to Cr(VI) and dissolution as $Cr_2O_7^{2-}$ with a further increase at $E > 1.7$ V due to oxygen evolution. For these reasons, Cr is the alloying element for stainless steel since it protects the alloy against corrosion even at negative potentials. In addition, it is very protective against aggressive anions such as chloride, so that pitting corrosion is inhibited.

According to XPS studies, the very protective passive layer grows linearly in 0.5 M H_2SO_4 with the potential from $d = 1$ nm at $E = 0.2$ V to $d = 3.0$ nm at $E = 1.2$ V (Figure 5.28) [82]. No smaller thickness than $d = 1$ nm can be measured because, after sputter cleaning, the reactive metal forms an anodic film by water decomposition. The extrapolation of d to vanishing values gives a value $E = -0.2$ V for the potential of passivation and formation of a Cr_2O_3 layer. The anodic charge Q increases linearly with potential in agreement with the thickness but with a larger slope for $E > 0.6$ V. The inverse capacity increases linearly with potential corresponding to a thickening oxide layer. It shows a decrease for $E > 0.7$ V, which matches the steeper increase of the anodic charge Q. Apparently, the approach to transpassive behavior with oxidation of Cr(III) to Cr(VI) causes a change of the electronic properties of the passive layer with an increase of the dielectric constant and the related decrease of $1/C$. XPS studies show a significant outer layer of Cr(III) hydroxide to the passive film [82,83]. Aging of the passive film under polarization was found to be critical for dehydration and buildup of the inner barrier layer of the passive bilayer film. This is accompanied by a change from an amorphous to a crystalline structure as shown by in situ STM investigations discussed in Section 5.8.4.

5.6.2.8 Passivity of Nickel

XPS investigations of the Ni surface agree well with the electrochemical behavior of the metal. In acid solution (0.5 M H_2SO_4), the passive layer consists of NiO with the contribution of ca. 0.5 nm of $Ni(OH)_2$ on top according to ARXPS studies [84,85]. Layer formation starts at $E = 0.6$ V with 2 nm and a small linear increase up to 2.5 nm at $E = 1.5$ V (Figure 5.29) [85]. In the transpassive range (Figure 5.4), the layer disappears at $E > 1.5$ V due to oxidation of Ni(II) to soluble Ni(III), which cannot be detected due to its instability. In alkaline solution (1 M NaOH) [85], the measured current density is much smaller because the components of the passive layer NiO and $Ni(OH)_2$ are insoluble. The anodic peaks of active

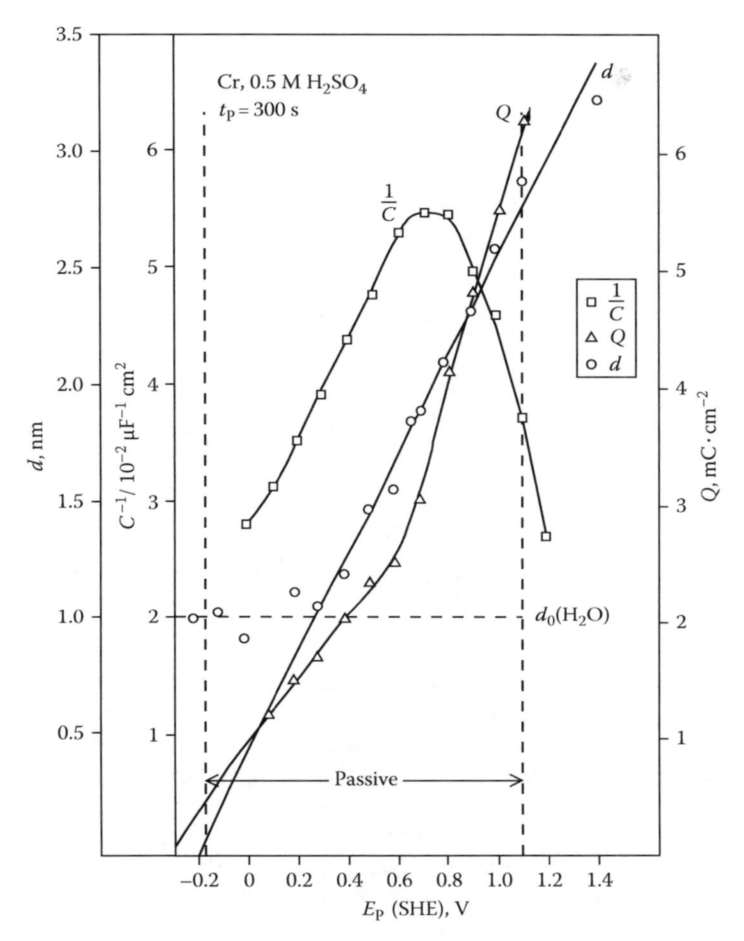

FIGURE 5.28
Oxide thickness (d), anodic charge for passivation (Q), and inverse capacity ($1/C$) of Cr in 0.5 M H_2SO_4 after 300 s passivation. d_0 is the minimum layer thickness that was measured by XPS due to water decomposition. (From Haupt, S. and Strehblow, H.-H., *J. Electroanal. Chem.*, 228, 365, 1987.)

and transpassive dissolution are shifted by ca. −0.9 V between both solutions as expected from the pH difference. NiO starts growing at $E = -0.5$ V with a linear increase to 3.5 nm at $E = 1.0$ V. This oxide sublayer is covered with an almost potential independent top layer of ca. 1.5 nm $Ni(OH)_2$ thus leading to a maximum layer thickness of 5 nm.

Some remarkable changes have been observed and have been attributed to a valence change of Ni within the layer going from the passive to the transpassive potential range at $E = 0.5$–0.6 V. All XPS signals related to the passive layer as $Ni2p_{3/2}$, O1s and C1s from the carbon contamination are shifted by −0.8 V to lower binding energy. UV photoelectron spectroscopy (UPS) studies show a shift of the cutoff edge that leads to an increase of the threshold energy by the same value of 0.8 eV. This result has been explained by a decrease of the Fermi level of the anodic oxide by −0.8 eV, which increases the threshold energy and decreases all binding energies by the same energy because they are all referred to the Fermi level. NiO is accepted as a p-type semiconductor. With increasing potential, the downward bending of the bands changes at the flat band potential

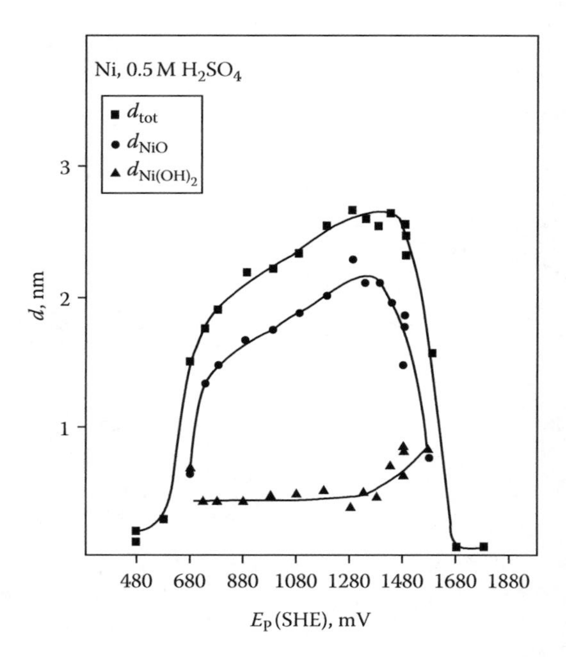

FIGURE 5.29
Thickness of passive layer formed on Ni after 300 s in 0.5 M H_2SO_4 as deduced from XPS measurements. (From Hoppe, H.W. and Strehblow, H.-H., *Surf. Interf. Anal.*, 14, 121, 1989.)

and is directed upward. This upward bending of the valence band finally causes its approach to and crossing with the Fermi level and the accumulation of positive holes at the layer surface. The chemical interpretation of this situation is the formation of higher valent cations, i.e., Ni(III) ions. Further potential increase will be located at the layer–electrolyte interface. It thus becomes kinetically effective to increase transpassive dissolution especially in acidic electrolytes with soluble corrosion products, the onset of oxygen evolution and NiOOH formation. At $E > 0.64$ V, changes of the $Ni2p_{3/2}$ signal are interpreted by a change of the outer $Ni(OH)_2$ film to NiOOH ($Ni(OH)_2 \rightarrow NiOOH + H^+ + e^-$). Quantitative evaluation of the O1s signal supports the formation of an oxy-hydroxide. In the range of oxygen evolution at $E = 1.64$ V, equal amounts of hydroxide and oxide are detected, which correspond to a thicker NiOOH film. Time-resolved XPS measurements show the linear growth of the NiO film with log t in the range of 1 ms to 1000 s. A less regular film growth is observed for transients in the transpassive range for $E > 0.7$ V. Here again, the NiO/$Ni(OH)_2$ layer is formed, with change of $Ni(OH)_2$ to NiOOH for $t > 0.1$ s.

5.6.2.9 Valve Metals

Valve metals like Al, Bi, Ce, Nb, Ta, Ti, Zr, or Hf may be oxidized up to very high electrode potentials [86–90]. No higher valent soluble species cause a transpassive behavior and oxygen evolution is suppressed due to the isolating properties of the oxide films. Al and Ta may be anodized to more than 100 V with thick anodic layers of Al_2O_3 or Ta_2O_5. The film thickness increases linearly with the potential and with the time of polarization with a constant current. In these cases, a field strength of 10^6–10^7 V cm^{-1} is found indicating a high field mechanism for film growth and migration of cations and anions through the layer. Al containing 1% of Cu may grow film thicknesses of 200 nm or more with potentials of 150 V. The thickness was deduced from the electrochemical charge and RBS investigations [91].

5.6.3 Passivity of Binary Alloys

The simple model of a homogenous passive layer of Figure 5.6 becomes more complicated if a second alloy component is present as shown in Figure 5.30. The composition of the passive layer is then determined by the oxidation rates of the components A and B at the metal surface, their transfer rates through the film, and their transfer across the passive layer–electrolyte interface, i.e., their individual corrosion rates in the passive state. The reaction rates at both interfaces may be decisive for the layer composition. One example is the preferential dissolution of Fe^{3+} ions due to the extremely slow cation transfer of Cr^{3+} ions at the surface of the film, which leads to an accumulation of Cr(III) within the film for FeCr alloys. Another example is the preferential oxidation of Al of an Al alloy containing 1% Cu. Cu does not enter the film and is accumulated at the metal surface while an Al_2O_3 film is formed. These examples are discussed in detail in the following.

Figure 5.30 shows how many parameters have to be taken into account, the metal fractions X_A and X_B at the metal surface, the cationic and anionic fractions within the film, and the dissolution rates of A and B at the film surface. Furthermore, the different transfer rates of the cations may cause a gradient in the layer composition. Finally, the cations A^{a+} and B^{b+} may be further oxidized at sufficiently positive potentials causing a distribution of lower and higher valent species within the film. This in turn requires the knowledge of the semiconducting properties that are involved in the oxidation of cations as well as the reactions of redox systems at the film surface, which require electron conduction across the layer. All these details show that the semiconductor properties and the chemical composition and structure have to be studied with appropriate tools. The complexity of these systems requires the application of surface analytical methods in order to understand the properties of these films and their influence on the corrosion properties of alloys.

5.6.3.1 Passivity of FeCr Alloys

Fe-Cr and Fe-Cr-Ni alloys are of high technical importance, the main benefit for ferritic and austenitic stainless steels resulting from the excellent corrosion resistance of Cr_2O_3 layers. Figure 5.31 shows the polarization curve of Fe-15 Cr in 0.5M H_2SO_4 [92]. Its characteristic features are determined by the electrochemical properties of the pure alloy components. Hydrogen evolution (with cathodic currents) is observed up to $E = -0.2$V followed by the potential range of active dissolution of Cr^{2+} and Fe^{2+} up to 0V where passivity starts due to

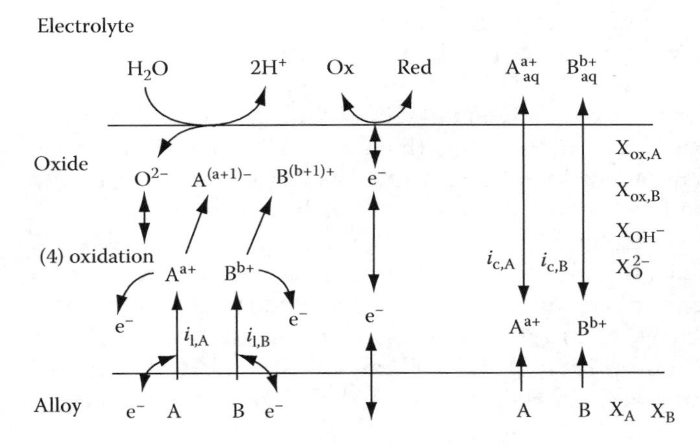

FIGURE 5.30
Model of passive layer for a binary alloy and related variables. (From Strehblow, H.-H., in *Passivity of Metals*, R.C. Alkire, D.M. Kolb, eds., Wiley-VCH, Weinheim, Germany, pp. 271–374, 2003.)

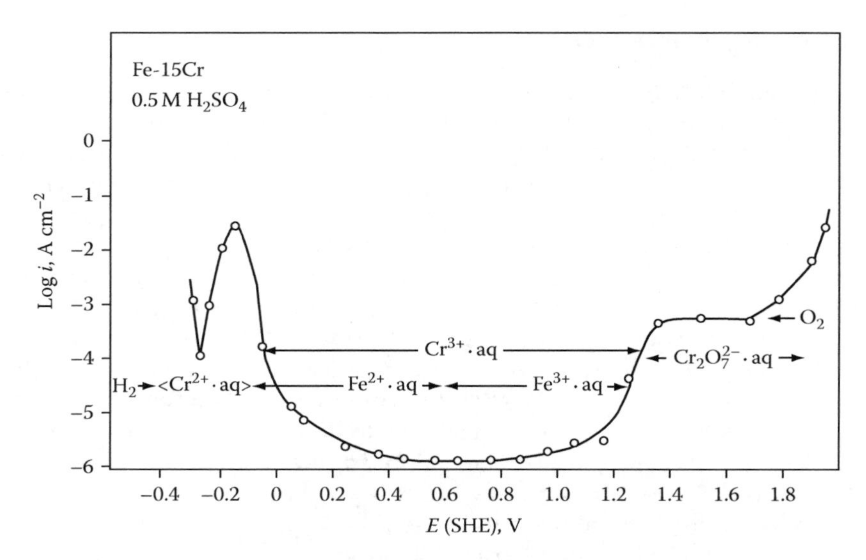

FIGURE 5.31
Polarization curve of Fe-15Cr in 0.5 M H_2SO_4 with indication of corrosion products and potential ranges of hydrogen evolution, active dissolution (Cr^{2+}), passivity (Cr^{3+}), transpassivity ($Cr_2O_7^{2-}$), and oxygen evolution. (From Haupt, S. and Strehblow, H.-H., *Corros. Sci.*, 37, 43, 1995.)

the formation of a protecting Cr(III) oxide film. The dissolution rate of Cr^{3+} is extremely small and thus excellent passive behavior of the alloy is obtained even at fairly negative potentials where pure Fe is still dissolving actively. Above the passivation potential of pure Fe, the Flade potential $E_{P2} = 0.58$ V, Fe^{3+} is dissolved at a slow rate. In the transpassive range, i.e., at $E > 1.0$ V, Cr is dissolved as $Cr_2O_7^{2-}$, however with a slower rate due to some protection by a passive layer of Fe. Fe(III) oxide may now serve as a good protecting alloy component due to the otherwise intense dissolution of Cr(VI) with current densities of several 100 μA cm^{-2}. Above $E = 1.6$ V, the current density increases again due to oxygen evolution.

This interpretation is supported by the XPS results presented in Figure 5.32a [92] and other analytical studies all pointing to Cr enrichment of the passive film in passivated FeCr [13,16,93–105] and FeCrNi [106–110] stainless steels. In the pre-passive range, i.e., at the transition from active to passive behavior at $E = 0.0$–0.2 V, an almost pure Cr(III) layer is found at the surface, which slightly decreases to a potential independent composition of 80% Cr and 20% Fe in the passive range from $E = 0.2$ to 0.9 V. The transpassive dissolution of Cr as Cr(VI) at $E > 0.9$ V causes a gradual increase of the still passivating Fe content within the film with increasing potential. The strong increase of Cr(III) in the passive range is a consequence of the extremely small dissolution rate of Cr^{3+} ions as also found for pure Cr (Section 5.6.2.7). Fe dissolves with current densities of 7 μA cm^{-2} in 0.5 M H_2SO_4, which is more than one order of magnitude larger than that of Cr^{3+}. As a consequence, Cr is enriched by a factor of 4 within the passive layer. This is seen even better with the ISS depth profiles presented in Figure 5.32b [92]. The enrichment of Cr is observed already after a short passivation time of 1 min and increases with time even up to $t_P = 1$ week. At the passive layer surface, there is still some Fe, which disappears after a short sputtering time and Cr increases up to 70% in the center of the passive film. After removal of the passive layer, the composition of the bulk metal underneath is reached with 15% Cr content. The layer thickness on Fe-15 Cr in 0.5 M H_2SO_4 increases linearly

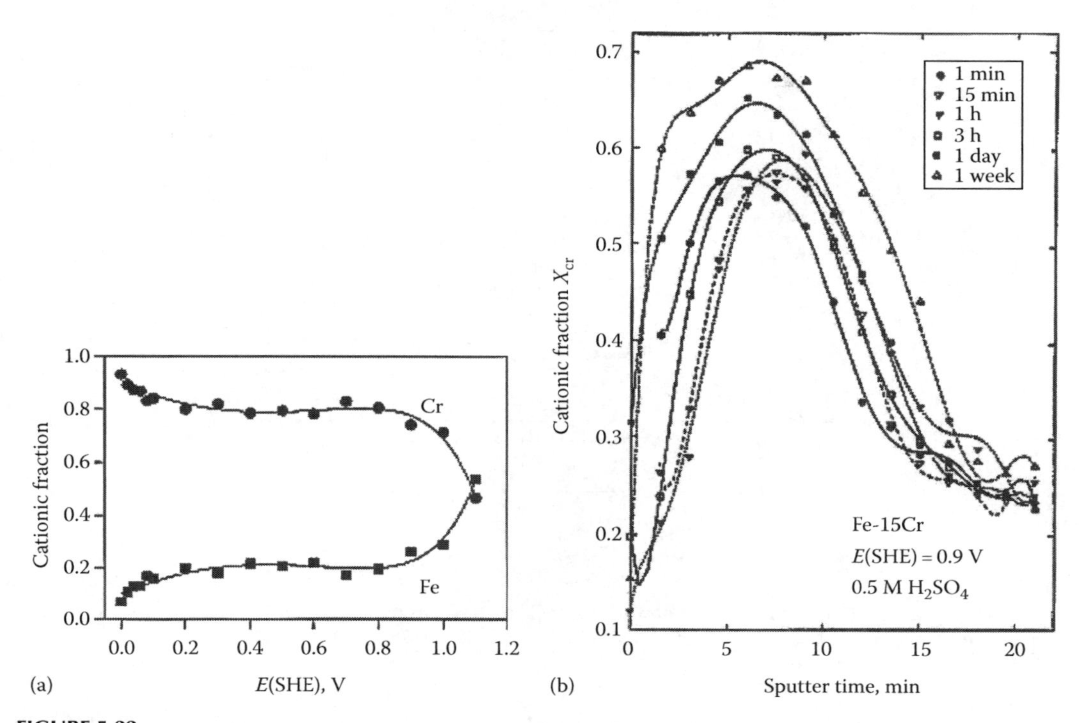

FIGURE 5.32
Composition of the passive layer formed in 0.5M H_2SO_4: (a) Fe-20 Cr, t_P = 300s as a function of electrode potential from XPS examinations. (From Keller, P. and Strehblow, H.-H., *Corros. Sci.*, 46, 1939, 2004.) (b) Fe-15Cr ISS depth profiles after passivation at E = 0.90V for different times t_P. (From Haupt, S. and Strehblow, H.-H., *Corros. Sci.* 37, 43, 1995.)

with potential from 1 to 2nm within the passive range (E = 0.0–1.0V) and decreases to 1nm in the transpassive range (E = 1.1–2.2V) when Fe becomes the passivating alloy component [92].

In alkaline solutions, the structure of the passive film is different as shown by ISS depth profiling (Figure 5.33) [98]. Due to insolubility, Fe(III) oxide preferential dissolution cannot take place and, consequently, Cr(III) is accumulated more moderately with a maximum of ca. 50% compared to 70% in 0.5M H_2SO_4. (Figure 5.33 shows the atomic ratio Cr/Fe instead of the cationic fraction in %.) Furthermore, a large outer part of the passive layer is rich in iron. This part thickens with increasing potential. Transpassive behavior of Cr is expected in 1M NaOH at $E > 0.3$V with oxidation of Cr(III) to soluble CrO_4^{2-}. This is in agreement with the ISS depth profile that shows a Cr-free tail of the passive layer formed at E = 0.34V. The layer grows up to ca. 5nm in the potential range of E = −0.86 to 0.34V. At E = −0.86V, Fe(II) shows some solubility as found from RRD studies (see Section 5.6.2.1). As a consequence, no Fe-rich part of the passive layer is found and Cr is accumulated at the surface to 55%. This process is slow with a negligibly small peak of Cr enrichment to 20% after 20s of passivation that increases to 55% after 2min.

In 0.5M H_2SO_4, XPS studies of Fe-20 Cr treated 300s in the pre-passive range of E = 0.0–0.2V show cationic fractions of 70% Cr(III) (hydroxide), 20% Cr(III) (sulfate), and 10% Fe(II). In the passive range, the composition becomes 50% Cr(III) (hydroxide), 25% Cr(III) (oxide), 10% Cr(III) (sulfate), and 20% Fe. These results were confirmed by XPS analysis of the anionic fractions showing a major contribution of 60% of OH^- in the pre-passive range

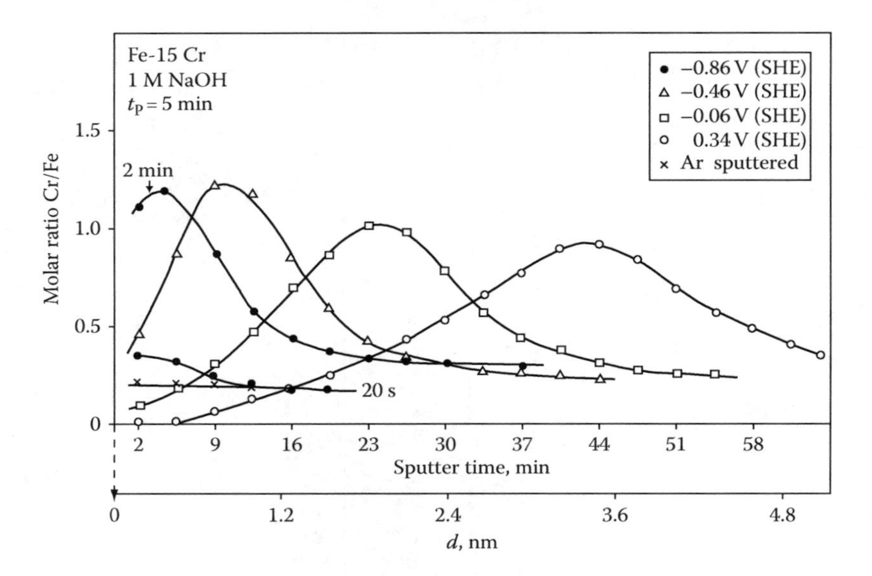

FIGURE 5.33
ISS depth profiles of passive layers formed on Fe-15Cr in 1 M NaOH, t_P = 300 s at potentials as indicated. (From Calinski, C. and Strehblow, H.-H., *J. Electrochem, Soc.* 136, 1328, 1989.)

with some contribution of O^{2-}, sulfate and water, and a slight decrease to 50% OH^- within the passive range with a contribution of 35% of O^{2-} and ca. 7% of water and 7% of sulfate [93].

In alkaline solution, the passive layer on Fe-5Cr is not completely reduced even at potentials of $E = -1.0$ V in contrast to the passive film on pure Fe [61]. Cr(III) cannot be reduced and remains at the surface. Fe(III) is reduced to Fe(II) but not to Fe-metal as shown in Figure 5.18 in comparison to pure Fe. Fe is reoxidized at appropriate positive potentials. Repetitive potentiodynamic anodic and cathodic scans yield growing anodic and cathodic peaks associated with the Fe(II)-to-Fe(III) oxidation reaction and the reverse reduction process, respectively. Repetitive cycling of Fe-5Cr specimens leads to accumulation of Fe species within the film with an increasing charge of oxidation and reduction.

5.6.3.2 Passivity of FeNi Alloys

Ni is a frequent alloying component, e.g., for stainless steels. Polarization curves of Fe-10Ni and Fe-53Ni alloys show the characteristic features of pure Ni shown in Figure 5.4 only to a small extent [111,112]. Small anodic peaks suggest passivation at $E = -0.6$ V in 1 M NaOH and at $E = 0.2$ V in 0.005 M H_2SO_4 + 0.495 M K_2SO_4. XPS shows that the outer part of the passive film is less than 1 nm thick and consists mainly of $Ni(OH)_2$ oxidizing to NiOOH in the transpassive range with some additions of FeOOH. The inner part of the film has been explained as a compound oxide of Fe_2O_3 and NiO. Its thickness increases linearly with potential from 0 to 5 nm in the range of $E = -0.6$ to 0.8 V. It levels off to 5 nm at $E = 0.7$ V in the transpassive potential range. The composition of the oxide and the metal surface corresponds to that of the bulk alloy with only negligible deviations for Fe-10Ni and Fe-53Ni in alkaline and weakly acidic electrolytes (pH 12.7 and 2.9).

Figure 5.34 gives an example for the Fe-53Ni alloy in 1 M NaOH. ISS depth profiles yield a more detailed chemical structure of the 4 nm thick layer due to the monolayer sensitivity of the technique. The Fe content increases from ca. 50% at the extreme surface to 100% at

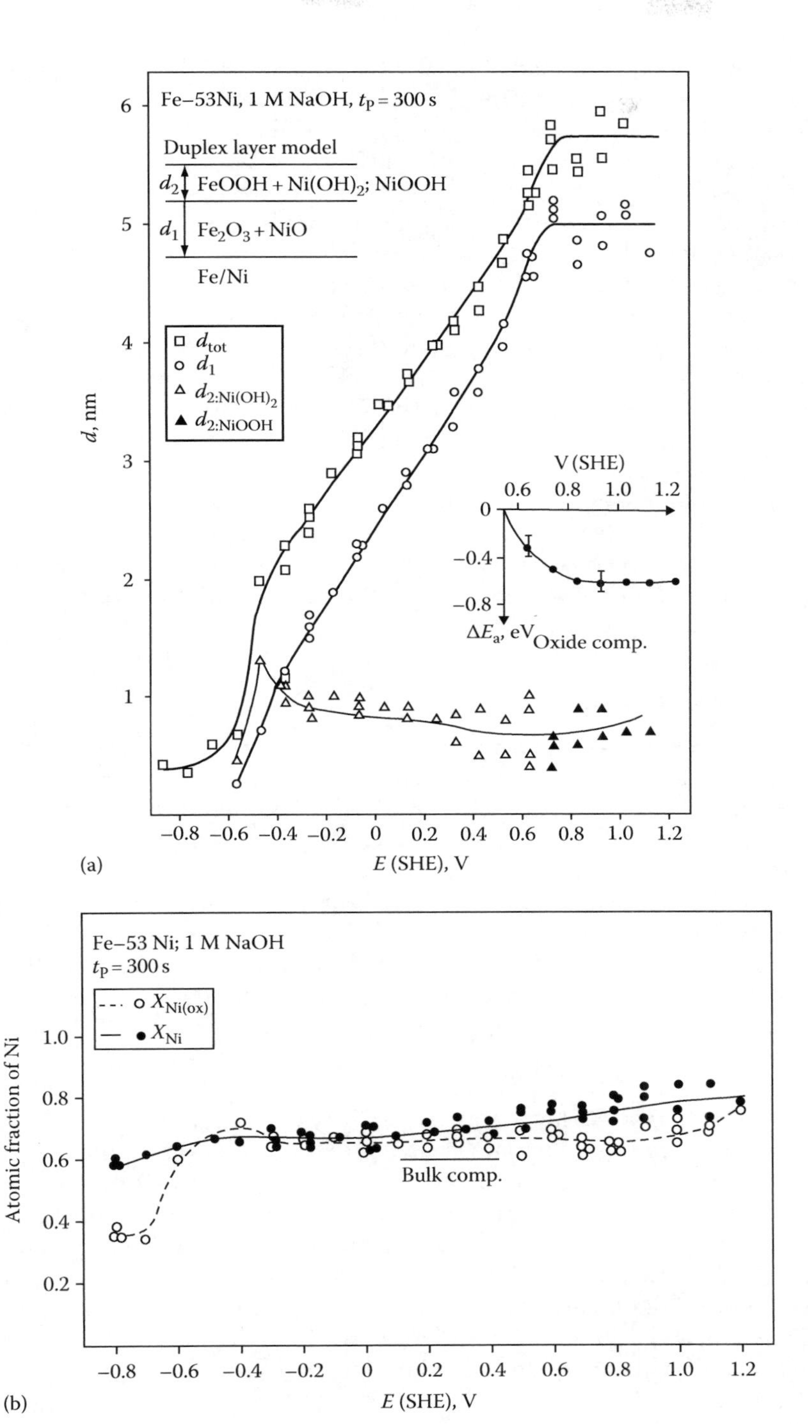

FIGURE 5.34
(a) Thickness of the outer hydroxide part (d_2), the inner oxide part (d_1) and total layer ($d = d_1 + d_2$) of the passive film on Fe-53Ni formed for 300 s in 1 M NaOH as a function of the electrode potential, (b) cationic fraction X_{Ni} (open circles) of the passive layer and atomic fraction X_{Ni} of the metal surface underneath (dark circles). (From Hoppe, H.W. et al., *Surf. Interf. Anal.*, 21, 514, 1994.)

a depth of 1 nm for a layer formed at $E = 0.440$ V for 300 s in 1 M NaOH. Apparently, this Fe enrichment zone refers to the hydroxide layer. It is followed by a steep decrease to a slight depletion of Fe of 30% at a depth of 2.5 nm with a final increase to the bulk composition close to the metal surface. This profile mirrors a duplex layer with an outer Fe-rich hydroxide and an inner Fe-depleted oxide. The same structure has been found throughout the whole passive range from $E = -0.46$ to 0.8 V. In the transpassive range, at $E = 0.84$ V, the ISS profile is qualitatively still the same with only a minor decrease to 90% of the maximum Fe enrichment. The thickness deduced from these depth profiles confirm the evaluation of the XPS data with a small 1 nm thick potential independent hydroxide film and an inner oxide, which increases linearly with potential up to ca. 6 nm.

Time-resolved XPS studies show again similar effects as on pure Ni. In 1 M NaOH, Fe-53Ni forms an outer hydroxide 0.3 nm thick and an inner oxide 4 nm thick at $E = 0.74$ V within 0.1 s. The inner part grows further to 5 nm of oxide within 300 s of passivation. Only slight changes of the composition are observed by XPS. Also similar to pure Ni, the binding energies of all XPS signals related to the passive layer decrease by −0.6 eV for FeNi alloys at the transition from the passive to transpassive range, i.e., at $E = 0.6–0.8$ V in 1 M NaOH, indicating a threshold energy increase by 0.6 eV. This is explained again by the semiconductor model with a negative shift of the Fermi level due to the formation of Ni(III) OOH as discussed in Section 5.6.2.8.

5.6.3.3 Passivity of NiCr Alloys

NiCr alloys with a sufficient Cr content are very resistant to corrosion and thus used for long lifetime construction as, i.e., for nuclear waste containers for underground repositories in the United States. Like for FeCr alloys, Cr enrichment is a characteristic of the passive film for Ni-base stainless steels passivated in acidic and alkaline electrolytes [113–115], as well as in high-temperature water [116].

Passivity is determined at negative potentials by the properties of Cr. Active Ni dissolution is suppressed to a residual current density of 25 μA cm^{-2} in 0.5 M H_2SO_4 due to the presence of a Cr_2O_3 film. In 1 M NaOH, the polarization curves indicate the oxidation of Ni at ca. −0.3 V by a shallow anodic peak corresponding to the −0.059 V/pH dependence of the passivation potential when compared to $E_P = 0.5$ V in 0.5 M H_2SO_4. At $E = 0.6$ V, a sharp anodic peak indicates the transpassive dissolution of CrO_4^{2-} and the formation of NiOOH in 1 M NaOH.

In 1 M NaOH, the passive layer consists mainly of hydroxide up to 0 V. It has a maximum thickness of 4 nm at 0.1 V. ARXPS investigations confirm a bilayer structure with outer hydroxide and inner oxide layers. At $E > 0.1$ V, the thickness of the hydroxide layer decreases down to 2 nm at $E = 0.9$ V in parallel to a linear increase of the inner oxide up to 4 nm. The total layer thickness has a constant value of 6 nm in the range of $E = 0.6–0.9$ V [113]. The outer hydroxide part consists mainly of Cr(III) up to $E = -0.3$ V where Ni(II) enters the layer and becomes the major (up to 90%) component at higher potentials. A small amount (10% at the most) of Cr(VI) is found in the transpassive range for $E = 0.5$ V where NiOOH is also detected. The inner oxide layer consist of pure Cr(III) oxide with a steep change at $E = 0.4$ V when approaching the transpassive potential range in which Ni becomes the major component. The presence of large amounts of $Ni(OH)_2$ and NiO is a consequence of their insolubility in 1 M NaOH.

In acidic electrolytes like 0.5 M H_2SO_4, the inner and outer sublayers of the passive film consist of pure Cr_2O_3 and $Cr(OH)_3$, respectively. In the transpassive potential range, up to 10% of Cr(VI) is found in the outer hydroxide part of the film. Thus the excellent corrosion properties of NiCr alloys in strongly acidic electrolytes are a consequence of the formation of a pure Cr(III)-oxide and -hydroxide layer with the well-known extremely slow metal dissolution in the passive state.

5.6.3.4 Passivity of FeAl Alloys

Al is a very reactive metal that shows however excellent passive behavior in weakly acidic and alkaline solutions because the solubility of Al_2O_3 is minimum at pH 6. The passive behavior of Fe with 4, 8, 15, and 22 at% of Al has been studied by polarization curves, potentiostatic transients, and XPS and ISS measurements with well-controlled transfer of the specimens from the electrochemical cell to the spectrometer. Passivity of these alloys has been investigated in a wide range of weakly acidic and alkaline solutions like H_2SO_4 with additions of Na_2SO_4, pH 2.5 and 3.8, phthalate buffer pH 5.0, and borate buffer pH 9.3 [117,118]. Alloying 4 at% Al reduces the maximum of Fe dissolution in acidic electrolytes of pH 3.8 and 5.0 by one order of magnitude. At pH 3.8, alloying 12 or 22 at% of Al suppresses the peak of active dissolution almost completely.

In all solutions, Al(III) is accumulated within the passive layer as shown by XPS and ISS sputter depth profiles presented in Figure 5.35 [117,118]. The XPS depth profiles of specimens with 8, 15, and 22 at% Al passivated at E = 1.0 V in phthalate buffer pH 5.0 yield a peak of Al enrichment in the center of the passive layer with a maximum of the cationic fraction X_{Al} = 20%, 40%, and 70%, respectively. These profiles are relatively broad because of the mean free path of the photoelectrons limiting the depth resolution. The top layer sensitivity of ISS resolves a sharper and higher maximum corresponding to an Al content of 40%, 60% and 95% for the 8, 15, and 22 at% Al alloys, respectively. ARXPS clearly indicates, an outer position of the hydroxide layer above the oxide sublayer. The oxide thickness grows linearly with potential in phthalate buffer pH 5.0 from a minimum value of 1 nm up to 7 nm at E = 1.2 V for all three alloys.

Potentiostatic transients show a similar sequence of layer formation as obtained for the variation of potential. Passivation of Fe-22Al at E = 1.0 V in phthalate buffer pH 5.0 forms first Fe(II) with a maximum at t_P = 10 s and a fall to negligible values at t_P = 100 s. Fe(III) starts to grow continuously after t_P = 1 s up to t_P = 1000 s. As a very reactive metal, Al forms already Al(III) at the very beginning of the transient. Its content decreases with time when Fe(III) enters the film between t_P = 1 and 10 s. Up to t_P = 0.1 s, the film contains large amounts of OH^- and water. After this initial time, water disappears completely (after 1 s) and OH stabilizes at 40%. ARXPS shows that OH^- is located in the outer part of the passive layer.

XPS studies of the reduction of a film formed at E = 0.96 V in phthalate buffer pH 5.0 on Fe-15Al show characteristic compositional changes. Galvanostatic reduction with $i = -20\,\mu A\ cm^{-2}$ yields a decrease of Fe(III) with an increase of Fe(II) to a maximum after 20 s and a decrease to a constant value of 12% after 40 s [117]. The Al(III) content stays constant till 40 s and drops afterwards to a constant value of 15%. Apparently, the Al(III) oxide remaining at the surface protects the remaining Fe(II) oxide against dissolution. The Al enrichment in the center of the passive layer is displaced to the surface due to dissolution of iron after its reduction to Fe(II). The oxidation of Fe(II)-to-Fe(III) and its reduction with appropriate changes of the potential remain the same compared to that of the passive layers formed on pure iron and FeCr alloys as described above.

5.6.3.5 Passivity of FeSi Alloys

Si is a relatively noble element, interesting as additive to Fe because of the protecting properties of its oxide especially in acidic electrolytes. High Si alloys resists corrosion

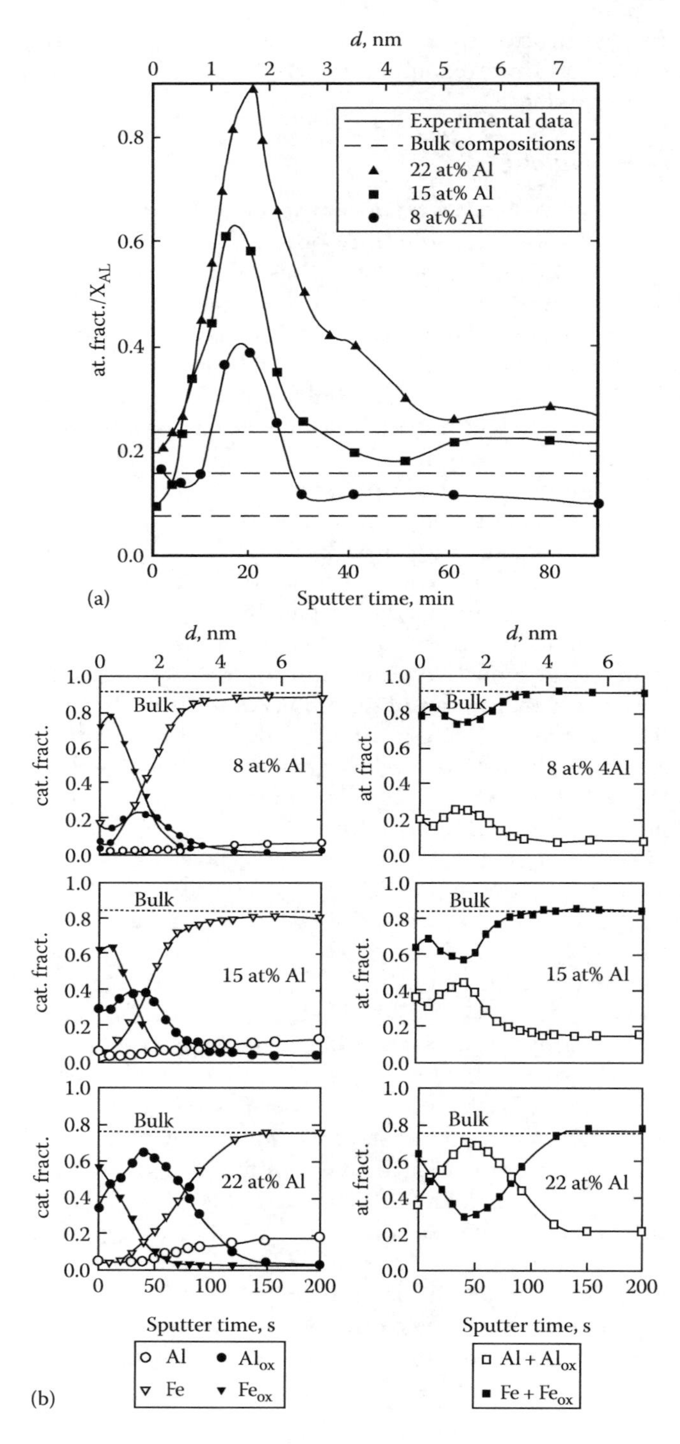

FIGURE 5.35
(a) ISS depth profiles of passive layers formed on three Fe-Al alloys passivated for 300s at 1V in 0.1M phthalate buffer pH 5.0, (b) XPS depth profile of the same specimens. (From Schaepers, D. and Strehblow, H.-H. *J. Electrochem. Soc.*, 142, 2210, 1995.)

even in strongly acidic electrolytes like sulfuric acid, nitric acid, and phosphoric acid. Alloys with small Si content up to 7.7 at% are less protected in phthalate buffer pH 5.0 compared to pure iron. Active dissolution is 20, 25, and 40 mA cm^{-2} and the passivation potential is $E_P = 0.3$, 0.5, and 0.8 V for 0, 2, and 4 at% Si, respectively [119,120]. FeSi alloys with 7.7 at% Si cannot be passivated in 0.001 M H_2SO_4 but are well protected with higher Si content like 21 and 25 at%. The polarization curve for Fe-25Si in 0.001 M H_2SO_4 + 0.5 M Na_2SO_4 shows only a negligible active dissolution peaking at ca. 0.1 mA cm^{-2} and a large passive range from 0.2 to 1.7 V [119,120].

XPS and ISS depth profiles show a pronounced accumulation of Si in the center of the passive layer for alkaline solutions. In acidic electrolytes, like acetate buffer pH 5.0, an almost pure SiO_2 film is found on Fe-21.3Si and Fe-33Si throughout the whole passive range with only minor contributions of a few percent of Fe oxide at $E > 1.0$ V [119,120]. Most likely, Fe is dissolved and the high Si content of the bulk alloy allows the formation of a continuous protective layer of SiO_2 with a minor content of Fe(III) at more positive potentials. For Fe-25Si, Fe enters the film for $E > 0.6$ V. At $E = 0.9$ V, a maximum of 100% Si is found at the center of the passive layer at a depth of 0.7 nm with the stabilization of some Fe(III) at the film surface. A similar profile is found for borate buffer pH 9.0 with a maximum of 80% Si.

Potentiostatic studies show that Si is oxidized first and forms SiO_2 films during the first millisecond of anodic transients. Fe enters these layers as Fe(II) and is oxidized to Fe(III) within the first seconds. In acidic electrolytes (pH = 5.0), Fe(III) is lost slowly with time leaving a permanent protective SiO_2 layer. A very characteristic behavior is found for low Si-containing alloys. In acetate buffer pH 5.0, even SiO_2 is lost on Fe-2Si when Fe(III) dissolves. As a consequence, the whole film disappears within 1000 s and the dissolution current starts to increase again after a fall to ca. $i = 10\,\mu A\,cm^{-2}$ within 3 min [120]. Apparently, the Si content is too small to form a continuous SiO_2 layer protecting the alloy surface for a bulk content of less than 7.6 at%.

5.6.3.6 Passivity of CuNi Alloys

CuNi alloys are important for many applications due to the combination of the chemical properties of both elements. Ni provides passivity in acidic electrolytes whereas Cu is protective in more alkaline solutions. The polarization curve is dominated by the anodic oxidation peaks A1 and A2 of Cu and the transpassive peak of Ni. In acidic electrolyte, the two anodic Cu peaks merge to give one active Cu dissolution peak. Alloys with a low Ni content cannot be passivated in acidic electrolytes because they cannot form a continuous Ni oxide layer at the surface.

The passive behavior and the composition of the passive layers have been studied in a wide range of alloy compositions and pH of the solutions using XPS and ISS [121,122]. The structure of the passive film is complex with an outer hydroxide and inner oxide layers, and a further subdivision due to both metal components and their two oxidation states. At low potentials, a film of mainly $Ni(OH)_2$ is formed with some NiO underneath. This inner oxide layer grows with potential. The outer hydroxide consists of pure $Ni(OH)_2$, which is oxidized to NiOOH in the transpassive range, i.e., at $E > 0.5$ V in phthalate buffer pH = 5. Copper enters the inner oxide layer at sufficiently positive potentials, i.e., at $E = 0$ V in phthalate buffer pH 5.0 with some NiO entering at $E > 0.4$ V. The metal surface is enriched in Cu due to the preferential oxidation of the less noble Ni. ARXPS shows that Cu_2O is located directly at the metal surface whereas CuO is in the upper oxide part. This complicated layer structure is also found by ISS depth profiling. However, the method cannot distinguish between the different valences of the layer components. The outer hydroxide film has only a negligible Cu content whereas the inner oxide is enriched in Cu, i.e., 60 at% for a Cu-50 Ni alloy. The ISS depth profiles also show a Cu enrichment of ca. 10 at% above the bulk value at the metal surface.

5.6.3.7 Passivity of CuSn Alloys

For CuSn alloys, both metal components influence the passive behavior as shown by studies of three alloys with 4, 12, and 19 at% Sn content [123]. In alkaline solutions (0.1 M NaOH), the polarization curves show the characteristic current peaks of the pure elements with the related reactions discussed before. As expected, for high Cu alloys like Cu-4Sn and Cu-12Sn, Cu is dominating with well-pronounced anodic CuA1 and CuA2 and cathodic CuC1 and CuC2 current peaks. With increasing Sn content, the contribution of its anodic SnA1 and SnA2 peaks increases (Figure 5.36). Pure Sn is less corrosion resistant in 0.1 M NaOH at least in the active range as shown by current peaks, which are more than one order of magnitude higher.

The chemical structure of the passive layers formed in 0.1 M KOH has been evaluated with XPS and ISS for Cu-4Sn and Cu-19Sn alloys as a function of potential and by potentiostatic transients [123]. At $E = -0.3$ V, Cu-4Sn forms a thin film of SnO and $Sn(OH)_2$ without Cu. When oxidation of Cu becomes possible at $E = 0.2$–0.8 V, the passive layer consists of Cu oxides with only minor contributions of a few percents of Sn oxides decreasing with increasing potential. One observes a duplex film of Cu_2O/CuO, $Cu(OH)_2$ like on pure copper. Only a small Sn content of less than 10% was observed in the center of the passive layer. At $E = 0.5$ V, this result was obtained for $t_P = 0.2$ s to 90 min.

At $E = -0.3$ V, the passive layer on Sn-19Cu also consists of a thin layer of pure Sn oxide, but a much larger contribution of Cu oxides is observed in the passive range, i.e., at $E = 0.2$–0.8 V, where all four Cu and Sn oxides are present. Cu oxides locate preferentially at the surface and the inner oxide layer is Sn-rich. The oxides show the usual distribution with the higher valent cations in the outer part as for the passive layers on the pure metals, i.e., SnO_2 on top of SnO and CuO on top of Cu_2O. At $E = 0.5$ V, the Sn oxide content increases with time up to $t_P = 30$ min, whereas the Cu oxide content decreases. CuO increases with time at the expense of Cu_2O.

In weakly acidic electrolyte (phthalate buffer, pH 5.5), both alloys form pure thin (0.2–0.5 nm) Sn oxide layers in the active range at $E = 0.3$ V. Within the passive range, Cu is

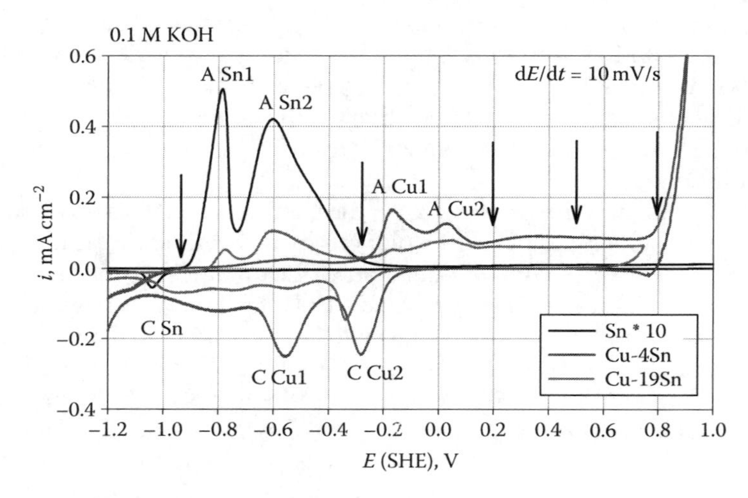

FIGURE 5.36
Polarization curves of pure Sn and Cu-4Sn and Cu-19Sn alloys in 0.1 M KOH. The current densities for Sn are reduced by a factor 0.1 to fit into the figure. (From Keller, P. and Strehblow, H.-H., *Z. Phys. Chem.*, 219, 1481 2005.)

dissolved preferentially at $E = 0.8–1.6$ V. The Cu-4Sn alloy forms an oxide layer containing both metal components with a Sn-rich inner part layer and with SnO directly at the metal surface and SnO_2 (containing some $Cu(OH)_2$ on top). The Cu-19Sn alloy is better protected due to a higher Sn content. Within the passive range, a pure duplex Sn oxide layer is formed with a 0.5-nm-thick SnO part also at the metal surface topped by a 2 nm thick outer part of SnO_2. The SnO inner part of the film contains residual CuO and $Cu(OH)_2$ although they should dissolve at this pH, suggesting that they are stabilized by SnO_2 and do not dissolved.

5.6.3.8 Passivity of AlCu Alloys

Al alloys often contain some percents of Cu. Vapor-deposited films of this composition range have been used in the semiconductor industry for conductive connections in integrated circuits. The Cu content is used to reduce electromigration, which leads to failure of the metallic connections. Cu can cause corrosion problems on a small scale. Electrolyte residues together with moisture and voltage supply of 5–18 V may lead to corrosion phenomena. Cu-rich inclusions are widely known to be detrimental to the corrosion resistance of Al alloys used in the ship building and airplane industry.

The anodic oxidation of thin Al films with Cu contents of up to 10 at% has been investigated [124]. Due to vapor deposition, these relatively high Cu contents are still homogeneously distributed within the Al films. Galvanostatic experiments with current densities of $i = 1$ mA cm^{-2} have been applied to films vapor deposited on quartz or sapphire substrates up to potentials of 100 V and more. The potential increases linearly with time but reaches a maximum where breakdown of the oxide layer occurs with local destruction and perforation of the whole metal film. This breakdown potential decreases with increasing Cu content within the film (Figure 5.37). For 0.5 at% Cu films, a potential of 150 V can be achieved during anodization in citrate buffer pH 6.0, whereas for 11.7 at% Cu only 18 V are reached. In the case of the higher Cu content, the continuous anodization after breakdown leads to a brittle surface due to numerous breakdowns of the film.

AlCu alloys containing a very reactive element and a seminoble metal of extremely different electrochemical properties, anodic oxidation causes the preferential oxidation of Al, which forms passivating aluminum oxide films whereas Cu remains at the metal surface. Although

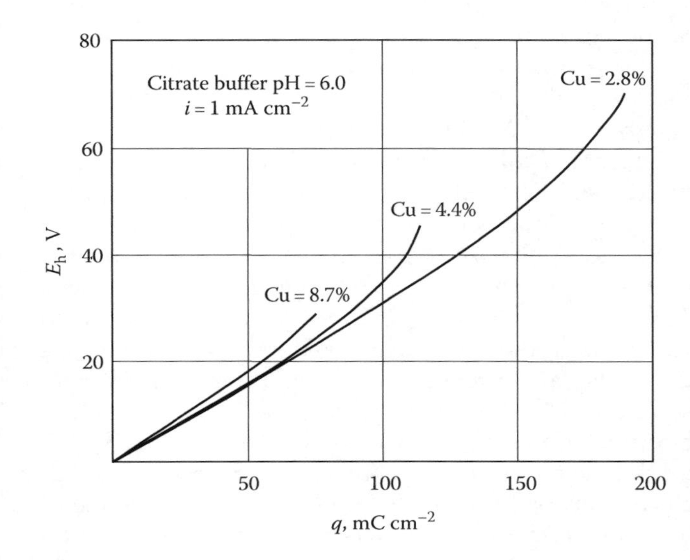

FIGURE 5.37
Galvanostatic oxidation of AlCu films in citrate buffer pH 6.0 with $i = 1$ mA cm^{-2} with breakdown at the maximum potential reached. (From Strehblow, H.-H. and Doherty, C.J., *J. Electrochem. Soc.*, 125, 30, 1978.)

the potential is increasing to several 10 V, the potential drop at the metal–layer interface is relatively small and the potential is too negative to oxidize Cu metal. The major part of the applied large potential difference is located within the passive film and thus does not contribute as a driving force for Cu oxidation. Consequently, Cu is accumulated at the metal surface underneath the anodic Al oxide layer. After sufficient Cu accumulation, Al oxidation is hindered at least locally with an increase of the potential drop at the metal surface and a possible Cu oxidation, which may cause local breakdown of the film and the observed localized dissolution.

RBS studies support this description [125]. RBS can be applied to study these anodic layers that are thick enough to allow investigation with the poor depth resolution of the method (ca. 5 nm) but with a high information depth of more than 1 μm for backscattered He ions with a primary energy of 4 MeV. The signals of Cu and Al are well separated due to their mass difference (Figure 5.38a). The Al signal of the anodized specimen shows a step with lower Al content corresponding to the Al_2O_3 layer. Its width is a measure of the layer thickness. An oxygen signal superimposed to that of the quartz substrate refers to the oxide and reflects again its thickness. Very important is the Cu signal at high energy of the backscattered He-ions. Its leading edge shows a shift for the anodized specimen in comparison to the untreated metal film, which again corresponds to the oxide layer. Figure 5.38b depicts the Cu signal for different stages of galvanostatic oxide formation corresponding to the various potentials that have been reached. The gap to the untreated metal specimen demonstrates clearly that the oxide layer does not contain any Cu. It also shows a well-pronounced Cu enrichment peak, which piles up with increasing potential and thus increasing oxide thickness. For thick oxide layers corresponding to high potentials, the width of this peak changes with the tilt of the specimens and mirrors the depth profile of the Cu enrichment, indicating that Cu accumulation is not completely homogeneously distributed across the metal surface and thus causes a roughening of the metal–oxide interface. At sites of high Cu accumulation, the film supposedly breaks due to insufficient supply of Al during anodic oxidation. Cross section of anodized AlCu specimens and further RBS studies have confirmed these findings [126,127].

5.6.4 Conclusion on the Chemical Composition of Passive Layers

Almost all pure metals and binary alloys form passive layers with a multilayered structure. Even for pure metals with only one oxidation state of the cations, a bilayer structure is observed with an inner oxide topped by an outer hydroxide. The situation becomes more complicated with several oxidation states and for alloys.

Figure 5.39 shows a schematic diagram for pure metals and binary alloys that have been investigated in detail with electrochemical and surface analytical methods. The most complicated chemical structure is found at sufficiently positive potentials where the higher valent species can be formed. The lower valent species are present within the layer as well, however usually inside and covered by the oxide formed by the higher valent cations. Typical examples for pure metals are the duplex films Cu_2O/CuO, $Cu(OH)_2$ on Cu, $NiO/Ni(OH)_2/NiOOH$ film on Ni and $SnO/SnO_2/Sn(OH)_4$ on Sn. In some cases, lower valent cations are found in the outer layer when their oxidation fails despite the positive potential because they are trapped in an insulating environment. A typical example is the passive layer on FeAl alloys with an accumulation of Al_2O_3 in the center of the layer. Fe(II) is found on top of an inner oxide containing Fe(III) and Al(III). Apparently, the Al_2O_3-rich layer has a poor conductivity not allowing the full oxidation of iron that requires the transport of electrons through the layer to the metal surface. The situation is complicated when two metal components of the alloy enter the

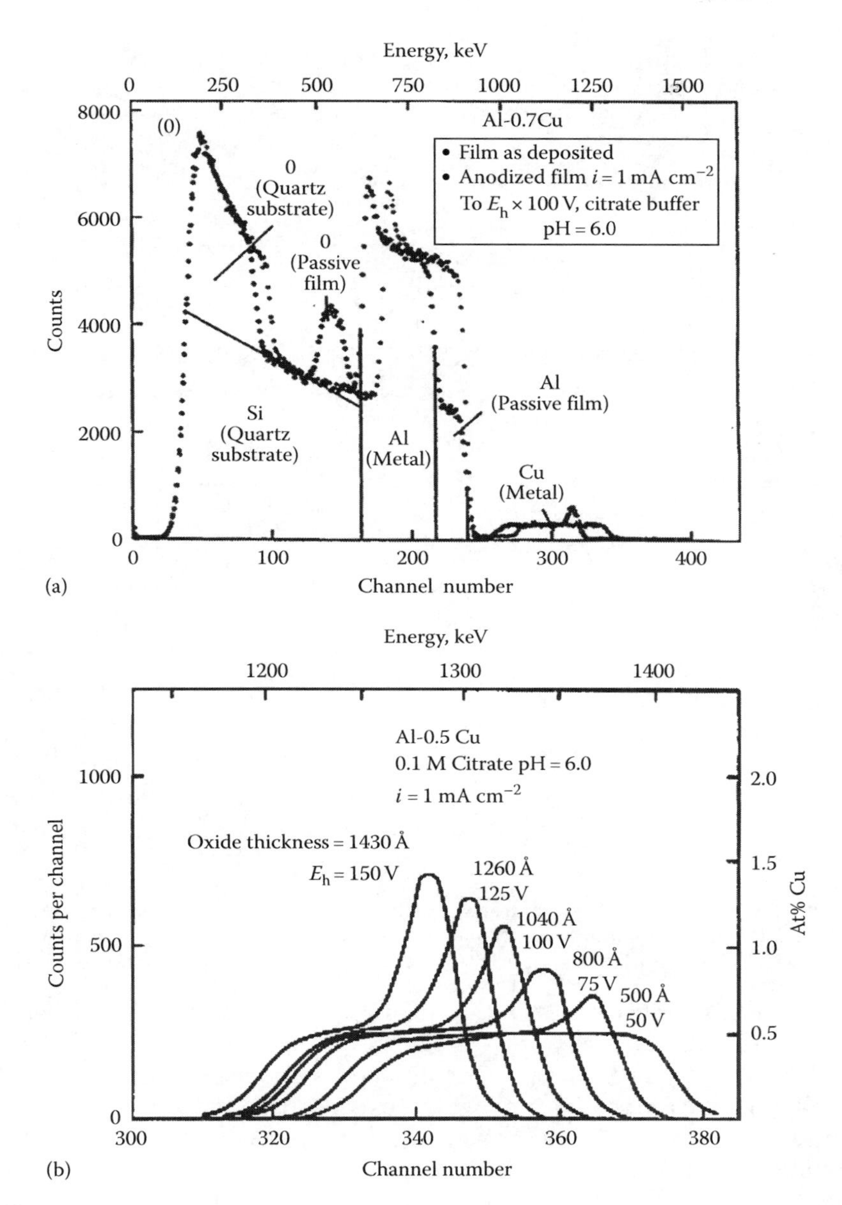

FIGURE 5.38
(a) RBS data for a Al-1Cu alloy film galvanostatically anodized with i = 1 mA cm^{-2} up to E = 100 V in citrate buffer pH 6.0, (b) Cu RBS signal for Al-0.5Cu anodized to different potentials. (From Strehblow, H.-H., et al., *J. Electrochem. Soc.*, 125, 915, 1978.)

passive film with different oxidation states like in the case of the passive layers formed on CuNi alloys with NiO in the inner part and additions of Cu_2O/CuO and $Cu(OH)_2$ to the $Ni(OH)_2$/NiOOH outer part.

The multilayer model of the passive film is the result of ARXPS studies combined to XPS and ISS depth profiles. The interpretation by a multilayered structure is not necessarily the only possibility. An alternative is a continuous change of the composition without phase

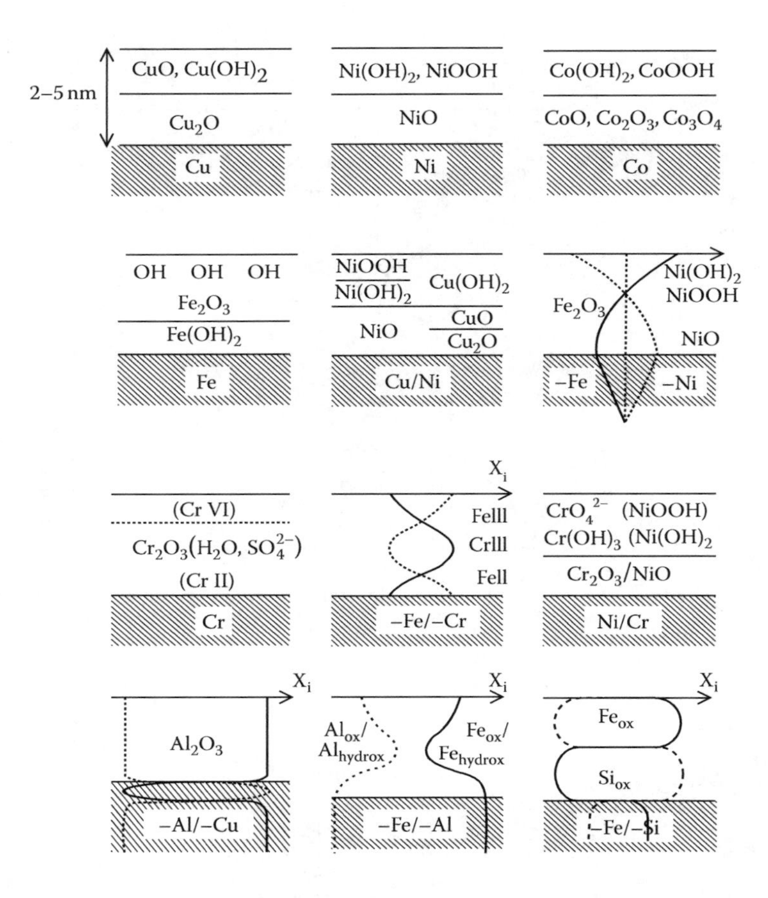

FIGURE 5.39
Schematic diagram of the chemical composition and structure of passive layers on pure metals and binary alloys as obtained from XPS and ISS investigations. (From Strehblow, H.-H., in *Passivity of Metals*, R.C. Alkire, D.M. Kolb, eds., Wiley-VCH, Weinheim, Germany, pp. 271–374, 2003.)

boundaries. This is strongly suggested by ISS depth profiles of passive layers on FeNi, FeCr, FeAl, and FeSi alloys.

Time-resolved studies are also important for a better understanding. It is a rule that the lower valent species are formed first and will be oxidized later. The layer structure develops within milliseconds or hours. This time scale depends on potential and thus on the driving force for the electrochemical reactions but also on the system. The duplex film on iron forms within a few seconds and it takes ca. 100s to get the final structure, which gets faster with increasing potentials. On the contrary, the accumulation Cr within passive layers on FeCr alloys and its change from a hydroxide to an oxide will get to its final stage after hours or days.

5.7 Electronic Properties of Passive Layers

Passive layers are electronic insulators or semiconductors depending on the band gap of their constituents. Valve metals like Al, Ta, Zr, Hf, Ti, and even pure Sn may be anodized to potentials much above $E = 1.5\,V$ without oxygen evolution, some like Al or Ta even to

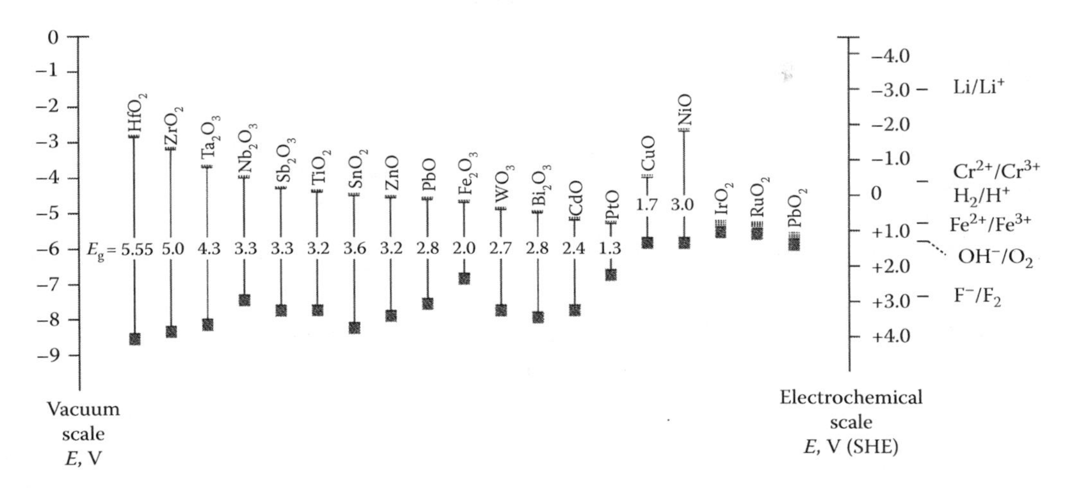

FIGURE 5.40
Band gap and energy of the upper valence band and the lower conduction band edges of several oxides relative to the SHE and the vacuum level including some redox systems. (From Schmickler, W. and Schultze, J.W., in *Modern Aspects of Electrochemistry*, Vol. 17, J.O'M. Bockris, B.E. Conway, R.E. White, eds., Plenum Press, New York, p. 357, 1986.)

much more than 100 V. Other metals like Fe, Co, Ni, Cr, Cu, and Ag form semiconducting oxides with a sufficiently small band gap to allow electron transfer reactions with redox systems including oxygen evolution. Some are even metallic conductors like IrO_2, RuO_2, and PbO_2.

Figure 5.40 compiles the band gap of some bulk oxides and the absolute energy of their band edges with reference to the standard hydrogen electrode and the vacuum level, as well as the energy of some redox systems of interest for electron transfer at passivated metal surfaces [128,129]. The band gap of insulating anodic oxides like Al_2O_3 or Ta_2O_5 exceeds 3 eV and thus electron transfer between a redox system within the electrolyte and the passive layer cannot take place. As discussed in Chapter 1, Section 1.18.1, this process requires electronic levels within the layer at the same energy as the involved redox system (horizontal energy transfer). For various oxides, such levels are not available and thus oxides at the left side of Figure 5.40 do not show redox processes even at very positive potentials. TiO_2 is at the limits of semiconducting properties. It requires a high band bending and thus a very positive potential for this n-type semiconductor in order to meet the condition of electron tunneling through the space charge layer to states in the conduction band. As a consequence, such an n-type semiconductors show diode characteristics with small anodic and large cathodic current densities as described in Chapter 1, Section 1.18.2. Band bending also dictates electron transfer for other semiconducting oxides in the center and right part of Figure 5.40. This is described in detail below.

5.7.1 Electron Transfer at Passivated Metallic Surfaces

Semiconducting passive layers present band bending relative to the surface of the substrate metal and to the electrolyte. The metal substrate and the electrolyte cannot take over the changes of the applied potential and thus the potential drops are located within the oxide film and at the metal–oxide and oxide–electrolyte interfaces. For a thick film

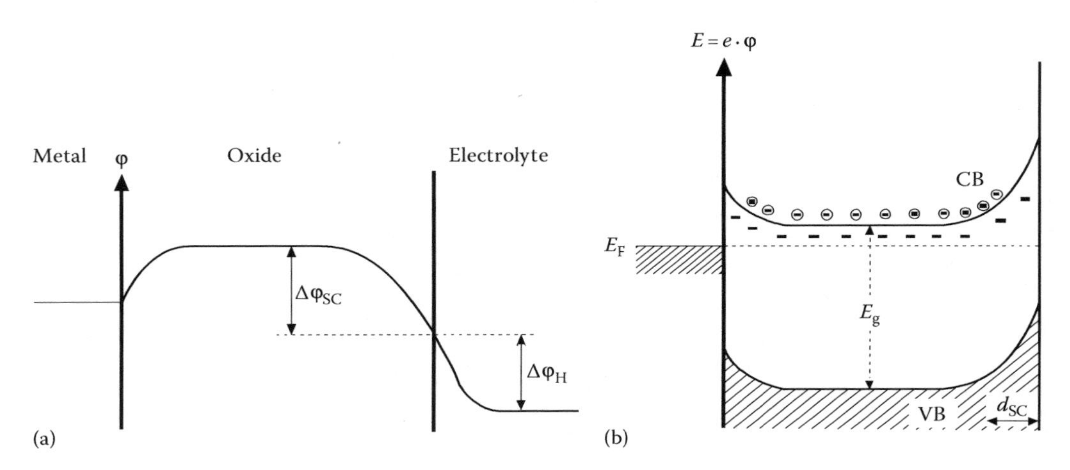

FIGURE 5.41
Potential profile at metal surface covered by a passive oxide layer of band gap E_g with (a) no potential drop within the center of the anodic oxide film and (b) band bending of the valence band and conduction edges at the interfaces with the metal and electrolyte. Electronic equilibrium with a constant Fermi level for the metal substrate throughout the film is assumed. (From Strehblow, H.-H., in *Passivity of Metals*, R.C. Alkire, D.M. Kolb, eds., Wiley-VCH, Weinheim, Germany, pp. 271–374, 2003.)

and not too high applied potentials, this leads to the situation depicted in Figure 5.41 with no potential drop in the center of the film and drops located at the metal–oxide and the oxide–electrolyte interface. At the metal surface, one usually assumes a fast transfer of metal ions, which stabilizes the potential drop at the metal–oxide interface [128]. At the oxide–electrolyte interface, a fraction of the potential drop, $\Delta\varphi_{SC}$, is located within the semiconducting oxide and another fraction, $\Delta\varphi_H$, within the Helmholtz layer. For stationary conditions, i.e., at constant oxide thickness, $\Delta\varphi_H$ is determined usually by the pH of the electrolyte. Here OH^- adsorption or O^{2-} formation at the layer surface (phase 2) is in equilibrium with the electrolyte (phase 3) as discussed for passive iron in Section 5.5.1. There is common agreement that further potential increase is located within the semiconducting film. This is a consequence of its small concentration of charge carriers, which determines the Debye length β (Equation 5.28) (Chapter 1, Section 1.18.2, Equations 1.179 and 1.180). Although the dopant concentration within passive layers is usually high in the range of 10^{17}–10^{20} cm^{-3}, it remains small compared to the electron concentration in metals (5×10^{23} cm^{-3}) and the ion concentration of electrolytes (1.2×10^{21} cm^{-3} for a 1 M solution). The depth of the space charge layer d_{SC} is given by Equation 5.29, which includes the Debye length β:

$$\beta = \sqrt{\frac{\varepsilon\varepsilon_0 kT}{e_0^2 N_D}} \tag{5.28}$$

$$d_{SC} = \beta\sqrt{\frac{2e_0(E - E_{FB})}{kT}} \tag{5.29}$$

d_{SC} increases with the square root of the potential drop $E - E_{FB}$. E_{FB} is the flat band potential, i.e., the electrode potential for which there is no potential drop in the semiconducting layer and thus with a vanishing space charge layer and a horizontal situation of the valence and conduction band. This refers to the situation when the potential is stepped to E_P after layer formation at a higher potential. Thus any increase of the potential to $E > E_P$ causes the formation of a space charge layer with a voltage drop $E - E_P$ within the layer close to its surface. For thin passive layers with $d < d_{SC}$, the voltage drop $E - E_P$ occurs across the total film causing a higher electrical field strength.

Similar to the discussion of thick layers in Chapter 1, Section 1.18.2, the capacity of a passivated metal surface is determined by two capacities in series, the capacity of the Helmholtz layer (C_H), and the capacity of the space charge layer of the semiconductor (C_{SC}). The addition of the reciprocal capacities yields the total capacity C given by Equation 5.30. Usually C_{SC} (~2 μF cm^{-2}) is about one order of magnitude smaller than C_H (~20 μF cm^{-2}) so that the total measured capacity equals approximately that of the space charge layer C_{SC}. Usually the capacity is measured as a function of the electrode potential E and is evaluated by a Schottky–Mott plot (Equation 5.31), which yields the flat band potential for $E - E_{FB} = 0$ if one neglects the small quantity kT/e. Another possibility is the determination of E_{FB} from photocurrent measurements as discussed in Chapter 1, Section 1.18.3. The photocurrent disappears for a vanishing band bending, i.e., for $E = E_{FB}$, when the photogenerated charge carriers are no longer separated due to a missing electrical field strength and thus recombine,

$$\frac{1}{C} = \frac{1}{C_H} + \frac{1}{C_{SC}} \approx \frac{1}{C_{SC}} \tag{5.30}$$

$$\frac{1}{C^2} = \frac{2}{eN_D}\left(E - E_{FB} - \frac{kT}{e_0}\right) \tag{5.31}$$

In order to determine the energy levels, the potential drop within the semiconductor $\Delta\varphi_{SC} = E_S - E_B$ with the potential E_S at the surface and E_B within the bulk of the semiconductor and the potential drop in the Helmholtz layer $E_{Sol} - E_S$ (Equation 5.32) have to be taken into account. At the flat band potential, i.e., for $E = E_{Fb}$ and a vanishing potential drop within the semiconductor $e\Delta\varphi_{SC} = e(E_S - E_B) = 0$, one obtains the relation for the Fermi energy within the passive layer, $eE_{F,Ox}$ (Equation 5.33). It is given by the energy at the flat band potential, eE_{Fb}, and the energy of the potential drop $e\Delta\varphi_H = e(E_S - E_{Sol})$ between the surface and the electrolyte solution:

$$\Delta\varphi_H = E_{Sol} - E_S = \Delta\varphi_H(0) - 0.059\,\text{pH} \tag{5.32}$$

$$e_0 E_{F,Ox} = e_0[-4.6 + E_{Fb} - (E_B - E_{Sol})] \tag{5.33}$$

The value of −4.6 eV is the difference between the hydrogen and the vacuum potential scales [130–133]. Usually the lower conduction band edge is ca. 0.25 eV above $E_{F,Ox}$ for an n-type semiconductor or the upper valence band edge is ca. 0.25 eV below $E_{F,Ox}$ for a p-type semiconductor. Thus, one knows approximately the energy of the lower conduction band edge or the upper valence band edge. The energy of the related other band is given by the band gap energy, which may be deduced from light absorption measurements. The threshold energy determined by UPS measurements gives a more reliable and exact value of the upper valence band edge in the potential scale.

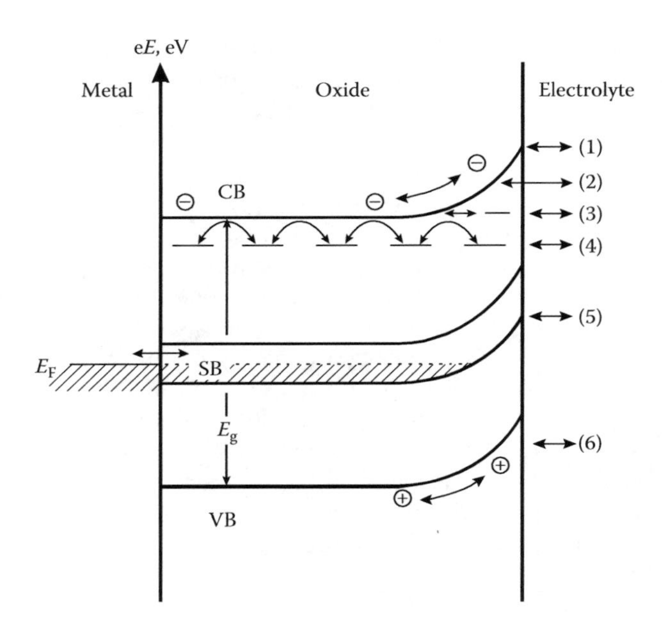

FIGURE 5.42
Electron transfer between redox systems within the electrolyte and the metal surface covered by a passivating oxide layer with band gap (E_g), conduction band (CB), and valence band (VB): (1) Direct tunneling into free states of the CB, (2) tunneling through the space charge layer to the conduction band CB, (3) hopping via surface states to CB, (4) multiple hopping via interband states, (5) transfer via a subband SB, and (6) transfer into positive holes of valence band VB. (From Strehblow, H.-H., in *Passivity of Metals*, R.C. Alkire, D.M. Kolb, eds., Wiley-VCH, Weinheim, Germany, pp. 271–374, 2003.)

Electron transfer between the redox systems within the electrolyte and in the metal or vice versa has to occur via the semiconducting oxide layer. This can proceed in several ways as depicted schematically for a n-type semiconducting oxide in Figure 5.42. The paths (1) and (2) have been already described in Chapter 1, Section 1.18.2. The transfer from the occupied states of the redox system occurs via tunnel processes through the space charge layer directly into empty states of the conduction band. The larger the band bending and thus the more positive the electrode potential the smaller the tunnel distance through the oxide. This increases the tunnel probability T via the exponential decay with the tunnel distance d_T as given by Equation 5.34 where E is the energy level of the tunneling electron, V_0 the total energy barrier, $V_0 - E$ the local energy barrier, d the distance the electron has to overcome (as shown in Figure 5.43), and m^* is the reduced mass of the electron. The coefficient $\kappa = \sqrt{2m^*(V_0 - E)}$ describes the dependence on the energy barrier height. Figure 5.43 also depicts the wave functions of the electron on both sides of a gap and in the gap. If one metal is replaced by the redox system, a qualitatively similar situation is obtained for the electron transfer into the conduction band. The same description will hold for STM, an imaging method based on electron tunneling used to study the structure of passive layers (see Section 5.8). The charge transfer at the oxide surface is proportional to the density of occupied states of the redox system D(Red) within the electrolyte and to that of the empty states within the conduction band D(Ox) and has to be multiplied with the probability of the tunnel process as given in Equation 5.35. The integration has to be on all energies above the Fermi level where occupied and empty states may interfere at the same energy level, which is a requirement for an adiabatic electron exchange. The change of oscillation

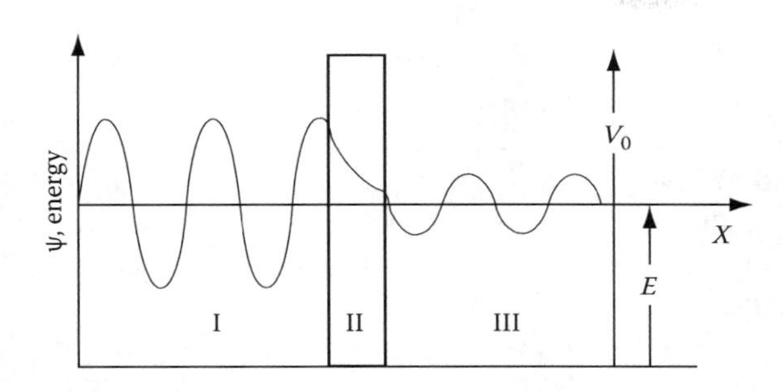

FIGURE 5.43
Schematic drawing for the explanation of electron tunneling between two metals I and III through a gap II of width d and energy height $V_0 - E$, and wave functions of the electron in the different phases.

modes and changes within the hydration shell are too slow to follow the rapid electron exchange within typically 10^{-15} s (Born–Oppenheimer principle):

$$T = \frac{16E(V_0 - E)}{V_0^2} \exp\left\{-2\sqrt{\frac{2m^* d_T^2 (V_0 - E)}{\hbar^2}}\right\} = \frac{16E(V_0 - E)}{V_0^2} \exp\left\{-\frac{2\kappa d_T}{\hbar}\right\} \tag{5.34}$$

$$i_+ = -F \int T D(Ox) D(\mathrm{Red}) \mathrm{d}E \tag{5.35}$$

Surface states may mediate the transfer of electrons from the redox system to the conduction band. The critical tunneling distance is then reduced so that deeper states of the redox system may contribute to charge transfer (Figure 5.42, case (3)). Similarly a high concentration of states within the forbidden band gap may contribute to charge transfer via multiple tunneling in sequence, i.e., a hopping mechanism (Figure 5.42, case (4)). Some oxides like Cu_2O have a large concentration of interband states within the band gap with a distribution of a certain energy width. This situation corresponds to the formation of a subband, which can mediate electron transfer where otherwise it should be excluded (Figure 5.42, case (5)). A subband has been already used to explain the conductivity of crystalline Cu_2O [134,135] and is required to explain redox currents on Cu covered by thin Cu_2O layers [136] as discussed in the following section.

Most localized states in the forbidden band gap are however close to the edges of the valence or conduction bands [137]. They are related to a high defect concentration of passive layers and suggest a smaller gap, called mobility gap, compared to the band gap (E_g) determined by light absorption measurements. They cause an extension of small photocurrents to photon energies $h\nu$ with $e_0E < e_0E_g$, which is known as Urbach tailing. Such effects are discussed in Section 5.7.3.

Finally electrons can be transferred to positive holes in the valence band and thus contribute to charge transfer (Figure 5.42, case (6)). Defect electrons can migrate from the inner parts to the surface of the oxide especially for positive potentials and an appropriate band bending. This is a preferred mechanism for p-type oxides, where positive holes are the majority carriers. For very thin oxides of one or two monolayers, electron transfer may occur by direct tunneling through the oxide into the conduction band of the metal substrate. Electron transfer can occur in both directions. The discussion above has treated the case of anodic current with electron transfer from the electrolyte to the metal substrate. Of course, the charge transfer may occur also in the opposite direction, via paths (1) to (6), leading to a cathodic current.

5.7.2 Photoelectrochemical Measurements

When light is shining on passivated metal surfaces a photocurrent is induced. Due to the ultra-thin layer, the effect is not very large and thin films of nanometer thickness must be examined with lock-in techniques. The light from a high-pressure xenon lamp then passes a mechanical chopper and a monochromator and hits finally the specimen within an electrolyte vessel. It passes usually to the electrochemical vessel via a quartz window to reduce absorption at short wavelengths. The alternating photocurrent is sorted out form the background by a lock-in amplifier. If the potential is fixed by the potentiostat, an alternating photocurrent is observed. It may be examined also directly. The cathodic scan of the polarization curve of Cu in Figure 5.5 shows current oscillations due to illumination with a chopped light beam.

The photoelectrochemical setup allows to measure photocurrent densities as a function of parameters like the wavelength of the light and the electrode potential, the latter inducing variations of the thickness and composition of the passive layer. Another possibility is the measurement of photopotentials, i.e., potential changes caused by the chopped light beam when the potential is not fixed as, e.g., for open circuit conditions. Well-defined photoelectrochemical studies add a redox system to the electrolyte adjusted to the band structure of the passive layer. This allows conclusions concerning the mechanism for the charge transfer reactions. Reference 137 described the technique in details.

5.7.3 General Electronic Properties of Passive Layers

Table 5.4 compiles some electronic property data for anodic oxide layers [138]. A compilation can also be found in Ref. [137]. A group of passivating oxides have large band gaps of $E_g > 3$ eV ranging from 7 to 9.5 eV for Al_2O_3, 4.6–8 eV for ZrO_2, 5.1 eV for HfO_2, and 4.0–4.6 eV for Ta_2O_5. They absorb light in the UV range and thus do not show color in the visible range. Other passivating oxides have poor or good semiconducting properties with $E_g =$ 3.2–3.8 eV for TiO_2, $E_g = 3.5$ eV for SnO_2, $E_g = 1.9$ eV for Fe_3O_4/Fe_2O_3, $E_g = 1.6$–2.2 eV for Cu_2O, and $E_g = 0.6$–1.7 eV for CuO. Some have properties of conductors with vanishing band gaps like IrO_2, RuO_2, and PbO_2. Most insulating and semiconducting oxides grow by anodic oxidation of the metal with a high field mechanism. The formation of porous oxides does not permit the exponential dependence of oxide growth on the electrical field strength and thus a less pronounced dependence on it is obtained. The anodic layers usually have high concentrations of dopants in the range of 10^{17}–10^{20} cm^{-3}.

Most anodic oxides are n-type semiconductors with a steep increase of cathodic currents with decreasing potential and a smaller increase of anodic currents with potential. Figure 5.44 shows a compilation of Tafel plots for passivated iron with different reacting redox systems [128]. The reactions of $Fe^{2+/3+}$ and $Fe(CN)_6^{4-/3-}$ follow an outer sphere mechanism with the exchange of an electron and some slight arrangement of the hydration shell of the ions or complexes. They are preferentially used to learn about the properties of the charge transfer reactions. A steeper cathodic and a less steep anodic branch of the redox current are observed, which is expected for an n-type semiconductor and a potential of the redox system in the vicinity of the conduction band. Others reactions follow a much more complicated mechanism like H_2O/O_2, which requires the breaking of bonds and the formation of new ones with intermediate products and adsorbates at the electrode surface. Figure 5.45 shows Tafel plots of the outer sphere $Fe^{2+/3+}$ system for several passivated metals [128]. Again most the cathodic branches are steep whereas most of the anodic ones are flat. The relatively high doping of passive layers on Fe and Ni causes much steeper anodic branches compared to those of SnO_2 layers. The increase of the slope with the concentration of dopants has been already discussed for SnO_2 in Chapter 1, Section 1.18.2.

TABLE 5.4

Density, ρ; Dielectric Permittivity, ε; Band Gap Energy, E_g; Flat Band Potential, U_{FB}; Equilibrium Potential of Oxide Electrode, U_{ox}; Donor Concentration, N; Difference of Electronegativity, ΔX; Transference Number of Cations, t_+; Formation Factor, dd/dU; and Initial Oxide Thickness, d_0 for Passive Films on Pure Metals

Metal	Oxide	ρ (g cm^{-3})	ε	E_g (eV)	U_{FB} (V)	U_{ox} (V)	N (cm^{-3})	ΔX	t_+	dd/dU (nmV^{-1})	d_0 (nm)
Al	Al_2O_3	3–3.4	7.5–15	7–9.5		−1.35		1.5	0.4–05	0.75–2	2–4.3
Au	$Au_2O_3/Au(OH)_3$	12.7	16–120		1.3	1.45		2.4		0.7–1	
Be	BeO	3	6.8			−1.76		1.5	0.75	0.76	0.65
Bi	Bi_2O_3	7–9	18–40	2.7		0.38		1.9			25
Cr	Cr_2O_3			3.5		−0.6		1.6			
Cu	Cu_2O		7–12	1.8	−0.23	0.42		1.9			
	CuO	6	18	0.6–1.7	0			2			
Fe	Fe_2O_3/Fe_3O_4	5.2		1.9	0.4	−0.08	10^{20}	1.9		0.6	
Hf	HfO_2	10	14–34	5.1	0	−1.57	10^{18}	1.3	0	1.8–2.4	4.1–5.1
Ir	IrO_2					0.93		2.2			
Mo	MoO_3	4.5				−0.04	1.3			3.5	3.1
Nb	Nb_2d_5	4.4–5	41–46	3.4–5.3	−0.08	−0.65	10^{18}–10^{19}	1.6	0.33	2.1–3.7	1.5–3.6
Ni	NiO/NiOOH			3.5		0.08		1.8			
Pb	PbO_2							1.8			
Pt	PtO			1.3	0.9	0.98		2.2		0.6	
Sb	Sb_2O_5	5.2	20	3	−0.22	0.15	10^{17}	1.9			
Si	SiO_2	2.1	3.8	9		−0.86		1.8	0	0.4–0.7	
Sn	SnO_2	6.7		3.5	0.5	−0.11	10^{19}–10^{20}	1.8		1	
Ta	Ta_2O_5	8–8.7	12–27	4–4.6	−0.8	−0.81		1.5	0.3–0.5	1.3–2.4	1–2
Te	TeO_2	5						2.1			
Ti	TiO_2	3.4–4.2	7–114	3.2–3.8	0–0.3	−0.86	10^{20}	1.6	0.4	1.3–3.3	1.3–5.4
V	V_2O_5	3.4–4.3				−1.02		1.6	0.28	6.5	5
W	WO_3	6.5	23–57	2.7–3.1	0.6		10^{17}–10^{18}	1.7	0.37	1.8	1.5
Zn	ZnO			3.2		−0.42		1.6			
Zr	ZrO_2	5.5–5.8	12–31	4.6–8	−0.82	−1.43		1.4	0	1.7–3	1.7–5.3

Source: From Schultze, J.W. and Lorhengel, M.M. *Electrochim. Acta*, 45, 24999, 2000.

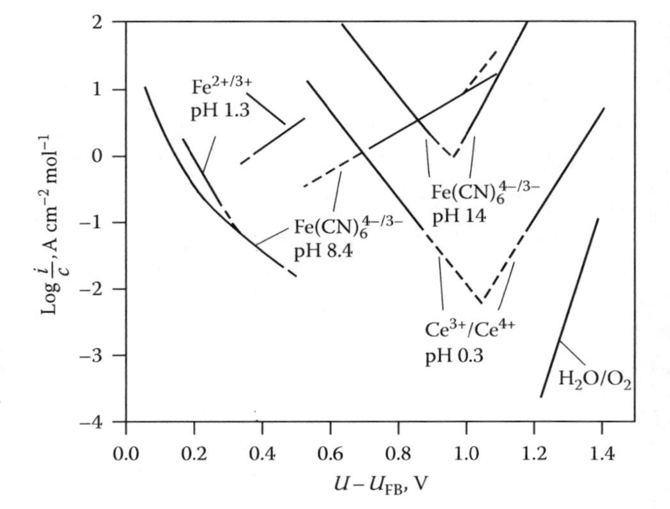

FIGURE 5.44
Tafel plot for redox currents on passivated Fe with different redox systems. i/c = current density normalized for the concentration of the redox system c. (From Schmickler, W. and Schultze, J.W., in *Modern Aspects of Electrochemistry*, Vol. 17, J.O'M. Bockris, B.E. Conway, R.E. White, eds., Plenum Press, New York, p. 357, 1986.)

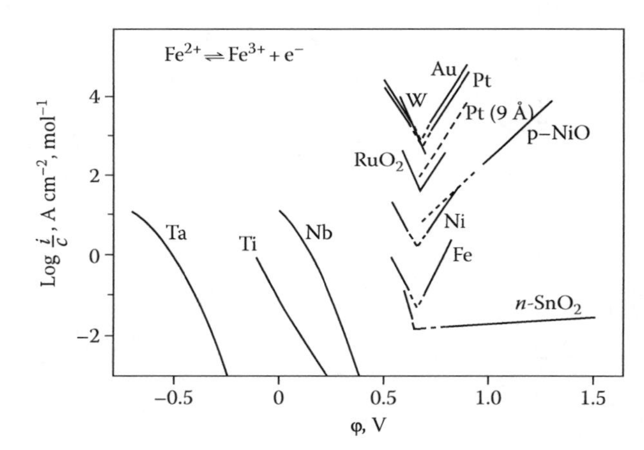

FIGURE 5.45
Redox currents for metals passivated with high (Ta, Ti, Nb) and low band gap (Sn, Fe, Ni) and metallic (Ru) oxides including metals with very thin oxides (Pt, Au) allowing direct tunneling through the oxide. (From Schmickler, W. and Schultze, J.W., in *Modern Aspects of Electrochemistry*, Vol. 17, J.O'M. Bockris, B.E. Conway, R.E. White, eds., Plenum Press, New York, p. 357, 1986.)

5.7.4 Photoeffects on Passivated Cu

Copper is an interesting metal that forms p-type passive layers. The cathodic photocurrents can be investigated in the potential range of $E = -0.25$ to $-0.50\,V$ without reduction of the Cu_2O layer. Care is thus taken that the metal surface is covered by a pure Cu_2O layer only. A duplex film forms at $E > 0\,V$ as discussed above. The thickness of Cu_2O sublayer can be varied in a wide range. Besides direct anodic growth, a duplex film formed at higher potentials can be reduced completely at potentials of the cathodic peak C2 (Figure 5.5). Furthermore one can grow additional Cu_2O by reduction of alkaline CuO_2^{2-} solutions on top of a Cu substrate covered already by a passivating Cu_2O film. Such cathodically thickened films have been used for photoelectrochemical studies and EXAFS investigations. They have similar properties as thinner passive layers formed by direct anodic oxidation [70]. For photocurrent measurements on Cu/Cu_2O, $[Co(NH_3)_5Cl]^{2+}$ has been added as a redox system to the electrolyte in order to establish well-defined electrochemical conditions for electron transfer. Similar photocurrents have been found with other redox systems like AsO_4^{2-}/AsO_3^{3-}, H_2O_2, Cu^{2+} ions. The solid CuO outer part of the duplex film can also serve as a redox system.

For Cu covered with a Cu_2O layer, electron transfer occurs at the oxide surface via surface states that accept the photoelectrons and transfer them at the same energy level to the empty states of the redox system (horizontal electron transfer). The photocurrent of passive Cu is diffusion limited for small redox concentrations of a few 0.0001 M and increases linearly with the concentration for small complex concentrations up to a constant level. It also increases with the thickness of the Cu_2O film due to the generation of more charge carriers for thicker light absorbing oxide layers.

Figure 5.46a shows an i_{Ph}^2 vs. E plot of photocurrents measured on a Cu electrode covered with a Cu_2O passive layer in a 0.1 M KOH + 0.002 M $[Co(NH_3)_5Cl]Cl_2$ electrolyte [69]. This plot allows to determine the flat band potential E_{FB} using to Equation 1.189 (Chapter 1, Section 1.18.3). $E_{FB} = -0.28$ V is found by extrapolation to $i_{Ph}^2 = 0$. Measurements for thicker Cu_2O films in the range of 4–40 nm including those obtained by reduction of alkaline CuO_2^{2-} solutions show only slightly smaller values of $E_{FB} = -0.23$ V [70]. Duplex films show two straight lines when they are examined with light of $\lambda = 300$ nm (Figure 5.46b). The extrapolation yields two values $E_{FB,I} = -0.23$ nm and $E_{FB,II} = 0$ V [69]. When the same passive layer is studied with $\lambda = 450$ nm, only one line is found with $E_{FB,II} = 0$ V. This result shows the presence of both oxides in the inner Cu_2O and the outer CuO, $Cu(OH)_2$ layers. The outer Cu(II) layer has a smaller band gap and generate charge carriers when illuminated with $\lambda = 450$ nm in which condition the inner Cu_2O does not contribute to the photocurrent. However, if illuminated with $\lambda = 300$ nm, both partial layers contribute depending on their flat band potentials. For $E > E_{FB,I} = -0.23$ V, the inner Cu_2O layer cannot contribute any more whereas the outer Cu(II) part contribute to the photocurrent until $E = 0$ V.

The band gap has been determined from the wave length dependence of the photocurrent as described by Equations 1.193 and 1.194 (Chapter 1, Section 1.18.3). The related $(\eta_{eff}\, h\nu)^2$ vs. $h\nu$ and $(\eta_{eff}\, h\nu)^{1/2}$ vs. $h\nu$ plots have been presented in Figure 1.56, leading to

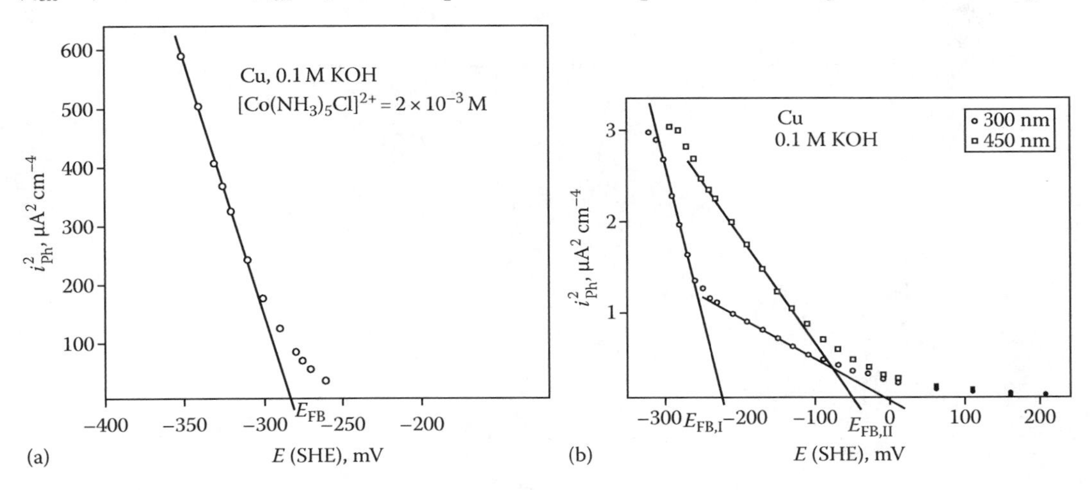

FIGURE 5.46
i_{Ph}^2 vs. E plot for passivated Cu electrodes in 0.1 M KOH with determination of E_{FB} by extrapolation. (a) Cu_2O passive film, $\lambda = 300$ nm, (b) Cu_2O/CuO, $Cu(OH)_2$ duplex film, $\lambda = 300$ and 450 nm. (From Collisi, U. and Strehblow, H.-H., *J. Electroanal. Chem.*, 210, 213, 1986.)

a band gap for direct and an indirect transition of $E_g = 3.0\,eV$ and $E_g = 2.3\,eV$, respectively. The values for thicker films formed by reduction from alkaline CuO_2^{2-} solutions are smaller by 0.5 and 0.2 eV, respectively [70].

With known values of E_{FB} and E_g, one only misses the energy of the bands relative to the hydrogen or vacuum scale to complete the energy diagram. The assumption of an energy distance of 0.25 eV between the Fermi level and the lower conduction or the upper valence band edges for n-type or p-type semiconductors, respectively is a rough estimate. A reliable experimental result is obtained by the application of UPS to these systems [17,136]. Figure 5.47 shows the UP spectra of Cu electrodes taken with the HeI line (21.2 eV). A negative bias of −5 V of the specimen relative to the energy analyzer of the spectrometer avoids cutoff of the photoelectrons with small kinetic energies [44]. All binding energies of solid specimens are referred to E_F. The spectrum of a sputter cleaned Cu surface starts at the Fermi edge with a binding energy $E_B = 0$. It has a width ΔE between the Fermi edge and the cutoff edge, which refers to photoelectrons with vanishing kinetic energy. In addition, Figure 5.47a shows the Cu 3d signal and Figure 5.47b and c the O2p signal of the oxide layers. The width of the spectrum ΔE of Figure 5.47 allows calculating the work function of Cu metal using Equation 5.36, which describes the energy balance of the photoionization process. ΔE is the kinetic energy of photoelectrons ejected by a photon of energy $h\nu$ from the highest energy level, i.e., the Fermi level (E_F). Φ_W is the work function, i.e., the energy difference between E_F and the vacuum

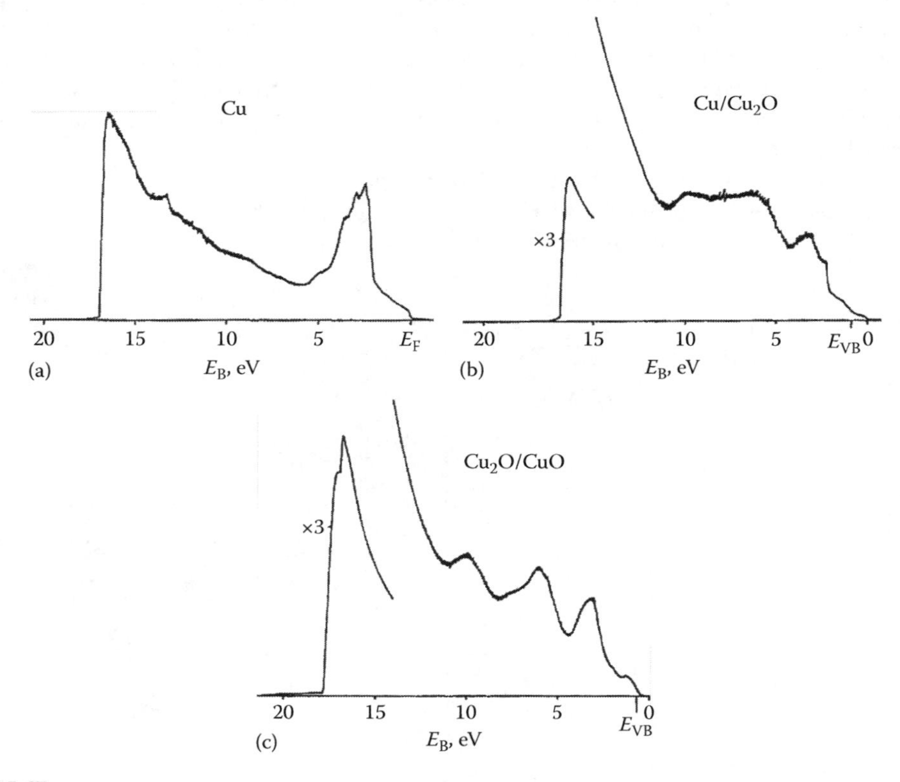

FIGURE 5.47
HeI-UP spectra of (a) sputter cleaned Cu metal, (b) Cu passivated with a Cu_2O single layer film, and (c) Cu passivated with a Cu_2O/CuO duplex film. (From Strehblow, H.-H. et al., *International Symposium on Control of Copper and Copper Alloys Oxidation*, Rouen 1992, Edition de la Revue de Métallurgie, Paris, France, p. 33, 1993.)

TABLE 5.5

Work Function Φ_W and Threshold Energies Φ_{Th} of Cu Specimens, Ar-Ion Sputtered and/or Prepared in 1 M NaOH at Appropriate Potentials to Produce Single Thin Film or Thick Duplex Layers of Anodic CuO and Cu_2O Oxides and Partially Reduced Surfaces

Specimen	Work Function Φ_W (eV) or Threshold Energy Φ_{Th} (eV)	Literature Data [139]
Cu sputtered	4.5	4.48
Cu reduced	4.3	—
Cu_2O layer	5.05	5.15
CuO layer	5.1	5.34
Cu/Cu_2O thin film	4.35–5.15	—
Cu_2O/CuO duplex film	4.3–4.55	—

level. With ΔE taken from the spectrum and the excitation energy $h\nu$, one obtains the work function of $\Phi_W = 4.5$ eV for the sputter-cleaned Cu. For the passivated surfaces, the spectra show the Fermi edge of the metal substrate at $E_B = 0$ eV and a further increase at $E = E_{VB}$ due to the semiconducting passive layer. The latter yields ΔE for the calculation of the threshold energy $e_0\Phi_{Th}$ using Equation 5.37. Slightly different values of E_{VB} are obtained for a Cu electrode covered with single or duplex layers, i.e., without or with the outer CuO film. Some specimen preparations included a partial reduction of the oxide, i.e., Cu_2O to Cu or CuO to Cu_2O in the case of a duplex layer, in order to obtain a clear step in the spectrum for both E_F and E_{VB} for both oxides:

$$\Delta E = h\nu - e_0\Phi_W \tag{5.36}$$

$$\Delta E = h\nu - e_0\Phi_{Th} \tag{5.37}$$

Table 5.5 presents the work functions and threshold energies for the electrochemically prepared Cu specimens. For comparison, data from the literature [139] are included that agree reasonably well with the results of passivated specimens. Some difference is found for a sputter-cleaned Cu surface dipped in water. There is still a residual O1s signal in the UPS, which has been attributed to a water layer at the surface, which has been also found by XPS on emersed Cu electrodes. It gets smaller with exposure time in the vacuum but does not disappear completely. It can also result from an OH adsorption layer, which has been found by STM studies (see Section 5.8.2.1). The calculated work function of the reduced Cu surface is slightly smaller compared to that of sputtered Cu, $\Phi_W = 4.3$ eV instead of $\Phi_W = 4.5$ eV. This should be seen as a consequence of a water dipole layer at the surface or an OH adsorption layer.

Combining the results from the electrochemical and photoelectrochemical investigations and the XPS and UPS studies yields the band structure model shown in Figure 5.48 for passivated copper [136]. At $E = E_{FB} = -0.25$ V, Cu_2O has a flat band structure (Figure 5.48a), which builds a space charge layer for $E > E_{FB}$ (Figure 5.48b). At $E = 0.55$ V, the valence band edge touches the Fermi level. Further potential increase leads to the growth of CuO with the corresponding formation of a high concentration of defect electrons at the surface, which takes over some of the potential increase (Figure 5.48c and d). At $E = 0.95$ V, the valence band touches again the Fermi level. Any further potential increase ($\Delta\varphi_H$) is located in the Helmholtz layer at the oxide–electrolyte interface. This increase of $\Delta\varphi_H$ is a

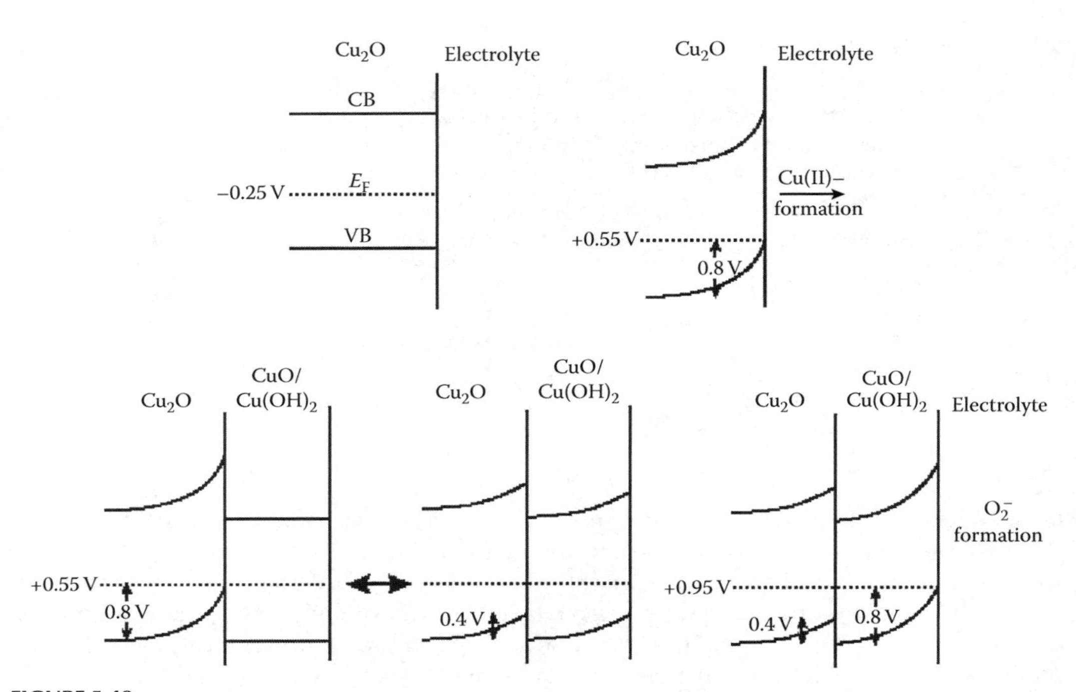

FIGURE 5.48
Band structure model of passivated Cu and its dependence on electrode potential. (From Strehblow, H.-H. et al., *International Symposium on Control of Copper and Copper Alloys Oxidation*, Rouen 1992, Edition de la Revue de Métallurgie, Paris, France, p. 33, 1993.)

driving force for oxygen evolution and an increased Cu^{2+} ion transfer at the oxide–electrolyte interface. Figure 5.49 depicts the potential diagram within the duplex passive layer and explains the two flat band potentials, which are obtained by photoelectrochemical measurements [70]. The band gap for CuO is slightly smaller causing a flat band situation for the valence and the conduction band. For $E < 0.23$ V, both oxides contain a space charge layer and contribute to i_{PH} (Figure 5.49a). At $E = -0.23$ V, the conduction band of Cu_2O is flat and only CuO contributes to i_{PH} (Figure 5.49b). At $E = 0$ V, the conduction band of CuO is flat too so that this layer cannot contribute any more to the photocurrent that yields $i_{PH} = 0$ A (Figure 5.49c).

Figure 5.50 summarizes the electronic properties with charge transfer reaction in the dark and under illumination for Cu_2O layers on Cu including the Co(III) complex as a redox system [136]. Cu_2O is a special semiconductor as its Fermi edge is in the center of the band gap due to many localized states forming a subband. Electronic equilibrium is assumed from the metal through the oxide with a horizontal position of E_F to the oxide surface. Cathodic charge transfer can occur throughout the oxide layer via the subband to the empty states of the redox system within the electrolyte. Similarly the photogeneration of pairs of charge carriers leads to their separation within the oxide layer. Electrons migrate to the surface and are transferred to the empty states of the redox system. In this case, the surface states can help to locate the electrons at the surface till they go to the empty states of the Co(III) complex. These surface states have been characterized by phototransient measurements. A sudden light pulse creates charge carriers that fill the empty states with time, causing a characteristic overshoot of the photocurrent with a half-life period of relaxation of 0.05 s. The detailed evaluation of the kinetics yields a concentration

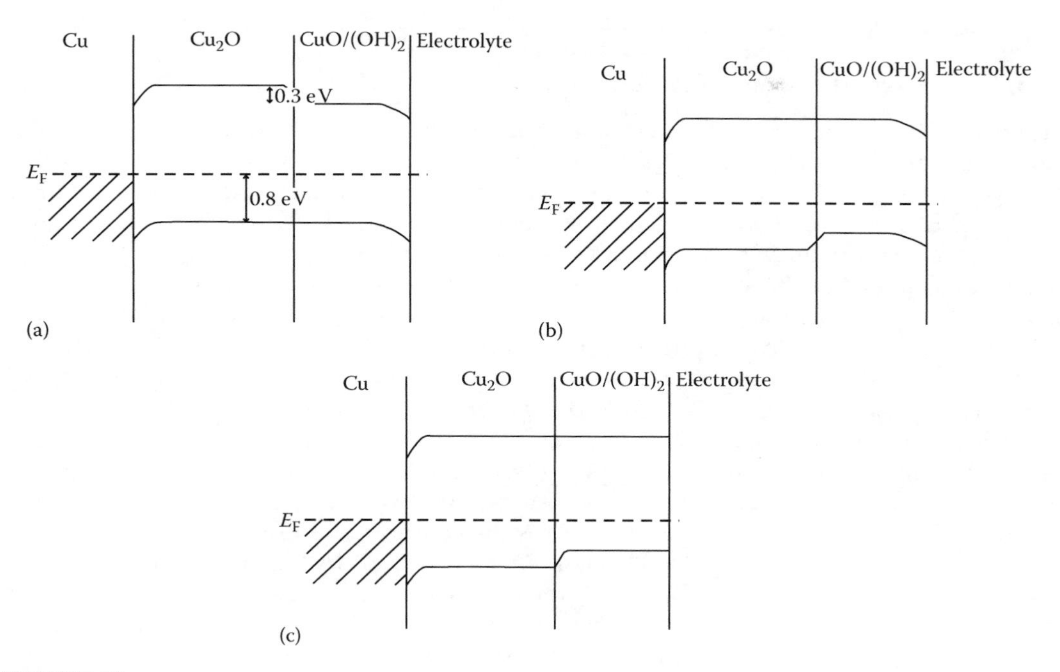

FIGURE 5.49
Potential diagram for the duplex passive layer on Cu for (a) $E < E_{FB,I}$, (b) $E = E_{FB,I}$ for Cu_2O, and (c) $E = E_{FB,II}$ for Cu_2O/CuO. (From Collisi, U. and Strehblow, H.-H., *J. Electroanal. Chem.*, 284, 385, 1990.)

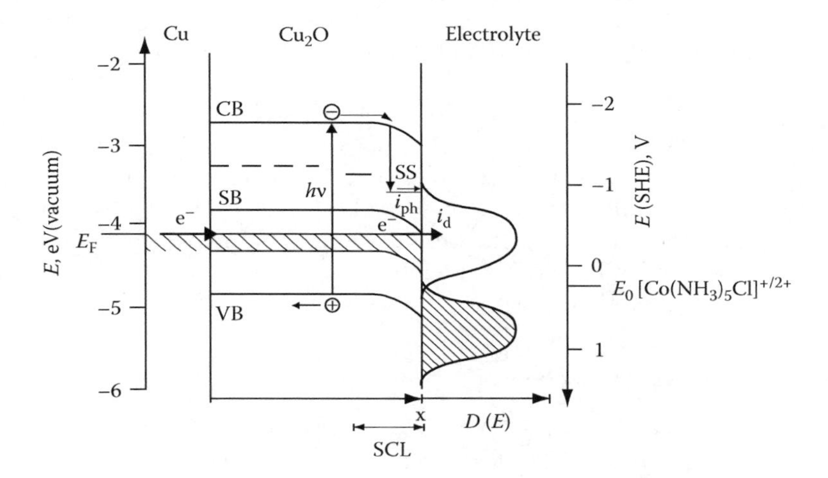

FIGURE 5.50
Band scheme for Cu/Cu_2O with the hydrogen and vacuum scale, the $[Co(NH_3)_6Cl]^{2+/3+}$ redox system, the subband and the path for transfer of photoelectrons via surface states and of electrons in the dark via the subband. (From Strehblow, H.-H. et al., *International Symposium on Control of Copper and Copper Alloys Oxidation*, Rouen 1992, Edition de la Revue de Métallurgie, Paris, France, p. 33, 1993.)

of $N_{SS} = 10^{18}$ cm^{-3} independent on the oxide thickness. In conclusion, one has to assume a constant concentration of these interband states throughout the whole anodic film, which agrees with the existence of the subband postulated for the electronic properties of the layer, and also required for the explanation of the electron transfer for STM during imaging of the surface.

5.8 Structure of Passive Films

There has been a long debate about the structure of passive layers with some authors claiming a crystalline structure and others an amorphous structure for a well-protecting film. Most conclusions were drawn indirectly from ex situ studies. Electron diffraction methods (TEM, LEED) were applied, which however work in vacuum with possible changes to the layer structure. In addition, the impact of a high-energy electron beam in TEM may also change considerably the structure. Therefore, there has been a strong need to study the structure of passive films with analytical methods working in the natural environment of the passivated metals, i.e., under applied potential and in contact with the electrolyte. In situ methods, based on XRD, x-ray absorption, or on scanning probes, have been developed and applied during the last 15–20 years to solve this issue.

Difficulties arise when applying x-ray analytical methods to characterize oxide films a few nanometer thick only. XRD, which gives information on the atomic structure and nanostructure, can be used at grazing incidence of the beam in order to get mainly an interaction with the thin oxide layer and suppress the contributions of the substrates (grazing incidence x-ray diffraction GIXD). This holds also for XAS, which gives information on the near range order of surface films even if they are amorphous. Absorption by the electrolyte being very high for a few keV energy of the incident photons, GIXD and XAS require high-intensity beams of cyclotron storage rings. The use of magnetic insertion devices like undulators or wigglers increases the intensity of the beam by two orders of magnitude compared to magnetic deflection. Diffractions experiments can also use higher energy radiation for which absorption is decreased. However, absorption spectroscopy relies on the absorption edge of the metal under study and thus does not allow a free choice of the energy. Any details about this method and the required electrochemical cells are given in Refs. [140,141].

Other applicable in situ methods are scanning probe microscopies, i.e., scanning tunneling microscopy (STM) and scanning force microscopy (SFM) [142,143]. Results obtained with these methods for the most part are presented in the following. A detailed account for the passive film growth and structure is presented for Cu. Other systems will be presented in a more condensed form. Not only the growth of passivating oxides but also their reduction is of interest. Some details about the related changes will be given for thin hydroxide films on Co as an example.

5.8.1 STM Experimental Details

Fundamental SPM studies (mostly STM but also AFM) on metals are performed with well-oriented single crystals surfaces in order to relate the observed effects to the structure of the substrate and not have the influence of the microstructure of the metal or alloy. The structural changes related to passivation and their mechanisms may vary with the substrate structure as illustrated below for oxide formation on Cu(111) and Cu(001). The growth of adsorption layers and oxides are best followed by in situ investigations on seminoble metals that can be easily reduced, so that one may start with an oxide-free surface, which is not always the case with more reactive metals. Hydrogen evolution sets limits to surface reduction at negative potentials in order to avoid water decomposition in the small electrolyte volume of in situ SPM setups and to prevent oxygen evolution at the counterelectrode.

After an appropriate cut of the single crystal, the surfaces are mechanically polished down to a grain size of less than $1\,\mu m$ and subsequently treated according to special procedures depending on the metal under study. For example, Ni and Cu are electropolished in sulfuric [144] and orthophosphoric [145] acid, respectively, and Ag and Co are chemically polished in a $HClO_4/HCl/CrO_3$ [146] or lactic acid/HCl/HNO_3 [147] solutions, respectively, in order to get atomically smooth surfaces. Annealing in ultrahigh purity (6N) hydrogen overnight (Cu and Ag at 1000 K, Ni at 1273 K, Co at 653 K) increases the terrace width and heals out imperfections. Cathodic reduction pretreatments are required to reduce the air-formed oxide grown during installation of the electrochemical cell.

For AFM, commercial tips and cantilevers are used whereas for STM tips are prepared in various ways. W tips are easily obtained by electrochemical etching in NaOH. They have to be isolated with wax or varnish for electrochemical in situ studies to reduce electrochemical currents at the tip to less than 0.05 nA. Small electrochemical cells made of PCTFE (polycholorotri fluoroethylene) are pressed with a Viton O-ring onto the crystal surface [142]. They contain a Pt counterelectrode and a small reference electrode. Often a Pt pseudoreference electrode is used, which yields a reproducible potential that can be calibrated using the characteristic features of the polarization curve for a given system. A simple Pt wire may be cleaned much better and does not introduce impurities to the small electrolyte volume. With oxygen present in the electrolyte, the Pt-wire potential is in the domain of the O_2/OH^- reaction.

5.8.2 Growth and Structure of Passive Films on Cu

5.8.2.1 OH Electrosorption on Cu(111) and Cu(001)

The polarization curve of Cu in alkaline solution has been already shown in Figure 5.5. In situ STM evidences marked structural modifications related to the adsorption of OH^- ions on Cu(111) and Cu(001) surfaces at potentials much more negative than 3D Cu(I) oxide formation at peak A1 [145,148–150]. Small anodic (a) and cathodic (c) peaks are observed on the polarization curves (Figure 5.51). For Cu(111), the peaks show hysteresis and are at more positive with (a) at $E = -0.45$ V and (c) at $E = -0.7$ V. On Cu(001), (a) and (c) are at the same potential $E = -0.8$ V. The charge density is 55 and 30 μC cm^{-2} for the Cu(111) and the Cu(001) orientations, respectively. These charges have been related to the electrosorption with complete (dis)charge of OH^- according to the following equation:

$$Cu + OH^- \leftrightarrow Cu{-}OH_{ads} + e^- \quad (5.38)$$

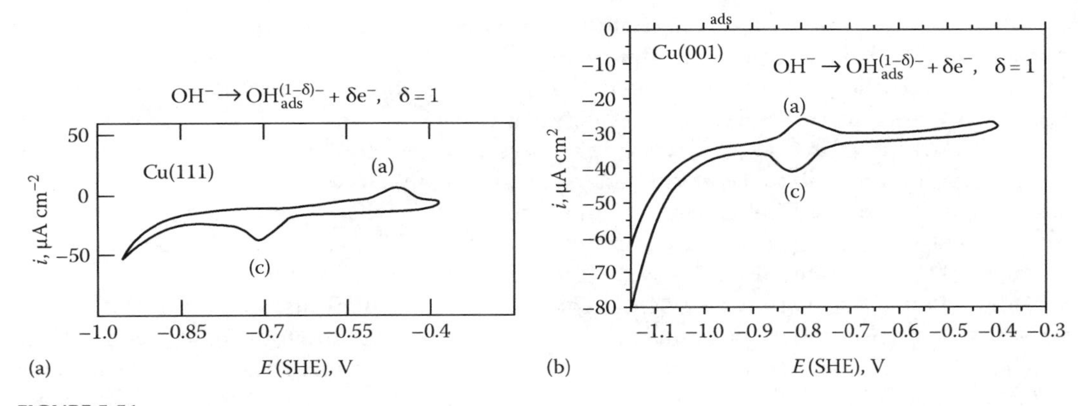

FIGURE 5.51
Polarization curve of (a) Cu(111) and (b) Cu(001) in 0.1 M NaOH showing anodic OH^- adsorption and cathodic OH desorption peaks. (From Maurice, V. et al., *Surf. Sci.*, 458, 185, 2000; Kunze, J. et al., *J. Electroanal. Chem.*, 554, 113, 2003.)

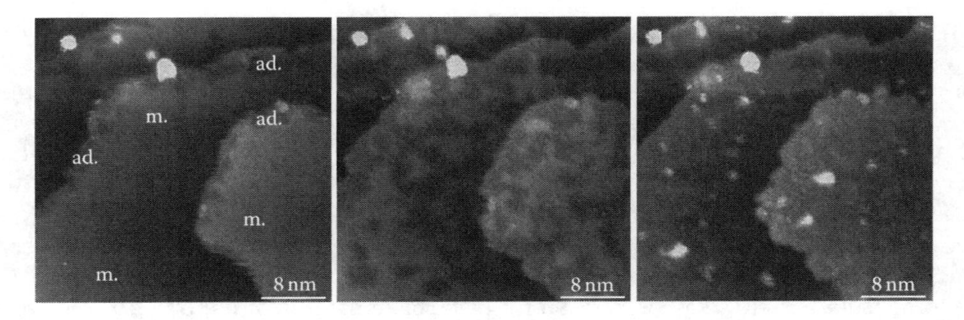

FIGURE 5.52
Sequence of in situ STM images showing the growth of the Cu–OH adlayer on Cu(111) in 0.1 M NaOH at −0.6 V. (From Maurice, V. et al., *Surf. Sci.*, 458, 185, 2000.)

as supported by the STM data. Besides EXAFS measurements show the development of a first coordination shell with a radius of 0.15 nm at these potentials [151]. The related peak in the real space distance plot increases in intensity with anodic oxidation and decreases reversely with cathodic reduction of the surface, supporting the assignment of peaks (a)/(c) to OH adsorption/desorption.

The Cu(111) surface shows characteristic changes when OH^- electrosorption occurs [145]. If one approaches peak (a) stepwise, the adsorption process may be studied in details with the relatively slow STM imaging of ca. one image per minute. At $E = -0.6$ V, dark islands (marked ad) appear preferentially at the step edges and then grow in size and merge together (Figure 5.52). Some islands also form within the terraces. At the end of the adsorption process, the darker islands cover the terraces completely with only some bright spots a few nanometer-wide protruding. The step edges are frizzy at the beginning, which is an indication of surface mobility of the Cu atoms at these sites. They become clearer during the adsorption process, indicating a loss of mobility associated with the formation of the Cu–OH adlayer. One can also see that the step edges are displaced during OH adsorption. This can be explained if the density of the Cu atoms in the topmost atomic plane decreases and the excess atoms travel across the terraces to the step edges where they accumulate. Thus OH adsorption induces the reconstruction of the Cu(111) topmost atomic plane to a less densely packed structure in the Cu–OH adlayer. The bright spots protruding from the terraces are assigned to clusters of excess Cu atoms. They correspond to Cu atoms ejected from the topmost plane by reconstruction and whose mobility on the Cu–OH adlayer is reduced so that they aggregate before reaching the step edges. The described reaction is reversible. If the potential is stepped to $E = -0.70$ V, the surface returns to its original appearance with reversible changes. The rate of these surface reactions is increasing with the related driving force, i.e., with the applied overpotential for OH adsorption/desorption.

At higher magnification, two orientations of a hexagonal structure of the adsorbed layer rotated by ±10° and ±20° relative to the substrate are revealed [145,149]. The coincidence cell with the substrate contains 4 and 9 unit cells of the Cu–OH adlayer and 21 and 49 unit cells of the Cu(111) substrate, respectively. Figure 5.53 shows the STM image of the 20°-rotated Cu–OH adlayer structure with a schematic drawing to explain its relation to the Cu(111) substrate. Each unit cell of the adlayer contains one OH group and four Cu atoms. The dark spots in the hexagonal structure correspond to the hexagonal OH lattice with a parameter of 0.6 nm. The brighter spots correspond to the Cu atoms forming a hexagonal sublattice with a parameter of 0.3 nm, larger than in the unreconstructed Cu(111) plane (0.256 nm) and thus less dense. The OH sits in a threefold hollow site of the reconstructed Cu_R sublattice.

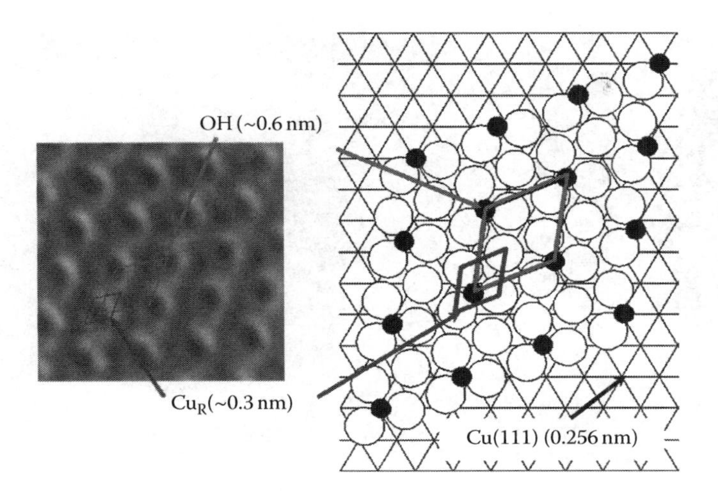

FIGURE 5.53
In situ STM image and model of the ordered structure of the 2D Cu–OH adlayer on Cu(111) in 0.1 M NaOH in the potential region below oxidation. The large and small cells mark the lattice of OH_{ads} species and of reconstructed Cu_R plane, respectively. (From Kunze, J. et al., *Electrochim. Acta*, 48, 1, 2003.)

The OH:Cu ratio is 1:4 with respect to the reconstructed Cu_R plane and 9:49 with respect to the unreconstructed Cu plane. Thus about every five Cu atom has to be excluded from the topmost plane by the adsorption-induced reconstruction and is relocated at step edges or clustered on the terraces. The coverage of 9:49 (0.18 ML) is in agreement with the charge density of 55 μC cm^{-2} of the anodic and cathodic peaks (a) and (c) in Figure 5.51a. Indeed, this charge corresponds to 3.4×10^{14} OH cm^{-2} according to Equation 5.38, which is 19% of 1.75×10^{15} Cu cm^{-2} of the unreconstructed Cu(111) planes.

The atomic structure in the Cu–OH adlayer is similar to that found in the (111)-oriented cuprite compound. In this orientation (depicted in Figure 5.56c), the fcc cuprite structure consists of the ABC stacking of trilayers made of one Cu^+ cation plane embedded between two O^{2-} anion planes. The O^{2-}(111) sublattice forms a (2 × 2) superstructure relative to the Cu^+(111) sublattice with parameters of 0.604 and 0.302 nm, respectively. The number of atoms in the Cu^+ middle plane is four times that of O^{2-} in the upper and lower planes, thus forming an electrically neutral Cu_2O(111) trilayer. The adlayer structure can thus be seen as a precursor of the three-dimensional layer of cuprite, which forms at potential $E = -0.2$ V in 0.1 M NaOH. The dark appearance of the adsorbed islands is not a topographic but an electronic effect (i.e., increase of tunneling barrier height) well-known for oxygen adsorption on Cu, Ag, Ni surfaces from gas-phase studies [152]. Oxygen atoms adsorbed from the gas phase also sit in the threefold hollow site of the Cu(111) surface and the formation of a 2D surface oxide leads to local ordering with a structure similar to Cu_2O(111) [153–157].

The detailed mechanism of OH adsorption is different on the more open atomic structure of the Cu(001) surface [150]. This is already suggested from the lower potential of the (a) and (ċ) peaks in the polarization curve (Figure 5.51b). STM images of the Cu(001) surface show terraces with monoatomic step edges oriented along the close packed <110> directions (Figure 5.54). Adsorption-induced reconstruction occurs already at $E = -0.81$ V. The step edges are rotated by 45°, and thus reoriented along the <100> directions. Higher magnification shows an atomic structure consisting of trimers attributed to Cu-OH-Cu units. These trimers terminate the terraces at the step edges and self-arrange in zigzag lines on the terraces along the <110> directions (Figure 5.54b and c). The units are rotated

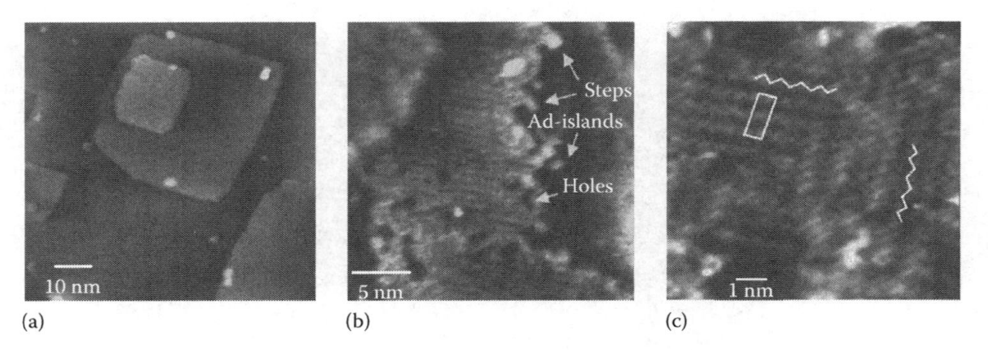

FIGURE 5.54
In situ STM images of the Cu(001) surface (a) prior and (b,c) after formation of the Cu–OH adlayer. (b) Local reorientation of the step edges is pointed. (c) The zigzag lines formed by the alternating Cu–OH–Cu trimers are marked as well as the $c(2 \times 6)$ superstructure. (From Kunze, J. et al., *J. Electroanal. Chem.*, 554, 113, 2003.)

at 45° relative to the line direction (i.e., parallel to <100>) and alternate at 90°. One trimer is 0.35 nm long and the repeating period is 0.53 nm along the zigzag line. Ordering of the zigzag lines in a $c(2 \times 6)$ superstructure with two variants at 90° is observed in domains limited in width. In the $c(2 \times 6)$ domains, the zigzag lines are separated by 0.76 nm.

The schematic diagram of Figure 5.55 has been proposed to explain the mechanism of surface reconstruction [150]. Two Cu atoms are ejected from the topmost Cu(001) plane and located above it with the OH presumably adsorbed in bridge sites, thus forming a Cu–OH–Cu trimer. The chemisorption of OH is thought to be the driving force for this reaction. By a collective reaction on the surface, multiple trimers are produced on the terraces and the zigzag lines and $c(2 \times 6)$ domains are formed by self-ordering. At the step edges, the trimers terminate the terraces and their orientation along the <100> directions dictate the reorientation of the step edges along these directions. Assuming a complete coverage of the surface

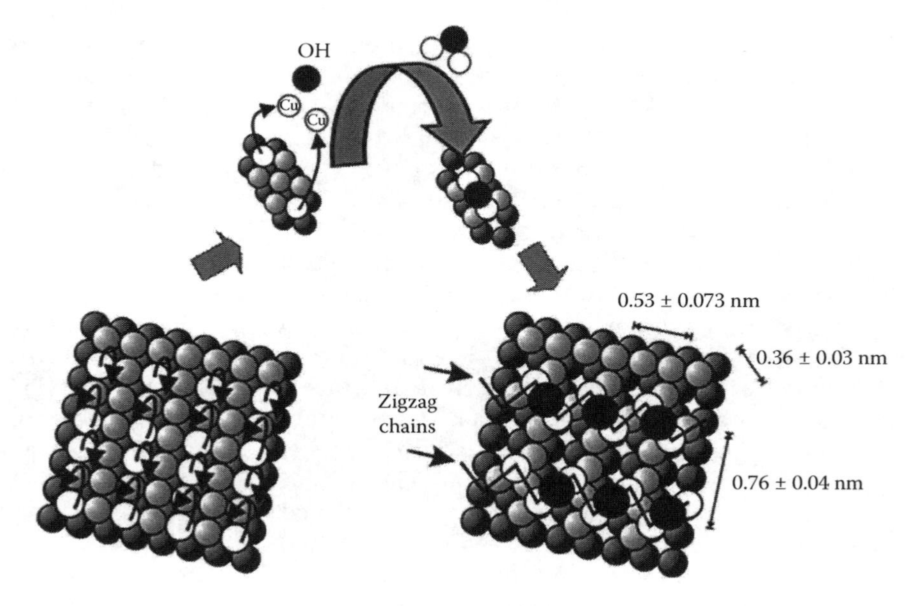

FIGURE 5.55
Scheme of OH adsorption-induced reconstruction of Cu(001) with formation of the Cu–OH–Cu trimers and self-arrangement in zigzag lines and $c(2 \times 6)$ superstructure. (From Kunze, J. et al., *J. Electroanal. Chem.*, 554, 113, 2003.)

with the $c(2 \times 6)$ superstructure, one obtains a coverage of 2.56×10^{14} OH cm^{-2} (0.16 ML), in reasonably good agreement with the value of 1.8×10^{14} O·cm^{-2} that can be deduced from the charge density (30 μC cm^{-2}) of the (a) and (c) peaks in the polarization curve assuming reaction according to Equation 5.38. The difference suggests a possibly less dense formation of trimers in the nonordered parts of the adlayer and/or a partial charge transfer.

As on Cu(111), the local structure of the adlayer on Cu(001) is similar to that encountered in the (001)-oriented cuprite. In $Cu_2O(001)$ (depicted in Figure 5.57), the O^{2-} ions also sit in bridge positions between two Cu^+ ions of the plane below and form Cu–O–Cu trimers that alternate at 90° to form zigzag lines. The structure of the Cu–OH adlayer formed on Cu(001) can thus also be seen as a 2D surface precursor of the structure of the 3D $Cu_2O(001)$ layer formed when the driving force for oxidation is increased as described below. On Cu(001), the OH adsorption-induced reconstruction is a faster process than on Cu(111) and cannot be followed with the same details by the relatively slow STM imaging. The Cu atoms ejected from the topmost plane build up the adlayer by a collective on-site reaction. The more open atomic structure of Cu(001) is thought to promote this process that can occur at lower potential. On Cu(111), the adlayer is also built up collectively on-site but by the Cu atoms remaining in the reconstructed plane, and thus more strongly bonded to the atomic plane below. Accordingly, a higher degree of structural ordering is obtained in the 2D Cu–OH adlayer formed on the Cu(111) surface than on Cu(001).

5.8.2.2 Cu_2O Formation on Cu(111) and Cu(001)

On both Cu surfaces, the growth of Cu_2O at peak A1 starts with the formation of an amorphous layer consisting of small grains ca, 3 nm wide and 0.25–0.3 nm high [148–150,158,159]. No atomic structure could ever be resolved on these grains. They nucleate preferentially at step edges but also on the terraces and cover them completely with time. Increasing the potential accelerates their nucleation. For Cu(111) in 0.1 M NaOH (pH 13) [149,159] and borate buffer solution (pH 9.3) [158], one observes grain growth at $E = -0.2$ V and $E = 0$ V, respectively, which agrees reasonably well with the −0.059 V/pH shift expected from thermodynamics. At pH 13 and $E = -0.25$ V, the amorphous oxide grains coexist with islands of the 2D Cu–OH adlayer formed at lower potential and described above (Figure 5.56a). At $E = -0.2$ V, the oxide layer rapidly grows up to seven monolayers in thickness as measured by subsequent cathodic reduction. STM shows that a crystalline layer of Cu_2O with a faceted surface is then developed (Figure 5.56b). These observations suggest that the change from essentially 2D amorphous grains or islands to a 3D crystalline layer occurs during the 3D growth of the oxide and possibly involves Ostwald ripening. This process becomes faster with increasing potential. For pH 9.3, one observes at $E = 0$ V some dissolution with a related step flow, which stops when oxide grains form.

The terraces of the 3D crystalline Cu(I) oxide layer grown on Cu(111) are 2.7 nm wide and 0.2–0.25 nm high, which agrees with the thickness of 0.246 nm of a (111)-oriented cuprite trilayer. At atomic resolution, one observes an atomic lattice on the facets. Its hexagonal symmetry and parameter of 0.3 nm (Figure 5.56b) correspond to the unreconstructed structure of the Cu sublattice of $Cu_2O(111)$ described above (Figure 5.56c). Model DFT calculations have shown that the Cu^+ plane of the oxide lattice may be terminated by an OH adlayer at the interface with the electrolyte [160]. The OH adlayer stabilizes an otherwise unstable termination if the Cu plane structure remains unreconstructed. This interpretation is supported by ARXPS data, which show a hydroxide layer on top of inner oxide. Similar results are obtained for passive layers on other metals, i.e., Ni as described in the following.

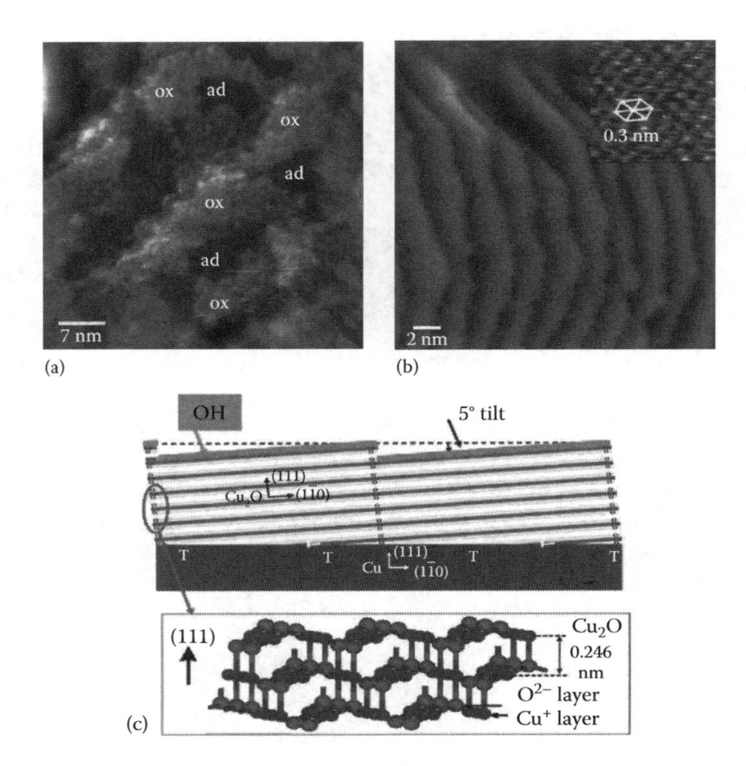

FIGURE 5.56
In situ STM images of the Cu(I) oxide grown on Cu(111) at (a) −0.25 V and (b) −0.20 V in 0.1 M NaOH. (a) At low overpotential, noncrystalline 2D oxide islands (ox.) separated by the Cu–OH adlayer (ad) cover partially the substrate. (b) At higher overpotential, a 3D crystalline oxide layer fully covers the substrate (atomic lattice in inset). (c) Model (section view) of the 7 ML thick Cu(I) oxide grown in tilted epitaxy on Cu(111). (From Kunze, J. et al., *J. Phys. Chem. B*, 105, 4263, 2001.)

The respective orientation of the oxide and substrate lattices indicates a parallel (Cu(111)[1$\bar{1}$0]//Cu_2O(111)[1$\bar{1}$0]) or antiparallel (Cu(111)[1$\bar{1}$0]//Cu_2O(111)[1$\bar{1}$0]) epitaxial relationship. The coincidence length in the <110> directions is 1.787 nm and corresponds to 6 and 7 periods within the Cu_2O(111) and Cu(111) planes, respectively. However, the width and height of the facets at the oxide layer surface indicate a tilt of 5° of the oxide lattice relative to the metal lattice (Figure 5.56c). The 5° tilt between the two lattices is expected to lead to a better fit with an allowance of 1% of contraction of the oxide lattice or expansion of the metal lattice. Faceting is therefore a possibility for stress relief for the epitaxial growth of the oxide lattice on top of the metal lattice with a large lattice mismatch. Stress relief may also occur at grain boundaries when small grains predominate in the layer structure.

3D Cu(I) oxide layers grown by anodic oxidation on Cu(001) surfaces in 0.1 M NaOH also present a facetted topography shown in Figure 5.57a [150]. The facets are 5 nm wide and 0.25 nm high, which agrees with the thickness of one (001)-oriented cuprite monolayer. Atomic resolution reveals a square lattice with a parameter of 0.3 nm corresponding to the Cu sublattice of the (001)-oriented Cu_2O structure. The STM images of the lattice agree with AFM data [161]. As on Cu(111), it is assumed that the Cu sublattice is terminated by a OH^- layer at the interface with the aqueous electrolyte in order to stabilize the otherwise polar surface structure if unreconstructed. The epitaxial relationship to the substrate is Cu(001)[100]//Cu_2O(001)[1$\bar{1}$0] (Figure 5.57b). It indicates a 45° rotation of the close packed direction of the Cu sublattice of the oxide relative to that in the Cu

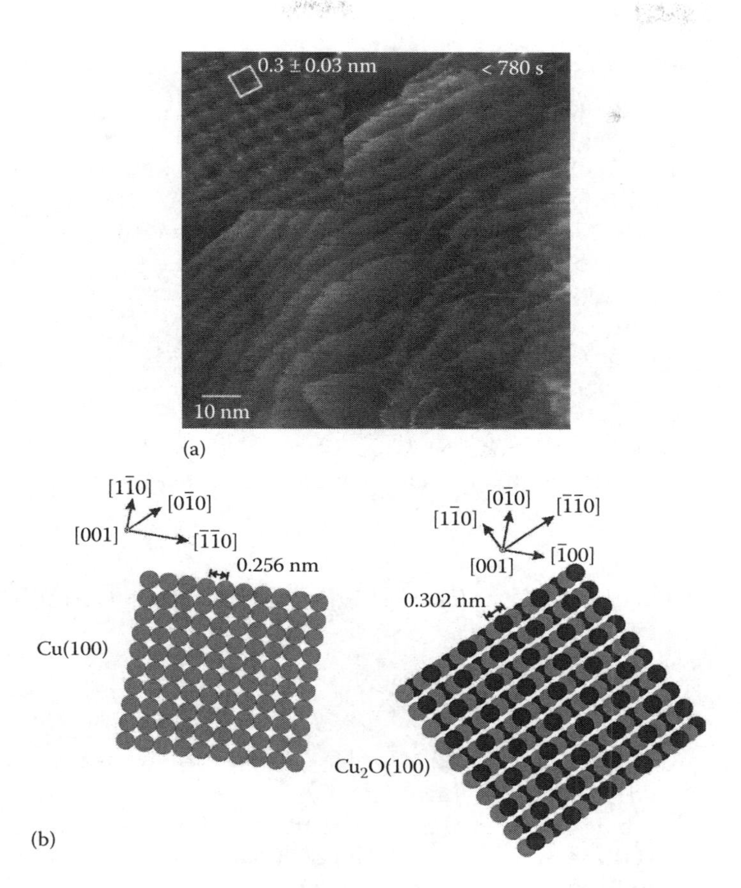

FIGURE 5.57
(a) In situ STM image of the 3D $Cu_2O(001)$ crystalline layer formed on Cu(001) in 0.1 M NaOH at −0.11 V (atomic lattice in inset). (b) Model (top view) of the $Cu_2O(001)$ surface with O^{2-}-Cu^{+}-O^{2-} trimers rotated by 45° with respect to the close-packed <110> directions of the Cu(001) substrate. (From Kunze, J. et al., *J. Electroanal. Chem.*, 554, 113, 2003.)

metal substrate, already present in the 2D Cu–OH adlayer formed at lower potential. From the width and height of the terraces at the oxide layer surface, the tilt between the oxide layer and metal substrate lattices is estimated to be 3°. The coincidence length of 3.32 nm between the two lattices corresponds to 11 periods of the Cu sublattice of the oxide and 9 periods in the metal lattice. The tilt contributes to accommodate the large misfit of 19.5% between the [110] direction in the oxide lattice and the [100] direction of the metal substrate.

On both Cu substrates, the cathodic reduction of 3D oxide films leaves the substrate surface irreversibly facetted. For example, after reduction of a 3.4 ML thick $Cu_2O(001)$ layer formed at $E = -0.18$ V for 3 min in 0.1 M NaOH on Cu(001) the surface still shows the imprint of the anodic oxide [150]. This indicates that the mobility of the reduced Cu atoms is too poor to recover the original surface with smooth and large terraces. A similar effect is observed for the 2D Cu–OH adlayer on Cu(001). Its reduction leads to a fragmented surface covered with monoatomic Cu islands. However, on Cu(111), similar treatments of the 2D Cu–OH adlayer allow to recover large and atomically smooth terraces, suggesting a higher mobility on this orientation.

5.8.2.3 Duplex Layer on Cu(111) and Cu(001)

Growth of the duplex Cu(I)/Cu(II) layer at $E = 0.75\,V$ yields the formation of 3D oxide crystals [162]. Their topography consists of terraces with a width of 5–20 nm and step edges having a linear or kinked morphology. The atomic structure of the terraces is consistent with a (001)-oriented CuO lattice characterized by a distorted hexagonal unit cell (Figure 5.58). This structure is found for both substrate orientations, i.e., Cu(111) and Cu(001). The step height of the terraces is 0.29 nm, which is close to the value 0.25 nm expected for 1 ML of CuO(001) from the crystal structure. Thus STM shows that the outer layer of the duplex film has a crystalline CuO(001) structure. This oxide structure may be terminated by O^{2-} or Cu^{+} planes but an unreconstructed structure would be polar and unstable. Here again, a termination of a Cu layer by an OH layer could explain the observed structure.

Combining the STM data obtained on the single 3D Cu(I) oxide layers and duplex Cu(I)/Cu(II) layers, the following epitaxial relationships with the two Cu substrates have been deduced:

$$\mathrm{CuO(001)[\bar{1}10]//Cu_2O(111)[1\bar{1}0]//Cu(111)[1\bar{1}0]} \quad \text{or} \quad [1\bar{1}0] \tag{5.39}$$

$$\mathrm{CuO(001)[\bar{1}10]//Cu_2O(001)[1\bar{1}0]//Cu(001)[100]} \tag{5.40}$$

Duplex layers on Cu(001) often reveal a surface covered by grains 3 nm wide and not showing any ordered atomic structure. However, changing the STM imaging conditions with a smaller bias between the tip and the specimen and a larger tunnel current yields the crystalline structure described above. This is explained by a closer approach of the tip to the surface. The crystalline oxide film could be covered by a thin amorphous layer, possibly corresponding to the Cu(II) hydroxide observed by XPS (see Section 5.6.2.2).

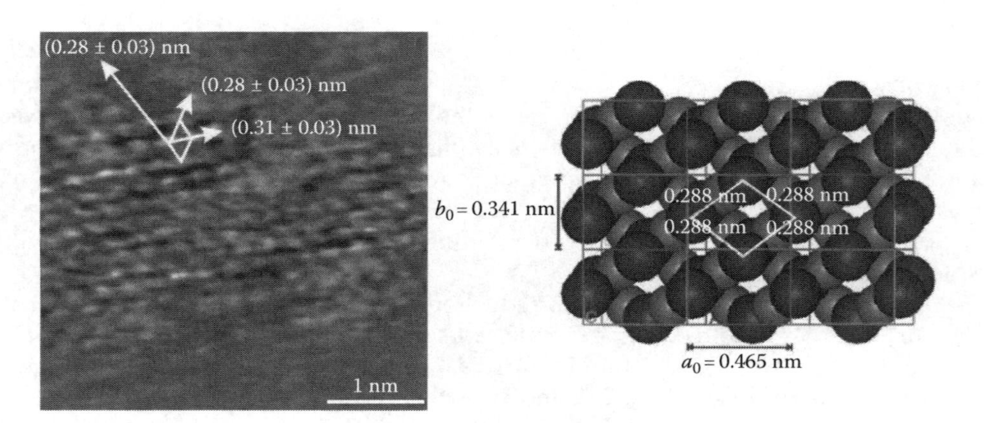

FIGURE 5.58
In situ STM image of the atomic lattice recorded on the terraces of the 3D Cu(II) anodic oxide formed on Cu(111) in 0.1 M NaOH at 0.83 V, and top view of the (001) face of CuO. The dark disks represent the O^{2-}, the bright disks Cu^{2+}. (From Kunze, J. et al., *Corros. Sci.*, 46, 245, 2004.)

5.8.3 Growth and Structure of Passive Films on Ag

5.8.3.1 STM Studies

The formation of 2D Ag(I) anodic oxide layers has been studied on Ag(111) with in situ STM [146,74]. As on Cu, the Ag surface shows OH adsorption and the formation of a 2D surface oxide in 0.1 M NaOH at potentials below the formation of the 3D oxide film at $E = 0.4$ V (Figure 5.59). A well-prepared Ag(111) surface shows large terraces at $E = -0.65$ V on which the hexagonal atomic structure of the metal surface with a parameter of 0.29 nm can be resolved. At $E = -0.05$ V, the major part of the surface is modified by OH adsorption [74]. It appears darker due to an increase of the tunneling barrier height like observed on Cu. With increasing potential ($E = 0.05$ V), the small ordered domains (marked D in Figure 5.59) protrude from the terraces. These protruding domains are assigned to the subsurface incorporation of OH^- (or O^{2-}) below the topmost Ag plane [74]. The hexagonal lattice measured locally has a period of 0.32 nm fitting reasonably well the Ag–Ag distance in the (111)-oriented Ag_2O lattice (0.334 nm), and thus the small ordered domains are assigned to 2D islands having a Ag_2O(111)-like structure. In addition, grains, ca. 3 nm wide and 0.38 nm high, assigned to 3D oxide nuclei are formed. The protruding domains grow larger when the film is formed more slowly, i.e., by a potentiodynamic scan rather than a step at a predefined potential anodic vertex, to form extended islands (marked I in Figure 5.59). However, the 2D surface oxide remains partially ordered with the larger ordered islands surrounded by a highly defective matrix formed by the smaller domains.

As on Cu(111), due to the larger Ag–Ag distance in the 2D oxide islands than in the Ag substrate plane, the Ag topmost plane must be reconstructed with the ejection of Ag atoms. The accumulation of the ejected atoms at the step edges after surface diffusion and the related displacement of the step edges is not observed as on Cu(111). Part of them cluster to form the 3D oxide nuclei observed on the terraces. A larger fraction has to be dissolved into the electrolyte because there are too much ejected Ag atoms compared to the number of grains found on the surface.

Further increase of the potential to $E = 0.15$ V causes the formation of two dimensional pits resulting from dissolution (Figure 5.60a) [74]. EXAFS studies in transmission mode have shown the formation of $Ag(OH)_2^-$ complexes in solution [163]. STM shows that the

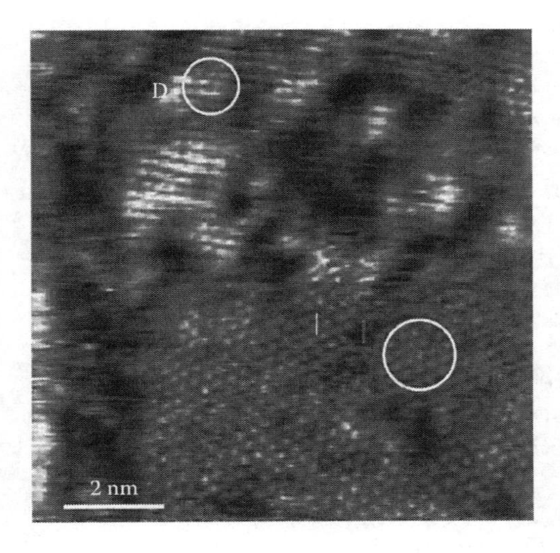

FIGURE 5.59
In situ STM image of the partially ordered 2D surface oxide layer formed on Ag(111) in 0.1 M NaOH at −0.05 V. D and I mark the highly defective matrix and extended ordered islands, respectively. Circles indicate the atomic lattice. (From Maurice, V. et al., *J. Phys. Chem. C*, 111, 16351, 2007.)

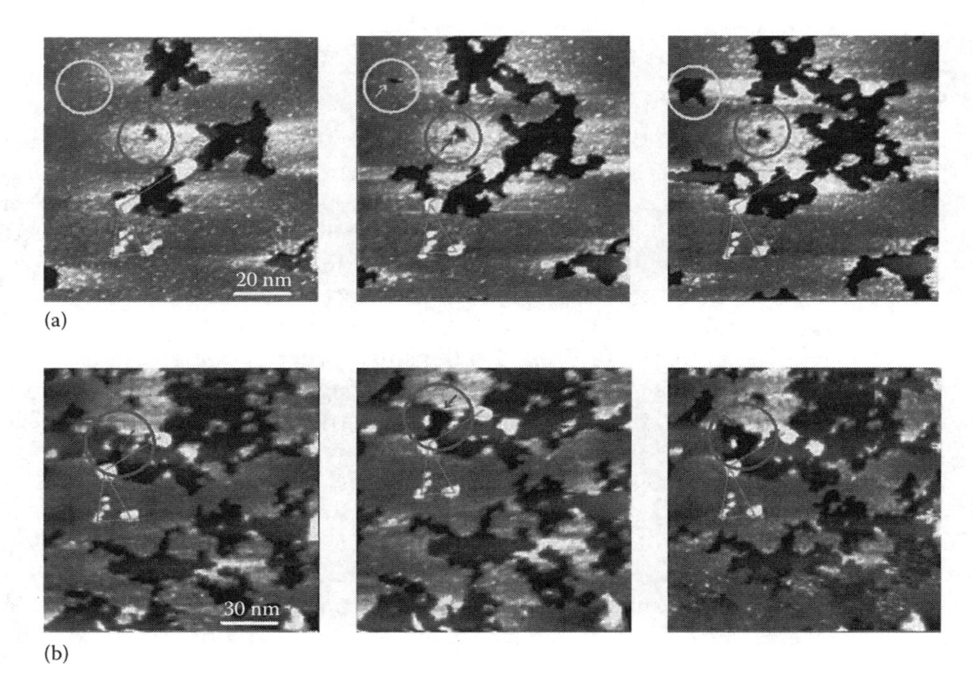

FIGURE 5.60
Sequences of in situ STM images of the Ag(111) surface in 0.1 M NaOH at +0.15 V, showing the initiation (pointers) and propagation/blocking of (a) 2D pits and (b) 3D pits. (From Maurice, V. et al., *J. Phys. Chem. C*, 111, 16351, 2007.)

resulting pits grow irregularly as a result of the surface structure. Lateral growth occurs in the less resistant parts of the 2D surface oxide layer made of the highly defective matrix. It is inhibited or blocked in the more resistant 2D extended ordered island and by the 3D oxide nuclei. Pit nucleation occurs in the second Ag plane also made of a 2D partially ordered surface oxide layer after dissolution of the first plane and exposure to the electrolyte, thus starting a slow in depth (i.e., 3D) pit growth (Figure 5.60b). As for the topmost layer, the 2D ordered islands inhibit or block dissolution that only propagates in the weaker highly defective matrix areas. As a consequence, 3D dissolution propagates locally.

5.8.3.2 XAS Studies

XAS is an interesting tool to obtain in situ results on the near range order of passive layers on metals even for highly disordered or amorphous films. Experimental requirements have been shortly described above. From the reflectivity data, one may calculate with the Fresnel equations the absorption spectra $\mu(E)$ of the film (see Refs. [140,141] for details). The oscillation of $\mu(E)$ above the absorption edge, the so-called extended absorption fine structure (EXAFS), leads to the EXAFS function $\chi(E) = (\mu - \mu_0)/\mu_0$ with the background value μ_0. The Fourier transform of $\chi(E)$ to the real space yields the structural parameters, i.e., the radius of the coordination shells R_i, their coordination numbers N_i, i.e. the number of neighbors for a given coordination shell and the Debye–Waller factor σ_i as a measure of the structural (and thermal) disorder.

The anodic film on Ag in 1 M NaOH formed at the potential of beginning oxide formation at $E = 0.40$ V is still too thin (2.4 nm) to avoid the contribution of the Ag substrate even

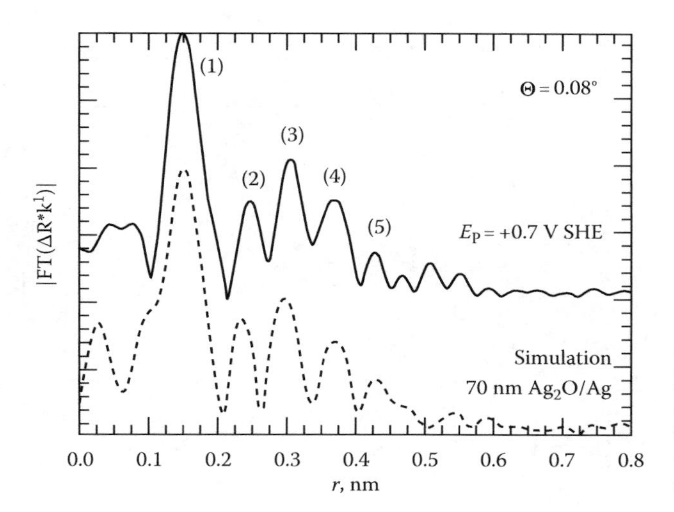

FIGURE 5.61
Fourier transform of reflectivity FT(R*k) for Ag covered with 70 nm Ag_2O formed at 0.70 V in 1 M NaOH, solid line experimental, dashed calculated data. (From Hecht, D. et al., *Surf. Sci.* 365, 263, 1996.)

for total reflection of the x-ray beam. Besides the first Ag–O coordination shell, one obtains the Ag–Ag shell of the substrate. A stronger Ag–Ag contribution of the oxide requires a thicker film (4 nm) formed at $E = 0.41$ V. Five nanometer thick films formed at $E = 0.43$ V show a good coincidence for calculated and measured peaks for several coordination shells. The data suggest the change of a primary less ordered Ag(I) oxide to crystalline Ag_2O with increasing electrode potential [73]. For thick oxide films, as e.g., 70 nm formed at $E = 0.70$ V, no substrate contribution is seen in the Fourier transform and up to five coordination shells are visible, which agrees with the calculation for a crystalline Ag_2O layer. Figure 5.61 shows an example for the Fourier transform of experimental and calculated reflectivity data [72]. Detailed evaluation of the reflection data for the first coordination shell of anodic Ag(I) oxide films formed at $E = 0.60$ and 0.70 V agree very well with transmission EXAFS data of polycrystalline bulk Ag_2O ($R_1 = 0.195$ nm, $N_1 = 2.01$, $\sigma = 0.008$). The lattice constant deduced from diffraction data in transmission of anodic oxide (0.70 V, 1 M NaOH) and a bulk Ag_2O reference agree well with those from literature ($a = 0.47$ nm) [164]. The evaluation of time-resolved EXAFS and diffraction studies show the formation of small crystallites, which grow from 10 to 25 nm in the range of $t = 20$–600 s.

For potentials $E > 0.7$ V, the oxide layer is oxidized in 1 M NaOH to AgO. Reflectivity data of these thick anodic layers in the range of 100 nm do not contain contributions of the metal substrate. The Fourier transform of EXAFS data deduced from absorption measurements in reflectivity and transmission show up to eight coordination shells that can be related to crystallographic data, i.e., to the Ag–O and Ag–Ag distances of the oxide. The first two shells related to the closest Ag–O distance have been evaluated in detail [75]. AgO consists of equal amounts of Ag^+ and Ag^{3+} ions. Therefore the first two coordination shells correspond to a planar quadratic orientation of Ag^{3+} with 4O and a linear arrangement of Ag^+ with 2O with Ag–O distances of $R_1 = 0.203$ nm and $R_2 = 0.22$ nm, respectively, in close agreement with the crystallographic data. This interpretation is supported by the broad shape of the Ag $3d_{5/2}$ XPS signal consisting of two Ag species [75] (see Section 5.6.2.3). The detailed evaluation of both the XAS and XPS data points to a highly oxygen-deficient AgO oxide. The coordination number for the closest coordination shell (Ag^{3+}–O) increases with the potential of oxide formation approaching $N_1 = 3.8$ for $E = 1.2$ V, close to the expected value 4. Figure 5.62 presents the results for the structural parameters $R_{1,2}$, $N_{1,2}$, and $\sigma_{1,2}$ of anodic silver oxide as a function of the potential of oxide formation [75].

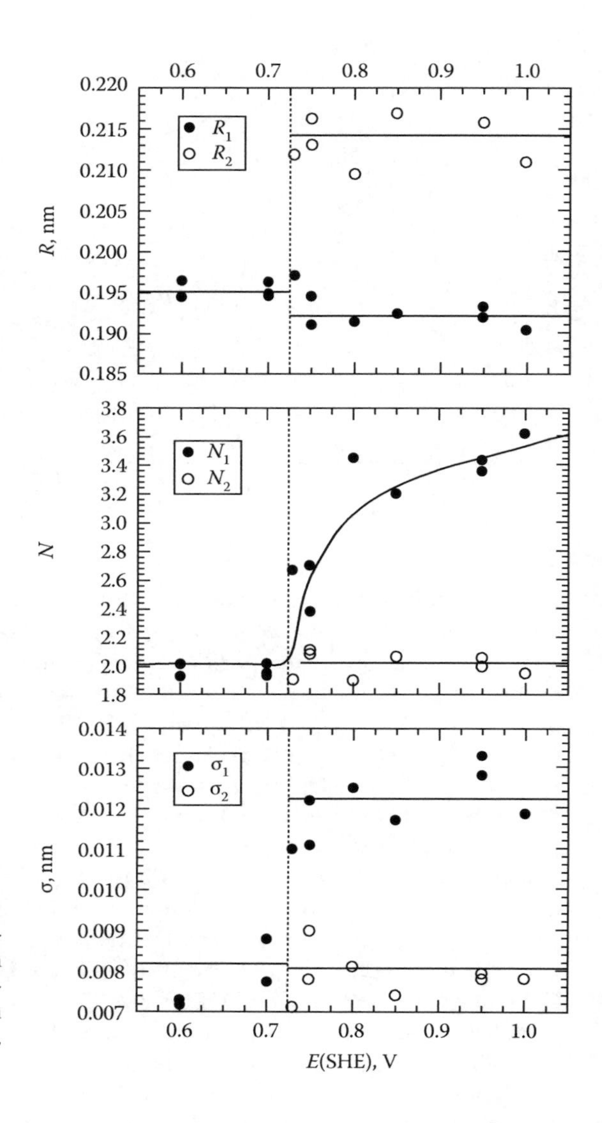

FIGURE 5.62
Parameters for near range order (first two coordination shells) of anodic oxide layers grown on Ag in 1 M NaOH in dependence of the electrode potential, $E < 0.725$ for Ag_2O, $E > 0.725$ for AgO. (From Lützenkirchen-Hecht, D. and Strehblow, H.-H., *Surf. Interf. Anal.*, 41, 820, 2009.)

These examples show that EXAFS in reflection mode may provide good data on the in situ structure of even highly disordered or even amorphous anodic oxide films. It yields data for the whole film and not only of the surface and is applicable also for rough films. Therefore, it is a good alternative and complimentary to scanning probe methods like STM and AFM. However, the experimental effort with respect to equipment (synchrotron radiation source) and specimen preparation is relatively large.

5.8.4 Structure of Passive Films on Fe, Ni, Cr, and Their Alloys

For iron, in situ STM studies [165] and in situ and ex situ GIXD measurements with synchrotron radiation [166,167] have shown that the passive layer formed in borate buffer solution (pH 8.4) is crystalline. In situ STM investigations reveal an atomic lattice consistent with the O sublattice of the inverse spinel structure of Fe_3O_4 or γ-Fe_2O_3. More details of the atomic structure have been found by in situ and ex situ GIXD studies. On Fe(001) and Fe(110), the oxide grows parallel to the (001) and (111) planes, respectively. In both cases, the [110] direction of the oxide is parallel to the [100] direction of the metal surface. The lattice parameter of the oxide layer is 0.839 nm within the surface plane and 0.83 nm perpendicularly, in better agreement with the parameters of Fe_3O_4 (0.8394 nm) [168] than γ-Fe_2O_3 (0.93396 and 0.3221 nm) [169]. The measured structure (so-called LAMM phase) has the same sublattice of O anions as in γ-Fe_2O_3 and Fe_3O_4 (thus confirming the STM data) but with different occupancies of the cation sites. In both reference compounds, the unit cell contains 32 oxide anions and there are 16 octahedral and 8 tetrahedral cation sites. The tetrahedral sites are fully occupied and there are no interstitial sites. For γ-Fe_2O_3, 33% of the octahedral sites are occupied. However, in the LAMM structure of the passive film, 80% ± 10% of the 16 octahedral cation sites are occupied by trivalent cations and 66% ± 10% of the tetrahedral cation sites are occupied by divalent cations. The occupancy of the octahedral interstitial by cations is 12% ± 4% and that of the tetrahedral interstitial is 0%. The Debye–Waller factor (0.01–0.028 nm) is an indication of a strong static disorder with varying bond lengths caused by interstitials and vacancies. The evaluation of the width of the Bragg peaks indicate a nanocrystalline structure with a grain size of 8 and 5 nm for the passive layers grown on Fe(001) and Fe(110), respectively. A value of 5 nm was obtained from ex situ STM data on passivated sputter deposited pure iron films [170]. The oxide layer on Fe thus consists of iron oxide nanocrystals having a spinel structure and with many 1D defects related to the cation site occupancy. A high linear density (1–2×10^6 cm^{-1}) of 2D defects (grain boundaries/twins) characterizes the nanostructure.

Figure 5.63a illustrates the typical topography observed by in situ STM of a Ni(111) single-crystal surface polarized in the middle of the passive domain in a sulfuric acid solution (pH ~ 2.9). The crystalline structure of the 3D Ni(II) passive film, first observed ex situ [144,171], has been later confirmed in situ by STM [172–175], by AFM [176] and by GIXD [177,178]. The film thickness increases up to 2 nm with increasing potential in the passive range. As on Cu, the passivated surface is faceted exhibiting terraces and steps. The facetted topography of the surface of the crystallized passive film is indicative of a slightly tilted epitaxy (~3°) between the NiO lattice forming the barrier oxide layer, and the Ni(111) substrate terraces. It has been proposed that this tilt partly relax the interfacial

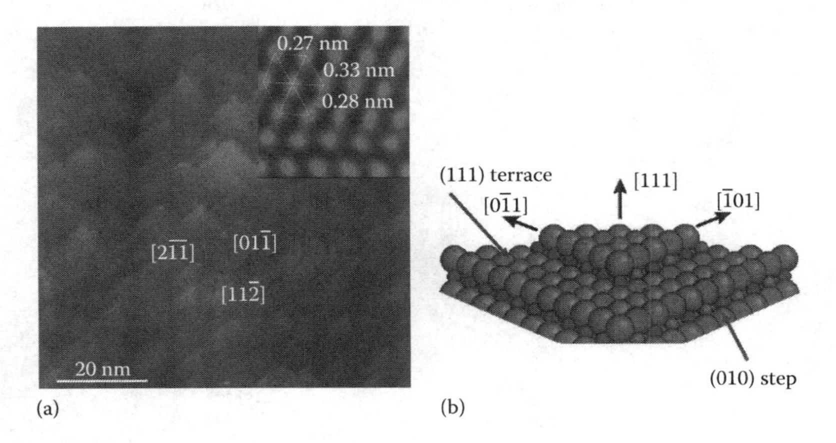

FIGURE 5.63
(a) In situ STM image of the 3D crystalline passive film formed on Ni(111) in 0.05 M H_2SO_4 + 0.095 M NaOH (pH ~ 2.9) at 0.95 V (atomic lattice in inset). (b) Model of the passive film surface with (111)-oriented facets delimited by edges oriented along the close-packed directions of the NiO lattice. Oxygen (surface hydroxyls) and Ni are represented by large and small spheres, respectively. (From Zuili, D. et al., *J. Electrochem. Soc.*, 147, 1393, 2000.)

stress associated with the mismatch of 16% between the two lattices. This tilt has been confirmed by in situ GIXD on Ni(111) [177,178]. The presence of surface crystalline defects (i.e., step edges) resulting from the slightly tilted epitaxy plays a key role in the dissolution in the passive state. Indeed dissolution is a 2D process leading to a step flow process as evidenced by STM sequence imaging [179,180]. The 2D step flow process leads to the stabilization of the facets with edges oriented along the close-packed directions of the oxide lattice and produces steps that are oriented along the {100} planes, the most stable orientations of the NiO structure (Figure 5.63b).

The lattice measured by STM on the terraces of passivated Ni(111) surfaces is hexagonal with a parameter of 0.3 ± 0.02 nm assigned to NiO(111) [144,171,174]. This result was also confirmed by GIXD that showed an antiparallel epitaxial relationship between the orientations of the oxide and metal lattices [177,178]. It must be pointed out that the (111) surface of NiO, which has the NaCl structure, is normally polar and unstable. It is however the surface that is obtained by passivation. The reason for this is that the surface is stabilized by adsorption of a monolayer of hydroxyl groups as confirmed by density functional theory (DFT) calculations [181,182]. These data show that the direction of growth of the oxide film is governed, at least in part, by the minimization of the oxide surface energy by the hydroxyl/hydroxide groups. The presence of water is thus a major factor for the structural aspects of the growth mechanism.

For Ni(111) in NaOH solutions (pH 11 and 13), the 3D growth of the crystalline passive film with a faceted topography is also observed [183,184]. Its thickness can be estimated to increase from 1.5 to 4.5 nm with increasing potential based on data obtained in 0.5 M NaOH [85]. The hexagonal lattice has a parameter of 0.32 ± 0.01 nm, in better agreement with the value of 0.317 nm expected for a 3D layer assigned to β-$Ni(OH)_2$(0001). This is consistent with a crystalline 3D outer hydroxide layer of the passive film formed in alkaline electrolytes, as opposed to the 2D outer hydroxide layer existing on the inner oxide layer in acid electrolytes (see Section 5.6.2.8). For Ni(001) at pH 14, a crystalline passive film with an hexagonal structure consistent with an outer layer of $Ni(OH)_2$(0001) is also formed [172].

The lateral dimensions of the crystalline grains forming the passive films on nickel are relatively well documented. Values determined from the morphology observed by STM and AFM (Figure 5.64) have been reported to range from ~2 nm in the initial stages of 2D growth

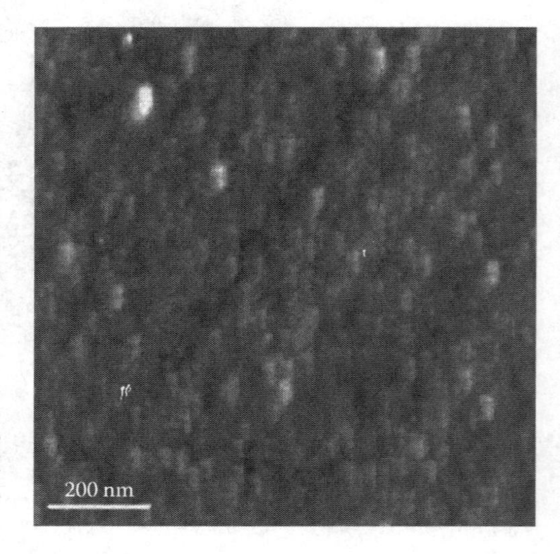

FIGURE 5.64
AFM image showing the granular nanostructure of the 3D crystalline passive film formed on Ni(111) in 0.05 M H_2SO_4 + 0.095 M NaOH (pH 2.9) at 0.55 V. (From Marcus, P. et al., *Corros. Sci.*, 50, 2698, 2008.)

to 30–230 nm for 3D films in stationary conditions of passivity [171,173,174,180,183,184]. A large dispersion could be found on the same sample, suggesting a varying degree of advancement of the coalescence of the oxide grains during the nucleation and growth mechanism. A lower average value of ~8 nm has been measured from the NiO(111) Bragg peaks by GIXD [177,178]. This difference may be assigned to the fact that STM and AFM measurements are unable to resolve the multiple twin or grain boundaries that can characterize the inner part of the passive film without markedly affecting the surface topography. The linear density (~1–2 × 10^6 cm^{-1}) of 2D defects (grain boundaries/twins) in the passive film is thus similar to that on Fe. Both AFM [180] and STM [185,186] data have evidence that breakdown events are located at the boundaries between the oxide crystals of the passive film, giving evidence that the grain boundaries of the passive film are preferential sites of passivity breakdown.

On chromium, potential and aging favor the development of the inner oxide layer of the passive film formed in acid solution (0.5 M H_2SO_4), as shown by XPS [83]. However, the film is mostly disordered at low potential where the oxide inner layer is not fully developed and where the passive film is highly hydrated and consists mainly of hydroxide [83,187]. This supports the view that the passive film on chromium can consist of small Cr_2O_3 nanocrystals buried in a disordered chromium hydroxide matrix. At high potential where the inner part of the passive film is dehydrated and consists mostly of chromium oxide, larger crystals having a faceted topography and extending over several tens of nanometers have been observed by in situ STM [187]. The nanocrystals have a lattice consistent with the O sublattice in α-Cr_2O_3(0001). The basal plane of the oxide is parallel to Cr(110). A special feature of the passive film on chromium is that the oxide nanocrystals can be cemented together by the chromium hydroxide–disordered outer layer. This anodic Cr(III) oxide/hydroxide film provides an excellent corrosion protection to the Cr surface with a strong resistance to dissolution although it might be far from dissolution equilibrium in strongly acidic electrolytes.

FeCr alloys and Fe-Cr-Ni alloys are dominated by the Cr enrichment within the anodic films as found by XPS [103,106]. For specimens passivated in 0.5 M H_2SO_4, STM yields again a nanocrystalline structure with a hexagonal lattice consistent with the O sublattice of Cr_2O_3. A similar result is found for Ni-based stainless alloys passivated in high-temperature (325°C) water [116]. The grains, assigned to α-Cr_2O_3, have their basal plane (0001) oriented parallel to the Fe-22Cr(110) and Fe-18Cr-13Ni(100) substrate planes [103,106]. It has been observed that, for short polarization times (≤2 h), the crystallinity of the passive films decreases with increasing Cr content of the alloy. Structural changes also occur during aging under anodic polarization (Figure 5.65). The major modification is an increase of the crystallinity of the film and the coalescence of Cr_2O_3 nanocrystals in the inner barrier oxide layer as observed over time periods of up to 65 h parallel to the increase of Cr(III) as found by XPS and ISS as described in Section 5.6.3.1. The comparison of the rates of crystallization of Fe-22Cr(110) and Fe-18Cr-13Ni(100) revealed that the rate of crystallization is more rapid on the austenitic stainless steel than on the ferritic one. This was tentatively explained by a regulating effect of Ni on the supply of Cr on the alloy surface, a lower rate of Cr enrichment being in favor of a higher degree of crystallinity.

5.8.5 Passive Monolayer on Co(0001) and Its Reduction

Figure 5.66 shows the structure of the Co(0001) surface in 0.1 M NaOH after stepping the potential from $E = -1.0$ to -0.65 V at the beginning of the passive range [147]. Oxidation starts at the step edges and spreads over the surface. The insertion of OH ions between

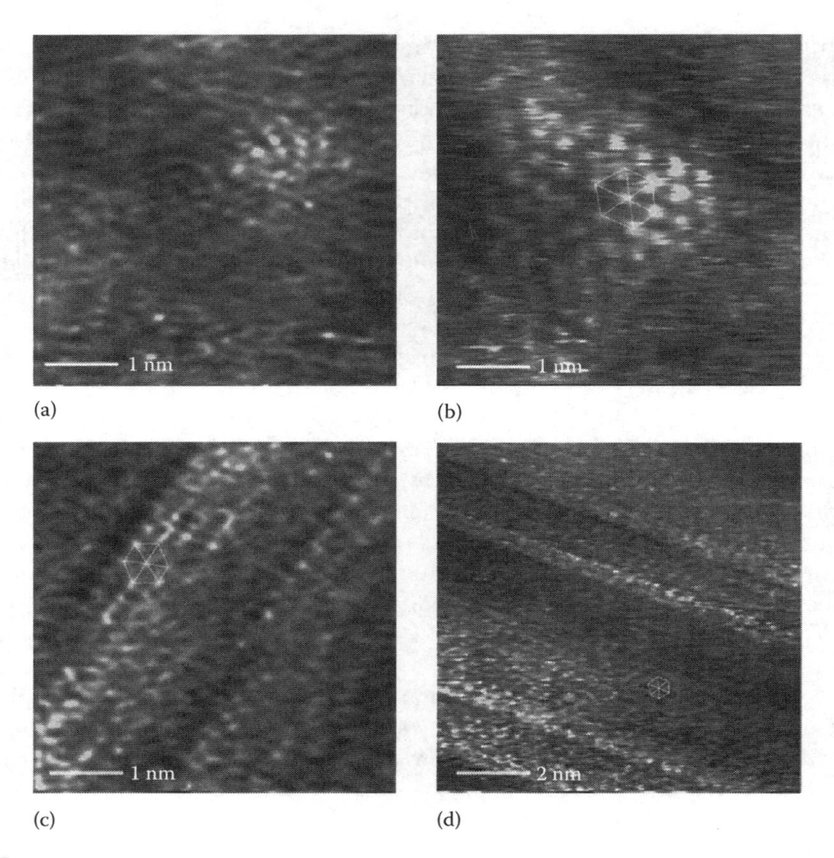

FIGURE 5.65
STM images of the Fe-22Cr(110) (left) and Fe-18Cr-13Ni(100) (right) surfaces recorded after passivation in 0.5 M H_2SO_4 at +0.5 V for (a,b) 2 h and for (c,d) 22 h. The nearly hexagonal lattice is marked. The effect of ageing under polarization is evidenced by the extension of the observed crystalline areas. (From Maurice, V. et al., *J. Electrochem. Soc.*, 143, 1182, 1996; Maurice, V. et al., *J. Electrochem. Soc.*, 145, 909, 1998.)

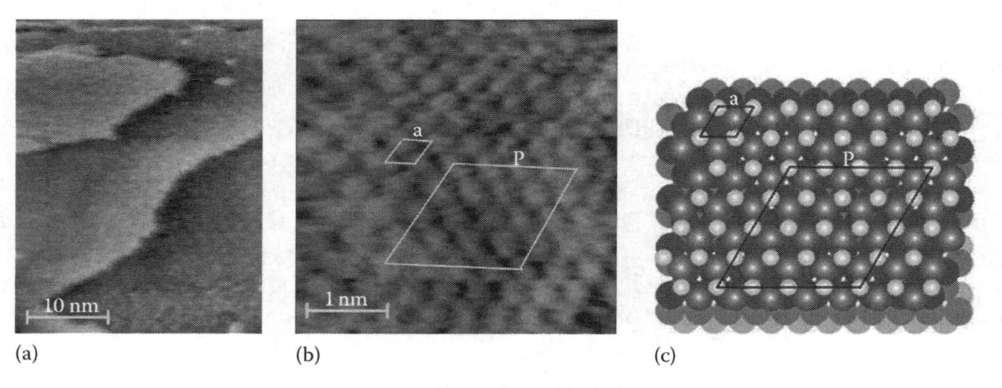

FIGURE 5.66
In situ STM images of one monolayer of $Co(OH)_2(0001)$ formed on Co(0001) in 0.1 M NaOH at −0.65 V. (a) Modified terraces with hexagonal Moiré pattern. (b) Unit cells of Moiré pattern (P) and $Co(OH)_2(0001)$ monolayer (a). (c) Model of the $Co(OH)_2(0001)$ monolayer on Co(0001) showing the unit cells (P) and (a) forming a (4 × 4) coincidence cell. (From Foelske, A. et al., *Surf. Sci.*, 554, 10, 2004.)

the topmost Co plane and the second Co plane at the step edges induces an increase of step height from 0.25 to 0.28 nm. The surface structure is characterized by a Moiré pattern with a parameter $P = 1.25$ nm and a corrugation of 0.04 nm (Figure 5.66a and b). The adlayer unit cell has a parameter $a = 0.33$ nm (Figure 66b), consistent with the structure of a (0001)-oriented monolayer of $Co(OH)_2$ ($a = 0.317$ nm from the bulk structure [164]) oriented parallel to the Co(0001) substrate surface. Potentiodynamic reduction after oxidation at $E = -0.65$ V confirms an equivalent thickness of 0.8 monolayer of $Co(OH)_2$ [77]. The misfit of 20% between the metal substrate ($a = 0.25$ nm) and the adlayer leads to a 4 × 4 coincidence superstructure forming the Moiré pattern as shown in the model of Figure 5.66c. This model shows one layer of OH and one layer of Co^{2+} on top of the Co(0001) metal. The final OH layer is omitted for clarity. The Co^{2+} ions presumably sit in the threefold hollow site of the OH layer.

Interesting details were observed by STM during the slow reduction of this hydroxide layer at $E = -1.0$ V in 1 M NaOH [147]. Reduction causes the formation of terraces, 3D Co metal clusters and OH–Co–OH trimers. At an intermediate stage, 2D islands grow and merge together to form terraces with clusters at the step edges (Figure 5.67a). The trimers (resolved as single protrusions) form transient local triangular and hexagonal patterns according to the hexagonal structure of the surface (Figure 5.67). Some domains formed by trimers display a rectangular superstructure assigned to a missing row structure (Figure 5.67b). The fuzzy edges of these domains suggest that surface diffusion is involved in the reaction consuming them. At the end, smooth terraces decorated at their edges by trimers are obtained. 3D Co clusters remain on the terraces and step edges (Figure 5.67c).

Figure 5.68 tentatively rationalizes the observations made during the reduction process. The $Co(OH)_2$-ordered monolayer presumably rearranges into a transition state via a flat alignment of the $Co(OH)_2$ trimers. The reduced Co atoms arrange in 2D islands and 3D clusters. The nonreduced trimers are mobile and get trapped and cluster at the edges of these islands where they are stabilized and remain as stable chemical units. This corresponds to the often observed situation that oxide or hydroxide forms already at relatively negative potentials at the energetically favored step edge sites. In parallel, the trimers that are not trapped at the island edges can self-arrange locally to form triangular and hexagonal local patterns imposed by the symmetry of the substrate or domains of a missing row

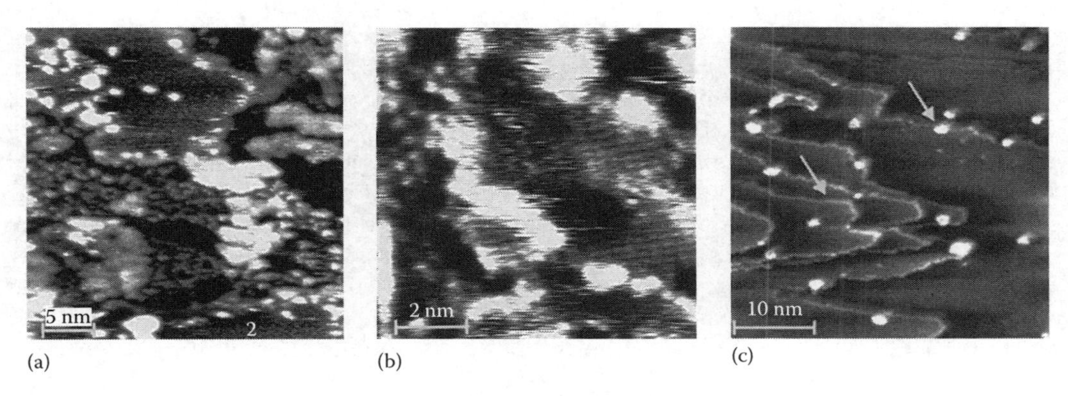

FIGURE 5.67
In situ STM images of the reduction of the monolayer of Co(OH)(0001) grown on Co(0001) in 0.1 M NaOH at −1.0 V. (a) Merging 2D islands, 3D clusters predominantly at edges and trimers in-between, (b) transient rectangular superstructure formed by trimers with fuzzy borders indicating atomic displacements, (c) Final structure with decoration of the step edges by OH–Co–OH trimers and trapped clusters (pointed). (From Foelske, A. et al., *Surf. Sci.*, 554, 10, 2004.)

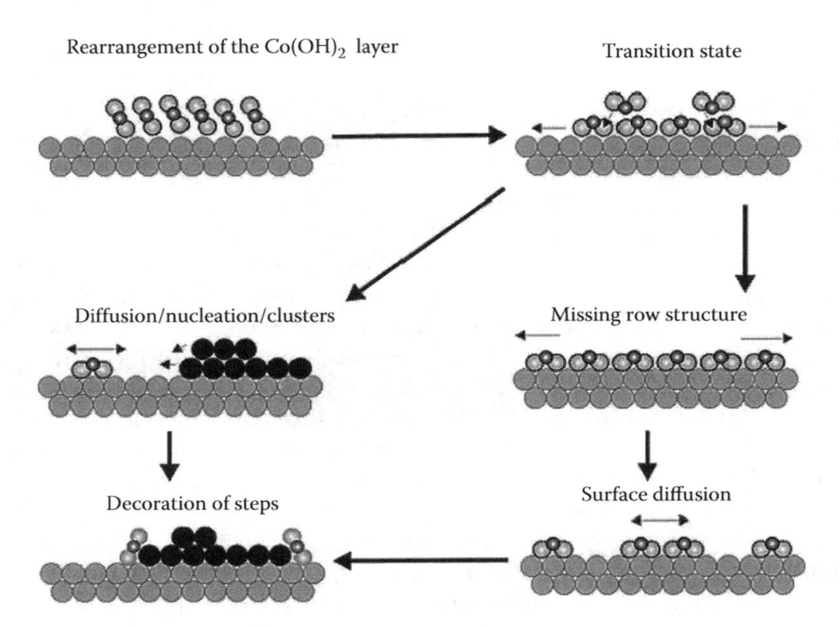

FIGURE 5.68
Schematic diagram of the $Co(OH)_2$ reduction process. (From Foelske, A. et al., *Surf. Sci.*, 554, 10, 2004.)

structure. However, even these locally ordered domains become eventually destabilizes by surface mobility imposing back and forth displacement at their edges. The trimers get then reduced and/or diffuse to the island edges. At the end, all transient local structures have been reduced and the 2D Co islands have grown and merged to form terraces decorated with unreduced trimers at their edges and with 3D Co clusters.

5.8.6 Conclusions on the Structure of Passive Films

The structure of passivated surfaces of Cu, Ag, Co, Fe, Ni, Cr, and stainless steels has been reviewed with emphasis on results obtained on metal surfaces by in situ STM and complements from AFM, XAS, and GIXD studies. The reviewed data demonstrate the adsorption of hydroxide ions in the potential range below the one for oxide growth, and the role of structural precursors of the 2D adsorbed layers in the growth of passive films, inducing surface reconstruction of the topmost metal plane to adopt the structural arrangement found in 3D passivating oxides. The corrosion resistance of metal surfaces covered by 2D passivating oxide/hydroxide layers in a transient stage of passivation was shown to be sensitive to the local structure. Dissolution is then promoted at the weakest nonordered sites, initiating 2D nanopits on the substrate terraces and the growth of 3D nanopits by a repeated mechanism on the newly exposed terraces after prolonged polarization at low anodic overpotential. Nucleation of passive films occurs preferentially at preexisting defects (i.e., step edges) of the metal surface. The 3D passive films on numerous substrates are nanocrystalline or become nanocrystalline with time. Their surface is hydroxylated and the orientation of the oxide crystals can be modified by the substrate structure. Aging under polarization is critical to the crystallization of chromium-rich passive films. Surface faceting of the passivating oxide results from a few degree tilt of the oxide lattice with respect of the metal lattice and leads a step flow mechanism of dissolution in the passive state.

Numerous grain boundaries result from the nanometer dimensions of the single-crystal grains in the passive films. These nanostructure-related defects play a preferential role in passivity breakdown.

5.9 Conclusion

Anodic oxide films on metal surfaces are very important for corrosion protection and determine decisively the electrochemical properties of electrode surfaces. They are the main reason that reactive metals may be used as materials in the various fields of construction and products for every day application. In order to understand the unique protecting properties of passive layers only a few nanometers thick, their chemical composition as well as their structure and electronic properties should be known. They determine how protective a passive film is, i.e., how long and in which environment a metal may be used. Passive layers have been investigated with numerous ex situ and in situ methods over more than a century.

The thermodynamic properties and the application of electrode kinetics have been described. As corrosion and passivity are in principle determined by electrochemistry at metal surfaces, a good understanding of the equilibria and the kinetics of electrode surfaces is a necessary requirement for any further study with more sophisticated methods. This involves complicated transients studies as well as electrochemical methods like the RRD electrode with or without hydrodynamical modulation. Here a systematic research is still needed for a better understanding of pure metals and especially alloys. Results for simplified conditions give answers for the often more complicated situation of corroding systems in a real environment.

If the electrochemistry is understood, one needs the application of surface analytical methods to learn about the chemical properties and chemical structure of passive layers. Then one has to take care that electrochemical specimen preparation has to occur with optimum control in order to get reliable results. This permits to draw clear mechanistic conclusions on the properties of the layers like their growth, reduction, changes of their composition, reactivity, degradation, and stability including realistic environmental conditions. Application of XPS, ISS, and RBS to a wide variety of pure metals and binary alloys has been described in Section 5.6. These techniques provide valuable results especially when applied together with a systematic change of the experimental parameters like the potential and time of passivation, the composition of the electrolyte, and alloy and conditions for layer degradation.

Electronic properties can be studied by photoelectrochemistry using redox reactions at passivated metal surfaces in the dark or with illumination. Their knowledge is very valuable to understand layer formation and corrosion phenomena for open circuit conditions, i.e., under the influence of redox systems within the electrolyte. Furthermore layer formation includes often oxidation of intermediates, i.e., oxidation of lower valent to higher valent cations, which requires electronic conduction through the already existing film. Similarly layer reduction needs electron transfer across the film.

The section on the structure of passive layers has shown the application of in situ STM and AFM and synchrotron methods like XAS and GIXD for the analysis of the structure of passivated surfaces at the atomic scale and at the nanoscale. These techniques provide a clearer view and understanding on the relation between structure and protecting properties and the influence of defects on the reactivity in the passive state and passivity breakdown.

Anodic films have usually a complex structure with two or more films on top of each other. They show a characteristic distribution of cations of different valencies and in most cases an inner oxide and an outer hydroxide part. STM provides details like OH adsorption in the initial stages of growth, the formation of oxide films often with an amorphous grain structure, which changes to a nanocrystalline oxide layer during aging and the loss of water. Crystallization may be fast as in the case of Cu in less than a minute or slow as for Cr, which needs hours. At sufficiently positive potentials, the oxide layer gets more complicated due to the formation of a duplex film with lower valent cations inside and higher valent cations at the surface of the passive layer.

The study of surface layers on electrodes has considerately improved during the last 40 years. Many details are known by investigations with new advanced techniques, which have been developed and applied to these systems. Numerous mechanisms have been studied and the models have been revised or improved and are the result of detailed investigations with much less or no speculation. All this helps to get to a better understanding and a detailed insight to the mechanisms of the formation of passive layers and their change with the environmental parameters.

References

1. H.H. Uhlig, in *Passivity of Metals*, R.P. Frankenthal, J. Kruger, eds., *Corrosion Monograph Series*, The Electrochemical Society, Princeton, NJ (1978), pp. 1–28.
2. H.-H. Uhlig, *Corrosion Handbook*, p. 21, Wiley, New York (1948).
3. N. Tomashov, *Theory of Corrosion and Protection of Metals*, Macmillan, New York (1966), p. 7, 325.
4. W. Ostwald, *Elektrochemie*, Ihre Geshichte and Lehre, Veit & Comp., Leipzig, 1151 p. (1896).
5. J. Keir, *Phil. Trans.* 80, 359 (1790).
6. Gmelin, *Handbuch der Anorganischen Chemie*, Teil 59 A p. 313, Winter, Heidelberg (1929–1932).
7. M. Schönbein, Letter to M. Faraday, in *M. Faraday, Experimental Researches in Electricity*, Vol. 2, Dover Publications, New York (1965), 238 pp.
8. M. Faraday, *Handbuch der Anorganischen Chemie*, Teil 59 A Berlin (1929–1932), pp. 234–250.
9. F. Gorn, K.J. Vetter, *Z. Phys. Chem.* 77, 317 (1972).
10. K.E. Heusler, *Ber. Bunsen Ges. Phys. Chem.* 72, 1197 (1968).
11. K.J. Vetter, F. Gorn, *Electrochim. Acta* 18, 321 (1973).
12. S. Haupt, C. Calinski, U. Collisi, H.-W. Hoppe, H.-D. Speckmann, H.-H. Strehblow, *Surf. Interf. Anal.* 9, 357 (1986).
13. P. Marcus I. Olefjord, *Corros. Sci.* 28, 589 (1988).
14. N. Sato, *Corros. Sci.* 31, 1 (1990).
15. K.E. Heusler, in *Passivity of Metals*, R.B. Frankenthal, J. Kruger, eds., The Electrochemical Society, Princeton, NJ (1978), p. 77.
16. P. Marcus, V. Maurice, in *Passivity of Metals and Alloys*, M. Schütze, editor, Materials Science and Technology, Vol. 19, Wiley-VCH, Weinheim, Germany (2000), pp. 131–169.
17. H.-H. Strehblow, in *Passivity of Metals*, R.C. Alkire and D.M. Kolb, eds., Advances in Electrochemical Science and Engineering, Vol. 8, Wiley-VCH, Weinheim, Germany (2003), pp. 271–374.
18. J.A. Bardwell, G.I. Sproule, M.J. Graham, *J. Electrochem. Soc.*140, 50 (1993).
19. M.J. Graham, *Corros. Sci.* 37, 1377 (1995).

20. M.J. Graham, J.A. Bardwell, G.I. Sproule, D.F. Mitchell, B.R. Macdougall, *Corros. Sci.* 35, 13 (1993).
21. C. Courty, H.J. Mathieu, D. Landolt, *Electrochim. Acta* 36, 1623 (1991).
22. A.J. Davenport, J.A. Bardwell, C.M. Vitus, *J. Electrochem. Soc.* 142 (1995) 721.
23. L.J. Oblonsky, A.J. Davenport, M.P. Ryan, H.S. Isaacs, R.C. Newman, *J. Electrochem. Soc.* 144, 2398 (1997).
24. P. Schmuki, S. Virtanen, *Electrochem. Soc. Interf.* 6, 38 (1997).
25. V. Maurice, P. Marcus, Structure, corrosion and passivation of metal surfaces, in *Modern Aspects of Electrochemistry*, Vol. 46, S.-I. Pyun, editor, Springer, New York (2009), pp. 1–58.
26. M. Pourbaix, *Atlas d'Equilibres Electrochimiques*, Guthiers Villars+ Cie, Paris, France (1963).
27. M. Pourbaix, *Atlas of the Electrochemical Equilibria in Aqueous Solutions*, Pergamon, Oxford (1966).
28. H.-H. Strehblow, U. Collisi, P. Druska, *International Symposium on Control of Copper and Copper Alloys Oxidation*, Rouen, 1989, Edition de la Revue de la Métallurgie, Paris, France (1993), p. 33.
29. H.-H. Strehblow, H.-D. Speckmann. *Werkst. Korr.* 35, 512 (1984).
30. J.A. Davies, B. Domeij, J.P.S. Pringle, F. Brown, *J. Electrochem. Soc.* 112, 575 (1965).
31. J.P.S. Pringle, *Electrochim. Acta* 25, 1403 (1979).
32. J.P.S. Pringle, *Electrochim. Acta* 25, 1423 (1979).
33. W.J. Albery, M.L. Hitchman, *Ring Disc Electrodes*, Clarendon, Oxford (1971).
34. K.J. Vetter, F. Gorn, *Electrochim. Acta* 18, 321 (1973).
35. F.P. Fehlner, N.F. Mott, *Oxid. Met.* 2, 59 (1970).
36. D.A. Vermilyea, *J. Electrochem. Soc.* 102, 655 (1955).
37. K.J. Vetter, *Z. Elektrochem.* 58, 230 (1954).
38. K.G. Weil, *Z. Electrochem.* 59, 711 (1955).
39. C. Wagner, *Z. Phys. Chem.* 25, B21 (1933).
40. C.Y. Chao, L.F. Lin, D.D. Macdonald, *J. Electrochem. Soc.* 128 (1981) 1187.
41. D.D. Macdonald, S.R. Baggio, H. Song, *J. Electrochem. Soc.* 139 (1992) 170.
42. D.D. Macdonald, *Pure Appl. Chem.* 71, 951 (1999).
43. P. Marcus, F. Mansfeld, eds., in *Analytical Methods in Corrosion Science and Engineering*, CRC Press, Taylor & Francis, Boca Raton, FL (2006).
44. H.-H. Strehblow, P. Marcus, X-ray photoelectron spectroscopy in corrosion research, in *Analytical Methods in Corrosion Science and Engineering*, P. Marcus, F. Mansfeld, eds., CRC Press, Taylor & Francis, Boca Raton, FL (2006), pp. 1–37.
45. J.E. Castle, Auger electron spectroscopy, in *Analytical Methods in Corrosion Science and Engineering*, P. Marcus, F. Mansfeld, eds., CRC Press, Taylor & Francis, Boca Raton, FL (2006), pp. 39–64.
46. D. Schmaus, I. Vickridge, MeV ion beam analytical methods, *in Analytical Methods in Corrosion Science and Engineering*, P. Marcus, F. Mansfeld, eds., CRC Press, Taylor & Francis, Boca Raton, FL (2006), pp. 103–132.
47. H. Beinert, K.F. Bonhoeffer, *Z. Elektrochem.* 47, 536 (1941).
48. F. Flade, *Z. Phys. Chem.* 76, 513 (1911).
49. U.F. Franck, *Z. Naturforsch.* 4a, 378 (1949).
50. K.G. Weil, K.F. Bonhoeffer, *Z. Phys. Chem.* (NS) 4, 175 (1955).
51. K. Heusler, K.G. Weil, K.F. Bonhoeffer, *Z. Phys Chem.* (NS) 15, 149 (1958).
52. B. Kabanov, D. Leikis, *Acta Physicochim. USSR* 21, 769 (1946).
53. B.B. Losev, B.N. Kabanov, *Zh. Fiz, Khim.* 28, 824, 914 (1954).
54. G.H. Cartledge, R.F. Sympson, *J. Phys. Chem.* 61, 973 (1957).
55. G.H. Cartledge, *Z. Elektrochem.* 62, 684 (1958).
56. K.J. Vetter, *Z. Elektrochem.* 62, 642 (1958).
57. H. Göhr, E. Lange, *Naturwiss.* 43, 12 (1956).
58. G. Hagg, *Z. Phys. Chem.* B29, 95 (1935).
59. K.J. Vetter, *Electrochemical Kinetics*, Academic Press, New York (1967), 753pp.
60. K.J. Vetter, G. Klein, Thesis, Technical University of Berlin, Berlin, Germany (1959).
61. S. Haupt, H.-H. Strehblow, *Langmuir* 3, 873 (1987).
62. G. Engelhardt, T. Jabs, D. Schaepers, H.-H. Strehblow, *Acta Chim. Hungarica* 129, 551 (1992).
63. G. Engelhardt, D. Schaepers, H.-H. Strehblow, *J. Electrochem. Soc.* 139, 2170 (1992).

64. G. Engelhardt, T. Jabs, H.-H. Strehblow, *J. Electrochem. Soc.* 139, 2176 (1992).
65. J.W. Schultze, U. Stimming, *Z. Phys. Chem.* 98, 285 (1975).
66. H.D. Speckmann, S. Haupt, H.-H. Strehblow, *Surf. Interf. Anal.* 11, 148 (1988).
67. H.-H. Strehblow, B. Titze, *Electrochim. Acta*, 25, 839 (1980).
68. M.M. Lohrengel, J.W. Schultze, H.D. Speckmann, H.-H. Strehblow, *Electrochim. Acta* 32, 733 (1987).
69. U. Collisi, H.-H. Strehblow, *J. Electroanal. Chem.* 210, 213 (1986).
70. U. Collisi, H.-H. Strehblow, *J. Electroanal. Chem.* 284, 385 (1990).
71. H.D. Speckmann, M.M. Lohrengel, J.W. Schultze, H.-H. Strehblow, *Ber. Bunsen Ges. Phys. Chem.* 89, 392 (1985).
72. D. Hecht, P. Borthen, H.-H. Strehblow, *Surf. Sci.* 365, 263 (1996).
73. D. Lützenkirchen-Hecht, H.-H. Strehblow, *Surf. Interf. Anal.* 38, 686 (2006).
74. V. Maurice, L.H. Klein, H.-H. Strehblow, P. Marcus, *J. Phys. Chem. C* 111, 16351 (2007).
75. D. Lützenkirchen-Hecht, H.-H. Strehblow, *Surf. Interf. Anal.* 41, 820 (2009).
76. P. Keller, PhD thesis, Elektrochemische und oberflächenanalytische Untersuchungen zur anodischen Deckschichtbildung auf Zinn und Kupfer/Zinn Legierungen, University of Düsseldorf, Dusseldorf, Germany (2005).
77. A. Foelske, H.-H. Strehblow, *Surf. Interf. Anal.* 34, 125 (2002).
78. K. Heusler, in *Passivity of Metals*, R.P. Frankenthal, J. Kriger, eds., Electrochemical Society, Princeton NJ (1978), p. 790.
79. W. Feitknecht, W. Lotmar, *Helv. Chim. Acta* 18, 1369 (1935).
80. A. Drexle, H.-H. Strehblow, *Proceedings Kurt Schwabe Symposium*, Helsinki, Espoo, Finland, (2004), 433pp.
81. S. Haupt, H.-H. Strehblow, *J. Electroanal. Chem.* 216, 229 (1987).
82. S. Haupt, H.-H. Strehblow, *J. Electroanal. Chem.* 228, 365 (1987).
83. V. Maurice, W. Yang, P. Marcus, *J. Electrochem. Soc.* 141 (1994) 3016.
84. P. Marcus, J. Oudar, I. Olefjord, *J. Microsc. Spectrosc. Electron.* 4, 63 (1979).
85. H.W. Hoppe, H.-H. Strehblow, *Surf. Interf. Anal.* 14, 121 (1989).
86. A. Güntherschulze, H. Betz, *Z. Phys.* 91, 70 (1934).
87. A. Güntherschulze, H. Betz, *Z. Phys.* 92, 367 (1934).
88. D.A. Vermilyea, *Acta Metall.* 1, 283 (1953).
89. D.A. Vermilyea, *Acta Metall.* 2, 476 (1954).
90. C. Doherty, H.-H. Strehblow, *J. Electrochem. Soc.* 125, 30 (1978).
91. H.-H. Strehblow, C.M. Melliar Smith, W.M. Agustyniak, *J. Electrochem. Soc.* 125, 915 (1978).
92. S. Haupt, H.-H. Strehblow, *Corros. Sci.* 37, 43 (1995).
93. P. Keller, H.-H. Strehblow, *Corros. Sci.* 46, 1939 (2004).
94. C.-O.A. Olsson, D. Landolt, *Electrochim. Acta* 48, 1093 (2003).
95. I. Olefjord, B. Brox, in *Passivity of Metals and Semiconductors*, M. Froment, editor, Elsevier, Amsterdam, the Netherlands (1983), p. 561.
96. D. F. Mitchell, M. Graham, *Surf. Interf. Anal.* 10, 259 (1987).
97. S. Mishler, H.J. Mathieu, D. Landolt, *Surf. Interf. Anal.* 11, 182 (1988).
98. C. Calinski, H.-H. Strehblow, *J. Electrochem. Soc.* 136, 1328 (1989).
99. R. Kirchheim, B. Heine, H. Fischmeister, S. Hofmann, H. Knote, U. Stolz, *Corros. Sci.* 29, 899 (1989).
100. J.E. Castle, J.H. Qiu, *Corros. Sci.* 29, 591 (1989).
101. C. Hubschmid, H.J. Mathieu, D. Landolt, *Proceedings of the 12th International Congress*, NACE, Houston, TX (1993), p. 3913.
102. W.P. Yang, D. Costa, P. Marcus, *J. Electrochem. Soc.* 141, 2669 (1994).
103. V. Maurice, W. Yang, P. Marcus, *J. Electrochem. Soc.* 143, 1182 (1996).
104. L.J. Oblonsky, M.P. Ryan, H.S. Isaacs, *J. Electrochem. Soc.* 145, 1922 (1998).
105. D. Hamm, K. Ogle, C.-O.A. Olsson, S. Weber, D. Landolt, *Corros. Sci.* 44, 1443 (2002).
106. I. Olefjord, B.-O. Elfström, *Corrosion* 38, 46 (1982).
107. B. Brox, I. Olefjord, *Surf. Interf. Anal.* 13, 3 (1988).

108. E. De Vito, P. Marcus, *Surf. Interf. Anal.* 19, 403 (1992).
109. C.-O.A. Olsson, S.E. Hörnström, *Corros. Sci.* 36, 141 (1994).
110. V. Maurice, W. Yang, P. Marcus, *J. Electrochem. Soc.* 145, 909 (1998).
111. H.W. Hoppe, S. Haupt, H.-H. Strehblow, *Surf. Interf. Anal.* 21, 514 (1994).
112. H.W. Hoppe, H.-H. Strehblow, *Corros. Sci.* 31, 167 (1990).
113. T. Jabs, P. Borthen, H.-H. Strehblow, *J. Electrochem. Soc.* 144, 1231 (1997).
114. P. Marcus, J.-M. Grimal, *Corros. Sci.* 33, 805 (1992).
115. L.J. Oblonsky, M.P. Ryan, *J. Electrochem. Soc.* 148, B405 (2001).
116. A. Machet, A. Galtayries, S. Zanna, L. Klein, V. Maurice, P. Jolivet, M. Foucault, P. Combrade, P. Scott, P. Marcus, *Electrochim. Acta* 49 (2004) 3957.
117. D. Schaepers, H.-H. Strehblow, *J. Electrochem. Soc.* 142, 2210 (1995).
118. D. Schaepers, H.-H. Strehblow, *Corros. Sci.* 39, 2193 (1997).
119. C. Schmidt, H.-H. Strehblow, *J. Electrochem. Soc.* 145, 834 (1998).
120. C. Schmidt, H.-H. Strehblow, *Surf. Interf. Anal.* 27, 984 (1999).
121. P. Druska, H.-H. Strehblow, S. Golledge, *Corros. Sci.* 38, 835 (1996).
122. P. Druska, H.-H. Strehbow, *Corros. Sci.* 38, 1369 (1996).
123. P. Keller, H.-H. Strehblow, *Z. Phys. Chem.* 219, 1481 (2005).
124. H.-H. Strehblow, C.J. Doherty, *J. Electrochem. Soc.* 125, 30 (1978).
125. H.-H. Strehblow, C.M. Melliar Smith, W.M. Augustyniak, *J. Electrochem. Soc.* 125, 915 (1978).
126. H. Habazaki, M.A. Paez, K. Shimizu, P. Skeldon, G.E. Thomson, G.C. Wood, X. Zhou, *Corros. Sci.* 38, 1033 (1996).
127. X. Zhou, G.E. Thompson, H. Habazaki, K. Shimizu, P. Skeldon, G.C. Wood, *Thin Solid Films* 293, 327 (1997).
128. W. Schmickler, J.W. Schultze, in *Modern Aspects of Electrochemistry*, Vol. 17, J.O'M. Bockris, B.E. Conway, R.E. White, eds., Plenum Press, New York (1986), p. 357.
129. J.W. Schultze, M.M Lohrengel, *Electrochim. Acta* 45, 3193 (2000).
130. S. Trasatti, *J. Electroanal. Chem.* 52, 313 (1974).
131. R. Gomer, S. Tryson, *J. Chem. Phys.* 66, 4413 (1977).
132. D.M. Kolb, *Z. Phys. Chem. NF* 154, 179 (1987).
133. D. Lützenkirchen-Hecht, H.-H. Strehblow, *Electrochim. Acta* 43, 2957 (1998).
134. C. Noguet, M. Papiero, J.P. Zierlinger, *Phys. Stat. Sol. (a)* 24, 565 (1974).
135. J.P. Zielinger, M. Tapiero, C. Noguet, *Phys. Stat. Sol (a)* 33, 155 (1985).
136. H.-H. Strehblow, U. Collisi, P. Druska, International Symposium on Control of Copper and Copper Alloys Oxidation, Rouen 1992, Edition de la Revue de Métallurgie, Paris, France (1993), p. 33.
137. F. Di Quarto, M. Santamaria, C. Sunseri, Photoelectrochemical techniques in corrosion science, in *Analytical Methods in Corrosion Science and Engineering*, P. Marcus, F. Mansfeld, eds., CRC Press, Taylor & Francis, Boca Raton, FL (2006), pp. 697–732.
138. J.W. Schultze, M.M. Lorhengel, *Electrochim. Acta* 45, 24999 (2000).
139. M. von Ardenne, *Tabellen zu Angewandten Physik*, 4th edn, VEB Deutscher Verlag der Wissenschaften, Berlin, Germany (1979).
140. D. Lützenkirchen-Hecht, H.-H. Strehblow, Synchrotron methods for corrosion research, in *Analytical Methods in Corrosion Science and Engineering*, P. Marcus, F. Mansfeld, eds., Taylor & Francis, Boca Raton, FL (2006), pp. 169–235.
141. H.-H. Strehblow, D. Lützenkirchen-Hecht, Spectroscopies, scattering and diffraction techniques, in *Schreir's Corrosion*, Vol. 2: *Corrosion in Liquids, Corrosion Evaluation*, T. Richardson, R.A. Cottis, J.D. Scantlebury, S.B. Lyon, P. Skeldon, G.E. Thompson, R. Lindsay, M. Graham, eds., Elsevier, Amsterdam, the Netherlands (2010), pp. 1374–1404.
142. V. Maurice, P. Marcus, Scanning tunneling microscopy and atomic force microscopy, in *Analytical Methods in Corrosion Science and Engineering*, P. Marcus, F. Mansfeld, eds., Taylor & Francis, Boca Raton, FL (2006), pp. 133–168.

143. V. Maurice, P. Marcus, Scanning probe microscopies, *Schreir's Corrosion*, Vol. 2: *Corrosion in Liquids, Corrosion Evaluation*, T. Richardson, R.A. Cottis, J.D. Scantlebury, S.B. Lyon, P. Skeldon, G.E. Thompson, R. Lindsay, M. Graham, Elsevier, Amsterdam, the Netherlands (2010), pp. 1430–1442.
144. V. Maurice, H. Talah, P. Marcus, *Surf. Sci.* 284 (1991) L431.
145. V. Maurice, H.-H. Strehblow, P. Marcus, *Surf. Sci.* 458, 185 (2000).
146. J. Kunze, H.-H. Strehblow, G. Staikov, *Electrochem. Commun.* 6, 132 (2004).
147. A. Foelske, J. Kunze, H.-H. Strehblow, *Surf. Sci.* 554, 10 (2004).
148. H.-H. Strehblow, V. Maurice, P. Marcus, *Electrochim. Acta*, 46, 3755 (2001).
149. J. Kunze, V. Maurice, L.H. Klein, H.-H. Strehblow, P. Marcus, *Electrochim. Acta* 48, 1 (2003).
150. J. Kunze, V. Maurice, L.H. Klein, H.-H. Strehblow, P. Marcus, *J. Electroanal. Chem.* 554–555, 113 (2003).
151. P. Borthen, B.J. Hwang, H.-H. Strehblow, D.M. Kolb, *J. Phys. Chem. B* 104, 5078 (2000).
152. F. Besenbacher, J.K. Norskov, *Prog. Surf. Sci.* 44, 5 (1993).
153. L.H. Dubois, *Surf. Sci.* 119, 399 (1982).
154. H.H. Niehus, *Surf. Sci.* 130, 41 (1983).
155. J. IHaase, H.-J. Kuhr, *Surf. Sci.* 203, L695 (1988).
156. Y. Xu, M. Mavrikakis, *Surf. Sci.* 494, 131 (2001).
157. F. Wiame, V. Maurice, P. Marcus, *Surf. Sci.* 601, 1193 (2007).
158. V. Maurice, L.H. Klein, H.-H. Strehblow, P. Marcus, *J. Electrochem. Soc.* 146, 524 (1999).
159. J. Kunze, V. Maurice, L.H. Klein, H.-H. Strehblow, P. Marcus, *J. Phys. Chem. B*, 105, 4263 (2001).
160. M.M. Islam, B. Diawara, V. Maurice, P. Marcus, *J. Mol. Struct. THEOCHEM* 903, 41 (2009).
161. N. Ikemiya, T. Kobo, S. Hara, *Surf. Sci.* 323, 81 (1995).
162. J. Kunze, V. Maurice, L.H. Klein, H.-H. Strehblow, P. Marcus, *Corros. Sci.* 46, 245 (2004).
163. D. Lützenkirchen-Hecht, C.U. Waligura, H.-H. Strehblow, *Corros. Sci.* 40, 1037 (1998).
164. R.W.G. Wyckoff, *Crystal Structures*, Vol. 1, 2nd edn, Interscience Publishers, New York (1963), 331pp.
165. M.P. Ryan, R.C. Newman, G.E. Thompson, *J. Electrochem. Soc.* 142, L 177 (1995).
166. M.F. Toney, A.J. Davenport, L.J. Oblonsky, M.P. Ryan, C.M. Vitus, *Phys. Rev. Lett.* 79, 4282 (1997).
167. A.J. Davenport, L.J. Oblonsky, M.P. Ryan, M.F. Toney, *J. Electrochem. Soc.* 147, 2162 (2000).
168. R. C. Weast, *Handbook of Chemistry and Physics*, Vol. 61, CRC Press, Boca Raton, FL (1981).
169. C. Graeves, *J. Solid State Chem.* 49, 325 (1983).
170. E.E. Rees, M.P. Ryan, D.S. MacPhail, *Electrochem. Solid-State Lett.* 5, B21 (2002).
171. V. Maurice, H. Talah, P. Marcus, *Surf. Sci.* 304, 98 (1994).
172. S.L. Yau, F.-R. Fan, T.P. Moffat, A.J. Bard, *J. Phys. Chem.* 98, 5493 (1994).
173. T. Suzuki, T. Yamada, K. Itaya, *J. Phys. Chem.* 100, 8954 (1996).
174. D. Zuili, V. Maurice, P. Marcus, *J. Electrochem. Soc.* 147, 1393 (2000).
175. M. Nakamura, N. Ikemiya, A. Iwasaki, Y. Suzuki, M. Ito, *J. Electroanal. Chem.* 566, 385 (2004).
176. N. Hirai, H. Okada, S. Hara, *Transaction JIM* 44, 727 (2003).
177. O.M. Magnussen, J. Scherer, B.M. Ocko, R.J. Behm, *J. Phys. Chem. B* 104, 1222 (2000).
178. J. Scherer, B.M. Ocko, O.M. Magnussen, *Electrochim. Acta* 48, 1169 (2003).
179. V. Maurice, L.H. Klein, P. Marcus, *Surf. Interf. Anal.* 34, 139 (2002).
180. P. Marcus, V. Maurice, H.-H. Strehblow, *Corros. Sci.* 50, 2698 (2008).
181. N. Pineau, C. Minot, V. Maurice, P. Marcus, *Electrochem. Solid-State Lett.* 6, B47 (2003).
182. A. Bouzoubaa, B. Diawara, V. Maurice, C. Minot, P. Marcus, *Corros. Sci.* 51, 2174 (2009).
183. A. Seyeux, V. Maurice, L.H. Klein, P. Marcus, *J. Solid State Electrochem.* 9, 337 (2005).
184. A. Seyeux, V. Maurice, L.H. Klein, P. Marcus. *J. Electrochem. Soc.* 153, B453 (2006).
185. V. Maurice, V. Inard, P. Marcus, in *Critical Factors in Localized Corrosion III*, P.M. Natishan, R.G. Kelly, G.S. Frankel, R.C. Newman, eds., The Electrochemical Society Proceedings Series, PV 98-17, Pennington, NJ (1999), 552pp.
186. V. Maurice, L.H. Klein, P. Marcus, *Electrochem. Solid-State Lett.* 4, B1 (2001).
187. D. Zuili, V. Maurice, P. Marcus, *J. Phys. Chem. B* 103, 7896 (1999).

6

Passivity of Austenitic Stainless Steels

Clive R. Clayton
State University of New York

Ingemar Olefjord
Chalmers University of Technology

CONTENTS

6.1 Introduction 327
6.2 Barrier and Deposit Layers 328
6.3 Boundwater in Passive Films 331
6.4 Roles of Molybdenum in the Passivity of Austenitic Stainless Steel 332
6.5 Roles of Nitrogen in the Passivity of Austenitic Stainless Steel 335
6.6 Alloy Surface Layers 339
6.7 Summary 346
Acknowledgments 346
References 346

6.1 Introduction

Austenitic stainless steels appear to have significantly greater potential for aqueous corrosion resistance than their ferritic counterparts. This is because the three most commonly used austenite stabilizers, Ni, Mn, and N, all contribute to passivity. As in the case of ferritic stainless steel, Mo, one of the most potent alloying additions for improving corrosion resistance, can also be added to austenitic stainless steels in order to improve the stability of the passive film, especially in the presence of Cl ions. The passive film formed on austenitic stainless steels is often reported to be duplex, consisting of an inner barrier oxide film and outer deposit hydroxide or salt film.

Passivation is generally believed to take place by the rapid formation of surface-adsorbed hydrated complexes of metals, which are sufficiently stable on the alloy surface that further reaction with water enables the formation of a hydroxide phase that, in turn, rapidly deprotonates to form an insoluble surface oxide film. Failure in any of these stages would lead to continued active dissolution. The passivation potential is critical to this process, in part because it governs the oxidation state of the metal, which in turn governs its solubility. In addition, the electric field strength has to be sufficient to cause deprotonation of the surface hydroxide phase in order to enable the oxide barrier film to become established. No evidence has been found by surface studies that passivity of austenitic stainless steels is possible by formation of a simple hydroxide film. It is perhaps surprising that the passive film formed on austenitic stainless steels does not always contain each of the alloying elements added to stabilize the austenitic phase, even when such additions appear to improve the chemical stability of the steel. Ni exemplifies this behavior.

To understand the influence of alloying elements on the passivity of stainless steels, researchers have combined electrochemical and surface analysis. Polarization diagrams provide the first indication of the overall influence of alloy additions on the active-passive transition, passivity, and pitting resistance. However, surface analysis by x-ray photoelectron spectroscopy (XPS) of prepassivated surfaces provides a direct observation of the location and the chemical state of an alloying element. Such information is invaluable for the development of a working model of the passivation process. The following surface layers have been attributed to the passivation process: barrier oxide layer, salt deposit layer, and alloy surface layer.

Alloying of transition elements, not surprisingly, often results in the formation of several different oxidation states. Some oxidation states are apparent at only a trace level and may be excluded from the general discussion. The main parameters that determine the oxidation state of an element in a passive film are:

- The passivation potential or oxidizing power of the solution
- The pH of the test electrolyte
- Local pH within highly hydrated passive films
- The age of the passive film
- The location of the ion relative to the external surface

For the purpose of a general introduction to the nature of the passive films formed on stainless steels, we shall consider one of the media most corrosive to stainless steels, i.e., acidic chloride solutions. In this discussion we shall consider several studies that include a range of stainless steel compositions. Trends in the nature of the passive film formed on these alloys will be emphasized.

It is common to find that the passive film is duplex, consisting of an inner oxide barrier film that is considerably less hydrated than the often thinner, outer deposit film. The outer film often contains salts or hydroxides of the alloy constituent metals. Although the inner oxide is the primary diffusion barrier against egressing cations and ingressing anions such as Cl^-, there is some evidence that in some cases the deposit film may serve to control ion transport in the passive film by the influence of fixed charges. In other cases it has been considered to be a relatively insoluble film that serves to protect the metal surface from aggressive ions while a barrier film develops underneath.

It will also be shown that the thickness of the passive films, while varying with the potential of passivation, is commonly only a few nanometers. Finally, we shall see that in austenitic stainless steels the active stage of repassivation often results in considerable change in composition at the alloy surface. The possible role of this modified alloy layer in the overall passivation process will be discussed.

6.2 Barrier and Deposit Layers

Surface analysis has provided the basis of our knowledge of the duplex nature of the passive film formed on austenitic stainless steels. Destructive depth profiling using inert gas ion sputter etching and nondestructive variable-angle electron spectroscopy are commonly used to probe the composition of these layers. Both types of analysis will be illustrated.

TABLE 6.1

Composition of AVESTA 832SL (wt%)

C	Mn	P	S	Si	Cr	Ni	Mo	Cu	N
0.040	1.64	0.020	0.003	2.62	16.7	15.0	4.28	0.10	0.27

The first example involves the use of XPS with inert gas ion sputter depth profiling. This study [1] involved the active dissolution and passivation of the stainless steel, AVESTA 832SL (Table 6.1).

The test electrolyte was a deaerated solution of 0.1 M HCl + 0.4 M NaCl. The polarization behavior of the alloy is compared in Figure 6.1 with the alloy constituent metals. The corrosion potential is seen to be more noble than that of Cr and Fe but close to that of Ni and Mo. This is quite typical for the corrosion potential of austenitic alloys. It will be shown that Ni and Mo are enriched on the surface in the metallic state during anodic dissolution. As a consequence, the corrosion potential becomes close to the corrosion potential of these elements. From the polarization data it is suggested that both Cr and Mo are more likely to contribute to passivity, especially the barrier layer, than Fe or Ni and that Mo will contribute only in a narrow range of potential before it undergoes transpassive dissolution. These simple indications will be shown to be only partially correct. In Figure 6.2 are typical XPS spectra of the outer region of the surface films obtained in the same study for the alloy polarized at passive potentials (–100 and 500 mV vs. SCE).

The thicknesses of the films formed at the passive potentials –100 and 500 mV (SCE) were calculated to be 1.0 and 1.5 nm, respectively. It can be seen from Figure 6.2 that the Ni spectra contain almost no detectable oxidized Ni. Therefore, Ni does not contribute directly to the structure of the barrier or deposit films. This is surprising because Ni is well known to contribute to pitting resistance [2]. In addition, it can be seen that the film formed at 500 mV (SCE), which is several hundred millivolts above the transpassive potential of Mo, contains contributions from Mo^{4+} (as MoO_2), which has been shown to be the main cation in the passive film formed on Mo in 0.1 M HCl [3]. Therefore, it would appear

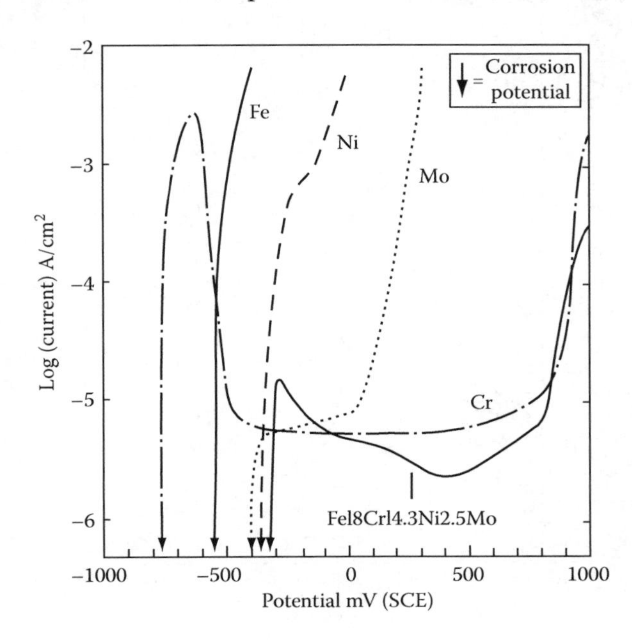

FIGURE 6.1
Anodic polarization curves of the pure metals Fe, Cr, Mo, and Ni and of an austenitic stainless steel, Fel8Crl4.3Ni2.5Mo (at%) (Fe16.7Cr15.0Ni4.28Mo wt%) exposed to 0.1 M HCl + 0.4 M NaCl at 25°C. Sweep rate, 3 mV/s. (From Olefjord, I. et al., *J. Electrochem. Soc.*, 132, 2854, 1985.)

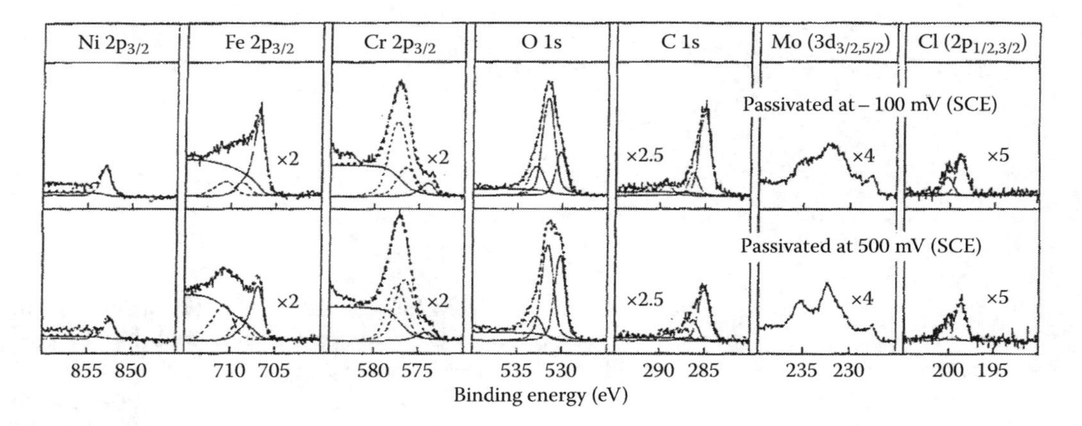

FIGURE 6.2
XPS spectra recorded from Fel8Crl4.3Ni2.5Mo (at%) after passivation at –100 mV (SCE) and 500 mV (SCE). (From Olefjord, I. et al., *J. Electrochem. Soc.*, 132, 2854, 1985.)

that Mo contributes directly to the passive film formed on austenitic stainless steel. The general spectra of Figure 6.2 provide no direct information concerning the duplex nature of the passive film. However, in Figure 6.3 is a depth profile produced by argon ion sputter depth profiling of the passive film formed at 500 mV (SCE) for the same alloy. From this figure it can be seen that there exist two layers, an inner oxide-based layer (barrier layer)

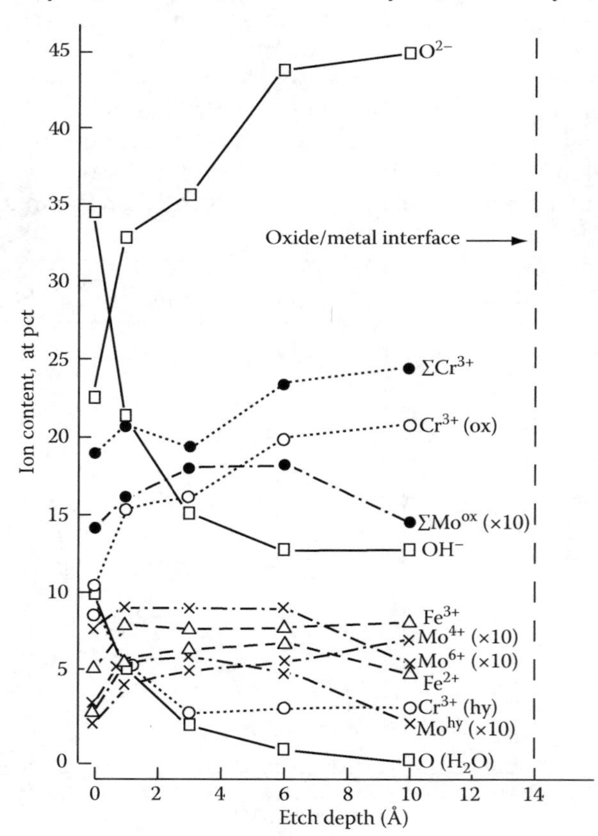

FIGURE 6.3
Ion content vs. etch depth for the austenitic stainless steel Fe18Cr14.3Ni2.5Mo polarized at 500 mV (SCE). (From Olefjord, I. et al., *J. Electrochem. Soc.*, 132, 2854, 1985.)

constituting the majority of the film and an outer hydrated layer (deposit layer), which is also richer in Mo^{6+} than Mo^{4+}. We shall discuss the role of Mo in the passivation process later. A further striking observation is that Cr^{3+} is found in greater abundance in the passive film in both the inner and outer layers than Fe, despite the fact that Fe is the majority element in the alloy. Clearly, Cr is the main passivating species in stainless steels.

6.3 Boundwater in Passive Films

A series of reported studies provide some insights into the nature and the role of bound water in the passivity of austenitic stainless steels [4–6]. The studies primarily focused on the passivity of type 304 stainless steel, passivated in deaerated 0.5 M H_2SO_4. Radiotracer studies were conducted with tritiated water in order to determine the quantity of water bound into the lattice following the formation of the passive film. In addition, the rate of desorption of bound water from the film was determined by dioxane solvent extraction. Dynamic rupture and self-repair of the passive film were seen to be critically influenced by the nature of the bound water. Two classes of bound water were determined from XPS, radiotracer, Coulombic titration, pitting incubation, and noise analysis [4–6]. The two classes of bound water were of the following types: (a) M-H_2O and M-OH (aquo and olation groups) and (b) M-O or M-OOH with oXo and olation bridges.

In Figure 6.4 the proposed structure of the passive films is presented schematically [4]. The bound water was proposed to be associated largely with Cr but with its nature governed by the applied potential, temperature, and time at potential. It was established that the water content in the passive films formed on 304 stainless steel in deaerated 0.5 M H_2SO_4 changed abruptly at the critical potential of 400 mV (SCE) as shown in Figure 6.5 [5]. Later reflection high-energy electron diffraction (RHEED) studies based on the same alloy composition and test electrolyte [7] revealed that corresponding structural changes with potential and time at potential were associated with chromium and iron products composing the barrier and deposit layers, respectively. Passive films that were formed after 1 min were found to be amorphous. At longer times of passivation some crystallization was revealed. The inner barrier film, which was found by XPS to consist mostly of Cr_2O_3, was seen by RHEED to be amorphous at all potentials studied in the passive range, except

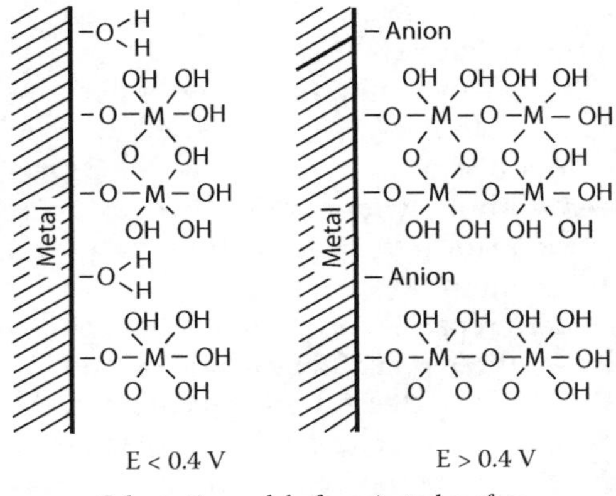

FIGURE 6.4
The Okamoto model of the structure of the passive film. (From Okamoto, G., *Corros. Sci.*, 13, 471, 1973.)

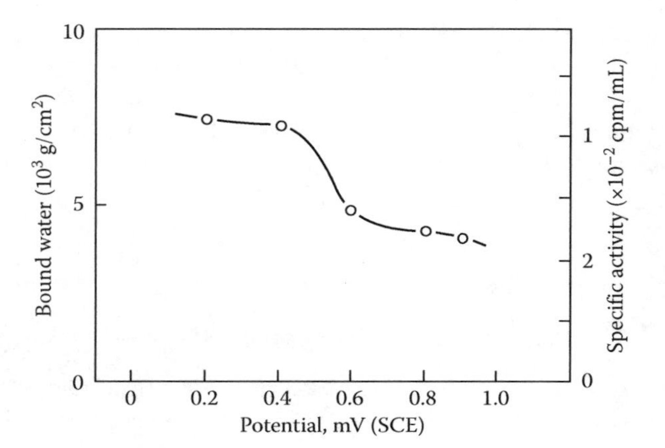

FIGURE 6.5
Amount of bound water in the passive film vs. the potential. (From Saito, H. et al., *Corros. Sci.*, 19, 693, 1979.)

at the transpassive potential, where it crystallized. Low-potential passive films (<400 mV) were seen to convert to the same structure as higher potential passive films (>400 mV), given sufficient time for field-induced deprotonation to take place. The chromium content of the deposit layer was observed to be $Cr(OH)_3$ in the presence of acidic SO_4^{2}/Cl and CrOOH in the presence of 0.5 M H_2SO_4. Highly hydrated Fe compounds (including green rusts) were also observed, which rapidly deprotonated with corresponding charge-balancing oxidation of Fe^{2+} at the critical potential of 400 mV (vs. SCE).

6.4 Roles of Molybdenum in the Passivity of Austenitic Stainless Steel

One of the most effective elements added to austenitic stainless steel, and for that matter even ferritic stainless steel, in order to improve pitting resistance is Mo [8]. Molybdenum, however, is a highly versatile element, existing in the passive film in a number of oxidation states. In the case of the hexavalent state it has been observed in both the cationic and anionic states, namely as molybdenum trioxide and ferrous molybdate. It has most commonly been reported to exist in the quadrivalent state as molybdenum dioxide and oxyhydroxide.

The majority of the studies reported agree that Mo^{4+} is incorporated into the inner region of the passive film, whereas Mo^{6+} is present in the outer layer. However, the outer layer is variously defined as a salt layer or an extension of the barrier layer.

Numerous studies have attempted to elucidate the role of Mo in the passivity of stainless steel. It has been proposed from XPS studies that Mo^{6+} forms a solid solution with CrOOH with the result that Mo is inhibited from dissolving trans-passively [9]. Others have proposed that active sites are rapidly covered with molybdenum oxyhydroxide or molybdate salts, thereby inhibiting localized corrosion [10]. Yet another study proposed that molybdate is formed by oxidation of an Mo dissolution product [11]. The oxyanion is then precipitated preferentially at active sites, where repassivation follows. It has also been proposed that in an oxide lattice dominated by three-valent species of Cr and Fe, ferrous ions will be accompanied by point defects. These defects are conjectured to be canceled by the presence of four- and six-valent Mo species [1]. Hence, the more defect-free film will be less able to be penetrated by aggressive anions. A theoretical study proposed a solute vacancy interaction model in which Mo^{6+} is assumed to interact electrostatically with oppositely charged cation vacancies [12]. As a consequence, the cation vacancy flux is gradually reduced in the passive film from the solution side to the metal–film interface, thus hindering vacancy condensation at the metal–oxide interface, which the authors postulate acts as a precursor for localized film breakdown [12].

The XPS spectra (Figure 6.2) show that the passive film formed in Cl^--containing solutions contains Cl^-. It has been shown [13,14] that Cl^- is present in both the barrier layer and the hydroxide layer. It is believed that chloride ions substitute oxygen in the passive film. Thereby the stability of the film is reduced because the number of cation-oxygen bonds decreases and dissolution of the oxide is enhanced. In one study [14] high-alloyed steels, Fe20Cr20Ni, Fe20Cr20Ni6Mo, and Fe20Cr20Ni6Mo0.2N were exposed to 0.1 M HCl + 0.4 M NaCl at –75 mV (SCE) for 10 min and 2 h. It was found that the Cl^- content of the Mo-free alloy was considerably higher than the Cl^- content of the Mo-containing alloy. Figure 6.6 [15] shows the experimental result. It was suggested [14,15] that the mechanism for lowering the Cl^- content in the film is due to the ability of Mo to form soluble stable oxo-chloro complexes in diluted HCl solutions. Reference [16] gives details about the Mo complexes. Table 6.2 summarizes some of the possible soluble stable Mo-oxo-chloro complexes. Formation of these complexes, during the passivation process, lowers the Cl^- content in the passive film and thereby make the film more resistant for pitting corrosion.

It has been suggested that molybdate ions may act as cation-selective species in the deposit layer, thus producing a bipolar film with the anion-selective component

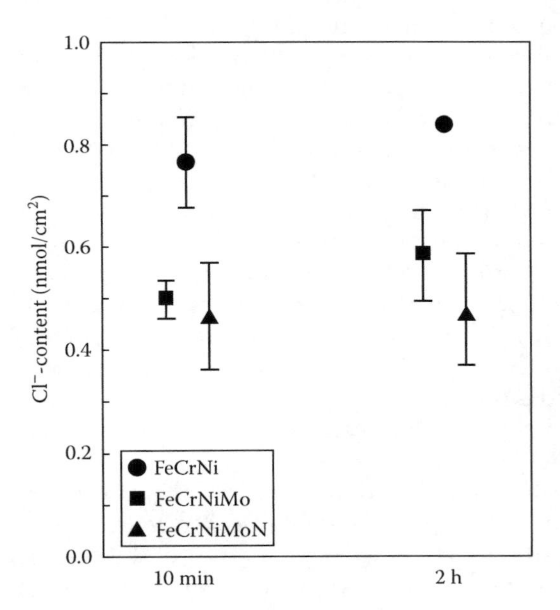

FIGURE 6.6
Total amount of Cl^- in the passive film formed at –75 mV (SCE) in 0.1 M HCl + 0.4 M NaCl. (From Falkenberg, F. and Olefjord, I., *Proceedings of the International Symposium in Honor of Professor Norio Sato: Passivity and Localized Corrosion*, M. Seo, B. MacDougall, H. Takahashi, and R.G. Kelly, eds., Electrochemical Society, Pennington, NJ, p. 404, 1999.)

TABLE 6.2

Soluble Stable Mo-Oxo-Chloro Complexes in Aqueous HCl Solutions

Mo(II)	$[Mo_6Cl_8]^{4+}$
Mo(III)	$[Mo_2(OH)_2Cl_n]\,(H_2O)_{8-n}]^{(4-n)+}$, $n < 3$ $[MoCl_5(H_2O)]^{2-}$
Mo(IV)	$[Mo_3O_4Cl_n]\,(H_2O)_{9-n}]^{(4-n)+}$, $n < 4$
Mo(V)	$[Mo_2O_4Cl_4(H_2O)_2]^{2-}$ $[Mo_2O_3Cl_8]^{4-}$
Mo(VI)	$Mo(OH)_5Cl$

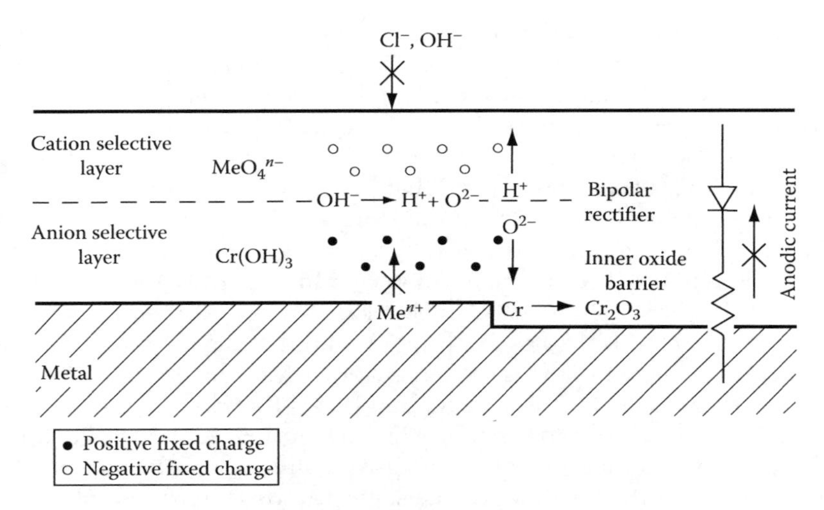

FIGURE 6.7
Schematically illustration of the bipolar model of the passive film. (From Clayton, C.R. and Lu, Y.C., *J. Electrochem. Soc.*, 133, 2465, 1986.)

represented by the barrier layer, which in acidic media is predicted to be anion selective [17]. The bipolar model of passivity is represented schematically in Figure 6.7.

Evidence in support of several of these models has been reported. XPS studies of the passive and transpassive films formed on Mo in deaerated 0.1 M HCl [3] established that molybdate was absent from both surface films. In a later study the same authors used a twin potentiostat arrangement, with a second working electrode of either Fe, Cr, or Ni that was polarized near an Mo electrode at the same potential (–180 mV vs. SCE) [18]. At this potential Mo and Cr are passive, while Ni and Fe are active. In this work it was shown that for the Fe-Mo and Ni-Mo electrode couples, iron or nickel molybdate was observed on the passive Mo surface. In the case of the Cr-Mo couple, molybdate was observed only on the passive film of Cr. This work was also able to show that transpassivity of Mo at 250 mV (SCE) was suppressed in the presence of Fe, which formed a molybdate salt on the surface of Mo. This indicated evidence of a possible mechanism by which Mo can remain passive in stainless steels at higher potentials than the transpassive potential of Mo. In addition, this work supported the idea that soluble molybdate anions can redeposit at active sites.

In other work, a major prediction of the bipolar model of passivity [19] was tested. This was that in the presence of a bipolar passive film with the cation-selective layer (molybdate anions, for example) in the outer layer or deposit film, there would be a tendency to increase O–H bond stretching and eventual deprotonation due to the conjoint effect of the electric field associated with the surface charge on the metal and the strong negative fixed charge on the oxyanion. As a result of this work it was shown by XPS analysis that when a passive film formed on an Fe-19Cr9Ni alloy was doped in solution with molybdate anions, there was a significant decline in the hydroxyl concentration in the film in favor of oxide anions. In addition, the concentration ratio of chromium oxide to hydroxide was increased. This work has several ramifications. First, it provides some support for the bipolar model of passivity, which would suggest that molybdate can rectify the transport of ions through the passive film in favor of inhibiting egress of cations other than protons as well as inhibiting the ingress of chloride ions that aid in the dissolution of the film. Second, the promotion of deprotonation of the passive film would favor oxidation of ferrous ions to ferric in order to establish charge neutrality. This would also have the effect of reducing the defect concentration in the trivalent-dominated oxide lattice in accordance with a previous suggestion [1].

So far, the role of molybdenum in the passive film has been discussed. However, several workers have postulated that Mo may exclusively or additionally control the kinetics of the active dissolution process. These models can be separated into two kinds: (a) insoluble salt models and (b) surface alloy models. One of the earliest reports to indicate the importance of insoluble salts in the repassivation kinetics of ferrous alloys involved studies of Fe-Mo alloys [20]. It was shown that when the alloy content exceeded 5% a protective chloride salt film stabilized by ferrous molybdate precipitated from the test solution after the initial dissolution of molybdenum. This work suggests that a similar mechanism may occur in Mo-bearing austenitic stainless steels in chloride solutions. The stability of the molybdenum-bearing passive film in chloride solutions has also been attributed to the ability of Mo to form insoluble chloride complexes at the base of pits, thereby arresting chloride ions in the pit solution, which in turn enables subsequent repassivation to occur. This has been shown by Auger electron spectroscopy (AES) of pit surfaces [21]. XPS analysis has revealed [15,22] the Mo(II) valent compound β-$MoCl_2$ formed in the passive film on high-alloyed stainless steels passivated in hydrochloric acid. β-$MoCl_2$ is the only one of the nonsoluble chloride compounds that can easily be detected by XPS. Other insoluble high-Mo-valency chloride compounds such as α-$MoCl_3$, MoOCl, $MoOCl_2$ and MoO_2Cl [16] are not easily detected due to the overlap of their Mo signals with Mo-oxide peaks.

The activity of Cl ions in the pits of Mo-bearing stainless steels was also shown by AES to be significantly reduced by precipitation of an insoluble chloride complex containing Mo. This behavior is also reflected in the passivation of high-purity Mo in a deaerated solution of 4 M HCl [23]. It has been shown that the passive range of potential for Mo in 4 M HCl is significantly greater than in 0.1 M HCl. XPS analysis of the surface of Mo polarized in the passive range for deaerated 4 M HCl solution provided evidence of a molybdenum oxyhydroxy chloride film. This product was not observed in the exclusively oxide-based passive film formed on Mo in 0.1 M HCl solution. The presence of such an insoluble film on Mo in 4 M HCl indicates that such a film may be formed on the bottom of pits formed on Mo-bearing stainless steels.

It has been shown that the low pH value of pitting solutions is caused by high concentrations of metal chlorides [24]. Therefore the formation of insoluble chloride complexes in pit bottoms suggests that lowering the concentrations of soluble metal chlorides in the pitting solution would cause the pH to shift toward that of the bulk solutions. The conjoint effects of raising pH and deposition of salt films provide conditions more conducive to repassivation [25].

The effect of surface alloy modification by anodic dissolution on the passivation of Mo-bearing austenitic stainless steels is discussed later in a separate section.

6.5 Roles of Nitrogen in the Passivity of Austenitic Stainless Steel

Nitrogen may be added to austenitic stainless steel (commonly 0.2–0.7 wt%) in order to stabilize and strengthen the austenitic phase. However, in the presence of Mo additions very significant improvements in general and localized corrosion have been demonstrated by several workers [26–30]. This early work indicated a strong synergistic effect of Mo and N in the corrosion resistance of austenitic stainless steel.

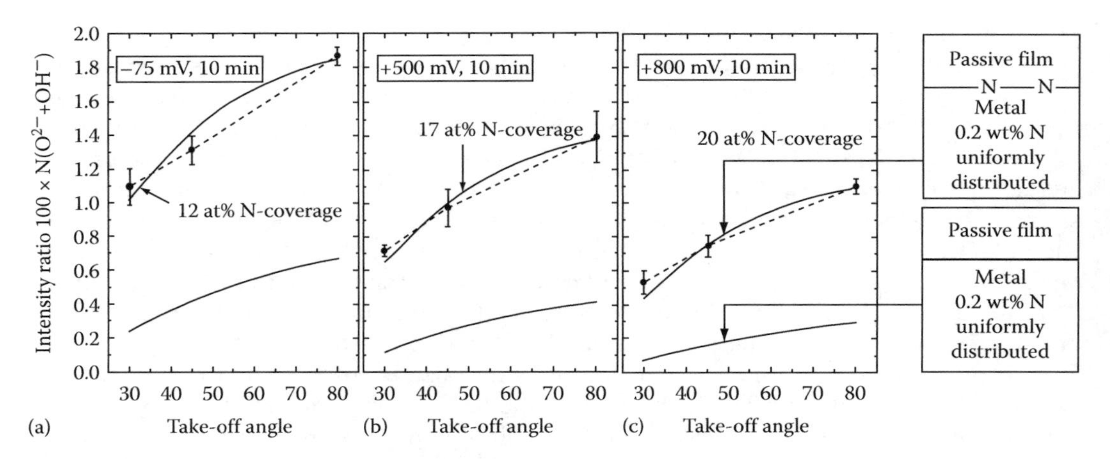

FIGURE 6.8
Measured intensity ratios, 100 × $N/(O^{2-} + OH^{-})$, at the take-off angles 30°, 45°, and 80° after passivation of the Fe20Cr20Ni6Mo0.2N alloy at (a) –75, (b) 500, and (c) 800 mV (SCE). The lower thick line and the upper thin line represent the estimated intensity ratios from the N-distributions shown in the models to the right of the figure. (From Falkenberg, F. and Olefjord, I., *Proceedings of the International Symposium in Honor of Professor Norio Sato: Passivity and Localized Corrosion*, M. Seo, B. MacDougall, H. Takahashi, and R.G. Kelly, eds., Electrochemical Society, Pennington, NJ, p. 404, 1999.)

More recently, it has been demonstrated [31] that N anodically segregates to the oxide–metal interface during passivation. The dots with the error bars in Figure 6.8 [15] show the measured intensity ratios, $N/(O^{2-} + OH^{-})$, recorded by XPS as a function of the take-off angle for an alloy (Fe20Cr20Ni6Mo0.2N) polarized in a 0.1 M HCl + 0.4 M NaCl solution at –75, 500, and 800 mV for 10 min. The distribution of N was estimated by utilizing quantitative analysis described in Ref. [32]. The thick solid lines at the bottom of the three figures represent the expected intensity ratios one should obtain if the nitrogen atoms (concentration 0.2 wt%) are uniformly distributed in the phase below the passive film. However, the measured intensity ratios are significantly higher, which implies that N is enriched at the surface. By assuming a model shown in the upper right corner of the figure it was possible to find a distribution of N that satisfies the measured data. The thin solid lines in Figure 6.8 are theoretical intensity ratios calculated by assuming that N is enriched at the metal–oxide interface and that the bulk concentration is 0.2 wt% N. The theoretical ratios are obtained for N coverage of 12, 17, and 20 at% at the interface.

It has been demonstrated [33] that a strong Mo-Ni-N interaction is likely to occur. In a series of papers on experimental N-bearing alloys #30 and #30c, which were compared with alloy AL6X, it was shown that nitrogen additions strongly improved corrosion resistance [29,30] (Table 6.3). In tests conducted in several acidic chloride solutions it was shown

TABLE 6.3
Chemical Composition of Main Elements of Steels (wt%)

	Fe Balance				
Steel	**Cr**	**Ni**	**Mo**	**N**	**C**
AL6X	20.5	24.7	6.7	0.04	0.02
#30	24.3	19.9	6.1	0.44	0.02
#30c	22.0	20.5	6.0	0.49	0.005

that nitrogen alloying was responsible for lowering the critical current density and passive current density at room temperature and at elevated temperature [29]. Alloy #30 was also tested at 80°C in a solution of 4% NaCl + 1% $Fe_2(SO_4)_3$ + 0.01 M HCl in which it showed no evidence of pitting. This result was all the more significant because it had previously been shown that for austenitic alloys a linear relationship existed between the critical pitting temperature and the compositional factor (wt% Cr + 2.4 wt% Mo) in the same solution [34]. The inference therefore remains that the N-bearing alloy may protect against pitting via an alternative overall mechanism.

In further studies of alloy #30 it was shown by Auger depth profiling [30] (see Figure 6.9) that nitrogen segregated to the alloy surface at the metal–oxide interface during passive dissolution at 500 mV (SCE) for 24 h in deaerated 0.5 M H_2SO_4. In later studies it was shown that N additions strongly influence the alloy composition at the metal–oxide interface [35]. Thus segregation of N was found to coincide with the enrichment of Ni and Cr in 304(N) stainless steel and Ni, Cr, and Mo in Mo-bearing austenitic stainless steels such as 317LX(N), 904L(N), and AL6X(N) (Table 6.4).

It was determined from the N 1s photoelectron spectra that the form in which nitrogen was segregated is a surface nitride [33]. XPS studies of surface nitrides formed on Fe, Cr, Ni, Mo,

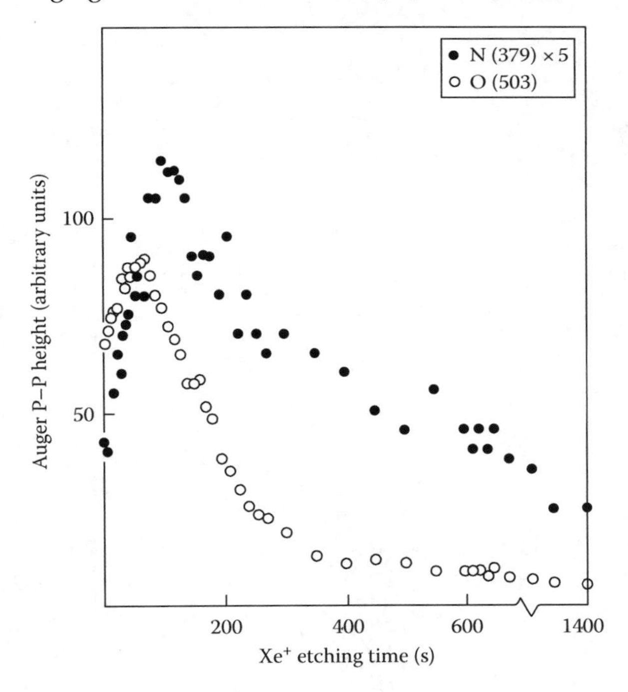

FIGURE 6.9
Auger depth profile for alloy 30 after passivation for 24 h in 0.5 M HCl + 2 M NaCl. Approximate etching rate was 0.01 nm/s. (From Lu, Y.C. et al., *J. Electrochem. Soc.*, 130, 1774, 1983.)

TABLE 6.4

Chemical Composition of Steels (wt%)

	Fe Balance									
Steel	**C**	**Mn**	**P**	**S**	**Si**	**Cr**	**Ni**	**Mo**	**Cu**	**N**
304	0.053	1.77	0.031	0.008	0.41	19.27	8.49	0.36	0.16	0.040
317LX	0.022	1.52	0.018	0.015	0.41	18.43	13.13	3.34	0.01	0.050
AL6X	0.020	1.47	0.030	0.002	0.50	20.45	24.65	6.30	0.19	0.040
904L	0.019	1.50	0.023	0.002	0.44	20.46	24.40	4.51	1.48	0.050

and the stainless steels 304, 317LX, 904L, and AL6X show that the nitride anodically formed on nitrogen-bearing austenitic stainless steels was a mixed nitride. For this work and for the purpose of studying the interaction of anodically segregated N with the individual alloying constituents, a room temperature electrochemical nitriding process was developed [33]. The process involved the cathodic reduction of nitrate ions. The outcome of the treatment was that the same surface nitrides were formed on the stainless steels as formed by anodic segregation. The benefit of the technique was that, unlike the case of ion implantation and plasma or thermal nitriding, a room temperature surface phase could be formed without expected surface damage or sensitization. The advantage of studying surface nitrides over the nitrogen-bearing alloys was reported to be that, whereas anodic segregation is a continuously regenerating process, surface alloying provides a finite quantity of nitrogen that can be monitored by surface analysis after anodic dissolution under active and passive conditions.

As seen in Figure 6.10, the potentiodynamic behavior in deaerated 0.1 M HCl of austenitic stainless steels is beneficially modified by the surface nitriding treatment, in

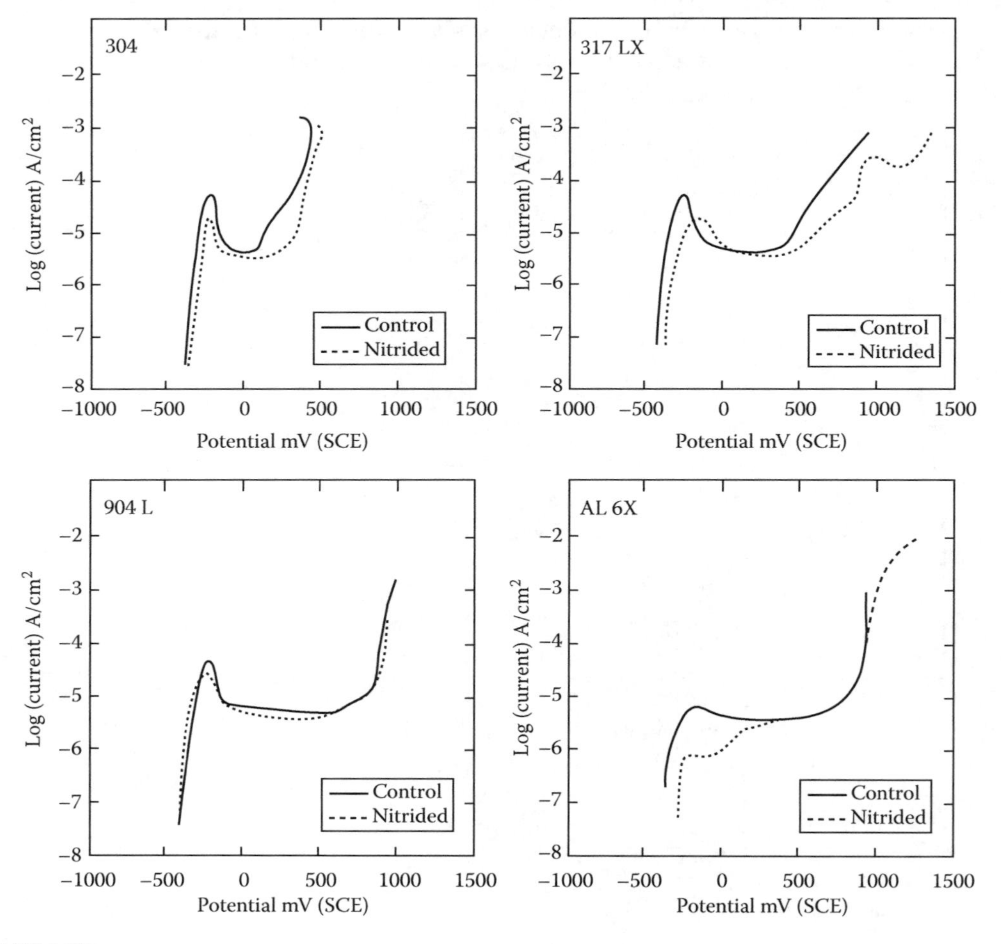

FIGURE 6.10
Polarization curves for surface-nitrided and untreated austenitic stainless steels in deaerated 0.1 M HCl where the specimens were permitted to float to the open-circuit potential before polarization. Sweep rate: 1 mV/s. (From Willenbruch, R.D., *Proceedings of the Sixth International Symposium on Passivity, Part 1*, N. Sato and K. Hashimoto, eds., *Corros. Sci.*, 31, 179, 1990.)

agreement with previous studies of the same steels alloyed with nitrogen [33]. Each of the steels showed that the surface nitride stifles active dissolution and that this effect is more pronounced as the addition of Mo is increased. In the lower alloyed steels, 304 and 317LX, the greatest influence of nitrogen was reported to be in raising the pitting potential. As shown by Figure 6.10, the pitting resistance of surface-nitrided 317LX steel is markedly better than that of type 304 due to the higher Mo content. It is also apparent that the greatest anodic inhibition was achieved by the AL6X alloy, which has the highest Mo content of the alloys studied and in particular is very similar in composition to 904L except for an additional 1.8 wt% Mo. These observations imply that N and Mo support the same mechanism of corrosion inhibition.

The authors also noted that following potentiodynamic polarization from the corrosion potential to 0 mV at a scan rate of 1 mV/s, XPS analysis was still able to detect significant quantities of surface nitride. This is illustrated in Figure 6.11. The most active of the alloys studied, type 304, was determined to have dissolved approximately 20 monolayers. This suggested that the nitride may form a kinetic barrier that is protected by the oxide passive film from rapid protonation to ammonia and ammonium in the active range of potential. In the same study the nitride phase formed on Ni had little effect on anodic behavior in 0.1 M HCl, whereas Fe became anodically activated by nitriding. Chromium exhibited complete removal of the active nose following nitriding. This was attributed to suppression of Cr^{2+} by direct chemical reaction of surface N and Cr forming a Cr^{3+} state as CrN, a precursor of Cr_2O_3, which was identified by XPS. Molybdenum, however, showed some ennoblement of the corrosion potential and transpassive potential. This effect on Mo was later found to be even more pronounced following thermal nitriding [36].

It has been reported that nitrogen alloying and surface nitriding result in a higher metal oxyanion content in the outer layers of the passive films of stainless steels formed in acidic solutions. By inspection of the *E*-pH diagrams of Mo and W, for instance, it is seen that MoO_4^{2-} and WoO_4^{2-} are both more likely to form in the middle to high pH range [37]. The presence of the metal oxyanions in the outer region of passive films formed on stainless steels indicates that the products are developed during the active stage of repassivation, precipitating as salt films. The well-known pitting inhibition derived from metal oxyanions such as molybdate suggests that the presence of such anions in the outer layers of the passive films would tend to reduce the probability of pit initiation.

6.6 Alloy Surface Layers

A stainless steel is normally used at potentials corresponding to the passive range. In this condition only a fraction of the oxidized atoms remain in the passive film; most are dissolved into the solution. The oxidation rate of the metal corresponds to the passive current density. As established earlier, the alloying elements Cr and Mo are enriched in the passive film. The other two main alloying elements, Fe and Ni, could in principle be found in the solution due to selective dissolution of these elements. XPS spectra [1,38–40] recorded from stainless steels after polarization to potentials in the active and passive regions show that the measured Fe content in the metal phase is lower than the composition of the bulk alloy. Fe must therefore be selectively dissolved and can thereby be found enriched in the solution. Ni and Mo, on the other hand, are enriched in the metal phase underneath the oxide. It has been proposed [1,38–40] that during active dissolution and passivation of

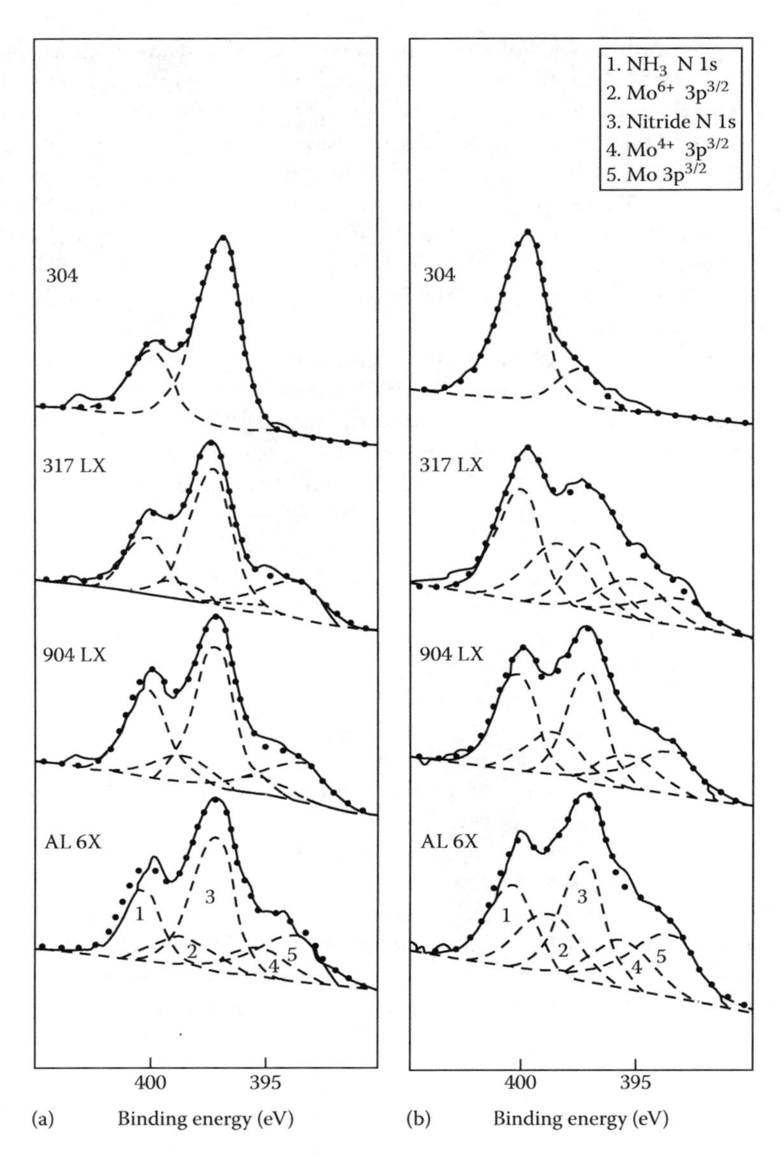

FIGURE 6.11
N 1s spectra for austenitic stainless steels: (a) As nitrided; (b) follow polarization from open circuit to 0 mV. Sweep rate: 1 mV/s. (From Willenbruch, R.D., *Proceedings of the Sixth International Symposium on Passivity, Part 1*, N. Sato and K. Hashimoto, eds., *Corros. Sci.*, 31, 179, 1990.)

stainless steel a thin layer of an intermetallic compound is formed in the outermost layers of the metal phase. Inhibition of the active dissolution of stainless steel by elemental Mo has been considered [41]. It was proposed that Mo serves to decrease active dissolution by binding to active surface sites such as kinks and thus increasing the coordination of more active species. More recently, it has been shown [42] by cyclic polarization and pit propagation rate tests on a series of Fe-Ni-Cr-Mo alloys having Mo contents of 3, 6, and 9 wt% for three Fe/Ni ratios that active dissolution was governed by an Ni-Mo surface complex. These observations led us to the conclusion that in describing the corrosion properties of

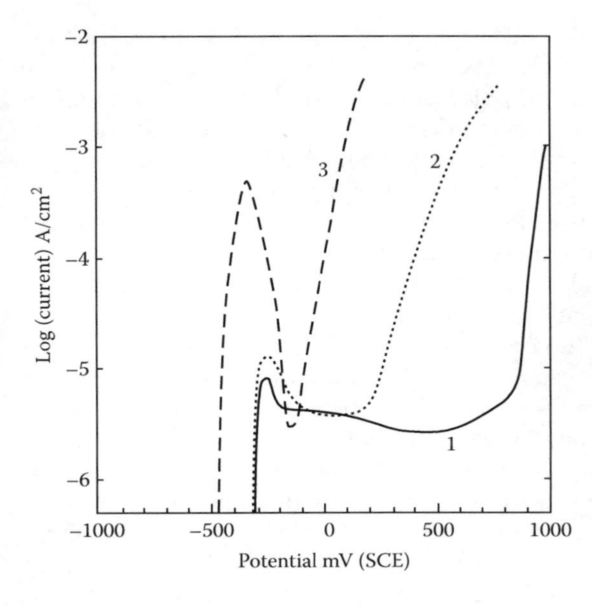

FIGURE 6.12
Polarization diagrams obtained during exposure to 0.1 M HCl + 0.4 M NaCl of the following alloy steels: (1) Fe-200Cr-18Ni-6.1Mo-0.2N; (2) Fe-18Cr-13Ni-2.7Mo; (3) Fe-18Cr-9Ni. (From Olefjord, I., *Surface Characterization*, E. Brune, R. Hellborg, H.J. Whitlow, and O. Hunderi, eds., Wiley-VCH, Weinheim, Germany, eds., p. 291, 1997.)

stainless steel it is not enough to describe the reactions leading to formation of the barrier layer; it is also necessary to describe the reactions taking place in the outermost layers of the metallic phase.

The influence of the alloy composition on the polarization diagrams of austenitic stainless steels polarized in 0.1 M HCl + 0.4 M NaCl is shown in Figure 6.12. The curves represent the following steels: curve 1, 20Cr-18Ni-6.1Mo-0.2N (wt%); curve 2, AISI 316, 18Cr-13Ni-2.7Mo; curve 3, AISI 304, 18Cr-9Ni. The polarization curves show that the high-alloyed steel is passive within a broad potential range. The other two alloys are sensitive to pitting corrosion in this solution. The critical current density is about the same for the two Mo-containing alloys, and it is an order of magnitude larger for the non-Mo-containing steel. The conclusion from the polarization diagrams is that the composition of the metal influences markedly the passivation and the pitting behavior of the steels. It will be shown in the following that synergistic effects exist between the alloying elements.

The spectra in Figure 6.13 were recorded from a high-alloyed austenitic stainless steel (16.7Cr-15Ni-4.3Mo) after polarization in 0.1 M HC1 + 0.4 M NaCl at −320 mV (SCE). The potential represents the active dissolution potential slightly above the corrosion potential. The result is from the same study as Figure 6.2. It appears from the spectra of Figure 6.13

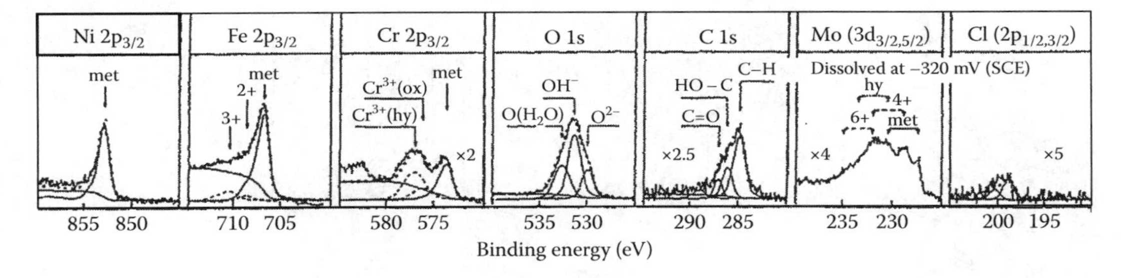

FIGURE 6.13
XPS spectra of an Fe-20Cr-18Ni-6.1Mo-0.2N (wt%) alloy recorded after polarization to the active potential −320 mV (SCR). (From Olefjord, I. et al., *J. Electrochem. Soc.*, 132, 2854, 1985.)

that the intensities of the metallic states are much higher after polarization to the active potential compared with the passive potentials of Figure 6.2. This is, of course, due to the fact that the oxide film formed on the sample polarized at the active potential is much thinner than on the passivated samples. One should in principle expect no oxide or at most a very small amount of oxidized species on the alloy polarized to the active region. However, during the rinsing and the transfer of the sample from the electrochemical cell to the XPS analyzer, the surface is slightly oxidized. The thickness of the oxide formed during handling of the sample polarized to the active region is about 0.5 nm.

The composition of the metal phase can be estimated from the recorded intensities. The dots in Figure 6.14a [1,40] show the apparent compositions of the metal phases of the alloy after polarization to the active potential. The composition is calculated from the measured peak intensities in Figure 6.13, taking into account the yields and the attenuation lengths of the photoelectrons. The horizontal lines in the up and down triangles in Figure 6.14 show the composition of the alloy. It appears from Figure 6.14a that Ni and Mo are enriched in the metal phase, while the elements Fe and Cr are depleted. The depth to which the composition changes from the bulk composition can be only a few atomic planes because of the very low diffusion rate at room temperature. The depth of analysis, which is about three times the attenuation length, is larger than the enriched zone. Hence, the bulk of the alloy contributes to the recorded spectral intensity. Thus, the composition in Figure 6.14a is an apparent composition. The actual surface composition can be found only if the distribution of the elements in the surface region is known. Unfortunately, this is not the case.

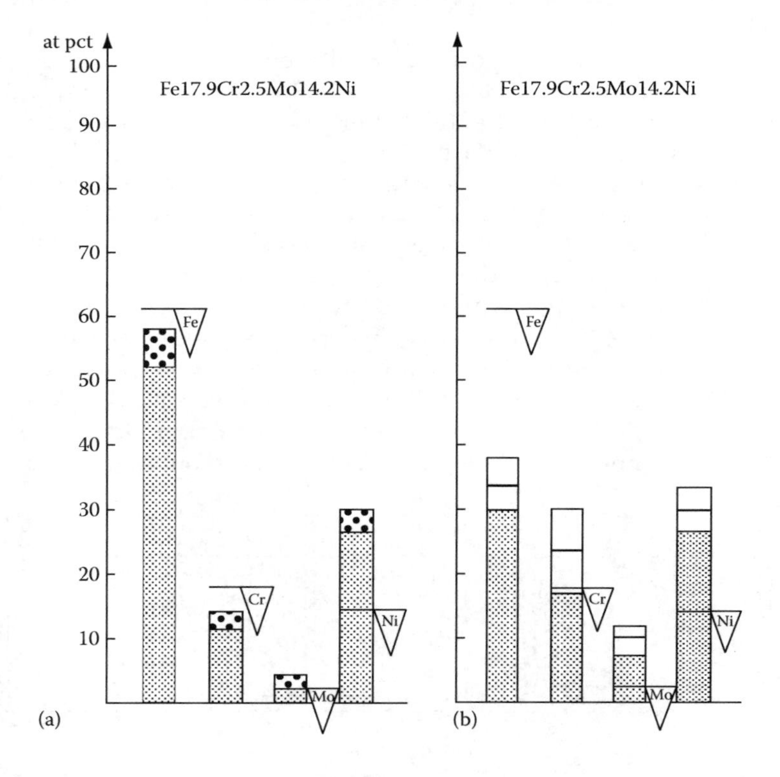

FIGURE 6.14
Composition of the metal phase during dissolution at an active potential. (a) Apparent metal content; (b) estimated metal content in the outermost atomic layer. (From Olefjord, I. et al., *J. Electrochem. Soc.*, 132, 2854, 1985; Brox, B. and Olefjord, I., *Proceedings of Stainless Steel 84*, Institute of Metals, London, p. 134, 1985.)

It is therefore necessary to assume a realistic distribution function of the elements in order to be able to calculate the surface composition of the metal phase.

The simplest distribution function describes enrichment of the elements Ni and Mo in only one atomic plane. If this model is applied to the spectra shown in Figure 6.13, it becomes apparent that it is not possible to obtain mass balance. However, most of the oxide present on the surface during the analysis of the sample polarized to the active potential is formed during transfer of the sample from the cell. Therefore, to assess the composition of the surface during anodic dissolution, the cations detected have to be converted to their metallic states and added to the contribution from the metallic spectra. By assuming a model for the distribution of the elements, the surface composition can be calculated. The details of the procedure have been published [1,40]. Figure 6.14b shows the quantitative analysis of the metal composition during anodic dissolution. The distribution function used was an error function where the composition varies over three atomic planes. The composition of the second plane is 50% of the difference between the bulk composition and the composition of the surface. The values shown with the coarse solid line represent the composition of the outermost atomic layer during anodic dissolution. As already pointed out, it was assumed that most of the oxide observed on the surface was due to the transfer of the sample. It appears from Figure 6.14 that Ni and Mo are markedly enriched on the surface of the metal during anodic dissolution. Even the Cr concentration in the outermost layer is significantly higher than the bulk concentration. Because of the enrichment of Cr, it is easy to understand that the oxide formed during handling of the sample was mainly Cr oxide. Furthermore, Figure 6.14b shows that the Fe content in the outer surface layer is markedly lower than the alloy composition.

It was shown before that the Ni content of the passive film is low. One can therefore expect that the Ni content at the metal–oxide interface is still high after passivation. Figure 6.15 [1] illustrates the measured apparent metal content of the metal phase after polarization to the active potential, −320 mV (SCE), and the passive potentials, −100 and 500 mV (SCE). It appears from the figure that Ni largely remains in the alloy phase after passivation.

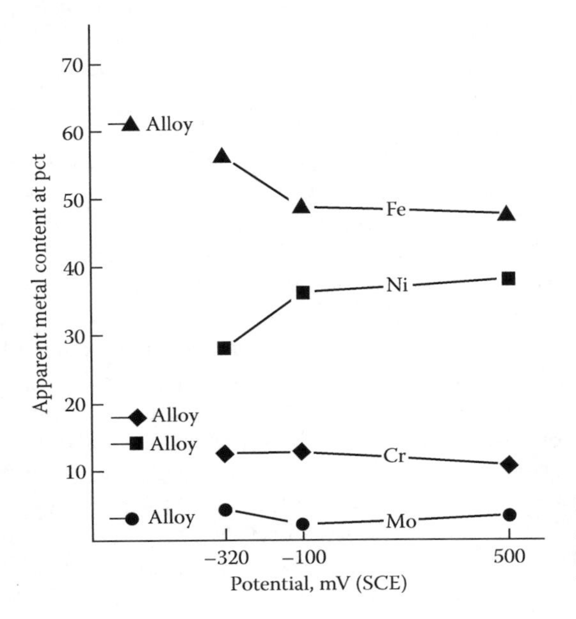

FIGURE 6.15
Apparent metal content of the alloy vs. the potentials (SCE). (From Olefjord, I. et al., *J. Electrochem. Soc.*, 132, 2854, 1985.)

It has been pointed out that the passive currents for all three alloys whose polarization diagrams were shown in Figure 6.12 are about the same. Even the passive current for pure Cr (Figure 6.1) is about the same as for the alloys. The noticeable difference in electrochemical behavior between the alloys is in the passivation current and the pitting potential. As shown earlier, the alloying elements are enriched on the surface in their metallic states during active dissolution. It has been suggested [1,40] that an intermetallic surface phase is formed during dissolution.

The formation of an intermetallic surface phase can be understood from the Engel–Brewer [43] valence bond theory of metallic bonding by considering the ground state electronic configuration of the elements and the nature of the possible bonding processes. The model predicts intermetallic bonding between "hyper" and "hypo" *d*-electron transition metals, resulting in very strong bonding. Such bonding, in principle, should result in low dissolution rates due to a higher activation energy for anodic dissolution. Such systems are formed between transition metals from the left of the periodic table having more vacant *d* electrons and those to the right having fewer *d*-electron vacancies. The intermetallic bonding between elements suitably separated in the transition row may result from penetration of an electron pair from the hyper *d* electron into the *d* orbital of the hypo *d*-electron metal. XPS analysis [44] of thin Ni-Mo intermetallic layers has confirmed the charge transfer. Compared with those of pure metal lattices, the binding energies of Ni and Mo from the intermetallic layer are shifted higher and lower, respectively. It has been pointed out that the strength of the *d*-electron bonding increases from *Zd* to *4d*, indicating that the stability of the intermetallic bond between Ni and Cr should be lower than that between Ni and Mo [33]. Therefore, it has been suggested that the overall lattice energy increases during dissolution of Mo-alloyed austenitic stainless steel in the metal layers close to the passive film due to formation of intermetallic bonds between Ni and Mo atoms [45]. Consequently, the activation energy increases for anodic dissolution and the dissolution rate decreases. According to this model, bonding between Mo and Fe is predicted to be weaker than between Mo and Ni and Mo and Cr, and therefore Fe is selectively dissolved.

The Engel–Brewer model [43] of intermetallic bonding would explain how Ni can lower the critical current density and elevate the pitting potential of austenitic alloys, while playing no direct role in the construction of the passive film. It is evident that Mo not only plays several direct roles in the formation and stability of the passive film but also enhances Ni's role by further enhancing the anodic segregation of Ni. Therefore, it is proposed that through sluggish dissolution kinetics alone, Ni, when bonded with Cr and more strongly with Mo, will lower the rate of metal dissolution during the pitting process and thereby reduce the maximum metal chloride concentration in the pit solution [45].

In the N-bearing stainless steels, N has a pronounced influence on the pitting corrosion properties. It has already been pointed out [15,26,31] that the positive effect of N is obtained for the Mo-alloyed stainless steels. The lower dissolution rate of the Mo-containing alloys due to formation of intermetallic layer during active dissolution provides a model for the synergistic effect between Mo and N. Figure 6.16 shows the proposed mechanism. Three alloys with and without Mo and N are assumed to be exposed to an acid chloride-containing solution. Pits are initiated on all three alloys. The pits formed on the Mo-containing alloys become smaller than the pits on the Mo-free alloy because the Mo lowers the dissolution rate of the alloy by formation of an intermetallic surface layer. If the acidity and/or the chloride content of the solution is high enough, the pits formed on the Mo-free alloy will become critical and grow. The dissolved ions are hydrated and the hydrolysis causes increased acidity in the pits. The pH value in the small pits formed on the Mo-containing alloys becomes low due to the outlet diffusion of H^+ ions. The environment in the large pits

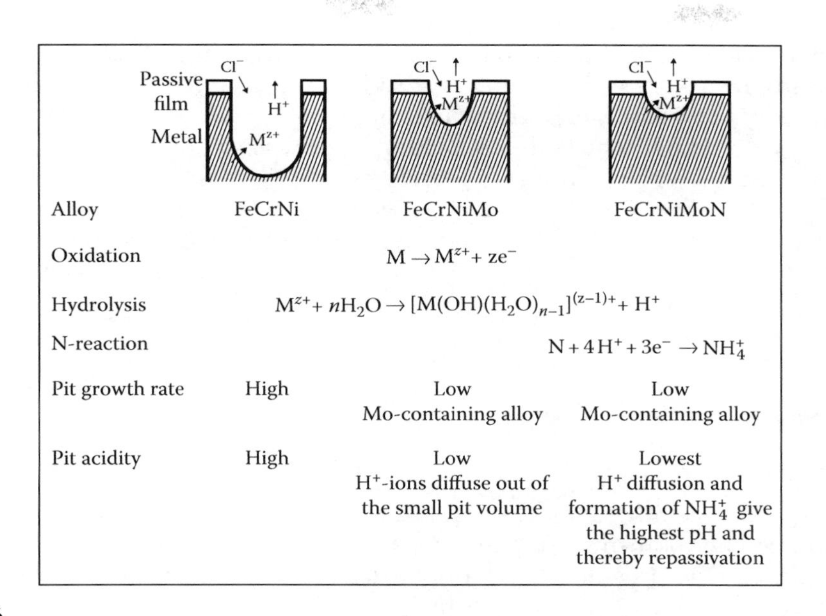

FIGURE 6.16
A model for the growth and repassivation of pits in Mo-free, Mo and Mo-N containing alloys. (From Falkenberg, F. and Olefjord, I., *Proceedings of the International Symposium in Honor of Professor Norio Sato: Passivity and Localized Corrosion*, M. Seo, B. MacDougall, H. Takahashi, and R.G. Kelly, eds., Electrochemical Society, Pennington, NJ, p. 404, 1999.)

formed on the Mo-free alloy changes, on the other hand, noticeably because the production of the solvated H^+ ions is high, the drop in pH at the pit surface becomes more pronounced, and the ingress migration of Cl^- ions is more effective. For the N-bearing steels the lowering of pH due to hydrolysis is compensated by formation of NH_4^+, as suggested by Osozawa and Okato [26]. The pH value of the solution in the pits will be above a critical value and thereby repassivation will occur in an early stage of the pitting process. Hence, this model [31] concludes that the role of Mo is to lower the dissolution rate of the initiated pits and thereby limit the increase in acidity in the pitted volume.

It seems therefore that Mo plays a multifold role in corrosion protection of stainless steels: (a) it lower the chloride content in the passive film during passivation and repassivation and thereby makes the film more resistant to breakdown; (b) it lowers the dissolution rate of the alloy during initiation of pitting; (c) it forms insoluble salt at the bottom of pits, which is favorable to the formation of passive film; and (d) during the repassivation process it forms molybdate, which governs the deprotonation reaction of the poorly protective hydroxide and thereby a more protective barrier layer is formed. According to the thick-membrane studies [17], molybdate has been shown to promote ionic rectification, which retards chloride ingress and promotes deprotonation of the hydroxide. The synergistic effect of N, Ni, and Mo can be understood from the work of Jack [46], who reported that whereas single nitrides of Mo and Ni are thermodynamically relatively unstable, with Ni being the less stable of the two, a mixed Mo and Ni nitride of the form Ni_2Mo_3N exhibited very high thermodynamic stability. Although this suggests the likely involvement of Ni in such mixed surface nitrides formed during anodic segregation of N-bearing steels, it also suggests strong intermetallic bonding. This is not contrary to the nature of bonding in interstitial nitrides formed by transition elements. For example, although N is the electron acceptor in the mixed nitride phase, one would expect strong interaction between Mo and Ni and somewhat weaker interaction between Cr and Ni.

Finally, it is evident that less than one monolayer of surface nitride is capable of increasing the pitting resistance of 304 stainless steel [33] and therefore it is probable that this mixed nitride phase is situated at previously high-energy sites such as grain boundaries and surface dislocations, which are preferential sites for both pitting and surface nitride formation. In the case of Ni-bearing steels these sites would be initially the most active and therefore the sites where anodic segregation would tend to be first developed. Therefore, as a consequence of anodic segregation at high-energy sites during the active stage of repassivation, a more homogeneous passive film may be generated that would be less vulnerable to localized breakdown.

6.7 Summary

The passivity of austenitic stainless steels is a field of study that will continue to develop as new service environments are found for this important class of steels. Many of the studies reviewed here provide a basis for new testing and analytical strategies that must be applied to these and similar materials to develop new stainless steels for such new markets as the bioprocessing industry and facilities for atmospheric scrubbing of industrial emissions. In this chapter we have emphasized the importance of barrier oxide layers, salt deposit layers, and alloy surface layers in the overall process of passivity and breakdown of passivity. Finally, it has been shown that by combining electrochemical and surface analytical methodologies, both corrosion parameters and structure-property relationships can be combined to model passivation processes in a way that provides new insights into alloy performance and new alloy design concepts.

Acknowledgments

This work has been supported by the National Swedish Board for Technical and Industrial Development (NUTEK), U.S.O.N.R. (A. J. Sedriks, Contract Officer) contract N0001485K0437, and the U.S.N.S.F. International Program Division (Christine Glenday, Contract Officer).

References

1. I. Olefjord, B. Brox, and U. Jelvestam, *J. Electrochem. Soc. 132*:2854 (1985).
2. J. Horvath and H. H. Uhlig, *J. Electrochem. Soc. 115*:791 (1968).
3. C. R. Clayton and Y. C. Lu, *J. Electrochem. Soc. 133*:2465 (1986).
4. G. Okamoto, *Corros. Sci. 13*:471 (1973).
5. H. Saito, T. Shibata, and G. Okamoto, *Corros. Sci. 19*:693 (1979).
6. K. Tachibana, K. Miya, K. Furuya, and G. Okamoto, *Corros. Sci. 31*:521 (1990).
7. A. R. Brooks, C. R. Clayton, K. Doss, and Y. C. Lu, *J. Electrochem. Soc. 133*:2459 (1986).
8. A. J. Sedriks, *Corrosion of Stainless Steels*, Wiley, New York, 1979.
9. K. Sugimoto and Y. Sawada, *Corros. Sci. 17*:425 (1979).

10. K. Hashimoto, K. Asami, and K. Teramoto, *Corros. Sci. 19*:3 (1970).
11. H. Ogawa, H. Omata, I. Itoh, and H. Okada, *Corrosion 34*:53 (1978).
12. M. Urquidi and D. D. MacDonald, *J. Electrochem. Soc. 132*:533 (1985).
13. L. Wegrelius and I. Olefjord, *Mater. Sci. Forum 185–188*:347 (1995).
14. L. Wegrelius, F. Falkenberg, and I. Olefjord, *J. Electrochem. Soc. 146*:1391 (1999).
15. F. Falkenberg and I. Olefjord, *Proceedings of the International Symposium in Honor of Professor Norio Sato*: *Passivity and Localized Corrosion* (M. Seo, B. MacDougall, H. Takahashi, and R. G. Kelly, eds.), Electrochemical Society, Pennington, NJ, 1999, p. 404.
16. F. Falkenberg, Metastable pits on austenitic stainless steels, Thesis for the degree Licentiate of Engineering, Chalmers University of Technology, Gothenburg, Sweden, 1998.
17. M. Sakashita and N. Sato, *Corros. Sci. 17*:473 (1977).
18. C. R. Clayton and Y. C. Lu, *Corros. Sci. 29*:881 (1989).
19. Y. C. Lu, C. R. Clayton, and A. R. Brooks, *Corros. Sci. 29*:863 (1989).
20. J. R. Ambrose, *Passivity of Metals* (R. P. Frankenthal and J. Kruger, eds.), Electrochemical Society, Pennington, NJ, 1978, p. 740.
21. A. Schneider, D. Kuron, S. Hofmann, and R. Kircheim, *Proceedings of the Sixth International Symposium Passivity, Part I* (N. Sato and K. Hashimoto, eds.), *Corros. Sci. 31*:191 (1990).
22. F. Falkenberg and I. Olefjord, Passivity of metals and semiconductors, eds. M. B. Ives, J. L. Luo, J. R. Roda, *The Electrochemical Society*, p. 570 (1999).
23. G. P. Halada, C. R. Clayton, H. Herman, S. Sampath, and R. Tiwari, *J. Electrochem. Soc. 142*:74 (1995).
24. Y. Hisamatsu, *Passivity and Its Breakdown on Iron Based Alloys, USA–Japan Seminar* (R. Staehle and H. Okada, eds.), National Association of Corrosion Engineers, Houston, TX, 1976, p. 99.
25. Z. Szklaska-Smialowska, *Pitting Corrosion of Metals,* National Association of Corrosion Engineers, Houston, TX, 1986, p. 358.
26. K. Osozawa and N. Okato, *Passivity and Its Breakdown on Iron and Iron Based Alloys, USA–Japan Seminar*, National Association of Corrosion Engineers, Houston, 1976, p. 135.
27. J. E. Truman, M. J. Coleman, and K. R. Pirt, *Br. Corros. J. 12*:236 (1977).
28. J. Eckenrod and C. W. Kovack, *ASTMSTP 679,* ASTM, Philadelphia, PA, 1977, p. 17.
29. R. Bandy and D. van Rooyen, *Corrosion 39*:227 (1983).
30. Y. C. Lu, R. Bandy, C. R. Clayton, and R. C. Newman, *J. Electrochem. Soc. 130*:1774 (1983).
31. I. Olefjord and L. Wegrelius, *Corros. Sci. 38*:1203 (1996).
32. I. Olefjord, *Surface Characterization* (E. Brune, R. Hellborg, H. J. Whitlow, and O. Hunderi, eds.), Wiley-VCH, Weinheim, Germany, 1997, p. 291.
33. R. D. Willenbruch, C. R. Clayton, M. Oversluizen, D. Kim, and Y. C. Lu, *Proceedings of the Sixth International Symposium on Passivity, Part 1* (N. Sato and K. Hashimoto, eds.), *Corros. Sci. 31*:179 (1990).
34. J. Kolts, J. B. C. Wu, and A. I. Asphahani, *Met. Prog.* September *125*:25 (1983).
35. C. R. Clayton, G. P. Halada, and J. R. Kearns, *Proceedings, US–Japan Seminar*, Timberline Lodge, Mount Hood, OR, June 1994 (R. Latanision and K. Hashimoto, eds.), *Mater. Sci. Eng. A 198*:135 (1995).
36. Y. C. Lu, M. B. Ives, C. R. Clayton, and D. Kim, *Corros. Sci. 35*:89 (1993).
37. G. P. Halada and C. R. Clayton, *J. Vac. Sci. Technol. A 11*, *4*:2342 (1993).
38. I. Olefjord, *Mater. Sci. Eng. 42*:161 (1980).
39. I. Olefjord and B.-O. Elfström, *Corrosion (Houston) 38*:46 (1982).
40. B. Brox and I. Olefjord, *Proceedings of Stainless Steel 84*, Institute of Metals, London, 1985, p. 134.
41. R. C. Newman, *Corros. Sci. 25*:331 (1985).
42. B. E. Clark, S. J. Thorpe, and K. T. Aust, *Corros. Sci. 31*:551 (1990).
43. L. Brewer, *Science 161*:115 (1968).
44. S. Börjesson and I. Olefjord, Private communication.
45. I. Olefjord and C. R. Clayton, *ISIJ Int. 31(2)*:134 (1991).
46. K. H. Jack, *High Nitrogen Steels* (J. Fact and A. Henry, eds.), Institute of Metals, London, 1989, p. 117.

7

Mechanisms of Pitting Corrosion

Hans-Henning Strehblow
Ecole Nationale Supérieure de Chimie de Paris,
Centre National de la Recherche Scientifique
and
Heinrich Heine University Düsseldorf

Philippe Marcus
Ecole Nationale Supérieure de Chimie de Paris
and
Centre National de la Recherche Scientifique

CONTENTS

7.1 Introduction ... 349
7.2 Some Basic Details of Passivity ... 350
7.3 Pitting Potentials and Inhibition Potentials ... 352
7.4 Breakdown of Passivity ... 353
  7.4.1 Penetration Mechanism ... 355
  7.4.2 Film-Breaking Mechanism ... 359
  7.4.3 Adsorption Mechanism ... 361
  7.4.4 Comparison of the Different Nucleation Mechanisms ... 366
7.5 Mechanistic Consequences for Pit Nucleation from Studies with Nanometer Resolution ... 366
7.6 Special Conditions for Localized Corrosion ... 370
7.7 Transition from Pit Nucleation to Pit Growth: Microscopic Observations ... 372
7.8 Pit Growth ... 376
  7.8.1 Fundamentals of Stable Pit Growth ... 376
  7.8.2 Precipitation of Salt Films ... 378
  7.8.3 Potential Drops ... 382
  7.8.4 Composition of Pit Electrolyte ... 383
7.9 Repassivation of Corrosion Pits ... 386
7.10 Factors Stabilizing Pit Growth ... 387
7.11 Conclusion ... 390
References ... 391

7.1 Introduction

Pitting corrosion occurs at passivated metal surfaces during the access of so-called aggressive anions. Halides very effectively attack passivating thin oxide layers, leading to an intense localized dissolution of the metal surface, which is otherwise protected by the passive layer against general dissolution. Chlorides cause the most serious problems due to their presence in many environments such as seawater and salt on roads, in food, and in the chemical industry. Many metals and their alloys are subject to this type of corrosion (e.g., iron, nickel, copper, aluminum, steels), whereas chromium is one of the few exceptions that resists pitting in aggressive environments. The restriction of the dissolution to pits within a large passivated metal surface, which may serve as a large cathode for the reduction of oxidants such as dissolved oxygen, leads to fast perforation of the metal, which weakens the construction and thus causes large economic losses and safety problems.

Pitting at passivated metal surfaces is a complex process with a sequence of steps. For each stage of the development and growth of a corrosion pit, one has to study the mechanistic details in order to understand the process as a whole. For a detailed mechanistic discussion, the following stages are usually distinguished:

1. Processes leading to breakdown of passivity
2. Early stages of pit growth
3. Late stages of pit growth
4. Repassivation phenomena

The details of the mechanisms also depend on the metal or the composition of the alloys as well as on the electrolyte and other environmental conditions. Metallic and nonmetallic inclusions often play a decisive role in the start of a corrosion pit. In most cases, the presence of aggressive anions is necessary for breakdown of passivity and stable pit growth. The discussion in this chapter explains the effect of these anions by their tendency to form complexes with metal ions. It concentrates on the behavior of some pure metals, such as pure Fe and Ni, in simple electrolyte solutions. Some basic concepts are the center of interest, although it is known from the technical applications of the different materials that the appearance and the details of the corrosion mechanisms may involve many complicating factors. Thus, this contribution intends to describe the main mechanistic aspects rather than the details of complicating factors.

7.2 Some Basic Details of Passivity

In order to get a better basis for the discussion and understanding of the corrosion mechanisms, a brief summary of the structure and the properties of passive layers is given. For more detailed information one should refer to the chapter on passivity in this book. Passive layers form on many reactive metals. If these films have semiconducting properties, as for Fe, Cr, Ni, and Cu, they will grow only up to a few nanometers in thickness to the potential for oxygen evolution. Valve metals such as Al, Nb, Ta, Ti, Zr, and Hf form insulating oxide films that may grow up to more than 100 V without oxygen evolution with thicknesses reaching several tens of nanometers. These films generally show a potential drop at the metal–oxide and the oxide–electrolyte interface as well as within the passive layer (see Figure 7.2a). The electrical field strength within the passive layer of some nanometers is in the order of some

10^6 V/cm, which enables the migration of ions through the film at room temperature at a measurable level in a region of corresponding current densities of some few $\mu A/cm^2$ or less. According to detailed studies using surface analytical methods such as x-ray photoelectron spectroscopy (XPS), the passive film has a complicated chemical structure that is described by a multilayer or at least by a bilayer model. Oxides and hydroxides are present, sometimes separated in different layers, sometimes as hydroxy-oxides. The size and structure of the related sublayers change with the electrode potential and with other factors such as the composition of the bulk metal and the electrolyte. At least one part of the passive layer acts as a barrier to corrosion, to which most of the details of the following discussion refer.

At appropriate pH the oxide films may be in equilibrium with the electrolyte according to the data of pH-potential diagrams of M. Pourbaix. In strongly acidic electrolytes the passive layer on metals such as Fe, Ni, Cr, and steel is far from its dissolution equilibrium, and passivity for these conditions is a consequence of the slow dissolution kinetics of the oxide. As this situation is not related to thermodynamics of layer formation, it is in contradiction to the results of Pourbaix diagrams. In other words, Pourbaix diagrams are not applicable to these situations. The phase diagram of Figure 7.2a shows the reactions occurring on a passive metal that are influenced by the potential drops $E_{1,2}$ at the metal–oxide, $E_{2,3}$ at the oxide–electrolyte interface, and $\Delta\varphi$ within the oxide layer. $E_{2,3}$ influences the kinetics of the dissolution reaction (1) of the metal ions and the related passive corrosion current density, as well as the kinetics of O^{2-} or OH^- formation of reaction (2) corresponding to the growth of the layer. Both reactions require in addition a field-assisted ion transport through the passive layer. Last but not least, electron transfer occurs as a third reaction (3) that is of interest for corrosion reactions under open-circuit conditions, where the reduction of oxidants as dissolved oxygen serves as a cathodic compensation for the anodic corrosion process. Redox reactions, of course, do not occur on insulating passive layers. For technically applied valve metals with insulating passive layers, inclusions with high electronic conductivity may provide sufficient electronic transfer from the passivated metal to a redox system within the electrolyte. They may serve as cathodic sites within the passivating oxide film on the metal, which enables a slow corrosion rate in the passive state or even the intense metal dissolution during localized corrosion. Copper inclusions in aluminum are a classical example. Here the differences in the electronic properties at the surface have been visualized with the scanning Kelvin probe, which tests the change of the work function with the material at the metal surface with lateral resolution [1].

A high electrical field is established at least in the part of a more complicated layer structure that serves as a barrier for metal dissolution. This in turn leads to logarithmic or inverse logarithmic growth of the film thickness and linear decay of the logarithm of the current density with time for potentiostatic passivation transients. The transfer of metal and O^{2-} ions through the passive layer increases exponentially with the electrical field strength. The transfer of aggressive anions through the film according to reaction (4) (Figure 7.2a) is controlled similarly by the electrical field strength, which is important for the discussion of the penetration mechanism for breakdown of passivity. The potential drop $E_{2,3}$ influences the adsorption of anions including aggressive anions, the formation of O^{2-} and OH^- anions from the electrolyte, and the transfer of metal cations at the oxide–electrolyte interface. It thus rules layer formation and the passive corrosion current density. $E_{2,3}$ changes when the composition of the oxide changes at the surface in contact with the electrolyte, especially when approaching the transpassive potential range and intermediately for nonstationary conditions when the electrode potential is altered. It also changes with the pH *via* reaction (2) of Figure 7.2a. The electrochemical equilibrium for this reaction leads to a -0.059 V/pH dependence of the potential drop $E_{2,3}$ at the oxide–electrolyte interface and thus to a decrease of the passive current density with increasing pH. The transfer of metal ions into the electrolyte is also influenced by complexing anions, which will be discussed in detail in the chapter on the adsorption mechanism of pit nucleation.

7.3 Pitting Potentials and Inhibition Potentials

Pitting occurs when the potential exceeds a critical value E_p in the passive range of a polarization curve. E_p decreases with the logarithm of the concentration of the aggressive anions [A] according to Equation 7.1. The presence of inhibiting anions I may shift it to more positive values. An interesting observation is the inhibition potential E_I for the additional presence of an inhibitor that causes an upper limit to pitting in the passive potential range. E_I increases with the logarithm of the concentration of the aggressive anion [A] and decreases with increasing logarithm of the concentration of the inhibitor [I] according to Equation 7.2 [2]:

$$E_p = a - b\log[A] \tag{7.1}$$

$$E_I = a + b\log\frac{[A]}{[I]} \tag{7.2}$$

Nitrate and perchlorate may act as inhibitors. Figure 7.1 gives an example of the dependence of E_p and E_I on the electrolyte composition for Ni in chloride- and nitrate-containing solutions. The gap between the two lines is the potential range of pitting corrosion. Similar results have been obtained for ClO_4^- and Fe [2]. A very peculiar property is observed for

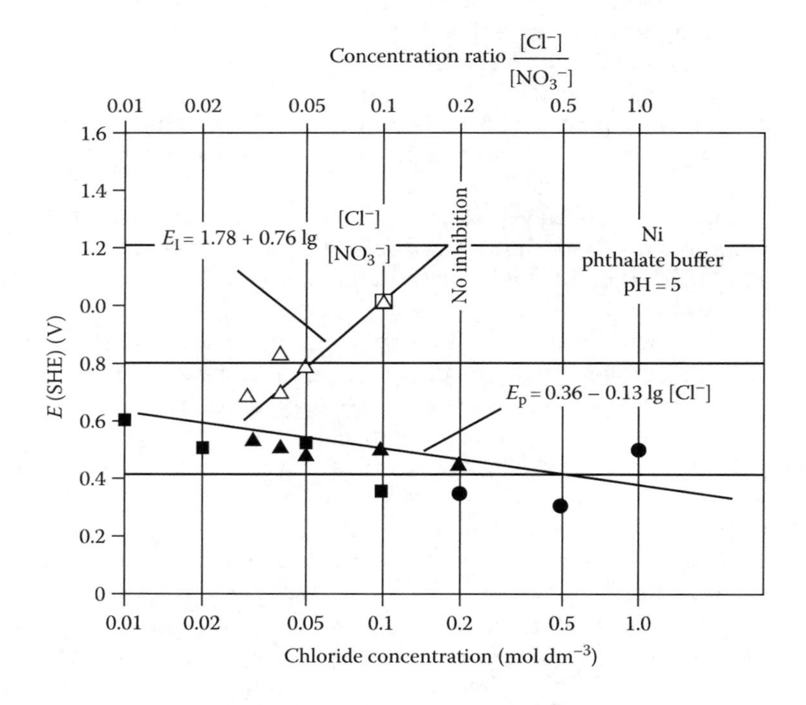

FIGURE 7.1
Pitting potential E_p and inhibition potential E_I as a function of the chloride concentration and the concentration ratio of chloride to nitrate in phthalate buffer, pH 5.0: (Δ) E_I, (▲) E_p, 1 M KNO_3; (□) E_I, (■) E_p, 0.1 M KNO_3; (●) galvanostatic experiments, 1 M KNO_3. (From Strehblow, H.-H. and Titze, B., *Corrosion Sci.*, 17, 461, 1977.)

Fe in ClO_4^--containing solutions. Although this anion is an inhibitor for chloride pitting at more negative potentials, it also causes pitting of Fe by itself with a very positive critical potential [2,3]. Its E_p follows Equation 7.1, is pH independent, and may be measured with an accuracy of some few mV. Usually critical potentials may be measured with an uncertainty of several 10 mV only. Surface analytical studies have shown that the aggressive properties of perchlorate are related to its decomposition close to the pitted surface, which yields aggressive chloride [3]. This is very special as one expects a ClO_4^- reduction at more negative potentials only where it is kinetically stable. As a possible explanation, the decomposition of ClO_4^- in the increased electrical field of the oxide–electrolyte interface has been proposed when approaching the transpassive potential range [3].

7.4 Breakdown of Passivity

Three main mechanisms are discussed by most authors for the processes leading to breakdown of passivity: the penetration mechanism, the film-breaking mechanism, and the adsorption mechanism [4]. Figure 7.2a through c presents diagrams for their explanation.

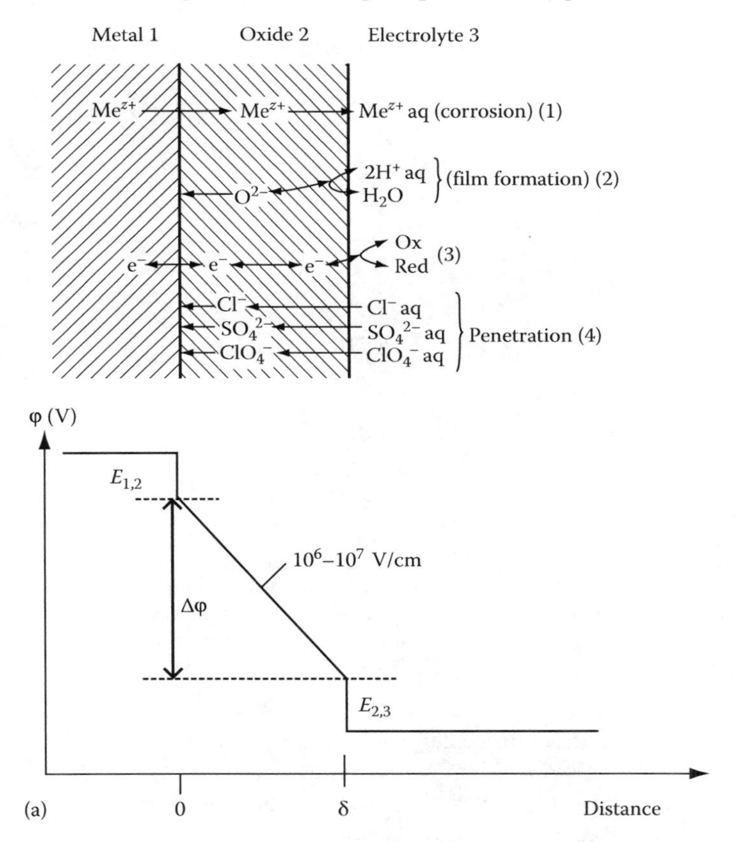

FIGURE 7.2
Phase diagram of a passive metal demonstrating the processes leading to pit nucleation. (a) Penetration mechanism and phase diagram of a passive layer with the related processes of ion and electron transfer within the oxide and at its phase boundaries including schematic potential diagram (φ),

(continued)

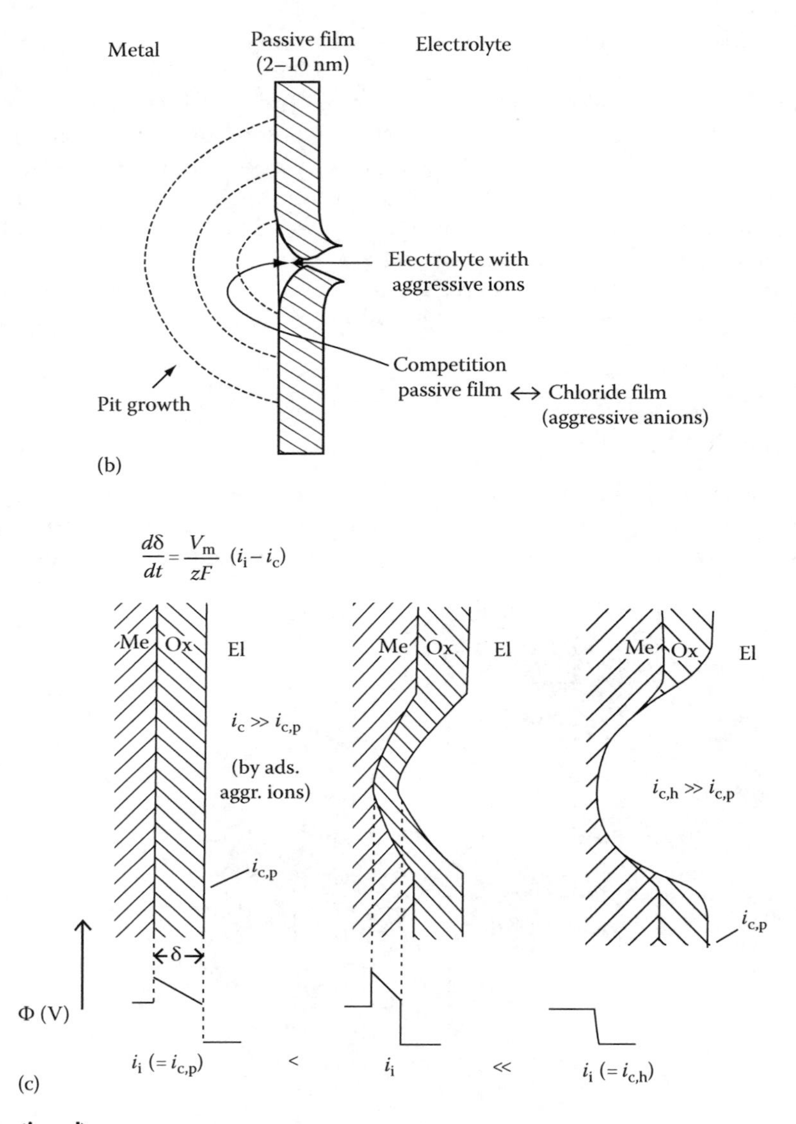

FIGURE 7.2 (continued)
(b) film-breaking mechanism and related competing processes, (c) adsorption mechanism with increased local transfer of metal ions and related corrosion current density i_c caused by complexing aggressive anions leading to thinning of the passive layer and increases in field strength and final free corrosion current density $i_{c,h}$ within the pit. (From Strehblow, H.-H., *Werkst. Korros.*, 27, 792, 1976.)

The penetration mechanism (Figure 7.2a), first discussed by Hoar et al. [5], involves the transfer of anions through the oxide film to the metal surface, where they start their specific action. The film-breaking mechanism (Figure 7.2b), proposed by Vetter and Strehblow [6] and Sato et al. [7,8], requires breaks within the film that give direct access of the anions to the unprotected metal surface. The adsorption mechanism, discussed first by Kolotyrkin [9] and Hoar and Jacob [10], starts with adsorption of aggressive anions at the oxide surface, which enhances catalytically the transfer of metal cations from the oxide to the electrolyte (Figure 7.2c). This effect leads to thinning of the passive layer with possible final total removal and the start of intense localized dissolution. Strict separation of these mechanisms might not always be appropriate, as penetration of a passive layer *via* some very small defect is not necessarily very different from the occurrence of fissures in the film that permit easy access of the aggressive anions to the metal surface as in the film-breaking mechanisms. In the following discussion of the mechanism of breakdown of passivity, a distinction is proposed between stationary and nonstationary conditions of the passive layer. These different experimental or environmental conditions may have a strong influence on the effective mechanism leading to breakdown of passivity.

7.4.1 Penetration Mechanism

The penetration mechanism (Figure 7.2a) requires transfer of the aggressive anions through the passive layer to the metal–oxide interface, where they cause further specific action. The high electrical field strength and a high defect concentration within the presumably severely disordered structure of the passivating oxide layer may explain this transfer. Many authors even postulate an amorphous structure of the oxide layer [11]. Direct evidence is given by a few extended x-ray absorption fine structure (EXAFS) measurements [12]. Systematic studies with this method are, however, difficult and still missing. For a highly disordered structure it is not clear why the breakdown event occurs locally.

On the basis of these ideas, Macdonald and coworkers [13,14] developed their model of passivity and its breakdown involving the action of vacancies within the passive layer. It is assumed that cation vacancies migrate from the oxide–electrolyte to the metal–oxide interface, which is equivalent to the transport of cations in the opposite direction. If these vacancies penetrate into the metal phase at a slower rate than their transport through the oxide, they accumulate at the metal–oxide interface and finally lead to a local concentration. The related voids lead to stresses within the passive film and its final breakdown. The inward diffusion or migration of cation vacancies is affected by the incorporation of Cl ions at the oxide–electrolyte interface according to the following mechanism: The concentration c of metal ion V_{M^+} and O^{2-} vacancies of $V_O{}^{2-}$ are determined by the equilibrium of the Schottky pair formation at the oxide–electrolyte interface (Equation 7.3), which causes an inverse dependence of their concentrations (Equation 7.4):

$$\text{Null} = V_{M^+} + \frac{x}{2} V_O{}^{2-} \tag{7.3}$$

$$c_{v,M} = \text{const}\, c_{v,O}^{-x/2} \tag{7.4}$$

In the presence of Cl^-, its incorporation in O^{2-} vacancies occurs according to the equilibrium of Equation 7.5, which is affected by the potential drop $E_{2,3}$ at the interface according to Equation 7.6 with the concentration of oxygen vacancies $c_{v,O^{2-}}$ at the surface, the activity of chloride a_{Cl^-} within the solution, and the potential-independent part of the equilibrium constant K:

$$V_{O^{2-}} + Cl^-_{(aq)} = Cl_O \tag{7.5}$$

$$c_{v,O^{2-}} a_{Cl^-} = K \exp\left(\frac{\alpha F E_{2,3}}{RT}\right) \tag{7.6}$$

According to Equation 7.6, $c_{O^{2-}}$ depends on a_{Cl^-} and the potential drop $E_{2,3}$; $c_{O^{2-}}$ in turn affects $c_{v,M}$, which is the driving force of the diffusion of the cation vacancies to the metal–oxide interface. Thus the interdependence of the concentrations of cation and anion vacancies within the oxide and the incorporation of Cl^- determine the concentration gradient of cation vacancies and their transport through the oxide layer that will cause a critical concentration for breakdown at the pitting potential. According to this outline, the further discussion yields a semilogarithmic dependence of the pitting potential E_p and the chloride activity a_{Cl^-} within the electrolyte similar to Equation 7.1 (Equation 7.7). The constant B contains the diffusivity constant of the cation vacancies within the oxide layer:

$$E_p = A - B \log a_{Cl^-} \tag{7.7}$$

Objections to the point-defect model are that in its original form it assumes linear transport equations with diffusion of the different species within the passive layer, whereas migration with an exponential dependence on the high electrical field strength of some 10^6 V/cm should be dominating as a driving force. It also concentrates the changes in electrode potential ΔE to the potential drop $E_{2,3}$ at the oxide–electrolyte interface, so that it fully enters the equilibrium of Cl^- incorporation of Equations 7.5 and 7.6. However, large parts of ΔE contribute to $\Delta\varphi$, i.e., are located within the barrier part of the passive layer (Figure 7.2), and $E_{2,3}$ changes only with the composition of the oxide surface, if at all, and the pH of the solution but not with the applied electrode potential, at least for stationary conditions. An overpotential $\eta_{2,3}$ for the potential drop $E_{2,3}$ is mainly expected for nonstationary conditions of the passive layer that will directly influence the adsorption equilibrium of Cl^-. The specific role of chloride and other halides in breakdown of passivity is still not sufficiently understood in light of this theory. Further refinements might improve this interesting view.

Surface analytical methods unfortunately do not always give a clear answer about the penetration of aggressive anions. Some authors found chloride within the film with XPS, Auger electron spectroscopy (AES) [15,16], and secondary ion mass spectroscopy (SIMS) [17]; others could not find it within the film [18–22]. The contradictory results may be explained in terms of sample preparation and the sensitivity of the methods. Very careful XPS studies with Fe–Cr alloys show incorporation within the outer hydroxide part of the duplex passivating film [23,24]. The inner oxide layer remains free of Cl^- if prepared within chloride-free electrolytes before exposure to the aggressive anions. Similarly, incorporation of Cl^- within the outer hydroxide layer was found for pure Ni [25]. It was found in the inner oxide part only when the passive layer was formed in solutions already containing Cl^-. If the electrode potential was above the critical value for breakdown, Cl^- penetrated into the preexisting oxide with possible lateral fluctuations of its concentration, leading finally to the formation of pits. According to these studies, the accumulation within the hydroxide overlayer serves as an accumulation of a sufficiently large amount to cause breakdown in the following step. Further discussions of the surface analytical investigations of passive layers that have been exposed to chloride-containing solutions may be found in the chapter on passivity in this book.

A detailed bilayer or even multilayer structure is observed for passive films on many metals and alloys [26,27]. The outer part is usually a hydroxide, whereas the main inner part is an oxide [23,25–27]. The hydroxide structure may well act as an ion exchanger or at least absorb anions, as has been proved for some systems. Although the access of aggressive anions leads to changes of the passive layer detected by ellipsometry [28] and reflection spectroscopy [29], it is still unclear what conclusion may be drawn from these observations. If the penetration of aggressive anions leads to weak channels where intense dissolution may start, it is unclear why the film does not re-form specifically at this site, and why these defects do not re-passivate. The self-healing mechanism of the passive layer is essential for its excellent protecting property. The specific role of the aggressive anions is missing in this mechanism of breakdown. Any explanation should involve the characteristic chemical properties of anions such as the halides.

Experiments with well-prepassivated specimens show that the formation of a corrosion pit may be an extremely fast process, in the time range of <1 s or even <1 ms [4,30]. Figure 7.3 depicts as an example the increase of the current density within less than 1 ms as a consequence of growing pits for an Ni specimen prepassivated for 1 s before a potential shift above the critical pitting potential. A simple comparison of the small stationary passive current densities in the range of μA/cm² with these short times leads to contradictions with the penetration mechanism [30,31]. If anions migrate inward as cations migrate outward during stationary dissolution, these fast nucleation times cannot be understood. Furthermore, it seems questionable that large anions such as SO_4^{2-} and ClO_4^- migrate sufficiently fast in the electrical field through the passive layer to cause nucleation in times $t < 1$ s [30,31]. These anions may also cause pitting of iron at the lower potential part of the passivity range and ClO_4 also at potentials close to the transpassive range [2,3]. However, the inward migration of aggressive anions and the outward migration of cations may be facilitated at local defects within the passive layer. Even for those situations, one needs the special chemical properties of the anions to understand why the defects do not repassivate but develop to a corrosion pit. Any mechanism of nucleation and growth of corrosion pits has to include the specific role of the aggressive anions that causes the formation of a pit instead of repassivating the defect site. These details will be discussed in one of the following sections.

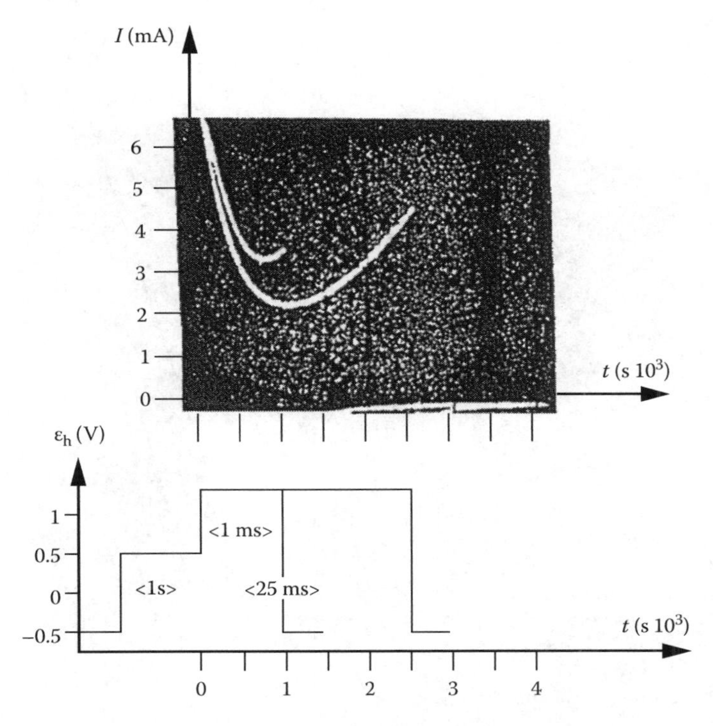

FIGURE 7.3
Pit nucleation of Ni ($A = 0.02\,cm^2$) at $E = 1.30$ V (SHE) in phthalate buffer, pH 5.0 + 0.1 M KCl after 1 s prepassivation at $E = 0.5$ V. (From Strehblow, H.-H., *Werkst. Korros.*, 35, 437, 1984.)

In this regard, one experiment with well-passivated Fe electrodes should be mentioned [6]. If the penetration of aggressive anions through the passive layer was the rate-determining step, a high electrical field strength should favor their migration and accelerate pit nucleation. For nonstationary conditions, for passive Fe, exactly the contrary has been observed (Figure 7.4). After passivation for 1 h at 1.18 V (SHE, standard hydrogen electrode) in phthalate buffer pH 5.0 plus 0.1 M K_2SO_4, the nucleation rate was severely increased when the potential was stepped to 0.78 V immediately before Cl^- addition to 0.01 M. This pretreatment resulted in a much faster increase of the current density with time and a slope of 3 for a double logarithmic current–time plot was found, which is an indication of an increased nucleation rate that is constant with time. These observations were confirmed by direct microscopic examination of the specimen's surfaces. If one waits at the lower potential before chloride addition, the nucleation is reduced again with a related change of the current increase from a slope 3 to 2 for the double logarithmic plot, which indicates a variation from a constant nucleation rate to a constant number of pits, i.e., no further nucleation. Apparently the nucleation is severely increased for nonstationary conditions and is reduced again when approaching a new stationary state. An explanation of this effect with the penetration mechanism leads to contradictions. When the potential is stepped to lower values the electrical field strength is decreased and one expects reduced rather than increased migration of the aggressive anions through the passive layer (Figure 7.2). Remaining at the lower potential before Cl^- addition should decrease the thickness of the passive layer because of an excess of film dissolution over its formation. The approach to the new stationary state with time at a lower potential should yield the original field strength

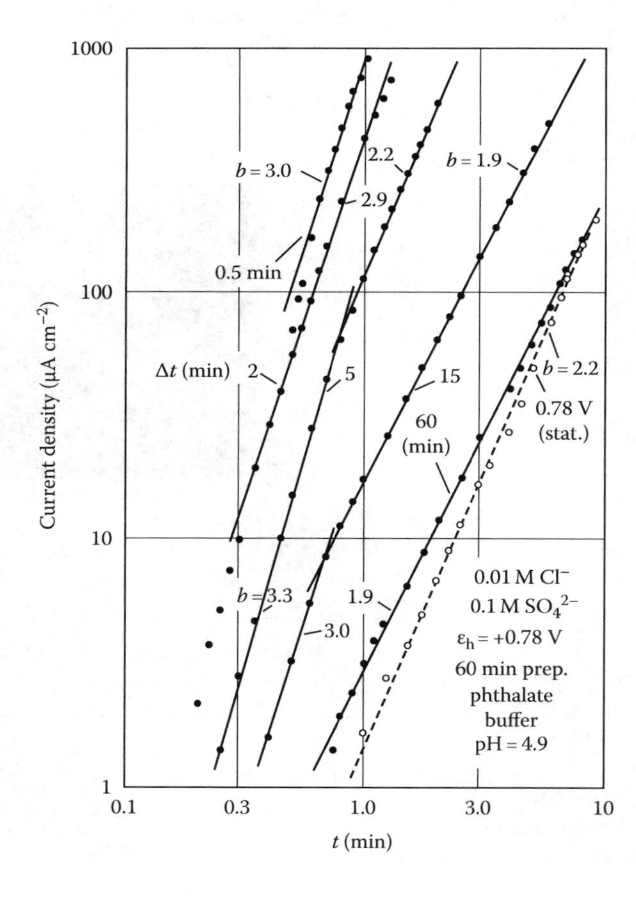

FIGURE 7.4
Double logarithmic plot of the increase of the geometric current density with time for electropolished iron during pitting corrosion, 1 h prepassivated at 1.18 V and potential drop to 0.78 V (solid line), Δt = time between potential change and chloride addition, 1 h prepassivated at 0.78 V (dashed line), phthalate buffer, pH 4.9, 0.1 M SO_4^{2-}, 0.01 M Cl^-. (From Vetter, K.J. and Strehblow, H.-H., *Ber. Bunsen. Ges. Phys. Chem.*, 74, 1024, 1970.)

with an increase of penetration. Again, the opposite effect was found [6]. Even the adsorption mechanism cannot explain these observations. If the potential is decreased, the voltage drop $E_{2,3}$ at the oxide–electrolyte interface will decrease intermediately with a negative overvoltage $\eta_{2,3}$ for O^{2-} formation from the water, i.e., the dissolution of the oxide, until a new stationary state with $E_{2,3} = E_{2,3,s}$ is achieved (Figure 7.2). This situation should disfavor the adsorption of Cl^- at this interface, which again should slow down the penetration instead of increasing it.

7.4.2 Film-Breaking Mechanism

The occurrence of fissures within the passive layer is a possible explanation for the observations mentioned last, especially for a nonstationary state of the passive layer. A sudden change of the electrode potential even in a negative direction will cause stresses within the film. Chemical changes [32] or electrostriction [7,8] is a reasonable explanation. Chemical changes, i.e., a reduction of Fe(III) to Fe(II), have been detected with XPS when negative potentials were applied to electrodes passivated at positive potentials well within the passive range [32]. There also exists direct evidence of film-breaking events for these nonstationary conditions according to observations with a rotating Fe-disk–Pt-ring electrode [33]. Even in the absence of aggressive anions, a sudden potential decrease from 1.3 to 0.7 V causes release of Fe^{2+} ions after a few seconds into the electrolyte, as detected by a temporary peak of the analytical ring current detecting Fe^{2+} ions (Figure 7.5). This is a

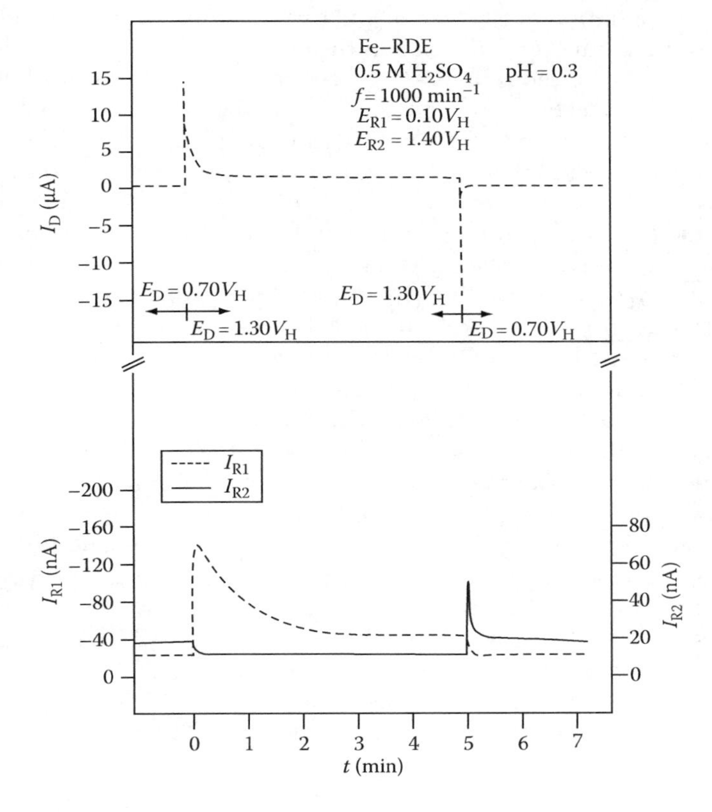

FIGURE 7.5
Disk current I_D and ring currents I_{R1} for Fe^{3+} analysis and I_{R2} for Fe^{2+} analysis for a potential variation from E_D = 0.7 to 1.30 V and back. E_{R1} = 0.10 V, E_{R2} = 1.40 V, 0.5 M H_2SO_4, pH 0.3, f = 1000 min^{-1}, disk area A = 0.09 cm^2. (From Löchel, B.P. and Strehblow, H.-H., *Werkst. Korros.*, 31, 353, 1980.)

consequence of film breaking and repair events. Any crack within the passive layer will lead to direct contact of small parts of the metal surface with the electrolyte and thus to dissolution of Fe as Fe^{2+} with current densities that refer to the relatively positive potentials, but without any protecting oxide at these defect sites. The self-healing mechanism of the passive layer causes only temporary Fe^{2+} formation of a few seconds duration. These potential changes should cause numerous defects of nanometer dimensions within the passive layer in order to get a sufficient amount of unprotected metal surface so that the temporary release of Fe^{2+} becomes measurable. In the presence of aggressive anions, their direct access to the metal surface will prevent repassivation and pits can form. The serious increase of the nucleation rate (Figure 7.4) after these potential changes is further evidence for the suggested formation of a large number of defects in the nanometer range.

As already mentioned, pit nucleation is a very fast process. For well-passivated Fe specimens, the time for the occurrence of a first pit is less than a second for potentials well above the critical value E_p, as determined by the increase of the current density after the addition of chloride to the solution and direct microscopic observation with a camera [4]. Most of the time is used to get the aggressive anions to the specimen surface by convection. If chloride is already present in the electrolyte, a short passivation below the pitting potential for approximately 1 s and a subsequent potential increase cause pit formation within less than a millisecond [4]. Results obtained with these electrochemical pulse techniques suggest that nonstationary conditions of the passive layer favor the film-breaking mechanism. The penetration of aggressive anions would require much longer times than observed during these pulse experiments. If the experimental conditions are not in favor of pitting, i.e., for small concentrations of aggressive anions ($<10^{-3}$ M) or low potentials in the vicinity of E_p, pit nucleation requires times longer than these extremely short times.

Long passivation times suggest that film breaking is a good explanation for the pit nucleation for nonstationary conditions too. The current increase caused by pitting slows down when the passivation time is increased up to more than 10 h [6,31]. The passive layer is formed during the first seconds and does not grow much after that time. This is a direct consequence of the barrier character of these films, which leads to logarithmic or inverse logarithmic growth with time. Further decrease of the passive current density and especially of the pit nucleation after hours may easily be explained by constantly occurring film breaking and repair events. The related stresses will heal with passivation time so that these events become rare with time. Figure 7.6 shows an example of a decrease of pitting with passivation time and the related reduction of the current increase [31].

Measurement of the electrochemical current noise has the aim of correlating the observed current fluctuations with breakdown and repair events that might lead to the formation of stably growing pits [34,35]. In the view of these theories, the application of statistical methods to the occurrence of current spikes and the observed probability of pit formation lead to a stochastic model for pit nucleation. The evaluation of current spikes in the time and frequency domains yields parameters such as the intensity of the stochastic process and the repassivation rate [34]. They depend on parameters such as the potential, state of the passive layer, and concentration of aggressive anions.

An interesting discussion of current measurements on microdimensional electrodes of stainless steel wires is given by Mattin and Burstein [36]. Their analysis of current transients at a very low level in chloride-containing 0.075 M $HClO_4$ leads to the distinction of metastable and stable pits. According to their discussion, the remaining passive layer protects the pit analyte from being diluted from the bulk solution. Only when the film breaks off too small pits will repassivate, whereas a few larger ones are deep enough to keep their local environment undiluted so that they survive.

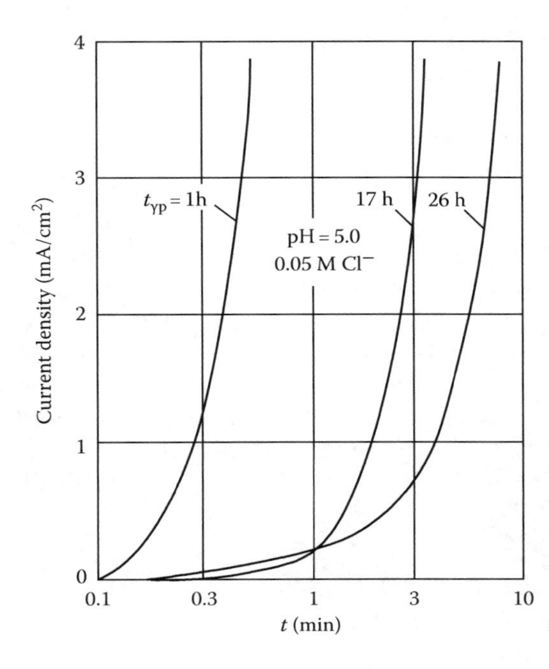

FIGURE 7.6
Increase of current density for passive Fe with pitting corrosion in phthalate buffer, pH 5.0 and 0.05 M Cl^- at $E = 1.38$ V for different prepassivation times t_{pp}. (From Vetter, K.J. and Strehblow, H.-H., in *Localized Corrosion*, NACE, Houston, TX, 1974, p. 240.)

7.4.3 Adsorption Mechanism

The passive current density is influenced by the action of anions even for conditions where pitting does not occur. Passive iron in 0.5 M H_2SO_4 shows stationary current densities (7 μA/cm²) that are higher by approximately one order of magnitude than for 1 M $HClO_4$ (<1 μA/cm²) [37]. This observation has been explained by a catalytic effect of SO_4^{2-} anions on the transfer of Fe^{3+} from the oxide to the electrolyte [38]. This cation transfer is the rate-determining step for the passive corrosion reaction. In the presence of SO_4^{2-}, an $FeSO_4^+$ complex forms. It is reasonable that this complex, with only one positive charge, requires less activation energy to be transferred from its O^{2-} ligands within the oxide matrix to the electrolyte than the highly charged Fe^{3+} ion. The complexation of cations by organic reagents causes a similarly enhanced dissolution in the passive state. The dissolution of Ni^{2+} from passive nickel and nickel-based alloys is enhanced by organic acids such as formic acid, which may lead to the removal of NiO in the passive layer [39]. Similarly, additions of citrate to the electrolyte cause thinning of the passive layer on stainless steel and increase of the Cr content within the oxide layer. This is a consequence of accelerated transfer of Fe^{3+} from the film surface to the electrolyte [40]. Apparently the complexation and transfer of Cr^{3+} are not enhanced.

It is well established that metals that show passivity within strong acidic electrolytes are far from the dissolution equilibrium of the oxide. The barrier character of the passive layers for these conditions requires slow dissolution kinetics at the oxide–electrolyte interface. Similarly to the catalytic effect of SO_4^{2-} and complexing organic agents, halides enhance the transfer of Fe^{3+} and Ni^{2+} from the oxide to the electrolyte. The adsorption mechanism for pit nucleation starts with the formation of surface complexes that are transferred to the electrolyte much faster than uncomplexed Fe^{2+} ions (Figure 7.2c). These details have been studied for Fe in Cl^-- and F^--containing solutions [21,41–43]. Similarly, Ni in solutions containing F^- has been studied [30,45–47]. Fe is a good metal with which to analyze the processes leading to passivity breakdown because of the characteristic changes of the valence of the dissolving ions, which may be measured with the rotating-ring-disk (RRD) technique. Fe^{3+} ions are dissolved if the metal surface is still covered completely with Fe(III) oxide, whereas Fe^{2+} is detected when bare metal comes into contact with the electrolyte at a pit surface or at a defect site, similar to Fe dissolution in the active state. The release of Fe^{3+} ions immediately after Cl^- has access to a prepassivated Fe electrode and was taken as a measure of locally enhanced dissolution of the oxide film. This leads to a local thinning of the passive layer and finally to its complete breakdown and the formation of a pit. In the case of F^- in acidic electrolytes the attack is more general; i.e., the passive layer is subject to a general attack. As a consequence, the measured passive current density is increased by orders of magnitude (Figure 7.7) [21,42,43]. Unfortunately, the strong $FeF_5H_2O^{2-}$ complex that finally forms in solution prevents its detection by its reduction to Fe^{2+} at the analytical ring. The current increases during the intermediate stage of attack with the first order of the HF concentration. Therefore, a proposed mechanism should yield an electrochemical reaction order of one for HF. The following sequence of reactions has been proposed for the processes that lead to thinning of the passive layer and enhanced passive dissolution [21,42–44]:

$$Fe^{3+}(ox) + HF(aq) \leftrightarrow FeF^{2+}(ad) + H^+(aq) \tag{7.8}$$

$$FeF^{2+}(ad) \rightarrow FeF^{2+}(aq) \tag{7.9}$$

$$4HF + H_2O + FeF^{2+}(aq) \rightarrow FeF_5H_2O^{2-}(aq) + 4H^+(aq) \tag{7.10}$$

The adsorption (7.8) is assumed to be fast and in a quasi-equilibrium. Reaction (7.9) is the rate-determining step, and (7.10) is a fast following reaction step in solution leading to the stable fluoro-aquo complex. These assumptions lead to a first-order process for HF

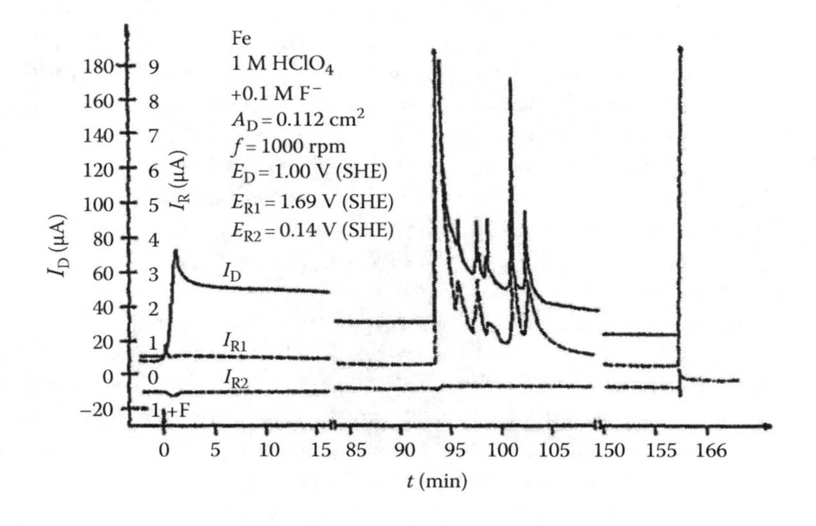

FIGURE 7.7
Current-time dependence of a rotating Pt-split-ring-Fe-disk electrode in 1M $HClO_4$ after HF addition to 0.1 M, 2 h prepassivation at 0.1 V in 1M $HClO_4$, Fe^{2+} detection at ring 1. (From Löchel, B.P. and Strehblow, H.-H., *Electrochim. Acta*, 28, 565, 1983.)

according to the rate equations (7.15) and (7.16). The Butler–Volmer equation (7.11) will hold for the rate-determining step (7.9) with a charge transfer coefficient α and with the potential drop $E_{2,3,s}$ at the oxide–electrolyte interface for stationary conditions. $E^0_{2,3,s}$ refers to a solution of pH 0. According to the equilibrium of O^{2-} formation of Figure 7.2a and Equation 7.13, one obtains Equation 7.14 for the pH dependence of the potential drop $E_{2,3,s}$ for stationary conditions of the passive layer:

$$i_{cs} = 3kF\left[FeF^{2+}_{ad}\right]\exp\left(\frac{2\alpha FE_{2,3,s}}{RT}\right) \quad (7.11)$$

Introducing the electrochemical equilibrium at the oxide/electrolyte interface of reaction (7.8) with its constant K according to the following equation:

$$\left[FeF^{2+}_{ad}\right] = \frac{K\left[HF_{aq}\right]\left[Fe^{3+}_{ox}\right]}{[H^+]}\exp\left\{\frac{FE_{2,3,s}}{RT}\right\} \quad (7.12)$$

and the equilibrium of O^{2-} formation and the related pH dependence of the potential drop $E_{2,3,s}$

$$H_2O \leftrightarrow O^{2-} + 2H^+ (aq) \quad (7.13)$$

and

$$E_{2,3,s} = E^0_{2,3,s} - 0.059\,pH \quad (7.14)$$

yields

$$i_{c,s} = 3kKF\frac{\left[HF_{aq}\right]\left[Fe^{3+}_{ox}\right]}{\left[H^+_{aq}\right]}\exp\left\{\frac{FE_{2,3,s}(2\alpha+1)}{RT}\right\}$$

$$i_{cs} = i^0_{c,s}\exp\left\{\frac{FE^0_{2,3,s}(2\alpha+1)}{RT}\right\} \quad (7.15)$$

and

$$i^0_{c,s} = 3kKF\left[HF_{aq}\right]\left[Fe^{3+}_{ox}\right][H^+]^{2\alpha} \quad (7.16)$$

The effect of HF is a model for pitting insofar as it involves the total passive surface so that the measured effects are much more pronounced and refer to a surface of known size instead of an unknown actual pit surface that even changes with time. The later stages of the attack of the passive layer lead to current peaks that go along with equivalent Fe^{2+} formation (Figure 7.7). Finally, a general breakdown of the passive layer is observed with a steep increase of the dissolution current density and Fe^{2+} formation [21,42,43]. Apparently, one may observe the different steps of breakdown of passivity for fluoride directly, which are difficult to follow for the local events in the case of the other halides. The difference in the action of fluoride and the other halides may be explained by a comparison of the

TABLE 7.1

pK Values of Stability Constants of Metal Ion Complexes with Halides for the Reactions $Me^{z+} + X^- \rightarrow MeX^{(z-l)+}$ Including Constants Referring to HF (*) in Acidic Solutions

Anion	Fe^{3+}	Ni^{2+}	Cr^{3+}
F^-/HF(*)	5.17/2.26(*)	0.66	4.36 sol; 1.42(*)
Cl^-	0.62	−0.25	−0.65
Br^-	−0.21	−0.12	−2.65
I^-	1.30	—	−5.0

Source: Sillén, L.G. and Martell, A.E., *Stability Constants of Metal Ion Complexes*, Suppl. 1, Special Publ. 25, The Chemical Society, London, 1971, pp. 167, 171, 217.

stability constants of their iron complexes as presented in Table 7.1 [30,45–47]. The stability constants K_1 with one anion are given, which refers to the reaction order one that has usually been found for the dissolution of passive layers under the influence of the aggressive anions. The reaction of Fe^{3+} with HF to form FeF^{2+} and H^+ yields the more realistic value of $\log K_1 = 2.28$ due to the small dissociation constant of HF ($pK = -\log K_d = 2.98$). Table 7.1 also contains the constants K_1 for Ni^{2+}– and Cr^{3+}–halide complexes. Their falling values from fluoride to iodide and Fe^{2+} to Ni^{2+} support the decreasing tendency for enhanced dissolution of the passive layer and localized corrosion. These data can be referred to the situation at the oxide surface. The fluoro complexes are very stable and form in high concentrations at all surface sites. Therefore their much faster transfer to the electrolyte yields enhanced general dissolution, whereas the attack of the other halides is locally restricted and much less pronounced. Besides the thermodynamically based values, i.e., the stability constants, the kinetics of complex formation and of the complex transfer to the electrolyte are another decisive factor for the attack of the passive layer. In this sense, the situation of Cr^{3+} is very special and will be discussed separately.

XPS measurements of passivated Fe and Ni electrodes that have been exposed to aggressive anions (Ni and Fe to F^-; Fe to Cl^-, Br^-, and I^-) but have not already formed corrosion pits support this mechanism. The quantitative evaluation of the data clearly shows a decrease of the oxide thickness with time of exposure [22,48]. Not only F^- but also the other halides cause thinning of the passive layer (Figure 7.8) [48]. The catalytically enhanced transfer of cations from the oxide to the electrolyte leads to a new stationary state of the passive layer. Its smaller thickness yields an increased electrical field strength for the same potentiostatically fixed potential drop, which in turn causes faster migration of the cations through the layer to compensate for the faster passive corrosion reaction (7.1) at the oxide–electrolyte interface (Figure 7.2a). Statistical local changes of this dissolution process will finally result in breakdown of the passive layer and exposure of bare metal surface to the electrolyte. For F^- this occurs after intermediate breakdown and repair events at the total surface. For the other halides, local effects and the formation of pits are found corresponding to their less pronounced complexing tendency. They will form complexes and cause their effect only at special active sites of the oxide surface. It should also be mentioned that fluoride causes a reduced local attack in more alkaline solutions such as phthalate buffer pH 5 [42], which finally leads to the formation of corrosion pits and not a general attack as observed in strongly acidic electrolytes. The thinning of the oxide layer is also much less pronounced or hardly detected for pH > 5. For a higher pH, the passive layer gets to or close to its dissolution

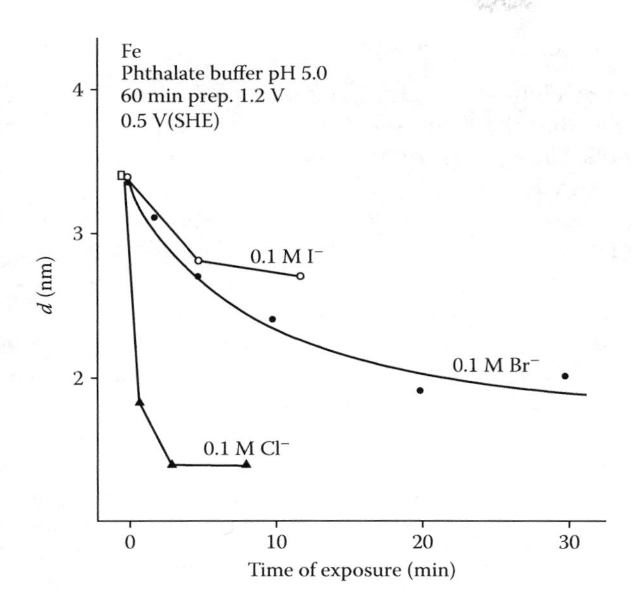

FIGURE 7.8
Decrease of oxide thickness with time of halide exposure at 0.5 V after 60 min prepassivation at 1.20 V for different halides, deduced from XPS measurements. (From Khalil, W. et al., *Werkst. Korros.*, 36, 16, 1985.)

equilibrium in the sense of the thermodynamically deduced potential–pH diagrams. In this situation, the tendency for the formation and transfer of soluble fluoride complexes will be reduced so that again only local effects are observed.

In this regard, the special situation of Cr should be discussed. This metal does not show pitting, and the passive layer is not attacked in the presence of chloride. The stability constants of CrX^{2+} complexes are smaller than 1. Besides that it is well known that the exchange of Cl^- ligands and H_2O between the inner and the outer sphere is an extremely slow process with a half-time of several hours [49]. This is a consequence of the large ligand field stabilization of the Cr(III) complexes with an octahedral coordination shell with three electrons in the lower t_{2g} and none in the e_g level. This situation causes "insolubility of $CrCl_3$ and Cr_2O_3 in cold water" as reported in literature [50]—however, it is rather the slow dissolution rate of the Cr^{3+} salt or oxide than their low solubility. The exchange of a ligand of the inner with one of the outer coordination shell requires a large activation energy, which makes the dissolution process even via complex formation extremely slow. Therefore, one has to conclude that the presence of a $CrCl^{2+}$ complex at the surface will not increase the dissolution rate because it will form and dissolve very slowly by itself. In contrast to this situation, the exchange is rapid for Fe^{3+} complexes. Thus, a chemical change of Cr^{3+} ions from a part of the oxide matrix to a $CrCl^{2+}$ complex will not increase the dissolution rate. Besides these circumstances, the smaller stability constants of the Cr^{3+} complexes are also in favor of the stability of the passive layer. In consequence, the tendency for Cr^{3+}–halide complexes to form is negligibly small, and once they have formed their dissolution rate is not increased relative to Cr^{3+} within an oxide matrix. Therefore, the halides will not attack the passive layer of chromium and pitting cannot occur, in agreement with the experimental findings.

For similar reasons, the dissolution rate of Cr(III) oxide is extremely slow in the passive state. Additions of Cr therefore stabilize the passive behavior of Fe–Cr alloys and stainless steel. Fe–Cr alloys are more resistant to pitting in chloride-containing electrolytes with more positive pitting potentials compared with pure Fe. The Cr concentration is increased within the passive layer relative to the composition of the bulk metal. Thus, Fe–Cr alloys are more protected against the attack of aggressive anions and pitting by the beneficial effect of Cr.

7.4.4 Comparison of the Different Nucleation Mechanisms

The discussion of the different nucleation mechanisms on the basis of experimental results for iron and nickel leads to the conclusion that the film breaking and adsorption mechanisms are very effective. As usual in kinetics, the fastest reaction path is dominating. This, however, depends on the experimental or environmental conditions. For a stationary state of the passive layer the adsorption mechanism seems to be most effective, as demonstrated for Fe in weakly acidic electrolytes. If a nonstationary state is attained by a fast change of the potential, film breaking is most probable. Of course, other nucleation mechanisms may contribute as well, such as mechanical damage of the surface, dissolution of inclusions, and last but not least the penetration mechanism. Penetration is believed to be the leading mechanism for pitting of Ni in Cl^--containing electrolytes. A conclusive critical experiment to determine whether penetration of Cl^- is an initial step for breakdown of Ni passivity is still missing. The role of inclusions is the subject of another chapter in this book and will not be discussed in detail here. It was the aim of this chapter to discuss effects related to pure or at least single-phase metals. In the technical world, however, these other effects are very important. As chemistry plays a decisive role in the pitting process, one should discuss the tendency of the different cations to form complexes with halides. Thus, one has to include the properties of the aggressive anions and of the specific metals under study as well. This idea is often neglected but seems to be a key question for breakdown of passivity and the stable growth of corrosion pits, and it will be discussed again in Sections 7.4.3 and 7.10.

7.5 Mechanistic Consequences for Pit Nucleation from Studies with Nanometer Resolution

The more general discussions on pit initiation allow only limited conclusions on the local nature of the responsible mechanisms. The previous discussion of Section 7.4 on the three main mechanisms does not necessarily allow conclusions on the local confinement of the attack. This is characteristic for the older models of the passive layer, which could not take care of its atomistic structure at the nanometer scale. More recent investigations are discussed in Chapter 5. They show that passive layers consist of a multilayer structure with at least one part acting as a barrier that in many cases is crystalline. The outer part of the passive film is often amorphous and not believed to be a barrier for the corrosion process. The inner part is in most cases the barrier. Its crystallinity improves usually with increasing potential and passivation time. This may be related to the loss of water, i.e., a change from a hydroxide or oxyhydroxide to an oxide. Even adsorption layers as precursors of the passive films consist of large and well-ordered domains and may give protection to the metal surface underneath. However, when three-dimensional layers are formed they consist mostly of nanocrystals sometimes even with amorphous parts in between. They frequently are faceted to overcome stresses between the layer and the metal substrate surface. These facets are usually only a few nm wide. Well-ordered although faceted oxide layers are obtained on Cu and Ni, nanocrystals on Fe, Cr, and Fe–Cr alloys. Crystallization takes long time for Cr and Fe–Cr alloys and the amorphous part of the film acts as a cement between the nanocrystals. On Co and Ag ordered domains are obtained for monolayers, whereas Ag forms thick porous layers at higher potentials. Details are discussed in Chapter 5 of this book. In summary, protecting passive layers are continuous but nanocrystalline with many defects. Especially at grain boundaries of the nanocrystals,

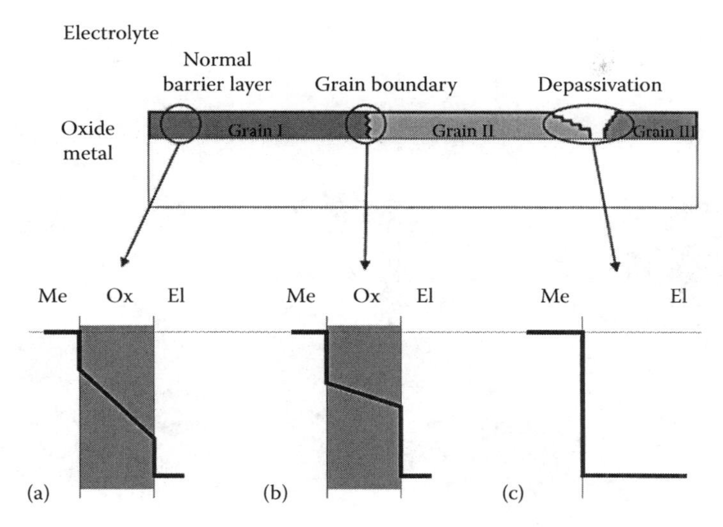

FIGURE 7.9
Potential distribution at sites protected by a crystalline barrier-type oxide: (a) at defect-free barrier layer, (b) at defect (grain boundary), and (c) at depassivated site (pit nucleus). (From Marcus, P. et al., *Corrosion Sci.*, 50, 2698, 2008.)

numerous disordered sites suggest to be weak regions where an attack may easily take place. Indeed, repassivated pits are found in vicinity of the pitting potential on passivated Ni(111) surfaces at step edges that seem to be the preferential site for the attack of chloride [51]. Therefore, the mechanism of breakdown of passivity and pit nucleation have been discussed on the basis of the structure of ordered passive layers with many defects like grain boundaries of nanocrystals of oxides, disordered or amorphous parts of the surface layers, step edges, and other defects [52].

These defects correspond to more reactive sites of the passivating oxide. As a consequence, cations are less bound to the oxide matrix and may dissolve faster corresponding to a smaller free activation enthalpy $\Delta G^{\ddagger}$. Therefore the passive layer gets thinner at these sites. In addition, the potential drop is redistributed (Figure 7.9). The defect sites of the passive layer are less resistive to ion migration and take over less potential drop. With a total potential difference between the substrate metal and the electrolyte given by the applied electrode potential a larger drop will be located at the oxide–electrolyte and/or the metal–oxide interface which accelerates the electrochemical reactions at these sites.

The effect of a local increase of the potential drop at the oxide surface is similar to the situation discussed for the adsorption mechanism of Section 7.4.3 and Figure 7.2c. Apparently, the defects lead to locally enhanced dissolution, however, with still maintained passivity (Figure 7.10a). Statistical fluctuations may cause even an intermediate localized activation, which, however, will not last due to the self-healing effect of passive surfaces. However, they cause local depressions at the specimen surface. Temporary local breakdown should therefore be accepted also for electrolytes free of aggressive anions. This concept is consistent with an increased general corrosion rate during the early stages of passivation that will decrease with time when the film structure becomes more crystalline with a related decrease of its defects.

A second mechanism is related to increased cation formation with a larger potential drop at defects of the film at the metal–oxide interface. The migration of these cations through the defects within the film should be fast and they get dissolved into the electrolyte. If oxygen ions cannot enter fast enough, cation vacancies will accumulate at the metal surface and form voids. These voids finally lead to a local collapse of the passive layer and

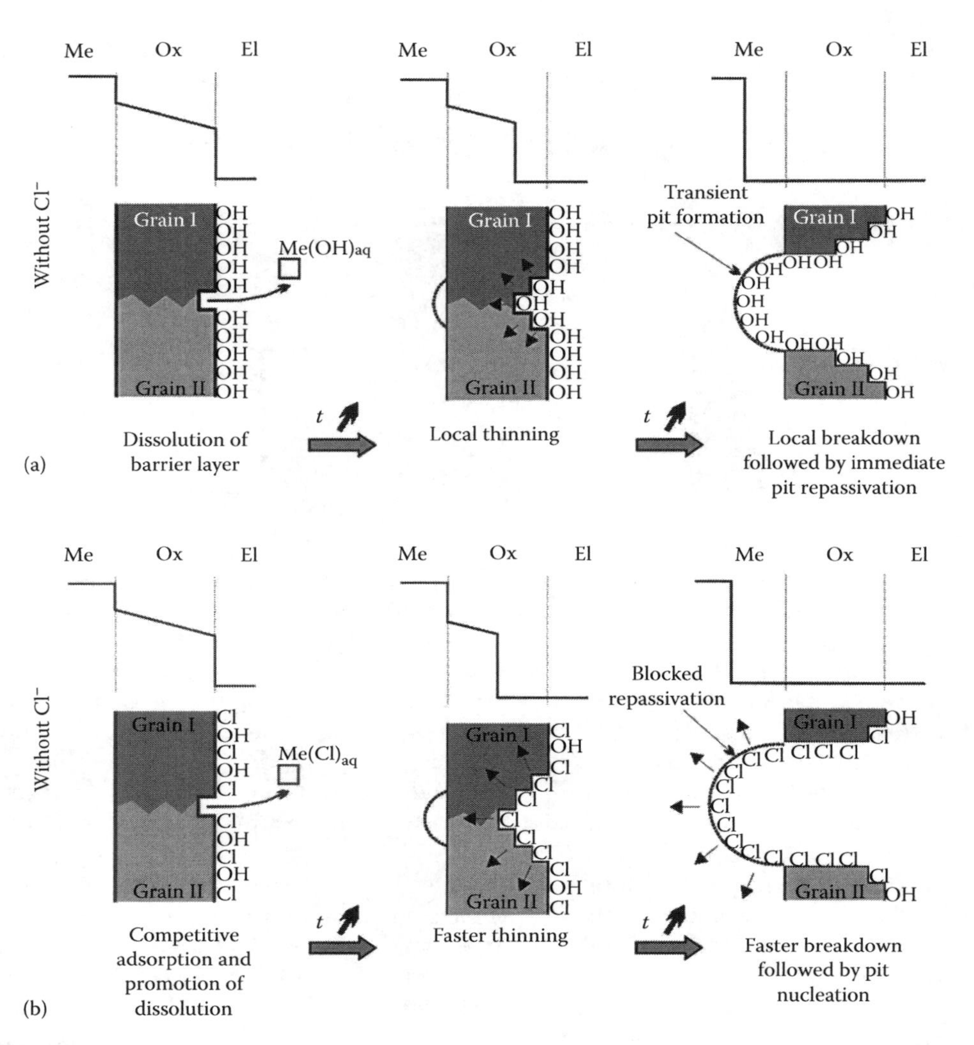

FIGURE 7.10
Enhanced dissolution at defect of the passive layer (a) without halides forming depression with repassivation and (b) in the presence of halides forming pit nucleus growing to corrosion pit. (From Marcus, P. et al., *Corrosion Sci.*, 50, 2698, 2008.)

a temporary local breakdown which again creates depressions (Figure 7.11a). This situation is similar to the proposed mechanism described by the point-defect model [13,14].

The predicted depressions have been found indeed by in situ scanning tunneling microscope (STM) investigations of Ni(111) surfaces in weakly acidic electrolytes [53]. Depressions of 0.4–1.4 nm are found between crystalline oxide grains in sulfate-containing electrolyte at $E = 0.55$ V. They are growing by stepping the potential to more positive values up to $E = 1.05$ V. Depressions 2.2–30 nm wide and 2.2–3.8 nm deep are found for these conditions, which have to be related to repassivated pit nuclei. They could not grow further due to the absence of aggressive anions.

If halides like Cl^- are present they will compete at the surface of the passive layer with OH^- ions. Chemisorbed halides may help to remove locally the passive layer due to their complexing properties similar to the already discussed adsorption mechanism.

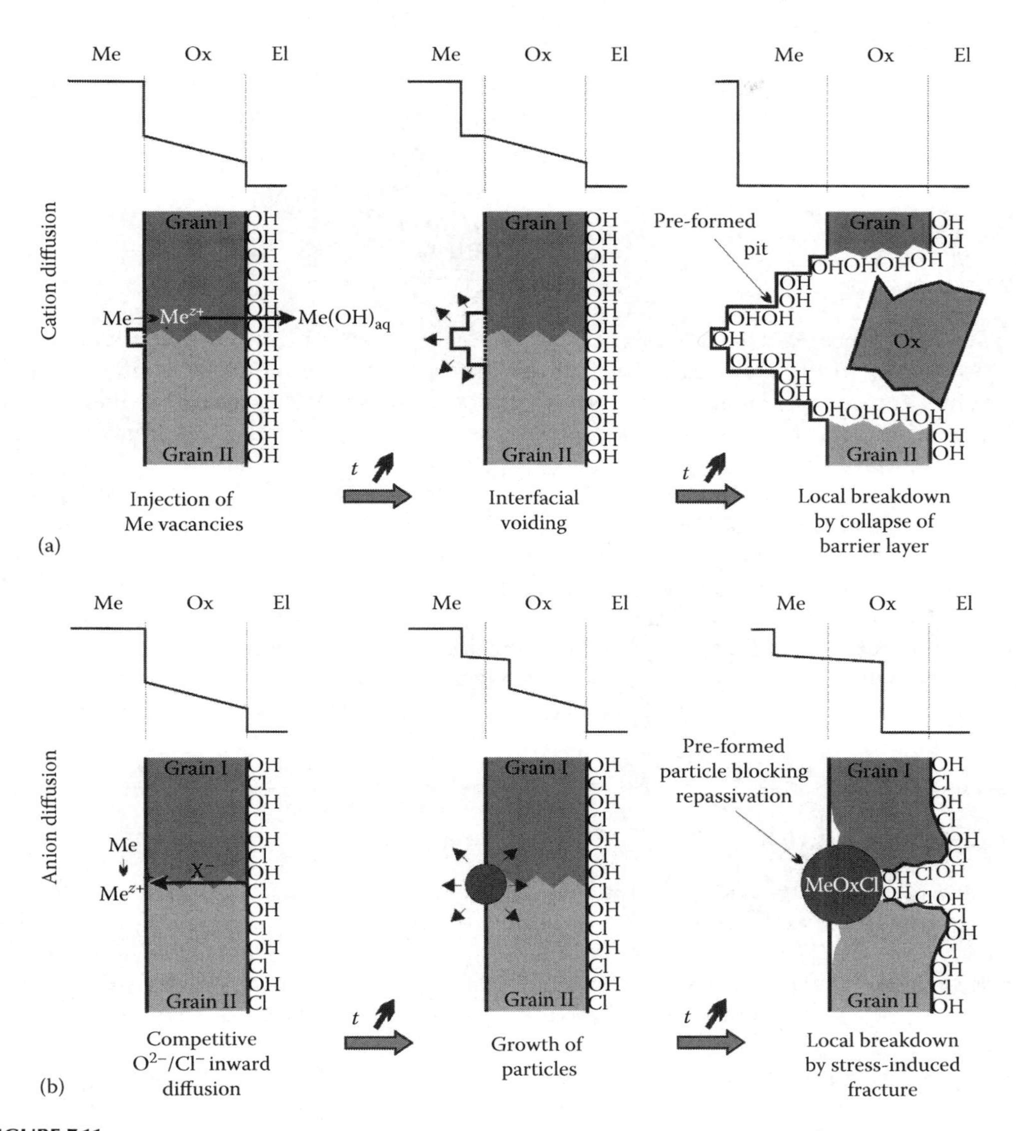

FIGURE 7.11
Accumulation of vacancies to voids at film defects leading to (a) the intermediate local collapse of passive layers and depression in the absence of halides (b) to pit nucleus which may grow to corrosion pit. (From Marcus, P. et al., *Corrosion Sci.* 50, 2698, 2008.)

The formation of $MeCl^{n+}$ and $MeOHCl^{(n-1)+}$ complexes at the surface of the passive layer leads to species which are more soluble and faster dissolving ($n = 3$ for Fe and $n = 2$ for Ni). The exchange of OH^- by Cl^- especially at defects like step edges is a exothermic process ($\Delta G < 0$) and leads to a decrease of the free activation enthalpy $\Delta G^{\ddagger}$ for the dissolution of the halide complexes to the electrolyte compared to the cations within the unchanged oxide surface (Figure 7.10b). The accelerated transfer of the halide complexes yields a local thinning of the layer and its breakdown due to fluctuations again at the more reactive defect sites. This has been confirmed by theoretical ab initio calculations, which show a preferential replacement of OH and chemisorption of chloride at defects, especially at step edges as shown on Figure 7.12 [54]. As a consequence several factors

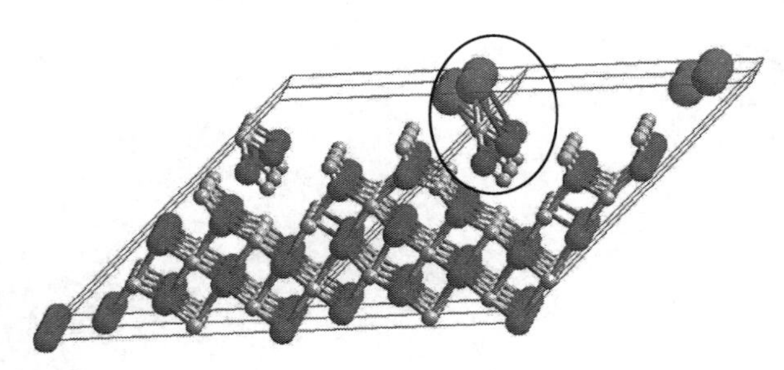

FIGURE 7.12
DFT calculation of the replacement of OH by Cl at a stepped passive film surface (hydroxylated NiO) and the detachment of ahydroxychloride complex. (From Bouzoubaa, A. et al., *Corrosion Sci.*, 51, 941, 2174, 2009.)

facilitate local breakdown at defects, the increased potential drop at the oxide–electrolyte interface, and the formation of fast-dissolving halide complexes. Similarly halides may enter the passive layer at defects and form halides or oxyhalides at the metal–oxide interface that prevents a reformation of the passive layer and causes the nucleation of a pit (Figure 7.11b).

Once local breakdown has occurred at these defects by local dissolution of the passive layer or by local accumulation of voids and mechanical collapse, halides from the electrolyte have direct access to the bare metal surface. They compete with OH^- and it is a matter of the electrode potential and the bulk concentration of the halides whether these nuclei will repassivate or continue their growth. The total potential drop concentrates at the metal–electrolyte interface at these pit nuclei. If the applied potential is large enough a related intense local metal dissolution leads to an accumulation of corrosion products and thus of halides that prevents their repassivation, i.e., the nucleus will grow to a pit.

7.6 Special Conditions for Localized Corrosion

Localized corrosion is not always related to aggressive anions. Pits may form on metal surfaces due to imperfect passivation. This may be a consequence of a poor film quality or the action of metallic inclusions. Two examples have been already mentioned in Chapter 5. One example is the localized corrosion of silver at potentials corresponding to early film formation (Chapter 5), another one is the effect of Cu inclusions on passivity breakdown of Al (Chapter 5).

Ag shows localized corrosion in alkaline solutions such as 0.1 M NaOH at low potentials where passivity starts. At $E = 0.15\,V$, a monolayer of Ag(I) oxide/hydroxide protects a Ag(111) surface with irregular and small ordered domains (Chapter 5, Figure 5.59). The latter have the lattice of (111) oriented Ag_2O. Protrusions with subsurface oxide and small grains of Ag(I) oxide are found. Dissolution starts at the nonordered parts of the surface oxide with pits that are one monolayer deep and grow laterally [55]. Their boundaries are random and thus dissolution follows weak and poorly structured parts of the surface oxide (Chapter 5, Figure 5.60). Apparently pitting occurs at weakly ordered sites of the surface oxide whereas ordered oxide domains and oxide grains stop the pit growth. Figure 7.13 shows schematically the competing processes of the formation of ordered protecting two-dimensional oxide layers and three-dimensional oxide grains, and pit initiation at weakly protected disordered two-dimensional hydroxide layers. Pit nucleation and growth at the second atomic plane is observed and apparently follows the same mechanism leading to a slow in-depth pit growth. The second plane does not undercut the first plane. This example shows that weak parts of the passive layer are prone to attack even in the absence of so-called aggressive anions. It should be mentioned that Ag tends to form $[Ag(OH)_2]^-$ anions in NaOH as a soluble corrosion product.

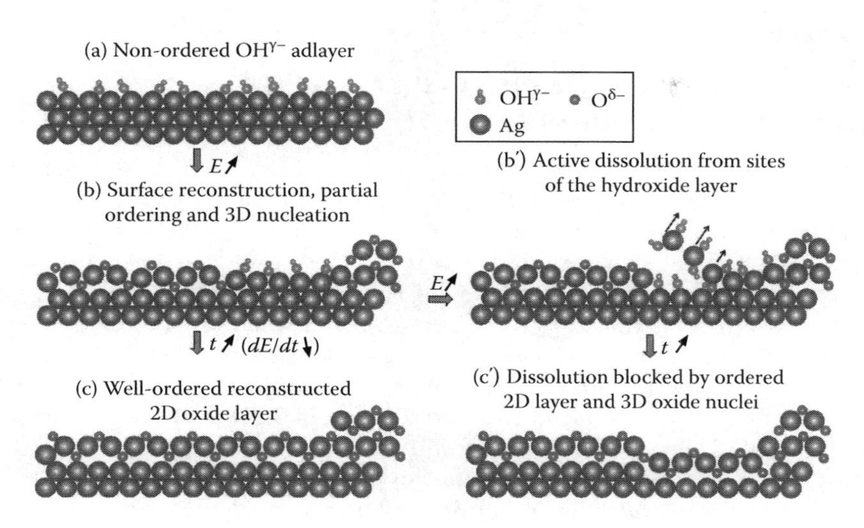

FIGURE 7.13
Schematic diagram for the formation of (a) a disordered hydroxide adsorption layer, (b), (c) an ordered oxide layer and (c) three-dimensional oxide grains on Ag(111) in 1 M NaOH in the underpotential rage of oxide formation ($E \leq 0.15$ V), (b′) with two-dimensional pit formation at disordered hydroxide layer with $Ag(OH)_2^-$ dissolution and (c′) its prevention at ordered oxide layers and oxide grains. (From Maurice, V. et al., *J. Phys. Chem. C*, 111, 16351, 2007.)

The formation of passive layers on pure aluminum leads to several hundred nm thick films of Al_2O_3 for potentials of more than 150 V (Chapter 5) [56]. According to RBS studies additions of Cu cause its accumulation at the metal surface underneath a growing pure Al_2O_3 layer with a final breakdown of passivity and intense localized metal dissolution. With increasing additions of Cu breakdown occurs at decreasing potentials [57]. These Cu additions are well distributed within the vapor-deposited films at the beginning and form even stable alloys at low concentration. Annealing of Al alloys with a few percent Cu causes the formation of small Cu-rich inclusions (such as the Al_2Cu Θ phase) that are known to cause localized attack. The Cu-rich particles protrude in the Al_2O_3 surface film, which is locally thinner. Dissolution of the oxide, enhanced at high pH and/or in the presence of Cl^- causes the exposure of the Cu-rich particles sooner than the Al matrix because of the thinner film. Anodic Al_2O_3 may even break down due to insufficient Al supply during oxide growth at these inclusions with an increased local potential drop or the inclusions may even not be covered completely by the passivating Al_2O_3 layer. These particles are then accessible to redox systems of the electrolyte and thus serve as local cathodes. This is an important condition for open-circuit corrosion of these materials because thick films of Al_2O_3 are electronic insulators. Aluminum may dissolve at the boundary of these surface inclusions due to a crevice between the two metal phases. As a consequence the crevice supports Al dissolution with pH shifts and ohmic drops, causing the propagation of localized corrosion. This might be further supported by aggressive anions like halides within the electrolyte. They may accumulate within these crevices and thus prevent their repassivation. Only those Cu inclusions at the metal surface induce this kind of localized corrosion. Those within the bulk of the material are not involved as long as they are not exposed to the electrolyte. Al alloys with Cu inclusions are widely used in industry and localized attack is a significant issue. The corrosion protection of Al alloys containing Cu is presented in details in the chapter on Al alloys.

7.7 Transition from Pit Nucleation to Pit Growth: Microscopic Observations

In recent literature the transition of nucleation to a stable growth of corrosion pits is examined with the STM [51,58]. These studies try to follow the development of corrosion pits from a diameter of some few nanometers to several micrometers using besides the STM the scanning electron microscope (SEM). The results thus give an extension from previous investigations with the SEM and the light microscope [6] from the μm to the nm range. STM studies may trace corrosion pits down to a size of some few nm. In principle, these effects have been followed in two different ways. Maurice, Inard, and Marcus [51] passivated Ni(111) single crystals in ~0.1 M sulfate solution of pH 3 for 30 min at $E = 0.85$ and exposed them finally to this solution after the addition of 0.05 V NaCl at the same potential (pitting potential 0.75 V). Numerous small pits ($5–9 \times 10^{10}\,cm^{-2}$) with a preferential round or some with a triangular shape aligned at steps of terraces of the (111) plane could be found (Figure 7.14). They have an average lateral size of 20 nm and a depth of 2 nm, which is in the thickness range of the passive layer of ~1 nm. A comparison of the total area of these pits (number and size) leads to a charge of metal dissolution of ~0.7 mC/cm^2, which is about one order of magnitude less than the directly measured charge of 1.6–8 mC/cm^2. This result leads to the conclusion that increased metal dissolution in the passive state occurs, which is expected according to the adsorption mechanism for pit nucleation. These small pits are interpreted as pit nuclei that apparently did not grow further. About 1% of these pit nuclei grows larger (70 nm) but with the same depth and a triangular shape oriented still parallel to the step edges at the Ni surface. They presumably are repassivated pits. About 0.1% of the pits grow to a size of some few 100 nm with an elongated irregular shape parallel to the steps. These pits are seen as repassivated, i.e., metastable pits and grow when current transients are observed in the potentiostatic current–time curves.

Another approach is the growth of corrosion pits at potentials several hundred mV more positive than the pitting potential for short times in the ms range. Kunze and Strehblow prepassivated Ni(111) surfaces by changing the potential from −0.1 V to values below the pitting potential ($E = 0.39$ V) for a few seconds in a 0.2 M NaCl solution of pH 5.6 to prepassivate the surface [58]. Then the electrode was pulsed to potentials in the range 1.0–1.5 V for some few ms to grow pits and finally stepped back to −0.1 V to stop pit growth. These pulse

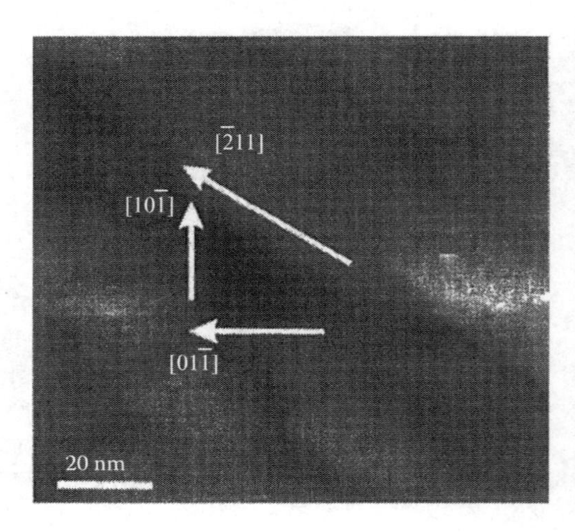

FIGURE 7.14
STM topographic image of a pit formed on Ni(111) passivated for 1800 s at 0.85 V (SHE) in 0.05 M Na_2SO_4, pH 3 and subjected to pitting in the same solution with additions of 0.05 M NaCl for 4500 s at the same potential. (From Maurice, V. et al., *Critical Factors in Localized Corrosion III* (P.M. Natishan, R.G. Kelly, G.S. Frankel, and R.C. Newman, eds.), The Electrochemical Society Proceedings Series, PV 89-17, Pennington, NJ, 1999, p. 552.)

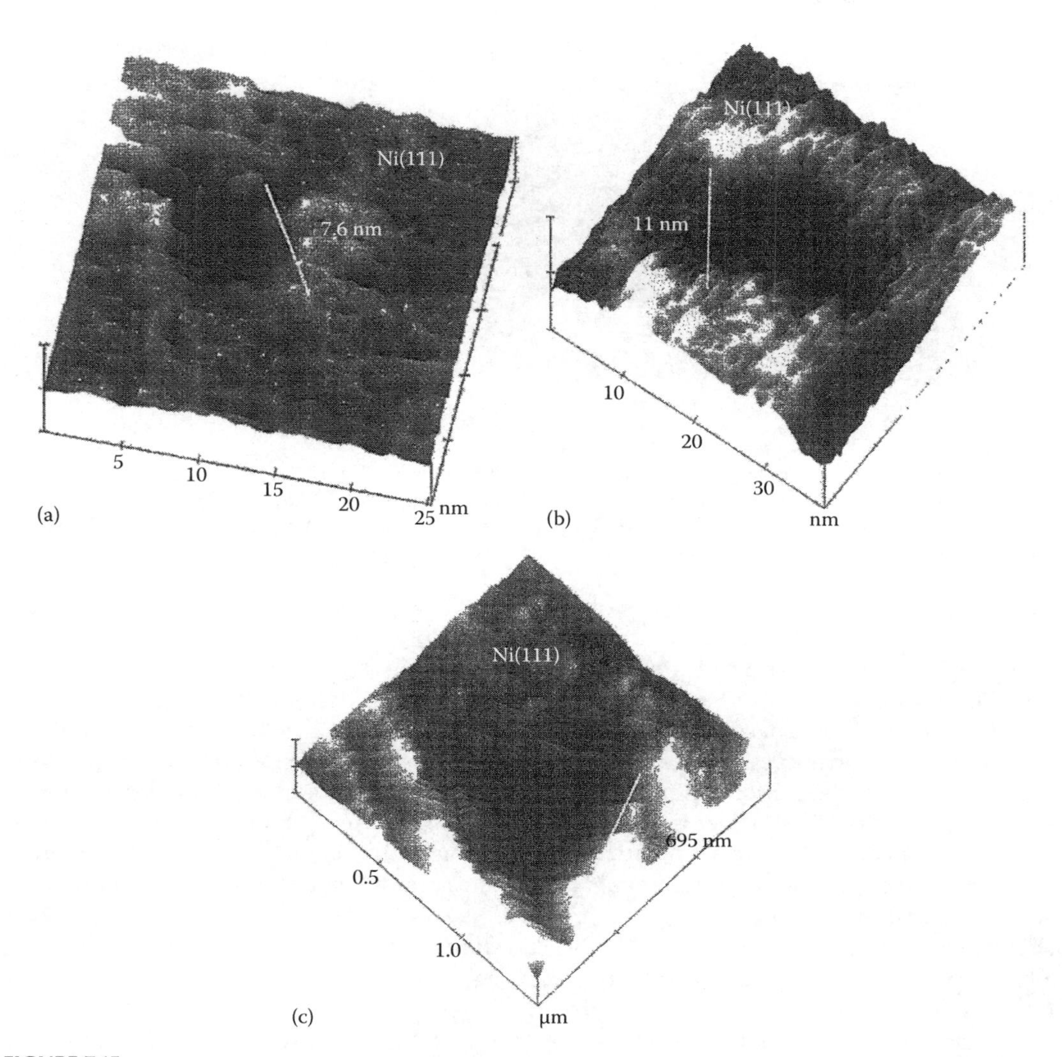

FIGURE 7.15
Sequence of in situ STM topographic images of growing pits formed on Ni(111) by potentiostatic pulses in 0.2 M NaCl, pH 5.6 during 50 ms at 1.1 V (SHE) after 3 s prepassivation at 0.39 V. (a) through (c) with increasing time of pit growth at 1.1 V. (From Marcus, P. et al., *Corrosion Sci.*, 50, 2698, 2008.)

measurements facilitated pit nucleation and thus caused a large density of very small corrosion pits [30,59] that could be studied in situ. The sulfate-free NaCl solution enhanced pit nucleation and hindered the early change of their shape from a polygon to a hemisphere [6]. Figure 7.15 presents a sequence of typical corrosion pits in different stages of development [58]. Small pits of less than 10 nm lateral size and 1.5 nm depth have an irregular slightly elongated shape. They change to triangular pits during further growth. This shape is easily explained by (111) crystal planes forming the pit surface and their intersection with the (111) Ni surface. They are the most densely packed crystal planes of face-centered cubic (fcc) Ni and thus have the largest resistance to metal dissolution. As a consequence, these planes remain during the dissolution of an actively corroding surface site and thus form the inner pit surface. They still grow further as shown on the SEM micrograph of Figure 7.16a for a similar experiment on polycrystalline Ni [58]. Depending on the orientation of

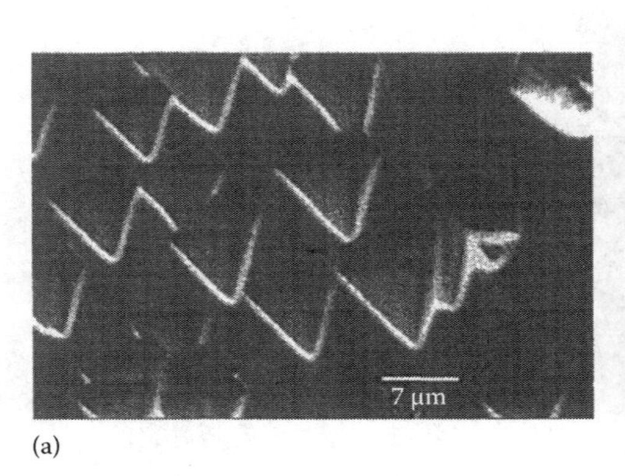

(a)

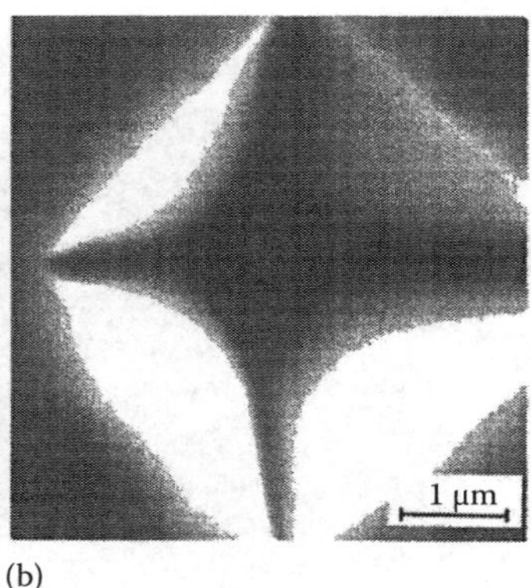

(b)

FIGURE 7.16
SEM images of pits formed on polycrystalline Ni electrodes: (a) Formed in 0.2 M NaCl, pH 5.6 for 5 ms at 1.5 V (SHE) after 10 s prepassivation at 0.5 V. (From Marcus, P. et al., *Corrosion Sci.*, 50, 2698, 2008.) (b) Formed in 0.05 M phthalate buffer, pH 5.0 with addition of 0.1 M KCl for 200 ms at 1.00 V (SHE) after preactivation at −0.5 V and 1 s prepassivation at 0.50 V. (From Strehblow, H.-H., *Habilitationsschrift*, Freie Universität, Berlin, 1977.)

the crystalline surfaces, other shapes of the pit orifice also appear such as squares (Figure 7.16b) [60] or triangles with cutoff edges approaching a hexagon.

Figure 7.17 presents examples of the combination of (111) and (100) planes for pits on crystallites of body-centered cubic (bcc) Fe grown with the potentiostatic pulse technique described earlier [6]. These polygonal pits change their shape to hemispheres when the increasing concentration of corrosion products leads to the precipitation of a salt film. This situation causes electropolishing of the pit surface by diffusion-limited metal dissolution within the pit. It starts at the pit bottom, where the electrolyte is concentrated first, and finally includes the whole pit surface, changing the polygonal form to a hemisphere [61]. These electropolished hemispheres keep their form when crossing a grain boundary, whereas the polygons may change their shape because of the different orientation of the adjacent crystallite (Figure 7.18a and b) [6]. The change of the pit surface from a polygon to a hemisphere will be discussed later in detail together with the calculation of the electrolyte composition for later stages of pit growth.

The change of the pit shape with its size and thus its age gives support to the film-breaking mechanism. A rapid change of the electrode potential causes small cracks within the

(a)

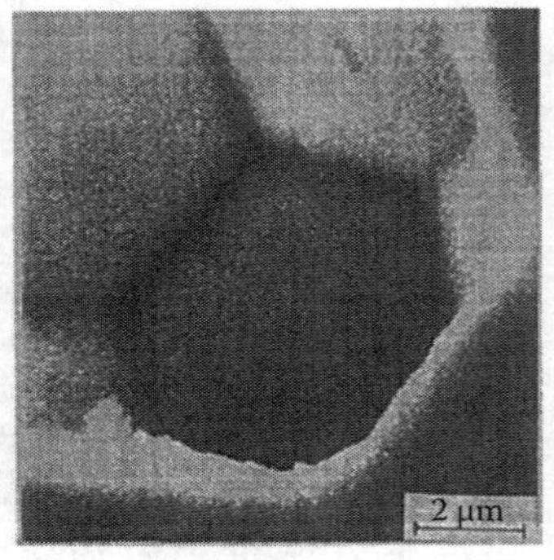

(b)

FIGURE 7.17
SEM images of pits on polycrystalline Fe formed in phthalate buffer, pH 5.0 with addition of 0.01 M KCl for 3 s at 1.10 V on crystallites with (111) orientation, (a) (110) planes at pit surface; (b) combination of (110) and (100) planes at pit surface. (From Vetter, K.J. and Strehblow, H.-H., *Ber. Bunsen. Ges. Phys. Chem.*, 74, 1024, 1970.)

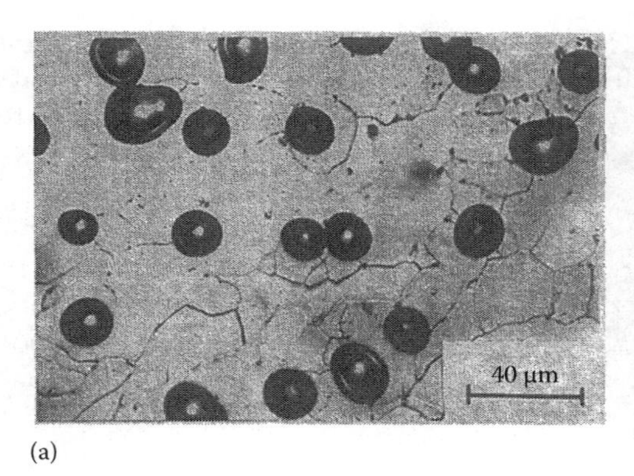

(a)

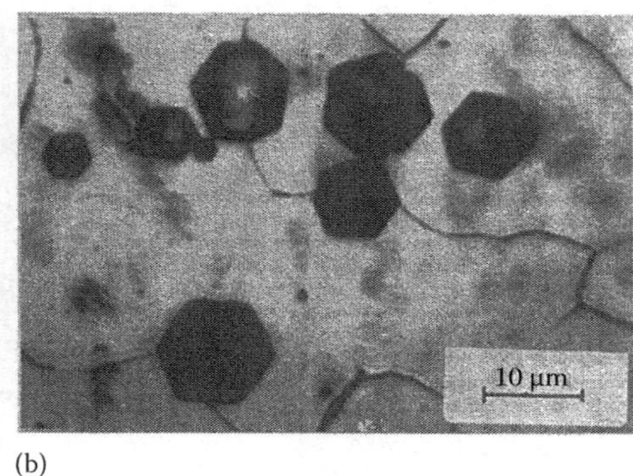

(b)

FIGURE 7.18
Microscopic images of pits on polycrystalline Fe formed in phthalate buffer pH 5.0. (a) With 0.01 M KCl and 0.1 M K_2SO_4 for 30 s at 1.10 V (SHE); (b) with 0.01 M KCl only, for 3 s at 1.10 V. (From Vetter, K.J. and Strehblow, H.-H., *Ber. Bunsen. Ges. Phys. Chem.*, 74, 1024, 1970.)

passive layer with a size similar to the thickness of the passive layer of 1–2 nm. This leads to small elongated pits of some few nm, which then change to polygonal forms during their growth to some 10 nm (Figure 7.15a through c). If the potential is high and thus the local current density at the pit surface is large enough, the increasing concentration of corrosion products finally leads to a round and electropolished hemisphere. Low potentials with smaller pit current densities do not reach this stage at all or after a longer time so that large polygonal pits may be observed for these conditions. In the vicinity of the pitting potential, pit growth becomes instable. As a consequence, a metastable pit growth with a stop-and-go sequence will lead to the observed irregular shape.

The presence of large amounts of nonaggressive anions in the bulk electrolyte also causes their accumulation within the pits. This competition leads to smaller halide concentrations with a less stable condition for pit growth. The smaller dissolution current in the presence of these nonaggressive anions (see Figure 7.20) further supports unstable growth and possible repassivation. As a consequence, discontinuous growth of pits with irregular shapes is observed. Figure 7.19 gives an example of a pit grown for 10 s on Ni in 0.05 M

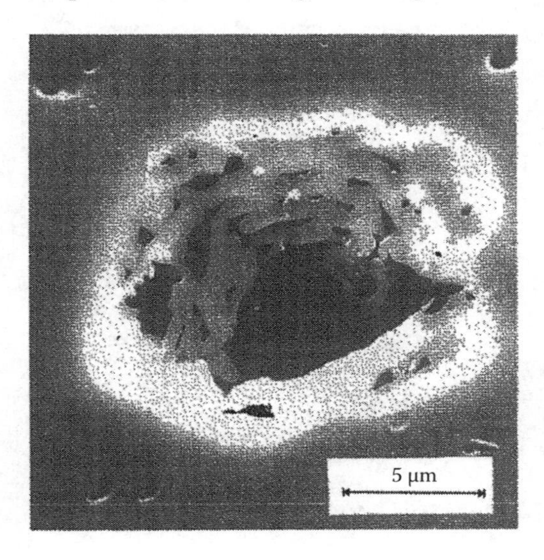

FIGURE 7.19
Microscopic image of a metal-covered pit with irregular shape on polycrystalline Ni formed in 0.05 M phthalate buffer, pH 5.0 with 0.1 M K_2SO_4, and 0.1 M KCl for 10 s at 0.90 V (SHE), small triangular pit at inner pit surface. (From Strehblow, H.-H., *Habilitationsschrift*, Freie Universität, Berlin, 1977.)

phthalate buffer, pH 5 with 0.1 M KCl and 0.1 M K_2SO_4. For these conditions pits grow into the metal leaving a metal cover that is perforated from their inner surface. Apparently, this irregular shape compensates for the lower current density. The covered pit keeps the accumulated halide inside and also causes a large ohmic drop that stabilizes its growth under less favorable conditions. At the inner surface of this larger pit a small triangular pit is observed, which supports the explanation of continuous passivation and nucleation for irregular growth. The formation of small corrosion tunnels with (100) surfaces during pitting of Al in the vicinity of its critical pitting potential further supports the discontinuous irregular growth at microscopic and submicroscopic sites [62]. At more positive potentials, repassivation no longer occurs and pits with a hemispherical and electropolished surface are obtained.

7.8 Pit Growth

7.8.1 Fundamentals of Stable Pit Growth

As visualized in the previous section, different stages of pit growth have to be distinguished, which are closely related to the question of the stability of localized corrosion. A pit will run through these different stages, each of which has its own characteristic conditions of stability. Once a pit has nucleated, its local current density refers to the applied electrode potential. For very positive potentials well within the passive range, extremely high dissolution current densities of several tens of A/cm^2 up to more than 100 A/cm^2 have been deduced from the growth of the corrosion pits using Faraday's law and taking into account their size and form (Figure 7.20) [59,61]. Equivalent current densities are measured directly on small wire electrodes of <1 mm diameter embedded in resin when the passivation of the whole electrode surface is prevented in solutions of high chloride content (>1 M) starting with a preactivated oxide-free surface [59,63]. Extrapolating the dissolution kinetics of free metal corrosion to potentials in the passive range of the polarization curve

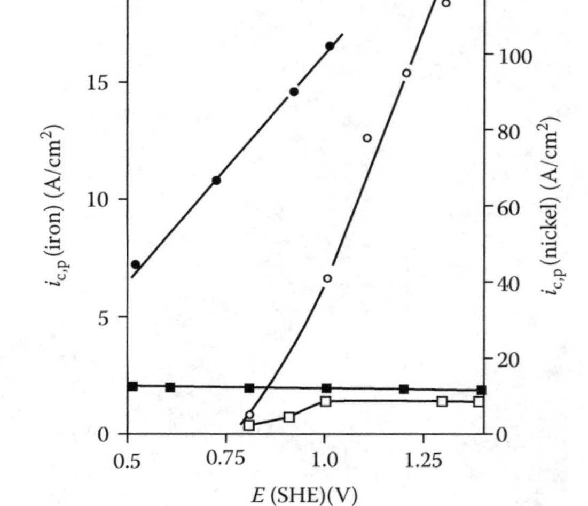

FIGURE 7.20
Local pit current density $i_{c,p}$, as a function of potential E deduced from pit growth on polycrystalline Ni and Fe electrodes in 0.05 M phthalate buffer, pH 5.0 Ni: (□) 0.1 M KCl + 0.1 M K_2SO_4, (○) 0.1 M KCl; Fe: (■) 0.01 M KCl + 0.1 M K_2SO_4, (●) 0.01 M KCl. (From Strehblow, H.-H., *Habilitationsschrift*, Freie Universität, Berlin, 1977.)

leads to extremely high local current densities of 10^3–10^6 A/cm^2 [64]. These large current densities would cause precipitation of a salt film within 10^{-4}–10^{-8} s. Current densities of more than 10^3 A/cm^2 are possible in principle but only up to 120 A/cm^2 has been observed for a free corroding metal surface [59,61].

The local current density increases with the chloride concentration, whereas the critical pitting potential decreases with increasing chloride concentration. If, however, $i_{c,p}$ is presented relative to E_p, the data fit a single curve. Apparently an increasing chloride concentration shifts the whole $i_{c,p}/(E - E_p)$ curve in the negative direction [60, p. 55]. The addition of nonaggressive anions such as sulfate or nitrate shifts the exponential current increase to positive potentials with a large potential-independent plateau at low potentials [59,60, p. 57]. These plateau values are in the range of a few A/cm^2, i.e., 5 A/cm^2 for Ni in 0.1 M KCl + 0.1 M K_2SO_4 and 2 A/cm^2 for Fe in 0.01 M KCl + 0.1 M KSO_4 (Figure 7.20) [59]. Apparently the potential-independent plateau values of $i_{c,p}$ are related to the formation of a salt layer that provides electropolishing conditions at the pit surface (Figure 7.18a). If, however, the current becomes still smaller such as for Ni below 1.0 V, the growth of the pit becomes unstable and irregular shapes are observed as shown in Figure 7.19.

A special geometry for localized corrosion studies has been applied by Frankel and coworkers [65,66]. They subjected vapor-deposited thin Ni–20% Fe and Al films on an inert substrate as working electrodes of a potentiostatic circuit to the local attack of chlorides, which causes fast perforation of the film and subsequent growth of two-dimensional pits with actively corroding cylindrical side walls. With this method the authors avoid the usual deepening of three-dimensional pits on bulk materials with the related increasing diffusion length for the transport of corrosion products and ohmic drops. Thus, the geometry for metal dissolution and transport of corrosion products were kept constant with time of growth of the two-dimensional pits. For this simpler and constant geometry, the local current density and the potential of repassivation of pits could be measured more accurately. The local current densities $i_{c,p}$ could be directly deduced from the growth of the radius of the two-dimensional pits applying Faraday's law without any assumptions for the geometry of the actively corroding surface. Usually, $i_{c,p}$ increases linearly with the electrode potential up to 100 A/cm^2, where it is controlled by mass transport and ohmic resistance. It finally gets to a diffusion-controlled potential-independent limiting value. There exist a critical local current density i_{cr} and a repassivation potential E_R that are closely related to each other. The accumulation of corrosion products is interpreted by the authors as a stabilizing factor for pit growth. If the local current density goes below a critical value i_{cr}, localized corrosion will stop due to a less aggressive environment corresponding to a smaller local concentration of the corrosion products that may be maintained. Apparently, a sufficiently high concentration of chloride is required to prevent repassivation. Estimates have shown that ~50% of the concentration of saturation is required to prevent repassivation [65]. As will be shown later for a hemispherical pit, the locally increased concentration of corrosion products and thus chlorides is proportional to the product of the local current density and the pit radius $i_{c,p}r$ (Equation 7.18). For a two-dimensional circular pit this relation simplifies to a proportionality to $i_{c,p}$ that yields critical conditions that are independent of the radius of the two-dimensional pit [65]. In this sense, these studies may give valuable information on the stability of polygonal and hemispherical pits. With increasing metal film thickness and consequently larger side walls of the actively corroding cylinder surface, i_{cr} decreases due to a larger corroding surface and hence to an increased ohmic resistance and higher accumulation of corrosion products. For similar reasons, di/dE decreases with increasing film thickness.

As already discussed in relation to the adsorption mechanism of pit nucleation, the halides cause catalytically enhanced dissolution of the passivating oxide. Any attempt of the bare metal surface of a pit to passivate by oxide formation will fail in the presence of high concentrations of aggressive anions, similar to the galvanostatic transient behavior of preactivated specimens. Thus, there is a close correspondence of small iron and nickel electrodes in solutions of high Cl^- content (>1 M) and an intensively corroding pit surface that will be discussed in the next section. In both cases, the formation of a passive layer is suppressed. In the case of Al a sufficiently low pH by hydrolysis of the corrosion products will stabilize pitting additionally. Passivity of this metal requires a dissolution equilibrium of the anodic oxide with the electrolyte that requires weakly acidic to alkaline solutions. Localized acidification is therefore a good additional stabilizing factor for corrosion pits on Al. This special situation, however, does not contradict the necessary active chemical role of halides, i.e., their complexing properties.

7.8.2 Precipitation of Salt Films

After a sufficiently long time of high dissolution current densities, the accumulation of corrosion products will lead to the precipitation of a salt film. This precipitation is expected in an early stage of pit growth if sufficiently large potentials and as a consequence large current densities are applied. Galvanostatic transients of preactivated small electrodes in solutions of high chloride content are not subject to passivation, and the dissolution at potentials within the passive range leads to a related steep increase of the measured electrode potential after the transition time τ (Figure 7.21). For the linear diffusion problem of the completely dissolving electrode surface it was found that Sand's equation is valid; this relates the current density $i_{c,p}$ and time t of corrosion to the difference $\Delta c = c_m - c_{m,b}$ of the

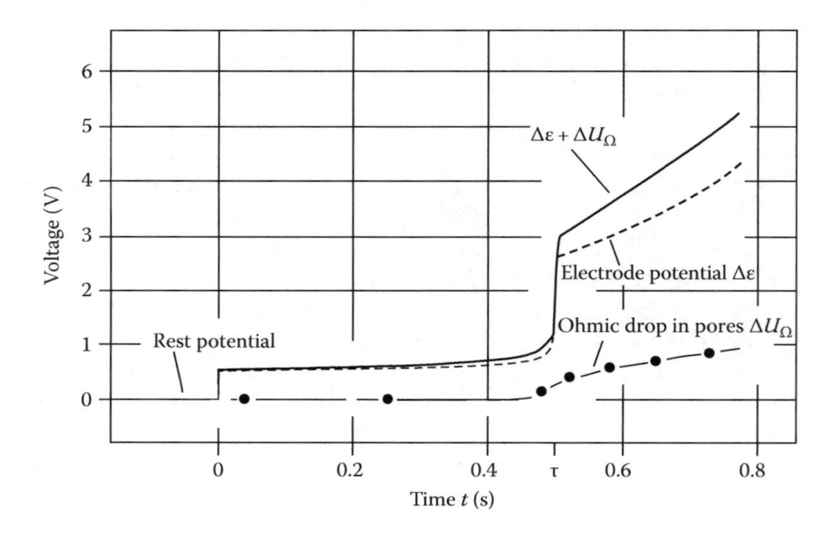

FIGURE 7.21
Galvanostatic transient of Ni in saturated $NiCl_2$ solution with $i = 1.5\,A/cm^2$ and the related voltage drop $\Delta E + \Delta U_\Omega$ across the total salt layer and ohmic drop ΔU_Ω across the porous film. (From Strehblow, H.-H. et al., *J. Electrochem. Soc.*, 125, 915, 1978.)

metal ion concentration with c_m at the electrode surface and its bulk value $c_{m,b}$, their charge z_m, and diffusion constant D of the metal ions [63].

$$i\sqrt{t} = \tfrac{1}{2} z_m F \sqrt{\pi D} \Delta c \tag{7.17}$$

In the absence of supporting electrolyte, i.e., in a solution containing the salt of the dissolving metal, a diffusion constant $D_{eff} = D(1 + z_m/|z_a|)$, which includes the contribution of migration to the transport, has to be used instead D. Here z_m and z_a are the valences of the metal ions and anions. Figure 7.22 presents the accumulation of metal ions Δc according to Equation 7.17 during dissolution of a small nickel electrode in solutions of high chloride content as a function of $i\sqrt{t}$. When c_m reaches its value of saturation $c_{m,s}$ for $t = \tau$, precipitation of a salt film will occur. This salt film has been examined further with galvanostatic pulses superimposed on the main transient to determine ohmic drops (Figure 7.21). The results lead to a model for the forming salt film with a spongy, porous outer part and a barrier-type inner part [63]. Besides experimental examinations [63], numerical calculations [67] have shown that Equation 7.17 may also be applied for concentrated electrolytes. Different solutions with and without aggressive anions have been tested using galvanostatic transients with small Fe and Ni electrodes, such as 0.5 M H_2SO_4, 1 M $HClO_4$, 1 M $NaClO_4$, and 1 M HNO_3. Figure 7.23 gives an example for Fe. In the absence of aggressive anions, no potential plateau but only a small shoulder was found for galvanostatic transients; i.e., only passivation with no large free metal dissolution was observed. The product $i\sqrt{\tau}$ starts at small values and decreases with increasing current density i. Only in the presence of >1 M Cl^- (or possibly other halides) were much larger values independent of i found for $i\sqrt{\tau}$. These studies also prove that the presence of aggressive anions is necessary to prevent passivation of Fe and Ni surfaces.

For a hemispherical pit the diffusion-limited local current density $i_{c,p}$ changes according to Equation 7.18, which is a simple combination of Fick's first diffusion law and Faraday's law for the hemispherical electrode geometry with the same variables as used for Equation 7.17:

$$i_{c,p} = \frac{z_m F D_{eff} \Delta c}{ar} \tag{7.18}$$

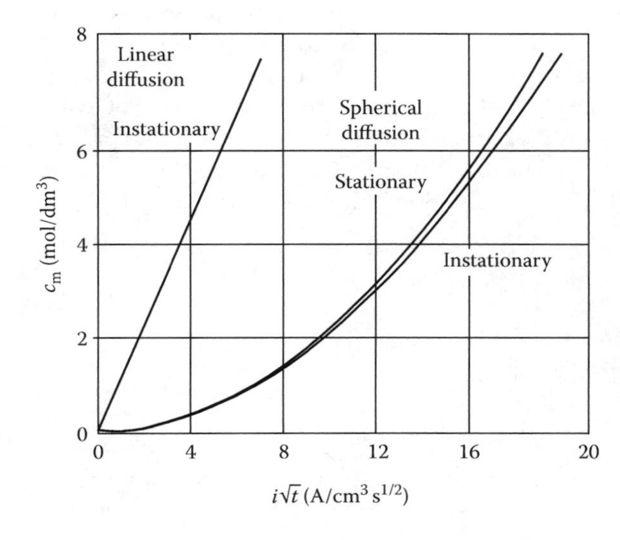

FIGURE 7.22
Concentration of $NiCl_2$ at a corroding Ni surface c_m as a function of the product $i\sqrt{t}$ for linear and spherical transport (including nonstationary conditions) calculated with Equations 7.17 through 7.20. (From Strehblow, H.-H., *Werkst. Korros.*, 35, 437, 1984.)

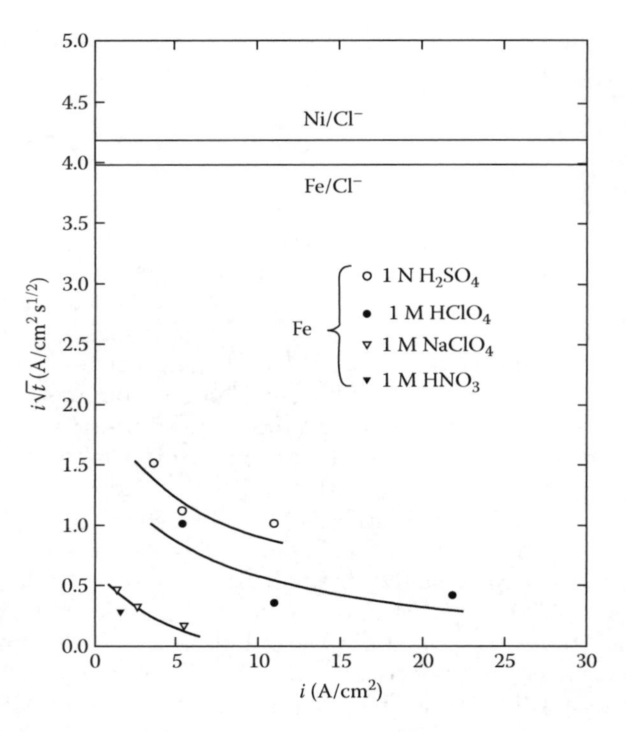

FIGURE 7.23
Plot of $i - i\sqrt{\tau}$, τ = transition time, for galvanostatic transients of small Fe electrodes in different solutions without and with 1M Cl^- depending on the applied current density. (From Strehblow, H.-H., *Habilitationsschrift*, Freie Universität, Berlin, 1977.)

This equation contains a constant a that takes into account the change from the convex to the concave pit geometry. For the simple analytical solution of the convex hemispherical transport problem $a = 1$ will hold. For a concave hemispherical pit a has been estimated to be 3 [6,30,31]. Computer-assisted numerical calculations yield a result that is very close to this value for the pit bottom [68,69]. With the simple stoichiometric relation between the local current density $i_{c,p}$ of a hemispherical pit and its growth rate of the radius dr/dt and Faraday's law (Equation 7.19), one obtains Equation 7.20, which is the equivalent relation for the transport from a hemispherical pit to the linear transport from a small electrode dissolving totally at its surface (Equation 7.17). V_m is the molar volume of the metal under study. Figure 7.21 compares the calculated results for $NiCl_2$ for both electrode geometries. Precipitation of corrosion products is expected for the transition times $t = \tau$ when saturation is achieved at the pit surface with $c_{m,s} = 4.8\,M$ and 4.2 M for $NiCl_2$ and $FeCl_2$, respectively. This condition is achieved first at the pit bottom for $a = 3$. The calculated data agree well with the experimentally observed precipitation within pits [30]. For Ni, $i\sqrt{\tau} = 14.8\,A/cm^2\,s^{1/2}$ is obtained. With the assumption of nonstationary transport conditions one yields almost identical results (Figure 7.21). Some deviation is obtained only for values $i\sqrt{t} > 12\,A/cm^2\,s^{1/2}$ [30].

$$i_{c,p} = \frac{(z_m F/V_m)\mathrm{d}r}{\mathrm{d}t} \tag{7.19}$$

$$i_{c,p}\sqrt{t} = z_m F\sqrt{\frac{\Delta c D_{eff}}{V_m a}} \tag{7.20}$$

The precipitation of a salt film is closely related to a change in the growth of a corrosion pit. When the salt film is formed, the growth rate slows down and the pit morphology

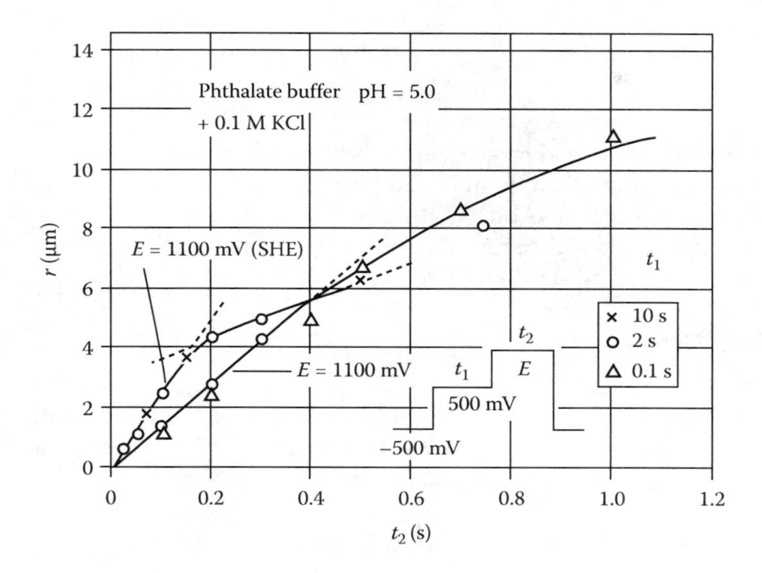

FIGURE 7.24
Deviation from linear pit growth with time when precipitation of a salt layer occurs, Ni, E = 1.00 and 1.30 V phthalate buffer, pH 5.0 + 0.1 M KCl with inserted diagram of applied potential pulses. (From Strehblow, H.-H., *Habilitationsschrift*, Freie Universität, Berlin, 1977.)

changes from a polygonal to a hemispherical structure. This change starts at the pit bottom, where it is expected first, and often is combined with an electropolishing effect [61]. Direct microscopic in situ and ex situ examinations of growing pits for potentiostatic conditions show a decrease of the growth of the radius when a salt film precipitates (Figure 7.24) [64]. The time of change of the growth rate coincides with the electropolishing effect at the pit bottom [61]. Heimgartner and Böhni [70] discussed the growth of corrosion pits for longer times under diffusion-limited conditions in more detail. If the local current density $i_{c,p}$ in a pit is not constant and decreases inversely with the pit depth r according to Equation 7.18a, one deduces with Equation 7.19 a reciprocal relation of its growth rate dr/dt with its depth r (Equation 7.21). Integration leads to a parabolic dependence of the pit radius on time t (Equation 7.22) and combination with Equation 7.18a to Equation 7.23 with an inverse change of the local pit current density $i_{c,p}$ with $\sqrt{t}$. This discussion assumes a saturated solution with a change of the metal ion concentration $\Delta c = c_{m,s}$, i.e., a salt film at the actively corroding pit surface and a vanishing bulk concentration $c_{m,b}$. Therefore a later stage of pit growth is discussed and not the initial situation of free metal dissolution as for the beginning of pit growth of Figure 7.20. Equations 7.22 and 7.23 assume a minimum pit radius r_0 where a saturated solution at its surface is achieved. Equation 7.23 is simply the combination of Equations 7.18a and 7.22.

$$i_{c,p} = \frac{z_m F D c_{m,s}}{ar} \tag{7.18a}$$

$$\frac{dr}{dt} = \frac{i_{c,p} V_m}{z_m F} = \frac{D c_{m,s} V_m}{ar} \tag{7.21}$$

$$r - r_0 = \sqrt{\frac{D V_m t c_{m,s}}{a}} \tag{7.22}$$

$$i_{c,p} = \frac{z_m F D c_{m,s}}{a r_0 + \sqrt{2 D V_m c_{m,s}^{1/a}}} \tag{7.23}$$

Enhanced convection may lead to a situation where ohmic resistance instead of diffusion is rate determining. Convection by an increasing flow rate of the electrolyte or the evolution of gas bubbles in corrosion pits will increase the local dissolution and finally cause ohmic current control. Like diffusion control, ohmic control will lead to a parabolic dependence of the pit radius or pit depth on time.

7.8.3 Potential Drops

There exist detailed studies for the potential drop ΔU, the composition of the pit electrolyte, and the related changes of the pH. All these factors have been used to explain the stability of a corroding pit. The potential drop within the electrolyte for an open hemispherical pit can be estimated by the following simple equation:

$$\Delta U = \frac{a i_{c,p} r}{\kappa} \tag{7.24}$$

with specific conductivity κ, local current density $i_{c,p}$, radius r, geometric factor $a = 3$ [6]. It depends on the specific situation whether ΔU is large enough to shift the potential below the Flade potential, i.e., in the active range of the polarization curve (Figure 7.25). For 0.5 M H_2SO_4 with $\kappa = 0.22\,\Omega^{-1}\text{cm}^{-1}$ and $i_{c,h} = 0.3\,\text{A/cm}^2$, which is a reasonable value for moderately high pit current densities for Fe, one obtains $\Delta U > 1$ V for a pit radius $r > 2.4$ mm, but only $\Delta U = 0.41$ mV for $r = 1\,\mu$m. This short calculation demonstrates that it is absolutely necessary to define well which state of the growth of a pit is discussed. Potential drops may stabilize localized corrosion for large pits but never for small ones, i.e., for initial stages. The conductivity of the electrolyte is, of course, another factor that has to be taken into account. The accumulation of ions within a corroding pit may increase κ and thereby reduce ΔU compared with the value obtained with the bulk conductivity. Evidence for potential control of pit growth is given for studies on Al by Hunkeler and Böhni [71]. They found a linear dependence of the product $i_{c,p} r$ on the voltage drop ΔU.

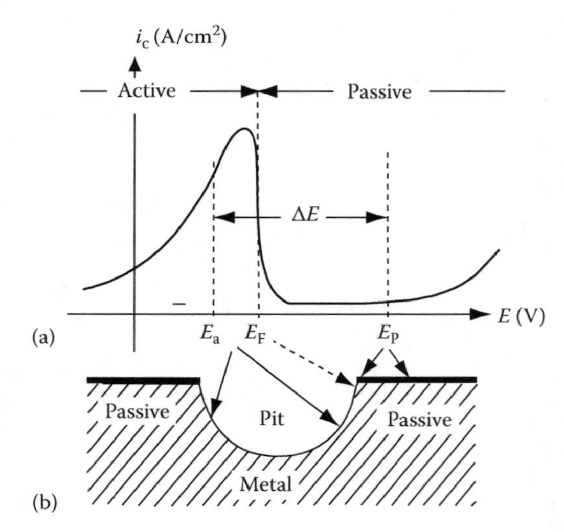

FIGURE 7.25
(a) Typical current density potential curve of a passive metal; (b) hemispherical pit with potentials E_a inside and E_p outside the pit according to the rejected theory frequently used for small pits. (From Vetter, K.J. and Strehblow, H.-H., in *Localized Corrosion*, NACE, Houston, TX, 1974, p. 240.)

7.8.4 Composition of Pit Electrolyte

The accumulation of electrochemically nonactive electrolyte components is closely related to the potential drop ΔU within the electrolyte, similar to a Boltzmann dependence of the concentration c_j of these species on the electrical energy ΔUzF [6,30]. Together with the electroneutrality equation, one obtains a dependence of the concentration of the corroding metal ions c_m at the metal surface on ΔU with $c_m - c_{m,b} = c_m$ for a vanishingly small bulk concentration $c_{m,b}$ [6,30]. The corresponding bulk values are $c_{j,b}$ and $c_{m,b}$.

$$c_j = c_{j,b} \exp\left(\frac{-\Delta U z_j F}{RT}\right) \tag{7.25}$$

$$\Delta c = c_m - c_{m,b} = c_m = -\left(\frac{1}{z_m}\right)\sum c_{j,b} z_j \exp\left(\frac{-\Delta U z_j F}{RT}\right) \tag{7.26}$$

Equation 7.25 requires that only one electrochemical process occurs with no subsequent reactions within the bulk electrolyte. The concentration of the metal cations c_m may also be calculated from Equation 7.18 or 7.18a. For vanishingly small, bulk concentration of the cations, Δc equals the surface concentration c_m. The accumulation of cations is proportional to the local pit current density and the pit radius (Equation 7.18). One may therefore take $i_{c,p}r$ as a measure of the cation concentration at the pit surface, which depends on the age of the pit. Therefore this scale is included in Figures 7.26 and 7.27.

$$\Delta c = c_m = \frac{a i_{c,p} r}{z_m F D_{eff}} \tag{7.18b}$$

The concentrations c_j at the pit surface of species that are not involved in the electrode process and the cation concentration have been calculated according to Equation 7.25 for

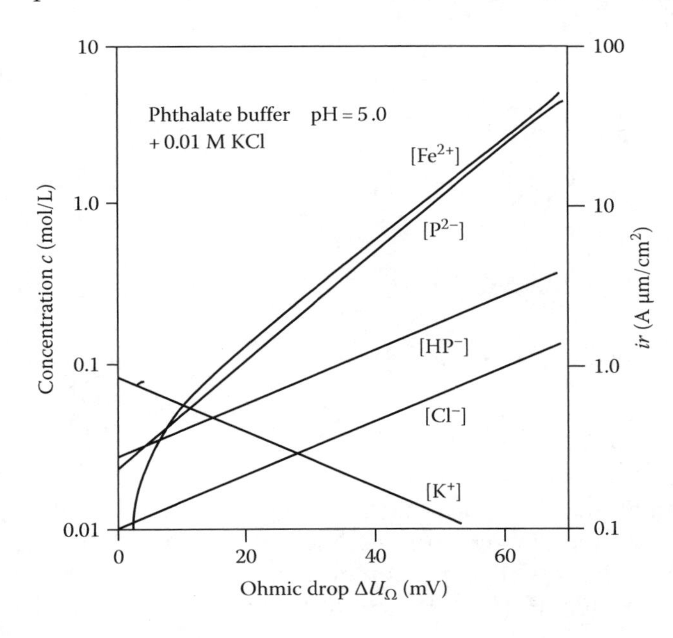

FIGURE 7.26
Ionic concentration at the metal surface depending on the ohmic drop ΔU_Ω within the electrolyte according to Equations 7.25 and 7.26. (From Strehblow, H.-H., *Werkst. Korros.*, 35, 437, 1984.)

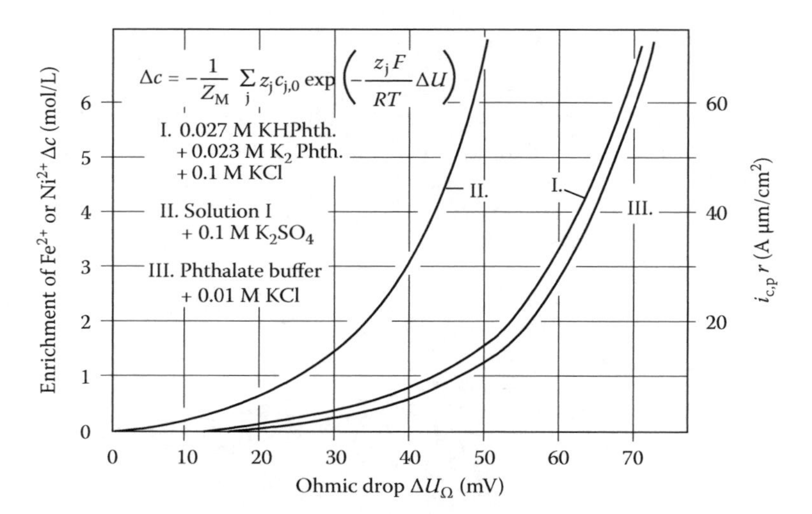

FIGURE 7.27
Concentration of metal ions at the dissolving metal surface depending on the ohmic drop ΔU_Ω and the parameter $i_{c,p}r$ according to Equation 7.26. (From Strehblow, H.-H., *Werkst. Korros.*, 35, 437, 1984.)

some bulk electrolytes frequently used for pitting studies (Figure 7.26). Metal ion concentrations c_m go up to saturation of a few moles per liter when the potential drop increases to several tens of millivolts, which is usually the case for a pit radius of several micrometers. Equation 7.26 was applied for the calculation of the concentrations c_m at the pit surface as given in Figure 7.27.

pH shifts have often served to explain the stability of a corroding pit. Equation 7.25 permits the calculation for buffered solutions when the pH is not affected by hydrolysis of the corrosion products [30]. Buffer ions change in concentration in agreement with the pH shift as given for the phthalate buffer in Figure 7.26. An accumulation of dissolving metal ions leads to accumulation of the (buffer) anions and to depletion of the other cations and consequently to a positive pH shift according to the following equation:

$$\text{pH} - \text{pH}_b = \frac{\Delta U}{0.059} \tag{7.27}$$

In nonbuffered solutions the usual hydrolysis equilibrium of the metal ions leads to acidification according to

$$\text{Me}^{z+} + \text{H}_2\text{O} \leftrightarrow \text{MeOH}^{(z-1)+} + \text{H}^+,$$

With $[\text{MeOH}^{(z-1)+}] = [\text{H}^+]$ for nonbuffered electrolytes, one obtains

$$[\text{H}^+] = \sqrt{K_h[\text{M}^{z+}]}, \tag{7.28}$$

With hydrolysis constants of $K_h = 10^{-10}$ and 10^{-7}, one obtains for the saturation concentrations 4.8 and 4.2 M, pH values of 4.3 and 2.4 for saturated $NiCl_2$ and $FeCl_2$ solutions, respectively. These calculations show that the precipitation of hydroxide within a corrosion pit may be prevented for these metals by acidification. However, as discussed before, passivation should nevertheless be possible for many technically important metals. For Fe, Ni, and other metals, especially in acidic media, passivity cannot be explained on the basis of thermodynamic equilibria and Pourbaix diagrams. In these cases passivity is a kinetic phenomenon. Otherwise these metals should not show passivity in a strongly acidic environment. Thus, different explanations are required for the stability of a corrosion pit.

The pH shift of the passivation potential E_{pa} may not give an explanation either. Its change usually by −0.059 V/pH is much too small to cause a shift above the potentiostatically applied value E well within the passive range, so that the metal surface might reach the active range of the polarization curve. In 0.5 M H_2SO_4, the passivation potential for Fe or Ni is $E_{pa} = 0.58$ or 0.35 V, respectively, with a shift of −0.059 V per pH unit. If the potential is set in the passive range up to 1.5 V one has to explain a shift of the passivation potential by 0.9–1.15 V, which is absolutely out of range of any reasonable explanation. If, however, the potential E is set to values close to E_{pa}, as often realized in technical corrosion situations with open-circuit conditions, acidification may cause its shift above E with $E_{pa} > E$, which especially in weakly acidic and alkaline unbuffered solutions may fulfill these requirements. In addition, the much more negative passivation potential of Fe for neutral and alkaline solutions, such as $E_{pa} = -0.1$ to 0.059 pH for Fe, will not be effective as a consequence of acidification by hydrolysis. The passivation potential becomes more negative because of the insolubility of lower valent Fe oxides such as Fe_3O_4 or $Fe(OH)_2$ in neutral and alkaline solutions. However, these oxides dissolve very quickly in acidic electrolytes, so they do not provide passivity for these conditions. In conclusion, one may say that local negative pH shifts due to hydrolysis of corrosion products may support and stabilize pitting at negative electrode potentials for neutral and weakly acidic and alkaline bulk electrolytes. This will hold especially for metals that cannot be passivated at low pH, e.g., Cu and Al, because of their fast dissolution or high solubility in this environment. However, this interpretation does not hold for Fe, Ni, and steels in strongly acidic electrolytes at sufficiently positive electrode potentials. Thus, pitting needs another, more general explanation. Furthermore, anions of other strong acids such as perchlorates and sulfates should act as aggressive anions as well if hydrolysis and pH shifts are the essential causes of localized corrosion. However, ClO_4^- may even act as an inhibitor if the potential is not too positive and most metals may be passivated in sulfuric and perchloric acid, which demonstrates that the specific chemical properties of the aggressive anions have to be taken into account in the explanation of pitting corrosion.

An attempt has been made to explain the requirement for a minimum concentration of aggressive anions to maintain stable pit growth [30,31]. With the assumption that a salt film of thickness δ has to be maintained at the growing pit surface, one obtains

$$c_{min} = \frac{aV_m i_{c,p} \delta}{DFV_s} = 0.5 i_{c,p} \delta \tag{7.29}$$

With this assumption, the experimentally found value of $c_{min} = 3 \times 10^{-4}$ M has been confirmed with a local current density $i_{c,p} = 1\ A/cm^2$, a salt layer thickness $\delta = 5$ nm, geometric factor $a = 3$, molar volume $V_m = 3.55\ cm^3/val$ for the equivalent volume for Fe metal and $V_s = 21.25\ cm^3/val$ for $FeCl_2$, and diffusion constant $D = 10^{-5}\ cm^2/s$ for Cl^- ions.

The accumulation of aggressive anions has been found at the surface even of small polygonal pits of some few μm diameter when their change to a hemisphere by the precipitation of a salt film was not already achieved. A special preparation technique of pulling the electrode with actively corroding pits into a benzene layer above the electrolyte preserved the special situation at the corroding pit surface even after rinsing with acetone and dry storage in air [72]. These pits remained active and continued their growth immediately when reintroduced into the electrolyte at the same potential; however, they became inactive on rinsing with water or stepping the potential below the critical value. With electron microprobe analysis, chloride that corresponded to layer of ~5 nm $FeCl_2$ could be found when the pit remained active but was lost completely on rinsing with water or repassivation [72]. These studies show clearly that the formation of a thin layer of aggressive anions even in pits of some few μm is responsible for stable pit growth. The later precipitation of much thicker salt films in larger pits of several tens of μm provides further stabilization but is not a necessary condition for localized corrosion. The presence of nonaggressive anions in the bulk electrolyte also caused their accumulation but did not suppress the additional presence of aggressive anions that is absolutely necessary for a stable pit growth. The electropolishing conditions in sulfate-containing solution caused large amounts of these anions corresponding to several hundred nm of $FeSO_4$ within the hemispherical pits but did not avoid the additional and necessary accumulation of ~5 nm $FeCl_2$ [72].

7.9 Repassivation of Corrosion Pits

The kinetics of repassivation of small pits in an early stage of their development seem to be related to the transport of aggressive anions, such as the chloride accumulated locally during the intense dissolution process from the pit to the bulk [4,30]. If this transport is the rate-determining step, one expects the repassivation time to increase with the depth of a corrosion pit. If we simply apply the Einstein–Smoluchowski relation for the transport time t_r (=repassivation time t_r) out of a pit of radius r (Equation 7.30), and if the radius r is given by the local current density i_{cp} and the lifetime t_p of the pit by Equation 7.31, we obtain for the repassivation time t_r (Equation 7.32)

$$t_r = \frac{r^2}{2D} \tag{7.30}$$

$$r_r = \frac{i_{c,p} t_p V_m}{z_m F} \tag{7.31}$$

$$t_r = \frac{i_{c,p}^2 t_p^2 V_m^2}{2D z_m^2 F^2} = k t_p^2 \tag{7.32}$$

The t_r increases with the square of the lifetime t_p of a pit. The constants $k = k_i$ have been estimated from $D = 5 \times 10^{-6}$ cm²/s, and $z_m = 2$ for both cations and $zF/V_m = 2.72 - 10^4$ A s/cm³ for Fe and 2.93×10^4 A s/cm³ for Ni to $k_{Fe} = 1.35 \times 10^{-4}\, i_{c,p}^2$ and $k_{Ni} = 1.16 \times 10^{-4}\, i_{c,p}^2$ for Fe and Ni, respectively. To check the $i_{c,p}^2$ and $t_{c,p}^2$ dependence of t_r, potentiostatic pulse measurements have been performed. The potential was pulsed for some 10 ms up to 1 s to values above the pitting potential E_p and back to $E \le E_p$ for increasing time Δt. When the potential was finally

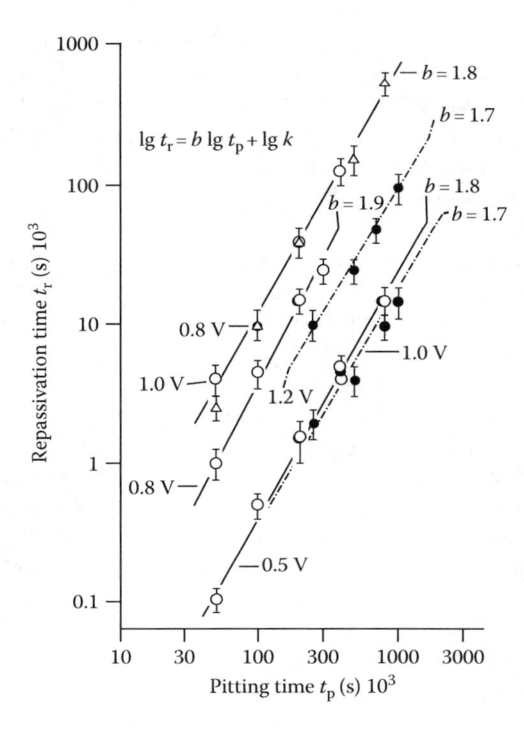

FIGURE 7.28
Repassivation time t_r vs. pitting time t_p for Fe (A = 0.13 cm²) and Ni (A = 0.2 cm²) at different potentials for pitting as indicated, t_r leading to deviations of continuous current increase for one potential sequence. (□) Fe 0.01 M Cl^-; (○) Fe 0.01 M Cl^-, (●) Ni 0.1 M Cl^- Δ Fe 0.01 M Cl^- t_r leading to continuous increase of current for many subsequent potential cycles, phthalate buffer, pH 5.0. (From Strehblow, H.-H., *Werkst. Korros.*, 27, 792, 1976.)

pulsed back to the potential in the pitting region, the current density started to increase further at the same level reached at the end of the previous pitting period if Δt was short enough. When Δt was too long, at least some of the pits repassivated and the subsequent pitting period had to start again with the nucleation of new pits and a smaller value of the average current density of the electrode. The interval Δt of a fast deviation from a continuous increase of the current density for a set of pulse experiments is taken as the repassivation time t_r. A double logarithmic plot of t_r and t_p shows a slope of approximately 2 as required by Equation 7.32. Also, t_r changes proportionally with $i_{c,p}^2$. The estimated values agree sufficiently well with the experimental data of Figure 7.28.

One may therefore conclude that the rate-determining step for the repassivation is the transport of accumulated aggressive anions out of small pits. This result coincides well with the explanation that localized corrosion is stabilized at the initial stage by the accumulation of aggressive anions that prevent the formation of a passive layer at the active pit surface.

7.10 Factors Stabilizing Pit Growth

The factors that stabilize the growth of a pit depend on the stage of its development. Ohmic drops cannot serve as an explanation for small pits with micrometer dimensions. The ohmic drops are <1 mV for strongly acidic, well-conducting electrolytes such as 0.5 M H_2SO_4 and a pit radius of 1 μm. This is far too small to shift the potential from its potentiostatically fixed value below the passivation potential by more than 1 V (Figure 7.25). However, in later stages for pits of several millimeters radius, the ohmic drop might become high enough to satisfy this condition. The extremely high local current densities are a further proof of the absence of large ohmic drops. The local potential at the pit surface is in the vicinity of the potentiostatically fixed value in order to permit these high dissolution currents. Ohmic drops have already been corrected in Figure 7.20, presenting these high current densities for Ni.

pH shifts also cannot explain the stable pit growth for metals such as Fe, Ni, and steel that are also passive in strongly acidic electrolytes. For these systems passivity is a kinetically determined phenomenon that is even in disagreement with the Pourbaix diagram. However, there are metals with passivation potentials in close agreement with their Pourbaix diagrams that form passivating oxide layers which dissolve at a high rate in strongly acidic electrolytes, such as Al and Cu. For these systems, local destabilization of passivity may occur on acidification of the pit electrolyte with pitting in neutral or weakly acidic bulk electrolytes. However, even for these metals there still remains the question why halides are necessary for pit nucleation and pit growth. The pH dependence of the Flade potential of −0.059 V/pH as obtained for metals such as Fe and Ni cannot serve as a stabilizing factor. The pH shift required to obtain active conditions at the pit surface is much too large in comparison with the applied electrode potentials.

The requirement for aggressive anions and their influence on the repassivation kinetics demonstrate that their accumulation stabilizes localized corrosion in an early stage. The autocatalytically enhanced dissolution of the passive layer by these anions prevents any repassivation once a pit has formed. It should be mentioned again that pure Cr is one of the few metals that is not subject to pitting by chloride. In agreement with this finding, Cr(III) salts dissolve very slowly in water. Fast transfer of the complexed ions to the electrolyte is necessary for the attack on the passive layer, which provides a good explanation for the pitting resistance of this metal compared with Fe and Ni.

For sufficiently positive potentials, the intense dissolution process leads to severe accumulation of corrosion products and finally, after a transition time, to the formation of a salt film, which is a further stabilizing factor for a pit. It may act as a reservoir to maintain a high concentration of the aggressive anions. These explanations are supported by the logarithmic dependence of the critical pitting and inhibition potential on the concentration of aggressive (A) or inhibiting (I) anions according to Equations 7.1 and 7.2.

These equations with large *b* factors have been found experimentally for Fe and Ni and various aggressive and inhibiting anions [2,73]. They have been deduced on the basis of a thermodynamic theory of adsorption of the anions at the metal surface of a pit in competition with each other and with the formation of the passive layer [37,73]. This explanation agrees well with the concept of repassivation prevented by accumulation of aggressive anions. According to this approach, a potential-assisted Langmuir type of adsorption isotherm for the aggressive anions is assumed at the parts of the bare metal surface where the passive layer has been removed by any of the mechanisms that have been discussed. For simplicity, homogeneous adsorption at the metal surface is assumed with no lateral interaction between the adsorbed ions. The accumulation of corrosion products and consequently of the aggressive anions due to the extremely high dissolution rates supports their adsorption and increases their surface coverage Θ, which also increases with the electrode potential E. Stable pit growth will be achieved when Θ exceeds a critical value Θ_{crit}. It should be large enough to prevent the formation of a continuous oxide film, i.e., the repassivation of the pit surface. In the light of this concept, a given composition of the bulk electrolyte requires a minimum electrode potential E_p to exceed Θ_{crit} and to achieve stable pit growth. Detailed evaluation of this adsorption equilibrium yields

$$E_p = \frac{a-(RT2.303)}{(\gamma F)\log\{[A]\}} = a + b\log\{[A]\}$$
$$a = \frac{RT(2.303)}{(\gamma F)\log\{\Theta_{A,crit}/(1-\Theta_{A,crit})\}} + \frac{\Delta G^o_{ad,A}}{(\gamma F)} + \Delta\varphi_h \quad (7.33)$$

where

$\Delta G^o_{ad,A}$ is the standard Gibbs free energy (free enthalpy) of adsorption of the aggressive anions A

$\Delta\varphi_h$ is the potential drop at the standard hydrogen electrode

γ is the electrosorption valence, which takes into account the closest approach of the anions to the metal surface because of their own dimensions

If inhibiting anions I are present as well, they compete with the aggressive anions A for the adsorption sites according to the same type of adsorption isotherm as given in Equation 7.33 with their specific ΔG^o_{ad} values and bulk concentrations. Combining the equations for the inhibiting and aggressive anions leads to the critical conditions of inhibition of localized corrosion. The detailed discussion [2] yields for the inhibition potential E_I

$$E_I = \frac{a+(RT2.303)}{(\Delta\gamma F)\log\{[A]/[I]\}} = a + b\log\left\{\frac{[A]}{[I]}\right\}$$
$$a = \frac{(RT2.303)}{\Delta\gamma F\log\{\Theta_{A,cirt}/\Theta_{I,cirt}\}} + \frac{\Delta G^o_{ad}}{(\Delta\gamma F)} + \Delta\varphi_h \quad (7.34)$$

where

ΔG^o_{ad} is the difference in Gibbs free energies of species A and I

$\Theta_{A,ad}$ and $\Theta_{I,ad}$ are the surface coverages of both ions

$\Delta\gamma$ is the difference in their adsorption valences

The negative $\Delta\gamma$ value causes a positive shift of E_I with increasing [A] and decreasing [I], which is an upper potential limit of the pitting range. The small values of $\Delta\gamma$ corresponding to small differences for the approach of ions A and I to the electrode surface cause large b factors for Equation 7.34, which is found experimentally for nitrate and perchlorate [2,64].

Pits may also be stabilized by other complicating effects. In some cases, the pit mouth is covered by precipitated corrosion products such as hydroxides or even a metal cover when the dissolution occurs irregularly (Figure 7.19). In these situations, an ohmic drop can lead to a potential shift of the pit surface below the passivation potential. The same will hold for large and especially deep and narrow pits. Crevice corrosion may be seen as a special case of localized corrosion. In these cases, ohmic drops act as stabilizing factors for pitting. Last but not least, the presence of a gas bubble within a pit will cause a situation similar to that of a crevice with related ohmic drops. Hydrogen bubbles have been observed by Pickering and Frankenthal [74] within pits and served as an explanation for stable pitting.

In this situation, it is still not understood how the potential becomes sufficiently negative within a small pit despite a potentiostatically fixed positive value to cause hydrogen evolution. Apparently gas bubbles may be an important stabilizing factor for some special systems and later stages of pit growth.

A further factor in the stability of a growing pit is alloy components that prevent or reduce pitting. They usually accumulate within the passive layer due to their slow dissolution kinetics. The special situation of Cr has already been discussed in detail. Fe–Si alloys with more than 20 at% Si are other examples. A sufficiently large Si content causes the formation of a continuous SiO_2 layer especially in acidic electrolytes [75], which prevents localized corrosion. The dissolution of SiO_2 is not enhanced by halides, which apparently is a consequence of the missing complexation of Si(IV) so that the passive layer remains protective even in the presence of a large chloride concentration. If a pit starts, a large bulk concentration of Si will lead to Si accumulation at the pit surface, which then may form a protective Si(IV)-rich film. This situation will lead to repassivation of the corrosion pit in an early stage as observed experimentally [76].

7.11 Conclusion

The discussion of various experimental results has shown that different effects may serve as stabilizing factors for localized corrosion. It depends on the stage of development of a corrosion pit and the environmental conditions, which one is the most effective. Therefore, one has to refer to the special situation under discussion. Often, the controversial points of view of authors depend on the different situations that are examined and discussed. It was in the aim of this contribution to sort out the different factors and to describe the conditions in which they are effective. In this sense, localized corrosion is a very complicated process involving different effects.

The same statement holds for the nucleation of corrosion pits. Here, three main mechanisms have been discussed. Again, it depends on the experimental or environmental conditions, which one is the most effective and dominates the nucleation step.

Although there are many common features in localized corrosion of the different systems, one should also take into account the specific chemistry of a system. In this sense, the tendency of halides to complex with metal cations is very important and seems to be the key for an understanding of the stabilization of a corrosion pit by prevention of the repassivation of a defect site within the passive layer. Enhancement of the transfer of metal cations from the oxide to the electrolyte by halides, especially the strongly complexing fluoride, holds for many metals. The special situation of Cr, which is resistant to localized corrosion, may be understood in terms of the slow dissolution kinetics of the Cr(III) salts. One should also consider the different passivation mechanisms. Metals such as Al and Cu are not passive in strongly acidic electrolytes. In these cases, localized acidification by the hydrolysis of corrosion products may well serve as a stabilizing factor for pitting. Other metals such as Fe, Ni, and steels are passive even in strongly acidic electrolytes, in disagreement with the predictions of the Pourbaix diagrams. Local acidification cannot be a stabilizing effect for pitting in these cases. In summary, all explanations of stable pit growth should include chemical arguments to explain why halides are so essential for localized corrosion. Their complexing and catalytic properties for the dissolution of the passive layers should not be ignored.

Besides the possible differences and the special influence of the chemical properties of the various systems, many common features are found for localized corrosion. Electrochemical and surface analytical methods for the study of passive layers and their breakdown have been very successful in the past, and closer insight into the effective mechanisms has been achieved. STM and SFM (scanning force microscopy) will give valuable information about the nucleation and very early stages of growth of corrosion pits. The application of these methods for fundamental studies of pitting corrosion has just started. Further investigations using these methods and extension to other metals will broaden the experience and knowledge and will give better insight into the mechanisms of this technologically important type of corrosion.

References

1. P. Schmutz and G. S. Frankel, *J. Electrochem. Soc.* 145:2285 (1998).
2. H.-H. Strehblow and B. Titze, *Corrosion Sci.* 17:461 (1977).
3. H. Prinz and H.-H. Strehblow, *Corrosion Sci.* 40:1671 (1998).
4. H.-H. Strehblow, *Werkst. Korros.* 27:792 (1976).
5. T. P. Hoar, D. C. Mears, and G. P. Rothwell, *Corrosion Sci.* 5:279 (1965).
6. K. J. Vetter and H.-H. Strehblow, *Ber. Bunsen. Ges. Phys. Chem.* 74:1024 (1970).
7. N. Sato, *Electrochim. Acta* 16:1683 (1971).
8. N. Sato, K. Kudo, and T. Noda, *Electrochim. Acta* 16:1909 (1971).
9. Ya. J. Kolotyrkin, *Corrosion* 19:261t (1964).
10. T. P. Hoar and W. R. Jacob, *Nature* 216:1299 (1967).
11. A. G. Revez and J. Kruger, *Passivity of Metals* (R. P. Frankenthal and J. Kruger, eds.), The Electrochemical Society, Princeton, NJ, 1978, p. 137.
12. J. Kruger, G. G. Long, M. Kuriyama, and A. J. Goldman, *Passivity of Metals and Semiconductors* (M. Froment, ed.), Elsevier Science Publishers, Amsterdam, the Netherlands, 1983, p. 163.
13. C. Y. Chao, L. F. Lin, and D. D. Macdonald, *J. Electrochem. Soc.* 128:1191 (1981).
14. C. Y. Chao, L. F. Lin, and D. D. Macdonald, *J. Electrochem. Soc.* 128:1194 (1981).
15. J. Augustynski, in *Passivity of Metals* (R. P. Frankenthal and J. Kruger, eds.), The Electrochemical Society, Princeton, NJ, 1978, p. 285.
16. S. Meitra and E. D. Verink, in *Passivity of Metals* (R. P. Frankenthal and J. Kruger, eds.), The Electrochemical Society, Princeton, NJ, 1978, p. 309.
17. T. E. Pou, O. J. Murphy, J. O. M. Bockris, L. L. Tongson, and M. Monkowski, *Proceedings of the 9th International Congress on Metallic Corrosion*, Toronto, Canada, 1984, Vol. 2, p. 141.
18. G. C. Wood, J. A. Richardson, M. F. Abd Rabbo, L. P. Mapa, and W. H. Sutton, in *Passivity of Metals* (R. P. Frankenthal and J. Kruger, eds.), The Electrochemical Society, Princeton, NJ, 1978, p. 973.
19. S. Szlarska-Smialowska, H. Viefhaus, and M. Janik-Czachor, *Corrosion Sci.* 16:644 (1976).
20. M. Janik-Czachor and K. Kaszczyszyn, *Werkst. Korros.* 33:500 (1982).
21. B. P. Löchel and H.-H. Strehblow, *Electrochim. Acta* 28:565 (1983).
22. B. P. Löchel and H.-H. Strehblow, *J. Electrochem. Soc.* 131:713 (1984).
23. W. P. Yang, D. Costa, and P. Marcus, *Oxide Films on Metals and Alloys* (B. MacDougall, R. S. Alwitt, and T. A. Ramanarayanan, eds.), *Proceedings*, Vols. 92-22, The Electrochemical Society, Princeton, NJ, 1992, p. 516.
24. S. Mischler, A. Vogel, H. J. Mathieu, and D. Landolt, *Corrosion Sci.* 32:925 (1991).
25. P. Marcus and J. M. Herbelin, *Corrosion Sci.* 34:1123 (1993).
26. H.-H. Strehblow, *Proceedings Corrosion 91*, NACE, Houston, TX, 1991, p. 76/1.

27. H.-H. Strehblow, *Proceedings Corrosion Prevention*, Australasian Corrosion Association, Brisbane, Australia, November 1997.
28. C. L. McBee and J. Kruger, *Localized Corrosion*, NACE, Houston, TX, 1974, p. 252.
29. W. Paatsch, *Ber. Bunsen. Ges. Phys. Chem.* 77:895 (1973).
30. H.-H. Strehblow, *Werkst. Korros.* 35:437 (1984).
31. K. J. Vetter and H.-H. Strehblow, in *Localized Corrosion*, NACE, Houston, TX, 1974, p. 240.
32. S. Haupt and H.-H. Strehblow, *Langmuir* 3:873 (1987).
33. B. P. Löchel and H.-H. Strehblow, *Werkst. Korros.* 31:353 (1980).
34. U. Bertocci, *Advances in Localized Corrosion, NACE Proceedings of the 2nd International Conference on Localized Corrosion*, Orlando, FL (H. S. Isaacs, U. Bertocci, J. Kruger, and S. Smialowska, eds.), NACE, Houston, TX, 1987, p. 127.
35. J. Steward and D. E. Williams, in *Advances in Localized Corrosion, NACE Proceedings of the 2nd International Conference on Localized Corrosion*, Orlando, FL (H. S. Isaacs, U. Bertocci, J. Kruger, and S. Smialowska, eds.), NACE, Houston, TX, 1987, p. 131.
36. S. P. Mattin and G. T. Burstein, *Progress in the Understanding and Prevention of Corrosion* (J. M. Costa and A. D. Mercer, eds.), Institute of Materials, Houston, TX, 1993, p. 1109.
37. H.-H. Strehblow, Dissertation, Freie Universität, Berlin, Germany, 1971.
38. K. J. Vetter and F. Gorn, *Electrochim. Acta* 18:321 (1973).
39. S. Benamar, Thesis, Université Pierre et Marie Curie, Paris, France, 1998.
40. I. Milosev and H.-H. Strehblow, *J. Biomed. Mater. Res.*, submitted.
41. K. E. Heusler and L. Fischer, *Werkst. Korros.* 27:551 (1976).
42. H.-H. Strehblow, B. Titze, and B. P. Löchel, *Corrosion Sci.* 19:1047 (1979).
43. H.-H. Strehblow and B. P. Löchel, *Passivity of Metals and Semiconductors* (M. Froment, ed.), Elsevier Science Publishers, New York, 1983, p. 379.
44. B. P. Löchel and H.-H. Strehblow, *J. Electrochem. Soc.* 131:522 (1984).
45. L. G. Sillén and A. E. Martell, *Stability Constants of Metal Ion Complexes*, Suppl. 1, Special Publ. 25, The Chemical Society, London, 1971, p. 171.
46. L. G. Sillén and A. E. Martell, *Stability Constants of Metal Ion Complexes*, Suppl. 1, Special Publ. 25, The Chemical Society, London, 1971, p. 167.
47. L. G. Sillén and A. E. Martell, *Stability Constants of Metal Ion Complexes*, Suppl. 1, Special Publ. 25, The Chemical Society, London, 1971, p. 217.
48. W. Khalil, S. Haupt, and H.-H. Strehblow, *Werkst. Korros.* 36:16 (1985).
49. F. A. Cotton and G. Wilkinson, *Anorganische Chemie*, Verlag Chemie, Weinheim, Germany, 1980, p. 920.
50. A. R. Cheyne, T. L. Coombs, and P. T. Grant, *Handbook of Chemistry and Physics*, CRC Press, Boca Raton, FL, 1978, p. B105.
51. V. Maurice, V. Inard, and P. Marcus, *Critical Factors in Localized Corrosion III* (P. M. Natishan, R. G. Kelly, G. S. Frankel, and R. C. Newman, eds.), The Electrochemical Society Proceedings Series, PV 89-17, Pennington, NJ, 1999, p. 552.
52. P. Marcus, V. Maurice, and H.-H. Strehblow, *Corrosion Sci.* 50 (Letter):2698 (2008).
53. V. Maurice, T. Nakamura, L. Klein, and P. Marcus, *Local Probe Techniques for Corrosion Research*, (R. Oltra, V. Maurice, R. Akid, and P. Marcus, eds.), EFC Publications No 45, Woodhead Publishing Ltd., Cambridge, U.K., 2007, p. 71.
54. A. Bouzoubaa, B. Diawara, V. Maurice, C. Minot, and P. Marcus, *Corrosion Sci.* 51:941, 2174 (2009).
55. V. Maurice, L. H. Klein, H.-H. Strehblow, and P. Marcus, *J. Phys. Chem. C* 111:16351 (2007).
56. H.-H. Strehblow, C. M. Melliar-Smith, and W. M. Augustyniak, *J. Electrochem. Soc.* 125:915 (1978).
57. H.-H. Strehblow and C. J. Doherty, *J. Electrochem. Soc.* 125:30 (1978).
58. J. Kunze, Diploma Thesis, Heinrich-Heine-Universität, Düsseldorf, Germany, 1998, p. 64.
59. H.-H. Strehblow and J. Wenners, *Z. Phys. Chem. NF* 98:199 (1975).
60. H.-H. Strehblow, *Habilitationsschrift*, Freie Universität, Berlin, Germany, 1977, p. 53.
61. H.-H. Strehblow and M. B. Ives, *Corrosion Sci.* 16:317 (1976).

62. H. Kaesche, *Die Korrosion der Metalle*, Springer, Berlin, 1979, Germany, p. 268.
63. H.-H. Strehblow and J. Wenners, *Electrochim. Acta* 22:421 (1977).
64. Th. R. Beck and R. C. Alkire, *J. Electrochem. Soc.* 126:1662 (1979).
65. G. S. Frankel, J. R. Scully, and C. V. Jahnes, *J. Electrochem. Soc.* 143:1834 (1996).
66. A. Seghal, D. Lu, and G. S. Frankel, *J. Electrochem. Soc.* 145:2834 (1998).
67. G. R. Engelhardt, A. D. Davydov, and T. B. Zhukova, *Sov. Electrochem.* 26:888 (1990).
68. J. Newman, D. N. Hansen, and K. J. Vetter, *Electrochim. Acta* 22:829 (1974).
69. G. Engelhardt and H.-H. Strehblow, *J. Electroanal. Chem.* 365:7 (1994).
70. P. Heimgartner and H. Böhni, *Corrosion* 41:715 (1985).
71. F. H. Hunkeler and H. Böhni, *Werkst. Korros.* 34:593 (1983).
72. H.-H. Strehblow, K. J. Vetter, and A. Willgallis, *Ber. Bunsen. Ges. Phys. Chem.* 75:823 (1971).
73. K. J. Vetter and H.-H. Strehblow, *Ber. Bunsen. Ges. Phys. Chem.* 74:449 (1970).
74. H. W. Pickering and R. P. Frankenthal, in *Localized Corrosion*, NACE, Houston, TX, 1974, p. 261.
75. C. Schmidt and H.-H. Strehblow, *J. Electrochem. Soc.* 145:834 (1998).
76. U. Wolff, Dissertation, Technische Universität Dresden, Dresden, Germany, 1999.

8

Sulfur-Assisted Corrosion Mechanisms and the Role of Alloyed Elements

Philippe Marcus

Ecole Nationale Supérieure de Chimie de Paris

and

Centre National de la Recherche Scientifique

CONTENTS

8.1 Introduction ... 395
8.2 Fundamental Aspects of Sulfur-Induced Corrosion ... 396
8.2.1 Activation (Acceleration) of Anodic Dissolution by Adsorbed Sulfur ... 396
8.2.2 Blocking or Retarding Effect of Adsorbed Sulfur on Passivation ... 399
8.2.2.1 Effect on the Growth of the Passive Film ... 399
8.2.2.2 Effect on the Passivation Kinetics ... 399
8.2.2.3 Effect on the Structure and Properties of the Passive Film ... 400
8.2.3 Anodic Segregation of Sulfur ... 400
8.2.4 Passivity Breakdown Caused by Interfacial Sulfur ... 402
8.2.5 The Joint Action of Sulfur and Chloride Ions ... 403
8.3 Counteracting the Sulfur Effects with Alloyed Elements ... 404
8.3.1 The Role of Molybdenum ... 404
8.3.2 The Role of Chromium ... 406
8.4 Implications in Areas of Practical Importance ... 407
8.5 Thermodynamics of Sulfur Adsorption on Metal Surfaces in Water ... 409
8.5.1 Principle of *E*-pH Diagrams for Adsorbed Species ... 410
8.5.2 *E*-pH Relations for the Equilibria between Dissolved and Adsorbed Species ... 411
8.5.3 Potential-pH Diagrams ... 413
8.6 Conclusion ... 416
References ... 416

8.1 Introduction

Sulfur is an element which is found in the bulk of most materials (in saturated solid solution and often in inclusions) as well as in many environments (in the form of, e.g., HS^-, H_2S, SO_2, $S_2O_3^{2-}$).

The role of sulfide inclusions in corrosion has been recognized in early works. The fact that sulfur-containing species are detrimental to the resistance of metals and alloys to localized corrosion has been established for a long time but the mechanisms have remained unclear until recently. In the area of passivity breakdown, where substantial research effort has been expended for several years, the effects of chloride ions have been investigated much more than the effects of sulfur. The aim of this chapter is to review the fundamental aspects of the mechanisms of S-induced corrosion, with special emphasis on the role played by adsorbed (chemisorbed) sulfur.

In the first section of this chapter, the basic effects of adsorbed sulfur on anodic dissolution, on passivation, and on the breakdown of passive films are presented. In a subsequent section the effects of alloyed elements (essentially Cr and Mo) are presented and the way in which they can counteract the detrimental influence of sulfur is emphasized. In the next section, implications of the mechanisms of sulfur-induced corrosion for different areas of practical importance are given, with connections to the related chapters of this book. In the last section of this chapter, the thermodynamic predictions of the conditions of adsorption of sulfur on metal surfaces in water are given.

8.2 Fundamental Aspects of Sulfur-Induced Corrosion

The basic concepts and the associated mechanisms underlying the effects of adsorbed sulfur on the corrosion of metals and alloys are presented here. They have been derived to a large extent from studies performed on well-defined metal and alloy surfaces. In many cases, oriented single crystals have been used to control the structure of the surface on which the corrosion reactions were investigated. Controlled amounts of sulfur in the range of a monolayer were produced prior to the electrochemical measurements by dosing the surface with gaseous H_2S. Surface science techniques, including Auger electron spectroscopy (AES), electron spectroscopy for chemical analysis (ESCA), also called x-ray photoelectron spectroscopy (XPS), low-energy electron diffraction (LEED), and a more specific radiochemical technique utilizing the ^{35}S radioisotope, were combined to analyze the surface prior to and after the electrochemical tests, in order to relate the electrochemical and corrosion behavior with the chemical composition and the structure of the surface. This approach has allowed us to determine the mechanisms of the sulfur-assisted corrosion reactions on an atomic scale.

8.2.1 Activation (Acceleration) of Anodic Dissolution by Adsorbed Sulfur

The first studies were conducted on Ni [1–3] and Ni-Fe [4] single crystals with well-controlled coverages of radioactive sulfur. They clearly demonstrated that adsorbed sulfur accelerates the anodic dissolution of the metal and the alloys. This effect is seen in Figure 8.1, where the polarization diagrams for a nickel-iron single-crystal alloy [Ni-25Fe(100)(at %)] with and without adsorbed sulfur are shown. Two major findings are apparent from the measurements: (a) a significant increase in anodic current density is observed when the surface is covered by a monolayer of adsorbed sulfur, showing that the dissolution is accelerated by the presence of sulfur, and (b) the surface sulfur is highly stable, since it remains adsorbed during dissolution of the metal. This result was established by the measurements of the coverage, denoted by θ in Figure 8.1. This gives evidence of the catalytic nature of the accelerating effect of adsorbed sulfur on the dissolution. In this way, a microscopic amount of adsorbed sulfur can stimulate the dissolution of a significant amount of material (about 1 μm of alloy is dissolved in the active region of Figure 8.1 under the catalytic action of a monoatomic layer of adsorbed sulfur). The catalytic effect of a monolayer of

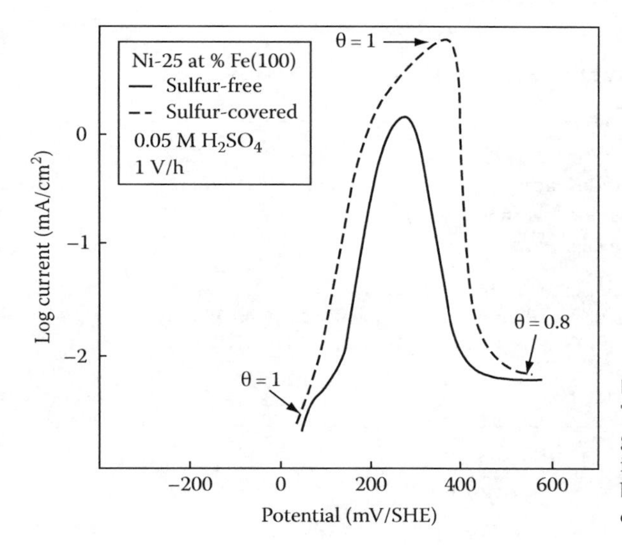

FIGURE 8.1
The influence of adsorbed sulfur on the *i-E* diagram for Ni-25Fe (at %) single crystal [(100) face] in 0.05 M H_2SO_4. θ is the coverage of the surface by adsorbed sulfur, measured with the radiochemical technique (^{35}S radiotracer).

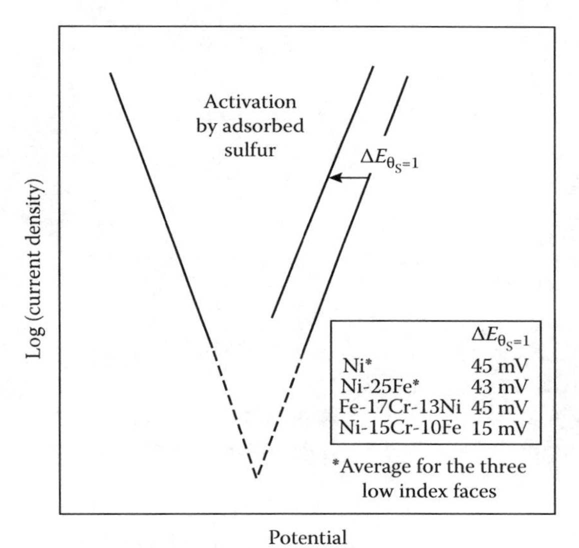

FIGURE 8.2
Schematic *i-E* plots showing the effect of adsorbed sulfur on anodic dissolution. Inset: Potential shift ($\Delta E_{\theta_S=1}$) induced by a complete monolayer of adsorbed sulfur on different metals and alloys ($\Delta E_{\theta_S=1}$ is measured at half-maximum of the active peak of the sulfur-free surface).

adsorbed sulfur on anodic dissolution of metals and alloys is summarized in the schematic *i-E* diagram of Figure 8.2 with, in the inset, potential shifts measured with adsorbed sulfur on different metals and alloys. Since the first direct demonstration of the accelerating effect of dissolution by adsorbed sulfur, which was for nickel and nickel-iron alloys, the more general nature of this effect has been revealed by similar experiments on nickel alloy 600 [5,6] and on austenitic stainless steels [7]. The acceleration factor R (ratio of the current densities at the maximum of the active peak with and without adsorbed sulfur) for nickel-base and iron-base alloys are reported in Table 8.1. Values in the range 5–12 have been observed. This striking effect has been attributed to a weakening of the metal–metal bonds induced by adsorbed sulfur, which leads to a lowering of the activation energy barrier for the passage of a metal atom from the surface to the solution. This mechanism is

TABLE 8.1

Ratio (*R*) of the Current Densities at the Maximum of the Active Peaks with and without Adsorbed Sulfur[a]

Metal or Alloy	*R*	Reference
Ni-S_{ads}/Ni (average on the three low-index faces)	7.7	[2]
Ni-25Fe-S_{ads}/Ni-25Fe (average on the three low-index faces)	5.5	[4]
Fe-17Cr-13Ni-S_{ads}/Fe-17Cr-13Ni [(100) face]	7	[7]
Ni-15Cr-10Fe-S_{ads}/Ni-15Cr-10Fe	12	[5]

[a] Coverages of the surfaces by preadsorbed sulfur are (in 10^{-9} g/cm^2): 42 on Ni(100), 47 on Ni(111), 44 on Ni(110), 40 on Ni-25Fe (100), 42 on Ni-25Fe(111), 41 on Ni-25Fe (110), 40 on Fe-17Cr-13Ni(100), and 40 on Ni-15Cr-10Fe.

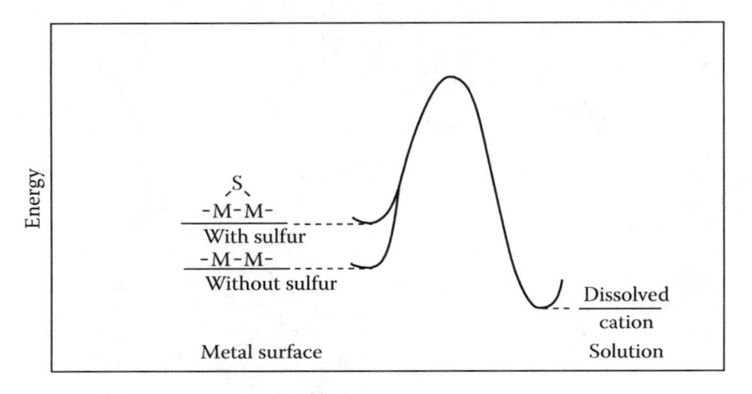

FIGURE 8.3
Weakening of the metal–metal bond by adsorbed sulfur and acceleration of the dissolution of the metal. (From Ando, S., Suzuki, T., and Itaya, K., *J. Electroanal. Chem.*, 412, 139, 1996.)

shown schematically in Figure 8.3 [8]. The existence of a dipole $M^{\delta+}$—$S^{\delta-}$ associated with the presence of sulfur can also promote the anodic dissolution.

The influence of sulfur can be strongly localized if the sulfur is not homogeneously distributed over the surface but adsorbed in specific sites, such as surface defects, where sulfur atoms are more tightly bonded.

Direct observation of the dissolution of sulfur-modified Ni(100) [9] using in situ scanning tunneling microscopy (STM) is in agreement with the previously shown effects of adsorbed S on dissolution (including the enhanced dissolution and the stability of the S layer). It also provides structural data at the atomic scale, indicating that the dissolution of nickel atoms is taking place at step edges, with a displacement of adsorbed S atoms from the edge of the upper terrace to the adjacent lower terrace, as shown in Figure 8.4.

On Cu, it was also shown that the anodic dissolution, in a mildly alkaline borate buffer solution, is catalyzed by a submonolayer of sulfur and that S remains adsorbed on the copper surface [10].

The anodic dissolution of iron in acidic medium is also accelerated by H_2S [11].

The major conclusion of the results reviewed in this section is that a mono-atomic layer of sulfur can promote the dissolution of macroscopic amounts of material. This gives evidence of a direct link between the atomic-level interactions of adsorbed sulfur atoms with metal surfaces and their macroscopic manifestations in corrosion.

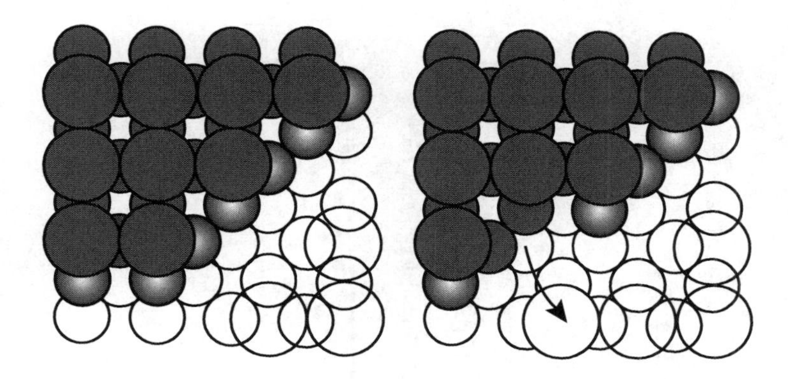

FIGURE 8.4
Model of the anodic dissolution process of sulfur-covered Ni(100) electrodes, showing dissolution of two Ni atoms from the step edge and displacement of an S atom from the upper terrace to a fourfold hollow site on the lower terrace (small circles denote Ni atoms, large circles denote S atoms, the gray color indicates atoms of the upper terrace). (Adapted from Ando, S. et al., *J. Electroanal. Chem.*, 412, 139, 1996.)

8.2.2 Blocking or Retarding Effect of Adsorbed Sulfur on Passivation

8.2.2.1 Effect on the Growth of the Passive Film

Another important effect of adsorbed sulfur that was observed first on Ni [1–3] and later on Ni-Fe alloys [4] is the poisoning of passivation. This effect takes place above a critical coverage of the surface by sulfur that was found to be 0.7–0.8 monolayer. This effect of sulfur is observed in the *i-E* diagram shown previously in Figure 8.1, where the passivation potential of the surface covered by a complete monolayer of sulfur ($\theta = 1$) is shifted (by ~100 mV) to a more anodic value with respect to the sulfur-free surface. In using the ^{35}S radiotracer, it was demonstrated that the desorption (or electro-oxidation) of ~20% of the full monolayer is necessary to allow the passive film to be formed. The retarding effect of sulfur on the growth of the passive film is caused by blocking the sites of adsorption of hydroxyl ions, which are the precursors in the formation of the passive layer. X-ray photoelectron spectroscopy (XPS) measurements have revealed that in the presence of adsorbed sulfur the adsorption of OH groups is not totally inhibited, but the OH groups on the surface are more diluted and thus the disproportionation reaction of adjacent OH groups is prevented. A schematic representation of this mechanism is given in Figure 8.5. In this way the oxide film, which would normally passivate the surface, is not formed.

8.2.2.2 Effect on the Passivation Kinetics

Measurements of the passivation kinetics by means of potential steps applied to surfaces without or with adsorbed sulfur have revealed that the time to complete passivation

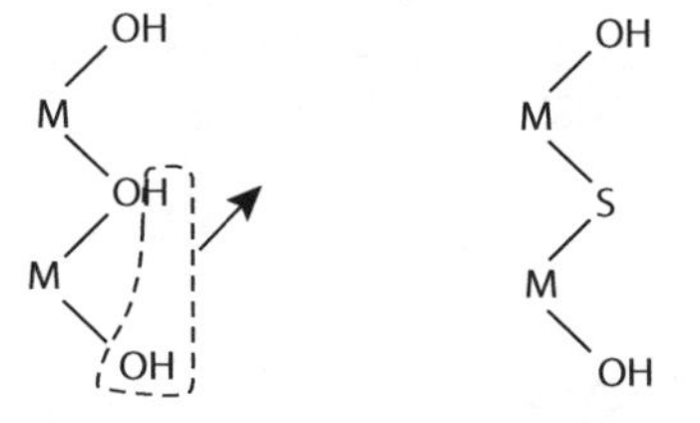

In this configuration, passivation is favored

In this configuration, passivation is precluded

FIGURE 8.5
The mechanisms of the blocking effect of adsorbed sulfur on passivation.

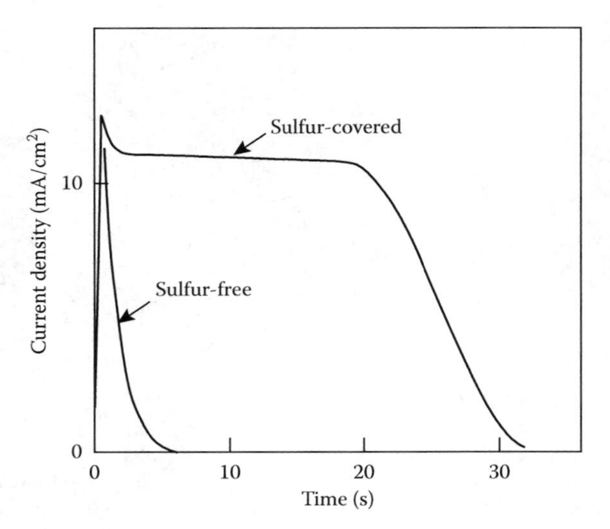

FIGURE 8.6
The effect of adsorbed sulfur on the passivation kinetics of nickel [Ni(111), 0.05 M H_2SO_4, 540 mV/SHE].

increases when the surface is covered by adsorbed sulfur. This effect is represented in the *i-t* diagram of Figure 8.6. A rapid decrease of the current immediately after the potential step is observed for the clean surface, whereas for the sulfur-covered surface the current density decreases slowly, indicative of a slow process of lateral growth of the film on top of the remaining fraction of a monolayer of adsorbed sulfur. This is consistent with the view that a complete monolayer of sulfur inhibits the nucleation of the oxide, as observed on nickel, whereas a fraction of a monolayer does not inhibit the nucleation but may diminish the density of nucleation sites for the oxide and make the lateral growth of the film more difficult. These two effects can slow down the passivation kinetics.

Such kinetic effects have been observed on sulfur-contaminated nickel [8], on nickel alloy 600 [6], and on copper [12].

8.2.2.3 Effect on the Structure and Properties of the Passive Film

Below the critical sulfur coverage of 0.7–0.8 monolayer on Ni, the passive film grows on top of the remaining adsorbed sulfur. However, the structure and the properties of the film formed on the sulfur-contaminated surface are modified. A study [2] by reflection high-energy electron diffraction (RHEED) of the structure of the passive films formed on nickel single crystals [Ni(111)] has shown that, in the absence of sulfur, a crystalline film epitaxial with the substrate is formed (a fact that has been confirmed by both ex situ and in situ atomic resolution imaging of the film by scanning tunneling microscopy [13–15]), whereas in the presence of sulfur the epitaxial growth is disrupted and a more defective polycrystalline film is obtained. Accordingly, the current density in the passive state is about four times larger for a film formed with sulfur at the metal–oxide interface. XPS measurements of the binding energy of sulfur (S 2*p*) provide evidence that it remains at the interface because the binding energy remains close to that of sulfur bonded to the metallic surface (~162 eV).

8.2.3 Anodic Segregation of Sulfur

The concept of anodic segregation was first proposed after the discovery that sulfur present as an impurity in the bulk of Ni and of Ni-25Fe alloys is enriched on the surface during anodic dissolution [2,4,16]. This phenomenon was also observed on Fe-50Ni [17] and on alloy 600 [5,18]. The experiments were performed on Ni and Ni-Fe alloys that had been doped with sulfur prior to the electrochemical treatments. The introduction of radioactive S into the bulk was performed by exposing the samples to an H_2S-H_2 gas mixture (with ^{35}S-labeled H_2S) at high temperature (1000°C–1200°C) under thermodynamic conditions in which sulfur is in solid solution in the metal or the alloy. Rapid quenching to room temperature after the high-temperature treatment avoids sulfide precipitation. The *i-E* curves recorded on a sulfur-free (i.e., annealed in pure hydrogen) Ni electrode and on Ni containing 50 ppm sulfur are shown in Figure 8.7a. The sulfur-free Ni sample exhibits the normal active–passive transition observed in 0.05 M H_2SO_4. The sample with bulk sulfur exhibits a completely different behavior: there is no active–passive transition, and

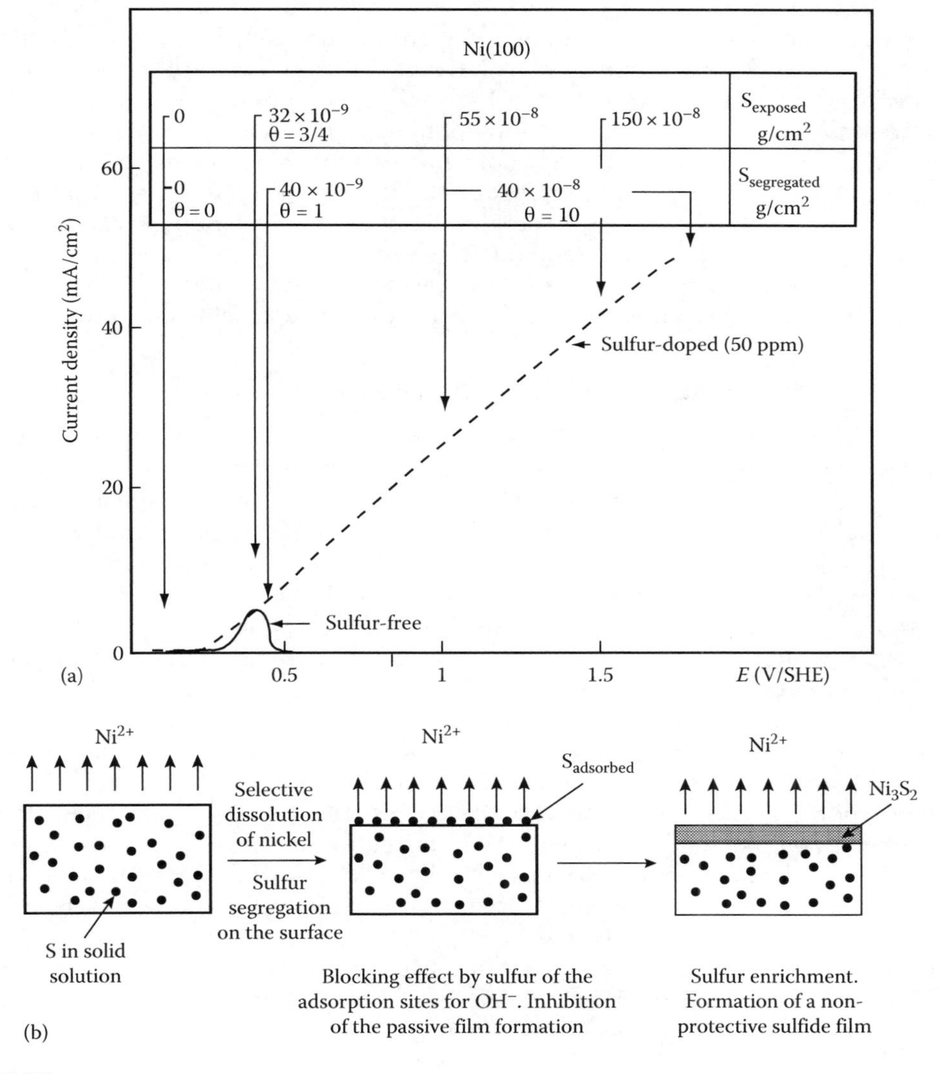

FIGURE 8.7
(a) Surface enrichment of sulfur during anodic dissolution: *i-E* curves and radiotracer measurements of the sulfur concentration on the surface of Ni(100) containing 50 ppm of sulfur (0.05 M H_2SO_4, 1 V/h). (b) The mechanism of anodic segregation.

the current density increases with increasing potential in the whole range of potentials corresponding to the passive state of sulfur-free Ni. The results of the measurements of the sulfur concentration on the surface, using radioactive sulfur, are shown in Figure 8.7a. They demonstrate that sulfur segregates on the surface during dissolution of the electrode with bulk sulfur (the measured values of the surface sulfur concentration are denoted $S_{segregated}$ in Figure 8.7a). The values denoted $S_{exposed}$ in Figure 8.7a represent the integrated amount of sulfur that became available on the surface due to the dissolution of the metal matrix. Comparison of the data for $S_{segregated}$ and $S_{exposed}$ revealed that all the sulfur present in the bulk remains on the surface during dissolution, up to a surface concentration of $40 \times 10^{-9} g/cm^2$, which corresponds to a complete monolayer of sulfur on Ni(100). Above this surface coverage, nickel sulfide precipitates and then a thin layer of nickel sulfide is formed. The amount of sulfur in this layer is $\sim 40 \times 10^{-8}$ g/cm^2, which corresponds to ~25 Å of Ni_3S_2. Sulfur in excess of this amount is dissolved. The formation, before reaching the passivation potential, of a complete monolayer of sulfur and the subsequent growth of a sulfide layer preclude the formation of the passive oxide film. The sulfide layer is not protective and thus high corrosion rates are obtained in the range of potentials in which pure Ni is normally passivated. This phenomenon was called anodic segregation [3,4]. The anodic segregation rate depends on the sulfur content of the material and on the rate of anodic dissolution. Because of differences in dissolution rates for single-crystal surfaces of different crystallographic orientation, the sensitivity to sulfur increases in the order (111) < (100) < (110), which is the order of increasing dissolution rates observed for Ni in 0.05 M H_2SO_4 [2]. The mechanism of anodic segregation is represented in Figure 8.7b.

8.2.4 Passivity Breakdown Caused by Interfacial Sulfur

After the formation of the passive film on a bare metal surface, slow dissolution of the metal cations continues, involving dissolution at the passive film surface and transport of ions through the oxide (see Chapter 6). The question then arises as to where the bulk impurities, e.g., sulfur, go during this process. This question was addressed in a detailed investigation of Ni and Ni-Fe alloys that were doped with radioactive sulfur in order to trace the path taken by the sulfur atoms during the slow dissolution in the passive state [19]. The radiochemical results gave direct evidence that sulfur present in the bulk accumulates at the metal-passive film interface. The enrichment rate was shown to be proportional to the sulfur content in the metal and to the dissolution current density. Above a critical concentration of sulfur at the interface (metal–oxide), breakdown of the passive film was observed. This critical concentration of interfacial sulfur was measured with radioactive sulfur and found to be close to one monolayer of sulfur (i.e. $\sim 40 \times 10^{-9}$ g/cm^2). The loss of adherence (or the decohesion) at the interface for this critical coverage has been attributed to a weakening of the bonding of the oxide to the substrate caused by sulfur. The nucleation of the sulfide that is expected to take place when the sulfur concentration exceeds a complete monolayer may be responsible for the observed local breakdown of the film and pitting. The defects that are likely to exist at the interface could serve as specific sites for the nucleation of the sulfide, once a quasi-complete monolayer of sulfur has been accumulated at the interface. A schematic representation of the proposed mechanism of the sulfur-induced breakdown of the passive film is shown in Figure 8.8.

It is to be noted that the three following factors are in favor of sulfur remaining at the metal-passive film interface: (a) the sulfur–metal chemical bond is very strong, (b) the solubility of S in nickel oxide (which constitutes the inner part of the passive film on Ni and Ni-Fe alloys) is very low, and (c) the electric field across the passive film, which assists the

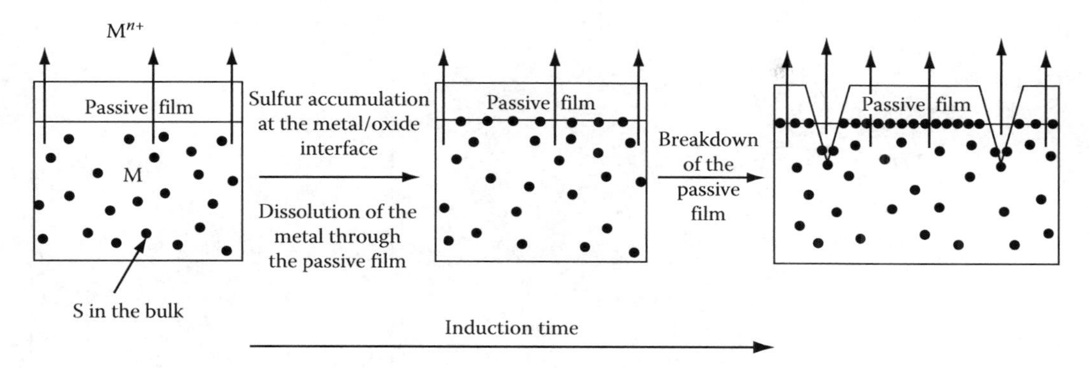

FIGURE 8.8
Mechanism of the breakdown of the passive film induced by enrichment of sulfur at the metal-passive film interface.

passage of cations from the metal-passive film interface to the passive film-solution interface, should impede the transport of sulfur, which would be negatively charged.

8.2.5 The Joint Action of Sulfur and Chloride Ions

The passivity breakdown and pitting caused by Cl^- ions has been the subject of an enormous amount of work. This area has been reviewed in Chapter 7.

In the preceding parts of this chapter we have seen that S alone can enhance the dissolution, block or retard the growth of the passive film, and cause passivity breakdown and pitting by enrichment at the metal–film interface. All these effects can evidently also take place in the presence of Cl^-. A major difference between the mechanisms of action of Cl^- and S is that S does not seem to interact directly with the oxide surface as strongly as Cl^-.

The effect of sulfur species and chloride ions has been studied in some detail in the case of corrosion of stainless steels in solutions containing thiosulfate ($S_2O_3^{2-}$) and chloride (Cl^-) ions [20–27].

It has been demonstrated [27] that $S_2O_3^{2-}$ is not reduced on the surface of the passive film formed in neutral solution on an Fe-17Cr alloy, whereas the reduction to adsorbed sulfur or sulfide does take place on the nonpassivated alloy surface. In this study, the combined action of chloride and sulfur was also investigated. The conclusion was that with 30 ppm thiosulfate added to the Cl^--containing solution (0.02 M) at pH 7, the initial breakdown of the passive film is caused by Cl^- and the thiosulfates do not play a major role in this step, whereas the presence of thiosulfate has a drastic effect on the subsequent step, the occurrence of stable pits. With Cl^-, and without $S_2O_3^{2-}$, unstable pits (i.e., pits that are repassivated) are formed over a wide range of potentials below the classical pitting potential corresponding to the formation of stable pits, whereas with the thiosulfates, the reaction on the bare metal surface produces reduced sulfur (adsorbed sulfur or sulfide), which precludes re-passivation and causes a marked increase of stable pits at low potential. This effect is shown in Figure 8.9. The formation of adsorbed sulfur by reaction of $S_2O_3^{2-}$ with the bare metal surface was demonstrated by XPS measurements of the binding energy of the sulfur core level electrons (S 2*p*), which was found to be 169 eV for adsorbed thiosulfate on the passive film surface and 162 eV after reduction of the thiosulfate on the Fe-17% Cr alloy surface. In previous works on the effects of addition of $S_2O_3^{2-}$ on corrosion of stainless steels [21,25], such a mechanism had been hypothesized, but direct experimental evidence was lacking. A synergistic effect of Cl^- and $S_2O_3^{2-}$ on pitting corrosion was also observed on 310 stainless steel [28]. Type 304 L stainless steel suffers stress corrosion

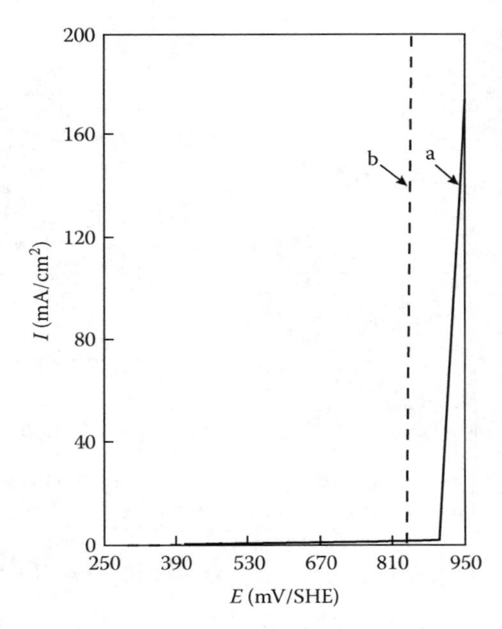

FIGURE 8.9
The joint action of $S_2O_3^{2-}$ and Cl^- on passivity breakdown and localized corrosion of an Fe-17Cr alloy, (a) 0.02 M Cl^-; (b) 0.02 M Cl^- + 30 ppm $S_2O_3^{2-}$. $S_2O_3^{2-}$ is reduced on the alloy surface, as evidenced by the difference in the binding energy of S 2*p* for $S_2O_3^{2-}$ (169 eV) and S_{ads} (162 eV).

cracking in chloride solutions with 10^{-3} to 10^{-1} M $S_2O_3^{2-}$ (at 353 K). It was shown that the cracks initiated at pits [29] and the authors concluded that $S_2O_3^{2-}$ affects the initiation process of pitting and cracking.

To summarize, a likely mechanism for the combined action of chlorides and sulfur is the local breakdown of the film by Cl^-, which then permits all the effects of S shown above on metal and alloy surfaces to take place. A major consequence is the stabilization of otherwise unstable (or metastable) pits. This mechanism is consistent with the lower pitting potential and/or shorter incubation time observed experimentally.

Sulfur species in aqueous solution may also cause directly the breakdown of the film if the sulfide is thermodynamically more stable than the oxide, a case that would require a high concentration of the sulfur species in solution to provide a reaction of the type $M_xO_y + zH_2S \leftrightarrow M_xS_z + yH_2O + 2(z - y)H^+ + 2(z - y)e^-$.

Sulfide inclusions have been known for many years to be detrimental to the corrosion resistance of steels and extensive studies have been performed on stainless steels ([30–33] and references in Chapter 9). Preferential adsorption of Cl^- on sulfide inclusions has been suggested to be an important step in the corrosion mechanism. The precise mechanism of the joint action of Cl^- and the sulfide is not fully elucidated.

8.3 Counteracting the Sulfur Effects with Alloyed Elements

8.3.1 The Role of Molybdenum

Numerous electrochemical and surface analytical studies have been performed to understand the effect of alloyed molybdenum on the corrosion resistance of stainless steels (see, e.g., the section on the role of Mo in Mo-containing austenitic stainless steels in Chapter 6).

In order to reach a better understanding of the role of molybdenum in the presence of sulfur, studies were undertaken on simple systems, i.e., binary and ternary single-crystal alloys on which the surface concentrations of sulfur and of Mo could be precisely measured by radiochemical (^{35}S) and spectroscopic (XPS) techniques. The experiments performed on nickel-molybdenum alloys [Ni-2Mo and Ni-6Mo (at %)] provided the first direct evidence of a specific surface interaction between molybdenum and adsorbed sulfur leading to the removal of sulfur from the surface and thereby the attenuation or the disappearance of the detrimental effects of adsorbed sulfur [34,35].

The most conclusive experiments were done in (a) preadsorbing, in a gas mixture of H_2S and H_2, a known amount of sulfur on the well-defined surface of an Ni-2 at % Mo alloy [the (100) crystallographic orientation of this alloy was used] and (b) measuring the variation of the sulfur coverage during anodic dissolution in the active state (at 320 mV/SHE) in 0.05 M H_2SO_4. The first and essential observation was that the sulfur coverage decreases, whereas it had been found to remain constant on pure Ni (discussed earlier in this chapter). The precise measurement of the desorption kinetics is shown in Figure 8.10 in terms of sulfur coverage (weight per cm^2 and number of sulfur atoms per metal atom on the surface) versus the amount of dissolved material (expressed in number of dissolved atom layers). The initial coverage corresponds to saturation of the surface by adsorbed sulfur (42 ng/cm^2, one S atom for two M atoms on the surface of the (100) face). Because sulfur was initially present only on the surface, the results demonstrate that a surface reaction between Mo and sulfur is the cause of the loss of the surface sulfur. This is a dynamic process in which the dissolution of the alloy brings Mo to the surface, which bonds to

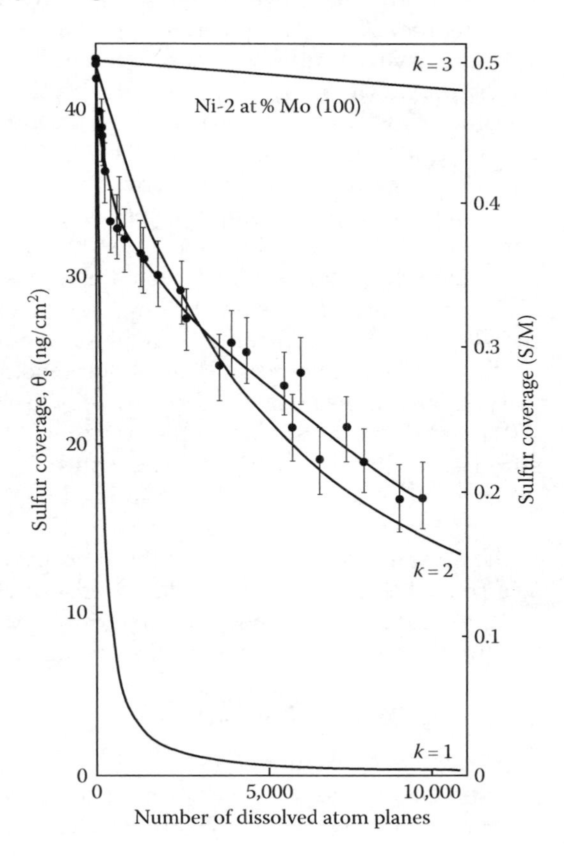

FIGURE 8.10
The loss of surface sulfur induced by molybdenum (experimental and theoretical curves).

adsorbed sulfur and is then dissolved with it. A theoretical model was proposed for the mechanism of the effect of Mo. In this model, one S atom adsorbed on the surface reacts with k Mo atoms of the surface plane and the cluster that is formed (in which O or OH may participate) is dissolved. It was shown that the sulfur coverage on the surface after dissolution of $(n + 1)$ atom planes of the alloy, denoted $\theta_s(n + 1)$, has the following expression [34]:

$$\theta_s(n+1) = \theta_s(n) - [(\theta_s(n))^2 (C^s_{Mo})^k]$$

where

C^s_{Mo} is the surface concentration of Mo, which was assumed to be identical to the bulk content of Mo

k is the number, defined above, of Mo atoms required to form with sulfur a cluster that is dissolved

This equation was used to fit the experimental variation of the sulfur coverage. The results for $k = 1, 2, 3$ are computed in Figure 8.10. The curve with $k = 2$ provides a very good fit to the experimental curve. Similar findings were obtained with Ni-6% Mo (100). These results strongly support the proposed mechanism in which, on a (100) surface, two Mo atoms bond to and remove one adsorbed S atom. It is to be noted that in this mechanism the passive film is not directly involved, but of course the major consequence of the removal of sulfur by Mo is that the passive film may be formed on an otherwise blocked surface. The same model has been used successfully to interpret in a quantitative manner the fact that on other Ni-Mo alloys the surface enrichment of sulfur by anodic segregation of bulk sulfur was very limited [36] compared with what had been reported previously for Ni [2]. On Ni-6%Mo (100) with 32 ppm sulfur in the bulk, the coverage by sulfur measured after different times of anodic polarization in the active region (in 0.05 M H_2SO_4) was always less than a complete monolayer. The theoretical amount of sulfur, calculated using an equation similar to that given above for the Mo-induced desorption of adsorbed S, but in which a term accounting for the supply of sulfur by the dissolution of the alloy was added, gave a good fit to the experimental values when the value of k was taken equal to 2. This confirmed the validity of the mechanism in which two Mo atoms bond to one sulfur atom and remove it from the surface by dissolution of the cluster formed. More recently, the molybdenum-induced removal of adsorbed sulfur has been observed in measurements performed on single-crystal alloys with chemical compositions close to the compositions of type 304 and 316 stainless steels, i.e., without and with molybdenum, respectively [7]. After preadsorbing sulfur on the surfaces of the two alloys, the sulfur coverage was measured as a function of time of exposure of the alloys to 0.05 M H_2SO_4 at the corrosion potential. The results are shown in Figure 8.11. The monolayer of sulfur adsorbed on the 304-type stainless steel is stable in 0.05 M H_2SO_4 at the corrosion potential, whereas, owing to the presence of molybdenum, the desorption of sulfur is observed on the 316-type stainless steel. The model used above gives a fit of the experimental data for $k \sim 1$, instead of $k = 2$ for the Ni-Mo alloys. This suggests that chromium lowers the number of Mo atoms required to promote the dissolution of sulfur.

8.3.2 The Role of Chromium

The role played by alloyed Cr in the presence of sulfur has been investigated in detail on Ni-based alloys with various concentrations of Cr in the range 8%–34% (a range of concentration within which alloys 600 and 690 are situated) [5,6,18]. The combined

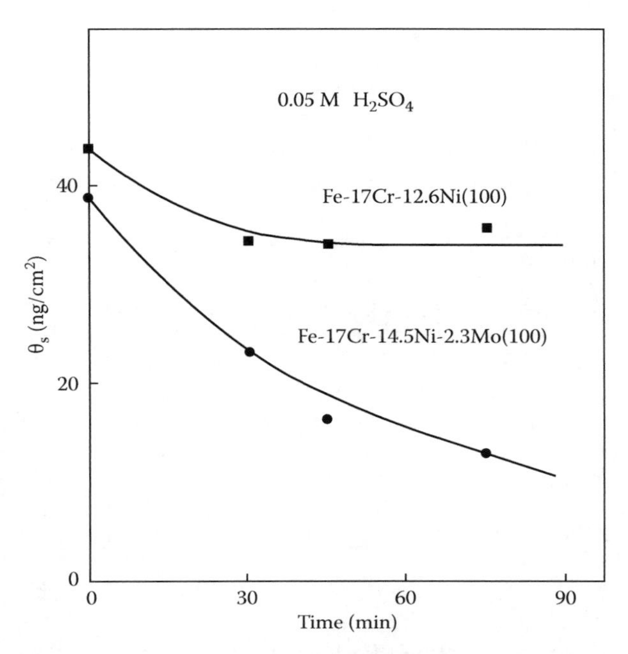

FIGURE 8.11
Mo-induced removal of sulfur adsorbed on a single-crystal alloy of composition close to 316 stainless steel. The measurements on an alloy of composition close to 304 stainless steel are shown for comparison. The coverage of sulfur was measured with radioactive sulfur (^{35}S).

electrochemical, radiochemical (^{35}S), and spectroscopic (XPS) measurements performed with sulfur pre-adsorbed on the surface and with bulk sulfur (introduced in the alloy in controlled amounts by bulk diffusion of S at high temperature) revealed that Cr allows the surface to be passivated under conditions in which the passivation would be precluded by sulfur if there were no Cr. The surface analysis by XPS [18] revealed the coexistence of nickel sulfide and chromium oxide on the surface. The mechanism of this counteracting effect of Cr is totally different from the mechanism of the Mo effect discussed earlier. With Cr the desorption of sulfur is not the key factor. What Cr does is to react selectively with oxygen (i.e., with OH or H_2O) to give Cr oxide, whereas Ni reacts with sulfur. This leads to the coalescence of S in small nickel sulfide islands that can be covered by the lateral growth of the chromium oxide islands. In other words, the antagonistic effects of Cr and S are based on the fact that, on the considered alloys, Cr acts as an oxide former whereas Ni acts as a sulfide former. This competitive mechanism is represented schematically in Figure 8.12. In agreement with this mechanism, it has been observed that H_2S, which accelerates the anodic dissolution of chromium, has no influence on passivated chromium [37].

8.4 Implications in Areas of Practical Importance

There is much evidence for the damaging effect of sulfur species in a wide range of corrosion-related service failures. The relation between the sulfur-induced corrosion mechanisms presented in this chapter and the implications in areas of practical importance can be rationalized on the basis of (a) the source of sulfur, (b) the transport process to the metal surface, and (c) the conditions of the reduction (or oxidation) of the sulfur species into the harmful chemical state of sulfur, i.e., the adsorbed (chemisorbed) sulfur (or the sulfide if the concentration of the sulfur species is high). Table 8.2 summarizes some of these

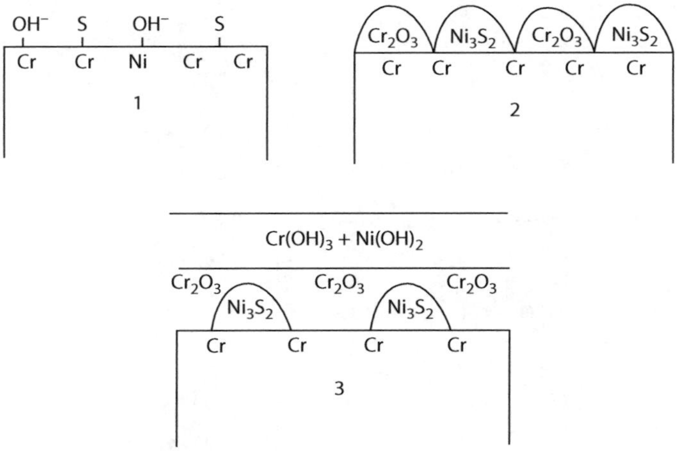

FIGURE 8.12
The mechanism of the antagonistic effects of chromium and sulfur on the corrosion resistance of Ni-Cr-Fe alloys.

TABLE 8.2

Sulfur Species that May Be Present in the Environment or the Material, with the Process of Transport to the Surface, Nature of the Surface Reaction Producing Adsorbed Sulfur, and Examples of Areas of Practical Importance

Sulfur Species	Transport Process to the Surface	Surface Reaction Producing Adsorbed Sulfur	Examples of Areas of Practical Importance
Environmental			
H_2S		Oxidative reaction	Oil and gas
HS^-	Diffusion in liquid	Oxidative reaction	
$S_2O_3^{2-}$	Diffusion in liquid	Reductive reaction	Pulp and paper
SO_2	Diffusion in the atmosphere and in liquid	Reductive reaction	Atmospheric
HSO_4^-	Diffusion in liquid	Reductive reaction	
SO_4^{2-}	Diffusion in liquid	Reductive reaction	High-temperature water
Material			
Sulfur in solid solution	Solid-state diffusion (at high temperature) or anodic segregation		
Sulfur segregated at grain boundaries	Solid-state diffusion (at high temperature) or anodic segregation		All areas
Sulfide inclusions	Dissolution and readsorption		

considerations and gives a few examples of practical areas in which detrimental effects of sulfur have been identified, e.g., oil and gas, pulp and paper industries, power plants (high-temperature water reactors), and atmospheric corrosion. Further details can be found in the other chapters of this book.

Two categories of sulfur species may be considered according to the origin of the species: (a) sulfur species (molecules or ions) in the environment, e.g., SO_2 (gaseous), H_2S (gaseous or aqueous), HS^-, $HS_2O_3^-$, $S_2O_3^{2-}$, $S_4O_6^{2-}$, HSO_4^-, SO_4^{2-}, and (b) sulfur in the material: sulfur in solid solution, sulfur segregated at the surface or in the grain boundaries, and sulfur in sulfide inclusions. After considering the source of sulfur, the process by which the sulfur species arrive at the surface (transport process) has to be identified. For environmental sulfur species it is generally diffusion in the liquid (e.g. aqueous solution). For sulfur present in the bulk metal or alloy, it may be surface segregation by solid-state diffusion at high temperature or anodic segregation (a process that has been defined in a preceding section). Higher anodic dissolution rates of grain boundaries, further accelerated by sulfur (according to the mechanism of sulfur-enhanced dissolution described earlier), result in rapid surface enrichment of sulfur at grain boundaries exposed to the electrolyte. In the case of sulfide inclusions, e.g., MnS, although their detrimental effect on corrosion of steels and stainless steels has been extensively studied [30–33,38–41], the exact mechanism is not fully understood. The important case of sulfide inclusions in stainless steels is discussed in detail in Chapter 9. On the basis of the data reviewed here, other authors have invoked the role of adsorbed sulfur to explain the detrimental effects of sulfide inclusions in stainless steels [38–40]. It must be pointed out that early studies of the effects of H_2S addition on corrosion of stainless steels [41] have revealed the equivalence of the effects of sulfur added in the form of H_2S or present in sulfide inclusions. The effects of $S_2O_3^{2-}$ observed in the pulp and paper industry [23] and of sulfur species in the steam generators of pressurized water reactors [42] have also been interpreted on the basis of the data presented in this chapter. Attempts to investigate the transport of sulfur from the inclusions to the surrounding surface led to the conclusion that sulfur is cathodically deposited after dissolution of the sulfide, producing rings of sulfur around the sulfide inclusion [38]. In a study of the mechanisms of pitting corrosion of stainless steels using submicron resolution scanning electrochemical and photoelectrochemical microscopy [43] it was concluded that the electrodissolution of certain MnS inclusions in stainless steel is chloride catalyzed. The enhancement of the dissolution of sulfide inclusions by Cl^- had also been suggested in an earlier work [40].

After the process by which sulfur is transported to the surface, the nature of the surface reaction producing adsorbed sulfur must be considered. Sulfur may be adsorbed by electro-oxidation of sulfides [H_2S (aqueous) or HS^-] or by electro-reduction of sulfates (HSO_4^- or SO_4^{2-}) or thiosulfates ($HS_2O_3^-$, $S_2O_3^{2-}$).

The oxidative or reductive reactions and the conditions of potential and pH in which they may take place are detailed in the following. For SO_2, the surface reactions are presented in the chapter on atmospheric corrosion (Chapter 15).

Once sulfur is present on the surface in the active chemical state (i.e. adsorbed), it has the same effects (which have already been described) irrespective of its origin.

The stage at which the surface reactions become localized is not exactly known. It is probably strongly related to the stage at which surface defects are crucial, which is the case for the surface reactions that produce adsorbed sulfur.

8.5 Thermodynamics of Sulfur Adsorption on Metal Surfaces in Water

The principle of potential-pH (Pourbaix) diagrams has been extended to the case of bidimensional layers of elements adsorbed on metal surfaces [44,45]. It allows us to predict the conditions of stability of adsorbed sulfur on metals immersed in water containing dissolved sulfur species such as S, H_2S, HS^-, $HS_2O_3^-$, $S_2O_3^{2-}$, HSO_4^-, and SO_4^{2-}. The calculated *E*-pH diagrams show that the stability domain of adsorbed sulfur extends beyond the usually predicted range of stability of metal sulfides, and thus adsorbed sulfur layers can exist under conditions in which no bulk sulfide is stable. Potential-pH diagrams have been calculated for sulfur adsorbed on surfaces of iron, nickel, and chromium, in water containing sulfides or sulfates [44–46], and in water containing sulfides and thiosulfates [47]. They are also available for copper [48].

8.5.1 Principle of *E*-pH Diagrams for Adsorbed Species

An element A (which may be O, S, N, or H) is adsorbed on a metal M in the form of a monoatomic layer, with a valence state equal to zero [denoted $A_{ads}(M)$]. The adsorption of A from a species dissolved in aqueous solution may be an electro-oxidation or an electro-reduction reaction, depending on the valence state of A in the dissolved species.

In contrast to adsorption in ultrahigh vacuum (UHV) and in gas phase, the adsorption of an atom or a molecule on a metal surface in water involves replacement of adsorbed water molecules [$H_2O_{ads}(M)$] and competition with the adsorption of oxygen [$O_{ads}(M)$] or hydroxyl [$OH_{ads}(M)$] resulting from the dissociation of H_2O molecules on the metal surface.

In a Langmuir model for adsorption, it is assumed that the two-dimensional phase is an ideal solution, where water, oxygen or hydroxyl, and sulfur adsorb competitively on the same surface sites and there are no interactions between adsorbed species.

Under these conditions the chemical potential of each element A in the phase adsorbed on a metal M can be expressed as follows:

$$\mu_{\text{Aads(M)}} = \mu^{\circ}_{\text{Aads(M)}} + RT\ln\theta_{\text{A}} \tag{8.1}$$

where

θ_A is the relative coverage of the surface of M by adsorbed A ($0 \le \theta_A \le 1$; $\theta_A = 1$ for the complete monolayer of A)

$\mu^{\circ}_{\text{Aads(M)}}$ is the standard chemical potential of A, corresponding to saturation of the surface by A

The chemical potential of adsorbed water is also given by

$$\mu_{\text{H}_2\text{Oads(M)}} = \mu^{\circ}_{\text{H}_2\text{Oads(M)}} + RT\ln\theta_{\text{H}_2\text{O}} \tag{8.2a}$$

with

$$\theta_{\text{H}_2\text{O}} = \left(1 - \sum_i \theta_{\text{A}i}\right) \tag{8.2b}$$

As an example of the method of calculation of the *E*-pH relations, let us consider the adsorption on a metal of oxygen from water:

$$H_2O_{ads}(M) = O_{ads}(M) + 2H^+ + 2e^-$$

The equilibrium potential of this half-reaction is obtained by applying the Nernst law with the chemical potentials of adsorbed oxygen and water expressed as in Equations 8.1 and 8.2:

$$E = E^\circ + \left(\frac{RT \ln 10}{2F}\right)\log\left[\frac{\theta_O}{(1-\theta_S-\theta_O)}\right] - \left(\frac{RT \ln 10}{F}\right)\text{pH} \tag{8.3a}$$

with the standard potential E° given on the standard hydrogen scale by

$$E^\circ = \frac{[\mu^\circ_{Oads(M)} + \mu^\circ_{H+} + \mu^\circ_{e-} - \mu^\circ_{H_2Oads(M)}]}{2F} \tag{8.3b}$$

with $\mu^\circ_{H+} + \mu^\circ_{e-} = \frac{1}{2}\mu^\circ_{H_2(g)}$.

Electrochemical experiments and surface analyses show that the adsorbed oxygen species in solution on most transition metals at 25°C are likely to be hydroxyls. Thermodynamic data obtained from electrochemical experiments are available only for OH_{ads} on copper [49]. Therefore O_{ads} is considered here on Fe, Ni, and Cr and OH_{ads} on Cu. The standard Gibbs energies of formation (chemical potentials) for sulfur and oxygen adsorbed on metal surfaces can be calculated [44–46] from literature thermodynamic data for reversible chemisorption at the metal–gas interface.

8.5.2 *E*-pH Relations for the Equilibria between Dissolved and Adsorbed Species

In water containing thiosulfates, the sulfur species to be considered are the thermodynamically stable sulfide species $H_2S_{(aq)}$ and HS^-, and the metastable thiosulfate species $HS_2O_3^-$ and $S_2O_3^{2-}$. The *E*-pH relations associated with the various equilibria between water, the dissolved sulfur species and adsorbed sulfur, and oxygen or hydroxyl adsorbed on Fe, Ni, Cr, and Cu at 25°C are as follows:

$$S_{ads}(M) + H_2O_{(l)} = O_{ads}(M) + H_2S_{(aq)}$$

$$\log\left(\frac{\theta_O}{\theta_S}\right) = -\text{pK} - \log m_S \tag{8.4}$$

$$\text{pK(Fe)} = 18.8;\quad \text{pK(Ni)} = 27.4;\quad \text{pK(Cr)} = 13.6$$

$$S_{ads}(M) + H_2O_{ads}(M) + H_2O_{(l)} = 2OH_{ads}(M) + H_2S_{(aq)}$$

$$\log\left[\frac{\theta^2_{OH}}{\theta_S(1-\theta_S-\theta_{OH})}\right] = -\text{pK} - \log m_S \tag{8.5}$$

$$\text{pK(Cu)} = 19.7$$

$$S_{ads}(M) + H_2O_{(l)} = O_{ads}(M) + HS^- + H^+$$

$$\log\left(\frac{\theta_O}{\theta_S}\right) = \text{pH} - \text{pK} - \log m_S \tag{8.6}$$

$$\text{pK(Fe)} = 25.8;\quad \text{pK(Ni)} = 34.4;\quad \text{pK(Cr)} = 20.6$$

$$S_{ads}(M) + H_2O_{ads}(M) + H_2O_{(l)} = 2OH_{ads}(M) + HS^- + H^+$$

$$\log\left[\frac{\theta_{OH}^2}{\theta_S(1-\theta_S-\theta_{OH})}\right] = pH - pK - \log m_S \tag{8.7}$$

$$pK(Cu) = 26.7$$

$$H_2O_{ads}(M) = O_{ads}(M) + 2H^+ + 2e^-$$

$$E^{25} = E^{\circ 25} + 0.030\log\left[\frac{\theta_O}{(1-\theta_S-\theta_O)}\right] - 0.059\,pH \tag{8.8}$$

$$E^{\circ 25}(Fe) = -0.11\,V; \quad E^{\circ 25}(Ni) = 0.04\,V; \quad E^{\circ 25}(Cr) = -0.52\,V$$

$$H_2O_{ads}(M) = OH_{ads}(M) + H^+ + e^-$$

$$E^{25} = E^{\circ 25} + 0.059\log\left[\frac{\theta_{OH}}{(1-\theta_S-\theta_{OH})}\right] - 0.059\,pH \tag{8.9}$$

$$E^{\circ 25}(Cu) = 0.095\,V$$

$$H_2O_{ads}(M) + H_2S_{(aq)} = S_{ads}(M) + H_2O_{(l)} + 2H^+ + 2e^-$$

$$E^{25} = E^{\circ 25} - 0.030\log m_S + 0.030\log\left[\frac{\theta_S}{\theta_{H_2O}}\right] - 0.059\,pH \tag{8.10}$$

$$E^{\circ 25}(Fe) = -0.67\,V; \quad E^{\circ 25}(Ni) = -0.77\,V$$

$$E^{\circ 25}(Cr) = -0.92\,V; \quad E^{\circ 25}(Cu) = -0.488\,V$$

$$H_2O_{ads}(M) + HS^- = S_{ads}(M) + H_2O_{(l)} + H^+ + 2e^-$$

$$E^{25} = E^{\circ 25} - 0.030\log m_S + 0.030\log\left[\frac{\theta_S}{\theta_{H_2O}}\right] - 0.030\,pH \tag{8.11}$$

$$E^{\circ 25}(Fe) = -0.88\,V; \quad E^{\circ 25}(Ni) = -0.97\,V; \quad E^{\circ 25}(Cr) = -1.12\,V;$$

$$E^{\circ 25}(Cu) = -0.695\,V$$

$$2S_{ads}(M) + 5H_2O_{(l)} = 2O_{ads}(M) + HS_2O_3^- + 9H^+ + 8e^-$$

$$E^{25} = E^{\circ 25} + 0.007\log\left(\frac{m_S}{2}\right) + 0.015\log\left(\frac{\theta_O}{\theta_S}\right) - 0.067\,pH \tag{8.12}$$

$$E^{\circ 25}(Fe) = 0.58\,V; \quad E^{\circ 25}(Ni) = 0.71\,V; \quad E^{\circ 25}(Cr) = 0.50\,V$$

$$2S_{ads}(M) + 5H_2O_{(l)} = 2OH_{ads}(M) + HS_2O_3^- + 7H^+ + 6e^-$$

$$E^{25} = E^{\circ 25} + 0.010\log\left(\frac{m_S}{2}\right) + 0.020\log\left(\frac{\theta_{OH}}{\theta_S}\right) - 0.069\,pH \quad (8.13)$$

$$E^{\circ 25}(Cu) = 0.76\,V$$

$$2S_{ads}(M) + 5H_2O_{(l)} = 2O_{ads}(M) + S_2O_3^{2-} + 10H^+ + 8e^-$$

$$E^{25} = E^{\circ 25} + 0.007\log\left(\frac{m_S}{2}\right) + 0.015\log\left(\frac{\theta_O}{\theta_S}\right) - 0.074\,pH \quad (8.14)$$

$$E^{\circ 25}(Fe) = 0.60\,V; \quad E^{\circ 25}(Ni) = 0.72\,V; \quad E^{\circ 25}(Cr) = 0.52\,V$$

$$2S_{ads}(M) + 5H_2O_{(l)} = 2OH_{ads}(M) + S_2O_3^{2-} + 8H^+ + 6e^-$$

$$E^{25} = E^{\circ 25} + 0.010\log\left(\frac{m_S}{2}\right) + 0.020\log\left(\frac{\theta_{OH}}{\theta_S}\right) - 0.079\,pH \quad (8.15)$$

$$E^{\circ 25}(Cu) = 0.78\,V$$

8.5.3 Potential-pH Diagrams

The preceding equations have been used to construct the potential-pH diagrams for sulfur and oxygen (hydroxyl) adsorbed in water containing sulfides (H_2S or HS^-) or thiosulfates ($HS_2O_3^-$ or $S_2O_3^{2-}$) on Fe, Ni, Cr, and Cu [44–48]. The diagrams are shown in Figures 8.13 through 8.16 at 25°C for different sulfur coverages (θ_s = 0.01; 0.5; 0.99) and a molality of dissolved sulfur $m_s = 10^{-4}$ mol/kg. The diagrams are superimposed on the S-M-H_2O diagrams (M = Fe, Ni, Cr, Cu), calculated for a molality of dissolved metal $m_M = 10^{-6}$ mol/kg.

These diagrams allow us to predict the *E*-pH conditions in which sulfur is adsorbed on a surface of Fe, Ni, Cr, or Cu in water by oxidation of sulfides or reduction of thiosulfates. In aqueous solution, when the potential is increased in the anodic direction, the adsorbed water molecules are replaced by sulfur atoms adsorbed from $H_2S_{(aq)}$ and HS^-. At higher potentials, sulfur is oxidized in $HS_2O_3^-$ or $S_2O_3^{2-}$ and replaced by adsorbed oxygen or hydroxyl. The replacement reaction is completed within a very narrow range of potential (~ 0.06 for O_{ads} and 0.08 V for OH_{ads}). The domains of stability of the adsorbed S monolayer (for $m_s = 10^{-4}$ mol/kg and $m_M = 10^{-6}$ mol/kg) overlap the domains of stability of the metals (Fe, Ni, Cr, Cu), the dissolved cations (Fe^{2+}, Ni^{2+}, Cr^{2+}, Cr^{3+}, Cu^+, and Cu^{2+}), and the oxides or hydroxides (Fe_3O_4, Fe_2O_3, $Ni(OH)_2$, NiO, Cr_2O_3, Cu_2O, CuO). The stability domains of S_{ads} are significantly wider than the stability domains of the bulk metal sulfides, and thus S_{ads} is stable in *E*-pH regions where there is no stable metal sulfide, which reflects the excess of stability of the chemisorbed state with respect to the corresponding 3D compound. On this basis, detrimental effects of sulfur on the corrosion resistance of metals are predicted even under potential and pH conditions where the metal sulfides are not thermodynamically stable. The comparison of the behaviors of the different metals in the presence of dissolved sulfur species shows that the extent of the overlap of S_{ads} with the stable metals decreases in the sequence Ni, Fe, Cr, so the effect

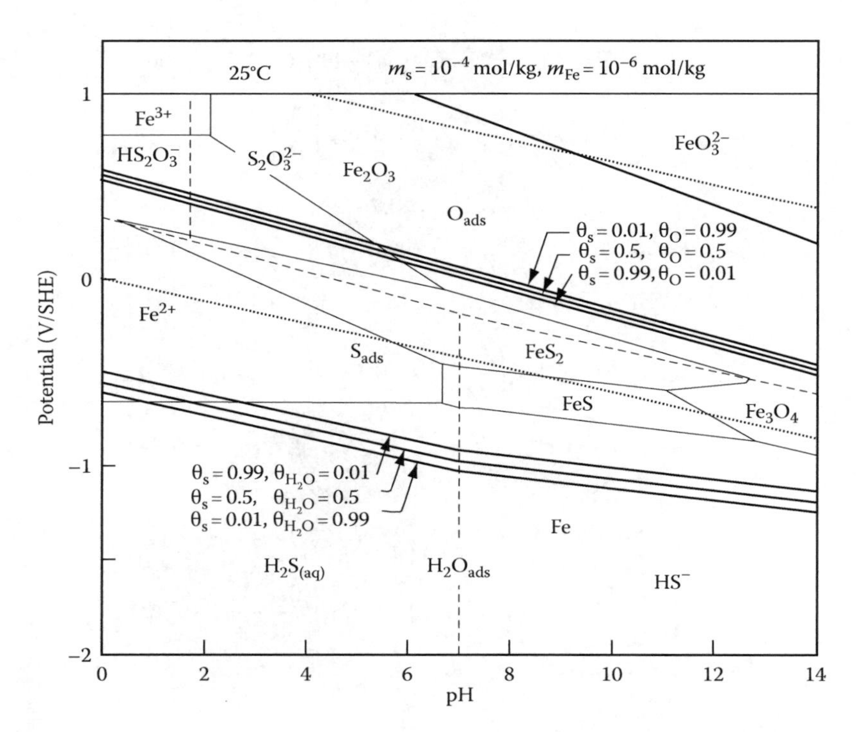

FIGURE 8.13
Potential-pH diagram for sulfur adsorbed on Fe (25°C, $m_s = 10^{-4}$ mol/kg).

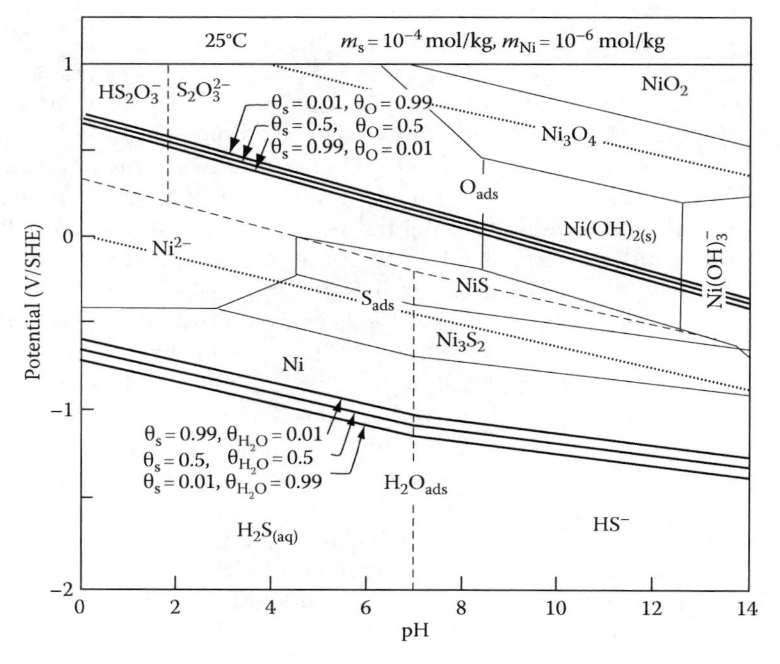

FIGURE 8.14
Potential-pH diagram for sulfur adsorbed on Ni (25°C, $m_s = 10^{-4}$ mol/kg).

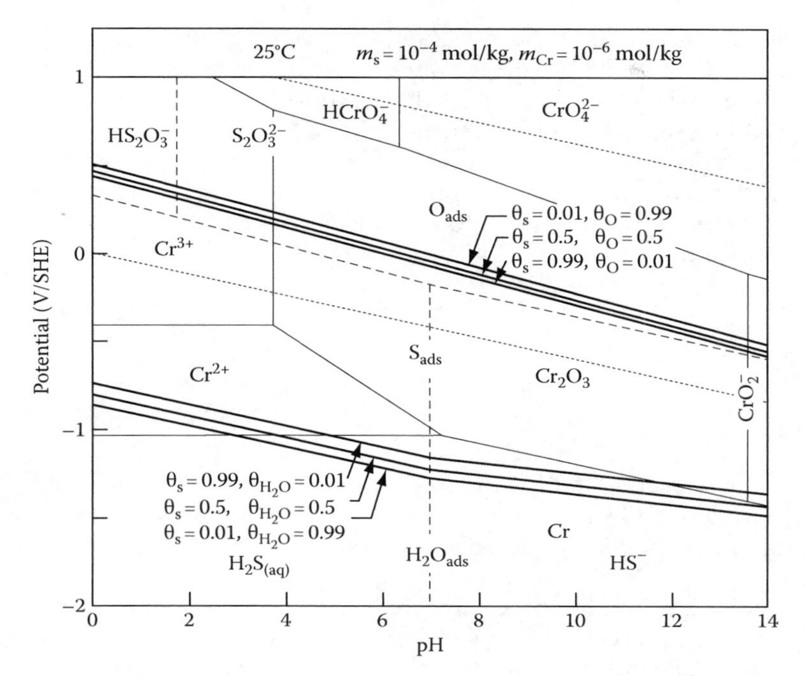

FIGURE 8.15
Potential-pH diagram for sulfur adsorbed on Cr (25°C, $m_S = 10^{-4}$ mol/kg).

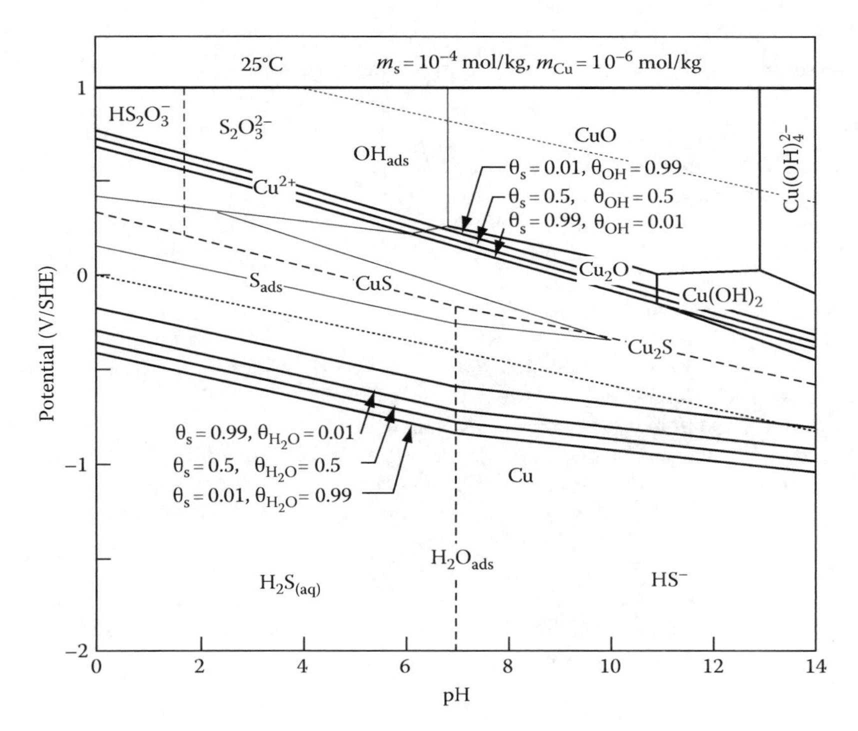

FIGURE 8.16
Potential-pH diagram for sulfur adsorbed on Cu (25°C, $m_S = 10^{-4}$ mol/kg).

of thiosulfates on corrosion of these metals is expected to decrease in the same order Ni > Fe > Cr. Prediction of S adsorption in the active domains (i.e., during anodic dissolution) is also important because this is the condition in which S-enhanced dissolution is experimentally observed. Another effect of thiosulfates expected from the stability of S_{ads} in the passive domains is the blocking or retarding of passivation (or repassivation) of stainless steels. Such diagrams are useful to assess the risk of corrosion of metals and alloys induced by adsorbed sulfur produced by electro-oxidation of sulfides or electro-eduction of thiosulfates.

8.6 Conclusion

The fundamental aspects of sulfur-induced corrosion have been reviewed. The mechanisms have been derived from data obtained on chemically and structurally well-defined surfaces using electrochemical and surface analysis techniques (^{35}S radiotracer and surface spectroscopies).

The data obtained show the direct link that exists between atomic-scale surface reactions of sulfur and macroscopic manifestations (enhanced dissolution, blocking or retarding of passivation, and passivity breakdown). The data provide the fundamental basis required to rationalize the detrimental effects of sulfur species encountered in a large number of service conditions.

References

1. P. Marcus, N. Barbouth, and J. Oudar, *C. R. Acad. Sci. Paris 280*:1183 (1975).
2. J. Oudar and P. Marcus, *Appl. Surf. Sci. 5*:48 (1979).
3. P. Marcus and J. Oudar, *Fundamental Aspects of Corrosion Protection by Surface Modification* (E. McCafferty, C.R. Clayton, and J. Oudar, eds.), The Electrochemical Society, Pennington, NJ, p. 173, 1984.
4. P. Marcus, A. Teissier, and J. Oudar, *Corros. Sci. 24*:259 (1984).
5. P. Combrade, M. Foucault, D. Vancon, J.M. Grimal, and A. Gelpi, *Proceedings of the 4th International Symposium on Environmental Degradation of Materials in Nuclear Power Systems—Water Reactors* (D. Cubicciotti, ed.), NACE, Houston, TX, p. 5, 1990.
6. D. Costa and P. Marcus, *Proceedings of the European Symposium on Modifications of Passive Films* (P. Marcus, B. Baroux, and M. Keddam, eds.), The Institute of Materials (EFC 12), London, p. 17, 1994.
7. A. Elbiache and P. Marcus, *Corros. Sci. 55*:261 (1992).
8. P. Marcus, *Advances in Localized Corrosion* (H.S. Isaacs, U. Bertocci, J. Kruger, and S. Smialowska, eds.), NACE, Houston, TX, p. 289, 1990.
9. S. Ando, T. Suzuki, and K. Itaya, *J. Electroanal. Chem. 412*:139 (1996).
10. G. Seshadri, H.-C. Xu, and J.A. Kelber, *J. Electroanal. Chem. 146*:1162 (1999).
11. H. Ma, X. Cheng, G. Li, S. Chen, Z. Quan, S. Zhao, and L. Niu, *Corros. Sci. 42*:1669 (2000).
12. H.C. Xu, G. Seshadri, and J.A. Kelber, *J. Electrochem. 747*:558 (2000).
13. V. Maurice, H. Talah, and P. Marcus, *Surf. Sci. 284*:L431 (1993).
14. V. Maurice, H. Talah, and P. Marcus, *Surf. Sci. 304*:98 (1994).
15. D. Zuili, V. Maurice, and P. Marcus, *J. Electrochem. Soc. 147*:1393 (2000).

16. P. Marcus, I. Olefjord, and J. Oudar, *Corros. Sci. 24*:269 (1984).
17. P. Marcus and I. Olefjord, *Corrosion (NACE) 42*:91 (1986).
18. P. Marcus and J.M. Grimal, *Corros. Sci. 57*:377 (1990).
19. P. Marcus and H. Talah, *Corros. Sci. 29*:455 (1989).
20. D. Tromans and L. Frederick, *Corrosion (NACE) 40*:633 (1984).
21. R.C. Newman, H.S. Isaacs, and B. Alman, *Corrosion (NACE) 38*:261 (1982).
22. A. Garner, *Corrosion (NACE) 41*:587 (1985).
23. R.C. Newman, *Corrosion (NACE) 41*:450 (1985).
24. R.C. Newman, K. Sieradski, and H.S. Isaacs, *Met. Trans. 13A*:2015 (1982).
25. R.C. Newman and K. Sieradski, *Corros. Sci. 25*:363 (1983).
26. S.E. Lott and R.C. Alkire, *J. Electrochem. Soc. 136*:973, 3256 (1989).
27. C. Duret-Thual, D. Costa, W.P. Yang, and P. Marcus, *Corros. Sci. 39*:913 (1997).
28. H.-S. Kuo, H. Chang, and W.-T. Tsai, *Corros. Sci. 41*:669 (1999).
29. T. Haruna, T. Shibata, and R. Toyota, *Corros. Sci. 39*:1935 (1997).
30. Z. Szklarska-Smialowska, *Corrosion (NACE) 28*:388 (1972).
31. G. Wranglen, *Corros. Sci. 14*:331 (1974).
32. G.S. Eklund, *J. Electrochem. Soc. 121*:467 (1974).
33. A. Szummer and M. Janick-Czachor, *Br. Corros. J. 9*:216 (1974).
34. P. Marcus, *C. R. Acad. Sci. Paris Ser. II 305*:615 (1987).
35. P. Marcus and M. Moscatelli, *J. Electrochem. Soc. 136*:1634 (1989).
36. P. Marcus and M. Moscatelli, *Mem. Et. Sci. Rev. Met. 85*:561 (1988).
37. X. Cheng, H. Ma, S. Chen, L. Niu, S. Lei, R. Yu, and Z. Yao, *Corros. Sci. 41*:773 (1999).
38. J.E. Castle and R. Ke, *Corros. Sci. 30*:409 (1990).
39. J. Stewart and D.E. Williams, *Corros. Sci.* 33:457 (1992).
40. M. Janik-Czachor, *Modifications of Passive Films* (P. Marcus, B. Baroux, and M. Keddam, eds.), The Institute of Materials (EFC 12), London, p. 280, 1994.
41. J.L. Crolet, L. Seraphin, and R. Tricot, *Met. Corros. Ind.*, 616 (1976).
42. P. Combrade, O. Cayla, M. Foucault, D. Vancon, A. Gelpi, and G. Slama, *Proceedings of the 3rd International Symposium on Environmental Degradation of Materials in Nuclear Power Systems—Water Reactors* (G.J. Theus and J.R. Weeks, eds.), The Metallurgical Society, Traverse City, MI, p. 525, 1987.
43. D.E. Williams, T.F. Mohiuddin, and Ying Yang Zhu, *J. Electrochem. Soc. 145*:2664 (1998).
44. P. Marcus and E. Protopopoff, *J. Electrochem. Soc. 137*:2709 (1990).
45. P. Marcus and E. Protopopoff, *J. Electrochem. Soc. 140*:1571 (1993).
46. P. Marcus and E. Protopopoff, *J. Electrochem. Soc. 144*:1586 (1997).
47. P. Marcus and E. Protopopoff, *Corros. Sci. 39*:1741 (1997).
48. E. Protopopoff and P. Marcus, *Corros. Sci. 45*:1191 (2003).
49. V. Maurice, H.-H. Strehblow, and P. Marcus, *Surf. Sci. 458*:185 (2000).

9

Further Insights on the Pitting Corrosion of Stainless Steels

Bernard Baroux
Grenoble Institute of Technology

CONTENTS

9.1 Introduction ... 419
9.2 From Pit Nucleation to Stable Pitting ... 421
 9.2.1 Pit Nucleation and the Effect of Passive Film ... 422
 9.2.2 Stochasticity ... 423
9.3 Electrochemical Aspects ... 423
 9.3.1 The Pitting Potential ... 423
 9.3.2 Pitting Potential Measurement ... 424
 9.3.3 Probabilistic Behavior ... 425
 9.3.4 The Repassivation Potential ... 426
 9.3.5 Critical Pitting Temperature ... 427
 9.3.6 Current Fluctuations under Potentiostatic Control ... 427
 9.3.7 Rest Potential Fluctuations ... 429
9.4 Metallurgical Aspects ... 429
 9.4.1 Effect of the Alloying Elements ... 429
 9.4.2 Metallurgical Processing ... 430
 9.4.3 The Sulfide Inclusions ... 432
9.5 Some Instances of the Effect of the Inclusions ... 434
 9.5.1 Titanium and Manganese Sulfides ... 434
  9.5.1.1 pH Effects and Pitting Sites ... 436
  9.5.1.2 Inhibitive Effect of Sulfate Ions ... 439
  9.5.1.3 Prepitting Events ... 439
  9.5.1.4 Aging Effects ... 441
 9.5.2 The Effect of the Sulfide Solubility ... 443
 9.5.3 The Effect of the Non-Soluble Inclusions ... 444
9.6 Concluding Remarks ... 445
References ... 446

9.1 Introduction

Fundamentals of the pitting mechanisms have been discussed in this book by Strehblow and Marcus and the whole question of pitting corrosion was reviewed by Szlarska-Smialowska[1] and Frankel.[2] It is intended here to shed light on some points investigated on industrial stainless steels exposed to neutral or weakly acidic chloride-containing aqueous

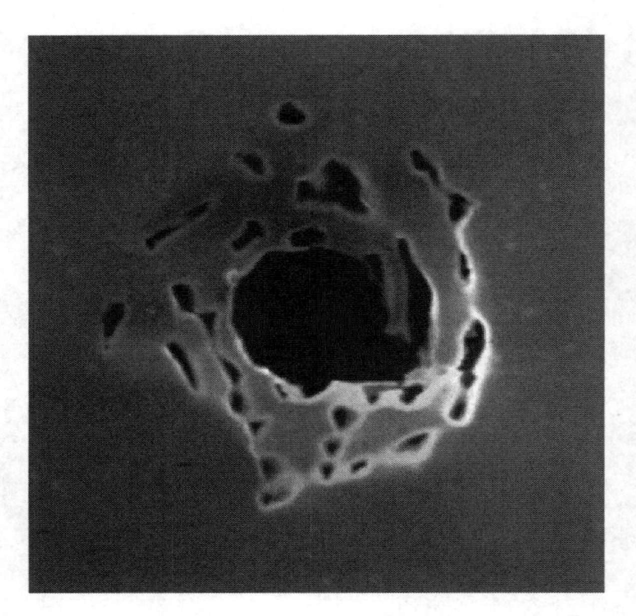

FIGURE 9.1
Typical view of a semi-developed pit (scanning electron microscopy ×1000). A thin metallic film is still present and covers part of the pit. Secondary pits are visible all around the main hole. The pit may either go on or repassivate if the thin "top" film breaks down.

media, i.e., whose pH is larger than the so-called depassivation pH, so that the passive film remains stable except in the pit itself.

Figure 9.1 shows a semi-developed pit observed at the scale of 10 μm on a 17% Cr stainless steel immersed in a NaCl-containing solution. The pit consists of an undermining hollow, covered by a thin metallic cap, leading to the formation of an occluded zone. Some secondary pits are visible all around the main hole. The possible collapse of the metallic cap can suppress the occlusion and provoke pit repassivation.

The onset of a stable pit is the result of several successive stages, from passive film breakdown (pit nucleation) which acts on the scale of some nanometers to irreversible pit growth giving macroscopic damages, which in practical situations are observed on the scale of some 100 μm. In between, pit repassivation competes with pit growth, which features metastable pitting. The larger the solution chloride content, the smaller the pitting resistance of the steel, but the precise mechanisms of action of chloride ions likely depend on the pitting stage under consideration. On the irreversible growth stage, chloride ions accelerate the anodic dissolution inside the pit. At the metastable pit stage, they prevent pit repassivation. At the nucleation stage they favor the formation of cation vacancies inside the film (which are likely responsible for pit nucleation) by extracting Cr and Fe cations from the passive film at the film electrolyte interface. Other anions (e.g., sulfate ions), increase the pitting resistance due either to competitive adsorption with chloride ions or to inhibition of the nucleation sites.

How to rank pitting corrosion resistance of industrial stainless steels remains a serious challenge, depending on the pitting stage which is considered as critical. Pit nucleation is hardly detected by standard electrochemical techniques whereas metastable pitting produces some observable potential and current transients (measuring 1 μA during 1 s corresponds to a pit size of the order of some micrometers). As far as stable pitting is concerned the corrosion resistance can be characterized by the so-called pitting potential. Furthermore, pitting initiation is recognized as being a stochastic process at the nucleation and repassivation stage and possibly also during pit growth. The result is that the pitting potential itself exhibits probabilistic behavior.

The theories proposed for accounting for pitting corrosion mainly deal with film breakdown and often fail to predict the behavior of industrial alloys, the pitting resistance of which strongly depends on their metallurgical properties. For instance, the final treatment of annealed sheets determines the properties of the passive film and then the resistance to pit nucleation. It was also shown that the pitting resistance of cold rolled sheets depends on the cold work ratio. Last, but not least, the inclusions contained in the steel (mainly the sulfides) are often more determining for the pitting resistance than the addition of expensive alloying elements since they may act as pitting sites. Their effect was evidenced both on the pitting potential and on the pitting transients observed at the metastable pitting stage. In the last section, some instances drawn from our industrial experience in these concerns will be given.

9.2 From Pit Nucleation to Stable Pitting

For simplicity, the elementary pitting process can be divided into three stages (Figure 9.2):

1. Pit nucleation, which leads to the formation of a small area of bare, un-filmed, metal (pit "nucleus") and is considered as a random process.[3] We note G (cm^2 s^{-1}) the nucleation probability per unit of time and area.
2. The development of a metastable pit, leading to the local dissolution of a significant part of the underlying metal. The dissolving metal area is referred to as a pit "embryo." The result is either the repassivation of the embryo or the formation of a stable pit (pit stabilization). The whole process leading to the formation of a stable pit (nucleation + pit stabilization) is referred to as the initiation stage. Its rate g (the pit generation rate, cm^{-2} s^{-1}) can be measured experimentally.[4] In some conditions g is proportional to the nucleation rate G and its measurement provides some information on the nucleation stage.[5]
3. The stable pit growth.

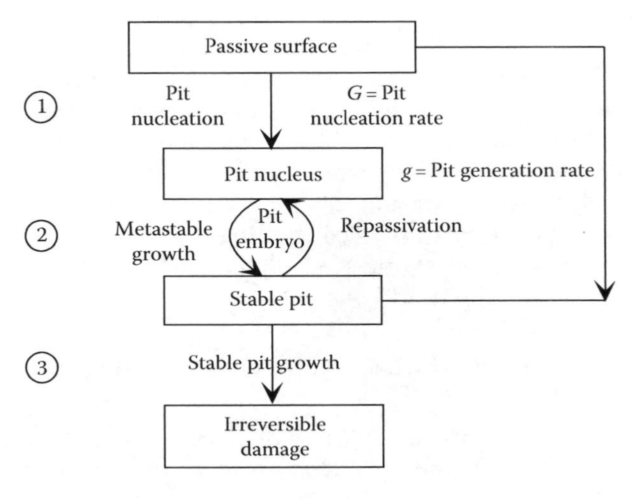

FIGURE 9.2
Multistep mechanism for the onset of a stable pit.

9.2.1 Pit Nucleation and the Effect of Passive Film

Many routes[6] may lead to passive film breakdown, e.g., local film thinning, voiding and collapse of the film, or local stress gradient.[7] It is generally accepted that some atomic sites at the film surface are more prone to pit nucleation than others, for instance at the emergence of inter-granular boundaries (as evidenced by nano-scale observations) the presence of which is thought to modify the distribution of potentials from the metal to the electrolyte. According to the point defect model, nucleation may also result from the injection of cation vacancies at the film electrolyte interface as a consequence of the formation of some MCl^{n-1} type complex, M^{+n} being either a chromium or iron cation. Due to the electric film prevailing trough the film, these vacancies are transported toward the metal film interface where they eliminate. The film breakdown may occur either from the accumulation of vacancies either at the metal film interface[8] or inside the passive film itself,[9] where they may form some voids which irreversibly grow after reaching a critical size.[10]

The passive film composition and thickness are expected to play an important role in the resistance of passive films to local breakdown but other parameters such as semiconducting properties (as discussed below) or crystalline order have to be considered (it was found for instance that charging a specimen with hydrogen[11] increases both pitting susceptibility and atomic disorder in the film). However, as these properties vary generally together from a passive film to another, it is rather difficult to determine their specific effect.

As far as film composition is concerned, the pitting resistance was found to increase with the Cr/Fe ratio in the film. It is generally admitted that the film is made of a Cr enriched oxide close to the metal film interface and a Fe enriched hydroxide close to the film electrolyte interface. Then, the larger the inner film with respect to the outer one, the larger the total Cr/Fe ratio, suggesting than the inner film likely plays the major role in the resistance to pit nucleation.

Rather short treatments (~1 h) in nitric acid drastically increase the further pitting resistance of stainless steels. It can be the result either of some modifications of the passive film (such as a chromium enrichment, as evidenced by AES and XPS measurements) or of the dissolution of harmful pitting sites.[12] In the same way, long term (some weeks) water aging[13] in non corroding conditions were found to increase both pitting resistance and average Cr/Fe ratio inside the film (which could be the result of a growth of the Cr rich inner film at the expense of the outer one).

Some workers[14] have correlated the pitting resistance and the number of pre-pitting transients (see further) to the electronic properties of the passive film considered as a n-type semiconductor and have found that the number of transients increased with the concentration in deep localized electronic traps in the band gap. Since the electronic defects likely result from the presence of atomic defects (such as cation vacancies), it suggests that the accumulation of these defects is responsible for the film breakdown. It was also found that UV irradiation increases the resistance to pit nucleation, which may result from some modifications induced in the electric field controlling the transport of vacancies.[15]

In the industrial processing of stainless steel sheets, two final annealing treatments (after cold rolling) are commonly used:

1. The so-called 2B treatment, which is performed in oxidizing atmospheres followed by an acidic pickling.
2. The bright annealing treatment ("BA"), performed in hydrogen containing atmosphere and requiring no further pickling. However, due to the presence of a small amount of residual water vapor in the annealing furnace, a thin superficial oxide (some nm), resembling to a passive film, is formed.[12]

The pitting resistance after a BA treatment is known to be better than after the 2B one.[16] Moreover, Mott–Schottky analysis shows[17] that 2B passive films are more donor-doped than BA ones, and that the density of donor states increases with chloride concentration. This is consistent with the detrimental influence of the donor levels in the band gap suggested above.[13] Considering that these donor levels are likely generated by negatively charged cation vacancies and that these vacancies are produced by the chloride ions at the passive film interface, the better resistance of the BA sheet could result from a less easy chloride adsorption than on the 2B one.[15,16]

9.2.2 Stochasticity

Nucleation and repassivation are now considered as the two steps of a birth and death initiation process.[3] It was proposed[5] that the repassivation probability (per unit of time) μ is constant up to a critical time τ_c, then nullifies (which means that a sufficiently developed pit can no longer repassivate). The pit generation rate writes then: $g = G\exp(-\mu\tau_c)$.

The question arises however to know whether the whole initiation process is Poissonian[3] in time and space, nearly Poissonian[18] or not Poissonian at all.[19,20] As concerns spatial correlations, it is expected that the probability of new pitting events in the vicinity of an existing pit[21] may be different from the pitting probability in a not pitted region.

Interactions between early dominating pits and the adjacent electrode surface were found to develop as regions of enhanced or suppressed pitting susceptibility[22,23] (triggered by concentration and potential fields developed during growth of large pits). Transient aggressive species accumulation around active pits may also cause alterations that persist for long periods after primary pits have been repassivated.

As concerns time correlation, we noted[15] that the cumulative numbers of metastable pits increase almost linearly with the logarithm of time for BA or 2B industrial surfaces (see above), which shows that the pit nucleation rate decreases with time, revealing an improvement of the pitting resistance to pit nucleation in the course of time (at least in not too severe environments).

9.3 Electrochemical Aspects

9.3.1 The Pitting Potential

Formally,[24] if J represents the anodic current and X the concentration in corrosion products in the pit, one writes: $dX/dt = KJ(V, X) - DX$, where V is the metal–electrolyte potential difference (J increases with V) and parameters K and D represent the production in corrosion products by the anodic dissolution and their dilution into the electrolyte by a diffusion process, respectively. A steady state is reached when $dX/dt = 0$ and $(\partial/\partial X)(dX/dt) < 0$, which also writes $(\partial/\partial X)J(V, X) < (D/K)$. This condition is fulfilled when V is smaller than a critical potential V_p (the "pitting potential") which depends on the bulk electrolyte concentration and on the nature of the pitting site. Finally, the pitting potential acts as a bifurcation parameter beyond which a pit grows irreversibly and below which it repassivates (as far as the diffusion parameter D does not decrease).

Beyond this formal presentation of the pitting potential, it should be noted that all factors increasing the corrosion products production (K) or decreasing their dilution in the electrolyte (D) should favor pit stabilization:

1. An increase in the K factor may come from a local increase in the active dissolution current density and should generate a local pH decrease, due to cation hydrolysis, and a local Cl^- concentration increase, as a consequence of electro-migration to support the current. The combination of the pH decrease and the Cl^- concentration increase leads then to a dissolution current increase, and so on.
2. A decrease in the D factor may result from the presence of a confined zone, for instance at the boundary between the steel matrix and an inclusion, or under a cap of primary solid corrosion products (the further breakdown of which will lead to pit repassivation); we note that a metastable pit which develops perpendicularly to the steel surface produces a confined geometry as well, so that elongated metastable pits which lived a sufficient time will not repassivate later.

It should be kept in mind that the "pitting potential" is larger than the "nucleation potential" which is the potential below which no passive film breakdown can occur and is not directly reachable by standard electrochemical measurements.

9.3.2 Pitting Potential Measurement

The pitting potential being defined as the potential below which no pit can stabilize, the question arises to define what is considered as a "stable" pit. In the most frequently used techniques, pitting is considered as irreversible when the anodic current increases up to some 10 μA, which corresponds rather to a macro-pit.

Both potentiostatic and potentiokinetic procedures can be used for measuring pitting potential.

1. In the potentiostatic method, the sample is polarized at a potential V and the induction time $\tau(V)$ for a pit occurrence is recorded. The pitting potential is the highest potential for which $\tau = \infty$. In practice, a large, but finite, value of τ is chosen. Should the pit initiation be a purely deterministic process, τ is considered as the "incubation time" necessary for the conditions for irreversible passive film breakdown to occur.
2. In the potentiokinetic method, the electrode potential is raised from the rest potential up to pitting. The measured pitting potential depends on the scanning rate s. Using a deterministic model, one finds[25]

$$\frac{1}{s}\int_{V_{rest}}^{V_{pit}} \frac{dV}{\tau(V)} = 1$$

and then the pitting potential is expected to increase when the scanning rate increases. However, for low scanning rates, the passive film may reinforce during the immersion time resulting in an increase of the incubation time ($[\partial\tau/\partial t]_V > 0$) and of the pitting potential. In practice, it is better to use large enough scanning rates (typically 100 mV min^{-1}) to avoid any interference with aging effects.

Whatever the test procedure, one must be aware of some important points:

- Pitting potential does not feature pit nucleation but irreversible pitting, including pit growth up to a size sufficient for being detected by the experiment, which in some circumstances can be rather different.
- The measured values strongly depend on the experimental procedure. For the examples given later in this chapter, and when no other conditions are specified, the tests were performed at constant temperature (23°C by default) in carefully de-aerated NaCl aqueous solution (distilled water) whose pH (6.6 by default) was adjusted to the test value by adding either NaOH or HCl. The samples were cut from thin cold-rolled sheets of thickness 0.5–1 mm as 10 mm diameter disks and were first mechanical polished under water with SiC paper grade 1200 then, in order to ensure better reproducibility of the results, aged 24 h in air before being immersed in the corrosive solution, where they are left at rest potential for 15 min before testing.
- Last, even in strictly defined experimental conditions, the pitting potential exhibits an intrinsic probabilistic behavior, as discussed below.

9.3.3 Probabilistic Behavior

Since pit nucleation and repassivation can be analyzed in terms of birth and death processes, the time τ before the occurrence of the first pit (the induction time) is a probabilistic value. Therefore, the pitting potential measured in potentiokinetic conditions is probabilistic as well.[4] For a set of N specimens tested under strictly identical conditions (potentiostatic or potentiokinetic), the pitting probability at a time t (or at a potential V) is defined as n/N, and the survival probability P (probability for no pit to occur on a single specimen) as $(N - n)/N$, with n being the number of pitted specimens. Noting $\varpi\delta S$ the pitting probability of an infinitesimal area δS of a single specimen, the survival probability of δS is $\delta P = 1 - \varpi\delta S$, ϖ being referred to as the elementary pitting probability (EPP) and its time derivative $g = d\varpi/dt$ as the pit generation rate (PGR). The survival probability $P = (1 - \varpi\delta S)^{S/\delta S}$ of the whole sample tends to exp $(-\varpi S)$ when δS tends to zero. The EPP can then be obtained from $\varpi = -LnP/S$, P being approximated by $P = (N - n)/N$. Strictly speaking, the intrinsic pitting potential is given by $g = 0$ and should be deduced from the transition from $P = 1$ to $P = 0$ for infinite surface areas S. However, this condition is not easily fulfilled in laboratory experiments, and it is often more convenient to use small surface areas and to characterize the pitting resistance by the probabilistic functions ϖ or g.

Potentiostatic measurements on a sufficient number N of samples ($N = 24$ in the following examples) or simultaneous measurements using a multichannel device, give the survival probability P and then the elementary pitting probability $\varpi(V,t)$ as a function of the polarization time t. The probability density for the random function τ is—$dP/P\,dt = Sg(V,t)$. Potentiokinetic measurements give the EPP $\varpi(V)$ and the PGR $g(V) = d\varpi/dt = s \cdot d\varpi/dV$. Figure 9.3a and b shows the EPP time dependence measured on AISI 304 type specimen for several electrode potentials. For low enough potentials, the PGR decreases with time, showing that when no corrosion occur potentiostatic holding in the corrosive solution improves the further pitting resistance. Last, the PGR for $t = 0$ shown in Figure 9.3c is the same as the PGR determined from potentiokinetic measurements with high enough scanning rates.

A conventional pitting potential V_p can be defined, for which $g_o(V_p)$ has a small, but measurable, arbitrary value (for instance 0.1 cm^{-2} s^{-1}). For practical reasons it is often more

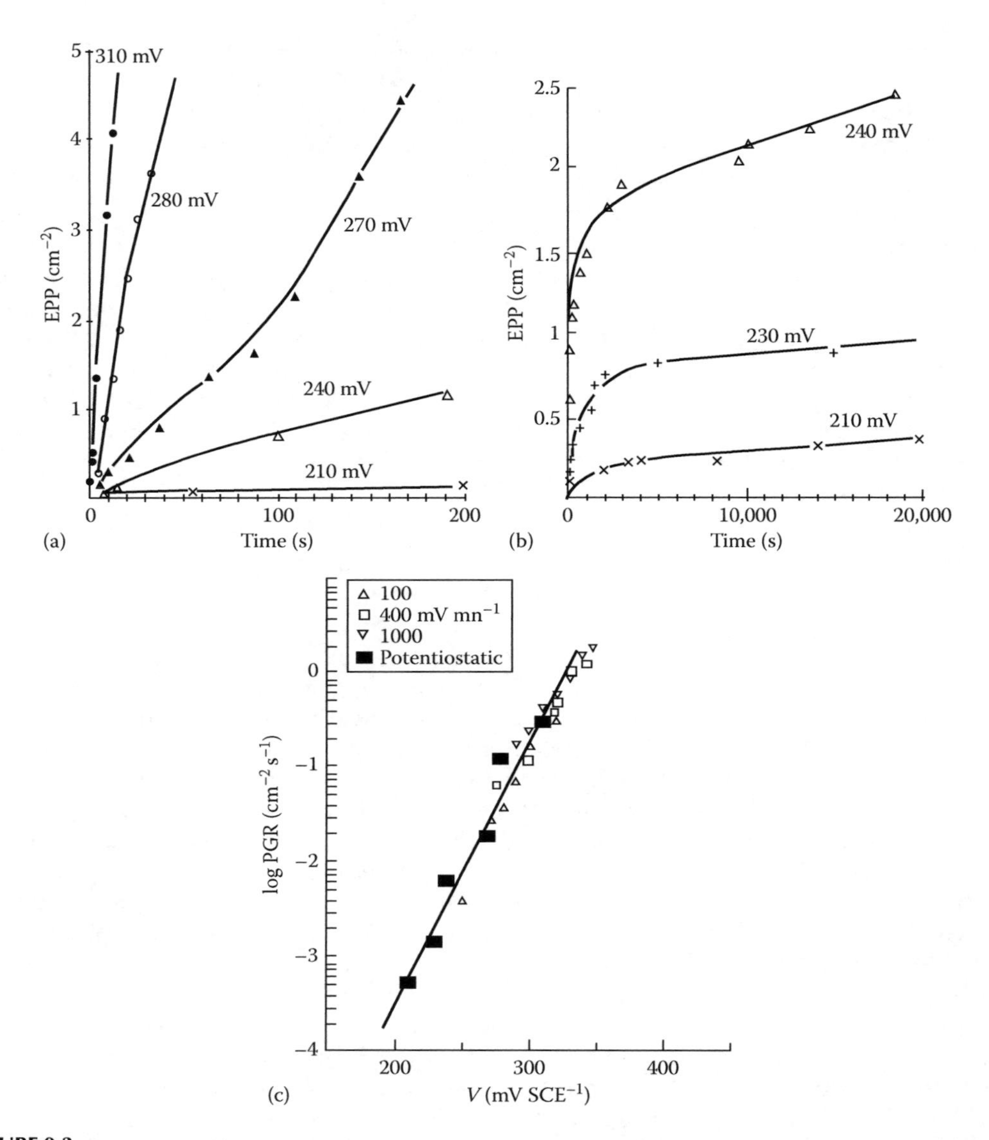

FIGURE 9.3
(a) Elementary pitting probability in NaCl (0.5 M) for an AISI 304 type stainless steel in potentiostatic conditions for different electrode potentials V (vs. SCE) and short holding times. (b) Same for long holding times. (c) Pit generation rate for $t = 0$ determined from either potentiostatic or potentiokinetic measurements.

convenient to consider the potential $V_p' \neq V_p$ for which the EPP (and not the PGR) has a conventional value ϖ_1 (for instance $\varpi_1 = 0.1\,\text{cm}^{-2}$). For high enough scanning rates, V_p' increases with s following: $\int^{V_p} g_0(V)dV = s\varpi_1$.

9.3.4 The Repassivation Potential

Another characteristic potential can be drawn from potentiokinetic measurements, namely the "repassivation" potential" V_{rep}. The electrode potential is first raised up to pitting; then, once the anodic current has reached a predetermined value i_1, the scanning is reversed toward the cathodic potentials. The anodic current then decreases and the so-called repassivation potential is obtained when the current measured during the backward scanning equals the one which was found during the forward scanning. For potentials larger than V_{rep} the pit keeps active but for smaller values it repassivates. One should keep in mind that the repassivation potential depends on the degree of pit development when the scanning is reversed, thus on the arbitrary current density i_1.

9.3.5 Critical Pitting Temperature

The temperature aids in stabilizing pitting and hindering repassivation.[26] For any steel grade, a critical pitting temperature (CPT) can be defined[27–29] below which pits cannot stabilize and then no pitting potential can be measured. The CPT can be defined within ±1°C and seems then to be a deterministic value. The more alloyed the steel is (particularly with molybdenum) the higher the CPT. In practice, CPT is an appropriate criterion to rank the pitting resistance of very resistant stainless steel grades, such as duplex grades, the CPT of which are much higher than room temperature (on the contrary of AISI 304 type grades for instance). Following some authors,[27] the anodic salt film present in a metastable pit is stable above the CPT but not below. However, the transition between the two situations does not require the properties of the salt itself to change suddenly with temperature but would rather behave as a bifurcation between two possible evolutions of a complex dynamical system.

9.3.6 Current Fluctuations under Potentiostatic Control

The existence of anodic current fluctuations under potentiostatic control at potentials lower than the pitting potential has been recognized for a long time[30–34] then revisited using some signal analysis[35,36] or microelectrodes[37] techniques.

The anodic current fluctuations are made of successive individual transients (pre-pitting events). The most frequently observed (Figure 9.4, type 1) shows a slow current increase corresponding to the pit growth, followed by a sharp decrease, likely resulting from the breakdown of a diffusion barrier (for instance the dissolution of the salt film or the breakdown of the pit cap formed by a vestige of the outer metallic surface at the pit mouth). A second type of transient (type 2) is sometimes observed, showing a fast current increase followed by a slow decrease. This situation may occur when no efficient diffusion barrier takes place and the pit repassivates shortly after its nucleation. It will be shown in the next section that the type of transient depends on the presence or not of the presence of soluble sulfide inclusions[24] acting as pitting sites in the industrial stainless steels. For the earlier workers[31,32] the observed transients were mainly of type 1, likely because they worked on MnS containing industrial steels (see below).

It was also shown that the frequency of the fluctuations decreases with the steel chromium content and the polarization time and increases with the solution chloride content. In potentiokinetic tests[32] this frequency increases with the scan rate but the dependence is not consistent with the assumption of a nucleation frequency depending only on the electrode potential frequency.

Some workers[34] approached the problem in a different way and suggested the existence of a Lotka–Volterra oscillator acting between some active and passive regions at the steel surface, the nature of which being not clearly identified. Others[38] considered the appearance of the current transients as a result of the potential-assisted formation of a salt film

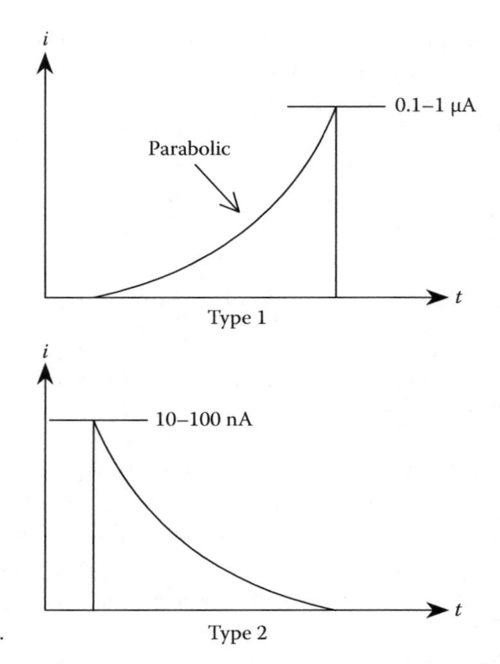

FIGURE 9.4
The two types of anodic transients in potentiostatic conditions.

(instead of the protective oxide film). This is not far from the idea[39,40] following which the pre-pitting noise is associated with halide island formation on the passive film. Another idea[41] is that this noise is the result of a two-step initiation mechanism, involving a competition between the OH^- and Cl^- adsorption and the formation of a transitional halide complex which promotes the dissolution. However, it is likely that such a mechanism does not act at the same scale as the pitting transients described above, since these transients clearly involve the dissolution of a significant part of the metallic substrate in the course of metastable pit development.

Cao et al.[42] analyzed the power spectral density (PSD) of the pre-pitting noise on some AISI 304 (MnS containing) and 321 (Ti-bearing, then MnS free) steels but made no reference to the difference in the nature of the inclusions present in the two steels. For 321 steel, the pre-pitting transients were found to be basically of type 2 and for 304 of type 1. The PSD was found to vary as f^{-n} at the high-frequency limit, with n = 2–4 following the electrode potential. A white noise (no frequency dependence) is found at very low frequency (some 0.1 Hz). From this work, it is also inferred that the solution chloride content affects the nucleation frequency but not the growth or the repassivation kinetics of the micro-pit.

Working on 304 type steels in acidic media, other workers[37,43] confirmed that the current increased as the square of time, up to some 100 nA. The use of 50 μm diameter electrodes[44] evidenced some spikes with heights of the order of 10–100 pA after either a slow current increase followed by a sharp decrease or a quick current increase followed by a relatively slow decay. For these authors a pit does not repassivate if the product of the pit depth and current density exceeds a minimum value (for maintaining a sufficiently aggressive solution at the dissolving surface), which in agreement with former theoretical works.[45] However, this minimum is generally not achieved in the first stages of pitting, and the pit growth requires the presence of a barrier to diffusion at the pit mouth, which is thought to be either a remnant of the passive film or a vestige of the outer surface of the metal itself. Rupture of this "pit cap" leads to repassivation of the metastable pit.

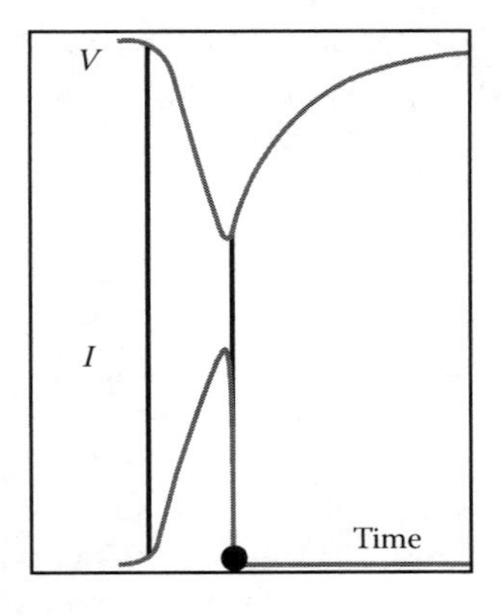

FIGURE 9.5
Schematic potential (*V*) and intensity (*I*) transients for a 304 type steel at open circuit. (From Berthome, G. et al., *Corros. Sci.*, 48, 2432, 2006.)

9.3.7 Rest Potential Fluctuations

In open-circuit conditions, metastable pitting results in rest potential fluctuations as well, although this question is less documented in the literature than the potentiostatic current transients. Analyzing the PSD of the rest potential fluctuations on a 304 type steel, some authors[46] evidenced however two results of great metallurgical importance: (a) The potential fluctuations are more pronounced for cold-rolled specimens than for solution-annealed ones, and (b) in the 0.05–2 Hz frequency range, the PSD amplitude increases with the steel sulfur content.

The individual rest potential transients consist of a rather sharp potential fall followed by a rather slow return to the stationary value.[47] Following Hashimoto and colleagues[48] the minimum of potential would correspond to repassivation and the potential recovery would reflect the repassivation kinetics. On the contrary Isaacs stated[49,50] that the potential recovery is the effect of the recharging of the capacitance of the surrounding passive surface. This last interpretation was confirmed by simultaneous measurements of the current and potential time evolution at rest potential,[16] using a specifically designed three electrodes experimental setup. Figure 9.5 shows schematically what happens for a AISI 304 type steel in situation of metastable pitting at rest potential. The onset of metastable pitting provokes both a potential fall and a corrosion current increase. After a time, the corrosion current decreases and finally nullify. At almost the same time, the potential reaches its minimum value. After this minimum, the corrosion current is null (the pit is then repassivated) and the potential slowly increases up to its stationary value, reflecting the capacitance recharging of the neighboring surface. The time constant for the potential recovery is determined by the transfer resistance and the capacitance of the passive film. The role of the impedance of the passive surface around the pit was studied in details following another way,[51] suggesting that a micro-pit growth is in fact under the control of the electrode area.

9.4 Metallurgical Aspects

9.4.1 Effect of the Alloying Elements

Chromium is obviously the major element which controls pitting of stainless steels,[52] but it is not the one. Other alloying elements (such as Ni, Mo, and Cu) or impurities (such as S) are known to play important roles, but these roles are often complex and not always fully identified. One should consider differently the elements which enter the passive film (and then may act on pit nucleation) and those which do not. In the first category one finds the minor oxidizable elements (such as silicon and aluminum) present at the level of some tenths of a weight % in the industrial steels steel, but able to concentrate much more in the passive films formed in appropriate conditions (e.g., after bright annealing). The elements which do not enter the passive film can lower the ionic current throughout the metal–passive film interface at the nucleation stage but they are believed to act mainly through their effect on the dissolution rate (a decrease of which favoring the pit repassivation in its competition with anodic dissolution) and the chemistry inside the pit (by preventing too large a pH decrease, for instance).

The sulfur content is a determining parameter due to the formation of sulfides, which may act as pitting sites (see later). The free machining steels, such as AISI 303 type austenitic steels, contain 0.2%–0.3% sulfur, which forms some numerous and large manganese sulfides. These inclusions are known to dramatically improve the steel machinability, but also to decrease a lot the pitting resistance. AISI 430 or 304 grades contain less than 0.03% S, which leads to less numerous and smaller sulfides than in AISI 303 grade. Among them, the grades used for long products (bars and wires) typically contain 0.02% S to ensure a sufficient machinability. In contrast, the grades used for flat products (sheets and plates) generally contain less than 30 ppm S, which leads to a better pitting resistance. The precise role of Mn sulfides in pit initiation has been extensively studied and will be detailed further. The sulfide dissolution provides some harmful sulfur species which are thought to play a major role in pit initiation. It was shown[53] that some alloying elements as Cu, Mo, and even Ni may combine with these sulfur species and decrease their noxious effect.

The role of Mo is a matter of discussion and may be multiple. First, this element do not enter the passive films formed on 434 or 316 type steels but its effects on the active dissolution rate (and possibly the pit chemistry) are recognized. Secondly, it can counteract the harmful effect of sulfur species by favoring[54] sulfur species desorption (possibly through combined state with Mo), the adsorbed uncharged S reducing finally to soluble H_2S.

The effect of Copper as alloying element is either beneficial or detrimental depending on the pitting stage into consideration[55] and no relevant information can be drawn from only considering its effect on the pitting potential:

1. When Cu ions are dissolved into solution, they inhibit the detrimental effect of dissolved sulfur species on pit initiation, via the formation of insoluble copper sulfides.
2. During the acidic dissolution of the steel, a residual copper film is formed,[56] accelerating the cathodic reaction and inhibiting the dissolution reaction. Copper addition decreases then the steel dissolution rate at the pit bottom and therefore also the pit propagation rate.
3. When the potential at pit bottom is more cathodic than that needed for the copper dissolution in the pit electrolyte, metallic copper hinders pit repassivation. A consequence is that the effect of copper depends indirectly on the Ni content: since Ni increases the potential at the pit bottom, it favors the dissolution of copper ions rather than their precipitation in the metallic state.

9.4.2 Metallurgical Processing

The metallurgical processing of the steel (hot and cold rolling, heat treatments, welding, surface treatments) may dramatically modify the pitting resistance of a steel grade as well. The effect of heat treatments on the phase equilibrium, recrystallization, and minor element segregation will not be discussed here, but some known or less known evidences must be recalled. First, the steel generally contains some nonmetallic phases such as precipitates (Cr carbides in ferritic steels, for instance) or inclusions (oxides, sulfides, etc.). The difference between precipitates and inclusions is that the former are produced in the metal solid state and the latter during (or at the end of) the melting process, in the metal liquid state. Then, at the contrary of precipitates (e.g., the chromium carbides), the inclusions do not produce significant segregation at their interface with the metallic matrix. However, they can be unstable in the corrosive medium, which is observed for sulfides and even for oxides in some cases.[57] Furthermore, because their ductility is generally not the same as that of the metallic matrix, hot and cold rolling may produce some micro-decohesions at the metal–inclusion interface, which behave as microcrevices forming preferential pitting sites, depending on the cold rolling ratio. Anyway, it is now well established that, in industrial steels, inclusions may act as pitting sites. On the other hand, chromium carbides may provoke chromium depletion in the metal around the precipitate, after a welding treatment or even after the metal cooling at the end of a hot transformation process. Even when these local negative segregations are too weak to produce inter-granular corrosion, they may lead to a decrease in the pitting resistance. For this reason, the final annealing must be carefully controlled, mainly for ferritic steels (due to the low carbon solubility in bcc structures).

In order to optimize their drawability, the stainless steels sheets are the more often delivered in as annealed conditions. It was seen above that the annealing treatment (or solution treatment for the austenitic grades) which takes place at the end of the industrial process is of major importance for the further pitting resistance. When this treatment is performed in oxidizing atmosphere, some Cr and Fe oxides are produced and a Cr depleted zone is formed under the oxide, which have to be removed by a final pickling treatment (generally in acidic conditions). When it is performed in hydrogen-based atmospheres (hydrogen or most often cracked ammonia), no further pickling is needed but the dew-point of the annealing atmosphere markedly influences the passive film properties, as discussed above.[13] The composition and the properties of the resulting passive films depend then strongly on the applied treatment which, together with a change in the metallic inclusion distribution at the metal surface, noticeably affects the pitting resistance.[16,17] In the same way, some nitric acid decontamination treatments are sometimes performed after at the end of the process and modify the surface properties as well. Last, the surface condition in the delivery state also depends on the surface roughness, which is itself controlled by the rolling conditions.

When large ultimate or yield strengths are needed, cold rolled stainless steels can be used without any final annealing. The pitting resistance in the as rolled condition is found lower than after annealing, but the effect depends on the pitting stage under consideration. On 304 type steel, the pit initiation frequency shows a maximum[58] after 20% cold-rolling reduction. This maximum is observed on both austenitic and ferritic grade, contradicting the hypothesis of a direct effect of strain induced martensite, the number of stable pits and the pit propagation rate increase monotonously with cold rolling reduction, while pit repassivation ability decreases (leading to a larger number of stable pits). The pitting potential decreases when the cold rolling rate increases, characterizing then stable pitting more than pit nucleation. It was suggested that pit propagation and the onset of stable pitting are controlled by the overall dislocation density whereas the nucleation frequency is related to dislocations piling-up (which could in turn generate a highly defective structure[59] in the passive film, favoring pit nucleation). The addition of alloying elements[60] (such as copper) would then affect the pitting resistance not only for chemical but also for mechanical reasons (such as their effect on the staking fault energy and then dislocations piling up), confirming the decisive role of the microstructure in pitting corrosion of industrial alloys.

9.4.3 The Sulfide Inclusions

Sulfide inclusions play a major role in pitting initiation even for the low sulfur 430, 434, 304, 316 type grades in which sulfur is considered as an impurity. For all these standards and more generally for all Ti-free stainless steels, sulfur is combined with Mn to form manganese sulfides (MnS) which act as pitting sites in chloride containing corrosive media. Rapid solidification or laser surface melting[5] provokes the removal or at least the redistribution of the Mn sulfides, resulting in a strong improvement in pitting resistance, but is not an industrial process. It is worthy to note that the so-called passivation treatments performed in nitric acid not only not only improve the passive film resistance and decontaminate the industrial surfaces from possible ferric pollution but also partially remove the MnS inclusions, improving then the pitting resistance for another reason.[12]

Following the sulfur content of the steel and the metallurgical processing, MnS is found either isolated or stuck to other inclusions (mainly oxides) or even combined in complex inclusions. Chromium may partly substitute for Mn to form some (Mn,Cr) S sulfides, whose Cr content depends on the Mn content of the steel[53] (currently 0.4% in ferritic steels and 1% or more in austenitic ones). The higher the chromium content in the sulfide, the higher the pitting resistance of the steel, resulting in a better pitting resistance for low Mn steels, at least when other metallurgical parameters are not modified by lowering the Mn content.

The case of free machining steels, as for instance the AISI 303 standard, whose the sulfur content can reach 0.3% or more is worthy to mention since their improved machinability is generally paid by a poor corrosion resistance, with however the advantage to form easy systems for the experimental study of the effect of sulfides! Standard stainless steels sheets contain commonly less than 30 ppm S and their sulfides are then much smaller and less numerous, remaining however a key factor for pitting resistance. The steels used for long products industry (bars and rods) are sometimes slightly enriched in sulfur (less than 0.03%) in order to improve a little bit their machinability. Their pitting resistance is then intermediate between the resistance of the low sulfur containing steels used for thin sheets and the one of high sulfur free machining steels.

Although being polarized together with the surrounding metal, sulfide inclusions should be unstable in the potential range where the metal is passive,[61] and then readily dissolve, providing some local chemical and electrochemical conditions which differ from ones which prevail on other parts of the surface. Several models have been proposed for the dissolution of MnS inclusions. The first idea[65,68] is that in acidified media MnS forms hydrogen sulfide, HS-ions or elemental sulfur according to:

$$MnS + 2H^+ \rightarrow H_2S + Mn^{+2}$$

$$MnS + H^+ \rightarrow HS^- + Mn^{+2} + e^-$$

$$MnS \rightarrow S + Mn^{+2} + 2e^-$$

$$H_2S \leftrightarrow H^+ + HS^-$$

$$HS^- \leftrightarrow H^+ + S + 2e^-$$

Direct action of water to form sulfates, sulfites and particularly thiosulfates have also be considered[62,63]

$$MnS + 4H_2O \rightarrow Mn^{+2} + SO_4^{-2} + 8H^+ + 8e^-$$

$$MnS + 3H_2O \rightarrow Mn^{+2} + HSO_3^- + 5H^+ + 6e^-$$

$$2MnS + 3H_2O \rightarrow 2Mn^{+2} + S_2O_3^{-2} + 6H^+ + 8e^-$$

$$S_2O_3^{-2} + H_2O \leftrightarrow S + SO_4^{-2} + 2H^+ + 2e^-$$

The last two reactions involving the formation and then dismutation of $S_2O_3^{2-}$ and resulting in the formation of adsorbed sulfur are confirmed by some photo-electrochemical investigations and photo-electrochemical microscopy[35] which shows the deposition around the inclusions of a ring of material[64] deduced to be sulfur. All these reactions are potential and/or pH dependent and then can occur or not according to the actual conditions in, or close to, the pit embryo. It is worth noting that they do not explicitly take into account the effect of chloride ions for the dissolution of Mn^{2+}. Cl^- ions are thought to adsorb preferentially on the inclusion[65] due to its higher electronic conductivity, giving stronger electrostatic image forces than on the surrounding oxide film.[53] In our opinion, Cl^- adsorption is followed by the potential-assisted formation of the adsorbed complex $MnCl^+$, which dissolves in the aqueous solution according to $(MnCl)^+_{ads} + Cl^- \rightarrow Mn^{2+} + 2Cl^-$. Because there is some evidence that pitting initiation on MnS containing steels is strongly dependent on the solution pH (see next section), it is suggested that Cl^- and OH^- adsorption compete on Mn sulfide surfaces, resulting in pH-dependent dissolution kinetics. The MnS dissolution results then[66] in local electrolyte enrichment in Mn^{2+} which, beyond a critical level, can form a $MnCl_2$ salt layer which may prevent repassivation. The question of what happens at the metal–salt layer interface has been discussed[67] and it was suggested that a FeCr oxychloride could form, whose properties and also possible remnance once the salt layer has dissolved, could play a determining role in repassivation mechanisms.

The recognition that manganese sulfides are the least resistant pit initiation sites in standard industrial stainless steels has motivated severe control of the steelmaking process either by lowering the sulfur content (when machinability is not the major criterion) or the Mn content or, better, by adding small amounts of titanium (Ti > some 0.1% by weight or more). In Ti bearing steels the MnS formation is prevented since Ti sulfides (Ti_xS) are formed at higher temperature during the melting process. Ti sulfides are considered less harmful than MnS for pitting corrosion but are also able to act as pitting sites in more severe conditions.

The earlier theories[68] assumed that MnS dissolution locally provokes the formation of a virgin metal surface. Following other workers, complete MnS dissolution is not needed, as the pits preferentially initiate at the metallic matrix–MnS interface. It was once suggested[69] that a Cr depleted zone could take place around sulfides (since Mn sulfides also incorporate Cr) but this idea is now rejected.[71] Following modern approaches, sulfide solubility seems to be a key factor with the dissolved sulfur containing species[71–75] species re-depositing around the inclusion and sensitizing the passive film to pit nucleation. Figure 9.6 summarizes this view, which leaves open the question of the exact nature of these sulfur dissolved species (elemental sulfur, thiosulfate ions, or others) and of their effect on pit initiation. The first idea is that they adsorb onto the passive film close to the inclusion, which would favor pit nucleation following a mechanism still under discussion. An alternative idea is that $S_2O_3^-$ would prevent the repassivation after the pit nucleation events (local film

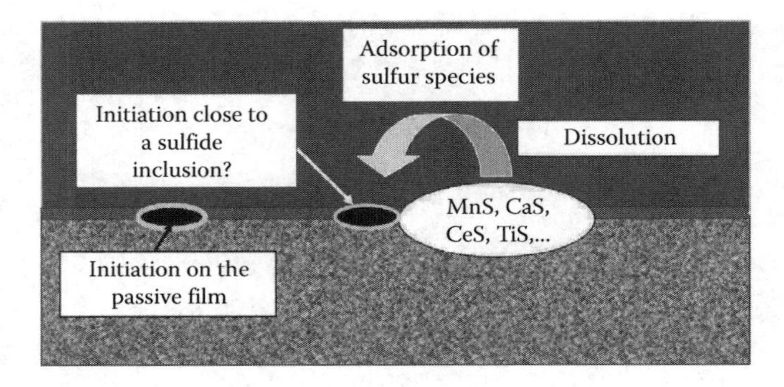

FIGURE 9.6
Schematic modeling of the effect of sulfides: First the sulfides dissolve more or less quickly, depending on their stability and of the solution corrosivity. Then dissolved sulfur containing species redeposit on the passive film close to the sulfide inclusion, promoting local film breakdown or preventing film healing after pit nucleation.

breakdown) which normally occur when the electrode potential is high enough but are generally followed by repassivation when far from dissolving sulfides.

However, whatever the details of the mechanisms, pitting initiation occur more easily close to a sulfide inclusion than far from it, and the pitting sensitivity increases with the inclusion solubility.[71,76] Such a model explains why Ti containing steels (then MnS free) are more resistant that MnS containing ones, since the Ti sulfide solubility is smaller than the one of MnS. The effect of other type of sulfides confirms this view, as illustrated in the last section of this chapter.

9.5 Some Instances of the Effect of the Inclusions

9.5.1 Titanium and Manganese Sulfides

In the following, the steels under consideration (Table 9.1) are mainly some annealed AISI 430-type FeCr ferritic alloys containing also either Nb (steels A and A′) or Ti (steels B and B′) additions. Also, some solution-treated 304 and 321 AISI-type steels were considered (C and D), in order to separate the matrix and inclusion effects. These two steels are austenitic and Ni bearing but the latter contains Ti and is MnS free, whereas the former contains MnS inclusions.

Except for steel A, Al additions are used as deoxidizing agents during the melting process, which leads to Al ~0.030% and induces the presence of Al_2O_3 inclusions. Note the sulfur content, which is easily attained with modern steelmaking techniques, and the presence of

TABLE 9.1
Steel Composition in wt% or (ppm)

	Cr	Ni	Si	Mn	Ti	Nb	(S)	(O)	(N)
A	16.4		0.4	0.45		0.5	40	240	n.d.
A′	15.7		0.4	0.45		0.7	50	340	340
B	16.8		0.4	0.45	0.4		30	260	140
B′	17.4		0.4	0.45	0.4		30	250	110
C	17.6	8.2	0.4	1.4			20	530	470
D	17.8	9.3	0.4	1.3	0.4		30	310	170

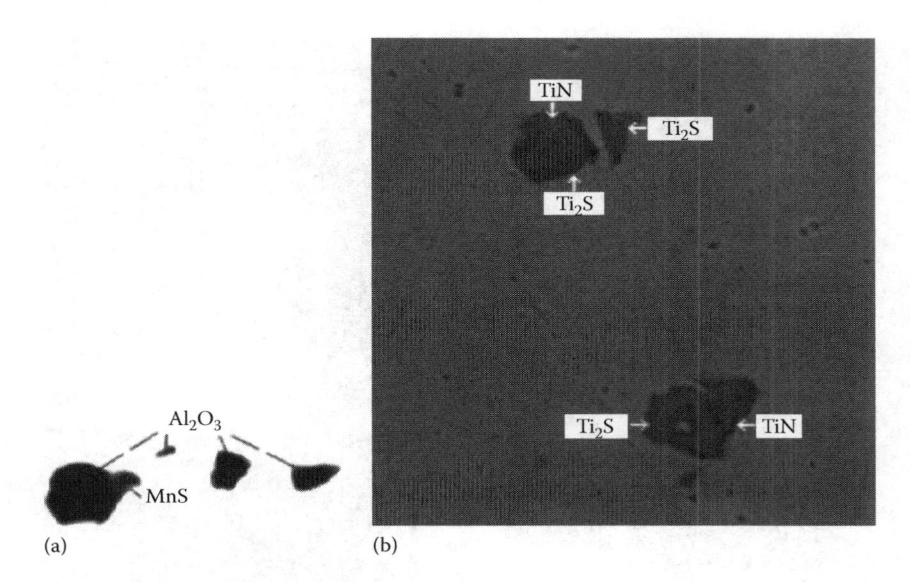

FIGURE 9.7
Non metallic inclusions (before pitting). Scanning electron microscopy (×3000). (a) Steel A′: MnS nucleated on an aluminum oxide and (b) steel B: Ti nitride surrounded by sulfur compounds.

some stabilizing elements (Ti or Nb), which trap carbon and avoid the formation of chromium carbides. For steels A and A′, sulfur is trapped under the form of manganese sulfide. The difference between steels B and B′ is their Cr content; both contain Ti, which also traps nitrogen in the form of titanium nitrides and sulfur in the form of Ti sulfides; this trapping occurs during the steelmaking process before the steel solidification. Concerning steel A′, some Mn sulfides are found around aluminum oxides (Figure 9.7a). Because these oxides have a poorer ductility than the metallic matrix, the cold-rolling process provokes some micro-decohesions around the inclusion, where some manganese sulfides are often located, leading to the possible formation of noxious microcrevices. This phenomenon is not observed on Al-free steels, where oxides are mainly malleable silicates. However, the art of the steelmaker consists in avoiding the formation of Cr oxides, which could produce a similar but worse effect than Al oxides. In every case, some Mn sulfides are also found closely stuck to Nb carbonitrides (Figure 9.8a). Because these carbonitrides precipitate at ~1200°C, this shows that (for the sulfur content considered), the high-temperature sulfur solubility is sufficient for MnS precipitation to occur in the solid steel (lower than 1200°C) while for higher sulfur contents MnS is generally considered to precipitate at the end of the steel solidification. Note that very few isolated MnS precipitates are found, probably because Nb(C,N) act as precipitation sites. This situation contrasts with the behavior of steel C (304), for which (1) no carbides may nucleate the MnS precipitate and (2) sulfides can precipitate in austenite at the delta→gamma transformation temperature.

Figures 9.7b and 9.8b show the location of Ti sulfides on steel B, i.e., around the titanium nitrides, embedded in a Ti carbide belt. Detailed examination suggests that at high temperature a homogeneous Ti carbosulfide belt precipitates around the Ti nitrides (which formed in the liquid steel). Then, lowering the temperature, titanium sulfides and carbides separate at some points as shown in Figure 9.7b, producing the Ti carbide TiC and a Ti sulfide, identified in some cases (using electron diffraction) as the hexagonal phase Ti_2S. As another result, note that Al oxides have been identified in the core of the Ti nitrides (which is not known in the figures), playing the role of nuclei for TiN precipitation.

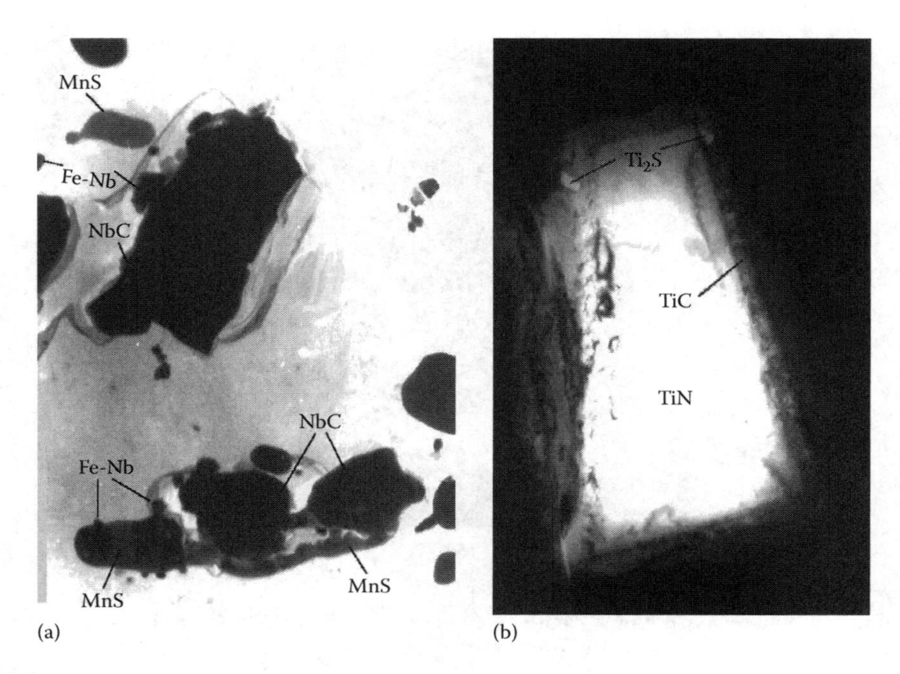

FIGURE 9.8
Non metallic inclusions (before pitting). STEM (×18,000) on thin foils. (a) Steel A′. MnS nucleated on a Nb carbonitride. Intermetallic (Fe, Nb) phases are also observed. (b) Steel B. Ti nitride surrounded by a Ti carbide, in which some Ti sulfides are embedded.

9.5.1.1 pH Effects and Pitting Sites

The pitting potentials were measured for all these steels using the potentiokinetic method in NaCl aqueous media whose pH was varied from 3 to 6.6. The results for steels A′ and B in NaCl (0.02 M) are shown in Figure 9.9a. No pH dependence is observed for steel B. In contrast, for steel A′, a sharp pitting potential decrease is evident when the pH is lowered below a critical value pH_c ranging between 4.5 and 5. Since the main difference between the two steels is the presence or absence of MnS, and Ti sulfides are known to have better stability in aqueous electrolytes than MnS, one can assume that this decrease is due to the pH-assisted MnS dissolution. Note that the same phenomenon is observed (Figure 9.9b) when comparing the two steels C and D, which are respectively MnS containing the MnS free, showing that the observed effect is related to the nature of sulfides present in the steel and not to other metallurgical factors. Figure 9.9b also shows the results obtained for steels, A and B′, which confirm the above findings. The slight difference between steels B and B′ may be attributed to the difference in Cr concentration. The strong difference between steels A and A′ may be due to the difference in Cr concentration and sulfur contents but also to the Al oxides present in steel A′ and around which MnS are found. The effect of the solution chloride content was also investigated between 0.02 and 0.5 M. Figure 9.9c shows that for high enough chloride concentrations, a pitting potential pH dependence is found even for Ti-containing steels, suggesting that Ti sulfides are not so stable in such electrolytes. The discontinuity of the pitting potential versus pH variations should then be related to the dissolution of sulfur species, which occurs easily for MnS-containing steels (whatever the steel matrix composition) but only for high enough chloride contents for Ti-bearing steels. Note that the critical pH which is found ($4.5 < pH < 5$) is the same for Ti-free and Ti-bearing steels and is close to the one deduced from the potential-pH equilibrium

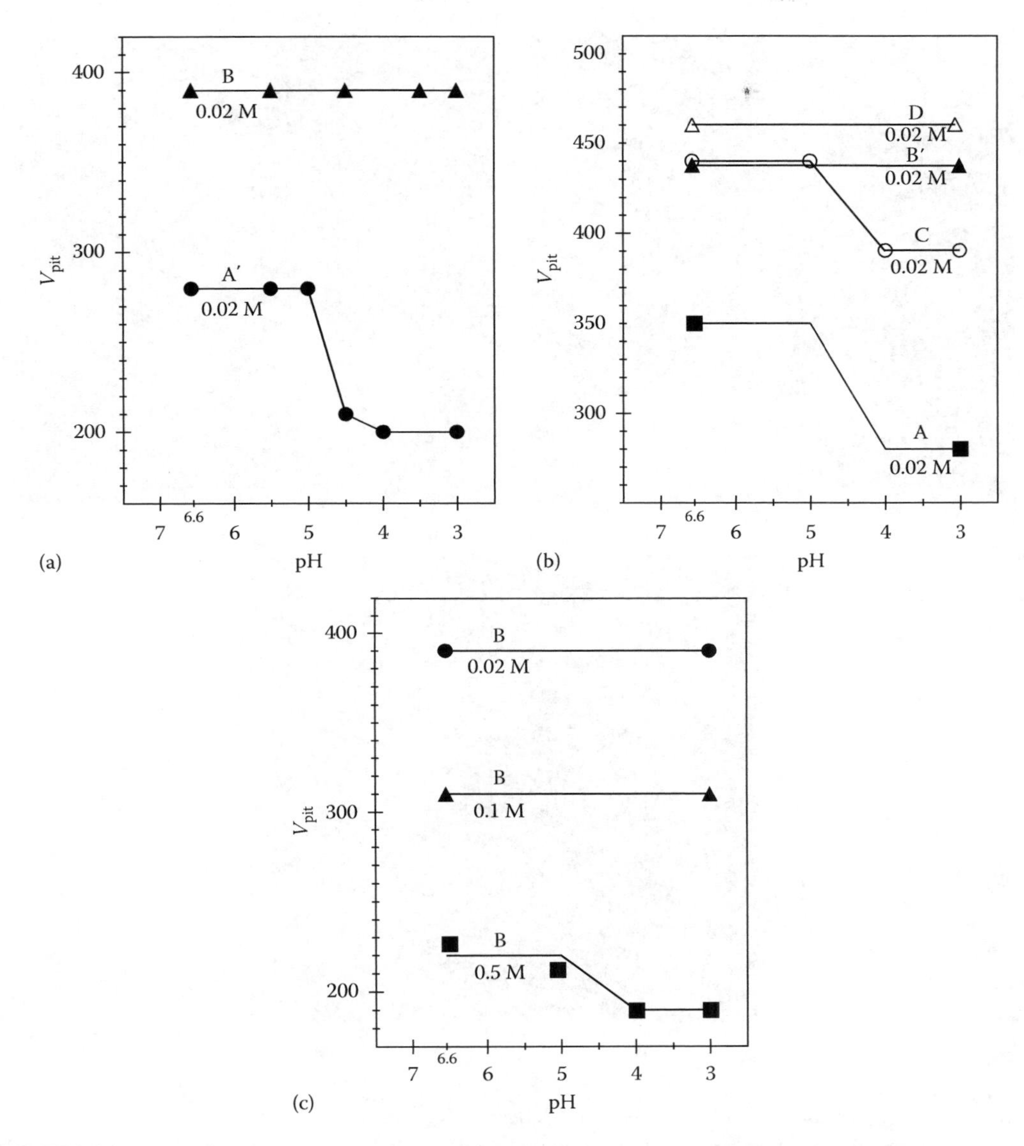

FIGURE 9.9
Effect of the solution pH on the pitting potentials. (a) Steels A′ and B in NaCl (0.02 M). (b) Steels A, B′ (FeCr steels) and C, D (FeCrNi steels) in NaCl (0.02 M). (c) Effect of the solution chloride content dependence for a Ti containing steel. For low chloride contents (0.02 and 0.1 M), there is no pH effects. For 0.5 M, lowering the pH decreases the pitting resistance.

diagrams in some chloride-containing aqueous solutions.[53,72] This supports the idea of a critical pH in the sulfur species-pH equilibria, regardless of the nature of the dissolved cation (manganese or titanium), once the conditions are reached for the sulfide to dissolve.

Figure 9.10 shows the typically observed pit initiation sites. For steel A′, whatever the pH or the chloride content, pits initiate either around Al oxides (Figure 9.10a), where some MnS is located, or on MnS inclusions (Figure 9.10b) around Nb(C,N) or (seldom) isolated. This confirms the preceding hypothesis. For steel B in NaCl (0.5 M) solution, pitting generally occurs at the TiN boundary (Figure 9.10c), where Ti sulfides are present. This shows

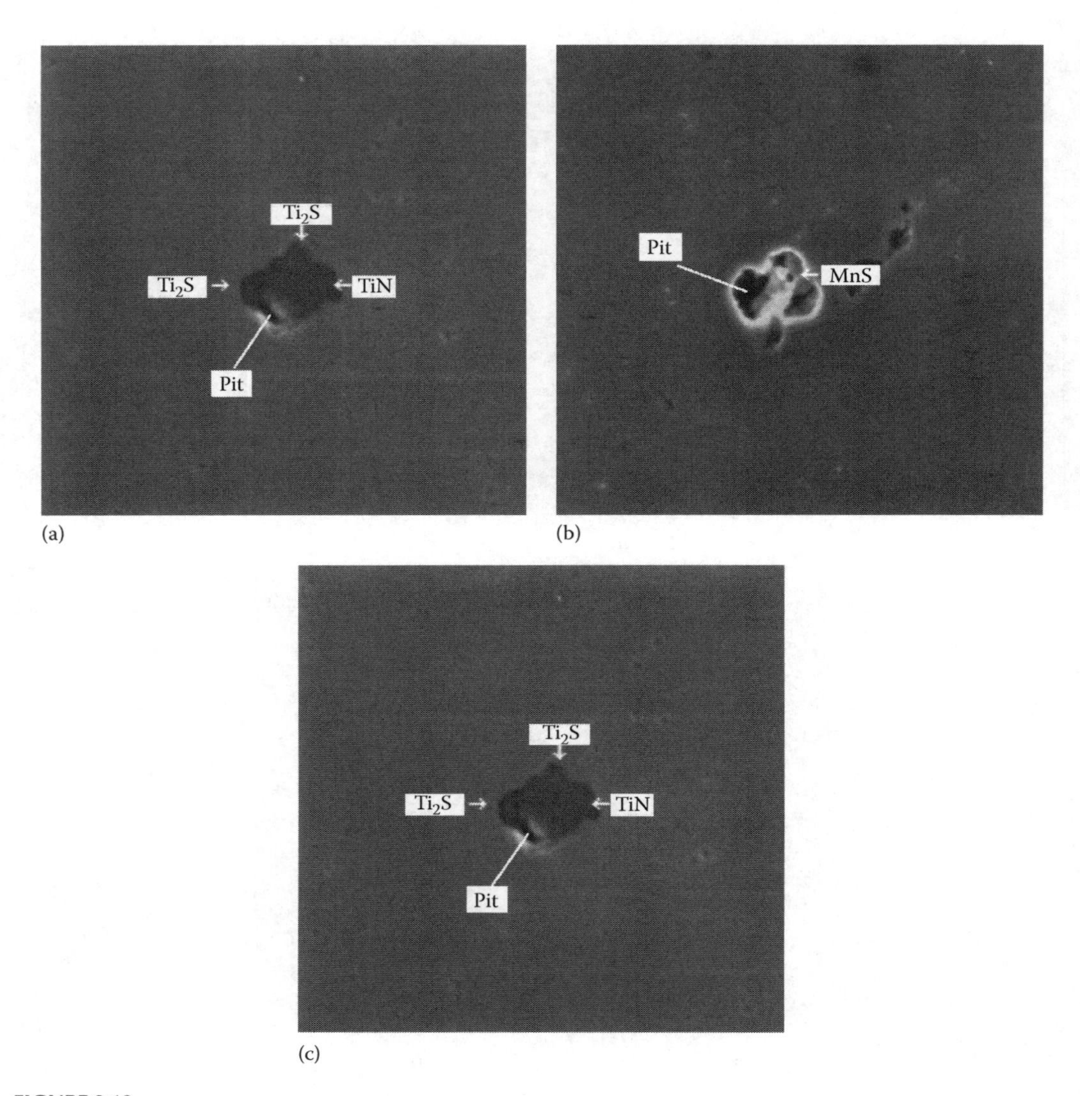

FIGURE 9.10
Scanning electron microscopy (×3000) on pitted samples. (a) Steel A′. Pitting at the boundary of an alumina inclusion. (b) Steel A′. Pitting on a manganese sulfide. (c) Steel B. Pitting at the boundary of a Ti nitride.

clearly that for such chloride concentrations, Ti sulfides act as pitting sites and becomes unstable when the potential increases. For lower chloride concentration (0.2 M NaCl), the situation is not so clear and pits could initiate directly on the metallic matrix, with no direct relation to nonmetallic inclusions, or in some cases on titanium nitrides. In the two situations, however, Ti sulfides do not seem to act as pitting sites, which is consistent with the absence of any pitting potential pH dependence.

A practical consequence can be drawn from these results: Ti-stabilized steels exhibit better pitting resistance than Ti-free ones, provided that the corrosive medium is not too severe, i.e., the chloride content does not exceed a critical value, depending on the solution pH, which could cause the Ti sulfides destabilization. In the case of crevice corrosion, the situation is more complex. Inside a crevice, the pH decreases and the chloride concentration increases with time. It is an accepted idea that corrosion occurs when the pH becomes lower than a critical value, which is referred to as the depassivation pH. However, at least for the less resistant steels, pitting corrosion may occur in the crevice before this general depassivation. From these results, it is concluded that MnS-containing steels are much more sensitive to this pitting-induced crevice corrosion than Ti-bearing ones. For very severe crevices, however, the chloride content can drastically increase with time Ti-stabilized steels are no longer different from Ti-free ones.

9.5.1.2 Inhibitive Effect of Sulfate Ions

Sulfate additions in 0.1 M NaCl were found to decrease the pitting probability for the two steels steels A and B but this effect is more pronounced for steel A than for steel B, even when this difference becomes smaller when the chloride concentration increases. Furthermore sufficient sulfate additions result in a deviation from the standard exponential $\omega(V)$ law, which tends to become bimodal (two different exponential $\omega(V)$ laws for high and low electrode potentials), for high enough sulfate amounts on steel B. MnS containing steels are then clearly less sensitive to the inhibitive action of sulfate ions than MnS free ones.

9.5.1.3 Prepitting Events

Recording the anodic current $i(t)$ during a polarization below the conventional pitting potential (Figure 9.11a) provides two types of information. First, the average anodic current

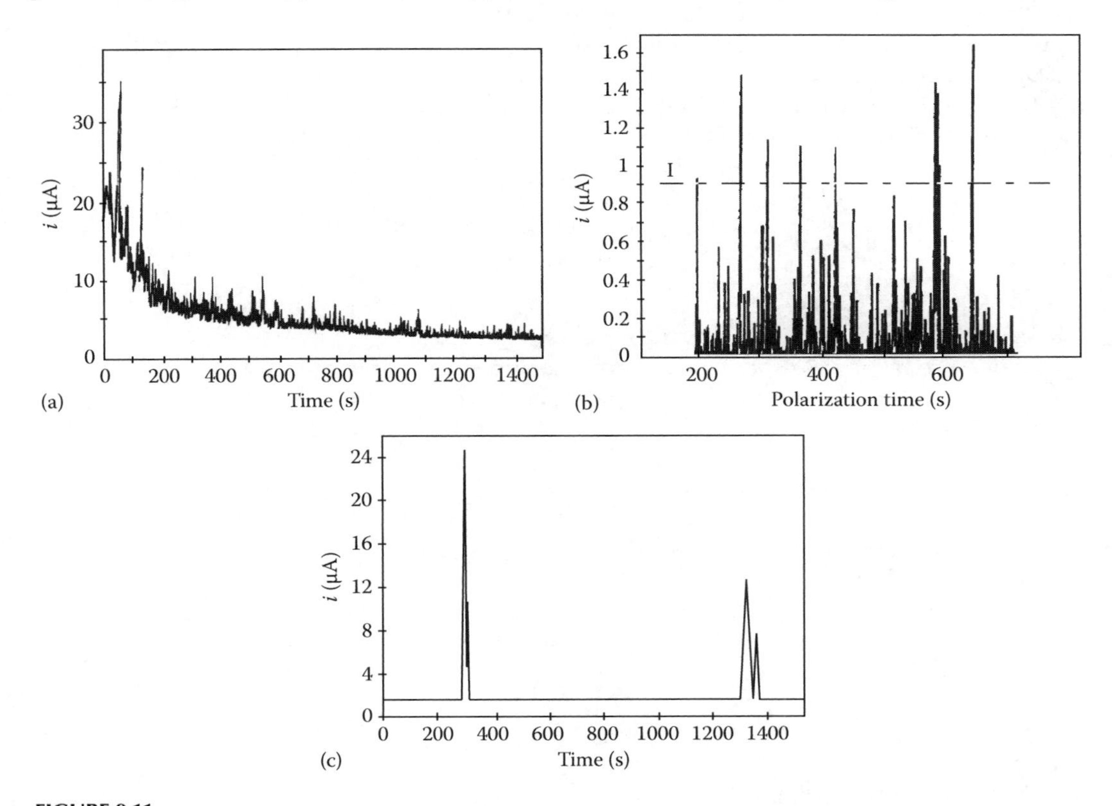

FIGURE 9.11
Prepitting events during a polarization at 200 mV SCE^{-1} in NaCl 0.02 M pH 6.6. (a) Typical anodic current variations for MnS containing steels. (b) Typical pre-pitting noise for MnS containing steels after subtracting the anodic current baseline. (c) Typical transients for Ti bearing steels.

decreases with time, probably corresponding to the onset of improved passivity. Second, some oscillations of this anodic current are observed, which, through the same SEM observations as above, were shown to correspond to some initiated and then repassivated pits. Figure 9.11b shows the typical reduced signal obtained for steel A′ by subtracting the "anodic current baseline" which likely corresponds to some passive film modifications in the course of time. For MnS-containing steels (A or A′), the prepitting events are numerous and produce what we call a pre-pitting noise. In contrast, for MnS-free steels the anodic events are much less numerous and well separated, as shown in Figure 9.11c. It is therefore logical to relate the so-called prepitting noise to the dissolution of manganese sulfides. Additionally, we note that the current time signature of the individual prepitting events is of type 1 (Figure 9.4a) for the MnS-containing steels (with a current increase varying approximately as t^2, followed by a sharp current decrease) and of type 2 (Figure 9.4b) for the MnS free ones (as expected for local passive film breakdown followed by passive film healing).

A simple way to analyze the reduced signal corresponding to the prepitting is to build up a "distribution function" $N(I)$ which counts, for a given time period (~8 min), the number of events for which the anodic intensity i is larger then I. Using this technique, one sees that the prepitting noise decreases with the polarization time and increases with the electrode potential and the solution chloride content. It also increases markedly when the pH decreases from 5 to 4 pH 4 and 5 is once more evidenced (Figure 9.12) providing then the same information that the pitting potential.

Another way to analyze the prepitting noise is to calculate the power spectral density (PSD) associated with the anodic current fluctuations. Let us consider a Poissonian series of birth and death events (birth frequency λ, death frequency μ), with a parabolic growth law between birth and death. This situation can be modeled as follows:

$$i(t) = \sum_n i_n(t) \quad \text{with } i_n(t) = \frac{\alpha}{2}(t - t_n)^2 H(t - t_n)[1 - H(t - t_n - \theta_n)]$$

where

t_n and θ_n are respectively, the induction time and the lifetime of the event number
H is the step function [$H(x) = 0$ when $x < 0$ and 1 when $x > 0$]
α is a constant

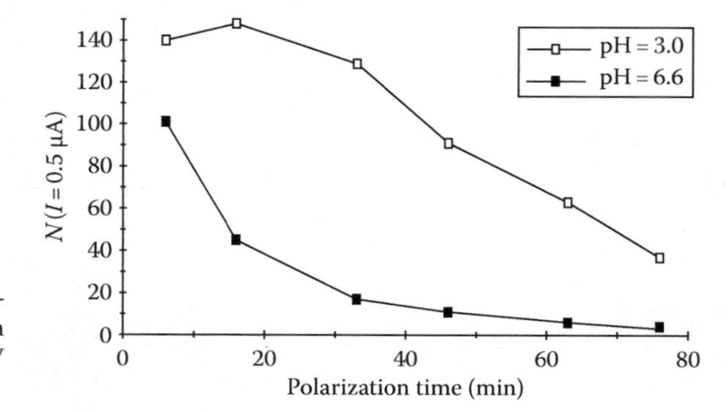

FIGURE 9.12
Effect of the polarization time t_p and the solution pH on the distribution function $N(I)$ for a MnS containing steel (A′) polarized at 200 mV SCE^{-1}) in NaCl 0.02 M.

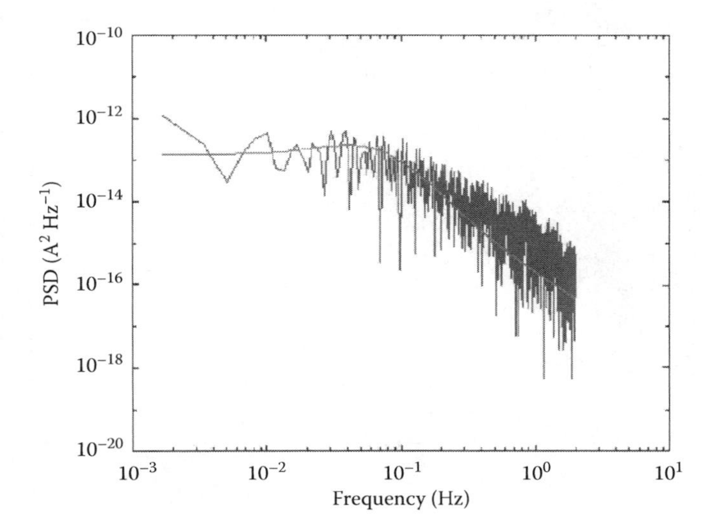

FIGURE 9.13
MnS containing steel: typical PSD vs. frequency diagram for an unitary acquisition period (512 s), $f < 2$ Hz. Polarization at 200 mV SCE^{-1} in NaCl 0.02 M, pH 6.6. $\lambda = 0.023$ Hz. $\mu = 0.43$ Hz. PSD(0) $= 1.33 \times 10^{-13}$ A^2 s.

The average current $\langle i \rangle$ and the PSD $\Psi(f)$ are found[36] to be $\langle i \rangle = \alpha\lambda/\mu^3$ and

$$\Psi(f) = \frac{\langle i \rangle^2}{\lambda} 8\mu^2 \frac{24\pi^4 f^4 + 18\pi^4 f^2 \mu^2 + 5\mu^4}{(4\pi^2 f^2 + \mu^2)^3}$$

Applying the fast Fourier transform (FFT) technique to the reduced signal $i(t)$ allows one to obtain the PSD and then the birth and death frequencies λ and μ by fitting these equations. The result evidences good agreement with the model (Figure 9.13) but the values obtained for λ and μ should be considered with caution, since they may depend on the method chosen for determining the $i(t)$ baseline.

9.5.1.4 Aging Effects

It has been shown above that the current fluctuations density for MnS-containing steels was a decreasing function of the polarization time. The first idea is that the pitting sites, having initiated an unstable pit, become inactive after pit repassivation, leading to a decrease of the available pitting sites and then of the further pits generation rate. However, it was observed that aging potentiostatically a MnS-free steel decreases the number of further prepitting events as well. Because in this case the prepitting events are very rare, the explanation above does not hold. It is believed that aging rather increases the resistance of the passive film. Figure 9.14 shows the effect of a prepolarization treatment (1 h at various potentials in the test solution itself) on the further pitting resistance. One can see that such a prepolarization increases the further pitting potential ($dV_{pit}/dV_{po1} \sim 0.5$ in any case, for $V_{po1} = -200$ to $+200$ mV SCE), regardless of the type of sulfide present in the steel. The main conclusion is that prepolarizing a sample in the corrosive solution under conditions in which pitting does not occur improves the further corrosion resistance, which is clearly related to passivity reinforcement. This shows that not only the nonmetallic inclusions but also the passive film plays a role in the pitting initiation. However, increasing the prepolarization time from 1 to 16 h shows that the effect of the aging time is not the same for the two types of steels (Figure 9.15). A 16 h potentiostatic aging is more beneficial for MnS-free

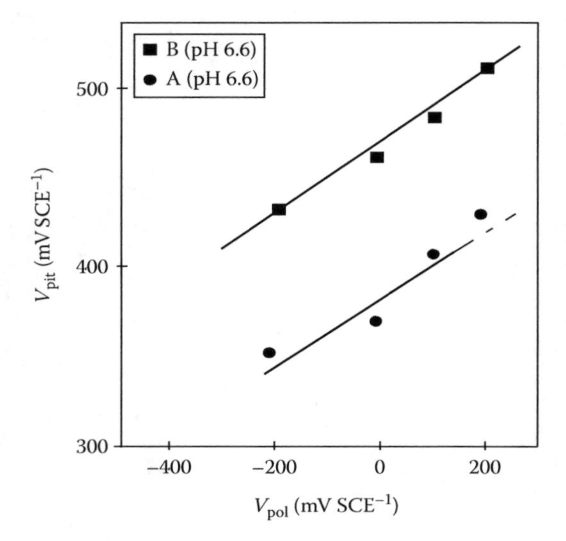

FIGURE 9.14
Effect of polarizing the samples 1 h at various potentials V_{pol} lower than the initial pitting potential, on the further pitting potential V_{pit} (potentials vs. SCE). NaCl 0.02 M, pH 6.6.

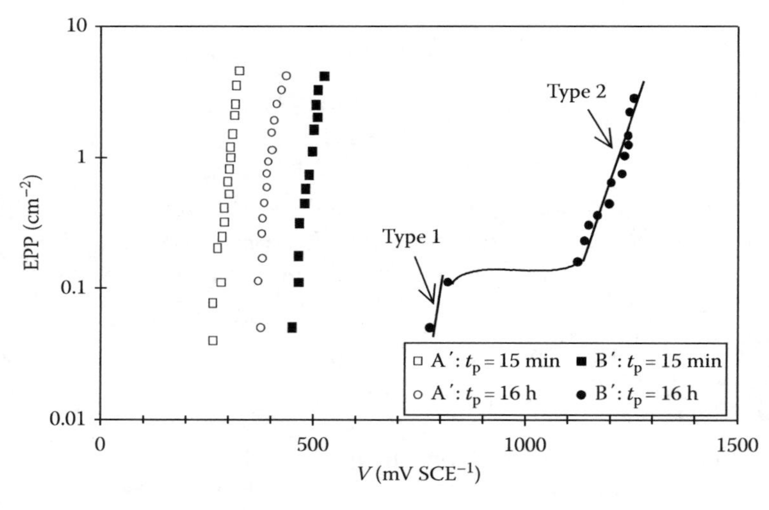

FIGURE 9.15
Effect of a 16 h aging at constant potential (steel A′: 150 mV, steel B′: 200 mV) on the elementary pitting probabilities $\varpi(V)$. NaCl 0.02 M, pH 6.6.

steels than for MnS-containing ones. It is believed that the dissolution of sulfur-containing species counteracts in the latter case the beneficial effect of the passive film reinforcement. Furthermore, in the first case (MnS-free steel), the pitting probability law $\omega(V)$ exhibits a bimodal behavior, as already observed in the case of sulfate-containing medium.

Figure 9.16 shows the effect of aging the samples for 24 h at rest potential before the pitting potential measurement in NaCl (0.02 M), pH 6.6, instead of for 15 min in the standard procedure. The rest potential evolution was recorded and found to be the same for MnS-containing and MnS-free steels (from nearly −330 mV SCE^{-1} at time zero to nearly −150 mV after 24 h). Rest potential aging does not produce any measurable prepitting events (which would result in some rest potential fluctuations), at least in these experimental conditions. However, it slightly increases the further pitting potential, a little more for the MnS-free steels than for

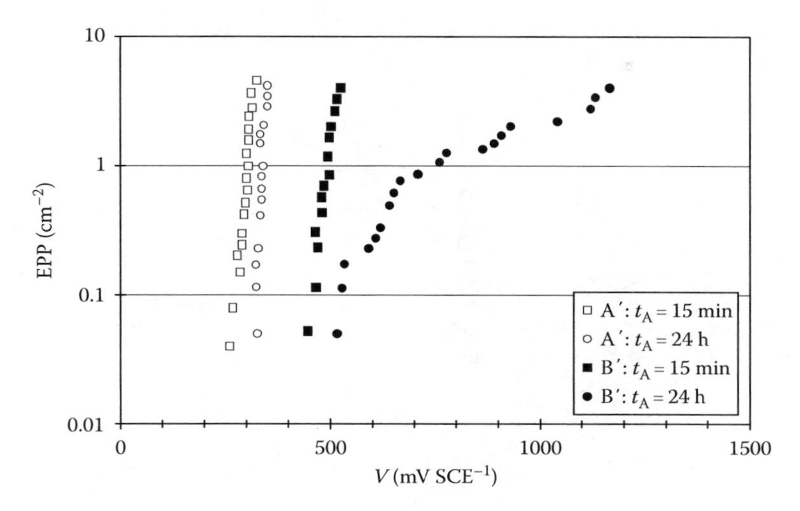

FIGURE 9.16
Effect of rest potential aging on the elementary pitting probabilities $\varpi(V)$. NaCl 0.02 M, pH 6.6.

the MnS-containing ones. This rules out the first assumption, following which the pitting potential improvement could be the consequence of the decrease of active pitting sites.

9.5.2 The Effect of the Sulfide Solubility

Due the emergence in the industrial world of some low cost, high manganese and low nickel austenitic steels (FeCrMnLNi) for which no Ti additions could be considered for some cost reasons, the question of the sulfide inclusions solubility has been recently revisited. In order to explore different possibilities (excluding titanium additions) we compared several experimental grades obtained from laboratory heats (25 kg ingots, melt under vacuum) after hot and cold processing on the form of 1 mm thick solution treated stainless sheets. All these steels had the same matrix base composition (16.4% Cr, 1.6% Ni, 7.3% MN, 0.07% C, 9 ppm S) but different additions of sulfide former elements, such as Ce and Ca, were performed. For comparison, a laboratory heat of 304 type grade was also considered.

The sulfides of the base steel (FeCrMnLNi) were almost pure MnS. Those of the 304 type grade were (Mn,Cr)S with about 5% Cr substituted to Mn in the sulfide. FeCrMnLNi + Ce steel shows only Ce sulfides and FeCrMnLNi + Ca steel shows Ca sulfides (in shell around the oxides). Thermodynamics predicts that Sulfide solubility ranges as:

$$\text{CeS} < (\text{Mn,Cr})\text{S} < \text{MnS} < \text{CaS}.$$

This point was verified by SEM examinations on diamond polished specimen before and after immersion in aerated NaCl (0.0.02 M) neutral aqueous solutions. The results are shown on Figure 9.17a and confirm the solubility ranking for the different steels under investigation. Last, pitting potentials were measured on polished specimen by the potentiokinetic method (scan rate 100 mV min^{-1}) in deaerated NaCl (0.02 M) pH 6.6 electrolyte at 23°C. The results show (Figure 9.17b) that the pitting potential increases when the sulfide solubility decreases except for the 304 grade where the beneficial influence of Ni interferes with the effect of sulfides.

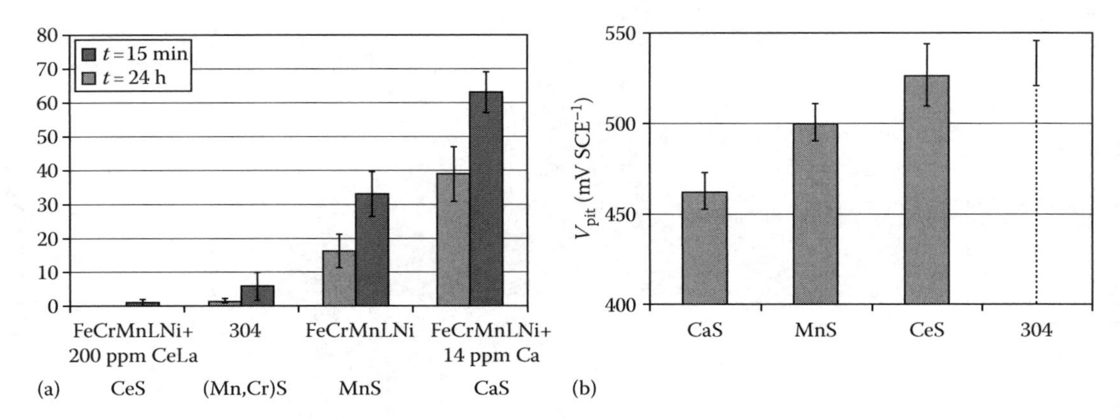

FIGURE 9.17
(a) Percentage of dissolved inclusions in a FeCrMnLNi steel after immersion 15 min or 24 h in aerated NaCl (0.0.02 M) neutral aqueous solutions (SEM examination). (b) Effect of the type of sulfide inclusions on the pitting potential in NaCl 0.02 M.

The conclusion is that the larger the sulfide solubility in the corrosive solution is, the lower the pitting resistance. The pitting corrosion resistance increases then in the same way that the solubility of sulfides: Ca < Mn < Cr < Ti < Ce. Finally, the role of the sulfides in pit initiation could be simply to provide sulfur species more or less easily (following the sulfur stability), which deposit then on the passive film around the sulfide, resulting in a locally poorer pitting resistance.

9.5.3 The Effect of the Non-Soluble Inclusions

Even non-soluble inclusions may have an influence on the pitting resistance of stainless steels. Let us consider for instance the effect of zirconium additions in a Fe17Cr AISI 430 type steel. We note the so-called Zr excess, i.e., the amount of Zr which is not trapped by C and N during the steelmaking process and is then available to form hard Zr containing intermetallic compounds at the end of the process. Figure 9.18a and b shows the correlation between δZr, the number of intermetallic compounds and the pitting potential measured on polished specimen by the potentiokinetic method (scan rate 100 mV min^{-1}) in deaerated

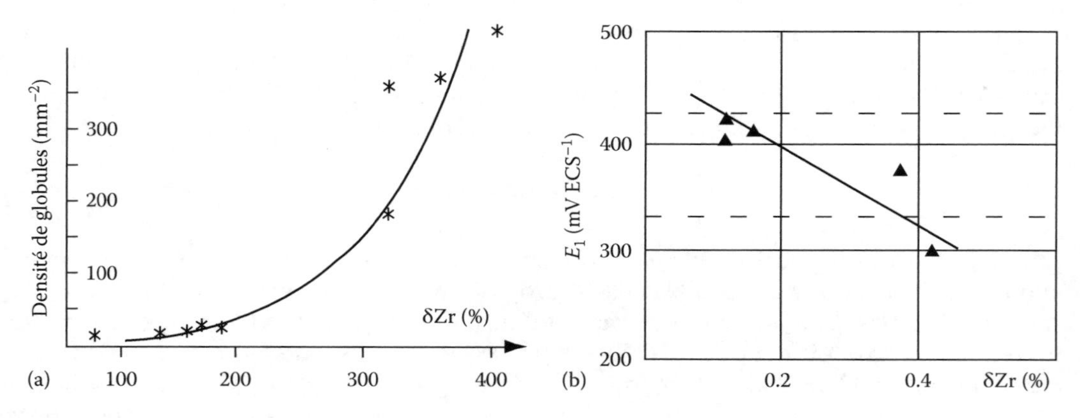

FIGURE 9.18
(a) Effect of the Zr excess in a Fe17CrZr alloy on the number of intermetallic compounds observed per unit area (mm^{-2}). (b) Effect of the Zr excess on the pitting in NaCl 0.02 M.

NaCl (0.02 M) pH 6.6 electrolyte at 23°C. The detrimental effect of these compounds on the pitting resistance is likely due to the existence of microcrevices at the metal inclusion interface, likely formed during the cold rolling due to the difference of plasticity between the inclusion and the matrix.

9.6 Concluding Remarks

The results presented in this chapter show that pit initiation on industrial stainless steels depends on a combination of electrochemical and metallurgical factors and exhibits a stochastic behavior. Moreover the pitting process acts on several successive size scales: pit nucleation (the passive film breakdown) on the nanoscopic scale (some nm), metastable pitting on the microscopic scale (some μm), which is also the size scale of the inclusions present in the steel, and pit growth on the macroscopic scale (some 100 μm when irreversible damages are observed).

As far as nucleation is concerned, pit may nucleate directly on the passive film, far from any nonmetallic inclusion. This is the case generally considered in most theoretical models, but there is no evidence that it is the prevailing situation in practice, except for ultra-pure stainless steels.[77] In industrial steels, pitting mainly occurs close to some metallurgical flaws. The first idea which comes in mind is that the passive film is less resistant in the neighborhood of the flaws than far from them. When these flaws are soluble sulfide inclusions (such as MnS), the dissolved sulfur species redeposit on the passive film around the sulfides, either favoring passive film breakdown or preventing the repassivation of pit nuclei. This phenomenon is a low-probability event, explaining that only a small number of sulfides act as pitting sites in practice.

Few things are known on the repassivation probability of metastable pits, except that this probability should nullify after a certain time, or after the pit has reached a certain size. The time required for a pit embryo to reach this critical size may depends on the metallurgical properties of the steel, such as the density of dislocations after cold working.[58] However, this time can be considered as a random variable as well[78] and therefore the statistics of pits growth should also contribute to the total probabilistic behavior.

Numerical simulation of metastable pitting is currently in progress, using for instance Monte Carlo and Cellular Automata techniques,[79] but further work is needed for some significant results to be obtained in this way. At the opposite, several significant works have been performed on numerical simulation of stable pitting. Derived from the pioneering work by Galvele,[46] the reactions between chemical species in the pit and their transport under electric field control have been formalized, give rise to a set of master equations which can be solved using the finite element method.[80–82] The conditions required for a pit to stabilize, the effect of the pit shape and its evolution, the dissolution kinetics in the near saturated pit environment prevailing inside the pit can be addressed and sometimes correlated to some experiments carried out on artificial pit electrodes.[80] Applying numerical simulation to real industrial steel compositions shows[82] that the pitting potential is controlled by both (1) the dissolution kinetics in the acidic and chloride enriched electrolyte which develops inside the pit and (2) the critical pH above which repassivation occurs.

Thanks to the development of computer tools, a motivating aim for future works could be to develop numerical simulation of pitting from pit nucleation to metastable and stable pitting, taking into account the metallurgical properties of the steel through some simple but sufficiently realistic models.

References

1. Z. Szlarska-Smialowska, *Pitting Corrosion of Metals*, Nace, Houston, TX, 1986.
2. G.S. Frankel, *J. Electrochem. Soc.*, 145 (1998) 2186–2198.
3. T. Shibata and T. Takeyama, *Corrosion (Nace)*, 33 (1977) 243 et seq.
4. B. Baroux, *Corros. Sci.*, 27 (1988) 969–986.
5. J. Steward and D.E. Williams, *Corros. ScL*, 33 (1992) 457.
6. P. Marcus, V. Maurice, and H.-H. Strehblow, *Corros. Sci.*, 50 (2008) 2698–2704.
7. V. Vignal, N. Mary, R. Oltra, and J. Peultier, *J. Electrochem. Soc.*, 153 (2006) B352–B357.
8. F. Lin, C.Y. Chao, and D.D. Macdonald, *J. Electrochem. Soc.*, 128 (1981) 1194–1107.
9. B. Baroux, *Corros. Sci.*, 31 (1990) 447–452.
10. G. Martin and B. Baroux, *Europhys. Lett.*, 5 (1988) 629–634.
11. Q. Yang and J.L. Luo, *J. Electrochem. Soc.*, 148 (2001) B29–B35.
12. J.S. Noh, N.J. Laycock, W. Gao, and D.B. Wells, *Corros. Sci.*, 42 (2000) 2069–2084.
13. D. Gorse, J.C. Joud, and B. Baroux, *Corros. Sci.*, 33 (1992) 1455–1478.
14. P. Schmuki and H. Bohni, *J. Electrochem. Soc.*, 139 (1992) 1908–1913.
15. S. Fujimoto, T. Yamada, and T. Shibata, *J. Electrochem. Soc.*, 145 (1998) L79–L81.
16. G. Berthome, B. Malki, and B. Baroux, *Corros. Sci.*, 48 (2006) 2432–2441.
17. J. Amri, T. Souier, B. Malki, and B. Baroux, *Corros. Sci.*, 50 (2008) 431–435.
18. T. Lunt, S.T. Pride, J.R. Scully, J.L. Hudson, and A.S. Mikhailov, *J. Electrochem. Soc.*, 144 (1997) 1620–1629.
19. T.T. Lunt, J.R. Scully, V. Brusamarello, A.S. Mikhailov, and J.L. Hudson, *J. Electrochem. Soc.*, 149 (2002) B163–B173.
20. T. Lunt, S.T. Pride, J. R. Scully, J. L Hudson, and A.S. Mikhailov, *J. Electrochem. Soc.*, 144 (1997) 1620–1629.
21. M. Dornhege, C. Punckt, J.L. Hudson, and H.H. Rotermund, *J. Electrochem. Soc.*, 154 (2007) C24–C27.
22. N.D. Budiansky, J.L. Hudson, and J.R. Scully, *J. Electrochem. Soc.*, 151 (2004) B233–B243.
23. N.D. Budiansky, L. Organ, J.L. Hudson, and J.R. Scully, *J. Electrochem. Soc.*, 152 (2005) B152–B160.
24. J.L. Crolet, L. Seraphin, and R. Tricot, *Mem. ScL Rev. Met.*, 74 (1977) 647 et seq. (in French).
25. J.L. Crolet, L. Seraphin, and R. Tricot, *C.R. Acad. ScL Paris*, 280C (1975) 333 et seq. (in French).
26. J.O. Park, S. Matsch, and H. Bohni, *J. Electrochem. Soc.*, 149 (2002) B34–B39.
27. N.J. Laycocq, M.H. Moayed, and R.C. Newman, *J. Electrochem. Soc.*, 145 (1998) 2622–2628.
28. N.J. Laycocq and R.C. Newman, *Corros. Sci.*, 40 (1998) 887–902.
29. N.J. Laycock, M.H. Moayed, and R.C. Newman, *Corros. Sci.*, 48 (2006) 1004–1018.
30. J.M. Defranoux, *Corros. Sci.*, 3 (1963) 75.
31. U. Bertocci and Y. Yang-Xiang, *J. Electrochem. Soc.*, 131 (1984) 1011.
32. T. Tsuru and M. Saikiri, *Corros. Eng.*, 39 (1990) 401.
33. M. Keddam, M. Krarti, and C. Pallotta, *Corrosion* (*Nace*), 43 (1987) 454.
34. J.J. Podesta, R.C.V. Piatti, and A.J. Arvia, *Corros. Sci.*, 22 (1982) 193.
35. A.R.J. Kucernak, R. Peat, and D.E. Williams, *J. Electrochem. Soc.*, 139 (1992) 2337.
36. G. Gabrielli, F. Huet, M. Keddam, and R. Oltra, *Corrosion* (*Nace*), 46 (1990) 266.
37. P.C. Pistorius and G.T. Burstein, *Corros. Sci.*, 33 (1992) 1885.
38. F. Hunkeler, G.S. Frankel, and H. Bohni, *Corrosion* (*Nace*), 43 (1987) 189.
39. R. Doelling and K.E. Heusler, *Z. Phys. Chem.*, 139 (1984) 39.
40. T. Okada, *J. Electrochem. Soc.*, 132 (1985) 537.
41. T. Okada, *Corros. Sci.*, 26 (1986) 839.
42. C. Cao, Q. Shi, and H. Lin, *Bull. Electrochem.*, 6 (1990) 710.
43. P.C. Pistorius and G.T. Burstein, *Mater. Sci. Forum*, 111–112 (1992) 429.
44. G.T. Burstein and S.P. Martin, *Philos. Mag. Lett.*, 66 (1992) 127.
45. J.R. Galvele, *J. Electrochem. Soc.*, 123 (1976) 464–474.

46. S. Magaino, *J. Electroanal. Chem.* 258 (1989) 227.
47. H.S. Isaacs, *Corros. ScL,* 29 (1989) 313.
48. M. Hashimoto, S. Miyajima, and T. Murata, *Corros. Sci.*, 33 (1992) 885.
49. H.S. Isaacs, *Corros. ScL,* 34 (1993) 525.
50. H.S. Isaacs and Y. Ishikawa, *J. Electrochem. Soc.*, 132 (1985) 1288.
51. R. Oltra, G.M. Indrianjafy, and R. Roberge, *J. Electrochem. Soc.*, 140 (1993) 343.
52. B. Baroux, F. Dabosi, C. Lemaitre, *Stainless Steels* (P. Lacombe, B. Baroux, and G. Beranger, eds.), EDP, Les Ulis, France, 1993, pp. 314 et seq.
53. G. Wranglen, *Corros. Sci.*, 14 (1974) 331 et seq.
54. A. Betts and R.C. Newman, *Corros. Sci.*, 34 (1993) 151 et seq.
55. T. Sourisseau, E. Chauveau, and B. Baroux, *Corros. Sci.*, 47 (2005) 1097–1117.
56. K. Ogle, J. Baeyens, J. Swiatowska, and P. Volovitch, *Electrochim. Acta*, 54 (2009) 5163–5170.
57. M.A. Baker and J.E. Castle, *Corros. Sci.*, 33 (1992) 1295 et seq.
58. L. Peguet, B. Malki, and B. Baroux, *Corros. Sci.*, 49 (2007) 1933 et seq.
59. A. Barbucci, G. Cerisol, and P.L. Cabot, *J. Electrochem. Soc.*, 149 (2002) B534–B542.
60. L. Peguet, B. Malki, and B. Baroux, *Corros. Sci.*, 51 (2009) 493–498.
61. J.L. Crolet, L. Seraphin, and R. Tricot, *Mem. Sci. Rev. Met.*, 73 (1976) 703 (in French).
62. S.E. Lott and R.C. Alkire, *J. Electrochem. Soc.*, 136 (1989) 973.
63. E.G. Webb, Richard, and C. Alkire, *J. Electrochem. Soc.*, 149(6) (2002) B272–B295.
64. D.E. Wilhiams, T.F. Mohiuddin, and Y.Y. Zhu, *J. Electrochem. Soc.*, 145(8) (August 1998) 2664–2672.
65. R. Ke and J.E. Castle, *Corros. Sci.*, 30 (1990) 409.
66. M.A. Baker and J.E. Castle, *Corros. Sci.*, 34 (1993) 667.
67. U. Steinsmo and H.S. Isaacs, *J. Electrochem. Soc.*, 140 (1993) 643.
68. E.G. Webb, T. Suter, and R.C. Alkire, *J. Electrochem. Soc.*, 148(5) (2001) B174–B195.
69. D.E. Williams and Y.Y. Zhu, *J. Electrochem. Soc.*, 147(5) (2000) 1763–1766.
70. P. Schmuki, H. Hildebrand, A. Friedrich, and S. Virtanen, *Corros. Sci.*, 47 (2005) 1239–1250.
71. B. Baroux, D. Gorse, and R. Oltra, Critical factors in localized corrosion IV, *Proceedings of ECS 202nd Meeting*, Salt Lake City, UT, 2002, pp. 335 et seq.
72. G.S. Eklund, *J. Electrochem. Soc.*, 121 (1974) 467.
73. C.H. Paik, H.S. White, and R.C. Alkire, *J. Electrochem. Soc.*, 147(11) (2000) 4120–4124.
74. I. Muto, Y. Izumiyama, and N. Hara, *J. Electrochem. Soc.*, 154(8) (2007) C439–C444.
75. H. Krawiec, V. Vignal, O. Heintz, R. Oltra, and J.M. Olive, *J. Electrochem. Soc.*, 152(7) (2005) B213–B219.
76. I. Muto, D. Ito, and N. Hara, *J. Electrochem. Soc.*, 156(2) (2009) C55–C61.
77. C. Boulleret, J.-L. Pastol, J. Bigot, B. Baroux, and D. Gorse, C7-415, *J. Phys.* IV (Nov. 1995) 5.
78. E.E. Mola, B. Mellein, E.M. Rodriguez de Schiapparelli, J.L. Vicente, R.C. Salvarezzo, and A.J. Arvia, *J. Electrochem. Soc.*, 137(5) May (1990) 1384–1390.
79. B. Malki and B. Baroux, *Corros. Sci.*, 47 (2005) 171–182.
80. N.J. Laycock and S.P. White, *J. Electrochem. Soc.*, 148 (2001) B264–B275.
81. S. Scheiner and C. Hellmich, *Corros. Sci.*, 49 (2007) 319–346.
82. B. Malki, T. Souier, and B. Baroux, *J. Electrochem. Soc.*, 155 (2008) C583–C587.

10

Crevice Corrosion of Metallic Materials

Pierre Combrade
Framatome

CONTENTS

10.1 Introduction ... 450
10.2 Basic Mechanisms of Crevice Corrosion ... 451
10.2.1 Crevice Corrosion due to *IR* Drop ... 452
10.2.2 Crevice Corrosion due to Local Changes of Anodic Behavior of the Material ... 452
10.2.3 Initiation and Propagation ... 452
10.2.4 Parametric Effects ... 453
10.2.4.1 Crevice Geometry ... 453
10.2.4.2 External Surfaces ... 454
10.2.4.3 *IR* Drop ... 454
10.3 Phenomenology of Crevice Corrosion of Passive Alloys in Aerated Chloride Environments ... 455
10.3.1 Effect of Potential: Critical Potentials for Initiation and Repassivation ... 456
10.3.2 Effect of the Environment ... 458
10.3.3 Geometric Factors ... 458
10.3.4 Alloy Composition ... 460
10.3.5 Propagation of Crevice Corrosion ... 461
10.4 Processes Involved in Crevice Corrosion of Passive Alloys in Aerated Chloride Environments ... 461
10.4.1 Environment Evolution in the Crevice Gap ... 461
10.4.1.1 Step 1: Deaeration of the Crevice Environment ... 461
10.4.1.2 Step 2: pH Evolution and Concentration of Dissolved Species ... 462
10.4.1.3 Thermodynamics of the Crevice Environments for Fe-Ni-Cr Alloys ... 465
10.4.1.4 Other Environmental Evolutions in the Crevice Gap ... 468
10.4.2 Breakdown of Passivity Inside the Crevice ... 468
10.4.2.1 "Classical Mechanism": General Breakdown of Passivity in Low pH, High Chloride Solutions ... 469
10.4.2.2 "Microcrevices" inside the "Macrocrevice" Gap ... 471
10.4.2.3 Pitting in the Crevice Gap ... 471
10.4.2.4 Role of Sulfide Inclusions on Stainless Steel ... 473
10.4.2.5 Passivity Breakdown due to High (Metal) Chloride Concentration ... 473
10.4.2.6 The Role of *IR* Drop in the Initiation of the Crevice Corrosion ... 474
10.4.3 Propagation of Crevice Corrosion ... 474
10.4.4 Crevice Repassivation ... 476

10.4.4.1 Spontaneous Crevice Arrest 476
10.4.4.2 Repassivation Potential 476
10.5 Modeling Crevice Corrosion 478
10.5.1 Surface Reactions 479
10.5.2 Solution Chemistry 479
10.5.2.1 Solution Models: Activity Coefficients 479
10.5.2.2 Hydrolysis and Complexation Products and Reactions 480
10.5.3 Transport Processes 481
10.5.4 Mass Balance 482
10.5.5 Electrical Charge Balance 482
10.5.6 Crevice Initiation Criteria 483
10.5.7 Results of the Models 483
10.6 Experimental Characterization of the Alloys versus Crevice Corrosion 487
10.6.1 Crevice Former Devices 488
10.6.2 Exposure Tests in Representative Environments 488
10.6.3 Exposure Tests in Conventional Environments: The Ferric Chloride Test 489
10.6.4 Electrochemical Tests 489
10.6.4.1 Determination of Critical Potentials for Initiation and Protection 489
10.6.4.2 Determination of the Material Behavior in Simulated Crevice Environments 490
10.7 Prevention of Crevice Corrosion 493
10.8 Conclusion 494
References 495

10.1 Introduction

Crevice corrosion is a form of localized corrosion that occurs in zones of restricted flow where a metallic material surface is in contact with a small volume of confined, stagnant liquid whereas most of the material surface is exposed to the bulk environment.

Crevice zones may result from the design of the component (see Figure 10.1) or from the formation of deposits during service, shutdown, or even fabrication. These deposits may come from suspended solids in the environment, corrosion products, or biological activity. Low-flow areas are prone to the formation of such deposits.

Crevice corrosion occurs mainly (but not exclusively) on passive materials. The most important problem is the crevice corrosion of stainless steels, nickel-base alloys, aluminum alloys, and titanium alloys in aerated chloride environments, particularly in sea or brackish water, but also in environments found in chemical, food, and oil industries. Other cases of crevice corrosion are also known such as the so-called *corrosion by differential aeration* of carbon steels, which does not require the presence of chloride in the environment. Also mentioned in the literature is the crevice corrosion of steels in concentrated nitric acid and inhibited cooling water and of titanium alloys in hot sulfuric environments.

In this chapter, the basic mechanisms of crevice corrosion are briefly presented but most of the text is devoted to the crevice corrosion of passive alloys, particularly Fe-Ni-Cr-... alloys in aerated chloride environments. Phenomenological aspects, the mechanisms of initiation, the conditions of propagation, the modeling, the experimental techniques, and the possibility of prevention are successively described.

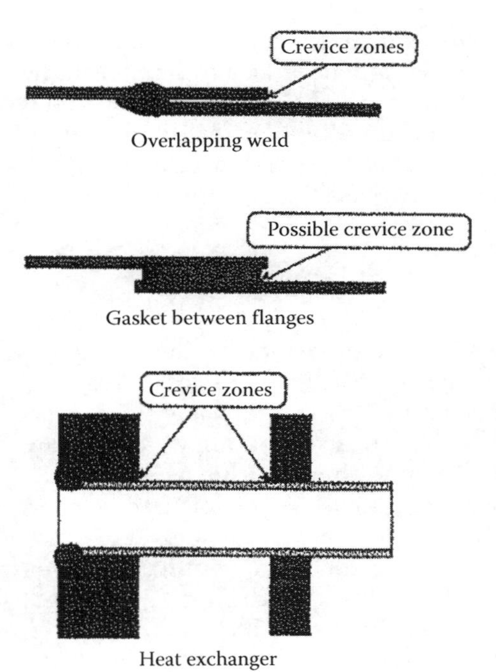

FIGURE 10.1
Examples of crevice zones due to design.

10.2 Basic Mechanisms of Crevice Corrosion

The specificity of the crevice zones arises (Figure 10.2) from (a) the limited mass transport by diffusion and convection between the inside of the crevice and the bulk environment, (b) the high (metal surface area)/(solution volume) ratio inside the crevice, (c) the presence of large external surfaces exposed to the bulk environment, and, in many cases, (d) the significant solution resistance between the inside and the outside of the crevice.

Crevice corrosion is caused by a change of the local environment inside the crevice zone into more aggressive conditions. On passive materials, the local conditions must become severe enough to cause passivity breakdown and the onset of active dissolution.

Usually the local change of environment occurs in two steps. *The first step* is due to the local exhaustion of reactants, usually the dissolved oxygen, and/or the local accumulation of corrosion products as a consequence of the restricted transport kinetics

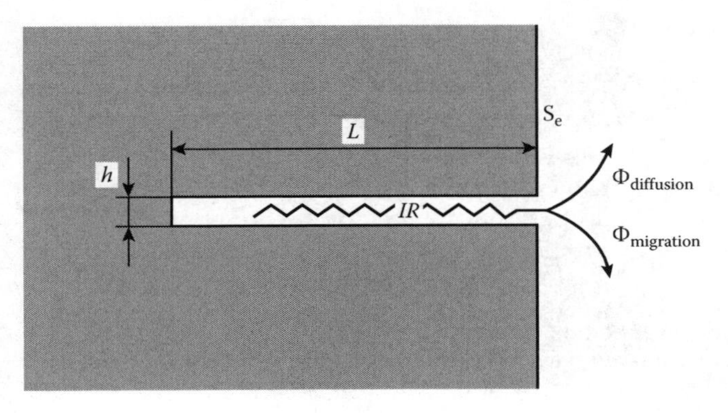

FIGURE 10.2
Characteristics of a crevice zone: reduced diffusional exchanges with the bulk solution, low $\Phi_{diffusion}$; small solution volume in contact with large surface area, L/h large; large external surfaces S_e; solution resistance between crevice gap and bulk solution, *IR* drop.

between the crevice and the bulk environment. This leads to the buildup of a galvanic cell between crevice and external surfaces with two consequences: (a) a difference in the corrosion potential inside and outside the crevice due to the solution resistivity* and (b) an additional source of environment modification caused by the electrolytic migration between the crevice and the bulk solution. This *second step* is self-accelerated if the local environment becomes more severe and tends to increase the galvanic current and the electrolytic migration.

10.2.1 Crevice Corrosion due to *IR* Drop

In some instances, the increase of the corrosion rate in the crevice may be due only to the *IR* drop. The more classical case where the *IR* drop is responsible for crevice corrosion is the corrosion by differential aeration of unalloyed steels in near-neutral water.

Unalloyed steels exhibit an active-passive transition in near-neutral water (Figure 10.3). In aerated environments, the availability of a cathodic current from oxygen reduction allows steel passivity to be stable, whereas active dissolution prevails in deaerated conditions (Figure 10.3a). Inside a crevice, the local change of environment is due to the rapid consumption of the dissolved oxygen. However, a cathodic current from oxygen reduction may be still provided by the external surfaces. If the *IR* drop is low enough, the surface may remain passive inside the crevice (curve 1 in Figure 10.3b). If the *IR* drop is too large (more efficient crevice effect), however, the corrosion potential inside the crevice is too low for the passivity to be stable (curve 2 in Figure 10.3b) and the surfaces inside the crevice gap become active. In this case, the corrosion rate in the crevice is usually much higher than the corrosion rate of the free surfaces in a deaerated environment because of the supply of cathodic current by the external surfaces.

10.2.2 Crevice Corrosion due to Local Changes of Anodic Behavior of the Material

In most cases, crevice corrosion also requires a local change of the environment that results in a change of the material anodic behavior. For example, crevice corrosion of stainless steels in aerated chloride environments requires a local depletion of the oxidizing species, but it is triggered by a second step of environment modification that causes a pH drop and an increase of chloride concentration in the crevice (see later).

Local pH drop is not the single cause of crevice corrosion. For example, local depletion of passivating species may be the origin of crevice corrosion as is the case for steels in cooling water with anodic inhibitors such as nitrite or chromate ions.†

10.2.3 Initiation and Propagation

Because of the general mechanisms described previously, crevice corrosion always includes two steps: (a) an *initiation or incubation period* required for the local environment to undergo a critical change during which no significant dissolution damage occurs in the crevice and (b) a *propagation period* during which the corrosion damages occur. Depending on the cause of the localized corrosion, the initiation time may or may not be a significant part of the lifetime of a structure suffering crevice corrosion.

* This potential difference is usually referred to as *IR* drop.

† Note that cathodic inhibitors do not cause such severe crevice corrosion because their local depletion does not significantly increase the cathodic current available in a deaerated crevice.

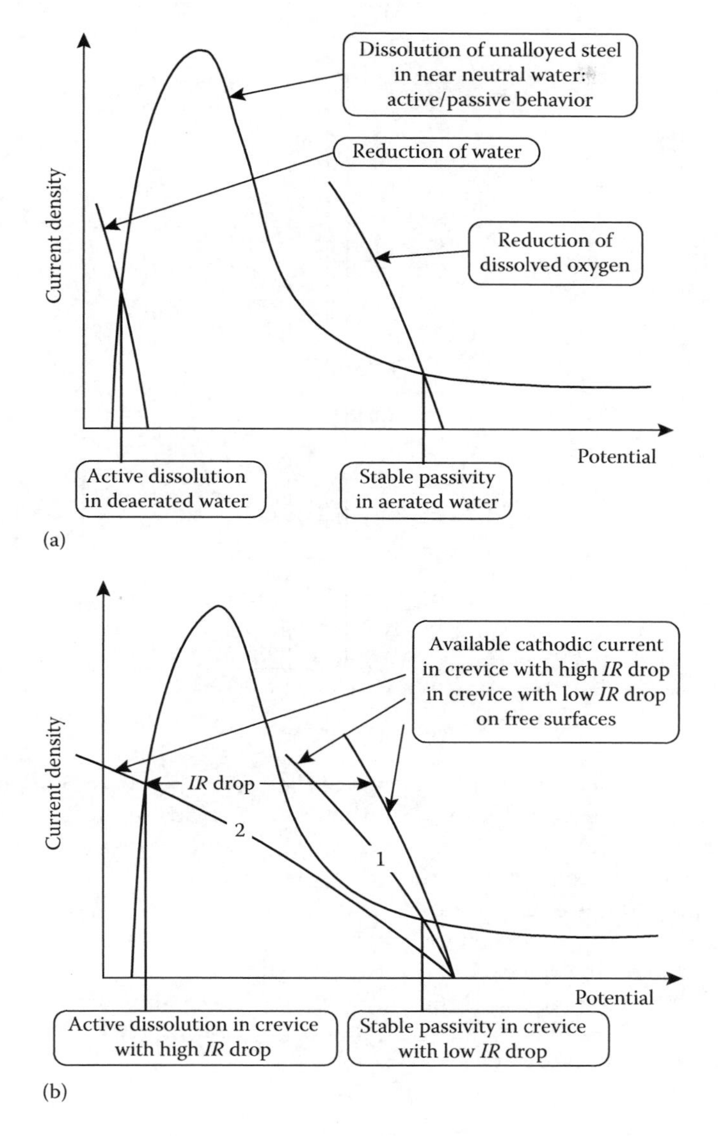

FIGURE 10.3
Crevice corrosion of unalloyed steel in near-neutral water. (a) Behavior of "un-creviced" steel surface in aerated and deaerated water. (b) Behavior in presence of crevice—role of *IR* drop.

10.2.4 Parametric Effects

10.2.4.1 Crevice Geometry

The severity of the crevice regarding the initiation stage strongly depends on its geometry, i.e., the crevice gap (h) and the crevice depth (L).

The crevice gap h (Figure 10.2) controls the volume/surface ratio of the crevice: the smaller this ratio, the faster the environment changes due to surface reactions. However, the crevice gap must be wide enough for the environment to enter but it becomes inefficient if it is too wide. Figure 10.4 [1] shows the extent of crevice corrosion of unalloyed steel in nitric acid as a function of the gap. For stainless steels in chloride environments, short

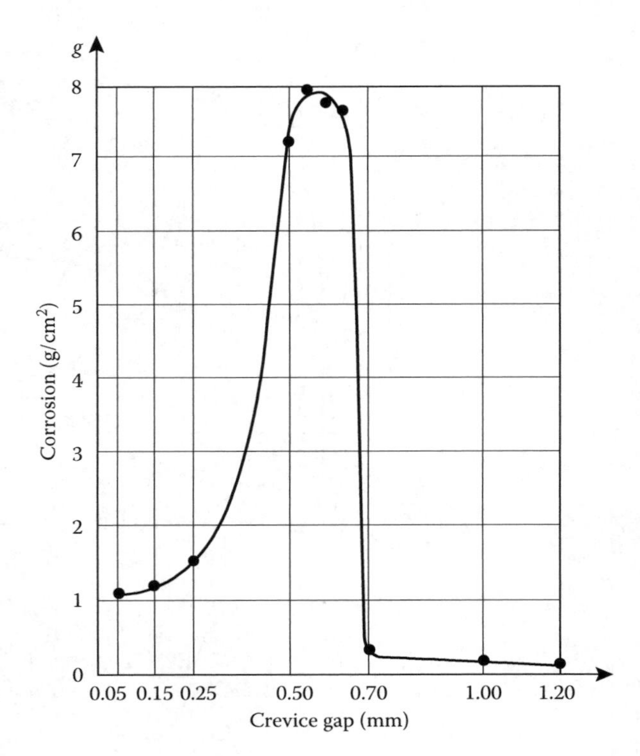

FIGURE 10.4
Effect of crevice gap on the amount of corrosion of unalloyed steel in nitric acid. (From Rosenfeld, I.L., *5th International Congress on Metallic Corrosion*, Tokyo, Japan, p. 53, 1972.)

incubation times are obtained for crevice gaps of the order of a micrometer or less [2,3] and the incubation time increases rapidly with the crevice width. As a consequence, surface roughness is of major importance, and, for example, crevice corrosion can be avoided by polishing the surfaces in contact with seals to a smooth finish.

The crevice depth L (Figure 10.2) controls the transport processes: the deeper the crevice, the slower the transport kinetics between the crevice tip and the bulk environment. A minimum crevice depth is often required for corrosion to occur [2].

Bernhardsson et al. [4] introduced the parameter L^2/h as a geometric criterion of crevice severity. A similar r_0^2/h ratio is included in one of the nondimensional parameters introduced by Alkire et al. [5].

10.2.4.2 External Surfaces

The free surfaces outside the crevice gap constitute the cathode, which provides cathodic current to balance the anodic dissolution in the crevice gap. Thus, the larger the free surfaces and the higher the bulk solution conductivity, the higher the available cathodic current and the higher the corrosion potential in the crevice. Except in the cases where the *IR* drop is the driving factor for crevice corrosion (see earlier), this usually shortens the incubation period and increases the corrosion rate in the propagation stage.

10.2.4.3 IR *Drop*

The *IR* drop depends on both the crevice geometry and the solution conductivity. Most of the *IR* drop usually occurs inside the crevice gap, but the *IR* drop in the bulk solution may be of major importance in dilute solutions and limit the effective area of the cathode.

It is worth noticing that the *IR* drop and the diffusion fluxes between the crevice and the bulk solution are controlled by the same restricted transport path. Thus, the more limited the diffusion transport, the larger the *IR* drop.

The *IR* drop is responsible not only for the buildup of a potential difference between the crevice and the free surfaces but also for potential gradients inside the crevice. If large enough, the *IR* drop could make possible water reduction near the crevice tip and this may create a more complex situation, the crevice tip becoming an additional source of cathodic current and reversing the environmental evolution [6].

10.3 Phenomenology of Crevice Corrosion of Passive Alloys in Aerated Chloride Environments

The first point to emphasize is that *crevice corrosion of passive alloys does not occur in environments deprived of oxidizing species other than water.* Generally, it does not occur in deaerated solutions. On the contrary, oxidizing agents such as many biocides (hypochlorite, chlorine, chlorine dioxide, etc.) may cause crevice corrosion.

As the other crevice corrosion phenomena, the development of crevice corrosion on passive alloys in aerated (and more generally in oxidizing) chloride environments involves two steps:

First, there is an *incubation period,* which may be very long for the more resistant materials. During this period, short corrosion potential transients can possibly be observed (Figure 10.5), but, when examining the surface inside the crevice, no significant traces of corrosion are observed.

Then, *initiation* of rapid corrosion occurs inside the crevice and the *propagation period* begins during which the corrosion proceeds quite rapidly. The formation of deposits outside the crevice allows visual detection of the corrosion. Initiation of corrosion in the

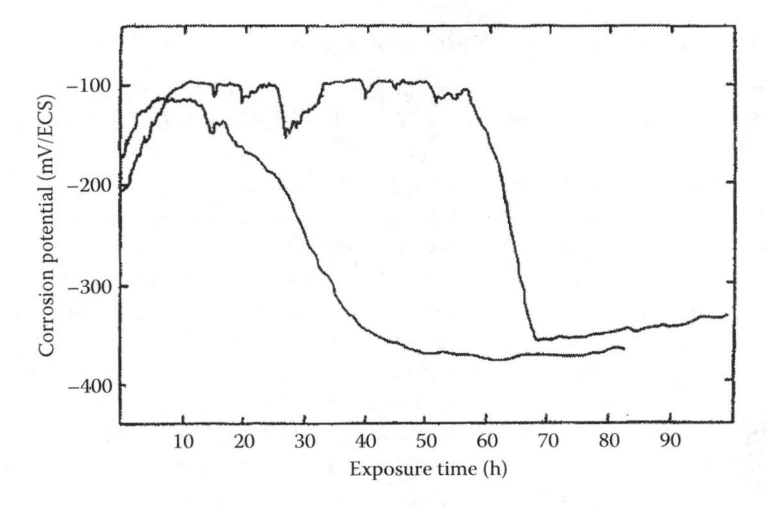

FIGURE 10.5
Potential drop due to initiation and propagation of crevice corrosion (crevice area ratio 1:1) in two different tests on a 316 stainless steel in 1 M NaCl, pH 6 solution at room temperature. (From Oldfield, J.W., *19th Journées des Aciers Spéciaux—International Symposium on Stainless Steels,* Saint Etienne, France, 1980.)

FIGURE 10.6
Severe crevice corrosion of stainless steel bolts in seawater. (From Audouard, J.P., Désestret, A., Catelin, D., and Soulignac, P., *CORROSION/88*, paper 413, St. Louis, MO, March 21–25, 1988.)

crevice is also marked by a significant drop in the corrosion potential of the material [3] (Figure 10.5). Thus, monitoring of the corrosion potential can be an efficient way to detect crevice corrosion (as well as other types of localized corrosion) in its early stages or even during the incubation period.

Figure 10.6 shows an example of severe crevice corrosion of stainless steel bolts in seawater.

10.3.1 Effect of Potential: Critical Potentials for Initiation and Repassivation

For a given bulk environment, crevice initiation appears to be dependent on the corrosion potential of the free surfaces: initiation occurs when this potential exceeds a critical value E_{crer}. The propagation may occur at potentials significantly lower than the potential that caused the initiation to occur, and indeed, the potential of a freely corroding material is generally lower during the propagation compared with the incubation period (Figure 10.5). However, propagation stops below a critical potential for the crevice repassivation E_{rp}.

As a consequence, the polarization curves drawn on a specimen equipped with a crevice former device exhibit (Figure 10.7) features very similar to those typical of pitting corrosion. During the upward potential scan, crevice corrosion initiates quite abruptly at a critical potential E'_{crev} that is generally lower than the pitting potential. Then, during a subsequent backward potential scan, corrosion stops at a critical potential E'_{rp} significantly lower than E'_{crev}.

However, both crevice and repassivation potentials are dependent on the experimental conditions and particularly on the scan rate when determined by using potentiokinetic techniques. Nevertheless, long-term experiments tend to show that there is a minimum value of E_{crev} above which the incubation time decreases as the potential increases (Figures 10.8 [8] and 10.9 [7]). Thus, adding an oxidizing agent (for example, biocides) in a chloride environment strongly enhances the crevice initiation (as well as the pitting probability).

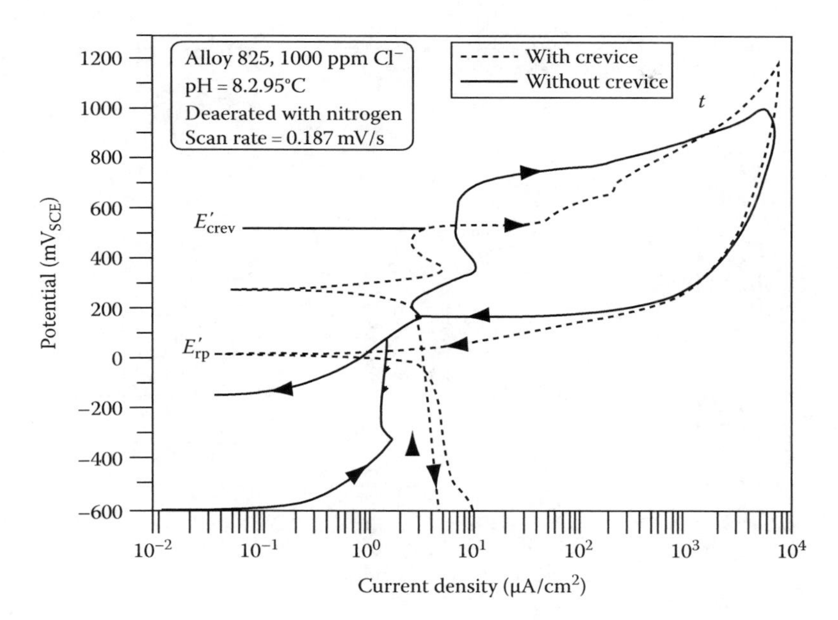

FIGURE 10.7
Polarization curves of alloy 825 in chloride solution: comparison of curves for pitting and crevice corrosion. (From Dunn, D.S. et al., *Corrosion*, 52, 115, 1996.)

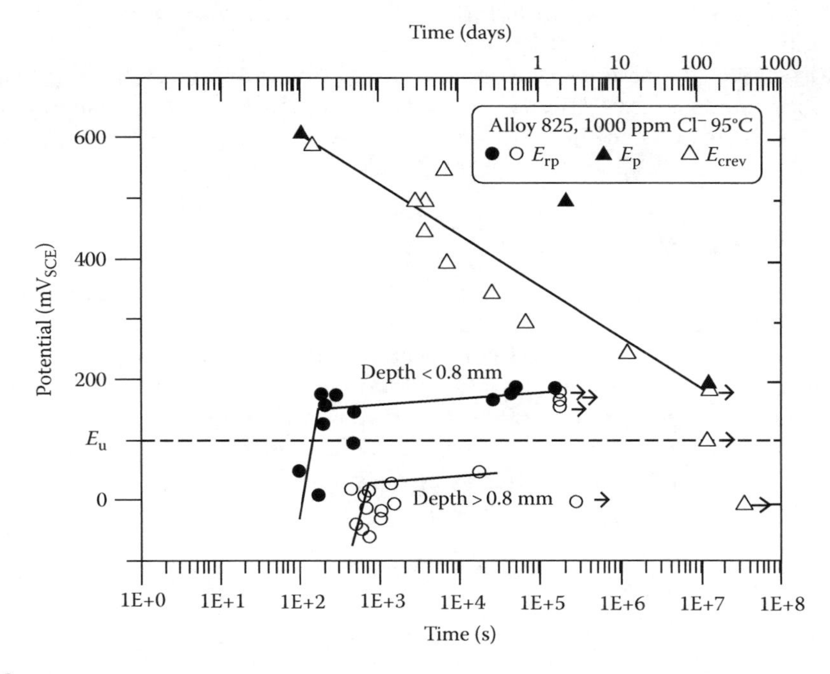

FIGURE 10.8
Effect of time on the crevice and repassivation potentials of alloy 825 in chloride solutions. (From Dunn, D.S. et al., *Corrosion*, 52, 115, 1996.)

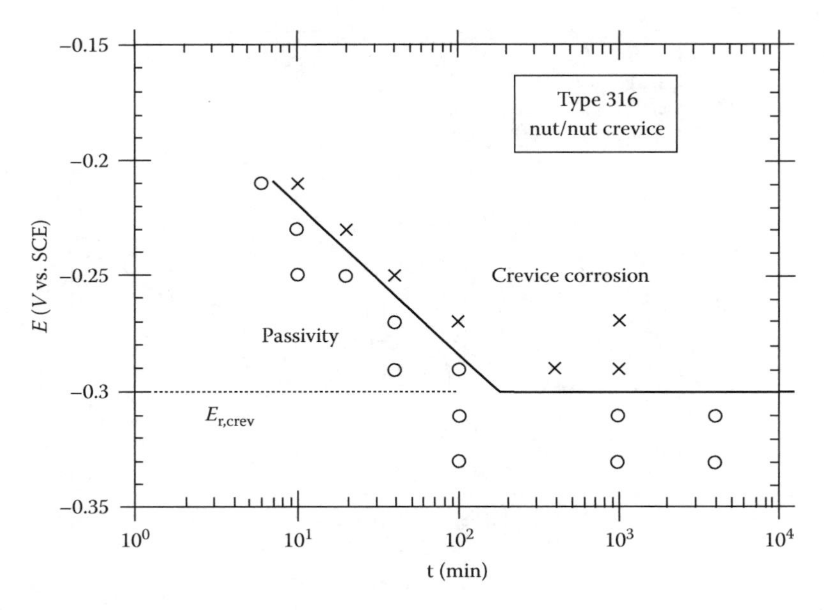

FIGURE 10.9
Effect of potential on the initiation time of crevice corrosion on 316 stainless steel (SS) in 3% NaCl solution. (From Akashi, M. et al., *CORROSION/98*, paper 158, NACE International, Houston, TX, 1998.)

Conversely, there is some agreement that a critical value E_{pr} (usually referred to as protection potential) exists that is the maximum possible value of E_{rp} below which no crevice corrosion may propagate significantly. This is consistent with the fact that crevice corrosion of passive alloys in chloride environments does not occur in deaerated environments.

Figure 10.8 suggests that long-term crevice and pitting potentials tend to the same value [8], but this still needs confirmation. In particular, it remains to be established whether this common value is intrinsic to the alloy-environment system or whether it depends on the crevice geometry.

10.3.2 Effect of the Environment

For a given crevice geometry, the critical potentials for crevice initiation and repassivation decrease with increasing chloride content (Figure 10.10) and increasing temperature (Figure 10.11) of the bulk solution. This means that the susceptibility to crevice corrosion of passivated alloys increases with the chloride content and the temperature. For example, titanium alloys become sensitive to crevice corrosion only in hot concentrated chloride solutions around 100°C–150°C [9,10]. Propagation rates also increase with temperature.

The presence of sulfate ions or buffering species may retard crevice initiation, at least in dilute chloride solutions [13].

10.3.3 Geometric Factors

The incubation time is strongly dependent on the crevice geometry (see Section 10.2.4). The initiation time decreases with decreasing crevice gap. By changing the crevice gap from about 0.2–5 μm, Oldfield [3] raised the initiation time from 6.5 to 350 h on AISI 430 stainless steel.

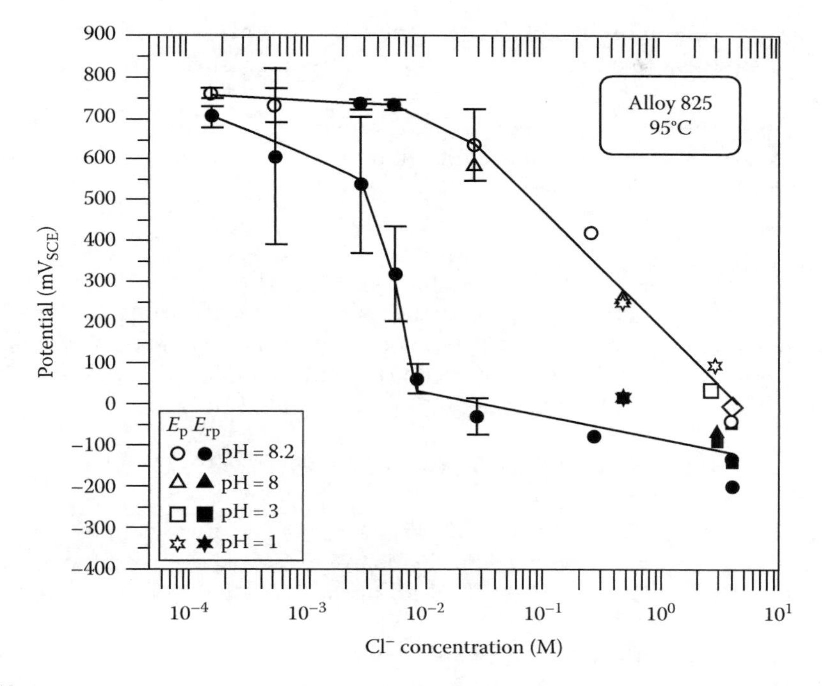

FIGURE 10.10
Effect of chloride concentration on the crevice and repassivation potential of alloy 825 at 95°C. (From Sridhar, N. and Cragnolino, G.A., *Corrosion*, 49, 885, 1993.)

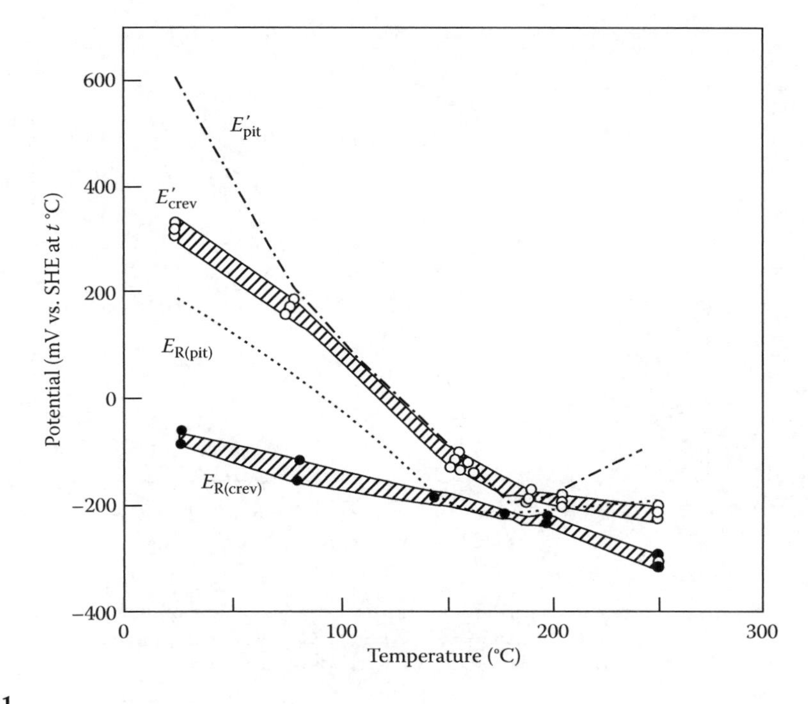

FIGURE 10.11
Effect of the temperature on the critical potentials for crevice and pitting corrosion on 304 SS in 0.5 M NaCl solution. (From Yashiro, H. et al., *Corrosion*, 46, 727, 1990.)

The crevice gaps that exhibit the greater efficiency in crevice initiation are of the order of tenth of micrometers to a few micrometers. At this scale, surface roughness is of major importance.

The area of the free surface surrounding the crevice zone is also of importance: a large free surface area causes shorter crevice initiation time, faster propagation rate, and smaller potential drop at the crevice initiation.

10.3.4 Alloy Composition

Chromium, molybdenum, and nitrogen are the most efficient alloying elements for increasing the resistance to crevice (and to pitting) corrosion initiation of stainless alloys (Fe-Ni-Cr-Mo-N-... alloys). A pitting index has been derived that characterizes the resistance of stainless alloys to these forms of corrosion [14] (Figure 10.12 [15]):

$$\text{PREN} = (\text{wt\% Cr}) + 3.3\,(\text{wt\% Mo}) + 16 \text{ to } 30\,(\text{wt\% N})$$

As an example, PREN values of about 40–45 are requested for a stainless alloy to resist crevice corrosion in natural seawater. This leads to the use of sophisticated alloys containing usually about 23%–25% Cr, up to 7% Mo, up to 0.45% N, and very often Cu and W additions.

The harmful effect of sulfur [16] as an alloy impurity has been mainly attributed to the presence of reactive sulfide inclusions (usually MnS), which constitute very efficient sites for initiation of localized corrosion (see later). Low sulfur or Ti alloying* improves the alloy resistance to crevice corrosion [16].

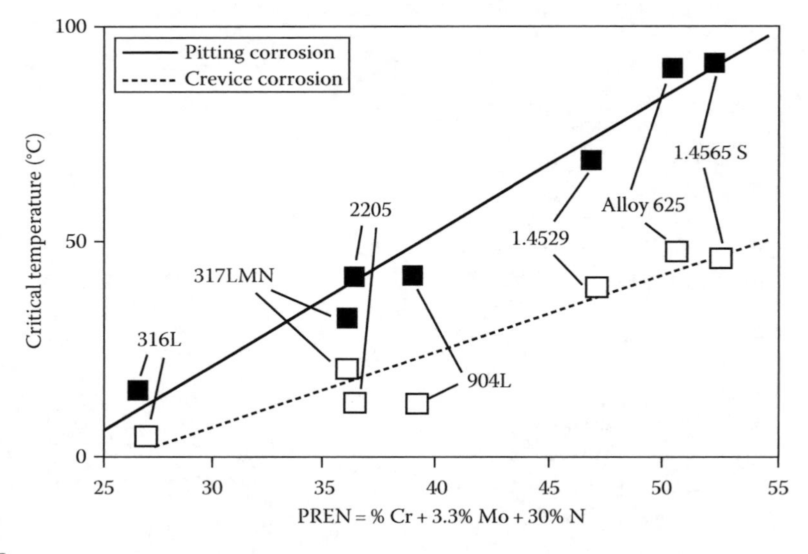

FIGURE 10.12
Effect of composition on the resistance to crevice and pitting corrosion of stainless alloys in chloride environments—the PREN index. (From *Remanit 4565S-Nitrogen Alloyed Austenitic Stainless Steels with Maximum Corrosion Resistance and High Strength*, Thyssen Stahl AG, Krefeld, Germany, 1995.)

* The beneficial effect of Ti alloying is probably due to the formation of sulfide compounds more stable than manganese sulfide.

The propagation rate also depends on the composition of the alloy, but Oldfield [3] observed that the ranking of alloys may be different because the effects of elements are different for initiation and propagation. For example, he claimed that Cr, which is very efficient in retarding the initiation, favors fast propagation rates Mo definitely has a positive effect on both stages [17,18], but Ni seems to have a positive effect only on the propagation rate [16,18].

10.3.5 Propagation of Crevice Corrosion

Crevice corrosion occurs at very high rates, which make this form of corrosion very difficult to manage as the lifetime of a corroding structure may be determined by the initiation period.

Laboratory studies showed that for most alloys, the corrosion rate is higher near the mouth of the crevice [19,20]. However, alloys such as ferritic stainless steels may corrode faster near the dead end of the crevice [20].

An interesting observation on long-term crevice tests of Fe-Ni-Cr-Mo-N alloys [21] is that well-developed crevice corrosion may stop spontaneously even though the environmental conditions do not change. If confirmed, such behavior could be of major importance because it would mean that crevice corrosion may not necessarily cause the failure of structures, particularly in applications involving thick materials such as the containers for long-term storage and disposal of radioactive wastes.

10.4 Processes Involved in Crevice Corrosion of Passive Alloys in Aerated Chloride Environments

The generally accepted scenario is that the environment in the crevice suffers a progressive evolution that leads to the breakdown of passivity and to a propagation stage during which the metal inside the crevice undergoes active dissolution. The following paragraphs briefly describe the current understanding of the different processes, i.e., environment evolution, passivity breakdown, and active dissolution in the crevice gap.

10.4.1 Environment Evolution in the Crevice Gap

The following mechanism is generally accepted to describe the evolution of the crevice environment on passivated alloys exposed to aerated chloride environments.

10.4.1.1 Step 1: Deaeration of the Crevice Environment

On the free surfaces, the cathodic reaction is the reduction of oxygen. However, the environment in the crevices becomes deaerated after periods of time that can be very short, at least compared with the lifetime of an industrial apparatus. As an example, a passive current of 10 nA/cm^2 will cause the deaeration of a crevice with a surface/volume ratio of 10^3 cm^{-1} (i.e., a crevice gap of 20 μm between two metal surfaces) in about 3 h.

The lack of oxygen causes the inhibition of the cathodic reaction inside the crevice. Thus, the local anodic reactions must be balanced by cathodic reactions occurring on the surfaces exposed to the bulk solution. This builds up a galvanic cell between the inside and the outside of the crevice. As a consequence, the *IR* drop between the crevice and the bulk is responsible for the buildup of a corrosion potential difference between the inside and outside surfaces, the former becoming more anodic. However, the contrary to what happens on unalloyed steels (see earlier), the stainless alloys exposed to neutral chloride solutions are passive in both aerated and deaerated solutions and the local oxygen consumption has no significant effect of the dissolution kinetics. Thus, this step does not induce any corrosion damage but it is essential to trigger the following process.

10.4.1.2 Step 2: pH Evolution and Concentration of Dissolved Species

The anodic dissolution of the metal inside the crevice produces metallic cations (and/or oxides that contribute to thickening the passive layer, at least before initiation of the propagation stage).

Production of solid oxides or oxyhydroxides

$$x\mathrm{M} + (y+z)\mathrm{H_2O} \rightarrow \mathrm{M}_x\mathrm{O}_y(\mathrm{OH})_z + (2y+z)\mathrm{H}^+ \quad (10.1)$$

Production of solvated metallic cations

$$\mathrm{M} \rightarrow [\mathrm{M(H_2O)}_n]^{z+} + ze^-, \quad \text{where } n \text{ is usually 6} \quad (10.2)$$

These metallic cations cannot concentrate indefinitely and they undergo hydrolysis.

The earlier theories assumed the formation of simple hydroxides:

$$[\mathrm{M(H_2O)}_n]^{z+} \rightarrow [\mathrm{M(H_2O)}_n]^{(z-h)} + h\mathrm{H}^+ \quad (10.3)$$

It is now recognized that metallic cations may also form metallic chlorides and/or complex oxy-hydroxy-chlorides according to reactions such as

$$\mathrm{M(H_2O)}_n^{z+} + m\mathrm{Cl}^- \rightarrow [\mathrm{M(H_2O)}_{(n-m)}\mathrm{Cl}_m]^{(z-m)+} + m\mathrm{H_2O} \quad (10.4)$$

$$[\mathrm{M(H_2O)}_{(n-m)}\mathrm{Cl}_m]^{(z-m)+} + h\mathrm{H_2O} \rightarrow [\mathrm{M(H_2O)}_{(n-m-h)}(\mathrm{OH})_h\mathrm{Cl}_m]^{(z-m-h)+} + h\mathrm{H_3O}^+ \quad (10.5)$$

The formation of metallic polycations should also be considered [4].

These anodic processes result in a local excess of cations inside the crevice. Because there is no cathodic reaction inside the crevice, the restoration of electrical neutrality requires the onset of a galvanic current in the base metal and a migration flux in the liquid. At least in the early stages of crevice evolution, most of the ions conveyed by this migration are chloride ions that are transported from the bulk solution into the crevice [22].

All these processes are summarized in Figure 10.13, where hydrolysis reaction (10.3) is assumed. As a consequence, *the crevice environment becomes more and more acidic and more and more concentrated in metallic ions (including metal chloride complexes) and free chloride ions.* We will see later that, in some instances, this pH evolution, combined with an *IR* drop, can make possible the reduction of water in the crevice.

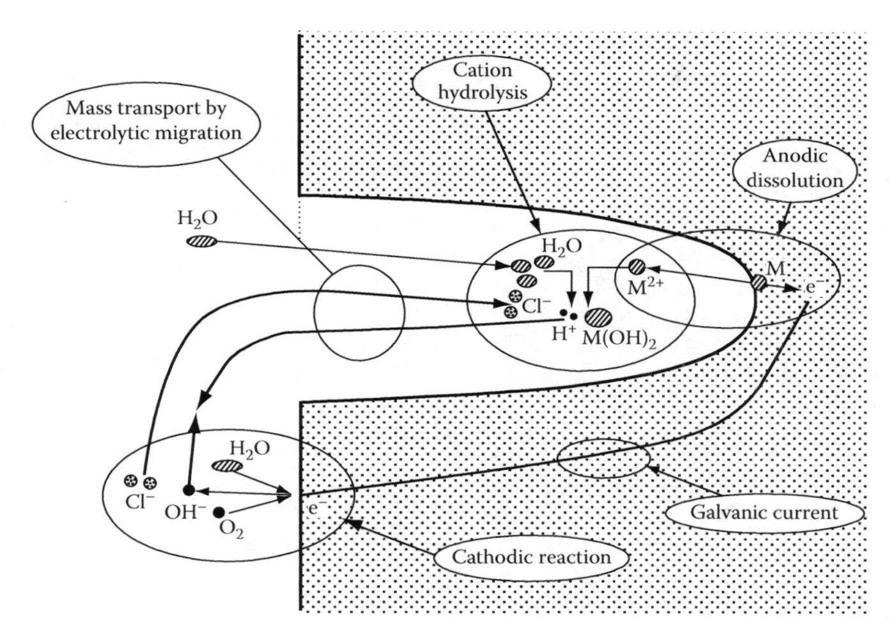

FIGURE 10.13
Sketch of the main processes leading to environment evolution in a crevice.

Indeed, numerous experimental studies have been performed to study the evolution of the environment in a crevice. Most data were obtained in actively corroding artificial crevices or pits, either by sampling the solution or by direct pH and Cl^- concentration measurements. Table 10.1 summarizes significant results that confirm the foregoing trends. In crevice solutions, the drop of pH depends on the hydrolysis constants of the metal cations. On stainless alloys, chromium and molybdenum are considered to be the cause of the very low, sometimes negative, pH observed. Iron, nickel, and aluminum exhibit much less acidic hydrolysis reactions and the pH values in the crevices are higher: values of 3–5 are reported for iron, values of 3–4 for the aluminum alloys.

It must be noted that the pH drop is due to the dissolution of the base material: thus, the more resistant the material, the lower is the pH required for dissolution to occur. Thus, the more acidic crevice pH values are observed on the more resistant materials. This is shown, for example, in the studies of Bogar and Fujii (Figure 10.14 [24]) on Fe-Cr ferritic stainless steels and Suzuki et al. [28] on 304L, 316L, and 18Cr-16Ni-5Mo austenitic stainless steels (Table 10.1).

The concentration of the ions in the crevices can reach very high values of the order of several mol/L [28,33], and the solution may eventually become saturated, which possibly limits the dissolution rate. In such concentrated solutions, salt films may form on the dissolving metal surfaces: $NiCl_2$ has been identified in nickel [34] and $FeCl_2$ in iron [35] and stainless steel [35] pits or crevices.

Several studies also outline the strong effect of the potential of the free surfaces on the pH drop inside an actively corroding crevice: in particular, early work of Pourbaix [36] on unalloyed steel showed (Figure 10.15) that the higher the potential of the external surfaces, the higher the potential and pH drops in the crevice. Similar results were obtained by Zuo et al. [37] for stainless steel actively corroding in an artificial crevice. Conversely, Pourbaix [36] and Turnbull and Gardner [38] on unalloyed steels and

TABLE 10.1

Some Experimental Measurements of Local pH and Composition in Crevices[a]

Material	Bulk Solution	pH	Local Composition	References and Notes
Unalloyed steel	0.06–0.6 M NaCl	3–4.7		Data collected in Ref. [23]
	Seawater	3.3		
Mild steel	0.5 N NaCl[b]	~4.3	Up to 2.5 N Fe, 5.5 N Cl	[22]
Fe-10% Cr	0.6 M NaCl	3.2	0.36 M Fe, 0.06 M Cr	[24]
Fe-15% Cr	0.6 M NaCl	3.0	0.21 M Fe, 0.17 M Cr	[24]
Fe-20% Cr	0.6 M NaCl	2.4	0.14 M Fe, 0.33 M Cr	[24]
Fe-25% Cr	0.6 M NaCl	1.8	0.11 M Fe, ?? M Cr	[24]
Fe-12% Cr	0.1 M NaCl + 0.01 M $NaHCO_3$	2.5		[25]
304SS	Seawater	2.3 1.5–2.2		Data collected in Ref. [23]
304SS	0.6 M NaCl	1.7–2.7		[26]
304SS	1 M NaCl[b]	0		[27]
304SS	0.5 N NaCl[b]	1.6	Up to 0.5 N Cr, 6 N Cl	[22]
Fe	0.5 N NaCl, 70°C[b]	4.7		[28], Artificial pit
Cr	0.5 N NaCl, 70°C[b]	0.09		[28], Artificial pit
Ni	0.5 N NaCl, 70°C[b]	2.93		[28], Artificial pit
304L SS	0.5 N NaCl, 70°C[b]	0.6–0.8	2.3 N Fe, 1.06 N Cr, 0.34 N Ni, 3.87 N Cl^-	[28], Artificial pit
316L SS	0.5 N NaCl, 70°C[b]	0.06–0.17	4.3 N Fe, 1.5 N Cr, 0.68 N Ni, 0.14 N Mo, 6.47 N Cl^-	[28], Artificial pit
18Cr-16Ni-5Mo	0.5 N NaCl, 70°C[b]	0.13–0.08	3.26 N Fe, 1.79 N Cr, 0.95 N Ni, 0.31 N Mo, 6.2 N Cl^-	[28], Artificial pit
7075 Al alloy	3.5% NaCl	3.45	1600 ppm Al + Zn + Mg + Cu	[29]
7474	3% NaCl	3.6–4.2		[30]
Pure Al	1 M NaCl–pH 11	3–4		[31], Natural pit

[a] For more details see Turnbull [32].

[b] Anodic polarization.

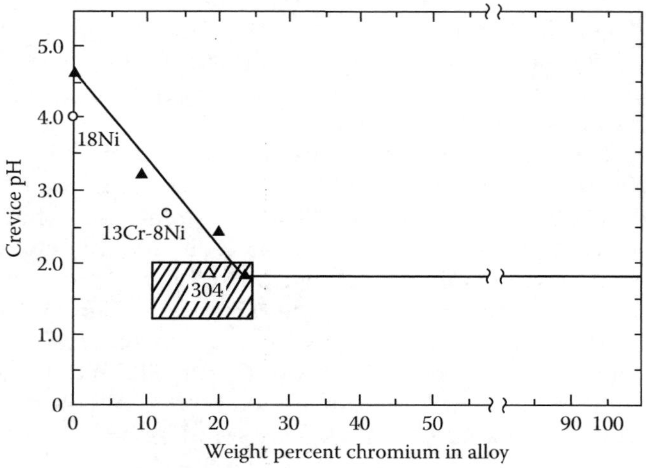

FIGURE 10.14
Effect of chromium content of the alloy on the pH in the crevice. (From Bogar, F.D. and Fujii, C.T., NRL Report 7690, AD 778 002, 1974.)

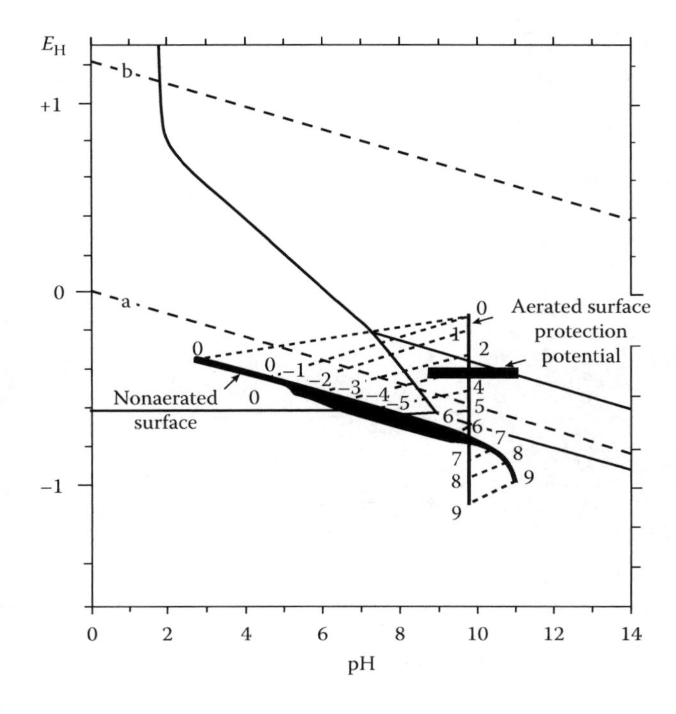

FIGURE 10.15
Effect of the potential of the external surfaces on the pH and potential inside an artificial C-steel crevice. Same numbers refer to the corresponding conditions on the free surface and inside the crevice. (From Pourbaix, M., *Corrosion*, 26, 431, 1970.)

Peterson and Lennox [39] on 304 stainless steel showed that cathodic polarization of the free surfaces induced an increase of pH and a cathodic polarization of the surfaces inside the crevice (Figures 10.15 and 10.16).

In the artificial crevice experiments, several other features are interesting. As an example, for the same 304 stainless steel:

The relationship between the total amount of anodic charge passed in an artificial crevice and the pH is not unique (Figure 10.17).

The relationship between the chromium content (and more generally the composition) of the crevice solution and the pH is not unique and does not agree with any simple hydrolysis reaction (Figure 10.18).

The pH of a synthetic solution made of reagent grade transition metals and sodium chlorides does not have the same pH as the crevice solution of the same composition (Figure 10.19). In addition, this pH spontaneously decreases with time [33,40] (Figures 10.18 and 10.20) for periods of time of several weeks which should not be neglected when modeling the crevice evolution. Indeed, Zhen [41] showed, by ultraviolet (UV) spectroscopy, the presence of at least two different Cr compounds in stainless steel crevice solutions and observed a spontaneous evolution with time toward one of these species.

10.4.1.3 Thermodynamics of the Crevice Environments for Fe-Ni-Cr Alloys

Thus it appears that the detailed chemistry of the crevice environments is not very well known for the following reasons:

The ions and solid species formed may be more complex than the simple metal hydroxides assumed in most models. For example, Tsuru et al. [22] and Ogawa [42] have shown, in the case of stainless steel, the formation of chloro-hydroxy complexes involving up to 20 chloride ions, although the coordination number of chromium ions should not exceed 6. Tsuru et al. [22]

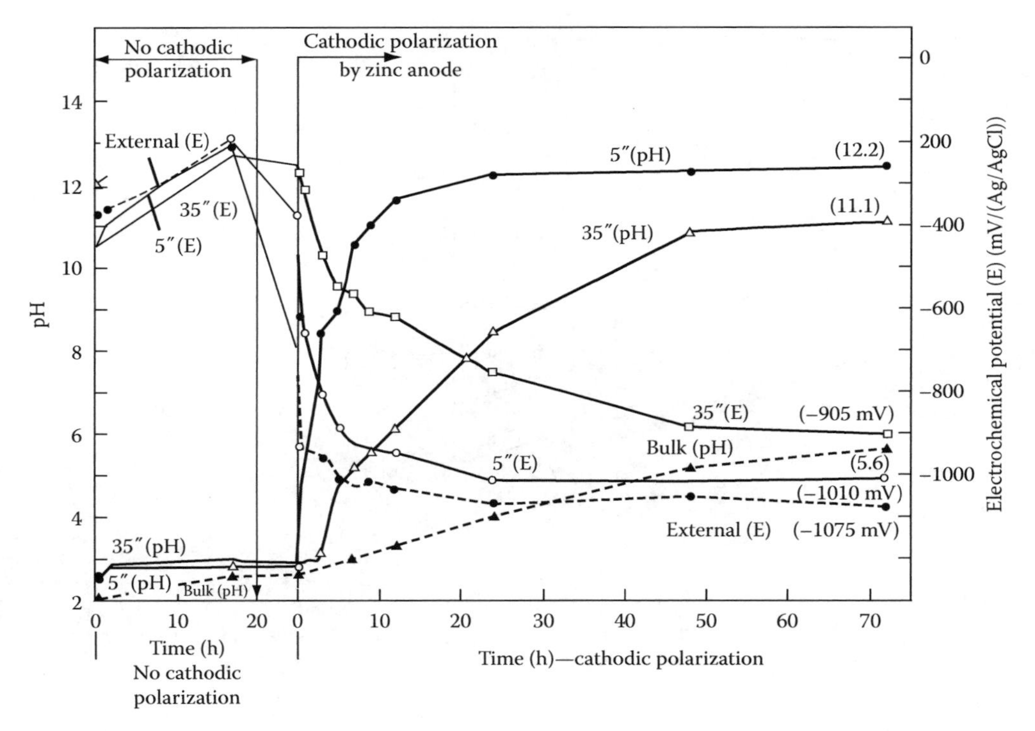

FIGURE 10.16
Effect of cathodic polarization by zinc anode on the pH and potential inside a 304 SS crevice. (From Zuo, J.Y. et al., *9th International Congress on Metallic Corrosion*, Toronto, Ontario, Canada, p. 336, 1984.)

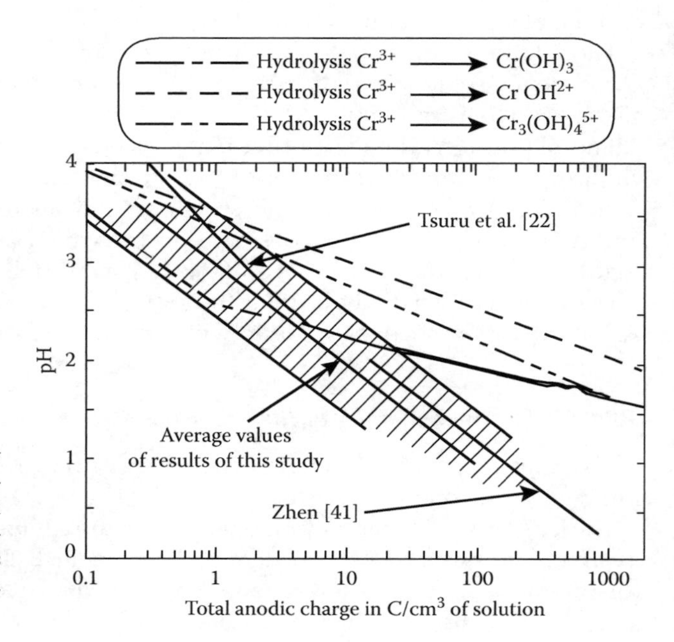

FIGURE 10.17
Experimental relationships obtained by different authors between pH and dissolution current 304 SS artificial crevice and comparison with simple hydrolysis equilibria. (From Combrade, P. et al., *Proceedings of Innovation Stainless Steels*, Florence, Italy, 1993, p. 215.)

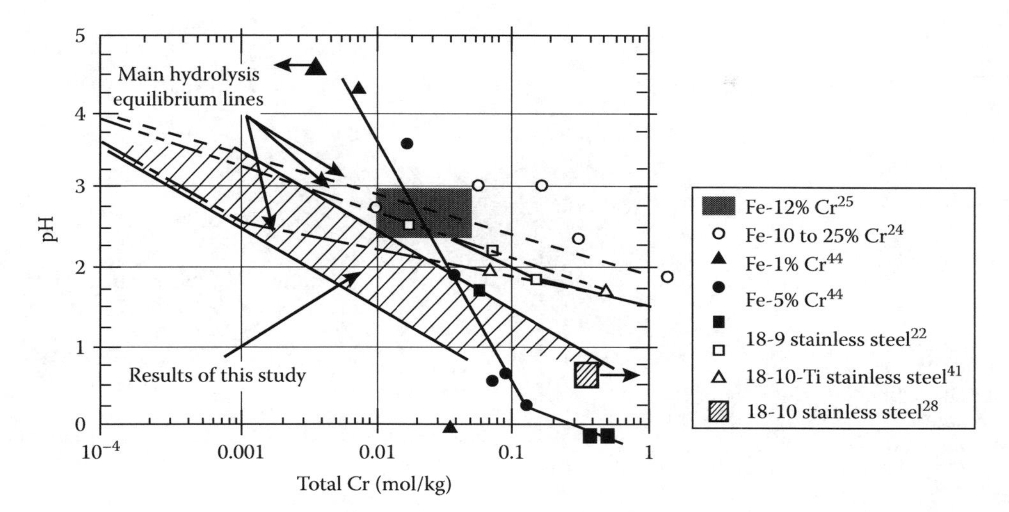

FIGURE 10.18
Experimental relationship obtained by different authors between pH and Cr content of artificial crevice solutions. (From Combrade, P. et al., *Proceedings of Innovation Stainless Steels*, Florence, Italy, 1993, p. 215.)

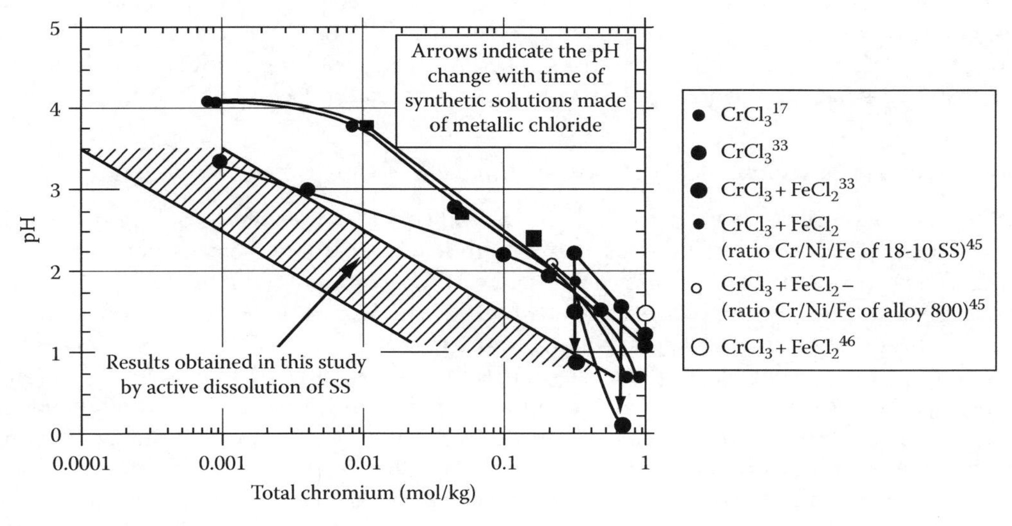

FIGURE 10.19
Comparison of artificial crevice solution with synthetic solution of the same composition. (From Combrade, P. et al., *Proceedings of Innovation Stainless Steels*, Florence, Italy, 1993, p. 215.)

also showed that in iron and stainless steel crevices, iron is likely to be complexed to chloride as $[FeCl(H_2O)_5]^+$ (i.e., $FeCl^+$), but Brossia et al. [35] found that $CrCl^{2+}$ is the most likely metallic chloride in stainless steel crevices.

In concentrated solutions, the presence of large amounts of chloride has been shown to decrease the pH [22,33,42] by increasing the activity coefficient of the hydrogen ions.

Some reactions of hydrolysis and complexation are not as fast as usually assumed (see earlier).

This may suggest that the nature of the cationic species produced by the anodic dissolution may depend on the surface conditions of the alloy and on the local environment.

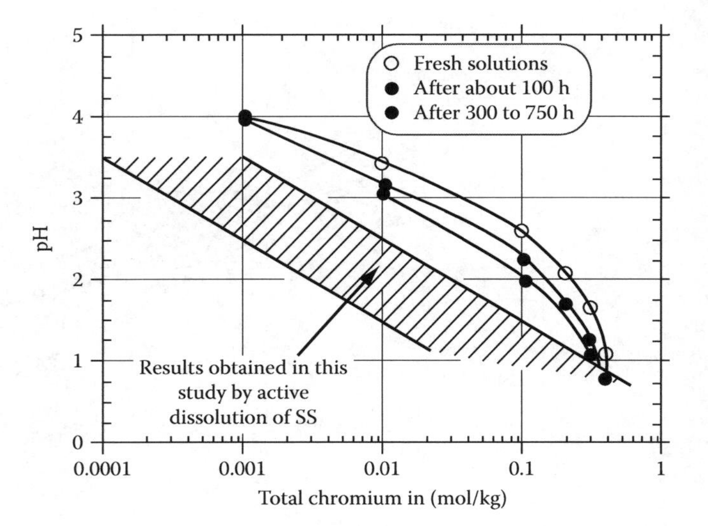

FIGURE 10.20
Evolution with time of the pH of synthetic solution made of metallic chloride salts. (From Combrade, P. et al., *Proceedings of Innovation Stainless Steels*, Florence, Italy, 1993, p. 215.)

In particular, the analyses of passive films on stainless steels showed [43] that the external layers are very hydrated and contain chloride. On the other hand, the number of hydroxyl groups and chloride ions adsorbed on active surfaces is likely to be dependent on the local pH, chloride content, and surface potential. Thus, dissolution from a passive surface may produce metallic hydroxy-chlorides different from those resulting from active dissolution, and active dissolution may produce cations depending on the local conditions.

10.4.1.4 Other Environmental Evolutions in the Crevice Gap

In the case of stainless steels, Lott and Alkire [47] consider that the dissolution of sulfide inclusions (MnS) inside a crevice is the main environment change before passivity breakdown. They showed by direct analysis of the local environment in an artificial crevice that manganese sulfides were oxidized to thiosulfates, which are known to promote pitting in aerated chloride environments [48–50] and, more generally, to inhibit film repair on stainless alloys. More recent work of Brossia and Kelly [51,52] showed that the dissolution of manganese sulfides actually produced sulfide rather than thiosulfate ions. Whatever the nature of the reactive sulfur species produced by the degradation of MnS inclusions, they can be deleterious for the passive layer, particularly because they may form on the bare metal surfaces an adsorbed layer of sulfur that enhances the anodic dissolution and inhibits (or retards) the formation of the passive film (Chapter 8 and [53]). Indeed Crolet et al. [54] showed that in acidic environments representing crevice solutions an increase in the sulfur content of the stainless steels favored the onset of active dissolution.

In the case of dissimilar crevices where one of the crevice walls is made of a nonmetallic material, such as a gasket, the possibility of dissolution of foreign elements must also be considered.

10.4.2 Breakdown of Passivity Inside the Crevice

The detailed mechanism of passivity breakdown inside a crevice is not clearly established and different possibilities are still discussed. Indeed, it is quite possible that different mechanisms may be involved depending on the material, bulk environment, crevice geometry, and, possibly, surface condition of the metal.

10.4.2.1 "Classical Mechanism": General Breakdown of Passivity in Low pH, High Chloride Solutions

The more "classical" mechanism involves general film breakdown in a *critical environment* characterized by a low pH and a high chloride content. Low pH and high chloride concentrations are known to be deleterious for the passivity of most alloys. Thus, the progressive evolution of the crevice environment causes a degradation of the passive film that may result in the following successive situations (Figure 10.21a):

1. An increase of the passive current
2. The onset of an active dissolution domain with a peak current that increases with decreasing pH and increasing chloride content (and, thus, increasing with time)
3. A complete inhibition of passivity

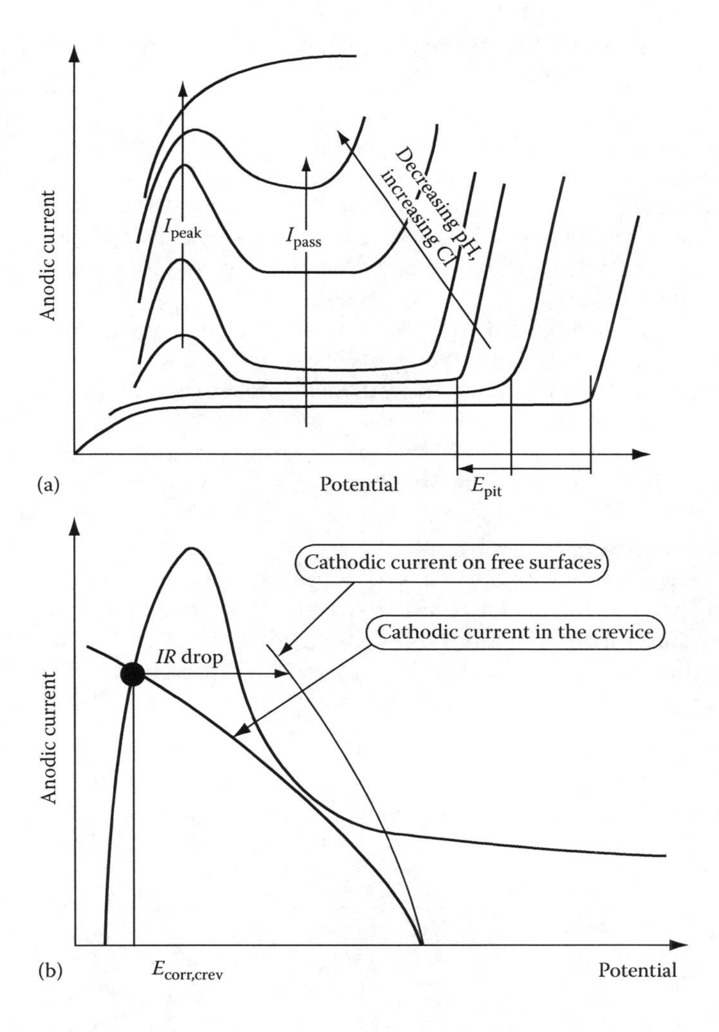

FIGURE 10.21
(a) Evolution of the anodic characteristic of passivated alloy when decreasing the pH and increasing the chloride content. (b) Activation inside a crevice when the corrosion potential is located in the activity peak due to ohmic drop in the crevice.

Passivity breakdown is thought to occur either when the corrosion potential of the crevice surfaces is located in the active peak due to ohmic drop (Figure 10.21b) or when there is no more active-passive transition on the anodic curve.

This mechanism of initiation gave rise to the notion of critical pH for film breakdown [55], which was extensively used as a criterion to rank the resistance of stainless alloys to crevice corrosion. For Crolet et al. [55] the critical pH is the pH corresponding to the onset of an active peak in the polarization curves, while Ogawa [42] considered the pH of spontaneous film breakdown under free corrosion conditions. These differences changed somewhat the critical pH values but usually not the alloy ranking. Table 10.2 gives some results of critical pH evaluated on stainless alloys. Okayama et al. [56] performed measurements of the depassivation pH of over 50 stainless steels and nickel-base alloys and found that the effect of alloying elements of austenitic materials was given by

$$pH_D = 4.09 - 0.248\,wt\%\,Cr - 0.219\,wt\%\,Mo + 1.29\,\log(wt\%\,Ni) + 75\,wt\%\,S + 2.66\,wt\%\,C \quad (10.6)$$

This relation showed the well-known beneficial effect of Cr and Mo and confirmed the very deleterious effect of sulfur (see previous section).

On very resistant stainless alloys, the criterion of depassivation pH tends to be replaced by a criterion of critical crevice solution (CCS) [19,57] and Gartland assumes that crevice corrosion occurs when the solution is so aggressive that passivity is no longer possible over the whole potential range.

However, this classical mechanism appears not to be consistent with different observations, at least on less resistant materials:

Several experiments [58–60] with in situ local measurement of pH and/or chloride concentration with microelectrodes showed that the large changes in crevice pH and chloride content mentioned in the former paragraph occurred not before but after passivity breakdown. In these experiments [58,61], the potential difference between crevice and free surfaces (i.e., the ohmic drop) is often very low before initiation (few mV) and becomes larger only after crevice initiation.

This mechanism does not explain clearly why the initiation time is so dependent on the potential of the external surfaces [60,62], particularly if the passive current is not significantly dependent on the potential as is assumed in most crevice models. This strong dependence on the applied potential would support the idea that the pH drop rate in the crevice is controlled by the migration process and/or by the available cathodic current.

TABLE 10.2

Examples of Depassivation pH Measured according to Crolet et al. [47] and Oldfield and Sutton [54] Criteria

Stainless Alloy	pH of Depassivation	Reference
AISI 430SS	2.9	[47]
304L	19.5–2.25	[47]
316L	1.7	[47]
904L	1.4	[47]
	1.25	[54]
Alloy 625	0	[54]

This is hardly understandable at the very low current densities involved during the initiation stage, when considering that dissolution rates several order of magnitude higher can be sustained during the propagation stage.

The following points have been raised to account for these observations:

The corrosion initiates by local passivity breakdown in the crevice as a result of (a) microcontacts due to wall asperity or (b) pitting inside the crevice.

The critical environment involves the buildup of deleterious species such as (c) sulfur or (d) chloride species rather than low pH.

10.4.2.2 "Microcrevices" inside the "Macrocrevice" Gap

Several authors [1,14,60] raised the point that at the scale of the crevice gap, the surface roughness may play a major role in the environment evolution. Local contact between the opposite surfaces may constitute very tight local "microcrevices" inside the "macrocrevice" (Figure 10.22).

Due to their very narrow gap, these microcrevices could be very efficient in promoting a very local pH change (see Section 10.5) and a local passivity breakdown by the mechanism presented in the former paragraph. As soon as active dissolution occurs somewhere inside the macrocrevice, the breakdown of passivity spreads out progressively to the whole crevice provided the potential inside the crevice is high enough.

Thus, according to this view, the classical mechanism could operate at the scale of the microcrevices because of asperity contacts rather than at the scale of the whole crevice. Most of the "apparent" incubation period would be a period of very local corrosion extending progressively into the whole crevice zone, and thus, it would be sensitive to the potential because of the locally active dissolution.

This interpretation is also consistent with the observation that pH and chloride microelectrodes detect only a significant environment evolution after a time lag following the current increase because the size of the sensitive area of these electrode is much larger than the size of the crevice nuclei inside the crevice. Thus, these microelectrodes detect only the spreading of the corrosion through the whole crevice (Figure 10.23).

10.4.2.3 Pitting in the Crevice Gap

The initiation of crevice corrosion by pitting inside the crevice gap is mainly invoked for stainless steels. Indeed, several investigators [23,46,63,64] have observed pitting and further coalescence of the pits inside stainless steel crevices. Eklund [64] showed that pitting on MnS inclusions near the mouth of the crevice may be the initial stage of crevice corrosion of stainless steels. Oldfield and Sutton [46] have given a detailed description of

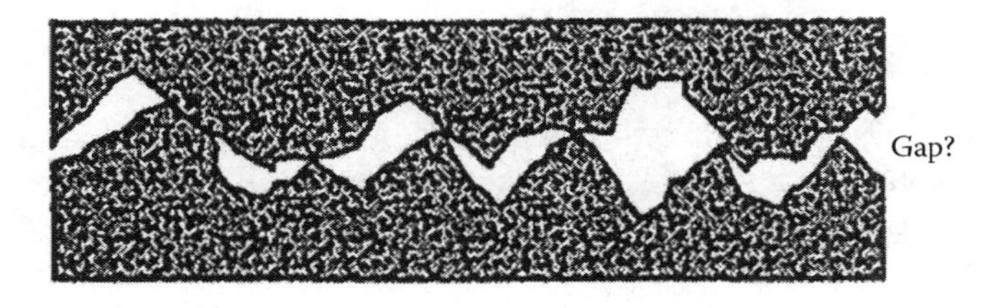

FIGURE 10.22
Crevice formed by asperity contact. (From Sedriks, A.J., *CORROSION'96—Research Topical Symposia. Part 2. Crevice Corrosion: The Science and Its Control in Engineering Practice*, NACE International, Houston, TX, pp. 279–310, 1996.)

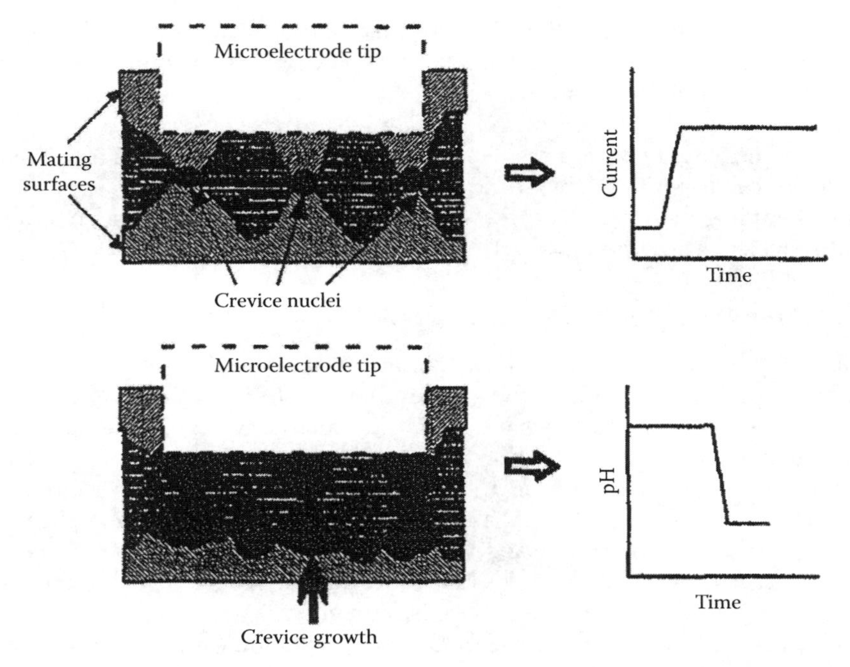

FIGURE 10.23
Development of localized corrosion inside a crevice gap formed by asperities and corresponding response of current. (From Sridhar, N. and Dunn, D.S., *Corrosion*, 50, 857, 1994.)

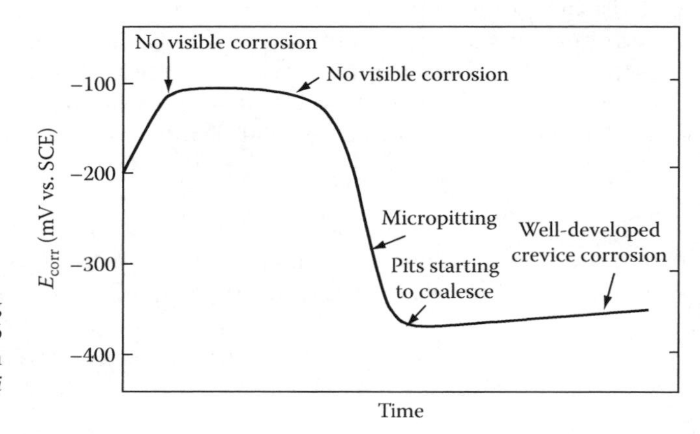

FIGURE 10.24
Crevice initiation by micropitting observed by Oldfield and Sutton on 316 SS in 1 M NaCl, pH 6 solution. (From Oldfield, J.W., and Sutton, W.H., *Bri Corros. J.*, 13, 104, 1978.)

the sequence of micropitting, pit coalescence, and crystallographic etching that results in a rapid dissolution stage in a stainless steel crevice in 1 M NaCl solution (Figure 10.24). In this case, early corrosion damages progressively spread through the crevice, suggesting an effect of the environment evolution in the crevice. By using a moiré fringe technique to observe directly the metal dissolution in a crevice, Shinohara et al. [65] confirmed the localized character of the early dissolution events and their lateral extension and coalescence.

It is worth noting that the crevice may favor pit initiation in two ways: (a) the pitting potential of a stainless alloys decreases with increasing chloride content (Figure 10.21a); (b) the restricted transport conditions may contribute to stabilize pit embryos at potentials lower than on the free surfaces [66]. This last point is supported by the experimental and modeling work of Laycock et al. [62]. Indeed, the deleterious effect of anodic polarization on the initiation time is obviously consistent with a pitting process.

10.4.2.4 Role of Sulfide Inclusions on Stainless Steel

Besides their possible role in pit initiation (see earlier), the dissolution of the MnS inclusions can generate thiosulfates [47,67] or more likely sulfide [51,52] ions, which are known to favor passivity breakdown. Indeed, Crolet et al. [54] observed higher depassivation pH on stainless steels with high sulfur content. According to Brossia and Kelly, the initiation involves a critical $[Cl^-]/[HS^-]$ ratio. This effect of MnS is unlikely in modern stainless alloys with very low sulfur contents.

10.4.2.5 Passivity Breakdown due to High (Metal) Chloride Concentration

On pure aluminum, Hebert and Alkire [68] proposed that passivity breakdown occurred when a critical concentration of aluminum chloride is attained in the crevice environment. Indeed, they showed convincingly (Figure 10.25) that the presence of aluminum cations in the solution caused passivity breakdown for a pH and chloride concentration at which solutions containing only sodium cations did not.

Based on an experimental work on alloy 7475 crevices and propagating cracks, Holroyd et al. [30] agree with this assumption. The critical aluminum concentration required decreases as chloride content increases [68] but low pH and high chloride content are not a prerequisite for passivity breakdown.

The possibility of the existence of a critical concentration of chloride rather than a critical pH has been proposed for stainless steels, for example, by Zakipour and Leygraf [69] in relation to a critical composition of the passive film (chromium content of about 48%).

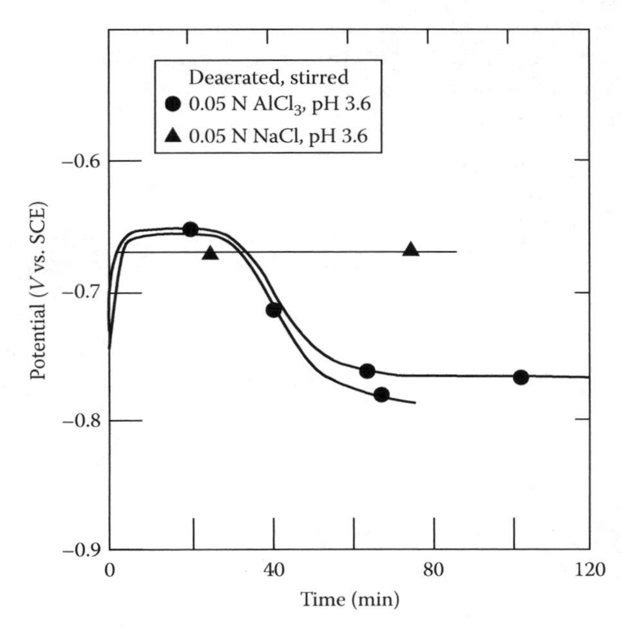

FIGURE 10.25
Effect of dissolved Al on the activation of pure Al in pH 3.6 solution. Test under galvanostatic conditions. (From Hebert, K.R. and Alkire, C., *J. Electrochem. Soc.*, 130, 1001, 1983.)

On stainless steels, several authors [70–73] showed that a minimum concentration of metallic chloride is required for stable propagation to occur. As mentioned previously, the external layer of the passive films on stainless steels contains hydroxyl groups and chloride. Thus, it may be possible that the presence in the solution of metal complexes modifies the external layers of the passive films and contributes to lowering their stability.

10.4.2.6 The Role of IR *Drop in the Initiation of the Crevice Corrosion*

The role of *IR* drop in crevice initiation is not clear. Different authors [58,60,61] observed crevice initiation on stainless steels at a very low *IR* drop level. It is clear that initiation processes can be separated into two classes: (a) those that operate at relatively high potentials (pitting in the crevice gap), which cannot be enhanced by large ohmic drops, and (b) those that occur at low potential (general passivity breakdown), which are favored by large *IR* drops. However, on stainless steels and nickel-base alloys, there is at present no direct evidence to support the last type of process, mainly because high free surface potential always enhances crevice initiation of passive alloys.

On titanium alloys, however, the metal still has an active-passive transition in the crevice environment and the surface potential must be low enough for the active corrosion to occur. Thus, *IR* drop is required to stabilize the active dissolution [9,74], particularly when large cathodic currents are available.

10.4.3 Propagation of Crevice Corrosion

Propagation of crevice corrosion occurs by active dissolution in an occluded cell and it is generally considered analogous to the propagation stage of pitting corrosion.

At least in a first stage, corrosion may not affect the whole crevice gap: it has been shown that on austenitic stainless steels, dissolution is mainly located near the mouth of the crevice [3,19] (Figure 10.26), whereas on ferritic stainless steels the maximum of dissolution rate is located deeper in the crevice [3].

As a consequence of rapid corrosion inside the crevice, the rate of evolution of the environment toward more acidic and more concentrated solutions increases, and indeed several experiments confirmed this point. However, the dissolution rate and the environment evolution may be limited by the following factors:

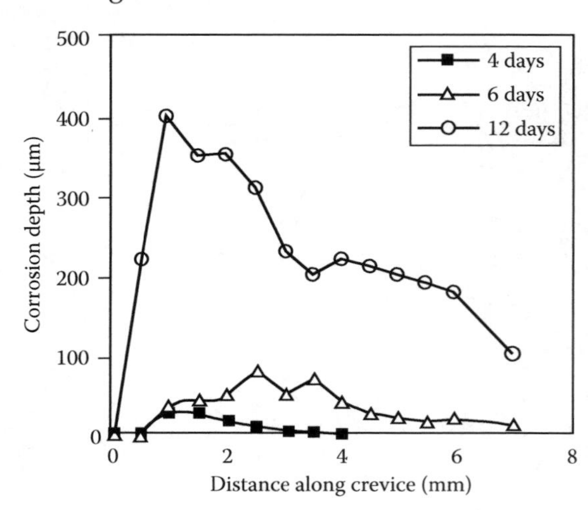

FIGURE 10.26
Preferential propagation near the crevice mouth. (From Valen, S., Thesis, Department of Materials and Processes, Norwegian Institute of Technology, Trondheim, Norway, 1991.)

Immediately after passivity breakdown, the dramatic increase of the dissolution current has two consequences that contribute to limit the dissolution rate by decreasing the corrosion potential of the active surfaces in the crevice:

1. The cathodic reaction kinetics increase on the free surfaces to balance the increased anodic dissolution and this induces a drop of their corrosion potential. Potential drops of several hundreds of mV can be observed on the free surfaces due to the initiation of crevice corrosion (see Figure 10.5) Large cathodic areas, high solution conductivity, and cathodic reaction depolarizers (such as carbon deposits) can increase significantly the amount of available cathodic current and thus the crevice propagation rate.
2. The migration current from the crevice gap increases dramatically and this also increases the ohmic drop: a role of the ohmic drop in the control of the propagation rate is consistent with a maximum of corrosion rate located near the crevice mouth as observed by several author [3,19]. One must notice that ohmic drop is not limited to the solution inside the crevice. Significant ohmic drops may also occur in the bulk solution near the crevice mouth, particularly in dilute environments.

Later, several other events may also occur to limit the environment evolution and the dissolution rate:

The solution inside the crevice reaches saturation and a salt film is formed on the active surface. The corrosion rate is then limited by mass transport trough the salt layer. Recent results [34,35,73,76–78] confirm this possibility, which was not as clearly established as it is for pit propagation.

The pH and the potential in the crevice may become low enough for local water reduction to be possible. This stifles further pH decrease as the local cathodic reaction is able to balance the production of cations by anodic dissolution.

Balance between production and outward transport of cations and dilution due to geometric changes of the crevice [79] are also invoked as possible limiting processes. However, on the simple basis of the difference of density between the dissolved metal and the saturated crevice solution, Hakkarainen [80] pointed out that at least 97% of the dissolved cations must be evacuated out of the active crevice. This weakens the preceding point. But, on the contrary, the possible limitation by the diffusion of water through a saturated salt layer has not been considered.

Thus, even though the corrosion occurs by active dissolution, its rate is not controlled by an activation process and, generally, it remains limited to current densities of the order of tens of mA/cm^2 in the saturated solutions, which prevail on well-developed crevices [19,33,57,70,71]. Figure 10.27 shows how the dissolution rate decreases when saturation of the solution is approached.

When discussing the propagation of crevice corrosion, it is worth noting that the analogy with the propagation of pits may have limitations. Most of the experimental work and many simplified calculations refer to a geometry of the corroding crevice with an aspect ratio L/h close to 1. This may have consequences for conclusions regarding the distribution of the dissolution current inside the crevice gap and the rate-controlling processes because an almost uniform corrosion potential on the corroding surface and an almost uniform transport path for the diffusion and migration processes can be assumed in pits. In very deep crevices, the potential and the transport path vary along the crevice profile; thus, the evolution of the local environment may vary. Consequently, the saturation and the formation of salt films leading to a dissolution process controlled by diffusion may not occur simultaneously on the whole crevice surfaces. In addition, the increasing *IR* drop near the tip of the crevice gap may allow the reduction of water to occur, changing the local pH evolution before any saturation has been reached.

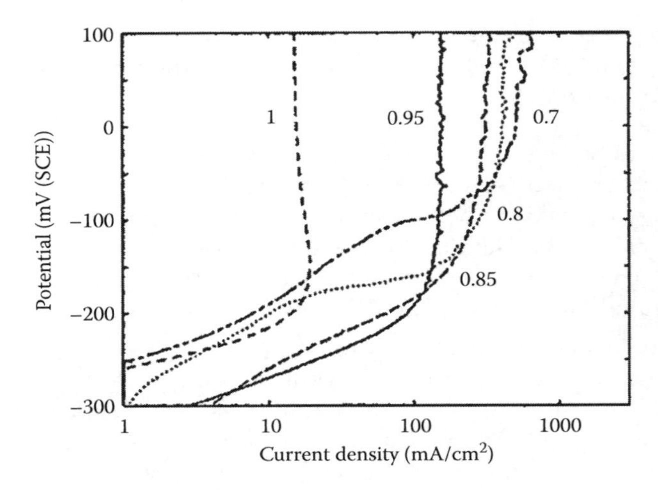

FIGURE 10.27
Anodic dissolution of 304 SS in simulated crevice solutions. The numbers indicate the degree of saturation of the solution. (From Hakkarainen, T., *Corrosion Chemistry in Pits, Crevices and Cracks,* A. Turnbull, ed., Her Majesty's Stationery Office, London, 1987.)

10.4.4 Crevice Repassivation

10.4.4.1 Spontaneous Crevice Arrest

Observations (see Section 10.4.3) suggest that, under constant bulk environment conditions, crevice corrosion may spontaneously cease. This has been taken into account in a model developed by Gartland [19]. Gartland attributed this behavior to an environment dilution caused by the geometric change of the crevice, but we have seen that this argument is not always likely. It also seems appropriate to invoke salt precipitation and/or local dryout due to the slow diffusion of water through a salt film of increasing thickness. This suggests that mass transport could not be efficient enough to eliminate such a large amount of corrosion products and this could lead to the precipitation of solid products that would fill the crevice gap. Another cause of self-arrest, or at least of kinetic limitation, could be lack of sufficient water in the crevice gap. Further studies are required to establish the conditions and the mechanisms involved in this spontaneous arrest, in order to take advantage of it in practical situations.

10.4.4.2 Repassivation Potential

The propagation of crevice corrosion can also be arrested by decreasing the potential of the outside surfaces below a critical value (see earlier). The existence of a "repassivation" or "protection" potential was recognized very early, in particular by Pourbaix et al. [81] for pitting corrosion. From a practical point of view, the existence of a protection potential below which no crevice corrosion is possible is of major importance because it guarantees the immunity of passivated alloys in near-neutral chloride solutions in the absence of oxidizing species and because it makes possible the cathodic protection of structures.

However, the significance and the intrinsic nature of this potential are still under discussion. There are at least two possible causes for active corrosion inside crevices to stop below a critical potential:

First, the critical potential may be a "deactivation potential," i.e., a potential that corresponds to a cancellation of the overpotential required for dissolution in the crevice. According to Starr et al. [82], if active corrosion stops by deactivation with decreasing potential, a further increase of the corrosion potential would cause an immediate reactivation of the crevice. Indeed, Starr et al. [82] and Dunn and Sridhar [83] observed such a reactivation on low-grade stainless steels in acidic environments.

Second, the critical potential may be a repassivation potential. It has been shown in artificial active crevices that lowering the potential of the free surfaces causes the local environment to become less aggressive (see, for example, Pourbaix [36]). Thus, at some point, the environment becomes not aggressive enough for active dissolution to be sustained and the metal surface in the crevice becomes passive. In this case, a subsequent increase of the corrosion potential does not produce immediate reactivation inside the crevice. Starr et al. [82] observed this situation on 12% Cr stainless steels in near-neutral environments; Dunn and Sridhar [83] observed the same behavior on alloy 825. However, the repassivation may be attributed to different environment changes: an increase of local pH in the crevice [82,84], a destabilization of the salt film that controls the dissolution kinetic in the crevice [77,85,86], or a decrease of the chloride concentration below a critical value [34,35,71–73,77,78,83]. Recent experimental observations [34,35,78] show that corrosion may be stable even after dissolution of the salt film.

In fact, studies of Cragnolino et al. [8,11] suggest that, depending on the conditions of the experimental determination, the "arrest" potential measured by backscan polarization may be close to either potential: using a relatively high scan rate, which does not allow time for the local environment to be significantly modified, the measured arrest potentials are close to the deactivation potentials and they are significantly lower than those measured under a low potential scan rate. The latter are more likely to be due to a modification of the local environment toward less aggressive conditions. We have already seen that the maximum repassivation potential seems to be almost identical to the minimum initiation potential (Figure 10.8). This observation has been reported for a relatively resistant stainless alloy (alloy 825) and it would support the idea of Gartland [19,57] (see earlier) that there is a critical environment for passivity breakdown and that the local environment required for repassivation is not very different. This could be an argument against the mechanism of initiation by pitting in the crevice and in favor of a mechanism involving a critical environment.

A pending question is still to know whether this repassivation potential has a unique value or depends on corrosion damages and crevice geometry. The following arguments are in favor of such a unique value:

1. For a given bulk environment and a given crevice configuration, the repassivation potential becomes independent of the corrosion extent as soon as some critical damage size is exceeded [8] (Figure 10.28).
2. At least for high chloride concentrations, the repassivation potential exhibits low values which are poorly dependent on the chloride content of the environment [87] (Figure 10.29). Dunn et al. expressed the average values of the repassivation potential as follows:

$$E_{rp} = E_{rp}^0(T) + B(T)\log[Cl^-] \quad (10.7)$$

with

$$E_{rp}^0(T) = 181.8 - 481(T) \quad \text{and} \quad B(T) = -64 - 0.8(T) \quad \text{for alloy 825} \quad (10.8)$$

However, Figure 10.29 shows that the effect of chloride content is low on the re-passivation potential, particularly when looking at the lower bound of the scatter b and of the measured value which is almost constant for chloride contents in excess of 10^{-1} M. This is consistent with the presence of a nearly saturated solution and/or precipitated salt film in well developed crevice corrosion, and indeed such environments should be almost independent on the bulk solution chemistry. Thus, a re-passivation potential which,

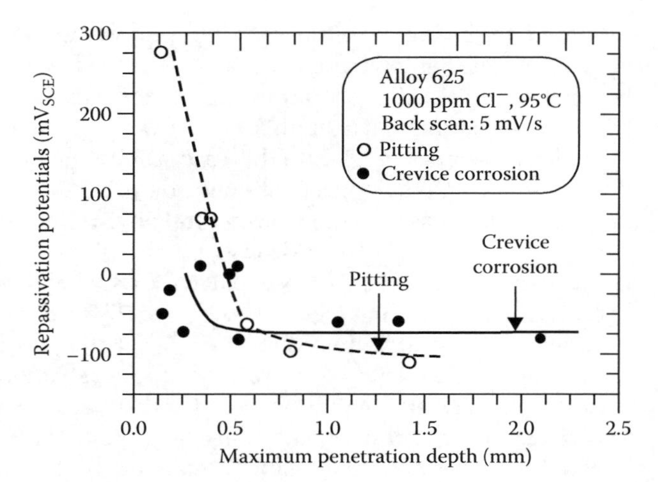

FIGURE 10.28
Influence of the corrosion damage on the repassivation potential of an alloy 825 in a 1000 ppm Cl^- solution at 95°C. (From Dunn, D.S., *Corrosion*, 52, 115, 1996.)

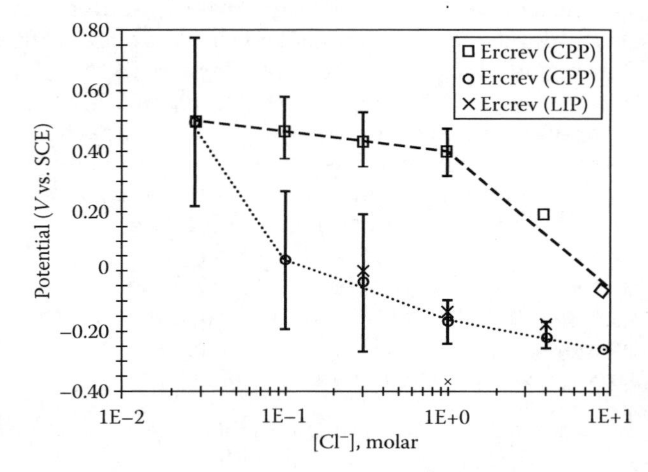

FIGURE 10.29
Influence of the chloride content on the repassivation potential of an alloy 625 at 95°C. (From Grass, K.A. et al., *CORROSION/98*, paper 149, NACE International, Houston, TX, 1998.)

according Pourbaix [36], is close to the local potential of an actively corroding crevice should be poorly or no dependent on the bulk environment and crevice geometry.

Nevertheless, this point still requires more work, particularly regarding the exact meaning of the high values of the repassivation potentials measured in low chloride and/or low temperature environments (left side of the Figures 10.10, 10.11, and 10.29). In this case, the local environment inside the crevice is likely to be controlled by mass transport and thus the repassivation potential could be dependent on the crevice geometry.

10.5 Modeling Crevice Corrosion

In order to try to predict the behavior of passive materials in chloride environments and particularly stainless alloys in seawater, many attempts have been made to model the environment in a crevice, mostly for stainless Fe-Ni-Cr-Mo alloys. Two kinds of model have been developed:

Steady state model [2,89–92], which try to calculate the environment in a propagating crevice or, more frequently, a pit, and

Transient models [2,4,19,57,79,93–99], which intend to describe the environment evolution in a crevice and, in most instances, to predict the passivity breakdown by using a criterion usually based on a critical environment resulting from experimental data.

Modeling the environment or the environment in a crevice requires solving a set of equations including:

The kinetics of surface reactions in the crevice, mainly the metal dissolution because the cathodic reaction is generally assumed to occur only on the outside surfaces

The kinetics of the reactions within the crevice solution, mainly the formation of metal chlorides and the hydrolysis of cations, which is the fundamental cause of pH drop within the crevice

The transport processes including electrolytic migration and chemical diffusion, as most of the models assume no convection inside the crevice gap

The mass and electrical charges balance in the crevice solution, across the metal-solution interfaces, and at the mouth of the crevice

10.5.1 Surface Reactions

With few exceptions [19], the passive anodic current of metal dissolution is usually taken as a constant and potential independent during the initiation stage and/or it is a parameter of the calculations.

During the propagation stage, different models are used, from the Tafel law [100] to experimentally fitted laws [19]. To take into account the strong current limitation due to the formation of a saturated salt layer on the dissolving surface, Gartland [19] introduced a damping coefficient that increases exponentially as the local concentration approaches saturation.

The possibility of water reduction inside the crevice gap is taken into account by Gartland [19] as a possible situation when the local pH and potential become low enough. Only one model [99] includes the cathodic reaction outside the crevice gap. It is more or less implicitly assumed in most models that this reaction is not a controlling process for the environment evolution inside the gap.

10.5.2 Solution Chemistry

The description of the crevice solution chemistry is probably the most difficult step (at least from the corrosion or chemistry standpoint if not the computer programming) as a result of the lack of reliable data. The main problems are:

The thermodynamic description of the solution inside the crevice, which cannot be done by the dilute solution models, at least during the propagation stage

The nature of the complex ions formed in the crevice and their thermodynamic properties, which are required to calculate the hydrolysis reactions governing the environment evolution

10.5.2.1 Solution Models: Activity Coefficients

Most models use dilute solution assumptions and former models [93] assumed unit values [2,19,93,94] of all activity coefficients (except in some cases for H^+ ions). Bernhardsson et al. [4] reached a compromise by using two sets of equilibrium constants, one for dilute solutions and the other for concentrated solutions. Other authors used Debye–Hückel, the truncated Davies model [96,98], or the B-dot Debye–Hückel model [98] to derive the activity coefficients in concentrated solutions.

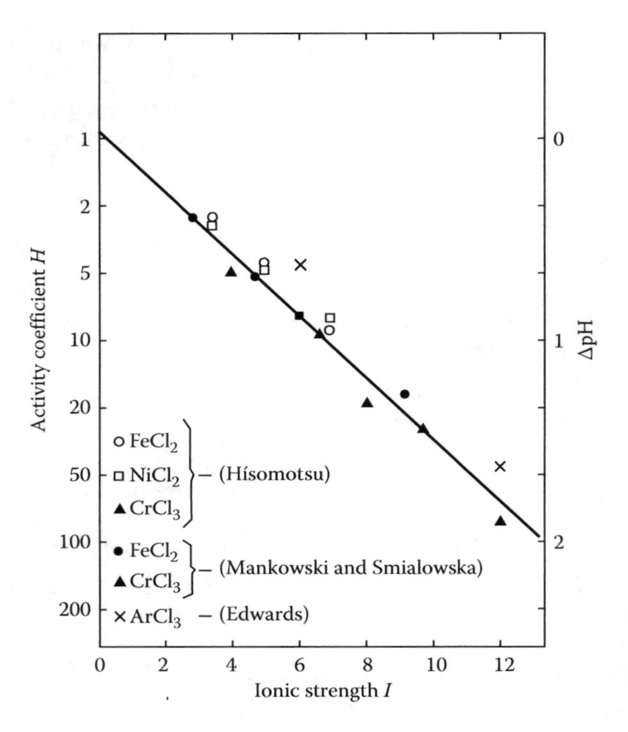

FIGURE 10.30
Influence of the ionic strength on the H^+ activity in chloride solutions. (From Gartland, P.O., *CORROSION'96—Research Topical Symposia. Part II. Crevice Corrosion: The Science and Its Control in Engineering Practice*, NACE International, Houston, TX, pp. 311–339, 1996.)

A particular case is the activity coefficient of H^+ ions. It has been shown experimentally (Figure 10.30) that the presence of metal ions strongly decreases the pH of a chloride solution. Many models [2,19,94,97] include a pH correction derived from the results of Mankowski and Smialowska [33].

More recently, Brossia et al. [35] introduced an approach based on new correlations for activity coefficients in concentrated solutions. This approach [101] is based on the Helgeson–Kirkham–Flowers equation for standard-state properties with a nonideal solution model based on the activity coefficient expression developed by Bromley and Pitzer. Using specific software, Brossia et al. were able to predict the dominant ionic species and the salt precipitation in Ni, Fe, Cr, and 308 stainless steels crevice solutions. Fair agreement was observed with the in situ analyses of these crevice solutions by Raman spectroscopy.

10.5.2.2 Hydrolysis and Complexation Products and Reactions

These species and reactions include the products of the metal cation hydrolysis and the formation of metallic chlorides.

In one of the first models [93], the formation of metal chlorides was not taken into account and the cation hydrolysis was assumed to produce solid hydroxides. In the case of stainless steels, the pH drop was assumed to be driven by the hydrolysis of chromium cations according to the following reaction:

$$Cr^{3+} + 3H_2O \rightarrow Cr(OH)_3 \tag{10.9}$$

In the case of aluminum alloys, the hydrolysis reaction was assumed to be

$$Al^{3+} + 3H_2O \rightarrow Al(OH)_3 \tag{10.10}$$

TABLE 10.3

Solubility and Hydrolysis Equilibria in the Model of Gartland

$Cr^{3}+H_2O \rightleftharpoons Cr(OH)^{2+}+H^+$	$pK = 4.84$
$2Cr^{3}+2H_2O \rightleftharpoons Cr_2(OH)_2^{4+}+2H^+$	$pK = 4.64$
$Cr^{3+}+Cl^- \rightleftharpoons CrCl^{2+}$	$pK = 0.65$
$Ni^{2+}+Cl^- \rightleftharpoons NiCl^+$	$pK = 0.25$
$Ni^{2+}+2Cl^- \rightleftharpoons NiCl_2$	$pK = 0.05$
$Fe^{2+}+Cl^- \rightleftharpoons FeCl^+$	$pK = -0.36$
$Fe^{2+}+2Cl^- \rightleftharpoons FeCl_2$	$pK = -0.40$
$Mo^{3+}+2H_2O \rightleftharpoons MoO_2+4H^++2e$	$E_0 = 0.061$ V SCE

Source: Gartland, P.O., *CORROSION'96—Research Topical Symposia. Part II. Crevice Corrosion: The Science and Its Control in Engineering Practice*, pp. 311–339, 1996.

Actually, the drop of pH is related to more complex reactions and species. Thus, in more sophisticated models, several hydrolysis reactions and metal chloride formation are taken into account but the selection of species and reactions is somewhat different from model to model. Oldfield and Sutton [94] and Watson and Postlethwaite [2] considered only hydroxides as the product of cation hydrolysis. Sharland [96] introduced simple metallic chlorides. The most complete set of species and reactions has been used by Bernhardsson et al. [4], which made available the thermodynamic data of a large number of species, including several iron, nickel, chromium, and molybdenum polycations as well as metal chlorides and hydroxychlorides. Gartland [19] used a more limited set of species (Table 10.3) selected among the Bernhardsson data. According to their experimental results, Hebert and Alkire [95] included $Al(OH)_n^{3-n}$ as the hydrolysis product in their model of the crevice corrosion of aluminum alloys.

A common feature of all the models is that all the chemical reactions of complex formation and cation hydrolysis are fast enough for thermodynamic equilibrium to be assumed. This may also be an incorrect assumption because we have already mentioned that the pH of synthetic crevice solutions can evolve for significant periods of time (up to several weeks) before reaching a stable value [33,40].

10.5.3 Transport Processes

Most models use transport equations for dilute solutions including diffusion, electrolytic migration, and convection such as:

$$J_i = -D_i \nabla C_i - z_i F u_i \nabla \varphi + v C_i \tag{10.11}$$

with

$$D_i = RTu_i \text{ (Nernst–Einstein relation)} \tag{10.12}$$

However, different attempts have been made to take into account the deviation from ideality of the crevice solutions:

Walton et al. [98] wrote the transport equation (10.11) in a form using the activity instead of the concentrations.

$$N_n = \frac{-D_n C_n}{RT} \nabla \tilde{\mu}_n = \frac{-D_n C_n}{RT} \nabla(\mu_n + z_n F \Theta_s) \tag{10.13}$$

However, this formulation assumes the validity of the Nernst–Einstein equation for any ionic strength.

Gartland [19] made an attempt to use corrected ionic mobility coefficients: he used the diffusion coefficients collected by Bernhardsson [4] regardless of the solution concentrations but, to calculate the ion mobility coefficient, he introduced in the Nernst–Einstein equation an ionic strength-dependent correction factor based on a limited set of experimental data.

Walton [102] also included a porosity tortuosity factor in the diffusion coefficient to take into account the presence of porous solids and/or gas bubbles in the crevice gap.

The ionic fluxes are generally used to derive the current flowing in the crevice:

$$i = \sum_i z_i J_i \tag{10.14}$$

One common assumption of most, if not all, the available models is that the transport is purely unidimensional; i.e., the crevice is assumed to be narrow enough for the transverse concentration and potential gradients to be negligible. In fact, this assumption is required in order to simplify the numerical solutions.

10.5.4 Mass Balance

The mass balance equation:

$$\frac{\partial C_i}{\partial t} = -\nabla J_i + R_i \text{ (=0 in steady-state propagation models)} \tag{10.15}$$

includes the production of species by chemical reactions in the liquid phase R_i. These reactions are the complexation and hydrolysis reactions as well as dissolution reactions on the crevice walls because models assume unidimensional transport fluxes.

10.5.5 Electrical Charge Balance

The electrical charge balance inside (and in one instance outside) the crevice is modeled in two ways:

Many models assume the electrical neutrality in any point [4,93,95,96]:

$$\sum_i z_i C_i = 0 \tag{10.16}$$

The potential distribution is obtained by using the Ohm law applied to the currents calculated in Equation 10.14 and solution resistivity calculated from mobility coefficients.

Others [2,79,97,99] solve the Poisson equation, which also gives the potential distribution along the crevice gap:

$$\nabla^2 \Theta = \frac{F}{\varepsilon} \sum_i z_i C_i = 0 \tag{10.17}$$

The solution of the Poisson equation dramatically increases the complexity and duration of the numerical simulations and requires numerical shortcuts. In addition, Watson and Postlethwaite [2] mentioned that the time required for electrical neutrality to be reached appears to be less than 10^{-10} s. This is very short compared with the time required for chemical equilibrium, and it is not clear whether the simple assumption of electrical neutrality could significantly change the results of such models.

10.5.6 Crevice Initiation Criteria

All criteria rely on a critical environment for passivity breakdown:

Critical pH [93]

Critical crevice solution [2,19,46,94] taking into account the pH and the chloride content but not the effect of metal cations

Critical dissolved Al content [95]

Despite the experimental observations, no attempt has been made to model crevice initiation by local pitting.

10.5.7 Results of the Models

As expected, the models predict the drop of local pH and the increase of chloride and metal ion content in the crevice gap. The predictions of several models were compared, with variable success, with the environment modifications observed in artificial crevices. Figure 10.31 shows a typical example of results indicating that the transport process is

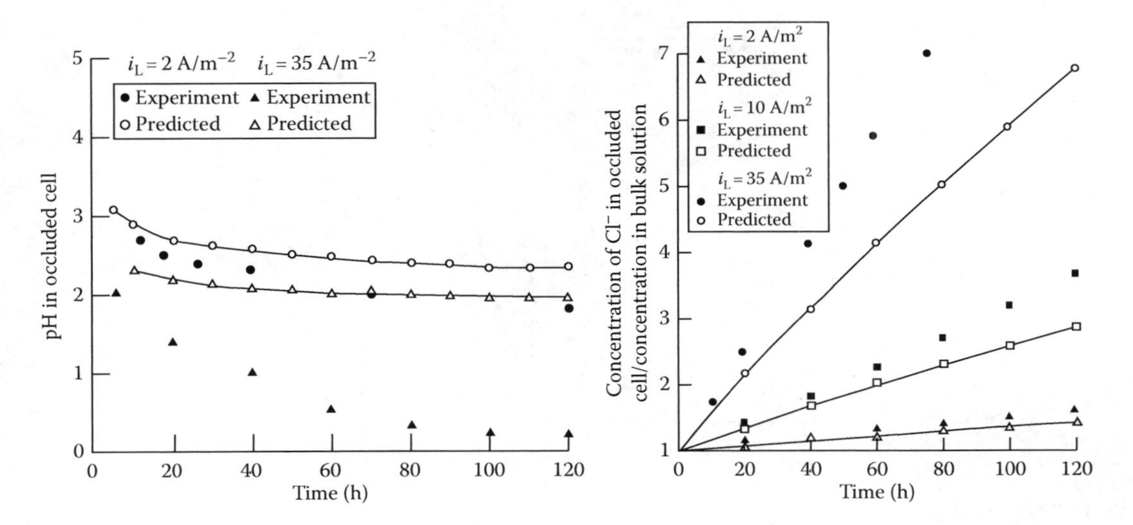

FIGURE 10.31
Comparison of the results of Zuo et al. [37] with the predictions of the model of Sharland [96]. Note that the chloride content is fairly well predicted, whereas the pH drop is underestimated. (From Zuo, J.Y., Jin, Z., Zhang, S., Xu, Y., and Wang, G., 9th International Congress on Metallic Corrosion, Toronto, p. 336/96, 1984; Sharland, S., *Corros. Sci.*, 33, 183, 1992.)

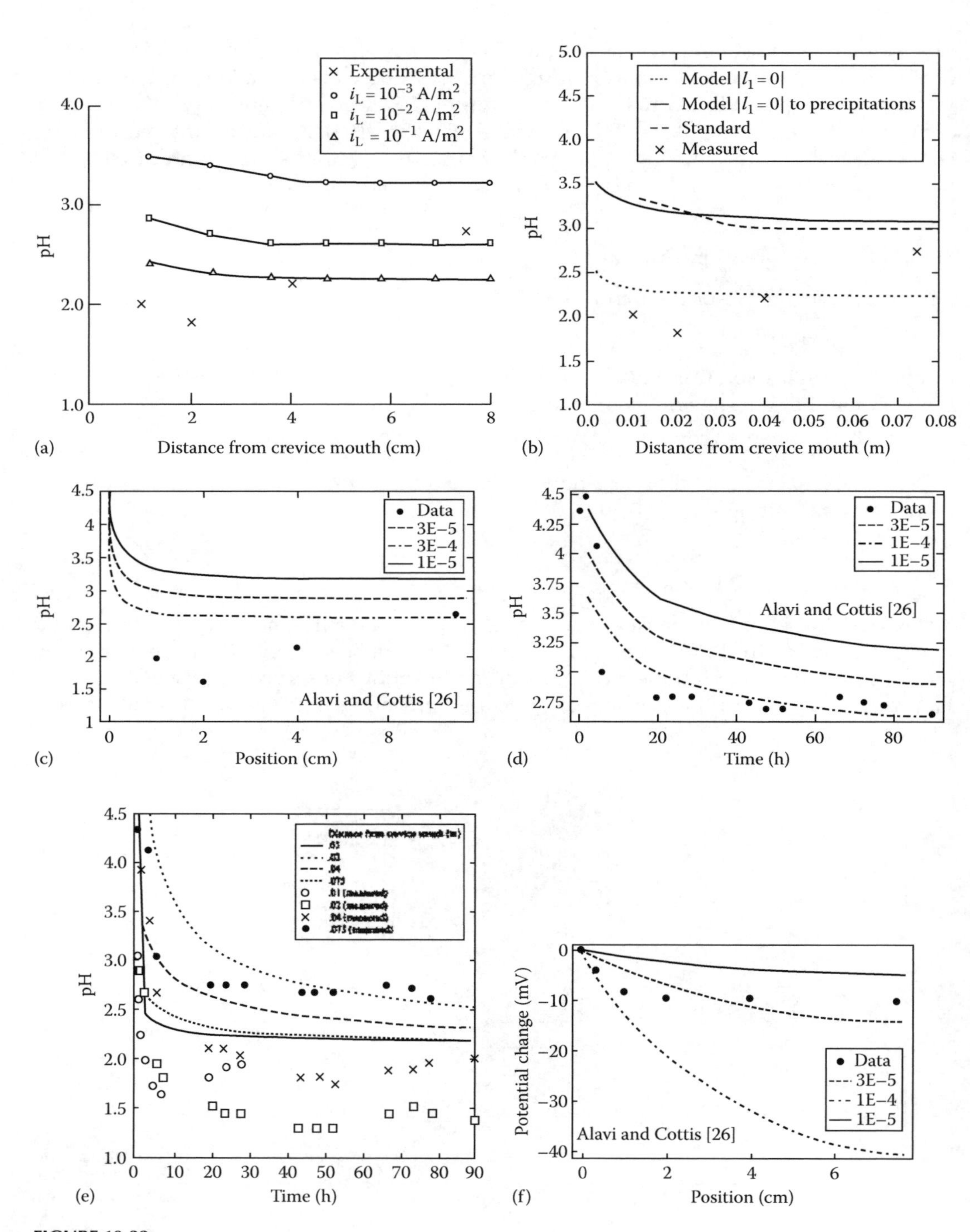

FIGURE 10.32
Comparison of the experimental results of Alavi and Cottis [26] with the predictions of the models of White et al. [99], Sharland [96], and Walton et al. [98] assuming various current densities. (a) pH as a function of position after 90 h after Sharland. (b) pH as a function of position after 90 h after White. (c) pH as a function of position after 90 h after Walton. (d) pH a 7.5 cm of the mouth after White. (e) pH after 90 h after White. (f) Potential drop as a function of the position after 90 h, after Walton.

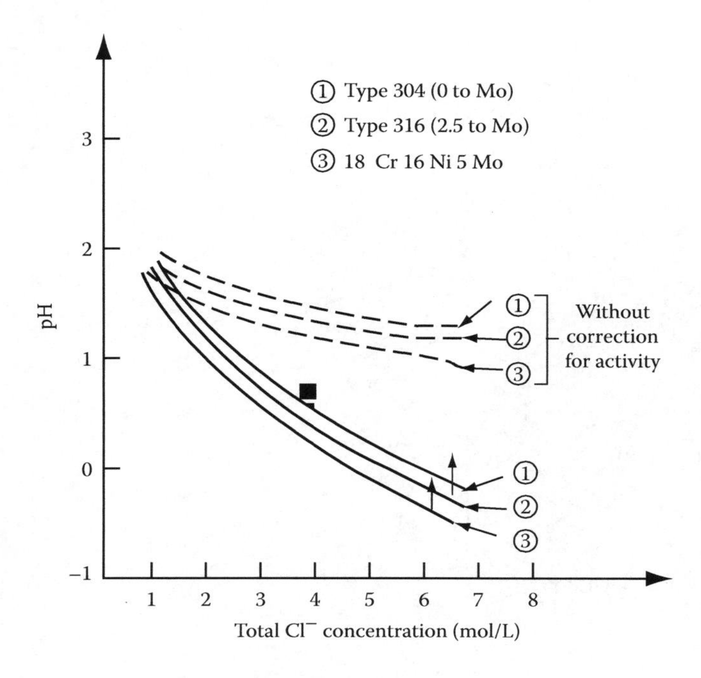

FIGURE 10.33
Effect of H^+ activity correction on the pH calculated by the Gartland model.

probably well enough modeled as indicated by the fair prediction of the chloride content but that the pH calculations are far less accurate. In Figures 10.32 and 10.33, the set of data of Alavi and Cottis [26] has been compared with several models: the pH drops are generally underestimated (Figure 10.32a through e), the potential gradient inside the crevice is not well described (Figure 10.32f), and no model is able to predict the pH increase observed near the crevice tip (Figure 10.3a and b).

Gartland [19] showed (Figure 10.33) that the use of a correction factor to evaluate the H^+ activity is necessary to approach the actual pH, regardless of the fact that the equilibria involving solubility and hydrolysis of metal cations are taken into account. This may be one of the major causes of the underestimation of pH drop observed in the preceding models.

Beside the quantitative results, which are impaired by the lack of basic knowledge of the crevice chemistry, the models provide interesting parametric results:

All models confirm that the crevice geometry is of major importance. As previously mentioned, Bernhardsson et al. [4] introduced an L^2/h geometric factor (in fact, a severity factor $S = i_a L^2/h$, Figure 10.34) that appears to control the environment evolution. The trends shown in Figure 10.34 are one of the main reasons to study in more detail the effect of crevice depth/width ratio on the crevice repassivation (see Section 10.4.4).

More precisely, the results of the Watson model clearly indicate (Figure 10.35) that corrosion can start only if the crevice is tight and deep enough and support the fact that a critical crevice size may exist. This is consistent with one conclusion of White et al. [99] which indicates that their model is not able to predict pH low enough for crevice initiation in gaps of the order of several tens of micrometers. These results, as well as results of Gartland [19], also show that efficient crevice gaps are of the order of micrometers or less, in accordance with experimental results of Oldfield [3]. This supports the assumption that microcrevices due to asperity may play a major role in crevice initiation, as suggested by Sridhar and Dunn [60].

Gartland [19] also predicted, in agreement with observation (Figure 10.26), that initial crevice damages are mainly located near the crevice mouth (Figure 10.36), where the *IR* drop is low, which makes possible higher dissolution rates.

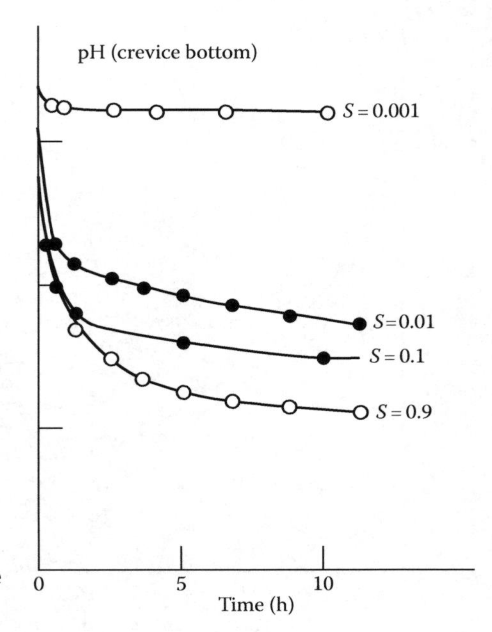

FIGURE 10.34
Effect of severity factor on pH drop calculated by the Bernhardsson model.

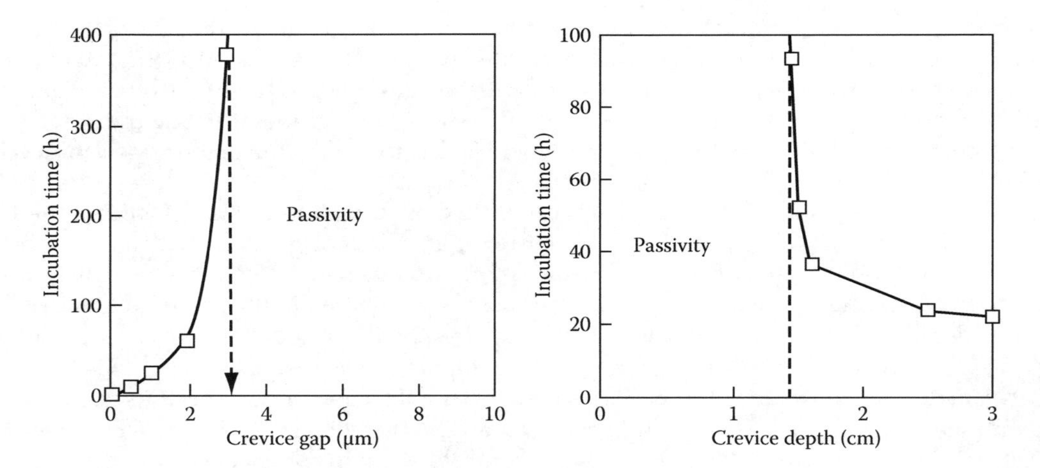

FIGURE 10.35
Effect of crevice geometry on crevice initiation time calculated by the Watson model.

Finally, these models also show that the initial pH drop may be fast but that, rapidly, the rate of evolution becomes very slow (Figures 10.34 and 10.37).

In summary, these models supply interesting information but still rely on experimental fitting to predict the initiation times correctly. Among their weaknesses, there is lack of precise data on the local chemistry of concentrated solutions and lack of prediction of the effect of the potential of the free surfaces. The use of a unidimensional transport equation and the assumption of instantaneous equilibrium of the hydrolysis and solubility reactions are also questionable.

Finally, whatever their accuracy, the use of crevice initiation models will always meet with some limitation because of the uncertainty in the actual crevice geometry.

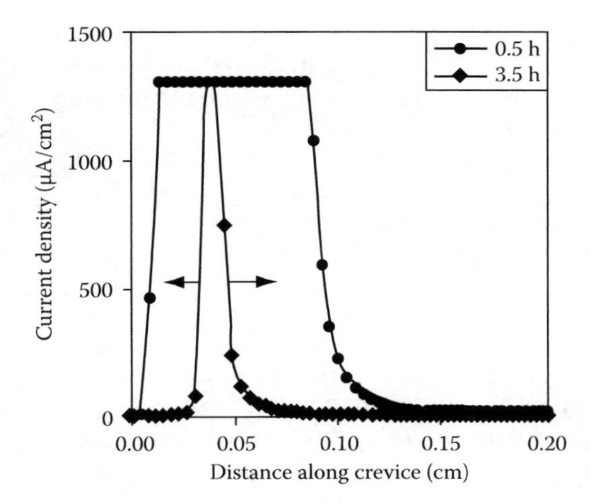

FIGURE 10.36
Prediction of the initial corrosion damage near the crevice mouth by the Gartland model (see Figure 10.26 to compare with experimental data).

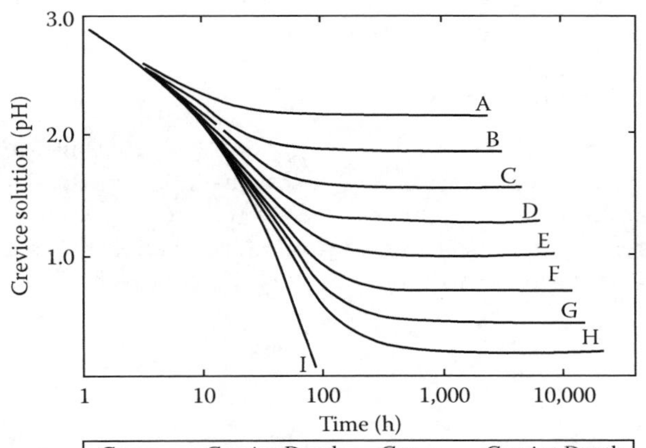

Curve	Crevice Depth	Curve	Crevice Depth
A	0.3 cm	F	0.8 cm
B	0.4 cm	G	0.9 cm
C	0.5 cm	H	1.0 cm
D	0.6 cm	I	∞
E	0.7 cm		

FIGURE 10.37
Drop of pH versus time as predicted by the Oldfield model.

However, a very useful model to be developed would be a propagation model, particularly in an attempt to obtain theoretical support to describe the repassivation potentials and, if possible, to predict self-arrest of crevice corrosion.

10.6 Experimental Characterization of the Alloys versus Crevice Corrosion

In order to select alloys for safe use in aerated chloride environments, a series of tests and criteria has been designed. They can be divided in several classes:

Exposure tests in environments representative of the service environment

Exposure tests in conventional environments

Electrochemical tests, which can be divided into two classes: (a) those intended to measure critical potentials for crevice initiation or protection and (b) those used to derive initiation criteria from the behavior of materials in conventional solutions supposed to simulate the crevice environment

For a more detailed description of crevice testing, the reader may consult the papers of Oldfield [103] and Ijsseling [104].

10.6.1 Crevice Former Devices

All the exposure and electrochemical tests that involve the use of crevice former devices encounter the problem of the geometry of the crevice. Many different designs of crevice former have been used, some examples of which are given in Figure 10.38a. The ASTM standard G78 defines a crevice former (Figure 10.38b) that involves 20 crevice zones and gives a series of recommendations, for example, for the setup of the specimens in the test vessels. This device allows more reproducible results because of the statistical effect of the large number of crevice zones. However, the standard does not specify important points such as the condition of the contact surface of the crevice former.

10.6.2 Exposure Tests in Representative Environments

These tests are designed to rank the behavior of different alloys, to get orders of magnitude of initiation times in service, and/or to select alloys that should guarantee long-term resistance to crevice initiation. Thus, they can be very long (months or even years) with carefully selected crevice former devices, materials, and surface conditions. A problem is the test monitoring to detect crevice initiation without any perturbation of the local processes.

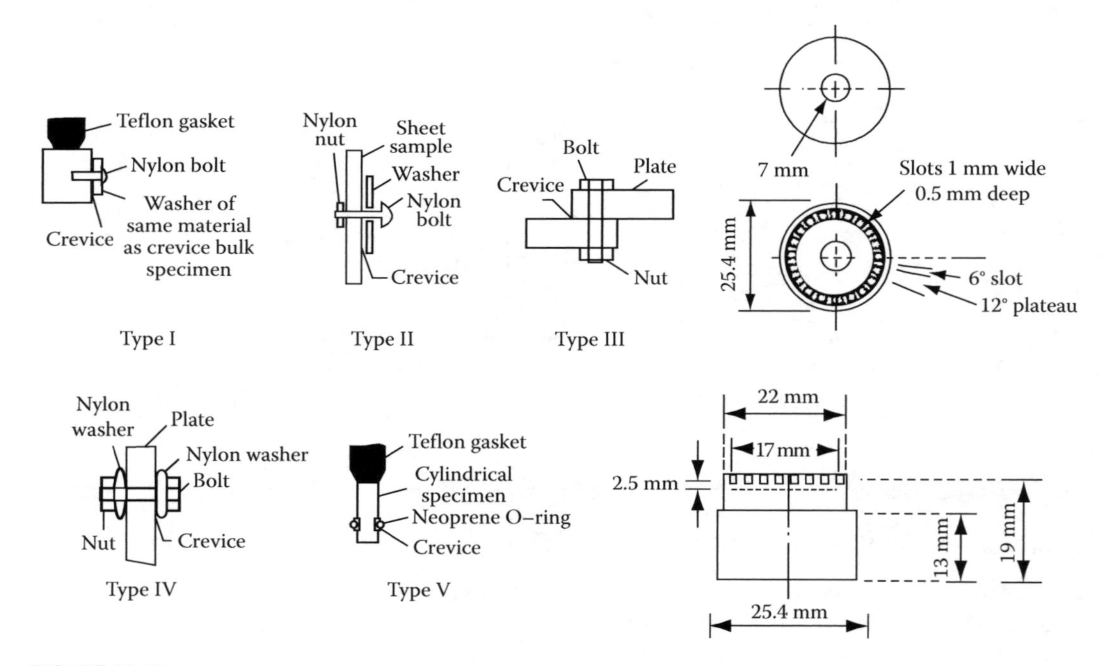

FIGURE 10.38
Crevice former devices for crevice corrosion testing: (a) Different types of crevice former. (From Wilde, B.E., *Corrosion*, 28, 283, 1972.) (b) ASTM G 78 crevice former device.

Apart from the selection of the crevice former device, it is important to have in mind that the severity of the test may depend on the availability of cathodic current. Thus, it is important to use specimens with large free surface area, particularly if the environment has high electric conductivity that allows long-distance galvanic coupling (seawater, for example).

There are different monitoring possibilities for exposure tests:

Periodic examination of the specimens is the simplest technique but it may dramatically modify the incubation period if the crevice zone becomes more or less dried during the examination. In particular, opening the crevice former to observe the inner surfaces may modify completely the local environment and cause the effective test time to go back to almost zero.

Monitoring of the corrosion potential allows one to detect any increase of the corrosion rate due to the polarization of the cathodic reaction on the free surfaces. Thus, the initiation of active corrosion in the crevice may be detected. In addition, electrochemical noise (i.e., potential fluctuations) indicating some dissolution transient may be interpreted as precursor events of passivity breakdown or at least of a low degree of passivity. The analysis of potential noise in terms of stochastic versus chaotic features has been shown [106] to allow early detection of pit initiation. This type of analysis should be checked for crevice corrosion.

Periodic measurements of polarization resistance may also be used to detect the onset of active corrosion.

10.6.3 Exposure Tests in Conventional Environments: The Ferric Chloride Test

This test is defined by the ASTM standard G48 and it is used for stainless steels and nickel-base alloys. Coupons with a crevice former device are exposed to a 6% $FeCl_3$ solution. Two criteria are used to characterize the tested materials:

The weight loss after a given time of exposure at a given temperature.

A critical crevice temperature obtained by periodically increasing the temperature by steps of 2.5 or 5°C until passivity breakdown occurs in the crevice. However, because the passive film becomes more stable with increasing exposure time, the critical temperature may depend to some extent on the temperature selected to start the test.

An experimental correlation shows that this test can be used to rank the behavior of stainless alloys in seawater. In practice, it is also used as a control test to guarantee the constant quality of a product or to check the effect of fabrication parameters such as thermal treatment, welding, and surface condition.

10.6.4 Electrochemical Tests

10.6.4.1 Determination of Critical Potentials for Initiation and Protection

These tests are the same as those used to determine the pitting resistance except that a crevice former device is present on the test coupons. The measured parameters are:

The crevice potential and the repassivation potential obtained on polarization curves (potentiokinetic technique) or, in some instances, by using potential steps or potentiostatic tests

The critical temperature for the crevice at a given potential

10.6.4.1.1 Potentiokinetic Techniques

Figure 10.7 represents typical polarization curves of a passivated alloy in a near-neutral chloride solution with and without a crevice former device. The shapes of the two curves are identical except that the crevice potential is definitely lower than the pitting potential. In addition, it may depend on the crevice geometry if not properly measured.

As already discussed, the main difficulty of this technique, beside the problem of the crevice former geometry and reproducibility, is that the results are strongly dependent on the potential scan rate both because of the time-dependent stability of the passive films and because of the time-dependent evolution of the environment inside a crevice. In particular, the repassivation potential may be overestimated if corrosion is not well developed in the crevice and it can be underestimated if the potential backscan is too fast to allow the evolution of the local environment to be in "quasi-steady" conditions. It is generally admitted that the scan rate has to be very low, which causes the two critical potentials to become closer. But the appropriate scan rate must be determined on each system because it may depend on the alloy and on the environment.

10.6.4.1.2 Potentiostatic Techniques

To determine the critical potential of crevice initiation, coupons in a crevice former device are exposed for a fixed period of time under potentiostatic control and monitoring of the anodic current is used to detect the onset of active corrosion. Several experiments are performed at different potentials and the crevice potential is the threshold potential that corresponds to an infinite initiation time (see Figures 10.8 and 10.9).

To measure the repassivation potential, severe crevice corrosion must be initiated on a "creviced" specimen and allowed to propagate for a fixed period of time or, more usually, to a definite amount of anodic charge. Then the potential is stepped to a lower value and current monitoring is used to check the propagation or repassivation. As shown in Figure 10.8, the repassivation time increases with increasing applied potential and the repassivation potential is the potential corresponding to an infinite repassivation time. Apparently, the repassivation potential may depend on the development of corrosion before the potential is stepped down only if the amount of accumulated charge is lower than a critical value.

The potentiostatic techniques should be preferred to the potentiokinetic ones, but they require more specimens and longer experiments.

A more sophisticated repassivation test has been designed [7,12] to minimize the number of specimens by using a stepwise backscan potential (Figure 10.39). During each step, the evolution of the current is analyzed and time is allowed to determine the trend. If, after decreasing, the anodic current increases again, indicating that propagation is not inhibited, the potential is stepped down again until repassivation occurs.

10.6.4.2 Determination of the Material Behavior in Simulated Crevice Environments

The tests that have been widely used in the past years are intended to determine the conditions of passivity breakdown in the crevice. Alloy coupons with no crevice area are tested in solutions that are supposed to simulate the local conditions in crevices. This is usually done by using acidic solutions with an increased chloride content compared with the bulk environment. For example, to perform tests to characterize the behavior in sea water, solutions containing 30–150 g/L of NaCl were used, the pH being lowered by HCl additions. Different measurements can be performed.

Activation pH: This pH is the pH below which an active peak is present on the polarization curves, (Figure 10.40). This criterion was mainly developed by Crolet et al. [55], The lower the activation pH, the more resistant the alloy to crevice corrosion.

pH of spontaneous film breakdown: It is measured by continuously lowering the pH of the solution until a freely corroding test coupon suffers a sudden potential drop due to passivity

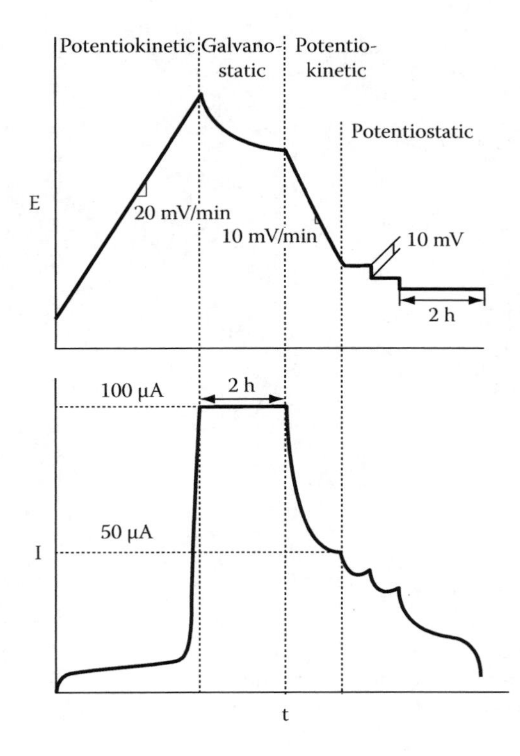

FIGURE 10.39
Stepwise potential backscan and current response to determine the repassivation potential. (From Yashiro, H. et al., *Corrosion*, 46, 727, 1990.)

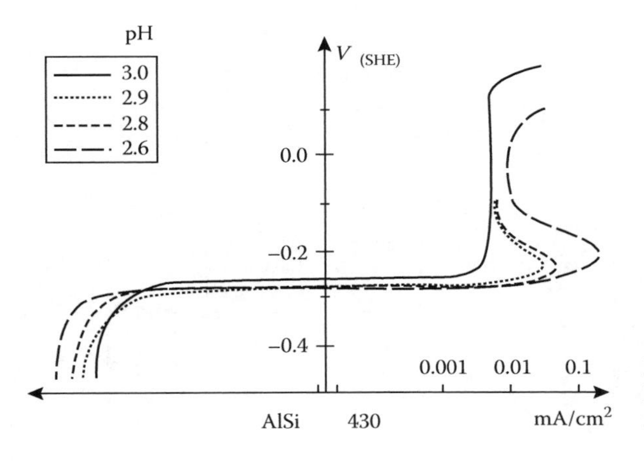

FIGURE 10.40
Measurement of activation pH on 430 SS (2 M NaCl, 23°C). (From Crolet, J.L. et al., *Mater. Tech.*, 355, 1978.)

breakdown (Figure 10.41). This criterion, which leads to a critical pH somewhat lower than the former, was used by Ogawa [42].

Active peak current as a function of pH: Several authors [3,103,108] use the measurement of the active peak current below the activation pH to compare the propagation kinetics of crevice corrosion on different materials. Figure 10.42 shows an example of this kind of measurement. Notice that the peak current of the duplex stainless steels (UR 47, 31803, 31603) is much higher at low pH values than that of austenitic grades (304 and 904L). This is consistent with the beneficial effect of nickel on the crevice propagation rate mentioned earlier.

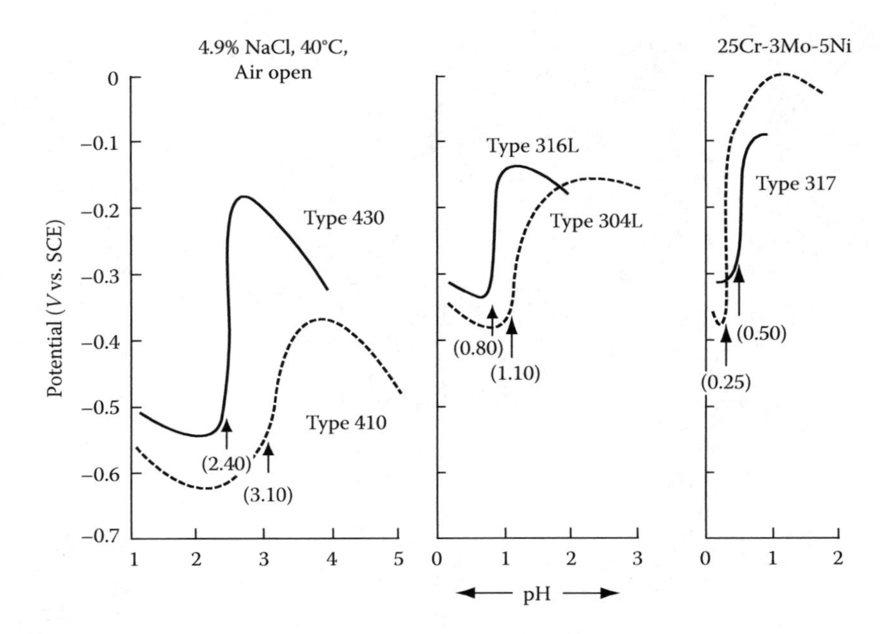

FIGURE 10.41
Measurement of spontaneous film breakdown pH. (From Ogawa, H., *A Mechanistic Analysis of Crevice Corrosion on Stainless Steels and the Calculation of the Incubation Time of the Crevice Corrosion Nucleation*, Nippon Steel Corporation, Japan, 1987.)

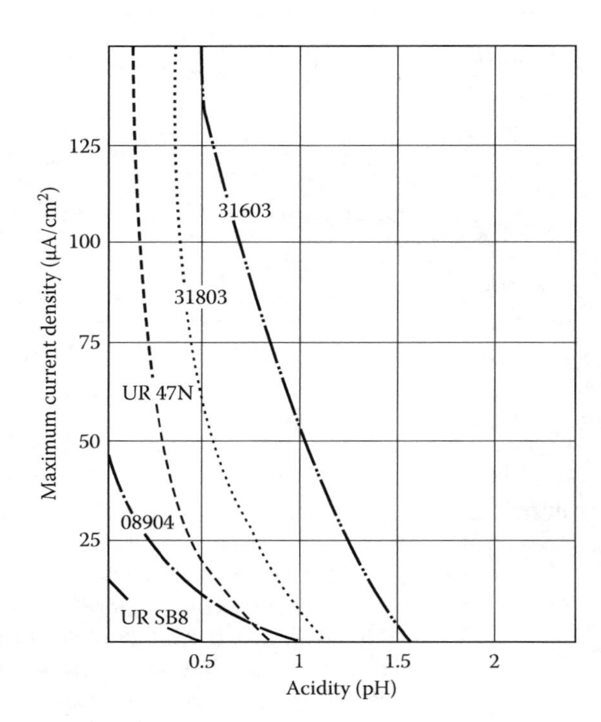

FIGURE 10.42
Evolution of peak current as a function of pH on different stainless steels. (From Crolet, J.L. et al., *Mater. Tech.*, 355, 1978.)

All these electrochemical criteria suffer from several drawbacks:

1. It is assumed that passivity breakdown in a crevice is due to pH drop but, as discussed, the actual mechanisms may be different.
2. Conventional solutions that are not representative of crevice environment are used. In particular, the presence of metal cations in the crevice solution is not considered.

However, these criteria probably allowed a correct ranking of the stainless alloys because the various methods are sensitive to the alloying elements that improve passivity (mainly Cr and Mo).

10.7 Prevention of Crevice Corrosion

Prevention of crevice corrosion must be taken into account at different stages of the life of an industrial apparatus:

Material selection: The more resistant the materials to crevice corrosion (and more generally to localized corrosion) in chloride environments, the more expensive they are. Thus, the selection of a material exhibiting a very high resistance to crevice corrosion should not be the only way considered to prevent crevice corrosion: improved design, construction procedures, surface finish, maintenance procedures, and process management can be of a major importance.

Cathodic protection: Because crevice corrosion occurs above a critical potential, cathodic protection is an efficient way to prevent crevice corrosion by maintaining the potential of the free surfaces below the protection potential. This is the case of 17-4-PH stainless steel, which is not resistant to crevice corrosion in seawater but is commonly used because it performs very well under cathodic protection (this protection is often due to galvanic coupling to unalloyed steel).

Design: An improved design should minimize the crevice areas, for example, by avoiding overlapping welded parts and by polishing seals bearing. However, crevice zones can be created by the formation of deposits during service and/or shutdown periods. Deposits can be minimized, for example, by avoiding low flow rate areas and by allowing complete draining of tanks.

Fabrication procedure and surface finish: The intrinsic resistance of a material to crevice corrosion may be dramatically hampered by inappropriate fabrication procedures. The most sensitive points are welding procedures, which must be optimized by an appropriate energy input and, if necessary, by the use of more alloyed filler metal; thermal treatments, particularly if stress relief has to be performed; and surface conditions, which may suffer from fabrication events (scratches with iron incrustation, drops of welding filler metal, etc.) and which should be restored at the end of the fabrication process, for example, by a passivation treatment in the case of stainless alloys.

In service, transient, and shutdown conditions—maintenance procedures: Care has to be taken of the possible change of conditions during shutdown and/or transient periods, which are not necessarily less prone to the development of crevice corrosion. An example is the case of desalination plants where concentrated chloride solutions may become very dangerous during shutdown if strict deaeration is not maintained. The possibility of formation of deposits during transient and shutdown periods and the occurrence of surface damage during maintenance operations are among the risks that have to be considered. Generally speaking, the best possible cleanliness of the apparatus must be maintained during all the apparatus lifetime.

10.8 Conclusion

Crevice corrosion occurs in restricted transport zones where the access of reactants and the elimination of corrosion products can be very slow. The detrimental effect of a crevice is related to a very small volume of solution in contact with very large metal surfaces. The restricted transport path between crevice and bulk solution is responsible not only for low diffusion exchanges but also for the buildup of a potential difference between the free surfaces and the crevice that becomes more anodic (ohmic drop effect).

These conditions trigger an evolution of the local environment in the crevice. Crevice corrosion may occur when the environment becomes more aggressive in the crevice than in the bulk environment.

On passivated alloys, and particularly on stainless alloys (Fe-Ni-Cr-Mo-... alloys), crevice corrosion occurs when the local conditions in the crevice cause passivity breakdown and the onset of active dissolution. As a consequence, crevice corrosion exhibits two different stages: an incubation period during which the local environment in the crevice evolves toward critical conditions and a propagation period that starts at the initiation time corresponding to the passivity breakdown.

The most widespread cases of crevice corrosion of passivated alloys are caused by aerated (or more generally oxidizing) chloride solutions such as sea or brackish water. In these chloride solutions, the environment in the crevice becomes progressively more acidic and more concentrated in chloride anions and metal cations. There are several possible causes of passivity breakdown, including low pH, high chloride content, presence of metallic chloride complexes, and pitting inside the crevice gap. Passivity breakdown occurs only if the corrosion potential of the free surface exceeds a critical value, but the relationship between the potential of the external surfaces and the evolution of the environment in the crevice is not completely understood.

Whatever the initiation mechanism, propagation causes further evolution of the local environment, which eventually becomes saturated (with concentrations estimated to be several mol/L of chloride and metal cations). At this point, a salt layer may form on the corroding surfaces. Thus, active corrosion in the crevice is rarely controlled by an activation process but rather by the ohmic drop and, for well-developed crevices, by cation transport through the salt layer, the thickness of which determines the corrosion rate.

An interesting feature that can be used to avoid crevice corrosion is that propagation stops if the potential of the free surfaces becomes lower than a critical value, referred to as repassivation potential. At least for alloys exhibiting a significant resistance to crevice corrosion, it is believed that this corrosion arrest is due to a repassivation of the metal surface inside the crevice. It is quite possible that, after long exposure periods, the critical potential for crevice initiation and the protection potential have a common value, but this point must be confirmed. The "protection" or "repassivation" potential can be a safe design criterion to avoid localized corrosion in long-term service. In particular, it provides a criterion independent of the surface condition because it guarantees against propagation of corrosion of the bulk material.

As the basic mechanisms of crevice corrosion have been understood to some extent, many models have been developed either to describe the steady state of an actively corroding crevice or to calculate the environment evolution and predict the passivity breakdown. These models are able to reproduce some major features of crevice corrosion, particularly geometric effects, but they all suffer from the lack of knowledge of the complex concentrated solutions that are formed in the crevice.

Crevice corrosion is a typical example of a corrosion problem in which the solution involves more than just the selection of fully resistant alloys. Design, fabrication procedure, and surface finish must be optimized and attention paid to the management of service conditions, as well as transient and shutdown periods. Optimization of maintenance procedures to maintain the best cleanliness of the surfaces may also contribute to prevent crevice corrosion.

References

1. I. L. Rosenfeld, *5th International Congress on Metallic Corrosion*, Tokyo, Japan, 1972, p. 53.
2. M. K. Watson and J. Postlethwaite, *Corrosion 46*:522–530 (1990).
3. J. W. Oldfield, *19th Journées des Aciers Spéciaux—International Symposium on Stainless Steels*, Saint Etienne, France, 1980.
4. S. Bernhardsson, L. Eriksson, J. Oppelstrup, I. Puigdomenech, and T. L. Wallin, *8th International Congress on Metallic Corrosion*, Mainz, Germany, 1981, pp. 193–198.
5. R. C. Alkire, T. Tomasson, and K. Hebert, *J. Electrochem. Soc. 132*:1027–1031 (1985).
6. D. F. Taylor and C. A. Caramihas *J. Electrochem. Soc. 129*:2458–2464 (1982).
7. M. Akashi, G. Nakayama and T. Fukuda, *CORROSION/98*, paper 158, NACE International, Houston, TX, 1998.
8. D. S. Dunn, G. A. Cragnolino, and N. Sridhar, *Corrosion 52*:115–124 (1996).
9. P. Mckay and D. B. Mitton, *Corrosion 41*:52–62 (1985).
10. A. Takamura, *Corrosion 23*:306–313 (1967).
11. N. Sridhar and G. A. Cragnolino, *Corrosion 49*:885–894 (1993).
12. H. Yashiro, K. Tanno, H. Hanayama, and A. Miura, *Corrosion 46*:727–733 (1990).
13. J. W. Oldfield and R. N. Kain, *12th International Corrosion Congress*, Houstan, Texas, 1993, p. 1876.
14. A. J. Sedriks, *CORROSION'96—Research Topical Symposia. Part 2. Crevice Corrosion: The Science and Its Control in Engineering Practice*, NACE International, Houston, TX, 1996, pp. 279–310.
15. *Remanit 4565S-Nitrogen Alloyed Austenitic Stainless Steels with Maximum Corrosion Resistance and High Strength*, Thyssen Stahl AG, Krefeld, Germany, 1995.
16. T. Sydberger, *Werkst. Korros. 32*:119–128 (1981).
17. W. Yang, R. Ni, H. Hua, and A. Pourbaix, *Corros. Sci. 24*:691–707 (1984).
18. F. Hunkeler, and H. Boehni, *Passivity of Metals and Semiconductors* (M. Froment, ed.), Elsevier, Amsterdam, the Netherlands, 1983.
19. P. O. Gartland, *CORROSION'96—Research Topical Symposia. Part II. Crevice Corrosion: The Science and Its Control in Engineering Practice*, NACE International, Houston, TX, 1996, pp. 311–339.
20. J. W. Oldfield, T. S. Lee, and R. N. Kain, *Corrosion Chemistry in Pits, Crevices and Cracks* (A. Turnbull, ed.), Her Majesty Stationery Office, London, 1987.
21. S. Valen, P. O. Gartland, and U. Steinsmo, *CORROSION/93*, paper 496, NACE International, New Orleans, LA, 1993.
22. T. Tsuru, K. Hashimoto, A. Nishikata, and S. Haruyama, *Materials Science Forum*, Vols. 44 and 45, Trans Tech Publications, Switzerland, 1989.
23. W. B. A. Sharp and L. H. Laliberté, *CORROSION/78*, paper 16, NACE International, Houston, TX, 1978.
24. F. D. Bogar and C. T. Fujii, NRL Report 7690, AD 778 002, 1974.

25. E. D. Verink, K. K. Starr, and J. M. Bowers, *Corrosion 32*:60–64 (1976).
26. A. Alavi and R. A. Cottis, *Corros. Sci. 27*:443–451 (1987).
27. B. E. Wilde and D. E. Williams, *Electrochim. Acta 16*:1971–1985 (1971).
28. T. Suzuki, M. Yamabe, and Y. Kitamura, *Corrosion 29*:18–22 (1973).
29. M. Marek, J. G. Rinker, and R. F. Hochman, *6th International Congress on Metallic Corrosion*, Sydney, Australia, 1975, p. 502.
30. N. J. H. Holroyd, G. M. Scamans, and R. Hermann, *Corrosion Chemistry in Pits, Crevices and Cracks* (A. Turnbull, ed.), Her Majesty's Stationery Office, London, U.K., 1987.
31. K. P. Wong and R. C. Alkire, *J. Electrochem. Soc., 137*:3010–3015 (1990).
32. A. Turnbull, *Corros. Sci. 23:* 833–870 (1983).
33. J. Mankowski and Z. Szklarska-Smialowska, *Corros. Sci. 15*:493–501 (1975).
34. N. Sridhar and D. S. Dunn, *J. Electrochem. Soc. 144*:4243 (1997).
35. C. S. Brossia, D. S. Dunn, N. Sridhar, *Critical Factors in Localized Corrosion III* (R. G. Kelly, G. S. Frankel, P. M. Natishan, and R. C. Newman, eds.), The Electrochemical Society Proceedings, Pennington, NJ, Vol. 98–17, 1998, p. 485.
36. M. Pourbaix, *Corrosion 26*:431–438 (1970).
37. J. Y. Zuo, Z. Jin, S. Zhang, Y. Xu, and G. Wang, *9th International Congress on Metallic Corrosion*, Toronto, Ontario, Canada, 1984, p. 336.
38. A. Turnbull and M. K. Gardner, *Br. Corros. J. 16*:140 (1981).
39. M. H. Peterson and T. J. Lennox, Jr., *Corrosion 29*:406–410 (1973).
40. P. Combrade, M. C. Bonnet, and H. Pages, *Proceedings of Innovation Stainless Steels*, Florence, Italy, 1993, p. 215.
41. L. Y. Zhen, *9th International Congress on Metallic Corrosion*, Toronto, Ontario, Canada, 1984, p. 350.
42. H. Ogawa, *A Mechanistic Analysis of Crevice Corrosion on Stainless Steels and the Calculation of the Incubation Time of the Crevice Corrosion Nucleation*, Nippon Steel Corporation, Japan, 1987.
43. P. Marcus and V. Maurice, *Corrosion and Environmental Degradation* (M. Schütze, ed.), Wiley, New York, 2000.
44. N. Lukomski and K. Bohnenkampf, *Werkst. Korros. 30*:482 (1979).
45. W. Battista, A. M. T. Louvisse, O. R. Mattos, and L. Sathler, *Corros. Sci. 28*:759–768 (1988).
46. J. W. Oldfield and W. H. Sutton, *Br. Corros. J. 13*:104–111 (1978).
47. S. E. Lott and R. C. Alkire, *J. Electrochem. Soc. 136*:973–979 (1989).
48. A. Garner, *Corrosion 41*:587–591 (1985).
49. R. C. Newman, *Corrosion 45*:282–287 (1989).
50. A. Garner and R. C. Newman, *CORROSION/91*, paper 186, NACE International, Houston, TX, 1991.
51. C. S. Brossia and R. G. Kelly, *Critical Factors in Localized Corrosion II*, The Electrochemical Society Proceedings Volume, Pennington, NJ, 1995, pp. 201–217.
52. C. S. Brossia, R. G. Kelly, *Corros. Sci. 40*:1851–1871 (1998).
53. J. Oudar and P. Marcus, *Applied Surface Science 3*:48 (1979).
54. J. L. Crolet, L. Séraphin, and R. Tricot, *Rev. Metall.* 937–947 (1975).
55. J. L. Crolet, J. M. Defranoux, L. Séraphin, and R. Tricot, *Mem. Sci. Rev. Metall. 71*:797–805 (1974).
56. S. Okayama, S. Tsujikawa, and K. Kikuchi, *Corros. Eng. 36*:631–638 (1987).
57. P. O. Gartland and S. Valen, *CORROSION/91*, paper 511, NACE International, Cincinnati, OH, 1991.
58. S. E. Lott and R. C. Alkire, *Corros. Sci. 28*:479–484 (1988).
59. K. Sugimoto and K. Asano, *Advances in Localized Corrosion—NACE 9* (H.S. Isaacs, U. Bertocci, J. Kruger, and S. Smialowska, eds.), NACE, Houston, TX, 1990.
60. N. Sridhar and D. S. Dunn, *Corrosion 50*:857–872 (1994).
61. G. Karlberg and G. Wranglen, *Corros. Sci. 11*:499–510 (1971).
62. N. J. Laycock, J. Stewart, and R. C. Newman, *Corros. Sci. 39*:1791–1809 (1997).
63. Z. Szklarska-Smialowska and J. Mankowski, *Corros. Sci. 18*:953–960 (1978).

64. G. S. Eklund, *J. Electrochem. Soc. 123*:170–173 (1976).
65. T. Shinohara, N. Masuko, and S. Tsujikawa, *Corros. Sci. 35*:785–789 (1993).
66. L. Stockert and H. Boehni, *Mater. Sci. Forum 44/45*:313 (1989).
67. S. E. Lott and R. C. Alkire, *J. Electrochem. Soc. 136*:3256–3262 (1989).
68. K. Hebert, R. C. Alkire, *J. Electrochem. Soc. 130*:1001–1007 (1983).
69. S. Zakipour and C. Leygraf, *Corrosion 37*:363–368 (1981).
70. T. Hakkarainen, *Electrochemical Methods in Corrosion Research—Materials Science Forum* (M. Duprat, ed.), Vol. 8, Trans Tech Publications, Switzerland, 1986.
71. T. Hakkarainen, *Corrosion Chemistry in Pits, Crevices and Cracks* (A. Turnbull, ed.), Her Majesty's Stationery Office, London, 1987.
72. G. T. Gaudet, W. T. Mo, T. A. Hatton, J. W. Tester, J. Tilly, and H. S. Isaacs, *AIChE J. 32*:949–958 (1986).
73. U. Steinsmo and H. S. Isaacs, *J. Electrochem. Soc. 140*:643–653 (1993).
74. J. C. Griess, *Corrosion 24*:96–109 (1968).
75. S. Valen, Thesis, Department of Materials and Processes, Norwegian Institute of Technology, Trondheim, Norway, 1991.
76. H. S. Isaacs, J.-H. Ho, M. L. Rivers, and S. R. Sutton, *J. Electrocherm. Soc. 142*:1111–1118 (1995).
77. N. J. Laycock and R. C. Newman, *Corros. Sci. 39*:1771–1790 (1997).
78. N. Sridhar, D. S. Dunn, C. S. Brossia, and G. A. Cragnolino, *CORROSION 2001—Research Topical Symposium on Localized Corrosion*, NACE International, Houston, TX, 2001.
79. R. W. Evitts, J. Postlethwaite, and M. K. Watson, *CORROSION/92*, paper 75, NACE International, Nashville, TN, 1992.
80. T. Hakkarainen, *CORROSION 96—Research Topical Symposia. Part II. Crevice Corrosion: The Science and Its Control in Engineering Practice*, Houston, TX, 1996, pp. 355–366.
81. M. Pourbaix, L. Klimzack-Mathieu, C. Martens, J. Maunier, C. Vanlengenhaghe, L. D. Munch, J. Laureys, L. Nellmans, and M. Warzee, *Corros. Sci. 3*:239 (1963).
82. K. K. Starr, E. D. Verink, and M. Pourbaix, *Corrosion 32*:47–51 (1976).
83. D. S. Dunn and N. Sridhar, *Critical Factors in Localized Corrosion*, The Electrochemical Society, Pennington, NJ, 1995, pp. 79–92.
84. J. R. Galvele, *Corros. Sci. 21*:551 (1981).
85. H. Boehni and F. Hunkeler, *Advances in Localized Corrosion—NACE 9* (H. S. Isaacs, U. Bertocci, J. Kruger, and S. Smialowska, eds.), NACE, Houston, TX, 1990.
86. T. R. Beck, *Advances in Localized Corrosion—NACE 9* (H. S. Isaacs, U. Bertocci, J. Kruger, and S. Smialowska, eds.), NACE, Houston, TX, 1990.
87. K. A. Grass, G. A. Cragnolino, D. S. Dunn, and N. Sridhar, *CORROSION/98*, paper 149, NACE International, Houston, TX, 1998.
88. D. S. Dunn, G. A. Cragnolino and N. Sridhar, *Corrosion 56*:90–104 (2000).
89. A. Turnbull and J. G. N. Thomas, *J. Electrochem. Soc. 129*:1412–1422 (1982).
90. S. M. Gravano and J. R. Galvele, *Corros. Sci. 24*:517–534 (1984).
91. S. M. Sharland and P. W. Tasker, *Corros. Sci. 28*:603–620 (1988).
92. S. M. Sharland, *Corros. Sci. 28*:621–630 (1988).
93. J. L. Crolet and J. M. Defranoux, *Corros. Sci. 13*:575–585 (1973).
94. J. W. Oldfield and W. H. Sutton, *Br. Corros. J. 13*:14–22 (1978).
95. K. Heber and R. C. Alkire, *J. Electrochem. Soc. 130*:1007–1014 (1983).
96. S. Sharland, *Corros. Sci. 33*:183–201 (1992).
97. N. Sridhar, G. A. Cragnolino, H. Pennick, and T. Y. Torng, *Life Prediction of Corrodible Structure* (R. N. Parkins, ed.), Vol. 1, NACE, Houston, TX, 1994.
98. J. C. Walton, G. A. Cragnolino, and S. K. Kalandros, *Corros. Sci. 38*:1–18 (1996).
99. S. P. White, G. J. Weir, and N. J. Laycock, *Corros. Sci. 42*:605–629 (2000).
100. S. Tsujikawa, Y. Soné, and Y. Hisamatsu, *Corrosion Chemistry within Pits, Crevices and Cracks* (A. Turnbull, ed.), Her Majesty's Stationery Office, London, 1987.
101. A. Anderko, S. J. Sanders, and R. D. Young, *Corrosion 53*:43–53 (1997).
102. J. C. Walton, *Corros. Sci. 30*:915–928 (1990).

103. J. W. Oldfield, *Int. Met. Rev. 32*:153–170 (1987).
104. F. P. Ijsseling, *Br. Corros. J. 15*:51–69 (1980).
105. B. E. Wilde, *Corrosion 28*:283 (1972).
106. B. Baroux and H. Mayet, *ECS Meeting, Proceedings*, Vol. 95-15, Chicago, IL, 1995.
107. J. L. Crolet, L. Séraphin, and R. Tricot, *Mater. Tech.* 355–359 (1978).
108. J. P. Audouard, A. Désestret, D. Catelin, and P. Soulignac, *CORROSION/88*, paper 413, NACE International, St. Louis, MO, 1998.

11

Stress-Corrosion Cracking Mechanisms

Roger C. Newman
University of Toronto

CONTENTS

11.1 Definition and Scope of Stress-Corrosion Cracking 499
11.2 Candidate Models for SCC 500
 11.2.1 Slip-Dissolution Model 500
 11.2.2 Film-Induced Cleavage Model 501
 11.2.3 Hydrogen Embrittlement Model 502
11.3 The Rate of SCC: Effects of Stress Intensity Factor 502
11.4 The Appearance of SCC 503
11.5 Initiation of SCC: Threshold Stresses and Role of Localized Corrosion 504
11.6 Metallurgy of SCC 507
11.7 The Electrochemistry and Mechanisms of SCC 515
11.8 Stress-Corrosion Testing in Relation to Mechanisms of Cracking 530
11.9 Notes on Other Proposed SCC Mechanisms 534
11.10 Summary and Recommendations for the Future 537
References 539

Several testable models for stress-corrosion cracking (SCC) of metals are discussed in terms of the main experimental variables: stress, metallurgy, and environment. Slip-dissolution, film-induced cleavage, and hydrogen embrittlement models are all shown to be consistent with experimental data in particular systems. Other models that cite effects of corrosion (without a film) or adsorption on crack tip deformation, leading to microcleavage or plastic microfracture, are less easy to test. No model can be "universal" in view of the demonstrable multiplicity of mechanisms. In many cases the atomistic mechanism is unknown, yet cracking can be controlled or predicted via the localized corrosion process that precedes SCC.

11.1 Definition and Scope of Stress-Corrosion Cracking

SCC is defined as the growth of cracks due to the simultaneous action of a stress (nominally static and tensile) and a reactive environment. For metals, "reactive" excludes gaseous hydrogen, cathodic polarization, and liquid metals but includes aqueous and nonaqueous electrolytes and reactive atmospheres (H_2O, I_2, Cl_2). Related phenomena occur in inorganic glasses and ceramics, especially in water, and are thought to be involved in major geological processes including earthquakes and midocean vulcanism. Static stress does not exclude slow monotonic straining or low-amplitude cycling ("ripple loading"), which accelerate SCC in many metallic systems—for example, by promoting oxide film rupture at the crack tip. Stresses arise in practice from applied loads or from residual stress due to welding or in-homogeneous plastic deformation.

This review deals mainly with metals. The literature on their SCC is immense, and there are many useful texts and conference proceedings [1–12]. A brief history was published by Newman and Procter [13] and should be consulted for further historical details, especially the metal-environment combinations known to exhibit SCC. Plasticity plays a key role in SCC of metals, in contrast to inorganic glasses and ceramics, which are brittle solids and crack via reaction of the corrodent with highly stressed bonds at an atomically sharp crack tip [14]. Metal-induced fracture [15], formerly called liquid-metal embrittlement, remains a tantalizing phenomenon and there is no agreement on the mechanism or the relationship to SCC. Several authors have proposed universal models of SCC and corrosion fatigue [16–18] or even universal models of environment-assisted fracture, encompassing aqueous, gaseous, and liquid-metal environments [19,20]. We shall conclude that although these models may apply to particular systems, they are *not* universal, and there remain several valid mechanisms of SCC. Finally, hydrogen effects play an important part in SCC but are also the subject of a large separate literature [21,22]. The atomistic action of hydrogen will be covered quite briefly here but can be investigated further by referring to Refs. [1–12,21,22].

11.2 Candidate Models for SCC

We have selected three testable cracking models that may account for most known cases of SCC in metals: slip-dissolution, film-induced cleavage, and hydrogen embrittlement. Other important models involve microcleavage [16,17] or plastic microfracture [19] induced by dissolution or adsorption, respectively. A sixth model, surface mobility [20], will be discussed only briefly, as there appear to be serious problems with the basic physics of this model [23], although not necessarily with the general concept. A model based on vacancy injection during dissolution, even of pure metals, was proposed by Jones [18]; this deserves attention but has not yet made any successful new predictions, and it relies on processes that might be discounted in the future using modern techniques such as scanning tunneling microscopy.

11.2.1 Slip-Dissolution Model

According to this model [24–33], crack growth occurs by extremely localized anodic dissolution. The sides of the crack are protected by a film, usually an oxide, which is fractured as a result of plastic strain in the metal at the crack tip. Crack growth proceeds by a cyclic process of film rupture, dissolution, and film repair (Figure 11.1). This model was once favored for both intergranular and transgranular SCC in a wide variety of systems [25] but is now applied mainly to the intergranular cracking of ferritic steels in passivating environments such as carbonate-bicarbonate solutions [26,27] and to sensitized stainless steels [28–32].

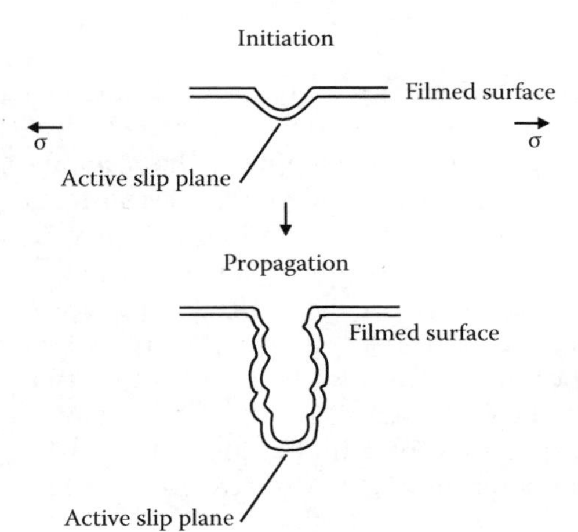

FIGURE 11.1
Schematic drawing of crack initiation and growth by slip-dissolution. (Adapted from Staehle, R.W., in *Stress-Corrosion Cracking and Hydrogen Embrittlement of Iron-Base Alloys* (R.W. Staehle, J. Hochmann, R.D. McCright, and J.E. Slater, eds.), NACE, Houston, TX, 1977, p. 180.)

11.2.2 Film-Induced Cleavage Model

Film-induced cleavage originated with the suggestion by Edeleanu and Forty [34] that superficial dezincification of α-brass in ammonia solution could trigger a brittle fracture through several micrometers of the unattacked substrate; they believed that the material through which this crack was propagating (between slip bands) was embrittled by short-range ordering. Their optical microscopy was elegant and came near to proving the reality of the crack-jumping phenomenon, which was later confirmed by Pugh and others using the scanning electron microscope (SEM) [35–39]. In the modern theory of film-induced cleavage [40–45] there is no special property of the face-centered cubic (fcc) substrate through which the crack jumps; the special properties lie in the film itself, which must be able to trigger a crack with a velocity of hundreds of meters per second within less than 100 nm (Figure 11.2). Nanoporous metallic layers produced by dealloying seem to be very effective, as they can be both brittle and epitaxial, or at least strongly bonded to the substrate. Tests of the film-induced cleavage model [33,46–51] have relied on producing single-shot brittle events through thin alloy foils and have shown that *intergranular* fracture can also be triggered by dealloyed layers in gold alloys. It may turn out that film-induced fracture involves a slower and more plastic process than originally envisaged by Sieradzki and Newman [40], but the reality of the phenomenon is now well established.

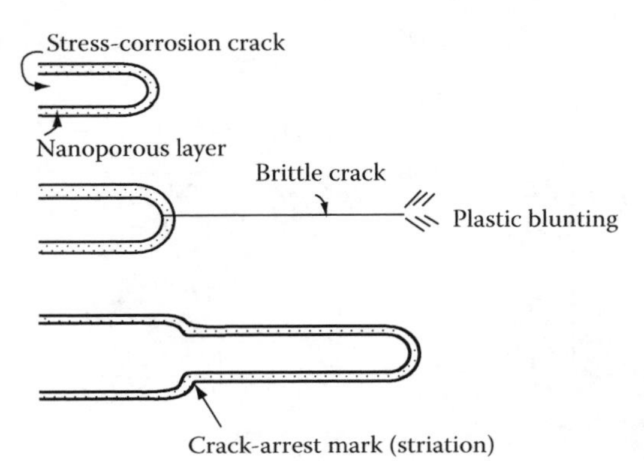

FIGURE 11.2
Schematic drawing of crack growth by film-induced cleavage in an fcc metal or alloy. (Adapted from Beavers, J.A. and Pugh, E.N., *Metall. Trans. A*, 11A, 809, 1980.)

11.2.3 Hydrogen Embrittlement Model

Hydrogen absorption is clearly responsible for SCC of high-strength steels in aqueous environments [52], notably in the presence of H_2S, where the fraction of discharged hydrogen that is absorbed by the metal can approach 100% owing to poisoning of the recombination reaction ($H + H = H_2$) by sulfur adsorption [53]. The role of hydrogen in fcc systems remains controversial, with positive indications in Al-Zn-Mg-Cu alloys [54,55] but waning evidence in austenitic stainless steels and Ni alloys [56]. Trans-granular SCC of stable austenitic steels at high temperatures such as 300°C does not correlate with their susceptibility to hydrogen embrittlement. Hydrogen discharge and absorption are favored when there is acidification of the local environment by cation hydrolysis [54], especially if this acidification damages or destroys a passive film in the crack. Once absorbed in the metal, hydrogen can promote cleavage, inter-granular separation, or a highly localized plastic fracture. In hydride-forming metals such as titanium, formation of brittle hydrides can be part of the fracture mechanism.

11.3 The Rate of SCC: Effects of Stress Intensity Factor

"Classical" SCC phenomena, such as the cracking of brass in aqueous ammonia, are classical because they occur at low stresses and at high rates: 10^{-9} to 10^{-6} m/s or 0.1 to 100 mm/day. The crack velocity varies with the mode I stress intensity factor (K_I) as shown in Figure 11.3 [$K_I \simeq \sigma(\pi a)^{1/2}$], where σ is the stress and a is the crack length [57].

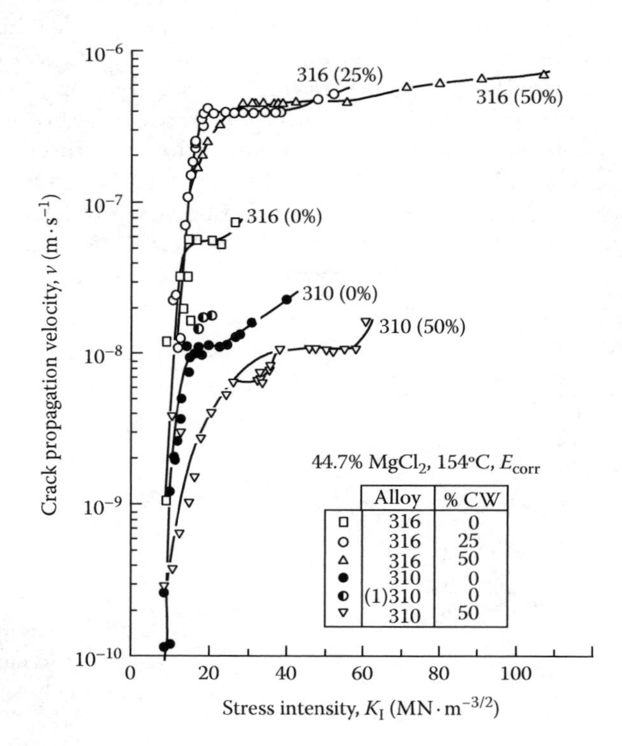

FIGURE 11.3
Examples of crack velocity-stress intensity curves for SCC, showing the effects of alloy composition and cold work on SCC of austenitic stainless steels in a hot chloride solution. (From Dickson, J.I. et al., *Can. Met. Q.*, 19, 161, 1980; Courtesy of Pergamon Press, Oxford, U.K.)

The plateau at intermediate K_I values indicates that something chemical rather than something mechanical is controlling the crack velocity; this might be dissolution, diffusion, or adsorption. SCC has been detected growing with very low velocities such as 10^{-12} m/s (0.1 μm/day, or 1 mm every 30 years) [58]. Such processes, occurring in piping or steam generator tubes, threaten the long-term integrity of nuclear power plants [59]. At these low rates and relatively high temperatures (300°C), solid-state transport of substitutional elements becomes a feasible part of the SCC mechanism, hence the aqueous SCC phenomenon starts to blend with high-temperature oxidation and might be controlled by processes such as grain boundary diffusion of atomic oxygen [60,61].

The important quantities in Figure 11.3 are the *threshold stress intensity*, K_{th} or K_{ISCC}, and the *region II crack velocity*, υ_{II}. In strong materials that can suffer fast fracture, the *critical stress intensity* or fracture toughness, K_{IC}, terminates the life of a component catastrophically, whereas ductile alloys fail by leakage or loss of cross section. The classical SCC systems have low K_{ISCC} values, especially austenitic stainless steel in hot chloride, where values as low as 1 Mpa $m^{1/2}$ have been reported [62,63]; there is a wide range of values in the literature, up to 12 Mpa $m^{1/2}$. The superior performance of duplex (austenoferritic) stainless steels in these environments is associated with a much higher (3–10 times) K_{ISCC} value [62]; thus duplex components must be highly stressed or defective in order to fail by SCC in hot chloride solutions. Incidentally, all the K_{ISCC} values mentioned, even the lowest, are high enough to cause yielding at the crack tip.

11.4 The Appearance of SCC

SCC failures are macroscopically brittle in the sense that the ductility and load-bearing capacity of the material are impaired. Microscopically, the cracks are either intergranular (sometimes along old boundaries such as those of prior austenite in martensitic steels) or transgranular and cleavage-like (Figure 11.4a and b). In duplex microstructures, one phase often cracks more easily than the other, leading to characteristic cracking patterns (Figure 11.4c). Plastic deformation always accompanies crack growth and plays a key role in most cracking mechanisms. Sometimes the fracture surfaces show evidence for stepwise (discontinuous) crack advance, especially in transgranular SCC [35–39] (Figure 11.5). It is possible to produce similar crack front markings by periodic over- or underloading [37,38,64,65] and to show that the natural crack arrest marks do indeed correspond to positions of the crack front. The average orientations of transgranular cracks are specific, e.g., {110} in α-brass [35], but such surfaces are sometimes composed of very fine, alternating {111} microfacets [66]. Fracture of the ligaments between the {110} crack facets, which produces the river lines on the fracture surface, is usually a ductile process such as necking (in pure Cu) or crystallographic shear (in α-brass). The latter favors crack propagation at low stresses owing to the small displacement involved [39]. The presence of intact ligaments behind the crack tip shields the tip from some of the applied stress intensity and explains why cracks are often much sharper [34] than one would calculate using elastic-plastic fracture mechanics. Ligament fracture, or its absence, can play a role in the formation of the crack arrest markings shown in Figure 11.5.

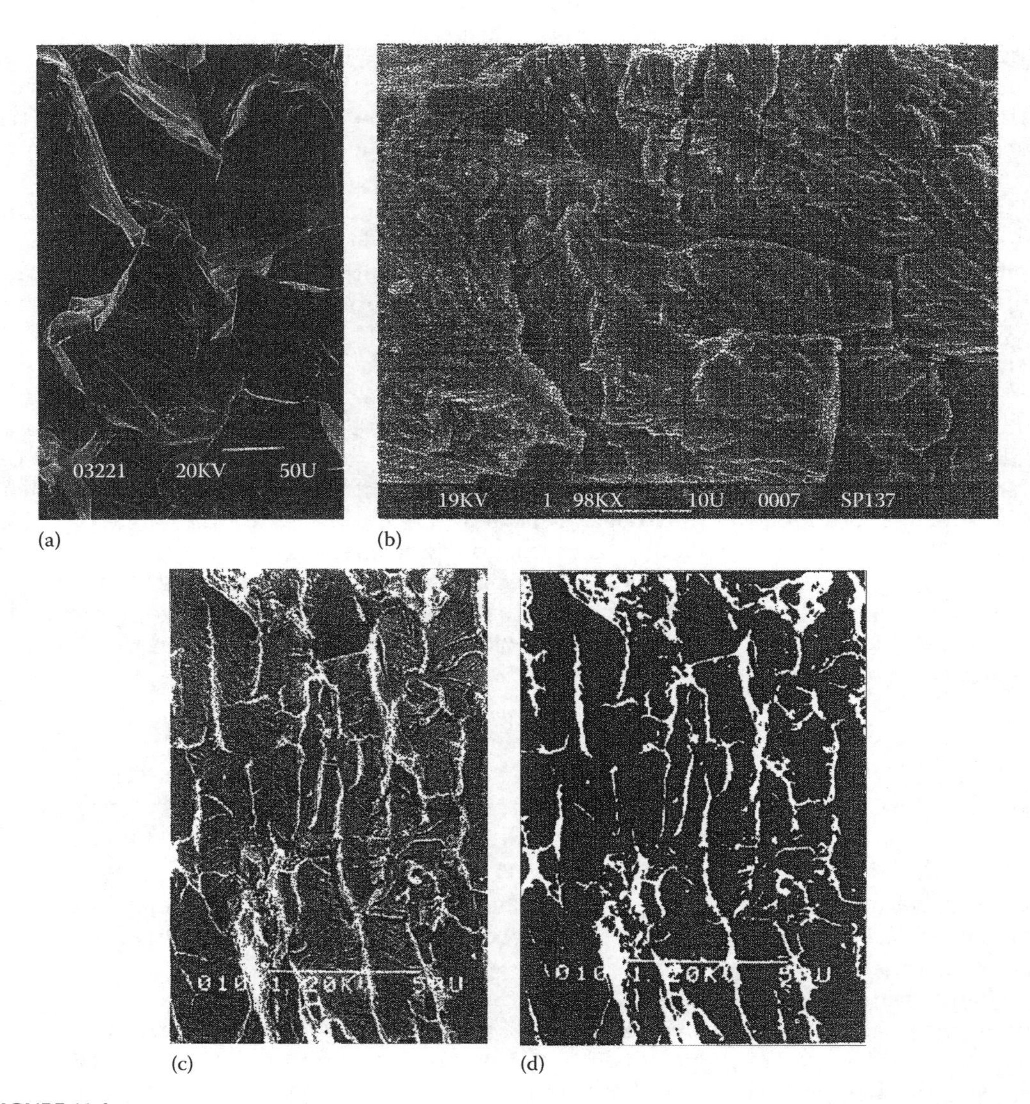

FIGURE 11.4
Examples of SCC fracture surfaces, all obtained with the slow strain rate test. (a) Intergranular SCC of α-brass in Mattsson's [109] solution, pH 7.2. (b) Transgranular SCC of α-brass in 15 M cuprous ammonia solution equilibrated with Cu and Cu_2O. (Courtesy of Shahrabi, T.) (c) SCC of 25Cr duplex stainless steel plate (50% ferrite) in NaCl-H_2S solution at 80°C, showing flat, cleavage-like cracking of ferrite and mainly ductile fracture of austenite. (Courtesy of Salinas Bravo, V.M.) (d) High-contrast schematic of (c), showing necked austenite grains as light (high) areas.

11.5 Initiation of SCC: Threshold Stresses and Role of Localized Corrosion

SCC initiation at smooth surfaces shows a threshold stress (σ_{th}) that varies from 20% to more than 100% of the yield stress (σ_y), depending on the metal and the environment. Thermal stress relief of welded or cold-worked components can prevent SCC if σ_{th}/σ_y is fairly high, e.g., 0.7. In some systems, σ_{th} can be lowered by low-amplitude dynamic

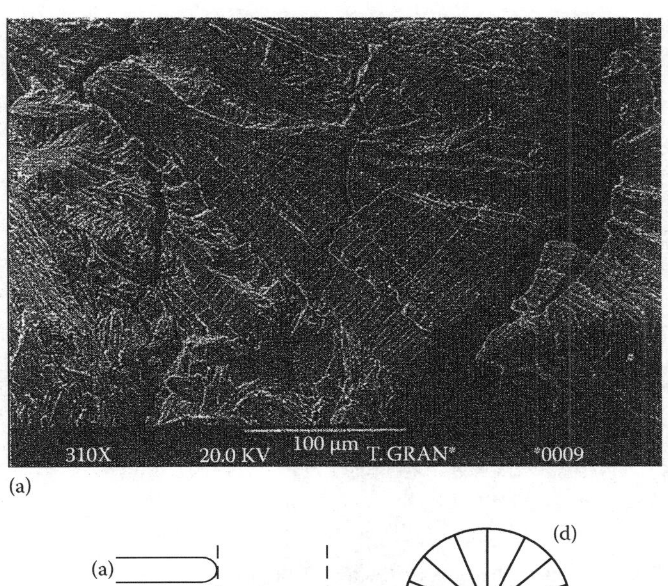

(a)

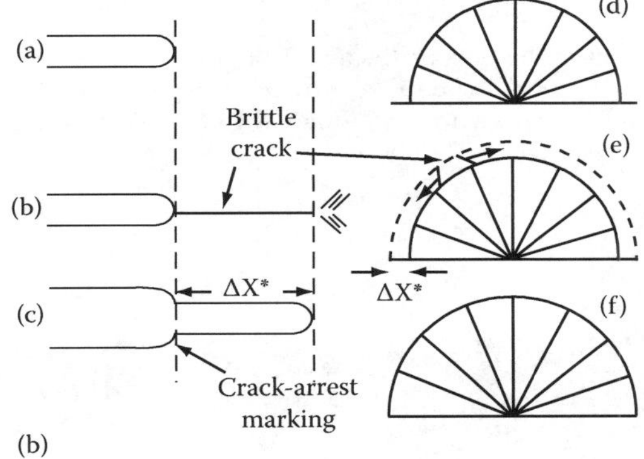

(b)

FIGURE 11.5
(a) Crack arrest marks on a transgranular SCC surface, observed near the overload region on a slow strain rate specimen ($\dot{\varepsilon} = 10^{-5} s^{-1}$) of 70–30 brass tested in pH 7.2 Mattsson's [109] solution with 0.05 M NaCl; (b) schematic drawing of the transgranular SCC process and the mechanism of striation formation. (After Beavers, J.A. and Pugh, E.N., *Metall. Trans. A*, 11A, 809, 1980; Pugh, E.N., *Atomistic of Fracture* (R.M. Latanision and J.R. Pickens, eds.), Plenum, New York, p. 997, 1983; Hahn, M.T. and Pugh, E.N., *Corrosion*, 36, 380, 1980; Beggs, D.V., et al., *Proceedings of the A. R. Troiano Honorary Symposium on Hydrogen Embrittlement and Stress-Corrosion Cracking* (R. Gibala and R.F. Hehemann, eds.), ASM, Metals Park, OH, 1984; Pugh, E.N. *Corrosion*, 41, 517, 1985; Courtesy of NACE, Houston, TX.)

loading (Figure 11.6), which is very important in practice [67]. SCC is usually nucleated by some form of localized corrosion; in chloride-SCC of stainless steels [68] or aluminum alloys [54], cracks start from areas of pitting, intergranular corrosion, or crevice corrosion that create the stress concentration and the acidity required for cracking (Figure 11.7). In SCC of C-Mn or low-alloy steels, intergranular corrosion occurs along segregated zones rich in carbon, nitrogen, or phosphorous and provides a stress concentration (helps to achieve K_{ISCC}) [27] (Figure 11.8). Pure iron shows no intergranular corrosion or SCC in this kind of solution. In mechanistic studies, one often comes across references to rupture of passive films, but in fact SCC rarely occurs in a true (local) passive state. There is usually some form of intergranular corrosion, pitting, crevice corrosion, or dealloying, or else SCC occurs in a conventional active state via a hydrogen embrittlement

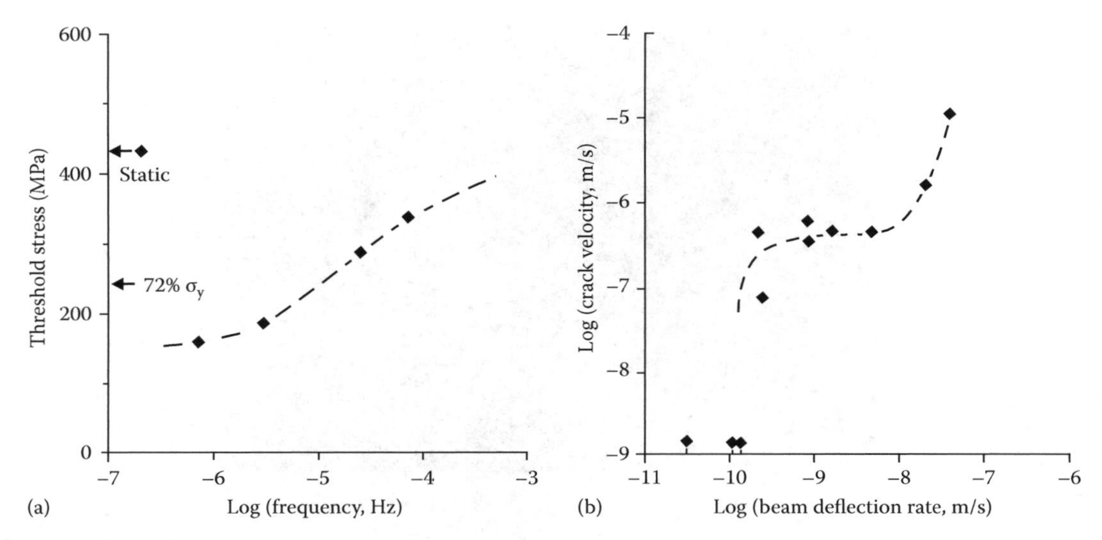

FIGURE 11.6
(a) The effect of low-amplitude cyclic loading on the threshold mean stress for SCC of C-Mn steel in carbonate-bicarbonate solution. (b) The corresponding data for growth of single cracks. (From Parkins, R.N. and Fessler, R.R., *Mater. Eng. Appl.*, 1, 80, 1978; Redrawn from Procter, R.P.M., *Cathodic Protection* (V. Ashworth and C.J.L. Booker, eds.), Ellis Horwood, Chichester, U.K., 1986.)

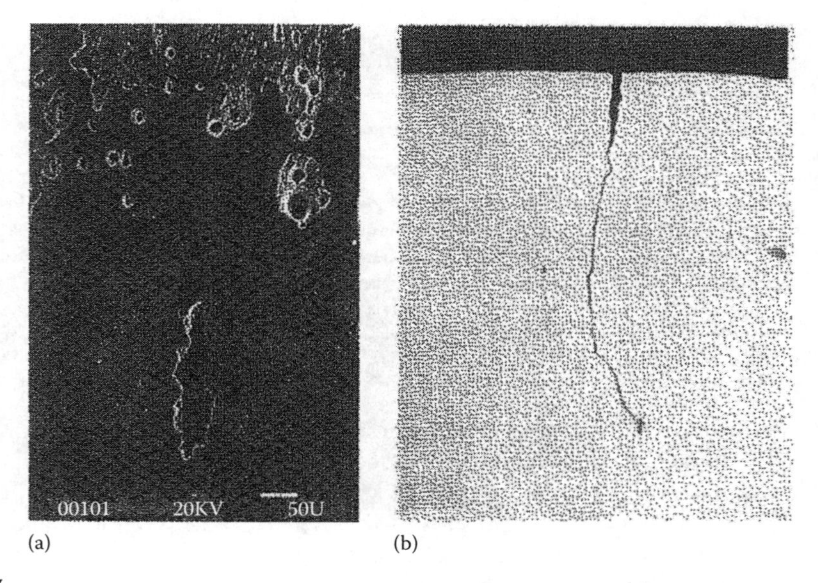

FIGURE 11.7
Aspects of the nucleation of SCC by localized corrosion. (a) Peak aged Al-Li-Cu-Mg alloy 8090 after unstressed preexposure in aerated 3.5% NaCl for 7 days. (b) SCC initiated from one of the fissures shown in (a), following removal of the solution and continued exposure to laboratory air under a short transverse tensile stress. (Courtesy of Craig, J.G., unpublished data.) (c) Creviced region of 316 L stainless steel after a slow strain rate test in 0.6 M NaCl + 0.03 M $Na_2S_2O_3$ at 80°C and an applied anodic current of 25 μA, showing unstable pitting leading to crevice corrosion and SCC initiation. (Courtesy of Suleiman, M.I.)

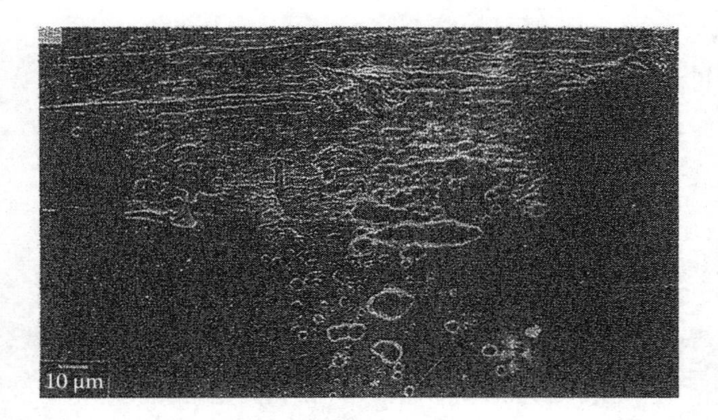

FIGURE 11.8
Caustic SCC of an Fe-3Ni-1Mo pure alloy at −400 mV (Hg-HgO) in a slow strain rate test in 9 M NaOH at 98°C, showing the intergranular corrosion that initiated SCC. (From Bandyopadhyay, N. et al., *Proceedings of the 9th International Congress on Metallic Corrosion*, NRC, Ottawa, Ontario, Canada, Vol. 2, 1984, p. 210.)

mechanism (as in high-strength, low-alloy steels). There are surface films in cracks, and these need to be ruptured to expose bare metal, but in many cases (such as carbon steel in sodium hydroxide), the film has a precipitated or gel-like nature rather than conforming to the conventional idea of a compact passive film a few nanometers in thickness.

11.6 Metallurgy of SCC

The main metallurgical variables in SCC are

Solid solution composition
Grain boundary segregation
Phase transformations and associated solute-depleted zones
Duplex structures
Cold work

In the case of hydrogen-induced SCC of high-strength steels, one could add strength, no matter how this is achieved. Environmental criteria for SCC are necessary but not sufficient: SCC will not occur without a susceptible metallurgy. The only exceptions to this rule are transgranular cracking processes in pure metals, e.g., iron in anhydrous ammonia [69] or copper in sodium nitrite solution [65,70].

"Advanced" materials such as metal-matrix composites also show SCC phenomena, but these are not distinctive and research has been quite qualitative in this area.

Solid solution composition classically controls the SCC of brasses [71,72], austenitic stainless steels in hot chloride solutions [73,74], and noble-metal alloys [75]. In all these systems there is evidence that *dealloying* dominates the SCC mechanism, although this remains controversial for stainless steels. Transgranular SCC ceases above a critical content (parting limit or dealloying threshold) of the most noble alloying element, either 80–85% in Cu-Zn or Cu-Al or 40% in Ni-Cr-Fe or Au-X alloys (Figures 11.9 and 11.10). These values

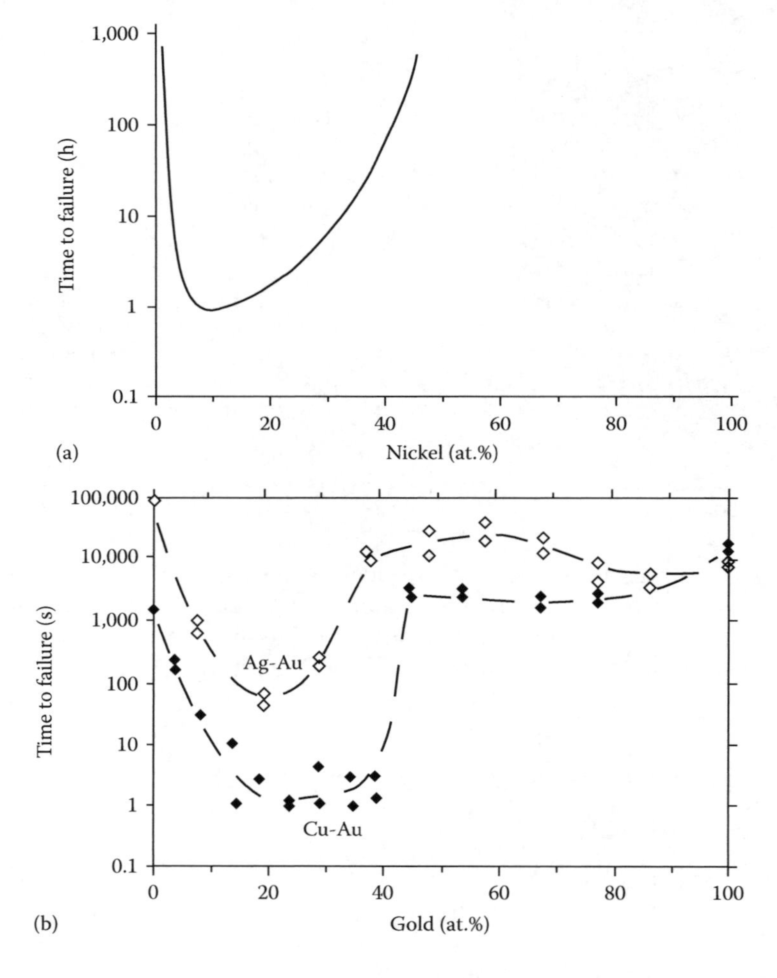

FIGURE 11.9
Direct and indirect evidence for a role of dealloying in SCC. (a) The Copson curve showing the dependence of SCC in austenitic stainless steels on their nickel content. (From Copson, H.R., in *Physical Metallurgy of Stress Corrosion Fracture* (T.N. Rhodin, ed.), Interscience, New York, 1959, p. 247.) (b) Similar behavior for gold alloys in aqua regia, where dealloying is known to control SCC. (From Graf, L., in *Fundamental Aspects of Stress-Corrosion Cracking* (R.W. Staehle, A.J. Forty, and D. van Rooyen, eds.), NACE, Houston, TX, 1969, p. 187.)

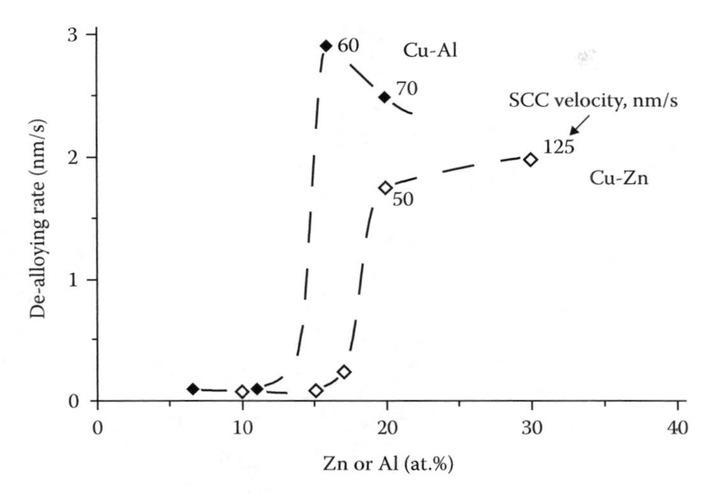

FIGURE 11.10
Direct and indirect evidence for a role of dealloying in SCC: SCC and dealloying of Cu-Zn and Cu-Al monocrystals in cuprous ammonia solution as a function of solute content. (From Sieradzki, K. et al., *J. Electrochem. Soc.*, 134, 1635, 1987.)

have been interpreted using percolation theory [72,76,77]. In all these systems the crack walls are either oxide free or else have thick porous oxides that allow contact of metal and electrolyte [78,79]. Flangan and coworkers [80,81] have shown that SCC of Au-Cu alloys occurs below the critical potential for macrodealloying (E_c) and consequently reject film-induced cleavage for this system. However, because dealloying in Au-Cu occurs in a localized mode resembling pitting [82,83], this conclusion may be premature. The concentration of Cl^- that occurs in the pit lowers E_c locally by complexing Cu(II), and any transient electrochemistry should be done in a $CuCl_2$-rich solution, simulating the crack environment, if it is to reproduce the behavior of a crack tip.

Minor alloying elements (C, P, N, As,…) can have several distinct effects on the electrochemistry of SCC. In caustic intergranular SCC of carbon steel, carbon or nitrogen segregation (and/or precipitation) at grain boundaries interferes with passive film formation and may affect plasticity [27], and phosphorus segregation introduces SCC in a new, more oxidizing range of potentials [84,85] (Figure 11.11). In chloride-SCC of austenitic stainless steels, the group 5A elements are so influential on *transgranular* cracking that it is difficult to crack high-purity ternary alloys in the laboratory [86]. This is not understood in detail, but one possibility was proposed by the author [87,88]: adsorption of group 5A elements is known to reduce surface self-diffusivity in electrolytes, and dealloyed layers need extremely fine porosity to cause SCC by film-induced cleavage, so if the 5A elements are absent there is a rapid coarsening of the porosity and a reduced susceptibility to SCC (Figure 11.12). This idea unifies the well-known effect of arsenic in brass with that of N or P (or even As) in stainless steel.

Phase transformations are used in strong alloys to provide dispersion (precipitation) strengthening and also occur in ductile alloys during welding or simply as a by-product of the traditional metallurgy of the system. In strong alloys, there is a different sequence of nucleation and growth at grain boundaries compared with the matrix [54]. Depending on the phase(s) formed at the grain boundary, enhanced reactivity may result, either of the phase itself or of an associated solute-depleted zone [89]. The most important of these systems are the Al-Zn-Mg (7000 series) aerospace alloys, which

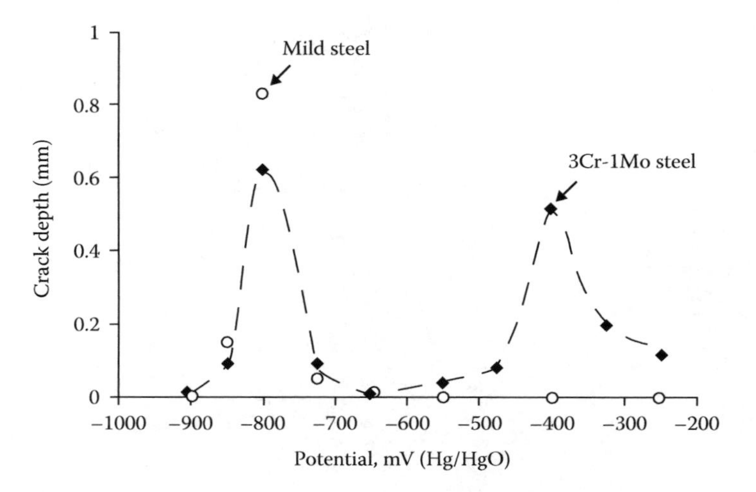

FIGURE 11.11
SCC penetration as a function of potential for carbon steel and low-alloy Cr-Mo steel tested in hot NaOH solution, showing a cracking regime at higher potential induced by P segregation to grain boundaries. See also Figure 11.22. (From Li Shiqiong et al., *Mater. Sci. Eng.*, A119, 59, 1989.)

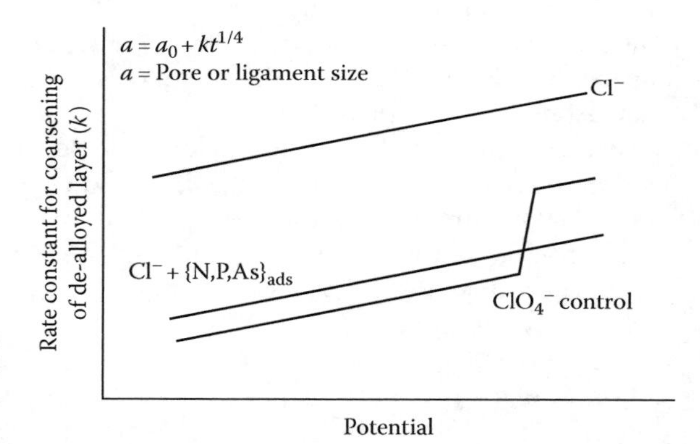

FIGURE 11.12
Schematic drawing of the proposed mechanism of action of group 5A elements on chloride-SCC of austenitic stainless steels. (From Newman, R.C., *Corros. Sci.*, 33, 1653, 1992.)

become especially susceptible to SCC when further alloyed with Cu. Grain boundary segregation of Cu has been detected in susceptible conditions using analytical electron microscopy [90]. Hydrogen plays a role in this cracking [54], and it is difficult to separate the roles of hydrogen and dissolution, let alone the contributions of the reactive $MgZn_2$ phase, the segregated Cu, the solute-depleted zone, and the grain boundary strength. Precipitation of grain boundary phases containing Cu-depleted (less noble) regions; this is easily demonstrated in Al-Cu (2000 series) systems, but the narrow and deep Cu depletion occurs only in underaged, noncommercial heat treatments, so practical problems are relatively rare, especially as these alloys are normally used in sheet rather than extrusion form (the arrangement of grain boundaries in unrecrystallized sheet inhibits transverse SCC). In the 7000-series alloys the high SCC susceptibility is shifted to the peak strength condition—a major inconvenience to users, as it denies them the use of the strongest condition of the alloy (it must be overaged for most applications).

Progress is being made in complex thermomechanical treatments to improve the toughness and SCC resistance at peak strength [54].

Intergranular corrosion does not necessarily correlate with SCC in aluminum alloys. The 6000-series (Al-Mg-Si) alloys such as 6061 can suffer intergranular corrosion but never fail by SCC. Examination of fractured specimens following intergranular corrosion shows crystallographic tunneling attack on either side of the grain boundary.

Precipitation strengthening of steels, such as the PH grades of ferritic stainless steel containing Cu, affects SCC when the operative mechanism is hydrogen embrittlement [91]. *Precipitation*, without strengthening, can occur at grain boundaries and is responsible for the most expensive form of SCC: the cracking of weld-sensitized austenitic stainless steel in oxygenated high-temperature water (boiling-water reactor pipe cracking) [59]. The sensitization process in AISI type 304 stainless steel consists of the precipitation of chromium carbides at the grain boundaries in the zone either side of a weld where the temperature has reached about 600°C–800°C. This leaves narrow Cr-depleted zones that repassivate more slowly than the matrix and therefore provide an active crack path. Effective remedies are the reduction of carbon content to about 0.03% (the 304 L grade) or the addition of alternative carbide formers (Ti, Nb, e.g., the 321 grade). The fraction of sensitized grain boundaries required to provide a continuous crack path in slow strain rate tests has been shown to conform to a percolation model with a percolation threshold of about 23% [92] (Figure 11.13). Under static loads the required fraction of sensitized boundaries may be higher.

Other phase transformations occurring in austenitic and duplex stainless steels (σ phase, α′ phase) generally produce *discontinuous* Cr-depleted zones around these Cr-rich phases, so crack growth is not necessarily enhanced, although pitting resistance is greatly degraded by σ phase, so SCC is initiated much more easily [91]. The austenite–ferrite transformation will be considered later.

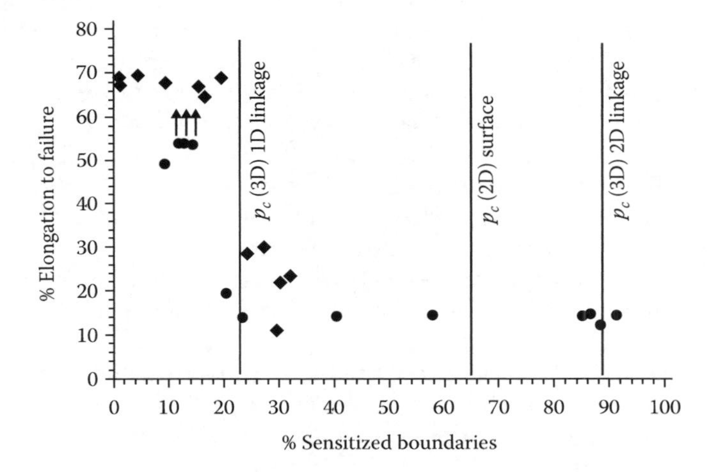

FIGURE 11.13
Dependence of SCC on the fraction of sensitized grain boundaries for slow strain rate tests of type 304 stainless steel in sodium thiosulfate solution, showing behavior in accordance with a 3D percolation model for linkage of sensitized grain boundary facets with a sharp cracking threshold at 23%. The thresholds at 65% and 89%, representing growth of the crack as a continuous rumpled sheet, may be relevant in constant-load tests where ductile ligaments can arrest the cracks. (From Wells, D.B. et al., *Corrosion*, 45, 649, 1989.)

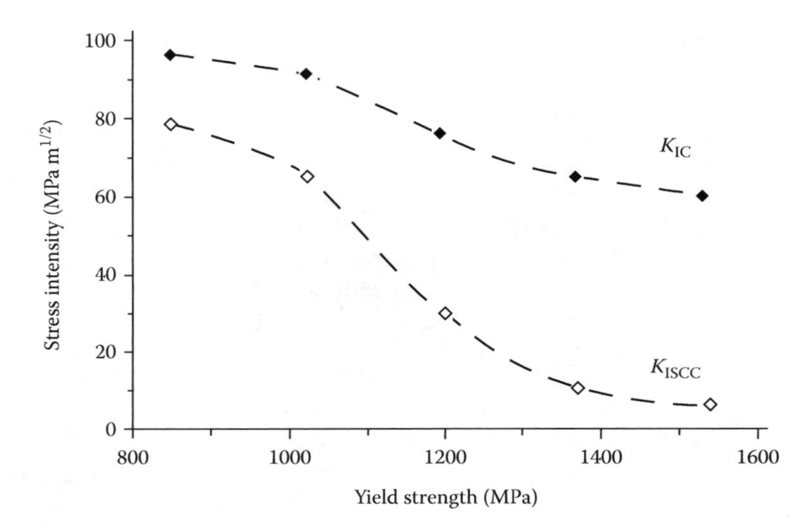

FIGURE 11.14
Dependence of K_{ISCC} and K_{IC} on yield strength for AISI 4340 high-strength steel tested in flowing seawater. (Redrawn from Brown, B.F., in *Theory of Stress Corrosion Cracking* (J.C. Scully, ed.), NATO, Brussels, Belgium, 1971, p. 186; Brown, B.F., in *Stress-Corrosion Cracking and Hydrogen Embrittlement of Iron-Base Alloys* (R.W. Staehle, J. Hochmann, R.D. McCright, and J.E. Slater, eds.), NACE, Houston, TX, 1977, p. 747.)

Martensite is very susceptible to hydrogen embrittlement, and strong low-alloy steels based on tempered martensite have an inescapable tendency for hydrogen-induced SCC in saltwater or other neutral environments [93]. *Bainite* can be the basis of strong steels (although not as strong as lightly tempered martensites) that are slightly less susceptible. Depending on the flux of hydrogen into the steel (highest in acidic H_2S, lower in NaCl or fresh water, lower still in moist air), curves of K_{ISCC} versus yield strength can be drawn for martensitic steels as shown in Figure 11.14. A current challenge is to improve the hydrogen embrittlement resistance of quenched and tempered steels in the 700–850 MPa range of yield strengths so that they can be used in wet H_2S environments; one strategy is to exploit the greater resistance of recrystallized ferrite that forms in some of these steels. Some specialized high-strength steels can show superior performance, such as cold-drawn pearlite [94] (prestressing wire or tire cord) and especially work-hardened, high-nitrogen austenite [58], which is virtually immune to hydrogen embrittlement. Neither of these materials is suitable as a general-purpose structural steel. Cold-drawn steel wire derives its SCC resistance from its highly directional microstructure; the resistance of such microstructures to transverse, intergranular SCC is well known in many systems including aluminum alloys and steels (Figure 11.15).

Duplex structures are defined as alloy microstructures in which two phases are present in comparable proportions. Duplex (austenoferritic) stainless steels have become very popular materials because they combine strength and SCC resistance and can be welded with good microstructural control [91,95]. Their pitting and crevice corrosion resistance compare favorably with those of austenitic steels of similar Mo content.

The SCC resistance of duplex stainless steels is manifested as a high value of the threshold stress intensity, K_{ISCC}, compared with austenitic steels [62] (Figure 11.16). This arises from the different chemistry of the individual phases, which gives them, *individually*, different optimal ranges of potential for SCC (higher for γ, lower for α).

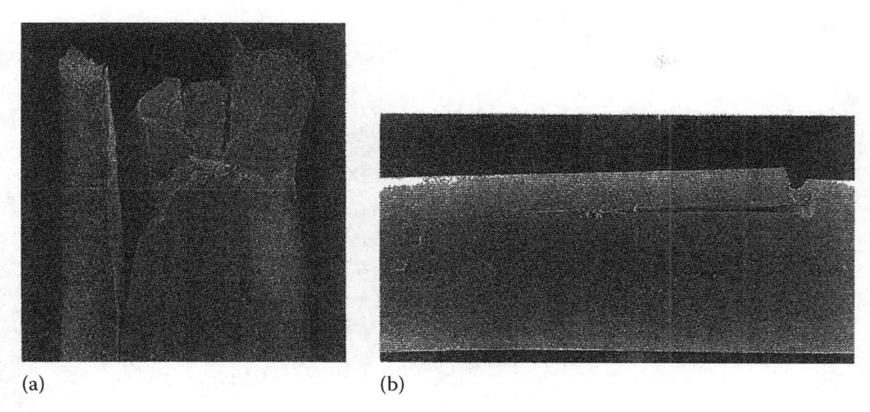

FIGURE 11.15
Example of transverse SCC resistance in a highly directional microstructure: 5-mm-thick wires of martensitic prestressing steel, notched and tested at a constant load [higher in (a) than (b)] in 0.6 M NaCl solution. (Courtesy of Doughty, A. S., unpublished data.)

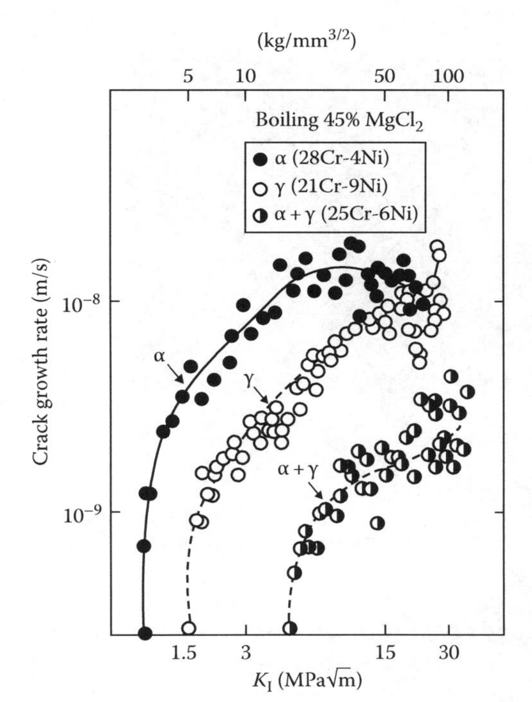

FIGURE 11.16
υ-K curves for a duplex stainless steel and its individually synthesized phases in hot chloride solution. (From Nagano, H. et al., *Met. Corros. Ind.*, 56, 81, 1981; *Boshoku Gijutsu* (*Corrosion Engineering*), 30, 218, 1981.)

Because there can only be one electrode potential at the tip of the crack, at least one of the phases must be below or above its best cracking potential (Figure 11.17). If the individual phases are synthesized and tested as bulk materials, they both crack easily with very low K_{ISCC} values [62,95], but in modern duplex steels SCC occurs in the ferrite and is arrested by the austenite (Figures 11.4c and 11.18a). However, if there is a strong oxidant present, or if oxygen is supplied through a thin boiling layer (as in evaporating

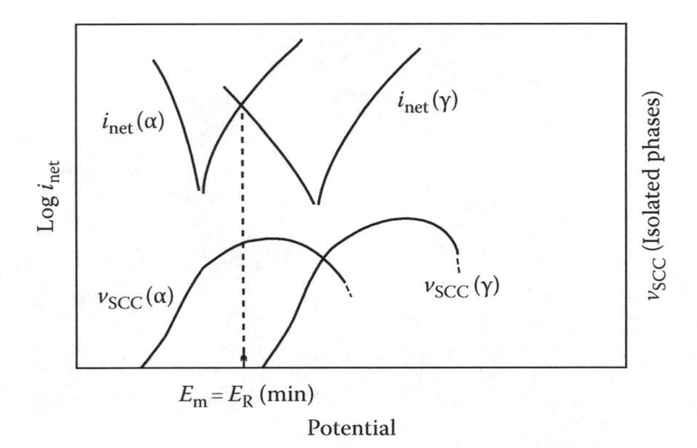

FIGURE 11.17
Schematic electrochemistry of chloride-SCC of duplex stainless steel.

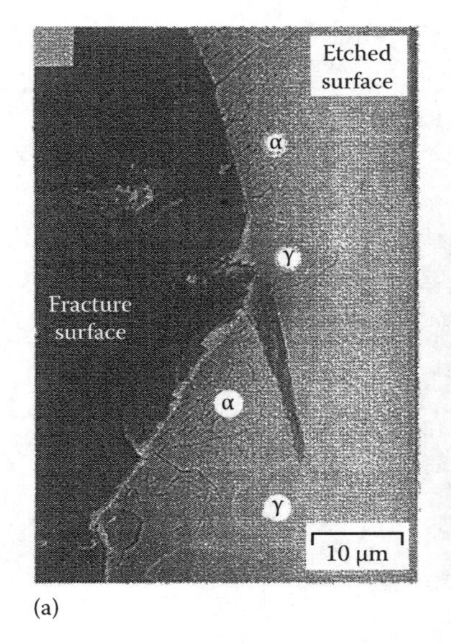

(a)

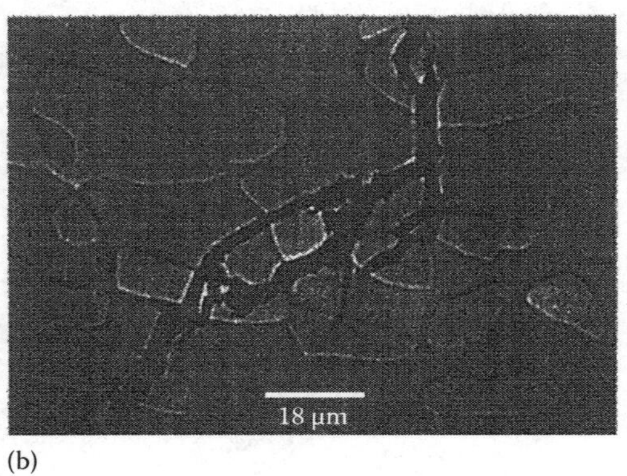

(b)

FIGURE 11.18
(a) Typical cracking pattern of chloride-SCC in a duplex stainless steel, showing cracking of ferrite and partial arrest by austenite. (Courtesy of Nisbet, W.J.R.) (b) Cracking of a similar steel in evaporating seawater, showing little or no restraint of the crack by the austenite under this more oxidizing condition. (Courtesy of Ezuber, H.)

sea water, which leaves behind a layer of saturated $MgCl_2$ solution), then the austenite is no longer protected by the ferrite (Figure 11.18b) and K_{ISCC} drops significantly. Such results are consistent with the reported effect of anodic polarization in deaerated $MgCl_2$ solution [91].

Cold work reduces the ductility and fracture toughness but does not necessarily increase the SCC velocity or reduce K_{ISCC}. In austenitic stainless steels that transform partially to martensite, the region II crack velocity in hot chloride solution is increased by cold work, and the crack path becomes partly intergranular [57,96,97]. This may be associated with a change of mechanism. In alloy 600 exposed to reducing high-temperature water, the velocity of intergranular SCC increases dramatically with increasing cold work [59,60,98]; according to Scott and Le Calvar [60], this may be related to enhanced intergranular oxidation, a familiar effect in high-temperature gaseous environments.

11.7 The Electrochemistry and Mechanisms of SCC

There are at least five electrochemical conditions that can lead to SCC in electrolytes, given that the material has a susceptible metallurgy; these are shown schematically in Figure 11.19:

A: A state of imperfect passivity near an active–passive transition, e.g., carbon steel in aqueous hydroxide, nitrate, or carbonate–bicarbonate solutions [26,27].

B: A state of slow, chloride-induced localized corrosion in stainless steels, aluminum alloys, or titanium alloys [54,68].

C: A state of surface dealloying with no continuous oxide, e.g., gold alloys in many aqueous solutions [75,99]. Dealloying may also occur within localized corrosion sites in passive alloys such as austenitic stainless steels.

D: Formation of unusual surface films that may play a casual role in SCC, e.g., nitrides formed on steel in anhydrous ammonia [69,100]. There are a number of related, room-temperature gaseous systems such as Zr/I_2 [101] and high-strength steel/Cl_2 [102,103]. Where such films cause only intergranular cracking and this occurs only at elevated temperatures, it is almost certainly due to penetration of atoms of the gaseous species along grain boundaries [60] and not to surface mobility as proposed by Bianchi and Galvele [104].

E: An active state leading to hydrogen-induced SCC, usually in high-strength steel [52,93,105,106] or medium-strength steels in H_2S media [53,107].

The most extensive mechanistic investigations have been carried out on passive systems showing cracking of type A or type B. All SCC mechanisms rely on the exposure of bare metal to the environment, and if this is too brief (owing to immediate repassivation), none of the likely cracking agents, such as dissolution or hydrogen, can operate. This explains the importance of a slow *repassivation rate* [26] (Figure 11.20). The kinetics of repassivation on bare metal surfaces have been measured by numerous potentiostatic techniques such as scratching, rapid straining, and laser illumination [108]. As well as active-passive

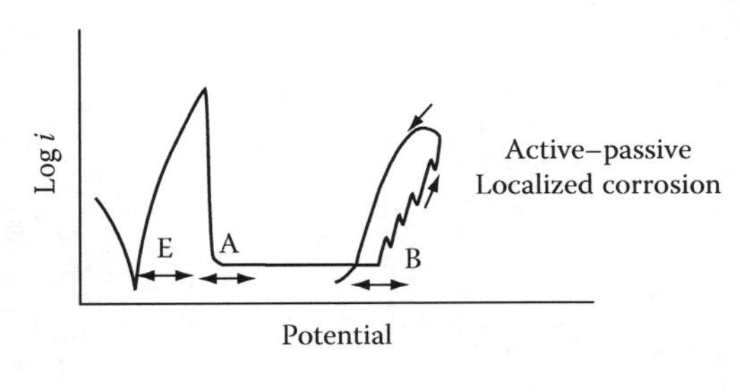

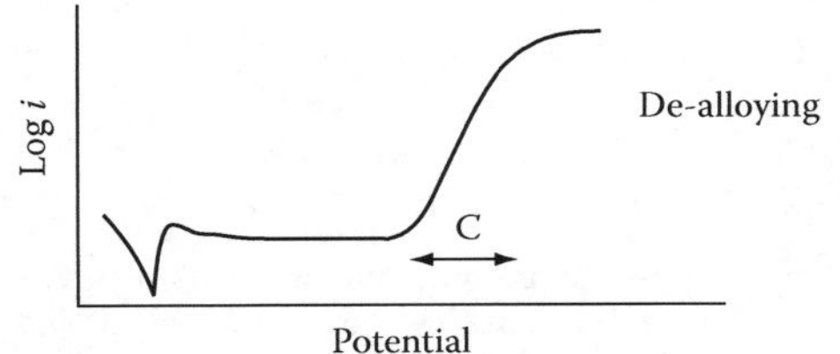

FIGURE 11.19
Correlation between SCC and various types of electrochemical behavior. The letters A, B, C, and E refer to the types of SCC defined in the text. The arrows indicate possible potential ranges of SCC.

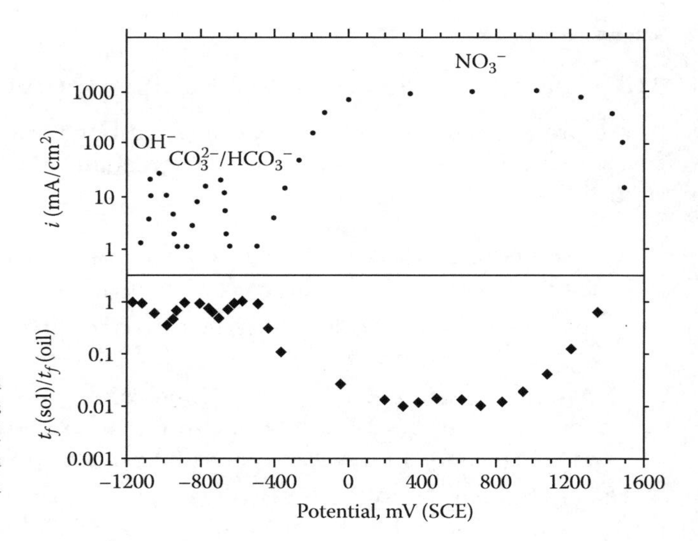

FIGURE 11.20
Correlation between inverse passivation rate (difference between current densities in fast and slow polarization scans) and slow strain rate SCC behavior for carbon steel in three SCC environments. (From Parkins, R.N., *Corros. Sci.*, 20, 147, 1980.)

transitions with potential, similar transitions occur with pH under free-corrosion conditions and may lead to critical pH requirements for SCC, notably for α-brass in near-neutral ammoniacal solutions [109,110].

This classification of SCC is useful in practice, as it can lead to remedial measures or material developments without necessitating a complete understanding of the atomistic mechanism. For example, in type B SCC it is sufficient to prevent the localized corrosion process or to ensure that it occurs too rapidly to allow crack nucleation [68,111,112]. This topic will be discussed in detail later.

Type A SCC normally occurs by intergranular slip-dissolution and is related to grain boundary segregation or precipitation. Most of our present understanding has been developed by studying the passivation characteristics of bulk material [26,27], but alloys or compounds simulating the grain boundary composition can be useful for assessing the effect of segregation or precipitation on the repassivation kinetics [25,84,85] (Figure 11.21).

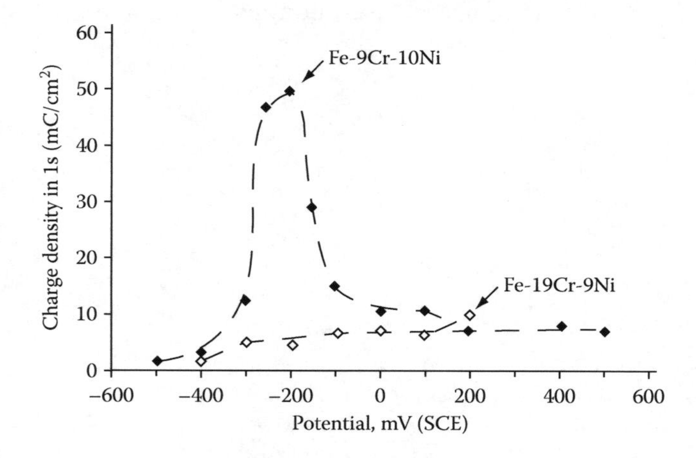

FIGURE 11.21
Delayed repassivation of material simulating the chromium-depleted grain boundary zone in sensitized stainless steel, measured in sodium thiosulfate solution using a scratch method. (From Newman, R.C. et al., *Metall. Trans. A*, 13A, 2015, 1982.)

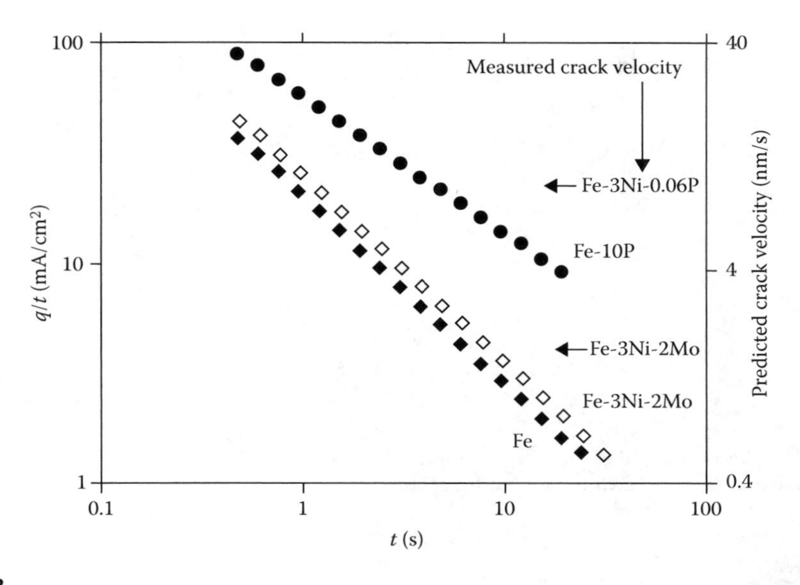

FIGURE 11.22
Anodic current decays for Fe, Fe-3Ni-2Mo, and an Fe-10P amorphous alloy (simulating a grain boundary) in 9 M NaOH solution at 98°C and −400 mV (Hg-HgO). (From Bandyopadhyay, N. et al., *Proceedings of the 9th International Congress on Metallic Corrosion*, NRC, Ottawa, Ontario, Canada, Vol. 2, p. 210, 1984.) The original $i(t)$ decays have been converted into $q(t)$ (charge density versus time) and then into $q(t)/t$ for comparison with a slip-dissolution model of SCC where the crack velocity is proportional to $q(\tau)/\tau$ with τ being the interval between film rupture events at the crack tip. The indicated crack velocities measured for Fe-Ni-Mo and Fe-Ni-P alloys are consistent with a single value of $\tau(6 \pm 3\,\text{s})$, giving some validity to the use of a simulated grain boundary alloy.

Rapid solidification can be used to incorporate large concentrations of segregants such as phosphorus [85] (Figure 11.22). In a number of systems it has been shown that the region II SCC velocity (υ_{II}) correlates with the maximum (reasonably sustained) anodic current density on a bare surface (i_{max}), via

$$\upsilon_{II} = \frac{i_{max} A}{nFs} \tag{11.1}$$

where

A is the atomic weight (relative atomic mass) of the metal
s is its density—that is, crack growth occurs by strain-enhanced intergranular corrosion [26]

The role of stress is to produce plastic strain in the metal, which fractures a brittle oxide layer.

SCC data are presented as velocity versus *static* stress parameter such as K_I, yet it is the *dynamic* plasticity at the crack tip that actually features the newly formed oxide film [67,113]. This problem was considered by Vermilyea [24], who showed that under a static stress the corrosion process could advance the crack tip into a region that had reached an equilibrium distribution of plastic strain (ε_p), and produce a *strain transient* ($\Delta\varepsilon_p$) that could fracture a newly formed oxide (Figure 11.23). Later Sieradzki [44] showed that the depth of transient corrosion required in Vermilyea's original model was unreasonably large (~μm), confirming that static slip-dissolution models can apply only to *intergranular* SCC, where there is a directional active path with a very low repassivation rate [33], It is not clear how the shielding of the crack tip stress intensity by intact metal ligaments [39] enters into this argument—it certainly makes real cracks much sharper than one would expect from fracture mechanics.

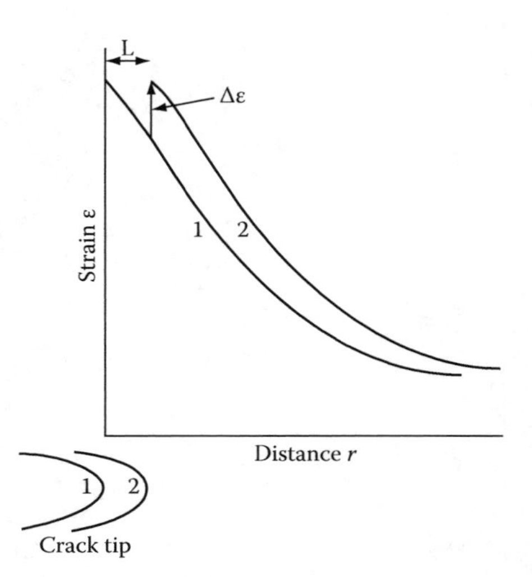

FIGURE 11.23
Plastic strain distribution ahead of a stress-corrosion crack tip and the effect of corrosion, according to Vermilyea [24].

The introduction of a *crack tip strain rate* ($\dot{\varepsilon}_c$) enables a wide range of SCC information to be rationalized [30]. For example, Equation 11.1 can be generalized to

$$\upsilon = \frac{\bar{i}A}{nFs} \tag{11.2}$$

where i is the time-averaged anodic current density at the crack tip. For a cyclical process in which the bare surface is exposed by film rupture, partially repassivated, and then exposed again, the interval t^* between the film rupture events is given by

$$t^* = \frac{\varepsilon_f}{\dot{\varepsilon}_c} \tag{11.3}$$

where ε_f is the fracture strain of the newly formed oxide film; thus $\bar{i}$ is given by

$$\bar{i} = \frac{\int_0^{t^*} i(t)dt}{t^*} \tag{11.4}$$

Writing the integral as $q(t)$, this gives

$$\upsilon = \frac{q(t^*)A\dot{\varepsilon}_c}{nFst^*\varepsilon_f} \tag{11.5}$$

The film rupture interval t^* decreases with increasing strain rate until eventually $i = i_{max}$ and υ approaches the maximum or region II velocity υ_{II}. For cyclic loads or slow strain rate tests, $\dot{\varepsilon}_c$ is relatively easy to estimate, and one can *test* the slip-dissolution model as a function of strain rate and potential [30–33]. More generally, one can include a continuum version of Vermilyea's concept [114]:

$$\dot{\varepsilon}_c = \frac{A\dot{\varepsilon}}{N} + \left[B \ln\left(\frac{C}{N}\right) \right] \upsilon \tag{11.6}$$

where
- $\dot{\varepsilon}$ is the bulk strain rate
- N is the number of identical circumferential cracks per unit length
- A, B, and C are constants

The disappearance of SCC above a critical applied strain rate (mechanical blunting) is rationalized by recognizing that, to maintain a sharp crack.

$$\upsilon(\dot{\delta}) \geq \dot{\delta} \tag{11.7}$$

where $\dot{\delta}$ is the rate of increase of the crack tip opening displacement (δ) due to the applied strain rate [33]. Sometimes SCC also vanishes *below* a particular strain rate, especially in systems that show extremely slow repassivation (chemical blunting); now a possible condition for crack stability is

$$u(\dot{\delta}) \leq \frac{\delta_{max}}{2} \tag{11.8}$$

where
- $u(= t^*\upsilon)$ is the depth of material dissolved between film rupture events
- δ_{max} is the maximum value of the crack tip opening displacement consistent with continued crack growth

This might correspond to the transition from small-scale to general yielding, given $\delta_{max} \simeq 2 - 3\,\mu m$ for carbon steel in typical test geometries (see Equation 11.9). That is, the amount of corrosion between film rupture events must not enlarge the mechanical crack opening to such an extent that the crack degenerates into a blunt slot. Because t^* is a function of δ this leads to another equation similar to Equation 11.7; the regions of SCC, chemical blunting, and mechanical blunting can then be mapped for various functions $i(t)$ (Figure 11.24). The main uncertainties in such a procedure are the composition of the material that is reaching at the crack tip [this obviously affects $i(t)$] and the restricted crack geometry, which alters the environment and increases both the solution resistance and the effective diffusion length. However, the wedge-shaped geometry of a real crack [33,115,116] significantly reduces the solution resistance and dissolved cation concentration compared with a one-dimensional slot. The crack tip opening displacement w_0 ($\equiv \delta$, see Figure 11.25) is given by

$$w_0 \equiv \delta = \frac{\alpha K_1^2}{E\sigma_0} \tag{11.9}$$

where
- E is Young's modulus
- σ_0 is the flow stress (roughly the average of the yield and ultimate stresses)
- α is a constant that varies with the stress state and strain hardening coefficient and is about 0.3 for carbon steel under plane strain [117]

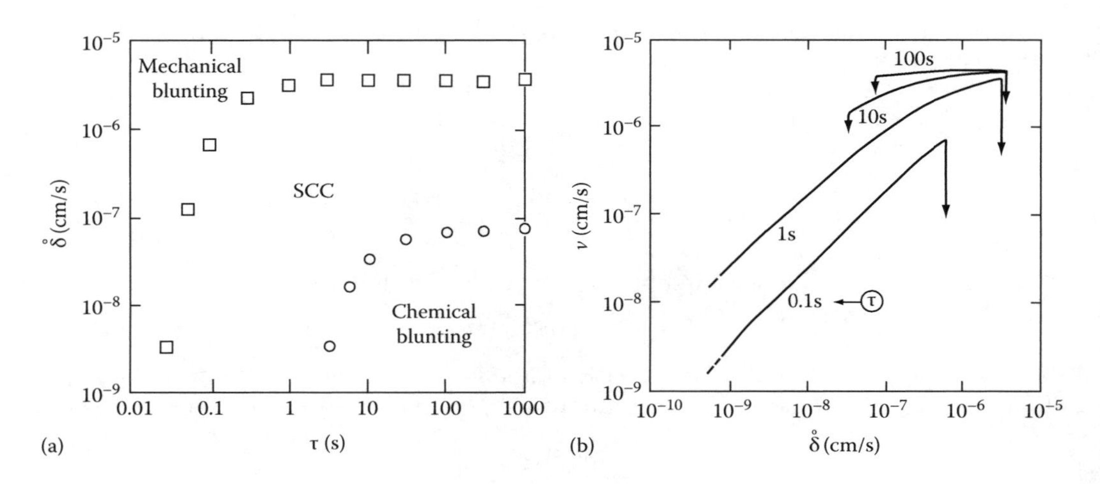

FIGURE 11.24
The effect of strain rate on SCC velocity by slip-dissolution for a system showing a t^{-1} current decay on a fresh surface, for various values of the decay time constant. (From Newman, R.C. and Saito, M., Anodic stress-corrosion cracking: Slip-dissolution and film-induced cleavage, in *Corrosion–Deformation Interactions* (T. Magnin and J.-M. Gras, eds.), Les Editions de Physique, Les Ulis, France, 1993, p. 3.)

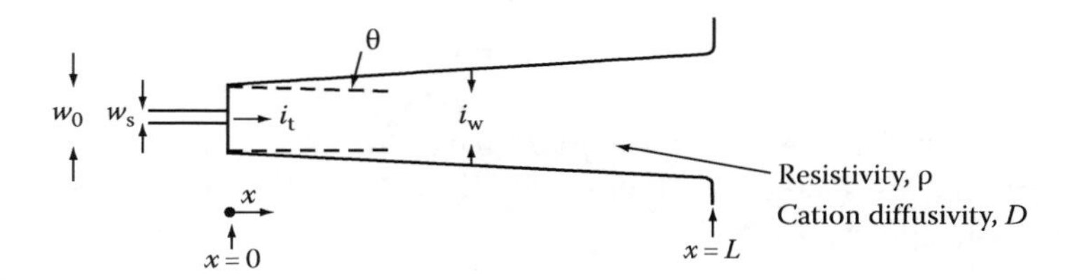

FIGURE 11.25
Crack geometry for calculation of the potential drop and metal ion concentration in a crack. (From Newman, R.C. and Saito, M., Anodic stress-corrosion cracking: Slip-dissolution and film-induced cleavage, in *Corrosion–Deformation Interactions* (T. Magnin and J.-M. Gras, eds.), Les Editions de Physique, Les Ulis, France, 1993, p. 3.)

For small-scale yielding, the crack half-angle θ is given by the elastic solution:

$$\theta = \frac{2(1-\upsilon^2)\sigma}{E} \tag{11.10}$$

where υ is Poisson's ratio, about 0.33. The *IR* potential drop or dissolved metal cation concentration in a growing crack (with small-scale yielding, passive walls, and an active path of width w_s embedded in w_0) was shown to depend on the stress and the crack velocity but not on the crack length [115]:

$$\Delta\phi = \frac{\rho E w_s i_t}{4\sigma} \ln\left(1 + \frac{4\sigma_0}{\sigma}\right) \tag{11.11}$$

where
- ρ is the electrolyte resistivity
- i_t is the anodic current density on the width w_s of active material at the crack tip (i.e., $i_t \equiv i$ in Equation 11.2)

This provides an interesting similitude argument for constant-load SCC tests of for practical SCC failures: no decrease in crack velocity with increasing crack length should be expected under constant load. The preceding argument is only slightly affected by the observation that real stress-corrosion cracks are sharper than would be expected from fracture mechanics [16,34,38,66,116], owing to the existence of uncracked ligaments that shield the crack tip from some of the applied stress intensity—the region over which this occurs (about one grain) is small compared with the crack length as a whole.

One can also analyze the strain-rate dependence of SCC for a slow strain rate test with general rather than small-scale yielding, assuming that plastic stretching at the crack sites dominates the specimen extension from shortly after crack initiation [33,115]. The crack half-angle θ is now given by

$$\theta = \frac{\dot{\varepsilon}}{2N\nu} \tag{11.12}$$

and the crack electrolyte resistance by

$$R = \frac{\rho N\nu}{\dot{\varepsilon}} \ln\left(1 + \frac{L\dot{\varepsilon}}{N\nu w_0}\right) \tag{11.13}$$

Values of θ at least 10 times those obtained in static tests are readily achieved in a slow strain rate test, with a large corresponding decrease in the *IR* potential drop or crack tip metal cation concentration for a given value of i_t. Such arguments show that concerns about the saturation of metal ions or very high *IR* potential drops within cracks are unfounded for slow strain rate tests, even if there is no active path w_s. Historically, this is an important issue, as these concerns were partly responsible for the abandonment of the slip-dissolution model by Galvele [20,118].

For a resistive system, the crack-tip anodic current density i_t can be written

$$i_t = \frac{E - E_m}{w_s R} \tag{11.14}$$

where

E is the potential at the external surface
E_m is the mixed (corrosion) potential of the straining crack tip
R is the electrolyte resistance in the crack

The crack velocity can be derived on this basis as

$$\nu = \left[\frac{\dot{\varepsilon}(E - E_m)A}{nFsw_s\rho N}\right]^{0.5} \left[\ln\left(1 + \frac{\dot{\varepsilon}L}{Nvw_0}\right)\right]^{-0.5} \tag{11.15}$$

As the log term is nearly a constant, this equation explains the commonly observed relationship [29]

$$\upsilon \sim \dot{\varepsilon}^{0.5} \tag{11.16}$$

without considering any of the details of the *i*(*t*) transient. It would be valuable to incorporate some of these crack geometry considerations into more rigorous electrochemical crack models [119].

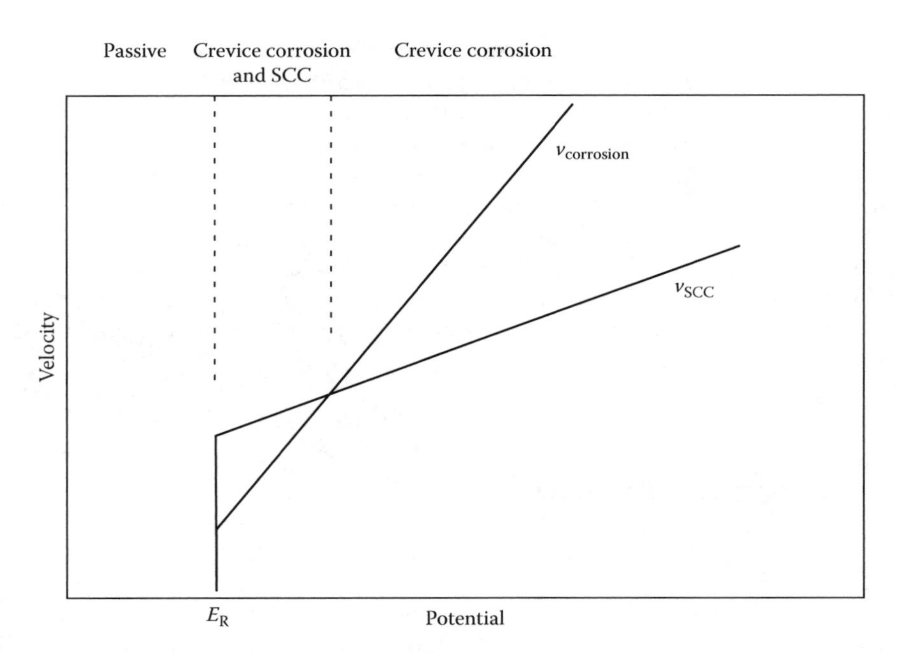

FIGURE 11.26
Rationale for occurrence of type B (localized corrosion-related) SCC over a range of potentials. (From Tamaki, K. et al., *Advances in Localized Corrosion* (H.S. Isaacs, U. Bertocci, J. Kruger, and S. Smialowska, eds.), NACE, Houston, TX, p. 207, 1991; Adachi, T., in *Proceedings of Stainless Steels'91*, ISIJ, Tokyo, Japan, 1991, Vol. 1, p. 189; Tsujikawa, S. et al., in *Proceedings of Stainless Steels'91*, ISIJ, Tokyo, Japan, 1991, Vol. 1, p. 196.)

Type B SCC includes a large number of passive systems in which cracking is induced by chloride ions. All these systems share an acid crack environment and an active or semiactive state at the crack tip. For a while it was thought that Al-Li-Cu-Mg alloys might be an exception to this rule because alkaline crack chemistries are possible under special conditions (little or no cathode area outside the crack), but it soon became clear that rapid cracking in this system is associated with acidification in the crack [120,121].

A fundamental requirement in type B SCC is that the crack velocity must be able to exceed the localized corrosion velocity; otherwise the crack will degenerate into a blunt fissure [68] (Figures 11.26 and 11.27). This leads us to expect SCC over a *range* of potentials between the repassivation potential of the localized corrosion process (E_R) and a higher potential (E_u) at which the localized corrosion velocity equals the crack velocity at the prevailing stress intensity value. By manipulating the composition of a susceptible austenitic stainless steel (Ni, P, Mo, Cu, Al), these potentials can be made to coincide, guaranteeing immunity to SCC [111,112]. An increase in temperature is then required to increase E_u sufficiently for SCC to occur; i.e., the manipulation of alloy composition increases the *critical temperature* for SCC initiation from a crevice. All the applicable alloying elements alter both the SCC velocity and the crevice corrosion kinetics, so a complete analysis would be quite complicated.

It is evident that type B SCC rarely occurs by the slip-dissolution mechanism. The metal in the crack is already in an essentially active state, although there may be a porous oxyhydroxide layer and possibly a dealloyed metallic layer [78,79]. If SCC does occur by anodic dissolution in such a system, it is of the trivial kind seen in sensitized stainless steel or underaged Al-Cu alloys [122] in chloride solutions: the kinetics of an intergranular pitting process are simply enhanced by the opening of the crack, reducing the ohmic resistance according to Equation 11.13.

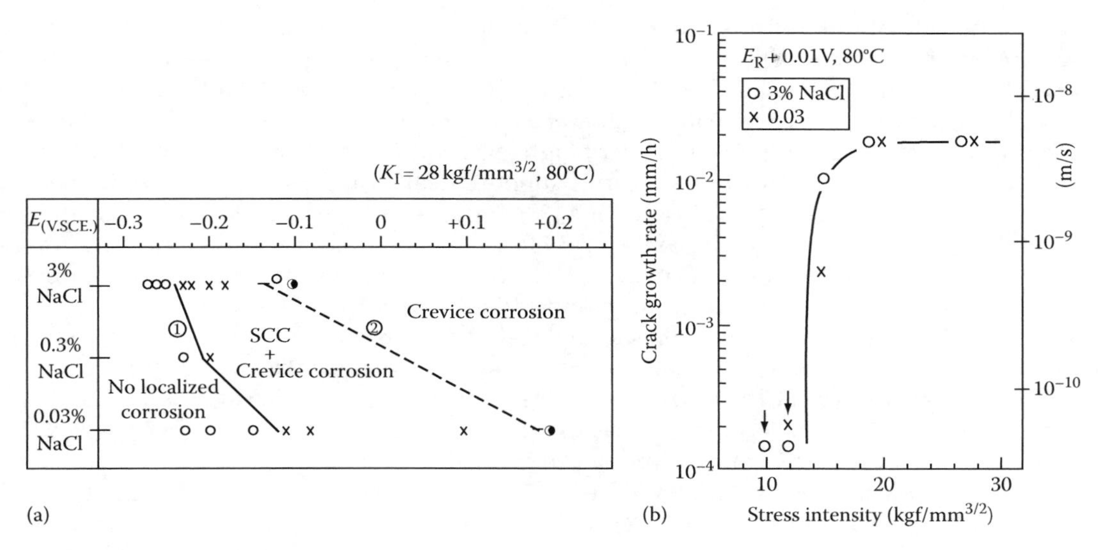

FIGURE 11.27
(a) Localized corrosion and SCC regimes from Tamaki et al. [68] using potentiostatic tests on creviced samples of 316L stainless steel in NaCl solution at 80°C. (b) υ-K curves measured in two different concentrations of NaCl just above the respective re-passivation potentials (E_R), showing that SCC is controlled by local chemistry. (Courtesy of NACE, Houston, TX.)

All type B systems generate hydrogen within cracks, and in some strong alloys, such as martensitic stainless steels [91], the cracking is obviously due to hydrogen embrittlement. High-strength Al-Zn-Mg-Cu (7000 series) alloys also have a proven role of hydrogen in SCC [54,55] (Figure 11.28). Both the formation and the entry of hydrogen are favored by the active state in the crack, but the *presence* of hydrogen does not guarantee a *role*, especially in fcc systems. A further complication is that in some Al alloys, slow crack growth occurs in the absence of an environment, due to creep-like processes or solid-metal embrittlement by lead particles in the alloy [54,123].

The mechanism of chloride-SCC of austenitic stainless steels is still debatable. We may discount the slip-dissolution model for crack growth at 80°C [68]: the required anodic

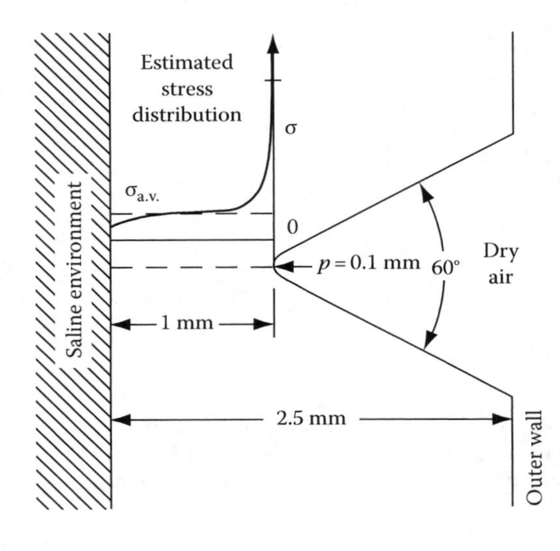

FIGURE 11.28
Notched hollow tube experiment of Ratke and Gruhl [55], used to show that SCC of high-strength Al-Zn-Mg alloys could involve internal crack initiation by hydrogen. (From Holroyd, N.J.H., in *Environmental-Induced Cracking of Metals* (R.P. Gangloff and M.B. Ives, eds.), NACE, Houston, TX, 1990, p. 311; Courtesy of NACE, Houston, TX.)

current density of some 50 mA/cm^2 (a velocity of about 2×10^{-8} m/s) is not available in the crevice environment, except during the first milliseconds of exposure of a fresh surface, and t^* (Equation 11.3) could never be this low. Hydrogen may play a role, especially in unstable austenites at relatively low temperatures, but it is difficult to rationalize a hydrogen effect in stable austenitic steels at very high temperatures such as 300°C [56]. The film-induced cleavage model [40,41] has indirect support: dealloying (Ni enrichment) has been demonstrated in simulated crack solutions [78] and observed on the sides of cracks [79], and the dependence of SCC on Ni content [73,74] is very similar to that seen in Au-Cu alloys as a function of Au content [75] (Figure 11.9). However, there is no direct evidence for film-induced substrate cleavage of the kind found in Cu-Zn, Au-Ag, and Au-Cu alloys [46–51]. The austenitic steel cracks only from sites of localized corrosion, so "single-shot" cleavage experiments are much more complicated than in systems that show uniform dealloying.

The electrochemistry of Cl-SCC in duplex (α-γ) stainless steels was discussed earlier and displayed in Figure 11.17. The ferrite phase has a higher E_R value, due to its higher Cr content, but dissolves more rapidly (due to its lower Ni content) when both phases are in the active state [124]. The crack approaches (from above, due to *IR* limitations) a mixed potential where the austenite is a net cathode and the ferrite a net anode. The ferrite is thus polarized above its normal (isolated) cracking potential and cracks rapidly, while the austenite is below its isolated cracking potential and cracks slowly or not at all. Cracks propagate in the ferrite and tend to be arrested by the austenite. Another situation is possible if the ferrite can remain passive in the crack while the austenite corrodes; now cracking occurs in the austenite and is hindered by the ferrite.

Addition of H_2S or its oxidation product $S_2O_3^{2-}$ (thiosulfate) greatly enhances localized corrosion and Cl-SCC of austenitic and duplex stainless steels [91,125,126]. The effect of H_2S is slight up to a critical concentration around 0.01 M because below this concentration it is depleted in deep pits and cracks [127]. "Nonpropagating" cracks found around this H_2S concentration have run out of H_2S at the crack tip (Figure 11.29). Enhancement of SCC by H_2S is

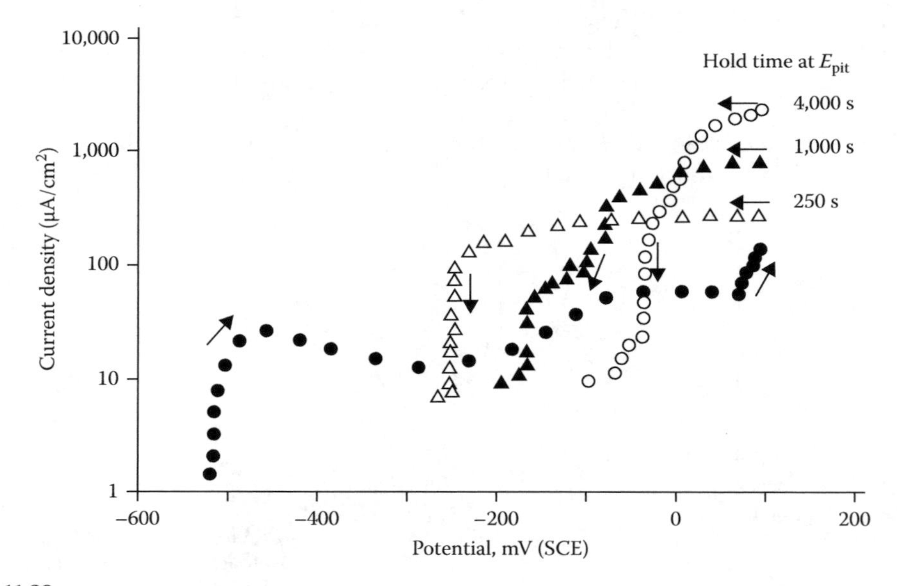

FIGURE 11.29
Potentiodynamic pitting scans for 25Cr duplex stainless steel in 20% NaCl + 0.03 M H_2S, pH 3.4 (acetate-buffered) at 80°C, with different hold times just above the pitting potential, showing the paradoxical raising of the repassivation potential (E_R) with increasing pit (crack) depth, due to depletion of H_2S at the dissolving surface. (Courtesy of Mat, S. B.)

often ascribed to a hydrogen effect, by analogy with low-alloy steels, but for stable austenites such as 310SS this argument is not tenable. Adsorption of sulfur on the active surface in the pit or crack changes the dissolution and passivation kinetics of the material, specifically by catalyzing Fe and Ni dissolution, and possibly by hindering passivation (but the passivation potential is not necessarily altered because no interaction between Cr and S is expected in aqueous systems) [128–130]. Austenite, with its lower Cr and Mo content, will tend to show more sulfur-induced activation than ferrite [130,131]; thus the kinetic difference between the phases will be reduced when adsorbed sulfur is present provided that both phases are in the active state. This may be one aspect of the increased susceptibility of duplex steels in Cl^-/H_2S systems, but obviously there is also an effect of S_{ads} on the atomistics of SCC.

SCC of austenitic or duplex stainless steels in Cl^-/H_2S environments usually occurs from pits, even at moderate temperatures such as 80°C, whereas crevice or underdeposit corrosion is normally needed to nucleate Cl^--SCC at such temperatures. This may be understood from Figure 11.26: pits in Cl^- solutions at 80°C are normally growing faster than cracks (so no SCC occurs), whereas in Cl^-H_2S solutions the crack velocity is increased *and* the pitting velocity can be lower without repassivation, owing to the stabilization of pit dissolution by adsorbed sulfur and possibly also the resistive effect of the black corrosion product that forms in the pits.

In Cl^-H_2S systems, the effect of Ni content is basically the same as in Cl-SCC, except that there is easier cracking around 20%–30% Ni, and a new form of mainly intergranular cracking appears at high T and $[H_2S]$; this latter cracking does not respect the Ni content [132].

Molybdenum alloying is not particularly beneficial for SCC resistance of austenitic stainless steels; nickel is the most important beneficial element. Referring to Figure 11.26, Mo increases E_R and E_u by about the same amount, through its effect on the localized corrosion velocity, while Ni (above 10%) decreases the crack velocity, thus lowering E_u [68,111,112].

Type C (dealloying-related) SCC has been mentioned in the context of austenitic stainless steels, but direct analytical evidence of dealloying is lacking in that case. However, a number of systems definitely show dealloying and can crack in an oxide-free state or with a loose, precipitated tarnish film, such as Cu-Zn, Cu-Al, Au-Cu, and Au-Ag. The original SCC system (brass in ammonia) clearly comes into this category [133], and the ease of dealloying in brass explains why it cracks in so many different environments, especially in slow strain rate tests [134,135]. In brass the dealloying is quite superficial (less than 100 nm) but correlates with SCC and can trigger foil fractures in cuprous $[Cu(NH_3)_2^+]$ solutions that simulate the crack environment [46,72,87,136,137] (Figures 11.10 and 11.30). In gold alloys, the correlation can be demonstrated more easily, as the electrochemistry is much simpler without any competing anodic reaction on the gold [47–51]. The attempt of Flanagan, Lichter, and others [80,81] to account for SCC of Au-Cu alloys without dealloying is premature, as rapid dealloying may well occur in the chloride-rich environment of a crack.

In several systems, including Au and Cu alloys, gross macrodealloying in chloride solutions has been shown *not* to favor SCC or microcleavage [33,47,48,87,135,136]. One view is that this represents a kind of crack blunting (in a slip-dissolution context) [135], but we have proposed a more detailed interpretation based on the film-induced cleavage model [47–49,87,88]. Essentially, gross dealloying is indicative of coarsening of the pores within the dealloyed material (by surface self-diffusion), to the point that the pores are too coarse to nucleate a brittle crack in the unattacked substrate. This aging phenomenon has been demonstrated directly in gold alloys (Figures 11.31 and 11.32) and in brass it has been shown that alloyed arsenic causes a transition from macro (μm) to micro (nm) dealloying [87,136], favoring SCC in chloride environments but having no effect in ammonia, where microdealloying occurs irrespective of the presence of arsenic; ammonia and arsenic

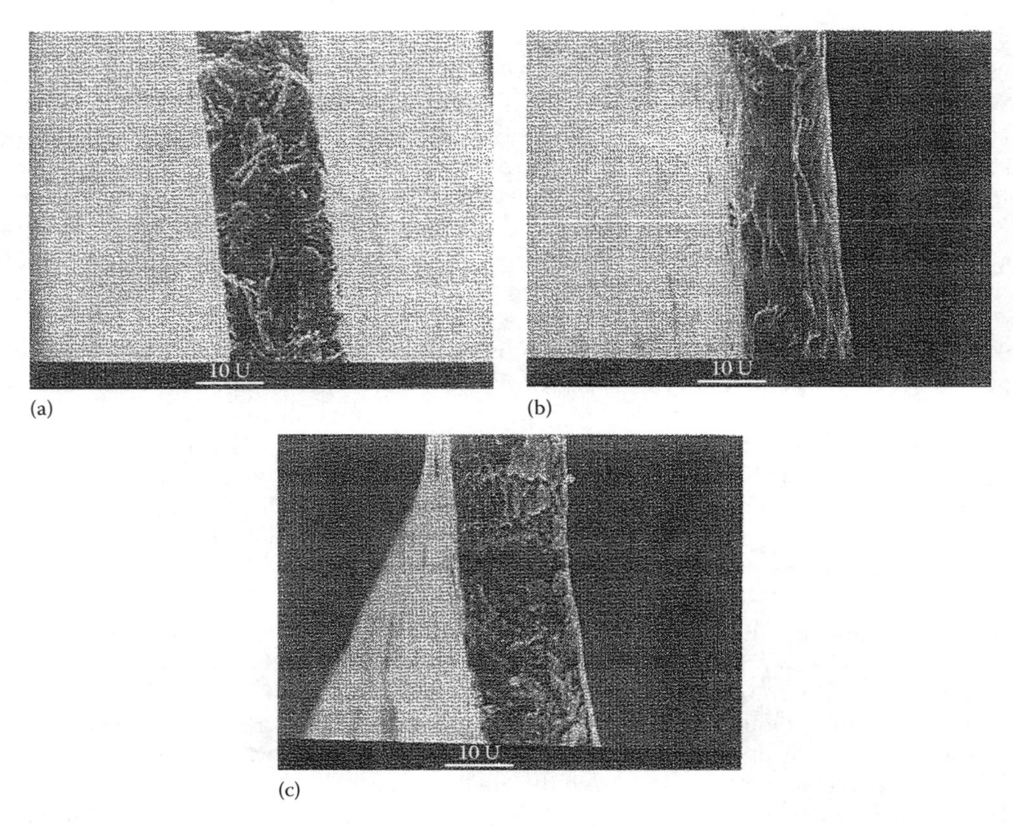

FIGURE 11.30
(a) Single-shot cleavage failure of α-brass rapidly strained in a cuprous ammonia solution after 100 min immersion. (b) Identical specimen rinsed in deionized water and dried with cool air before fracturing, showing aging of the dealloyed layer to a harmless form. (c) Identical specimen quenched in liquid nitrogen without rinsing, then fractured near 77 K, confirming that cleavage can occur with no liquid environment present. (From Newman, R.C. et al., *Scr. Metall.*, 23, 71, 1989.)

are both believed to hinder surface mobility of Cu, by specific adsorption in cationic or uncharged form [87,88,136–138]. The dealloyed layers formed on brass in ammoniacal solutions age very rapidly when out of contact with the solution, losing the ability to trigger cleavage in thin foils (Figure 11.30).

Cracking of silver alloys has been reported by Galvele and others in a series of papers [139–141]. Interesting SCC phenomena were reported in both aqueous and gaseous environments, including halogen vapor. Clearly, all these systems involve selective reaction of the silver in the alloy, which in halide solutions or atmospheres generates a composite of silver halide and nearly pure noble metal (Au or Pd); this layer should behave mechanically much like a dealloyed layer formed under conditions of free silver dissolution, so the mere occurrence of SCC does not help to distinguish between SCC mechanisms. Foil-breaking experiments [33,46–51] should be performed in some of these environments to test the generality of the film-induced intergranular fracture mechanism.

Type D SCC involves surface films that are neither oxides nor dealloyed layers. The cracking of high-strength 4340 steel in dry Cl_2 [102,103] or dissociated N_2 [142] appears to be a film-induced process; Sieradzki [103] proposed that $FeCl_2$ acted as a stiff layer, causing a ductile-to-brittle transition due to the modulus mismatch with the substrate.

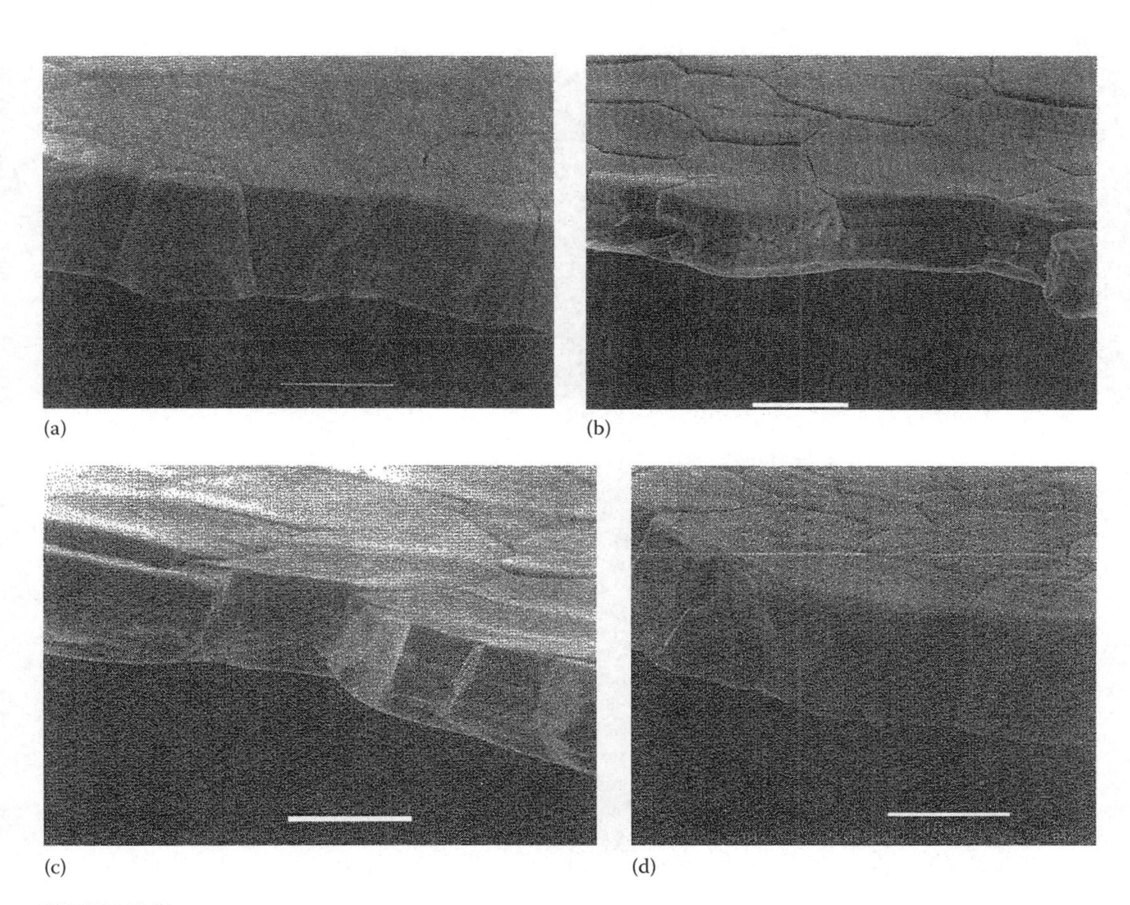

(a) (b) (c) (d)

FIGURE 11.31
(a) Ag-20 at % Au foil specimen, dealloyed to a depth of 3μm in perchloric acid without stress, then fractured after stepping the potential to 0V (SCE) for 5min (brittle intergranular failure). (b) Dealloyed and stepped to 700mV for 30min (ductile failure—the coarsening of the porosity in the dealloyed layer is more rapid at higher potentials). (c) As in (a) but with 10mM NaCl added after stepping the potential to 0V (ductile failure due to rapid pore coarsening). (d) As in (c) but with a further addition of 10mM pyridine (brittle failure due to counteraction of chloride effect by pyridine). Horizontal lines (scale markers) represent 10 microns. (From Newman, R.C. and Saito, M., Anodic stress-corrosion cracking: Slip-dissolution and film-induced cleavage, in *Corrosion–Deformation Interactions* (T. Magnin and J.-M. Gras, eds.), Les Editions de Physique, Les Ulis, France, 1993, p. 3; Saito, M. et al., *Corros. Sci.*, 35, 411, 1993.)

Later the lattice parameter was emphasized as a casual factor [143]. A detailed rationalization of SCC was made for iron exposed to anhydrous ammonia-methanol, where anodic oxidation of ammonia leads to interstitial penetration of nitrogen [69,100]. According to the film-induced cleavage model, such very thin, brittle layers are sufficient to allow cleavage through several μm of a body-centered cubic (bcc) material, whereas in fcc systems a nanoporous metallic layer 10–100nm thick is a specific requirement as it is the only way of epitaxially coupling a brittle reaction product of sufficient thickness to the fcc substrate. A special case may be the SCC of pure copper or silver, where a micropitted or tunneled zone is a possible brittle layer that could nucleate cracking [144,145].

There are several other film-induced SCC processes, such as cracking of Zr alloys in gaseous iodine [101]. In no case has monolayer *adsorption* been validated as the cause of SCC—every known case involves a 3D film. However, liquid metal embrittlement often occurs, even in fcc systems, without alloy or compound formation [15]. Possibly there is a

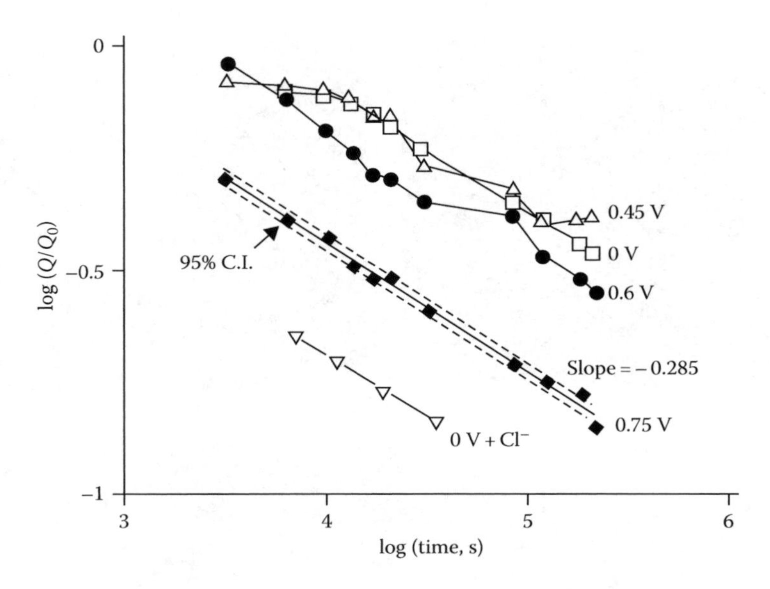

FIGURE 11.32
Measurements of double-layer capacitance (proportional to pore surface area; inversely proportional to pore radius) as a function of time during potentiostatic aging of dealloyed layers on Au–Ag, showing the effects of potential and chloride ions. (From Kelly, R.G. et al., *Electrochemical Impedance Spectroscopy* (J.R. Scully, D.C. Silverman, and M.W. Kendig, eds.), ASTM, ASTM STP 1188, Philadelphia, PA, 1991, p. 94.)

regime of very rapid adsorption-induced cracking that is rarely accessed in aqueous systems as it requires the adsorption process to be faster than plastic relaxation processes at the crack tip; hence in aqueous environments this is likely to occur only in very reactive metals (Al, Ti, Mg).

Type E SCC (embrittlement of steels by cathodic hydrogen in the active state) is quite well understood except for the atomistic action of the hydrogen, which is still debatable. Briefly, hydrogen atoms are produced by water reduction during corrosion in neutral solutions, and some of them enter the steel, especially if their recombination to H_2 is poisoned by adsorption of S, As, P, or Sb. Once in the steel, the hydrogen can move with an effective diffusivity (D_{eff}) of 10^{-8} to $10^{-4} cm^2/s$, depending on the microstructure as well as temperature [21] (e.g., small carbides in tempered martensite act as traps and greatly reduce D_{eff}). The lattice (interstitial) hydrogen solubility is elevated ahead of stressed cracks or notches, and these sites accumulate higher hydrogen concentrations. The same hydrostatic tension that increases the H solubility in the lattice also enhances hydride phase precipitation in metals such as Ti and Nb [101]. Hydrogen segregates to interfaces, including grain boundaries, and may weaken these in combination with other segregants such as P and Sb [146]. Once it reaches its site of action, and provided it is in a strong microstructure, the hydrogen causes local fracture by cleavage [147], intergranular separation [146], or enhanced microplasticity [148], or some combination of these processes. In low-strength steels, dynamic loading is required for cracking. If the input fugacity of hydrogen is high enough (hundreds of atmospheres in media such as acidic H_2S), recombination at internal sites such as nonmetallic inclusions may cause blistering or cracking (hydrogen-induced cracking, or HIC). This is minimized in pipeline steels by careful microstructural control.

The altered solution chemistry within cracks is an important aspect of SCC and corrosion fatigue in structural and high-strength steels. Turnbull [149–151] has shown that crack acidification, as proposed by Brown [152], is not the norm for steels corroding freely in NaCl solutions or seawater. Only in Cr-containing steels with very short cracks could any acidification be predicted [153]. Possibly some of Brown's historic measurements were flawed by oxidation of Fe^{2+} by atmospheric oxygen; more recent work shows that deep cracks normally become net cathodes and reach a pH of about 9 [154] (Figure 11.33). Nevertheless, with assistance from the *IR*-induced isolation of the crack enclave, the potential can remain below that of a reversible hydrogen electrode, and in sufficiently strong steels embrittlement will occur above some K_{ISCC} value. An excellent correlation exists between hydrogen uptake on the walls of simulated cracks and SCC of high-strength steels in the same solution [154] (Figure 11.34). Cathodic protection, within limits, can reduce the hydrogen uptake from cracks by increasing the local pH [149–151]; however, this is rarely used for high-strength

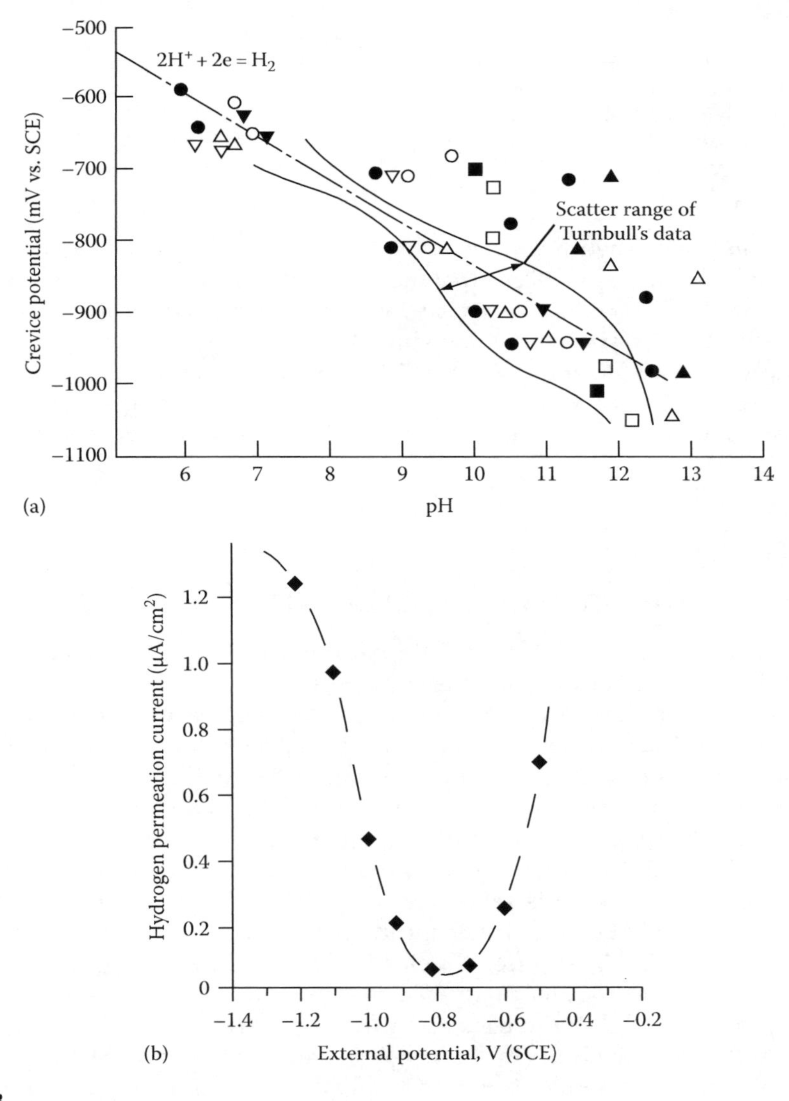

FIGURE 11.33
(a) Crevice potential and pH and (b) hydrogen permeation rate, as a function of potential for carbon steel crevices in NaCl solution. (From Cottis, R.A. and Taqi, E.A., *Corros. Sci.*, 33, 483, 1992; Courtesy of HMSO, London.)

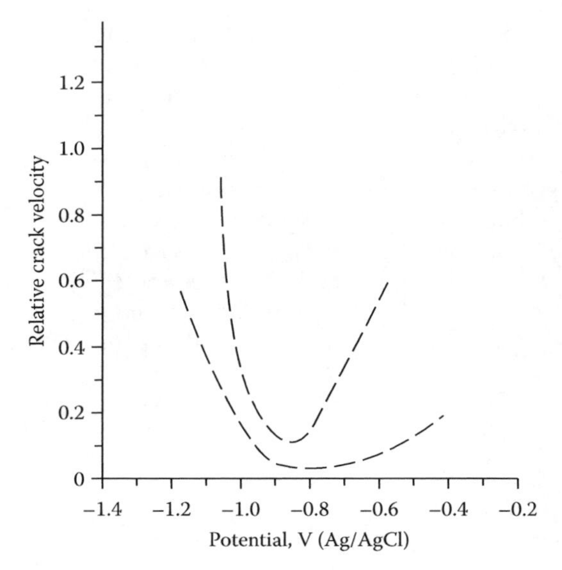

FIGURE 11.34
Band of results showing the effect of potential on the rate of SCC, for a range of high-strength steels in seawater; compare with Figure 11.33b. (From Brown, B.F., in *Stress-Corrosion Cracking and Hydrogen Embrittlement of Iron-Base Alloys* (R.W. Staehle, J. Hochmann, R.D. McCright, and J.E. Slater, eds.), NACE, Houston, TX, 1977, p. 747.)

steels that are sensitive to hydrogen, even though major benefits can be demonstrated in the laboratory. The problem in practice is that most cathodic protection systems locally reduce the potential to values that are dangerously low for high-strength steels.

Hydrogen effects often show the same kind of strain rate sensitivity as slip-dissolution processes. Low-strength steels are immune to SCC under static loads in salt water, but steels of all strengths can suffer hydrogen-assisted fatigue crack growth [105,155]. High strength continues to be detrimental, but relatively less so than under static loading.

11.8 Stress-Corrosion Testing in Relation to Mechanisms of Cracking

Because standard methods of SCC testing have been excellently reviewed by Sedriks [156], we focus on the implications of mechanisms for testing and vice versa.

To rationalize or predict service performance, and to test models of SCC, the slow strain rate test [26,27,113,157] is convenient as it enables a large number of tests to be conducted at a large number of potentials within a defined time; the maximum duration is simply the failure time in air (Figure 11.35). The maximum load, elongation to failure, and percent reduction in area at fracture are measured in the test environment and normalized to values measured in an inert environment. The disadvantages are that the mechanical condition is relatively undefined when there are multiple cracks, the crack velocity cannot be measured continuously, and failures occur in materials that would never fail by SCC in service. An elastic slow strain rate or ultraslow cyclic test [67] is a useful compromise that maintains the dynamic loading without gross plastic straining (Figure 11.36), but these tests are difficult to carry out in large numbers. It has been suggested that slow strain rate and cyclic-loading tests are part of a continuum [28–30,113,158] and that in some systems SCC and corrosion fatigue are also a continuum with a single mechanism (Figure 11.37); certainly the fractography of slow strain rate SCC and low-frequency fatigue crack growth can be very similar, e.g., in low-strength

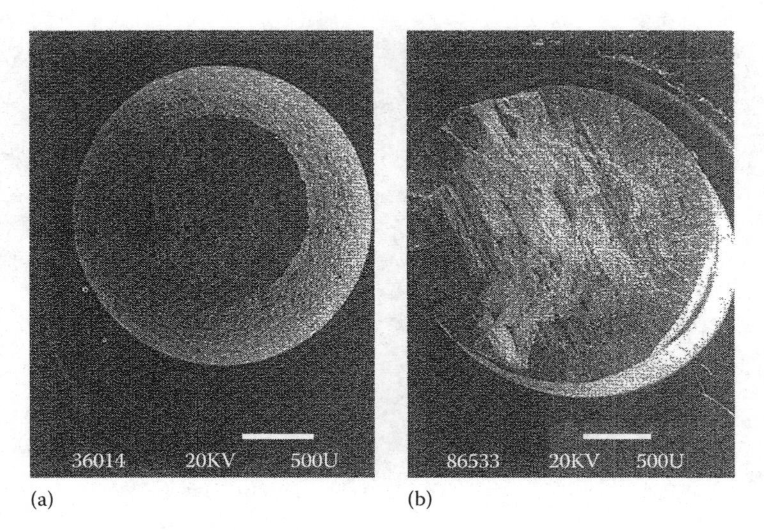

FIGURE 11.35
Typical fractures of C-Mn steel obtained in slow strain rate tests: (a) in air; (b) in anhydrous ammonia-methanol. (Courtesy of Hannah, I.M. and Zheng, W.)

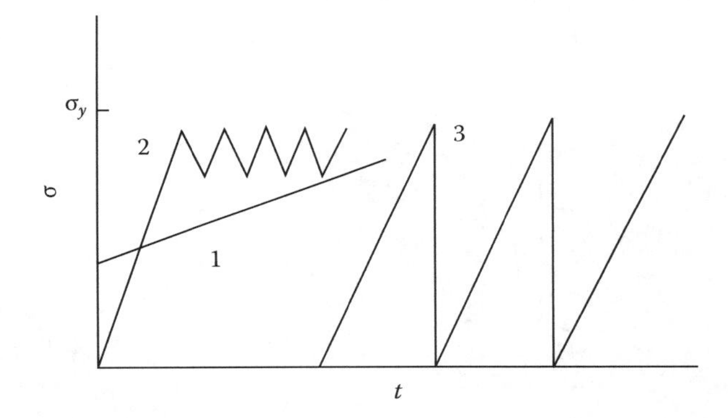

FIGURE 11.36
Various types of elastic slow strain rate test.

steels. The most notorious cases in which the slow strain rate test overestimates the susceptibility of a material are the hydrogen-induced cracking of low-strength steels under cathodic protection [159], the SCC of commercial-purity titanium in salt water [160], and the hydrogen-induced cracking of duplex stainless steels [161]. In every case the K_{ISCC} value, if it exists, is extremely high, at least 50 MPa m$^{1/2}$. Sometimes this hydrogen-induced cracking, which would not occur in practice, overshadows a "genuine"-SCC phenomenon that occurs at a lower velocity but has a much lower K_{ISCC} value, e.g., SCC of carbon steel in CO–CO_2–H_2O solutions [162,163]. One approach to this kind of problem (without changing the test) is to study the distribution of secondary cracks on the failed tensile specimen. Cathodic hydrogen embrittlement is confined to the necked region [159,164], but the genuine SCC is distributed as secondary cracks along the whole length of the tensile specimen [162] (Figures 11.38 and 11.39).

Having classified CO_2-induced cracking as possibly an artifact, we note that this is a favored mechanism for a transgranular cracking phenomenon seen in high-pressure gas transmission pipelines. Clearly, the hydrogen uptake in CO_2 or $NaHCO_3$ solution must be high compared with cathodic protection in salt water, where similar steels do not crack, or

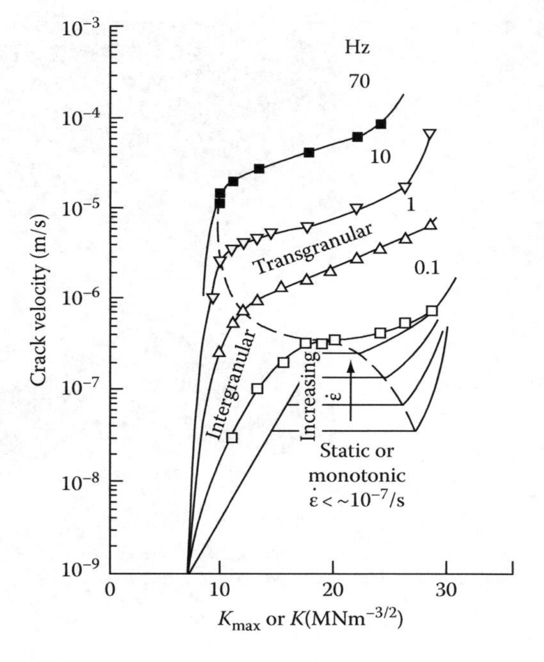

FIGURE 11.37
The effect of crack tip strain rate, via loading rate, on crack growth in a high-strength aluminum alloy. (From Holroyd, N.J.H., in *Environmental-Induced Cracking of Metals* (R.P. Gangloff and M.B. Ives, eds.), NACE, Houston, TX, 1990, p. 311; Holroyd, N.J.H. and Hardie, D., *Corros. Sci.*, 23, 527, 1983; Courtesy of NACE, Houston, TX.)

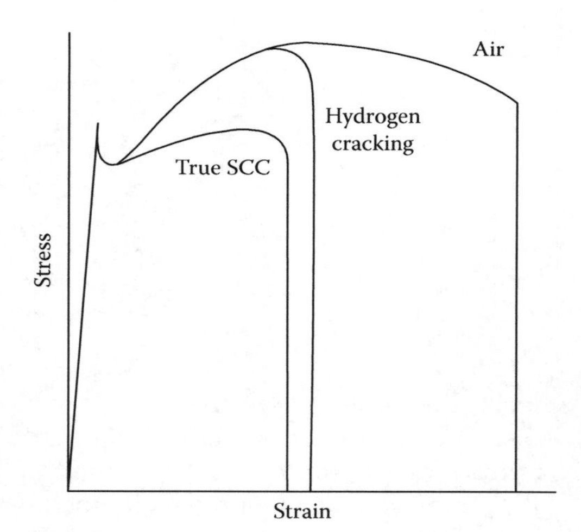

FIGURE 11.38
Stress–strain curves for "real" SCC (cracks initiated soon after yield) and cathodic hydrogen embrittlement (where cracking is confined to the necked region), for low-strength steels.

else the coexistence of localized corrosion and hydrogen entry helps to maintain the crack tip strain rate. Dynamic service stresses are also an important factor.

Having reviewed some of the flaws of slow strain rate testing, we must stress that this is an outstanding useful test. It is always desirable to have the slow strain rate information, even if it is not used for direct prediction o service performance. Neither the pipe cracking in boiling water reactors [59] nor the SCC of high-pressure gas transmission lines [27,67] (both type A systems) could have been predicted by static-load tests on smooth specimens.

Loading rate, or dK_I/dt, affects SCC initiation and growth in precracked specimens [113] (Figure 11.37) and a good appreciation of the cracking mechanism is essential to avoid

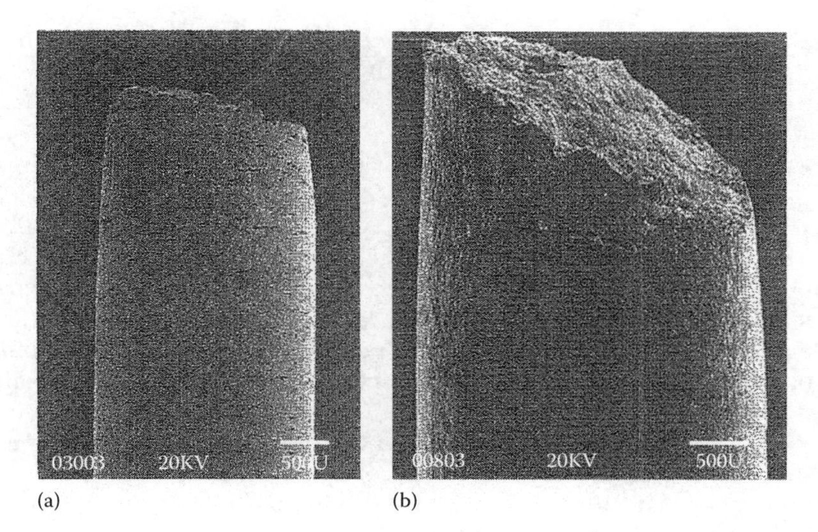

FIGURE 11.39
Fractography corresponding to Figure 38 (from Hannah, I.M. et al., in *Hydrogen Effects on Material Behavior* (N.R. Moody and A.W. Thompson, eds.), TMS, Warrendale, PA, 1990, p. 965), for C-Mn steel in $CO–CO_2–H_2O$ solution: (a) genuine SCC; (b) cathodic hydrogen embrittlement. (Courtesy of Hannah, I.M.)

surprises in service, as not all variants of $K_I(t)$ can be tested in the laboratory. Attempts to rationalize all such data with the slip-dissolution model (for high-temperature aqueous environments) have been more in the nature of a sophisticated fitting exercise than a first-principles approach [29], and there is a great divergence of views on the cracking mechanisms, especially in pressurized water reactor (PWR) pressure vessel steels, where everything from slip dissolution [28–30] to hydrogen embrittlement [165] to film-induced cleavage [42] to surface mobility [166] has been suggested.

In some systems, SCC is initiated at a very large number of sites, and the crack tip strain rate is greatly reduced at each individual site, leading to near arrest of the cracks (another reason for crack arrest might be discontinuous segregation or precipitation; recall Figure 11.13). Coalescence of two or more cracks is then required to produce a dominant crack that causes final failure [27,167] (Figure 11.40). Statistical physicists are very interested

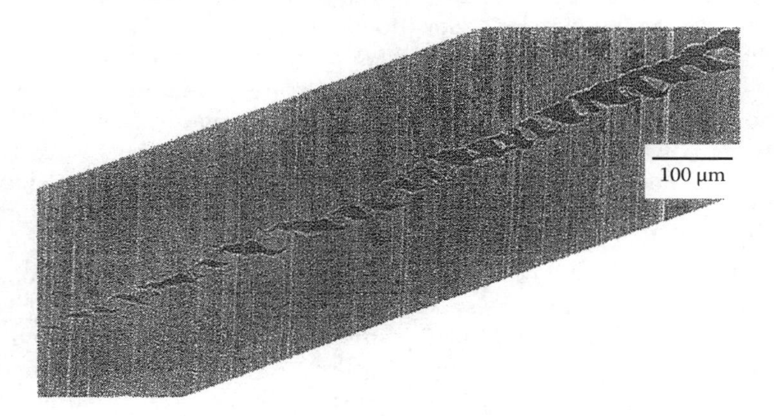

FIGURE 11.40
Crack coalescence of {110} microcracks leading to a dominant slip-band crack for a Cu monocrystal dynamically strained in $NaNO_2$ solution. (From Sieradzki, K. et al., *Metall. Trans. A*, 15A, 1941, 1984.)

in similar problems [168], ranging from mud cracking to fracture of random composite media. *Multiple crack interactions* are a major growth area in fracture mechanics. Such phenomena are peculiar to smooth specimens and provide a strong argument for carrying out several kinds of laboratory test.

A major area of development in SCC testing is the role of crevices in crack initiation. We have discussed the essential role of localized corrosion in Cl-SCC of austenitic stainless steels [68,111,112], and topical problems with duplex stainless steels necessitate extension of the same approach to these more resistant materials. The incorporation of crevices into slow strain rate specimens is not at all trivial, as we discovered during 2 years' work by Suleiman at UMIST [169]. Tamaki et al. [68] showed that a fine notch or precrack could be used to establish a reproducible condition in a fracture mechanics specimen, but this too is difficult work, especially if one attempts to carry out the work potentiostatically. Tamaki et al. found that, for 316L stainless steel at 80°C, SCC sometimes occurred only within a few tens of mV above the repassivation potential of the crevice (E_R) (Figures 11.25 and 11.26). Without accurate prior knowledge of E_R, or in a complex service environment, such work can be very time consuming, so we have proposed a galvanostatic variant in which a low anodic current is applied that automatically establishes a steady potential just above the lowest possible value of E_R before starting the tensile machine (see Figure 11.7c). This galvanostatic technique is also applicable to the rapid screening test used by Shinohara and colleagues [112], in which two sheets of austenitic steel are spot welded together and polarized in a chloride solution. All these procedures are recommended for further development except perhaps the use of the creviced slow strain rate test.

11.9 Notes on Other Proposed SCC Mechanisms

Magnin's model of SCC and corrosion fatigue [16,17] proposes that the motion of crack-tip dislocations is impeded by an obstacle ahead of the crack tip, which in the purest case might be a Lomer-Cottrell lock. Dissolution down a slip band toward the obstacle leads to the achievement of K_{Ic} at the obstacle and the propagation of a cleavage crack on the slip plane back toward the crack tip; in addition (or alternatively), this decohesion of the microfacet can be facilitated by hydrogen adsorption on the fracture plane (Figure 11.41). Such a model is consistent with certain observations in corrosion fatigue cracking, and with fractographic studies of Dickson and others [66,170] on transgranular SCC surfaces, but is not yet adapted to explain the multitude of special environmental factors in SCC. Application of this model to alloy 600 in hydrogenated high-temperature water [17] might be considered arbitrary in comparison with Scott and Le Calvar's [60] highly focused investigations of intergranular oxidation. Magnin's model is elegant and plausible but can be tested only by fractography, which is bound to be inconclusive.

The model of environmental cracking proposed by Lynch [19] will be considered only briefly as it accounts for none of the special environmental effects in SCC. This does not mean that the model is wrong, only that it is untestable except by microscopic means. The concept of enhanced local plasticity leading to plastic microfracture is an important aspect of hydrogen embrittlement [148] and has interesting implications in liquid metal embrittlement. Most authorities reject Lynch's adsorption-based approach to hydrogen effects preferring to appeal to effects of *internal* hydrogen on plasticity, yet Lynch has a powerful

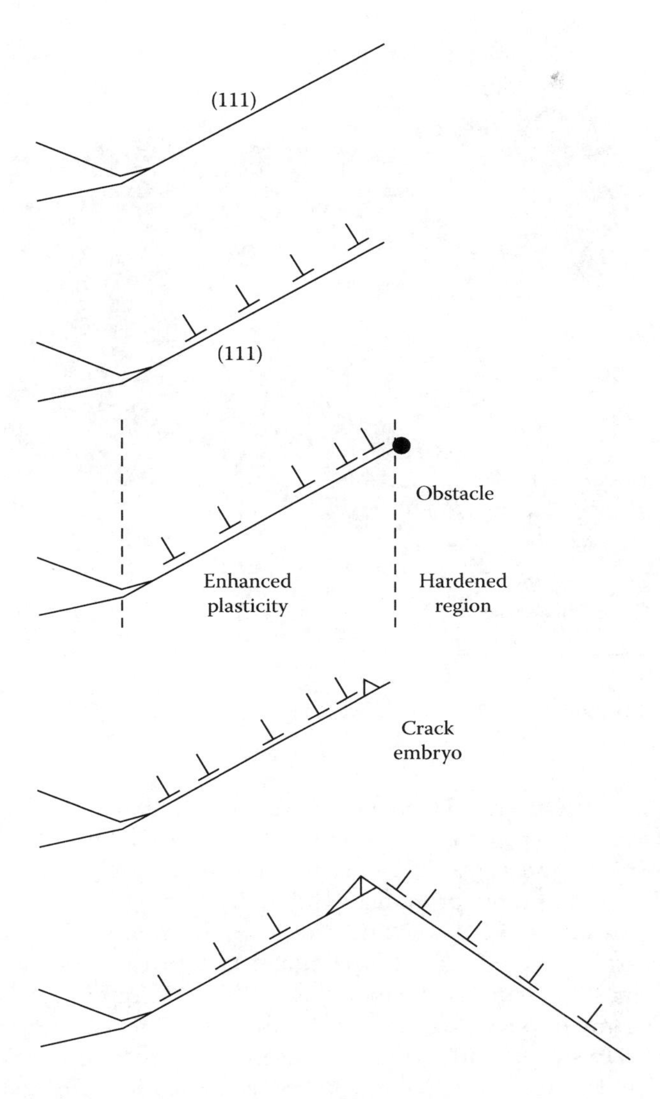

FIGURE 11.41
Magnin's model of SCC. (From Magnin, T. and Lepinoux, J., Metallurgical aspects of the brittle SCC in austenitic stainless steels, in *Parkins Symposium on Fundamental Aspects of Stress-Corrosion Cracking* (S.M. Bruemmer, E.I. Meletis, R.H. Jones, W.W. Gerberich, F.P. Ford, and R.W. Staehle, eds.), TMS, Warrendale, PA, 1992, p. 323; Magnin, T., in *Corrosion–Deformation Interactions* (T. Magnin and J.-M. Gras, eds.), Les Editions de Physique, Les Ulis, France, 1993, p. 27.)

argument based on the similarity between liquid metal–induced and hydrogen-induced fractures in certain systems. There is no doubt that plastic microfracture will form part of the complete SCC spectrum, should this ever be elucidated, but we must reject the notion that all SCC can be explained by such mechanisms.

"Universal" models of environmental fracture rarely survive for long, and SCC specialists such as Parkins [27,171] have devoted much effort to promoting a spectrum of mechanisms from mainly chemical (intergranular slip-dissolution) to mainly mechanical (hydrogen-induced SCC of high-strength steel). Nevertheless, universal models continue to appear from time to time and must be tested seriously. Galvele's surface mobility model

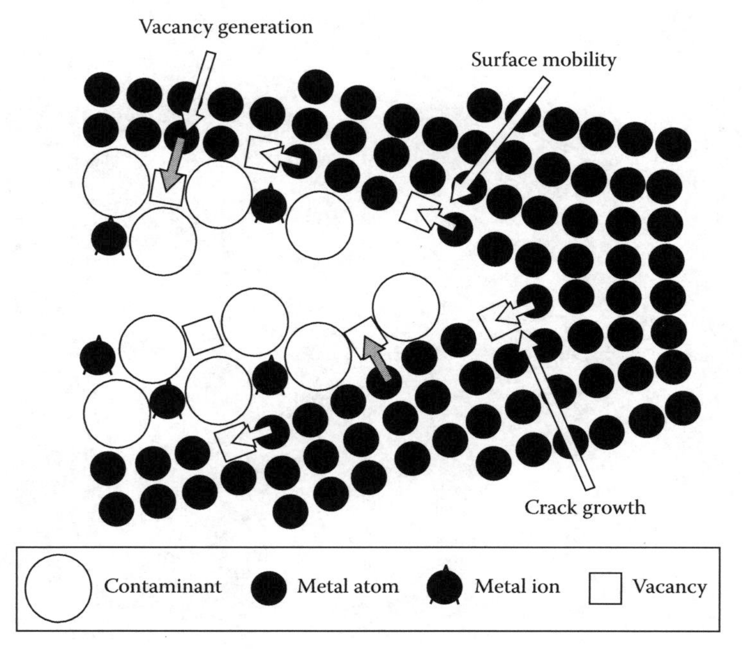

FIGURE 11.42
Galvele's surface mobility model for SCC. (From Galvele, J.R., *Corros. Sci.*, 27, 1, 1987.)

[20,166] proposes that cracks grow by surface diffusion of metal atoms (possibly combined with ions or molecules from the environment) from a very sharp crack tip to the crack walls, i.e., capture of vacancies by the crack tip (Figure 11.42). Formation of surface compounds with low melting points promotes SCC by increasing the surface self-diffusivity of the metal, D_s. The details of the crack tip and its interaction with the environment (i.e., electrochemistry) are considered to be secondary factors that contribute by maintaining the critical surface compound. The model is applied to liquid metal embrittlement as well as SCC and seems to have considerable predictive power. One of the key factors in Galvele's conversion to surface mobility was his conviction that slip-dissolution could not work in the restricted geometry of the crack, but, as explained earlier, the crack is not a one-dimensional slot and the crack opening angle permits very high anodic current densities in cracks without saturation of metal salts or very high *IR* potential drops.

According to Galvele's model, the crack velocity (υ) is given by

$$\upsilon = \frac{D_s}{L}\left[\exp\left(\frac{\sigma a^3}{kT}\right) - 1\right] \tag{11.17}$$

where

D_s is the surface self-diffusion coefficient
L is a characteristic diffusion length
σ is the elastic surface stress at the crack tip
a^3 is the volume of a vacancy

The treatment of the diffusion process is greatly simplified, but this is not one of the main flaws in the analysis. These are, according to Sieradzki and Friedersdorf [23]:

1. The equilibrium vacancy concentration in a plane can be increased by a *normal* tensile stress [172], but Galvele applied the relevant equation to the crack tip where this stress is zero. Lacking this effect, one must appeal to small, second-order effects.
2. Proper accounting of the chemical potential of the vacancies must include a capillary term of the form $+\gamma a^3/r$ (γ = surface energy, r = crack-tip radius) [173]. This prevents the crack from attaining very low r values, yet the surface mobility mechanism assumes an atomically sharp crack.

Sieradzki and Friedersdorf end their analysis by concluding that the crack velocities attainable by surface mobility (with the blunt crack tip just mentioned) are 10^8 to 10^{14} times lower than those calculated by Galvele. Their expression for the crack velocity is

$$\upsilon = \frac{D_s N_s \Omega^2}{kT} \frac{2}{\pi} \frac{1}{r^2} \left(\frac{\sigma^2_{yy}}{2E} - \gamma\kappa \right) \tag{11.18}$$

where
N_s is the number of lattice sites per unit area
Ω is the atomic volume (a^3)
κ is the curvature, which is just $1/r$ for the crack tip

11.10 Summary and Recommendations for the Future

Several models of SCC have been tested successfully in particular systems. It would be premature to regard these tests as definitive (there is a particular need for direct studies of crack dynamics in Au-Cu or similar alloys showing film-induced cleavage), but they encourage us to look to other, unsolved systems for a more complete picture of the range of SCC phenomena. Some of these systems are:

1. Creep-like phenomena that may involve cavity formation induced by *internal oxidation* [60,61]. These will normally occur at elevated temperatures, and the outstanding example is the SCC of alloy 600 in PWR primary water or steam-hydrogen mixtures. The species being internally oxidized may include both Cr and C [174]; hence Ni-C alloys may also crack under some conditions. Some progress has been made by pre-exposing thin foils to the SCC environment, then allowing hydrogen to escape and straining the foils to failure at room temperature [61]; significant irreversible embrittlement has been observed and is confined to the surface, suggesting an internal oxidation effect rather than an irreversible hydrogen effect such as methane bubble formation.
2. Gaseous embrittlement by species other than H_2 [101–103,141,142], where a thorough test of the applicable models, using filmed thin foils would be valuable.

3. SCC of iron or carbon steel under controlled potential conditions in conducting, high-temperature aqueous environments [29], where more than one mechanism may be operating, but the slip-dissolution and film-induced cleavage models are eminently testable using established methods with modern refinements.
4. Brittle SCC in iron exposed to CO-CO_2-H_2O [162,163] or anhydrous NH_3 [69]; indirect evidence for film-induced cleavage has been presented, but this is too indirect at present, and the special surface layers that form in these environments need to be studied.
5. Liquid metal embrittlement (LME) [15] (properly called metal-induced fracture, as the metal need not be liquid so long as it is at a high T/T_m) remains an outstanding problem for SCC specialists. Lynch [19] has stated that transgranular LME, in particular, involves only a surface interaction with the environment and hence must be classed as an adsorption-induced process even in the fcc materials; in situ electron microscopy, to observe the cracking process directly, seems to be one way forward (although not with such a crude environment as liquid mercury). The softening or local plastic fracture claimed by Lynch may be demonstrated or may resolve into a hardening or cleavage process at high enough resolution. Not enough is known about the total crack tip system in liquid metal embrittlement, and a proper accounting of all the forces and chemical potential gradients has not been made.
6. Systems of interacting cracks should be studied using techniques developed by statistical physicists [168].

Finally, there is room for more careful surface chemistry studies in SCC, utilizing recent advances in the characterization of the crack solution (pH, etc.) and the crack path (segregation, etc.). Although these studies may not reveal anything radically new, they are essential for a rigorous development of the subject, for example:

1. Interactions of anhydrous liquid ammonia with iron surfaces, to test the proposed role of nitride layers [69,100]. An experimental problem is that extremely low H_2O levels are required to produce the pure Fe-N interaction on a free surface, whereas in a crack the gettering action of the crack walls maintains a very low water content at the crack tip.
2. Surface analysis of films formed transiently on bare, e.g., laser-irradiated surfaces [175], should be undertaken. Such studies should include rapidly solidified alloys that incorporate segregant elements and should be integrated with mechanical studies on the same films.
3. The status of dealloying and film-induced cleavage in chloride-SCC of austenitic stainless steels is still unclear. Further studies should be undertaken in solutions that simulate the acid crack environment. The role of alloyed P [88] should be examined using these or simpler (Fe-Ni) alloys.
4. High-resolution Auger or secondary ion mass spectrometry (SIMS) analyses of SCC fracture surfaces, especially around crack-arrest features, have not kept pace with instrument development, and there is great scope for well-designed experiments of this type supported by cross-sectional transmission electron microscopy.
5. The scanning tunneling and atomic force microscopes (STM and AFM) have considerable potential for mapping fracture surface topography at very high resolution [176], provided one chooses an appropriate system with minimal corrosion and large flat areas on the fracture surface—this is illustrated every time one tests

an STM or AFM on cleaved graphite or mica. Such studies may resolve some of the micromechanics of film-induced cleavage—for example, by showing that crack-arrest markings are present everywhere on transgranular fracture surfaces but are simply too fine to be resolved in the SEM. One may also be able to use the new generation of these instruments, with their improved position control and microscope facilities, to carry out high-resolution side-surface studies *during* SCC.

Finally, while SCC of metals is important in our industrial society, SCC of other inorganic solids can be said to have shaped our whole world. It is now widely accepted that time-dependent fracture (stress corrosion) of rocks in contact with water is responsible for such phenomena as premature fault movement leading to earthquakes [177] and fracture of newly formed crustal rocks emerging on the ocean floor [178].

References

1. H. H. Uhlig, *Physical Metallurgy of Stress Corrosion Fracture* (T. N. Rhodin, ed.), Interscience, New York, 1959.
2. A. S. Tetelman, *Fundamental Aspects of Stress-Corrosion Cracking* (R. W. Staehle, A. J. Forty, and D. van Rooyen, eds.), NACE, Houston, TX, 1969.
3. E. N. Pugh, *Theory of Stress Corrosion Cracking* (J. C. Scully, ed.), NATO, Brussels, Belgium, 1971.
4. G. J. Theus and R. W. Staehle, *Stress-Corrosion Cracking and Hydrogen Embrittlement of Iron-Base Alloys* (R. W. Staehle, J. Hochmann, R. D. McCright, and J. E. Slater, eds.), NACE, Houston, TX, 1977.
5. A. J. Bursle and E. N. Pugh, *Mechanisms of Environment Sensitive Cracking of Materials* (P. R. Swann, F. P. Ford, and A. R. C. Westwood, eds.), The Metals Society, London, U.K., 1977.
6. D. E. Williams, C. Westcott and M. Fleischmann, *Corrosion Chemistry within Pits, Crevices and Cracks* (A. Turnbull, ed.), HMSO, London, U.K., 1987.
7. J. R. Rice, *Chemistry and Physics of Fracture* (R. M. Latanision and R. H. Jones, eds.), Martinus Nijhoff, Dordrecht, the Netherlands, 1987.
8. A. Turnbull, H. W. Pickering, and A. Valdes, *Embrittlement by the Localized Crack Environment* (R. P. Gangloff, ed.), TMS-AIME, Warrendale, PA, 1984.
9. *Environmental-Induced Cracking of Metals* (R. P. Gangloff and M. B. Ives, eds.), NACE, Houston, TX, 1990.
10. *Corrosion sous Contrainte* (D. Desjardins and R. Oltra, eds.), Les Editions de Physique, Les Ulis, France, 1992.
11. J. R. Galvele, *Parkins Symposium on Fundamental Aspects of Stress-Corrosion Cracking* (S. M. Bruemmer, E. I. Meletis, R. H. Jones, W. W. Gerberich, F. P. Ford, and R. W. Staehle, eds.), TMS, Warrendale, PA, 1992.
12. *Corrosion–Deformation Interactions* (T. Magnin and J.-M. Gras, eds.), Les Editions de Physique, Les Ulis, France, 1993.
13. R. C. Newman and R. P. M. Procter, *Br. Corros. J. 25*:259 (1990).
14. D. R. Clarke and K. T. Faber, *J. Phys. Chem. Solids. 48*:1115 (1987).
15. N. S. Stoloff, in *Environmental-Induced Cracking of Metals* (R. P. Gangloff and M. B. Ives, eds.), NACE, Houston, TX, 1990, p. 31.
16. T. Magnin and J. Lepinoux, Metallurgical aspects of the brittle SCC in austenitic stainless steels, in *Parkins Symposium on Fundamental Aspects of Stress-Corrosion Cracking* (S. M. Bruemmer, E. I. Meletis, R. H. Jones, W. W. Gerberich, F. P. Ford, and R. W. Staehle, eds.), TMS, Warrendale, PA, 1992, p. 323.
17. T. Magnin, in *Corrosion–Deformation Interactions* (T. Magnin and J.-M. Gras, eds.), Les Editions de Physique, Les Ulis, France, p. 27.

18. D. A. Jones, *Metall. Trans. A 16A*:1133 (1985).
19. S. P. Lynch, *Acta Metall. 36*:2639 (1988).
20. J. R. Galvele, *Corros. Sci. 27*:1 (1987).
21. R. W. Pasco and P. J. Ficalora, *Hydrogen Degradation of Ferrous Alloys* (R. A. Oriani, J. P. Hirth, and M. Smialowski, eds.), Noyes, Park Ridge, NJ, 1985.
22. H. K. Birnbaum, *Hydrogen Effects on Material Behavior* (N. R. Moody and A. W. Thompson, eds.), TMS, Warrendale, PA, 1990.
23. K. Sieradzki and F. J. Friedersdorf, *Corros. Sci. 36*:669 (1994).
24. D. A. Vermilyea, in *Stress-Corrosion Cracking and Hydrogen Embrittlement of Iron-Base Alloys* (R. W. Staehle, J. Hochmann, R. D. McCright, and J. E. Slater, eds.), NACE, Houston, TX, 1977, p. 208.
25. R. W. Staehle, in *Stress-Corrosion Cracking and Hydrogen Embrittlement of Iron-Base Alloys* (R. W. Staehle, J. Hochmann, R. D. McCright, and J. E. Slater, eds.), NACE, Houston, TX, 1977, p. 180.
26. R. N. Parkins, *Corros. Sci. 20*:147 (1980).
27. R. N. Parkins, in *Environmental-Induced Cracking of Metals* (R. P. Gangloff and M. B. Ives, eds.), NACE, Houston, TX, 1990, p. 1.
28. F. P. Ford, *Corrosion Processes* (R. N. Parkins, ed.), Applied Science Publishers, London, 1982.
29. F. P. Ford, in *Environmental-Induced Cracking of Metals* (R. P. Gangloff and M. B. Ives, eds.), NACE, Houston, TX, 1990, p. 139.
30. F. P. Ford, in *Corrosion sous Contrainte* (D. Desjardins and R. Oltra, eds.), Les Editions de Physique, Les Ulis, France, 1992.
31. P. L. Andresen, *Corrosion 47*:9–17 (1991).
32. R. C. Newman, K. Sieradzki, and H. S. Isaacs, *Metall. Trans. A 13A*:2015 (1982).
33. R. C. Newman and M. Saito, Anodic stress-corrosion cracking: Slip-dissolution and film-induced cleavage, in *Corrosion–Deformation Interactions* (T. Magnin and J.-M. Gras, eds.), Les Editions de Physique, Les Ulis, France, 1993, p. 3.
34. C. Edeleanu and A. J. Forty, *Philos. Mag. 46*:521 (1960).
35. J. A. Beavers and E. N. Pugh, *Metall. Trans. A 11A*:809 (1980).
36. E. N. Pugh, *Atomistic of Fracture* (R. M. Latanision and J. R. Pickens, eds.), Plenum, New York, 1983, p. 997.
37. M. T. Hahn and E. N. Pugh, *Corrosion 36*:380 (1980).
38. D. V. Beggs, M. T. Hahn, and E. N. Pugh, *Proceedings of the A. R. Troiano Honorary Symposium on Hydrogen Embrittlement and Stress-Corrosion Cracking* (R. Gibala and R. F. Hehemann, eds.), ASM, Metals Park, OH, 1984.
39. E. N. Pugh, *Corrosion 41*:517 (1985).
40. K. Sieradzki and R. C. Newman, *Philos. Mag. A 51*:95 (1985).
41. K. Sieradzki and R. C. Newman, *J. Phys. Chem. Solids 48*:1101 (1987).
42. R. C. Newman and K. Sieradzki, in *Chemistry and Physics of Fracture* (R. M. Latanision and R. H. Jones, eds.), Martinus Nijhoff, Dordrecht, the Netherlands, 1987, p. 597.
43. K. Sieradzki, in *Chemistry and Physics of Fracture* (R. M. Latanision and R. H. Jones, eds.), Martinus Nijhoff, Dordrecht, the Netherlands, 1987, p. 219.
44. K. Sieradzki, in *Environmental-Induced Cracking of Metals* (R. P. Gangloff and M. B. Ives, eds.), NACE, Houston, TX, 1990, p. 125.
45. T. B. Cassagne, in *Corrosion sous Contrainte* (D. Desjardins and R. Oltra, eds.), Les Editions de Physique, Les Ulis, France, 1992.
46. R. C. Newman, T. Shahrabi, and K. Sieradzki, *Scr. Metall. 23*:71 (1989).
47. R. G. Kelly, T. Shahrabi, A. J. Frost, and R. C. Newman, *Metall. Trans. A 22A*:191 (1991).
48. R. G. Kelly, A. J. Young, and R. C. Newman, *Electrochemical Impedance Spectroscopy*, ASTM STP 1188 (J. R. Scully, D. C. Silverman, and M. W. Kendig, eds.), ASTM, Philadelphia, PA, 1991, p. 94.
49. M. Saito, G. S. Smith, and R. C. Newman, *Corros. Sci. 35*:411 (1993).
50. J. S. Chen, M. Salmerón, and T. M. Devine, *Scr. Metall. 26*:739 (1992).
51. J. S. Chen, M. Salmerón, and T. M. Devine, *J. Electrochem. Soc. 139*:L55 (1992).

52. G. E. Kerns, M. T. Wang, and R. W. Staehle, in *Stress-Corrosion Cracking and Hydrogen Embrittlement of Iron-Base Alloys* (R. W. Staehle, J. Hochmann, R. D. McCright, and J. E. Slater, eds.), NACE, Houston, TX, 1977, p. 700.
53. C. S. Carter and M. V. Hyatt, in *Stress-Corrosion Cracking and Hydrogen Embrittlement of Iron-Base Alloys* (R. W. Staehle, J. Hochmann, R. D. McCright, and J. E. Slater, eds.), NACE, Houston, TX, 1977, p. 524.
54. N. J. H. Holroyd, in *Environmental-Induced Cracking of Metals* (R. P. Gangloff and M. B. Ives, eds.), NACE, Houston, TX, 1990, p. 311.
55. L. Ratke and W. Gruhl, *Werkst. Korros. 31*:768 (1980).
56. R. C. Newman and A. Mehta, in *Environmental-Induced Cracking of Metals* (R. P. Gangloff and M. B. Ives, eds.), NACE, Houston, TX, 1990, p. 489.
57. J. I. Dickson, A. J. Russell, and D. Tromans, *Can. Met. Q. 19*:161 (1980).
58. M. O. Speidel, *Proceedings of Stainless Steels'91*, ISIJ, Tokyo, Japan, 1991, Vol. 1, p. 25.
59. R. W. Staehle, in *Environmental-Induced Cracking of Metals* (R. P. Gangloff and M. B. Ives, eds.), NACE, Houston, TX, 1990, p. 561.
60. P. M. Scott and M. Le Calvar, Some possible mechanisms of intergranular stress corrosion cracking of alloy 600 in PWR primary water, *Proceedings of 6th International Symposium on Environmental Degradation of Materials in Nuclear Power Systems—Water Reactors*, San Diego, CA, 1993.
61. T. S. Gendron and R. C. Newman, Irreversible Embrittlement of Alloy 600 in Hydrogenated Steam at 400°C, *Corrosion/94*, paper 227, NACE, Houston, TX, 1994.
62. H. Nagano, T. Kudo, Y. Inaba, and M. Harada, *Met. Corros. Ind. 56*:81 (1981); also *Boshoku Gijutsu* (*Corrosion Engineering*) *30*:218 (1981).
63. M. O. Speidel, *Corrosion 33*:199 (1977).
64. H. Stenzel, H. Vehoff, and P. Neumann, in *Chemistry and Physics of Fracture* (R. M. Latanision and R. H. Jones, eds.), Martinus Nijhoff, Dordrecht, the Netherlands, 1987, p. 652.
65. K. Sieradzki, R. L. Sabatini, and R. C. Newman, *Metall. Trans. A 15A*:1941 (1984).
66. Li Shiqiong, J. I. Dickson, J.-P. Bailon, and D. Tromans, *Mater. Sci. Eng. A119*:59 (1989).
67. R. N. Parkins and R. R. Fessler, *Mater. Eng. Appl. 1*:80 (1978).
68. K. Tamaki, S. Tsujikawa, and Y. Hisamatsu, *Advances in Localized Corrosion* (H. S. Isaacs, U. Bertocci, J. Kruger, and S. Smialowska, eds.), NACE, Houston, TX, 1991, p. 207.
69. R. C. Newman, W. Zheng, and R. P. M. Procter, *Corros. Sci. 33*:1033 and 1009 (1992).
70. S. P. Pednekar, A. K. Agrawal, H. E. Chaung, and R. W. Staehle, *J. Electrochem. Soc. 126:*701 (1979).
71. E. N. Pugh, J. V. Craig, and A. J. Sedriks, in *Fundamental Aspects of Stress-Corrosion Cracking* (R. W. Staehle, A. J. Forty, and D. van Rooyen, eds.), NACE, Houston, TX, 1969, p. 118.
72. K. Sieradzki, J. S. Kim, A. T. Cole, and R. C. Newman, *J. Electrochem. Soc. 134*:1635 (1987).
73. H. R. Copson, in *Physical Metallurgy of Stress Corrosion Fracture* (T. N. Rhodin, ed.), Interscience, New York, 1959, p. 247.
74. M. O. Speidel, *Metall. Trans. A 12A*:779 (1981).
75. L. Graf, in *Fundamental Aspects of Stress-Corrosion Cracking* (R. W. Staehle, A. J. Forty, and D. van Rooyen, eds.), NACE, Houston, TX, 1969, p. 187.
76. K. Sieradzki, R. R. Corderman, K. Shukla, and R. C. Newman, *Philos. Mag. A 59*:713 (1989).
77. K. Sieradzki, R. C. Newman, and T. Shahrabi, in *Advances in Localized Corrosion* (H. S. Isaacs, U. Bertocci, J. Kruger, and S. Smialowska, eds.), NACE, Houston, TX, 1991, p. 161.
78. R. C. Newman, R. R. Corderman, and K. Sieradzki, *Br. Corros. J. 24*:143 (1989).
79. W. J. R. Nisbet, G. W. Lorimer, and R. C. Newman, *Corros. Sci. 35*:457 (1993).
80. R. M. Bhatkal, W. F. Flanagan, and B. D. Lichter, in *Corrosion–Deformation Interactions* (T. Magnin and J.-M. Gras, eds.), Les Editions de Physique, Les Ulis, France, 1993, p. 43.
81. L. Zhong, W. F. Flanagan, and B. D. Lichter, in *Corrosion–Deformation Interactions* (T. Magnin and J.-M. Gras, eds.), Les Editions de Physique, Les Ulis, France, 1993, p. 309.
82. M. Saito, M.Sc. thesis, UMIST, Manchester, U.K., 1993.
83. F. J. Friedersdorf, PhD thesis, Johns Hopkins University, Baltimore, MD, 1993.

84. R. P. Harrison, D. de G. Jones, and J. F. Newman, in *Stress-Corrosion Cracking and Hydrogen Embrittlement of Iron-Base Alloys* (R. W. Staehle, J. Hochmann, R. D. McCright, and J. E. Slater, eds.), NACE, Houston, TX, 1977, p. 659.
85. N. Bandyopadhyay, R. C. Newman, and K. Sieradzki, *Proceedings of the 9th International Congress on Metallic Corrosion*, NRC, Ottawa, Ontario, Canada, 1984, Vol. 2, p. 210.
86. V. Cihal, *Corros. Sci. 25*:815 (1985).
87. T. Shahrabi and R. C. Newman, *Mater. Sci. Forum, 44–45*:169, Trans. Tech., Zurich (1989) (*Proceedings of Electrochemical Methods in Corrosion Research III*, Zurich, 1988).
88. R. C. Newman, *Corros. Sci. 33*:1653 (1992).
89. J. R. Galvele and S. de Micheli, *Corros. Sci. 10*:795 (1970).
90. W. Hepples, M. R. Jarrett, J. S. Crompton, and N. J. H. Holroyd, in *Environmental-Induced Cracking of Metals* (R. P. Gangloff and M. B. Ives, eds.), NACE, Houston, TX, 1990, p. 383.
91. H. Spaehn, in *Environmental-Induced Cracking of Metals* (R. P. Gangloff and M. B. Ives, eds.), NACE, Houston, TX, 1990, p. 449.
92. D. B. Wells, J. Stewart, A. W. Herbert, P. M. Scott, and D. E. Williams, *Corrosion 45*:649 (1989).
93. B. F. Brown, in *Theory of Stress Corrosion Cracking* (J. C. Scully, ed.), NATO, Brussels, Belgium, 1971, p. 186.
94. R. N. Parkins, M. Elices, V. Sanchez-Galvez, and L. Caballero, *Corros. Sci. 22*:379 (1982).
95. J. Hochmann, A. Desestret, P. Jolly, and R. Mayoud, in *Stress-Corrosion Cracking and Hydrogen Embrittlement of Iron-Base Alloys* (R. W. Staehle, J. Hochmann, R. D. McCright, and J. E. Slater, eds.), NACE, Houston, TX, 1977, p. 956.
96. A. J. Russell and D. Tromans, *Metall. Trans. A 10A*:1229 (1979).
97. A. J. Russell and D. Tromans, *Metall. Trans. A 12A*:613 (1981).
98. M. O. Speidel and R. Magdowski, in *Corrosion–Deformation Interactions* (T. Magnin and J.-M. Gras, eds.), Les Editions de Physique, Les Ulis, France, 1993, p. 107.
99. T. B. Cassagne, W. F. Flanagan, and B. D. Lichter, *Metall. Trans. A 17A*:703 (1986).
100. M. Ahrens and K. E. Heusler, *Werkst. Korros. 32*:197 (1981).
101. D. Hardie, in *Environmental-Induced Cracking of Metals* (R. P. Gangloff and M. B. Ives, eds.), NACE, Houston, TX, 1990, p. 347.
102. G. E. Kerns and R. W. Staehle, *Scr. Metall. 6*:1189 (1972).
103. K. Sieradzki, *Acta Metall. 30*:973 (1982).
104. G. L. Bianchi and J. R. Galvele, *Corros. Sci. 34*:1411 (1993).
105. R. P. Gangloff, in *Environmental-Induced Cracking of Metals* (R. P. Gangloff and M. B. Ives, eds.), NACE, Houston, TX, 1990, p. 55.
106. R. P. Wei and M. Gao, in *Hydrogen Effects on Material Behavior* (N. R. Moody and A. W. Thompson, eds.), TMS, Warrendale, PA, 1990, p. 789.
107. M. Kimura, N. Totsuka, and T. Kurisu, *Corrosion 45*:340 (1989).
108. R. C. Newman, *Corrosion Chemistry within Pits, Crevices and Cracks* (A. Turnbull, ed.), HMSO, London, 1987, p. 317.
109. E. Mattsson, *Electrochim. Acta 3*:279 (1961).
110. T. P. Hoar and G. P. Rothwell, *Electrochim. Acta 15*:1037 (1970).
111. T. Adachi, in *Proceedings of Stainless Steels'91*, ISIJ, Tokyo, Japan, 1991, Vol. 1, p. 189.
112. S. Tsujikawa, T. Shinohara, and Chenghao Liang, in *Proceedings of Stainless Steels'91*, ISIJ, Tokyo, Japan, 1991, Vol. 1, p. 196.
113. R. N. Parkins, *Environment-Sensitive Fracture: Evaluation and Comparison of Test Methods*, ASTM STP 821 (S. W. Dean, E. N. Pugh, and G. M. Ugiansky, eds.), ASTM, Philadelphia, PA, 1984, p. 5.
114. J. Congleton, T. Shoji, and R. N. Parkins, *Corros. Sci. 25*:663 (1985).
115. R. C. Newman, *Corrosion 50*:682 (1994).
116. M. J. Danielson, C. A. Oster, and R. H. Jones, in *Corrosion Chemistry within Pits, Crevices and Cracks* (A. Turnbull, ed.), HMSO, London, 1987, p. 213.
117. J. R. Rice and R. Sorensen, *J. Mech. Phys. Solids 26*:163 (1978).
118. J. R. Galvele, *J. Electrochem. Soc. 133*:953 (1986).

119. D. D. Macdonald and M. Urquidi-Macdonald, *Corros. Sci. 32*:51 (1991).
120. N. J. H. Holroyd, A. Gray, G. M. Scamans, and R. Herrmann, *Proceedings of Aluminium–Lithium III*, Institute of Metals, London, U.K., 1986, p. 310.
121. J. G. Craig, R. C. Newman, M. R. Jarrett, and N. J. H. Holroyd, *J. Phys. (Orsay) 48*:825 (colloq. C3) (1987) [*Proceedings* of *Aluminium–Lithium IV*, Paris, 1987].
122. A. Rota and H. Boehni, in *Mater. Sci. Forum*, *44–45*:169, Trans. Tech., Zurich, Switzerland, (1989) (*Proceedings of Electrochemical Methods in Corrosion Research III*, Zurich, Switzerland, 1988), p. 177.
123. Y. S. Kim, N. J. H. Holroyd, and J. J. Lewandowski, in *Environmental-Induced Cracking of Metals* (R. P. Gangloff and M. B. Ives, eds.), NACE, Houston, TX, 1990, p. 371.
124. C. Edeleanu, *J. Iron Steel Inst. 173*:140 (1953).
125. T. Kudo, H. Tsuge, and T. Moroishi, *Corrosion 45*:831 (1989).
126. M. Barteri, F. Mancia, A. Tamba, and G. Montagna, *Corros. Sci. 27*:1239 (1987).
127. S. B. Mat and R. C. Newman, Local chemistry aspects of hydrogen sulfide–assisted SCC of stainless steels, *Corrosion/94*, paper 228, NACE, Houston, TX, 1994.
128. J. Oudar and P. Marcus, *Appl. Surf. Sci. 3*:48 (1979).
129. P. Marcus and J. Oudar, in *Chemistry and Physics of Fracture* (R. M. Latanision and R. H. Jones, eds.), Martinus Nijhoff, Dordrecht, the Netherlands, 1987, p. 670.
130. R. C. Newman, *Corros. Sci. 25*:341 (1985).
131. P. Marcus, *C. R. Acad. Sci. Paris 305*:675 (1987).
132. R. H. Jones and S. M. Bruemmer, in *Environmental-Induced Cracking of Metals* (R. P. Gangloff and M. B. Ives, eds.), NACE, Houston, TX, 1990, p. 287.
133. R. P. M. Procter and G. N. Stevens, *Corros. Sci. 15*:349 (1975).
134. A. Kawashima, A. K. Agrawal, and R. W. Staehle, *Stress-Corrosion Cracking—The Slow Strain Rate Technique*, ASTM STP 665 (G. M. Ugiansky and J. H. Payer, eds.), ASTM, Philadelphia, PA, 1977, p. 266.
135. R. N. Parkins, C. M. Rangel, and J. Yu, *Metall. Trans. A 16A*:1671 (1985).
136. T. Shahrabi, PhD thesis, University of Manchester, Manchester, U.K., 1989.
137. T. Shahrabi, R. C. Newman, and K. Sieradzki, *J. Electrochem. Soc. 140*:348 (1993).
138. R. C. Newman, T. Shahrabi, and K. Sieradzki, *Corros. Sci. 28*:873 (1988).
139. G. S. Duffo and J. R. Galvele, *Corros. Sci.34*:79 (1993).
140. G. S. Duffo and J. R. Galvele, *Metall. Trans. A 24A*:425 (1993).
141. G. L. Bianchi and J. R. Galvele, *Corros. Sci. 36*:611 (1994).
142. K. Sieradzki and P. Ficalora, *Scr. Metall. 13*:535 (1979).
143. A. Paskin, K. Sieradzki, D. K. Som, and G. J. Dienes, *Acta Metall. 31*:1258 (1983).
144. K. Sieradzki and J. S. Kim, *Acta Metal. Mater. 40*:625 (1992).
145. S. G. Corcoran and K. Sieradzki, *Scr. Metall. 26*:633 (1992).
146. J. Kameda and C. J. McMahon, *Metall. Trans. A 14A*:903 (1983).
147. W. W. Gerberich and S. Chen, in *Environmental-Induced Cracking of Metals* (R. P. Gangloff and M. B. Ives, eds.), NACE, Houston, TX, 1990, p. 167.
148. H. K. Birnbaum, in *Environmental-Induced Cracking of Metals* (R. P. Gangloff and M. B. Ives, eds.), NACE, Houston, TX, 1990, p. 21.
149. A. Turnbull, *Corros. Sci. 23*:833 (1983).
150. A. Turnbull, *Corros. Sci. 27*:1323 (1987).
151. A. Turnbull, in *Advances in Localized Corrosion* (H. S. Isaacs, U. Bertocci, J. Kruger, and S. Smialowska, eds.), NACE, Houston, TX, 1991, p. 359.
152. B. F. Brown, in *Stress-Corrosion Cracking and Hydrogen Embrittlement of Iron-Base Alloys* (R. W. Staehle, J. Hochmann, R. D. McCright, and J. E. Slater, eds.), NACE, Houston, TX, 1977, p. 747.
153. A. Turnbull and R. C. Newman, *Small Fatigue Cracks* (R. O. Ritchie and J. Lankford, eds.), TMS-AIME, Warrendale, PA, 1986, p. 269.
154. E. A. Taqi and R. A. Cottis, in Corrosion chemistry within pits, crevices and cracks, A. Turnbull, ed., N.P.L, London, 1984, p. 483.

155. S. Suresh, G. F. Zaminski, and R. O. Ritchie, *Metall. Trans. A 12A*:1435 (1981).
156. A. J. Sedriks, *Stress Corrosion Testing Made Easy*, NACE, Houston, TX, 1991.
157. See various references in *Stress-Corrosion Cracking—The Slow Strain Rate Technique*, ASTM STP 665 (G. M. Ugiansky and J. H. Payer, eds.), ASTM, Philadelphia, PA, 1977.
158. N. J. H. Holroyd and D. Hardie, *Corros. Sci. 23*:527 (1983).
159. B. R. W. Hinton and R. P. M. Procter, *Corros. Sci. 23*:101 (1983).
160. J. C. Scully and T. A. Adepoju, *Corros. Sci. 17*:789 (1977).
161. J. R. Valdez-Vallejo, R. C. Newman, and R. P. M. Procter, in *Hydrogen Effects on Material Behavior* (N. R. Moody and A. W. Thompson, eds.), TMS, Warrendale, PA, 1990, p. 1003.
162. I. M. Hannah, R. C. Newman, and R. P. M. Procter, in *Hydrogen Effects on Material Behavior* (N. R. Moody and A. W. Thompson, eds.), TMS, Warrendale, PA, 1990, p. 965.
163. A. Brown, J. T. Harrison, and R. Wilkins, in *Stress-Corrosion Cracking and Hydrogen Embrittlement of Iron-Base Alloys* (R. W. Staehle, J. Hochmann, R. D. McCright, and J. E. Slater, eds.), NACE, Houston, TX, 1977, p. 686.
164. R. P. M. Procter, *Cathodic Protection* (V. Ashworth and C. J. L. Booker, eds.), Ellis Horwood, Chichester, U.K., 1986.
165. H. Hanninen, H. Illi and M. Kemppainen, in *Corrosion Chemistry within Pits, Crevices and Cracks* (A. Turnbull, ed.), HMSO, London, 1987, p. 646.
166. J. R. Galvele, in *Parkins Symposium on Fundamental Aspects of Stress-Corrosion Cracking* (S. M. Bruemmer, E. I. Meletis, R. H. Jones, W. W. Gerberich, F. P. Ford, and R. W. Staehle, eds.), TMS, Warrendale, PA, 1992, p. 85.
167. R. N. Parkins, *Corrosion 46*:178 (1990).
168. P. Meakin, *Statistical Models for the Fracture of Disordered Media* (H. J. Herrmann and S. Roux, eds.), Elsevier, Amsterdam, the Netherlands, 1990, p. 291.
169. M. Suleiman, PhD thesis, University of Manchester, Manchester, U.K., 1993.
170. J. I. Dickson, D. Tromans, S. Li, S. El Omari, and X. Z. Wu, in *Corrosion–Deformation Interactions* (T. Magnin and J.-M. Gras, eds.), Les Editions de Physique, Les Ulis, France, 1993, p. 643.
171. R. N. Parkins, *Corrosion* (L. L. Shreir, ed.), 2nd edn., Newnes-Butterworths, London, 1976, Chapter 8.1.
172. F. C. Larche and J. W. Cahn, *Acta Metall. 33*:331 (1985).
173. R. J. Asaro and W. A. Tiller, *Metall. Trans. 3*:1789 (1972).
174. A. Pineau, in *Environmental-Induced Cracking of Metals* (R. P. Gangloff and M. B. Ives, eds.), NACE, Houston, TX, 1990, p. 111.
175. R. Oltra and M. Keddam, in *Advances in Localized Corrosion* (H. S. Isaacs, U. Bertocci, J. Kruger, and S. Smialowska, eds.), NACE, Houston, TX, 1991, p. 17.
176. J. Lankford and M. Longmire, *J. Mater. Sci. 26*:1131 (1991).
177. S. Das and C. H. Scholz, *J. Geophys. Res. 86*:6039 (1981).
178. P. G. Meredith (University College, London, U.K.), private communication.

12

Corrosion Fatigue Mechanisms in Metallic Materials

T. Magnin
Ecole Nationale Supérieure des Mines de Saint-Etienne

CONTENTS

12.1 Introduction 545
12.2 Corrosion Fatigue Crack Initiation 546
12.2.1 Classical Approaches to Corrosion Fatigue Damage 546
12.2.2 Influence of Cyclic Plasticity on Electrochemical Reactions 551
12.2.3 Softening Effect due to Anodic Dissolution 553
12.2.4 Influence of Corrosion on PSB Configurations 554
12.2.5 An Example of Mechanical and Electrochemical Coupling Effects: The CF Crack Initiation Mechanisms of a Two-Phase Stainless Steel in NaCl Solutions 555
12.3 Corrosion Fatigue Propagation Mechanisms 555
12.3.1 Phenomenology 555
12.3.2 Anodic Dissolution Effects 559
12.3.3 Hydrogen Effects 560
12.4 Monte Carlo-Type Simulations of Corrosion Fatigue Lifetimes from the Evolution of Short Surface Cracks 564
12.4.1 Introduction 564
12.4.2 Physical Description of the Fatigue Damage 565
12.4.3 Numerical Modeling 566
12.5 Interactions with Stress Corrosion Cracking (SCC) 569
12.6 Concluding Remarks 570
References 571

12.1 Introduction

The deleterious effect of an aqueous environment on fatigue crack initiation and propagation in metals and alloys has been observed for a long time [1–5]. The applied electrochemical potential has a large influence on localized corrosion reactions and, consequently, on corrosion fatigue (CF) damage. These complex effects are used for cathodic protection in offshore structures. Nevertheless, CF is influenced by various mechanical, chemical, and microstructural parameters that interact locally. Even if anodic dissolution versus hydrogen effects are known to occur during CF, the damage mechanisms are still controversial and fatigue lifetime predictions are still empirical.

The application of linear elastic fracture mechanics (LEFM) led to the determination of threshold stress intensity factors for long crack growth, which are quite useful for engineering material–environment systems. Nevertheless, it is now well established [6,7] that short cracks can develop and grow from smooth surfaces, due to localized corrosion–deformation interactions, at crack tip stress intensities below the previous threshold. CF damage models of short crack growth are then necessary to improve the fatigue damage predictions in aqueous solutions.

In the first part of this chapter, crack initiation mechanisms are reviewed and quantitative approaches are discussed, bearing in mind the limitations related to the evaluation of localized corrosion–deformation interactions. A particularly interesting example of electrochemical and mechanical coupling effects during CF in duplex stainless steels will illustrate the complexity of the interactions that lead to crack initiation.

Crack propagation models are then presented for both short and long cracks. Anodic dissolution and hydrogen effects at the crack tip are analyzed. Finally, the possible relation between stress corrosion cracking and CF is shown for crack initiation and crack propagation processes.

12.2 Corrosion Fatigue Crack Initiation

It is well known that slip bands, twins, interphases, grain boundaries, and constituent particles are classical sites for crack initiation. Moreover, persistent slip band (PSB)-grain boundary interactions are often observed to be preferential crack initiation sites during CF, as well as localized pits around metallurgical heterogeneities.

The main need in fatigue crack initiation modeling is related to the quantitative approach to local synergistic effects between environment and cyclic plasticity. In this section, quantitative approaches to corrosion fatigue crack initiation from different electrochemical conditions are presented. Then improvement of such models is given through corrosion–deformation interaction effects. Finally, an interesting example is given of the coupling effects between cyclic plasticity and corrosion that must be taken into account to improve the crack initiation resistance of duplex stainless steels in chloride solutions.

12.2.1 Classical Approaches to Corrosion Fatigue Damage

Electrochemical corrosion can be schematized as an "electronic pump or an electronic circuit" related to oxidation and reduction reaction:

$$M \rightarrow M^{n+} + ne^- \text{ anodic dissolution}$$

$$\left.\begin{array}{l} 2H_2O + O_2 + 4e^- \rightarrow 4OH^- \\ 2H^+ + 2e^- \rightarrow H_2 \end{array}\right\} \text{cathodic reactions}$$

together with a cation hydrolysis reaction: $M^{n+} + nH_2O \rightarrow M(OH)_n + nH^+$.

Here M^{n+} is a solvated ion, e^- is an electron, and η represents the ion state of charge. The electrons, liberated by the oxidation, must flow through the material M to be consumed in an appropriate cathodic reaction. Beyond a solubility limit, precipitates of hydroxide or hydra ted oxide are formed, and this surface film can provide a barrier to further dissolution. In fact, there are two film formation mechanisms: the dissolution-precipitation mechanism addressed before and also the solid-state oxidation process $M + H_2O \rightarrow MO + 2H^+ + 2e^-$.

Some films are termed "passive," for stainless steels or aluminum alloys, for instance. These films can play an important role in environment-sensitive crack initiation and fracture. Under thermodynamic equilibrium conditions, the film stability may be inferred from $E = f(\text{pH})$ diagrams, where E is the electrical potential related to the chemical free energy G by $G = -nEF$, and F is Faraday's number. At equilibrium, one can define the electrode potential (related to ΔG) and the current density I ($I \sim e^{-\Delta G/RT}$ where AG^* is the activation energy of dissolution).

Thus, the relation $E = f(I)$ gives different corrosion rates for a given metal in a given solution. Figure 12.1 shows such a relation (polarization curve) in the case of an austenitic stainless steel in an acidic Cl^- solution. Five domains can be considered for corrosion and corrosion fatigue damage:

1. In zone 1, $E > E_r$, pitting occurs by destabilization of the passive film. Pits act as stress concentrators during fatigue. During CF under pitting conditions, pits grow into the material. If such a pit reaches a critical depth d_{CL}, a fatigue crack can develop. The critical depth is then a function of the applied stress range [8].

   Let us suppose the following conditions:

   Constant corrosion conditions (pH, concentration of bulk solution)

   Constant alternating load, $d\Delta P/dt = 0$

   Constant loading frequency, $dv/dt = 0$

   It is well established that growth kinetics of corrosion pits are determined by a simple power law such as

$$d_L(t) = C(t - t_0)^{\beta}, \quad t > t_0 \tag{12.1}$$

   where t_0 is the incubation time for pit nucleation. If the pit depth reaches the critical value:

$$d_L(t) = d_{CL} \tag{12.2}$$

   corrosion fatigue crack initiation occurs. The critical pit depth d_{CL} depends on the applied stress range $\Delta\sigma_0$, the cyclic yield strength σ_{FC} (which can be different from the tensile yield strength), the fatigue crack growth threshold ΔK_0, and the geometry of the specimen, expressed in terms of a geometric factor G. It can be calculated

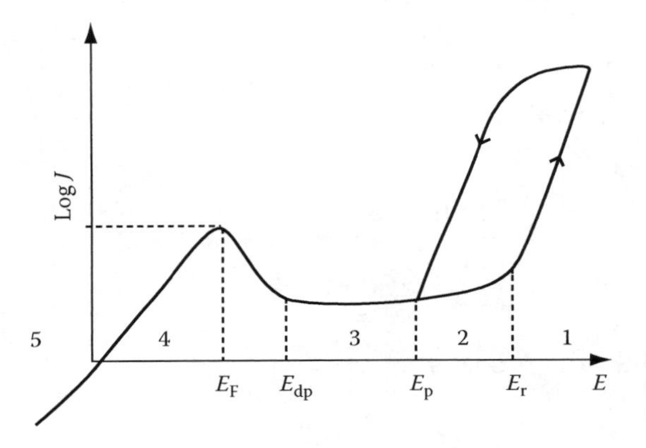

FIGURE 12.1
General polarization curve for an austenitic stainless steel in acidic Cl^- solutions.

by elastic–plastic fracture mechanisms based on the Dugdale model [7]. Then d_{CL} is, given by the following equation:

$$d_{CL} = \frac{\cos(\pi\Delta\sigma_0/4\sigma_{FC})\pi\Delta K_0^2}{32G^2\sigma_{FC}^2[1-\cos(\pi\Delta\sigma_0/4\sigma_{FC})]} \tag{12.3}$$

The number of cycles to initiate a corrosion fatigue crack under pitting conditions is, by combining the previous equations with $N = tv$,

$$N_i = v\left[t_0 + \left(\frac{d_{CL}}{C_2}\right)^{1/\beta}\right] \tag{12.4}$$

The dependence of N_i on the applied stress range $\Delta\sigma_0$, calculated according to the previous equation, is schematically represented in Figure 12.2a. Also, under pitting conditions no corrosion fatigue limit exists. For $N_i \leq vt_0$, the influence of corrosion on the fatigue crack initiation life disappears (only if pitting is the necessary effect). Then, the lifetime is determined by the air fatigue behavior. Figure 12.2b shows an example for which the proposed calculation of N_i seems quite appropriate. Nevertheless, the main problem is related to the fact that the coefficients C and β of the pit kinetics are often not constant during cycling; it is a clear example of a cooperative effect between plasticity and electrochemistry that needs finer analyses.

Figure 12.3 illustrates the fatigue crack initiation from a pit in a face-centered cubic (fcc) Fe-Mn-Cr alloy cyclically deformed at low strain rate in a Cl^- solution. In many multiphase engineering materials, the presence of constituent particles favors pitting and crack initiation.

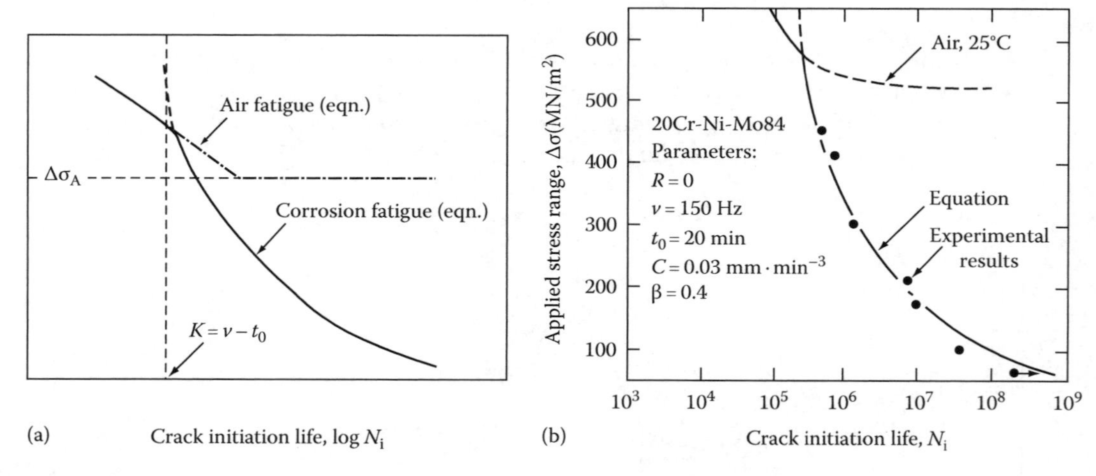

FIGURE 12.2
Corrosion fatigue crack initiation by pitting corrosion: (a) Schematic representation of a σ-N curve. (b) Comparison of experimentally and theoretically derived fatigue lives for the 20Cr-Ni-Mo alloy in 30 g/L NaCl solutions. (From Mueller, M., *Metall. Trans.*, 13A, 649, 1982.)

FIGURE 12.3
Crack initiation from a pit for a Fe-17Mn-13Cr alloy during CF in a 110°C Cl^- solution at a plastic strain amplitude $\Delta\varepsilon_p/2 = 4 \times 10^{-3}$ and a strain rate $\dot{\varepsilon} = 10^{-5}\ s^{-1}$.

2. In zone 2, $E_p < E < E_r$, pits are repassivated. If the plastic strain amplitude is too small for localized depassivation (by slip band emergence), pits will not grow and the CF behavior is then close to that in air. On the other hand, pits can grow if mechanical depassivation occurs and the CF behavior is then close to that of zone 1.
3. In the passive region 3, a competition between the kinetics of depassivation by slip and that of repassivation takes place. Thus the influence of the plastic strain amplitude and strain rate is quite obvious. In the same mechanical conditions as in (1), the quantity of matter dissolved per cycle (related to a distance) in the de-passivated slip bands can be expressed, using Faraday's law:

$$dN = \frac{M}{nF\rho} 2 \int_0^{1/V} i(\text{repassivation})dt \tag{12.5}$$

where i (repassivation) is generally of the form $i_{max}\exp(-\gamma t)$ with γ taken as a constant. Then $N_i = N$ (for $dN = d_c$). If d_c is taken as the grain size φ, for instance, then

$$N_i = \frac{\varphi}{dN} = \varphi \frac{nF\rho}{M} \frac{\gamma}{2i_{max}} \left[1 - \exp\left(\frac{\gamma}{\nu}\right)\right]^{-1} \tag{12.6}$$

A schematic representation of the equation is given in Figure 12.4a and corresponding experimental results are shown in Figure 12.4b.

Nevertheless, one of the main problems is that the repassivation law evolves during cycling as shown in Figure 12.5. This result also emphasizes the synergy in CF that leads to a complex predictive approach.

4. In zone 4, generalized dissolution occurs, which is generally quite dangerous for materials even without stress! In some cases, general corrosion can, however, blunt the cracks, which improves the fatigue life.

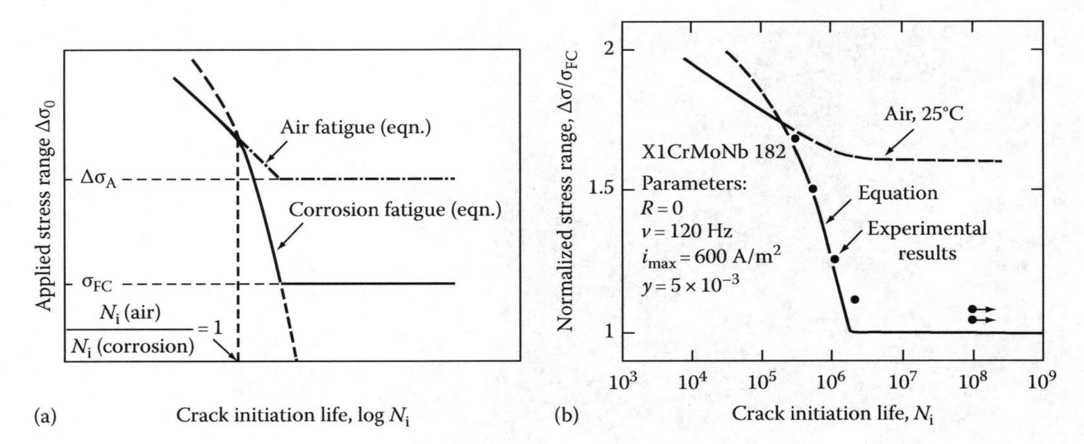

FIGURE 12.4
(a) Schematic representation of a σ-*N* curve for CF crack initiation under passive conditions, (b) Comparison of experimentally and theoretically derived fatigue lives for the X1CrMoNb 182 alloy in 30 g/L NaCl at 80°C. (From Mueller, M., *Metall. Trans.*, 13A, 649, 1982.)

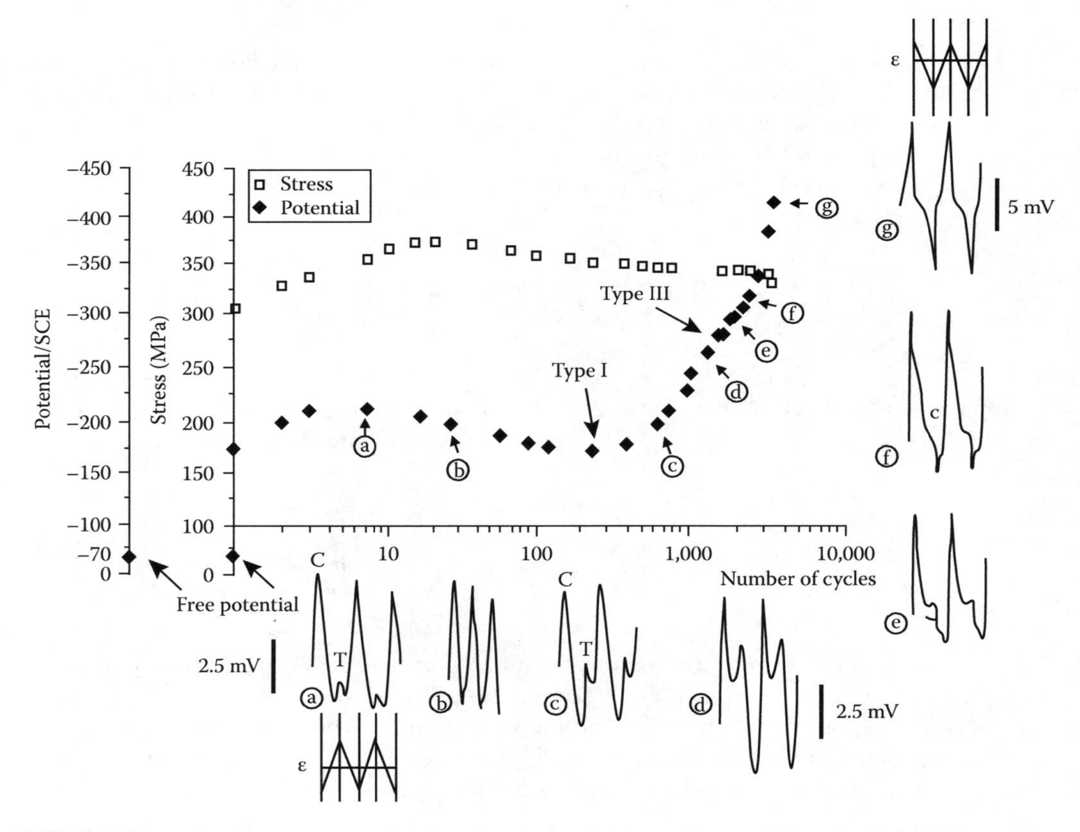

FIGURE 12.5
Simultaneous evolution of the cyclic stress σ, the average potential, and the shape of the cyclic potential transients for a 316 L alloy in 30 g/L NaCl ($\Delta\varepsilon_p/2 = 10^{-3}$, $\dot{\varepsilon} = 10^{-2}$ s^{-1}).

5. In zone 5, cathodic reactions are favored. If reduction of hydrogen occurs, we can have

$$2H^+ + 2e^- \rightarrow H_2$$

$$\rightarrow 2H_{adsorbed} \rightarrow H_2$$

$$H_{ads} \rightarrow H_{absored}$$

which can induce hydrogen diffusion and/or transport by dislocations, often leading to macroscopic brittle fracture under stress.

The electrochemical approach presented here has many limitations. First of all, the kinetics of the electrochemical reactions are closely dependent on the cyclic plasticity and the number of cycles [9,10]. Thus, predictive laws are very complex. Moreover, these laws use the local current densities, which are very difficult to model. Finally, this approach does not really take into account the local corrosion–deformation interactions (CDIs) and the effects of the corrosive solution on the deformation mode. Indeed, such synergetic effects between corrosion and deformation can be of prime importance; the following examples emphasize the role of CDI in crack initiation processes.

12.2.2 Influence of Cyclic Plasticity on Electrochemical Reactions

The evolution of dissolution current density transients during cycling of ferritic and austenitic stainless steels in NaCl solutions is shown in Figure 12.6, where curves $J_T - f(N)$ are plotted. Here J_T is the peak current density related to the depassivation process due to cyclic plasticity, and N is the number of cycles. During the first cycles the amount of dissolution increases, particularly at a high strain rate. One of the ferritic steels exhibits twinning, which induces a more marked depassivation during cycling, compared with the behavior of the second ferritic steel, which deforms by pencil glide [11].

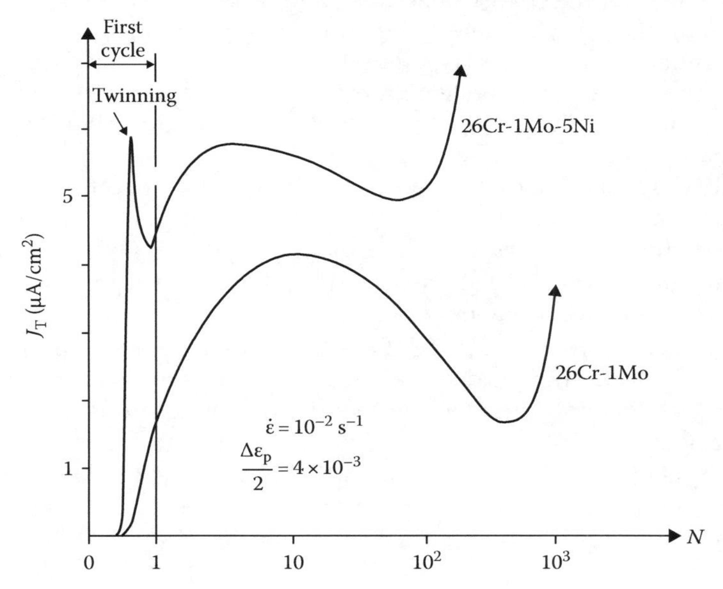

FIGURE 12.6
$J_T = f(N)$ curves for stainless steels during CF in a 3.5% NaCl solution.

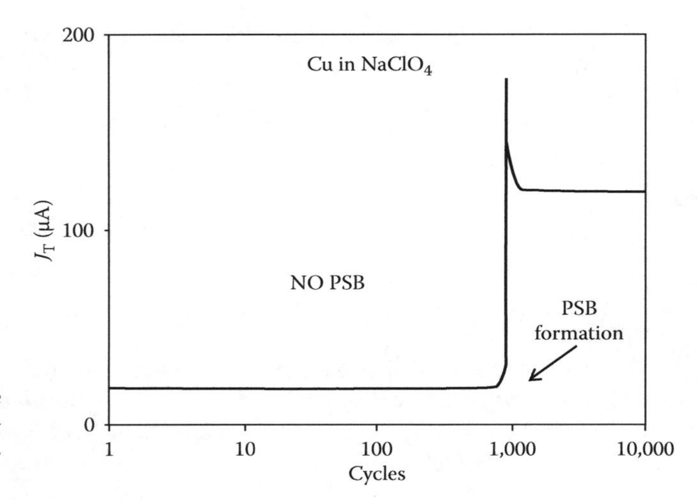

FIGURE 12.7
Influence of the PSB formation on the dissolution current for Cu single crystals in $NaClO_4$. (From Yan, B.D. et al., *Acta Metall.*, 33, 1593, 1985.)

Figure 12.6 clearly illustrates the influence of the deformation mode on the electrochemical reactions and the evolution of such electrochemical transients as a function of N (i.e., the localization of the plastic deformation with a decrease of J_T after the first cycles and then the formation of microcracks with a new increase of J_T until fracture due to a more difficult repassivation process).

PSBs and intense slip bands are very prone to specific dissolution, not only for passivated alloys but also in conditions of generalized dissolution as shown in Figure 12.7 for copper single crystals in $NaClO_4$ solution [9].

As soon as the PSBs form, the anodic current increases even though the applied plastic strain remains constant. This effect is related not only to the localization of the cyclic plasticity but also to the influence of the dislocation microstructure of PSBs on the free energy of dissolution ($-\Delta G$) and the energy of activation (ΔG^*) [10].

Moreover, cyclic plasticity has also been shown to promote localized pitting well below the pitting potential without stress [10]. Thus, for the ferritic Fe-26Cr-1 Mo stainless steel in 3.5% NaCl solution, a high strain rate $\dot{\varepsilon}$ promotes strain localization at grain boundaries, which induces intergranular pitting for an applied potential of about 400 mV below the pitting potential without stress effect.

The applied strain rate (or frequency) is a very sensitive parameter for CF damage. Figure 12.8 gives an interesting example for an Al-Li 8090 alloy in NaCl solutions. N_i is defined as the number of cycles to obtain a rapid 3% decrease of the saturation stress [10]. At high strain rate ($\dot{\varepsilon} > 5\times10^{-3}\ s^{-1}$), the anodic dissolution occurs at slip band emergence and induces an enhancement of the transgranular mechanical microcracking. At medium strain rate ($5\times10^{-5}\ s^{-1} < \dot{\varepsilon} < 5\times10^{-3}\ s^{-1}$), pitting is favored and responsible for crack initiation. So when the plastic strain decreases, pitting is more profuse (because of time) and the reduction in the fatigue life to crack initiation is more pronounced in comparison with air.

At low strain rate ($5\times10^{-6}\ s^{-1} < \dot{\varepsilon} < 5\times10^{-5}\ s^{-1}$), the fatigue time to initiation increased by blunting of the mechanically formed microcracks because of generalized pitting that acts as general corrosion. At very low strain rate ($\dot{\varepsilon} < 5\times10^{-6}\ s^{-1}$), CF crack initiation occurs by intergranular stress corrosion due to localized dissolution at grain boundaries. The rapid occurrence of stress corrosion cracking (SCC) induces a marked decrease of N_i.

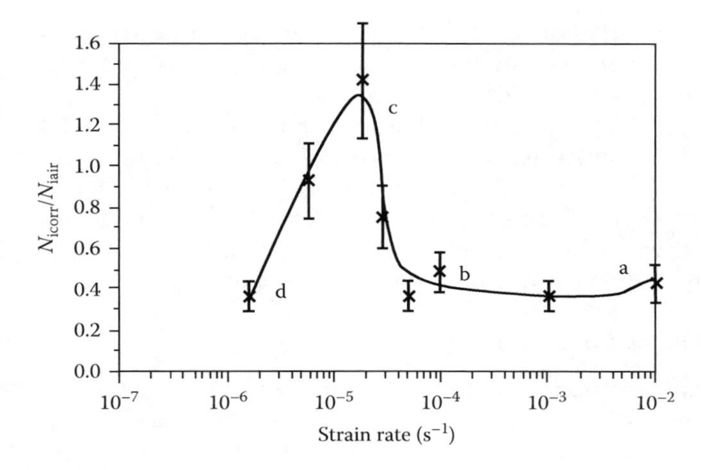

FIGURE 12.8
Influence of strain rate on 8090 Al-Li alloy fatigue life to crack initiation in a 3.5% NaCl solution for $\Delta\varepsilon_p/2 = 4 \times 10^{-3}$ at free potential.

12.2.3 Softening Effect due to Anodic Dissolution

CF tests on smooth specimens were performed at room temperature on a 316L austenitic stainless steel in a 0.5N H_2SO_4 solution at different electrochemical potentials and for a prescribed plastic strain amplitude of 4×10^{-3} ($\dot{\varepsilon} = 10^{-2}$ s^{-1}). The depassivation–repassivation process occurs in a very regular way, well before any microcracks can form [11]. It is of particular interest to follow the evolution of the maximum flow stress in the corrosive solution at free potential and at imposed cathodic potential and to compare this evolution with that observed in air (Figure 12.9). It clearly appears that (a) a cyclic softening effect occurs at the free potential in comparison with the behavior in air; (b) this softening effect disappears when the cathodic potential is applied (and the anodic dissolution is markedly reduced), after about 150 cycles; (c) the softening effect then occurs in the same way when

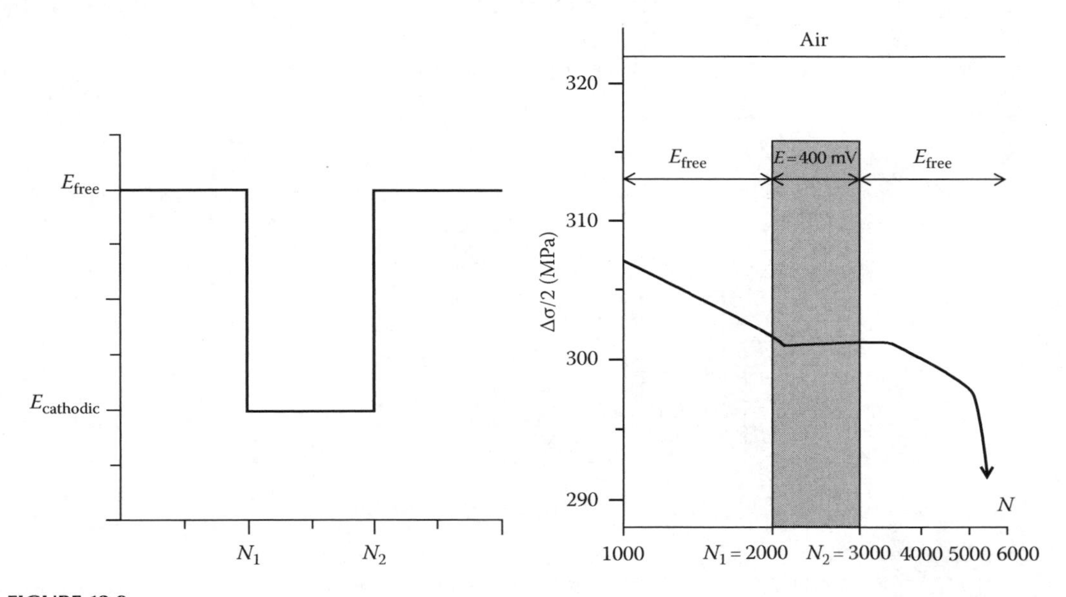

FIGURE 12.9
Evolution of the peak stress $\Delta\sigma/2$ during cycling in a 0.5N H_2SO_4 solution at free potential for $\Delta\varepsilon_p/2 = 4 \times 10^{-3}$ and ($\dot{\varepsilon} = 10^{-2}$ s^{-1}) compared with the air behavior.

the free potential is reestablished; and (d) a delay in the evolution of the flow stress with regard to the number of cycles for which a potential change is imposed can be observed for the free potential to the cathodic potential change (and vice versa).

This effect has also been observed during creep in corrosive solutions for copper [1]. It corresponds to the time during which vacancies due to dissolution are still acting on the dislocation mobility.

The macroscopic cycling softening effect observed in H_2SO_4 solution at room temperature (which is not due to microcracking) is very relevant to take into account quantitatively the local dissolution-deformation interactions that will lead to the fatigue crack initiation process.

12.2.4 Influence of Corrosion on PSB Configurations

Electrochemical control of corrosion has been shown to affect significantly the morphology of surface deformation. A modification of the number of persistent slip bands (PSBs) and of the slip offset height in PSB has been observed for Ni single crystals [4] in 0.5N H_2SO_4 and in copper single crystals [4,9] according to the applied potential, in comparison with air. It is easy to understand that such influences on PSB distribution will affect the crack initiation conditions. Figure 12.10 shows histograms of the PSB distribution produced on monocrystalline nickel in 0.5N H_2SO_4 at a constant strain amplitude.

Experiments conducted at the corrosion potential and at +160mV/SCE result in a reduction of the inter-PSB distance and a reduction of slip offset height.

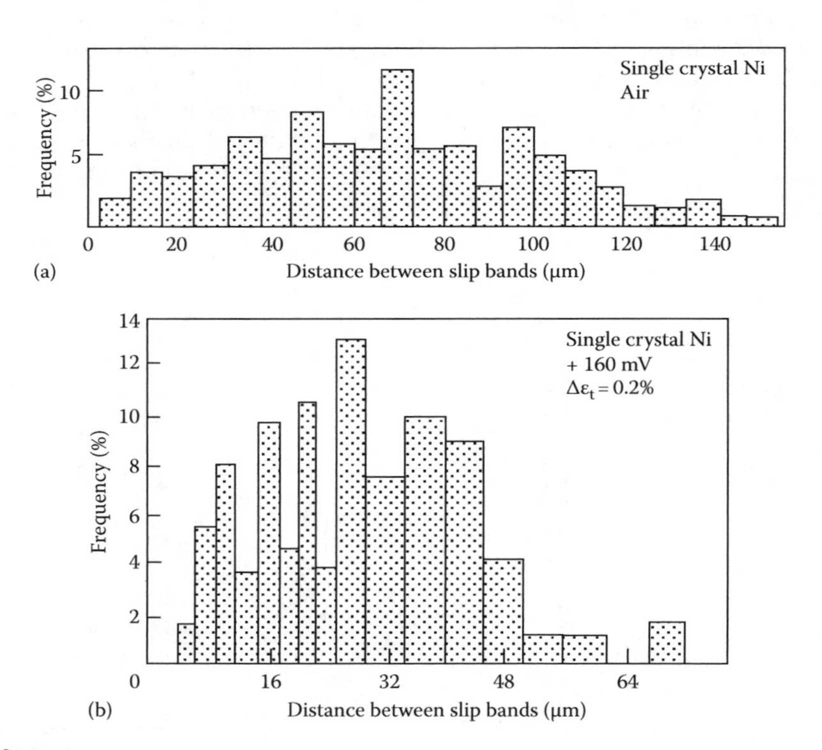

FIGURE 12.10
Histograms of distance between PSB clusters, from the replicate Ni single-crystal specimens fatigued for 1000 cycles under total strain control $\Delta\varepsilon_t$. Shear strain = 0.12%, (a) air and (b) +160mV/SCE in 0.5N H_2SO_4. (From Gangloff, R.P. and Duquette, D.J., *Chemistry and Physics of Fracture*, R. Latanision, ed., Nijhoff, Dordrecht, the Netherlands, p. 612, 1987.)

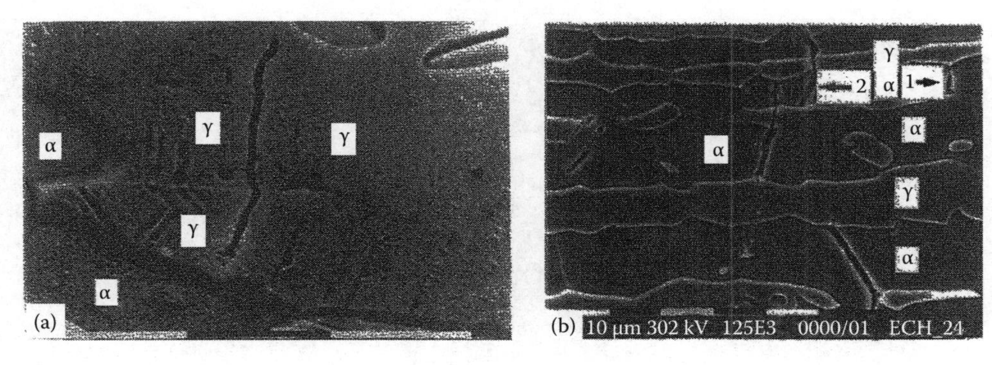

FIGURE 12.11
Crack initiation in a duplex stainless steel at $\dot{\varepsilon} = 2 \times 10^{-3}\ s^{-1}$. (a) In the γ phase at $\Delta\varepsilon_p/2 = 3 \times 10^{-4}$. (b) In the α phase at $\Delta\varepsilon_p/2 = 4 \times 10^{-3}$.

12.2.5 An Example of Mechanical and Electrochemical Coupling Effects: The CF Crack Initiation Mechanisms of a Two-Phase Stainless Steel in NaCl Solutions

Mechanical and electrochemical coupling effects are generally the key for understanding the crack initiation mechanisms in multiphase alloys. This is clearly illustrated for a duplex α/γ stainless steel (without nitrogen) in a 3.5% NaCl solution at pH 2 and free potential [11]. At low plastic strain amplitude, the softer γ phase is depassivated but this phase is cathodically protected by the non-plastically deformed α phase [11]. This coupling effect reduces the dissolution of the γ phase and delays CF damage, which is not the case at higher strain amplitude when the α phase is also depassivated by slip band emergence.

Observations of the crack initiation sites by scanning electron microscopy showed that at low plastic strain amplitudes ($\Delta\varepsilon_p/2 < 10^{-3}$) for which the fatigue resistance of the α-γ alloy is close to that of the γ alloy, cracks nucleate only in the austenitic phase (Figure 12.11a), but at higher strain amplitudes ($\Delta\varepsilon_p/2 < 10^{-3}$), the first cracks nucleate principally in the ferritic phase (Figure 12.11b). The excellent CF resistance of duplex stainless steels (for $\Delta\varepsilon_p/2 < 10^{-3}$) can then be understood through the electrochemical and mechanical coupling effects on crack initiation processes.

12.3 Corrosion Fatigue Propagation Mechanisms

12.3.1 Phenomenology

One can find an enormous amount of mechanistic work about corrosion fatigue crack growth in the literature (see, for instance, Refs. [4,6,7,12]). The aim of this section is not to review such studies but to analyze the possible mechanisms at the crack tip leading to crack advance. The question is, "What is the crack tip driving force that controls the corrosion fatigue crack growth?" among the different mechanical (stress, strain, strain rate) and chemical (dissolution, film formation, hydrogen production) effects.

Figure 12.12 shows the influence of different environments on the crack growth rate of a classical industrial steel [4]. Moist air is shown to have a marked influence on the crack growth rate in comparison with vacuum. Crack tip velocity is very sensitive to sodium chloride solution.

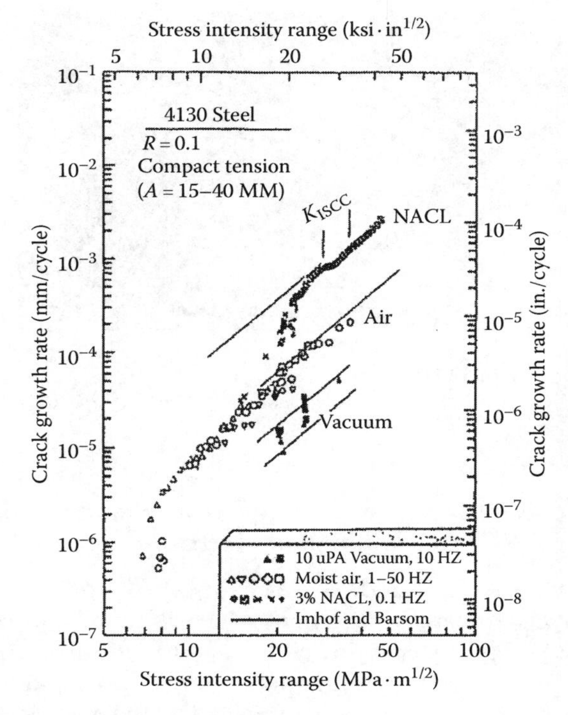

FIGURE 12.12
Influence of environment on the crack propagation velocity for a 4130 steel. (From Gangloff, R.P. and Duquette, D.J., *Chemistry and Physics of Fracture*, R. Latanision, ed., Nijhoff, Dordrecht, the Netherlands, p. 612, 1987.)

In fact, one must distinguish between long crack growth and short crack growth. For long crack growth, using linear elastic fracture mechanics (LEFM), typical characteristics of the corrosion fatigue behavior are given in Figure 12.13 as a function of the stress intensity factor ΔK. Interactions with SCC are emphasized.

The models then propose that the rate of crack growth corresponds either to the sum of pure mechanical fatigue and the rate of stress corrosion cracking or to the fastest available

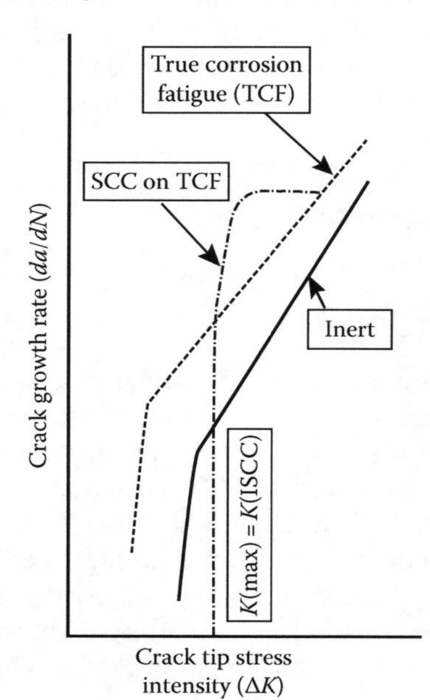

FIGURE 12.13
Schematic of fatigue and corrosion fatigue crack growth rate as a function of crack tip stress intensity.

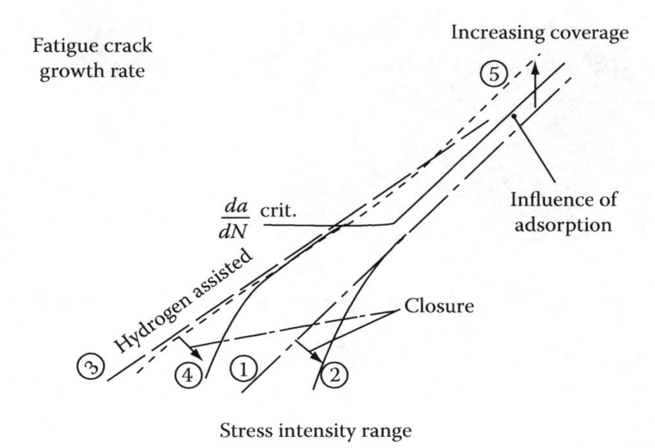

FIGURE 12.14
Schematization of the different CF crack growth mechanisms. (From Henaff, G. and Petit, J., *Corrosion Deformation Interactions CDI'92*, T. Magnin, ed., Les Editions de Physique, Paris, France, p. 599, 1993.)

mechanisms among those described in the following. Figure 12.14 summarizes the influence of corrosion on the fatigue crack growth characteristics [13–15], To the purely mechanistic mode I crack propagation behavior under vacuum (curve 1) must be added (a) the effect of closure [4,6] and mode II on the near-threshold (curve 2), (b) the hydrogen-assisted crack propagation behavior (curve 3), and (c) the influence of absorption and diffusion of hydrogen at low loading *R* ratio (curve 4) and at higher *R* ratio (curve 5). One must also take into account dissolution and film effects.

However, it has been shown that short cracks propagate at stress intensity factors well below the long crack threshold, as shown in Figure 12.15 for carbon steel in seawater.

The application of microstructural fracture mechanisms (MFMs) has been successfully used to predict the growth of short cracks during fatigue in air [16]. These models have been adapted to characterize and predict the uniaxial and multiaxial corrosion fatigue loading [17].

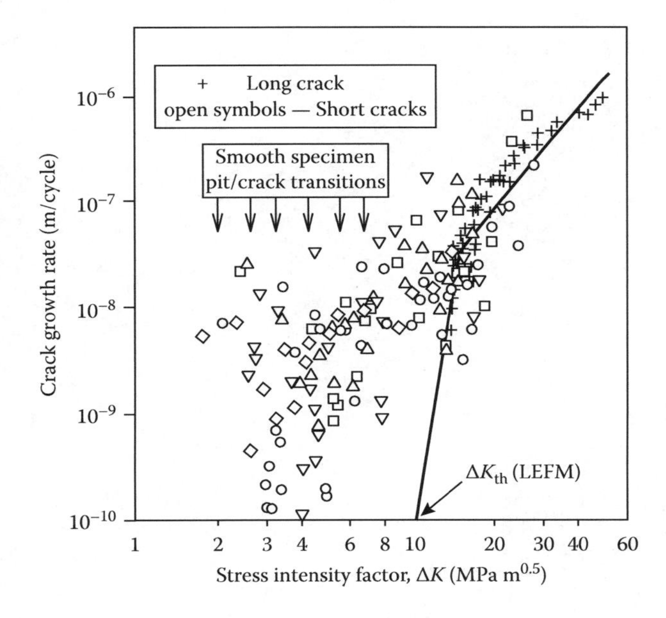

FIGURE 12.15
Comparison of long and short CF crack growth rates for carbon steels (σ_y = 500 MPa) in artificial water, 0.2 Hz, R = 0.1. (From Akid, R., *Environmental Degradation of Engineering Materials Vol. I*, Technical University of Gdansk and Université Bordeaux 1, p. 112, 1999.)

The two following equations provide the basis of the Brown–Hobson model [18], in which the fatigue crack growth rate is expressed as a function of an equivalent strain term γ_{eq} for stage I and stage II cracks, respectively.

$$\frac{da}{dN} = B_{I}\gamma_{eq}^{\beta_{I}}(d_i - a) \quad \text{(stage I, shear crack growth)} \tag{12.7}$$

$$\frac{da}{dN} = B_{II}\gamma_{eq}^{\beta_{II}}a - D \quad \text{(stage II, tensile crack growth)} \tag{12.8}$$

where
- a is the crack length
- d is a microstructural parameter such as grain size
- i is the number of grains through which the crack has traversed
- B_I, B_{II} and β_I, β_{II} are constant depending upon the material-environment system
- D is the long crack threshold

The determination of the crack tip environment is not easy because of the numerous electrochemical reactions and the associated mass transport and thermodynamic criteria that govern this environment. The nature of the solution (composition, pH, species, corrosion products that can induce roughness, etc.) can be very different at the crack tip and in the bulk solution. Figure 12.16 shows the possible electrochemical reactions at the crack tip. Both anodic dissolution effects and a coupled hydrogen reaction must often be taken into account. The mathematical modeling of the localized crack tip chemistry is therefore possible [19].

One can find in Ref. [20] a very significant effect of the transport properties of the crack solution on the CF crack growth rates of steel in seawater.

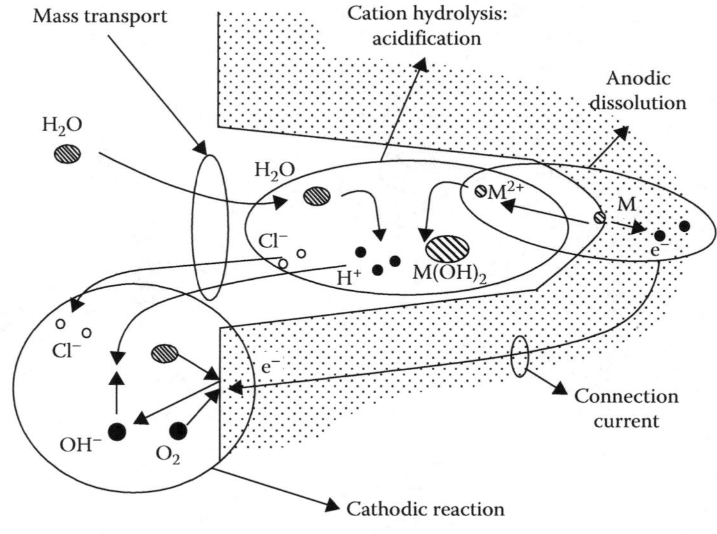

FIGURE 12.16
Possible electrochemical reactions at a corrosion fatigue crack tip.

12.3.2 Anodic Dissolution Effects

Anodic dissolution has been shown to occur preferentially in slip bands [11] very near the crack tip, which leads to many effects on the CF behavior as shown in Figure 12.17 [21]. A restricted slip reversibility model for stage I fatigue crack propagation has been proposed. The average crack growth increment per cycle, $\Delta\chi$, is determined by factors controlling slip reversibility on the forward slip planes (i.e., S1 and S4). For fee materials, it involves alternating slip on (111) planes near the crack tip. The slip reversibility is controlled by the degree of work hardening and recovery on a slip plane and the presence of corrosion products (oxides) on the slip steps. The rate of hydrated oxide nucleation plays a critical role in this model and therefore is expected to control stage I fatigue crack growth of an austenitic stainless steel in Cl^- solutions.

This localized dissolution process is not easy to take into account in numerical modeling of crack propagation. The slip dissolution model for CF is based on the fact that for many alloys in different solutions the crack propagation rate is proportional to the oxidation kinetics. Thus, by invoking Faraday's law, the average environmentally controlled crack

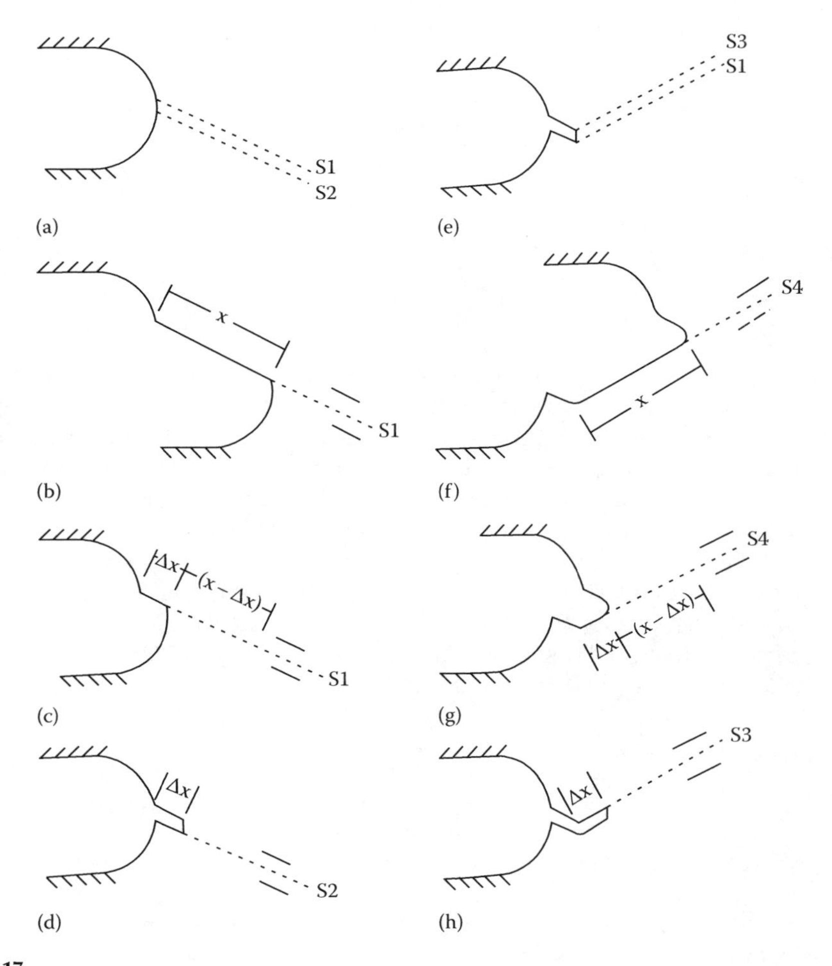

FIGURE 12.17
Schematic diagram of stage I corrosion fatigue crack propagation mechanism. (From Fong, C. and Tromans, D., *Metall. Trans.*, 19A, 2765, 1988.)

propagation rate $\bar{V}_t$ for passive alloys is related to oxidation charge density passed between film rupture events, Q_f:

$$\bar{V}_t = \frac{M}{n\rho F} Q_f \frac{1}{t_f} \tag{12.9}$$

where t_f is the film rupture period. Thus:

$$\bar{V}_t = \frac{M}{n\rho F} Q_f \frac{\dot{\varepsilon}}{\varepsilon_f} \tag{12.10}$$

where
$\dot{\varepsilon}$ is the strain rate
ε_f the strain for film rupture (about 10^{-3})

If we take a classical law for current transients at the crack tip,

$$Q_f = i_0 t_0 + \int_{t_0}^{t_f} i_0 \left(\frac{t}{t_0}\right)^{-\beta} dt \tag{12.11}$$

Then, for $t_f > t_0$, $\beta > 0$,

$$\bar{V}_t = \frac{M i_0 t_0 \dot{\varepsilon}}{n\rho F(\beta - 1)\varepsilon_f} \left[\beta - \left(\frac{\dot{\varepsilon} t_0}{\dot{\varepsilon}_f}\right)^{\beta - 1}\right] \tag{12.12}$$

Even if mechanical analyses give good approximations for $\dot{\varepsilon}$ at the crack tip, some problems still remain with the previous equation. In particular, the value of β evolves all along cycling. But the main effect is related to the localized corrosion–deformation described previously. It has been shown that vacancy generation at the crack tip due to localized dissolution can induce cyclic softening effects and that hydrogen absorption, which can also be coupled to localized dissolution, can also enhance the local cyclic plasticity. This is why improvement of CF predictive laws is needed even if V_t can be adjusted in the equation, which is still very useful. New models based on the description of Figure 12.17 but taking into account the localized corrosion deformation processes are under study.

Moreover, films related to solid-state oxidation ($M + H_2O \rightarrow MO + 2H^+ + 2e^-$) can also play a role in the crack advance. This needs further studies to be quantitatively more relevant [22].

12.3.3 Hydrogen Effects

Hydrogen-assisted cracking is often invoked, particularly for body-centered cubic (bcc) materials but also together with anodic dissolution for fee alloys. Figure 12.18 shows schematically a hydrogen-assisted cracking event. Interactions between a discretized dislocation array and the crack tip under an applied stress produce a maximum stress field from behind the tip. When the hydrogen concentration reaches a critical value, a microcrack is nucleated because either the local cohesive strength is reduced or dislocation motion is blocked in the hydrogen-enriched zone, or both. The microcrack arrests about 1 μm ahead of the original location of the tip and these processes then repeat, leading to discontinuous microcracking.

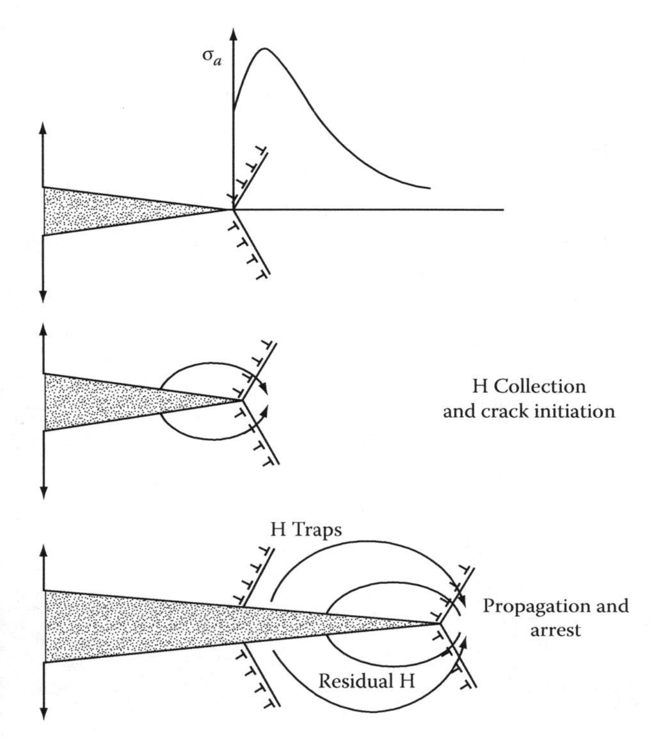

FIGURE 12.18
Schematic illustration of hydrogen-assisted cracking mechanism. (From Magnin, T., *Advances in Corrosion Deformation Interactions*, Trans Tech, Zurich, Switzerland, 1996.)

Other mechanisms have been proposed, particularly the hydrogen-induced plasticity model for precipitates containing materials such as Al-Zn-Mg alloys. This model is shown in Figure 12.19 [5].

Absorbed hydrogen atoms weaken interatomic bonds at the crack tip and thereby facilitate the injection of dislocations (alternate slip) from the crack tip. Crack growth occurs by alternate slip at crack tips, which promotes the coalescence of cracks with small voids nucleated just ahead of the cracks. In comparison with the behavior in neutral environments, the CF crack growth resistance decreases as the proportion of dislocation injection to dislocation egress increases. More closely spaced void nuclei and lower void nucleation strains should also decrease the resistance to crack growth in CF. This mechanism is proposed for Al-Zn-Mg alloys and is highly supported by observations that environmentally assisted cracking can occur at high crack velocities in materials with low hydrogen diffusivities and that the characteristics of cracking at high and low velocities are similar.

Other hydrogen-dislocation interactions at the corrosion fatigue crack tip must be taken into account. One can show, for instance, that hydrogen promotes planar slip in fee materials. Cross-slip ability can be discussed for the peculiar situation of the dissociation of a perfect screw dislocation into two mixed partíais separated by a stacking fault ribbon (Figure 12.20). The cross-slip probability depends on the work necessary for recombination.

Each partial is subjected to two forces: the repulsion due to the other partial and the attraction due to the stacking fault ($\Gamma \simeq 100\,\mathrm{mJ/m^2}$ in nickel, for instance). In the absence of an external stress, the partíais are in equilibrium at a distance where the two forces annihilate.

Hydrogen interacts only with the edge parts of the partíais (hydrostatic stresses). It migrates to the tensile zone of the edge parts, as shown in Figure 12.20 from an initial concentration C_0 (for calculations, see Ref. [23]).

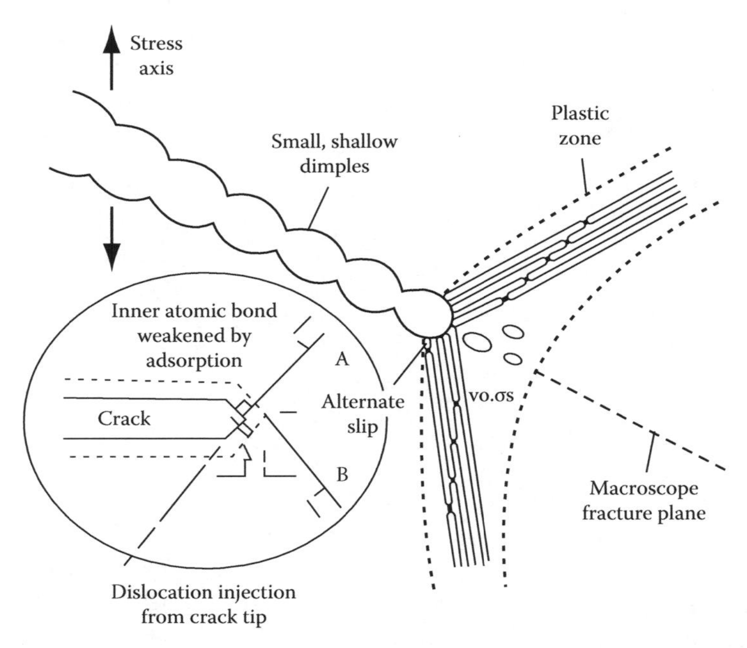

FIGURE 12.19
Schematization of the hydrogen-induced plasticity model for CF cleavage like cracking. (From Lynch, S.P., *Acta Metall.*, 36, 2639, 1988.)

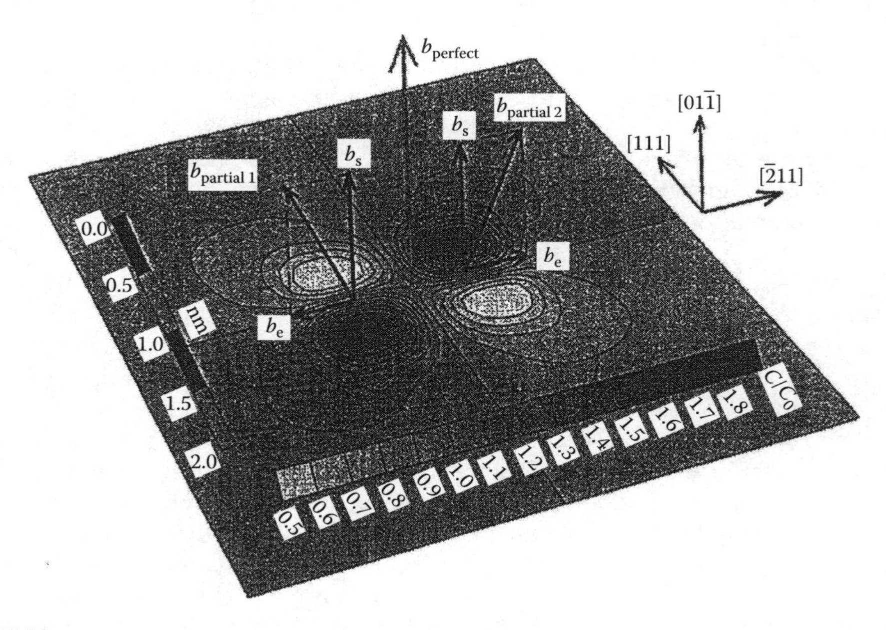

FIGURE 12.20
Equilibrium hydrogen distribution around a dissociated screw dislocation in nickel (C_0 is the initial concentration of hydrogen).

The dilatation strains due to the antisymmetric hydrogen distribution induce a net shear component along the dislocation line, whose sign is opposed to that of the edge dislocation. Because the Burger's vectors of the two edge parts of the partíais have opposite sign, hydrogen segregation will induce a supplementary effect, which is to reduce their normal attraction. From the viewpoint of the mixed partíais, the repulsion between the screw parts is the leading term, and hydrogen induces a supplementary repulsion by "screening" the attraction between the edge components.

This screening of the pair interactions between coplanar edge dislocations was shown to be independent to the first order on the distance between dislocations [23], provided that diffusion time and temperature allowed the hydrogen distribution to reach a configuration "in equilibrium with the local stress." Under these conditions, the relative screening of the pair interactions (i.e., the "hydrogen component" of the resolved shear stress normalized by that of the dislocation itself) was shown by numerical simulations to be given with excellent accuracy by the following simple expression [23]:

$$A(C_0, T) = \frac{75\%}{1+\beta(T/C_0)} \quad \text{with } \beta = \frac{9(1-v)RV_M}{2EV^{*2}} \tag{12.13}$$

where

C_0 is the remote hydrogen concentration (in atom fraction)
T is the absolute temperature
R is the gas constant
E, v, and V_M are Young's modulus, Poisson's ratio, and the molar volume of the host metal
V^* is the hydrogen partial molar volume

The coefficient β contains all the material parameters that influence the hydrogen screening effect as a function of temperature and concentration. In nickel, $\beta = 2.34 \times 10^{-4}\ K^{-1}$. Using this expression for the dissociated configuration of Figure 12.20, Figure 12.21 shows the forces

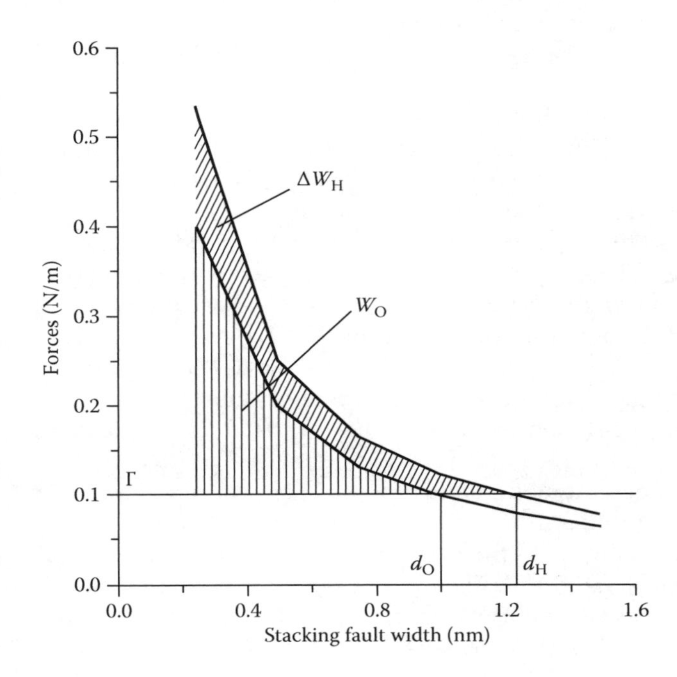

FIGURE 12.21
Effect of hydrogen on the work of recombination of partíais in nickel at 300 K.

on one partial versus the width of the stacking fault ribbon in nickel for an initial amount of hydrogen $n_H/n_{N_i} = 005$ at room temperature. The d_O and d_H are, respectively, the equilibrium distances in the absence and in the presence of hydrogen, neglecting a possible decrease of Γ. The recombination work is represented by the hatched areas between the curves. In the presence of hydrogen at concentration $C = 0.05$, calculations show that the recombination work is increased by 50%. The elastic effect of hydrogen is to increase the recombination work and then to decrease the cross-slip ability. This confirms experimental studies that show that hydrogen promotes planar glide [24].

Finally, it must be said that very often anodic dissolution and hydrogen effects must both be taken into account.

12.4 Monte Carlo-Type Simulations of Corrosion Fatigue Lifetimes from the Evolution of Short Surface Cracks

12.4.1 Introduction

The modeling of fatigue and corrosion fatigue damage is still not completely solved. Its definition is controversial, and it depends on the scale of interest in the study. The high-cycle fatigue regime was first investigated, and it has been summed up as due to the propagation from scratch of a single crack leading eventually to fracture [25,26]. Hence, an equivalence between the level of damage and the fatigue crack depth was considered [27]. In this case, fatigue damage accumulation corresponds to the bulk growth rate of one crack. Those speculations have proved to be both successful and satisfactory.

The next step aimed at extending the latter definitions from high- to low-cycle fatigue. However, as multiple cracks nucleate and propagate simultaneously in low-cycle fatigue, the previous definitions of both fatigue damage and fatigue damage accumulation are ambiguous. The majority of cracks are comparable to each other from a surface length viewpoint. Moreover, all crack lengths are comparable to grain size throughout all the fatigue lives.

On the one hand, the mechanical approach describes the behavior of long through-cracks in a macroscopically elastic field [7,25]. It has given interesting empirical laws but is unable to take into account the physical processes of fatigue damage. On the other hand, the microscopic approach was developed with the analysis of the dislocation behavior, the corresponding strain hardening or softening effects, and the formation of persistent and intense slip bands [28,29]. Nevertheless, this approach is generally limited to the formation of the first microcracks: the corresponding physical and numerical models concern only the first 10% of the fatigue lifetime for polycrystalline materials. Moreover, when applied to crack propagation, the microscopic approach does not take into account the influences of other formed cracks. Thus, the classical mechanical and microscopic approaches cannot be used to model the evolution of a population of surface cracks and the corresponding low-cycle fatigue damage.

In this section an investigation of the fatigue damage process at an intermediate scale (i.e., a mesoscopic scale corresponding to the grain size) is addressed. The importance of the latter scale to modeling low-cycle fatigue lifetimes is pointed out through the physical analysis of fatigue damage and of fatigue damage accumulation. Particular attention is paid to the evolution of populations of surface short cracks. Experimental results mainly correspond to push–pull low-cycle fatigue tests of 316L stainless steel in air and in 3% NaCl solutions. From these results and from assumptions concerning surface short crack behavior, a numerical model is developed. Monte Carlo principles are used to deal with random crack nucleation and crack interactions. It is finally shown that such an analysis is very relevant to prediction of corrosion fatigue lifetimes of austenitic stainless steels in chloride solutions.

12.4.2 Physical Description of the Fatigue Damage

The experimental results on the low-cycle fatigue damage of 316L stainless steel have been fully discussed elsewhere [30,31]. Transgranular cracking is generally observed in the 316L polycrystalline stainless steel for fully reversed push–pull tests at intermediate plastic deformation amplitudes ($\Delta\varepsilon_p/2$ in the range $[5 \times 10^{-5}, 5 \times 10^{-3}]$, the plastic deformation rate being 10^{-3} s^{-1}).

Multiple short cracks are observed at the specimen surface. Different surface short crack types are introduced according to both cracking behavior and surface lengths. Surface short cracks, the lengths of which are less than one grain size (i.e., less than 50 μm, which is the average grain size in the case of classical 316 alloy), are the more numerous ones. Every crack propagates first crystallographically, i.e., within the intense slip bands. The first main obstacles to their propagation are the grain boundaries. The closer to the grain boundary, the lower the crack growth rate [30,32]. Once the grain boundary is crossed, the propagation speeds up to the next grain boundary. It has been observed experimentally that two to three grain boundaries need to be crossed for the change of cracking behavior to take place. Surface crack propagation evolves from crystallographic to "mechanical" growth. The surface cracks are then observed to grow perpendicular to the specimen axis (likely to be related to the onset of the so-called stage II cracking). The preceding cracking process leads to distinction of three main categories of surface short cracks.

Type I cracks have a surface length less than one grain size. Type II cracks are longer than one grain size at the surface but smaller than three (two to three grain boundaries have been overcome by the surface short crack). Type III crack surface length goes from 3 up to 10 grain sizes. A fourth type concerns cracks longer than 10 grain sizes at the surface. The latter cracks are numbered in the range 1–3 per specimen and form during the last 10% of the fatigue lifetime, following completion of N_i cycles. N_i corresponds to the number of cycles to form the type IV short crack. It is about 90% of the fatigue lifetime [30,31].

The type I crack population evolves in three steps following a first hardening of samples, which needs 5 to 10% of the reduced lifetime (N/N_i). The first step consists of numerous nucleations of new cracks and leads to a surface density of approximately 50 type I cracks per mm^2 in less than 25% of the reduced lifetime. Crossing of the grain boundary corresponds to a second step during which the new crack nucleation rate decreases by two thirds. A few type II cracks form. This step extends up to 60%–70% of the reduced lifetime N/N_i. The type I crack density reaches about 70 cracks per mm^2. This is the highest type I crack density during the test. Stage III of the damage process corresponds to the remaining lifetime. There is a speeding up of damage accumulation. Numerous type II cracks form while the type I crack population decreases by a half (final density is about 40 cracks per mm^2). Their interactions and their surface propagation lead to generation of type III cracks. Two to three type IV cracks are eventually formed for the total specimen surface and propagate into the bulk, leading to rupture.

High short crack density and competitive growth of the multiple surface short cracks during more than 90% of lifetimes suggest a random process to form a fatal crack. High surface crack densities have been reported on other materials submitted to fatigue tests [33–35].

Randomness must occur during both surface crack nucleation and propagation. Consequently, crack interactions are likely to be unpredictable one by one. As fatigue damage results from a combination of these mechanisms (i.e., nucleation, propagation, and interactions), the usual approach of fatigue through a single crack growth relation is unsatisfactory. As a matter of fact, this approach does not take into account the effect of crack densities and hence the statistical aspect of fatal crack formation.

A result [33,34] has been achieved emphasizing both the importance of the statistical aspect of the fatigue damage accumulation and the role of type I surface short cracks. An accumulation of elementary damage events, the cracking of a single grain, was assumed and was shown to be successful in deducing the Coffin–Manson relationship using the statistical physics of disordered systems.

A few numerical models have been developed to take into account the developments and the interactions of surface short crack populations [34,36]. The next paragraph briefly presents results of such modeling.

12.4.3 Numerical Modeling

The simulation has been developed according to the statistical aspect of the physical damage accumulation process. Surface short cracks nucleate at random. Their subsequent propagation is not continuous from a time viewpoint. Hence, a random treatment was chosen to deal with both nucleation and propagation mechanisms. Moreover, only surface data are taken into account because surface fatigue damage dominates over 80% of the fatigue lifetime in low-cycle fatigue. The modeling therefore belongs to a 2D Monte Carlo type [36].

Figure 12.22 shows predictions of crack density evolutions with statistical scatter. Moreover, the simulations help to plot major crack developments. Figure 12.23a illustrates the fatal crack evolutions. Interestingly, for $\Delta\varepsilon_p/2 = \pm 4 \times 10^{-3}$, the curve clearly exhibits multiple coalescence even for crack length of the order of two grain sizes (i.e., 100 μm). The coalescences lead to significant increases of surface lengths. Note the absence of those sharp increases in Figure 12.23b for $\Delta\varepsilon_p/2 = \pm 4 \times 10^{-4}$. More results are detailed elsewhere [36]. For example, Table 12.1 compares a few experimental lifetimes with the corresponding simulation results.

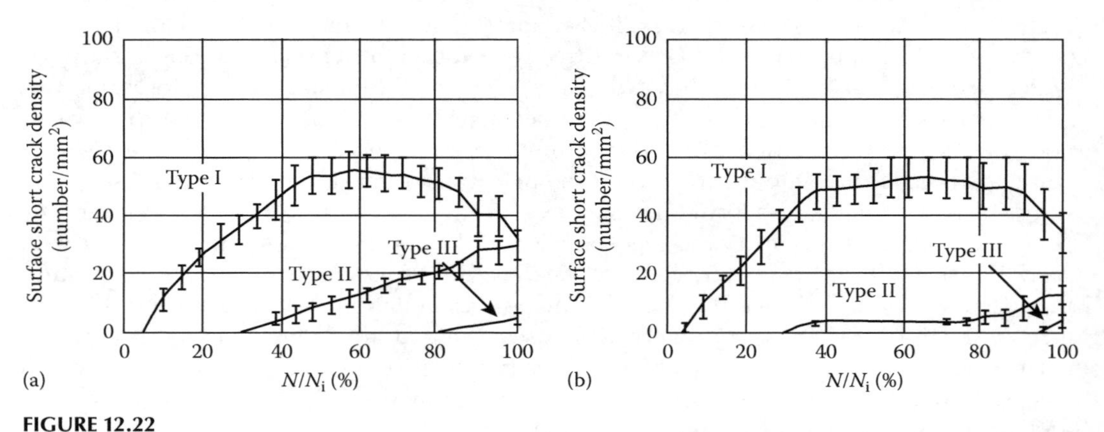

FIGURE 12.22
In-air test simulations. Surface short crack density vs. N/N_i at (a) a plastic strain amplitude of 4×10^{-3}, $N_i = 6800 \pm 800$ cycles and (b) a plastic strain amplitude of 4×10^{-4}, $N_i = 250{,}000 \pm 35{,}000$ cycles. Scatter is calculated from several computations.

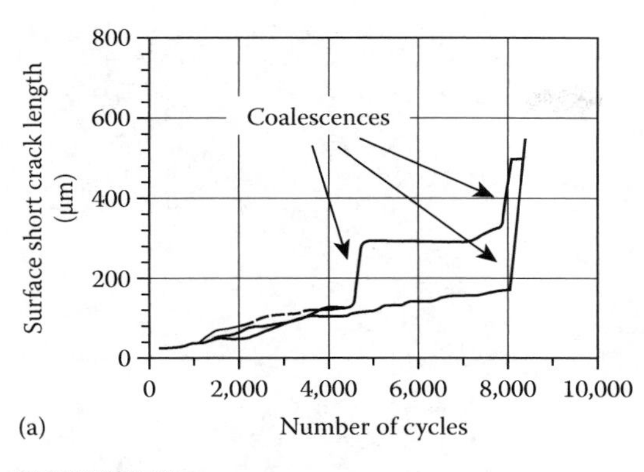

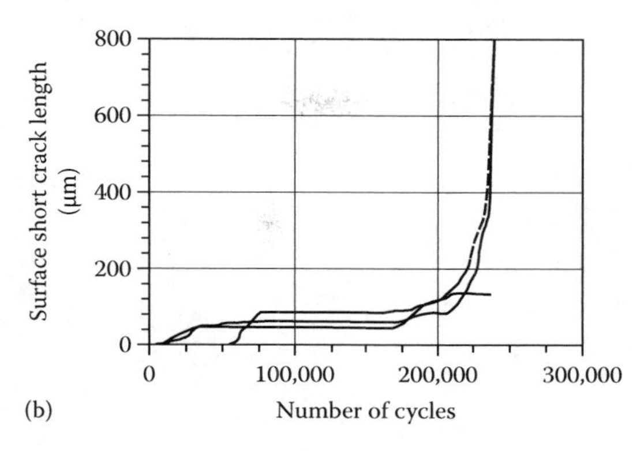

FIGURE 12.23
In-air test simulations. Major crack development vs. the number of cycles: (a) $\Delta\varepsilon_p/2 = \pm 4 \times 10^{-3}$ and (b) $\Delta\varepsilon_p/2 = \pm 4 \times 10^{-4}$. Coalescence mechanisms participate strongly in formation of a fatal crack at $\Delta\varepsilon_p/2 = \pm 4 \times 10^{-3}$.

TABLE 12.1

Comparison of Experimental and Fatigue Lifetimes (316 L Alloy at $\dot{\varepsilon} = 10^{-3}\ s^{-1}$)

	$\Delta\varepsilon_p/2 = \pm 4 \times 10^{-4}$		$\Delta\varepsilon_p/2 = \pm 2 \times 10^{-3}$		$\Delta\varepsilon_p/2 = \pm 4 \times 10^{-3}$	
Experimental lifetime (cycles)	268,000		19,250		7,500	
Modeling prediction (cycles)	250,000	38,000	16,440	1,800	6,800	950
		Standard scatter		Standard scatter		Standard scatter

Furthermore, the three short crack type densities correlate well. Also, the screen display of the simulated specimen surface shows similar patterns of crack spatial distribution (Figures 12.23 and 12.24).

Finally, the relevance of the mesoscopic scale considered in the present study (i.e., the grain size) can be emphasized when predicting the corrosion fatigue behavior of 316 L alloys in a 3.5% NaCl solution under free potential. In the corrosive solution the fatigue lifetime decreases by a factor of 2 for an applied plastic strain $\Delta\varepsilon_p/2 = \pm 4 \times 10^{-3}$ [11]. In this case the localized anodic dissolution at the slip band emergence increases mainly the number of nucleation sites, the surface micro-propagation threshold (particularly to cross the grain boundaries), the kinetics of surface propagation, and then the formation of the fatal crack. The experimental evolution of the short crack densities is shown in Figure 12.25a. It must be noted that the evolution of type I, II, and III crack populations is homologous to its evolution for in-air tests when referring to N/N_i. The simulation (Figure 12.25b) shows satisfactory agreement with experiments. A very interesting tool for corrosion fatigue prediction, based on the effect of the corrosive solution at the mesoscopic scale, is then available. This is new in this field. Other effects of a corrosive solution (such as pitting) can now be introduced in the software to simulate other possible corrosion fatigue mechanisms in metal and alloys.

Finally, such Monte Carlo treatment has been successfully applied to other free single-phase materials [37].

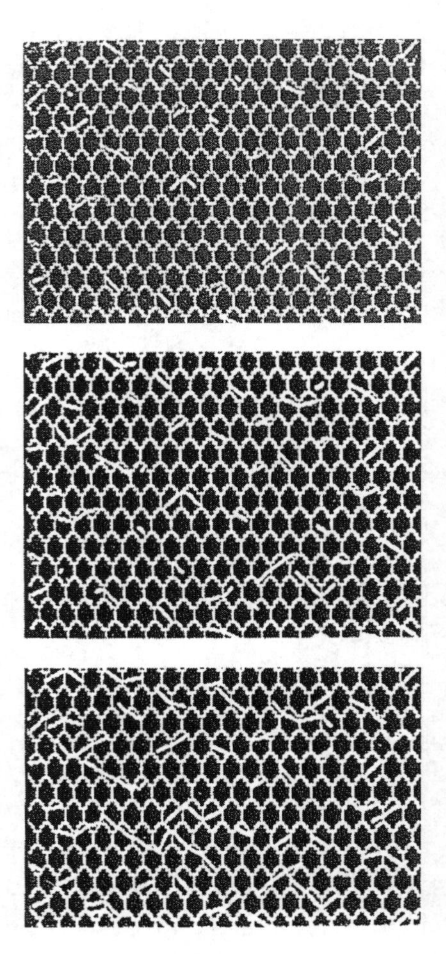

FIGURE 12.24
Screen display of modeling surface. $\Delta\varepsilon_p/2 = \pm 4 \times 10^{-3}$.

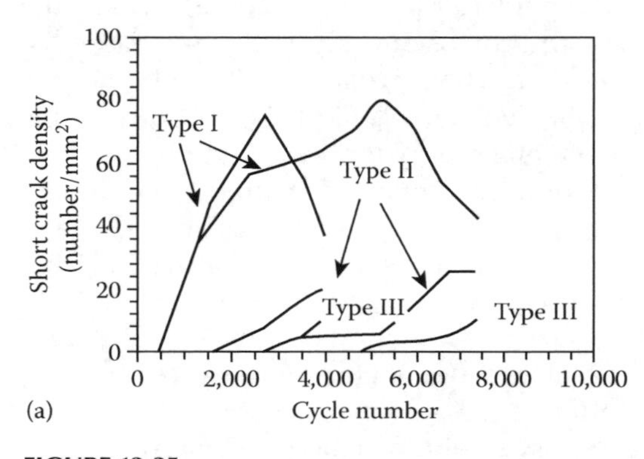

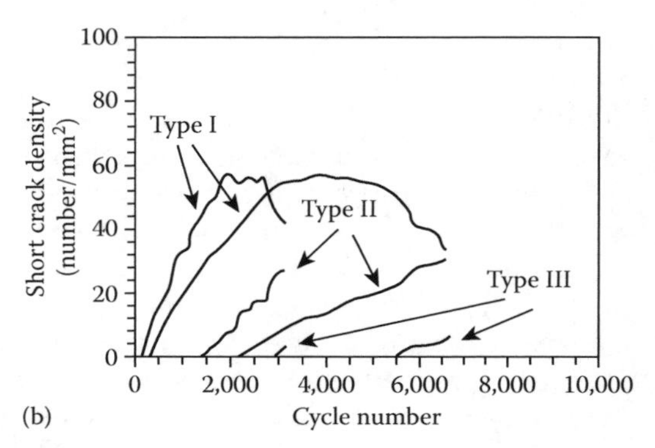

FIGURE 12.25
Influence of the environment on the evolution of type I, II, and III surface crack densities vs. the number of cycles (thick lines are for in-air test, thin ones for a 3.5% NaCl solution test): (a) experimental data and (b) simulation results.

12.5 Interactions with Stress Corrosion Cracking (SCC)

It is well known that SCC can interfere with corrosion fatigue during crack propagation (Figure 12.13). But this interaction also occurs for crack initiation and crack propagation near the threshold, as illustrated in Figure 12.3 for an fee Fe-Mn-Cr alloy cyclically deformed at low strain rate in a Cl^- solution and in Figure 12.8 for Al-Li alloys in 30 g/L NaCl. On the fracture surface of the Fe-Cr-Mn alloy, a crack initiates from a pit (which is favored by fatigue processes) by SCC with a cleavage-like process that has been described in detail elsewhere [22]. Then fatigue and corrosion fatigue take place with classical striations. Finally, it must be noted that the distinction between stress corrosion cracking and fatigue cracking is not always easy to make because SCC can also be discontinuous (see the mechanisms in Ref. [22]) with crack arrest markings very similar to fatigue ones. Thus, combination of pure SCC and fatigue mechanisms is often possible during corrosion fatigue damage.

CF can sometimes be more dangerous than SCC because it occurs whatever the strain rate is. This is illustrated in the case of a 7020 Al-Zn-Mg alloy for two different heat treatments, T4 (underaged) and T6DR (peak aged). Figure 12.26 compares the SCC behavior through the ratio between the elongation to fracture in 30 g/L NaCl and the elongation in air for tensile tests at an imposed strain rate with the CF behavior at a given applied plastic strain $\Delta\varepsilon_p/2 = 10^{-3}$, for different strain rates. One can see that, whatever the strain rate, CF always occurs, unlike SCC, which takes place only below a critical strain rate dependent on the heat treatment. Moreover, CF is affected

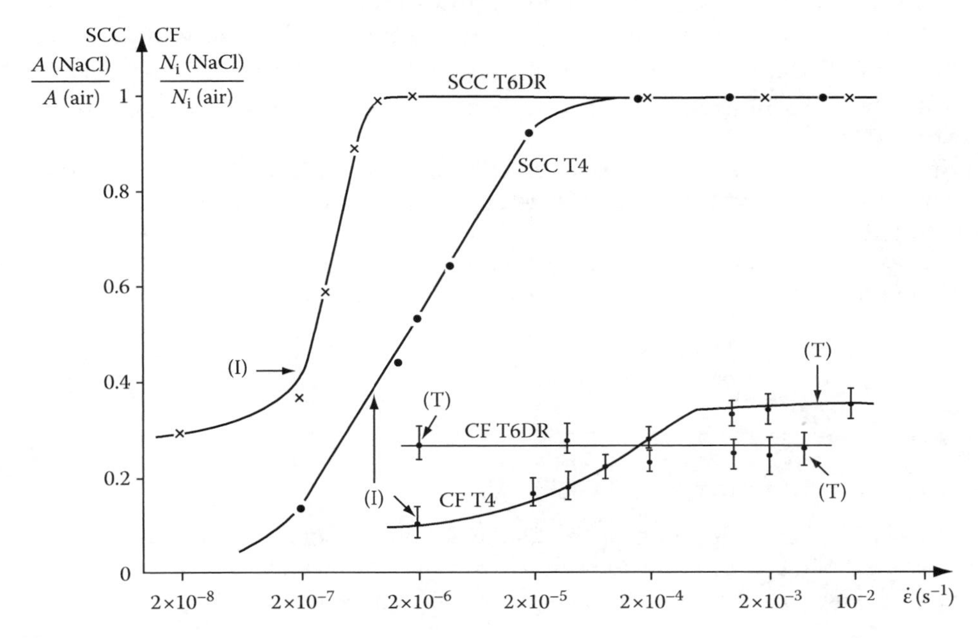

FIGURE 12.26
Comparison of SCC and CF (at $\Delta\varepsilon_p/2 = 10^{-3}$) behavior of a 7020 T4 and T6DR alloy at free potential in a 30 g/L NaCl solution. Crack initiation can be transgranular (T) or intergranular (I) according to the strain rate $\dot{\varepsilon}$.

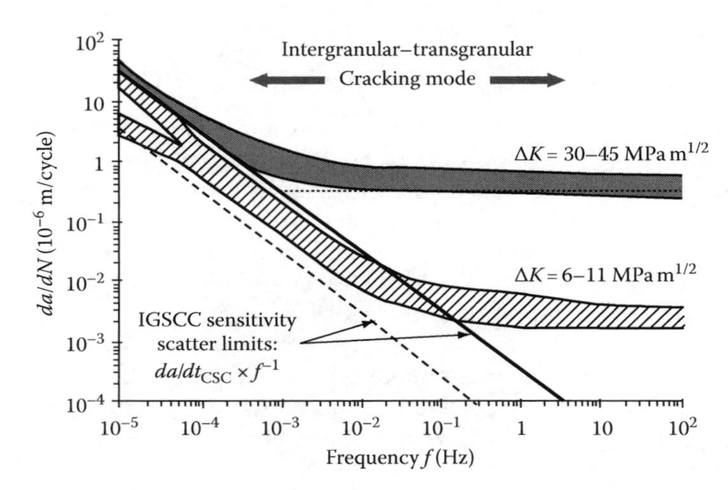

FIGURE 12.27
Crack growth rates versus frequency during CF and SCC on CT alloy 600 specimens in a PWR environment. (From Bosch, C. et al., *Corrosion Sci.*)

by SCC below the critical SCC strain rate, as shown by the fact that crack initiation is intergranular only when SCC occurs.

Nevertheless, cyclic loading is not always deleterious in regard to SCC. The example of cracking in alloy 600 for a pressurized water reactor (PWR) illustrates such behavior. Intergranular SCC is known to occur in the alloy 600/PWR system [38] when tests are performed on CT specimens for constant loading conditions with a stress intensity factor $K_I = 30\,\text{MPa}\sqrt{m}$. The effect of cycling loading at a load ratio $R = 0.7$ with a triangular wave form at $K_{max} = 30\,\text{MPa}\sqrt{m}$ is shown in Figure 12.27 according to the imposed frequency.

An increase in frequency induces a decrease of the crack growth rate and a transition from intergranular to transgranular cracking mode until air cyclic behavior becomes predominant.

This last example clearly shows that cycling can completely change the crack propagation mode when stress corrosion crack propagation is intergranular.

12.6 Concluding Remarks

The analysis of CF micromechanisms related to anodic dissolution and hydrogen effects is in progress but needs to be more quantitative through the localized corrosion–deformation interactions. The trends for future research are mainly related to:

1. The modeling of crack tip chemistry
2. The quantitative analysis of corrosion–deformation interactions at the CF crack tip (scale of 1 μm) according to the electrochemical conditions

3. A comparison between CF and SCC based on a detailed analysis of micromechanisms near the fatigue threshold
4. Development of numerical simulations at mesoscopic scales

These studies are needed to propose predictive laws for CF damage increasingly based on physicochemical controlling factors.

References

1. C. Laird and D. J. Duquette, *Corrosion Fatigue*, NACE, New York, 1972, p. 82.
2. C. Patel, *Corros. Sci. 21*:145 (1977).
3. (a) R. P. Wei and G. W. Simmons, *Int. J. Fract. 77*:235 (1981); (b) J. Congleton, J. H. Craig, R. A. Olich, and R. N. Parkins, *Corrosion Fatigue, Mechanics, Metallurgy, Electrochemistry and Engineering*, ASTM 801 (T. W. Crooker and B. W. Leis, eds.), ASTM, Philadelphia, PA, 1983, p. 367.
4. R. P. Gangloff and D. J. Duquette, *Chemistry and Physics of Fracture* (R. Latanision, ed.), Nijhoff, Dordrecht, 1987, p. 612.
5. S. P. Lynch, *Acta Metall. 36*, L10:2639 (1988).
6. R. O. Ritchie and J. Lankford, *Small Fatigue Cracks*, TMS-ASME, 1986, p. 33.
7. K. J. Miller and E. R. de Los Rios, *The Behaviour of Short Fatigue Cracks*, EGFl, Mechanical Engineering Publications, London, 1986.
8. M. Mueller, *Metall. Trans. 13A*:649 (1982).
9. B. D. Yan, G. C. Farrington, and C. Laird, *Acta Metall. 33*:1593 (1985).
10. T. Magnin and L. Coudreuse, *Acta Metall. 35*:2105 (1987).
11. T. Magnin and J. M. Lardón, *Mater. Sci. Eng. A 104*:21 (1988).
12. P. Scott and R. A. Cottis, *Environment Assisted Fatigue*, EGF/MEP, London, 1990.
13. G. Henaff and J. Petit, *Corrosion Deformation Interactions CDI'92* (T. Magnin, ed.), Les Editions de Physique, Paris, 1993, p. 599.
14. L. Hagn, *Mater. Sci. Eng. A 103*:193 (1988).
15. R. Akid, *Environmental Degradation of Engineering Materials Vol. I*, Technical University of Gdansk and Université Bordeaux 1, 1999, p. 112.
16. A. Navarro and E. R. de Los Rios, *Fat. Fract. Eng. Mater. Struct. 11*:383 (1988).
17. W. Zhang and R. Akid, *Fat. Fract. Eng. Mater. Struct. 20*:167 (1997).
18. P. D. Hobson, M. W. Brown, and E. R. de Los Rios, *The Behaviour of Short Fatigue Cracks* (K. J. Miller and E. R. de Los Rios, eds.), EGFl, London, 1986, p. 441.
19. A. Turnbull, *Corrosion sous contrainte* (D. Desjardins and R. Oltra, eds.), Les Editions de Physique, Paris, France, 1990, p. 79.
20. C. J. Van der Wekken, *Corrosion Deformation Interactions CDI'96* (T. Magnin, ed.), Institute of Materials, London, 1997, p. 76.
21. C. Fong and D. Tromans, *Metall. Trans. 19A*:2765 (1988).
22. T. Magnin, *Advances in Corrosion Deformation Interactions*, Trans Tech, Zurich, Switzerland, 1996 (see p. 126, with reference to Gerberich et al., Parkins Symposium, TSM, 191, 1992).
23. D. Delafosse, J. P. Chateau, and T. Magnin, *J. Phys. IV 9*:251 (1999).
24. J. P. Hirth and J. Lothe, *Theory of Dislocations*, Krieger Publications, Miami, FL, 1992.
25. P. C. Paris and F. Erdogan, *J. Basic Eng. Trans. AIME D 85*:528 (1963).
26. B. Tomkins, *Philos. Mag. 18*:104 (1968).
27. K. J. Miller, *Fatigue Eng. Mater. Struct. 5*:223 (1982).
28. T. Magnin, J. Driver, J. Lepinoux, and L. P. Kubin, *Rev. Phys. Appl. 19*:467 (1984).

29. H. Mughrabi, F. Ackermann, and K. Herz, *Fatigue Mechanisms*, ASTM STP 675 (J. T. Fong, ed.), ASTM, Philadelphia, PA, 1979, p. 69.
30. T. Magnin, L. Coudreuse, and J. M. Lardon, *Ser. Met. 19*:1487 (1985).
31. C. Ramade, Mécanismes de microfissuration et endommagement en fatigue oligocyclique de polycristaux monophasés, Thesis, Ecole des Mines St. Etienne, France, 1990.
32. K. J. Miller, H. J. Mohammed, and E. R. de Los Rios, *The Behaviour of Short Fatigue Cracks* (K. J. Miller and E. R. de Los Rios, eds.), Mechanical Engineering Publications, Sheffield, U.K., 1986, p. 491.
33. Y. Brechet, T. Magnin, and D. Sornette, *Acta Metall. Mater. 40*:2281 (1992).
34. A. Bataille, T. Magnin, and K. J. Miller, *Short Fatigue Cracks* (K. J. Miller, ed.), Esis 13, Mechanical Engineering Publications, Sheffield, U.K., 1997.
35. J. Weiss and A. Pineau, *Low-Cycle Fatigue and Elasto-Plastic Behaviour of Metals* (K. T. Rie, ed.), Elsevier Applied Science, Amsterdam, the Netherlands, 1993, p. 82.
36. A. Bataille and T. Magnin, *Acta Metall. Mater. 42*:3817 (1995).
37. A. Bataille, thesis, University of Lille, France, 1992.
38. C. Bosch, Etude de la relation entre la corrosion sous contrainte et la fatigue corrosion basse fréquence de l'alliage 600 en milieu primaire des réacteurs à eau sous pression, Thesis, Université Bordeaux 1, 1998.

13

High-Temperature Corrosion

Michael Schütze
Karl Winnacker Institute of DECHEMA

CONTENTS

13.1 Introduction 573
13.2 Thermodynamics of the Formation of Corrosion Products 574
13.3 Solid Corrosion Products 578
13.3.1 Formation and Growth Kinetics in "External Corrosion" 578
13.3.2 Role of Alloy Composition 584
13.3.3 Internal Corrosion 588
13.3.4 Metal Dusting as a Special Form of Solid Corrosion Product 591
13.4 Liquid Phases in High-Temperature Corrosion 594
13.4.1 General Aspects 594
13.4.2 Sulfate Corrosion (Hot Corrosion) 596
13.4.3 Liquid Chloride Corrosion 599
13.5 Formation of Volatile Phases in High-Temperature Corrosion 602
13.5.1 General Aspects 602
13.5.2 Gas-Phase Chlorine Corrosion 602
13.5.3 Formation of Volatile Phases from Water Vapor 610
13.6 Concluding Remarks 613
References 613

13.1 Introduction

High-temperature corrosion is often termed dry corrosion, which indicates that it is not aqueous electrolytes that are responsible for the corrosion mechanisms. As will be discussed later, liquid phases can also be involved in this kind of corrosion, which will, however, not be of an aqueous nature. Concerning the temperatures above which high-temperature corrosion is expected to occur, there is a definition in an ISO draft that states [1]: "Corrosion occurring when the temperature is higher than the dew point of aqueous phases of the environment but at least 373 K." This definition may still be a matter of controversial debate in the corrosion community as this temperature of 373 K is regarded as being too low, but it should be pointed out that the typical forms of high-temperature corrosion resulting from halogen attack (e.g., chlorine gas) can be observed at temperatures lower than 200°C (473 K). Nevertheless, the key criterion distinguishing high-temperature corrosion from wet or aqueous corrosion is the absence of an aqueous electrolyte. There are a number of textbooks addressing the issue of high-temperature corrosion in general with all significant aspects, which are recommended for further reading (see Refs. [2–10]).

The reactions occurring during high-temperature corrosion can lead to the formation of solid, liquid, and gaseous corrosion products. The discussion of high-temperature corrosion will be divided into three sections, which follow these three types of corrosion products. Generally, only the formation of solid corrosion products provides long-term stability of metallic components, which is required for technical use. In this case, protective surface layers can be formed as corrosion products that act as diffusion barriers and slow down the reaction between the environment and metallic material to an extent that it allows the use of the material for a longer time. As will be discussed in this chapter, such solid protective layers must have a low concentration of lattice defects resulting in low diffusion rates through the layers and must also be free from physical defects such as cracks or open porosity. Liquid phases generally do not provide any significant corrosion protection due to the high transport rates in these layers. The formation of gas-phase corrosion products prevents the formation of diffusion barriers on the surface. As the transport rates through the gas phase are the highest among the different states, this situation of the formation of volatile corrosion products has to be regarded as the most critical one from a technical point of view.

Depending on the temperatures, the reactants in the corrosion reaction, and the pressures and activities, a large variety of different corrosion mechanisms and features exists, which are characteristic for each specific situation. The features of these mechanisms will form the main part of the discussion in the present chapter. Table 13.1 summarizes the most common elements responsible for high-temperature corrosive attack in technical applications together with the potential corrosion products being formed. In the case of liquid and gaseous corrosion products, the temperatures above which these phases can be present in a significant extent are given. Whether these corrosion products form, depends on the composition of the corrosive environment and the alloy activities of the materials being attacked.

The two key aspects that characterize the behavior of materials under conditions of high-temperature corrosion consist on the one hand of a thermodynamic assessment providing information about the potential phases formed by the corrosion reaction and of a kinetic treatment that indicates the rate at which these phases are being formed yielding information about the consumption rates of the metallic alloy. Both aspects will be treated together with the mechanisms occurring and this is intended to provide a fundamental understanding of the field of high-temperature corrosion.

13.2 Thermodynamics of the Formation of Corrosion Products

The procedure for an assessment of which corrosion products can be formed in a certain environment by their reaction with a metallic alloy will be illustrated in the following by using an oxidation reaction. As a matter of course, all other types of reactions (sulfidation, carburization, chlorination, etc.) can also be addressed using this procedure. As a simplified general form of the oxidation reaction, the following equation can be written as follows:

$$\frac{2x}{y}\mathrm{M}+\mathrm{O}_2=\frac{2}{y}\mathrm{M}_x\mathrm{O}_y, \tag{13.1}$$

where

M stands for the metal component reacting with oxygen (O_2)

x and y are stoichiometric factors

TABLE 13.1

Potential Corrosion Products Found under High-Temperature Corrosion Conditions

Corrosion By	Solid	Liquid (m.p.)	Gaseous ($pX = 10^{-4}$ bar)
O_2	Al_2O_3	B_2O_3 (580°C)	CrO_3 (1000°C)
	Cr_2O_3	Bi_2O_3 (820°C)	MoO_3 (610°C)
	CoO	MoO_3 (795°C)	PtO_2 (1000°C)
	FeO, MnO, Fe_3O_4, Fe_2O_3	SiO_2 (ca. 1400°C)	SiO_2 (1250°C, low pO_2)
	NiO	V_2O_5 (674°C)	WO_3 (1000°C)
	SiO_2		
	Spinel phases		
	TiO_2		
H_2O	Like O_2	Like O_2	CrH_2O_4 (650°C)
			MoH_2O_4 (600°C)
			MnOH (800°C)
H_2			CH_4 (250°C)
N_2	AlN		
	CrN, Cr_2N		
	FeN		
C	CrC, Cr_7C_3, $Cr_{23}C_6$		
	Fe_3C		
	NbC		
	TiC		
	Metal dust		
S	FeS	Co–Co_4S_3 eutectic (880°C)	
		Fe–FeS eutectic (985°C)	
	CrS	Fe–FeO–FeS eutectic (925°C)	
	MnS_2	Ni–Ni_3S_2 eutectic (635°C)	
	NiS, Ni_3S_2		
	NbS		
	TiS, TiS_2		
	Sulfur spinels		
S + Alkali	FeS	Me(K, Na)SO_4 eutectic (530°C)	
		Me = Fe, Ni, Co	
	CrS		
	Ni_3S_2		
V	V_2O_5	Cu–Cu_2O–V_2O_5 eutectic (530°C)	
		Heat-resistant steel–V_2O_5 eutectic (600°C)	
Cl	$FeCl_2$, $FcCl_3$	$FeCl_2$ (676°C)	$AlCl_3$ (76°C)
	$AlCl_3$	$FeCl_3$ (303°C)	$CoCl_2$ (587°C)
	$CrCl_2$, $CrCl_3$	$AlCl_3$ (193°C)	$CrCl_2$ (741°C)
	$CoCl_2$	$CrCl_2$ (820°C)	$CrCl_3$ (611°C)
	$MoCl_4$, $MoCl_5$	$CrCl_3$ (1150°C)	$FeCl_2$ (536°C)
	$NiCl_2$	$CoCl_2$ (740°C)	$FeCl_3$ (167°C)
		$MoCl_4$ (317°C)	$MoCl_5$ (72°C)
		$MoCl_5$ (194°C)	$NiCl_2$ (607°C)

(continued)

TABLE 13.1 (continued)

Potential Corrosion Products Found under High-Temperature Corrosion Conditions

Corrosion By	Solid	Liquid (m.p.)	Gaseous (pX = 10^{-4} bar)
		$NiCl_2$ (1030°C)	$TiCl_4$ (38°C)
		$SiCl_4$ (70°C)	
		$TiCl_4$ (24°C)	
F	CrF_3	CrF_3 (1404°C)	CrF_3 (860°C)
	FeF_3	FeF_3 (1000°C)	FeF_3 (670°C)
	NiF_2	NiF_2 (1450°C)	NiF_2 (940°C)

Compiled from different sources.

The result of this reaction is the formation of the metal oxide M_xO_y if the free energy of formation ΔG^0 is negative. The law of mass action

$$K = \frac{a_{M_xO_y}^{2/y}}{a_M^{2x/y} \cdot pO_2} = \exp\left(-\frac{\Delta G^0}{RT}\right) \tag{13.2}$$

allows the definition of the boundary-separating conditions where the metal is still stable and unaffected and conditions where the oxidation product is being formed. In this law, K represents the equilibrium constant of the reaction, a is the activity of the different participants in the reaction, R is the general gas constant, and T is the absolute temperature. In the case of gaseous phases, a is replaced by the partial pressure p (in the present case pO_2 for oxygen) and for pure phases, the activity can be set to 1. In the case of oxidation, this boundary is defined by the so-called equilibrium oxygen partial pressure, which is the minimum oxygen pressure in the environment needed to keep an oxide formed by reaction (13.1) stable.

Based on this approach, so-called stability diagrams can be established that delineate areas of existence of the different corrosion products and of the unaffected material as a function of the activities and partial pressures of the components of the corrosive environment and provide information as to which corrosion mechanism or type of corrosion product can generally be expected under the respective environmental conditions. The principles of establishing such stability diagrams are based on equations of the type given by Equation 13.2 for the respective potential corrosion reactions with ΔG values taken from tables in the literature and are given in Ref. [6]. However, nowadays computer programs are often used that allow the assessment of corrosion reactions even in very complex technical environments and of materials consisting of a larger number of alloying elements [11,12]. An example of such a diagram is given in Figure 13.1 that shows the situation for a process gas environment based on carbon species and oxygen (coal gasification) [13]. Here it can be seen that under these conditions, the metallic elements tungsten and niobium would form carbides whereas all the other alloying elements that are commonly encountered in technical high-temperature materials, such as chromium, manganese, silicon, and aluminum, find sufficient oxygen in order to guarantee the stability of the respective oxides. It can be mentioned here that in particular, chromium, silicon, and aluminum can form slowly growing and dense oxide barriers (Al_2O_3 = alumina, Cr_2O_3 = chromia, SiO_2 = silica), which provide the function of a "passive layer," thus, protecting the underlying metal from increased high-temperature corrosion attack. Therefore, the diagram in Figure 13.1 also serves for an assessment on whether the process environment has to be regarded as critical with regard to high-temperature corrosion attack or whether resistance of the alloy in this environment can be expected. For the example of the process environment

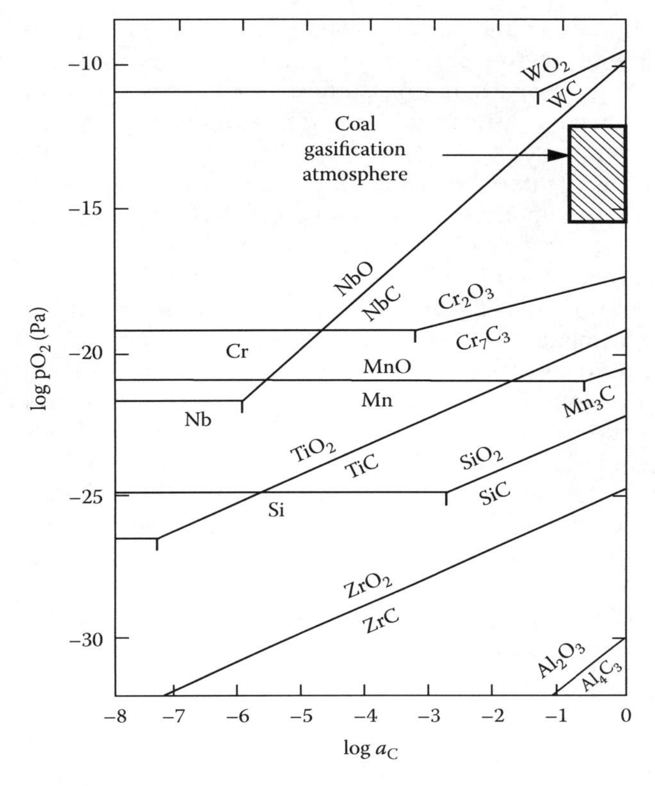

FIGURE 13.1
Stability diagram for different metals, oxides, and carbides at 950°C. The hatched area represents the composition of a coal gasification environment. (From Schendler, W., in *Corrosion Resistant Materials for Cool Conversion Systems*, eds., D.B. Meadowcroft and M.I. Manning, Applied Science, Barking, London, U.K., 1983, p. 201.)

in Figure 13.1, a protective situation can be assumed allowing the long-term use of metallic alloys by forming oxide phases like Al_2O_3, Cr_2O_3, or SiO_2 by the reaction between the corrosive environment and the respective alloy component on the metal surface. Another example of an environment containing sulfur and oxygen [14] is given in Figure 13.2 leading to similar results in that in particular aluminum-containing alloys should have a high

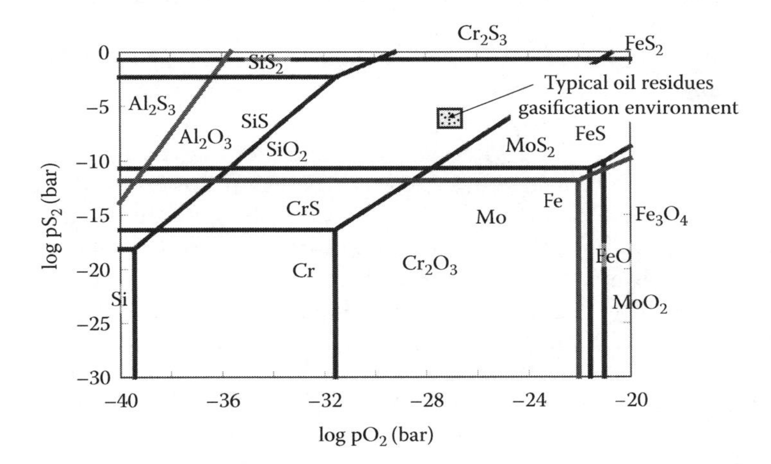

FIGURE 13.2
Stability diagrams for metal, oxides, and sulfides of Al, Cr, Fe, Mo, and Si at 700°C. The hatched area represents the composition of a refinery residues gasification process. (From Schütze, M. and Spiegel, M., *Korrosion in Anlagen zur thermischen Abfallverwertung, Proceedings of Jahrestagung 2004*, GfKORR, Frankfurt/Main, 2004.)

capability for forming a protective oxide scale acting as a diffusion barrier and, thus, providing good stability of the material in the respective process environment. Contrary to the situation in Figure 13.1, chromia-forming alloys are not resistant under the conditions of the example of an oil residues gasification environment as given in Figure 13.2, since Cr_2O_3 will not be stable and rather CrS will be formed.

It should, however, be mentioned at this point that these stability diagrams are only suitable to assess the stabilities of solid and liquid phases as corrosion products. As will be discussed later in the respective section, the formation of gaseous corrosion products requires a more sophisticated approach, since there is no fixed boundary concerning the partial pressure values in the environment between the existence of an unreacted metal and the reaction product. Rather, there is the formation of a gaseous corrosion product on top of either a metallic surface or a solid or liquid corrosion product layer on the metallic surface where this gaseous phase develops a certain vapor pressure leading to evaporation of a metal-containing gaseous corrosion product. The value of this vapor pressure is a function of the composition of the environment, and, thus, the definition of a fixed boundary between the existence of two reaction products is not possible in such a case. Therefore, a limiting vapor pressure or a limiting metal consumption rate has to be defined in this case for the characterization of the "stability" of the metal or alloy. Details on this aspect will, however, be given later.

In the following sections, the phenomena, mechanisms, and kinetics of the corrosion reactions are discussed for the three different cases of solid, liquid, and gaseous phases as corrosion products.

13.3 Solid Corrosion Products

13.3.1 Formation and Growth Kinetics in "External Corrosion"

The simplest form of a solid corrosion product on a metal surface is a continuous homogeneous surface scale consisting of one phase. Such a situation is encountered, e.g., in the oxidation of pure nickel where only nickel oxide is formed. Therefore, in many textbook examples, this type of reaction has been used to illustrate the principles of high-temperature corrosion. In the present chapter, the initial stages of the oxidation process are not addressed as this has been part of another chapter in this book.

If a dense surface scale is formed without any physical defects (e.g., cracks or spallation), the transport in this scale is determined by solid-state diffusion via chemical defects (vacancies, interstitials) in the lattice and at the grain boundaries. The growth rate of the surface scale depends on whether transport in the surrounding gas phase to the reacting surface, reaction at the scale surface or the scale/metal interface, or transport through the surface scale form the rate-limiting step. In most cases of technical applications, solid-state diffusion in the surface scale determines the scale growth rates, which is why this situation has been intensively investigated and modeled in the literature [2–10]. The corrosion products of such scales are ionic structures and diffusion in these structures requires lattice disorder, i.e., the corrosion product layer must have a nonstoichiometric composition.

The nomenclature of disorder in these corrosion products is given for the example of oxide formation in Table 13.2 for a divalent oxide [15]. As this table shows, defects may be present as interstitials or vacancies in either of the two sublattices and to maintain electronic neutrality, electronic defects have to be introduced that are either negatively charged

TABLE 13.2
Nomenclature of Lattice Disorder in a Divalent Oxide

Nomenclature	Meaning
Me_{Me}	Me^{2+} ion on a cation position
Me'_{Me}	Me^{+} ion on a cation position
$Me_i^{\cdot\cdot}$	Me^{2+} ion on an interstitial position
V''_{Me}	Me vacancy on a cation position
O_O	O^{2-} ion on an anion position
$V_O^{\cdot\cdot}$	O vacancy on an anion position
e'	e^- electron
$h^\bullet$	H^+ electron hole
•	+ (positive charge)
′	– (negative charge)

Source: Kröger, F. A., *The Chemistry of Imperfect Crystals*, North Holland, Amsterdam, the Netherlands, 1964.
The charges in the left column refer to a defect-free ionic lattice.

electrons or positively charged electron holes (defect electrons). As a result the corrosion products are electronic semiconductors. Generally two types of corrosion products can be expected. The first type shows a deficiency in the nonmetal sublattice or an excess in the metal lattice. An example for this situation is ZnO where zinc is present as an interstitial Zn_i. The other group of corrosion products shows either a metal deficit or an excess of the nonmetal species in the lattice. Staying with oxides it is NiO that can be mentioned as an example for this situation where there is a metal deficit due to nickel vacancies in the cation sublattice. The nickel vacancies are formed by taking in oxygen into the nickel oxide lattice at the outer part of the scale during the oxidation reaction (creation of vacancies in the metal sublattice of the oxide = high vacancy concentration):

$$\tfrac{1}{2}O_2 \rightarrow V''_{Ni} + 2h^\bullet + O_o. \quad (13.3)$$

At the oxide/metal interface the Ni vacancies are annihilated by taking Ni atoms from the metal substrate into Ni vacancy sites in the oxide lattice and conversion into Ni cations:

$$Ni + V''_{Ni} + 2h^\bullet = Ni_{Ni} \quad (13.4)$$

where Ni without subscript is the Ni atom in the metal lattice of the substrate.

Applying the law of mass action to Equation 13.3 yields Equation 13.5:

$$[V''_{Ni}] = const \cdot pO_2^{+1/6} \quad (13.5)$$

if $[h^\bullet]$ is replaced by $\tfrac{1}{2}[V''_{Ni}]$.

This equation indicates that the concentration of the nickel vacancies is strongly dependent on the partial pressure of the nonmetal species in the corrosive environment.

The complete oxidation reaction resulting from Equations 13.3 and 13.4 is

$$Ni + \tfrac{1}{2}O_2 + \left(V''_{Ni} + 2h^\bullet\right) = \left(Ni_{Ni} + O_O + V''_{Ni} + 2h^\bullet\right)_{NiO} \quad (13.6)$$

which again illustrates the role of the chemical defects in scale growth.

These considerations form the basis for the explanation as to why corrosion product layers on the surface can grow at high temperatures and what is the driving force for scale growth. Equation 13.5 applies to all sorts of corrosion product layers with lattice defects in its general form:

$$[\text{defects}] = \text{const} \cdot \text{pX}_n^{\pm 1/m}. \tag{13.7}$$

The exponent $1/m$ has a negative sign in the case of nonmetal vacancies and metal interstitials and a positive sign for metal vacancies and nonmetal interstitials.

Equation 13.7 also provides the explanation as to why the growth of corrosion product layers at high temperatures will never stop as long as the metallic material is exposed to the corrosive environment and not fully consumed.

Figure 13.3 shows a simplified schematic representation of the situation at a corrosion product layer. It is obvious that the partial pressure of the corrosive species leading to the growth of the reaction layer is higher in the environment than at the interface between the corrosion product layer and the metallic substrate. For the example of oxidation, the equilibrium partial pressure of oxygen at the interface between metal and the oxide layer corresponds to that resulting from the thermodynamic considerations in Section 13.2. When using the example in Figure 13.1, it becomes clear that this equilibrium oxygen partial pressure pO_2 can be extremely low (e.g., 10^{-32} bar for Al_2O_3 at a carbon activity of 0.1) whereas on the outside of the scale, again for this example of a coal conversion process in gas environment, the oxygen partial pressure would be in the range of 10^{-16} – 10^{-12} bar. For the case of oxidation in air the outside oxygen partial pressure would be even much higher ($pO_2 \approx 0.2$). As a result an extremely steep nonmetal partial pressure gradient is built up over the corrosion product layer that in turn results in a steep gradient concerning the lattice disorder $\Delta c = c_1 - c_2$ in Figure 13.3a. For the example of nickel oxide, there would be a high-concentration gradient in the number of nickel vacancies in this lattice with the result that at high temperatures significant transport of nickel cations from the substrate metal (reaction according to Equation 13.4) to the oxide surface (reaction according to Equation 13.3) takes place in order to decrease the slope of this gradient (Figure 13.3b). Once the nickel cations arrive at the oxide surface they react with the oxygen from the environment and form new NiO, i.e., the oxide scale grows. The transport of either the nickel cations or the nickel vacancies can be described by Fick's first law as shown in this schematic with the transport rate $-j$ being determined

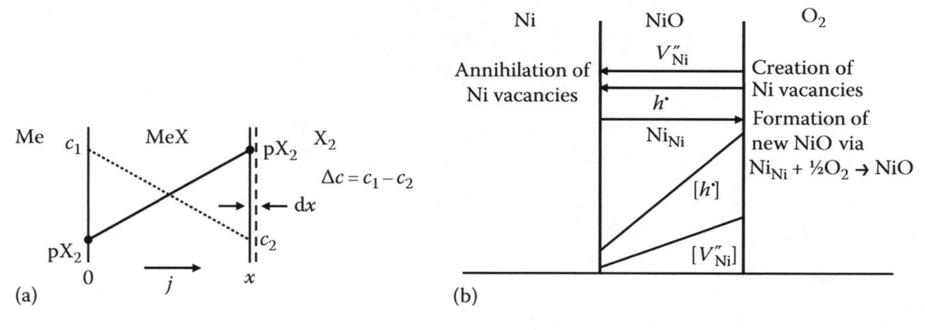

FIGURE 13.3
(a) Schematic of the partial pressure pX_2 of the oxidant and the metal concentration situation c in a nonstoichiometric corrosion product layer. (b) Schematic for the specific situation of the oxidation of nickel.

by the slope of the metal concentration gradient $\mathrm{d}c/\mathrm{d}x$ (or defect concentration gradient) with x being the thickness of the surface layer:

$$-j = \text{const}\,(1)\cdot D\cdot\frac{\mathrm{d}c}{\mathrm{d}x} \tag{13.8}$$

or in a simplified "linear" form as follows:

$$-j = \text{const}\,(1)\cdot D\cdot\frac{\Delta c}{x}, \tag{13.9}$$

where D is the diffusion coefficient of the diffusing species.

With regard to the concentration of the lattice defects (and by that of the metal in the oxide lattice), there are two fixed points, one at the interface between metal and corrosion product layer and the other one on the surface of the corrosion product. The one at the interface is defined by the constant equilibrium partial pressure at this point resulting from the thermodynamic equilibrium between corrosion product and metal. The one on the outer surface is defined by the concentration of the surrounding corrosive environment and is also constant, as long as the composition of the environment does not change. Therefore, Δc is constant in Equation 13.9 if the environment does not change (see Equations 13.5 and 13.7 and Figure 13.3a). In other words, this concentration gradient can never become zero since the values of c_1 and c_2 will always be different from each other. The gradients lead to transport of matter through the scale by diffusion that leads to growth of the surface layer (increase of x) but for this reason the growth of the surface layer can never end. Since the transport length x through the surface layer increases by layer growth the overall growth rate of this layer decreases with time. The respective kinetic equation for the surface layer growth can easily be deduced from Equation 13.9 by combining all the constants of this equation including the diffusion constant and the concentration difference between scale surface and metal scale interface Δc in the form a new constant k_p:

$$-j = \text{const}\,(2)\cdot\frac{1}{x} = k_p\cdot\frac{1}{x} \tag{13.10}$$

with const (2) = const (1) $\cdot D \cdot \Delta c = k_p$.

The growth of the oxide scale corresponds to the diffusion flux through the scale, which is $-j = \mathrm{d}x/\mathrm{d}t$ and, thus, the general equation for transport in diffusion-controlled growth of surface layers can be written in the form

$$\frac{\mathrm{d}x}{\mathrm{d}t} = k_p\cdot\frac{1}{x}. \tag{13.11}$$

This kinetics equation for diffusion-controlled surface scale growth can be converted into the integral form, which is better known and described in most textbooks:

$$x^2 = 2k_p\cdot t. \tag{13.12}$$

A detailed theoretical analysis of the electrochemical potential situation and the transport conditions in the scale [16] led to an equation of the rate constant k_p according to

$$k_p = 0.5 \int_{pO_2^i}^{pO_2^o} \left\{ \frac{z_c}{z_a} (D_M + D_O) \right\} d \ln pO_2, \tag{13.13}$$

where

D_M and D_O are the self-diffusion coefficients for random diffusion of the respective ions (metal and nonmetal)
z is the valence of the respective ion (anion and cation)
pO_2^o is the oxygen partial pressure at the outside of the surface scale
pO_2^i is the oxygen partial pressure at the interface between corrosion product and metal

The corrosion rate constant k_p is actually the most important parameter for describing the resistance of a material against high-temperature corrosion. If k_p is low, the overall high-temperature corrosion rate is low and the metal consumption occurs at a low rate. In this case the surface scale forms a good diffusion barrier and prevents access of the corrosive species from the reaction environment to the metal. If the value of k_p is high, this means that the consumption rates of the metal are high and a situation of nonprotective behavior exists.

Figure 13.4 shows a schematic of the course of high-temperature corrosion in the case of diffusion-controlled growth of a homogenous surface layer, which has parabolic

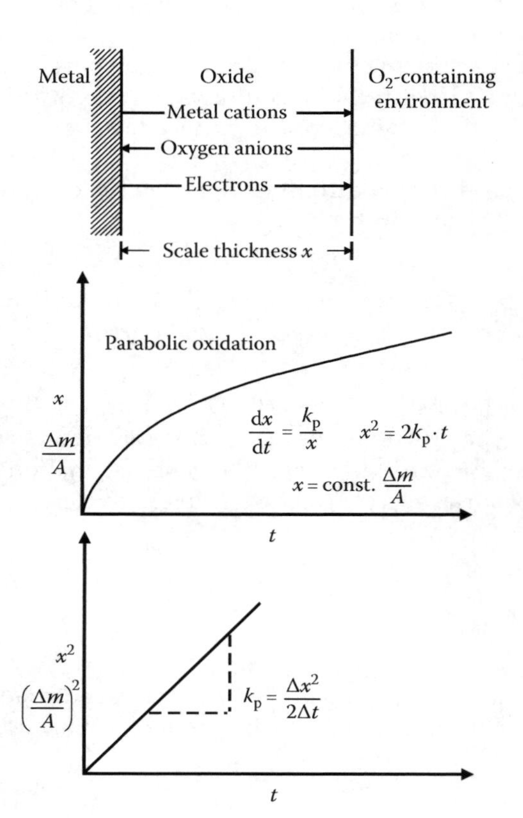

FIGURE 13.4
Schematic of the transport in the scale and the course of oxidation for diffusion-controlled (parabolic) scale growth. x and Δm are scale thickness and mass change due to corrosion, respectively. A is the surface area of the specimen in thermogravimetric measurements.

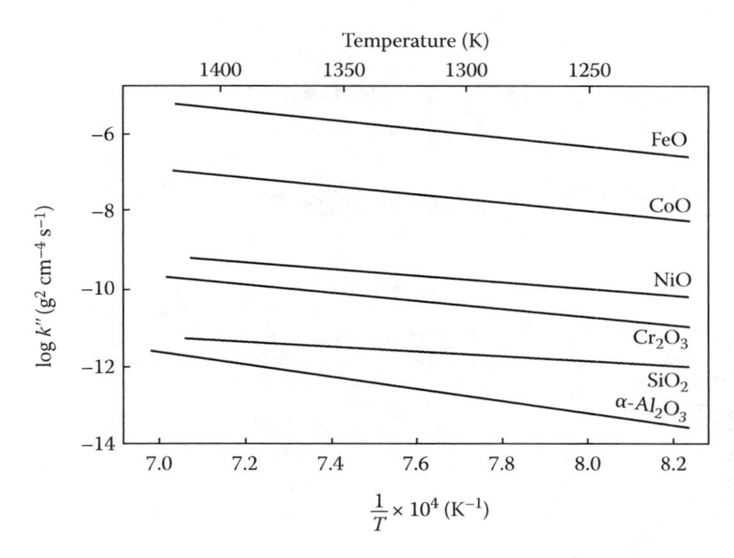

FIGURE 13.5
Order of magnitude of the parabolic rate constant k_p for several oxides as a function of temperature. (From Birks, N. and Meier, G.H., *Introduction to High Temperature Oxidation of Metals*, Edward Arnold, London, U.K., 1983.)

behavior as shown by Equation 13.12. The parabolic rate constant can be determined by measuring the weight change occurring during the corrosion reaction in a thermogravimetric analysis unit, which is described in Ref. [17]. The results are plotted in a model-oriented presentation as shown in the lower part of Figure 13.4, where the ordinate is the square of the measured mass change $\Delta m/A$ or scale thickness x, while on the abscissa the time is plotted in a linear way. The slope in such a plot corresponds directly to the rate constant k_p.

As can be concluded from the discussion so far it is indeed the amount of lattice disorder in the corrosion product layer that determines the growth rate in the form of the diffusion constant D. A lattice with a very high disorder (e.g., FeO) leads to extremely high scale growth rates as can be seen in Figure 13.5 whereas an oxide with an extremely low defect concentration such as α-Al_2O_3 has a very low scale growth rate (see again Figure 13.5). It should be mentioned at this point that the growth of an oxide scale can be dominated by either transport of a metallic species from the metal to the surface of the corrosion product layer and further reaction with the environment there or by transport of nonmetallic species from the environment through the corrosion product layer to the metal/scale interface. In the former case the corrosion product layer would grow on the outside whereas in the latter case the growth would occur at the scale/metal interface. This situation is schematically illustrated in Figure 13.6. In many technical situations there is also mixed transport through the surface scale, which is shown in Figure 13.6c and which can end up in the formation of new corrosion product not only at either side of the corrosion product layer but also inside of it.

It is often the case that one reacting species diffuses through the bulk lattice whereas the other reacting species moves in the counter direction along the grain boundaries of the scale. In this case formation of new corrosion product takes place at the scale grain boundaries leading to a new solid volume inside the scale and, thus, to compressive stresses that may end up in physical defects in the scale (cracks, microcracks, convolutions, buckles, etc.), thus, impairing the protective capability [7]. In Figure 13.7, doping

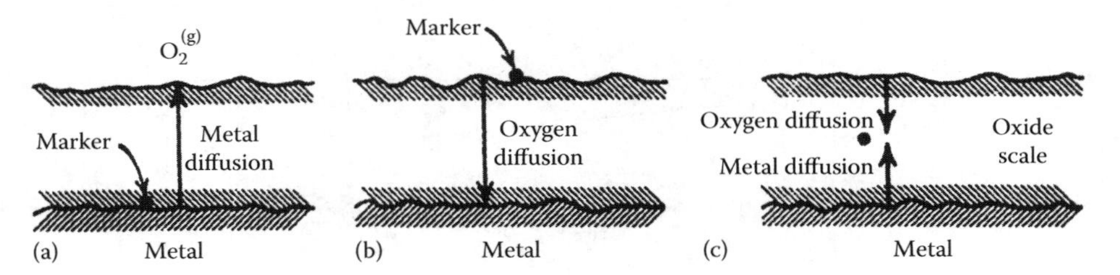

FIGURE 13.6
Position of inert markers (e.g., Pt or Al_2O_3 particles) after oxidation when the scale growth is diffusion controlled: (a) growth by outward metal cation diffusion, (b) growth by inward oxygen anion diffusion, and (c) growth by counterdiffusion of both species. (From Kofstad, P., *High Temperature Corrosion*, Elsevier Applied Science, London, U.K., 1988.)

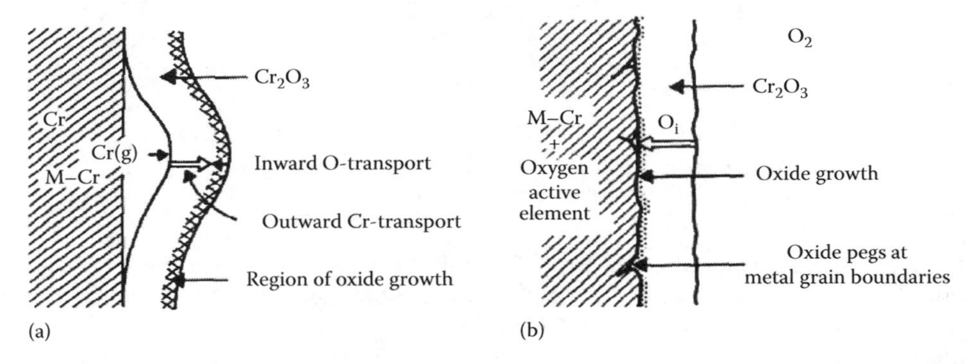

FIGURE 13.7
Schematic showing the effect of oxide scale growth by counterdiffusion leading to compressive intrinsic oxide growth stresses (a). If diffusion in one direction is suppressed by doping (e.g., REE) the scale grows only on one side and no intrinsic growth stresses are built up (b). (From Kofstad, P., *High Temperature Corrosion*, Elsevier Applied Science, London, U.K., 1988.)

effects are shown that can influence the diffusion situation in the surface scale. Transport in one direction is being suppressed by blocking the respective side exchange mechanism so that in the end only transport of one species and in one direction occurs with the result that again the formation of a new corrosion product is at one of the faces of the scale only. Therefore, the formation of corrosion products inside the scale and, thus, the buildup of compressive intrinsic growth stresses is suppressed. This doping effect is used in high-temperature technology to a large extent in that so-called active elements are added to the alloys, e.g., yttrium or cerium that suppress the outward metal cation diffusion along grain boundaries and, thus, change the growth mechanism to the formation of a new oxide at the oxide/scale interface solely by inward transport of oxygen (Reactive Element Effect, REE) [6].

13.3.2 Role of Alloy Composition

Thus far, the relatively simple situation of the formation of a homogenous continuous oxide scale formed on a pure metal has been regarded, which, however, is not representative of the situation encountered in complex technical alloys. For such alloys it is mainly their composition and by this the amount of the surface scale-forming elements that decides the formation and growth rates of the different corrosion products and thus which surface layer is formed. This is particularly important if a protective continuous surface oxide layer is to be formed providing sufficient corrosion resistance of the metallic substrate under operating conditions. As has already been discussed under the aspect of the thermodynamic considerations, a certain amount of the protective scale-forming alloying element is required together with a minimum nonmetal partial pressure in the corrosive environment, which is above the equilibrium partial pressure. For the case of oxidation this situation has been modeled in the theory by Wagner [18] and further developed by Whittle [19]. As a result of their considerations a minimum concentration of protective scale-forming alloying element can be calculated according to

$$N_c = \left(\frac{\pi k_i}{D_{int}}\right)^{1/2} \tag{13.14}$$

with

$$k_i = \left(\frac{2V_A}{\mu M_O}\right)^2 \cdot k_p, \tag{13.15}$$

where
- V_A is the molar volume of the alloy
- μ is the ratio of oxygen anions to metal cations in the oxide
- M_O is the molecular weight of oxygen
- D_{int} is the interdiffusion coefficient of the scale-forming element in the alloy

The ratio k_i/D_{int} is a characteristic value for an alloy indicating its ability to form a protective surface scale.

With k_i being the metal recession rate constant due to the corrosion process, the critical concentration of an alloying element for different situations of protective behavior can be calculated [19], as illustrated in the graph in Figure 13.8, as a function of the k_i/D_{int} ratio.

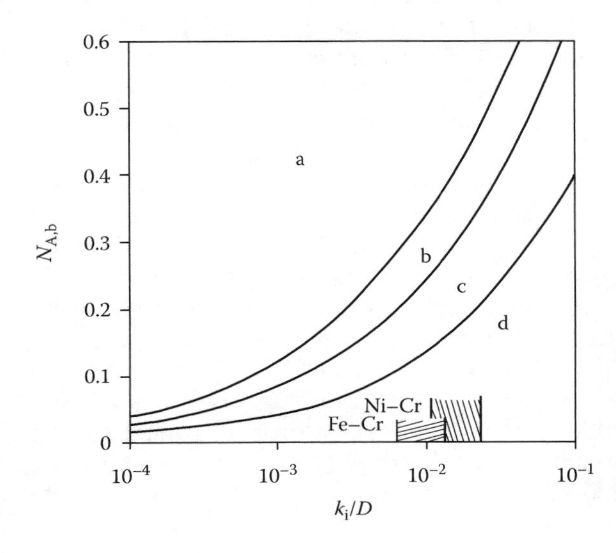

FIGURE 13.8
Critical values for the atom fraction N_c of the protective scale-forming element (e.g., Cr in the present case) as a function of the k_i/D_{int} ratio including columns for the ratios for Fe–Cr and Ni–Cr alloys. For explanation see text. (After Whittle, D. P., *Oxid. Met.*, 4, 171, 1972.)

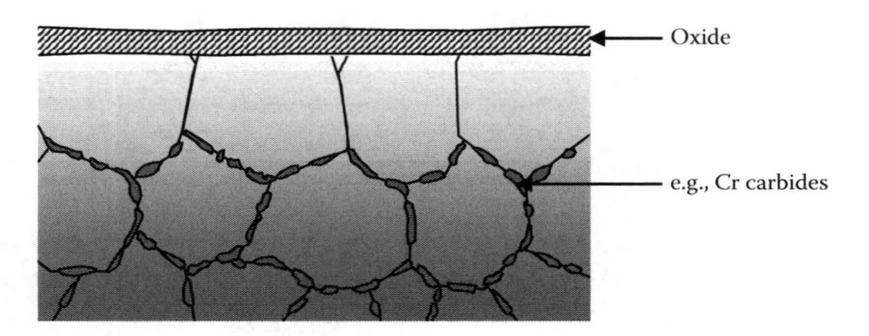

FIGURE 13.9
Schematic of alloy depletion by the corrosion process (in this example by oxidation, i.e., chromium-rich surface oxide scale formation). The dark areas are those of high (original) Cr content of the alloy. By Cr consumption due to scale growth also chromium carbides in the metal subsurface zone become dissolved (cross section of the surface area).

In field d of this figure, no protective oxide scale can be formed since the amount of the scale-forming element is insufficient and instead internal oxidation in the base material may occur, which will be discussed later. Field c corresponds to the situation where a protective scale can be formed but not reformed after spalling as the content of the scale-forming element in the alloy is too low. Field b stands for the situation where the scale can be reformed at least once after spalling whereas field a is characteristic for the case that even after multiple spalling the concentration of the scale-forming element is sufficiently high for reforming a protective scale.

A feature that is particular for the high-temperature corrosion of alloys is that the depletion of alloying elements can occur in the subsurface area of the metal underneath the corrosion product layer due to consumption of the respective alloying element for scale formation and growth. This situation is illustrated by the schematic in Figure 13.9. Subsurface zone depletion is of special importance for the formation of protective oxide scales acting as a diffusion barrier and, thus, of a corrosion protection layer since depletion of the respective alloying element may impair the protective effect in that either the amount of the alloying element falls below the chemical stability limit of the protective surface oxide scale or at least drops below the limit, which is needed for new formation or reformation of a protective scale. This situation is schematically illustrated in the form of the Cr concentration in the metal subsurface zone directly underneath the oxide scale in Figure 13.10. For this example of the formation of a protective chromium oxide scale, the different possible subsurface zone depletion kinetics for chromium in an alloy are given. In all cases a drop compared to the original chromium level will occur during the oxidation process as a function of time. If the kinetics follow characteristics given by case 1, the situation can be regarded as uncritical since the amount of chromium in the alloy subsurface area never falls below the limit needed for the formation of a protective chromia scale. The situation is more critical in case 2 where at least temporarily this amount drops below this chromium limit, however, recovers by rediffusion of chromium from the metal interior to the surface after a certain period of time. A much more critical situation arises in case 3 where only at the very beginning the chromium level stays above the critical limit but can never be recovered in the later stages of oxidation. The most critical case is encountered in case 4 where the chromium level drops continuously and in the end even falls below the thermodynamic instability limit for the protective chromia scale, so that after falling below this value the originally formed protective surface oxide

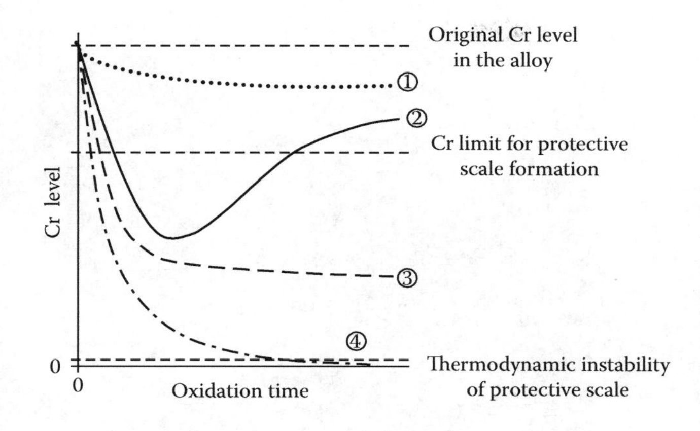

FIGURE 13.10
Schematic of the different possible types of Cr depletion kinetics in an alloy and consequences on the protective effect of the scale. (1) Uncritical. (2) Potential chemical–mechanical failure with repassivation. (3) Potential chemical–mechanical failure without repassivation. (4) True chemical failure.

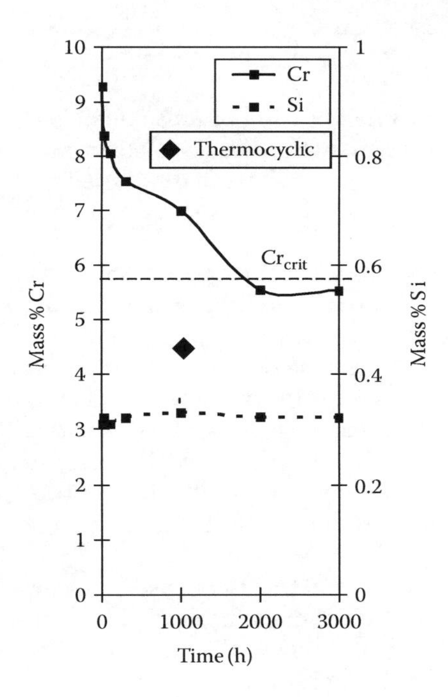

FIGURE 13.11
Si and Cr concentration in the metal subsurface zone directly underneath the oxide scale of a 9% Cr steel of the type P91 after oxidation at 650°C as a function of oxidation time. (From Vossen, J. P. T. et al., *Mater. High Temp.*, 14, 387, 1997.)

scale disintegrates in a chemical way. In all other cases the surface oxide scale is at least chemically stable, but in the case of a crack or another type of physical surface defect when the protective oxide scale is open for direct access of the corrosive environment to the metal, rehealing of these defects is not possible as long as the chromium concentration lies below the limit for protective scale formation.

As the example in Figure 13.11 shows, such depletion kinetics can be encountered under practical conditions for technical alloys so that considerations according to Figure 13.8 and Figure 13.10 are of significant importance for the technical use of materials under high-temperature corrosion conditions. Figure 13.12 nicely illustrates that at physical defects such as cracks in the protective surface oxide scales, fast growing nonprotective corrosion products can be formed if the original level of the protective scale-forming element (in this

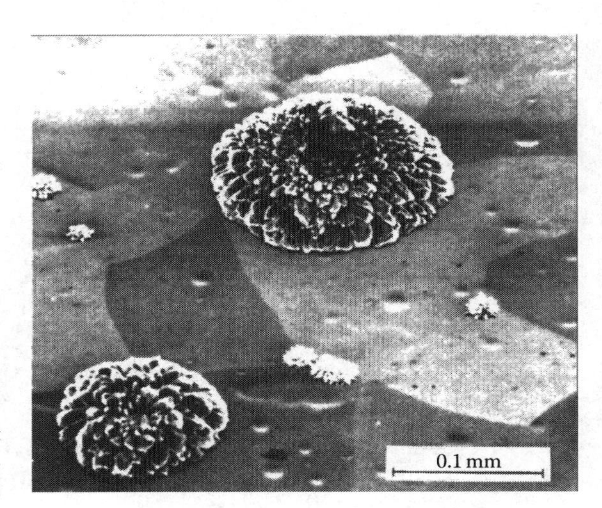

FIGURE 13.12
"Oxide roses" on an Fe–5Al alloy after oxidation at 800°C for 22h. (From Boggs, E., *J. Electrochem. Soc.*, 118, 908, 1971.)

case aluminum) falls below the critical value so that protective healing of these defects is no longer possible. In areas that are still covered by an extremely thin (optically transparent) alumina scale with high protective effect, the underlying metal grain boundaries are visible indicating the high efficiency of the protective effect keeping the oxidation rate at an extremely low level.

The question may arise as to why in case 2 in Figure 13.10 the chromium level can rise again after a certain amount of time. This question is answered by the fact that the oxidation rates decrease with continuing oxidation time (parabolic or similar behavior) so that the consumption rate of the scale-forming element also decreases and rediffusion from the metal interior can increasingly fill up the chromium deficit and thus lead again to a higher chromium level in the metal surface area. Usually, the situations 1–4 in Figure 13.10 follow an order of decreasing amount of protective scale-forming element in a material at the beginning of exposure to the environment. These situations can also be related to Figure 13.8 with situations a–d being similar to cases 1–4 in Figure 13.10.

Figure 13.13 schematically summarizes the consequences for the four cases if scale damage has occurred, based on Figure 13.10. In case 1 protective crack healing can occur (Figure 13.13a). If in case 2 cracking takes places in the period where the Cr subsurface content is below the limit for protective scale formation then first a fast growing oxide nodule is formed at the defect, which may, however, later be "sealed" by a repassivation layer when the Cr content has increased again by rediffusion (Figure 13.13c). Final loss of the protective effect (Figure 13.13b) takes place in cases 3 and 4.

13.3.3 Internal Corrosion

In particular, for alloys consisting of several alloying elements, internal corrosion is also possible, which means that instead of forming a continuous surface layer there are discrete corrosion product particles surrounded by metal in the metal subsurface zone, which result from the corrosion process. In particular, this situation is encountered if a protective surface scale shows cracks or other types of channels for direct gaseous transport through the scale to the metal subsurface zone or if no surface scale exists at all acting as a barrier to keep the reactive species away from the metal. Internal corrosion

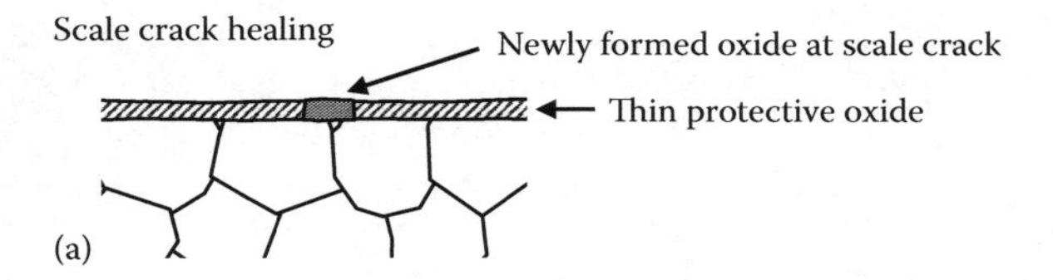

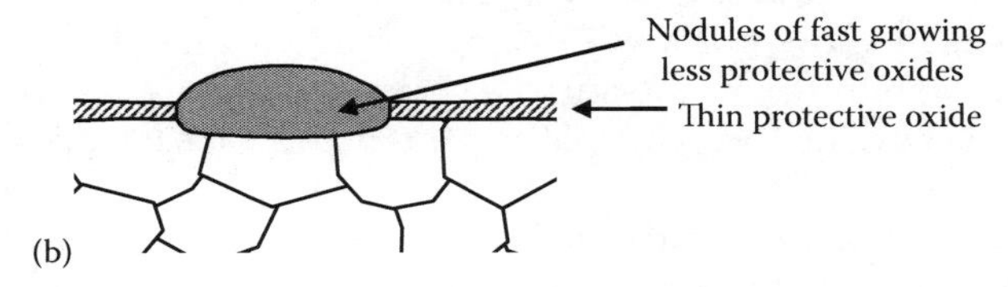

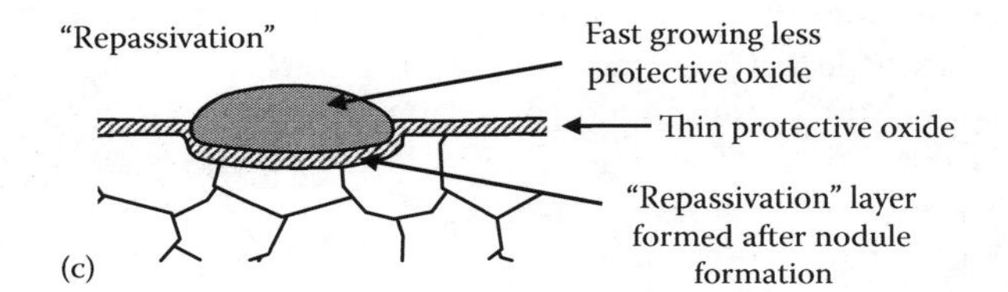

FIGURE 13.13
Schematic of the consequences of scale failure for the protective effect of the oxide scale (for explanation see text).

in an alloy with alloy basis A and an additional alloying element B, i.e., formation of the internal corrosion product BX, requires five prerequisites:

1. The base metal A and the alloy consisting of the elements A and B must be able to dissolve the oxidant X.
2. The free energies of formation must be lower for the corrosion product BX than for the corrosion product AX, i.e., $\Delta G_{BX} < \Delta G_{AX}$, which means that B must have a higher affinity to X than A.
3. The corrosive species X must diffuse faster into the alloy than the alloy component B from the inside of the alloy to the surface.
4. The alloy compound B must not exceed a minimum concentration, which was given by Equation 13.14, otherwise a continuous surface scale of BX would form.
5. No dense surface layer preventing the dissolution of X must exist.

The growth kinetics of the internal corrosion zone depend first of all on the partial pressure pX_2 of the corrosive species present at the metal surface and leading to internal corrosion. This dependence is described by Sievert's law [22]:

$$N_O^S \propto pX_2^{1/2}, \tag{13.16}$$

where N_O^S is the solubility of the oxidant in the metal. As for the growth of a surface scale, the kinetics of internal corrosion also depend on diffusion-controlled processes and can therefore again be described by a parabolic law:

$$x_i = \left(\frac{2N_O^S \cdot D_O^A}{y \cdot N_B^O} \cdot t \right)^{1/2}, \tag{13.17}$$

where

x_i is the penetration depth of the internal corrosion zone
N_B^O is the initial solute concentration
y is a stoichiometric factor
D_O^A is the diffusivity of the oxidant in the metal A
$N_O^S D_O^A$ is the solute permeability in the metal A

From a thermodynamic point of view even under a dense oxide scale BX, an internal oxide can be formed as discrete particles in alloys of the type A–B–C in the form of CX if the latter has a higher thermodynamic stability than the surface scale oxide BX. A schematic of a (metallographic) cross section showing the principle appearance of internal corrosion is given in Figure 13.14.

As already mentioned, internal corrosion is often encountered under protective oxide scales if the latter show physical defects. In this case the corrosive species can penetrate the oxide scale via the gas phase and react with the underlying metal-forming internal corrosion products. This has been encountered, e.g., in the case of mixed oxidizing/sulfidizing environments, where an important issue is also the composition of the gas phase of the environment. If oxygen is involved in the equilibrium between the different components of the surrounding corrosive environment, the activity of the other species may increase toward the metal/oxide interface as shown in Figure 13.15b. This can be explained by the fact that along the narrow transport channels, e.g., cracks, through the oxide scale the oxygen partial pressure decreases according to the thermodynamic situation as discussed earlier. Therefore, the equilibrium of the gas phase in these cracks is shifted toward the sulfur-containing compounds, e.g., the sulfur partial pressure in these cracks increases in the same direction as the oxygen partial pressure decreases. As a result the sulfur activity at the bottom of the cracks in the area of the oxide/metal interface is significantly increased compared to the original outer corrosive environment leading to severe internal sulfidation as shown schematically in Figure 13.15b. If oxygen is not involved in the equilibrium of the gas phase as in Figure 13.15a, no change in the sulfur activity occurs through the oxide scale inside the cracks. As a result the amount of internal sulfidation is significantly smaller than that in the case shown in Figure 13.15b.

Similar observations have been made in a large number of technical environments where oxygen and other corrosive compounds are present at the same time, e.g., oxidizing/carburizing environments. This situation also explains a phenomenon that is called

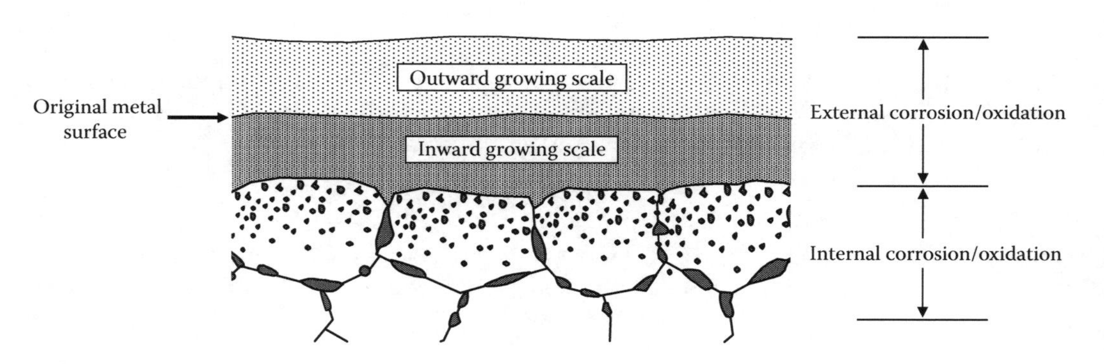

FIGURE 13.14
Schematic of a cross section showing the structures of internal corrosion underneath an outer oxide scale, at metal grain boundaries, and inside the metal grains.

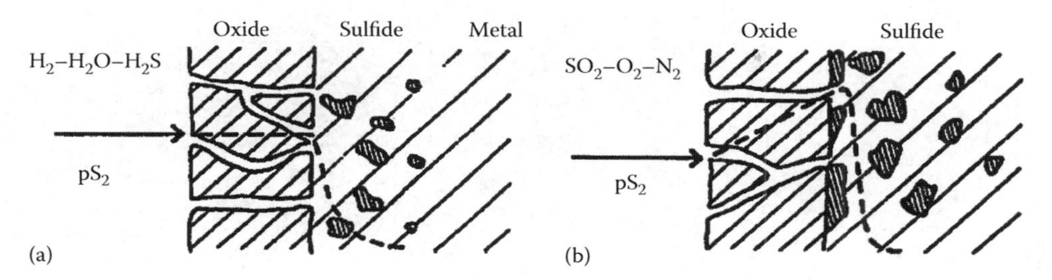

FIGURE 13.15
The role of the reaction equilibrium in the corrosive environment for internal corrosion (in this example internal sulfidation) under a defective (originally protective) oxide scale. (a) Oxygen not involved in the reaction equilibrium, no pS_2 gradient. (b) Oxygen involved in the reaction equilibrium, pS_2 increases through the oxide scale in inward direction. (From Grabke, H. J. et al., in *High Temperature Alloys for Gas Turbines and Other Applications*, eds., E. Betz et al., D. Reidel, Dordrecht, the Netherlands, 1986, p. 245.)

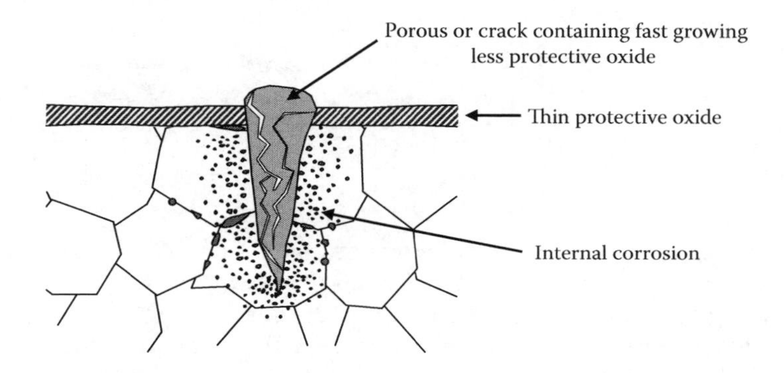

FIGURE 13.16
Schematic of the appearance of internal corrosion at a surface crack or crevice in the substrate material ("high-temperature crevice corrosion").

"high-temperature crevice corrosion" and which has been observed not only in surface cracks of materials, but also in areas of technical equipment where narrow crevices were present from design. Again due to a change in the composition of the corrosive environment along the crevice to the root of the crevice the situation may become more critical than what is to be expected from the composition of the original outside environment [24]. The result in such cases is usually a strong internal corrosion localized at such critical locations, which in the case of surface cracks in the metal under mechanical load cannot only accelerate corrosion at these spots but at the same time also lead to significant weakening of the component leading to premature failure [25]. The appearance of internal corrosion at a crevice or crack at the surface is schematically summarized in Figure 13.16.

13.3.4 Metal Dusting as a Special Form of Solid Corrosion Product

Metal dusting is a special form of catastrophic corrosion that takes place in carbon-containing environments where the carbon activity exceeds a value of 1 [26,27]. For Fe-based materials the principles are shown schematically in Figure 13.17. The catalytic effect of the metal or cementite surface leads to a dissociation of the carbon-containing gaseous species (CO_x, C_xH_y) with the effect of carbon uptake into the metal subsurface zone and carbon deposition (amorphous coke) on the surface. This mechanism is evidently influenced by the crystallographic situation of the metal grains on the material surface [28]. For carbon monoxide (CO) adsorbed on metal

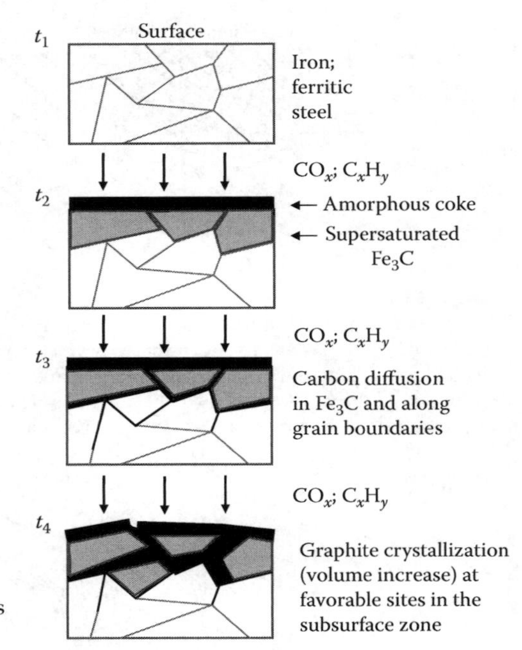

FIGURE 13.17
Initial stages of the metal dusting process of ferritic steels (for explanation see text).

surfaces, there exists a general qualitative trend between the stretch frequency, the adsorption energy, and the tendency for dissociation. The electron exchange from the d-orbitals of the iron to the antibonding π-orbitals of, e.g., CO increases the bond strength of carbon to the metal, lowers the C–O stretch frequency, and decreases the activation barrier for dissociation of the CO molecule [29]. For iron-based materials the formation of cementite (Fe_3C) takes place. Due to the high carbon activity in the gas phase the cementite becomes supersaturated. At sites that are epitaxially favorable, the supersaturated carbon may leave the cementite lattice and may precipitate as crystalline graphite [30,31]. The transition from adsorbed C on the metal surface to C in crystalline graphene (graphite) is energetically favored with a ΔG value of −72.4 kJ/mol for nickel surfaces (no data were found for iron and cementite surfaces, but the same tendencies exist there) [32]. In order to grow fully crystalline graphite without a nickel or cementite surface as a catalyst an annealing temperature of 2000°C would be required [33]. Apart from areas on the material surface preferred sites for such precipitation or crystallization of carbon are, for instance, grain boundaries with an optimum epitaxial relationship between metal or carbide lattice on the one side and graphite lattice on the other side. Furthermore, physical defects inside the cementite with surfaces of preferred crystallographic orientation may also serve as nucleation sites for the crystallization of carbon in the form of graphite. Formation of graphite leads to a significant volume increase in these areas and creates high local internal stresses in the material. For every graphite layer growing on a grain boundary additional space of 0.67 nm thickness is needed [34]. Accordingly, the formation of several graphite layers, n, on top of each other leads to a thickness increase at the metal or cementite grain boundary of $n \times 0.67$ nm. The resulting mechanical stresses lead to rupture with the formation of (further) physical defects, which means that a self-accelerating mechanism starts, which in the end leads to the formation of a powder consisting of graphite and cementite particles even down to nano size. In other words, the metal disintegrates into a dust of cementite particles and graphite that exhibits no mechanical strength at all and in many cases leads to rapid perforation of the original metallic wall. The result of this mechanism is summarized in the schematic in Figure 13.18.

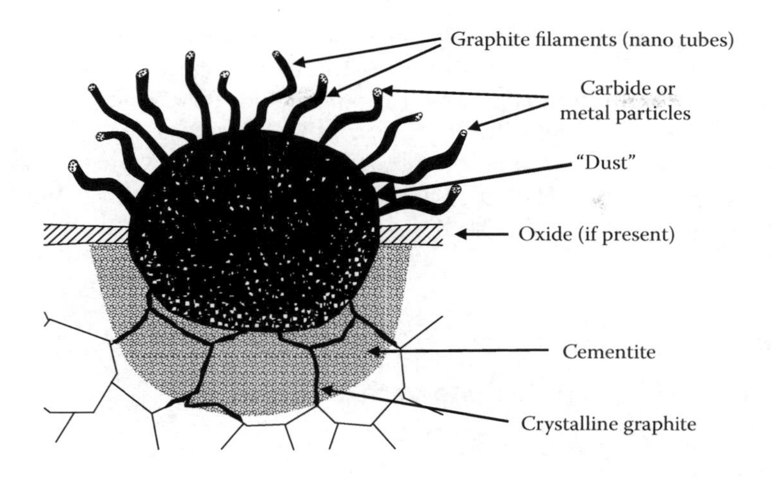

FIGURE 13.18
Schematic of the morphology of metal-dusting attack on high-alloy steels with pitting at unprotected (oxide free) sites. The pits are filled with a mixture of "metal dust" (white) and carbon powder (black). On the surface of the attacked site graphite filaments (carbon "nanotubes") grow catalyzed by the presence of carbide or metal particles at their tips.

A similar mechanism without the involvement of cementite formation may occur in the case of nickel-based alloys, which actually do not dissolve any significant amount of carbon in the metal lattice [34]. In this case it is the supersaturated grain boundaries of the metallic material where the carbon starts to crystallize in the form of graphite at preferred crystallographic orientations and physical defects, thus, rupturing the bonds at the grain boundaries due to the large volume increase from graphite formation. As shown in the literature metal dusting in these alloys in most cases indeed starts at grain boundaries [35]. Due to the formation of additional physical defects in the nickel-based metal lattice, disintegration in the end again leads to nano-size metallic particles and graphite powder. Such a mechanism also operates in the case of high-alloy austenitic steels.

The cementite particles formed by metal dusting in the case of iron-based materials and the metallic particles in the case of nickel-based materials can serve as a catalyst for crystallization of carbon in the form of graphite filaments. This is illustrated in Figure 13.18 where carbon nanotubes (graphite filaments) can be formed at the end of such carbide or metallic particles at preferred crystallographic orientations of the surface of these particles. At the "upper" side of the particle carbon uptake occurs whereas at the "lower" side of a preferred crystallographic orientation the carbon leaves the particle and converts into a crystalline graphite filament, thus, leading to a growth mechanism, which ends up in the formation of long whiskers or nanotubes of graphite [36]. Such structures have often been observed in the literature [37,38].

In all cases the actual driving force for the mechanism of metal dusting and for nanotube growth is the energy difference between the amorphous carbon and the crystalline carbon in the form of graphite. Recent investigations have shown that by blocking the catalytic activity of metal surfaces (catalytic poisoning) via doping with suitable elements, the mechanisms of carbon deposition on the surface and metal dusting can be completely suppressed [39,40]. This is confirmation that for the initiation of these mechanisms the catalytic activity of the metal surface is a major prerequisite. If the surfaces are covered by inert oxide scales the mechanism of metal dusting can also be suppressed as long as these oxide scales are dense and intact [41]. As soon as local physical defects are introduced into the protective oxide scale on the surface by oxide growth stresses or thermocycling conditions, metal dusting will start at these spots where the metal again comes into direct contact with the carbon-rich environment. Therefore, in particular for high-alloy metallic materials metal-dusting attack takes the form of localized pitting, e.g., at spots where local defects in the surface oxide scale were present [42]. Low alloy or unalloyed steels on the contrary usually show a continuous type of metal consumption due to metal-dusting attack since no protective oxide scale prevents the uptake of carbon into the material [43].

13.4 Liquid Phases in High-Temperature Corrosion

13.4.1 General Aspects

In high-temperature corrosion a number of situations can exist where the formation of liquid phases plays a decisive role for the corrosion process, see the schematic in Figure 13.19. These situations can comprise mechanisms where the corrosion product from the reaction between the environment and the metal is a liquid phase (Figure 13.19a), where a solid or a liquid deposit on the metal surface reacts with the metal directly or an oxide scale that had been previously formed (Figure 13.19b) or where from solid corrosion products eutectics can be formed with the metal (Figure 13.19c). Even among the oxides there may be liquid phases with examples given in Table 13.3 together with the respective melting points. Of course layers formed from such oxides do not offer any protective effect due to the high diffusion rates in liquids. In particular, the oxides formed from boron, vanadium, molybdenum, and lead can lead to severe problems since their melting points are far below the maximum operation temperatures for many metallic alloys.

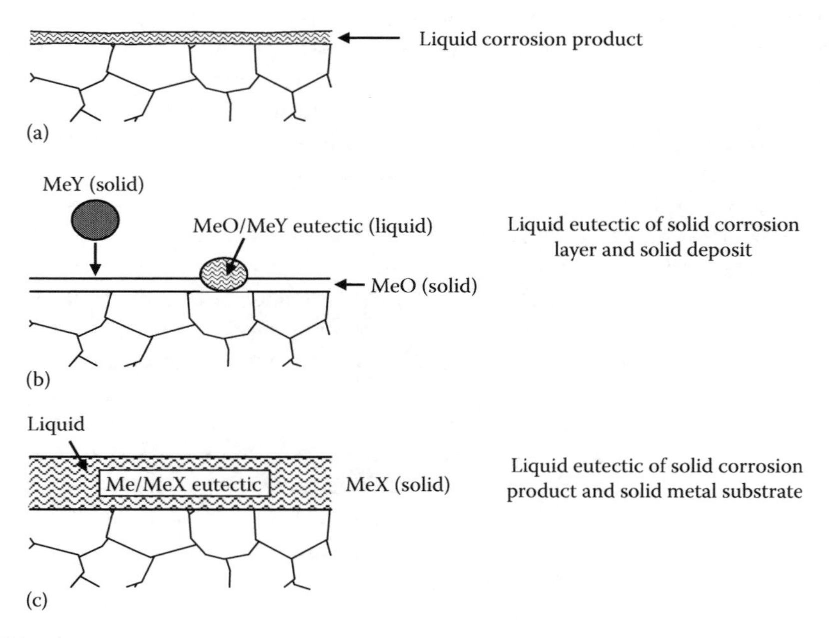

FIGURE 13.19
Schematic of the different situations in high-temperature corrosion where liquid phases play a key role.

TABLE 13.3

Melting Points of Some Technically Relevant Oxides

Oxide	Melting Point (°C)
B_2O_3	580
V_2O_5	674
MoO_3	795
Bi_2O_3	820
PbO	888
FeO	1370
Nb_2O_5	1490
WO_3	1490

Compiled from different sources.

Vanadate-induced corrosion most often results from the combustion of organic vanadium-containing compounds, e.g., in crude oil when burning the fuel with excess oxygen, leading to the formation of V_2O_5 and other vanadates. All the vanadate phases have a melting point below 650°C, in some cases even reaching temperatures of 550°C [22]. Such low melting temperatures are usually the result of the formation of eutectic phases between the oxide scale, the underlying metal, and the vanadium-containing compounds. As an example for the system Cu–Cu_2O–V_2O_5 a eutectic is observed that has a melting point of 530°C [44]. Even for heat-resistant steels vanadium-induced corrosion can be found at temperatures from 600°C upward [22]. Also in this case a eutectic is formed between the steels, the oxide, and the V_2O_5 deposits. Another eutectic that has an even lower melting point of 535°C is formed from Na_2O–V_2O_5 and in the case of the presence of Na_2SO_4 and sodium vanadates melting points as low as 400°C have been observed for slags on valves of diesel engines [45]. It is, however, not only deposits containing V_2O_5 that lead to severe corrosion by the destruction of the protective oxide scale but also high amounts of vanadium as an alloying element in the material itself, which can lead to rapid destruction of the material if the melting temperature of V_2O_5 is exceeded.

In a similar manner high-molybdenum-containing steels are prone to liquid phase corrosion, and it is reported in the literature [22] that liquid PbO seems to have played a role with regard to liquid-phase corrosion at high temperatures at times when lead compounds were used as additives in gasoline. In particular B_2O_3 but also other low-melting oxides or oxide eutectics have been reported to play a role in ceramics corrosion mainly at grain boundaries when they are added as a sintering aid [46]. As a result boron oxides often form as liquid layers together with other oxide formers of the material, which in particular for silicon-based ceramics is silicon.

However, it is not only oxides that may form liquid phases at high temperatures. The most well-known example of a corrosion product that is liquid at elevated temperatures and can lead to a catastrophic type of corrosion is sulfur in combination with nickel leading to the formation of a nickel/nickel sulfide eutectic that melts at around 635°C, see Table 13.4 [17]. For this reason it is often recommended not to use high-nickel-containing materials in high-sulfur environments above 600°C. On the other hand it should, however, be mentioned that if the oxygen partial pressure in the respective process environment is sufficiently high to form a protective oxide scale then even nickel-based alloys may have significant resistance against sulfidation attack. This, however, requires that the corrosive environment provides a sufficiently high oxygen partial pressure, which in many cases

TABLE 13.4

Melting Points of Some Eutectics Relevant to the High-Temperature Corrosion of Technical Materials

Eutectic	Melting Point (°C)
$Ni–Ni_3S_2$	635
$Co–Co_4S_3$	880
Fe–FeO–FeS	925
FeO–FeS	940
Fe–FeS	985

Compiled from different sources.

cannot be guaranteed, in particular in environments with high concentrations of H_2S. Other examples of low-melting eutectics formed from corrosion products and the underlying metal are given in Table 13.4. Most of these examples are due to the presence of sulfur that may cause significant problems for many of the technical materials.

13.4.2 Sulfate Corrosion (Hot Corrosion)

High-temperature corrosive attack by sulfate phases plays a significant role in technical conditions. This type of corrosion where the formation of liquid corrosion phases is always involved is commonly called "hot corrosion." (Note: High-temperature corrosion is the general term while "hot corrosion" is only used in the case of the presence of liquid phases.) Sulfate-induced corrosion requires the presence of either S_x or SO_2 or SO_3 [6]. As long as a dense and protective Cr_2O_3 or Al_2O_3 surface scale can be formed these materials are usually resistant against attack from these phases. However, in particular, in the presence of Na and a sufficiently high SO_2/SO_3 partial pressure, the formation of Na_2SO_4 can occur and also of mixtures of sulfates leading to liquid phases in the temperature range of 500°C–1000°C. In such cases a chemical dissolution of the Cr_2O_3 or Al_2O_3 surface scales can occur and often an incubation period is observed before a catastrophic type of corrosion commences. The reason for this incubation phase is that at the beginning the oxide scales still offer a protective effect; however, due to the chemical dissolution at a later stage this protection is no longer possible.

This type of chemical destruction of the protective oxide scale is often called fluxing [6]. This fluxing mechanism can have an acidic or a basic character depending on whether the reaction product is an acid or a base. An acid in this sense would be a compound that can take up electrons (electron acceptor) whereas a basic compound would be such that donates electrons (electron donor). In the case of acidic fluxing of the oxide scale, a positively charged metal cation would go into solution, which according to the definition as an acid would be, e.g., Ni^{2+} or Cr^{3+}. Basic fluxing would yield a negatively charged oxygen anion or other anions, e.g., NiO_2^- or CrO_4^{2-}.

Generally, two types of hot corrosion are distinguished in the literature talking about the corrosion situation induced by Na and S [6]:

- Low-temperature hot corrosion in the temperature range up to 800°C, which is also called type II hot corrosion
- High-temperature hot corrosion in the temperature range between 800°C and 1000°C, which is also called type I hot corrosion

TABLE 13.5

Main Substances Involved in the Corrosion Reaction of Combustion Processes

Intake	O_2, N_2, SO_2, NO, NO_2, CO, CO_2, H_2O, hydrocarbons, NaCl with parts of Na_2SO_4, and other chlorides and sulfates, other compounds specific for the operation site
Fuel	S, Na, Cl, V, Pb,..., in the form of compounds
Combustion environment	O_2, N_2, H_2O, S_2, SO_2, SO_3, CO, CO_2, NO_x, Na_2SO_4, NaCl, Na_2O, HCl, V_2O_4/V_2O_5, PbO

Source: Bürgel, R., *Handbuch Hochtemperatur-Werkstoftechnik*, Vieweg, Braunschweig, Germany, 2001.

TABLE 13.6

Melting Points of Salts and Eutectics That May Be Present in Combustion Processes

Salt	**NaCl**	**Na_2SO_4**	**K_2SO_4**	**$MgSO_4$**		**$CaSO_4$**	**$Na_3Fe(SO_4)_3$**	**$K_3Fe(SO_4)_3$**
T_m in °C	800	884	1069	1127		1397	620	620
				Na_2SO_4/				**$Na_3Fe(SO_4)_3$/**
Eutectic	**V_2O_5**	**$CoSO_4$**	**$MgSO_4$**	**$NiSO_4$**	**NaCl**	**K_2SO_4**	**$CaSO_4$**	**$K_3Fe(SO_4)_3$**
T_m in °C	525	565	668	671	790	830	913	555

Source: Bürgel, R., *Handbuch Hochtemperatur-Werkstoftechnik*, Vieweg, Braunschweig, Germany, 2001.

In the following these two types of hot corrosion shall be discussed in further detail. Tables 13.5 and 13.6 show that apart from Na_2SO_4, compounds formed from fuel and air can also lead to low-melting salt eutectics with sulfates on the one side and with elements from the alloy on the other side. This particularly concerns the elements nickel and cobalt that are present in high concentrations or even as base elements in most high-alloy high-temperature materials [6]. The sulfates formed from these metals result from the respective surface oxide and a reaction with the SO_3-containing environment according to

1. Oxide = Metal ions + O^{2-} (acidic fluxing)
2. $O^{2-} + SO_3 = SO_4^{2-}$
3. SO_4^{2-} + Metal ions = Sulfate

This reaction sequence that characterizes type II hot corrosion represents an acidic fluxing mechanism since positively charged metal cations are formed. As a summary reaction for nickel the following equation can be given:

$$NiO + SO_3 = NiSO_4. \tag{13.18}$$

This equation also illustrates that the incubation period is determined by the level of sulfate that has to be formed from nickel by the reaction with gaseous SO_3 for the formation of a significant amount of low-melting eutectic between $NiSO_4$ and Na_2SO_4. This incubation period decreases with increasing temperature since the amount of nickel sulfate needed for the formation of the eutectic decreases. In all cases, however, a critical SO_3 partial pressure has to be exceeded in order to initiate this type of corrosion. The formation of sulfates from Cr_2O_3 or Al_2O_3 is not observed at the commonly encountered SO_3 partial pressures in combustion environments [47]. In particular, sulfate formation from nickel can therefore

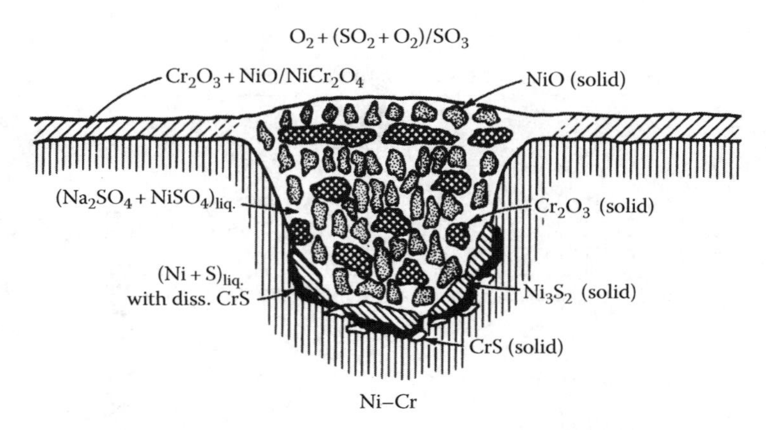

FIGURE 13.20
Schematic of type II hot corrosion (for explanation see text).

only occur if the surface scales of the alloys do not consist only of pure Cr_2O_3 or Al_2O_3 but also contain NiO or spinels containing nickel. Another explanation can also be physical defects in the protective oxide scale that explains why type II hot corrosion is usually observed commencing locally on the surface with nodule type corrosion products before a larger attack on the surface can take place. This is different from the large scale attack in type I hot corrosion, which is a general form of attack and does not occur only locally. Figure 13.20 shows an example of the appearance of type II hot corrosion in a schematic representation. In this representation it becomes clear that the liquid phase may be present at a relatively limited amount only, surrounding the solid phases (usually oxides of the type Cr_2O_3) that can exist to a very large amount together with NiO. In this temperature range Na_2SO_4 is a solid phase that only forms a liquid phase as a eutectic with $NiSO_4$ (see Figure 13.20). Another liquid phase may be the Ni- and S-containing eutectic, which has already been mentioned. From a corrosion protection point of view the question arises whether Cr_2O_3 or Al_2O_3 formers show better resistance under type II hot corrosion. There is no general and reliable information about this issue in the literature but tendentiously it seems as if higher chromium-containing materials can exhibit higher resistance than materials relying only on the formation of an aluminum oxide scale [47].

As already mentioned type I hot corrosion reveals a general form of attack that is different from the localized initiation of hot corrosion in the case of type II. In this case acidic fluxing of Cr_2O_3 and Al_2O_3 by the reaction with SO_3 can be excluded since the SO_3 partial pressure in combustion environments decreases with increasing temperature. However, due to the low SO_3 partial pressure and in combination with a relatively high O^{2-}-ion concentration in the salt melts basic fluxing of the oxides can occur according to

$$Al_2O_3 + O^{2-} = 2AlO_2^- \quad (13.19)$$

or

$$Cr_2O_3 + O^{2-} = 2CrO_2^- \quad \text{(at low oxygen partial pressures)} \quad (13.20)$$

or

$$Cr_2O_3 + 2O^{2-} + \tfrac{3}{2}O_2 = 2CrO_4^{2-} \quad \text{(at high oxygen partial pressures).} \quad (13.21)$$

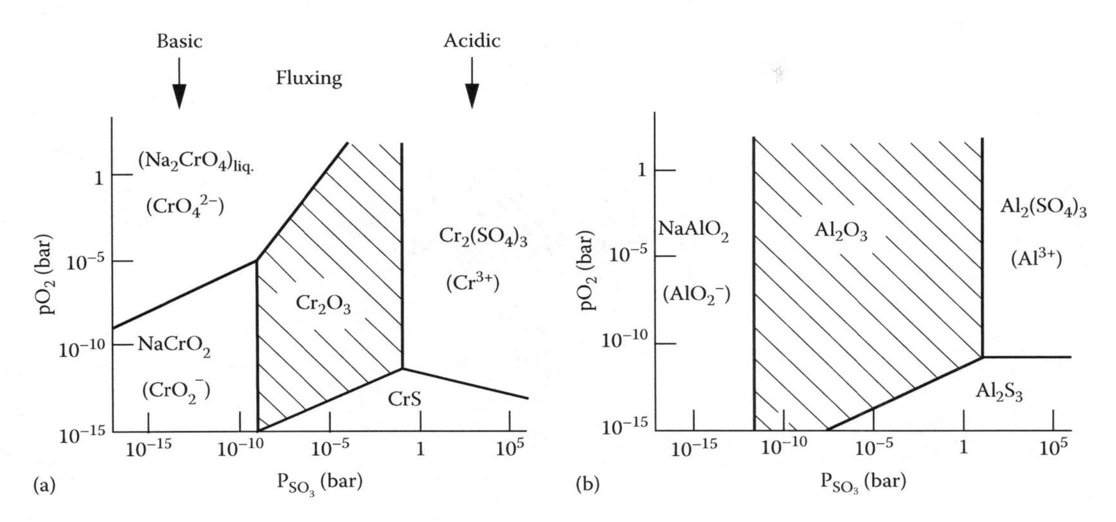

FIGURE 13.21
Stability diagrams for Cr- and Al-containing phases in Na_2SO_4 melt at 900°C: (a) Cr-containing phases and (b) Al-containing phases. (From Kofstad, P., *High Temperature Corrosion*, Elsevier Applied Science, London, 1988.)

In Figure 13.21, stability diagrams of the Cr- and Al-phases are shown as a function of the O_2 and SO_3 partial pressures at 900°C under the assumption that a liquid Na_2SO_4 film covers the surface. In this diagram it can be concluded that the stability field of Al_2O_3 is larger than that of Cr_2O_3. Therefore, the basic fluxing area is larger in the case of Cr_2O_3 than in the case of Al_2O_3 and, thus, the latter type of oxide scale should be more resistant under type I hot corrosion conditions. Generally, for this type of corrosion mechanism significant internal sulfidation is also observed since the corrosion reaction releases sulfur that can penetrate into the material and lead to internal sulfide formation. The general appearance of type I hot corrosion is shown in Figure 13.22. If the mechanism of chemical destruction of the protective oxide scale and subsequent sulfidation of the metallic alloy has started, then healing of the oxide scale is no longer possible. In other words, corrosion attack, once it has started can never be stopped under hot corrosion conditions.

The sulfur that is released from the sulfides by the oxidation reaction penetrates the material in front of the oxide, even if no additional sulfur is supplied. This combined sulfidation and oxidation reaction penetrates particularly fast if low-melting eutectics from Ni_3S_2 phases are formed, which exhibit high transport rates. The situation is less critical for the case of the formation of Cr sulfides, which are not liquid under such conditions. Figure 13.23 schematically explains the difference between hot corrosion and "dry oxidation" with regard to the temperature characteristics. For hot corrosion a temperature maximum is observed in the range between 700°C and 750°C since at higher temperatures evaporation and also disintegration of the sulfate salts can occur. The dry type of high-temperature corrosion (oxidation) usually exhibits a behavior that follows an Arrhenius function, which is also valid for vanadate-induced corrosion so that with increasing temperature, the oxidation or corrosion rates increase exponentially.

13.4.3 Liquid Chloride Corrosion

In particular in waste incineration environments liquid deposits of chlorides or mixed chlorides and sulfates can lead to extreme corrosion attack in the temperature range of

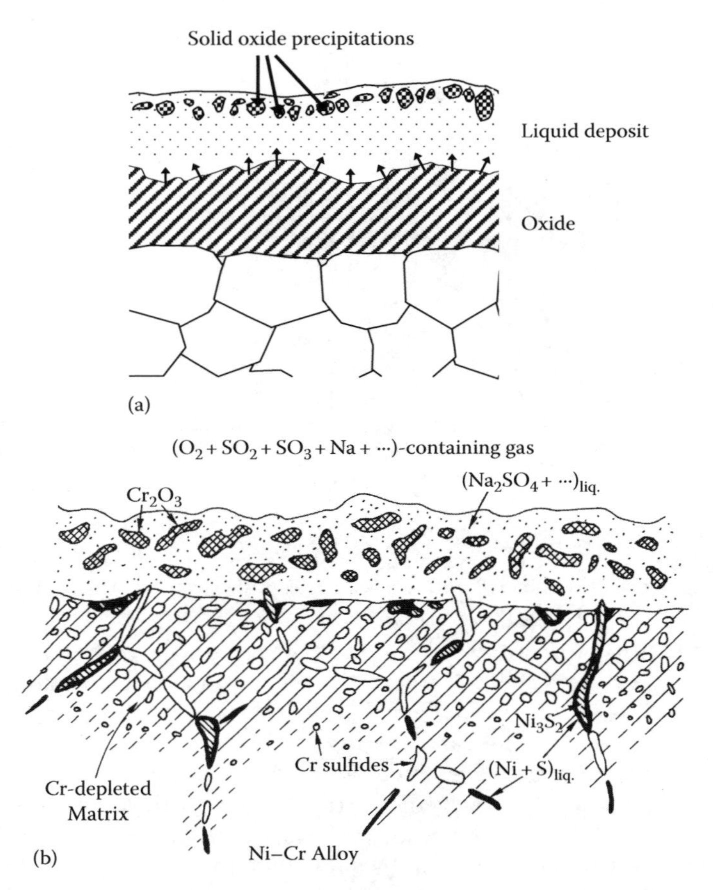

FIGURE 13.22
Schematic of type I hot corrosion (for explanation see text): (a) during the incubation period and (b) after the end of the incubation period.

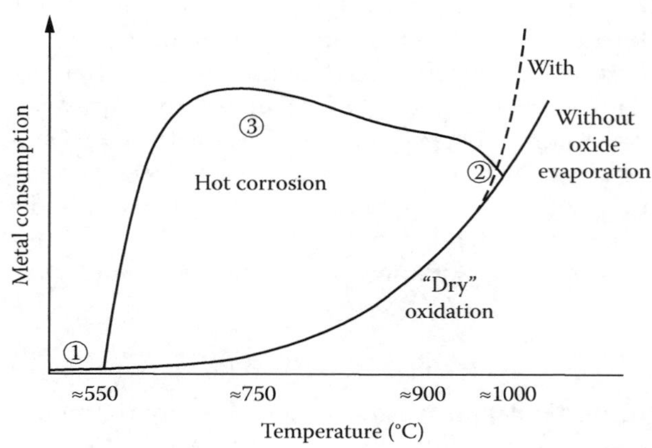

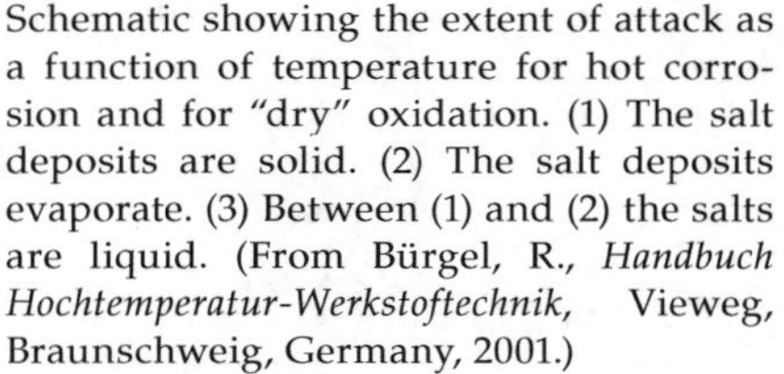

FIGURE 13.23
Schematic showing the extent of attack as a function of temperature for hot corrosion and for "dry" oxidation. (1) The salt deposits are solid. (2) The salt deposits evaporate. (3) Between (1) and (2) the salts are liquid. (From Bürgel, R., *Handbuch Hochtemperatur-Werkstoftechnik*, Vieweg, Braunschweig, Germany, 2001.)

TABLE 13.7

Composition and Melting Point of Several Low-Melting Chloride and Sulfate Eutectics

Composition (wt%)	Melting Point
$48ZnCl_2$–$52KCl$	250
$82ZnCl_2$–$18KCl$	262
$84ZnCl_2$–$16KCl$	262
$73ZnCl_2$–$27PbCl_2$	300
$31NaCl$–$69PbCl_2$	410
$21KCl$–$79PbCl_2$	411
$17NaCl$–$83PbCl_2$	415
$39ZnCl_2$–$50KCl$–$11PbCl_2$	275
$35ZnCl_2$–$48NaCl$–$17PbCl_2$	350
$16NaCl$–$40KCl$–$44PbCl_2$	400
K_2SO_4–Na_2SO_4–$ZnSO_4$	384
KCl–$ZnCl_2$–K_2SO_4–$ZnSO_4$	292

Source: Schütze, M. and Spiegel, M., *Korrosion in Anlagen zur thermischen Abfallverwertung, Proceedings of Jahrestagung 2004*, GfKORR, Frankfurt/Main, Germany, 2004.

250°C–400°C as given in Table 13.7. Although in gas phase chlorine-containing environments in this temperature range significant corrosive attack may not necessarily occur, the consumption of the metal by liquid chlorides can be increased compared to the gas-phase chlorine by more than one order of magnitude [14]. Examples of a eutectic mixture of KCl–$ZnCl_2$ at 250°C indicate extremely rapid corrosion in the first few hours, which is followed by a period of decreasing corrosion rate similar to what is observed under parabolic oxidation. Generally this type of corrosion is determined by dissolution of the metal, transport processes of the dissolved species through the molten film on the metal surface, and reprecipitation at the phase boundary among liquid, salt, and gas phases. At the metal/salt film interface the following reaction occurs for the example of iron due to the low oxygen partial pressure at this position:

$$Fe + Cl_2\,(\text{deposit}) = FeCl_2\,(\text{liquid}). \tag{13.22}$$

Due to this reaction iron is continuously dissolved as a chloride and diffuses in an outward direction through the melt to the phase boundary between liquid and gas phase. These diffusion processes are driven by the concentration gradient of dissolved $FeCl_2$ from the metal surface (high concentration of $FeCl_2$) to the phase boundary between salt melt and gas phase (low concentration of $FeCl_2$) (see also the schematic in Figure 13.24). The corrosion rate is very much dependent upon the solubility of the metal or the alloying compounds in the chloride melt. In this context iron has a significantly high solubility in KCl–$ZnCl_2$ salt melts compared to nickel and chromium with the concentration of dissolved chromium increasing after longer exposure times due to the formation of chromate. This explains the experimentally observed high corrosion rates for low-alloy steels and the relatively high resistance of nickel-based materials. These experimental investigations are also confirmed by thermodynamic considerations as shown in Ref. [14].

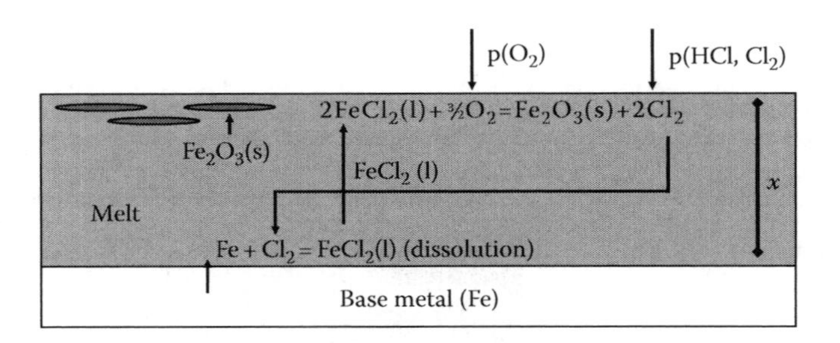

FIGURE 13.24
Corrosion mechanisms by molten chlorides. Fe is dissolved by $FeCl_2$ formation at the metal/melt interface and reprecipitated as solid Fe_2O_3 at the melt surface. (From Schütze, M. and Spiegel, M., *Korrosion in Anlagen zur thermischen Abfallverwertung, Proceedings of Jahrestagung 2004*, GfKORR, Frankfurt/Main, Germany, 2004.)

13.5 Formation of Volatile Phases in High-Temperature Corrosion

13.5.1 General Aspects

There are a number of situations under high-temperature corrosion conditions where volatile corrosion products can be formed. Examples of such phases are given in Table 13.1, which shows that mainly metal halides are volatile above certain temperatures. Interestingly, however, chromium oxide can also form a volatile phase in the form of CrO_3 with evaporation rates becoming significant at temperatures of 1000°C and higher. This is usually the reason why pure chromium oxide scale formers should not be used at temperatures above 1000°C if protection by the chromia scale is required. Another surprising observation is the formation of a volatile platinum oxide at temperatures of 1000°C and higher. This leads to the observation of a shiny metal surface on platinum metal at high temperatures, which, however, does not mean that platinum is inert in oxidizing environments. It has already been mentioned that molybdenum forms a low-melting oxide phase but as Table 13.1 shows the evaporation rates of this molybdenum oxide phase can also become significant at temperatures above 600°C. Silica has its problems at temperatures of 1250°C and higher at low oxygen partial pressures where the volatile SiO phase is formed. As will be discussed later, the presence of water vapor in the environment can also lead to the formation of gaseous metal hydroxyl phases leading to increased metal consumption through evaporation of these phases.

13.5.2 Gas-Phase Chlorine Corrosion

In this section the discussion will focus on the example of volatile metal chlorides that can also be used to understand the role of the formation of volatile corrosion products for other compounds. A general feature of high-temperature corrosion processes where volatile corrosion products are formed is that in stability diagrams, a coexistence of the volatile corrosion product phases and the metal or other solid and liquid phases of different composition is possible. Although in the traditional thermodynamic stability diagram, boundaries limit the fields of existence of the different phases, e.g., metals, oxides, and sulfides, there can be a superposition of gaseous halides on the metallic phase or the liquid or solid metal halide phase. Furthermore, a certain halide vapor pressure can be present above oxide surfaces. In other words, the traditional stability diagrams can no longer be used for defining areas of the formation of different corrosion products as a function of the respective activities or partial pressures of the corrosive environment. The first attempt to deal quantitatively with the situation of the formation of volatile corrosion products and to predict the role of the formation of gaseous species was to somewhat arbitrarily define a limiting vapor pressure of 10^{-4} bar for the corrosion products. This approach was used in conjunction with the Hertz–Langmuir equation, which is given by the following [48]

$$V\left(\frac{\text{g}}{\text{cm}^2\,\text{s}}\right)=\frac{\alpha_{\text{M}_x\text{Cl}_y}\,\text{p}_{\text{M}_x\text{Cl}_y}\,M_{\text{M}_x\text{Cl}_y}}{\sqrt{2\pi m_{\text{M}_x\text{Cl}_y}\,N^2kT}}, \tag{13.23}$$

where

V is the metal chloride evaporation rate

$\text{p}_{\text{M}_x\text{Cl}_y}$ is the metal chloride vapor pressure above the solid metal chloride phase given by the thermodynamic equilibrium

$m_{\text{M}_x\text{Cl}_y}$ is the mass of the evaporating species (M_xCl_y)

$M_{\text{M}_x\text{Cl}_y}$ is the metal chloride molecular weight

k is the Boltzmann constant

N is the Avogadro number

T is the temperature

The parameter $\alpha_{\text{M}_x\text{Cl}_y}$ is defined as representing the situation for the maximum possible evaporation rate resulting from a given set of environmental pressure, composition, temperature, and flow conditions.

This equation expresses the rate of metal halide evaporation as a function of the metal halide vapor pressure at the respective temperature and the respective halide activity. A key role is played by the factor α, which becomes 1 in the case of unimpeded evaporation into an infinite space but is around 10^{-4} if one assumes a metal consumption rate (metal halide evaporation rate) of 1 mm/a at a metal halide vapor pressure of 10^{-4} bar [48,49]. The latter sounds somewhat arbitrary and as will become clear later, this approach represents an idealized situation, which is usually not encountered under conditions of a flowing gas stream on the metal surface. Nevertheless, this approach seems to be supported by practical experience [17]. It can give an orientation about the stability of the different alloying elements, e.g., under chloridizing conditions and therefore has so far been used in almost all considerations on the role of chlorine corrosion for metal stability. In the literature, a criterion has been defined called T_4, which is the temperature at which the metal halide vapor pressure reaches a value of 10^{-4} bar. Such values are given in tables (see also Table 13.1 of this chapter) and this limiting value can also be taken from Figure 13.25.

Since under practical conditions and in most of the laboratory tests, oxygen can also be present to some extent, this approach has been extended in the form of the so-called "static" quasi-stability diagrams [50]. Again the criterion of a critical Me_xCl_y pressure of 10^{-4} bar has been used in order to define fields, where the corrosion rates should be low enough to guarantee stability of the metallic materials. An example of such a diagram is given in the middle of Figure 13.26 where at the same time a comparison is performed with the predictions from a traditional thermodynamic stability diagram (left) that indicates that the chlorine partial pressure at which the solid metal chloride phase appears is higher than the chlorine partial pressure where an equilibrium with the metal and the metal halide partial pressure of 10^{-4} bar exists. Diagrams of this type have been developed for most of the common high temperature material alloying elements and examples for 800°C are given in Figure 13.27.

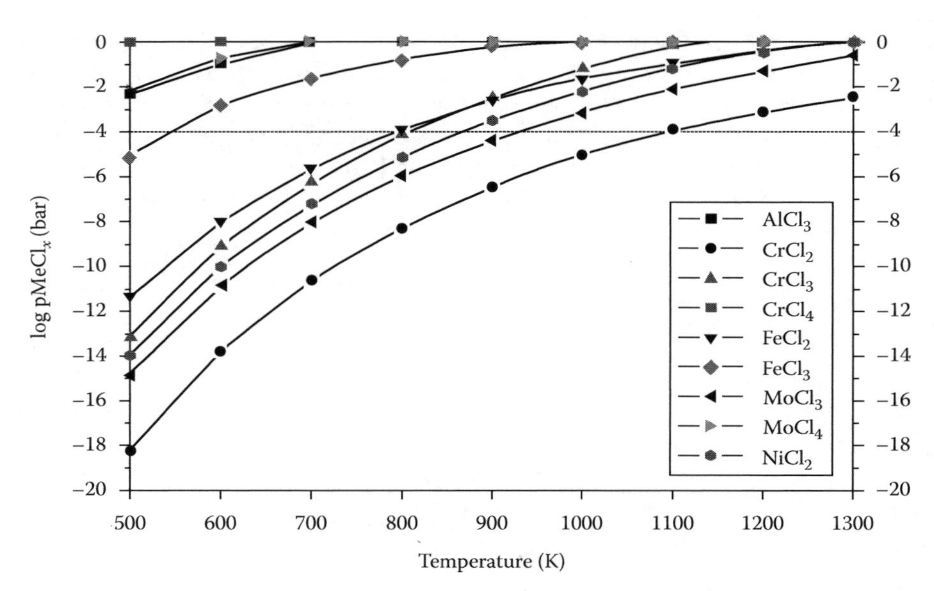

FIGURE 13.25
Vapor pressure pMe_xCl_y of several metal chlorides above the solid or liquid metal chloride phase as a function of temperature, together with the "limiting line" of $pMe_xCl_y = 10^{-4}$ bar. (From Latreche, H. et al., *Oxid. Met.*, 72, 1, 2009.)

When looking a little closer at the situation that stands behind the 10^{-4} bar criterion, it becomes clear that this does not correspond to the situation encountered in the common laboratory corrosion tests and in industrial plants. Figure 13.28 is a schematic that illustrates that the 10^{-4} bar criterion used for determining T_4 and for the establishment of the "static" quasi-stability diagrams describes the situation in a closed system. The approach of the T_4 temperature criterion corresponds to the schematic in the left part of this figure. In this case, it is assumed that inside a closed system, a solid or liquid metal halide is present and a "saturation" vapor pressure is being formed by volatilization of this halide inside the closed system. Indeed, metal halides with a high tendency for vaporization will develop the highest vapor pressures under such conditions so that this criterion definitely can be used for orientation in systems, where gaseous halides are formed on the surface of a metallic material or of solid or liquid metal halides. The situation in the middle of Figure 13.28 describes a closed system containing all the educts and products of the potential metal/oxygen/chlorine reactions in a closed system and this quantifies the Me_xCl_y partial pressure resulting from these reactions. This approach corresponds to the "static" quasi-stability diagram concept. The situation in a laboratory furnace or in an industrial plant must, however, be compared with an open system of a flowing gas above a metal surface. This situation is depicted on the right side of this figure. In this case, kinetic considerations play a role in that the rate-determining step for metal consumption is actually the transport rate of the corrosion product, e.g., the metal chloride, through the gas boundary layer formed on top of the metal surface. This situation has been dealt with in great detail in a more recent approach, which is called the "dynamic" quasi-stability diagram [49].

Before addressing this kind of diagram it is, however, necessary to discuss the different potential mechanisms of corrosion again for the example of a mixed chloridizing–oxidizing environment. The conclusions from this example can also be used for understanding other corrosion processes where volatile corrosion products are formed. The four different possible

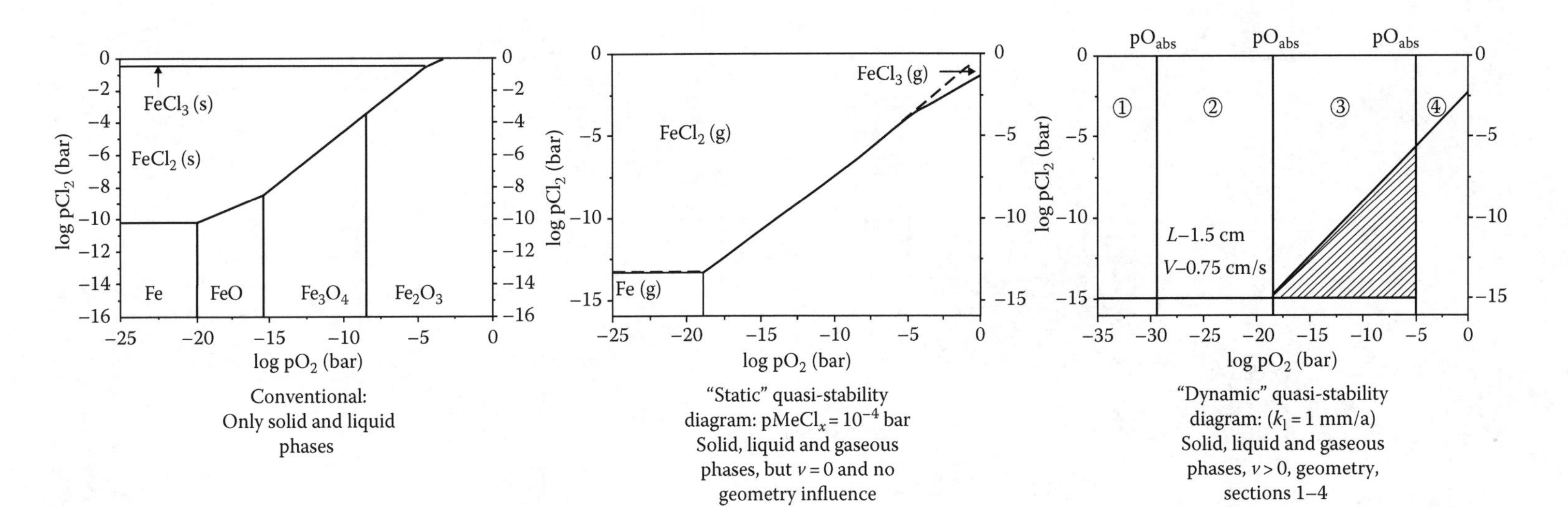

FIGURE 13.26
Comparison of the conventional stability diagram (left), the "static" quasi-stability diagram (middle), and the "dynamic" quasi-stability diagram (right) for iron at 800°C (for explanation see text).

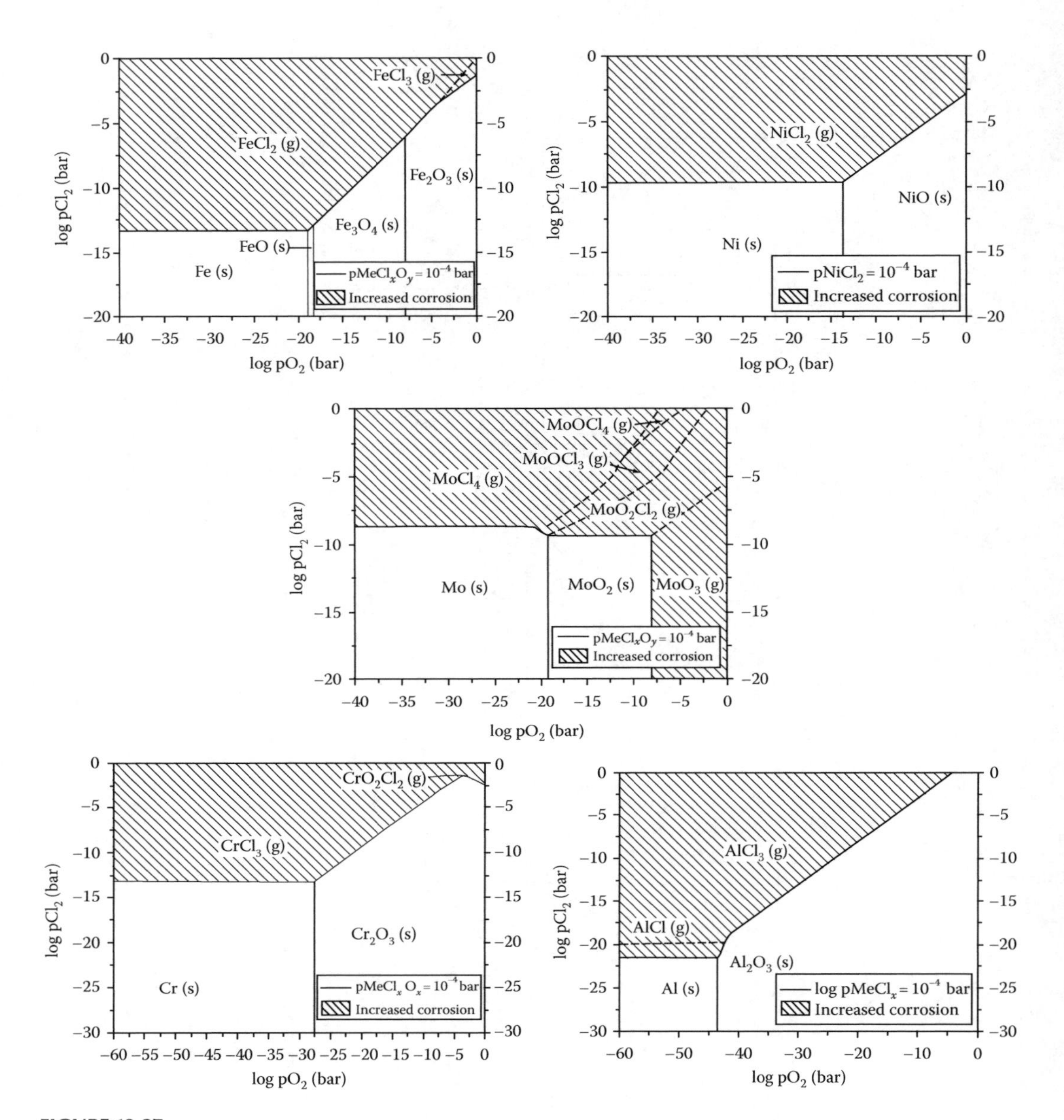

FIGURE 13.27
"Static" quasi-stability diagrams of Al, Cr, Fe, Ni, and Mo at 800°C based on the 10^{-4} bar criterion. The hatched areas denote conditions where increased corrosion rates are to be expected ($dx/dt > 1$ mm/a). (From Latreche, H. et al., *Oxid. Met.*, 72, 1, 2009.)

situations encountered in the case of halogen corrosion at high temperature under the presence of oxygen are summarized in Figure 13.29. All in all there may be four different cases:

- Case 1 describes the situation at extremely low oxygen partial pressures (oxygen partial pressures below the equilibrium partial pressure for oxide formation), where only metal halides can be formed and have to diffuse through the gas boundary layer, thus, determining the rate of metal consumption by the corrosion

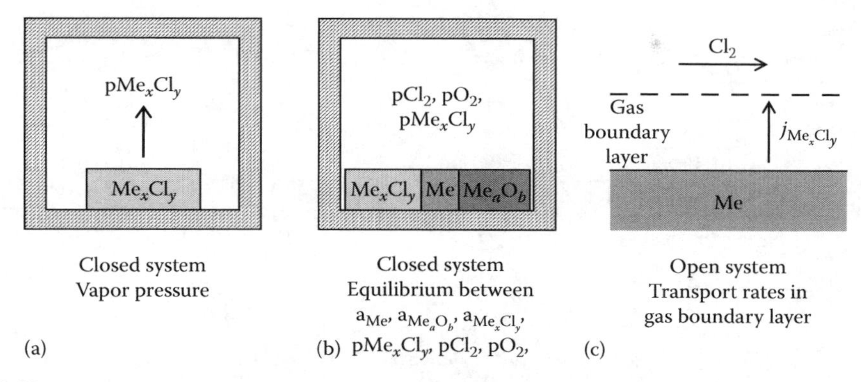

FIGURE 13.28
Schematic of the three situations that are used for the assessment of metal chloride evaporation. (a) Vapor pressure of the gaseous metal chloride above solid or liquid metal chloride in a closed system (value used in Equation 13.23 and for determining T_4). (b) Equilibrium partial pressure of pMe_xCl_y in a closed system containing O_2 and Cl_2 as gas phase and Me, Me_aO_b, and Me_xCl_y as solid or liquid phases (value used for the establishment of the "static" quasi-stability diagram). (c) Me_xCl_y transport rate in the gas boundary layer (metal consumption rate) as criterion for the amount of chlorine-induced corrosion in an open system with gas flow across the surface (value used for the establishment of the "dynamic" quasi-stability diagram).

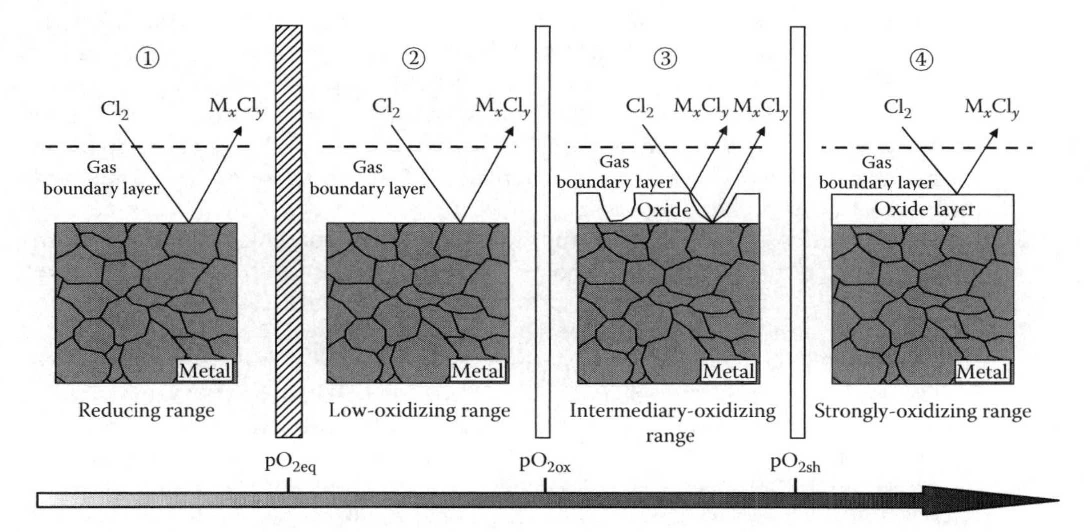

FIGURE 13.29
The four potential cases of high temperature chlorine corrosion in oxygen-containing Cl-environments (for explanation see text). (From Latreche, H. et al., *Oxid. Met.*, 72, 1, 2009.)

process. This is actually the easiest case because there is no interference with an oxide layer on top of the metal.

- Case 2 corresponds to a similar situation with the only difference in this case the oxygen partial pressure being higher than the equilibrium oxygen pressure for oxide formation. However, the oxygen pressure is still too low to overcome the impeding role of the presence of chlorine in the environment with regard to oxide formation and nucleation on the metal surface. Thus, the rate-determining step still is the evaporation of the metal halide with its transport through the gas boundary layer.

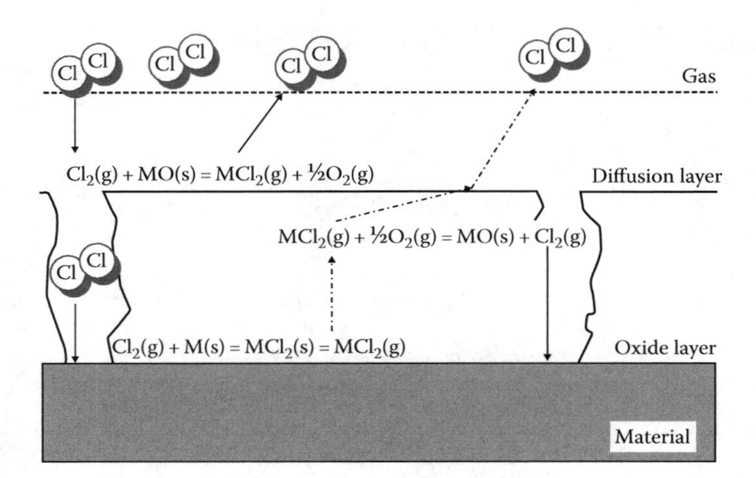

FIGURE 13.30
Schematic of the mechanism of active oxidation (case 3 in Figure 13.29) (for explanation see text). (From Latreche, H. et al., *Oxid. Met.*, 72, 1, 2009.)

- Case 3 is the so-called active oxidation process where a superposition of oxide formation (solid phase) and metal halide formation (volatile phase) takes place. This situation is further detailed in Figure 13.30. In active oxidation several different reactions can take place. The first is that halogen can come into direct contact with the metal by penetration of the oxide scale, which has formed on the surface. This oxide scale under such conditions is, however, pervious and porous so that the chlorine can migrate through this scale. As a result, volatile metal halides can be formed and, due to their vapor pressure, move in an outward direction through the oxide scale. However, these metal halides may become unstable at certain increased oxygen partial pressures (the oxygen partial pressure increases from the metal/oxide interface to the surface of the oxide layer) so that the volatile halides are converted into solid metal oxides. Through this chlorine molecules are released, which may diffuse back to the metal/oxide interface and react again forming new volatile metal halides. Another reaction that may take place is the direct attack of the oxide by chlorine again forming volatile metal chlorides. The circular process taking place inside the oxide scale and depositing solid metal oxide in the channels creates high compressive stresses that end up in cracking of the oxide and, thus, providing new transport paths of the gaseous molecular chlorine from the corrosive environment through the scale down to the metal/oxide interface. All in all a very porous and brittle oxide scale results from this mechanism that provides only very limited protection and, thus, explains the highly increased corrosion rates under such conditions.
- Case 4 can occur for certain oxides where the oxygen partial pressure is sufficiently high to form a dense and impervious protective oxide layer so that the attack by chlorine can only take place on the surface of the oxide scale but not by reaction with the underlying metal. This situation that is also called strongly oxidizing range is the only case where long-term stability of a material under chloridizing or halogenizing conditions can be expected.

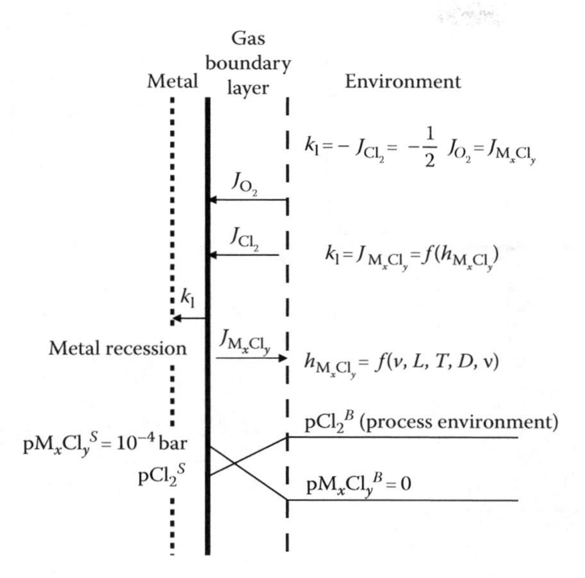

FIGURE 13.31
Basic considerations for establishing the dynamic quasi-stability diagram (for explanation see text). The situation corresponds to Figure 13.28c.

These four cases will be examined again in the so-called dynamic quasi-stability diagram later. The basis for the development of a quasi-stability diagram of the dynamic type is given in Figure 13.31. This schematic shows the role of the different values of the partial pressures pCl_2^B (chlorine partial pressure in the bulk gas atmosphere equals the process environment), pCl_2^S (chlorine partial pressure directly on the metal surface), and $pMe_xCl_y^S$ (metal chloride partial pressure on the metal surface). Since the situation at the surface underneath the gas boundary layer can be regarded as a quasi-closed "micro" system, $pMe_xCl_y^S$ can be set equal to a value that corresponds to a critical value for a certain metal consumption rate (e.g., 10^{-4} bar for 1 mm/a) and $pMe_xCl_y^B$ in the quasi-infinite process reaction space is taken as close to zero. Based on these assumptions, pCl_2^B in the process environment for a certain corrosion rate can be calculated. The corrosion rates in this case are determined by the concentration gradients inside the gas boundary layer resulting from these partial pressure values, which together with (among others) the viscosity define the transport rates inside the boundary layer. The thickness of the gas boundary layer (and by that the concentration gradients) is determined by the gas flow rate and the surface geometry. Entering all these parameters into a relatively complex equation system (a computer code has been developed for this [49]) allows the development of a dynamic quasi-stability diagram as shown in Figure 13.32, for iron for the different zones of oxygen partial pressure where the gas flow rates (v = 0.75 mm/s) and the geometry of the metal surface (length of a flat surface 1.5 mm) determine the results. This diagram illustrates that there can be a range (the very narrow area range no. 4′ on the right-hand side) where even at very high chlorine partial pressures the metal consumption rates do not exceed the given limiting value (in the example of this Figure 1 mm/a) since a dense and impervious oxide scale has formed, which resists direct attack by chlorine at these high oxygen partial pressure values.

In Ref. [49] the development of the dynamic quasi-stability diagram has been described in great detail and diagrams of this type have been developed for different flow rates and different surface geometries as well as for different alloying elements. From these investigations it is mainly the oxides Al_2O_3 and SiO_2 that allow for a type 4′ situation over a significant range of oxygen partial pressure values, whereas all other surface oxides are

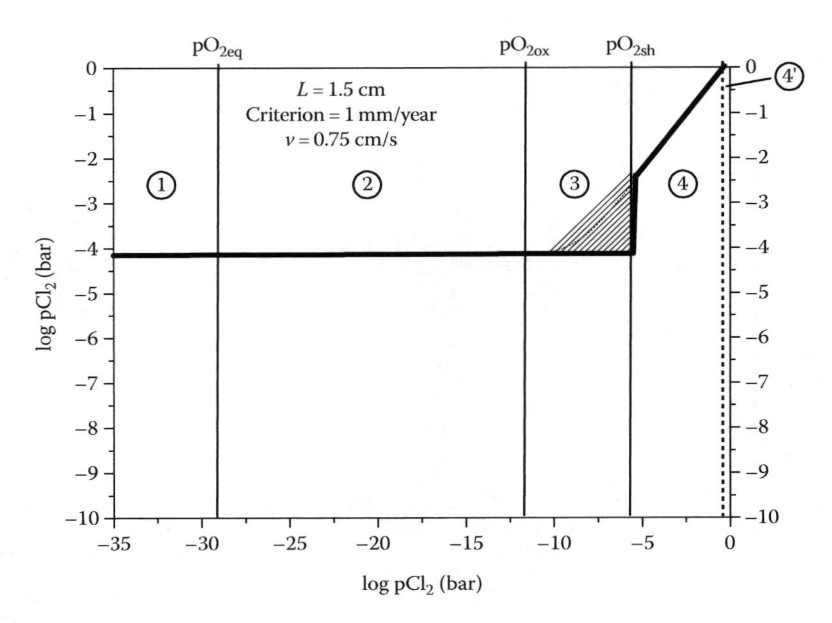

FIGURE 13.32
Dynamic quasi-stability diagram for Fe at 500°C. Zones 1–4 correspond to those in Figure 13.29. (From Latreche, H. et al., *Oxid. Met.*, 72, 1, 2009.)

not suitable for permanent protection of the metallic material against chlorine corrosion. In Ref. [49] this approach has also been applied to several commercial alloys and the initial results of this comparison look promising so that this approach should be preferred over the former approaches, which as described mainly serve for a certain orientation with regard to resistance under chloridizing environments.

13.5.3 Formation of Volatile Phases from Water Vapor

Another technically important situation where volatile corrosion products are formed is the case of the presence of water vapor in the process environment. Table 13.8 shows a number of potential reaction products in air with 10% H_2O at 650°C if conventional high temperature chromium steels are exposed to such an environment. In this case the vapor

TABLE 13.8
Partial Pressures of Gaseous Species in Air with 10% H_2O at 650°C in the Presence of Cr, Cr_2O_3, Mn, Mo, W, and Si as Solid Phases

Compound	Partial Pressure (bar)
MoH_2O_4	5.5×10^{-6}
MnH	4.6×10^{-8}
CrH_2O_4	1.1×10^{-9}
WH_2O_4	5.5×10^{-12}
CrO_3	9.8×10^{-12}
Others	$<10^{-13}$

Source: Schütze, M. et al., *Corrosion Eng. Sci. Tech.*, 39, 157, 2004.

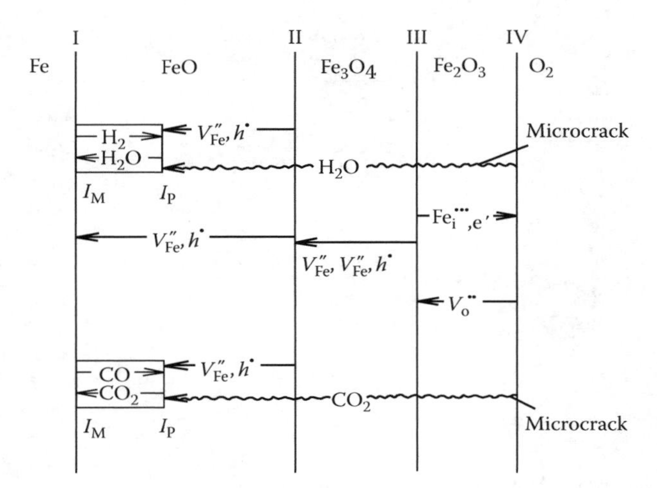

FIGURE 13.33
Transport processes in the oxide scale on iron in the presence of water vapor accelerating the oxidation rates. (From Rahmel, A. and Tobolski, J., *Corrosion Sci.*, 5, 333, 1965.)

pressures of the different species have again been calculated according to the approach described for halides in the former section and by applying the Hertz–Langmuir equation the evaporation rates have been received [51]. As the results of these calculations show, it is mainly the molybdenum and chromium hydroxyl species as well as the manganese hydride that exhibit measurable evaporation rates even at temperatures as low as 650°C.

However, even for pure iron it has been observed that the oxidation rates may increase significantly in the presence of water vapor [52]. This has been explained by the involvement of gas-phase species in the oxidation process in the form of gaseous H_2O and H_2. Figure 13.33 shows that the gaseous H_2O from the environment can penetrate iron oxide scales through microcracks that may have been formed due to oxide growth stresses and then react with the underlying metal directly forming new oxide but at the same time releasing gaseous hydrogen. This gaseous hydrogen may migrate in an outward direction and lead to a reduction reaction of the existing oxide, thus, "releasing" iron cations and at the same time forming gas-phase water vapor. The iron cations may migrate in an outward direction and then react with oxygen at the oxide surface, thus, leading to an increase of oxide scale thickness. The water vapor in a pore where this reaction had taken place may migrate in an inward direction and lead to direct attack of the metal starting the reaction cycle once again. As a consequence a circular process becomes active that includes gas-phase transport and, thus, increases the overall oxidation rate. A similar situation is also observed in the case of CO_2 high-temperature corrosion as shown in the lower part of this schematic.

The vapor pressures of metal oxides and hydroxyls potentially formed from low- or high-alloy steels are given in Figure 13.34. These results from thermodynamic calculations indicate that the vapor pressures can be highly dependent upon the pH_2O/pO_2 ratio in the environment and upon exposure temperature. While chromium hydroxyl formation and evaporation do not play a significant role in pure steam environments with low oxygen partial pressure, the formation and evaporation of manganese hydroxyl can become more pronounced under the same conditions, Figure 13.34c. On the other hand, significant chromium hydroxyl evaporation will take place at lower pH_2O/pO_2 ratios, Figure 13.34b.

In the literature also an increase of internal oxidation has been observed in the presence of water vapor [53]. Furthermore, increased acoustic emission activity took place in the oxide scales when water vapor was present indicating high compressive growth stresses in the oxide leading to microcracking and, thus, impairing the barrier effect of the scale [54]. This could possibly be one of the explanations for increased internal oxidation since

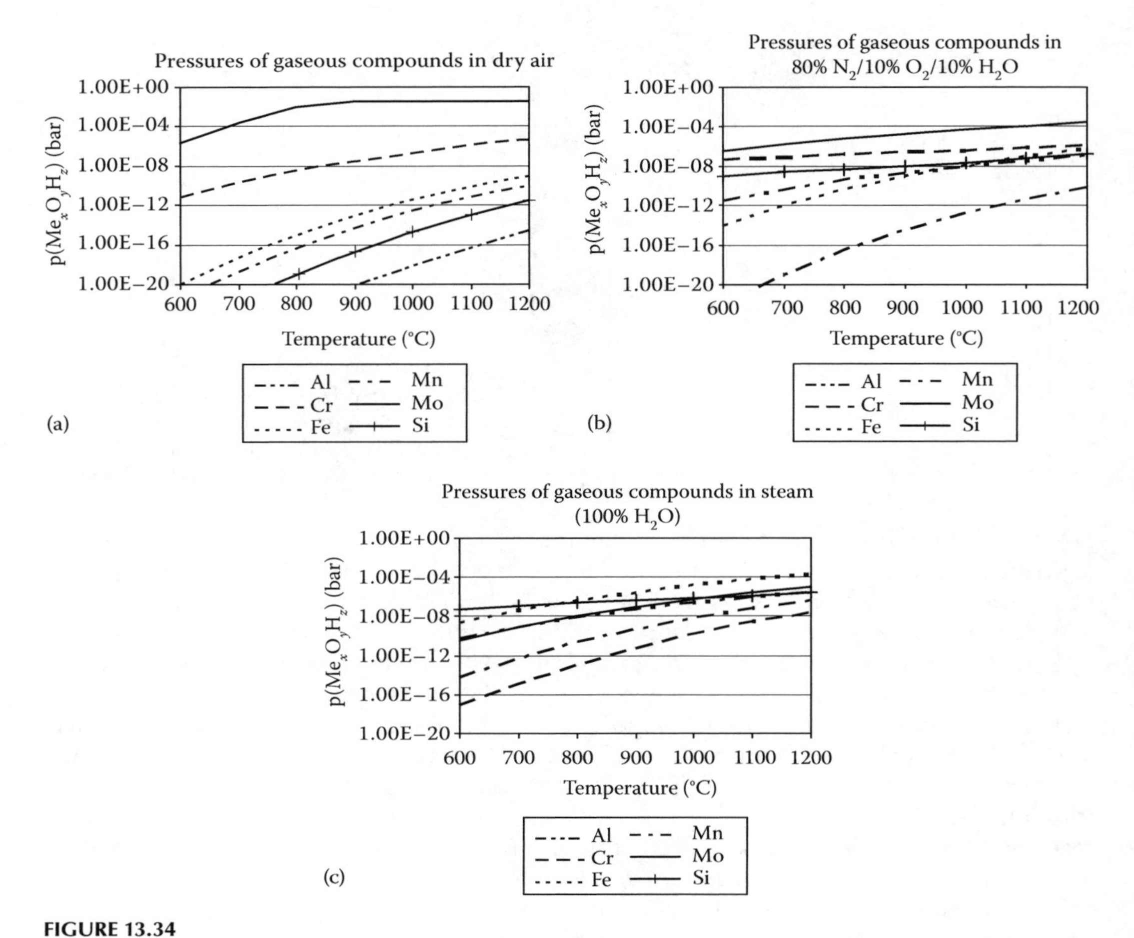

FIGURE 13.34
Vapor pressure values of the metal oxides and hydroxyls potentially formed from (low or high alloy) steels at 850°C as a function of temperature and for different oxygen and water vapor contents (ChemSage® calculations). (a) Dry air, (b) nitrogen-based environment with 10% oxygen and 10% water vapor, and (c) pure water vapor.

the oxygen partial pressure underneath such a microcracked oxide scale will have a higher value than underneath a dense scale. At the same time the penetration of H_2O and by this the supply of H_2 into the underlying metal is facilitated, which can further stimulate all the effects of increased internal oxidation.

It should also be mentioned for completeness of this chapter that pressurized hydrogen at high temperatures can induce a mechanism of internal attack, which can weaken in particular steels of high carbon content significantly (high temperature hydrogen embrittlement) [8]. In this case the cementite precipitates inside high carbon steels are directly attacked by hydrogen forming gaseous methane (cf. Table 13.1) and leaving behind pores at sites where the solid cementite particles had originally been. As a result the steel will exhibit a spongy structure due to the conversion of the solid cementite particles into pores whereas the methane escapes through the porous structure in an outward direction. This effect is not observed for steels with alloying elements that are strong carbide formers like molybdenum or niobium. In this case the carbon is tied up in stable carbides that can resist hydrogen attack and the steel is no longer endangered by this effect.

13.6 Concluding Remarks

As can be seen from the present chapter the phenomenon of high-temperature corrosion consists of a number of different mechanisms with sometimes complex interactions. A key role is played by the formation of a protective surface layer ("passive" layer), which usually is an oxide layer and serves as a barrier against the ingress of further corrosive species into the metal subsurface zone. Oxide scales with the "best" protective properties are those based on Al_2O_3 and Cr_2O_3, in some cases also on SiO_2, on spinel phases containing Al, Cr, and Si or at lower temperatures also on iron oxides. It should, however, be pointed out that these "passive" layers are living systems at high temperatures, i.e., they keep growing and, as a positive aspect, possess self-healing capabilities in case of scale damage (cracking, spalling, etc.). Concerning the latter, the alloy reservoir of protective scale-forming elements plays a decisive role in that protective scale healing is only possible if a sufficient supply from the alloy for the formation of a protective type of oxide can be guaranteed. An issue that has not been addressed in the present chapter but is of vital importance for the protective effect is the mechanical properties of the surface scales. For this a respective textbook is referred to [7].

An entirely different situation arises if liquid or gaseous corrosion products play a role. In this case the resistance of materials is only partially dependent upon the effect of barrier layers. In most cases it is, however, the kinetics of the metal consumption processes (dissolution, evaporation) that determine material resistance. As well as experimental measurements there are now tools for an assessment of material resistance based on the respective corrosion mechanisms and on thermodynamic considerations, allowing the production of diagrams which delineate critical and less critical situations.

As was also indicated in this chapter the development of an understanding of the mechanisms is still in a dynamic phase as recently significant steps forward have been made in the areas of metal dusting and chlorine corrosion. This shows that high-temperature corrosion research despite all the invaluable achievements in the past is far from being a static situation and in addition to textbook knowledge it is necessary to follow the most recent literature in order to stay up-to-date in the field.

References

1. ISO/TC 156 NP 21608: Corrosion of metals and alloys—Test method for isothermal exposure testing under high temperature corrosion conditions for metallic materials, International Organisation for Standardisation, Geneva, 2006.
2. O. Kubaschewski, B. E. Hopkins, *Oxidation of Metals and Alloys*, Butterworths, London, U.K., 1962.
3. K. Hauffe, *Oxidation of Metals*, Plenum Press, New York, 1965.
4. S. Mrowec, T. Werber, *Gas Corrosion of Metals*, National Bureau of Standards, Washington, DC, 1978.
5. N. Birks, G. H. Meier, F. S. Pettit, *Introduction to High Temperature Oxidation of Metals*, Cambridge University Press, New York, 2006.
6. P. Kofstad, *High Temperature Corrosion*, Elsevier Applied Science, London, 1988.
7. M. Schütze, *Protective Oxide Scales and Their Breakdown*, John Wiley, Chichester, U.K., 1997.
8. M. Schütze (ed.), *Corrosion and Environmental Degradation*, Wiley-VCH, Weinheim, Germany, 2000.

9. G. Y. Lai, *High Temperature Corrosion and Materials Applications*, ASM International, Materials Park, OH, 2007.
10. D. J. Young, *High Temperature Oxidation and Corrosion of Metals*, Elsevier, Oxford, U.K., 2008.
11. Fact Sage, GTT-Technologies, Herzogenrath (Germany).
12. Thermo-Calc, Thermo-Calc Software Inc., Stockholm (Sweden).
13. W. Schendler, in *Corrosion Resistant Materials for Cool Conversion Systems*, eds., D.B. Meadowcroft and M.I. Manning, p. 201, Applied Science, Barking, London, U.K., 1983.
14. M. Schütze, M. Spiegel, *Korrosion in Anlagen zur thermischen Abfallverwertung, Proceedings of Jahrestagung 2004*, GfKORR, Frankfurt/Main, Germany, 2004.
15. F. A. Kröger, *The Chemistry of Imperfect Crystals*, North Holland, Amsterdam, the Netherlands, 1964.
16. C. Wagner, *Z. Phys. Chem.* B21; 25 (1933).
17. G. Y. Lai, *High Temperature Corrosion of Engineering Alloys*, ASM International, Materials Park, OH, 1990.
18. C. Wagner, *J. Electrochem. Soc.* 99; 369 (1952).
19. D. P. Whittle, *Oxid. Met.* 4; 171 (1972).
20. J. P. T. Vossen, P. Gawenda, K. Rahts, M. Röhrig, M. Schorr, M. Schütze, *Mater. High Temp.* 14; 387 (1997).
21. E. Boggs, *J. Electrochem. Soc.* 118; 908 (1971).
22. A. Rahmel, W. Schwenk, *Korrosion und Korrosionsschutz von Stählen*, Verlag Chemie, Weinheim, Germany, 1977.
23. H. J. Grabke, J. F. Norton, F. G. Casteels, in *High Temperature Alloys for Gas Turbines and Other Applications*, eds., E. Betz et al., p. 245, D. Reidel, Dordrecht, the Netherlands, 1986.
24. M. Schütze, A. Rahmel, in *Corrosion Resistant Materials for Cool Conversion Systems*, eds., D.B. Meadowcroft and M.I. Manning, p. 439, Applied Science, Barking, London, U.K., 1983.
25. V. Guttmann, M. Schütze, in *High Temperature Alloys for Gas Turbines and Other Applications*, eds., W. Betz et al., p. 293, D. Reidel, Dordrecht, the Netherlands, 1986.
26. A. Kempen, H. van Wortel, *Stainless Steel World*; 23 (2003).
27. D. J. Young, J. Zhang, C. Geevs, and M. Schütze, *Materials and Corrosion*, 61; 1 (2010).
28. D. Röhnert, F. Phillipp, H. Reuther, T. Weber, E. Wessel, M. Schütze, *Oxid. Met.* 68; 271 (2007).
29. U. Seip, M.-C. Tsai, K. Christmann, J. Küppers, G. Ertl, *Surf. Sci.* 139; 29 (1984).
30. Z. Zeng, K. Natesan, V. A. Maroni, *Oxid. Met.* 58; 147 (2002).
31. C. H. Toh, P. R. Munroe, D. J. Young, *Mater. High Temp.* 20; 527 (2003).
32. F. Abild-Pedersen, J. K. Norskov, J. R. Rostrup-Nielsen, J. Sehested, S. Helveg, *Phys. Rev. B* 73; 1 (2006).
33. M. Nakamizo, R. Kammereck, J. Walker *Carbon* 12; 259 (1974).
34. J. Zhang, P. R. Munroe, D. J. Young, *Acta Mater.* 56; 68 (2008).
35. J. Zhang, D. J. Young, *Corrosion Sci.* 49; 1496 (2007).
36. S. Helveg, C. López-Cartes, J. Sehested, P. L. Hansen, B. S. Clausen, J. R. Rostrup-Nielsen, F. Abild-Pedersen, J. K. Norskov, *Nature* 427; 426 (2004).
37. P. E. Nolan, M. J. Schabel, D. C. Lynch, A. H. Cutler, *Carbon* 33; 79 (1995).
38. C. M. Chun, T. A. Ramanarayanan, *Oxid. Met.* 62; 71 (2004).
39. J. R. Rostrup-Nielsen, I. Alstrup, *Catal. Today* 53; 831 (2004).
40. C. Geers, M. Schütze, *Proceedings of NACE Corrosion 2009*, Atlanta, Paper No. 09149, NACE International, Houston, TX, 2009.
41. C. Rosado, M. Schütze, *Mater. Corrosion* 54; 831 (2004).
42. P. Szakalos, M. Lundberg, R. Pettersson, *Corrosion Sci.* 48; 1679 (2006).
43. C. M. Chun, T. A. Ramanarayanan, in *Corrosion by Carbon and Nitrogen: Metal Dusting, Carburisation and Nitridation (EFC 41)*, eds., H.J. Grabke and M. Schütze, p. 484, Woodhead, Cambridge, U.K., 2007.
44. E. Heitz, R. Henkhaus, P. Rahmel, *Corrosion Science—An Experimental Approach*, Ellis Harwood, New York, 1992.
45. I. Kvernes, M. Seiersten, in *High Temperature Corrosion*, ed., R.A. Rapp, p. 615, NACE International, Houston, TX, 1983.

46. R. J. Fordham (ed.), *High Temperature Corrosion of Technical Ceramics*, Elsevier Applied Science, London, U.K., 1990.
47. R. Bürgel, *Handbuch Hochtemperatur-Werkstoftechnik*, Vieweg, Braunschweig, Germany, 2001.
48. P. L. Daniel, R. A. Rapp, *Adv. Corros. Sci. Technol.* 5; 55 (1976).
49. H. Latreche, S. Doublet, M. Schütze *Oxid. Met.*, 72; 1 (2009).
50. C. Schwalm, M. Schütze, *Mater. Corrosion* 51; 34 (2000).
51. M. Schütze, D. Renusch, M. Schorr, *Corrosion Eng. Sci. Tech.* 39; 157 (2004).
52. A. Rahmel, J. Tobolski, *Corrosion Sci.* 5; 333 (1965).
53. E. Essuman, G. H. Meier, J. Zurek, M. Hänsel, W. J. Quadakkers, *Oxid. Met.* 69; 143 (2008).
54. A. Donchev, H. Fietzek, V. Kolarik, D. Renusch, M. Schütze, *Mater. High Temp.* 22; 137 (2005).

14

Corrosion Prevention by Adsorbed Organic Monolayers and Ultrathin Plasma Polymer Films

Michael Rohwerder
Max Planck Institute for Iron Research

Martin Stratmann
Max Planck Institute for Iron Research

Guido Grundmeier
University of Paderborn

CONTENTS

14.1 Introduction 617
14.2 Organic Monolayer Films 621
14.2.1 Corrosion Protection by Self-Assembled Films 621
14.2.1.1 Self-Assembly on Oxide-Free Metal Surfaces 622
14.2.1.2 Self-Assembly on Oxide-Covered Metal Surfaces 625
14.2.2 Stability of Defective Monolayer Films 637
14.3 Corrosion Protecting Ultrathin Films 641
14.3.1 Introduction 641
14.3.2 Conversion Film Formation from Aqueous Solutions 642
14.3.2.1 Formation of Inorganic Amorphous Conversion Films 644
14.3.2.2 Formation of Inorganic/Organic Hybrid Films 645
14.3.3 Adsorption of Thin Organosilane Films from Aqueous Solutions 647
14.3.4 Plasma-Enhanced Chemical Vapor Deposition of Ultrathin Protective Films 651
14.3.4.1 Plasma Modification of Passive Films 651
14.3.4.2 Plasma Polymer Deposition of Ultrathin Protective Films 652
14.3.5 Ultrathin Coatings Based on Conducting Polymers 654
14.3.5.1 How to Prepare Coatings That Work 654
14.3.5.2 Intelligent Corrosion Protection 658
References 661

14.1 Introduction

Traditionally, corrosion protection of many reactive materials is associated with organic coatings, which are usually applied as thick multilayer systems with thicknesses ranging from 10 to 100 μm. However, increasing interest exists in thin and ultrathin corrosion protecting films. This is particularly true for modern areas of materials research such as microelectronic devices or micromechanics. In both cases, any corrosion protecting film must be much less than 0.1 μm in thickness and frequently even monolayer films are wanted. However, even for conventional materials such as steel, galvanized steel, or Al-based alloys, thin film coatings are of considerable interest as frequently such films are generated with unusual properties by modern surface technology. For such applications, the timescale for film preparation must be in the seconds range and many techniques allow only the preparation of thin films under those restrictions. However, as the phase boundary between a metal and an organic coating plays a key role in determining corrosion resistance and adhesion, ultrathin films applied to the metal surface have the potential to help tailoring exactly these properties. Thicker coatings are then applied in later steps on the ultrathin films by conventional techniques. This chapter focuses on three different ultrathin coating systems: self-assembled monolayers, plasma polymer (PP) coatings, and thin coatings based on conducting polymers, capable of intelligent corrosion protection.

The corrosion rate of reactive metals can be reduced significantly by a modification of the metal surface by organic molecules or polymers. A well-known example is the corrosion protection by lacquers and other organic coatings [1–3], which is used in most practical applications to protect, for example, cars against atmospheric corrosion, pipelines against corrosion in humid soil, and ships against corrosion in seawater. It has long been believed that the corrosion protection is mainly due to the barrier properties of the coating, which impedes the penetration of water and oxygen [4] to the metal/polymer interface. However, many coatings are highly permeable for water and oxygen, and therefore it is not the barrier effect on the diffusion process that gives rise to the corrosion stability but the specific electrochemical properties of the metal/polymer interface, in particular the formation of an extended diffuse double layer and the related decrease of electrochemical reaction rates [5–7]. In the presence of defects (pores, pinholes, etc.), which may penetrate through the coating, the diffusion barrier is lowered, and the delamination rate of the polymer at the defect is determined by the formation of galvanic elements [8–12]. In many circumstances, the local anode of these elements is the defect and the local cathode, at which predominantly oxygen is reduced, is given by the delamination frontier. In other cases—in particular for Al-based alloys—the galvanic element is just the opposite and the delamination frontier is characterized by the local anode. In the first case, the stability of the metal/polymer interface is determined by inhibition of the oxygen reduction at this interface, as during the electrochemical reduction of oxygen very aggressive species are liberated ($OH\cdot$, OH^-), which will immediately attack and destroy chemical bonds within the polymer [13–16]. The corrosion inhibition of the coating, therefore, depends more on the composition, structure, and chemical bonds of the metal/polymer interface than on the thickness of the coating.

It is therefore desirable to improve the chemical interaction between the first monolayer of the coating and the substrate so that electrochemical reactions such as the reduction of oxygen are inhibited and the bonds may withstand the attack of water and other aggressive species such as $OH\cdot$. Modern concepts are aimed at the development of "molecular adhesion promoters" (Figure 14.1) [17–19], which will provide a link between the substrate and the organic coating. This model requires, however, that reactive centers are prepared on top of the metal substrate, which may serve as anchor sites for suitable functional groups of the organic molecules, as shown in Figure 14.1. The question is how to select reactive centers, organic molecules, and reaction conditions [20–25].

In order to do so, it is useful to study the relevant literature, which is concerned with organic molecules being used as corrosion inhibitors [26–29]. It is well known that many organic compounds are able to reduce the corrosion rate of metals significantly [30–34].

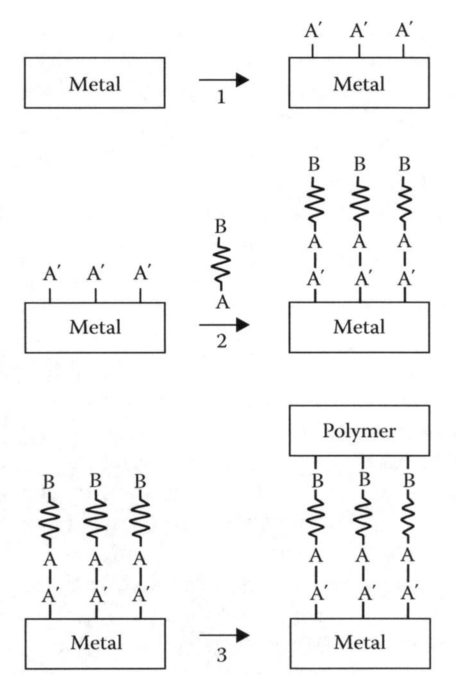

FIGURE 14.1
Principal scheme of the chemical modification of metal surfaces.

Inhibitors reduce the rate of either or both of the partial reactions of the corrosion process, the anodic metal dissolution and the cathodic oxygen reduction. Various inhibition mechanisms must be considered, taking into account the different situations created by changing, on the one hand, the chemical properties of the inhibitor and, on the other hand, the medium. The organic compounds may either be dissolved in the electrolyte (conventional inhibitors) or adsorbed onto the metal surface by condensation from the vapor phase (volatile corrosion inhibitors). From a mechanistic point of view [35], the inhibitors may change the composition of the medium (e.g., hydrazinium reduces the oxygen activity) or they may be enriched at the metal/medium interface. In the latter case, they may enhance the formation of passive films on top of the substrate, such as benzotriazole on Cu [30,31] or benzoates on Fe (passivators), or they may form monomolecular adsorption layers and prevent the dissolution of the substrate and the reduction of oxygen by changing the potential drop across the interface and/or the reaction mechanism (interface inhibitors) [36]. The interaction between organic ions or dipoles and electrically charged metal surfaces may be purely electrostatic, due to the electric field at the outer Helmholtz plane of the electric double layer. A competitive adsorption between water dipoles and the monomer has to be considered, and water adsorption always takes over the adsorption of nonpolar monomers if the electrode potential is shifted far from the potential of zero charge (PZC), as then the surface is highly charged [37]. The adsorption of the monomer prevails in the vicinity of the PZC. Cations are adsorbed at potentials negative to PZC and anions at potentials positive to the PZC. Because of the dependence on the charge of the metal surface, the adsorption is not specific for certain metals.

The charge of the anions or cations may be constant but also may change during adsorption. The number of electrons exchanged per adsorbed molecule is called the adsorption valence (typically less than 1) [38–40]. Because of the low activation energy, the adsorption based on pure electrostatic interaction is fast and reversible. If no electron transfer occurs, this kind of adsorption is frequently called physisorption. However, during adsorption, an electron transfer between the substrate and the adsorbed molecule may occur with orbital overlap between a single pair of electrons of the adsorbed molecule (N, S, P) and empty bands of the solid. Strong interaction is then observed (chemisorption) and the adsorption itself is highly irreversible [41–43]. Chemisorption is slower than physisorption, and it requires higher activation energy. In contrast to physisorption, it is specific for certain metals. Electron transfer is typical for transition metals having vacant low-energy d-electron orbitals. On the other hand, the inhibitor should have free single electron pairs or π electrons. The higher the polarizability of the involved heteroatom, the stronger is the adsorption (Lewis acid–base concept). The inhibitor is then the electron donor and the metal is the electron acceptor. This is in agreement with the principle of soft and hard acids and bases (SHAB).

It is clear from the situation just described that the latter situation is ideal for a stable chemical modification of the metal substrate and that molecules should be chosen that are able to interact strongly with the metallic substrate. However, there are some characteristic differences between corrosion inhibitors and molecular adhesion promoters: whereas for corrosion inhibition, the composition and structure of the metal surface are defined by the corrosive medium, the surface properties can be changed and adjusted to the structure of the adhesion promoter. Inhibitors must be soluble in the electrolyte (e.g., water) and can be applied only for well-defined reaction conditions. Adhesion promoters, however, may be applied from aqueous or nonaqueous solvents or even from the gas phase and the reaction conditions can be optimized for the given substrate. Based on the broad knowledge gained from decades of experience with inhibitors, it seems reasonable that some characteristic molecular features of inhibitors such as heteroatoms S, P, and O should be incorporated into the structure of the adhesion promoter; however, the molecule itself should show minimum solubility in water and the possibility to bind a polymer onto the adhesion promoter.

In this chapter, only very simple molecules are discussed. They are composed of one reactive center such as –SH, $-Si(OCH_3)_3$, or $-PO(OH)_2$, which should be able to bind to the metal surface [20,21] and a long aliphatic chain (e.g., $-C_{18}H_{37}$), which allows ordering of the individual molecules and the formation of a dense packing on top of the substrate (self-organization; see Figure 14.2). Ideally, for use not only for corrosion protection but also as adhesion promoters, these molecules should also have a terminal group that can react with functional groups within the organic coating that is applied on top of the monolayer film. In the following, we will discuss on a fundamental level how these modified surfaces may be prepared, what kind of structure is observed, and how stable the modification is in aggressive electrolytes.

The preparation technique must fulfill certain requirements: the surface properties of the substrate, for example, the density of chemisorbed OH groups, must be well defined, and the organic monomer must be allowed to bind to the reactive centers of the surface without destroying the defined surface structure; the monomer itself should form a dense structure with a high degree of ordering so that the substrate surface is not accessible to water molecules.

Another method for the modification of metal surfaces by ultrathin organic films is plasma polymerization. Plasma polymerization, as a process technology for corrosion-resistant thin-film deposition, has been explored during the last 20 years. PPs can be deposited from an electric discharge containing organic or metal–organic molecules [44,45]. A glow discharge is formed by exposing a gaseous monomer at reduced pressure to an electric field. The monomer is fragmented in the discharge and the reactive intermediates generated polymerize on a substrate according to a special reaction mechanism [46,47]. The resulting films can be highly cross-linked and, depending on process parameters, show more inorganic or more organic properties; moreover, adhesion is excellent to most metal

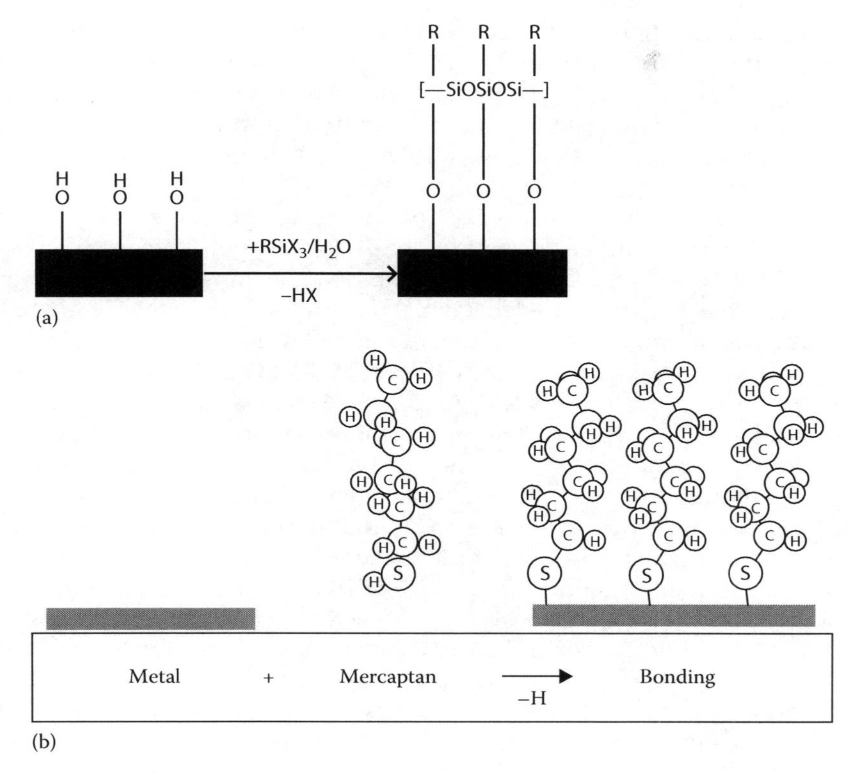

FIGURE 14.2
(a) Reaction of silanes with hydroxylated surface in humid atmospheres. (b) Reaction of mercaptans with metal surfaces.

surfaces, that is, the process is less specific to certain metals than is the case with molecular self-assembly (SA), and deposition of ultrathin films is fast. The main disadvantage is that until recently the process required a very low residual pressure; that is, vacuum equipment was needed. Lately, PPs have also been prepared under atmospheric pressure conditions.

The last ultrathin coatings systems to be discussed in this chapter are thin organic coatings based on conducting polymers. Conducting redox polymers such as polyaniline (PANI), polypyrrole (PPy), and polythiophene are controversially discussed for corrosion protection. In their conducting state, these polymers can play an active role in corrosion protection, not only a passive one as other organic coatings. However, under some conditions, these conducting polymers can also cause enhancement of corrosion. It will be discussed here how this can be prevented and how the positive properties can be optimized.

14.2 Organic Monolayer Films

14.2.1 Corrosion Protection by Self-Assembled Films

As outlined above, in order to improve the stability of the polymer/metal interface, it is of utmost importance to find ways to prepare interfaces that have better ability to inhibit oxygen reduction and are less vulnerable to the products of oxygen reduction. One way is to use monolayers of bifunctional molecules as adhesion promoters. Ideally, such a molecule should form a tight chemical bond to the metal or metal oxide surface with its head group and to the polymer with its tails groups. The monolayer should be as dense as possible with as few defects as possible, for optimum stability and inhibition capability. Also, for technical application, the formation of such monolayers should be quick, that is, finished within a few seconds.

The following paragraph will focus on electrochemical aspects of the self-organization and the resulting effect on the final defect structure. The discussion will distinguish between oxide-covered and oxide-free surfaces. The protective impact of the films will also be discussed.

In recent years, the process of molecular SA on solid surfaces, that is, the adsorption and self-organized formation of highly ordered monolayers from monomers in solution, has received increasing interest, especially the SA of thiols on gold, which is the focus of intense research since the late 1980s [48]. As could be shown, thiol monolayers proved to be excellent inhibitors of oxygen reduction and moreover are not easily destroyed by the radicals set free during the oxygen reduction [49]. First tests on iron also gave promising results [50,51], although the preparation is not easy on this substrate. In these studies, it was found that thiol molecules do not adsorb on iron oxide, the iron has to be electrochemically polarized to cathodic potentials in order to get rid of the oxide layer and then to keep it free of oxide during self-organization. As will be discussed in the following, since these early works, substantial progress has been made on many practical materials, such as iron and even zinc. Thiol SA on copper and silver is much easier to be achieved in good quality and is used, for example, as temporary corrosion protection for connectors in the electronics industry. Another important class of molecules used for SA of protective and adhesion promoting monolayers are phosphonates. These form monolayer films on many oxides and are already industrially applied on aluminum alloys. Many other classes of molecules could be listed here. Of great technical importance also are silanes and benzotriazoles, which also form ultrathin protective films, although these usually cannot be considered as two-dimensional monomolecular films as is the case for thiols and phosphonates. In the following, the main focus will be on the latter ones.

14.2.1.1 Self-Assembly on Oxide-Free Metal Surfaces

The best investigated example for surface modifications of metals are alkanethiols, $CH_3(CH_2)_nSH$, which consist of a long hydrocarbon chain with a sulfur head group with which they strongly adsorb onto the metal surface [48,52–54]; that is, a strong sulfur–metal bond is formed.

Besides on gold, thiols self-assemble also readily on the surfaces of silver and copper, and form well-ordered monolayers. These latter two materials find widespread technical application. However, both metals are subject to corrosion. In many cases, such as for use as electrical contacts in microelectronics, corrosion protection by thick, that is, insulating, organic coatings are no option. Here, ultrathin monolayer films are applied for temporary corrosion protection. These films are usually formed using corrosion-inhibiting molecules that are added to the cooling and process water to form thin diffusional barriers on the metal surface [55–58]. As in the case for gold, also on copper, the best investigated system is alkanethiols [52,59–70], such as on silver [52]. Reflection infrared spectroscopy indicates that monolayers on silver and on copper (when carefully prepared) have the chains in well-defined molecular orientations and in crystalline-like phase states, as has been observed on gold [52]. The formation of reproducible films on copper and silver is complicated as for both systems there is a certain sensitivity to air due to oxide formation, and to extended exposure to the solution containing the thiol. However, it is possible to produce reproducibly well-organized monolayer films, if established protocols are followed. For instance, if freshly evaporated copper samples are handled under nitrogen and anaerobically transferred to the thiol containing solution, well-defined monolayer films are obtained on the copper surface [52,71]. When less-stringent conditions are used, multilayer films form on the copper surface [72]. Also, extended exposure to alkanethiol solution results in reaction of the alkanethiol with the surfaces of silver and copper to form an interphase of metal sulfide on which an organized, oriented monolayer is supported.

In contrast, to function as inhibitors in a cooling or process stream, where inhibitor molecules can continuously adsorb on the surface and repair imperfections in the adsorbed protective film, the application of self-assembled monolayers as static protective films, such as for temporary corrosion protection of electric contacts, has to cope with the fact that there is no steady repair from a reservoir of inhibitors. Hence, high-quality monolayer films are required in order to ensure optimal protection. For copper and silver, self-assembled high-quality thiol monolayers efficiently inhibit corrosion of the underlying substrate in dry, humid, and wet oxidizing conditions. The performance of the SAMs (self-assembled monolayer) under these conditions was found to be insensitive to the concentration of water, suggesting that the layers are effective in screening the interaction between water and the metal substrate [73]. Over long-term exposures, the copper underneath the monolayer gets oxidized as shown in Figure 14.3.

The better the original quality of the monolayer, the better the protection. Hence, the optimal conditions for obtaining a monolayer of optimal performance are important to know. Usually, the concentration chosen for the SA of alkanethiols on gold is 1 mM in an ethanol solution. For gold, the suggested immersion time is 24 h [74], although it is found that better quality films are obtained from less-concentrated solutions, as thiol layers obtained from higher thiol concentrations are less structurally ordered even at similar overall thiol impingement [75]. Other factors such as temperature need to be controlled for high reproducibility. A usually neglected factor is the electrode potential of the sample. Even without active control of the electrode potential and even in the absence of intentionally added electrolyte-dissolved gases, impurities and dissociated solvent molecules will lead to a diffuse double layer and according to the balance between anodic and cathodic reaction to a certain electrode potential, which will sensitively depend on these uncontrolled factors. This is discussed in the following for the case of thiol SA on gold.

14.2.1.1.1 Thiol Self-Assembly on Gold

Now, even though self-organization is the subject of hundreds of publications, until very recently, nothing has been known about the influence of the electrode potential and surface charge on the SA process. For bare metal surfaces, the surface charge is controlled by the electrode potential and for oxide-covered samples by the pH of the solution.

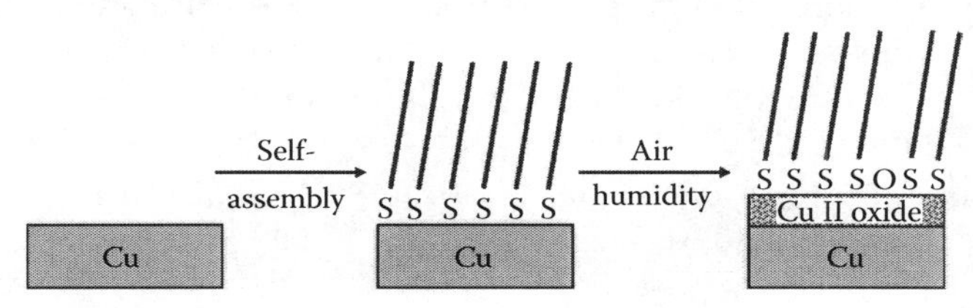

FIGURE 14.3
Sketch of thiol SA and of the changes due to oxidation. (After Jennings, G.K. and Laibinis, P.E., *Colloid. Surf. A Physicochem. Eng. Asp*, 116, 105, 1996.)

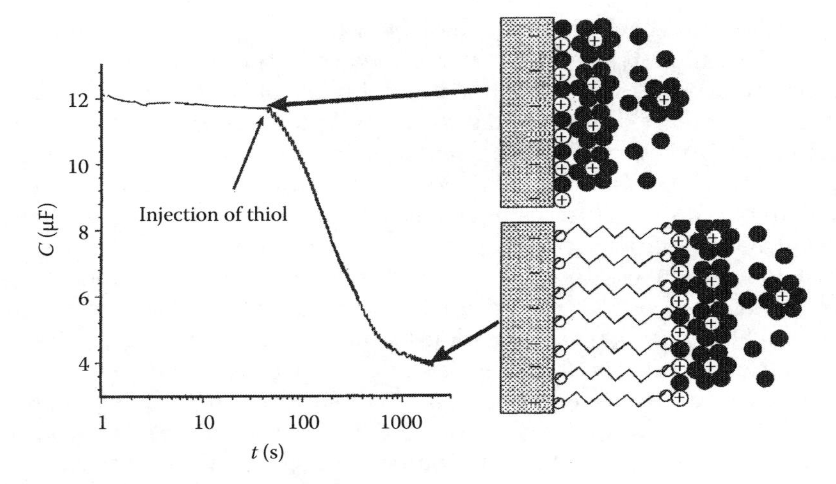

FIGURE 14.4
If the capacity of the gold surface is measured during thiol adsorption, a decrease in capacitance can be observed due to the pushing apart of the double layer by the adsorbing thiol.

The well-studied system thiol/Au is ideal for investigating the effect of the electrode potential on the kinetics of SA and on the resulting defect structure.

A direct way to monitor the SA *in situ* is to measure the decrease in capacitance of the immersed sample (see Figure 14.4). The double layer at the metal/electrolyte interface is pushed apart by the growing monolayer, which causes a decrease of capacitance. Because an electrode's capacitance depends on the electrode potential, it is useful to normalize curves obtained for adsorption at different potentials, for example, by referring the change in capacitance ΔC at the time t to the final capacity change ΔC_{max}. Such normalized curves for the adsorption at −400 and −800 mV versus a special Ag/AgCl reference electrode [19,76] in ethanol and typical nanoscopic structures of the films at different stages of the thiol SA are shown in Figure 14.5. A detailed analysis of the curves yields the distinction of three characteristic potential ranges for thiol SA [19,76].

An intermediate potential range from −400 to +200 mV versus the Ag/AgCl reference electrode where the rate of SA shows only a comparatively weak dependence on the electrode potential; scanning tunneling microscopy (STM) images of the completed films show basically the same features already known from the literature [77–85].

A cathodic potential range where the rate of SA decreases significantly with increasingly cathodic potentials; STM images of the completed films show that the average domain size also increases, so that for adsorption at −800 or −900 mV the thiol domains fill whole terraces, even if the lateral size of these exceeds 100 or even 200 nm (Figure 14.5).

An anodic potential range where there is a slight decrease in the rate of SA with increasingly anodic potentials.

Whereas the film is finished to 95% within 10 s at potentials of −400 to +200 mV, yielding a film composed of domains with a lateral size of 10–20 nm takes thousands of seconds at very cathodic potentials for the domains to cover the surface, but the domains are much

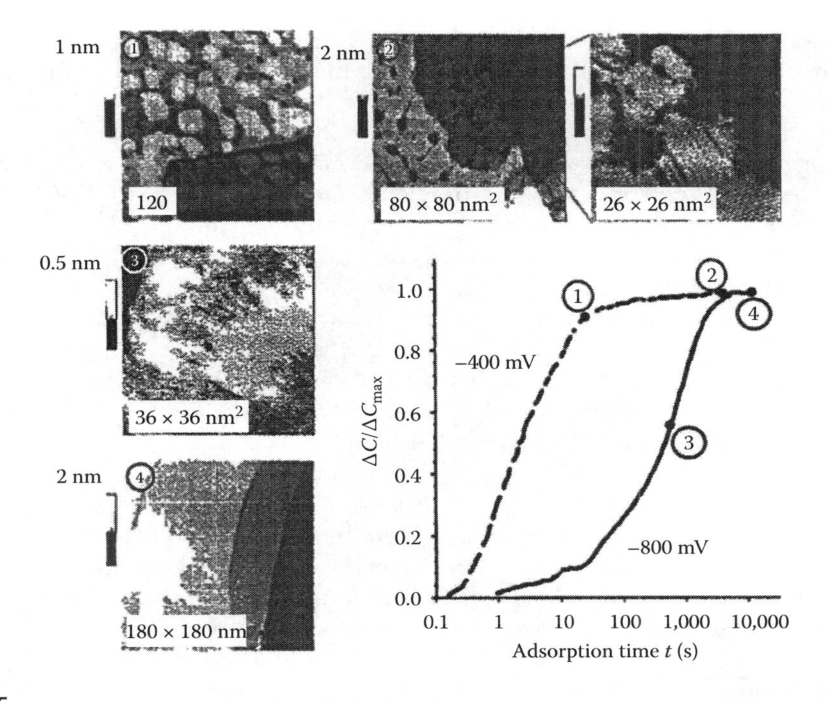

FIGURE 14.5
The normalized capacity curves clearly show that thiol (here decanethiol) SA is much faster at intermediate potentials (here −400 mV) than at cathodic potentials (−800 mV). Typical STM images of the film structure are shown for different stages of SA at the two potentials. At −400 mV, thiol SA is fast and occurs via small individual domains (1), which sets the foundation for the final domain structure (2). At −800 mV, SA is very slow and occurs via domains that are interconnected by molecular rows of chemisorbed thiol (3), and thus very large domains result in the final film (4).

larger. A detailed analysis shows that the thiol adsorption is at least a two-step process: physisorption followed by chemisorption [76a]:

$$R-SH_{sol} \xrightarrow{\text{Phys.}} R-SH_{ad}$$

$$R-SH_{ad} \xrightarrow{\text{Chem.}} R-S-Au+H^{+}+e^{-}\ \text{(oxidation)}$$

It is the electrode potential dependence of the latter step that is responsible for the extreme slowdown in adsorption velocity at cathodic potentials.

Assuming a symmetry factor $\beta = 0.5$ (because of the symmetry of the system) and a complete charge transfer, it can be shown from the measured capacity curves that about 36% of the overall potential difference between the electrode and the solution occurs over the alkane chains; that is, the longer the alkane chain, the smaller the slowdown effect at cathodic potentials. This could be shown for the adsorption of octadecylthiol as compared with decylthiol [76b].

14.2.1.2 Self-Assembly on Oxide-Covered Metal Surfaces

Since thiols are known to form a strong sulfur–metal bond and, as discussed above, it is also known that oxide formation is a main factor in lowering the thiol monolayer quality on copper and silver, it is surprising that thiol SA is also observed on the oxide-covered surfaces of reactive materials such as iron and zinc.

Presumably, thiol adsorption on these surfaces involves desorption of surface oxygen and hydroxide groups and the thiol seems to be mainly coordinated to the metal atoms in the oxide, as x-ray photoelectron spectroscopy (XPS) indicates, for instance, for thiol adsorption on zinc-terminated ZnO [86,87], where the thiol reduces the oxygen at the surface and forms a zinc–sulfur bond. O-polar ZnO, on the other hand, seems only to adsorb the thiols molecularly at room temperature [87].

Thiols adsorbed on oxidized metals have been studied, though not as extensively as gold, and include substrates such as tin [88], iron [51,89], nickel [90–94], zinc [95–101], and alloys such as CuNi [102] or ZnNi [103].

Thiols on oxidizable metals have been proven to act as good molecular corrosion inhibitors [97–99,104], provided the interface between the thiol group and the metal is well controlled. On nickel, this molecular barrier can resist several weeks to protect against corrosion [90]. However, mostly this happens by the formation of a thick and compact inhibitor film, which is not discussed further here.

14.2.1.2.1 Thiol Adsorption on Zinc

Since zinc is an active metal, it is more sensitive than gold to immersion time, concentration of thiols, and choice of the solvent. It has been proposed that a higher concentration of the thiol leads to a higher coverage [95]. It was found that alkanethiols form an ordered self-assembled adsorbate on both oxidized and reduced zinc [99].

According to Hedberg et al., alkanethiols adsorb on reduced and oxidized zinc surfaces through formation of Zn–S bonds [101] and three steps in the adsorption of octadecanethiol (ODT) on reduced zinc can be distinguished, as schematically depicted in Figure 14.6. For a 1 mM solution of ODT in ethanol, in the first step, the thiol forms ordered, highly oriented, domains, although the number of adsorbed molecules is far from saturation. This first step is very fast and just takes a few seconds. Despite the incomplete coverage, the ordered thiol regions measurably improve the corrosion protection performance [101]. The next step takes about 3 s to 1 h of immersion time. After 1 h, the number of molecules adsorbed on the reduced zinc surface reaches saturation, but the monolayer film becomes less ordered compared to that of 3 s immersion time. The corrosion protection properties of the thiol film, however, do increase during this period, despite the decreased order. In the third and last step, the thiol layer becomes even less ordered, which is accompanied also by a reduction of the protective properties. Hence, after 1 h adsorption, the optimal corrosion resistance is achieved.

This is in clear contrast to thiols on gold. For that system, longer immersion times continuously improve the ordering of the thiol. This is most likely due to the effect that during longer immersion time the corrosiveness of the ethanolic thiol solution overcomes the protective properties of the thiol film. Obviously, thiol adsorption on zinc is much more complicated than on gold and the procedure of film preparation becomes a by far more critical parameter.

On oxidized zinc, the adsorption process is slower, and only after 5 min immersion time an ordered, but again incomplete, film is formed. The reason for this slower rate of formation

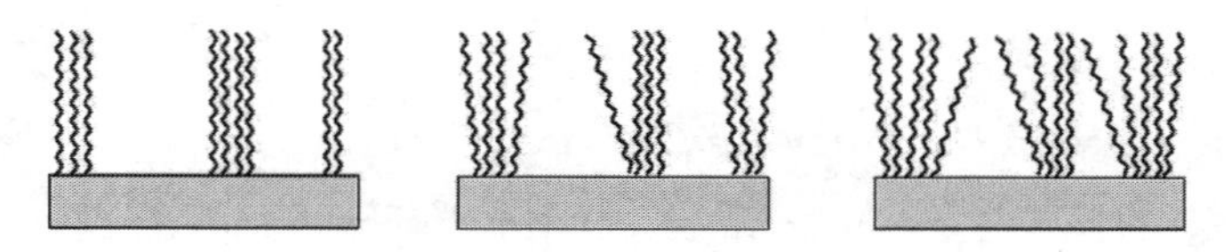

FIGURE 14.6
The three stages of thiol film formation on zinc oxide according to Hedberg et al. (From Leidheiser Jr., H., *Corrosion Control by Organic Coatings*, NACE, Houston, TX, p. 87, 1981.)

of ordered regions may be that on rougher surfaces the van der Waals forces between the molecules might be reduced due to an increased distance between the chains [101].

Another reason for this slower formation of ordered regions may be that the thiol molecules will have to desorb the oxide in order to gain access to the zinc atoms to which they adsorb.

During the second step, which takes from 5 min to about 2 h of immersion time, the number of adsorbed molecules reaches saturation, and most likely, similarly to reduced zinc, accompanied by an increased number of disordered molecules. Also here, an increase in corrosion protection is observed. In contrast to reduced zinc, not much changes during longer immersion, most likely due to the higher protective properties of the passivated zinc surface in the ethanol solution.

Higher concentration (20 mM) of ODT in the ethanol solution (compared to 1 mM) increases the protective properties of the thiol film on both the reduced and the oxidized zinc. For longer immersion times, the highest improvement in the impedance is seen for the reduced zinc. This indicates that there are competing reactions alongside the adsorption and organization of the ODT adsorbate.

In summary, the adsorbate formed by ODT on reduced and oxidized zinc is heterogeneous and becomes less ordered when immersed for longer times in 1 mM concentrated solution, a behavior that is distinctly different from that on gold.

14.2.1.2.2 Thiol on Iron

Also thiol adsorption on iron is possible. For this case, we will discuss how one of the main problems of modifying reactive, that is, oxide-covered, metals by thiol monolayers can be overcome: the long adsorption times necessary for obtaining good quality films. For technical application, this is a problem because adsorption times of the order of thousands of seconds are out of the question. In order to circumvent this problem, another coating technique can be applied: the sample (the working electrode in the figure) is polarized to cathodic potentials to reduce the oxide layer and to prevent oxide formation during the coating process; it is then pulled through a thiol film, floating on top of the aqueous electrolyte [50,51]. In this way, about 10-nm-thick multilayer films can be prepared, which show excellent blocking of oxygen reduction (see Figure 14.7). It is quite obvious that the oxygen reduction, a key reaction during corrosion, is drastically limited on the modified surface and a much higher cathodic overpotential is necessary to reduce oxygen at significant rates. If oxygen is reduced, however, the film is destroyed by the radicals formed during oxygen reduction [19].

For adhesion promotion only the first monolayer will be effective. Because it proved not to be possible to prevent iron surface reoxidation completely while pulling the sample through the floating thiol film, and thus parts of the sample are not covered by chemisorbed thiols but by oxide, it is necessary to get rid of such defects in a second preparation

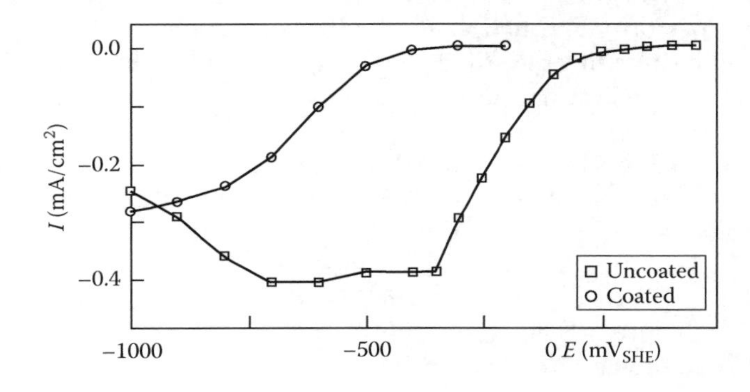

FIGURE 14.7
Rate of oxygen reduction on iron (○) and iron modified by one monolayer of *n*-decylmercaptan (□).

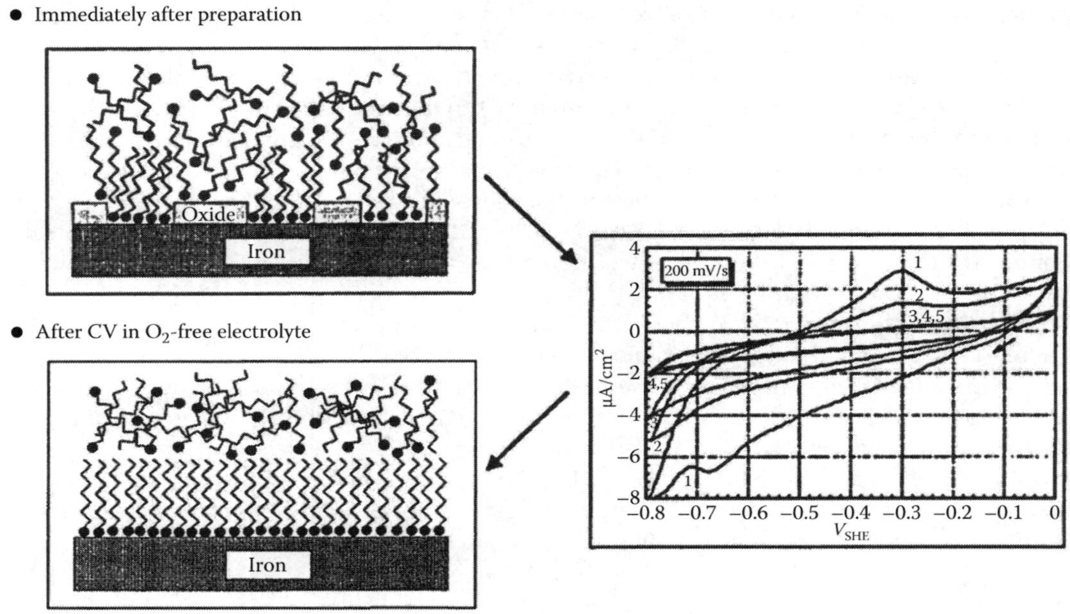

FIGURE 14.8
Model of a thiol multilayer on iron with remnants of oxide (upper left). After several cycles of the electrode potential (CV) in O_2-free electrolyte (right), the interface between thiol and iron is healed (lower left).

step in order to improve the quality of the chemisorbed monolayer beneath the multilayer film. Immersing the as-prepared sample (i.e., with the multilayer thiol film) in an aqueous solution and cycling the potential from the cathodic limit, where hydrogen evolution is beginning, through the potential range where iron is oxidized (peak in curves 1 and 2), and back for several times results in a healing of the disordered film (see Figure 14.8), as can be seen from the decrease of the oxidation peak with increasing number of cycles. With each sweep into the potential range where iron oxide is reduced, bare iron is exposed at the defect sites to the thiol in the multilayer, and in the subsequent anodic sweep the thiol can at least partly be chemisorbed before reoxidation of the remaining bare iron surface sets in. Finally, the entire surface is covered by chemisorbed thiol molecules. Then the modified sample is pulled out of the solution and the polymer coating is applied. Alternatively, the sample may be polarized at negative −800 mV. But then, the healing process takes much longer (see Figure 14.9a) because the thiol chemisorption at cathodic potentials is very slow (see earlier sections). Most of the multilayer thiol is dissolved into the polymer so that the first chemisorbed monolayer should perform its function as an adhesion-promoting layer. Scanning Kelvin probe (SKP) mappings of a thus modified iron sample coated with a polymer show more than three times slower delamination kinetics. Still, a more pronounced effect is desirable for future applications.

Closer investigation shows that the bond between the first thiol monolayer and the metal surface is destroyed by reoxidation of the iron surface beneath the multilayer film within tens of seconds. Of course, this results in diminished ability of the layers to promote adhesion compared with what would be expected from an intact monolayer. The faster the reoxidation the thinner the multilayer film. For a monolayer film, the reoxidation takes about 10 s. That is an impressive factor of 10^6–10^7 slower than for bare iron, achieved by an only 1-nm-thick monolayer film, as can be seen from Figure 14.9b.

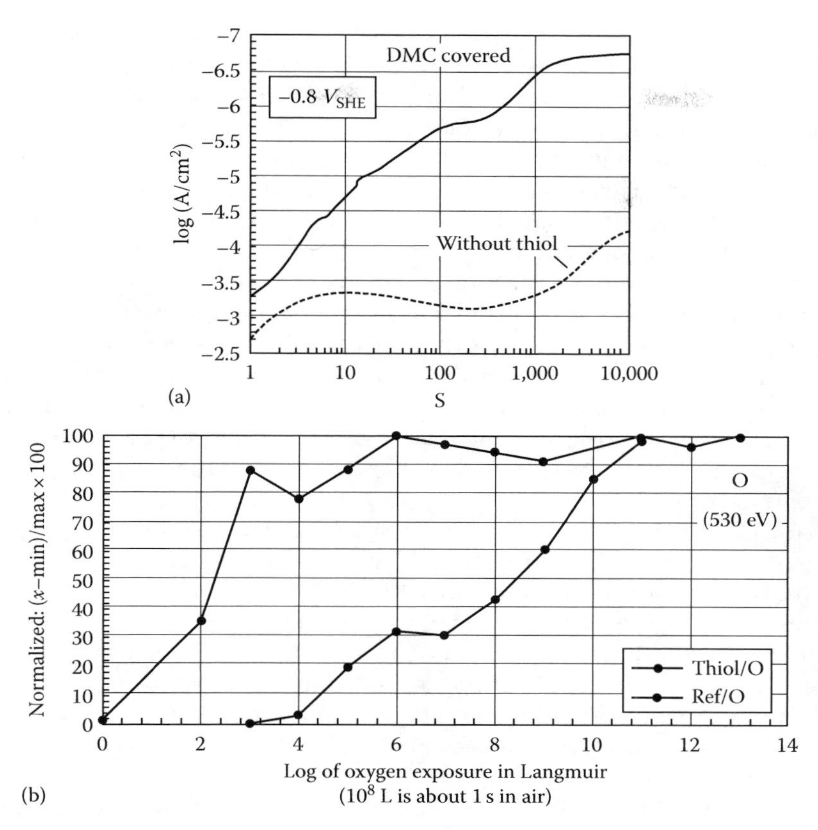

FIGURE 14.9
(a) Current observed during healing at −800 mV of a decanethiol monolayer compared with bare iron. (b) Normalized oxygen peak versus oxygen exposure for thiol-covered iron in comparison with bare iron.

A first step to find ways to improve this technique is an understanding of the reoxidation process. Other important parameters are chain length and functional groups of the thiol molecules. If the choice of more suitable thiols together with a more refined modification technique resulted in a thiol layer that survives even one order of magnitude longer in air, this could be an important breakthrough because then the polymer coating could be applied before most of the thiol film is destroyed.

14.2.1.2.3 Phosphonate Self-Assembly on Stable Oxide Surfaces

Surface modification by phosphonates or silanes plays an important role in technical corrosion protection. While organosilanes are widely discussed as adhesion promoters and for corrosion protection, less has been reported about organophosphonates. For this reason and because for the case of silanes the metal surface is usually not modified by a monolayer but by an ultrathin polymeric layer, the main focus of the following shall be on phosphonates. A good overview on silanes is given by Child and van Ooij [105,106]. Silane treatments of metals can increase their corrosion performance not only after painting but protection can also be obtained even without paint coatings, especially if nonorganofunctional silanes are used. Hence, silane treatments have the potential to replace currently used pretreatments by chromate and/or phosphate systems.

Whereas in the case of iron the adhesion layers have to be adsorbed directly on the metal because the oxides are unstable, the situation is different for metals that form stable oxides. Here it is better to have the films formed on the stable oxide. Phosphonates are suitable for real monolayer SA and hence allow well-controlled modification of the surface properties. Especially for aluminum phosphonate, SA plays already a considerable role in industrial application. On aluminum oxides, phosphonates, $X–R–PO_3^{2-}$, form stable and well-ordered monolayers. In $X–R–PO_3^{2-}$, X designates the functional end group, R the alkane chain, and PO_3^{2-} the head group. For long-chain organophosphonate monolayers, it was observed that surface hydroxyl groups promote the adsorption of the organophosphonate and that the adhesion of the phosphonate group is based on acid–base interactions [107,108]. The driving force is the formation of a surface salt as already described for long-chain carboxylic acids by Allara and Nuzzo [109] and recently by Terryn and coworkers [110]. Adolphi et al. [111] showed that by the adsorption of organophosphonic acids on aluminum oxide as well as on tantalum oxide surfaces, bidentates via the two hydroxyl functions of the phosphonic acid are formed, whereas the adsorption on titanium oxide surfaces leads to the formation of tridentates.

With technical samples, high-resolution STM, as it is the case for thiols on gold, is not applicable to obtain information about the molecular order. Other methods such as Fourier transform infrared (FTIR) spectroscopy and XPS have to be applied.

As mentioned above, the adsorption of phosphonates on aluminum surfaces is an acid–base reaction. Figure 14.10 shows FTIR spectra supporting this theory. The similarity in the spectra of a Zn-biphosphonate multilayer on gold and of an SA film of biphosphonic acid on aluminum oxide clearly shows the formation of the surface salt, sketched in Figure 14.11. The peaks at wave numbers 1100 and 950 cm^{-1} are attributed to vibrational modes of the RPO_3^{2-} anion.

This adsorption reaction requires the presence of OH groups on the aluminum oxide surface. This dependence becomes obvious from Figure 14.12 [112]: Different levels of OH on the oxide surface have been adjusted by gas-phase adsorption of different amounts of oxygen and water on aluminum in a special ultrahigh vacuum (UHV) chamber, which also allows electrochemical experiments and adsorption from solution without exposing

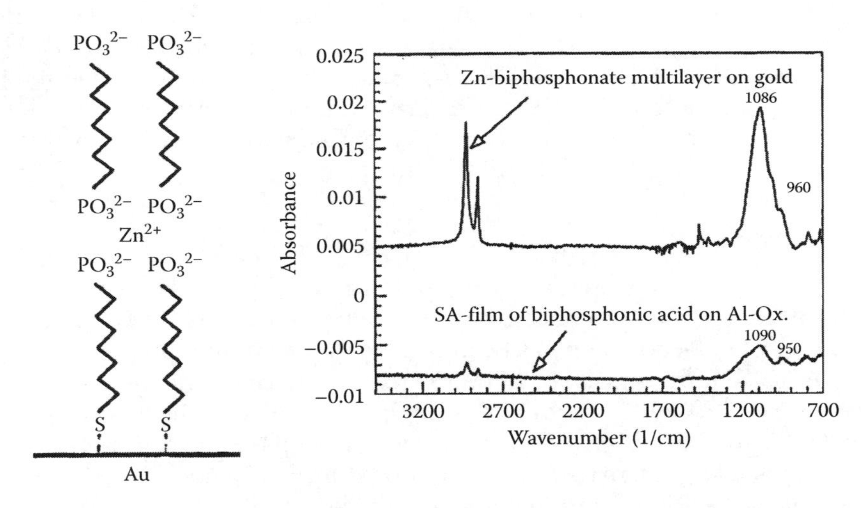

FIGURE 14.10
Infrared adsorbance spectra of a phosphonate film on an aluminum oxide surface and of Zn-biphosphonate multilayer on gold (see the figure on the left).

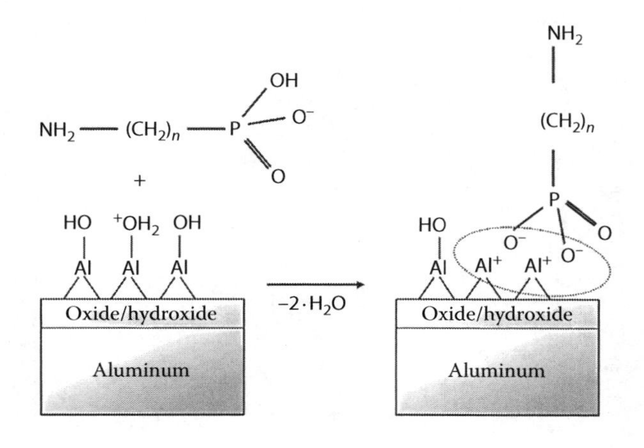

FIGURE 14.11
Sketch of the acid–base reaction during phosphonate adsorption.

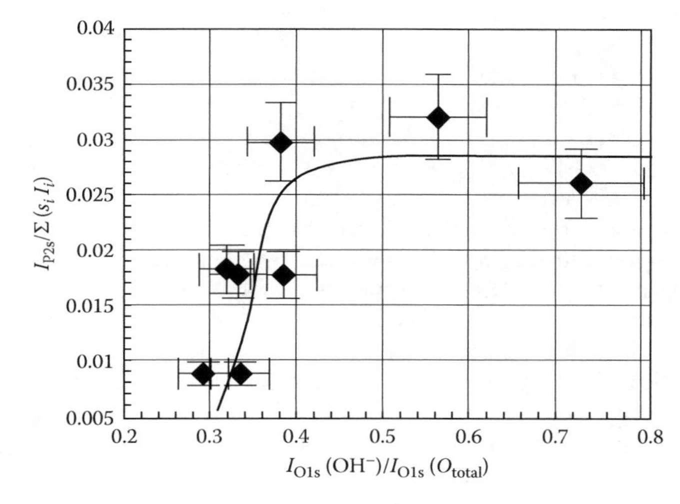

FIGURE 14.12
Intensity of the phosphorus 2s photoelectron signal as a function of the OH^- content in the oxide surface.

the samples to air. After the OH ratio in the oxide layer was determined with XPS, the sample was exposed to the phosphonate solution and then the amount of adsorbed phosphonates was determined with XPS. Figure 14.12 shows that for low OH content in the oxide surface no phosphonate adsorption occurs. This was confirmed by Giza et al. [113] who found that, with an increase in the surface hydroxyl density, the adsorption kinetics can be accelerated. Such acceleration can be explained by the adsorption of the phosphonic acid via surface hydrogen bonds prior to the condensation reaction, leading to the finally adsorbed phosphonate.

Because aluminum is only rarely used as the pure material, it is of crucial importance for successful application of protecting and adhesion-promoting monolayer films that not only the aluminum surface is covered by the molecules but also the inclusions, such as Al_2Fe, as they are typical of most aluminum alloys. Even though these inclusions of submicron to few microns size usually cover less than a few percent of the surface, they play a key role in the delamination of polymer films from aluminum alloys because they are not covered by an insulating passive oxide and thus reactions such as oxygen reduction can easily occur on their surface.

Cathodic water decomposition and especially oxygen reduction require considerable overvoltages on Al and Al alloys. Usually, potentials well below $-400\,mV_{SHE}$ are required

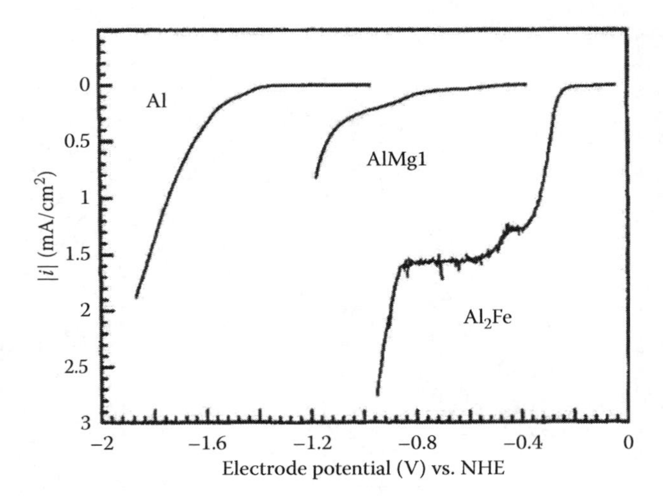

FIGURE 14.13
Current/potential curves for Al, AlMg1, and Al_2Fe in oxygen-purged 0.1 M Na_2SO_4 electrolyte.

to activate a cathodic current. Thus, both water decomposition and oxygen reduction occur together and both contribute to the overall current density. The effect of the inclusions on the current/potential curves can be seen in Figure 14.13 where the curves for aluminum, Al_2Fe, and AlMg1 in an oxygen-purged 0.1 M $NaSO_4$ electrolyte are shown [114]. The scans start at the open circuit potential (OCP) and are ramped with a velocity of 1 mV/s, while the samples are rotated at 200 rpm. Obviously, the inclusions in the AlMg1 act as active sites for the cathodic reactions and hence play the crucial role in the corrosion behavior.

Investigation of the inclusions in an AlMg1 surface that has been polarized to cathodic potentials for a short time typically shows a corrosive attack on the Al matrix around the inclusion. Obviously, the inclusions act as local cathodes and the matrix in their vicinity will dissolve anodically in the alkaline electrolyte around the inclusion. The extent of the attack on the Al matrix surrounding the inclusion, that is, the trenches around the inclusions, varies considerably. This means the oxygen reduction occurs to a different extent on the different inclusions, which can easily be explained by the concept of a microelectrode array: at low rotation velocities, the oxygen transport to the inclusions is limited and diffusion cones issuing from the inclusions will evolve and overlap at a certain distance from the surface, from which a planar diffusion zone is effective. Thus, the situation occurs that diffusion cones of inclusions activate earlier than others, use up all the oxygen available, and thus shield the other inclusions from oxygen. Figure 14.14 shows a scanning electron microscope (SEM) and a scanning Auger microscope (SAM) image of such an inclusion. Clearly visible is the corrosion trench in the matrix surrounding the inclusion; also interesting is the ring-like deposit of oxides around the trench. The surface of the inclusion is mostly Fe_3O_4, whereas before its activation, it showed a high content of aluminum oxide. The lower part of Figure 14.14 depicts a schematic drawing of the processes going on at the inclusions. Activation of the surface, that is, the dissolution of the aluminum oxides from the surface of the inclusion as it is induced by an increase in pH caused by oxygen reduction, results in an iron oxide–rich surface; oxygen reduction can take place with high reaction rates, aluminum is dissolved from the matrix surrounding the inclusion, and aluminum hydroxides precipitate in the vicinity of the trenches.

A good test of the protective properties of self-assembled monolayer films is to investigate to what extent these layers can inhibit the activation of the inclusions. This can be done by cathodic potential jumps and monitoring the current transients. Figure 14.15 shows such transients for an unmodified surface at different rotation velocities.

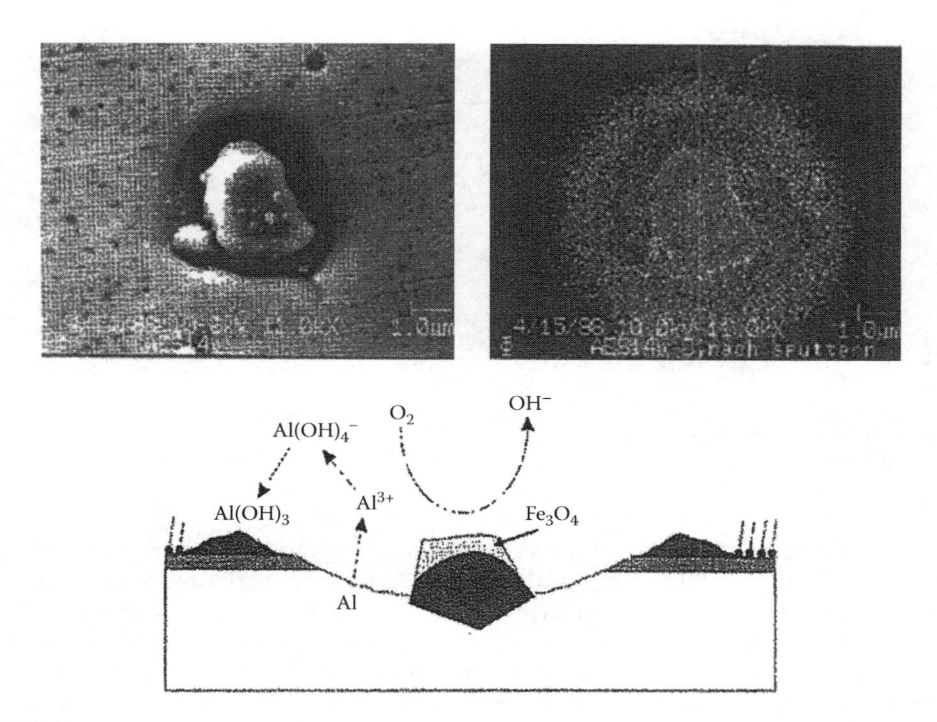

FIGURE 14.14
(Top) SEM image and oxygen SAM mapping of an Al_2Fe inclusion in AlMg1. (Bottom) Reaction scheme for corrosion at the inclusions.

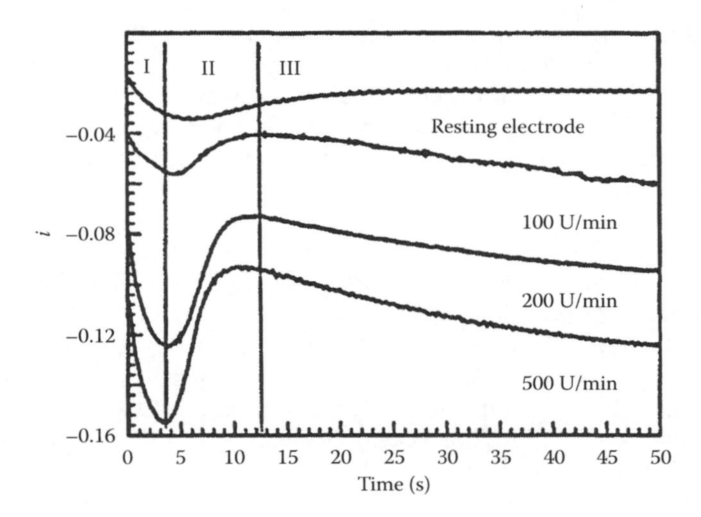

FIGURE 14.15
Current transients for AlMg1 samples after potential jumps from the OCP to −800 mV for different rotation velocities.

Three stages of the $i(t)$ curves can be distinguished. For the first few seconds, an onset of the cathodic currents can be seen. Here the activation of the inclusions occurs. The decrease in the cathodic current in stage II is most likely due to the onset of the anodic aluminum corrosion in the surrounding matrix. The behavior in stage III is more complicated to explain; it is assumed that the deposition of corrosion products plays an important role here.

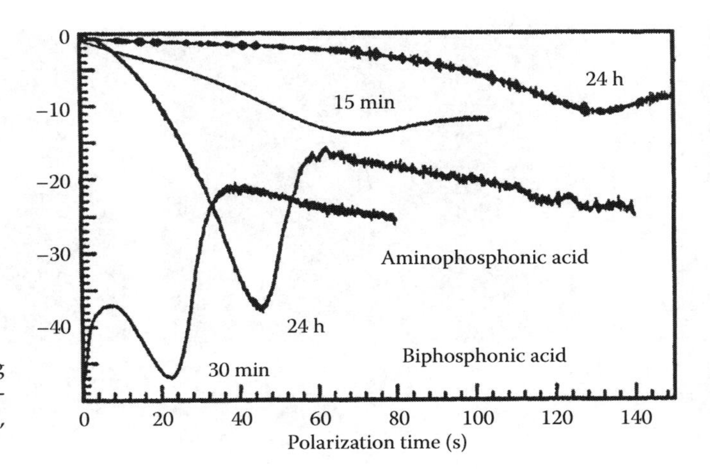

FIGURE 14.16
Current transients for the resting samples after modification with aminophosphonic or biphosphonic acid, respectively.

Figure 14.16 shows the transients for AlMg1 samples modified with biphosphonic or aminophosphonic acid; samples modified with octadecylphosphonic acid do not show much difference from the unmodified samples. Especially, the aminophosphonic acids show a significant improvement in the corrosion behavior. Sputter profiles on the matrix and inclusions show that similar films are formed on the Al matrix but that the aminophosphonic acid forms several-nanometer-thick multilayer films on the inclusions, much thicker films than the biphosphonic acid. In this way, the inclusions are effectively passivated with aminophosphonic acids. Similar observations have more recently been made for the effect of phosphonate modification of Al6016 alloy [115].

At a first look, the decrease of the peak current by a factor of 4–5 and the delay of the activation peak by a factor of 20–40, from 3–4 to 60–120 s, does not seem to be sufficient for an effective improvement. However, if an additional polymer coating is applied, a factor of 20 of increase in activation time might extend the lifetime of a product by several years. For this reason, phosphonate monolayers are suitable candidates for efficient adhesion promotion and corrosion protection. In fact, industrial application for this has already started and is spreading fast.

14.2.1.2.4 Adhesion Promotion by Phosphonate Films

Figure 14.17 shows the effect of the phosphonate monolayer (self-assembled *N*-ethylaminophosphonate, NEAP) on the delamination of an amine-modified epoxide ester from Fe–6.3Al–alloy surfaces. This ester is a basis for a number of industrial primers. As can be seen, the just polymer-coated alloy is delaminated quite rapidly. Delamination is much slower if the alloy is thermally treated in such a way that a thin insulating layer is formed by selective oxidation, and thus oxygen reduction is inhibited. If an additional phosphonate layer is built into the interface, the delamination is also significantly slowed down [116].

Besides inhibiting oxygen reduction, another role of the self-assembled monolayer is to improve the adhesion of the polymer film to the surface under mechanical strain, which of course also plays a role in slowing down coating delamination. Usually, it is assumed that it is the covalent bond of the head group of the molecules to the surface and the tail group to the polymer that is responsible for the improved mechanical stability. From many applications, it is known that this is not the main factor. For instance, in order to protect microelectronic devices, especially microchips, from harmful environmental influences, more than 90% of all semiconductor devices are encapsulated with polymeric compounds.

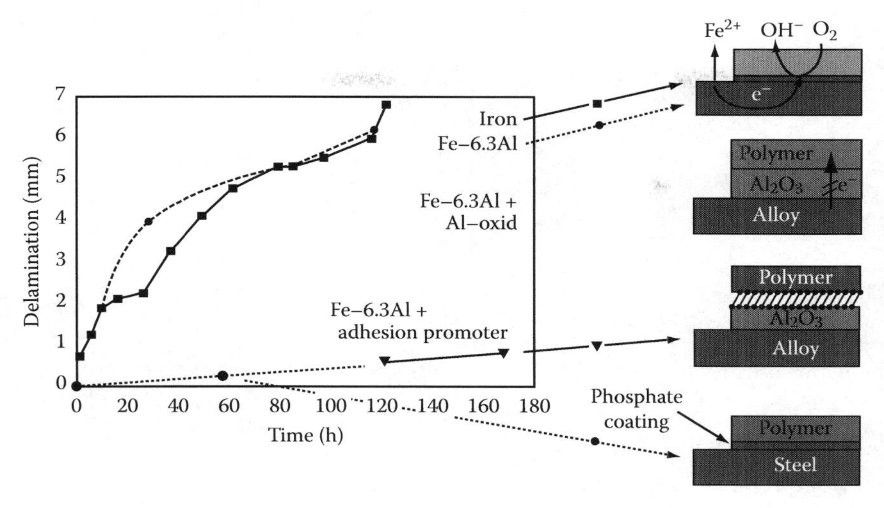

FIGURE 14.17
Delamination of an amino-modified epoxide ester from differently pretreated Fe–6.3Al-alloy surfaces and phosphated steel as a reference.

Often the polymers have to be briefly heated up to temperatures around 200°C–250°C to improve their protective properties. In this case, the explosive evaporation of adsorbed moisture can cause considerable mechanical strain at the polymer/substrate interface, which in turn might result in breakage of the polymer/substrate composite at the interface. Interestingly, in most cases, the failure occurs not directly at the interface but near the interface [117–124]. For this kind of adhesive breakage near the interface, several nanometers of polymer coating are usually still covering the substrate. It seems that the molecular forces between substrate and polymer are not crucial for the failure, but the transition zone from the substrate into the polymer bulk plays the key role.

In order to investigate the effects of an aminophosphonate monolayer on the adhesive properties, aluminum samples were modified with NEAP and then coated with a polycyanurate film [125], a prepolymer of the dicyanate of bisphenol A (DCBA), which consisted mostly of monomers and trimers of DCBA, and was dissolved in tetrahydrofuran (THF; 25 mg/mL) and then the prepolymer solution spin coated on the samples and exposed for 2 h at 220°C in laboratory air to complete the polymerization. The triazine rings in the DCBA monomer are IR inactive when parallel to the substrate surface and increasingly IR active with increasing normal component. Figure 14.18 shows the normalized intensity of the corresponding IR signal (at wave number 1380 cm^{-1}) as a function of the thickness of the polymer film for NEAP-modified and NEAP-unmodified samples. The normalization of the signal was to divide the signal obtained from thicker polymer films (>40 nm) by their thickness as obtained by ellipsometry and set this to one. It can be seen from Figure 14.18 that for the unmodified samples the orientation of the triazine rings in the monomers near the interface has, in comparison with the statistically oriented molecules in the polymer bulk, a distinct preferred component parallel to the surface. This preferred orientation extends up to about 10 nm from the interface. For the modified sample, this transition zone extents only over the first 2 nm from the surface. The reason for this is that the NEAP molecules are adsorbed on the surface as an oriented monolayer with the secondary amino group of the NEAP molecules at the surface. This amino group can react with the cyanate group of the polycyanurate prepolymer to an

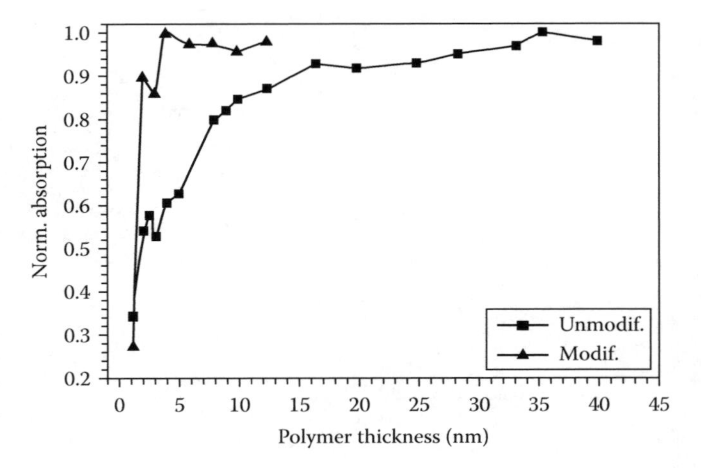

FIGURE 14.18
Normalized infrared adsorbance at the wave number 1380 cm^{-1} for polycyanurate films of different thickness for the unmodified and the NEAP-modified sample.

isourea bond, forcing it to a more or less upright orientation. Thus, the transition zone at the interface is reduced by a factor of 5. This is shown schematically in Figure 14.19.

Technical butt joint tests on samples coated with 400 nm polycyanurate polymer result, for the samples not modified with NEAP, in failures of the polymer/aluminum bond for stresses between 2 and 6 MPa, depending on the relative humidity. Interestingly, the failure occurred not directly at the interface but about 6–10 nm away from it, as could be derived from SEM and XPS measurements. This clearly shows that failure occurs within the transition zone. For samples modified with NEAP, no failure of the polycyanurate-polymer/NEAP/aluminum composite could be measured because of cohesive breakage in the coupling glue (XD4600) used for the test at about 7 MPa, independent of humidity; that is, much higher stability is to be expected.

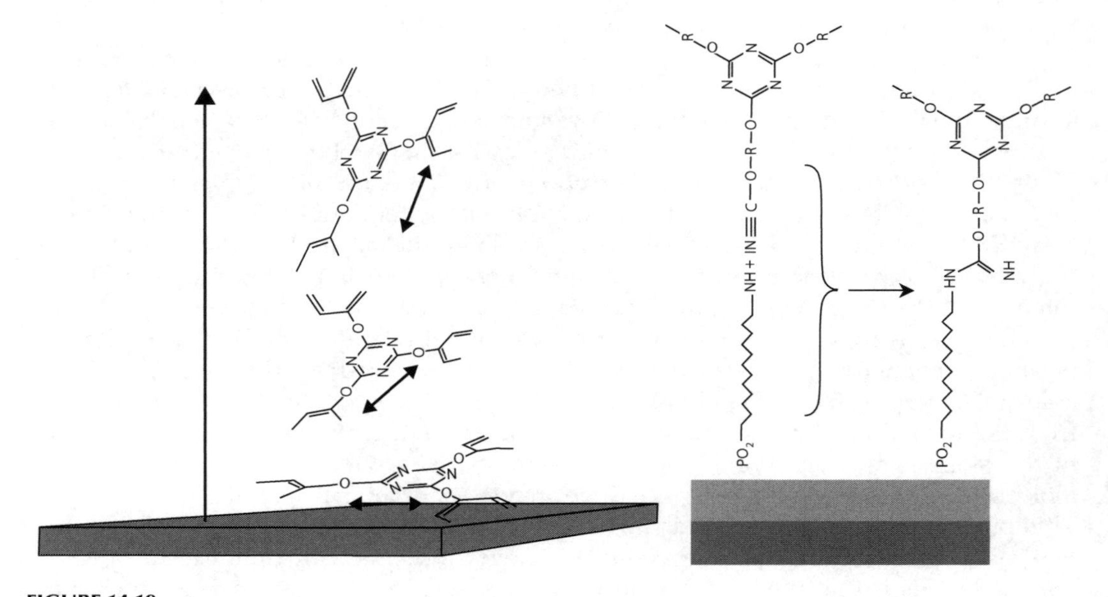

FIGURE 14.19
(Left) Supposed molecular structure within the polymer near the unmodified surface. The orientation of the molecules is parallel to the surface. (Right) Molecular structure at the interface in the case of NEAP modification.

14.2.2 Stability of Defective Monolayer Films

For the evaluation of the corrosion performance, it is standard to test the behavior of coating systems with artificially applied defects down to the metal. This is important because a coating system may show very good corrosion performance in the presence of only small defects, such as pinholes, but may break down very quickly in the presence of larger ones. As will be shown in the following, this is an often observed case for corrosion performance, for example, of coating systems based on conducting polymers.

Besides extrinsic defects also intrinsic defects, that is, defects in the coatings system itself, are of importance for the later performance. Hence, for ultrathin monolayer films applied between the metal and further organic coating systems for corrosion protection and adhesion promotion, defects in the monolayer film may play an especially critical role. For instance, it may be assumed that a monolayer film that is still quite disordered may show a significantly decreased corrosion performance as compared to a highly ordered one. However, practical experience from industrial application, for example, in electronics industry from modifying electric contacts or from industries modifying aluminum (i.e., alumina surfaces) with phosphonate self-assembled layers, shows that the degree of order is usually not a very critical factor as long as the saturation coverage is more or less reached. Such nearly complete monolayer films usually show a considerable degree of inhibition of typical key reactions of corrosion, such as oxygen reduction. More critical may be uncovered regions, where, for example, oxygen reduction is not inhibited and also the adhesion promotion to the additional polymer coating not effective. However, care has to be taken when evaluating measurements of the electrochemical activity of monolayer films. For instance, oxygen reduction on gold modified by thiols does not seem inhibited at all when microscopic regions of bare gold are present in the film in high density. Even though the total amount of thiol molecules per square centimeter may then be practically the same as for a fully intact monolayer, the limited transport of oxygen and the related formation of overlapping diffusion cones will cause similar oxygen reduction currents as for the bare surface (see, e.g., Ref. [49]).

As already mentioned, on the other hand, a monolayer film of saturation coverage but not yet achieved perfect order may already show very good inhibition, as long as no defect regions of bare gold are present. This is true for the monolayer film without additional coating applied and also with additional coating applied (see Figures 14.20 and 14.21).

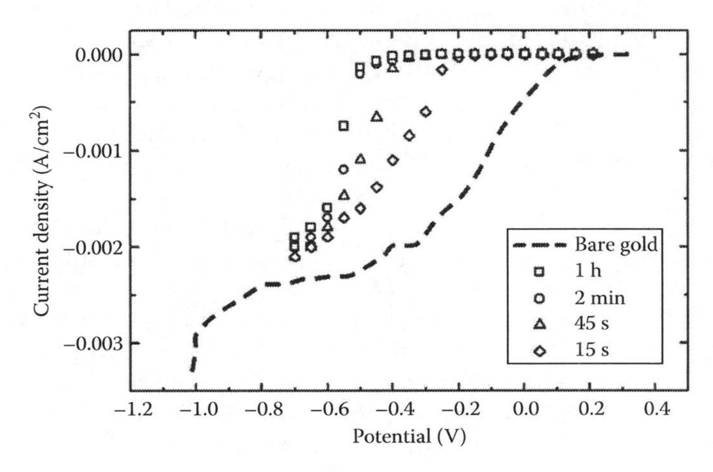

FIGURE 14.20
A C10 thiol SAM of 2 min has nearly the same inhibition efficiency as a film that adsorbed for hours.

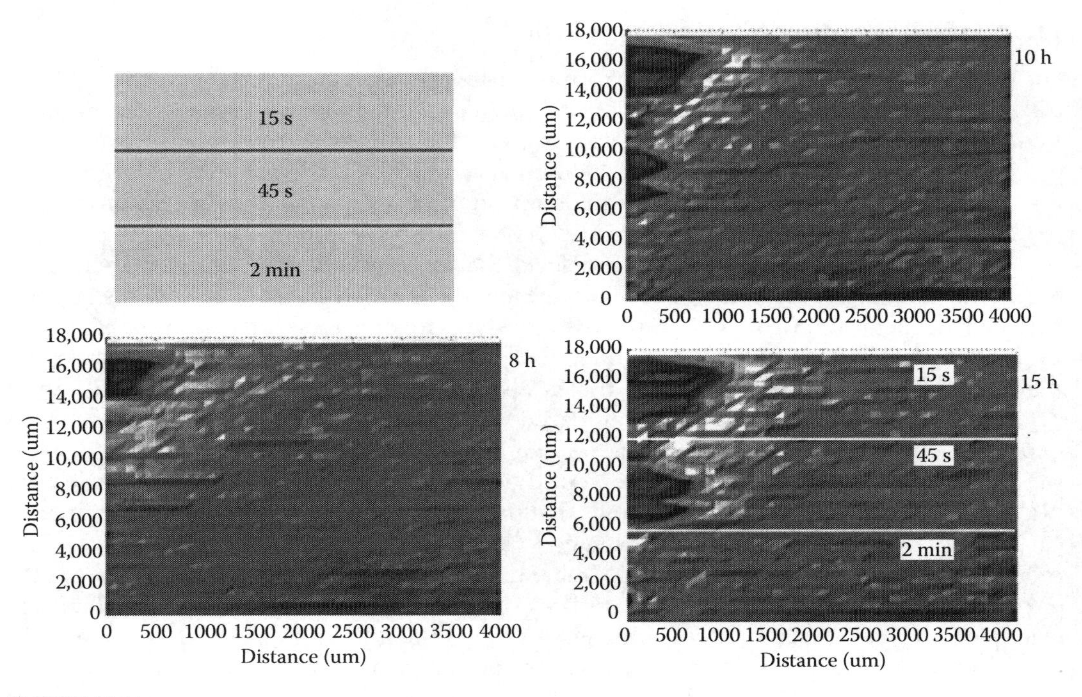

FIGURE 14.21
Sample modified stepwise for different adsorption times, that is, with different layer quality, and then covered by a simple clear coat. Delamination was initiated from the left-hand side by bringing the left edge of the sample in contact with electrolyte covered corroding iron (see also Ref. [126]).

Although a direct relation between the stability/reactivity of a self-assembled monolayer used for surface modification and the corresponding properties of the overall coating system, with applied additional coating layer(s) is to be expected, it is difficult to predict the behavior of the latter one, even if the monolayer itself is well characterized. This is easily explained by the fact that, for example, the concentrations of ions and oxygen at the interface between the monolayer-modified surface and the additional coating are not known. Understanding these factors is the aim of a current European Research Project "Numerical modeling and lifetime prediction of delamination, polymer coating disbonding & material degradation" [127]. For corrosion driven coating delamination, for instance, not only the electrochemical reactions at the interface play a role but also the mobility of ions at the interface, which seems to depend critically on amount and distribution of functional groups such as OH [127].

However, studying the effect of defects in the monolayer films in detail is difficult as they are mostly nanoscopic and can be studied only with atomic force microscopy (AFM) and STM, which require very time-consuming preparation and limit the flexibility of the experiments. Other operation modes such as scanning Kelvin probe force microscopy (SKPFM) [128,129] may play an important role in future work but still require preparation of suitable model samples with defects of suitable size. For thiol monolayer films, considerable control of the film quality can be gained by self-assembling under potential control. For negative potentials, much larger domains are obtained than for higher potentials [19,76]. However, much larger defects can be prepared in a very defined way by reparation with the Langmuir–Blodgett (LB) technique, which allows deposition of ordered monolayers of suitable (amphiphilic) organic molecules on a solid substrate [130].

Although thiols are classic SA molecules, it is possible to transfer thiol films onto gold by the LB technique [130]. Lösch et al. [130] studied the LB transfer of thiol molecules onto gold under electrochemical potential control, which allows adjusting the defect density. Cyclovoltammograms (CVs) of gold covered by one monolayer of *n*-octadecyl mercaptan were measured and it was found that the gold oxidation/reduction peaks were significantly suppressed for the LB-film modified gold electrodes. The residual peaks are determined by the part of the gold surface that is not blocked by chemisorbed thiol. The surface coverage by these defect sites is critically affected by the electrode potential on the stability as is clearly visible.

The LB technique can also be successfully used for transferring monolayers of phosphonate [131] or of silanol molecules onto a metal surface [132,133]. As mentioned above, silanols are widely used for corrosion protection (see also Ref. [134]). Now, in contrast to mercaptan, a dissociation of the head group takes place on the water surface [135] followed by the formation of silanols and polymerization. Hence, a two-dimensional polymer film results. The existence of the Si–O–Si bonding can be proven by infrared spectroscopy. The two-dimensional polymer film formed on the water aqueous subphase is rather stiff and viscous, so that during the transfer, a cleavage of the film occurs and an island structure of the monomolecular LB film results on the substrate [132,133]. This gives ideal samples for investigation of the role of defects in delamination. An ideal tool for such investigations is the SKP [19,128,129].

The dipole potential of the surface will change in the presence of polar organic molecules at the metal/air interface. This can be seen in Figure 14.22a, which shows an SEM picture of an iron surface partly coated by one monolayer of octadecylsilanol (ODS) [132]. In this situation, the individual molecules are ordered, as they have been deposited by the Langmuir–Blodgett–Kuhn method. Laterally resolved Auger analysis shows that the dark areas correspond to the noncoated substrate, whereas the bright areas are coated by one monolayer of the organic molecule. It is interesting that the same contrast is obtained with an SKP (Figure 14.22b). This is due to the fact that the covered part of the substrate has a different dipole potential in comparison with the noncovered areas and this also results in a different Volta potential [132].

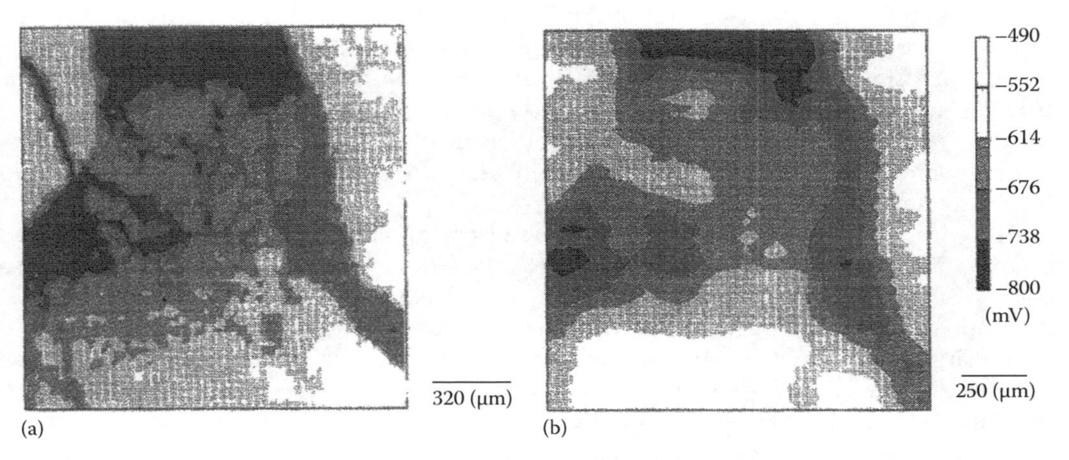

FIGURE 14.22
(a) SEM map of iron modified by approximately 0.7 monolayer of ODS. Bright areas: surface covered by ODS; dark areas: surface not covered. (b) Map of the Volta potential of the same surface as in (a). (From Stratmann, M. et al., *Ber. Bunsenges. Phys. Chem.*, 95, 1365, 1991.)

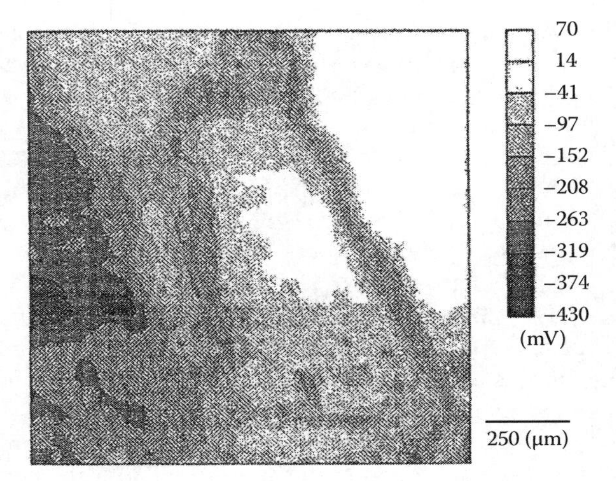

FIGURE 14.23
Map of the Volta potential of one monolayer of ODS on iron after exposure to 100% relative humidity for 343 h. Area: 5 × 5 mm. (From Wolpers, M. et al., *Thin Solid Films*, 210/211, 592, 1992.)

In humid atmospheres, a thin electrolyte layer is formed on top of the substrate surface. Under those circumstances, the Volta potential of the surface of the electrolyte layer is measured. It has been shown that this Volta potential is mainly determined by the Galvani potential difference of the metal/electrolyte interface and therefore by the electrode or corrosion potential [128,136].

Now the SKP can be used to monitor the local corrosion potential of a partly coated metal surface in a humid atmosphere. If, for example, the presence of the organic molecule changes the kinetics of the metal dissolution reaction, then this is reflected in a change of the corrosion potential; an acceleration of the metal dissolution will shift the corrosion potential cathodically and a retardation will shift the corrosion potential anodically. As an example, Figure 14.23 shows a map of the corrosion potential of an iron surface that is partly coated by one monolayer of ODS as measured with the SKP in a humid atmosphere [133]. The corrosion potential changes locally by several 100 mV due to the presence or absence of the film and it has been proven that the anodic potentials correspond to the areas that are coated by the polymer [133]. The potential plot of Figure 14.23 is therefore a representation of the inhibition of the anodic metal dissolution in a humid atmosphere. If the same specimen is exposed to a water-saturated atmosphere for longer corrosion times, it will be observed that starting from the uncoated areas of the iron surface, the corrosion (negative corrosion potentials) slowly proceeds into the coated areas [133]. The SKP is therefore an ideal tool for the investigation of inhibiting properties of organic molecules and it also allows analysis of the expansion of corrosion on partly inhibited surfaces.

Hence, for the first time, the delamination of the monolayer coating can be evaluated quantitatively and directly compared with the delamination of industrial paints. Such a comparison is shown in Figure 14.24, and it is quite obvious that the delamination rate is determined not by the thickness of the coating but by the bonds that prevail at the interface. Using the SKP in the future, the stability of other suitable metal/polymer bonds may be analyzed under corrosive conditions, and it is very likely that coatings will be developed that are considerably superior to the ones used up to now. It is essential, however, to understand the chemical or electrochemical reactions that take place at the metal/polymer interface and that give rise to the undermining. Such studies are performed preferentially on homogeneous electrode surfaces, as most spectroscopic and electrochemical techniques do not provide spatially resolved information. Exceptions are the scanning probe

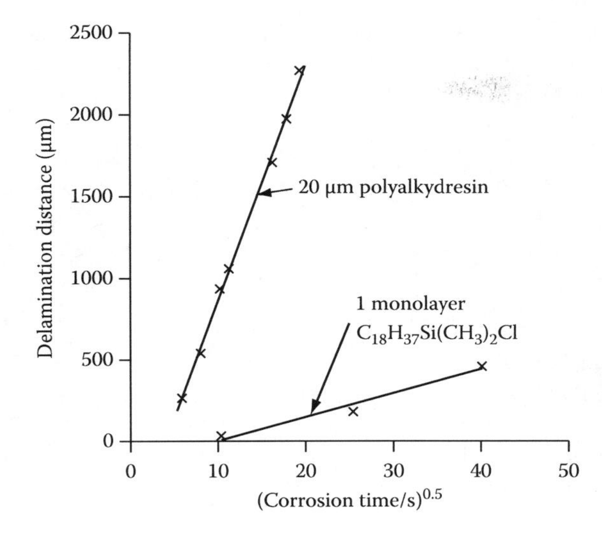

FIGURE 14.24
Comparison of the delamination rate of iron coated by 20 μm polyalkyd resin with iron coated by one monolayer of dimethylhexadecylsilanol. (Data from Brasher, D.M. and Mercer, A.D., *Br. Corros. J.*, 95,122, 1967.)

microscopy techniques such as STM or AFM and SKPFM. However, as already stated, the application of these techniques for investigation of the corrosion properties of chemically modified surfaces has only begun and few results have been obtained so far.

14.3 Corrosion Protecting Ultrathin Films

14.3.1 Introduction

Ultrathin inorganic and organic films on engineering metal substrates are of technical importance with regard to the desired functional materials properties such as

- Low friction of the metal sheet undergoing forming
- Temporary corrosion protection under atmospheric conditions
- Adhesion promotion to applied organic lacquers or adhesives
- Corrosion resistance of applied organic lacquers or adhesives

Micrometer-thick crystalline films of metal phosphates formed during a conversion treatment in aqueous solutions played a major role in the surface treatment of important engineering metals (such as Fe, Zn, and Al) and their alloys [137,138]. Such films are applied in continuous lines of metal sheet producers, in automotive paint shops and in many batch processes of metal parts producing industries. Additionally, chromating and chromate rinsing were applied to achieve self-repair properties based on the mobility of chromate ions and their redox potential and the possibility to form Cr(III) oxide films under reducing conditions [139].

Such thin films have to act as a barrier between corrosive ions and the metal and have to inhibit electron as well as ion transfer reactions. Moreover, they should be stable over a quite wide pH range since corrosive reactions especially under polymer films change the local pH at the front of deadhesion ranging from acidic pH values (e.g., during filiform corrosion) to very alkaline ones (e.g., cathodic delamination on steel) [140].

In recent years, the development of new environmentally friendly amorphous ultrathin films with a thickness of less than 100 nm became of increasing importance for the substitution of the traditional surface treatments. Studied surface technologies cover

- Alternative inorganic or inorganic–organic hybrid conversion treatments
- Organosilane ultrathin films
- PP film formation
- Sol-gel chemistry

In comparison to the self-assembled monolayers, they are three-dimensional in nature and are less sensitive to the chemistry of the substrate so that multimetal applications become possible.

In the following, selected modern surface ultrathin film technologies are described in detail.

14.3.2 Conversion Film Formation from Aqueous Solutions

The development of new advanced surface conversion coatings on metal substrates is of increasing importance for the steel and automotive industry. Classical surface treatments such as chromate passivation or Zn-phosphatation [137,138], which were standard industrial procedures are being replaced more and more by new thin-film conversion treatments with tailored chemistry for applications such as corrosion protection [141–143], adhesion promotion [144], or low friction coatings [145]. The newly developed systems should match or outperform standard technical systems with regard to efficiency, corrosion protection, adhesion, and formability. Amorphous thin films deposited from aqueous solutions have high technological potential since they make use of so-called smart green chemistries, which include phosphates [146–150], organosilanes [151–154], rare earth metal salts (REMS) [155–159], and transitional metals like molybdenum [160–163], zirconium [164,165], and manganese [147,166,167], while enabling the further use of established production routes. Ogle et al., for example, studied how the addition of different elements to the phosphate bath influences the corrosion protection of the phosphate layer [147]. These studies showed the importance of manganese in increasing the alkaline stability of phosphate layers more than the other tested elements. However, a detailed understanding of film-forming reactions at the interface between the water-based conversion solution and the metal surface has not been elucidated because of the complex structure of the deposited films.

In general, chemical conversion layers are insoluble crystalline or amorphous layers that are formed during a chemical reaction with the natural oxide layer on top of a metal substrate.

Table 14.1 gives an overview over the commonly used conversion chemistries for the different metal substrates.

As an example for the process of conversion layer formation on a metal substrate, the different steps of the formation of Zr-oxyhydroxide films from acidic hexafluorozirconate solutions on zinc substrate are schematically displayed in Figure 14.25.

In general, during a conversion process carried out in aqueous solutions, the pristine metal oxide surface layer is dissolved and even the base metal is partially etched. As a result, the ionic gradients in the vicinity of the surface initiate the growth of the designated conversion

TABLE 14.1

Overview of the Different Conversion Treatments for Metal Substrates

Substrate	Conversion Chemistry	References
Al	Zirconium	[166–168]
	Titanate	[167,168]
	Rare earth metals	[169–173]
Zn	Phosphates	[146–150,174]
	Rare earth metals	[173,155]
	Molybdate	[160–163]
	Zirconium	[164,165]
	Manganese	[147,175,176]
Fe	Phosphate	[177]
	Rare earth metals	[178]
Mg	Phosphates	[167,179]
	Molybdate	[179]
	Zirconium	[180]
	Rare earth metals	[180–182]

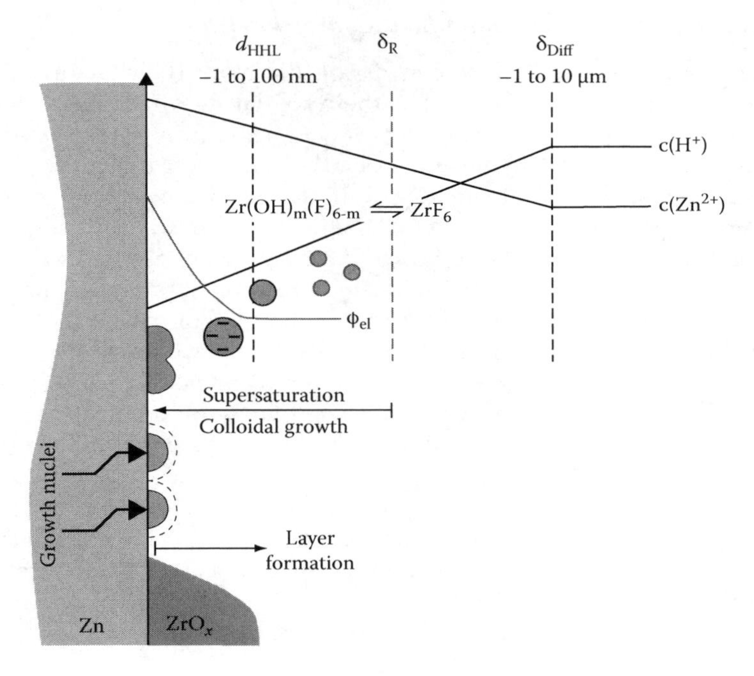

FIGURE 14.25
Schematic of the film formation process of a Zr-oxyhydroxide conversion layer formed from an acidic aqueous solution of hexafluorozirconates.

coating and the native metal oxide is replaced by an inorganic film, which is composed of anions and cations of the conversion solution. Depending on the conversion process, the dissolved metal ions of the substrate are essential for the formation of the protective inorganic coating, for example, phosphating and chromating process [183,184], or are just incorporated partially into the conversion film, for example, cerium-based coatings on aluminum [185].

For Zr-oxyhydroxide film formation, the exchange of fluoride ligands by hydroxides occurs during the pH increase at the interface between the metal and the electrolyte based on the cathodic reduction of protons to hydrogen and oxygen to hydroxide, and the anodic dissolution of metal ions into the electrolyte. The formed Zr-hydroxides are insoluble and nucleate in the interface region and directly on the dissolving metal surface. The formed surface Zr-hydroxide films partly directly condensate but mainly form hydrogen bonds leading to a continuous film on the surface that inhibits a further metal dissolution and proton or oxygen reduction.

Currently, there is only limited understanding on the correlation of the chemical structure and morphology of the conversion films with their electrochemical and corrosion protection properties. Grundmeier and coworkers studied the formation and properties of surface conversion layers on zinc-coated steel and could prove that organic macromolecules as additives in the conversion bath have crucial importance for the sealing of nanopores in the formed inorganic network of the conversion films [186,187]. However, still the establishment of a structure–property relationship is difficult to extract and hindered by the influence of the surface chemistry of the technical alloys on the film formation kinetics and the final film chemistry. Large numbers of technical samples have to be surface treated and evaluated for statistically assured results.

14.3.2.1 Formation of Inorganic Amorphous Conversion Films

The interesting structural aspects of conversion films are their chemical composition, film thickness, and morphology as well as their insulating properties. These film properties can then be correlated with the impedance of the film, its respective free corrosion potential and thereby finally with the deadhesion kinetics of an applied polymeric film. Following this approach, Stromberg et al. designed a new way of combining the synthesis of surface gradient films with localized surface analytical and electrochemical techniques [188]. The spatial resolution of all analytical tools applied to characterize the conversion coating enabled the authors to specify roughly 20 states of film formation on one surface gradient sample that is only 20 mm in length. For the synthesis of the surface gradient films, an automated dip coater was used. Spatially resolved surface analytical studies were performed by means of small spot XPS and FTIR reflection absorption microscopy (μ-IRRAS). These data were correlated with impedance data gained by the application of an electrochemical capillary cell and Volta potential surface mappings with the SKP.

A recent development in the treatment of zinc-coated substrates consists of a conversion process accomplished by zirconium acids. One of the first applications of the zirconium-based conversion chemistry was the sealing of phosphate deposits on zinc [189]. Alternatively, the surface is treated with $Zr(NO_3)_4$ first and subsequently immersed into an organosilane solution [190]. In current developments, pure zirconium acids or zirconium salts in combination with hydrofluoric acid are used for the pretreatment of zinc surfaces prior to the lacquer deposition [191].

Stromberg et al. [188] observed after the conversion treatment of hot dip galvanized zinc, a thin film of a zirconium oxyhydroxide with a homogeneous bulk composition during the conversion process. XPS sputter profiles showed that this oxide layer includes small amounts of zinc (~4 at%) from the base metal and fluorine (<9 at%) from the zirconium precursor. Due to the gradient preparation of the Zr layer via a dip coater, a linear increase of the conversion layer was observed. SKP measurements showed three areas of different Volta potentials. A region of high cathodic Volta potentials with values around −0.8 V with

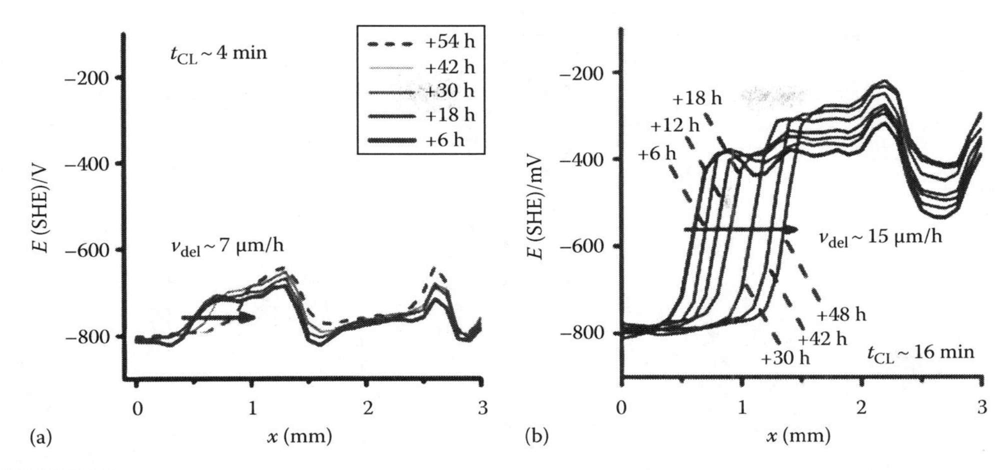

FIGURE 14.26
In situ Volta potential line scans of cathodic delamination taken at fixed conversion times: (a) t_{CL} = 4 min and (b) t_{CL} = 16 min. The horizontal arrows denote the predominant delamination velocity. (From Stromberg, Ch. et al., *Electrochim. Acta*, 52, 804, 2006.)

impeded oxygen reduction (3 min < t_{CL} < 13 min) was enclosed by areas of more anodic Volta potentials representing the initial stages of conversion layer buildup (t_{CL} < 3 min) and the region of "overetching" (t_{CL} > 13 min). Also the delamination velocities of the different areas differed as shown in Figure 14.26. The area of an intact conversion layer showed a delamination velocity of approximately 7 μm/h, while the "overetched" area showed higher velocities of around 15 μm/h.

14.3.2.2 Formation of Inorganic/Organic Hybrid Films

Beside the use of pure inorganic components as $Zr(OH)_x$ or cerates for the conversion film formation, inorganic/organic hybrid films have also been received considerable attention as new functional material in the recent years. Incorporation of organic functional groups like functional organosilanes improved the barrier and adhesive properties of the conversion films.

Figure 14.27 shows the changes in the open circuit potential during the formation of a crystalline phosphate layer and an amorphous inorganic/organic hybrid film. The phosphate layer formation results in a potential step of around 200 mV to more anodic potentials while the amorphous layer shows only little changes (~20 mV) to more cathodic values.

For the three-cation phosphate layer, the initial drop of the potential reflects the metal dissolution and the subsequent potential increase illustrates the formation of the inorganic barrier film with Ni deposited in nanoscopic noncovered interface defects [192].

For the ultrathin conversion layer that does not consist of Ni or Cu as noble metals that could be deposited in such defects, the potential remains very negative due to the small metal dissolution in nanoscopic defects.

Beside the use of organosilanes or silicates, complex prepolymers can also be added to inorganic conversion components to form an inorganic/organic hybrid film as shown by Grundmeier and coworkers [187,193]. An initial rapid buildup of the conversion layer was observed that slows with increasing immersion time due to the reduction in the availability of the reactive galvanized surface. The initial states of the conversion film are small

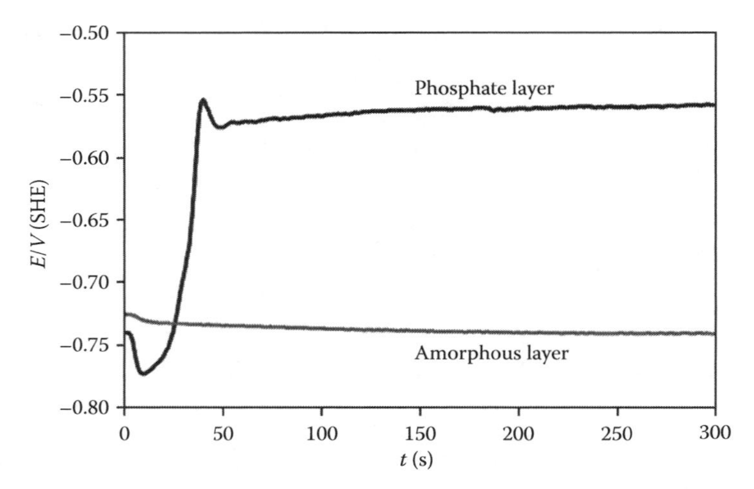

FIGURE 14.27
Detection of the open circuit potential during the formation of different conversion layers (crystalline, micrometer thick, three cation phosphate layer containing Ni in comparison to an about 20 nm thin amorphous conversion film containing organosilane, manganese, and zirconium).

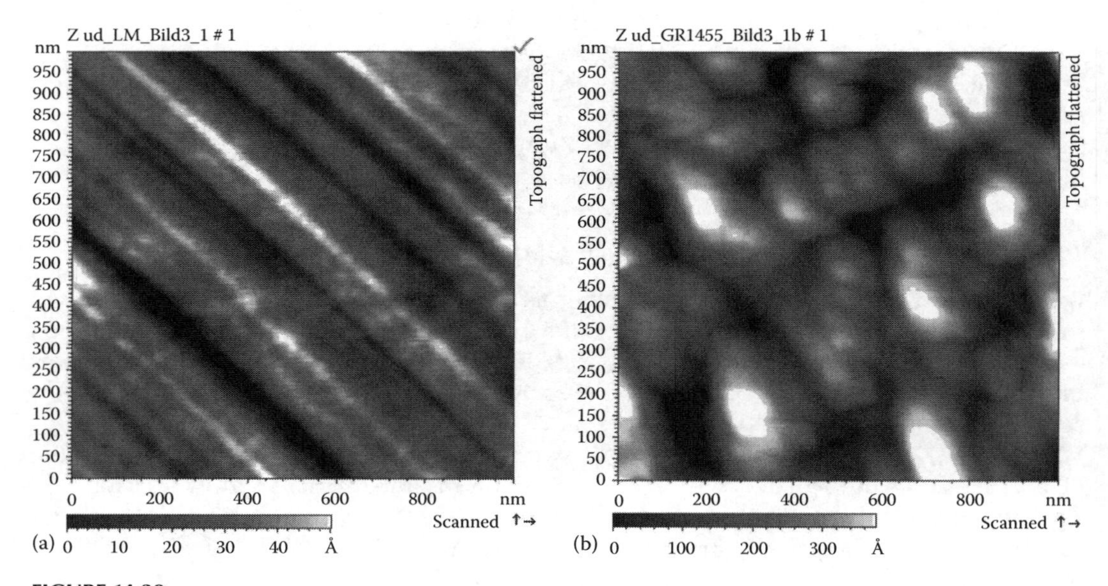

FIGURE 14.28
AFM topography image of a nonskin passed hot dip galvanised (HDG) sample: (a) without conversion layer and (b) with conversion layer (40 s immersion).

islands that grow and merge to produce an almost fully formed conversion layer with different terraces obscuring the surface of the zinc coating beneath (see Figure 14.28). An almost complete coverage of the mixed inorganic/organic hybrid layer comprising of metal hydroxides, phosphates, and zinc ions coordinated by the complex organic macromolecule was observed after approximately 60 s of immersion.

SKP measurements of the cathodic delamination process showed large differences in the potentials of the intact areas and the delamination velocities for a conversion-treated

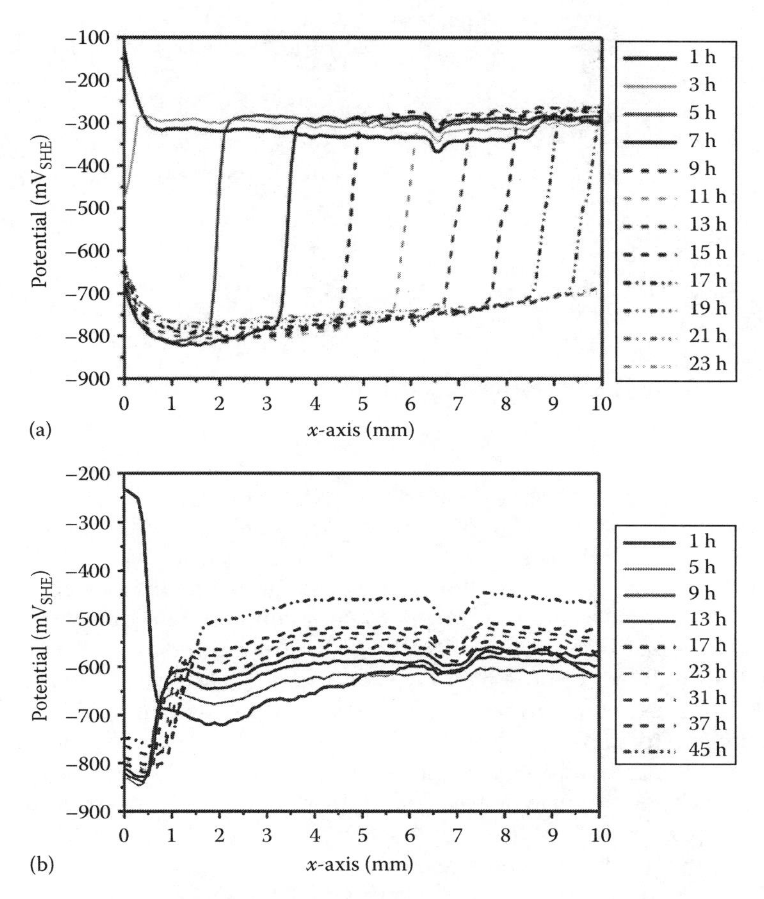

FIGURE 14.29
SKP line scans of polymer-coated electrogalvanized steel during the cathodic delamination process: (a) alkaline cleaned zinc surface without conversion layer and (b) conversion layer coated zinc surface. (From Klimow, G. et al., *Electrochim. Acta*, 53, 1290, 2007.)

sample and an electrogalvanized steel sample without conversion layer (see line scans for the delaminating lacquer film for both samples in Figure 14.29).

The potential difference between the actively corroding defect and the intact polymer/metal interface is 500 mV for the just oxide-covered zinc surface and only 150 mV for the conversion-coated surface. The delamination rate of the transparent polymer film in the conversion-coated area observed to be about 40 times smaller than for the just alkaline cleaned sample. This fact is in qualitative agreement with the measured inhibition of the oxygen reduction.

14.3.3 Adsorption of Thin Organosilane Films from Aqueous Solutions

Organosilanes are molecules that have four functional groups bonded to a silicon atom. Among these functional groups, the most common ones are alkyl chains, alkoxy groups, halogens, and hydroxyls [194,195]. The alkyl chains can be terminated with another

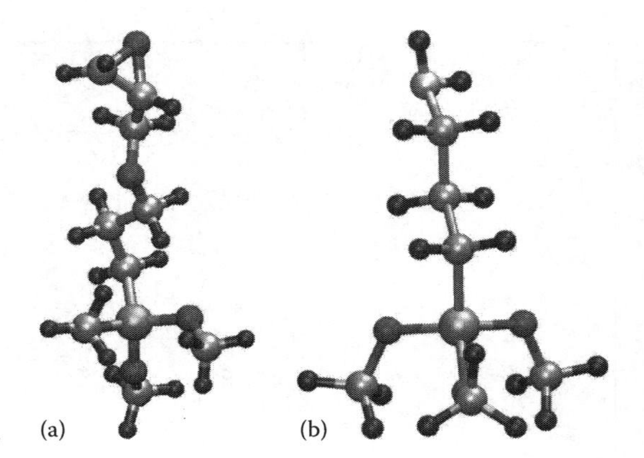

FIGURE 14.30
Molecular structures of (a) γ-APS and (b) GLYMO.

functional group that can further increase the number of the possible structures. Common functional groups are pyridine (–NC_5H_5), mercaptan (–SH), phenyl (–C_6H_5), methyl (–CH_3), amino (–NH_2), cyano (–CN), and vinyl (–CH=CH_2) groups [196].

The common and industrially relevant organosilanes share the common formula: F–Si–$(R)_n(R')_{3-n}$. Usually, "F" is in the form of a functionalized organic chain, which determines the hydrophobicity/hydrophilicity of the resulting film. It also determines how the organosilane film will bind to the next layer of organic coating. "R" presents the leaving group, which upon exposure to water can form silanols or can directly react with the surface hydroxyls to form metal–oxygen–silicon bonds. The choice of the leaving group is quite important since they will stay in the reaction medium. They can slowdown or accelerate the reactions; they can also be toxic or environmentally hazardous. "R′" is the group that in general does not take part in any binding (mostly CH_2 groups). The number of R′ groups is decisive when monolayers are desired.

The possibility of bond formation with the surface and the ability of cross-linking within the film increase among other factors with the number of the leaving groups. Therefore, a high percentage of the organosilanes used in industrial applications have three leaving groups. Two of very widely used examples, aminopropyltrimethoxysilane (γ-APS) and (3-glycidyloxypropyl)trimethoxysilane (GLYMO) are shown in Figure 14.30.

Although the mechanism leading to film formation is still a topic of research, the widely accepted route consists of the following steps [195,197–199]:

1. Reaction of alkoxy groups to silanols

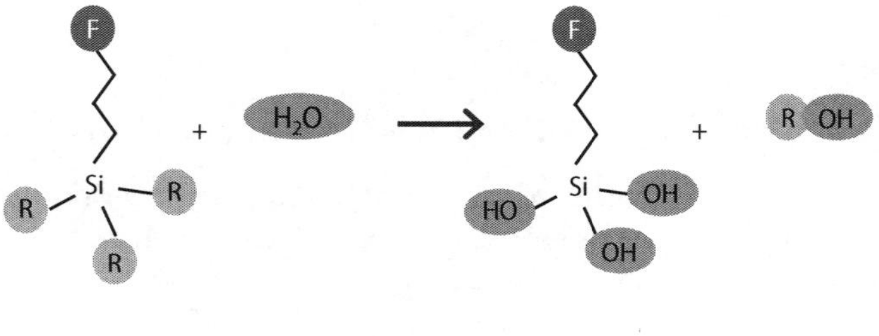

2. Simultaneous hydrogen bond formation between surface hydroxyls and silanols, and among silanols in the film

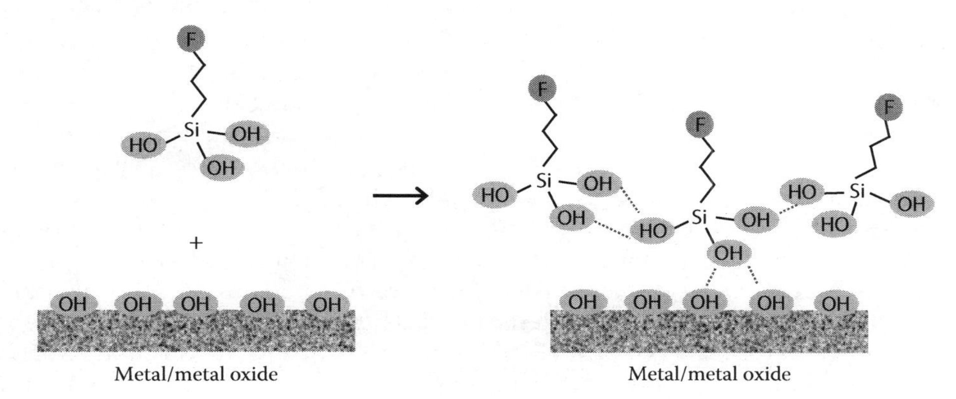

3. Conversion of the hydrogen-bonded structure to silicon–oxygen–metal and silicon–oxygen–silicon bonds via condensation reactions

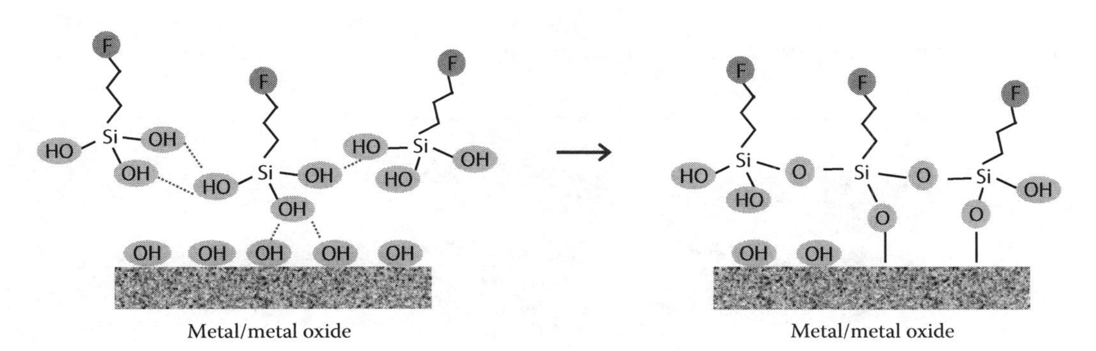

There are many methods for application of organosilanes on metals and metal oxide surfaces. Dip coating in aqueous solutions or alcohol solutions containing controlled amount of water, vapor deposition in atmosphere with controlled humidity, and spraying for large application areas are among the mostly utilized methods.

For aqueous solutions, the adjustment of the pH and the concentration of the solution are very important since they directly determine the degree of polymerization or cross-linking among the film and therefore the thickness of the resulting layer [195,200,201]. For industrial substrates like cold-rolled steel and glass, this may not be very crucial since the inherent roughness of the substrates is also very high. However, when working at the nanoscale, the thickness of the film has to be controlled more precisely. This is accomplished by using alcohol solutions with controlled water content. This limits the speed and degree of silanol formation and therefore enables a more precise control of the film thickness. For more delicate applications where distinctly one monolayer is desired, organosilanes with just one leaving group have to be used [196]. These molecules upon contact with water can form just one silanol functionality and therefore either stay in the solution in forms of dimers or react with the surface via a single bond and form monolayers.

Another decisive factor for the success of film formation on metal/metal oxide surfaces is the availability of surface sites for binding. For the formation of silicon–oxygen–metal bonds via condensation reactions, surface hydroxyls are required. One approach in the literature was to increase the surface hydroxyl density to promote the binding of silanols to the surface via chemical treatments or plasma modification [202]. On most metal/metal oxide surfaces, the conversion from hydrogen bonded to chemically linked organosilane layer requires an additional curing step at ~110°C. On industrial alloys, the precipitation of minor alloy elements, which do not readily adsorb organosilanes like copper in aluminum alloys, creates additional challenges for their applicability [203].

Organosilanes are not only used for corrosion protection but most often applied to enhance the adhesion of the upper polymer layers to metal/metal oxide surfaces. It is shown in the literature that the amino group of γ-APS can successfully link to epoxy coatings and the resulting films have superior corrosion resistance [204–206]. Wapner and Grundmeier [204] showed that the adhesion promoter γ-APS prepared from an aqueous solution on iron significantly decreases the rate of the cathodic delamination. Additionally, it allows a stable linkage of the adhesive to the metal substrate, thus the adhesive could only be removed in the delaminated area. The best stabilization was achieved by the successive deposition of a hexamethyldisilane (HMDS)/O_2 PP layer and the γ-APS layer (see Figure 14.31). Here, no cathodic delamination and also no wet deadhesion is observed even after 40 h of exposure.

The selection of the functionalized organic chain affects the macroscopic surface properties of organosilane film such as hydrophilicity and hydrophobicity. The end

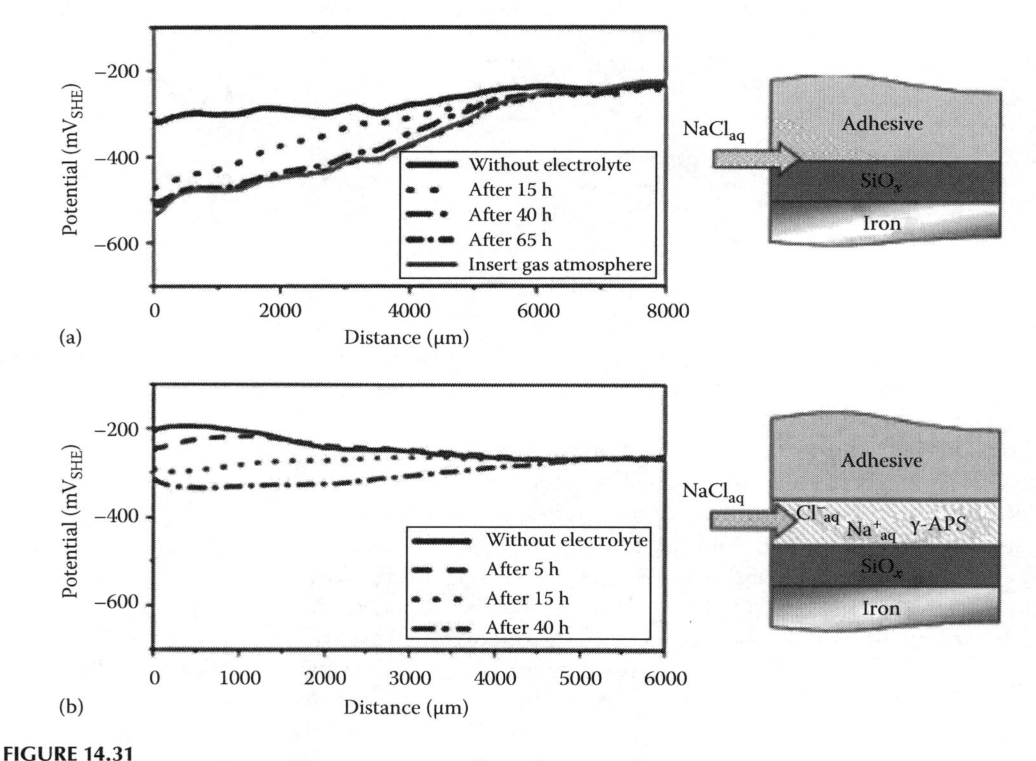

FIGURE 14.31
SKP line scan measurements of (a) HMDS/O_2-PP layer on iron surface, coated with epoxy resin, electrolyte 0.5 M NaCl and (b) HMDS/O_2-PP, then γ-APS layer from aqueous solution on iron surface, coated with epoxy resin, electrolyte 0.5 M NaCl. (From Wapner, K. and Grundmeier, G., *Adv. Eng. Mater.*, 6(3), 163, 2004.)

functionality and the length of the alkyl chain are of high importance. The functionality determines the mechanism of water adsorption and the nature of the formed film. The alkyl chain acts like a spacer or barrier for the diffusion of various species through the film toward the surface like water molecules and ions, which is of crucial importance for corrosion protection. The chain length and structure also contributes to the orientation of the forming film as expected from the studies on self-assembled monolayers.

Organosilanes are also very important monomers for sol-gel chemistry. Coatings obtained with sol-gel chemistry are known to provide dense structures with high barrier properties and good adhesion of organic paint systems. It is also possible to tune the properties of sol-gel-based coatings by changing the precursors, the reaction conditions, or by using different solvents. Moreover, it is possible to add purpose specific active species like corrosion inhibitors to benefit from their synergistic effects [207,208].

14.3.4 Plasma-Enhanced Chemical Vapor Deposition of Ultrathin Protective Films

14.3.4.1 Plasma Modification of Passive Films

Already in the early 1950s, Engell and Hauffe studied the chemisorption on nickel oxide [209] and found out that the chemisorption process was promoted by using an ionized gas [210]. The explanation of this behavior was given by the assumption of an increase in chemical potential of oxygen due to the excitation in the plasma and, associated to this, a significantly higher electron affinity that implicates the possibility of the buildup of a larger space charge.

According to the dependence of oxide formation on the oxidation state of oxygen and the electric field adjusted across the already formed layer, it becomes obvious to apply plasmas to control and to intensify the oxide modification processes. More recently, Grundmeier et al. have studied the modification of native oxide films on iron and zinc samples by oxygen and argon plasmas as pretreatment methods to adjust a certain surface chemistry, before further surface treatments were performed [211–213]. The complementary application of *in situ* (μ-IRRAS, quartz-crystal microbalance [QCM]) and *ex situ* (SKP, XPS) techniques allowed the observation and characterization of an oxide growth on iron during oxygen plasma treatment at room temperature as well as a further modification of the formed oxide layer by a subsequent argon plasma. The developed dependence of the Volta potential difference to the activities of the corresponding oxidation states after certain plasma treatment could be proved by SKP measurements [214]. The effects of the plasma pretreatment to the chemistry and electronic structure of the modified iron oxide as well as the cleaning effect of the discharge to the surface were used in other studies to obtain a strong adhesion of thin PPs to the ion surface and to demonstrate excellent properties of the system with regard to electrochemical corrosion resistance.

Comparable experiments on zinc surfaces have shown that all studied kinds of plasmas are able to remove the hydrocarbon contamination layer from a native zinc oxide, but water, oxygen, carbon dioxide, and hydrogen plasmas led to slightly different modifications of the oxide chemistry [215]. Further studies on plasma-modified and subsequent PP-coated zinc substrates have shown that delamination rates of coatings were significantly decreased. Therefore, these systems were identified to be promising for design of highly stable polymer/metal interfaces [216,217]. Bellakhal et al. investigated the influence of humid-air plasma in comparison to pure oxygen plasma treatment on brass and pure zinc samples and found by means of electrochemical methods a complex oxidation mechanism, involving the first formation of a precursor copper oxide Cu_xO and ZnO with a subsequent formation of Cu_2O and CuO. The resulted oxide composition was comparable to those resulting from thermal treatments [218–220].

The corrosion resistance of an industrial magnesium alloy (AZ31B) could be significantly increased by plasma oxidation in oxygen and water plasmas. Corresponding electrochemical studies, Rutherford backscattering spectrometry and x-ray diffractometry at grazing incidence results were published by Tian et al. [221].

Bertrand et al. have studied the influence of hydrogen plasma excited by means of microwaves at 2.45 GHz to native oxide films on stainless steel and aluminum. Using *in situ* IR-ellipsometry, it could be shown that the native oxide on stainless steel was almost completely reduced in a hydrogen plasma. In contradiction to this, the aluminum surface was more resistant and only the outermost hydroxylated part of the oxide layer could be reduced. It was also observed that a dense oxide layer could be grown on the pretreated steel surface by subsequent oxygen plasma [222].

The composition of the oxide layer covering the base metals and semiconductors is crucial for the behavior of the system during subsequent processing of the surface, in particular with regard to adhesion of any kind of coatings. All studies presented above demonstrate the ability of a significant modification of the oxide chemistry on the corresponding materials by means of low temperature plasmas. Due to the undefined influence of the ambient atmosphere on the often very sensible conditions on plasma-modified surfaces, the application of *in situ* techniques is essential for a deep understanding of the oxidation mechanism in a plasma.

Recently, Giza et al. showed that the increase of surface hydroxyls on oxide-covered aluminum by water plasmas substantially increased the kinetics of chemical binding of organophosphonic acids to the substrate [113].

14.3.4.2 Plasma Polymer Deposition of Ultrathin Protective Films

The plasma polymerization of organosilane precursors under various conditions is a topic that has received a great deal of attention by many researchers in recent years and has been extensively reported in various literatures. A first summarizing article about plasma polymerizations in glow discharges was published by Yasuda in 1981 summarizing the work of different authors, which have described PP films as by-products of phenomena associated with electric discharge, in the early twentieth century. By recognition of various advantages of PP films, more and more work was focused on the deposition and characterization of these films. However, many authors, especially those trying to deposit SiO_2-like films, did not classify the plasma polymerized siloxane films as organically modified ceramics despite the presence of carbon and hydrogen.

The plasma deposition processes can be separated into two classes, direct and remote plasma deposition by using different power supplies and experimental setups. When an organosilane precursor is introduced into the plasma zone, it is ionized and fragmented by high-energy electrons and ions inside the plasma. At low pressure and therefore a long mean free path, ionized molecular fragments and ions are more likely to collide with the chamber walls than with other gas-phase species and hence adsorb and react on surfaces within the reactor to grow thin films of similar chemical composition and structure in comparison with the injected precursor monomer. The schematic film-forming processes are shown in Figure 14.32.

This plasma polymerization process has been used to deposit hybrid films from a variety of organosilane precursors with applications ranging from corrosion protection [223–225], biomedical films for implants [226], barrier coatings for packaging [227–229] to dielectric layers in integrated electronic circuits [230].

For the deposition of organically modified ceramics, the most popular precursors are hexamethyldisiloxane (HMDSO) and tetramethoxysilane (TEOS) often in combination with oxygen to adjust cross-linking in the films. The chemical and physical properties of

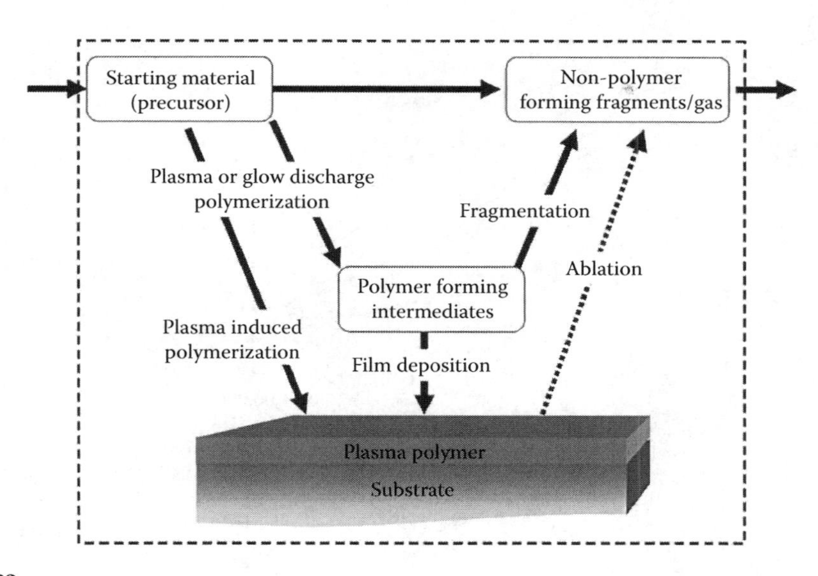

FIGURE 14.32
Schematic representation of the plasma polymerization mechanism according to Yasuda. (From Yasuda, H., Plasma Polymerization, Academic Press, New York, 1985.)

the resulting films depend strongly on the process parameters, such as the way of coupling the power into the gas, injection of the precursor, and the amount of oxidants.

Coatings produced from siloxane-based precursors have the general composition $SiO_xC_yH_z$, where $x \leq 2$, and y and z can vary up to the stoichiometry of the used precursor.

The chemistry of the siloxane plasma process is complicated by the wide variety of compounds formed by both carbon and silicon; however some general observations can be made.

The effect of plasma parameters in a microwave reactor on the growth and properties of HMDS-PP films was analyzed by Barranco et al. [231] with regard to the morphology and chemical composition at the inner and outer surface of the PP. According to XPS results, the composition of the different film was $SiC_{2.7}O_{0.6}$ for the film deposited at higher pressure (=HP-PP) and $SiC_{3.0}O_{1.5}$ at lower pressure (=LP-PP). Figure 14.33 shows

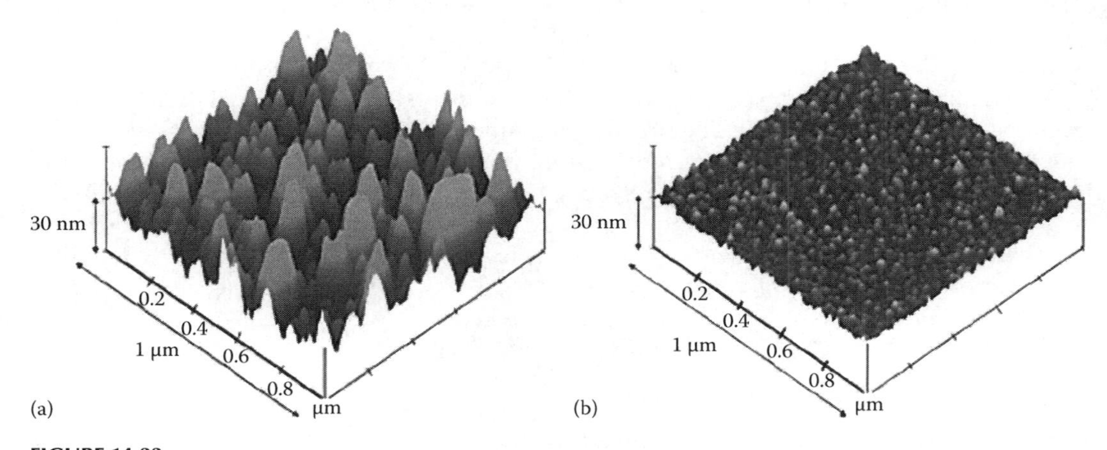

FIGURE 14.33
AFM maps (1 μm × 1 μm) of the deposited PP layers (D = 50 nm) at the outer surface (substrate: polished silicon wafer): (a) HP-PP and (b) LP-PP. (From Barranco, V. et al., *Electrochim. Acta*, 49, 1999, 2004.)

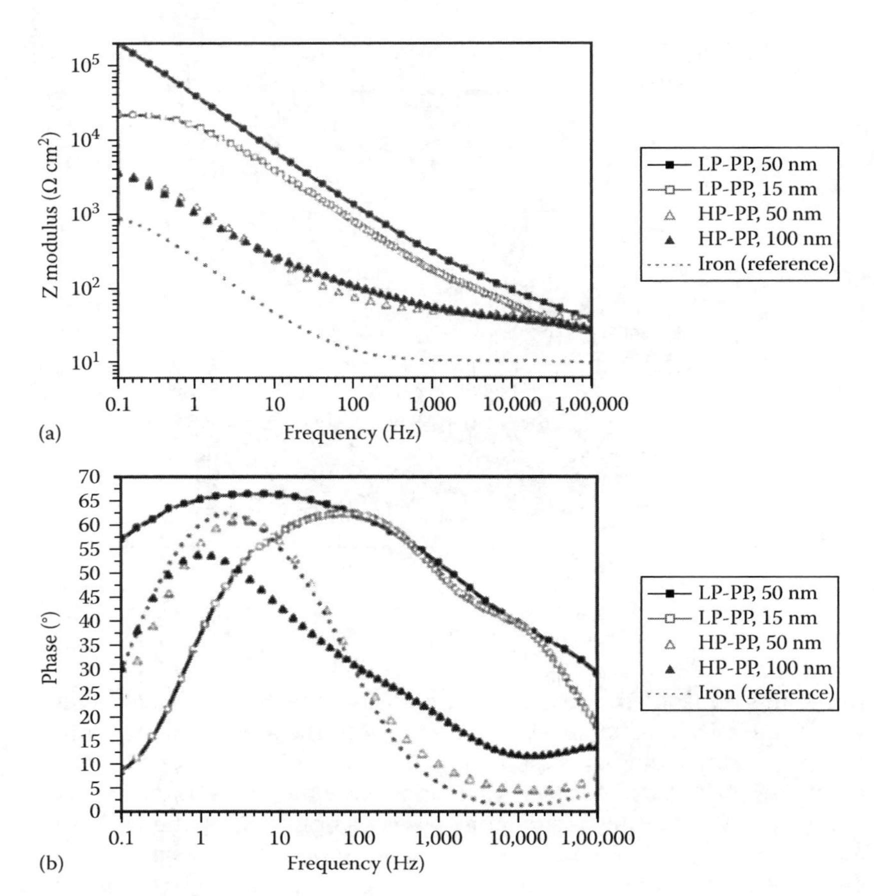

FIGURE 14.34
Bode plots of the reference iron sample and the PP layers after 30 min of immersion in borate buffer solution: (a) impedance modulus versus frequency and (b) phase angle versus frequency. (From Barranco, V. et al., *Electrochim. Acta*, 49, 1999, 2004.)

the topography of the two PP layers deposited under different conditions on a polished wafer surface. For the HP-PP, the roughness is significantly higher (root-mean-squared roughness [RMS] = 11 nm) than for the LP-PP (RMS = 2 nm).

Additionally, electrochemical impedance measurements of LP-PP and the HP-PP layers were performed (Figure 14.34). The LP-PP layers show higher impedance values at small frequencies than the HP-PP layers. Even the 15-nm-thick LP-PP shows impedance values with a factor of 5 higher than both HP-PP layers. Beyond that, the LP-PP layers showed a good correlation between the thickness and the barrier properties while the HP-PP layers do not show this behavior. This is a consequence of the high density of pinholes within the films deposited at higher pressure that result in the interfacial voids detected by AFM (see Figure 14.33).

14.3.5 Ultrathin Coatings Based on Conducting Polymers

14.3.5.1 How to Prepare Coatings That Work

Since the discovery of intrinsically conducting polymers (ICPs) in the late 1970s by Heeger, MacDiarmid, and Shirakawa, for which they were awarded with the Nobel prize [232–234], the unique combination of physical and chemical properties of ICPs has drawn the attention of scientists and engineers from many different fields of research and they were studied for various application possibilities. One idea that came up already quite early was to use them for corrosion protection [235,236].

However, although in most publications about conducting polymers for corrosion protection, only positive results are reported, it is generally known from industrial trials of conducting polymer based coatings that the practical performance usually is by far inferior to coatings not containing conducting polymers. This is due to the fact that in most published reports only simple immersion tests were performed, where the chance is good that due to its high potential in its oxidized state, the conducting polymer can passivate small defects [237,238] by partly being reduced, while under atmospheric corrosion conditions or in presence of too large defects, this does not work and fast coating breakdown is observed [238–240]. This is due to two reasons: first, under atmospheric corrosion conditions, the amount of active ICP is limited as only a small part of the coating is in direct contact with the electrolyte (see Figure 14.35), and second, when the ICP is not able to passivate a defect site, that is, to polarize the actively corroding metal over its passivation peak and providing the passive current density to keep it passive, it will on the contrary increase the corrosion by acting as a natural partner of the anodic metal oxidation by being itself completely reduced. This latter process was observed to be very fast. The reason for this fast reduction of the whole coating, even if the defects cover only a small part of the surface, lies in the transformation of the ICP to cation permselective behavior that seems to be a process that all ICPs undergo upon reduction over extended distances, as it is the case for coating that gets reduced starting from a defect site [239].

This is explained by the fact that in order to maintain charge neutrality in the ICP, its reduction has to be accompanied by either cation uptake or anion release (or both together). Now, if during reduction, any anions remain in the reduced polymer, these remaining anions (compensated by incorporated cations) can be regarded as fixed negative charges promoting cation mobility. While it is to be expected that large anions have a tendency to remain in the polymer, under the conditions of atmospheric corrosion; this is also the case for small anions. This may at first seem surprising. Although the size of an anion is usually not the only factor influencing its mobility and other factors such as the polymerization conditions, the thickness of the polymer, and the ions present in the electrolyte solution [241,242] play a role, usually it is expected that small dopant anions

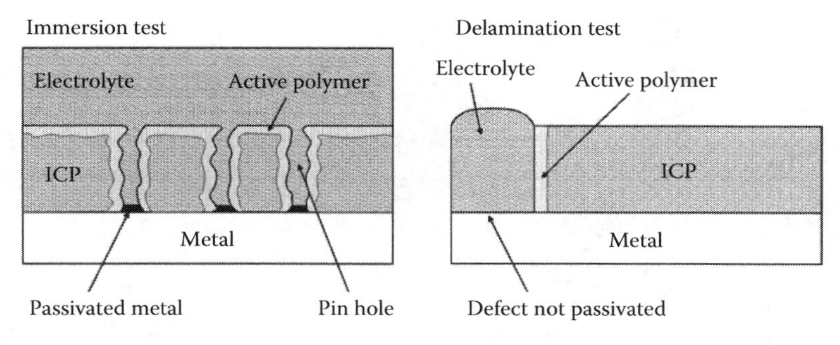

FIGURE 14.35
Immersion corrosion test versus atmospheric corrosion test (delamination of the coating starting at a corroding defect site). In the immersion case, the whole surface of the conducting polymer is active and can contribute to the passivation of the defect; in the delamination case, only the part of the polymer in direct contact with the electrolyte in the defect is active. In the first case, the reduction current of the conducting polymer may easily be higher than the current required to overcome the critical passivation current density in the defects. In this case the corrosion in the defects would be stopped. In the second case (the atmospheric corrosion case) the chances for this are much lower.

show good exchange properties or at least a mixed anion/cation exchange should occur. However, this is only true under standard electrochemical conditions where a thin film of at most a few micrometers thickness is reduced. During delamination of a conducting polymer film from a defect site, the ion exchange, that is, either release of anions or uptake of cations, takes place over hundreds of micrometers. Hence, even if at first, that is, near the border with the defect site, only a few anions remain in the reduced matrix, the cation mobility will steadily increase due to these fixed negative charges and thus a steady increase in cation incorporation and a correlated decrease of anion release will follow. After a certain distance, the reduced matrix will then have turned cation permselective (see also Ref. [239]).

Obviously, this hypothesis is based on the fact that the necessary ion exchange to compensate the electronic reduction of the ICP will never be 100% supported by exclusive anion release (only then the transition to cation permselectivity would not occur). Claimed 100% exclusive exchange of either anions or cations has so far never been proven, with the exception of experiments where exceptionally large anions or cations are used so that there incorporation is definitely impossible (during standard corrosion conditions always small cations such as sodium or potassium are present). With this exception, the incorporation of at least a few percent of cations cannot be excluded even under standard immersion electrochemistry of ICP films showing good anion release properties. This is the main reason why cation incorporation is always the dominating effect under delamination conditions. Of course, if anion release is enforced, for example, by excluding small cations from the defect, then no cations will be incorporated and the reduced ICP matrix will be free of any ions. In Figure 14.36, this is demonstrated by results obtained from a simple experiment: chloride-doped PPy was reduced from the side, that is, in a delamination setup, the progress of the reduction front was recorded for two different situations: in one case, the electrolyte was 0.1 M potassium chloride; in the other case, 0.1 M tetrabutylammonium chloride. As can be seen, in the first case the reduction proceeds much faster than in the latter case. If at a later stage, that is, when anions have been expelled from the ICP along a certain distance, for the case of the large cations in the defect, KCl is injected into the electrolyte, the reduction velocity is unaffected and remains slow.

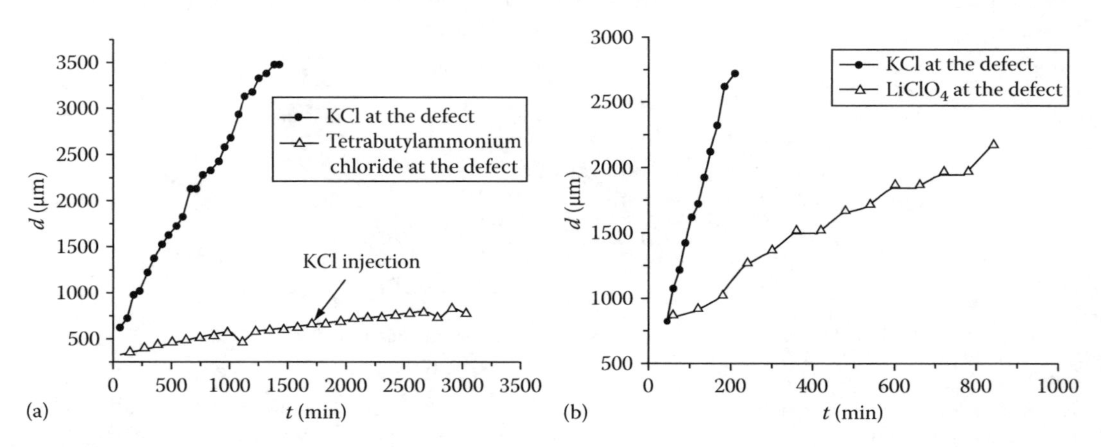

FIGURE 14.36
Comparison of the progress of the reduction front versus time for chloride doped PPy in the presence of a small cation in the defect (filled circles) and for the case of the big tetrabutylammonium cation in the electrolyte. (a) The injection of KCl has no effect on the slow curve started with only the big cation in the defect. (b) The effect of the cation size on reduction velocity for PPy doped with tosylate.

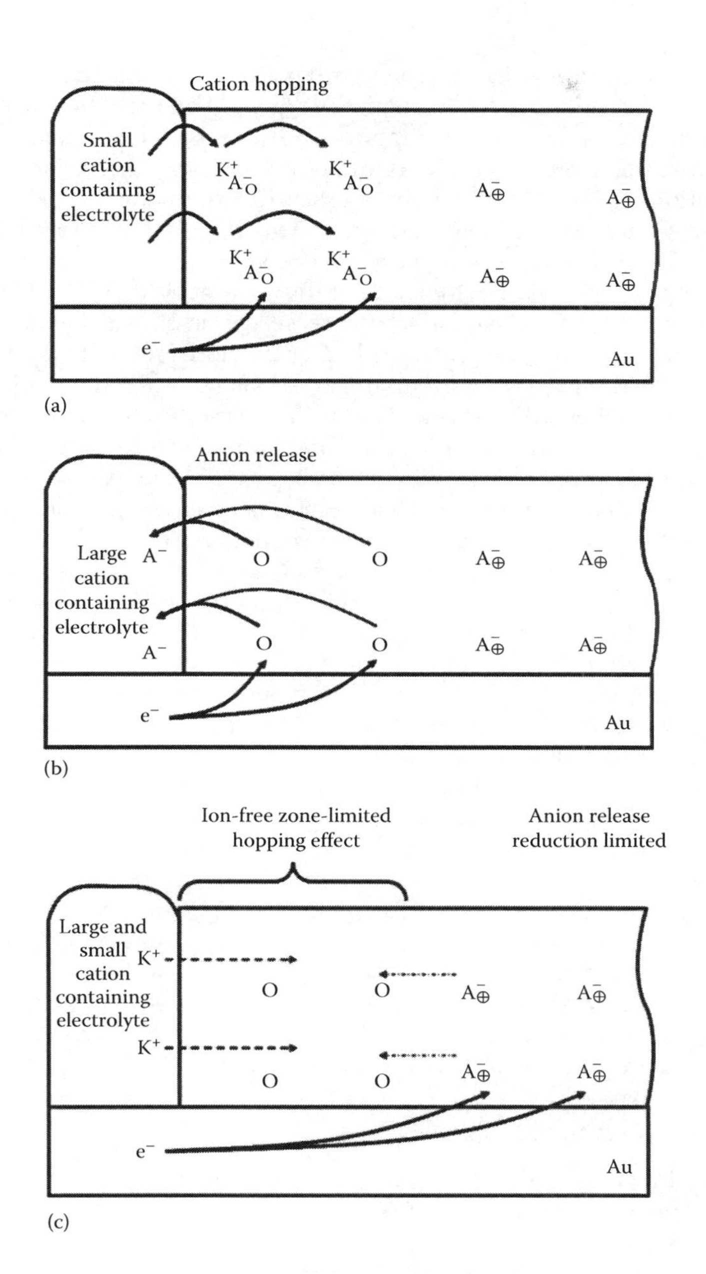

FIGURE 14.37
Schematic explanation of (a) the high mobility for cations in the presence of anions remaining in the polymer. (b) If anions are quantitatively expelled from the conducting polymer during reduction, then the fast cation hopping is inhibited in the ion-free zone (c).

This is schematically explained in Figure 14.37.

At the beginning of the reduction of the conducting polymer, both anion expulsion as well as cation incorporation occur, supposedly in the same way as they would do during a standard electrochemical reduction experiment. However, while in case of the immersed sample, the transport has to occur just over the thickness of the conducting polymer film, usually not more than 10 μm, in the delamination setup, the reduction proceeds laterally from the border with the electrolyte into the oxidized coating. Hence, the distance across which either transport of anions or cations has to take place is quickly one to two orders of magnitude larger. Now, even for a conducting polymer doped with a mobile anion, showing during standard electrochemical reduction predominant anion release behavior even in the presence of small cations in the electrolyte, some cation incorporation will occur in addition to the anion release. Hence, some negative charge will be immobilized in the polymer, neutralized by cations. This is the scenario of a typical cation transport membrane, where cations can hop from fixed negative charge to fixed negative charge. With each anion remaining in the polymer, the mobility of the cations will increase, that is, the further the reduction front progresses into the conducting polymer, the higher the contribution of cation incorporation to the charge neutralization. Finally, predominant cation incorporation is observed.

On the other hand, when all anions were expelled from the conducting polymer during reduction (by use of a very large organic cation that cannot be incorporated at all, see Figure 14.36a), the mobility of small cations is also very low in the reduced polymer, because the fast hopping from negative charge to negative charge cannot take place anymore. This is clearly shown by the fact that the progress of the reduction front remains unaffected when the KCl was injected after the reduction had already substantially advanced from the border with the electrolyte (see Figure 14.36a). In this case, the cations have to cross polymer free of anions. As no effect at all is observed, anion expulsion remains the dominating process, that is, the cations are not incorporated.

It should, however, be possible to avert the danger of sudden coating breakdown that is postulated to be an inherent property of all conducting polymers [239] by preventing too extended percolation networks of conducting polymer in composite coatings. Coatings that follow these design rules generally seem to perform well (see discussion in Refs. [239,243–247]) and especially to effectively inhibit fast delamination originating from external defects. For the case of the PANI films used by McMurray and coworkers [246,247], it is reported that no further enhancement of the performance was observed when the coating thickness was increased over about a few micrometers, which indicates that the average percolation length for the coatings is in that range too [248]. For the coatings used by Paliwoda et al. [243], Williams et al. [246], and Holness et al. [247], the percolation length seems to be about 1 μm. A general quantification of the maximum tolerable percolation length, however, cannot be given at this stage. This value will depend on a number of parameters and is subject of current research.

14.3.5.2 Intelligent Corrosion Protection

The most promising aspect of conducting polymers is that it is possible to use inhibitor anions as dopants (which are necessary for compensating the positive charge sitting on the polymer backbone of the oxidized ICP), which will be released when the ICP is reduced as a result of corrosive attack (see Figure 14.38). As the release will occur only when corrosion in a defect site will cause reduction of the ICP, this release is a true release on demand: only when the inhibitors are needed, the release will occur. This is different from the standard corrosion inhibitor pigments that are added to organic coatings and that continuously leach corrosion-inhibiting substances at high humidity, also when not needed.

Hence, such an intelligent approach would be highly desirable for future corrosion protection coatings. As discussed above, continuous networks of conducting polymer are to be avoided. If the conducting polymer forms macroscopic percolation networks in the coating, fast cation incorporation will result, preventing intelligent corrosion-triggered anion release for corrosion inhibition [238,240] and potentially causing fast

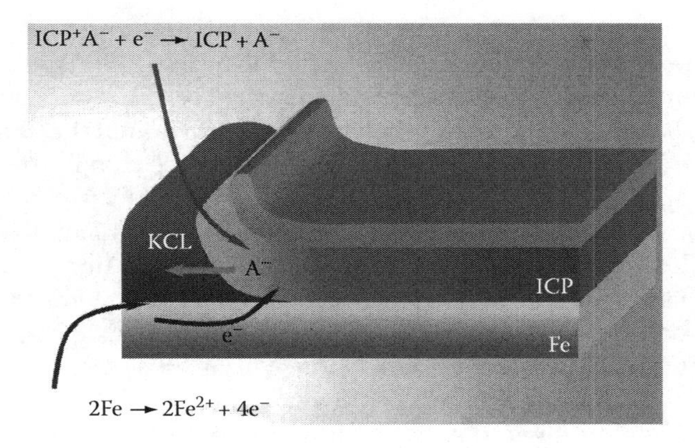

FIGURE 14.38
If cation incorporation does not occur, dopant anion release will take place when the ICP is reduced, for example, starting at a defect site where metal is corroding. The anions will then migrate to the corroding metal and if they have corrosion inhibiting properties, they will stop the corrosion. Then the release will also stop, as the ICP is not longer reduced. Hence, the release occurs, so to speak, only on demand.

coating breakdown [237–239]. If the particles of conducting polymer in a composite coating are dispersed in low density, no risk of fast cation incorporation is to be expected, since the particles are not in direct contact with the high concentration of small cations in the electrolyte. But only the few particles in direct contact with the metal can be electrochemically active, that is, the amount of possible inhibitor anion release is small. If however, the conducting polymer is organized in microscopic networks, but not macroscopic ones, fast cation incorporation should be prevented and significant anion release be enabled.

As an example, the work by Paliwoda et al. [243] shall be shortly discussed here.

In this work, a monolayer film of PPy nanoparticles dispersed into a nonconducting matrix was applied on iron. This monolayer film was then coated by a nonconductive top coat and a delamination experiment was started. The PPy was doped with polyphosphomolybdate (PMo) anions that decompose upon release into mainly molybdate and some phosphate, both inhibiting corrosion on iron. The nonconducting matrix polymer successfully prevented particles from forming macroscopic percolation networks [243] within the

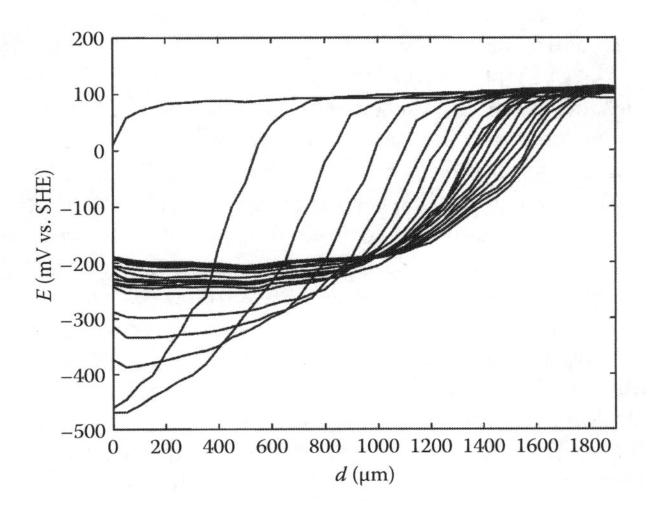

FIGURE 14.39
SKP delamination profiles obtained on a composite coating containing PPyPMo. Intervals between profiles are in case of all samples the same and equal, 20 min.

monolayer. In Figure 14.39, we can see that after some initiation time, the delamination slows down and the potential in the delaminated zone increases [243]. Parallel measurement of the corrosion potential in the defect revealed that the corrosion potential moved into the same direction, indicating an increasing passivation. Surface analytical investigations of the defect area clearly revealed the presence of molybdate on the iron in the defect area, that is, indeed release of the molybdate has occurred successfully. However, it took about 1 mm of delamination before enough inhibitor was released to effectively stop the corrosion process. This is easy to understand because the PPy is just present in an ultrathin layer on the metal surface, that is, the amount of PPy and hence the amount of inhibitor to be released is very low. A better approach would be to have a clustering of the ICPs in microscopic structures without the formation of macroscopic percolation. This is the subject of further research.

Concluding, pure conducting polymer coatings or composite coatings containing too high amounts of conducting polymer, that is, coatings where extended percolation networks of the ICP prevail, may show good corrosion protection under some conditions, especially in the absence of chloride and of larger defects. However, such coatings are prone to fast and disastrous breakdown as soon as a large defect is inflicted into the coating. Then the coupling between ICP and the exposed metal in the defect may even lead to an enhanced corrosion (see, e.g., Ref. [249]). This is the main reason why only very few works on electrodeposited ICP coatings have been included into this discussion, although excellent work has been carried out in depositing good quality ICP coatings onto reactive materials. This is far from easy and intense research was performed over the years on this topic. A more complete overview over the research on conducting polymers for corrosion protection that contains such information is given for instance in the reviews on ICP corrosion performance on nonferrous and ferrous metals by Dennis et al. and Spinks et al. [250,251]. This danger of pure ICP coatings may in future be lowered by tailoring the interface between ICP coating and metal toward low conductivity (see, for example, the work by Van Schaftinghen et al. [252]), but it is unclear yet whether this approach may ever work, especially because, when no conductivity between ICP and metal remains, no active function of the ICP will be available for intelligent protection strategies. Moreover, even in absence of larger defects, breakdown may occur. Tallman et al. [253], for instance, have observed PANI coatings on steel first performing well but then breaking down after about 50 days immersion.

Hence, the recommended approach for further research on ICP corrosion protection by coatings containing active ICP (i.e., not deactivated ICP that has been used up in a pretreatment process or which has been intentionally annealed for improved barrier properties, as discussed above) is to focus on composite coatings where the ICP particles do not form macroscopically extended percolation networks but just microscopic ones. As discussed above, such coatings are safe from disastrous breakdown in the presence of large defects. However, in order to optimize the performance of the coatings, two contradicting properties have to be optimized: in order to enhance the ability for inhibitor anion release and metal passivation, an as high as possible amount of ICP needs to be in electric contact with the metal at the defect site, that is, the percolation network needs to be as extended as possible. In order to prevent the transformation to cation permselectivity, this percolation network must not exceed a certain length, which will depend on the ICP used. How to find the best compromise is still subject of current research. The aim is good corrosion protection coatings with intelligent inhibitor release ability that will release inhibitors only when needed, that is, when corrosion occurs, and automatically stop the release when the corrosion stops.

However, problems with anodic corrosion at the interface between metal and composite coating still have to be solved. In principal, the high potential of the conducting polymer should ensure that the metal in direct contact with it is kept reliably within its passive range. Now, in a composite coating, not all the coated metal surface will be in direct contact with the conducting polymer particles, but just in their vicinity. Hence, over the time, the metal surface that is just close to the conducting polymer but not directly in contact with it might, due to water and ions diffused through the coating to the interface, form a weak galvanic element with the conducting polymer. The low conductivity at the interface then prevents the full polarization to the potential of the conducting polymer and hence, the metal may not be kept in its passive range. On the contrary, it may be polarized toward increased corrosion activity. This is in agreement with the findings of Williams and McMurray [254] and Michalik and coworkers [255] and constitutes a serious problem for the introduction of conducting polymers into technical corrosion protection.

The logical counter action would be to dope the conducting polymer with suitable anodic corrosion inhibitors. This again is subject of current research.

References

1. H. Leidheiser Jr., ed., *Corrosion Control by Organic Coatings*, NACE, Houston, TX, 1981.
2. R. A. Dickie and F. L. Floyd, eds., *Polymeric Materials for Corrosion Control*, American Chemical Society, Washington, DC, 1986.
3. H. Leidheiser Jr., *Coatings, Corrosion Mechanisms* (F. Mansfeld, ed.), Marcel Dekker, New York, 1987.
4. J. Kinloch, *Adhesion 3* (K. W. Allen, ed.), Applied Science, London, U.K., 1979.
5. R. Feser, PhD thesis, University of Clausthal-Zellerfeld, Germany, 1990.
6. R. Feser and M. Stratmann, *Werkst. Korros. 42*:187 (1991).
7. M. Stratmann, R. Feser, and A. Leng, *Electrochim. Acta 39*:1207 (1994).
8. H. Leidheiser Jr., W. Wang, and L. Igetoft, *Progr. Org. Coat. 11*:19 (1983).
9. H. Leidheiser Jr. and W. Wang, *Corrosion Control by Organic Coatings*, NACE, Houston, TX, p. 70, 1981.
10. H. Leidheiser Jr., *Corrosion Control by Organic Coatings*, NACE, Houston, TX, p. 124, 1981.
11. E. L. Koehler, *Proceedings of the 4th International Congress on Metallic Corrosion*, Amsterdam, the Netherlands, p. 736, 1972.
12. E. L. Koehler, *Corrosion Control by Organic Coatings*, NACE, Houston, TX, p. 87, 1981.
13. J. J. Ritter and J. Kruger, *Corrosion Control by Organic Coatings*, NACE, Houston, TX, p. 28, 1981.
14. H. Leidheiser Jr., L. Igetoft, W. Wang, and K. Weber, *Proceedings of the 7th International Conference on Organic Coatings*, Athens, Greece, 1981.
15. J. S. Hammond, J. W. Holubka, and R. A. Dickie, *J. Coat. Technol. 51*:45 (1979).
16. J. E. Castle and J. F. Watts, *Corrosion Control by Organic Coatings*, NACE, Houston, TX, p. 78, 1981.
17. M. Stratmann, *Adv. Mater. 2*:191 (1990).
18. M. Stratmann, B. Reynders, and M. Wolpers, *Haftung bei Verbundwerkstoffen und Werkstoffverbunden* (W. Brockmann, ed.), Deutsche Gesellschaft für Materialkunde, Oberursel, 1991.
19. M. Rohwerder and M. Stratmann, *MRS Bull. 24*:43 (1999).
20. E. P. Plueddemann, *Silane Coupling Agents*, Plenum Press, New York, 1982.
21. K. L. Mittal, ed., *Silanes and Other Coupling Agents*, VSP, Utrecht, the Netherlands, 1992.
22. K. L. Mittal, ed., *Adhesion Aspects of Polymeric Coatings*, Plenum Press, New York, 1983.
23. D. E. Leyden, ed., *Silanes, Surfaces and Interfaces*, Gordon & Breach Science Publishers, New York, 1985.
24. P. E. Cassidy and B. J. Yager, *J. Macromol. Sci. Rev. Polym. Technol. D1(1)*:1 (1971).
25. H. Ishida, *Polym. Compos. 5*:101 (1984).

26. G. Trabanelli, *Corrosion Inhibitors, Corrosion Mechanisms* (F. Mansfeld, ed.), Marcel Dekker, New York, 1987.
27. H. Kaesche, *Die Korrosion der Metalle*, Springer, Berlin, Germany, 1990.
28. *Proceedings of the (a) 1st, 1961 (b) 2nd, 1966 (c) 3rd, 1971 (d) 4th, 1976 (e) 5th, 1980 (f) 6th, 1985 (g) 7th, 1990, European Symposium on Corrosion Inhibitors*, Ferrara, Annali dell'universita di Ferrara, Sezione V.
29. J. Lipkowski and P. Ross, eds., *Adsorption of Molecules at Metal Electrodes*, Verlag Chemie, Weinheim, Germany, 1992.
30. M. E. Tarvin and B. A. Miksic, *Corrosion*, New Orleans, A, paper no. 344, 1989.
31. B. A. Miksic, M. Tarvin, and G. R. Sparrov, *Corrosion*, New Orleans, A, paper no. 607, 1989.
32. D. M. Brasher and A. D. Mercer, *Br. Corros. J. 95*:122 (1967).
33. G. Schmitt, *Br. Corros. J. 9*:165 (1984).
34. A. Leng, Diploma thesis, University of Dusseldorf, Germany, 1991.
35. H. Fisher, *Werkst. Korros. 23*:445 (1972).
36. W. J. Lorentz and F. Mansfeld, *International Conference on Corrosion Inhibition*, NACE, Houston, TX, 1987.
37. J. O. M. Bockris and R. A. J. Swinkels, *J. Electrochem. Soc. 111*:736 (1964).
38. K. J. Vetter and J. W. Schultze, *J. Electroanal. Chem. 53*:67 (1974).
39. J. W. Schultze and K. D. Koppitz, *Electrochim. Acta 21*:327 (1976).
40. K. J. Vetter and J. W. Schultze, *Ber. Bunsenges. Phys. Chem. 76*:920 (1972).
41. N. Hackermann and R. M. Hard, *1st International Congress Metal Corrosion*, Butterworth, London, U.K., 1962.
42. R. G. Pearson, *Science 151*:172 (1966).
43. L. Horner, *Chem. Ztg. 100*:247 (1976).
44. H. K. Yasuda, *J. Polym. Sci. Macromol. Rev. 16*:199 (1981).
45. R. d'Agostino, ed., *Plasma Deposition, Treatment, and Etching of Polymers*, Academic Press, New York, 1990.
46. H. Yasuda, *Plasma Polymerization*, Academic Press, New York, 1985.
47. N. Morosoff, *An Introduction to Plasma Polymerization, Plasma Deposition, Treatment, and Etching of Polymers* (R. d'Agostino, ed.), Academic Press, New York, 1990, pp. 1–95.
48. C. D. Bain, E. B. Troughton, Y.-T. Tao, J. Evall, G. M. Whitesides and R. G. Nuzzo, *J. Am. Chem. Soc. 111*:321 (1989)
49. E. Vago, K. de Weldige, M. Rohwerder, and M. Stratmann, *Fresen. J. Anal. Chem. 335*:316 (1995).
50. M. Volmer, M. Stratmann, and H. Viefhaus, *Surf. Interface Anal. 16*:278 (1990).
51. M. Volmer-Uebing and M. Stratmann, *Appl. Surf. Sci. 55*:19 (1992).
52. P. E. Laibinis, G. M. Whitesides, D. L. Allara, Y.-T. Tao, A. N. Parikh, and R. G. Nuzzo, *J. Am. Chem. Soc. 113*:7152 (1991).
53. H. Ron and I. Rubinstein, *Langmuir 10*:4566 (1994).
54. Z. Hou, N. L. Abbott, and P. Stroeve, *Langmuir 14*:3287 (1998).
55. G. Brunaro, F. Parmigiani, G. Perboni, G. Rocchini, and G. Trabanelli, *Br. Corros. J. 27*:75 (1992).
56. S. Gonzalez, M. M. Laz, R. M. Souto, R. C. Salvarezza, and A. J. Arvia, *Corrosion 49*:450 (1993).
57. K. T. Carron, M. L. Lewis, J. Dong, J. Ding, G. Xue, and Y. Chert, *J. Mater. Sci. 28*:4099 (1993).
58. F. M. A1-Kharafi, F. H. A1-Hajjar, and A. Katrib, *Corros. Sci. 30*:869 (1990).
59. C. M. Whelan, M. Kinsella, L. Carbonell, H. M. Ho, and K. Maex, *Microelectron. Eng. 70*:551 (2003).
60. F. Laffineur, J. Delhalle, S. Guittard, S. Geribaldi, and Z. Mekhalif, *Colloid Surf.* A *198–200*:817 (2002).
61. Z. Mekhalif, F. Sinapi, F. Laffineur, and J. Delhalle, *J. Electron Spectrosc. Relat. Phenom. 121*:149 (2001).
62. Y. Yamamoto, H. Nishihara, and K. Aramaki, *J. Electrochem. Soc. 140*:436 (1993).
63. M. Itoh, H. Nishihara, and K. Aramaki, *J. Electrochem. Soc. 141*:2018 (1994).
64. M. Itoh, H. Nishihara, and K. Aramaki, *J. Electrochem. Soc. 142*:1839 (1995).
65. M. Itoh, H. Nishihara, and K. Aramaki, *J. Electrochem. Soc. 142*:3696 (1995).
66. R. Haneda, H. Nishihara, and K. Aramaki, *J. Electrochem. Soc. 144*:1215 (1997).

67. R. Haneda, H. Nishihara, and K. Aramaki, *J. Electrochem. Soc. 145*:1856 (1998).
68. R. Haneda and K. Aramaki, *J. Electrochem. Soc. 145*:2786 (1998).
69. M. Ishibashi, M. Itoh, H. Nishihara, and K. Aramaki, *Electrochim. Acta 41*:241 (1996).
70. D. Taneichi, R. Haneda, and K. Aramaki, *Corros. Sci. 43*:1589 (2001).
71. P. E. Laibinis and G. M. Whitesides, *J. Am. Chem. Soc. 114*:1990 (1992).
72. H. Keller, P. Simak, W. Schrepp, and J. Dembowski, *Thin Solid Films, 244*:799 (1994).
73. G. K. Jennings and P. E. Laibinis, *Colloid. Surf. A Physicochem. Eng. Asp. 116*:105–114 (1996).
74. C. D. Bain, E. B. Troughton, Y.-T. Tao, J. Evall, G. M. Whitesides, and R. G. Nuzzo, *J. Am. Chem. Soc. 111*:321 (1989).
75. G. A. Edwards, A. J. Bergren, E. J. Cox, and M. D. Porter, *J. Electroanal. Chem. 622*:193 (2008).
76. (a) M. Rohwerder, K. de Weldige, and M. Stratmann, *J. Solid State Electrochem. 2*:88 (1998); (b) M. Rohwerder, dissertation, University of Düsseldorf, Shaker Verlag, 1997.
77. A. Widrig, C. A. Alves, and M. D. Porter, *J. Phys. Chem. 113*:2805 (1991).
78. A. Alves, E. L. Smith, C. A. Widrig, and M. D. Porter, *SPIE, Appl. Spectrosc. Mater. Sci. II 1636*:125(1992).
79. Li Sun and R. M. Crooks, *J. Electrochem. Soc. 138*:L23 (1991).
80. Y.-T. Kim and A. J. Bard, *Langmuir 8*:1096 (1992).
81. K. Edinger, A. Gölzhäuser, K. Demota, Ch. Wöll, and M. Grunze, *Langmuir 9*:4 (1993).
82. G. E. Poirier and M. J. Tarlov, *Langmuir 10*:2853 (1994).
83. E. Delamarche, B. Michel, Ch. Gerber, D. Anselmetti, H.-J. Güntherodt, H. Wolf, and H. Ringsdorf, *Langmuir 10*:2869 (1994).
84. J. A. M. Sondag-Huethorst, C. Schönenberger, and L. G. Fokkink, *J. Phys. Chem. 98*:6826 (1994).
85. C. Schönenberger, J. Jorritsma, J. A. M. Sondag-Huethorst, and L. G. J. Fokkink, *J. Phys. Chem. 99*:3259 (1995).
86. J. Dvorak, T. Jirsak, and J. A. Rodriguez, *Surf. Sci. 479*:155 (2001).
87. B. Halevi and J. M. Vohs, *J. Phys. Chem. B 109*:23976 (2005).
88. C. Yan, M. Zharnikov, A. Gölzhäuser, and M. Grunze, *Langmuir 16*:6208 (2000).
89. H. P. Zhang, C. Romero, and S. Baldelli, *J. Phys. Chem. B 109*:15520 (2005).
90. Z. Mekhalif, J. Riga, J.-J. Pireaux, and J. Delhalle, *Langmuir 13*:2285 (1997).
91. L. Tortech, Z. Mekhalif, J. Delhalle, F. Guittard, and S. Geribaldi, *Thin Solid Films 491*:253 (2005).
92. Z. Mekhalif, F. Laffineur, N. Couturier, and J. Delhalle, *Langmuir 19*:637 (2003).
93. Z. Mekhalif, J. Delhalle, J.-J. Pireaux, S. Noël, F. Houzé, and L. Boyer, *Surf. Coat. Technol. 100–101*:463 (1998).
94. S. Noël, F. Houzé, L. Boyer, Z. Mekhalif, R. Caudano, and J. Delhalle, *J. IEEE Trans. Compon. Packag. Technol. 22*:79 (1999).
95. F. Sinapi, L. Forget, J. Delhalle, and Z. Mekhalif, *Appl. Surf. Sci. 212–213*:464 (2003).
96. A. R. Noble-Luginbuhl and R. G. Nuzzo, *Langmuir 17*:3937 (2001).
97. F. Sinapi, T. Issakova, J. Delhalle, and Z. Mekhalif, *Thin Solid Films 515*:6833 (2007).
98. Z. Mekhalif, L. Massi, F. Guittard, S. Geribaldi, and J. Delhalle, *Thin Solid Films 405*:186 (2002).
99. H. Zhang and S. Baldelli, *J. Phys. Chem. B 110*:24062 (2006).
100. C. Nogues and P. Lang, *Langmuir 23*:8385 (2007).
101. J. Hedberg, Ch. Leygraf, K. Cimatu, and S. Baldelli, *J. Phys. Chem. C 111*:17587 (2007).
102. F. Laffineur, Z. Mekhalif, L. Tristani, and J. Delhalle, *J. Mater. Chem. 15*:5054 (2005).
103. F. Berger, J. Delhalle, and Z. Mekhalif, *Electrochim. Acta 53*:2852 (2008).
104. A. Srhiri, M. Etman, and F. Dabosi, *Werkst. Korros. 43*:406 (1992).
105. T. F. Child and W. J. van Ooij, *Trans. Inst. Met. Finish. 77*:64 (1999).
106. W. J. van Ooij and T. F. Child, *Chemtech 28*:26 (1998).
107. I. Maege, E. Jaehne, A. Henke, H.-J. P. Adler, C. Bram, C. Jung, and M. Stratmann, *Macromol. Symp. 126*:7 (1997).
108. I. Maege, E. Jaehne, A. Henke, H.-J. P. Adler, C. Bram, C. Jung, and M. Stratmann, *Progr. Org. Coat. 34*:1 (1998).
109. D. L. Allara and R. G. Nuzzo, *Langmuir 1(1)*:45 (1985).

110. J. van den Brand, S. Van Gils, P. C. J. Beentjes, H. Terryn, V. Sivel, and J. H. W. de Wit, *Progr. Org. Coat. 51*:339 (2004).
111. B. Adolphi, E. Jahne, G. Busch, and X. D. Cai, *Anal. Bioanal. Chem. 379*:646 (2004).
112. C. Bram, C. Jung, and M. Stratmann, *ECASIA 97: 7th European Conference on Applications of Surface and Interface Analysis*, Goteborg, Sweden, pp. 97–100 (1997).
113. M. Giza, P. Thissen, and G. Grundmeier, *Langmuir 24(16)*:8688 (2008).
114. M. Stratmann, M. Rohwerder, C. Reinartz et al., *ECASIA 97: 7th European Conference on Applications of Surface and Interface Analysis*, Goteborg, Sweden, pp. 13–20 (1997).
115. K. Wapner, M. Stratmann, and G. Grundmeier, *Int. J. Adhesion Adhesives 28*:59 (2007).
116. M. Dannenfeldt, Dissertation, University of Erlangen, Erlangen, 2000.
117. J. Marsh, L. Minel, M. G. Barthes-Labrousse, and D. Gorse, *Appl. Surf. Sci. 133*:270 (1998).
118. M. Taylor, C. H. McLean, M. Charlton, and J. F. Watts, *Surf. Interface Anal. 23*:342 (1995).
119. M. Taylor, J. F. Watts, J. Bromley-Barrat, and G. Beamson, *Surf. Interface Anal. 21*:697 (1992).
120. J. F. Watts, R. A. Blunden, and T. J. Hall, *Surf. Interface Anal. 16*:227 (1990).
121. G. M. Chritchlow and A. Madison, *Surf. Interface Anal. 17*:539 (1990).
122. D. M. Brewis, *Int. J. Adhesion Adhesives 13*:251 (1993).
123. De Neve, M. Delamar, T. T. Nguyen, and M. E. R. Shanahan, *Appl. Surf. Sci. 134*:202 (1998).
124. L. Olsson-Jaques, A. R. Wilson, A. N. Rider, and D. R. Arnott, *Surf. Interface Anal. 24*:569 (1996).
125. C. Schilz, Dissertation, University of Erlangen, Erlangen, 2000.
126. M. Rohwerder, E. Hornung, and X.-W. Yu, *MRS Proc. 734*:B2.8.1–B2.8.6 (2003).
127. Numerical modelling/lifetime prediction of delamination polymer coating disbonding and material degradation, Final Report, European Research Project, Degradation Modeling (STRP 01376).
128. M. Rohwerder and F. Turcu, *Electrochim. Acta 53*:290 (2007).
129. M. Rohwerder, E. Hornung, and M. Stratmann, *Electrochim. Acta 48(9)*:1235 (2003).
130. R. Lösch, M. Stratmann, and H. Viefhaus, *Electrochim. Acta 39*:1215 (1994).
131. T. Rigó, A. Mikó, J. Telegdi, M. Lakatos-Varsányi, A. Shaban, and E. Kálmán, *Electrochem. Solid State Lett. 8(10)*:B51 (2005).
132. M. Stratmann, M. Wolpers, H. Streckel, and R. Feser, *Ber. Bunsenges. Phys. Chem. 95*:1365(1991).
133. M. Wolpers, M. Stratmann, H. Viefhaus, and H. Streckel, *Thin Solid Films 210/211*:592 (1992).
134. K. L. Mittal, ed., *Silanes and Other Coupling Agents*, Vol. 2, VSP, 2000.
135. P. Nazarov, M. A. Petrunin, and Y. N. Mikhailovsky, *Proceedings of the 11th International Corrosion Congress*, Florence, Italy, 1990.
136. M. Stratmann and H. Streckel, *Corros. Sci. 30*:681 (1990).
137. T. Narayanan, *Rev. Adv. Mater. Sci. 9*(2):130 (2005).
138. A. Losch, E. Klusmann, and J. W. Schultze, *Electrochim. Acta. 39*:1183 (1994).
139. T. Prosek and D. Thierry, *Corrosion 60(12)*:1122 (2004).
140. P. Marcus, *Corrosion Mechanisms in Theory and Practice*, 3rd edn.
141. S. M. Cohen, *Corrosion 51*:71 (1995).
142. F. Mansfeld, C. B. Breslin, A. Pardo, and F. J. Perez, *Surf. Coat. Technol. 90*:224 (1997).
143. M. Bethencourt, F. J. Botana, M. J. Cano, and M. Marcos, *Appl. Surf. Sci. 189*:162 (2002).
144. M. Murase and J. F. Watts, *J. Mater. Chem. 8*:1007 (1998).
145. P. Hivart, B. Hauw, J. Crampon, and J. P. Bricout, *Wear 219*:195 (1998).
146. K. Ogle and R. Bucheit, *Conversion Coatings, Encyclopaedia of Electrochemistry* (A.J. Bard, M. Stratmann, eds.), Wiley-VCH, Weinheim, Germany vol. 5, p. 460, 2003.
147. K. Ogle, A. Tomandl, N. Meddahi, and M. Wolpers, *Corros. Sci. 46*:979 (2004).
148. H. Leidheiser Jr., *Coatings in Chemical Industries 28: Corrosion Mechanisms* (F. Mansfeld, ed.), Marcel Dekker, New York, p. 202, 1987.
149. L. Fedrizzi, F. Deflorian, S. Rossi, L. Fambri, and P. L. Bonora, *Progr. Org. Coat. 42(1–2)*:65 (2001).
150. J. F. Ying, B. J. Flynn, M. Y. Zhou, P. C. Wong, K. A. R. Mitchell, and T. Foster, *Progr. Surf. Sci. 50(1–4)*:259 (1995).
151. D. Zhu and W. J. van Ooij, *Electrochim. Acta 49*:1113 (2004).
152. A. Franquet, C. Le Pen, H. Terryn, and J. Vereecken, *Electrochim. Acta 48*:1245 (2003).

153. M. G. S. Ferreira, R. G. Duarte, M. F. Montemor, and A. M. P. Simões, *Electrochim. Acta 49*:2927 (2004).
154. M.-L. Abel, J. F. Watts, and R. P. Digby, *Int. J. Adhesion Adhesives 18*:179 (1998).
155. K. Aramaki, *Corros. Sci. 43*:1573 (2001).
156. B. R. W. Hinton, *J. Alloy. Compd. 180*:15 (1992).
157. A. J. Aldykiewicz Jr., A. J. Davenport, and H. S. Issacs, *J. Electrochem. Soc. 142(10)*:3342 (1995).
158. A. J. Aldykiewicz Jr., A. J. Davenport, and H. S. Issacs, *J. Electrochem. Soc. 143(1)*:147 (1996).
159. H. N. McMurray, S. M. Powell, and D. A. Worsley, *Br. Corros. J. 36*:42 (2001).
160. E. Almeida, T. C. Diamantino, M. O. Figueiredo, and C. Sá, *Surf. Coat. Technol. 106(1)*:8 (1998).
161. A. A. O. Magalhães, I. C. P. Margarit, and O. R. Mattos, *J. Electroanal. Chem. 572*(2):433 (2004).
162. C. G. da Silva, I. C. P. Margarit-Mattos, O. R. Mattos, H. Perrot, B. Tribollet, and V. Vivier, *Corros. Sci. 51*:151 (2009).
163. G. M. Treacy, G. D. Wilcox, and M. O. W. Richardson, *J. Appl. Electrochem. 29*:647 (1999).
164. J. H. Nordlien, J. C. Walmsley, H. Østerberg, and K. Nisancioglu, *Surf. Coat. Technol. 184(2–3)*:278 (2004).
165. O. Lunder, C. Simensen, Y. Yu, and K. Nisancioglu, *Surf. Coat. Technol. 153(1)*:72 (2002).
166. T. Scram, G. Goeminne, H. Terryn, and W. Vanhoolst, *Trans. Inst. Met. Finish. 70(3)*:91–95 (1995).
167. J. H. Nordlien, J. C. Walmsley, H. Osterberg, and K. Nisancioglu, *Surf. Coat. Technol. 153*:72–78 (2002).
168. O. Lunder, C. Simensen, Y. Yu, and K. Nisancioglu, *Surf. Coat. Technol. 184*:278–290 (2004).
169. M. Dabala, E. Ramous, and M. Magrini, *Mater. Corros. 55*(5):381–386 (2004).
170. M. F. Montemor, A. M. Simoes, and M. G. S. Ferreira, *Progr. Org. Coat. 44*:111–120 (2002).
171. M. A. Jakab, F. Presuel-Moreno, and J. R. Scully, *Corrosion 61*(3):246 (2005).
172. A. K. Mishra and R. Balasubramaniam, *Mater. Chem. Phys. 103*:385 (2007).
173. D. Weng, P. Jokiel, A. Uebleis, and H. Boehni, *Surf. Coat. Technol. 88*:147 (1996).
174. A. Neuhaus, E. Jumpertz, and M. Gebhardt, *Korrosion 16*:155 (1963).
175. B. R. W. Hinton and L. Wilson, *Corros. Sci. 34*:1773 (1983).
176. K. Z. Chong and T. S. Shih, *Mater. Chem. Phys. 80*(1):191 (2003).
177. A. Goldschmidt and H.-J. Steitberger, *BASF-Handbuch Lackiertechnik*, Vincentz Verlag, Hannover, p. 476, 2002.
178. Y. C. Lu and M. B. Ives, *Corros. Sci. 37(1)*:145–155 (1995).
179. Z. Yong, J. Zhu, C. Qiu, and Y. Liu, *Appl. Surf. Sci. 255*:1672 (2008).
180. H. Ardelean, I. Frateur, and P. Marcus, *Corros. Sci. 50*:1907 (2008).
181. M. F. Montemor, A. M. Simões, M. G. S. Ferreira, and M. J. Carmezim, *Appl. Surf. Sci. 254*:1806 (2008).
182. M. F. Montemor, A. M. Simões, and M. J. Carmezim, *Appl. Surf. Sci. 253*:6922 (2007).
183. Y. J. Warburton, D. L. Gibbon, K. M. Jackson, L. F. Gate, A. Rodnyansky, and P. R. Warburton, *Corrosion 55*:898 (1999).
184. Y. Yoshikawa and J. F. Watts, *Surf. Interface Anal. 20*:379 (1993).
185. M. Dabala, L. Armelao, A. Buchberger, and I. Calliari, *Appl. Surf. Sci. 172*:312 (2001).
186. N. Fink, B. Wilson, and G. Grundmeier, *Electrochim. Acta 51*:2956 (2006).
187. B. Wilson, N. Fink, and G. Grundmeier, *Electrochim. Acta 51*:3066 (2006).
188. Ch. Stromberg, P. Thissen, I. Klueppel, N. Fink, and G. Grundmeier, *Electrochim. Acta 52*:804 (2006).
189. N. Tang, W. J. van Ooij, and G. Gorecki, *Progr. Org. Coat. 30*:255 (1997).
190. M. F. Montemor, A. M. Simoes, M. G. S. Ferreira, B. Williams, and H. Edwards, *Progr. Org. Coat. 38*:17 (2000).
191. P. Puomi, H. M. Fagerholm, J. B. Rosenholm, and K. Jyrkas, *Surf. Coat. Technol. 115*:70 (1999).
192. D. Zimmermann, A. G. Munõz, and J. W. Schultze, *Surf. Coat. Technol. 197*:260 (2005).
193. G. Klimow, N. Fink, and G. Grundmeier, *Electrochim. Acta 53*:1290 (2007).
194. P. Matinlinna, S. Areva, L. V. J. Lassila, and P. K. Vallittu, *Surf. Interface Anal. 36*:1314 (2004).
195. P. Plueddemann, *Silane Coupling Agents*, 2nd edn. Plenum Press, New York, 1990.
196. Product Brochure, PCR *Prosil Organofunctional Silane Coupling Agents*, 7 pages.

197. H. Wang, T. Tang, S. O. Cao, K. Y. Salley, and Ng. Simon, *J. Colloid Interf. Sci. 291*:438 (2005).
198. W. J. van Ooij and D. Zhu, *Corrosion 57*:413 (2001).
199. M.-L. Abel, R. P. Digby, I. W. Fletcher, and J. F. Watts, *Surf. Interf. Anal. 29*:115 (2000).
200. A. Franquet, J. De Laet, T. Schram, H. Terryn, V. Subramanian, W. J. van Ooij, and J. Vereecken, *Thin Solid Films 384*:37 (2001).
201. J. Bouchet, G. M. Pax, Y. Leterrier, V. Michaud, and J.-A. E. Manson, *Compos. Interf. 13*:573 (2006).
202. I. Piwonski and A. Ilik, *Appl. Surf. Sci. 253*:2835 (2006).
203. P. E. Hintze and L. M. Calle, *Electrochim. Acta 51*:1761 (2006).
204. K. Wapner and G. Grundmeier, *Adv. Eng. Mater. 6*(3):163 (2004).
205. K. Wapner, M. Stratmann, and G. Grundmeier, *Electrochim. Acta 51*:3303 (2006).
206. G. Grundmeier, C. Reinartz, M. Rohwerder, and M. Stratmann, *Electrochim. Acta 43*:165 (1998).
207. S. V. Lamaka, M. F. Montemore, A. F. Galio, M. L. Zheludkevich, C. Trindade, L. F. Dick, and M. G. S. Ferreira, *Electrochim. Acta 53*:4773 (2008).
208. L. Zheludkevich, I. Miranda Salvado, and M. G. S. Ferreira, *J. Mater. Chem. 15*:5099 (2005).
209. H. J. Engell and K. Hauffe, *Z. Elektrochem. 57*(8):762 (1953).
210. H. J. Engell and K. Hauffe, *Z. Elektrochem. 57*(8):773 (1953).
211. G. Grundmeier, E. Matheise, and M. Stratmann, *J. Adhesion Sci. Technol. 10*(6):573 (1996).
212. G. Grundmeier and M. Stratmann, *Thin Solid Films 352*(1–2):119 (1999).
213. V. Barranco, P. Thiemann, H. K. Yasuda, M. Stratmann, and G. Grundmeier, *Appl. Surf. Sci. 229*(1–4):87 (2004).
214. G. Grundmeier and M. Stratmann, *Appl. Surf. Sci. 141*(1–2):43 (1999).
215. N. J. Ahirtcliffe, M. Stratmann, and G. Grundmeier, *Surf. Interf. Anal. 35*(10):799 (2003).
216. G. Grundmeier, M. Brettmann, and P. Thiemann, *Appl. Surf. Sci. 217*:223 (2003).
217. T. Titz, F. Hörzenberger, K. Van den Bergh, and G. Grundmeier, *Electrochim. Acta 52*(2):369 (2010).
218. K. Draou, N. Bellakhal, B. G. Cheron, and J. L. Brisset, *Mater. Chem. Phys. 51*(2):142 (1997).
219. N. Bellakhal, K. Draou, and J. L. Brisset, *Mater. Chem. Phys. 73*(2–3):235 (2002).
220. N. Bellakhal and M. Dachraoui, *Mater. Chem. Phys. 82*(2):484 (2003).
221. X. B. Tian, C. B. Wei, S. Q. Yang, R. K. Y. Fu, and P. K. Chu, *Nucl. Instr. Meth. Phys. Res. B. Beam Interact. Mater. Atoms 242*(1–2):300 (2006).
222. N. Bertrand, P. Bulkin, B. Drevillon, S. Lucas, and S. Benayoun, *Surf. Coat. Technol. 94–5*(1–3):362 (1997).
223. T. F. Wang and H. K. Yasuda, *J. Appl. Polym. Sci. 55*:903–909 (1995).
224. N. Shirtcliffe, P. Thiemann, M. Stratmann, and G. Grundmeier, *Surf. Coat. Technol. 142–144*:1121–1128 (2001).
225. G. Grundmeier, P. Thiemann, J. Carpentier, N. Shirtcliffe, and M. Stratmann, *Thin Solid Films 446*:61–71 (2004).
226. G. R. Prasad, S. Daniels, D. C. Cameron, B. P. McNamara, E. Tully, and R. O'Kennedy, *Surf. Coat. Technol. 200*:1031–1035 (2005).
227. H. Chatham, *Surf. Coat. Technol. 78*:1–9 (1996).
228. U. Moosheimer and C. Bichler, *Surf. Coat. Technol. 116–119*:812–819 (1999).
229. M. Deilmann, M. Grabowski, S. Theiss, N. Bibinov, and P. Awakowicz, *J. Phys. Appl. Phys. 41*: 135207 (2008).
230. V. S. Nguyen, J. Underhill, S. Fridmann, and P. Pan, *J. Electrochem. Soc. 132*:1925 (1985).
231. V. Barranco, J. Carpentier, and G. Grundmeier, *Electrochim. Acta 49*:1999 (2004).
232. A. J. Heeger, *Rev. Mod. Phys. 73*(3):681 (2001).
233. A. G. MacDiarmid, *Rev. Mod. Phys. 73*(3):701 (2001).
234. H. Shirakawa, *Rev. Mod. Phys. 73*(3):713 (2001).
235. G. Mengoli, M. T. Munari, P. Bianco, and M. M. Musiani, *J. Appl. Polym. Sci. 26*(12):4247 (1981).
236. D. W. Deberry, *J. Electrochem. Soc. 132*(5):1022 (1985).
237. T. D. Nguyen, M. Keddam, and H. Takenouti, *Electrochem. Solid State Lett. 6*:B2 (2003).
238. A. Michalik and M. Rohwerder, *Z. Phys. Chem. 219*:1547 (2005).
239. M. Rohwerder and A. Michalik, *Electrochim. Acta 53*:1300 (2007).

240. G. Paliwoda-Porebska, M. Stratmann, M. Rohwerder, U. Rammelt, L. Minh Duc, and W. Plieth, *J. Solid State Electrochem.* *10*:730 (2006).
241. J. Tamm, A. Alumaa, A. Hallik, and V. Sammelselg, *J. Electroanal. Chem.* *448*(*1*):25 (1998).
242. C. Ehrenbeck and K. Jüttner, *Electrochim. Acta* *41*(*4*):511 (1996).
243. G. Paliwoda, M. Stratmann, M. Rohwerder, K. Potje-Kamloth, Y. Lu, A. Z. Pich, and H.-J. Adler, *Corros. Sci.* *47*:3216 (2005).
244. P. J. Kinlen, V. Menon, and Y. Ding, *J. Electrochem. Soc.* *146*:3690 (1999).
245. M. Kendig, M. Hon, and L. Warren, *Progr. Org. Coat.* *47*:183 (2003).
246. G. Williams, A. Gabriel, A. Cook, and H. N. McMurray, *J. Electrochem. Soc.* *153*:B425 (2006).
247. R. J. Holness, G. Williams, D. A. Worsley, and H. N. McMurray, *J. Electrochem. Soc.* *152*:B73 (2005).
248. A. R. Bennet and D. A. Worsley, Spatially and temporally resolved optical and potentiometric studies of corrosion-driven redox processes in polyaniline coating, Presentation at the *ISE Conference in Edinburgh*, 2006.
249. J. M. Gustavsson, P. C. Innis, J. He, G. G. Wallace, and D. E. Tallman, *Electrochim. Acta* *54*:1483 (2009).
250. D. E. Dennis, G. Spinks, A. Dominis, and G. G. Wallace, *J. Solid State Electrochem.* *6*:73 (2002).
251. G. M. Spinks, A. J. Dominis, G. G. Wallace, and D. E. Tallman, *J. Solid State Electrochem.* *6*:85 (2002).
252. T. Van Schaftinghen, C. Deslouis, A. Hubin, and H. Terryn, *Electrochim. Acta* *51*:1695 (2006).
253. D. E. Tallman, Y. Pae, and G. P. Bierwagen, *Corrosion* *55*:779 (1999).
254. G. Williams and H. N. McMurray, *Electrochem. Solid State Lett.* *8*:B42 (2005).
255. M. Rohwerder, Le Minh Duc, and A. Michalik, *Electrochim. Acta* (2009), doi:10.1016/j.electacta.2009.02.103.

15

Atmospheric Corrosion

Christofer Leygraf
Royal Institute of Technology

CONTENTS

15.1 Introduction 669
15.2 The Atmospheric Region 670
15.2.1 Sulfur-Containing Compounds 672
15.2.2 Nitrogen-Containing Compounds 673
15.2.3 Chlorine-Containing Compounds 673
15.2.4 Other Atmospheric Compounds 674
15.3 The Aqueous Layer 674
15.3.1 Formation of the Aqueous Layer 674
15.3.2 Electrochemical Reactions 675
15.3.3 Acidifying Pollutants 676
15.3.4 Deposition of Pollutants 678
15.4 The Solid Phase 680
15.4.1 Acid-Dependent Dissolution 680
15.4.2 Metal–Anion Coordination Based on the Lewis Acid–Base Concept 682
15.4.3 Formation of Corrosion Products 682
15.4.4 Atmospheric Corrosion of Selected Metals 683
15.5 Atmospheric Corrosion Outdoors 687
15.5.1 General 687
15.5.2 Corrosion Rates 687
15.5.3 Field Exposures 689
15.5.4 Dose–Response Functions Describing the Multi-Pollutant Situation 690
15.5.5 The Effect of Temperature in Combination with Gases and Aerosol Particles 692
15.6 Atmospheric Corrosion Indoors 693
15.7 Dispersion of Metals 694
15.8 Concluding Remarks 697
Acknowledgments 698
References 698

15.1 Introduction

Atmospheric corrosion is the oldest type of corrosion recognized, and the atmosphere is the environment to which metals are most frequently exposed [1]. The impact on high-technology societies has been exemplified by the assumption that the cost of protection against atmospheric corrosion is about 50% of the total costs of all corrosion protection measures [2].

Atmospheric corrosion is the result of interaction between a material—mostly a metal—and its atmospheric environment. When exposed to atmospheres at room temperature with virtually no humidity present, most metals spontaneously form a solid oxide film. If the oxide is stable, the growth rate ceases and the oxide reaches a maximum thickness of 1–5 nm (1 nm = 10^{-9} m). Atmospheric corrosion frequently occurs in the presence of a thin aqueous layer that forms on the oxidized metal under ambient exposure conditions; the layer may vary from monomolecular thickness to clearly visible water films. Hence, atmospheric corrosion can be said to fall into two categories: "damp" atmospheric corrosion, which requires the presence of water vapor and traces of pollutants, and "wet" atmospheric corrosion, which requires rain or other forms of bulk water together with pollutants [3].

It was the pioneering work of Vernon [4] that developed the field of atmospheric corrosion to a scientific level. He discovered that corrosion rates increase rapidly above a "critical humidity" level, and he was also the first scientist to demonstrate the combined accelerating effects of SO_2 and relative humidity on atmospheric corrosion rates. Later, Evans [3], Rozenfeld [5], Barton [2], and others demonstrated that electrochemical processes play important parts, and it became evident that atmospheric corrosion of metals involves a broad range of electrochemical, chemical, and physical processes. The progress in the understanding of atmospheric corrosion has been reviewed in several books and review chapters [2,5–7] and updated in various conference proceedings [8–10].

There are obvious reasons why scientists' interest in atmospheric corrosion has become even more widespread during the past decades and why atmospheric corrosion has become a truly interdisciplinary field of science. Environmental effects on outdoor constructions—not least those that are part of our cultural and historical heritage—and on vehicles have triggered national and international exposure programs with comprehensive characterization of acidifying pollutant levels. Today the environmental effects not only concern what the environment does to metals but also what the metal does to the environment after running off any outdoor construction and reaching the biosphere. Frequently observed failures of highly miniaturized electronic components and equipment—even at low levels of pollutants—have focused technicians' and scientists' interest on indoor conditions too. Furthermore, recent progress in the development of new analytical techniques has resulted in a more detailed understanding of surface and interface phenomena as well as a better characterization of the corrosion products formed.

The purpose of this chapter is to outline briefly some of the new insights that have emerged as a result of basic studies of the atmospheric corrosion of metals. The chapter is entirely devoted to metallic materials, which means that the reader will have to turn to other sources for reviews of atmospheric corrosion of, for example, calcareous stones, glasses, plastics, painted materials, and textiles.

The complex interaction between the metal and the environment ranges from the atmospheric region over the thin aqueous layer to the solid surface region. Hence, the next three sections are devoted to various reactions and other phenomena in each of the regions involved. They are followed by three sections that deal with selected aspects of atmospheric corrosion outdoors and indoors, and the dispersion of metals induced by atmospheric corrosion.

15.2 The Atmospheric Region

The main constituents of the troposphere—the atmospheric region closest to the earth—are N_2, O_2, and rare gases (Ne, Kr, He, and Xe). They comprise more than 99.9% by weight of all molecules involved [11, p. 9]. Among these, N_2 and the rare gases are of no significant importance to atmospheric corrosion because of their inability to react with metal surfaces. O_2, on the other hand, is of paramount importance due to its participation as an electron acceptor in cathodic reactions and its involvement in chemical transformations of the atmosphere. Other tropospheric constituents conducive to atmospheric corrosion are H_2O and CO_2. H_2O is of critical importance in several respects (its role is discussed in more detail in the following sections). Its concentration in the troposphere varies over several orders of magnitude—from around 100 volume parts per million (ppm) to 10,000 ppm. CO_2, a naturally occurring molecule, has a concentration of about 330 ppm, high solubility in water, and contributes to the acidification of the aqueous layer.

The remaining constituents of importance in atmospheric corrosion are present only as trace gases with a total concentration of less than 10 ppm. Numerous investigations based on both laboratory and field exposures have proved at least the following gaseous constituents to be of significant importance [12]: O_3, H_2O_2, SO_2, H_2S, COS, NO_2, HNO_3, NH_3, HCl, Cl_2, HCHO, and HCOOH. Typical ranges for most of these constituents are given in Table 15.1 for outdoor and indoor conditions. Their presence may be the result of either natural or anthropogenic processes, and they may undergo a variety of chemical transformations during transport in the troposphere. Because all species are reactive, they have a certain average lifetime, which is often limited by the ability to react with important atmospheric oxidizers, primarily the hydroxyl radical, OH·, and O_3. OH· is generated by photoinduced dissociation of O_3 and the subsequent reaction of the electronically excited, energy-rich, oxygen atom $O(^1D)$ and water vapor:

$$O_3 + h\nu \rightarrow O(^1D) + O_2 \quad (\lambda < 310\,\text{nm}) \tag{15.1}$$

$$O(^1D) + H_2O \rightarrow 2OH\cdot \tag{15.2}$$

Although OH· molecules can oxidize several of the corrosion-stimulating species, such as SO_2, H_2S, and NO_2, a large fraction of OH· molecules are consumed through reactions with hydrocarbon molecules, whereby one of the end products is the hydroperoxyl radical, $HO_2\cdot$.

TABLE 15.1

Concentration of Selected Gaseous Air Constituents (ppb)

Species	Outdoor	Reference	Indoor	Reference
O_3	4–42	[17]	3–30	[17]
H_2O_2	10–30	[18]	5	[19]
SO_2	1–65	[17]	0.3–14	[17]
H_2S	0.7–24	[17]	0.1–0.7	[17]
NO_2	9–78	[17]	1–29	[17]
HNO_3	1–10	[18]	3	[19]
NH_3	7–16	[17]	13–259	[17]
HCl	0.18–3	[17]	0.05–0.18	[17]
Cl_2	<0.005–0.08[a]	[17]	0.001–0.005	[17]
HCHO	4–15	[18]	10	[19]
HCOOH	4–20	[18]	20	[19]

[a] Corresponding to 5 wt.% HCl according to Ref. [17].

This is another important component in gas-phase chemistry because it disproportionates into hydrogen peroxide, H_2O_2, and O_2 according to

$$HO_2\cdot + HO_2\cdot \rightarrow H_2O_2 + O_2 \tag{15.3}$$

H_2O_2, being highly soluble in aqueous environments, is another powerful oxidizer. The importance of oxidizing species in atmospheric chemistry and the multitude of reaction paths can be exemplified by considering different sulfur-, nitrogen-, and chlorine-containing compounds of significance in atmospheric corrosion.

15.2.1 Sulfur-Containing Compounds

SO_2 has long been recognized as the most important gaseous stimulant in atmospheric corrosion. It has many anthropogenic sources, such as combustion of sulfur-containing coal and oil, and emission from metal, petrochemical, and pulp and paper industries. Depending on the environment, there are several possible ways for SO_2 to oxidize, some of which are summarized below. In the gas phase, SO_2 can oxidize according to [13]

$$SO_2 + OH\cdot \rightarrow HSO_3\cdot \tag{15.4}$$

$$HSO_3\cdot + O_2 \xrightarrow{M} SO_3 + HO_2\cdot \tag{15.5}$$

$$SO_3 + H_2O \rightarrow H_2SO_4 \tag{15.6}$$

where M designates another gaseous molecule (e.g., N_2 or O_2) that collides with a generated $HSO_3\cdot$ molecule and removes some of the excess energy released. In the aqueous phase [14]:

$$SO_2 + xH_2O = SO_2\cdot xH_2O \tag{15.7}$$

$$SO_2\cdot xH_2O = HSO_3^- + H_3O^+ + (x-2)H_2O \tag{15.8}$$

$$HSO_3^- \xrightarrow{H_2O_2} SO_4^{2-} \tag{15.9}$$

The oxidation of S(IV) to S(VI) in the aqueous phase may also slowly proceed through

$$SO_2 + H_2O + \tfrac{1}{2}O_2 \rightarrow H_2SO_4 \tag{15.10}$$

or be promoted by O_3, Fe(III)- or Mn(II)-containing catalysts, or an increased pH [14].

H_2S, together with COS, is another principal atmospheric corrosion stimulant. It is emitted both from natural biological sources (e.g., volcanoes, moss, or swamp areas) and from anthropogenic sources (e.g., pulp and paper industries, catalyst converters in motor vehicles, sewage plants, garbage dumps, animal shelters, and geothermal power plants). H_2S can react with OH· to form SO_2 according to [15]

$$H_2S + OH\cdot \rightarrow HS\cdot + H_2O \tag{15.11}$$

$$HS\cdot + 2O_2 \rightarrow HO_2 + SO_2 \tag{15.12}$$

15.2.2 Nitrogen-Containing Compounds

NO and NO_2 are mainly formed through high-temperature combustion processes in power plants, vehicles, etc. The fraction of NO in the combustion gas is much higher than that of NO_2 but it is rapidly converted to NO_2 according to

$$2NO + O_2 \rightarrow 2NO_2 \tag{15.13}$$

Farther away from the emission source NO may also form NO_2 through

$$NO + O_3 \rightarrow NO_2 + O_2 \tag{15.14}$$

NO_2 plays a most important part in atmospheric chemistry by absorption of solar light and the subsequent formation of O_3 through

$$NO_2 + h\upsilon \rightarrow NO + O \quad (\lambda < 420\ \text{nm}) \tag{15.15}$$

$$O + O_2 \xrightarrow{M} O_3$$

NO_2 can also be oxidized to nitric acid according to

$$NO_2 + OH\cdot \xrightarrow{M} HNO_3 \tag{15.16}$$

Several of the reactions described have been identified in so-called photochemical smog formation [16]. This complex set of reactions involves an initial mixture of nonmethane organic compounds, NO, and NO_2, which is photochemically transformed into a final mixture including HNO_3 and O_3, aldehydes, and peroxy-acetylnitrate.

Principal sources of emission of NH_3 are animal shelters, fertilizer production, and cleaning detergents. In the aqueous phase, NH_3 establishes equilibrium with $NH_4{}^+$, which results in increased pH. An important role of NH_3 in atmospheric corrosion chemistry is to partly neutralize acidifying pollutants by forming particulate $(NH_4)_2SO_4$ and acid ammonium sulfates, such as NH_4HSO_4 and $(NH_4)_3H(SO_4)_2$. By increasing the pH of the aqueous phase, NH_3 also increases the oxidation rate of S(IV) to S(VI), as discussed earlier.

15.2.3 Chlorine-Containing Compounds

Chlorides participate in atmospheric corrosion reactions mainly as aerosols through transport from marine atmospheres. Other important sources are road deicers and dust binders on roads, coal burning, municipal incinerators, and fingerprints. Burning of high-chlorine coals may also result in emission of HCl, which is highly soluble in water and strongly acidifies the aqueous phase. Cl_2 is emitted from industrial processes, such as bleaching plants in pulp and paper industries and certain metal production industries, and from cleaning detergents. Cl_2 can also photodissociate into chlorine radicals, which react with organic compounds (RH) and form HCl according to

$$Cl_2 + h\upsilon \rightarrow Cl\cdot + Cl\cdot \quad (\lambda < 430\,\text{nm}) \tag{15.17}$$

$$RH + Cl\cdot \rightarrow R\cdot + HCl \tag{15.18}$$

15.2.4 Other Atmospheric Compounds

In addition to the gaseous species already commented on, Table 15.1 includes HCHO and HCOOH, which are important indoor corrosion stimulants (as discussed later) and which can originate from adhesives, tobacco smoke, combustion of biomass, and plastics, for example. A comparison between typical outdoor and indoor concentrations of the most important gaseous corrosion stimulants (Table 15.1) reveals, in general, lower levels indoors than outdoors. This is mostly due to enhanced indoor absorption of gases and particulates and also to the retardation and damping of outdoor variations by ventilating systems and air filtration. Exceptions are NH_3 and the organic species, which, as a rule, show higher levels indoors than outdoors as a result of anthropogenic activity.

Of utmost importance in atmospheric corrosion is the presence of particles and aerosols (an ensemble of particles suspended in the air) of mostly chlorides, sulfates, and nitrates. The size, shape, and chemical and physical properties of these particles and aerosols can vary widely. A more detailed description of particles and their role in atmospheric corrosion is given elsewhere [7,20].

15.3 The Aqueous Layer

15.3.1 Formation of the Aqueous Layer

The reaction of water vapor with a metal surface is of paramount importance to atmospheric corrosion. A large number of studies of initial water–metal interaction have been made on well-defined single-crystal surfaces of pure or oxidized metals (for a review, see Ref. [21]). Water may bond in molecular form to most clean and well-characterized metal surfaces. Through the oxygen atom it bonds to the metal surface or to metal clusters and acts as a Lewis base, as bonding is connected with a net charge transfer from the water molecule to the surface. Simple models of preferred adsorption sites are based on Lewis acid–base chemistry, in which water adsorbs on electron-deficient adsorption sites [22]. Water may also bond in a dissociated form, in which case the driving force is the formation of metal–oxygen or metal–hydroxyl bonds. The end products resulting from water adsorption are then adsorbed hydroxyl, atomic oxygen, and atomic hydrogen [21]. On metal oxides, water may also adsorb in either molecular or dissociative form. The tendency to dissociate seems to be facilitated by lattice defect sites, as observed, for instance, on monocrystalline TiO_2 [23], NiO [24], and α-Fe_2O_3 [25]. The monomolecular thick film of surface hydroxyl groups formed from dissociation of water is relatively protective and reduces the subsequent reaction rate of water [26]. The first monolayer of water adsorbed to the hydroxylated oxide substrate appears to be highly immobile, whereas the second and third monolayers are more randomly oriented and less immobile [27]. The water layer adjacent to the gas phase appears to have an ice-like structure [28]. The adsorption characteristics of water are strikingly similar for many different metals, and the exact nature of the metal oxyhydroxide seems to have only a minor influence on water adsorption phenomena. The quantity of reversibly adsorbed water increases with the relative humidity and with time. Table 15.2 presents the approximate number of water monolayers at 25°C and steady-state conditions, which have been experimentally determined by the quartz crystal microbalance method on a variety of metals [28,29].

TABLE 15.2

Approximate Number of Water Monolayers on Different Metals versus Relative Humidity

Relative Humidity (%)	Number of Water Monolayers
20	1
40	1.5–2
60	2–5
80	5–10

The bond strength between water and the hydroxylated metal surface is similar to the bond strength between neighboring, hydrogen-bonded, water molecules [21]. From this follows the possibility of water clusters on relatively homogeneous surfaces, which is further promoted on less well-defined surfaces with highly reactive sites, such as kinks or steps, thereby increasing the probability of anode–cathode area formation.

The aqueous phase formed acts as a solvent for gaseous constituents of the atmosphere. Preferred sites for corrosion attack may be related to sites where water adsorption is favored and gaseous molecules, such as SO_2 and NO_2, are easily dissolved. At aqueous films thicker than about three monolayers, the properties approach those of bulk water [30]. The relative humidity when this occurs is close to the "critical relative humidity" [31,32], above which atmospheric corrosion rates increase substantially and below which atmospheric corrosion is insignificant. The critical relative humidity for different metals in the presence of SO_2 has been reported to be between 50% and 90% [28].

15.3.2 Electrochemical Reactions

The aqueous phase also acts as a conductive medium for electrochemical reactions. Although atmospheric corrosion is largely dependent on electrochemical reactions, relatively few electrochemical studies have been focused on the elucidation of these basic mechanisms. The reason seems to be obvious if one considers the difficulties in reproducing a thin aqueous film in an electrochemical experiment. The thickness of such a film may vary with different outdoor exposure conditions, involving the complex chemistry and photochemistry produced by atmospheric pollutants together with solar light and resulting in the precipitation of representative corrosion products and changed transport properties. Nevertheless, systematic electrochemical experiments have been conducted with special emphasis on atmospheric corrosion (see, e.g., Refs. [5,33–37]). In the absence of pollutants, the most common anodic and cathodic processes in corrosion of metals exposed to a thin, neutral, aqueous phase are

$$\mathrm{Me} \rightarrow \mathrm{Me}^{n+} + n\mathrm{e}^- \quad \text{(anode reaction)} \tag{15.19}$$

$$\mathrm{O_2} + 2\mathrm{H_2O} + 4\mathrm{e}^- \rightarrow 4\mathrm{OH}^- \quad \text{(cathode reaction)} \tag{15.20}$$

Under most atmospheric corrosion conditions, the anode reaction rather than the cathode reaction is observed to be the rate-limiting step [2]. Upon evaporation of the aqueous layer, a film of corrosion products—consisting of metal hydroxides or metal oxyhydroxides—may precipitate. With repeated condensation–evaporation cycles, this film usually hinders the transport of ions through the corrosion product or the transport of Me^{n+} from the anodic site. Hence, the anodic reaction rate is lowered and, thereby, the atmospheric corrosion rate.

15.3.3 Acidifying Pollutants

In the presence of atmospheric acidifying pollutants, such as SO_2, the anode reaction is facilitated and, consequently, the total corrosion rate as well. Upon deposition of SO_2, interaction with the aqueous phase proceeds with the following reactions:

$$SO_2 + x\,H_2O = SO_2 \cdot x\,H_2O \tag{15.7}$$

$$SO_2 \cdot x\,H_2O = HSO_3^{\,-} + H_3O^+ + (x-2)H_2O \tag{15.8}$$

$$HSO_3^- = H^+ + SO_3^{\,2-} \tag{15.21}$$

Despite numerous studies of the important influence and role of SO_2, there is as yet no complete description of the interaction, on a molecular level, of SO_2 with an oxidized metal substrate in the presence of an aqueous phase. Recent in situ studies by means of infrared reflection-absorption spectroscopy have provided evidence that $HSO_3^{\,-}$ is coordinated with the hydroxylated metal oxide surface through a fast ligand exchange mechanism, in which the surface OH^- group is replaced by $SO_3^{\,2-}$ [38] (see Figure 15.1). Surface-bound H^+ released according to, e.g., reactions (15.8) and (15.21), may easily move from one functional group to another or from one surface site to another. In accordance with earlier proposed mechanisms for mineral dissolution in aquatic environments [39], the $SO_3^{\,2-}$-coordinated metal surface center may be detached from the oxide lattice when surrounded by two or more H^+-bonded functional groups. This results in the release of H^+ ions, which can participate in the detachment of another $SO_3^{\,2-}$-coordinated metal surface center. The net effects represent an acid-dependent dissolution rate (R_{diss}) of the oxide that has been written as

$$R_{diss} = C[a_{H^+}]^n \tag{15.22}$$

where

$[a_{H^+}]$ is the activity of the protons
C and n are constants [40]

When further elaborated, Equation 15.22 should include the influence of surface coordinated ligands.

If S(IV) is oxidized to S(VI), this will result in the formation of H_2SO_4 and the release of considerably more H^+ than if H_2SO_3 is involved. It follows that possible reaction paths for the oxidation of S(IV) to S(VI) are of crucial importance to atmospheric corrosion rates, and this has also been the subject of many investigations. Examples of reaction paths have been summarized in the preceding section.

The acid-dependent dissolution rate of the initially formed oxide or oxyhydroxide requires that all of the most important acidifying atmospheric pollutants be considered. Besides SO_2, these pollutants include CO_2, NO_2, and carbonyl compounds, such as HCHO and CH_3CHO, which can contribute to the total acidity in the aqueous film by forming H_2CO_3, HNO_3, HCOOH, and CH_3COOH.

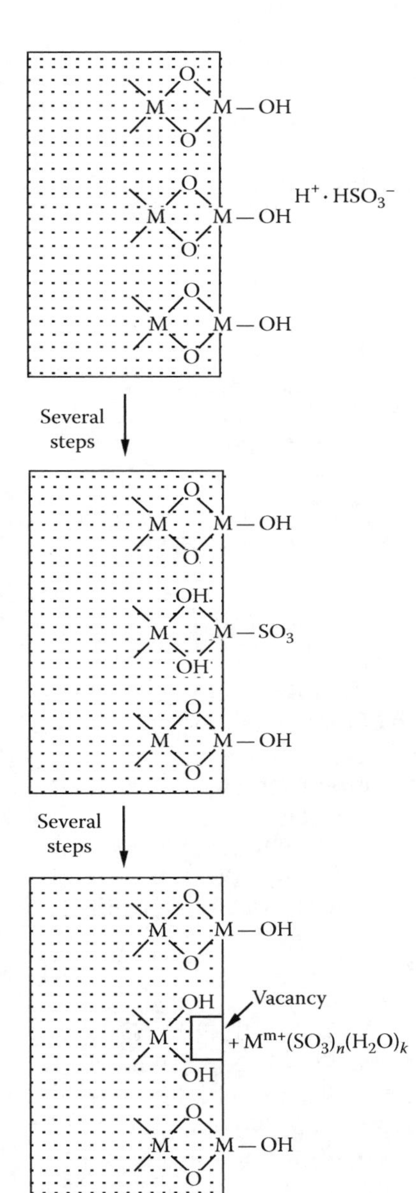

FIGURE 15.1
Fundamental processes in the initial stages of SO_2-induced atmospheric corrosion, including coordination of HSO_3^- with the hydroxylated metal oxide surface, replacement of surface OH^- group by SO_3^{2-}, and detachment of the SO_3^{2-}-coordinated metal surface center when surrounded by two or more H^+-bonded functional groups.

At equilibrium between the water film and the atmosphere, the concentration of dissolved atmospheric gases in the film is expressed according to Henry's law:

$$[X] = H(X)p(X) \tag{15.23}$$

where

[X] is the concentration in aqueous phase
H(X) is Henry's law constant for species X
p(X) is the partial pressure of X in atmosphere

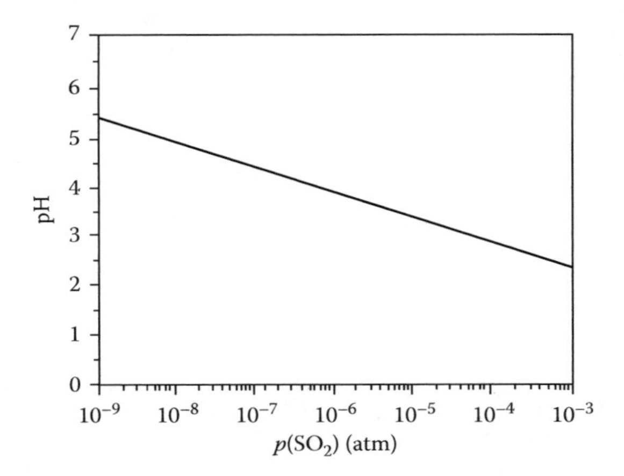

FIGURE 15.2
Variation of pH of the aqueous layer as a function of $p(SO_2)$.

For many gases, reversible ionization reactions occur and the ordinary Henry's law constant, H, is replaced by an effective Henry's law constant, H_e. In the case of SO_2 this can be written as

$$[H_2SO_3] + [HSO_3^-] + [SO_3^{2-}] = H_e(SO_2) \times p(SO_2) \tag{15.24}$$

Less dissociation takes place if the pH is decreased, which also results in a decrease of the effective Henry's law constant and a further slowdown of the dissolution of SO_2 in the aqueous layer [41].

The pH of the aqueous layer can be determined by formulating the general requirement equation that electrical neutrality must be maintained within the layer.

This means that the sum of products of molar concentration and charge of cationic species must equal the sum of products of molar concentration and charge of anionic species. If we consider only the interaction of SO_2 with the aqueous layer, the pH is determined by

$$[H^+] = [HSO_3^-] + 2[SO_3^{2-}] + [OH^-] \tag{15.25}$$

The calculated pH of the aqueous layer as a function of $p(SO_2)$, according to (15.24) and (15.25), is shown in Figure 15.2. A more exact estimate of the aqueous layer pH must involve the contribution of other acidifying pollutants, such as CO_2, NO_2, and HCHO, as well as the properties of the solid surface film, such as its coordination properties and dissociation product. A complicating factor is the high ionic strength of many species involved, especially if the aqueous film undergoes wetting and drying cycles. If so, the concentrations of species are replaced by their activities. Henry's law constants for different gases are given in Table 15.3. It should be noted that Henry's law applies only if equilibrium is maintained between the gas in the atmosphere and that in the aqueous phase. Consequently, it does not apply under nonequilibrium conditions, caused by mass transport restrictions in the atmosphere, in the aqueous layer, or across the gas–liquid interface.

15.3.4 Deposition of Pollutants

The incorporation of atmospheric species into the aqueous layer may occur through either dry or wet deposition. In dry deposition, there is no involvement of any precipitation, whereas wet deposition requires, e.g., rain, dew, fog, or snow for atmospheric pollutants to deposit. Indoors or in highly polluted areas close to emission sources, dry deposition

TABLE 15.3

Characteristics of Selected Gaseous Air Constituents[a]

	A	B	Eq. Conc. (μM)		Deposition Velocity (cm s^{-1})		Deposition Rate (ng cm^{-2} s^{-1})	
Species	H	(M atm^{-1})[b]	Outdoor	Indoor	Outdoor	Indoor[c]	Outdoor	Indoor
O_3	1.8	(−2)	2.3 (−4)	1.7 (−4)	0.05–1[d]	0.036	5.8 (−3)	6.8 (−4)
H_2O_2	2.4	(5)	4.2 (3)	1.2 (3)	—	0.07	—	5.0 (−4)
SO_2	1.4		1.1 (−2)	2.9 (−3)	0.01–1.2[d]	0.05	7.5 (−3)	2.7 (−4)
H_2S	1.5	(−1)	6.1 (−4)	4.0 (−5)	0.38[e]	0.03	2.2 (−3)	1.1 (−5)
NO_2	7.0	(−3)	1.9 (−4)	3.8 (−5)	0.02–0.8[d]	0.006	2.0 (−2)	6.2 (−5)
HNO_3	9.1	(4)	2.9 (2)	2.7 (2)	0.1–30[c]	0.07	1.4 (−2)	5.5 (−4)
NH_3	1.0	(1)	1.1 (−1)	5.8 (−1)	0.3–2.6[f]	0.05	6.6 (−3)	2.1 (−3)
HCl	2.0	(1)	1.5 (−2)	1.9 (−3)	—	0.04	—	5.8 (−6)
Cl_2	6.2	(−2)	1.2 (−6)	1.4 (−7)	1.8–2.1[e]	—	1.1 (−4)	—
HCHO	1.4	(4)	1.1 (2)	1.4 (2)	—	0.005	—	6.3 (−5)
HCOOH	3.7	(3)	3.3 (1)	7.4 (1)	—	0.006	—	2.3 (−4)

[a] The equilibrium solution concentration and deposition rate values were based on concentrations from Table 15.1 and using geometric mean values for the intervals. 1.8 (−2) means 1.8×10^{-2}.
[b] Henry's law constant, from Ref. [19].
[c] Ref. [19].
[d] Ref. [42].
[e] Ref. [43].
[f] Ref. [44].

is considered to be dominating but the relative importance of wet deposition may be difficult to establish because of the incidental nature of precipitation. Controlled field studies combined with extensive laboratory exposures have been undertaken within the National Acid Precipitation Assessment Program (NAPAP) to explore the relative contribution of wet and dry deposition to increased corrosion rates of a number of metals [45].

A useful parameter is the dry deposition velocity, which is defined as the ratio of deposition rate or surface flux per time unit of any gaseous compound and the concentration of the same compound in the atmosphere [46]. The concept of dry deposition velocity of SO_2 and its relevance to atmospheric corrosion rates is well established [47]. By examining data based on both field and laboratory exposures, it can be concluded that the factors controlling dry deposition fall into aerodynamic processes and surface processes. Aerodynamic processes are connected with the actual depletion of the gaseous constituent (e.g., SO_2) in the atmospheric region next to the aqueous phase and the ability of the system to mix new SO_2 into this region. This ability depends on, for instance, the actual wind speed, type of wind flow, and shape of the sample. Surface processes, on the other hand, are connected with the ability of the aqueous layer to accommodate SO_2. This ability increases with the thickness of the aqueous layer and, hence, with the relative humidity, the pH of the solution (as discussed earlier), and the alkalinity of the solid surface.

The dry deposition velocity (V_d) can be expressed as the inverse of the sum of two resistances, namely the aerodynamic resistance (R_a) and the surface resistance (R_s):

$$V_d = \frac{1}{(R_a + R_s)} \tag{15.26}$$

Under most exposure conditions, the dry deposition velocity will be the combined effect of both resistances [46]. At well-stirred and highly turbulent airflow conditions, however,

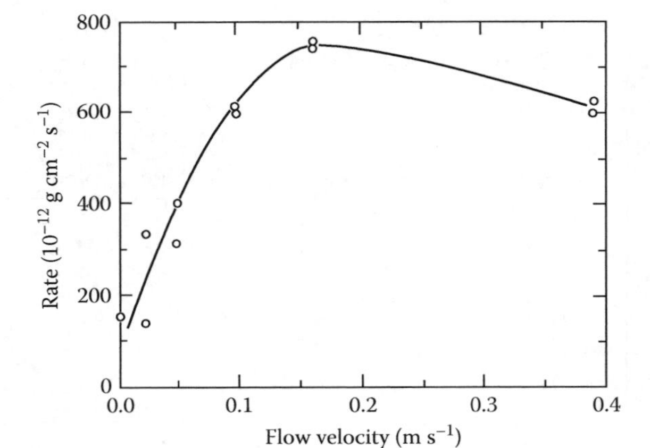

FIGURE 15.3
Influence of airflow velocity on atmospheric corrosion rate of Cu in a laboratory mixed-gas environment. (From Volpe, L., Mass-transport limitations in atmospheric corrosion, in *Proceedings of 1st International Symposium on Corrosion of Electronic Materials and Devices*, J. D. Sinclair, ed., p. 22, 1991, Courtesy of The Electrochemical Society, Pennington, NJ.)

$R_a \approx 0$ and the dry deposition velocity solely depends on surface processes. Ideal absorbers of SO_2 are alkaline surfaces such as lead peroxide or triethanolamine, for which $R_s \approx 0$. The dry deposition velocity in this case is determined mainly by aerodynamic processes. Typical ranges for dry deposition velocities onto various materials under outdoor and indoor conditions are summarized in Table 15.3.

A combined experimental and theoretical approach to mass transport limitations in atmospheric corrosion has been performed by Volpe [48,49] to obtain dry deposition velocities and probabilities of H_2S, NO_2, and O_3 reacting with Ag. This approach clearly shows the effect of airflow velocity on the corrosion rate and the significance of hydrodynamic and concentration boundary layers next to the metal surface, characterized by deviating tangential airflow velocity and concentration of air constituents (see Figure 15.3). An application of boundary layer theory to buildings and other objects has been reported aimed at postulating the economic assessment of corrosion damage [50].

An immediate consequence of dry deposition velocities is the kinetic constraints to obtain equilibrium between a gas in the atmosphere and the same gas in the aqueous layer. Especially in outdoor exposure conditions, characterized by wet–dry cycles, it is anticipated that the actual concentration of most corrosion-stimulating gases under many conditions may be far from equilibrium. Nevertheless, thermodynamic considerations have been most useful for predicting the formation of different corrosion end products and their stability domains [51,52]. A general and useful observation made by Graedel is the similarity between corrosion products found after prolonged exposure of metals and minerals formed by natural processes and containing the same metals (see, e.g., Ref. [53]).

Figure 15.4 is a schematic illustration of important processes occurring in or at the aqueous layer.

15.4 The Solid Phase

15.4.1 Acid-Dependent Dissolution

As discussed earlier, initial electrochemical reactions and repeated wet–dry cycles may result in precipitation of thin films of metal hydroxide, oxyhydroxide, or oxide. The precipitation processes of these compounds are complex and pass through the colloid state

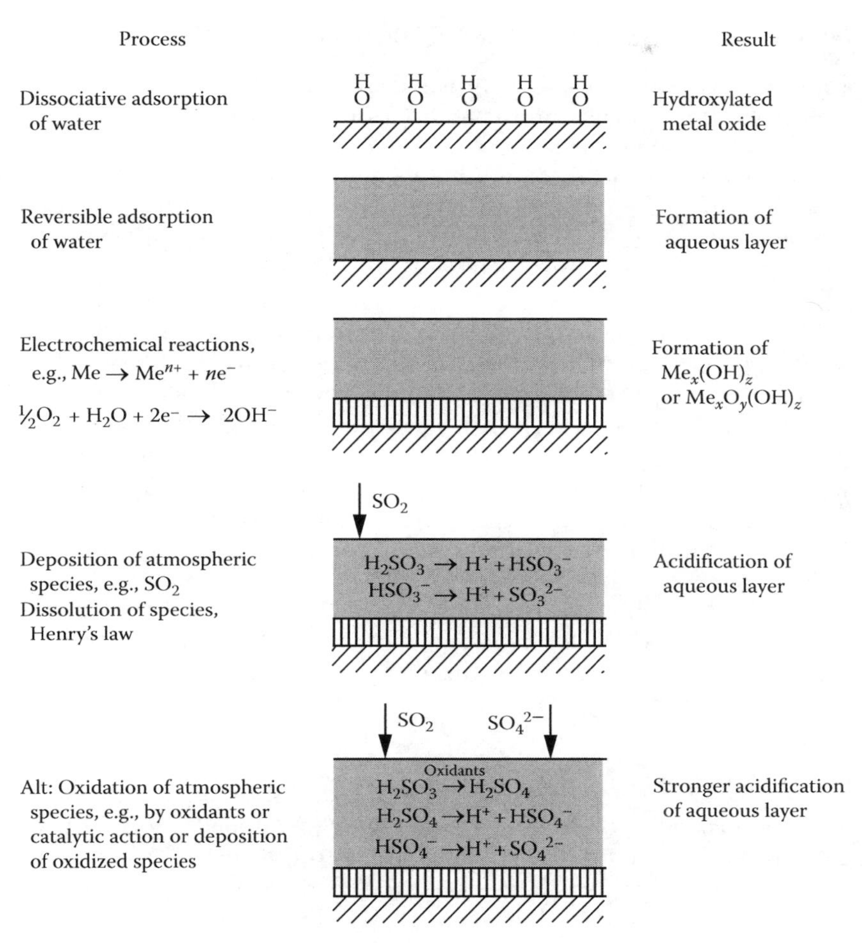

FIGURE 15.4
Schematic illustration of processes occurring in or at the aqueous layer.

before solid films are formed [54]. These films form on the metal surface normally by fast nucleation and spreading and by slower thickening, once the film has completely covered the surface. If the film is stable, it thickens through a high-field ion conduction mechanism; otherwise the film may thicken through a dissolution–reprecipitation mechanism [55]. One approach to predicting the ability to form a protective oxygen-containing film is by considering the heat of the adsorption of oxygen on the metal. Compilations of such data [56] suggest strong metal–oxygen bond formation and highly protective layers in various atmospheres in the case of, e.g., Ti, Al, or Cr, but only weak metal–oxygen bonds and practically nonexistent protective oxygen-containing films in the case of, e.g., Ag or Zn.

As a consequence of exposure to the aqueous layer, the oxygen-containing film can dissolve—the corrosion rate being normally controlled by the rate of dissolution of the film formed [55]. As mentioned in the preceding section, the dissolution rate of many oxides and other minerals is acid dependent and can be written in the form

$$R_{\text{diss}} = C[a_{\text{H}^+}]^n \tag{15.22}$$

15.4.2 Metal–Anion Coordination Based on the Lewis Acid–Base Concept

During prolonged exposure, a variety of different corrosion products may form, the exact nature of which largely depends on the ability of the dissolving metal ion to coordinate with other ions in the aqueous layer. A general treatment of classification of participating ions in coordination processes can be based on the Lewis acid–base concept. So far this theory has found only few applications in corrosion science [57]. It turns out to be most useful in rationalizing the specific behavior of different metals during atmospheric corrosion and therefore merits a short introduction.

Upon interaction of two species, a pair of electrons from one species can be used to form a covalent bond to the other one. The species that donates the electron pair is called a *Lewis base*, and the species that accepts the electron pair is called a *Lewis acid*. Acids or bases with valence electrons that are tightly held and not easily distorted are *hard* acids or bases. Acids or bases with valence electrons that are easily polarized or removed are *soft*. According to the hard and soft acid–base (HSAB) principle, hard acids are preferably coordinated with hard bases and soft acids are preferably coordinated with soft bases [58]. The hardness of an element usually increases with its oxidation state. Selected Lewis acids and bases and their classification into hard, intermediate, and soft acids and bases are listed in Table 15.4 [59].

In full agreement with experience from atmospheric corrosion, Table 15.4 suggests that hard acids such as Cr^{3+} and Ti^{4+} form oxygen-containing films, whereas soft acids such as Cu^{+} and Ag^{+} coordinate with reduced sulfur compounds. Intermediate acids, such as Fe^{2+}, Cu^{2+}, and Zn^{2+}, are expected to coordinate with a broader range of bases.

15.4.3 Formation of Corrosion Products

The atmospheric corrosion rate is influenced by many parameters, one of the most important being the formation and protective ability of the corrosion products formed. The composition of the corrosion products depends on the participating dissolved metal ions and the access to anions dissolved in the aqueous layer. The eventual thickening of the film of corrosion products can be described in a sequence of consecutive steps—dissolution–coordination–reprecipitation—where the dissolution step is acid dependent, coordination is based on the HSAB principle, and reprecipitation depends on the activities of the species involved.

Depending on the rate of crystallization and the rate of formation, the corrosion products may be amorphous or crystalline. If the former is rate determining, one expects

TABLE 15.4

Classification of Selected Lewis Acids and Bases

Hard	Intermediate	Soft
Acids		
H^{+}, Na^{+}, Mn^{2+}, Al^{3+}	Fe^{2+}, Ni^{2+}, Cu^{2+}	Cu^{+}, Ag^{+}
Cr^{3+}, Fe^{3+}, Ti^{4+}	Zn^{2+}, Pb^{2+}	
Bases		
H_2O, OH^-, O^{2-}	SO_3^{2-}, NO_2^-	R_2S, RSH
SO_4^{2-}, NO_3^-, CO_3^{2-}		RS^-

Source: Jolly, W. L., *Modern Inorganic Chemistry*, 2nd edn., McGraw-Hill, New York, p. 238, 1991.

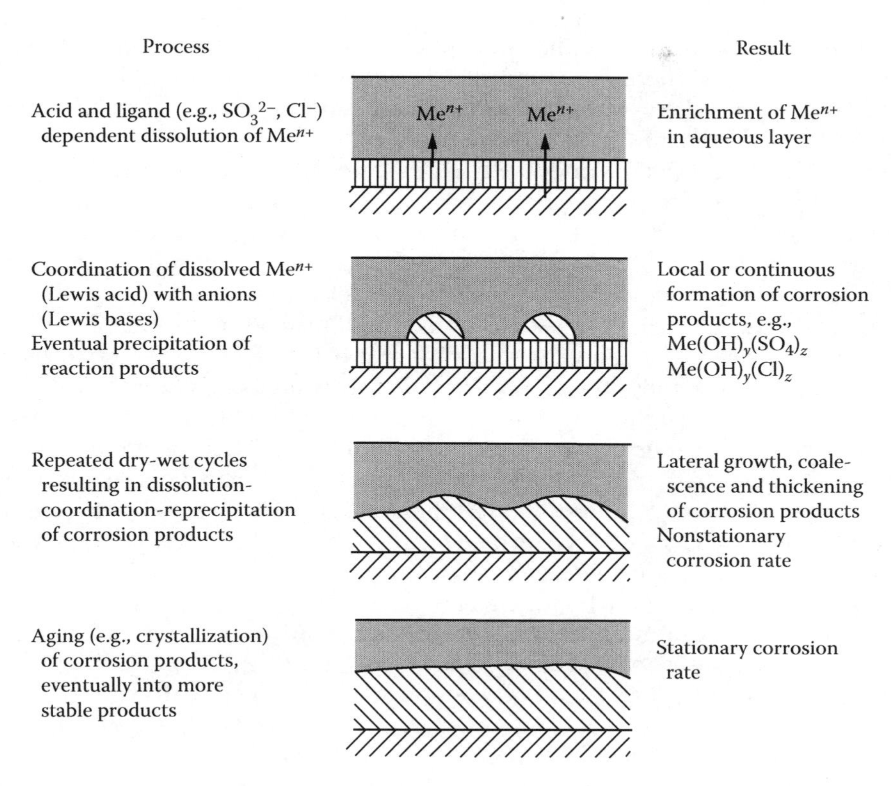

FIGURE 15.5
Schematic illustration of processes occurring in or at the solid phase.

amorphous phases to form. From colloid chemistry, it is known that aging or slow growth of amorphous phases may result in a transition from the amorphous to the crystalline state, a process that may occur through slow transformation in the solid state or through dissolution–reprecipitation processes [60]. This seems to be in agreement with findings of a transition from amorphous to crystalline basic nickel sulfates, the former being less corrosion resistant than the latter [61a].

Figure 15.5 is a schematic illustration of important processes occurring in or at the solid phase.

15.4.4 Atmospheric Corrosion of Selected Metals

Access to new and more sensitive analytical techniques has resulted in substantial progress in the characterization of corrosion products formed under both laboratory and field exposure conditions. These techniques permit the determination of, e.g., thickness, chemical composition, and atomic structure of corrosion products formed at both early and later stages of exposure. When combined with environmental data, such as deposition rates of corrosion-stimulating atmospheric constituents, relative humidity, temperature, and sunshine hours, the new techniques have resulted in a more comprehensive understanding of the complex processes that govern atmospheric corrosion. In a series of papers, Graedel has summarized the corrosion mechanisms of zinc [62], aluminum [63], copper [18], iron and low-alloy steel [64], and silver [19]. It is beyond the scope of this chapter to provide a detailed description of the specific atmospheric corrosion behavior of each individual metal. Nevertheless, a summary of atmospheric corrosion behavior of selected metals will be given in light of the general processes that have been discussed. The summary is based on the papers by Graedel, unless otherwise stated.

The atmospheric corrosion of zinc starts with the instant formation of a thin film of zinc hydroxide, which seems to occur in different crystal structures, and the subsequent formation of a protective layer of basic zinc carbonate, $Zn_5(CO_3)_2(OH)_6$. The pH of the aqueous layer controls the stability of initial corrosion products and results in the dissolution of Zn^{2+}. From the HSAB principle, one expects Zn^{2+}, classified as an intermediate acid, to coordinate with a number of different bases. In accordance with this, the prolonged exposure of zinc can proceed along a variety of different paths of reaction sequences depending on the actual deposition rates of atmospheric constituents. Among these Cl^- and SO_2 seem to be the most important. In a relatively benign rural atmosphere, the basic zinc carbonate may continue to grow slowly or may be followed by the formation of a protective basic zinc sulfate, e.g., $Zn_4SO_4(OH)_6 \cdot 4H_2O$. In a typical marine atmosphere, characterized by higher amounts of deposited Cl^- than of SO_2, islands of a less protective basic zinc chloride, $Zn_5Cl_2(OH)_8 \cdot H_2O$, are formed within days of exposure. These islands grow laterally and coalesce. Within weeks of exposure, a more protective basic zinc chlorosulfate, $NaZn_4Cl(OH)_6SO_4 \cdot 6H_2O$, may be observed [65]. In a typical urban environment, characterized by higher amounts of deposited SO_2 than of Cl^-, the basic zinc sulfate, $Zn_4SO_4(OH)_6 \cdot 4H_2O$, is observed within, typically, weeks of exposure. In highly polluted industrial environments, it is eventually followed by another basic zinc chlorosulfate, $Zn_4Cl_2(OH)_4SO_4 \cdot 5H_2O$. A feature common to most zinc-containing corrosion products observed is their strong structural resemblance, with sheets of Zn^{2+} in octahedral and tetrahedral coordination and with the main difference between the structures being the content and bonding between the sheets [66]. A generalized reaction sequence has been proposed that considers the evolution of corrosion products on sheltered zinc in a variety of atmospheric environments, including rural, marine, urban, and industrial [66]. Figure 15.6 displays schematically the reaction sequence in which sulfate deposition dominates over

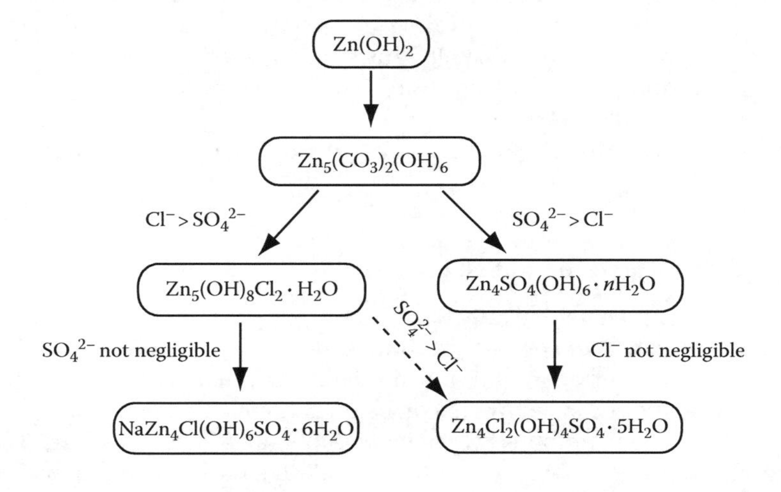

FIGURE 15.6
Formation sequences for compounds in zinc corrosion products formed under sheltered conditions dominated by chloride deposition (left sequence) and sulfate deposition (right sequence). (From Odnevall, I. and Leygraf, C., *Reaction Sequences in Atmospheric Corrosion of Zinc*, ASTM STP 1239, W. W. Kirk and H. H. Lawson, eds., p. 215, 1995, Courtesy of the American Society for Testing and Materials, Philadelphia, PA.)

chloride deposition (right part) and the reaction sequence in which chloride deposition dominates over sulfate deposition (left part). Reported corrosion rates (in μm $year^{-1}$) of zinc in outdoor atmospheres range from 0.2 to 0.3 (rural), from 0.5 to 8 (marine), and from 2 to 16 (urban, industrial).

Aluminum forms initially a few-nanometers-thick layer of aluminum oxide, γ-Al_2O_3, which after prolonged exposure in humidified air is covered by aluminum oxy-hydroxide, γ-AlOOH, and subsequently by various hydrated aluminum oxides and aluminum hydroxides. The stability of the compounds decreases with acidity and results in the dissolution of Al^{3+}. The ability of aluminum to form various oxygen-containing corrosion products is in agreement with the HSAB principle; this is also the case with the general observation of frequent formation of basic or hydrated aluminum sulfates and no detection of aluminum sulfides. The sulfates most frequently found on aluminum are poorly soluble, amorphous, and highly protective. As with other passivating metals, atmospheric corrosion rates of aluminum increase readily in the presence of Cl^-, resulting in local rather than uniform corrosion. Rates of uniform corrosion (in μm $year^{-1}$) of aluminum outdoors are substantially lower than for most other structural metals and are from 0.0 to 0.1 (rural), from 0.4 to 0.6 (marine), and ≈1 (urban).

In ambient air, copper initially forms a film with a total thickness of around 3 nm. It seems to contain at least two layers, both of which contain copper as Cu^+. The inner layer close to the metal consists of Cu_2O and the outer layer of bound hydroxyl, water, and adventitious hydrocarbon [67]. Cu dissolves into the aqueous layer as Cu^+, which readily oxidizes to Cu^{2+}. From the presence of both Cu^+ (soft acid) and Cu^{2+} (intermediate acid), a variety of corrosion products are expected. With reduced sulfur compounds present in significant concentrations, copper mainly forms Cu_2S. Without reduced sulfur compounds, the corrosion products are frequently complex mixtures of basic copper sulfates, chlorides, nitrates, and carbonates, with a composition that depends on the actual deposition rates of the corresponding air constituents. Abundant phases are $Cu_4SO_4(OH)_6$ and $Cu_3SO_4(OH)_4$ in urban areas and $Cu_2Cl(OH)_3$ and possibly basic sulfates in marine atmospheres. Other phases observed are $Cu_4SO_4(OH)_6 \cdot H_2O$, $Cu_2NO_3(OH)_3$, and $Cu_2CO_3(OH)_2$. Similar to that for zinc, a generalized reaction sequence has been proposed for sheltered copper, which integrates all existing knowledge of phases formed in copper patina during exposure in atmospheric environments with varying degree of sulfur and chloride pollution [67]. The phases observed bear structural resemblance and are mostly layered, a common observation in basic salts of divalent cations such as Cu^{2+} and Zn^{2+}. Representative corrosion rates (in μm $year^{-1}$) of copper exposed outdoors are ≈0.5 (rural), ≈1 (marine), 1–2 (urban), and ≤2.5 (industrial).

Initial stages of atmospheric corrosion of iron involve the incorporation of oxygen and water into a rich variety of different iron oxides or iron oxyhydroxides as rust layers. These processes are preceded by initial dissolution of Fe^{2+}, which only slowly oxidizes to Fe^{3+}. From this follows that corrosion products formed at earlier stages of exposure contain iron in both valence states but only as Fe^{3+} at later stages. A complex chemistry is anticipated with the participation of an intermediate acid slowly transforming into a hard acid. The oxidation of Fe^{2+} in the aqueous layer may involve several possible oxidants, such as the hydroxyl radical (OH·), the hydroperoxyl radical (HO_2·), and hydrogen peroxide (H_2O_2). Detailed discussions of the interaction between SO_2 and iron have been presented elsewhere [68]. It results in the formation of iron(II) sulfates, subsequent oxidation of Fe^{2+}, and concomitant formation of iron(III) oxyhydroxide, the release of SO_4^{2-}, the dissolution of more Fe^{2+}, formation of new iron(II) sulfate, etc. [69]. This acid regeneration cycle continues until basic iron(III) sulfate is formed as an end product. The complex interaction of SO_2

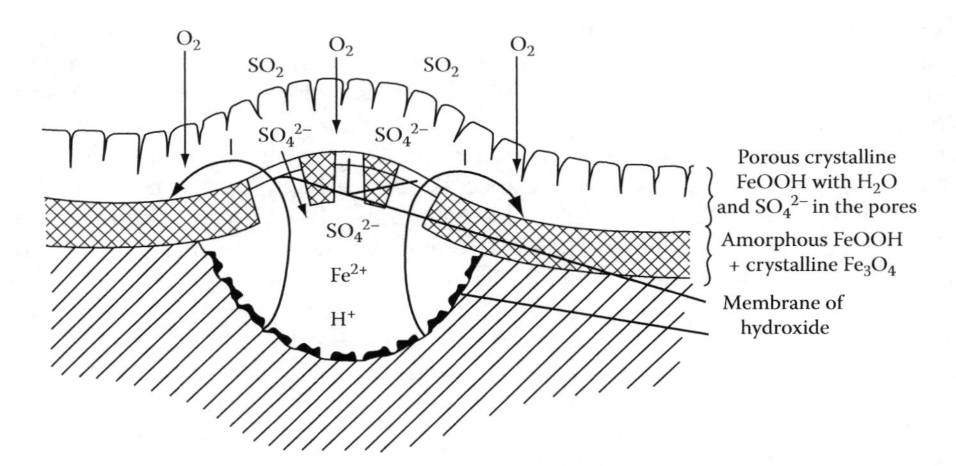

FIGURE 15.7
Sketch of sulfate "nest" on steel. (From Kucera, V. and Mattsson, E., Atmospheric corrosion, *Corrosion Mechanisms*, F. Mansfeld, ed., Marcel Dekker, New York, 1987.)

and iron is believed to take place in locally distributed sulfate "nests," which involve the assumed existence of semipermeable membranes of iron(III) oxyhydroxide maintaining a high activity of species involved at localized parts of the electrolyte (Figure 15.7). Evidence of nitrate- or carbonate-containing corrosion products is sparse, although coordination of these anions with Fe^{3+} is expected according to the HSAB principle. There is abundant evidence of accelerated atmospheric corrosion rates for iron caused by chlorides, which may result in the formation of basic iron(II,III) chlorides and β-FeOOH. Atmospheric corrosion rates for iron are relatively high and exceed those of other structural metals. They range (in μm $year^{-1}$) from 4 to 65 (rural), from 26 to 104 (marine), from 23 to 71 (urban), and from 26 to 175 (industrial).

In addition to the structural metals already discussed, results have been reported from studies of the atmospheric corrosion behavior of other metals, many of which are used as materials in electronic or electric equipment.

The atmospheric corrosion of nickel is similar to that of zinc. Nickel exists solely as Ni^{2+} and instantaneously forms nickel hydroxide and, subsequently, $NiSO_4 \cdot 6H_2O$, has been observed [61b]. After prolonged exposure, an amorphous basic nickel sulfate is formed, frequently mixed with small amounts of carbonate, with less protective ability. This phase can crystallize and form another basic nickel sulfate, with higher stability and protective ability [61a]. No evidence of other anions in the corrosion products has been found so far. The corrosion rates of nickel are comparable to those of copper [70].

Silver exhibits a corrosion behavior that hardly resembles that of any of the other metals described. Its unique behavior is to a large extent governed by the existence of Ag^+ upon dissolution of silver into the aqueous layer. Ag_2S is the most abundant component of the corrosion products formed. AgCl can form in environments with high chloride content, whereas no oxides, nitrates, sulfates, or carbonates have been reported in connection with atmospheric exposure. Silver exhibits corrosion rates comparable to those of aluminum, lower than those of zinc, and much lower than those of iron.

Tin forms both SnO_2 and SnO as dominating corrosion products in a variety of environments, resulting in a relatively high protective ability. Laboratory exposures involving mixtures of SO_2, NO_2, and H_2S suggest that tin is relatively unaffected by these pollutants [71]. Accordingly, efforts to correlate corrosion effects of tin in natural atmospheric environments with concentrations of SO_2 and NO_2 have so far been unsuccessful [72]. Tin exhibits corrosion rates comparable to those of silver [72].

15.5 Atmospheric Corrosion Outdoors

15.5.1 General

Early field exposure programs with selected metals were implemented in the United States and the United Kingdom in the beginning of the twentieth century. The environments were classified as "rural," "urban," "marine," and "industrial," and it was soon recognized that corrosion rates vary considerably between different types of environment. Over the decades, a large number of outdoor corrosion test programs have been implemented with the general aim of identifying and, possibly, quantifying the most important environmental parameters involved in atmospheric corrosion processes. The large body of data available has provided evidence that corrosion rates under outdoor exposure conditions are strongly influenced by SO_2 and Cl^- as well as by climatic factors (humidity and temperature). Despite this, the general goal of predicting the performance of a given metal in a given environment is far from being attained [73]. The reason is that many other factors influence corrosion rates: initial exposure conditions, sample mass and orientation, extent of sheltering, wind velocity, the nature of corrosion products formed, and pollutants that are not measured. The difficulties in making accurate predictions of corrosivity can be exemplified by the variations in corrosion rates between steel and zinc in a worldwide ASTM Site Calibration at 45 outdoor locations [74]. The mass loss between different sites for both steel and zinc after 2 years of exposure varied two orders of magnitude or more. Moreover, the ranking of test sites with respect to mass loss of zinc was considerably different from that of iron. The ratio of steel to zinc mass loss varied from around 10 to more than 350. Significant variations in mass loss between different exposure periods were also observed for both metals. Similar experience has been gained from several other exposure programs [75,76].

15.5.2 Corrosion Rates

According to Barton [2], a generalized description of corrosion rate versus time includes an induction period, a period for establishing stationary conditions, and a stationary period (see Figure 15.8). These periods can be associated with the present description of the aqueous layer and formation of corrosion products. During the induction period, the metal is covered with a spontaneously formed oxide, which affords a certain protection depending on the metal and the aggressiveness of the atmosphere and the aqueous layer. The transition period involves the transformation from the oxide layer into a fully developed layer of corrosion products via local formation and coalescence of corrosion products. The stationary period, finally, is characterized by full coverage of the surface by corrosion products, eventually reaching constant properties with respect to chemical composition and atomic structure, and stationary corrosion rates. The two initial periods are shorter the more aggressive the exposure conditions are. For steel, they typically last for years in very benign (indoor) atmospheres but only a few months in highly polluted industrial areas [2].

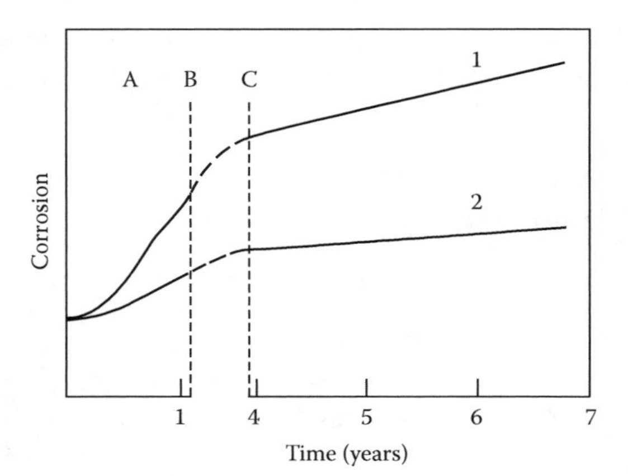

FIGURE 15.8
Schematic division of atmospheric corrosion into different periods, namely, an induction period (A), a period for establishing stationary conditions (B to C), and a stationary period (from C). (From Barton, K., *Protection against Atmospheric Corrosion*, 1976, Courtesy of John Wiley & Sons, London, U.K.)

A frequent observation during the two initial periods is the marked influence of initial exposure conditions on the subsequent corrosion rate [77,78]. Wet conditions—caused by rainfall or high relative humidity—are more severe than dry conditions during the first days of exposure; they result in higher corrosion rates when the exposure time extends to several months. In this respect, zinc seems to be more sensitive than steel. These observations are, at least partly, explained by the formation of corrosion products possessing different protective abilities. Especially on zinc, a diversity of structurally related corrosion products can be formed, the nature of which depends on initial exposure conditions [66]. An additional cause may be the seasonal dependence of the concentrations of H_2O_2 and O_3 [62].

During the third period of exposure, characterized by stationary corrosion rates, the interpretation of corrosion rates in terms of environmental data is more easily accomplished than during the two initial periods with varying corrosion rates. In reality, atmospheric corrosion is only quasistationary and regarded as a series of processes, when the surface is wet enough to allow rapid corrosion rates, and interrupted by periods of dryer conditions with negligible corrosion rates. An important concept here is the so-called time of wetness based on devices that monitor the actual time during which the surface is wet or the critical relative humidity exceeded [79]. In many cases, time of wetness estimates has been based on measurements of temperature and relative humidity. A common definition of time of wetness is the time during which, simultaneously, the temperature is above 0°C and the relative humidity ≥80%. However, this definition should not be taken too literally because the actual time when the surface is wet enough for rapid corrosion to occur depends on many other parameters, such as type, mass, orientation, and degree of sheltering of the metal; hygroscopic properties of corrosion products and of surface contaminants; type and level of pollutants; air velocity; and solar flux. An alternative way to estimate the time of wetness is by means of electrochemical cells, which consist of thin electrically separated metal electrodes and involve the detection of a current or an electromotive force when the thickness of the aqueous layer is sufficient to accelerate atmospheric corrosion [79–82]. In this case, too, the result will depend on many parameters, such as cell design, formation of corrosion products on metal electrodes, and criteria for defining time of wetness based on current or electromotive force responses.

It is also possible to measure quantitatively the amount of water on metal surfaces under in situ conditions by means of the quartz crystal microbalance [83]. This is illustrated in Figure 15.9, where the water content quotient on a gold surface, given as the mass of water in the aqueous adlayer divided by the mass of other deposited species on gold, is plotted as a function of relative humidity during consecutive 24 h cycles in an outdoor environment.

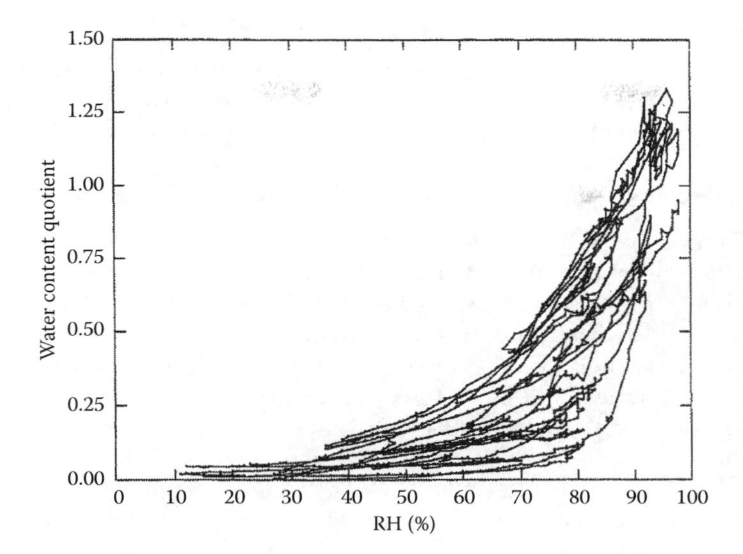

FIGURE 15.9
The water content quotient on gold, given as the mass of water in the aqueous adlayer divided by the mass of other deposits, as a function of relative humidity (RH) during 14 consecutive days of outdoor exposure in an urban environment. (From Forslund, M. and Leygraf, C., *J. Electrochem. Soc.*, 144, 105, 1997, Courtesy of The Electrochemical Society, Pennington, NJ.)

Each cycle is represented by one increase (during nighttime) and one decrease (during daytime) in water mass and relative humidity, respectively. From the figure, it is evident that significant amounts of water are present at relative humidity far below 80%. Analysis of the data shows that the actual time when the surface is wetted is significantly longer than the time when the temperature is above 0°C and the relative humidity ≥80%. Hence, this definition of time of wetness may be erroneous [83]. Despite these uncertainties in definition and experimental accuracy, the time of wetness exhibits a surprisingly good correlation with outdoor corrosion rates of steel in many types of environment [84].

15.5.3 Field Exposures

Because of the many parameters capable of influencing atmospheric corrosion rates, exposure programs have placed increased emphasis on standardized test procedures. The International Organization for Standardization (ISO) has formulated several corrosion testing standards [85–88] and also implemented a worldwide atmospheric exposure program known as ISOCORRAG [89]. This program aimed at generating a basis for the procedures used in the ISO classification system. The ISO classification system presents two different approaches to assessing the corrosivity of any environment. One approach is based on the exposure of standard specimens of aluminum, steel, copper, and zinc for 1 year and determining a *measured corrosivity class* from mass loss data. The other approach is based on SO_2 concentration, deposition of Cl^-, and time of wetness at the site, which results in an *estimated corrosivity class.* Having determined a corrosivity class with one of these methods, it is possible to estimate the extent of corrosion damage to aluminum, steel, copper, and zinc for either short- or long-term exposures. The ISO classification system provides adequate data for many practical purposes, including prediction of long-term corrosion behavior in different environments and the need for protective coatings. However, experience has shown that certain observations need further clarification. These observations include the frequent difference in corrosion rate between the top side and bottom side of flat standard specimens and between flat standard specimens and open helix standard specimens [90]. They also include the possible role of pollutants other than those measured.

15.5.4 Dose–Response Functions Describing the Multi-Pollutant Situation

The decreasing levels of SO_2 and increasing frequency of car traffic has resulted in a new multi-pollutant situation in many urban and industrial areas. In order to possibly quantify the corrosion effects caused by this new multi-pollutant situation an extended exposure program was performed that took place between 1997 and 2001 and involved some 30 test sites in 18 countries in Europe and North America [91]. Specimens of carbon steel, zinc, copper, bronze, limestone, paint-coated steel, and glass representative of medieval stained glass windows were exposed for up to four years. At each site, the environmental data measured included climatic parameters (temperature, relative humidity, and sunshine radiation), gaseous pollutants (SO_2, NO_2, HNO_3, and O_3), particles (presented as PM_{10}, i.e., concentration of particles with diameter $\leq 10\,\mu m$), and precipitation (total amount, conductivity, and concentration of, i.e., H^+, SO_4^{2-}, NO_3^-, Cl^-, NH_4^+).

A regression analysis of all data obtained resulted in dose–response functions, i.e., mathematical relationships that express the rate of deterioration of a material as a function of concentration of one or more corrosion stimulators. Table 15.5 shows the parameters included in the dose–response functions obtained for selected materials. For reason of space only the function for zinc will be used as an illustration of the new dose–response function from the multi-pollutant exposure:

$$ML_{zinc} = 3.53 + 0.471[SO_2]0.22 \cdot e^{0.018RH} \cdot e^{f(T)} + 0.041 \cdot Rain[H^+] + 1.37[HNO_3] \quad (15.27)$$

where

ML is the mass loss of zinc ($g\ m^{-2}$) after 1 year of exposure
$f(T) = 0.062(T - 10)$ when $T < 10°C$, otherwise $-0.021(T - 10)$
T is the temperature (°C)
RH is the relative humidity (%)
Rain is the amount of precipitation ($mm\ year^{-1}$)
SO_2 is the concentration ($\mu g\ m^{-3}$)
HNO_3 is the concentration ($\mu g\ m^{-3}$)
$[H^+]$ is the acidity of precipitation ($mg\ L^{-1}$)

TABLE 15.5

Materials, Parameters, and Their Inclusion in Dose–Response Functions as Obtained in an Extensive Multi-Pollutant Exposure Program

Material	T	RH	SO_2	O_3	HNO_3	PM_{10}	Rain	H^+
Carbon steel	X	X	X			X	X	X
Zinc	X	X	X		X		X	X
Copper	X	X	X	X			X	X
Cast bronze	X	X	X			X	X	X
Portland limestone		X	X		X	X	X	X

Source: Kucera, V. and Tidblad, J., Development of atmospheric corrosion in the changing pollution and climate situation, in *Proceedings of 17th International Corrosion Congress*, Las Vegas, NV, NACE, 2009.

Even though SO_2 has decreased substantially, it is still included in all functions, as is the effect of wet acid deposition (acid rain). NO_2 is not included in any of the functions directly but is closely related to HNO_3. The parameter HNO_3 was significant for zinc and limestone while PM_{10} was included in the functions for carbon steel, cast bronze and limestone.

In order to provide a more detailed understanding of the effect of HNO_3, an extensive comparison of laboratory and field corrosion effects of HNO_3 on copper, zinc, and carbon steel has been performed. A quantitative agreement between extrapolated laboratory results and field results could be established for copper and zinc and the total corrosion effect could be represented by the sum of two contributions; one from HNO_3 and one from remaining corrosion stimulators, including SO_2, O_3, precipitation characteristics and PM_{10}. In accordance with the dose–response function presented in Equation 15.27, the corrosion effects of zinc based on combined laboratory and field exposures could be graphically depicted (see Figure 15.10) [92].

However, an extrapolation of laboratory results of HNO_3-induced corrosion of carbon steel resulted in corrosion effects much lower than the total effect observed in the field. During identical laboratory exposures to HNO_3, the corrosion rate of carbon steel was nearly three times higher than that of copper or zinc. On the other hand, when comparing the corrosion effects induced by HNO_3 with those induced by SO_2 alone or in combination with either NO_2 or O_3, HNO_3 turned out to be far more aggressive than SO_2. This effect was mainly attributed to the high sticking coefficient, high solubility in water, and high chemical reactivity of HNO_3 in comparison to the other gaseous corrosion stimulators [93].

In all, dose–response functions based on field exposures can be used to predict corrosion rates, and in favorable cases also to provide some mechanistic insight, in particular if combined with laboratory exposures. Despite the fact that ambient SO_2 levels are still much higher than HNO_3 levels, these studies show that HNO_3 plays a significant role in the corrosion of copper and zinc, but not carbon steel.

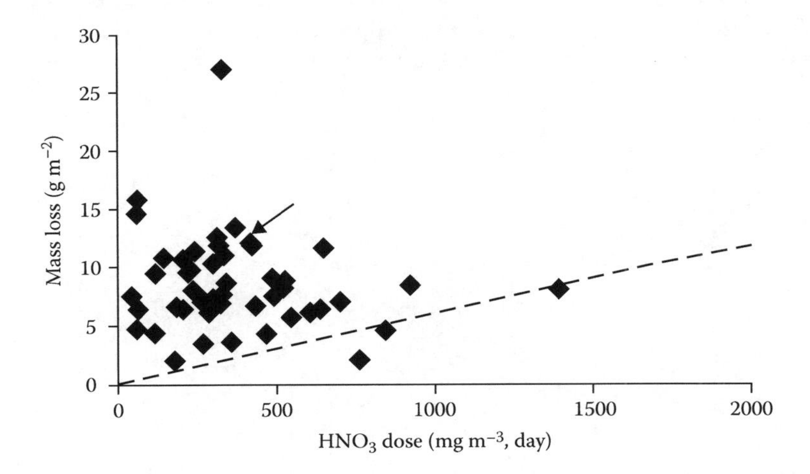

FIGURE 15.10
A plot of mass loss data of zinc after 1 year of exposure versus dose of HNO_3 in either field or laboratory exposures. The data points represent different exposure sites in an international field exposure program. The dashed line shows the results from laboratory exposures of zinc in humidified air at 70% relative humidity, to which sub-ppm concentrations of HNO_3 were added. For a given HNO_3-dose, the mass loss up to the dashed line represents the effect of HNO_3, while the mass loss above the dashed line includes the integrated effect of the remaining corrosion stimulators.

15.5.5 The Effect of Temperature in Combination with Gases and Aerosol Particles

It is generally agreed that time of wetness largely is governed by surface conditions, i.e., when the surface relative humidity exceeds the deliquescent relative humidity of any aerosol salt on the surface. In an international exposure program, it was found that sites with similar classifications according to ISO 9223 exhibit similar surface contaminants and wetting characteristics. The results furthermore showed that rain events usually clean the surface and thereby may change the wetting characteristics of the surface [94]. The complex interplay between deposition of chlorides and gases on surfaces of different metals has lately been explored by Cole et al. [95,96], as well by Cao et al. [97] and Morcillo et al. [98].

Detailed studies have been undertaken to explore the initial atmospheric corrosion of copper and zinc induced by particles, including NaCl and $(NH_4)_2SO_4$. They show a complex interplay between gases involved, such as SO_2 and CO_2, and the droplets around the deposited particles. The composition and lateral distribution of the corrosion products turn out to be governed by the location of anodic and cathodic sites, the transport of ions and other species between these sites, the deposition of gases onto the surface, and complicated spreading effects of the droplets formed [99].

Besides gases and aerosols, temperature is another critical factor that acts in combination with relative humidity to determine the wetness conditions on a metal surface. Hence, in order to obtain an overall picture of surface wetness, climate and pollution should be considered together. Based on dose–response functions obtained from field exposures [91], the most common temperature dependence for metals is illustrated by curve "a" in Figure 15.11, showing that the time of wetness has a maximum at round 10°C. The explanation is that at lower temperatures corrosion increases with temperature due to increase of time of wetness while, at higher temperatures corrosion decreases due to faster evaporation of moisture after rain or condensation periods.

The corrosion rate of most metals is strongly affected by the concentration of chlorides on the surface. Chlorides have hygroscopic properties and thus facilitate the formation of the aqueous adlayer, as discussed earlier in this chapter. This leads to a prolongation of the periods when the surface is wet, even at high temperatures. Therefore, the temperature decrease observed in curve "a" in Figure 15.11 is not observed when chlorides are

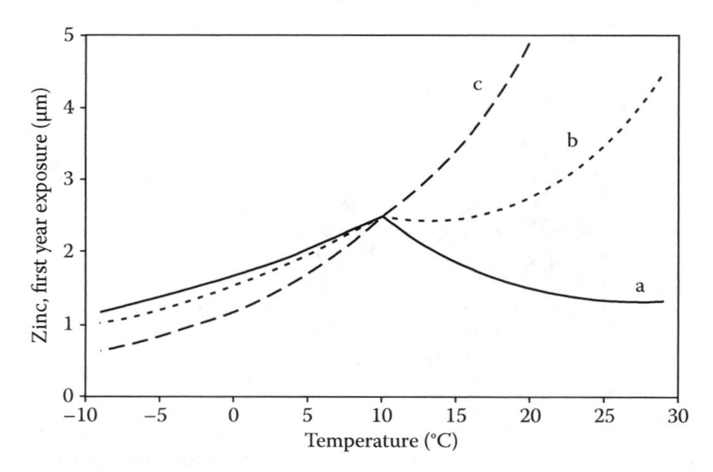

FIGURE 15.11
Calculated corrosion effect of zinc (in μm) vs. temperature using parameters value of (a) $SO_2 = 40\,\mu g\ m^{-3}$, $Cl = 3\,mg\ m^{-2}\ day^{-1}$; (b) $SO_2 = 20\,\mu g\ m^{-3}$, $Cl = 60\,mg\ m^{-2}\ day^{-1}$; (c) $SO_2 = 1\,\mu g\ m^{-3}$, $Cl = 300\,mg\ m^{-2}\ day^{-1}$. The data have been extracted from dose–response functions based on field exposures, as further described in Ref. [91].

present on the surface in sufficient amounts. Instead, the corrosion continues to increase with temperature also above 10°C, as shown for two different cases, illustrated by curves "b" and "c" in Figure 15.11. In addition, chlorides are corrosive and can cause pitting corrosion. The highest observed corrosion rates in the world today, with the exception of certain aggressive industrial atmospheres, can be found at warm marine sites in subtropical and tropical regions.

15.6 Atmospheric Corrosion Indoors

Whereas the accumulated knowledge of outdoor exposures has extended over many decades of corrosion tests, the history of indoor corrosion tests is short and mainly goes back to the growing interest in corrosion effects on electronic materials during the last two decades or so [100]. There are obvious differences in outdoor and indoor exposure conditions and, consequently, expected differences between outdoor and indoor corrosion behavior, some of which are summarized here [101].

The aqueous layer formed under outdoor exposure conditions is strongly influenced by seasonal and diurnal variations in humidity and by precipitation, dew, snow, or fog. Indoors, on the other hand, the aqueous layer is often governed by relatively constant humidity conditions. This means that there are practically no wet–dry cycles indoors and that the influence of indoor humidity can hardly be described by introducing a time of wetness factor.

Mostly due to enhanced indoor absorption of gases and particulates and also due to retardation and damping of outdoor variations by ventilating systems and air filtration, comparisons between outdoor and indoor concentrations of the most important gaseous corrosion stimulants in general reveal lower levels indoors than outdoors (Table 15.1). Exceptions are NH_3 and HCHO, which usually exhibit higher levels indoors than outdoors as the result of anthropogenic activity [17]. Another important difference is the decreased levels of indoor atmospheric oxidants, many of them photochemically produced as described earlier, and of Fe(III)- and Mn(II)-containing catalysts for promoting the oxidation of S(IV) to S(VI). Accordingly, the expected lifetime of S(IV) compounds is greater indoors than outdoors [17]. As discussed earlier, not only the concentration of pollutants but also the air velocity determines the dry deposition velocity of corrosion stimulants. With decreased indoor air velocities follow significantly lower dry deposition velocities indoors than outdoors.

Consistent with the differences in outdoor and indoor characteristics, the corrosion rate of many metals is significantly lower indoors than outdoors. By examining the corrosion rates of copper, silver, nickel, cobalt, and iron in eight indoor locations it was found that all metals except silver exhibited much lower rates indoors than outdoors [102]. Within the UN/ECE (United Nations Economic Commission for Europe) exposure program, a detailed study was made of the influence of sheltering nickel samples in a ventilated box, simulating exposure in unheated storage conditions. It was concluded that the deposition of SO_2 outside and inside the ventilated box differed by (approximately) a factor of 10, this difference being attributable to different airflow conditions, a factor of 2, and SO_2 concentrations, a factor of 5 [103].

As manifested by the ISO classification system, the outdoor corrosion behavior of structural metals can often be adequately predicted by only three parameters: SO_2 concentration, deposition of Cl^-, and time of wetness. With considerably less influence of these parameters indoors, it is reasonable to assume that the relative importance of other corrosion stimulants increases. Evidence for this can be exemplified in different ways. A classification system from the Instrument Society of America (ISA) for predicting corrosivity in indoor environments includes measurements of SO_2, NO_2, H_2S, and Cl_2 in addition to relative humidity and temperature [104]. Corrosion products formed on copper show much larger variations in chemical composition after indoor than after outdoor exposure [17]. Results from Fourier transform infrared reflection absorption spectroscopy analysis of corrosion products on copper, zinc, nickel, and iron formed in various benign corrosivity indoor environments exhibit large amounts of infrared bands from carboxylate ions, such as formate, acetate, and oxalate [105]. This forms evidence of a stronger influence of carboxylic acids and other organic species under benign indoor conditions than in more aggressive outdoor environments.

The deposition of particles on corroding surfaces is of significant importance in indoor conditions as well [106,107]. This is clearly seen when a comparison is made between the main sources of deposited SO_4^{2-} and Cl^- on aluminum and zinc surfaces during outdoor and indoor exposure. Outdoors, the accumulated SO_4^{2-} on both metal surfaces mainly originates from gaseous SO_2 and from SO_4^{2-} in precipitation. Indoors, surface-accumulated SO_4^{2-} is derived mainly from particles. Similarly, surface-accumulated Cl^- has three approximately equal outdoor sources—reactive gases, precipitation, and particles—and two approximately equal indoor sources: reactive gases and particles [107,108]. It is evident that future descriptions of the mechanisms governing indoor atmospheric corrosion processes must involve deposition of particles as a major source of corrosion-stimulating pollutants [109,110].

Lower corrosion rates indoors than outdoors will require the development of more sensitive techniques for corrosivity evaluation of indoor environments, and also an extended classification of indoor corrosivities [111].

15.7 Dispersion of Metals

So far, this chapter has treated various aspects of the impact of the atmospheric environment on metals. The opposite situation, in which a corroding metal can have detrimental influences on the surrounding environment, is now an important issue because of the increased concern regarding the amount of metals dispersed from roofs, facades, and other outdoor structures. Whereas long-term corrosion rates of metals, including copper and zinc, have been studied in many outdoor exposure programs, the dispersion of the same metals and their potential impact on the environment are less well understood. Risk assessments by, e.g., organizations and legislators have assumed that the runoff rate of a metal from a structure is equal to its corrosion rate. It has moreover been assumed that all metals dissolved and dispersed is bioavailable and therefore toxic to the environment. Both assumptions represent upper limits, and the risk assessments made therefrom are quite conservative. A realistic risk assessment of the corrosion-induced dispersion of metals in the environment needs answers to at least the following questions:

1. To what extent are metals released, and what is the ratio between the runoff rate of a metal and its corrosion rate?
2. Does the runoff rate of metals vary with time?

3. How does the chemical form of released metals change upon environmental entry, and how does this affect their bioavailability and potential adverse environmental effects?
4. To what extent are released metals retained by absorbing surfaces in the near vicinity of the point of release?

Copper and zinc have become the most frequently studied metals from this perspective. The copper corrosion rate in unsheltered outdoor environments is always highest in the beginning and slowly decreases with exposure time. The reason is the gradual formation of corrosion products, including Cu_2O (cuprite), $Cu_4SO_4(OH)_6 \cdot H_2O$ (posjnakite), and $Cu_4SO_4(OH)_6$ (brochantite), that act as a surface barrier and reduce the corrosion rate. The time for this sequence to happen varies considerably between different environments but brochantite, the end product in most patina formation, is often detected within 2 years also at unsheltered exposure conditions [67]. Zinc shows a similar behavior, in which case the main corrosion products formed in urban and rural environments of low chloride content are $Zn_5(CO_3)_2(OH)_6$ (hydrozincite) and $Zn_4SO_4(OH)_6 \cdot nH_2O$ [66].

From the large changes in copper corrosion product composition during the first 2 years and the concomitant reductions in corrosion rate, one would suspect that the copper runoff rate undergoes similar changes as a function of exposure period. Results from measurements on new and naturally aged copper have shown, however, that the copper runoff rate on an annual basis is relatively constant, or decreases slightly with time, during exposures up to 12 years. Runoff rates in the range between 0.7 and 1.6 g m^{-2} per year have been determined in relatively benign urban environments such as Stockholm, with an annual precipitation typically varying between 450 and 600 mm $year^{-1}$. The runoff rate is strongly correlated to parameters such as prevailing rain characteristics including amount, pH and intensity, also to gaseous pollutant concentrations and to surface inclination and orientation [112,113]. From a significantly decreasing copper corrosion rate with time, in particular during the first years of exposure, and a relatively constant copper runoff rate on an annual basis, follows that the quotient between copper runoff and copper mass loss initially is low and increases with time. Taking Stockholm as an example, the quotient was 7% after 1 month and 22% after 2 years [112].

As with copper, annual zinc runoff rates in atmospheric environments have turned out to remain relatively constant over exposure periods of a few years. In Stockholm, the runoff rate of zinc from zinc sheet varied between 1.6 and 2.3 g m^{-2} per year during a 10 year period [114]. Measured zinc runoff rates from zinc sheet exposed to different test sites in Europe during 1 year of exposure ranged from 3 and 5 g m^{-2} per year and could be correlated with the SO_2 concentration at sites with comparable annual amounts of precipitation [115,116]. For most sites, the quotient between zinc runoff rate and zinc corrosion rate varied between 50% and 60% during the first 2 years of exposure.

In order to predict runoff rates of copper from naturally patinated copper on buildings at specific urban or rural sites of low chloride influence, a general model was deduced based on both laboratory and field data [117,118]:

$$R = (0.37\,SO_2^{\,0.5} + 0.96\,\text{mm} \cdot 10^{-0.62\,\text{pH}})\cos\theta/\cos 45° \quad (15.28)$$

R denotes the copper runoff rate (g m^{-2} $year^{-1}$) through two contributions, both with a physical meaning. The first contribution originates from dry periods between rain events (the first-flush effect) and the second from the rain events. Input parameters are the SO_2 concentration (μg m^{-3}), pH and amount of precipitation (mm $year^{-1}$), and surface angle of inclination (θ). In 76% of all reported annual runoff rates, the predicted values are within 35% from the observed values. Further investigations have revealed that rain pH has a dominating effect on copper patina dissolution, nitrate in rainwater has a small inhibiting effect, whereas no significant effects were seen for chloride and sulfate in rainwater. Chlorides turn out to have a minor effect on the runoff rate of both copper and zinc, whereas the effect of chlorides on the corrosion rate is significant [119,120]. In another study an empirical model of copper runoff rates was deduced showing that 15% of the total copper mass loss from patinated copper was retained in the corrosion film after 100 years of exposure [121]. In a similar way, efforts have also been undertaken to model the runoff rate of zinc in terms of surface angle inclination, patina age, exposure direction, pH and amount of precipitation [122,123].

The runoff rate of the metal provides only the total metal release from any outdoor construction without considering the chemical speciation of the metal or its bioavailable fraction, i.e., the part of the metal that can interact with living organisms and cause toxic effects. The most bioavailable and therefore most toxic form is usually the free, hydrated, cation. Hence, in the case of copper or zinc, $Cu(H_2O)_6{}^{2+}$ and $Zn(H_2O)_6{}^{2+}$ are among the most bioavailable species. From this follows the importance to track the change in chemical identity of the released metal during transport from the source to the recipient. Such efforts have been undertaken for copper [124,125], zinc [126], and chromium and nickel from stainless steels [127]. Similar trends could be observed for all metals investigated. Immediately after release, the fraction of the free hydrated ion is very high, in the range from 60% to 90% depending on the conditions of the actual precipitation event. During its way to the environment the metal will react with, e.g., organic matter and the fraction of the most bioavailable forms of the metal will become successively smaller. Solid surfaces in the near vicinity of the location for metal release, such as limestone in buildings, soil surrounding the building, or concrete in pavement and storm water systems, act as efficient sinks for metal absorption whereby the fraction of bioavailable metal is significantly reduced. For zinc, a detailed study was performed in which the concentrations and chemical speciation of zinc from 14 zinc-based materials exposed during 5 years in Stockholm was determined [126]. Depending on surface conditions, the runoff rates of total zinc ranged from 0.07 to 2.5 g Zn m^{-2} $year^{-1}$ with zinc primarily released as the free ion for all materials investigated. Upon interaction of the zinc-containing runoff water with representative soil systems, which simulated a few decades of exposure, the total zinc concentration was reduced from milligram per liter to microgram per liter levels. Simultaneously, the most bioavailable fraction of zinc in runoff, $Zn(H_2O)_6{}^{2+}$, decreased from more than 95% to about 30%. Nearly all of the introduced zinc in the runoff water was retained within the soil. As long as the soil retention capacity was not reached, this resulted in zinc concentrations in the percolate water transported through the soil layer close to background values. To conclude, all studies performed to assess the environmental fate of corrosion-induced metal runoff show that surfaces near the location of metal release act as sinks for the released metal, whereby the absorbing material may transform into naturally occurring minerals [124–127].

These data, among many others, are important within REACH (Registration, Evaluation, Authorization and restriction of CHemicals), the new chemical policy of the European commission, which entered into force in 2007 [128]. An important issue within REACH is how to predict the environmental effects from metal alloys. The general approach so far has been to treat alloys as a mixture of the pure metals that constitute the alloys. Because of the passive behavior of many metals or alloys, but also because of other factors, this turns out to be an erroneous assumption. As an example, the release rates of copper and zinc from a brass (Cu20%Zn) surface exposed to rain water during 2 years of exposure in a benign environment differed significantly from their proportions in the bulk alloy [129]. These and other studies clearly show that alloys and the pure metals behave very differently when exposed to rainwater. Accordingly, release rates from pure metals cannot be directly used to predict release rates of individual constituents from their alloys.

15.8 Concluding Remarks

Atmospheric corrosion of metals comprises a broad range of electrochemical, chemical, and other processes in the interfacial domain, ranging from the atmospheric region over the thin aqueous layer to the solid metal region. Of utmost importance is the thin aqueous layer, which forms the medium for electrochemical reactions and accommodates gaseous, particulate, or other pollutants from the atmospheric region. Anode rather than cathode reactions are normally rate limiting and may be facilitated by acidifying pollutants, such as SO_2. The subsequent reaction sequences in the aqueous layer are governed mainly by kinetic constraints, including the concept of dry deposition velocity of pollutants and the nucleation, spreading, and coalescence of corrosion products. Atmospheric corrosion rates are strongly influenced by the formation and properties of corrosion products, where the formation process can be described in a sequence of consecutive steps: dissolution–coordination–reprecipitation. The dissolution step is usually acid dependent; possible coordination steps can be predicted by the HSAB principle, and the reprecipitation depends on the activities of species involved.

Extensive outdoor exposure programs have provided evidence that corrosion rates can be interpreted in terms of deposition of primarily SO_2 and Cl^- and time of wetness. The history of indoor exposure programs is much shorter. Atmospheric corrosion indoors is, in general, influenced by more constant humidity conditions and lower levels of SO_2 and Cl^-, from which follows an increased relative importance of other corrosion stimulants including organic gaseous species and particulate pollutants.

Despite all progress made in understanding individual processes, there is a need to develop more comprehensive models in the study of atmospheric corrosion. The first computational model developed in atmospheric corrosion [130] considers six distinct regimes, namely the gas phase (G), the gas–liquid interface (I), the liquid phase (L), the deposition layer (D), the electrodic regime (E), and the solid phase (S), and is designated GILDES. Chemical reactions within the regimes can occur to change the constituents of the regimes. Transport of chemical species of interest between the regimes is also considered together with condensation and dissolution processes and deposition and transport through the reaction products. For the six regimes, mathematical formulations have been specified to describe the transitions and transformations that occur. The conceptual framework used in the GILDES formulation is truly multidisciplinary and requires the incorporation of many scientific fields: gas phase—atmospheric chemistry; interface layer—mass transport engineering and interface science; liquid phase—freshwater, marine, and brine chemistry; deposition layer—colloid chemistry, surface science, and mineralogy; electrodic regime—electrochemistry; and solid phase—solid-state chemistry. GILDES model studies have so far included the initial SO_2-induced atmospheric corrosion of copper [131] and nickel [132], and also the SO_2+O_3- and SO_2+NO_2-induced atmospheric corrosion of copper [133]. Comparison with experimental data suggests that the model has been able to capture some of the most important processes.

Atmospheric corrosion involves a series of processes with periods of high corrosion rates interrupted by periods of negligible corrosion rate. For a deeper understanding of atmospheric corrosion, there is a need for more sophisticated techniques for measuring instant corrosion rates coupled with monitoring the deposition of the most important corrosion stimulants. Activities in these and other areas are presently being carried out, and it is anticipated that atmospheric corrosion will continue to develop into a multidisciplinary scientific field.

Acknowledgments

The author wishes to express his gratitude to Prof. I. Odnevall Wallinder at the Royal Institute of Technology (KTH) and Dr. J. Tidblad at Swerea KIMAB AB, Stockholm, Sweden, for most valuable review comments.

References

1. D. Fyfe, The atmosphere, in *Corrosion*, 2nd edn. (L. L. Shreir, ed.), Newnes-Butterworths, Sevenoaks, Kent, U.K., 1976, p. 2.26.
2. K. Barton, *Protection against Atmospheric Corrosion*, Wiley, London, U.K., 1976.
3. U. R. Evans, *The Corrosion and Oxidation of Metals*, Edward Arnold, London, U.K., 1960, p. 482.
4. W. H. J. Vernon, Effect of sulphur dioxide on the atmospheric corrosion of copper, *Trans. Faraday Soc.* 27:35 (1933).
5. I. L. Rozenfeld, *Atmospheric Corrosion of Metals*, National Association of Corrosion Engineers, Houston, TX, 1972.
6. V. Kucera and E. Mattsson, Atmospheric corrosion, in *Corrosion Mechanisms* (F. Mansfeld, ed.), Marcel Dekker, New York, 1987.
7. C. Leygraf and T. Graedel, *Atmospheric Corrosion*, John Wiley & Sons, New York, 2000.
8. (a) *Atmospheric Factors Affecting the Corrosion of Engineering Metals*, ASTM STP 646 (K. Coburn, ed.), American Society for Testing and Materials, Philadelphia, PA, 1978; (b) *Atmospheric Corrosion of Metals*, ASTM STP 767 (S. W. Dean and E. C. Rhea, eds.), American Society for Testing and Materials, Philadelphia, PA, 1982.
9. (a) *Degradation of Metals in the Atmosphere*, ASTM STP 965 (S. W. Dean and T. S. Lee, eds.), American Society for Testing and Materials, Philadelphia, PA, 1988; (b) *Atmospheric Corrosion*, ASTM STP 1239 (W. W. Kirk and H. H. Lawson, eds.), American Society for Testing and Materials, Philadelphia, PA, 1995; (c) *Symposium on Outdoor and Indoor Atmospheric Corrosion*, ASTM STP 1421 (H.E. Townsend, ed.), American Society for Testing and Materials, Philadelphia, PA, 2002.
10. W. H. Ailor (ed.), *Atmospheric Corrosion*, Wiley, New York, 1982.
11. T. E. Graedel, D. T. Hawkins, and L. D. Claxton, *Atmospheric Chemical Compounds, Sources, Occurrence, and Bioassay*, Academic Press, Orlando, FL, 1986, p. 9.
12. T. E. Graedel, D. T. Hawkins, and L. D. Claxton, *Atmospheric Chemical Compounds, Sources, Occurrence, and Bioassay*, Academic Press, Orlando, FL, 1986, p. 543.
13. J. G. Calvert and W. R. Stockwell, Mechanisms and rates of the gas-phase oxidations of sulfur dioxide and nitrogen oxides in the atmosphere, in *SO_2, NO, and NO_2 Oxidation Mechanisms: Atmospheric Considerations* (J. G. Calvert, ed.), Butterworth, Woburn, MA, 1984, p. 1.

14. L. R. Robbin, Kinetic studies of sulfite oxidation in aqueous solution, in *SO_2, NO, and NO_2 Oxidation Mechanisms: Atmospheric Considerations* (J. G. Calvert, ed.), Butterworth, Woburn, MA, 1984, p. 63.
15. T. E. Graedel, D. T. Hawkins, and L. D. Claxton, *Atmospheric Chemical Compounds, Sources, Occurrence, and Bioassay*, Academic Press, Orlando, FL, 1986, p. 68.
16. B. J. Finlayson-Pitts and J. N. Pitts Jr., *Atmospheric Chemistry: Fundamentals and Experimental Techniques*, Wiley, New York, 1986, p. 29.
17. S. K. Chawla and J. H. Payer, Atmospheric corrosion: A comparison of indoor vs. outdoor, in *Proceedings of 11th International Corrosion Congress*, Florence, Italy, 1990, p. 2.17.
18. T. E. Graedel, Copper patinas formed in the atmosphere—II. A qualitative assessment of mechanisms, *Corros. Sci.* 27:721 (1987).
19. T. E. Graedel, Corrosion mechanisms for silver exposed to the atmosphere, *J. Electrochem. Soc.* 739:1963 (1992).
20. Z. Y. Chen, S. Zakipour, D. Persson, and C. Leygraf, Effect of sodium chloride particles on the atmospheric corrosion of pure copper, *Corrosion* 60(5):479 (2004).
21. P. A. Thiel and T. M. Madey, The interaction of water with solid surfaces: Fundamental aspects, *Surf. Sci. Rep.* 7:211 (1987).
22. M. A. Barteau and R. J. Madix, The surface reactivity of silver: Oxidation reactions, in *Chem. Phys. Solid Surf. Heretog. Catalysis*, Vol. 4 (D. A. King and D. P. Woodruff, eds.), Elsevier, Amsterdam, the Netherlands, 1982, p. 95.
23. W. J. Lo, Y. W. Chung, and G. A. Somorjai, Electron spectroscopy studies of the chemisorption of oxygen, hydrogen and water on the titanium dioxide (100) surfaces with varied stoichiometry: Evidence for the photogeneration of titanium (3+) and for its importance in chemisorption, *Surf. Sci.* 71:199 (1978).
24. C. Benndorf, C. Nöbl, and T. E. Madey, Water adsorption on oxygen-dosed nickel (110): Formation and orientation of adsorbed hydroxyl, *Surf. Sci.* 138:292 (1984).
25. R. L. Kurtz and V. E. Henrich, Surface electronic structure and chemisorption on corundum transition-metal oxides: α-Ferric oxide, *Phys. Rev. B* 56:3413 (1987).
26. D. J. Dwyer, S. R. Kelemen, and A. Kaldor, The water dissociation reaction on clean and oxidized iron (110), *J. Chem. Phys.* 76:1832 (1982).
27. (a) E. Mc Cafferty, V. Pravdic, and A. C. Zettlemoyer, Dielectric behaviour of adsorbed water films on the α-iron(III)oxide surface, *Trans. Faraday Soc.* 66:1720 (1970); (b) P. B. P. Phipps and D. W. Rice, The role of water in atmospheric corrosion, in *Corrosion Chemistry, ACS Symp. Ser.* 89 (G.R. Brubaker and P.B.P. Phipps, eds.), American Chemical Society, Washington, DC, 1979, p. 235.
28. Y. R. Shen, A few selected applications of surface nonlinear optical spectroscopy, *Proc. Natl. Acad. Sci. USA* 99:12104 (1996).
29. Y. N. Mikhailovsky, Theoretical and engineering principles of atmospheric corrosion of metals, in *Atmospheric Corrosion* (W. H. Ailor, ed.), Wiley, New York, 1982, p. 85.
30. B. D. Yan, S. L. Meilink, G. W. Warren, and P. Wynblatt, Water adsorption and surface conductivity measurements on α-alumina substrates, in *Proceedings of 36th Electronic Components Conference*, Seattle, WA, 1986, p. 95.
31. W. H. J. Vernon, A laboratory study of the atmospheric corrosion of metals, Part I, *Trans. Faraday Soc.* 27:255 (1931).
32. W. H. J. Vernon, A laboratory study of the atmospheric corrosion of metals, Part II and III, *Trans. Faraday Soc.* 31:1668 (1935).
33. H. Kaesche, Elektrochemische Merkmale der atmosphärischen Korrosion, *Werkst. Korros.* 15:379 (1964).
34. F. Mansfeld, Electrochemical methods for atmospheric corrosion studies, in *Atmospheric Corrosion* (W. H. Ailor, ed.), Wiley, New York, 1982, p. 139.
35. C. Fiaud, Electrochemical behaviour of atmospheric pollutants in thin liquid layers related to atmospheric corrosion, in *Atmospheric Corrosion* (W. H. Ailor, ed.), Wiley, New York, 1982, p. 161.
36. M. Stratmann, The atmospheric corrosion of iron—A discussion of the physico-chemical fundamentals of this omnipresent corrosion process, *Ber. Bunsenges. Phys. Chem.* 94:626 (1990).

37. M. Stratmann, The investigation of the corrosion properties of metal surfaces covered by condensed or adsorbed electrolyte layers, in *Proceedings of 1st International Symposium on Corrosion of Electronic Materials and Devices* (J. D. Sinclair, ed.), The Electrochemical Society, Pennington, NJ, 1991, p. 1.
38. D. Persson and C. Leygraf, Initial interaction of sulfur dioxide with water covered metal surfaces: An in situ IRAS study, *J. Electrochem. Soc.* 142:1459 (1995).
39. W. Stumm, B. Sulzberger, and J. Sinniger, The coordination chemistry of the oxide-electrolyte interface: The dependence of surface reactivity on surface structure, *Croat. Chem. Acta* 63:277 (1990).
40. W. Stumm and G. Furrer, The dissolution of oxides and aluminium silicates: Examples of surface-coordination-controlled kinetics, in *Aquatic Surface Chemistry* (W. Stumm, ed.), Wiley, New York, 1987, p. 197.
41. S. E. Schwartz, Gas-aqueous reactions of sulfur and nitrogen oxides in liquid–water clouds, in *SO_2, NO, and NO_2 Oxidation Mechanisms: Atmospheric Considerations* (J. G. Calvert ed.), Butterworth, Woburn, MA, 1984, p. 173.
42. H. Rodhe, Luftföroreningars Sprinding (in Swedish), Compendium, University of Stockholm, Sweden, 1991.
43. G. A. Sehmel, Particle and gas dry deposition: A review, *Atmos. Environ.* 14:983 (1980).
44. P. J. Hanson and S. E. Lindberg, Dry deposition of reactive nitrogen compounds: A review of leaf, canopy and non-foliar measurements, *Atmos. Environ.* 25A:1615 (1991).
45. Effects of acidic deposition on materials, State-of-Science/Technology Report 19, National Acid Precipitation Assessment Program, 1990.
46. P. S. Liss and P. G. Slater, Flux of gases across the air–sea interface, *Nature* 247:181 (1974).
47. F. W. Lipfert, Dry depsosition velocity as an indicator for SO_2 damage to materials, *J. Air Pollut. Control Assoc.* 39:446 (1989).
48. L. Volpe, New studies using the tubular corrosion reactor, in *Proceedings of 11th International Corrosion Congress*, Florence, Italy, 1990, p. 2.25.
49. L. Volpe, Mass-transport limitations in atmospheric corrosion, in *Proceedings of 1st International Symposium on Corrosion of Electronic Materials and Devices* (J. D. Sinclair, ed.), The Electrochemical Society, Pennington, NJ, 1991, p. 22.
50. F. W. Lipfert and R. E. Wyzga, Application of theory to economic assessment of corrosion damage, in *Proceedings of Degradation of Materials due to Acid Rain* (R. Baboian, ed.), American Chemical Society, Washington, DC, 1986, p. 411.
51. E. Mattsson, The atmospheric corrosion properties of some common structural metals—A comparative study, *Mater. Perf.* 21:9 (1982).
52. M. Carballeira, A. Carballeira, and J. Y. Gal, Contribution to the study of corrosion phenomena in industrial atmosphere. Thermodynamic approach, in *Proceedings of 14th International Conference on Electric Contacts*, Société des Électroniciens, Paris, 1988, p. 239.
53. T. E. Graedel, K. Nassau, and J. P. Franey, Copper patinas formed in the atmosphere. I. Introduction, *Corros. Sci.* 27:639 (1987).
54. E. Matijevic, Preparation and characterization of model colloidal corrosion products, *Corrosion NACE* 35:264 (1979).
55. D. A. Vermilyea, Physics of corrosion, *Phys. Today* 9:23 (1976).
56. G. A. Somorjai, *Chemistry in Two Dimensions,* Cornell University Press, Ithaca, NY, 1981, p. 293.
57. N. Sato, Some concepts of corrosion fundamentals, *Corros. Sci.* 27:421 (1987).
58. R. G. Pearson, Hard and soft acids and bases, *J. Am. Chem. Soc.* 85:3533 (1963).
59. W. L. Jolly, *Modern Inorganic Chemistry*, 2nd edn., McGraw-Hill, New York, 1991, p. 238.
60. W. Feitknecht, Ordnungsvorgänge bei kolloiddispersen hydroxyden und hydroxy-salzen, *Kolloid Z.* 136:52 (1954).
61. (a) D. Persson and C. Leygraf, Analysis of atmospheric corrosion products of field exposed nickel, *J. Electrochem. Soc.* 139:2243 (1992); (b) I. Odnevall and C. Leygraf, The atmospheric corrosion of nickel in a rural atmosphere, *J. Electrochem. Soc.* 144:3518 (1997).

62. T. E. Graedel, Corrosion mechanisms for zinc exposed to the atmosphere, *J. Electrochem. Soc.* 136:193C (1989).
63. T. E. Graedel, Corrosion mechanisms for aluminium exposed to the atmosphere, *J. Electrochem. Soc.* 136:204C (1989).
64. T. E. Graedel and R. P. Frankenthal, Corrosion mechanisms for iron and low alloy steels exposed to the atmosphere, *J. Electrochem. Soc.* 137:2385 (1990).
65. I. Odnevall and C. Leygraf, Formation of $NaZn_4Cl(OH)_6SO_4 \cdot 6H_2O$ in a marine atmosphere, *Corros. Sci.* 34:1213 (1993).
66. I. Odnevall and C. Leygraf, *Reaction Sequences in Atmospheric Corrosion of Zinc*, ASTM STP 1239 (W. W. Kirk and H. H. Lawson, eds.), American Society for Testing and Materials, Philadelphia, PA, 1995, p. 215.
67. A. Krätschmer, I. Odnevall Wallinder, and C. Leygraf, The evolution of outdoor copper patina, *Corros. Sci.* 44:425 (2002).
68. L.-G. Johansson, SO_2-induced corrosion of carbon steel in various atmospheres and dew point corrosion in stack gases, Thesis, Chalmers University of Technology and University of Gothenburg, Göteborg, Sweden, 1982.
69. G. Schikorr, Über den mechanismus des atmosphärischen rostens des eisens, *Werkst. Korros.* 14:69 (1983).
70. J. Tidblad, C. Leygraf, and V. Kucera, Acid deposition effects on materials: Evaluation of nickel and copper, *J. Electrochem. Soc.* 138:3592 (1991).
71. S. Zakipour and C. Leygraf, Evaluation of laboratory tests to simulate indoor corrosion of electrical contact materials, *J. Electrochem. Soc.* 133:21 (1986).
72. J. Tidblad, C. Leygraf, and V. Kucera, UN/ECE international co-operative programme on effects on materials. Report No. 8: Corrosion attack on electric contact materials. Evaluation after 1 and 2 years of exposure, Swedish Corrosion Institute, Stockholm, Sweden, 1991.
73. H. Guttman, Atmospheric and weather factors in corrosion testing, in *Atmospheric Corrosion* (W. H. Ailor, ed.), Wiley, New York, 1982, p. 51.
74. S. K. Coburn, Corrosiveness of various atmospheric test sites as measured by specimens of steel and zinc, in *Metal Corrosion in the Atmosphere*, ASTM STP 435 (S. W. Dean, Jr. and E. C. Rhea, eds.), American Society for Testing and Materials, Philadelphia, PA, 1968, p. 360.
75. H. Guttman and P. J. Sereda, Measurement of atmospheric factors affecting the corrosion of metals, in *Metal Corrosion in the Atmosphere*, ASTM STP 435 (S. W. Dean, Jr. and E. C. Rhea, eds.), American Society for Testing and Materials, Philadelphia, PA, 1968, p. 326.
76. D. Knotkova, B. Bosek, and J. Vlckova, Corrosion aggressivity of model regions of Czechoslovakia, in *Corrosion in Natural Environments*, ASTM STP 558, American Society for Testing and Materials, Philadelphia, PA, 1974, p. 52.
77. O. B. Ellis, Effects of weather on the initial corrosion rate of sheet zinc, *ASTM Proc. Am. Soc. Test. Mater.* 47:152 (1947).
78. H. Guttman, Effects of atmospheric factors on the corrosion of rolled zinc, in *Metal Corrosion in the Atmosphere*, ASTM STP 435 (S. W. Dean, Jr. and E. C. Rhea, eds.), American Society for Testing and Materials, Philadelphia, PA, 1968, p. 223.
79. I. Odnevall and C. Leygraf, *Atmospheric Corrosion*, ASTM STP 1237 (W. W. Kirk and H. H. Lawson, eds.), American Society for Testing and Materials, Philadelphia, PA, 1994.
80. P. J. Sereda, Measurement of surface moisture, *ASTM Bull.* 53:228 (1958).
81. N. D. Tomashov, *Theory of Corrosion and Protection of Metals*, Macmillan, New York, 1966, p. 672.
82. V. Kucera and E. Mattsson, Electrochemical technique for determination of the instantaneous rate of atmospheric corrosion, in *Corrosion in Natural Environments*, ASTM STP 558, American Society for Testing and Materials, Philadelphia, PA, 1974, p. 239.
83. M. Forslund and C. Leygraf, Humidity sorption due to deposited aerosol particles studied in-situ outdoors on gold surfaces, *J. Electrochem. Soc.* 144:105 (1997).
84. M. Benarie and F. L. Lipfert, A general corrosion function in terms of atmospheric pollutant concentrations and rain pH, *Atmos. Environ.* 20:1947 (1986).

85. ISO 9223 Corrosion of metals and alloys. Classification of corrosivity of atmospheres, International Organization for Standardization, Geneva, Switzerland, 1991.
86. ISO 9224 Corrosion of metals and alloys. Guiding values for the corrosivity categories of atmospheres, International Organization for Standardization, Geneva, Switzerland, 1991.
87. ISO 9225 Corrosion of metals and alloys. Corrosivity of atmospheres. Methods of measurement of pollution, International Organization for Standardization, Geneva, Switzerland, 1991.
88. ISO 9226 Corrosion of metals and alloys. Corrosivity of atmospheres. Methods of determination of corrosion rate of standard specimens for the evaluation of corrosivity, International Organization for Standardization, Geneva, Switzerland, 1991.
89. D. Knotkova and L. Vrobel, ISOCORRAG—The International Testing Program within ISO/TC 156/WG 4, in *Proceedings of 11th International Corrosion Congress*, Florence, Italy, 1990, p. 5.581.
90. S. W. Dean, Corrosion testing of metals under natural atmospheric conditions, in *Corrosion Testing and Evaluation: Silver Anniversary Volume*, ASTM STP 1000 (R. Baboian and S. W. Dean, eds.), American Society for Testing of Materials, Philadelphia, PA, 1990, p. 163.
91. V. Kucera and J. Tidblad, Development of atmospheric corrosion in the changing pollution and climate situation, in *Proceedings of 17th International Corrosion Congress,* Las Vegas, NACE, 2009.
92. F. Samie, J. Tidblad, V. Kucera, and C. Leygraf, Atmospheric corrosion effects of HNO_3, *J. Electrochem. Soc.* 154:C249 (2007).
93. F. Samie, J. Tidblad, V. Kucera, and C. Leygraf, Atmospheric corrosion effects of HNO_3—Comparison of laboratory-exposed copper, zinc and carbon steel, *Atm. Environ.* 41:4888 (2007).
94. I. S. Cole, W. D. Ganther, J. D. Sinclair, D. Lau, and D. A. Paterson, A study of the wetting of metal surfaces in order to understand the processes controlling atmospheric corrosion. *J. Electrochem. Soc.* 151:B627 (2004).
95. I. S. Cole, D. A. Paterson, and W. D. Ganther, Holistic model for atmospheric corrosion Part 1—Theoretical framework for production, transportation and deposition of marine salts, *Corr. Sci. Eng. Techn.* 38:129 (2003).
96. I. S. Cole, T. H. Muster, D. A. Paterson, S. A. Furman, G. S. Trinidad, and N. Wright, Multiscale modeling of the corrosion of metals under atmospheric corrosion, *Mater. Sci. Forum 561–565*:2209 (2207).
97. O. Ou, C. Yan, Y. Wan, and C. Cao, Effects of NaCl and SO_2 on the initial atmospheric corrosion of zinc, *Corros. Sci.* 44:2789 (2002).
98. M. Morcillo, B. Chico, L. Mariaca, and E. Otero, Salinity in marine atmospheric corrosion: Its dependence on the wind regime existing in the site. *Corros. Sci* 42:91 (2000).
99. Zhuo Yuan Chen, The role of particles on initial atmospheric corrosion of copper and zinc—Lateral distribution, secondary spreading and CO_2-/SO_2-influence, Doctoral thesis, ISBN 91-7178-155-2, Royal Institute of Technology, Stockholm, Sweden, 2003.
100. G. Frankel, Corrosion of microelectronic and magnetic storage devices, in *Corrosion Mechanisms in Theory and Practice* (P. Marcus and J. Oudar, eds.), Marcel Dekker, New York, 1995.
101. C. Leygraf, Indoor atmospheric corrosion, *Proceedings of 15th International Corrosion Congress*, Granada, Spain, 2002, p. 547.
102. D. W. Rice, R. J. Cappell, W. Kinsolving, and J. J. Laskowski, Indoor corrosion of metals, *J. Electrochem. Soc.* 127:891 (1980).
103. J. Tidblad, C. Leygraf, and V. Kucera, Acid deposition effects on materials: Evaluation of nickel after 4 years of exposure, *J. Electrochem. Soc.* 140:1912 (1993).
104. ISA-s 71.04–1985 Standard: Environmental conditions for process measurements and control system: Airborne contaminants. Instrument Society of America, Research Triangle Park, NC, 1985.
105. D. Persson and C. Leygraf, Metal carboxylate formation during indoor atmospheric corrosion of Cu, Zn, and Ni, *J. Electrochem Soc.* 142:1468 (1995).
106. J. D. Sinclair, Corrosion of electronics: The role of ionic substances, *J. Electrochem. Soc.* 135:89C (1988).

107. J. D. Sinclair, L. A. Psota-Kelty, R. B. Comizzoli, R. P. Frankenthal, R. L. Opila, and G. R. Crane, Corrosion of electronics: The role of ionic substances, in *Proceedings of 11th International Corrosion Congress*, Florence, Italy, 1990, p. 2.95.
108. J. D. Sinclair, L. A. Psota-Kelty, and C. J. Weschler, Indoor/outdoor concentrations and indoor surface accumulations of ionic substances, *Atmos. Environ.* 19:315 (1985).
109. R. E. Lobnig, D. J. Siconolfi, L. Psota-Kelty, G. Grundmeier, R. P. Frankenthal, M. Stratmam, and J. D. Sinclair, Atmospheric corrosion of zinc in the presence of ammonium sulfate particles, *J. Electrochem. Soc.* 143:1539 (1996).
110. R. E. Lobnig and C. A. Jankoski, Atmospheric corrosion of copper in the presence of acid ammonium sulfate particles, *J. Electrochem. Soc.* 145:946 (1998).
111. E. Johansson, Corrosivity measurements in indoor atmospheric environments—A field study, licentiate thesis, ISBN 91-7170-267-9, Royal Institute of Technology, Stockholm, Sweden, 1998.
112. I. Odnevall Wallinder, and C. Leygraf, A study of copper runoff in an urban atmosphere, *Corros. Sci.* 39:209 (1997).
113. W. He, I. Odnevall Wallinder, and C. Leygraf, A comparison between corrosion rates and runoff rates from new and aged copper and zinc as roofing material, *Water Air Soil Poll.* 1:67 (2001).
114. I. Odnevall Wallinder, Private communication.
115. I. Odnevall Wallinder, P. Verbiest, W. He, and C. Leygraf, The influence of patina age and pollutant levels on the runoff rate of zinc from roofing materials, *Corros. Sci.* 40:1977 (1998).
116. P. Verbiest, H. Waeterschoot, R. Racek, and M. Leclercq, A study of runoff and corrosion rates of rolled zinc sheet in different exposures, *Prot. Coat. Eur.* 9:47 (1997).
117. I. Odnevall Wallinder, B. Bahar, C. Leygraf, and J. Tidblad, Modelling and mapping of copper runoff for Europe, *J. Environ. Mon.* 9:66 (2007).
118. I. Odnevall Wallinder, S. Bertling, X. Zhang, and C. Leygraf, Predictive models of copper runoff from external structures, *J. Environ. Mon.* 6:704 (2004).
119. J. Sandberg, I. Odnevall Wallinder, C. Leygraf, and N. Le Bozéc, Corrosion-induced zinc runoff from construction materials in a marine environment, *J. Electrochem. Soc.* 154:C120 (2007).
120. J. Sandberg, I. Odnevall Wallinder, C. Leygraf, and N. Le Bozéc, Corrosion-induced copper runoff from naturally and pre-patinated copper in a marine environment, *Corros. Sci.* 48:4316 (2006).
121. S. D. Cramer, S. A. Matthes, B. S. Covino Jr., S. J. Bullard, and G. R. Holcomb, Environmental factors affecting the atmospheric corrosion of copper, *ASTM Special Techn. Publ.* 1421:245 (2002).
122. I. Odnevall Wallinder, P. Verbiest, W. He, and C. Leygraf, Effects of exposure direction and inclination of the runoff rates of zinc and copper roofs, *Corros. Sci.* 42:1471 (2000).
123. W. He, I. Odnevall Wallinder, and C. Leygraf, A laboratory study of copper and zinc runoff during first flush and steady-state conditions, *Corros. Sci.* 43:127 (2001).
124. S. Bertling, I. Odnevall Wallinder, D. Berggren Kleja, and C. Leygraf, Long term corrosion-induced copper runoff from natural and artificial patina and its environmental fate, *Environm. Toxic. Chem.* 25:891 (2006).
125. B. Bahar, G. Herting, I. Odnevall Wallinder, K. Hakkila, C. Leygraf, and M. Virta, The interaction between concrete pavement and corrosion-induced copper runoff from buildings, *Environ. Mon. Assess.* 140:175 (2008).
126. S. Bertling, I. Odnevall Wallinder, C. Leygraf, and D. Berggren Kleja, Occurrence and fate of corrosion-induced zinc in runoff water from external structures, *Sci. Tot. Environ.* 367:908 (2006).
127. I. Odnevall Wallinder, S. Bertling, D. Berggren Kleja, and C. Leygraf, Corrosion-induced release and environmental interaction of chromium, nickel and iron from stainless steel, *Water Air Soil Poll.* 170:17 (2006).
128. Regulation (EC) No 1907/2006 of the European Parliament and of the Council of 18 December 2006 concerning the Registration, Evaluation, Authorisation and Restriction of Chemicals (REACH), OJ L 396:1–849 (2006).

129. G. Herting, S. Goidanich, I. Odnevall Wallinder, and C. Leygraf, Corrosion-induced release of Cu and Zn into rainwater from brass, bronze and their pure metals. A 2-year field study, *Environm. Monit. Asses.* 144:455 (2008).
130. L. A. Farrow, T. E. Graedel, and C. Leygraf, GILDES model studies of aqueous chemistry. II. The corrosion of zinc in gaseous exposure chambers, *Corros. Sci.* 38:2181 (1996).
131. J. Tidblad and T. E. Graedel, GILDES model studies of aqueous chemistry III. Initial SO_2-induced atmospheric corrosion of copper, *Corros. Sci.* 38:2201 (1996).
132. J. Tidblad and T. E. Graedel, GILDES model studies of aqueous chemistry V. Initial SO_2-induced atmospheric corrosion of nickel, *J. Electrochem. Soc.* 144:2676 (1997).
133. J. Tidblad, T. Aastrup, and C. Leygraf, GILDES model studies of aqueous chemistry VI. Initial SO_2/O_3- and SO_2/NO_2-induced atmospheric corrosion of copper, *J. Electrochem. Soc.* 152:B178 (2005).

16

Corrosion of Aluminum Alloys

Nick Birbilis
Monash University

T.H. Muster
CSIRO Materials Science and Engineering

Rudolph G. Buchheit
The Ohio State University

CONTENTS

16.1 Introduction....706
16.2 Metallurgical Aspects of Aluminum Alloys....707
16.2.1 Aluminum Production....707
16.2.2 Physical Metallurgy: Alloys and Tempers....707
16.2.2.1 Overview of Alloy Classes....710
16.2.3 Properties of Aluminum Alloys....712
16.2.4 Processing of Aluminum Alloys....713
16.2.4.1 Castings....713
16.2.4.2 Direct Chill Casting....713
16.2.4.3 Hot and Cold Rolling....714
16.2.4.4 Extrusion....714
16.2.4.5 Continuous Casting....715
16.3 Corrosion of Aluminum Alloys....715
16.3.1 Forms and Causes of Corrosion....716
16.3.1.1 General Corrosion....717
16.3.1.2 Pitting and Localized Corrosion....717
16.3.1.3 Bimetallic or Galvanic Corrosion....718
16.3.1.4 Dealloying....719
16.3.1.5 Crevice Corrosion....721
16.3.1.6 Filiform Corrosion....721
16.3.2 Effects of Microstructure on Corrosion....722
16.3.3 Intergranular Forms of Corrosion....724
16.3.4 Environmentally Assisted Cracking....725
16.3.4.1 Stress Corrosion Cracking....725
16.3.4.2 Liquid Metal Embrittlement....726

16.3.4.3 Corrosion Fatigue Cracking 727
16.3.4.4 Hydrogen Embrittlement 728
16.3.4.5 Environmental Influences upon Corrosion 728
16.4 Corrosion Preventions Strategies 729
16.4.1 Inorganic Coatings 729
16.4.2 Organic Coatings 730
16.4.3 Inhibitors 731
16.4.4 Specialized Coatings 731
Acknowledgments 732
References 732

16.1 Introduction

Aluminum (Al) has only recently completed its first century of commercial production [1]. The key to its extensive use today is its corrosion resistance, strength:density ratio, toughness, and its versatility that makes it suitable for a wide range of applications (from household foil and beverage cans to essential construction material for aircraft and space vehicles). Transportation, largely aerospace applications, has provided the greatest stimulus for alloy development and corrosion research that continues today [2]. Al and its alloys offer a diverse range of desirable properties that can be matched precisely to the demands of each application by the appropriate choice of composition, temper, fabrication, and processing mode [3]. It is increasingly the metal chosen for reducing the weight and hence emissions from the world's vast and rapidly expanding fleet of vehicles, making it the commodity "light metal."

Al as an engineering material ranks in tonnage use only behind ferrous alloys and growth in production has steadily increased. The global tonnage in 2007 was 60 million tons, of which 37 million tons was provided by primary production and 23 million tons by recycled scrap [4]. In regard to the history of Al and its production from ore, readers are directed to dedicated texts [5,6]. Al is produced using the Hall–Héroult process, involving electrolysis of aluminum oxide dissolved in cryolite, which in 1886 was used independently by Charles Martin Hall in the United States and Paul Héroult in France, and it remains the basis for production today.

From a corrosion perspective, Al has been a successful metal. In the past half century, major corrosion issues addressed include the localized corrosion of aluminum alloys containing magnesium, stress corrosion cracking (SCC) of alloys used in aerospace applications, galvanic corrosion of Al in architectural and automotive applications, and most recently the filiform corrosion of painted Al sheet in both architectural and automotive applications. Some of the present Al corrosion challenges are the ramifications from the elimination of chromates, the tolerance of increased impurity levels due to the increased use of recycled metal, the integration of Li into alloys while retaining corrosion resistance, and the sensitization of non-heat-treatable Al alloys. This chapter principally focuses on the material and surface electrochemistry of Al alloys, with lesser emphasis on the surface processing and coatings, which is a dedicated field in itself.

16.2 Metallurgical Aspects of Aluminum Alloys

16.2.1 Aluminum Production

Bauxite production has increased from 144 million tons worldwide in 2002 to 178 million tons in 2006. Recent growth in demand for Al has led to a predicted increase in annual demand by ~20% from 2006 to 2011. Typically four tons of bauxite is used to produce two tons of alumina, which then produces one ton of Al. The industry average emission associated with primary Al production is 9.73 kg CO_2e kg^{-1} with 55% of this from electricity generation [4]. Recycling of Al requires 95% less energy than that required for primary Al production and recycling of used Al products generates only 0.5 kg of CO_2e kg^{-1} of Al produced [7].

However, in order to meet the mechanical and corrosion performance requirements of many alloy and product specifications, much of the recycled metal must be blended or diluted with primary metal to reduce impurity levels. The result is that, in many cases, recycled metal tends to be used primarily for lower grade casting alloys and products [8]; however, with ~30% of Al being produced from recycled material, the future ramifications for corrosion will need to be addressed.

16.2.2 Physical Metallurgy: Alloys and Tempers

The properties of aluminum alloys (mechanical, physical, and chemical) depend on alloy composition and microstructure as determined by casting conditions and thermomechanical processing. While certain metals alloy with Al rather readily [9], comparatively few have sufficient solubility to serve as major alloying elements. Of the commonly used alloying elements, magnesium, zinc, copper, and silicon have significant solubility, while a number of additional elements (with less that 1% total solubility) are also used to confer important improvements to alloy properties. Such elements include manganese, chromium, zirconium, titanium, and scandium [2,10].

The low yield strength of pure Al (~10 MPa) mandates strength increase by alloying for subsequent engineering applications. The simplest strengthening technique is solution hardening, whereby alloying additions have appreciable solid solubility over a wide range of temperatures and remain in solution after any thermal cycles. Solid solution strengthening can lead to strength increases of about a factor of three.

The most significant increase in strength for aluminum alloys is derived from age hardening, which can result in strengths beyond 800 MPa. Age hardening requires a decrease in solid solubility of alloying elements with decreasing temperature. The age-hardening process can be summarized by the following stages: solution treatment at a temperature within a single phase region to dissolve the alloying element(s), quenching or rapid cooling of the alloy to obtain what is termed a supersaturated solid solution, and decomposition of the supersaturated solid solution at ambient or moderately elevated temperature to form finely dispersed precipitates.

The fundamental aspects of decomposition of a supersaturated solid solution are complex [11–13]. Typically, however, Guinier-Preston (GP) zones and intermediate phases are formed as precursors to the equilibrium precipitate phase [10]. (Figure 16.1 reveals a typical micrograph showing precipitate particles.)

Properties can be enhanced further by careful thermomechanical processing, which may include heat treatments like duplex aging and retrogression and reaging. Maximum

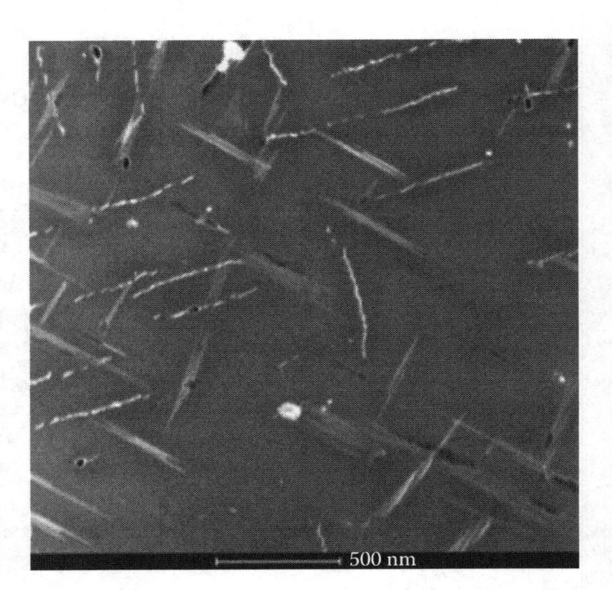

FIGURE 16.1
Dark field scanning transmission electron micrograph of coarse Al_2CuMg precipitate particles in an Al–Cu–Mg alloy. Imaged down the <100> zone axis. (Image courtesy of Nick Birbilis.)

hardening in commercial alloys is often achieved when the alloy is cold worked by stretching after quenching and before aging, increasing dislocation density, and providing more heterogeneous nucleation sites for precipitation.

Yield strength increases may also be achieved by grain refinement, exploiting the Hall–Petch relationship [2]. Grain refinement in aluminum alloys is achieved by additions of small amounts of low-solubility additions such as Ti and B to provide grain nuclei and by recrystallization control using dispersoids—trace alloying additions of Cr, Zr, or Mn to promote submicron-sized insoluble particles that subsequently can restrict or pin grain growth.

The microstructures developed in aluminum alloys are complex and incorporate a combination of equilibrium and nonequilibrium phases. Typically commercial alloys can have a chemical composition incorporating as many as 10 alloying additions. It is prudent, from a corrosion point of view, to understand the effect that impurity elements have on microstructures. While not of major significance to alloy designers, impurity elements such as Fe, Mn, and Si can form insoluble compounds. These constituent particles are comparatively large and irregularly shaped with characteristic dimensions ranging from 1 to ~10 μm. These particles are formed during alloy solidification and are not appreciably dissolved during subsequent thermomechanical processing. Rolling and extrusion tend to break up and align constituent particles within the alloy. Often constituents are found in colonies made up of several different intermetallic compound types. Because these particles are rich in alloying elements, their electrochemical behavior is often significantly different to the surrounding matrix phase. In most alloys, pitting is associated with specific constituent particles present in the alloy [14,15]. A range of alloying elements are found in constituent particles (typical example seen in Figure 16.2); examples include Al_3Fe, Al_6Mn, and Al_7Cu_2Fe.

The International Alloy Designation System (IADS) gives each wrought alloy a four-digit number, of which the first digit is assigned on the basis of the major alloying element(s), as is summarized in Table 16.1, along with the associated temper description.

FIGURE 16.2
SEM of constituent particles in AA7075-T651 imaged in backscattered electron mode. (Image courtesy of Nick Birbilis.)

TABLE 16.1

Wrought Aluminum Alloy and Temper Designations

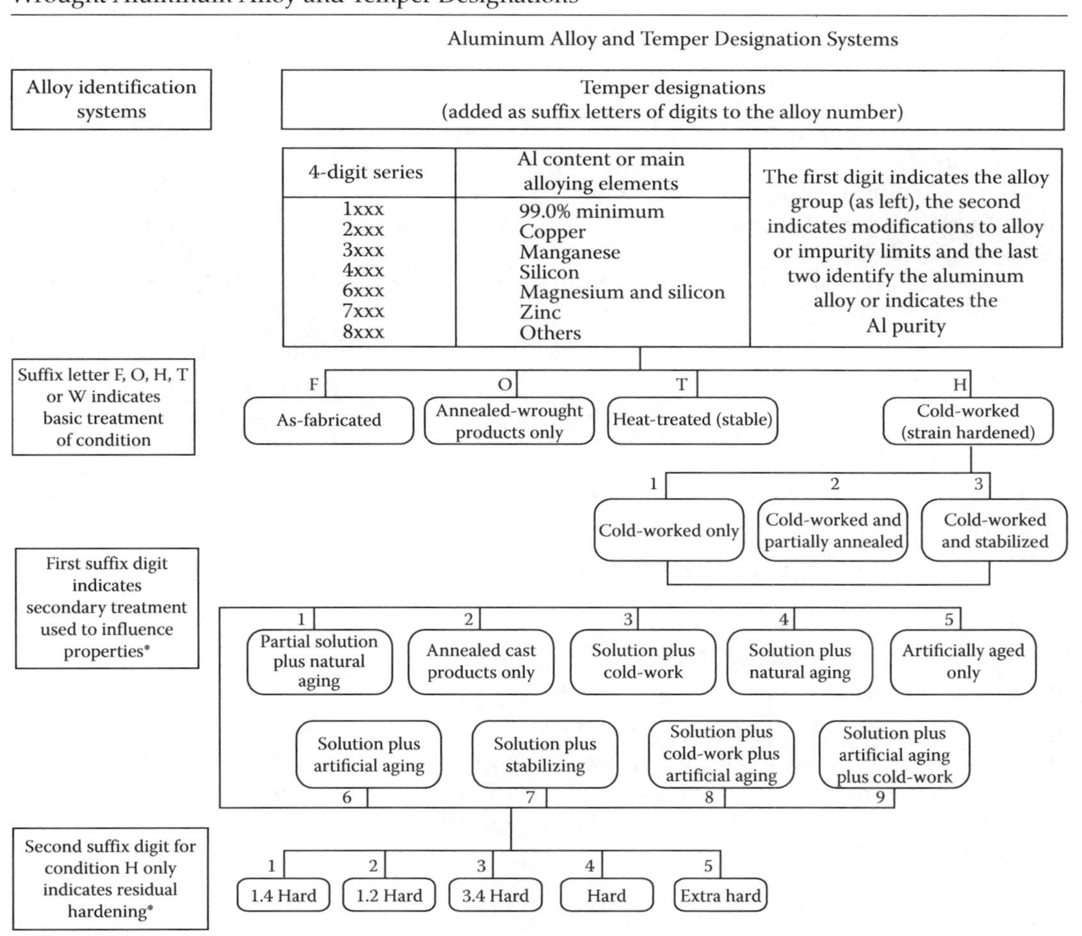

4-digit series	Al content or main alloying elements
1xxx	99.0% minimum
2xxx	Copper
3xxx	Manganese
4xxx	Silicon
6xxx	Magnesium and silicon
7xxx	Zinc
8xxx	Others

Sources: Adapted from Polmear, I.J., *Light Alloys: Metallurgy of the Light Metals*, 3rd edn., Arnold, London, U.K., 1995; Winkelman, G.B. et al., *Acta Mater.*, 55(9), 3213, 2007.

TABLE 16.2
Four-Digit System for Cast Aluminum Alloy Designations

Alloy Class	Designation
Al of =/>99.0%	1xx.x
Al + copper	2xx.x
+ Silicon (with copper and/or magnesium)	3xx.x
+ Silicon	4xx.x
+ Magnesium	5xx.x
(series is unused)	6xx.x
+ Zinc	7xx.x
+ Tin	8xx.x
+ Other elements	9xx.x

Sources: Adapted from Polmear, I.J., *Light Alloys: Metallurgy of the Light Metals*, 3rd edn., Arnold, London, U.K., 1995; Winkelman, G.B. et al., *Acta Mater.*, 55(9), 3213, 2007.

For cast aluminum alloys, alloy designations principally adopt the notation of the Aluminum Association System summarized in Table 16.2. The casting compositions are described by a four-digit system that incorporates three digits followed by a decimal. The .0 decimal indicates the chemistry limits applied to an alloy casting; the .1 decimal indicates the chemistry limits for ingot used to make the alloy casting and the .2 decimal indicates ingot composition but with somewhat different chemical limits. Some alloy designations include a letter. Such letters, which precede an alloy number, distinguish between alloys that differ only slightly in percentages of impurities or minor alloying elements (for example, 356.0, A356.0, and F356.0).

The temper designation system adopted by the Aluminum Association is similar for both wrought and cast aluminum alloys.

16.2.2.1 Overview of Alloy Classes

16.2.2.1.1 Pure Aluminum

Corrosion resistance of Al increases with increasing metal purity. The use of 99.8% and 99.99% grades is usually confined to those applications where very high corrosion resistance or ductility is required. General-purpose alloys for lightly stressed applications such as cooking utensils are approximately 99% pure Al. These alloys are known as the 1xxx series.

16.2.2.1.2 Copper and Copper–Magnesium-Containing Alloys

Copper is one of the most common of the alloying additions to Al, since it has both good solubility and a significant strengthening effect by its promotion of age-hardening response. Copper is added as a major alloying element in the 2xxx series alloys. These alloys were the foundation of the modern aerospace construction industry and, for example AA2024 (Al–4.4Cu–1.5 Mg–0.8 Mn), can achieve strengths of up to 520 MPa depending on temper [1].

16.2.2.1.3 Manganese-Containing Alloys

Manganese has a relatively low solubility in Al but improves its corrosion resistance in solid solution. Additions of manganese of up to 1% form the basis for an important series of non-heat-treatable (3xxx series) wrought alloys, which have good corrosion resistance, moderate strength (i.e., AA3003 tensile strength ~110 MPa), and high formability.

16.2.2.1.4 Magnesium-Containing Alloys

Magnesium has significant solubility in Al and imparts substantial solid solution strengthening and improved work hardening characteristics. The 5xxx series alloys (containing <7% Mg) do not age harden. Nominally, the corrosion resistance of these weldable alloys is good and their mechanical properties make them ideally suited for structural use in aggressive conditions. These alloys are used both for boat and ship building. Fully work hardened AA5456 (Al–4.7 Mg–0.7 Mn–0.12Cr) has a tensile strength of 385 MPa.

16.2.2.1.5 Silicon and Magnesium–Silicon-Containing Alloys

Silicon additions alone can lower the melting point of Al while simultaneously increasing fluidity, which is largely the basis of Al casting alloys and the associated shape casting industry. These alloys are increasing in importance in automotive applications for engine and drive train components. Corrosion issues with Al casting alloys are rare or at least rarely reported. The heat-treatable Al–Mg–Si are predominantly structural materials, all of which have a high resistance to corrosion, immunity to stress corrosion cracking (SCC) and are weldable. These 6xxx series alloys are mainly used in extruded form, although increasing amounts of automotive sheets are being produced. Magnesium and silicon additions are made in balanced amounts to form quasi-binary Al–Mg_2Si alloys, or excess silicon additions are made beyond the level required to form Mg_2Si. Alloys containing magnesium and silicon in excess of 1.4% develop higher strength upon aging.

16.2.2.1.6 Zinc and Zinc–Magnesium-Containing Alloys

The Al–Zn–Mg alloy system provides a range of commercial compositions, primarily where strength is the key requirement. Al–Zn–Mg–Cu alloys have traditionally offered the greatest potential for age hardening and as early as 1917 a tensile strength of 580 MPa was achieved; however, such alloys were not suitable for commercial use until their high susceptibility to SCC could be moderated. Military and commercial aerospace needs following World War II led to the introduction of a range of high strength aerospace alloys of which AA7075 (Al–5.6Zn–2.5 Mg–1.6Cu–0.4Si–0.5Fe–0.3 Mn–0.2Cr–0.2Ti) is perhaps the most well known, partly superseded by AA7150 in recent years. The high strength 7xxx series alloys derive their strength from the precipitation of η-phase ($MgZn_2$) and its precursor forms. The heat treatment of the 7xxx series alloys is complex, involving a range of heat treatments that have been developed to balance strength and SCC performance [16].

16.2.2.1.7 Lithium-Containing Alloys

Lithium is soluble in Al to about 4 wt% (which corresponds to about 16 at%). As these alloys of high specific strength and stiffness readily respond to heat treatment, research and development has intensified due to their potential for widespread usage in aerospace applications.

Recent studies have focused on Al–Cu–Li, Al–Li–Mg, and Al–Li–Cu–Mg alloys (including the use of minor Ag additions). These alloys derive their strength from age hardening, involving intermetallics such as Al_2CuLi and S′ phase, resulting in strengths in excess of 700 MPa [17]. Based on the impressive lightweight of such alloys, the new Airbus A380 and the Boeing Dreamliner are expected to be comprised of some portion of Al–Li based alloys (produced by Alcoa in the United States). These alloys also have potentially the lowest corrosion resistance of all aluminum alloys, where susceptibility to intergranular corrosion (IGC) is the most challenging present issue.

16.2.3 Properties of Aluminum Alloys

The basic physical properties for Al are given in Table 16.3.

The diverse and exacting technical demands made on aluminum alloys in different applications are met by the considerable range of alloys available, each of which has been designed to provide various combinations of properties. These include strength/weight ratio, corrosion resistance, workability, castability, high-temperature properties, etc., as seen in Table 16.4.

TABLE 16.3

Properties of Aluminum

Atomic number	13
Atomic weight	10.0
Atomic volume	26.97
Valency	3
Crystal structure	Face-centered cubic
Interatomic distance	2.863 Å
Electrochemical equivalent	0.3354 g Ah^{-1}
Density at 293 K	2700 kg m^{-3}
Melting point	931 K
Specific heat at 293 K	896 J kg^{-1} K
Latent heat of fusion	387 kJ kg^{-1}
Coefficient of linear expansion (293–393 K)	0.61×10^{-6} m K^{-1}
Thermal conductivity at 273 K	214 W m^{-1} K
Electrical volume resistivity at 293 K	2.7–3.0 μΩ cm
Electrical volume conductivity at 293 K	63%–57% IACS

TABLE 16.4

Properties of Selected Aluminum Alloys

Alloy	Temper	Wrought/Cast	Density (g cm^{-3})	Electrical Conductivity (% IACS)	Yield Strength (MPa)	Tensile Strength (MPa)	Elongation (%)
1199	O	W	2.71	60	10	45	50
3003	H14	W	2.73	50	145	152	8
5005	H38	W	2.70	52	200	186	5
5052	H38	W	2.68	35	290	255	7
2024	T861	W	2.77	30	490	517	6
6061	T6	W	2.7	43	276	310	12
7075	T6	W	2.80	22	503	572	11
	T73				434	503	13
201.0	T4	C (sand cast)	2.80	30	215	365	20
356.0	T51	C (sand cast)	2.69	41	140	175	2
413.0	F	C (die cast)	2.66	39	140	300	2.5

Sources: Adapted from Polmear, I.J., *Mater. Forum*, 28, 1, 2004; Grjotheim, K. and H. Kvande (eds.), *Introduction to Aluminium Electrolysis*, 2nd edn, Aluminium-Verlag, Dusseldorf, 1993; Winkelman, G.B. et al., *Acta Mater.*, 55(9), 3213, 2007.

16.2.4 Processing of Aluminum Alloys

The surface layers of aluminum alloys can be altered during processing and storage environments, which adds complexity to the surface finishing and corrosion performance of many aluminum alloys [18,19]. This section covers the major methods used in aluminum alloy production and comments on the main effects that influence corrosion performance. These effects include the formation of grain refined surface layers (GRSLs) during mechanical processing, the elongation of crystalline structure during rolling and extrusion, differences in surface roughness and porosity, and the segregation of specific alloying elements to the surface.

16.2.4.1 Castings

Cast products are usually produced in foundries from prealloyed metal supplied from secondary smelters (although certain high-performance castings are made from primary metal). The three most commonly use processes are sand casting, permanent mold casting, and die casting.

Sand molds are gravity fed whereas the metal moulds used in permanent mold casting are either gravity fed or by using air or gas pressure to force metal into the mold. Sand castings and permanent mold castings are made from alloys that respond to heat treatment.

In high-pressure die castings, parts up to approximately 5 kg are made by injecting molten aluminum alloy into a metal mold under substantial pressure using a hydraulic ram. High-pressure die castings contain high levels of porosity due to entrapped gas and are not easily welded or heat treated.

Despite their wide use, comparatively little work has been carried out to date to understand the corrosion behavior of aluminum casting alloys. Due to their high content of alloying elements like silicon, iron, and magnesium casting alloys may have a higher density of coarse intermetallic particles compared to wrought alloys. Processing parameters like cooling rate and pouring temperature, and even minor alloying element content variation, lead to significant changes in the microstructure of these alloys.

16.2.4.2 Direct Chill Casting

Direct chill (DC) casting is a semicontinuous process used for the production of rectangular ingots or slab for rolling to plate, sheet, foil, and cylindrical ingots or billet for extruded rods, bars, shapes, hollow sections, tube, wire, and rod. Most of production is from primary Al and process scrap or selected post consumer scrap. DC casting is the first step in the production of Al alloys prior to the value adding thermomechanical treatments, and while it may appear to be a topic not requiring discussion in such a chapter, it is important for the reader to realize that corrosion performance of an alloy is dictated by each step in the processing, from the very casting.

Before DC casting the melt is degassed, filtered, and grain refined. Casting starts by pouring molten metal into a water-cooled Al or copper mold. Accumulation of alloying elements at the surface can occur via the segregation processes where mobile elements diffuse from the bulk and from grain boundaries. Studies on binary alloys have provided insights into the kinetic and thermodynamic driving forces for segregation to surfaces, as enrichments have been correlated with the diffusion rates through molten aluminum [20]. In general, the surface-enriched elements have a high negative free energy for oxide formation and high diffusion coefficient through Al metal. Such elements include lithium, magnesium, and silicon. Other low melting point elements such as lead, indium, tin, and gallium can be problematic in recycled aluminum alloys. The segregation occurs during forming and heat treatments and has been demonstrated to influence the corrosion and wear properties of the alloys [21].

Thermodynamic considerations often fail to correctly predict the phase content and solid solution content of an as-cast microstructure because of the nonequilibrium nature of solidification during DC casting. This is important as alloy corrosion properties are controlled by solid solution levels and intermetallic phase crystallography and morphology, which depend on complex kinetic competitions for nucleation and growth. An understanding of the factors that govern phase selection in aluminum alloys under conditions of nonequilibrium solidification is important, since varying solidification conditions can lead to variations in secondary Al–Fe and ternary Al–Fe–Si phase contents at different positions in the casting, which in turn can lead to a degradation in the corrosion resistance of fabricated products [22]. Most importantly, the constituent particles do not appreciably dissolve during subsequent solution heat treatment and will hence persist.

16.2.4.3 Hot and Cold Rolling

Rolling blocks or slabs that are up to 30 ton in weight are heated to temperatures up to 400°C–500°C and passed through a reversing breakdown mill using heavy reductions per pass to reduce the slab gauge down to 15–35 mm. The slab surface undergoes intense shear deformation during this process and a GRSL is developed. The slab from the breakdown mill is then typically hotrolled on a multistand tandem mill down to a gauge of 2.5–8 mm. Hot rolling deforms the original cast structure and the as-cast grains are elongated in the rolling direction. The elongated microstructure developed during hot rolling can have a profound effect on corrosion properties like SCC and exfoliation corrosion. For example, the exfoliation corrosion of 7xxx alloys was shown to be due to manganese segregation during DC casting [23]. Milling and rolling processes have also been demonstrated to break up intermetallic particles, resulting in an increase in the number of phases that are susceptible to fracture [24,25].

The GRSL results from mechanical processing and is enhanced where rolling requires multiple passes. It contains a mixture of recrystallized fine metallic grains that incorporate oxide fragments and contain significant porosity [26,27]. Grain sizes in the range of 40 to 400 nm have been reported. The presence of GRSLs has been linked to the susceptibility of several aluminum alloys to filiform corrosion susceptibility. As a result, metal finishing and surface treatments are required to remove electrochemically active layers [28,29]. There are various reports on the depth of the modified surface region ranging from 1 to 8 μm. The GRSL is thought to be metastable as aging has been reported to lead to changes in the oxide composition and environment [30]. Subsequent heat treatment of the 3xxx alloys have been demonstrated to precipitate manganese-containing particles that renders the surface layer extremely surface active [31].

16.2.4.4 Extrusion

In Al extrusion, a hydraulic ram forces a preheated billet against a die and squeezes the metal through the die opening. Control of temperature and speed are important to avoid problems associated with overheating. However, most of these problems relate to surface defects seen during finishing operations such as anodizing rather than direct corrosion issues. Grain size and shape is important particularly if recrystallization does not occur and the as-cast grain structure is elongated in the extrusion direction. Partial recrystallization of an outer band of the extrusion can lead to a duplex structure with a predominately fibrous core and a thin recrystallized outer zone.

16.2.4.5 Continuous Casting

Al flat products may also be produced by twin roll casting. This effectively removes the requirement for DC casting, formal homogenization, and hot rolling. Molten Al is directly cast onto a water-cooled hollow steel roll. As solidification is more rapid in continuous casting, higher levels of supersaturation are achieved and the size of iron-bearing intermetallic phases is refined, which is largely beneficial from a corrosion perspective. However, surface quality is generally inferior to more conventionally cast and rolled strip, as the cast surface is not removed by scalping and variations in casting conditions can lead to surface streaks that persist at final gauge after cold rolling.

16.3 Corrosion of Aluminum Alloys

Al is a very reactive metal with a high affinity for oxygen. This is indicated from its position on the electromotive force series. The metal is nevertheless highly resistant to most atmospheres and chemicals. This resistance is due to the inert and protective character of the aluminum oxide film, which forms on the metal surface and which rapidly reforms if damaged. In most environments, therefore, the rate of corrosion of Al decreases rapidly with time.

The protective oxide film on Al attains a thickness of about 10 Å on freshly exposed metal in seconds. Oxide growth is modified by impurities and alloying additions and is accelerated by increasing temperature and humidity (or immersion in water). The protective oxide film inhibits corrosion because it is both resistant to dissolution and a good insulator that prevents electrons produced by oxidation of the metal.

The oxidation of Al at room temperature is reported to conform to an inverse logarithmic equation [32]. At elevated temperatures, oxidation studies over shorter periods illustrate conformity to parabolic, linear, and logarithmic relationships according to time and temperature. These kinetic variations are attributed to different mechanisms of film formation [33,34].

Corrosion of Al is an electrochemical process that involves the dissolution of metal atoms, only taking place once the oxide film has been dissolved or damaged. Al is amphoteric in nature, its oxide film being stable in neutral conditions, but soluble in acidic and alkaline environments. The thermodynamic stability of aluminum oxide films is expressed by the potential versus pH diagram seen in Figure 16.3 [35].

This diagram indicates the theoretical circumstances in which Al should corrode (forming Al^{3+} at low pH values and AlO_2^- at high pH values), passivity due to hydrargillite, i.e., $Al_2O_3 \cdot 3H_2O$ (at near-neutral pH values) and immunity (at very negative potentials). The nature of the oxide varies according to temperature, and above about 75°C, boehmite ($Al_2O_3 \cdot H_2O$) is the stable form. It should be noted that the potential–pH diagram does not

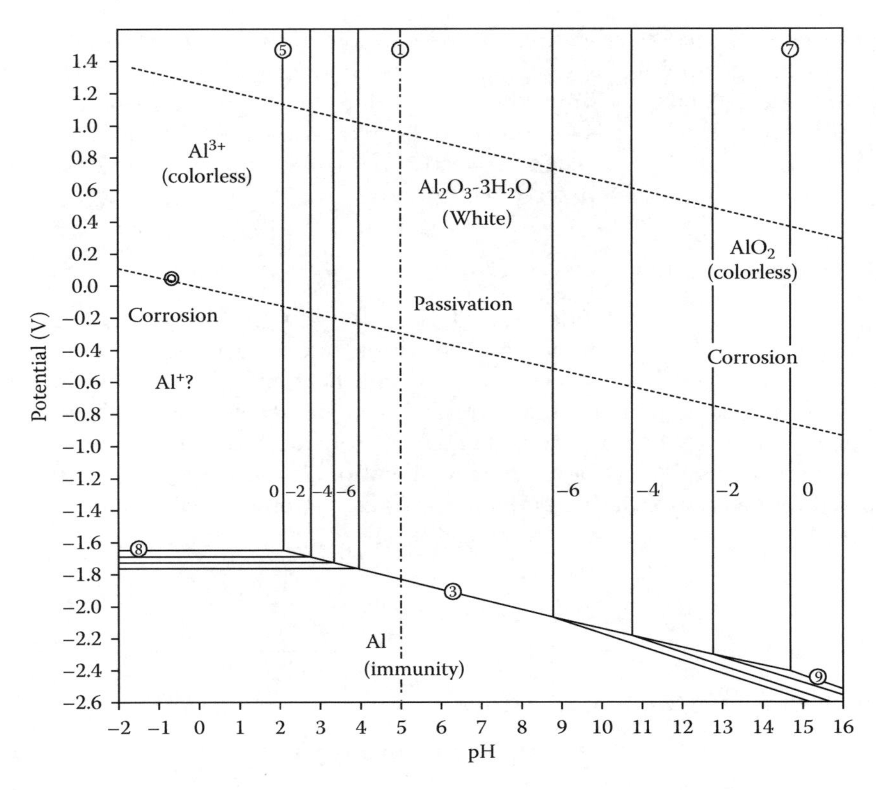

FIGURE 16.3
Potential–pH (Pourbaix) diagram for pure Al.

indicate an important property of Al, i.e., its ability to become passive in strongly acid solutions of high redox potential such as concentrated nitric acid, which can produce a pseudo passive Al-nitrate film that can protect the Al from further attack [36]. In addition, it must also be emphasized that in a practical sense, Al does not really display a region of immunity per se. This was classically shown by Perrault [37], who calculated a modified Pourbaix diagram accounting for the possibility of hydride formation. Such hydride formation will compromise immunity, and furthermore, the cathodic reaction upon the surface of Al at such negative potentials is also likely to increase pH such that a regime of Al dissolution evolves. As a result, in a practical context, the immunity region of Al would unlikely be able to be realized.

16.3.1 Forms and Causes of Corrosion

Corrosion is an electrochemical process and hence the corrosion potentials achieved at the surface of different aluminum alloys is of considerable importance. While the major forms of corrosion are covered elsewhere in this monograph, it is noted that Al alloys are susceptible to all forms of corrosion, and hence a brief mention of these manifestations is given in the context of Al. Also, the difference between the potential of aluminum alloys and other metals is important, as is the relationship between the potential of microstructural constituents of a single alloy. The major forms of corrosion of Al are as follows:

16.3.1.1 General Corrosion

As a general rule, dissolution occurs in strongly acid or strongly alkaline solutions (as per the Pourbaix diagram), but there are specific exceptions. For instance, in concentrated nitric acid, the metal is passive and the kinetics of the process are controlled by ionic transport through the oxide film. Also, inhibitors such as silicates permit the use of some alkaline solutions (up to pH 11.5) to be used with Al. Even where corrosion may occur, Al may be preferred to other metals because its corrosion products are colorless.

16.3.1.2 Pitting and Localized Corrosion

This is the most commonly encountered form of corrosion for aluminum alloys. In certain near-neutral aqueous solutions, a pit once initiated will continue to propagate into the metal due to acidification of the solution within the pit. The acidic conditions limit the formation of protective alumina films that usually prevent pit growth [38].

Pitting arises when localized/aggressive environments break down the nominally passive and corrosion-resistant film. Pits may form at scratches, mechanical defects, second phase particles, or stochastic local discontinuities in the oxide film. Since by definition pitting refers to local loss passivity, pitting only occurs in the near-neutral pH range for Al.

Solutions containing chlorides are very harmful and can create local corrosion potential "drops" between the metal surface and occluded regions (i.e., within a pit) where the chloride is concentrated or accumulated. Chlorides facilitate the breakdown of the film by forming $AlCl_3$, which is also usually present in the solution with the pits. When aluminum ions migrate away from pits, alumina precipitates as a membrane, further isolating and intensifying local acidity, and sustained pitting of the metal results. While the shape of pits can vary rather significantly depending on alloy type and environment, pit cavities are nominally hemispherical. This distinguishes pits from other forms of corrosion such as intergranular or exfoliation corrosion. Pitting is strongly influenced by alloy type and microstructure as discussed later in this chapter.

Overall, Al alloys display the hallmark/classical aspects of pitting; by this the authors refer to the following: (1) the Al-oxide is soluble in the presence of halide ions, namely chloride, and the formation of $AlCl_3$ is favored, causing localized attack of the oxide film. (2) When pitting occurs, the acidic conditions that evolve make the process autocatalytic, since Al readily dissolves at low pH (unlike other alloys, like say, the corrosion-resistant classes of stainless steel or nickel). (3) Regions of high cathodic activity will cause local alkalinity increase, which renders the Al susceptible to local dissolution where the pH is alkaline (this is obviously not the case for Fe or Zn where passivation occurs at high pH). As a result of this, Al is highly prone to all forms of localized attack with respect to other metal systems.

A review of certain aspects of pitting, specifically for Al, was given by Szklarska-Smialowska [39,40]. This review highlighted that from a detailed mechanistic point of view, the processes (on the atomic level) that lead to breakdown of the film due to halide interaction were presently not understood in sufficient detail. However, different analytical techniques including secondary ion mass spectrometry (SIMS), x-ray crystallography (XPS), and autoradiography have revealed an adsorption of Cl^- upon passive films of Al.

In near-neutral pH environments, pitting potentials for Al and its alloys have been readily and reproducibly measured by many investigators [38,41–43]. However, it is also noteworthy to mention that the potentiodynamically determined pitting potential yields no information regarding the number and size of pits that may form upon a given alloy.

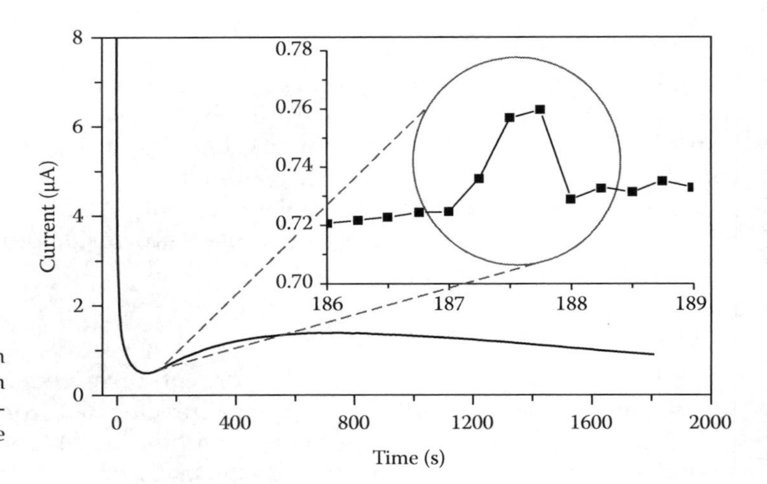

FIGURE 16.4
Current versus time record for an experimental Al–Cu–Mg alloy in deaerated 0.1 M NaCl held potentiostatically at $-0.615\,V_{SCE}$. (Image courtesy of Kevin D. Ralston.)

It is well known that Cu has the ability to ennoble the pitting potential; however, it is also capable of increasing the number of pit sites that ultimately occur.

In order to study Al pitting in more detail and to discriminate pitting statistics between alloys, certain researchers have recently studied current oscillations at a constant anodic potential below the pitting potential [40,44–48]. The occurrence of these oscillations is explained by the formation and repassivation of nano/micropits; termed "metastable" pits. These investigations have been largely carried out to understand the processes that lead to the formation of stable pits. A typical current versus time record is shown in Figure 16.4. The reason why this is mentioned herein is that this is an emerging area of research, and is differentiated to like measurement, on say, stainless steels [49] by the fact that significantly more pits per area (per unit time) are measured upon Al, which necessitates computer-assisted peak counting.

Recent studies have shown that the surface of rolled, ground, or machined aluminum alloys have deformed surface layers that range in thickness from 100 to 200 nm up to several microns [50–52] and that the presence of these layers has a strong effect on the initiation of pitting corrosion (and other forms of corrosion). These deformed layers are characterized by ultrafine grains formed due to high levels of shear strain at the aluminum alloy surface. The surface layer grains are 50–100 nm in diameter and are stabilized by oxide particles on their boundaries and are more susceptible to corrosion than the underlying bulk alloy. Deformed layers on AA3005 and similar architectural alloys are activated by the preferential precipitation of manganese-containing dispersoids, whereas the nucleation of precipitates during aging give rise to similar layers on AA6016 and AA6111 automotive and AA7075 aerospace alloys.

16.3.1.3 Bimetallic or Galvanic Corrosion

Al is anodic to most other metals when coupled in an electrolyte, with the exception of only Mg for the structural metals. While galvanic corrosion is largely an engineering design issue and covered elsewhere in this monograph, some small mention will be made here owing to the Al alloys being significantly less noble and unique compared to other systems. Table 16.5 shows the corrosion potentials for a range of nonheat-treatable (wrought), heat-treatable (wrought), and cast aluminum alloys with some other standard corrosion potentials based on measurements made according to ASTM G69 are shown

TABLE 16.5

Comparison of Measured Corrosion Potentials in 1 M NaCl Containing 9 ± 1 g L^{-1} H_2O_2

Alloy	Corrosion Potential (V_{SCE})
Al (99.999)	−0.75
Cr (99.9)	+0.23
Cu (99.999)	+0.00
Fe	−0.55
Mg	−1.64
Zn	−0.99
1100	−0.74
2014-T6	−0.69
2024-T3	−0.60
3003	−0.74
5052	−0.76
5154	−0.77
6061-T4	−0.71
6061-T6	−0.74
6063	−0.74
7039-T6	−0.84
7055-T77	−0.75
7075-T6	−0.74
7075-T7	−0.75
7079-T6	−0.78
8090-T7	−0.75

in Table 16.5. It is prudent, however, to consider this table only qualitative, as potential values may change with time and also between alloy manufacturers. Nonetheless, Table 16.5 is a useful guide for ranking of Al alloys.

Contact with steel can accelerate attack on Al, but in some natural waters and other special cases, Al can be protected at the expense of ferrous materials, particularly when the Al is "passive." Titanium appears to behave in a similar manner to steel. Stainless steel in contact with Al may increase attack on Al, notably in seawater or marine atmosphere, but the high electrical resistance of the two surface oxide films minimizes bimetallic effects in less aggressive environments. Where Al is coupled to copper, or exposed to metallic copper contamination (such as in water systems), corrosion of the Al is very rapid. This is because Cu is particularly efficient at supporting cathodic reactions (e.g., oxygen and water reduction). Limiting cathodic currents measured for pure copper are reported to be in the vicinity of 1.5 mA cm^{-2}, whereas limiting currents on pure Al are three orders of magnitude lower (0.5–1 μA cm^{-2}) [52].

16.3.1.4 Dealloying

Dealloying is analogous to galvanic corrosion but occurs where there is an accelerated/preferential loss of one or more elements from an intermetallic phase or from the wider alloy surface. Under electrochemical control, intermetallic phases have been demonstrated to dealloy as a result of both anodic and cathodic mechanisms. Anodic polarization can result in the generation of localized acidity, which is able to preferentially attack certain elements, removing them from a bulk phase. Conversely, cathodic polarization can remove elements susceptible to dissolution under alkaline conditions and noted in the previous section may also replate and concentrate noble metals on cathodic sites.

For aluminum alloys, copper enrichment at the interface has been of both scientific and industrial interest. One of the most studied examples of dealloying concerns the Al_2CuMg phase, which is commonly found in 2xxx series alloys and has been shown to exhibit less noble electrochemical potentials in comparison to the alloy matrix. Under neutral pH conditions magnesium and Al are preferentially dissolved from the Al_2CuMg phase, leaving a Cu-enriched and high surface area remnant, which then exhibits solution potentials noble to the matrix [53]. Figure 16.5 illustrates the enrichment of an Al_2CuMg particle in copper, its release into solution, and redeposition onto adjacent surface areas. Ultimately, the form that copper takes on the surface is thought to be important in determining the corrosion performance of alloys such as AA2024. The redistribution of copper has been demonstrated to enhance the kinetics of oxygen reduction processes and negatively affect corrosion. In dealloying from a bulk phase, the physical structure of a surface has been predicted by percolation theory to be dependent upon the dissolution rate and concentration of the noble elements in the phase [54,55]. Rapid dissolution rates lead to more porous network structures, where there is a possibility that unoxidized fragments enriched in the more noble metal will be released into solution, whereas slow dissolution allows surface diffusion and relaxation processes to maintain a stable surface structure. Also, if the noble metal content is sufficient, dealloying will not lead to an isolation of the percolation network. Theory suggests the copper concentration of 25 at% contained in the Al_2CuMg phase allows it to dealloy and form both porous copper-rich networks and also to release

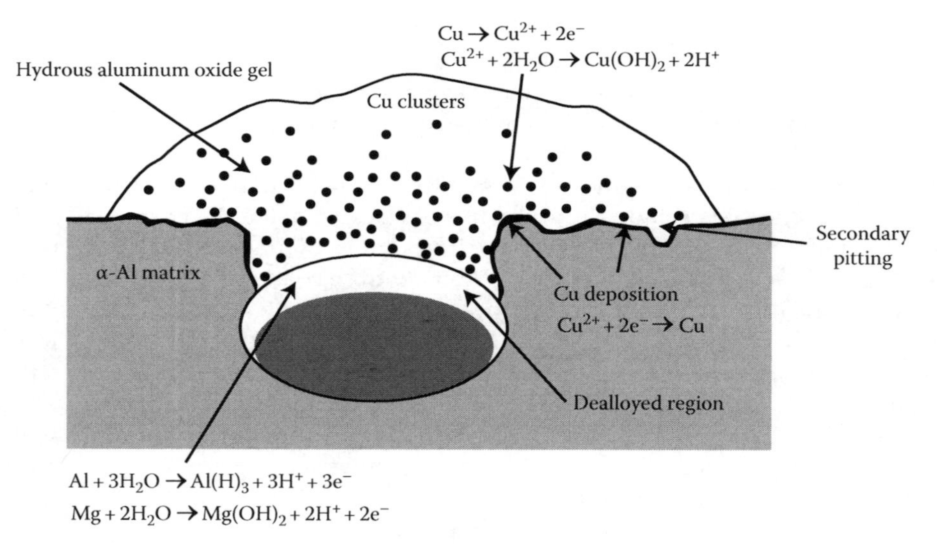

FIGURE 16.5
Schematic of dealloying of Al_2CuMg phase contained within an aluminum alloy matrix. Anodic polarization of particle results in preferential loss of Al and Mg. A Cu-rich network, which coarsens with age and is susceptible to breakup during hydrodynamic flow, releases small Cu-rich particles (diameter approximately 10–100 nm). The electrically isolated particles are dissolved, making Cu ions available for replating as elemental Cu onto cathodic sites. The replated cathodic sites serve as efficient local cathodes that stimulate secondary pitting. (After Buchheit, R.G. et al., *J. Electrochem. Soc.*, 147(1), 119, 2000.)

clusters of both oxidized and unoxidized copper into an electrolyte. It is also noted that hydrodynamic forces may assist in the release of fragments [56–59].

The accumulation of noble metals such as iron and copper at the surface of aluminum alloys is problematic even in the absence of specific intermetallic phases such as Al_2CuMg. Vukmirovic and coworkers showed that copper accumulation on the surface of AA2024 also arises from the copper in the aluminum solid solution. In a range of corrosive environments, Al has been shown to preferentially oxidize, resulting in the buildup of copper within a layer approximately 2–5 nm thick at the alloy surface [56,60,61]. The behavior of alloying elements in this sense is somewhat dependent upon the Gibbs free energy of oxide formation, which controls the enrichment of elements at the alloy surface and in the surface oxides during corrosion processes. Copper and other more noble elements (i.e., gold) have high Gibbs free energies of oxide formation per equivalent ($\Delta G°/n$) relative to that of alumina and therefore show extensive enrichment at the metal–oxide interface. In contrast, elements such as magnesium and lithium for example, have a lower $\Delta G°/n$ value than Al and, therefore, are more likely to appear in the oxide or electrolyte solution following corrosion processes [57].

In environments where aluminum alloys continually experience anodic dissolution, it has been suggested that alloys with a wide range of copper concentrations (0.06–26 at%) can display copper enrichment at the metal–oxide interface [62–64]. Once a certain level of enrichment occurs, copper atoms (and most likely other noble alloying elements) are thought to arrange themselves into clusters through diffusion processes and eventually protrude from the alloy surface due to undermining of the surrounding Al matrix [54,61]. These copper clusters may be released as elemental copper into the oxide layer by being undermined or copper ions may be oxidized directly from the protruding clusters. It has also been demonstrated that the level of copper enrichment is also influenced by grain orientation. In terms of general corrosion performance, the enrichment of copper at the alloy surface is also likely to increase the number of flaws that exist in the aluminum oxide.

16.3.1.5 Crevice Corrosion

If a crevice is formed between two Al surfaces, or between the surfaces of aluminum and a nonmetallic material (i.e., a polymer) localized corrosion can, and most often will, occur within the crevice in the presence of electrolyte. For Al alloys, mention is made here since crevice corrosion can be a very problematic form of damage in an engineering sense as real constructions that include welded lap joints, riveting, valve seats, or even deposits that arise in service [65,66] are common since Al is used extensively in the aerospace and automotive sectors. The general rules for the severity of crevice corrosion are presently under active research for several metal alloy systems, Al included [67]. Typically, in Al, tighter crevices lead to more rapid initiation of attack (owing to less electrolyte and a steeper oxygen concentration profile being achieved more rapidly).

16.3.1.6 Filiform Corrosion

Filiform corrosion may be considered as a specific type of differential aeration cell corrosion that occurs from defects where bare metal is exposed on painted or coated Al surfaces. It is worth of mention here, since as recently as the 1970s and 1980s, filiform corrosion was not considered as a damage mode upon Al. Much like crevice corrosion, filiform attack is driven by a differential aeration cell with an anodic head growing under a coating and a cathodic tail where oxygen is reduced. The filiform filaments are filled with corrosion products that can include alumina gel and partially hydrated corrosion products. Filiform corrosion is a corrosion issue for aluminum alloys in the aging aircraft sector. This is because aircraft are routinely painted for corrosion protection with polyurethanes and more complex coating systems [68–70]. Filiform corrosion however is not observed in cases where the Al has been anodized or conversion coated. Recent work has shown that susceptibility to filiform corrosion in architectural and automotive applications is due, in most if not all cases, to the presence of a deformed layer on the surface of the sheet.

16.3.2 Effects of Microstructure on Corrosion

Aluminum alloy microstructures are developed as a result of alloy composition and thermomechanical treatment. From a corrosion perspective, the dominant features of alloy microstructure are: grain structure and the distribution of second phase (intermetallic) particles as constituent particles, dispersoids, or precipitates. Such particles have electrochemical characteristics that differ from the behavior of the surrounding alloy matrix, making alloys susceptible to localized forms of corrosion attack that has been termed microgalvanic corrosion.

Several studies over the years have been carried out in order to assess the effect of intermetallic particles on the corrosion susceptibility of specific aluminum alloys [40,71–75]. In the 1990s, Buchheit collected the corrosion potential values for intermetallic phases common to aluminum alloys mainly in chloride-containing solutions [14]. More recently, various groups have focussed on the electrochemical properties of Fe-containing intermetallics [76,77], and Cu-containing intermetallics [78–80] and this has been expanded into a more comprehensive treatise covering a variety of common intermetallics present in commercial aluminum alloys (both wrought and cast) [38]. An abridged summary of the results of these studies is shown in Table 16.6. Again, such potentials are time-dependant; however, they serve as a very useful classification tool.

TABLE 16.6

Summary of Corrosion Potentials for Intermetallic Particles Common to Al Alloys

Phase	Corrosion Potential (mV_{SCE}) in 0.1 M NaCl
Al_3Fe	−539
Al_2Cu	−665
Al_6Mn	−779
Al_3Ti	−603
$Al_{32}Zn_{49}$	−1004
Mg_2Al_3	−1013
$MgZn_2$	−1029
Mg_2Si	−1538
Al_7Cu_2Fe	−551
Al_2CuMg	−883
$Al_{20}Cu_2Mn_3$	−565
$Al_{12}Mn_3Si$	−810
Al-2%Cu	−672
Al-4%Cu	−602

The identification and structural characterization of intermetallic particles present in aluminum alloys has been quantified by particle extraction techniques combined with electron probe microanalysis and by scanning and transmission electron microscopy combined with x-ray microanalysis.

Intermetallic particles in aluminum alloys may be either anodic or cathodic relative to the matrix. As a result, two main types of pit morphologies are typically observed. Circumferential pits appear as a ring around a more or less intact particle or particle colony and the corrosion attack is mainly in the matrix phase. This type of morphology arises from localized galvanic attack of the more active matrix promoted by the more noble (cathodic) particle as is shown in Figure 16.6, which also shows the phenomenon with an image collected via optical profilometry.

The second pit type of pit morphology is due to the selective dissolution of the constituent particle. Pits of this type are often deep and may have the remaining remnants of the particle in them (Figure 16.6 also shows this phenomenon). This morphology has been interpreted as particle fallout, selective particle dissolution in the case of electrochemically active particles, or in the case of some Cu-bearing particles, particle dealloying, and non-faradaic liberation of the Cu component.

Localized corrosion activity is however a complex phenomenon that is still under active research. Localized corrosion leads to local pH gradients as recently studied in detail

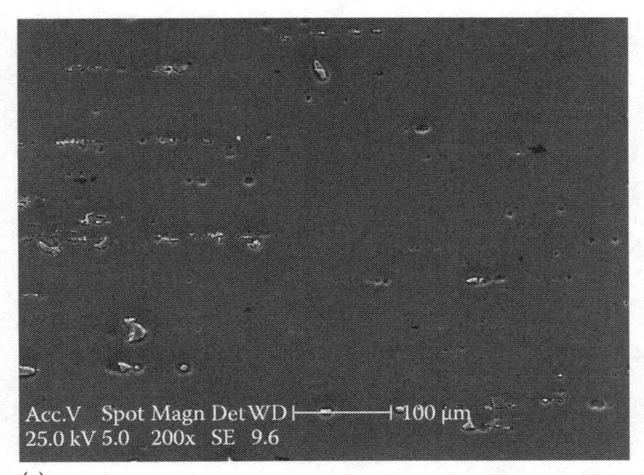

(a)

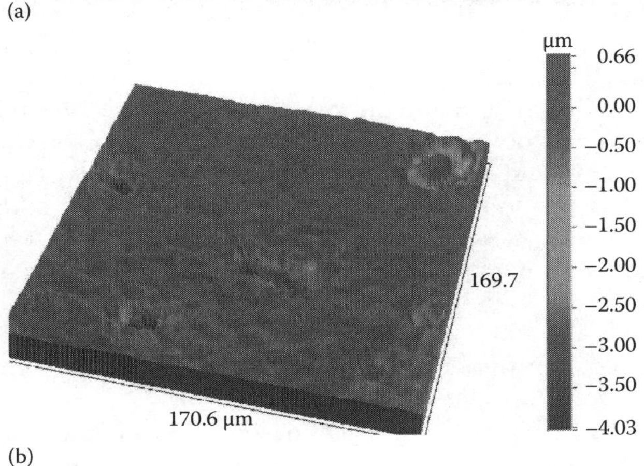

(b)

FIGURE 16.6
(a) SEM of early stage corrosion development upon AA7075-T651 immersed in quiescent 0.1 M NaCl for 1 day. (b) Optical profilometry image of AA7075-T651 immersed in quiescent 0.1 M NaCl for 1 day. (Images Courtesy of Mary K. Cavanaugh.)

by Ilevbare and Schneider [81,82]. Cathodic sites of enhanced oxygen reduction generate hydroxyl ions promoting local pH increase, which can then modify the subsequent rate and morphology of corrosion propagation. The precise morphology of particle-induced pitting is important for emerging damage accumulation models. For these models to be predictive, it is necessary to develop a comprehensive, self-consistent accounting of this type of pitting. In cases where the electrochemical characteristics of constituent particles have been rigorously characterized, they have been found to have much more complicated behavior than categorized by simple characterizations like "noble" or "active."

16.3.3 Intergranular Forms of Corrosion

IGC is a phenomenon and its precise mechanisms have been under debate for almost half a century [83]. While in a simple view we can consider IGC a special form of microstructurally influenced corrosion, IGC can be summarized as a process whereby the grain boundary "region" of the alloy is anodic to the bulk or adjacent alloy microstructure.

Corrosion activity may develop because of some heterogeneity in the grain boundary structure. In aluminum–copper alloys, precipitation of Al_2Cu particles at the grain boundaries leaves the adjacent solid solution anodic and more prone to corrosion. With aluminum–magnesium alloys, the opposite situation occurs, since the precipitated phase Mg_2Al_3 is less noble than the solid solution. Serious intergranular attack in these two alloys may however be avoided, provided that correct manufacturing and heat treatment conditions are observed.

In the case of the aluminum–magnesium system, almost all commercial alloys are supersaturated as the magnesium solubility at ambient temperatures is less than 1 wt%. This effectively means that for alloys with more than 3 wt% magnesium, elevated service temperatures can lead to grain boundary precipitation and sensitization of grain boundaries to an IGC susceptibility. The extent of this sensitization may be approximately deduced from the continuity of Mg_2Al_3 precipitation at the boundaries. Apparently continuous boundaries correspond to high levels of sensitization to IGC.

The level of sensitization to IGC of a fabricated sheet, plate, or extrusion can be easily determined by measurement of the NAMLT value (ASTM G67). This is a 24 h exposure to nitric acid and a measurement of weight loss due to IGC and loss of grains [84]. A NAMLT value of more than 30 g cm^{-2} of surface is necessary before an alloy has become sensitized to IGC or cracking. The test method used to evaluate susceptibility to IGC depends on alloy type. For 5xxx series alloys, the NAMLT method (ASTM G67) is adopted, while for 2xxx and 7xxx series alloys ASTM G110 is most common, employing testing in sodium chloride solutions containing hydrogen peroxide.

In the case of Al–Zn–Mg alloys, where the precipitated phase is the very anodic $MgZn_2$ (η) phase, IGC occurs rather readily. Again however susceptibility to IGC is strongly dependent on the heat treatment condition and its effect on grain boundary solute segregation and the morphology and composition of the grain boundary precipitate and the surrounding alloy matrix. [85]. The most resistant heat treatments are based on the use of overaging to the T7 treatment or more complex heat treatments that involve retrogression and reaging to minimize the trade-off between alloy strength and IGC resistance [2,3].

The images in Figure 16.7 help rationalize the origins of IGC using the Al–Zn–Mg system as a model.

IGC differs from pitting corrosion in that while IGC may initiate from a pit, its propagation penetrates more rapidly than pitting corrosion, and while both may have a deleterious

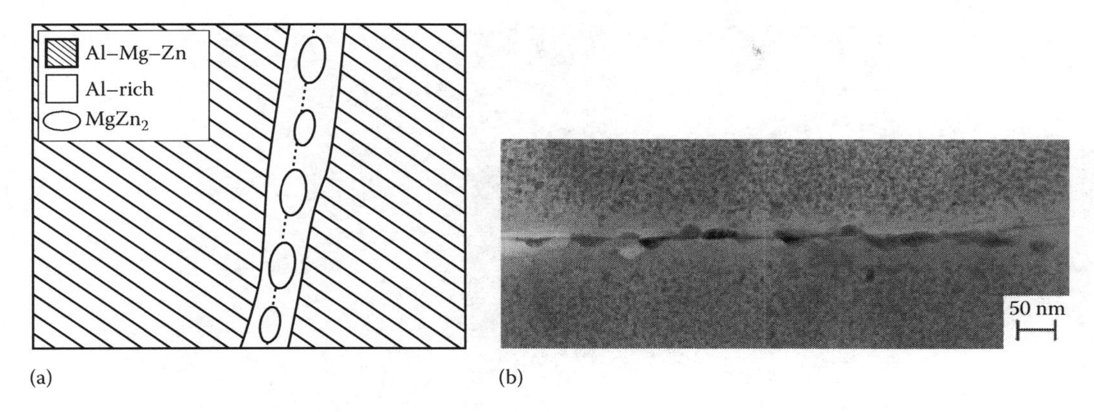

FIGURE 16.7
(a) Schematic of hypothetical grain boundary in an Al–Zn–Mg alloy. This schematic indicates the different chemistry that exists in the grain interior, solute depleted zone (precipitate free zone) and grain boundary precipitates, giving rise to electrochemical heterogeneity localized at the grain boundary region. (b) Conventional bright field TEM image of high angle grain boundary in AA7075-T651, revealing grain boundary precipitates ($MgZn_2$) and a distinguishable precipitate free zone. (Image courtesy of Steven P. Knight.)

effect on corrosion fatigue (CF), the sharper tips produced by intergranular attack are drastic stress concentrators, which may reduce the number of cycles to failure.

Exfoliation corrosion [3,86] of aluminum alloys is also frequently due to IGC. It generally occurs where the alloy microstructure has been heavily deformed (i.e., by rolling) and the grain structure has been flattened and extended in the direction of working. An IGC attack from transverse edges and pits then runs along grain boundaries parallel to the alloy Exfoliation is characterized by leafing off of layers of relatively uncorroded intragranular metal caused by the swelling of corrosion product in the layers of IGC. Exfoliation is observed on aircraft components, for example, around riveted or bolted components. Testing for exfoliation corrosion is carried out by a number of ASTM tests, including the acidified salt spray test (ASTM G85), the ASSET immersion test (ASTM G66), and most commonly the EXCO immersion test (ASTM G34).

16.3.4 Environmentally Assisted Cracking

Environmentally assisted cracking (EAC) is a term that incorporates SCC, liquid metal embrittlement (LME), CF, and hydrogen embrittlement (HE).

16.3.4.1 Stress Corrosion Cracking

SCC is a time-dependant intergranular fracture mode (in aluminum alloys) that requires the combined presence of a susceptible alloy, a sustained tensile stress and a corrosive environment.

The minimum tensile stress required to cause SCC in susceptible alloys is usually small and significantly less than the macroscopic yield stress [87,88]. Susceptibility to SCC has restricted the use of aluminum alloys, particularly 7xxx series alloys, to specific applications. There are several theories postulated for the mechanism of SCC. The main theories are either corrosion-dominated where cracking is due to preferential corrosion along the grain boundaries by anodic dissolution (i.e., analogous to IGC) or hydrogen dominated where cracking along grain boundaries is enhanced by absorbed atomic hydrogen.

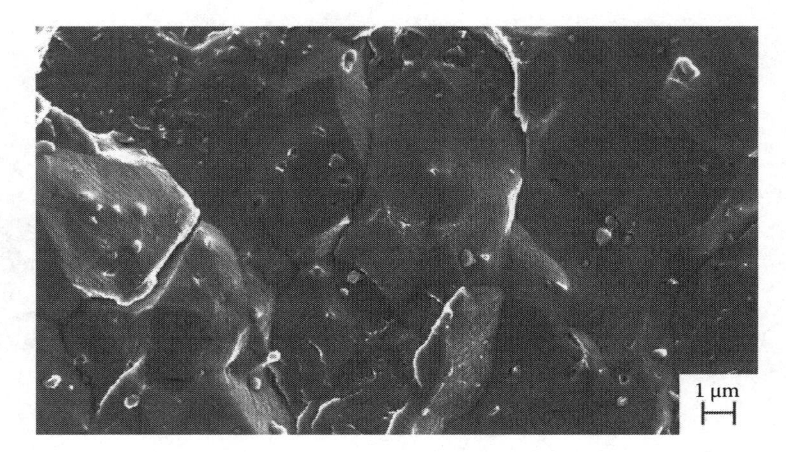

FIGURE 16.8
High-resolution SEM of an SCC fracture surface taken from an AA7079-T651 DCB specimen exposed to 75% relative humidity (specimen was from the T/6 position from 3 in. thick rolled plate). (Image courtesy of Steven P. Knight.)

The origin of this hydrogen is the result of IGC itself (so generated by corrosion in acidified cracks) and it is considered that the presence of absorbed hydrogen weakens grain boundaries.

Hence, SCC is a very important phenomenon for aluminum alloys, in order to do the topic justice, the reader is referred to specific monographs on the topic for a more detailed treatise of the SCC theories [89,90], along with the mechanical aspects of SCC. An example of intergranular SCC is seen in Figure 16.8.

SCC development depends on both the duration and magnitude of applied tensile stress. Fracture mechanics tools for the determination of crack growth rates are commonly used in the evaluation of SCC resistance for aluminum alloys [90]. Such tests suggest a minimum (threshold) stress intensity that is required for cracking to develop.

It is important to note that residual stresses in Al products that may arise as a result of quenching and cold working may play an important role in SCC should the level of residual stress be significant. SCC of 7xxx alloys occurs in water and water vapor in addition to chloride-containing electrolytes. Most other susceptible alloys fail only due to exposure to environments containing chloride ions.

Low strength and relatively pure aluminum alloys are not susceptible to SCC. The alloys most prone to SCC are the 7xxx, 2xxx, and higher strength 5xxx series alloys (i.e., those that have grain boundaries populated by precipitates). Most service failures involving SCC have occurred in the short transverse direction [91]. The relative resistance of certain aluminum alloys to SCC is tabulated in Table 16.7.

16.3.4.2 Liquid Metal Embrittlement

LME is not commonly observed in the general usage of aluminum alloys. It can be defined as a mode of attack that results in a complete loss of alloy ductility of a solid metal, well below the normal yield stress, as a result of the surface being wet by a liquid metal. LME-induced fractures are generally intergranular in aluminum alloys; an image of a LME fracture surface in mercury is shown in Figure 16.9.

TABLE 16.7

Qualitative SCC Resistance of Aluminum Alloys (in the Rolled Plate Form)

Alloy and Temper	SCC Resistance
2014-T3	Poor
2024-T3, T4	Poor
2024-T8	Good
2124-T851	Good
2219-T6, T8	Excellent
6061-T6	Excellent
7049-T73	Good
7x75-T6	Poor
7x75-T73	Excellent
7x75-T76	Intermediate

Sources: Adapted from Polmear, I.J., *Light Alloys: Metallurgy of the Light Metals*, 3rd edn, Arnold, London, 1995 (original source: U.S. Air Force Research Laboratory, Wright Patterson Air Force Base, OH); Davies, J.R. (ed.), *Corrosion of Aluminium and Aluminium Alloys*, ASM International, Materials Park, OH, 1999.

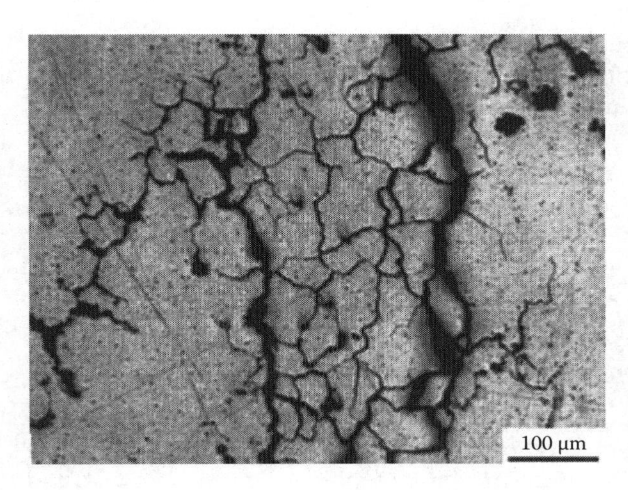

FIGURE 16.9
Intergranular failure of an Al alloy pressure vessel in a natural-gas plant due to LME by Hg. (Image courtesy of Stan P. Lynch (Australian Defence Force, not subject to copyright).)

It is well known that gallium in contact with Al can result in disintegration of the Al (or aluminum alloy) into individual grains, placing serious restrictions on Ga usage. In addition, mercury can also embrittle aluminum and its alloys. Al has also been noted as being embrittled appreciably by indium, sodium, tin–zinc, and lead–tin alloys.

16.3.4.3 Corrosion Fatigue Cracking

CF is the interaction of irreversible cyclic plastic deformation with localized corrosion activity. The combination of each of the mechanism, and the transition from initiation to propagation is a matter under present research and of considerable technological importance. Corrosion pits have been routinely observed to nucleate crack growth in structures subject to fatigue loading [92]. In fact, a recent paper reveals that in a fatigue study of aluminum alloy 7075-T6, all specimens in that study fractured from cracks associated with pitting [93]. Indeed the numerous studies referenced within Ref. [3] support the notion that pitting has a critical and detrimental effect on fatigue life. A fatigue crack can initiate from a corrosion pit or surface flaw when the flaw reaches a critical size at which the stress intensity factor reaches a threshold for fatigue cracking or conversely when the rate of fatigue crack growth exceeds that of pit growth. Again, since CF is a phenomenon that involves a mechanical component (and not corrosion alone), a significant amount of work exists in the mechanical area, including several technical publications, dedicated monographs, and handbooks [94–97].

CF phenomena are diverse and specific to the environment and particular application at hand; however, the variables that influence CF crack growth propagation are stress intensity, frequency of loading, stress ratio, alloy composition and microstructure (hence electrochemical potential), and the environment composition (and temperature).

16.3.4.4 Hydrogen Embrittlement

Hydrogen may readily dissolve in Al and all of its alloys, both in the molten state and during elevated temperatures close to the alloy melting temperature where the atmosphere may include water vapor or excessive hydrocarbons. Hydrogen from molten aluminum results in porosity in cast products but does not have an influence on corrosion performance or subsequent HE.

Recent works, however, are beginning to show there is experimental evidence that hydrogen generated during corrosion can penetrate into grain boundaries and lead to embrittlement [98–100]. The mechanism by which hydrogen causes the embrittlement is difficult to study and is posited to act by either facilitating enhanced dislocation emission ahead of an advancing crack or by bond weakening and enhanced plasticity ahead of an advancing crack.

Understanding HE is important in unraveling the mechanism and modes of failure processes for Al alloys; however, it thus far has not been a key factor that has restricted the usage of Al. No specific tests for assessing HE susceptibility of aluminum alloys exist.

16.3.4.5 Environmental Influences upon Corrosion

Aluminum alloys weather outdoors to a pleasant gray color. Superficial pitting occurs initially but gradually ceases, being least marked on high-purity Al. In a general sense, the atmospheric corrosion of Al is accelerated in the presence of moisture, sulfur dioxide, and ozone, where degradation models are alloy specific [101–103]. In urban areas, atmospheric fallout of carbon from partially burned fuel can cause severe localized pitting by galvanic action, although this is not commonly encountered. Indoors, aluminum retains its appearance well, and even after prolonged periods may show no more than slight dulling. This superficial deterioration can be accelerated by the presence of moist conditions and condensation that in extreme cases may lead to staining.

In order to enhance the durability of structural Al, anodizing provides a good level of protection. Anodized aluminum extrusions can survive even prolonged exposure in accelerated salt spray testing, indicating that the anodizing process where applicable is an industrially acceptable means of protection. With some alloys, including the copper-rich alloys and the medium-to-high-strength Al–Zn–Mg alloys, additional protection, e.g., painting, is desirable in the more aggressive atmospheres to avoid any risk of IGC.

Over many years, Alcan used the CLIMAT test classify industrial and marine atmospheres. In the late 1990s, it was estimated that Alcan had carried out over 13,000 individual CLIMAT tests globally and summarized versions of the test results have been published [104–106].

No corrosion occurs immediately upon the immersion of Al and its alloys in near-neutral freshwater and tap waters, and Al gives satisfactory service with all types of tap water provided regular cleaning and drying can take place. Selected aluminum alloys have excellent resistance to attack by distilled or pure condensate water, and are used in industry in condensing equipment and in containers.

Most acids are corrosive to aluminum-based materials. The oxidizing action of nitric acid at concentrations above about 80%, however, causes passivation of Al. Very dilute and very concentrated sulfuric acid dissolves Al only slowly.

Boric acid also exerts little attack on Al, while a mixture of chromic and phosphoric, or nitric acids can be used for the quantitative removal of corrosion products from Al without attacking the metal.

16.4 Corrosion Preventions Strategies

Where the natural passive oxide does not protect aluminum alloys in a sufficient manner for application, a range of technologies and multilayered protection schemes to confer corrosion protection exist. These employ the use of sacrificial, barrier, and inhibitive coatings. This section reviews the main approaches to give the reader completeness; however, the field is large in itself, and hence only an abridged account is given herein. Protection schemes nominally consist of an inorganic coating that can serve either as a coating itself, or function as a pretreatment or surface preparation step as part of a multilayered coating system (which would include an organic coating such as paint). In some environments, protection is provided by a solution phase inhibitor, or in advanced cases, with coatings that are functional (inhibitor doped).

16.4.1 Inorganic Coatings

The main two types of inorganic coatings are conversion coatings and anodized coatings.

Similar to steels, the conversion coating process for Al involves contacting the surface to be coated with an aqueous solution containing surface activators and coating-forming ingredients. Contact times in commercial processes range from seconds to minutes for certain aerospace applications, and solution contact with the surface can be applied by immersion, spraying, or rolling. In some instances, conversion coatings involve multiple steps (i.e., exposure to various chemicals in series) with commercial coating processes for Al well supported globally owing to the aircraft industry.

Conversion coatings are not as protective as anodized coatings. In most cases, conversion coated surfaces are subsequently primed or painted. Traditionally, pretreatment of Al has relied on chromate-based systems. There are two general classes of chromate conversion processes: the activated acidic formulations such as the chromium chromate processes that use a sodium fluoride-chromic acid chemistry, and alkaline oxide processes based on a sodium chromate-sodium hydroxide or carbonate chemistry.

Chromate conversion coatings are noted for their ability to self-heal. Self-healing refers to the ability of the coating to resist corrosion from scribes or defects in the coating. This phenomenon is attributed to the release of hexavalent chromium in the coating into an aggressive solution contacting the surface. As is now widely recognized, chromate is a human carcinogen and chromate-based coatings are banned in many countries. There continues a major worldwide initiative to develop chrome-free pretreatment systems for aluminum alloys. Such initiatives were reviewed in Refs. [107,108]; however, readers are directed elsewhere for a better account of emerging surface and coatings technology.

Anodizing can enhance corrosion protection of Al rather significantly (viz. anodized Al is a staple of the architectural industry) and can also improve adhesion of subsequently applied organic coatings. For this reason, it is a very widely used process. Anodized coatings are formed electrochemically in an aqueous solution that results in an aluminum oxide surface film. Anodized coatings on aluminum alloys are on the order of microns in thickness. Anodizing is carried out in an acidic bath that contains ingredients that promote formation of an adherent oxide film. To improve corrosion protection, anodized coatings can be sealed in a second step in the process.

16.4.2 Organic Coatings

A very broad range of organic coatings can be applied to aluminum alloys, and this is common when the application requires prolonged exposure to outdoor environments [109]. For many engineering applications, and certainly the most demanding ones (i.e., aerospace), a coating system consisting of multiple coating layers is used. Each layer contributes to the functionality of the overall coating system. Functions imparted may include adhesion promotion for subsequent layers, flexibility, impermeability, chemical resistance, abrasion resistance, electromagnetic signature control, and excellent corrosion protection.

The two strategies for imparting corrosion protection by organic coating are: barrier protection and active corrosion protection. Barrier protection is based on the concept of preventing contact of the underlying Al substrate with the environment. The vast majority of protective coatings systems used today are based on this concept. As elementary as such coatings may be, the concept is remarkably difficult to implement and sustain in practice. Coatings can contain defects that are formed during the application process. Defects can serve as corrosion initiation sites in service. Hence, a great effort is made to apply coatings on scrupulously clean surfaces under very well-controlled conditions. Coatings can, and often will, sustain damage in service. Damage can arise from mechanical action at the surface, viz. scratches. Damage can also arise by exposure to aggressive chemical agents. This can degrade the organic component of the coating destroying barrier properties. Ultraviolet radiation can also degrade the integrity of the organic coatings.

For active corrosion protection strategies, barrier properties of the coating are not necessarily important. Coatings that protect by this mechanism contain a releasable corrosion inhibitor and are often termed "smart-coatings." As moisture from the environment penetrates the coating, the corrosion inhibitor is released. In well-designed coatings, any moisture that reaches the coating–metal interface will have enough inhibitor in it so that the natural passivity of the underlying alloy is preserved. This protection strategy is not as widely employed as barrier protection, but strontium chromate–pigmented organic primer coatings used widely in aerospace applications depend critically on this protection mechanism.

16.4.3 Inhibitors

Corrosion inhibitors can be used to assist the corrosion resisting performance of aluminum and its alloys when exposed to aggressive aqueous environments. In most cases, an inhibitor is a chemical substance, soluble in water that slows corrosion. There are a range of chemicals that inhibit Al corrosion; however, the selection and delivery mechanism of inhibitors is in many cases application specific.

Soluble corrosion inhibitors act by limiting anodic reactions, cathodic reactions, or both. This gives rise to a useful scheme for classifying chemical inhibitors. For aluminum alloys, anodic inhibitors typically act to increase the pitting potential or suppress the onset of pitting in exposure testing. Even with good anodic inhibitors, pitting may occasionally occur if preexisting defects on the alloy surface exist. Additionally, anodic inhibitors may have no effect on slowing the growth of existing pits.

Cathodic inhibitors slow the rate of the oxygen reduction reaction, and hence the companion oxidation reaction must also slow. This results in an overall decrease in the corrosion rate, as well as a decrease in the free corrosion potential. For the best inhibitors, the decrease in the corrosion potential is usually to a value well below the alloy's pitting or repassivation potential. Cathodic inhibitors have the advantage of being able to improve corrosion resistance at low concentrations. For example, chromate added at micromolar concentrations to an aerated dilute chloride solution is enough to significantly reduce the rate of oxygen reduction.

Soluble inhibitors are usually ions in solution. Important inorganic anions that inhibit Al corrosion include: chromate, phosphate, permanganate, nitrate, vanadate, molybdate, tungstate, and silicate. Cations of strontium, cerium and the lanthanides, and zinc are inorganic cationic inhibitors. Organic substances that are inhibitors of Al corrosion include phosphonates, sulfonates, benzoates, thiols, azoles, and amines. Among these special attention is given to chromates. Chromate is a proven and powerful inhibitor of oxygen reduction and an excellent inhibitor of Al corrosion. Chromates are used across all industries as Al corrosion inhibitors. However, their use is becoming increasingly restricted over concerns for workplace safety and environmental pollution because chromates are human carcinogens.

Inhibitors can be incorporated into coatings in a variety of ways (i.e., inhibitor ions can be attached to reactive sites on coating resin polymers or directly applied to Al surfaces using an evaporable solvent).

16.4.4 Specialized Coatings

There are several "spray" coating technologies that can apply Al coatings onto other metals, or layers of other materials onto Al. These can be divided into two categories, thermal spray or cold spray, based on the need to heat the surface in order to ensure metallic bonding. Cold spray has the advantage over thermal spray methods that include pulsed thermal spray, plasma spray, and high-velocity oxygen fuel (HVOF) spray, as the particles are less likely to oxidize as heat treatment is avoided. Thermal methods have the advantage that the feedstock, usually $<20\,\mu m$ diameter particles, can melt during the process, quenching rapidly when impacting the substrate. Al substrate temperatures during such processes are reported to stay below 60°C.

Relatively thick Al coatings can be deposited onto metal and other selected substrates using cold gas sprayed (or cold spray) coatings. Cold spray accelerates small particles, typically $<50\,\mu m$ diameter to high velocity ($>700\,m\,s^{-1}$) using a supersonic gas jet. The impacting particles bond to the surface through plastic deformation, and it is thought that any previous oxides on the particle and substrate are broken up during impact. The morphology of the deposited coatings can vary significantly on the variables such as powder characteristics, impact velocity, etc. [110]. Properties must be optimized to achieve metallic bonding; otherwise, bonding may either be poor or reliant on mechanical interlocking and therefore suffer from adhesion problems. Alumizing layers can be applied using cold spray. Cold spray can be particularly useful for repairing damaged items, particularly high-cost components, which, for example, can be rebuilt back to original thicknesses.

Aluminum-based metallic glasses have been explored as a means to form coatings where high concentrations of desirable elements can be supported, where below the glass transition temperature (T_g) they form chemically homogeneous solid solutions. The coatings can potentially be designed to serve various functions; as a barrier to corrosion, to function as a sacrificial anode and to release soluble metal ions that act as corrosion inhibitors. Typical compositions might include a combination of transition elements and lanthanides, which have been demonstrated to have good ability to form aluminum-based metallic glasses with doping levels over 15% [108]. Such coatings may be deposited using HVOF, cold spray, and pulsed thermal spray.

Research into coating technologies and corrosion inhibition methodologies are shedding new light on the way that traditional protective systems work. This is allowing coatings to be developed to optimize functions such as galvanic protection, passivation, and to introduce self-healing mechanisms into coatings [111,112].

Acknowledgments

The authors gratefully acknowledge Julie Fraser (Monash University) for the careful production of diagrams, figures, and tables. Additionally, all those who provided material for the figures within the chapter are also very gratefully acknowledged.

References

1. I.J. Polmear, Aluminium alloys—A century of age hardening, *Materials Forum, 28*, 1–14 (2004).
2. I.J. Polmear, *Light Alloys: Metallurgy of the Light Metals*, 3rd edn., Arnold, London, 1995.
3. J.R. Davies (ed.), *Corrosion of Aluminium and Aluminium Alloys*, ASM International, Materials Park, OH, 1999.
4. http// www.world-aluminium.org (accessed August 2009).
5. K. Grjotheim and H. Kvande (eds.), *Introduction to Aluminium Electrolysis*, 2nd edn., Aluminium-Verlag, Dusseldorf, Germany, 1993.
6. K. Grjotheim and B.J. Welch, *Aluminium Smelting Technology*, 2nd edn., Aluminium-Verlag, Dusseldorf, Germany, 1988.
7. M. Schlesinger, *Aluminium Recycling*, CRC Press, Boca Raton, FL, 2006.
8. S.K. Das, Designing aluminum alloys for a recycling friendly world, Materials Science Forum 519–521, 1239–1244 (2006).
9. K.R. Van Horn (ed.), *Aluminium*, vol. 1, ASM International, Materials Park, OH, 1967.
10. J.E. Hatch, *Aluminium: Properties and Physical Metallurgy*, ASM International, Materials Park, OH, 1984, 424pp.

11. K. Raviprasad, C.R. Hutchinson, T. Sakurai et al., Precipitation processes in an Al–2.5Cu–1.5 Mg (wt. %) alloy microalloyed with Ag and Si, *Acta Mater. 51*(17), 5037–5050 (2003).
12. L. Kovarik, M.K. Miller, S.A. Court et al., Origin of the modified orientation relationship for S(S″)-phase in Al–Mg–Cu alloys, *Acta Mater. 54*(7), 1731–1740 (2006).
13. G.B. Winkelman, K. Raviprasad, and B.C. Muddle, Orientation relationships and lattice matching for the S phase in Al–Cu–Mg alloys, *Acta Mater. 55*(9), 3213–3228 (2007).
14. R.G. Buchheit, *J. Electrochem. Soc. 142*, 3994 (1995).
15. www.aluminium.org (The Aluminium Association) (accessed October 2008).
16. D.O. Sprowls, High strength aluminium alloys with improved resistance to corrosion and stress corrosion cracking, *Aluminum 54*(3), 214–217 (1978).
17. T.V. Rao and R.O. Ritchie, Fatigue of aluminum lithium alloys, *Int. Mater. Rev. 37*(4), 153–185 (1992).
18. M. Fishkis and J.C. Lin, Formation and evolution of a subsurface layer in metalworking process, *Wear 206*, 156 (1997).
19. M. Fishkis and J.C. Lin. *Wear 206*, 156–170 (1997).
20. T.J. Carney, P. Tsakiropoulos, J.F. Watts, and J.E. Castle, *Int. J. Rap. Sol. 5*, 189–217 (1990).
21. K. Nisancioglu, Ø. Sævik, Y. Yu, and J.H. Nordlien, *55th Annual Meeting of International Society of Electrochemistry*, Paper no. S7FP92, September 19–24, 2004, Thessaloniki, Greece.
22. C.M. Allen, K.A.Q. O'Reilly, B. Cantor, and P.V. Evans, Intermetallic phase selection in 1xxx Al alloys, *Prog. Mater. Sci. 43*, 89–170 (1998).
23. D.G. Evans and P.W. Jeffrey, Exfoliation corrosion of AlZnMg alloys, in: R.W. Stachle, B.F. Brown, J. Kruger, A. Agarwal (Eds.) U.R. Evans Conference on Localized Corrosion, NACE-3, Houston, Texas, 1974.
24. O. Lunder and K. Nisancioglu, *Corros. Sci. 44*, 414–422 (1987).
25. A.E. Hughes, A.P. Boag, L.M. Pedrina, L. Juffs, D.G. McCulloch, J. Du Plessis, P.J.K. Paterson, I.K. Snook, and B. O'Malley, *ATB Metallurgie 45*, 551–556 (2006).
26. A. Afseth, J.H. Nordlien, G.M. Scamans, and K. Nisancioglu. *Corros. Sci. 43*, 2093–2109 (2001).
27. H. Leth-Olsen, J.H. Nordlein, and K. Nisancioglu, *Corros. Sci. 40*, 2051–2063 (1998).
28. H. Leth-Olsen, J.H. Nordlein, and K. Nisancioglu, *J. Electrochem. Soc. 144*, L196–L197 (1997).
29. J.M.C. Mol, J.H. de Wit, and S. Van der Zwaag, *J. Mater. Sci. 37*, 2755–2758 (2002).
30. R.K. Viswanadham, T.S. Sun, and J.A.S. Green, *Corrosion 36*, 275–278 (1980).
31. G.M. Scamans, A. Afseth, G.E. Thompson, and X. Zhou, *ATB Met. 43*, 90–94 (2003).
32. H.P. Godard, *J. Electrochem. Soc. 114*, 354 (1967).
33. D.W. Aylmore, S.J. Gregg, and W.B. Jepson, *J. Inst. Met. 88*, 205 (1959–60).
34. R.W. Bartlett, *J. Electrochem. Soc. 111*, 903 (1964).
35. M. Pourbaix, *Atlas of Electrochemical Equilibria in Aqueous Solutions*, NACE International, Houston, TX, 1966.
36. D.D.N. Singh, R.S. Chaudhary, and C.V. Agarwal, *J. Electrochem. Soc. 129*(9), 1869–1874 (1982).
37. G.G. Perrault, *Electrochem. Soc. 126*(2), 199–204 (1979).
38. G.S. Frankel, Pitting corrosion of metals—A review of the critical factors, *J. Electrochem. Soc. 145*(6), 2186–2198 (1998).
39. Z. Szklarska-Smialowska, *Corros. Sci. 41*, 1743 (1999).
40. Z. Szklarska-Smialowska, *Pitting and Crevice Corrosion*, NACE International, Houston, TX, 2005, 650pp.
41. F.D. Wall and M.A. Martinez, A statistics-based approach to studying aluminum pit initiation—Intrinsic and defect-driven pit initiation phenomena, *J. Electrochem. Soc. 150*(4), B146–B157 (2003).
42. N. Birbilis and R.G. Buchheit, *J. Electrochem. Soc. 152*(4), B140–B151 (2005).
43. H. Kaesche, Investigation of uniform dissolution and pitting of aluminum electrodes, *Werkstoffe Korros. 14*, 557 (1963).
44. S.T. Pride, J.R. Scully, and J.L. Hudson, Metastable pitting of aluminum and criteria for the transition to stable pit growth, *J. Electrochem. Soc. 141*(11), 3028–3040 (1994).
45. Y. Kim and R.G. Buchheit, A characterization of the inhibiting effect of Cu on metastable pitting in dilute Al–Cu solid solution alloys, *Electrochim. Acta 52*, 2437–2446 (2007).

46. M.K. Cavanaugh, N. Birbilis, and R.G. Buchheit, A quantitative study on the effects of environment and microstructure on pit initiation in Al-alloys, *ECS Trans. 16*(52), 1–11, (2009).
47. A.R. Trueman, Determining the probability of stable pit initiation on aluminium alloys using potentiostatic electrochemical measurements, *Corros. Sci. 47*(9), 2240–2256 (2005).
48. K. Ralston, N. Birbilis, and C.R. Hutchinson, *Proceedings of the Conference on Corrosion and Prevention'09*, Coffs Harbour, Australia, pp. 1796–1804, (2009).
49. D.E. Williams, C. Westcott, and M. Fleischmann, *J. Electrochem. Soc.* 132, (1985).
50. G.M. Scamans, A. Afseth, G.E. Thompson, and X. Zhou, Ultra-fine grain sized mechanically alloyed surface layers on aluminium alloys, in *8th International Conference on Aluminium Alloys, Aluminium Alloys their Physical and Mechanical Properties*, Cambridge, U.K., pp. 1461–1466, (2002)
51. X. Zhou, G.E. Thompson, and G.M. Scamans, The influence of surface treatment on filiform corrosion resistance of painted aluminium alloy sheet, *Corros. Sci. 45*(8), 1767–1777 (2003).
52. J.C. Seegmiller, R.C. Bazito, and D.A. Buttry, *J. Electrochem. Soc. 7*(1), B1–B4 (2004).
53. R.G. Buchheit, R.P. Grant, P.F. Hlava, B. McKenzie, and G. L. Zender, *J. Electrochem. Soc. 144*, 2621 (1997).
54. K. Sieradzki, *J. Electrochem. Soc. 140*(10), 2868–2872 (1993).
55. R.C. Newman and K. Sieradzki, *Science 263*, 1708–1709 (1994).
56. M.B. Vukmirovic, N. Dimitrov, and K. Sieradzki, *J. Electrochem. Soc. 149*(9), B428–439 (2002).
57. T. Muster, A.E. Hughes, and G.E. Thompson, Copper distributions in aluminium alloys, in I.S. Wang (ed.), *Corrosion Research Trends*, Nova Science Publishers, New York, pp. 35–106 (2008).
58. R.G. Buchheit, M.A. Martinez, and L.P. Montes, *J. Electrochem. Soc. 147*(1), 119–124 (2000).
59. M.A. Jakab, D.A. Little, and J.R. Scully, *J. Electrochem. Soc. 152*(8), B311–B320 (2005).
60. D.Y. Jung, I. Dumler, and M. Metzger, *J. Electrochem. Soc. 132*, 2308–2312 (1985).
61. H. Habazaki, K. Shimizu, P. Skeldon, G.E. Thompson, G.C. Wood, and X. Zhou, *Corros. Sci. 39*(4), 731–737 (1997).
62. S.G. Garcia-Vergara, F. Colin, P. Skeldon, G.E. Thompson, P. Bailey, T.C.Q. Noakes, H. Habazaki, and K. Shimizu, *J. Electrochem. Soc. 151*(1), B16–B21 (2004).
63. E.V. Koroleva, G.E. Thompson, G. Hollrigl, and M. Bloeck, *Corros. Sci. 41*, 1475–1495 (1999).
64. C. Blanc, B. Lavelle, and G. Mankowski, *Corros. Sci. 39*(3), 495–510 (1997).
65. T.D. Burleigh, The postulated mechanisms for stress corrosion cracking of aluminum alloys: A review of the literature 1980–1989, *Corrosion 47*(2), 89–98 (1991).
66. P.R. Roberge, *Handbook of Corrosion Engineering*, McGraw Hill, New York, 1999.
67. Z.Y. Chen, F. Cui, and R.G. Kelly, Calculations of the cathodic current delivery capacity and stability of crevice corrosion under atmospheric environments, *J. Electrochem. Soc. 155*(7), C360–C368 (2008).
68. O. Schneider, G.O. Ilevbare, R.G. Kelly et al., In situ confocal laser scanning microscopy of AA2024-T3 corrosion metrology—III. Underfilm corrosion of epoxy-coated AA2024-T3, *J. Electrochem. Soc. 154*(8), C397–C410 (2007).
69. H.N. McMurray, A.J. Coleman, G. Williams et al., Scanning Kelvin probe studies of filiform corrosion on automotive aluminum alloy AA6016, *J. Electrochem. Soc. 154*(7), C339–C348 (2007).
70. A.J. Coleman, H.N. McMurray, G. Williams et al., Inhibition of filiform corrosion on AA6111-T4 using in-coating phenylphosphonic acid, *Electrochem. Solid State Lett. 10*(5), C35–C38 (2007).
71. M. Zamin, *Corrosion 37*, 627 (1981).
72. B. Mazurkiewicz and A. Piotrowski, *Corros. Sci.* 23, 697 (1983).
73. O. Seri, *Corros. Sci. 36*, 1789 (1994).
74. J.R. Scully, T.O. Knight, R.G. Buchheit, and D.E. Peebles, *Corros. Sci. 35*, 185 (1993).
75. K. Nişancioğlu, *J. Electrochem. Soc. 137*, 69 (1990).
76. M.J. Pryor and J.C. Fister, *J. Electrochem. Soc. 131*, 1230 (1984).
77. Afseth, J.H. Nordlien, G.M. Scamans, and K. Nişancioğlu, *Corros. Sci. 44*, 2543 (2002).
78. J.L. Searles, P.I. Gouma, and R.G. Buchheit, *Met. Mat. Trans. A 32A*, 2859 (2001).
79. N. Birbilis and R.G. Buchheit, An experiment survey of electrochemical characteristics for intermetallic phases in aluminium alloys, *J. Electrochem. Soc. 152*(4), pB140 (2008).

80. N. Birbilis, M.K. Cavanaugh, and R.G. Buchheit, Electrochemical response of AA7075-T651 following immersion in NaCl solution, *ECS Transactions*, *1*(4), 115 (2006).
81. O. Schneider, G.O. Ilevbare, R.G. Kelly, and J.R. Scully, *J. Electrochem. Soc. 151*, 465 (2004).
82. G.O. Ilevbare, O. Schneider, R.G. Kelly, and J.R. Scully, *J. Electrochem. Soc. 151*, 453 (2004).
83. M.S. Hunter, G.R. Frank, and D.L. Robinson, *2nd International Congress on Metallic Corrosion*, New York, vol. 66 (1963).
84. ASTM G67-99 Standard test method for determining the susceptibility to intergranular corrosion of 5XXX Series aluminum alloys by mass loss after exposure to nitric acid (NAMLT Test) (1999).
85. S.P. Knight, *A Review of Heat Treatments*, Australasian Corrosion Association, Melbourne, Victoria, Australia, 2003.
86. X.Y. Zhao and G.S. Frankel, Quantitative study of exfoliation corrosion: Exfoliation of slices in humidity technique, *Corros. Sci. 49*(2), 920–938 (2007).
87. G.M. Scamans, N.J.H. Holroyd, and C.D.S. Tuck, The role of magnesium segregation in the intergranular stress corrosion cracking of aluminium alloys, *Corros. Sci. 27*(4), 329–347 (1987).
88. G.M. Scamans and N.J.H. Holroyd, Stress-corrosion of aluminium aerospace alloys, *J. Electrochem. Soc. 133*(8), C308–C308 (1986).
89. A.J. Sedriks, *Stress Corrosion Cracking: Test Methods*, National Association of Corrosion Engineers, Houston, TX, 1990.
90. R.H. Jones (ed.), *Stress-Corrosion Cracking*, ASM International, Materials Park, OH, 1992.
91. T.J. Summerson and D.O. Sprowls, Corrosion behavior of aluminium alloys, *International Conference of the Hall-Heroult Process*, Conference Proceedings, Vol. III, Engineering Materials Advisory Services Ltd. (University of Virginia), pp. 1576–1662.
92. K. van der Walde et al., Multiple fatigue crack growth in pre-corroded 2024-T3 aluminum, *Int. J. Fatigue 27*, 1509–1518 (2005).
93. K. Jones and D.W. Hoeppner, Pit-to-crack transition in pre-corroded 7075-T6 aluminum alloy under cyclic loading, *Corros. Sci.* 47 3109-3122 (2005).
94. S. Suresh, *Fatigue of Materials*, Cambridge University Press, Cambridge, MA, 1998.
95. R.W. Hertzberg, *Deformation and Fracture Mechanics of Engineering Material*, 4th edn, Wiley, New York, 1995.
96. D.W. Hoeppner, Model for prediction of fatigue lives based upon a pitting corrosion fatigue process, in *Fatigue Mechanisms*, J.T. Fong (ed.), *Proceedings of ASMT-NBS-NSF Symposium*, Kansas City, MO, ASTM STP675, ASTM, 1979, p. 841.
97. Y. Kondo, Prediction of fatigue crack initiation life based on pit growth, *Corrosion 45*, 7 (1989).
98. A. Thakur, R. Raman, and S.N. Malhotra, Hydrogen embrittlement studies of aged and retrogressed-reaged Al–Zn–Mg alloys, *Mater. Chem. Phys. 101*(2–3), 441–447 (2007).
99. S. Osaki, H. Kondo, and K. Kinoshita, Contribution of hydrogen embrittlement to SCC process in excess Si type Al–Mg–Si alloys, *Mater. Trans. 47*(4), 1127–1134 (2006).
100. S.P. Lynch, S.P. Knight, N. Birbilis, and B.C. Muddle, Stress corrosion cracking of Al–Zn–Mg–Cu alloys: Effects of composition and heat treatment, in *Proceedings of the 2008 International Hydrogen Conference: Effects of Hydrogen on Materials*, Jackson Hole, WY, 2008.
101. F.W. Lipfert, M. Benarie, and M.L. Daum, Metallic corrosion damage functions for use in environmental assessments, in *Proceedings of the Symposia on Corrosion Effects of Acid Deposition and Corrosion of Electronic Materials*, Las Vegas, NV, 1985, F. Mansfeld, S.F. Haagenrud, V. Kucera, F.H. Haynie, and J.D. Sinclair (eds.), The Electrochemical Society, Pennington, NJ, vol. 86(6), p. 108 (1986).
102. S.E. Haagenrud, J.F. Henriksen, and F. Gram, Dose–response functions and corrosion mapping of a small geographical area. Presented at the *Electrochemical Society Symposium on Corrosion Effects of Acid Deposition*, Las Vegas, NV, October 1985, Norwegian Institute for Air Research, NILU F 53/85, (1985).

103. V. Kucera, in Baboian (ed.), *Influence of Acid Deposition on Atmospheric Corrosion of Metals: A Review*, ACS Symposium Series, vol. 318, 1985, pp. 104–118.
104. D.P. Doyle and H.P. Godard, A rapid method for determining the corrosivity of the atmosphere at any location, *Nature 200*, 1167–1168 (1963).
105. D.P. Doyle and T.E. Wright, Rapid methods for determining atmospheric corrosivity and corrosion resistance, in W. Ailor (ed.), *Atmospheric Corrosion*, John Wiley & Sons, NY, 1982, pp. 227–243.
106. G.A. King, Assessment of the corrosivity of the atmospheres in an intensive piggery using CLIMAT testers, Corrosion Australasia, vol. 14, 8–14 (1987).
107. B.R.W. Hinton, *Met. Finish. 89*(10), 15 (1991).
108. F. Presuel-Moreno, M.A. Jakab, N. Tailleart, M. Goldman, and J.R. Scully, *Mater. Today* 11(10), 14–23 (2008).
109. J.R. Vilche, E.C. Bucharsky, and C.A. Giudice, *Corros. Sci.* 44, 1287 (2002).
110. M. Jahedi, A. Oh, E. Gulizia, S. Gulizia, A.J. Malallah, M.A. Jallaf, N.A. Jabbri, and A.A. Zarouni, *Minerals, Metals and Materials Society*, Warrendale, PA, pp. 951–955 (2009).
111. A. Kumar, L.D. Stevenson, and J.N. Murray, *Prog. Org. Coat. 55*, 244–253 (2006).
112. A.E. Hughes, S.A. Furman, T.G. Harvey, S.G. Hardin, P. Corrigan, F. Scholes, T.H. Muster, P. White, and H. Fischer, Self-healing coating systems for corrosion protection, Presented at the *1st International Self-Healing Conference*, Noorwijk aan Zee, the Netherlands, April 18–20, 2007.

17

Microbially Influenced Corrosion

Dominique Thierry
French Corrosion Institute

Wolfgang Sand
University of Duisburg-Essen

CONTENTS

17.1 Introduction 738
17.2 Microorganisms 738
17.3 Microbially Influenced Geochemical Cycles 742
17.3.1 Carbon Cycle 743
17.3.2 Nitrogen Cycle 743
17.3.3 Sulfur Cycle 745
17.3.4 Iron Cycle 746
17.4 Mechanisms of Biodeterioration 747
17.4.1 Excretion of Acid 747
17.4.2 Chelatization 748
17.4.3 Organic Solvents 748
17.4.4 Other Metabolic Compounds 748
17.4.5 Biofilm 749
17.4.6 Salt Stress 749
17.4.7 Exoenzymes and Emulsifying Agents 750
17.5 Microbial Corrosion of Constructional Materials 750
17.5.1 Metallic Materials 750
17.5.1.1 Mechanisms of Microbially Induced Corrosion of Iron and Mild Steel 752
17.5.1.2 Mechanisms of Microbially Induced Corrosion of Stainless Steels and Titanium 757
17.5.1.3 Mechanisms of Microbially Induced Corrosion on Aluminum and Aluminum Alloys 763
17.5.1.4 Mechanisms of Microbially Induced Corrosion on Copper and Copper Alloys 763
17.5.2 Other Construction Materials 764
17.5.2.1 Mineral Materials 764
17.5.2.2 Organic Polymers of Natural and Synthetic Origin 764
17.6 Countermeasures 766
17.6.1 Changing or Modifying the Material 766
17.6.2 Modifying the Environment or Process Parameters 766
17.6.3 Organic Coatings 766

17.6.4 Cathodic Protection ... 767
17.6.5 Biocides ... 767
17.6.6 Microbiological Methods ... 767
17.6.7 Physical Methods ... 767
17.6.8 Simulation of the Biogenic Attack ... 768
17.7 Diagnosis of MIC ... 768
17.8 Concluding Remarks ... 769
References ... 770

17.1 Introduction

Microbially influenced corrosion (MIC) is, by definition, corrosion associated with the action of microorganisms present in the system. MIC is therefore an interdisciplinary subject that embraces the fields of materials science, chemistry, microbiology, and biochemistry.

The first report of MIC on metals was published in 1891 by Garrett [1], who found that the corrosion of covered cables could be attributed to biogenic ammonia, nitrite, and nitrate. Almost two decades later, it was shown by Gaines [2] that sulfate-reducing, sulfur-oxidizing, and iron bacteria were responsible for part of the corrosion of iron in soils. These observations were associated with aerobic conditions. In 1934, von Wolzogen Kühr and van der Vlugt [3] presented the first evidence that microorganisms play a direct role in the corrosion of metals under anaerobic conditions. They postulated that sulfate-reducing bacteria were able to pick up adsorbed hydrogen from the metal surface. From 1934 up to now, there have been an increasing number of reports of MIC on metallic and nonmetallic constructions and in various environments.

The cost of MIC is very significant. Several estimates have been made in the United States and in the United Kingdom. For instance, it has been reported by Iverson [4] that the annual cost of MIC of buried pipes in the United States was $500–$2000 million. In the United Kingdom, it was postulated that at least 50% of the corrosion occurring on buried metals was of microbial origin [5].

The main purpose of this chapter is to provide an overview of the present knowledge of the mechanisms of MIC on common constructional materials. This review includes both metallic and nonmetallic materials. The reason for including nonmetallic materials is that biodegradation occurs on these materials and that they are widely used for construction purposes. Because detailed knowledge in the fields of microbiology and biochemistry is necessary in order to understand the mechanisms of MIC, considerable space has been devoted to a general description of microorganisms and to a description of natural cycles of matter in which microorganisms play an important role. A later section briefly presents the most commonly used countermeasures for avoiding or minimizing the influence of microorganisms on the degradation of materials.

17.2 Microorganisms

The term "microorganism" covers a vast diversity of life forms. Bacteria, blue-green cyanobacteria, algae, lichens, and fungi together with protozoa are all classed as microorganisms. The main characteristic is the size of the individual organism. As a rule, it ranges between 0.5 and 10 μm, although organisms several meters in size are known (e.g., brown algae). Because of their small size, microorganisms have a large, catalytically active surface. A volume of 1 cm^3 can contain 10^{12} bacterial cells having a surface of approximately 1 m^2 (cell size 1 to 1 μm, length to width). Through this enormous surface area, agents may be excreted into the surrounding medium, causing detrimental effects on materials. The vast majority of microorganisms cannot be detected with the naked eye. Techniques such as light and/or electron microscopy (transmission and/or scanning) are required to magnify single cells so that they become visible [6]. The detection may be enhanced using dyes, which bind to cells and make them colored or fluorescent. Specific detection can be achieved using the polymerase chain reaction (PCR) technique with specific gene probes and/or immunolabeled antibodies binding to selected microorganisms. Thus, microorganisms with special characteristics may be detected in a mixed population.

Other indirect techniques measure the effect of the metabolism of microorganisms on the environment, such as acidification; alkalization; oxygen consumption; degradation of substrates, production of insoluble, dissolved, and/or gaseous intermediates and/or end products of catabolism; and heat production [7]. An indirect technique for visualization of "single" cells of microorganisms is transfer of a cell to a solidified nutrient solution and subsequent incubation at a given temperature, which enables the cell to grow and to multiply. After several multiplication cycles, a colony (consisting now of at least 10^7 daughter cells) becomes visible to the naked eye. The solidifying agent may be agar-agar, derived from algae, a polysaccharide forming a gel.

A powerful, recently developed tool is the PCR technique. The technique allows theoretically the detection of a single cell in 10^7 or more cells. A probe containing a counterpart of that piece traps a piece of genetic information, unique for a microorganism (a genetic fingerprint). Afterward, the double-stranded piece is multiplied by means of an enzyme, taq polymerase. After several multiplication cycles, the genetic information is sufficiently amplified to become detectable, e.g., by autoradiography (because of the use of radioactive components). However, so far only a few absolutely specific probes for detection exist.

Microorganisms exhibit a vast diversity with regard to their metabolism. Nevertheless, this diversity may be described by six main terms. Microorganisms obtaining their metabolic energy by use of light are called phototrophs. In contrast, those gaining energy through chemical reactions are known as chemotrophs. The source of reducing power is used for a further subdivision. In the case of inorganic hydrogen donators, the microorganisms are called lithotrophs, whereas those using organic compounds are called organotrophs. Finally, the source of cell carbon is important. If CO_2 is used, the organisms are called autotrophs. If carbon is derived from organic molecules, they are designated heterotrophs. By combing these six terms, bacteria can easily be described with regard to their nutritional requirements. For instance, if energy is derived from inorganic hydrogen donators and biomass is derived from organic molecules, they are called mixotrophs (=chemolithoorganotroph). *Escherichia coli*, for example, a bacterium occurring in the gut of animals, is a chemoorganoheterotroph because it grows by chemical oxidation of organic compounds such as glucose in the medium and derives cell carbon by assimilating part of the glucose. Another example is the chemolithotrophic bacterium *Acidithiobacillus thiooxidans*, which grows by the chemical oxidation of inorganic sulfur compounds using the energy of the oxidation for metabolism and deriving the cell carbon from CO_2 of the air.

Most microorganisms may be classified as chemoorganoheterotrophs. Exceptions are photosynthetic microorganisms (mainly a few bacterial genera), cyanobacteria, algae, and in part lichens because of the algal or cyanobacterial symbiont. Exclusively restricted to bacteria are chemolithotrophic life forms.

An important feature of microbial life is the possibility to degrade any naturally occurring compound. This is called microbial omnipotence. Exceptions to this rule are a few manmade compounds such as highly polymerized materials (resins, plastics, etc.) and halogenated compounds. These are called xenobiotica because they are strange to the living world and living organisms have had too little time to adapt to these compounds and develop degradative enzymes. Eventually, genetic engineering of microorganisms may help in the development of degradatively active ones.

Besides energy and carbon sources, microorganisms need nitrogen, phosphorus, and trace elements. Nitrogen compounds may be inorganic ammonium and/or nitrate (sometimes nitrite, too) as well as organically bound nitrogen (amino acids, mieleotides, etc.). Some microorganisms (bacteria and cyanobacteria) are able to fix nitrogen from atmospheric nitrogen with the help of an enzyme called nitrogenase. The end product is ammonia, which is incorporated in cell constituents.

Phosphorous is usually taken up as inorganic phosphate or as (organically bound) phosphorylated compounds such as phosphorus-containing sugars and lipids. Phosphate in the form of adenosine triphosphate (ATP) serves as the main energy storage compound. Whenever a reaction takes place generating metabolically useful energy, ATP is produced to conserve at least a part of it.

Trace elements are needed for many metabolic purposes. They make up only a negligible amount of the total cell weight, but they support vital functions. Iron as Fe^{2+} or Fe^{3+} is necessary for the electron transport system. It functions as an oxidizable/reducible central atom in cytochromes or in nonheme iron–sulfur proteins. Magnesium plays this role in the chlorophyll molecule. Cobalt functions in the transfer of methyl groups from/to organic or inorganic molecules (vitamin B_{12}, cobalamine, is involved in the methylation of heavy metals such as Hg). Copper is an integral part of a cytochrome (aa_3), which at the terminal end of the electron transport system mediates the reduction of oxygen to water (cytochrome oxidase). Further examples exist for other metals.

Microbial growth is influenced and sometimes restricted by several chemical and physical factors. Life generally cannot exist without water. Hence, water is a prerequisite for microbial life and growth. Microorganisms differ considerably with regard to the amount of water needed. In particular, fungi are able to live under extremely dry conditions. Three types of water surrounding a solid material need to be distinguished: hygroscopic, pellicular, and gravitational water. Hygroscopic water, a film 3×10^2 μm thick, directly surrounds the solid particle. It is not available for life and does not freeze or move. Pellicular water and gravitational water are biologically available and, thus, may be used by microorganisms. The biologically available water is usually measured as the water activity a_w of a sample:

$$a_w = \frac{\text{vapor pressure of a solution}}{\text{vapor pressure of pure water}} \text{(at the same temperature)}$$

Most bacteria are restricted to a_w values of more than 0.90 (equivalent to an osmotic pressure of –150 bar). Exceptions are bacteria living in salterns or salinas—halobacteria—that exhibit a_w values of 0.75 (equivalent to an osmotic pressure of –400 bar). The lowest known value for a (micro) organism has been found for the fungus *Xeromyces bisporus* [8]. The organism is able to grow at an a_w value of 0.61, equivalent to −681 bar. For comparison, egg powder has an a_w value of 0.40 (–1260 bar) and biscuits of 0.30 (–1660 bar). Lichens, because of the symbiosis of a photosynthetic partner (alga or cyanobacterium) with a fungus, may resemble fungi in their need for available water. All other microorganisms (and also higher organisms) are very sensitive to water stress. They barely tolerate water shortage. The tolerated a_w value is 0.99 (−15 bar). Physical factors may contribute to water shortage. In porous systems such as soil or sedimentary stone (sandstone), water activity is already reduced because of the capillary bonding in pores of low diameter (below 10 µm).

Another important factor is the hydrogen ion concentration. Microorganisms may be distinguished by their ability to grow under acidic, neutral, or alkaline conditions. Hence, they are called acidophiles, neutrophiles, or alkaliphiles. The bacterium *A. thiooxidans* has been detected in samples exhibiting a negative pH value, whereas in soda lakes life has been detected at pH values of 12 and above. Fungi are able to grow over a large range of pH values. Species of *Penicillium* have been found at pH 2 and up to pH 12. Most microorganisms, however, thrive in the neutral pH range from 6 to 8.

The redox potential is another factor determining microbial growth. If, under standard conditions, hydrogen is assumed to have a redox potential (E_h) of −420 mV and oxygen to have an E_h value of +820 mV, this gives the range in which metabolism can take place. The basic process of life is the reaction of hydrogen with oxygen to form water. Depending on oxidizing or reducing conditions, different types of metabolism are found. Under oxidizing conditions with E_h values of +500 mV and above, oxygen is usually used as terminal electron acceptor. E_h values of +400 mV favor the reduction of nitrate if oxygen is not available. If neither oxygen nor nitrate is available, Mn^{4+} may serve as an electron acceptor, being reduced to soluble Mn^{2+} species. At an E_h value of −180 mV, ferric iron starts to be used as terminal electron acceptor. At an even lower redox potential, sulfate is reduced to sulfide (−220 mV). The lowest potential (−250 mV) is necessary for the reduction of carbon dioxide to methane.

Closely linked to the redox potential is the available oxygen. Microbial life is possible under well-aerated as well as under totally oxygen-free conditions. The differences are used to characterize the oxygen needs of microorganisms. Aerobes are organisms living with the amount of oxygen contained in the air. They may tolerate shortage of up to 5% or 10% of the regular partial pressure of oxygen. Facultative aerobes are able to live with the oxygen partial pressure even under conditions of exhaustion. In this case, they convert their metabolism to the use of chemically bound oxygen (nitrate) or to fermentation. Anaerobes are organisms that perform their metabolism without any free oxygen. They are able to use only bound oxygen (sulfate, carbon dioxide) or to ferment organic compounds [9]. In the latter group, there are a few species that are called strict anaerobes because oxygen is highly toxic to these organisms. The process of utilization of chemically bound oxygen is called anaerobic respiration. It is also used for the reduction of inorganic metal ion species such as Mn^{4+} or Fe^{3+}.

Last but not least, the temperature needs to be discussed. Microbial life is possible between −5°C and +114°C. Again, the temperature needs of microorganisms are used for their classification.

Psychrophiles are organisms living in the range of −5°C up to 20°C. Psychotrophs live between 5°C and 30°C. Mesophiles thrive in temperatures between 20°C and 45°C. At higher temperatures, moderate thermophiles find their habitat (40°C–55°C). The next group consists of the thermophiles, living in the range from 55°C to 85°C. The extreme thermophiles grow at temperatures up to about 120°C. Above that, living organisms have not so far been detected. Most organisms live in the mesophilic range corresponding to the usual temperature on the surface of the earth. Interestingly, only a special group of bacteria is able to grow at elevated temperatures (above 70°C). They are called archaebacteria and are distinguished from eubacteria by many chemical and physiological traits.

Many of them seem to represent ancestors who are believed to have been active in the early history of this planet (high temperature, reducing atmosphere). Organisms are usually not restricted to grow at a certain temperature. Adaptation within a range of 20°C–30°C above or below their optimized growth temperature is possible. It is achieved by a modification of the fatty acid composition in the cytoplasmic membrane. At higher temperatures, unsaturated and/or branched fatty acids prevail, whereas, at lower temperatures saturated, unbranched ones are found.

In summary, microbial life is influenced by many parameters. Adaptation to a change is possible [9–11]. This allows microorganisms to be present wherever a substrate is available for metabolism. A variation in the parameters previously mentioned may be used to control microbial growth. However, care must be taken to ensure that merely adaptation of the microorganisms present and not a change in the composition of the microbial biocoenosis takes place.

Microorganisms grow predominantly attached to surfaces. It is estimated that more than 90% of all microorganisms live sessile on the surface of a solid material (called the substratum, in contrast to the nutrient, which is called the substrate). Attachment of microorganisms is mediated by a complex series of events by excretion of linking exopolymers, resulting in firmly substratum-bound cells.

The exopolymers consisting of lipopolysaccharides, lipoproteins, and sometimes nucleic acids [12] constitute the outer membrane of the cells and, because they are strain specific, may be used for identification purposes. Furthermore, they may contain metal cations, some of which are known to be necessary for substrate degradation [13]. If several cells are growing together on a substratum, the exopolymers fill the voids. This is called a biofilm. If a biofilm grows and increases in thickness, it becomes stratified or patchy. The oxygen supply of cells buried deeply in a biofilm may become limiting. Thus, anaerobic zones or layers develop [14]. This may result in the appearance of closely associated zones with aerobic and anaerobic types of metabolism. Localized biofilm (microcolonies) may have serious detrimental effects on materials [15]. Because of the oxygen consumption by a microcolony or a biofilm, localized aeration elements exist, causing in the case of metals an enhanced dissolution by stimulation or electrochemical effects. Another effect may be reduced heat transfer in heat exchange systems [16]. Furthermore, biofilms protect the organisms living there because of reduced penetrability of poisons such as heavy metals, biocides, and natural enemies such as viruses (phages) and grazing protozoa [17,18]. In addition, a biofilm protects against dryness. The microorganisms growing in a biofilm have, thus, created a microenvironment of their own favorable for their survival and growth. Another aspect is that in dilute solutions, due to ionic interactions, (organic) molecules tend to adsorb to surfaces and thus becoming increasingly available for attached organisms.

17.3 Microbially Influenced Geochemical Cycles

Most elements on earth are subject to cycling. During this process, the oxidation status may be changed from fully oxidized to totally reduced. All cycles are influenced by microorganisms, and some are controlled by them [19]. Four examples—the carbon cycle, the nitrogen cycle, the sulfur cycle, and the iron cycle—are selected here to illustrate the importance of microorganisms.

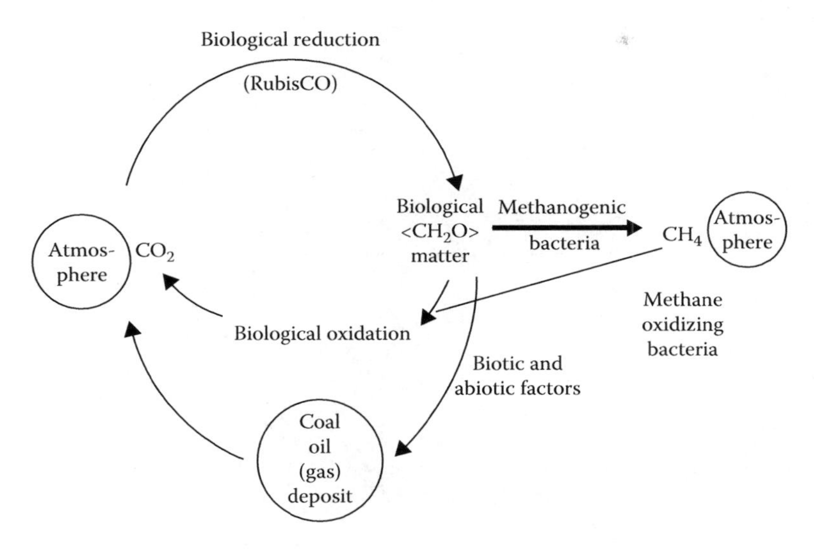

FIGURE 17.1
Carbon cycle.

17.3.1 Carbon Cycle

Carbon occurs in the biosphere in either the reduced (methane, fatty acid, carbohydrate) or the oxidized (alcohol, aldehyde, carbonic acid, carbon dioxide) form. The valence state 0 is found only in coal (a compound of biological origin).

(Micro)organisms control the carbon cycle (see Figure 17.1). By fixing CO_2 from the air, cell carbon is generated. Primary producers such as green plants, algae, cyanobacteria, and photosynthetic bacteria are responsible for this reaction. The enzyme mainly involved in this process is called ribulose-1,5-bisphosphate-carboxylase/oxygenase (RubisCO). It is the most abundant protein on earth. The resulting cell carbon is degraded by organisms, the end product being carbon dioxide, which may be fixed again. Another possibility is the incomplete degradation occurring in sediments, etc. This may give rise to accumulations of organic matter as observed in shallow marine areas, swamps, and marshes. Presumably, deposits of carbon such as oil, gas, and coal were formed in this way [20]. Another pathway is the fixation of CO_2 followed by its use in structural elements of organisms such as valves and bones. In this way, inorganic deposits such as limestone and shales are formed.

17.3.2 Nitrogen Cycle

Nitrogen occurs in the biosphere in either the reduced (to ammonium) or oxidized (nitrite, nitrate) form, together with intermediates such as hydroxylamine (NH_2OH), nitrogen (N_2), and nitrogen oxides (N_2O, NO, NO_2). In addition, organically bound nitrogen is found in many compounds such as proteins, nucleic acids, amines, and urea. Microorganisms mediate all and control most of the reactions in this cycle (see Figure 17.2). The cycle starts with an enzymatic reaction unique to bacteria and cyanobacteria. Atmospheric dinitrogen, N_2, is reduced and fixed by the action of the enzyme nitrogenase. The product NH_3 is incorporated in cell constituents, the amount of ammonia that becomes available for the biosphere being comparable to the amount produced in technical ammonia production.

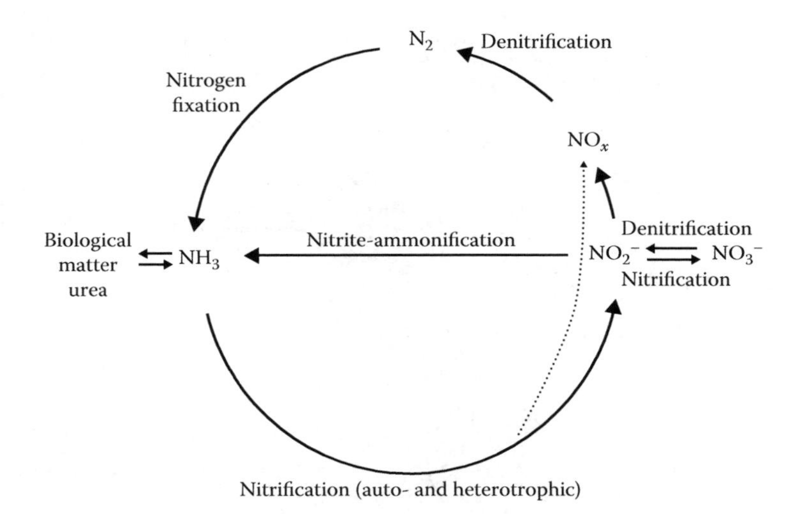

FIGURE 17.2
Nitrogen cycle.

When cell constituents are degraded, ammonia is liberated and may be used for the synthesis of other nitrogenous compounds. Because nitrogen is usually a limiting factor for life, organisms take care not to lose it. However, upon the death of cells, ammonia may become available. Usually, a process called nitrification starts. Nitrifying bacteria oxidize ammonia via nitrites to nitrate. The nitrifying bacteria consist of two groups: the ammonia oxidizers, which are responsible for the oxidation of ammonia to nitrite, and the nitrite oxidizers, which are responsible for converting nitrite to nitrate. Hence, a cation is converted to an anion plus protons (acidification). The process has positive and negative consequences. Whereas ammonia may be adsorbed to clay particles and, thus, may become unavailable for plant growth, nitrate is a mobile ion, which can be washed out of soil by rainfall and enter into ground water. Because of this, nitrification is responsible for nitrogen loss from fertilizers and groundwater pollution by nitrate. In the case of a large supply of organic matter, the availability of oxygen is often limiting for degradation. Under these circumstances, a reductive process called denitrification or nitrate (nitrite) respiration gains importance.

Nitrate and other oxidized nitrogen compounds may serve as electron acceptors and be reduced to nitrite, NO, N_2O, and finally N_2 [21]. The oxygen is released as water. Thus, denitrifiers can get rid of hydrogen resulting from the oxidation of organic matter. By this mechanism, a considerable amount of organic matter may be mineralized under anaerobic conditions. Growth yields with chemically bound oxygen are in the range of atmospheric oxygen usage (less than 10%). The reduction does not always proceed until dinitrogen is released. Often, NO or N_2O are the end products. These may either be reoxidized or be further reduced by microorganisms. In any case, the nitrogen cycle is closed by these microbiologically catalyzed reactions. It needs to be added that there is an enormous input of ammonia mainly from two processes: the degradation of manure from livestock breeding and loss of fertilizers from farmland. Compared with industrial processes such as the use of ammonia for flue gas desulfurization, these two processes dominate the ammonia emission. Nitrogen oxides are becoming increasingly important because of their influence on the global climate. In contrast to the emissions of dust particles and sulfur dioxide, the emission of nitrogen oxides is still increasing because of increasing use of combustion processes, e.g., for automobiles and heating systems. Microbiology will adapt to the increased supply of these compounds, which furthers the growth of ammonia-oxidizing and NO-oxidizing microorganisms.

17.3.3 Sulfur Cycle

Sulfur occurs in the biosphere in many compounds. It is essential for the formation of the sulfurylated amino acids methionine and cysteine/cystine. Other highly important compounds are those containing reactive thiol groups, such as coenzyme A or iron-sulfur redox centers involved in electron transfer reactions. The commercially most important sources are deposits of metal sulfides (production of valuable metals by mining) and sulfur (source of sulfur for sulfuric acid production). Metal sulfides can be attacked and degraded microbiologically by the action of highly specialized bacteria, which oxidize the metal sulfide to a metal sulfate [22]. This is accompanied by sulfuric acid production, which keeps the metal ions solubilized. The metals may be recovered by processes such as sedimentation, solvent extraction, or ion exchange and further purified for practical use. The oxidation of the sulfide moiety takes place in several steps including sulfur, polythionate, and sulfite (see Figure 17.3). The energy liberated by this oxidation is recovered by adenosine triphosphate (ATP) production. The microorganisms active in this process tolerate high heavy metal concentrations (up to several grams per liter) and low pH values (pH 1.5 and below). They may grow at temperatures from 4°C up to 90°C. Once sulfate has been produced, a process similar to denitrification may take place. If sufficient organic matter is available and anaerobic conditions prevail, sulfate may act as an electron acceptor for reduction equivalents, being reduced via sulfite and sulfur to sulfide [23]. The process is called sulfate reduction (sulfate respiration) and the bacteria are referred to as sulfate reducers or sulfate-reducing bacteria (SRB). This is a physiologically diverse group of microorganisms including photosynthetically active bacteria. Besides mesophilic eubacteria, very ancient life forms belong to this group. It contains archaebacteria, which are able to live at 110°C by sulfate reduction, conditions probably resembling early stages of life on this planet. If sulfide accumulates, two different pathways are selected depending on the oxygen availability. Under aerobic conditions, oxidation to sulfate occurs as described before. Under anaerobic conditions in the light, another type of oxidation takes place. Photosynthetic microorganisms oxidize sulfide to sulfur and, at least some of them, to sulfate using the electrons liberated to fill up their photosynthetic system.

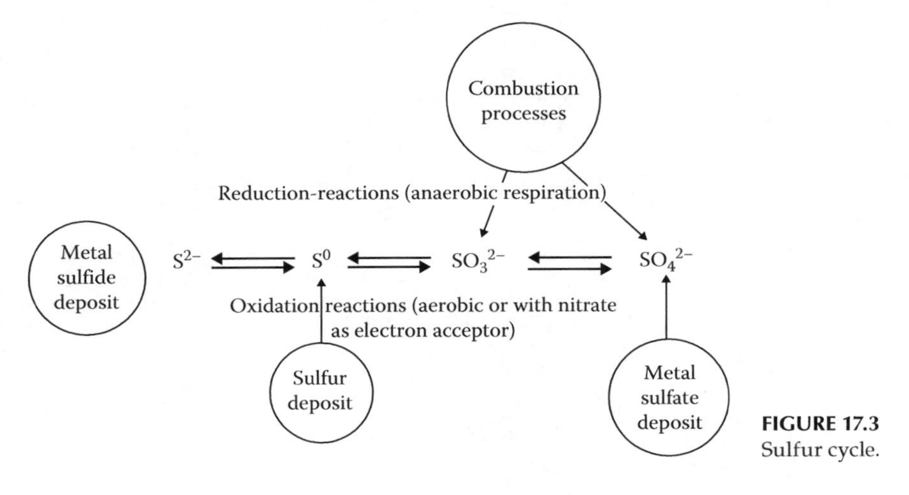

FIGURE 17.3
Sulfur cycle.

Without light, it has been discovered that sulfide oxidation coupled to nitrate reduction can take place. A small part of the sulfide is used for the production of sulfur-containing biomolecules. Thus, biologically, sulfur and its compounds may be oxidized and reduced by many reactions. Deposits of sulfur compounds resulting from this activity may include elemental sulfur, metal sulfides, and sulfate-like barites. The biological origin can be detected by measurement of the $^{32}S/^{34}S$ ratio. The enzymes involved in sulfur metabolism often prefer to handle the lighter isotope. As a consequence, a biological deposit will be enriched with ^{32}S compared with sulfur from meteorites which has a strictly defined ratio of 22.22/1.

17.3.4 Iron Cycle

Iron is an essential element for life. Although it belongs to the trace elements, a shortage endangers many vital functions. Generally, it exists as Fe^{2+} or Fe^{3+} mediating redox reactions combined with electron transfer (see Figure 17.4). For example, iron is the central atom of cytochromes (heme type) and is part of redox proteins having an iron-sulfur reaction center (nonheme). Ferrous iron may serve as energy source for many bacteria. Well known are the acidophilic lithoautotrophic bacteria *Acidithiobacillus ferrooxidans* and *Leptospirillum ferrooxidans* living in sulfidic, aerobic habitats by oxidizing ferrous to ferric iron (besides sulfide oxidation to sulfuric acid with *Acidithiobacillus ferrooxidans*). The energy of the reaction is used for metabolic purposes. Reaction products are usually ferric iron, sulfuric acid, and to a certain extent intermediary sulfur. Although these bacteria thrive best at a pH range of 1–3.5 [24], where ferrous iron is only slowly autoxidized by oxygen, organisms such as *Metallogenium* sp. (pH 3.5–6.5) and *Gallionella ferruginea* (pH 6–8) as well as *Leptothrix ochracea* (=*Sphaerotilus natans*, pH 6–8) are responsible for ferrous iron oxidation at higher pH values. As Fe^{2+} becomes increasingly autoxidizable at pH values above 3.5, these organisms must be adapted to live with low concentrations of ferrous iron and/or a reduced oxygen partial pressure (microaerophiles). Once ferric iron has entered the cycle, it will probably be precipitated in the form of ferric hydroxide, $Fe(OH)_3$, followed by several reactions resulting in the formation of hematite, Fe_2O_3. Under conditions where organic compounds are available, e.g., in anaerobic sediments with decaying organisms, ferric iron may be used as an electron acceptor and be reduced to ferrous iron [9]. Thus, an immobile iron form is transformed into a mobile one. This process becomes even more important because coprecipitated heavy metals also become solubilized, consequently endangering the environment. Many microorganisms, bacteria and fungi, are able to use

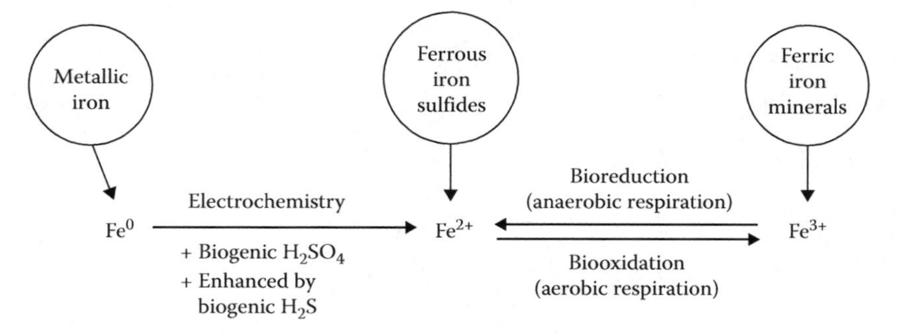

FIGURE 17.4
Iron cycle.

this reduction reaction. Even the ferrous iron–oxidizing bacteria may be able to perform this reaction in the event of oxygen shortage. The reactions are not restricted to mesophilic bacteria. Thermophilic bacteria are known to participate in these reactions (oxidatively as well as reductively).

As mentioned, sulfuric acid is produced concomitantly with the metal sulfide oxidation. Sulfuric acid may react with metallic iron, Fe^0, to give ferrous sulfate and hydrogen. Both compounds may be oxidized biologically. Furthermore, ferric iron may react with metallic iron to form ferrous iron, $Fe^0 + 2Fe^{3+} \rightarrow 3Fe^{2+}$, which may then be oxidized to ferric iron. Another reaction with iron results from the sulfur cycle and demonstrates the intimate connection between both cycles. Reduction of sulfate under anaerobic conditions, e.g., by SRB, results in the formation of hydrogen sulfide. This may react with ferrous iron to form a precipitate of ferrous sulfide and, if sulfide is in excess, finally pyrites (FeS_2).

In summary, it becomes obvious that most reactions of materials with the environment are influenced by microorganisms and are often even controlled by them [19]. Thus, in the case of MIC, the participation of microorganisms needs to be anticipated [25,26]. Furthermore, cases are known in which excreted metabolic products, e.g., exopolysaccharides (EPS) free from microbial cells, cause corrosion.

17.4 Mechanisms of Biodeterioration

Despite the vast diversity of microorganisms participating in various natural cycles, the biological mechanisms contributing to or causing biodeterioration may be summarized in a few main categories [27]. It needs to be pointed out, however, that one microorganism may exert multiple detrimental effects. In addition, under natural conditions, pure cultures do not exist. Thus, mixed cultures called biocoenoses are active by usually creating growth conditions for the corrosion-causing microorganisms. In most cases, microorganisms cause an attack resulting from a chemical compound produced and excreted by metabolism. Thus, the basic action will be a chemical reaction.

17.4.1 Excretion of Acid

Specialized bacteria are able to produce and excrete strong mineral acids. Usually under aerobic conditions, thiobacilli oxidize inorganic sulfur compounds and sulfur to sulfuric acid. The energy of this oxidation is coupled via special enzymes to growth. Besides sulfur and sulfur compounds, the bacteria need only carbon dioxide (cell mass). The genus *Thiobacillus* consists of several species that are able to grow at moderately alkaline down to strongly acidic pH values. The species able to grow and proliferate on alkaline materials are pioneers for the species growing only under acidic conditions. After the buffer substance of, e.g., concrete (lime) has been exhausted, the pH value in the surface water declines and the acidophilic species start to proliferate. This finally causes severe biogenic sulfuric acid corrosion [6,28–31].

The second group to be mentioned here is the nitrifying bacteria excreting nitric acid. The nitrifying bacteria consist of two groups. The ammonia oxidizers convert ammonia to nitrite, and the nitrite oxidizers subsequently form nitrate. Like sulfuric acid, nitric acid can react with alkaline materials forming highly soluble salts (in contrast, sulfates are much less soluble). Biogenic deterioration results [28,32–34]. Nitrifiers are also lithoautotrophs. The energy source is a nitrogen compound; cell carbon is derived from CO_2.

The third important acid is produced by all life forms. Carbon dioxide is excreted as the end product of metabolism. It reacts with water to form carbonic acid, which may dissolve, e.g., to carbonates forming soluble bicarbonates. In this way, the binding material of concrete, lime, may be dissolved.

The fourth group of microorganisms consists of those which during their metabolism excrete organic acids such as oxalic, citric, malic, lactic, or acetic acid, amino acids, uronic acids, etc. [35,36]. Usually, these acids are excreted because of an unbalanced metabolic state. However, they may thus be taken up again by the cells in later growth phases or be metabolized by other microorganisms present in the biotope. This means that organic acids are usually available only temporarily. Nevertheless, their presence may have caused transformations in the crystal lattice. Organic acids may be excreted by almost all bacteria, cyanobacteria, algae, lichens, and fungi. Sometimes the excretion is coupled to the uptake of limiting cations into the cell.

17.4.2 Chelatization

Besides acid attack on materials, organic acids are chelating cations. Because of the stability of the complexes, metals may be dissolved from a crystal lattice resulting in a weakening of the structure. The action is sometimes deliberate because it allows the cells to replenish limiting cation concentrations; e.g., pathogenic bacteria possess special iron chelators to replenish growth-limiting ferric iron at the expense of the host.

17.4.3 Organic Solvents

Many microorganisms are able to metabolize organic substances under anaerobic conditions. If an electron acceptor such as nitrate, Fe^{3+}, Mn^{4+}, or sulfate is not available, fermentation results [9]. This means that hydrogen (=redox equivalent) is transferred from one to another organic compound. By this transfer, growth may be possible because of a substrate phosphorylation. The result of a fermentation process is thus usually another organic compound, sometimes carbon dioxide. The organic compounds are in many cases organic acids, as mentioned before, or organic solvents such as ethanol, propanol, or butanol. These solvents may react with materials of natural and/or synthetic origin, causing swelling, partial or total dissolution, etc., which finally results in deterioration.

17.4.4 Other Metabolic Compounds

An important compound for MIC is hydrogen sulfide. It is produced under anaerobic conditions by the action of sulfate-reducing bacteria from sulfate, sulfite, and sometimes sulfur [37]. It has also been shown that thiosulfate can by used by SRB [38]. As mentioned before, the SRB use the not fully reduced sulfur compounds as electron acceptors in anaerobic respiration.

Hydrogen sulfide may act in several ways. Either it is reoxidized to sulfuric acid under aerobic conditions [39] or in the presence of compounds such as nitrate or it may react with cations to form metal sulfides that precipitate [40]. Further, H_2S may be oxidized anaerobically in the light by photosynthetic bacteria (mostly to sulfur, sometimes to sulfate). In the case of materials containing acid-reactive substances, the biogenic acid will become harmful.

In the case of metals, the presence of H_2S may cause MIC. Another source of H_2S is not restricted to anaerobiosis. Amino acids containing sulfur are degraded under aerobic conditions, giving rise to H_2S (cysteine and cystine).

Another important compound is ammonia. It may be generated as a result of amino acid or urea degradation by microorganisms. Furthermore, ammonia (or mainly ammonium salts such as sulfates or chlorides) is a major part of airborne gases (or dust particles). By dry or wet deposition, ammonia/ammonium salts reach the surface of materials and, in the case of porous materials, enter the pore system and become biologically available. Nitrifying bacteria as already described may degrade ammonia/ammonium compounds. However, chemical reactions with materials may also be possible.

Nitrogen oxides, N_2O, NO, NO_2, resulting from biological activity (nitrogen cycle—oxidation of ammonia or nitrite reduction to N_2) as well as from processes such as the burning of fuels (heat and power generation, traffic), are a group of compounds that may react with materials. NO_2 is a water-soluble gas dissociating upon dissolution into nitric and nitrous acid. These acids may attack susceptible materials. NO is less reactive but may be oxidized to NO_2 by light. N_2O is not known to interfere with materials.

17.4.5 Biofilm

Microorganisms tend to adhere to surfaces [27,41–43]. It is believed that more than 90% of all microorganisms grow in this way [12]. Concomitantly, exopolymeric substances are excreted. This may result in a slimy superficial layer on materials and/or, in the case of porous materials, in total clogging of the free pore volume [44]. The exopolymers are usually hydrated and may contain ionic groups favoring water inclusion. Thus, one result may be an increase in water content in the case of porous materials (a consequence is freeze-thaw attack; see the following).

Furthermore, the biofilm may act as insulation, reducing heat transfer. In the case of heating systems, the efficiency may be considerably reduced [16]. Another effect is the reduction of the cruising speed of ships due to increased friction caused by the surfacial biofilm [45]. Finally, biofilm development may cause technical processes to become susceptible to trouble. In paper machines, the inclusion of exopolymers in the pulp may cause the paper web to break in the paper machine. The coating of cans with resins may also be affected. Cans for food are coated on the inside with a resin to protect the metal from corrosion by foodborne acids. If the metal is covered at a few spots with a biofilm (microcolonies), the coating process will not be totally successful. Pores will remain in the resin at these spots, rendering the metal accessible to acids from the food. Corrosion from the inside combined with deterioration and spoilage of the contents will result.

17.4.6 Salt Stress

The biogenic reactions that have been mentioned result in the generation and, generally, accumulation (except in an aquatic environment) of salts as reaction products. Because salts are hydrophilic, they are usually hydrated, resulting in an increase in the water content of a porous material. This may increase the susceptibility to physical attack by freezing and thawing because of the volume change in water/ice crystals.

Furthermore, upon desiccation, salt crystals may develop, causing the surfacial removal of layers of material, e.g., deterioration destroying wall paintings on natural stone. The third detrimental effect of salts results from the formation of large crystals causing a swelling attack. A well-known example is the formation of ettringite from gypsum crystals destroying concrete and bricks.

17.4.7 Exoenzymes and Emulsifying Agents

Besides exopolymers of a lipopolysaccharidic nature, microorganisms excrete lipoproteins and proteins. The latter include exoenzymes for the degradation of high-molecular-weight compounds such as cellulose. Cellulases degrade the insoluble cellulose fibrils to soluble molecules such as cellobiose (dimer) and glucose, which may be taken up by the cells for metabolism. Similar exoenzymes exist for other materials, e.g., esters, amines, and waxes. Other molecules, although not as large as cellulose, may also be insoluble and, thus, only poorly degradable.

In some cases, microorganisms excrete emulsifying agents to increase the solubility of hydrophobic substances [46]. In the case of elemental sulfur, which is highly insoluble due to its hydrophobic nature, the excretion of such emulsifying agents causes an increase in dispersibility from 5 to 20,000 μg/L [47]. This holds for other hydrophobic materials as well. Water-insoluble hydrocarbons become water soluble because of microbial production of emulsifying agents and consequently become degradable.

These microbial deterioration mechanisms represent the main categories of action. Several mechanisms are usually active jointly, and because these mechanisms may also influence other types of physical or chemical attack, the microbial share of the total attack can rarely be determined. Much too often, studies of deterioration mechanisms suffer from inadequate microbiological analysis (if it is done at all). If corrosion occurs in the range where life is possible, a microbial contribution needs to be taken into account and, hence, a microbiologist should be consulted [25,48].

17.5 Microbial Corrosion of Constructional Materials

17.5.1 Metallic Materials

Metallic materials are an important group of construction materials. MIC may occur for these materials for many industrial applications, as listed in Table 17.1 [49]. MIC of metallic materials does not involve any new form of corrosion. Thus it is necessary to discuss the electrochemical nature of corrosion briefly before continuing with the mechanisms of MIC for different construction metals.

The corrosion of metals in the presence of water is of an electrochemical nature. This includes corrosion in water-containing solutions, atmospheric corrosion due to the presence of a humidity film at the metal surface, and corrosion in soils due to the humidity of the soil. A detailed discussion of electrochemically induced corrosion is outside the scope of this chapter. For more details, a large number of good comprehensive reviews may be consulted (for instance, Ref. [50]). Hence only some basics will be given here insofar as they may help in understanding MIC of metals.

The anodic reaction involves the oxidation of the metal into metal ions, whereas the cathodic reaction involves the reduction of some component in the corrosive environment:

$$\text{Anodic reaction: Me} \rightarrow \text{Me}^{n+} + ne^- \qquad (17.1)$$

TABLE 17.1

Industrial Applications for Which Microbial Corrosion Has Been Reported

Industry	Problem Areas
Chemical processing industries	Stainless steel tanks, pipelines, and flanged joints, particularly in welded areas after hydrotesting with natural river or well waters
Nuclear power generation	Carbon and stainless steel piping and tanks; copper-nickel, stainless, brass, and aluminum-bronze cooling water pipes and tubes, especially during construction, hydrotest, and outage periods
Onshore and offshore oil and gas industries	Mothballed and waterflood systems, oil and gas handling systems, particularly in environments sourced by sulfate-reducing bacteria (SRB)-produced sulfides
Underground pipeline industry	Water-saturated clay-type oils of near-neutral pH with decaying organic matter and a source of SRB
Water treatment industry	Heat exchangers and piping
Sewage handling and treatment industry	Concrete and reinforced concrete structures
Highway maintenance industry	Culvert piping
Aviation industry	Aluminum integral wing tanks and fuel storage tanks
Metalworking industry	Increased wear from breakdown of machining oils and emulsions

Source: Dexter, S.C., Localized corrosion, *Metals Handbook,* 13, 9th edn., ASM International, Metals Parb, OH, pp. 104–122, 1987.

Cathodic reactions may be summarized as follows:

$$O_2 + 2H_2O + 4e^- \rightarrow 4OH^- \quad \text{(acid, neutral or alkaline solutions)} \tag{17.2}$$

$$O_2 + 4H^+ + 4e \rightarrow 2H_2O \quad \text{(acidic solutions)} \tag{17.3}$$

$$2H^+ + 2e \rightarrow H_2 \quad \text{(acidic solutions)} \tag{17.4}$$

$$2H_2O + 2e^- \rightarrow H_2 + 2OH^- \quad \text{(neutral and alkaline solutions)} \tag{17.5}$$

The rates of the anodic and cathodic reactions must be balanced in order to preserve electroneutrality.

The nature of the cathodic reaction depends on the pH of the solution and on the presence of dissolved oxygen or other oxidizers such as dissolved CO_2. For instance, in the pH range between 4 and 10, the diffusion of dissolved oxygen to the metal surface controls the rate of corrosion of iron. In this pH range and in the absence of oxygen, very low corrosion rates should therefore be expected.

The term "uniform corrosion" implies that the anodic and cathodic sites are virtually inseparable, whereas "localized corrosion" implies that macroscopic anodic and cathodic sites are physically separable.

As already stated, MIC of metallic materials does not involve any new form of corrosion. The mechanisms involved in the biodeterioration of materials have already been given in detail. The main ways in which microorganisms may enhance the rate of corrosion of metals and/or the susceptibility to localized corrosion in an aqueous environment are given below in relation to known corrosion mechanisms (for each case some examples are also given).

Formation of concentration cells at the metal surface and in particular oxygen concentration cells. This effect may occur when a biofilm or bacterial growth develops heterogeneously on the metal surface. Concentration cells are also associated with the tubercles formed by iron-oxidizing bacteria, such as *Gallionella.* Certain bacteria may also trap heavy metals such as copper and cadmium within their extracellular polymeric substance, resulting in the formation of ionic concentration cells.

Modification of corrosion inhibitors. To this group belong microorganisms that may destroy corrosion inhibitors such as bacteria that transform nitrite (corrosion inhibitor for iron and mild steel) to nitrate, or nitrate (corrosion inhibitor for aluminum and aluminum alloys) to nitrite and ammonia and N_2.

Production of corrosive metabolites. Bacteria may produce different metabolites, such as inorganic acids (e.g., *Thiobacillus thiooxidans),* organic acids (almost all bacteria, algae, and fungi), sulfide (sulfate-reducing bacteria), and ammonia, which are corrosive to metallic materials.

Destruction of protective layers. Various microorganisms may attack organic coatings, and this may lead to corrosion of the underlying metal.

Stimulation of electrochemical reactions. An example of this type of action is the evolution of cathodic hydrogen from microbially produced hydrogen sulfide.

Hydrogen embrittlement. Microorganisms may influence hydrogen embrittlement on metals by acting as a source of hydrogen or/and through the production of hydrogen sulfide.

In summary, all known cases of microbial corrosion of metals can be attributed to known corrosion mechanisms. In the following sections, the mechanisms of MIC for different metals and alloys are discussed.

17.5.1.1 Mechanisms of Microbially Induced Corrosion of Iron and Mild Steel

17.5.1.1.1 Corrosion under Anaerobic Conditions

Considering the electrochemical reactions discussed earlier, very low corrosion rates are expected for iron and steel in near-neutral conditions and in the absence of oxygen. However, a large number of case histories in the literature, particularly for buried pipes and marine structures, report that the corrosion may be very severe and can be several orders of magnitude higher than that experienced under normal aerobic conditions [13]. It is well accepted that this is due to sulfate-reducing bacteria and their ability to produce sulfide. However, despite the large body of literature available, the exact mechanism(s) responsible for the corrosion in such environments is still the subject of discussion. Several good review articles present this subject in more detail than is attempted here [40,51–53].

The first attempt to explain the anaerobic corrosion of iron was that by Von Wolgozen Kühr and van der Vulgt [3]. These authors proposed as early as 1934 that sulfate-reducing bacteria were responsible for the pitting corrosion they observed on buried cast-iron pipes through their ability to remove the hydrogen from the metal surface via their enzyme hydrogenase for the dissimilar reduction of sulfate according to the following reactions:

$$4Fe \rightarrow 4Fe^{2+} + 8e^{-} \quad \text{(anodic reaction)} \tag{17.6}$$

$$8H_2O \rightarrow 8H^{+} + 8OH^{-} \quad \text{(dissociation of water)} \tag{17.7}$$

$$8H^+ + 8e^- \rightarrow 8H_{ad} \quad \text{(cathodic reaction)} \tag{17.8}$$

$$SO_4^{2-} + 8H_{ad} \rightarrow S^{2-} + 4H_2O \quad \text{(cathodic depolarization by sulfate-reducing bacteria)} \tag{17.9}$$

$$Fe^{2+} + S^{2-} \rightarrow FeS \quad \text{(corrosion products)} \tag{17.10}$$

$$3Fe^{2+} + 6OH^- \rightarrow 3Fe(OH)_2 \quad \text{(corrosion products)} \tag{17.11}$$

$$4Fe + SO_4^{2-} + 4H_2O \rightarrow 3Fe(OH)_2 + FeS + 2OH^- \quad \text{(overall reaction)} \tag{17.12}$$

Reaction (17.9) represents the ability of SRB to remove the adsorbed hydrogen from the metal surface through the enzyme hydrogenase. This reaction was referred to by the authors as "depolarization." This term was used only to underline that there was an undefined change in the electrochemical response of the system studied [54]. This is an alternative reaction path compared with the classical hydrogen evolution reaction:

$$H^+ + e^- \rightarrow H_{ad} \tag{17.13}$$

$$\nearrow \quad 2H_{ad} \rightarrow H_2 \quad \text{(chemical desorption)} \tag{17.14}$$

$$\searrow \quad H_{ad} + H^+ + e^- \rightarrow H_2 \quad \text{(electrochemical desorption)} \tag{17.15}$$

Reactions (17.14) and (17.15) are the rate-determining steps (i.e., a higher activation energy is required for these reactions compared with the discharge of hydrogen ions) for many metals and alloys, including iron in deaerated aqueous solutions. The rate-determining steps may, however, be different in the case of complex corrosion layers. Sulfate-reducing bacteria, by removing adsorbed hydrogen to form sulfide according to reaction (17.9), lower the activation energy for the desorption steps. As these steps are rate determining, this may lead to higher corrosion rates if hydrogen evolution is thermodynamically possible (i.e., if the immunity potential is lower than the redox potential for H_2/H^+).

This theory, also referred to in the literature as the classical depolarization theory, suggests that only sulfate-reducing bacteria that are hydrogenase-positive (i.e., bacteria in which the enzyme hydrogenase is present) are responsible for the anaerobic corrosion of iron. Even though there are data supporting this view [55,56], it has been shown in many other works that this is not the case and that a high corrosion rate may be obtained with hydrogenase-negative strains [57–59]. In addition, several other important factors are not taken into account in the classical depolarization theory: (a) corrosion is a kinetically controlled process that is globally irreversible [146]; (b) the effects of sulfide, bisulfide, and hydrogen sulfide produced from the sulfate reduction on the anodic reaction; (c) the effect of hydrogen sulfide on the cathodic reaction; (d) the effect of elemental sulfur from the biotic or abiotic oxidation of sulfur; (e) fluctuations in the environmental conditions between anaerobic and aerobic conditions [13]; and (f) the production of other corrosive metabolites [53].

TABLE 17.2

Alternative Mechanism to the Classical Depolarization Theory

Name Referred to in the Literature and References	Main Mechanisms	Role of Hydrogenase
Depolarization by iron sulfide [63]	Formation of an iron/iron sulfide galvanic cell, iron sulfide acting as the site for cathodic reduction of molecular hydrogen	Secondary through the regeneration of ferrous sulfide
Depolarization by hydrogen sulfide [64]	Cathodic reduction of microbially produced hydrogen sulfide: $H_2S + e^- \rightarrow HS^- + H_2$	Secondary through the production of hydrogen sulfide
Elemental sulfur [65]	Formation of a concentration cell with elemental sulfur acting as the reactant	Secondary through the production of elemental sulfur
Iverson's mechanism [66]	Production of a volatile and corrosive iron phosphite metabolite	Not defined
Local acidification of anodes [61]	Localized acidification of anodes due to the formation of iron sulfide corrosion products: $Fe^{2+} + HS^- \rightarrow FeS\downarrow + H^+$	None

It is therefore now widely accepted that it is rather unlikely that the cathodic reaction is affected directly by the enzymatic consumption of hydrogen as described in the "cathodic depolarization theory." This theory is therefore reported here more for historical interest than as a potential mechanism for pitting corrosion of iron in presence of SRB.

More recent theories have been proposed in which the role of the biogenic sulfide, the formation of a galvanic cell between the metal and the iron sulfide film formed, the role of elemental sulfur, the role of iron phosphites, and the local acidification of anodes have been discussed. The alternative theories to the classical depolarization theory are presented briefly in Table 17.2.

It has also been shown that the nature of the iron sulfide film plays an important role in the initiation of pitting corrosion. Figure 17.5 shows the different biogenically and chemically formed iron sulfide films [60]. When protective films such as mackinawite and siderite are formed, the corrosion rate may be very low. However, with changes in the environmental conditions, these films may be transformed to iron sulfides that are less protective, such as pyrrhotite, with the possible initiation of pits as a result. The importance of the environment for the initiation of localized corrosion by SRB as been mentioned by several authors [61,62]. Of particular importance is the ability of SRB to regulate pH, the ionic composition, the composition of Fe^{2+}, aerobic/anaerobic cycles, and the presence of other microorganisms, as they may lead to conditions resulting in passivation or uniform corrosion or localized corrosion. This is probably the main reason why the severe pitting corrosion experienced in the field is only seldom reproduced in the laboratory using growing cultures of SRB, where uniform corrosion is generally observed. Indeed, in practice, SRB are not found in isolation but in consortia of microorganisms or biofilms in which many physicochemical parameters such as pH and dissolved oxygen concentration change both with time and within the thickness of the biofilm. This dynamic picture of microbial corrosion is discussed in more detail later.

17.5.1.1.2 Corrosion under Aerobic Conditions

Corrosion may occur under aerobic conditions through the production of sulfuric acid by bacteria of the genus *Thiobacillus* [67]. The sulfuric acid is produced through the oxidation of various inorganic sulfur compounds, as illustrated in Table 17.3 [68]. Some of these

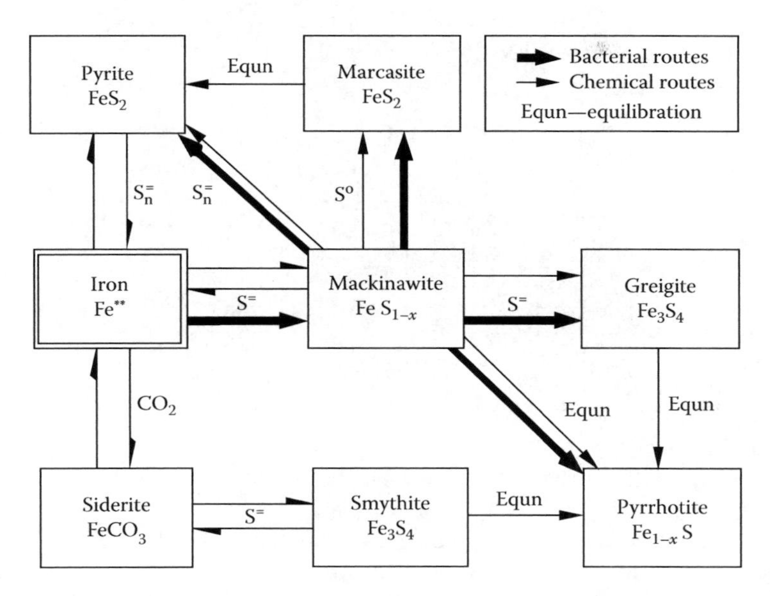

FIGURE 17.5
Chemical and biological interrelationships between iron sulfides. (From Rickard, D.T., The microbiological formation of iron sulphides, *Proceedings of Contribution to Geology*, Stockholm, Sweden, pp. 67–72, 1969.)

TABLE 17.3
Oxidation Reactions of Thiobacilli and Other Aerobic Sulfur-Oxidizing Bacteria

$S_2O_3^{2-}$	+	$2O_2$	+	H_2O	→	$2SO_4^{2-}$	+	$2H^+$		
2S	+	$3O_2$	+	$2H_2O$	→	$2SO_4^{2-}$	+	$4H^+$		
$4S_2O_3^{2-}$	+	O_2	+	$2H_2O$	→	$2S_4O_6^{2-}$	+	$4OH^-$		
$2S_4O_6^2$	+	$7O_2$	+	$6H_2O$	→	$8SO_4^{2-}$	+	$12H^+$		
2SCN	+	$4O_2$	+	$4H_2O$	→	$2SO_4^{2-}$	+	$2CO_2$	+	$2NH_4^+$
		H_2S	+	$2O_2$	→	SO_4^{2-}	+	$2H^+$		
		$2H_2S$	+	O_2	→	2S	+	$2H_2O$		
$2S_3O_6^2$	+	$4O_2$	+	$4H_2O$	→	$6SO_4^2$	+	$8H^+$		
		$5H_2S$	+	$8NO_3$	→	$5SO_4^{2-}$	+	$4N_2$	+	$4H_2O$
5S	+	$6NO_3$	+	$2H_2O$	→	$5SO_4^{2-}$	+	$3N_2$	+	$4H^+$
$5S_2O_3^{2-}$	+	$8NO_3$	+	H_2O	→	$10SO_4^{2-}$	+	$4N_2$	+	$2H^+$

Source: Cragnolino, G. and Tuovinen, O.H., *Int. Biodeterior.*, 20, 9, 1984.

bacteria can tolerate a concentration of sulfuric acid up to 10%–12%. Under these conditions, iron and mild steel are heavily corroded.

Corrosion of iron and steel may also occur in aerobic conditions due to the colonization of bacteria at the metal surface and the formation of an uneven patchy biofilm. Nonuniform biofilms or bacterial colonization results in the formation of differential aeration cells in which the anodic areas are found under respiring colonies or in the thick part of the biofilm (zones depleted in oxygen) whereas the cathodic areas are found in noncolonized areas or in the areas where the biofilm is thin (regions rich in oxygen). This is schematically illustrated in Figure 17.6. The iron-oxidizing bacteria such as *Gallionella fermginea, Crenothrix,* and *Leptothrix* are often associated in the literature with this form of

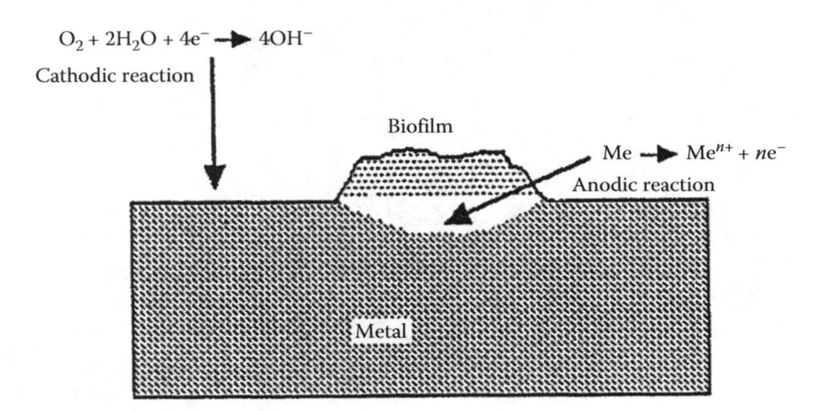

FIGURE 17.6
Schematic representation of the influence of the biofilm in the formation of differential aeration cells.

corrosion, particularly in the internal surfaces of water pipes. These bacteria are aerobes and are believed to oxidize ferrous ions into ferric ions with the resulting formation of heavy deposits of ferric oxide or hydroxide, also referred to as tubercles. However, even if iron-oxidizing bacteria are the best-documented case, many other organisms may be involved in this kind of MIC.

Finally, it should be mentioned that microorganisms may cause the degradation of corrosion inhibitors such as aliphatic amines, nitrite, and phosphate-based corrosion inhibitors used, for instance, for steel pipes in cooling water systems. This results in a higher demand for corrosion inhibitor and an increase in the bacterial population, which leads to high corrosion rates, if the system is not controlled [69,70].

17.5.1.1.3 Microbial Biofilms and the Interactions between Aerobic and Anaerobic Populations

In the preceding discussion, the mechanisms of microbial corrosion have been divided in the traditional way into anaerobic and aerobic mechanisms, which refer to the living conditions of the microorganisms involved in the corrosion processes. However, it is now well known that bacteria are not found in isolation but rather in consortia or biofilms in which many bacterial communities coexist. A good illustration of this is given in Figure 17.7 [71]. As already mentioned, a biofilm is composed mainly of bacterial cells, extracellular polymers, inorganic ions, and water. The biofilm structure allows the flow of nutrients, metabolates, waste products, and enzymes [72]. Although it may be difficult to make a general picture of the biofilm, some authors even consider the biofilm matrix as a system of immobilized enzymes [73] with dynamic exchange with the medium. The biofilm influences the mass transfer of different species and in particular dissolved oxygen from the solution to the metal, thereby creating a zone depleted in oxygen at the metal–biofilm interface (Figure 17.7a). In this part of the biofilm, facultative and obligate anaerobic microorganisms such as sulfate-reducing bacteria may find suitable conditions of growth (anaerobic conditions and low redox potential). The result of this is the formation of microcolonies of SRB, which, through their production of metabolite products, may generate conditions favorable for other anaerobic bacteria to grow (Figure 17.7b). This creates areas on the metal surface with local gradients in, for instance, hydrogen sulfide and hydrogen concentrations that may initialize localized corrosion on iron and steel (Figure 17.7c). Many authors [74–76], have reported similar synergistic effects of bacterial consortia on the rate of corrosion of mild steel.

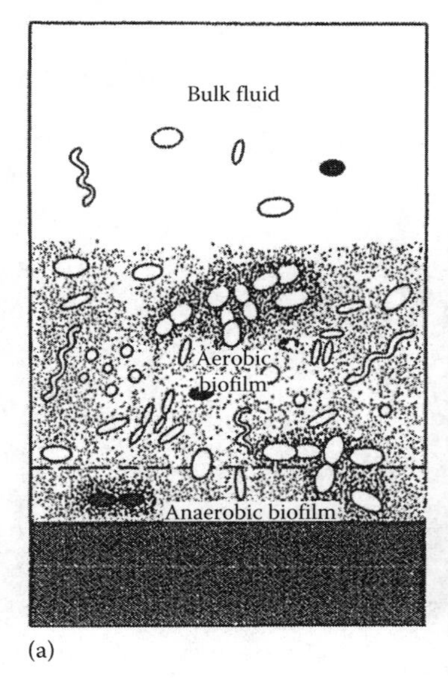

(a)

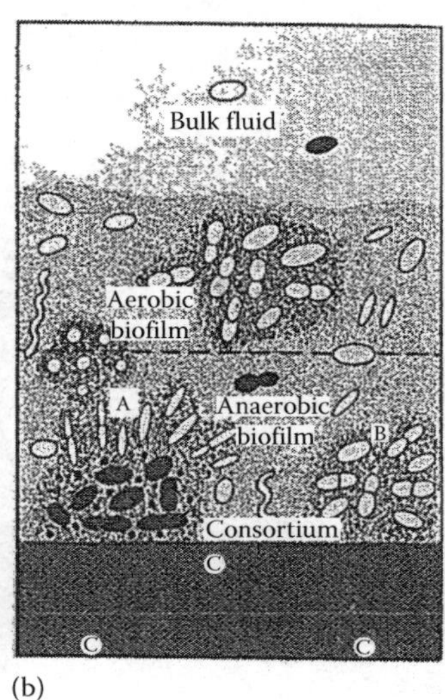

(b)

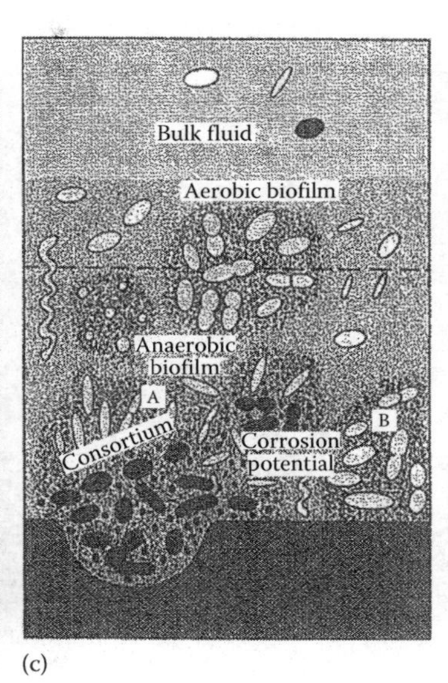

(c)

FIGURE 17.7
Diagrammatic representation of the formation of microbial consortia and their influence on the corrosion processes: (a) When the film becomes sufficiently thick, its inner part will be anaerobic with the possible development of SRB microcolonies (black cells). (b) The SRB attracts secondary colonizers by its metabolic products and forms a consortium with them. (c) The development of local areas with varying physiochemical parameters leads to pitting corrosion. (From Costerton, J.W. and Geesey, G.G., The microbial ecology of surface colonization and of consequent corrosion, *Proceedings of Biologically Induced Corrosion*, S. C. Dexter, ed., NACE-8, Houston, TX, pp. 223–232, 1986.)

In addition to the ways already mentioned in which microorganisms influence the corrosion of iron and mild steel, there are numerous reports that microorganisms may cause hydrogen embrittlement, stress corrosion cracking, and corrosion fatigue [77–79]. It has been shown that in pure cultures of hydrogen-producing bacteria (hydrogen is produced as an end product of the fermentation process) hydrogen embrittlement is enhanced. In natural microbial biofilms, the situation may be described according to Figure 17.8 [79]. Consortia of hydrogen-producing and hydrogen-consuming bacteria are formed at the metal surface, creating, for instance, gradients of hydrogen [80,81]. Hence, depending on the primary colonizers (hydrogen producers or hydrogen consumers) and their distribution on the metal surface, hydrogen embrittlement may or may not occur.

It is also obvious that the production of H_2S from SRB may accelerate hydrogen embrittlement, stress corrosion cracking, and corrosion fatigue.

From these studies, it seems clear that, even if studies of microorganisms in isolation (i.e., in pure cultures) give valuable information concerning the mechanisms of MIC of a particular species in a given environment, more work should be done on the effect of mixed bacterial populations on the mechanisms of MIC.

17.5.1.2 Mechanisms of Microbially Induced Corrosion of Stainless Steels and Titanium

There are large numbers of reported case histories of MIC on stainless steel in water and aqueous waste systems. They are related to different industrial applications such as freshwater storage and circulation systems in nuclear power plants [82–85] and cooling water

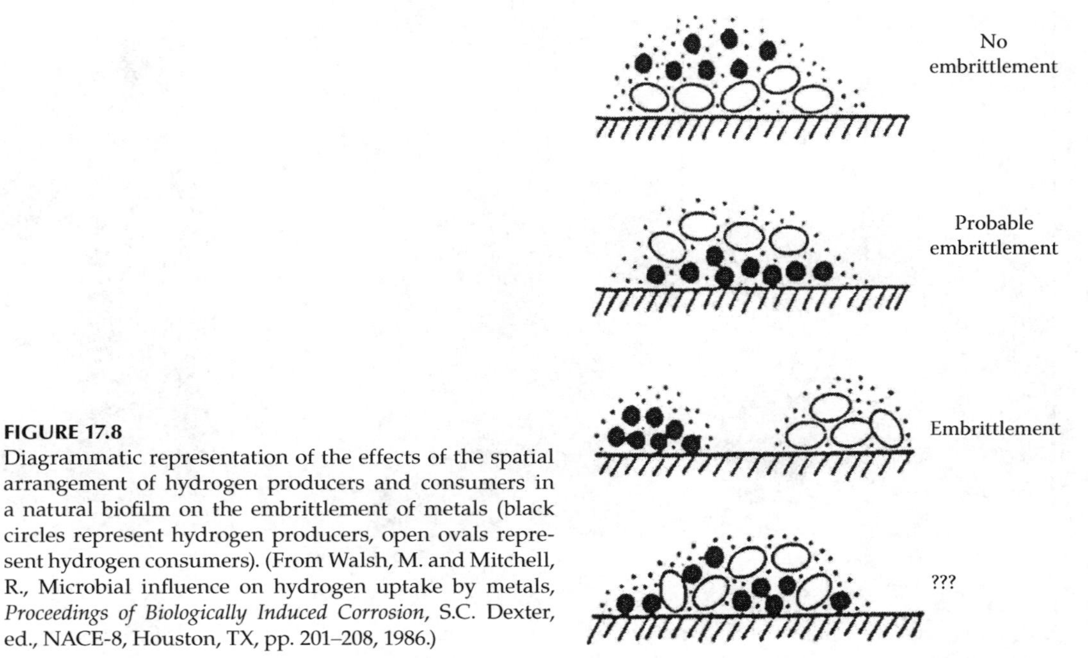

FIGURE 17.8
Diagrammatic representation of the effects of the spatial arrangement of hydrogen producers and consumers in a natural biofilm on the embrittlement of metals (black circles represent hydrogen producers, open ovals represent hydrogen consumers). (From Walsh, M. and Mitchell, R., Microbial influence on hydrogen uptake by metals, *Proceedings of Biologically Induced Corrosion*, S.C. Dexter, ed., NACE-8, Houston, TX, pp. 201–208, 1986.)

systems in chemical process industries [86,87]. There are basically three cases: (a) crevice corrosion under unexpected deposits, (b) sensitivity of pitting and crevice corrosion to trace of H_2S, and (c) crevice corrosion in natural seawater. Most of these reports are not well documented concerning the microorganisms involved in the process. However, some general features are common in almost all cases: (a) MIC is reported for the alloys with a relatively low content of molybdenum; (b) pitting or crevice corrosion generally occurs at or around welds (i.e., near the fusion line, or with sensitization in the heat-affected zones); (c) discrete localized deposits are often found in direct connection, with or close to the corrosion sites.

The higher sensitivity to MIC of welds compared with the unwelded metal may be explained by a difference in surface roughness or/and chemical composition that facilitates the colonization of the surface by microorganisms.

In some reports, a higher sensitivity to MIC of the delta-ferrite in duplex welds has been reported [88,89], and in others both the austenite and delta-ferrite phases have been shown to be sensitive to MIC [90]. There is also evidence that surface treatment of welds, such as solution annealing and pickling or polishing and grinding, may result in welds that are less susceptible to microbial corrosion [90,91].

The observation of localized deposits is often associated with the presence of slime-forming bacteria such as *Pseudomonas* and iron bacteria. These bacteria consume oxygen diffusing into the deposit and thus create a differential aeration cell that may initiate crevice corrosion. Under the deposit, the environment becomes anaerobic with the subsequent possibility of growing SRB in a scenario similar to that described for microbial biofilms. Besides the effect of concentration cells, the specific role of metal-concentration/oxidizing bacteria such as iron- and manganese-oxidizing bacteria in the initiation of localized corrosion is often mentioned [92]. By controlling the redox potential of Fe^{2+}/Fe^{3+} and/or Mn^{2+}/Mn^{4+}, these bacteria may polarize the metal surface at a potential at which Fe^{3+} and/or Mn^{4+} may exist. Together with the accumulation of chloride ion (due to the necessity of preserving electroneutrality), this leads to the formation of $FeCl_3$ and $MnCl_4$ solutions that are aggressive to stainless steels. It is well known that both situations facilitate the breakdown of the passive film with crevice corrosion and/or pitting corrosion as a result.

Sulfate-reducing bacteria have also been reported to be responsible for pitting corrosion on stainless steels in aqueous environments. The mechanisms proposed are mostly related to iron and steel. However, a different mechanism has been proposed in which the role of thiosulfate in the microbial pitting of stainless steel has been emphasized [93]. The same authors have also demonstrated clearly that SRB-induced pitting corrosion of stainless steel is unlikely to occur in a uniformly anaerobic SRB medium, whereas it will occur when the anaerobic sites are coupled to an oxygen cathode [94]. In a recent study, the electrochemical behavior of AISI 316 L has been studied in presence of hydrogenase and NAD^+ in phosphate buffer. There findings are summarized in Figure 17.9 [95,96]. Mechanism A is in fact the same as the one described previously and called "cathodic depolarization" with the same limitations in validity as discussed previously. Mechanisms B and C correspond to a direct electron transfer between the hydrogenase and the stainless steel. Mechanism B suggests a direct contact between the hydrogenase and the metal whereas this is not the case in for mechanism C. However, the question of the need of hydrogenase in direct contact with the stainless steel surface is still a matter of discussion.

Atomic force microscopy was used to quantify pitting corrosion on AISI 304 colonized by either aerobic *Pseudomonas* or anaerobic *Desulfovibrio desulfuricans* [97]. The results indicate that SRB exhibited more corrosive properties with deeper pits compared to slime formers such as *Pseudomonas*.

This discussion is related almost solely to alloys with a relatively low molybdenum content, but the high molybdenum alloy stainless steels may also be susceptible to MIC. 904 L, AL-6X, and 254 SMO were attacked by bacteria in pure cultures of *D. desulfuricans* and *Hyphomicrobium indicum*. The corrosion attack seems to be localized at the grain boundaries [98]. However, MIC on highly alloyed stainless steels has never clearly been reported in natural environments.

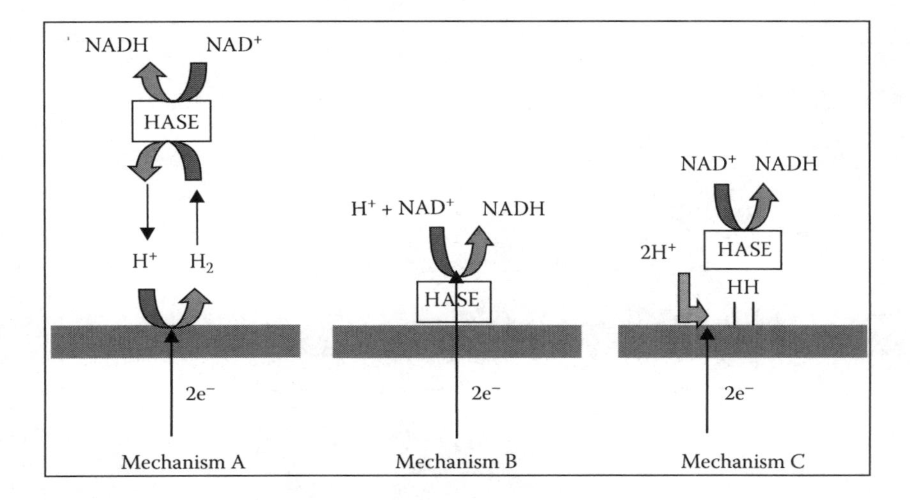

FIGURE 17.9
Possible mechanisms explaining cathodic depolarization on stainless steel. (From Da Silva, S. et al., *Bioelectrochemistry*, 56, 77, 2002.)

Besides the reports of MIC described above that are mainly related to water with relatively low salt content such as fresh water, an increase in the corrosion potential has also been reported for stainless steels exposed to natural seawater. The ennoblement of the corrosion potential has been observed for various stainless steel compositions (i.e., austenitic, ferritic, duplex, and superaustenitic stainless steels) exposed in natural seawater with different salinities and at different temperatures [99–104]. It is well known that in a given steel/environment system, localized corrosion occurs on stainless steel when the corrosion potential becomes greater than the pitting potential. Hence, the result of the potential ennoblement is that there is a higher probability of localized corrosion of stainless steel in a natural environment than in artificial seawater. This effect has been connected with the presence of a biofilm on the metal surface as shown in Figure 17.10. The corrosion potential of 21Cr-3Mo steel increases as a function of time in natural seawater to a value of about 400 mV/SCE, whereas it remains unchanged in sterilized seawater. When a specific inhibitor of respiratory activity (sodium azide) is added to both natural and sterilized seawater, a rapid decrease in the corrosion potential is observed in the case of natural seawater but it remains unchanged in sterile seawater [105]. The increase of the corrosion potential on stainless steel has been correlated with an increase of the rate of the cathodic reaction, e.g., the reduction of oxygen. This is shown in Figure 17.11. During the development of the biofilm at the stainless steel surface, an important depolarization of the oxygen reduction reaction occurs. This will lead to a higher rate of propagation of localized corrosion on stainless steel. Hence the effect of the biofilm settlement on stainless steel may be summarized as follows:

1. The corrosion potential of stainless steel increases in the noble direction, which increases the probability of localized corrosion.
2. A higher corrosion propagation rate is observed once localized corrosion develops.
3. Higher galvanic currents between stainless steel coupled to less noble materials are measured.

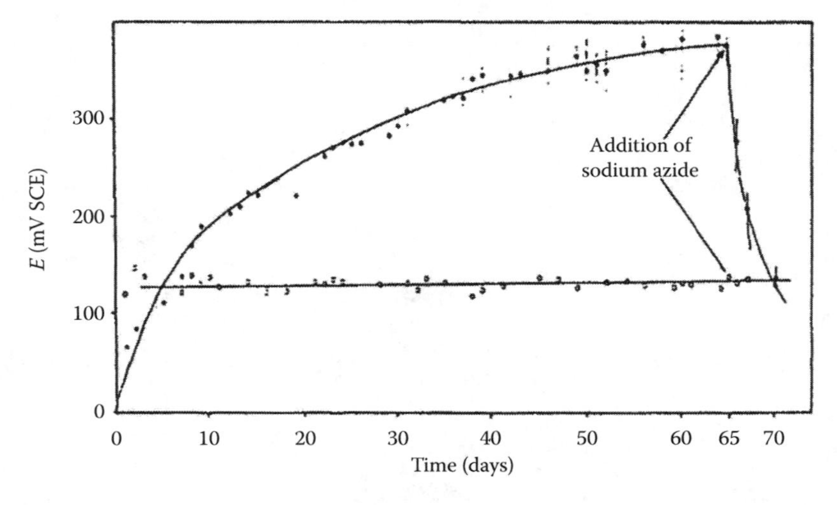

FIGURE 17.10
Corrosion potential as a function of exposure time for 21Cr-3Mo steel in natural seawater (black circles) and in artificial seawater (white squares). Time of addition of sodium azide is indicated by arrows. (From Scotto, V. et al., *Corros. Sci.*, 25, 185, 1985.)

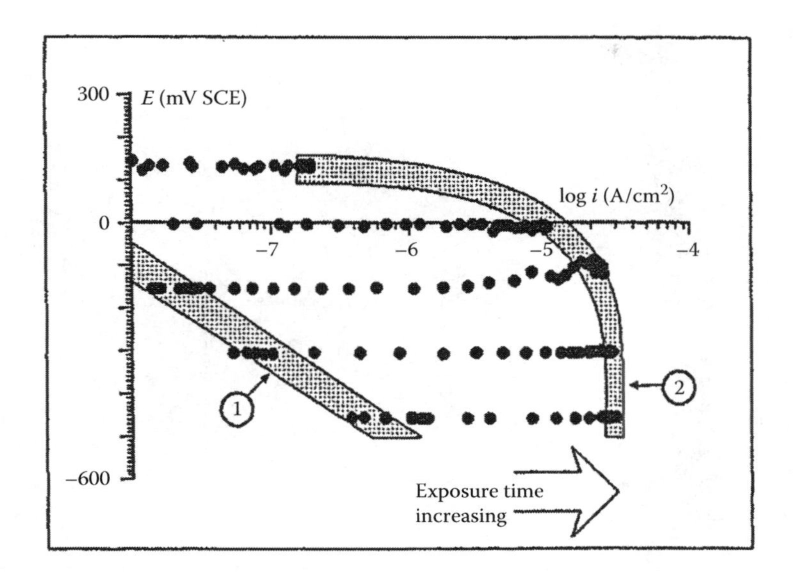

FIGURE 17.11
Oxygen reduction current density on stainless steel exposed to natural seawater and continuously polarized at fixed cathodic potentials. The envelopes of curves 1 and 2 correspond to the set of data on clean and already fouled stainless steel surfaces, respectively. (From Mollica, A. et al., *Aspects of Microbially Induced Corrosion*, D. Thierry, ed., University Press, Cambridge, U.K., 1996.)

The study of the mechanisms of the ennoblement of stainless steel is complex, as seawater constitutes an environment in which numerous parameters can act. For instance the role of passive films on the oxygen reduction of stainless steel has been demonstrated by several authors [106,107]. A literature review of the subject has been published [108]. It has been shown that the potential ennoblement in the presence of biofilms is probably due to an enhancement of the cathodic reduction of oxygen by (a) a decrease in pH at the metal–biofilm interface, (b) the formation of macrocyclic organometallic catalysts, such as porphyrins and phtalocyanins, (c) bacterially produced enzymatic catalysis, or (d) the production of hydrogen peroxide within the biofilm or the catalysis of the oxygen reduction by manganese. More recent literature on the subject seems to confirm the role of enzymes in the depolarization of oxygen reduction reaction on stainless steel [109,110]. A possible mechanism in which the potential ennoblement results from enzymatic competition between the production of hydrogen peroxide and acids by oxidases and the enzymatic consumption of these two chemicals has been proposed. This is schematically shown in Figure 17.12. Based on a large amount of biofilm growth on stainless steel at different European locations, it has been shown that enzymes entrapped in the EPS matrix of the biofilm could be responsible for the ennoblement of the corrosion potential on stainless steel [111]. However, other authors have shown that EPS alone were not able to reproduce the electrochemical effects observed on stainless steel and the ennoblement was more related to the enzymatic activity within the biofilm and the subsequent production of H_2O_2 [110]. The important role of the formation of H_2O_2 on the ennoblement was evidenced on AISI 316 in seawater by addition of H_2O_2 decomposing enzymes (e.g., catalase and peroxidase) resulting in an offset of the ennoblement potential [113]. In connection to these results, it has been shown that a solution containing glucose oxidase and its substrate D-glucose provided the same ennoblement and the same crevice corrosion results for different stainless steels compared to that obtained in natural seawater [114]. All these results support the hypothesis based

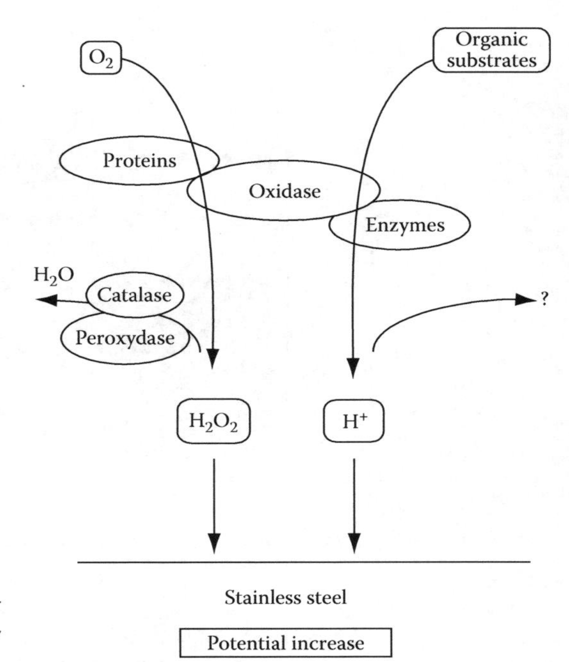

FIGURE 17.12
Hypothetical mechanism of the stainless steel ennoblement in natural seawater. (From Dupont, I. et al., *Int. Biodeterior. Biodegrad.*, 41, 13, 1998.)

on the importance of enzymatic catalysis in the potential ennoblement of stainless steel in natural waters. This is in good correlation with several reports showing that biofilms formed in natural seawater are able to generate hydrogen peroxide at a concentration in the range of mmol/L [115,116]. More information on the importance of the enzymatic approach on MIC and in particular on the ennoblement of stainless steel can be found in a recent review [117].

Finally it should be noted that potential ennoblement of stainless steel has also been reported for stainless steel exposed to river water (e.g., at low chloride content) [118–120]. In this case, pitting corrosion associated with the potential ennoblement occurred on low alloyed stainless steel as for instance AISI304 and AISI316. The ennoblement has often been explained by the deposition of manganese oxides. This mechanism also called manganese biomineralization has been evidenced by Dickinson and Lewandowski [121,122]. In this case, MnO_2 is formed on the stainless steel surface from Mn^{2+} by Mn-oxidizing bacteria and then it is reduced according to reactions (17.16) and (17.17). It should be emphasized that this mechanism implies a direct contact between the MnO_2 and the stainless steel surface as well as the presence of manganese oxidizing bacteria.

$$MnO_{2(s)} + H^+ + e^- \rightarrow MnOOH_s \tag{17.16}$$

$$MnOOH_s + 3H^+ + e^- \rightarrow Mn^{2+} + 2H_2O \tag{17.17}$$

According to a literature review, there is no reported case of MIC on titanium [123].

17.5.1.3 Mechanisms of Microbially Induced Corrosion on Aluminum and Aluminum Alloys

Case histories of MIC on aluminum and aluminum alloys have been mainly reported for aircraft fuel storage tanks and heat exchanger tubes using water with different salinities. In all cases, pitting corrosion occurred.

Even though these cases are not as well documented as MIC on iron and steel, the possible mechanisms and microorganisms involved have been identified for the microbial contamination of aircraft aluminum fuel tanks [124,125]. Some of the most predominant bacteria and fungi are *Pseudomonas aeruginosa, Aerobacter aerogenes, Desulfovibrio,* and *Cladosporium* species. The corrosion occurs mainly in the water phase of the fuel–water mixture at the bottom of the tanks and at the fuel–water interface. The mechanisms involved are the same as those already given for iron and steel: (a) the production of corrosive metabolites by bacteria and fungi (organic acids and hydrogen sulfide), (b) the creation of differential aeration cells, (c) the transformation of nitrate (corrosion inhibitor for aluminum) to nitrite, and (d) extracellular activity resulting in the removal of major or minor metallic atoms from the basic structure of the alloy [59].

17.5.1.4 Mechanisms of Microbially Induced Corrosion on Copper and Copper Alloys

As in the case of aluminum and aluminum alloys, the MIC of copper and its alloys is not very well documented. This is probably due to the general belief that copper is toxic to microorganisms, which is not, however, the case for all microorganisms. As an example, bacteria of the genus *Thiobacillus* may tolerate copper concentrations up to 6%.

Case histories of MIC on copper and its alloys have been reported for piping systems and heat exchangers. Whereas heat transfer problems are observed with growing biofilms, corrosion increases after the death of the microorganisms within the biofilm. This is believed to be due to the production of ammonia and carbon dioxide upon the death of cells that may result in pitting corrosion and/or stress corrosion cracking for copper and its alloys [59,123].

The production of hydrogen sulfide by sulfate-reducing bacteria is also believed to lead to pitting corrosion and stress corrosion cracking on copper and copper alloys. In this case, corrosion is due to the formation of a thick nonadherent layer of chalcocite (Cu_2S) or covellite (CuS_{1-x}). Pitting corrosion may occur where the copper sulfide film has been removed, with the cathodic reaction taking place on the intact copper sulfide film [126].

Microbially induced pitting corrosion has also been observed in isolated locations in Europe, the United States, Australia, and New Zealand in potable water reticulation systems plumbed by copper [127,128]. This results in a contamination of the water such that the water is in breach of the maximum copper levels permitted by the U.S. Environmental Protection Agency (EPA) lead-copper rule (i.e., 1.3 ppm) [126]. Generally, a biofilm is observed under a black deposit of copper (II) oxide and basic copper salts. Although the link between the pitting corrosion of copper and the microbial activity has been clearly established, the exact mechanism by which microorganisms increase the probability of initiation of pitting corrosion on copper is not known at the present time. However, it has been suggested that the role of the biofilm could be to oxidize copper (I) oxide to copper (II) oxide [129]. Finally, it should be noted that an increase in the bicarbonate level of water as well as a change in the chloride/sulfate ratio with increasing chloride content yielded a decrease in the initiation and propagation of pitting corrosion copper, respectively [127].

17.5.2 Other Construction Materials

17.5.2.1 Mineral Materials

Concrete is a mixture of a coarse-grained aggregate, hydraulic binding agents (cement, gypsum, lime, asphalt, sulfur, and resin), and water (except in the case of sulfur and resin, where water is not needed). For a limited period after preparation, the mixture can be molded into different shapes until hardening occurs as a result of chemical reactions between the components.

Ceramics are products made from fired clay–containing masses such as kaolin. The products are distinguished by the grain size of the clay and additives used and the firing temperature.

Glass is a transparent, sometimes colored, inorganic material, which is fragile and predominantly noncrystalline. It has no well-defined melting point, but with continuous heating, it changes from being viscous to a soft and finally thin fluid state (i.e., an undercooled melt without crystallization).

Natural stone originates from rocks obtained from quarries in natural deposits. Rocks are defined as natural, solid formations of the earth's crust. They consist of a mixture of minerals or a single mineral and may contain organic residues.

Characteristics are the composition of primary (main), secondary, and auxiliary minerals, the structure, and the stratification. Three main groups need to be distinguished. Magmatic rocks are molten masses in the earth's crust (basalt, granite, etc.). Sedimentary rocks contain grains (often quartz) held together by a binding material (often calcareous or siliceous, such as gypsum, limestone, sandstone, schist). Metamorphic rocks are formed by transformation from magmatic or sedimentary types by heat and pressure (quartzite, marble, crystalline schist).

All these materials are characterized as nonmetallic mineral materials. A biogenic attack can be caused by almost all microorganisms and by the mechanisms described above [28,29,35,36,130–139]. Only the action of exoenzymes, emulsifying agents, and organic solvents seems to be negligible for these materials. However, if mixtures of mineral materials with organic polymers are used to improve the properties, as in resin-modified mortars or in resin- or sulfur-bound concrete, the latter mechanisms may contribute to some extent.

17.5.2.2 Organic Polymers of Natural and Synthetic Origin

To the group of construction materials [25] belong, besides stone, those that have been used since humans began construction work. Wood is the oldest known construction material. Because of its unique characteristics, it is still in use today. The main components of wood are cellulose, polyoses (hemicelluloses), and lignin. The proportions of these components in wood are 40%–50%, 15%–35%, and 25%–35%, respectively. Extractable compounds such as terpenes, waxes, or tannins may account for 1%–3% and mineral compounds for 0.1%–0.5%. The tensile strength of wood is determined by the cellulose cell structure. The cellulose molecule contains from 10,000 up to 14,000 β-(1,4)-glycosidically bound glucose units and represents the valuable part of wood for further products such as paper, cardboard, foils, films, and fibers.

Polyoses are also polysaccharides, but with lateral chains and ramifications. Between 50 and 200 saccharide molecules are polymerized. Galactoglucomannan or 4-*O*-methylglucuronoxylan is the main constituent. The polyoses connect the polysaccharides with lignin in the cell wall. They determine the swelling and shrinkage of wood.

Lignin, either guaiacylpropane solely or combined with syringylpropane, is a complex molecule responsible for the pressure resistance of wood. The extractable substances such as terpenes provide the resistance against degrading microorganisms, e.g., bacteria and/or fungi. From these two groups and lichens, wood-degrading microorganisms originate. The main action is due to exoenzymes degrading the highly polymerized substances into water-soluble monomers or dimers. Similar mechanisms are involved in the degradation of wood-based materials such as laminated timber, blackboard, plywood, chipboard, and fiberboard. Plywood and chipboard may be especially endangered by biogenic attack because the binders used for these materials may increase their susceptibility. Binders may be urea-formaldehyde resins, phenol-formaldehyde resins, cresol-formaldehyde resins, isocyanates, etc., which may serve as a nitrogen source.

Besides wood, construction materials are made from plastics. These are materials consisting mainly of macromolecular organic compounds obtained synthetically or by modification of natural products. On application of heat and pressure, synthetics melt and become formable. Three main groups of synthetic polymers may be distinguished.

The term "polycondensates" covers phenolic, urea, thiourea, melamine, unsaturated polyester, alkydic, and allyl resins as well as silicones, polyimides, and polybenzimidazoles in the group of duroplasts. To the category of thermoplasts belong polyamides, polycarbonates, polyesters, polyphenylene oxides, polysulfones, and polyvinyl acetals.

Polymerizates are generally thermoplasts (polyethylenes, polypropylenes, poly-1-butenes, poly-4-methyl-1-penteneionomers, polyvinyl chlorides, polyvinylidene chlorides, polymethyl methacrylates, polyacrylonitriles, polystyrenes, polyacetals, fluorosynthetics, polyvinyl alcohols, polyvinyl acetates, and poly-*p*-xylylenes).

Polyadducts include duroplasts such as epoxidic resins and cross-linked polyurethanes as well as thermoplasts such as linear polyurethanes and chlorinated polyethers.

Although natural compounds are believed to be degradable by microorganisms in general—microbiological omnipotence—this does not always apply to synthetic polymers. These substances belong at least partially to the xenobiotica, which are known to be resistant to biological attack. For a few synthetic resins, experience of biodegradability has been available for nearly a century. Perspex has been demonstrated to be biologically nondegradable. The same holds for nonflexible PVC. The main groups of resins may be characterized with regard to their biodegradability as follows. Polyurethanes with molecular weights of 10,000 and above are generally believed to be easily biodegradable. However, introducing substituents into the molecule may modify this. Polyethylenes with molecular weights of 1500–7000 range from resistant to easily biodegradable. Polypropylenes (MW 25,000–500,000) also vary from resistant to easily biodegradable. Polymethyl methacrylates (MW 500,000–1,000,000) and polyamides (MW 10,000–100,000) are comparable to those already mentioned. Polystyrenes (MW 200,000–300,000) are generally believed to be resistant to biological attack. The same holds for polyvinyl chlorides (MW 30,000–520,000) and polytetrafluoroethylenes (MW 500,000–5,000,000).

Besides the basic polymers, the final products often contain a variety of additives to adapt a resin to the intended purpose. This includes softeners (may make up to 60% of the final product), stabilizers, antioxidants (0.01%–2%), antistatics, light-protecting agents (0.1%–1%), optical brighteners, anti-ignition agents (up to 30%), microbicides (0.3%–5%), lubricating agents, filling agents (up to 30%), accelerators, and pigments (0.02%–5%).

It is not possible in the available space to mention all the compounds used. It is important that the majority of these additives are organic compounds, which are often easily biodegradable. As a consequence, a synthetic polymeric material may not be biodegradable with respect to the polymeric backbone, the synthetic polymer, but it becomes biodegradable because of the additives used. A result may be an embrittlement of the material (e.g., flexible PVC after being used several times). Even if a synthetic material is not attacked structurally, depending on the use, microbial growth may cause deterioration associated with, e.g. aesthetically intolerable color changes. A variety of microorganisms produce pigments (mainly black, some also blue or red) that may diffuse into the resin, causing a change in appearance.

Rubber, of either natural or synthetic origin, also contains because of its method of production many biodegradable compounds. It is used for purposes in which it needs to maintain its elasticity for a long time. Generally, synthetic rubbers are more stable than natural ones. In any case, rubbers may be attacked by microorganisms.

17.6 Countermeasures

A large number of approaches may be used to prevent or minimize the microbiological degradation of construction materials. The choice of the proper approach depends on many factors, such as the nature of the environment in which MIC occurred (e.g., soil, cooling water, seawater), the type of microorganisms involved, and the nature of the material. In practice, several approaches are often used in order to increase their efficiency. Many reviews on this subject are available in the literature [51,140–142]. Thus, the different approaches are discussed here only briefly. In any case, it is important to note that in the case of MIC, the affected structure should not be cleaned prior to analysis in order to be able to trace the origin of the problem.

17.6.1 Changing or Modifying the Material

The choice of a material that is not susceptible or less susceptible to bacterial degradation may be one solution. However, this is often not possible for economic reasons.

17.6.2 Modifying the Environment or Process Parameters

The environment can be modified through, for instance, avoidance of anaerobic zones or modification of the pH in order to prevent acid accumulation. A good example of this is in the use of sand, gravel, or chalk as a nonaggressive backfill around buried steel pipes. This provides drainage and consequently better aeration of the soil. Chalk also provides an alkaline environment that may prevent acid accumulation. Similarly, stagnant conditions should be avoided in water systems. The process parameters may also be changed. For example, an increase in water velocity in heat exchanger tubes leads to partial detachment of biofilm. However, it is not always possible to change the process parameters. Besides, one must be careful not to induce other corrosion problems. For instance, a high water velocity may cause erosion corrosion on some metals and alloys.

17.6.3 Organic Coatings

Organic coatings are widely used for the protection of steel in environments such as buried pipes, exteriors of buildings, and marine structures. However, all paint coatings are more or less biodegradable (see plastic materials). Besides, care must be taken with the surface preparation before painting and application of organic coatings in order to avoid voids or pores, where microorganisms may grow with highly localized corrosion as a result.

17.6.4 Cathodic Protection

Cathodic protection is widely used in the case of buried steel pipes and marine structures. As a rule, in order to control the anaerobic corrosion of iron by SRB, the potential of the structure is depressed to at least –1 V versus $Cu/CuSO_4$ instead of –0.85 V versus $Cu/CuSO_4$, which is usually recommended in the absence of SRB [143]. Cathodic protection is often used in combination with organic coatings because the presence of the organic coating diminishes the need for protection current by several decades compared with that of the unpainted metals. In addition, cathodic protection will be efficient to protect coated steel at defects.

17.6.5 Biocides

Biocides are commonly used for many industrial applications such as industrial water systems. Biocides may be divided into two categories: (a) oxidizing agents such as chlorine, ozone, and chlorine dioxide and (b) nonoxidizing agents such as bisthiocyanate, isothiazolines, acrolein, dodecylguanidine hydrochloride, formaldehyde, glutaraldehyde, chlorophenols, and quaternary ammonium salts.

However, bacteria may adapt to biocides in different ways, such as the production of enzymes, changes in the internal structure of the cell, or changes in the composition of the cell wall. In order to minimize this problem, several biocides are generally used alternatively in practice. Furthermore, although all biocides are efficient on the planktonic population (i.e., microorganisms in the water phase), few are efficient in the case of microorganisms in a biofilm, so that an increase in the dosage up to 100 times is needed [78]. Besides, the use of biocides is more and more limited by environmental legislation due to their toxicity to higher organisms.

17.6.6 Microbiological Methods

As already mentioned, bacterial life depends on many factors. A variation in, for instance, pH, oxygen concentration, temperature, or light conditions may be used to control microbial growth. However, care must be taken to ensure that changes in these parameters do not induce changes in the bacterial population that result in the growth of other microorganisms that may cause MIC. Besides, a change in these parameters may increase the electrochemical corrosion of metals and alloys. A decrease or increase in pH, for instance, may cause corrosion on copper and aluminum alloys.

17.6.7 Physical Methods

Physical methods such as filtration of the water, mechanical removal of the biofilm, or the use of ultraviolet (UV) radiation can be used for some specific applications. For instance, UV light sterilization has been used in potable water systems and cooling water systems instead of chlorination, which is otherwise commonly used. However, in general, these methods are less efficient and more costly than the use of chemicals [69,70].

17.6.8 Simulation of the Biogenic Attack

Besides the previously mentioned countermeasures, a technique that has gained in importance reproduces the biogenic attack under conditions that are optimized for the deterioration-causing agents, the microorganisms. By simulation of the biogenic attack on materials, a quick-motion effect can be produced that allows materials to be tested for their suitability for given applications. The simulation may even be used for the development of resistant materials [136]. This has been shown in the case of biogenic sulfuric acid attack on cement-bound concrete, mortar, and bricks; on resin- or sulfur-bound concrete; on pipeline linings made of synthetics; and in the case of biogenic nitric acid attack on cement-bound concrete and natural stone. It must be emphasized that simulation of biogenic attack on materials requires profound knowledge of *all* the processes and participating microorganisms. If these are known, the situation may be remodeled under conditions optimized for the microorganisms resulting in a reduced time span for the corrosion to become detectable. The problem of biogenic sulfuric acid corrosion in sewage pipelines has been described extensively [6,28,29,144]. A countermeasure was a simulation that allowed remodeling of the microbially caused process within a period of 1 instead of 8 years. Corrosion is caused in this system by sulfuric acid–excreting thiobacilli, which thrive because of hydrogen sulfide emissions from the sewage. H_2S was increased if sewage became anaerobic and was stripped later on. Thiobacilli need water and oxygen and grow optimally at 30°C (instead of 16°C in the Hamburg sewage system). These conditions were created in a simulation apparatus constructed for materials testing. The results of the experiments confirmed that thiobacilli were the corrosion-causing agents [31]. However, differences were noted between the results of chemical testing (resistance of the same materials against pure sulfuric acid) and the results of these biotests. The chemical test was not able to detect significant differences in the resistance of the materials toward the acid attack (all materials corroded more or less similarly), whereas the biotest revealed severe differences. Some concretes were only slightly attacked (1% loss of substance after 1 year of simulation) whereas others lost 10% of their substance [136]. This finding demonstrates that microbial interactions between material and microorganism are of the utmost importance in the determination of resistance to a biogenic attack. The biotest cannot be replaced by chemical and/or physical tests. Because of these test results, the simulation apparatus is now used (internationally) for the development of new materials with increased resistance to biogenic attack.

17.7 Diagnosis of MIC

The diagnosis of MIC is a severe problem. In most cases, damage seems to be explainable by the well-known and from the practical work familiar mechanisms involving chemistry, physics, and materials sciences. This is a result of the fact that engineers and technicians were never introduced, while studying, to the possibility of microorganisms being a causative agent in corrosion. Thus, the often-observed sediments, deposits, slimy layers, etc. tend to be neglected, although they indicate at least the presence of microorganisms. Whereas it is well known that under natural conditions microorganisms are ubiquitous, this fact seems to be neglected in the case of technical environments.

However, life may also be possible in the latter case if several requirements are fulfilled. Microorganisms, like all living beings, need water. However, they are more versatile than other organisms, because their growth limit is the water activity value of $a_w = 0.6$ (–680 bar osmotic pressure). Active microorganisms have been detected in the temperature range from –10°C to 114°C; in the pH range from 0 to 12; at pressures above 1000 bar (deap sea vents); with and without oxygen, hydrogen sulfide, carcinogenic and mutagenic compounds; and even under conditions of high radiation (storage basins of nuclear power plants). Thus, if these requirements are met, participation of microbes in cases of corrosion should be anticipated.

Especially when the following additional markers are found, MIC damage becomes likely:

1. The occurrence of a slimy layer on the surface of a material. This can be tested with the finger. Besides, removal of some slime and burning with a lighter might cause the smell of burned hair = proteins.
2. The smell of mud or even rotten eggs (H_2S) indicates MIC.
3. Coloration or discoloration of a material is suspicious.
4. A high water content besides a soft nature of deposits is often indicative of MIC.
5. Finally, crevice or pitting corrosion is often the result of microbial activity.

In general, almost all materials are susceptible to MIC problems: metals such as mild and stainless steels, alloys, aluminum, copper; mineral materials such as concrete, stone, brick, marble, sandstone, ceramics, glass; natural organic materials such as wood, paper, leather, textiles, starch, polymers (lipids); synthetic organic materials such as plastics, paints, coatings, glues, polymers; and hydrocarbons such as mineral oils, waxes, tar, lubricants, cutting emulsions, fats, and greases. Only in the groups of plastics are there several compounds that are not biodegradable. Microorganisms do not have the enzymes necessary for a depolymerization. Besides, some nondegradable compounds resulting from chemical synthesis called xenobiotics exist (often pesticides, etc.).

Consequently, MIC may play an important role in many cases of (bio)deterioration/corrosion. For a thorough diagnosis, it is crucial that the preceding points are checked and that experts are consulted. The site of corrosion should be left untouched and under the same conditions as before the detection of the failure. Cleaning should be avoided in any case. Good documentation is needed, meaning photographs, videos, data from probes, materials, history, etc. If the diagnosis is not done properly, a good chance exists that there will soon be another failure at the same (or a neighboring) site.

17.8 Concluding Remarks

MIC is a ubiquitous phenomenon, but the participation of microorganisms and their importance are not fully realized. This is often the result of a lack of knowledge, because courses of study in the technical disciplines do not impart this knowledge to students [145]. Another reason may be the fact that, in contrast to a case of corrosion in which no biological factors are involved, in a case of MIC, only an interdisciplinary approach will give the knowledge necessary to understand the mechanisms, However, this is inevitable because the microbiology and the physicochemical processes causing MIC are usually too complicated to be solved by simple test methods [43,132,133,146]. Most processes involved belong to naturally occurring terrestrial cycles of matter. These cycles are the reason why life can persist. Thus, we shall only be able to modify and slow down these processes, never to inhibit them totally. Profound knowledge of all participating processes will make it possible to enhance the life span of materials considerably [147].

References

1. J. H. Garrett, *The Action of Water on Lead*, H. K. Lewis, London, U.K., 1891 (43).
2. R. H. Gaines, *N. Eng. Ind. Chem.* 2:128 (1910).
3. G. A. H. von Wolzogen Kühr and L. R. van der Vlugt, De graphiteering van gietijzer ais electrobiochemisch proces in anaerobe gronden, *Water 18*:147–165 (1934).
4. W. P. Iverson, Biological corrosion, *Advances in Corrosion Science and Technology* (M. G. Fontana and R. W. Staehle, eds.), Plenum, London, U.K., 1972, p. 2.
5. G. H. Booth, Bacterial corrosion, *Discovery 6*:524–527 (1964).
6. E. Bock and W. Sand, Applied electron microscopy on the biogenic destruction of concrete and blocks—Use of the transmission electron microscope for identification of mineral acid producing bacteria, *Proceedings of the Eighth International Conference on Cement Microscopy* (J. Bayles, G. R. Gouda, A. Nísperos, eds.), International Cement Microscopy Association, Duncanville, TX, 1986, pp. 285–302.
7. A. W. Schröter and W. Sand, Estimations on the degradability of ores and bacterial leaching activity using short-time microcalorimetric tests, *FEMS Microbiol. Rev. 11*:79–86 (1993).
8. H. Stolp, *Microbial Ecology: Organisms, Habitats, Activities*, Cambridge University Press, Cambridge, U.K., 1988, pp. 1–308.
9. D. R. Lovley, Dissimilatory Fe(III) and Mn(IV) reduction, *Microbiol. Rev. 55*:259–287 (1991).
10. S. R. de Sánchez and D. J. Schiffrin, Bacterial chemo-attractant properties of metal ions from dissolving electrode surfaces, *J. Electroanal. Chem. 403*:39–45 (1996).
11. J. P. Busalmen, S. R. de Sánchez, and D. J. Schiffrin, Ellipsometric measurement of bacterial films at metal-electrolyte interfaces, *Appl. Environ. Microbiol. 64*:3690–3697 (1998).
12. H. J. Busscher and A. H. Weerkamp, Specific and nonspecific interactions in bacterial adhesion to solid substrata, *FEMS Microbiol. Rev. 46*:165–173 (1987).
13. T. Gehrke, M. Drews, and W. Sand, Microbiological examinations of low-water corrosion on steel piling structures, *DECHEMA-Monographie 133, Biodeterioration* (W. Sand, ed.), VCH, Weinheim, Germany, 1996, pp. 101–106.
14. W. C. Lee and W. G. Characklis, Corrosion of mild steel under an anaerobic biofilm, *Corrosion, 49*:3 (1993).
15. E. C. Hill, J. L. Shennan, and R. J. Watkinson, eds., *Microbial Problems in the Offshore Oil Industry*, Institute of Petroleum, John Wiley, Chichester, Great Britain, 1987.
16. H.-C, Flemming and G. G. Geesey, eds., *Biofouling and Biocorrosion in Industrial Water Systems*, Springer, Berlin, Germany, 1991.
17. J. W. Costerton and E. S. Lashen, Influence of biofilm on efficacy of biocides on corrosion-causing bacteria, *Mater. Performance 23*:34–37 (1984).
18. T. E. Ford, J. S. Maki, and R. Mitchell, The role of metalbinding bacterial exopolymers in corrosion processes, *CORROSION 87*, NACE, Washington, DC, 1987.
19. H. L. Ehrlich, *Geomicrobiology*, Marcel Dekker, New York, 1990, pp. 1–646.
20. R. M. Atlas, Microbial degradation of petroleum hydrocarbons: An environmental perspective, *Microbiol. Rev. 45*:180–209 (1981).
21. M. Baumgärtner, A. Remde, E. Bock, and K. Conrad, Release of nitric oxide from building stones into the atmosphere, *Atmos. Environ. 24B*:87–92 (1990).
22. A. Schippers, T. Rohwerder, and W. Sand, Intermediary sulfur compounds in pyrite oxidation: Implications for bioleaching and biodepyritization of coal, *Appl. Microbiol. Biotechnol. 52*(1):101–104 (1999).

23. W. Dilling and H. Cypionka, Aerobic respiration in sulfate-reducing bacteria, *FEMS Microbiol. Lett. 71*:123–128 (1990).
24. A. R. Colmer and M. E. Hinkle, The role of microorganisms in acid mine drainage: A preliminary report, *Science 106*:253 (1947).
25. E. Heitz, H.-C. Flemming, and W. Sand, *Microbially Influenced Corrosion of Materials*, Springer, Berlin, Germany, 1996.
26. A. H. Rose, *Microbial Biodeterioration*, Academic Press, London, U.K., 1981, pp. 1–516.
27. P. S. Guiamet, S. G. Gómez de Saravia, and H. A. Videla, An innovative method for preventing biocorrosion through microbial adhesion inhibition, *Int. Biodeterior. Biodegrad. 43*:31–35 (1999).
28. E. Bock and W. Sand, A review: The microbiology of masonry biodeterioration, *J. Appl. Bacteriol. 74*:503–514 (1993).
29. K. Milde, W. Sand, W. Wolff, and E. Bock, Thiobacilli of the corroded concrete walls of the Hamburg sewer system, *J. Gen. Microbiol. 129*:1327–1333 (1983).
30. C. D. Parker, The corrosion of concrete. I. The isolation of a species of bacterium associated with the corrosion of concrete exposed to atmospheres containing hydrogen sulphide, *Aust. J. Exp. Biol. Med. Sci. 23*:81–90 (1947).
31. W. Sand, Importance of hydrogen sulfide, thio-sulfate, and methylmercaptan for growth of thiobacilli during simulation of concrete corrosion, *Appl. Environ. Microbiol. 53*:1645–1648 (1987).
32. M. J. Kauffmann, Rôle des bacteries nitrifiantes dans l'alteration des pierres calcaires des monuments, *C. R. Acad. Sci. Paris. 234*:2395–2397 (1952).
33. K.-O. Kirstein, W. Stiller, and E. Bock, Mikrobiologische Einflüße auf Betonkonstruktionen, *Beton-Stahlbetonbau 81*:202–204 (1986).
34. R. Mansch and E. Bock, Microbial deterioration of materials—Simulation, case histories and countermeasures: Testing of the resistance of ceremic materials, *Werkst. Korros. 45*:96–104 (1994).
35. W. Dannecker and K. Selke, Simultanbestimmung organischer und anorganischer Anionen aus verwitterten Natursteinoberflächen mittels Gradienten-Ionen-Chromatogra-phie, *Fresenius Z. Anal. Chem. 335*:966–970 (1989).
36. F. E. W. Eckhardt, Solubilization, transport, and deposition of mineral cations by microorganisms—Efficient rock weathering agents, *The Chemistry of Weathering* (J. I. Drever, ed.), Reidel, Dordrecht, the Netherlands, 1985, pp. 161–173.
37. I. P. Pankhania, A. N. Moosavi, and W. A. Hamilton, Utilization of cathodic hydrogen and *Desulfovibrio vulgaris* (Hildenborough), *J. Gen. Microbiol. 132*:3357–3365 (1986).
38. M. Magot, L. Carreau, J. L. Cayol, B. Oliver, and J. L. Crolet, Sulphide-producing, not sulphate reducing anaerobic bacteria presumptively involved in bacterial corrosion, *Proceedings of 3rd European Workshop on Microbial Corrosion*, Estoril, Portugal, 1994.
39. W. Sand, Microbial mechanisms of deterioration—A general mechanistic overview, *Int. Biodeterior. Biodegrad. 40*:183–190 (1998).
40. W. A. Hamilton, Sulphate-reducing bacteria and anaerobic corrosion, *Annu. Rev. Microbiol. 39*:195–217 (1985).
41. A. P. Hunt and J. D. Parry, The effect of substratum roughness and river flow rate on the development of a freshwater biofilm community, *Biofouling 12*:287–303 (1998).
42. G. Schmitt, Sophisticated electrochemical methods for MIC investigation and monitoring, *Mater. Corros. 48*:586–601 (1997).
43. F. Widdel, Mikrobielle Korrosion, *Jahrbuch Biotechnologie*, Band 3 (P. Präve, M. Schlingmann, W. Crueger, K. Esser, R. Thauer, and F. Wagner, eds.), Carl Hanset, Munich, Germany, 1990, pp. 277–318.
44. G. E. Jenneman, M. J. Mclnerney, and R. M. Knapp, Microbial penetration through nutrient saturated Berea sandstone, *Appl. Environ. Microbiol. 50*:383–391 (1985).
45. S. G. Berk, R. Mitchell, R. J. Bobbie, J. S. Nickels, and D. C. White, Microfouling on metal surfaces exposed to seawater, *Int. Biodeterior. Bull. 17*:29–37 (1981).
46. J. M. Shively and A. A. Benson, Phospholipids of *Thiobacillus thiooxidans*, *J. Bact. 94*:1679–1683 (1967).
47. R. Steudel and G. Holdt, Solubilization of elemental sulfur in water by cationic and an-ionic surfactants, *Angew. Chem. Int. Ed. Engl. 27*:1358–1359 (1988).

48. L. H. G. Morton, ed., *Biodeterioration of Constructional Materials*, Biodeterioration Society Occasional Publication No. 3, Publications Service, Lancashire Polytechnic, Great Britain, 1987, pp. 1–140.
49. S. C. Dexter, Localized corrosion, *Metals Handbook*, 13, 9th edn., ASM International, Warrenville, OH, 1987, pp. 104–122.
50. D. A. Jones, *Principles and Prevention of Corrosion*, Macmillan, New York, 1991, pp. 1–568.
51. A. K. Tiller, Aspects of microbial corrosion, *Corrosion Processes* (R. N. Parkins, ed.), Applied Science, London, U.K., 1983, pp. 115–159.
52. A. K. Tiller, A review of the European research effort on microbial corrosion between 1950 and 1984, *Proceedings of Biologically Induced Corrosion* (S. C. Dexter, ed.), NACE, Houston, TX, 1986, pp. 8–29.
53. H. A. Videla, Sulphate reducing bacteria and anaerobic corrosion, *Corros. Rev. 9*:103–141 (1990).
54. D. J. Duquette and R. E. Ricker, Electrochemical aspects of microbiologically induced corrosion, *Proceedings of Biologically Induced Corrosion* (S. C. Dexter, ed.), NACE-8, Houston, TX, 1986, pp. 121–130.
55. G. H. Booth and A. K. Tiller, Cathodic characteristics of mild steel by sulphate reducing bacteria—An alternative mechanism, *Corros. Sci. 8*:583–600 (1968).
56. W. P. Iverson, Direct evidence for the cathodic depolarization theory of bacterial corrosion, *Science 151*:986–988 (1966).
57. G. H. Booth, L. Elford, and D. S. Wakerley, Corrosion of mild steel by sulphate reducing bacteria: An alternative mechanism, *Br. Corros. J. 3*:242–245 (1968).
58. J. D. A. Miller and R. A. King, Biodeterioration of metals, *Microbial Aspects of the Deterioration of Materials* (D. W. Lovelock and R. J. Gilbert, eds.), Academic Press, London, U.K., 1975, pp. 83–103.
59. D. H. Pope, D. J. Duquette, A. H. Johannes, and P. C. Wayner, Microbiologically influenced corrosion of industrial alloys, *Mater. Performance 4*:14–18 (1984).
60. D. T. Rickard, The microbiological formation of iron sulphides, *Proceedings of Contribution to Geology*, Stockholm, Sweden, 1969, pp. 67–72.
61. J. L. Crolet, S. Daumas, and M. Magot, pH regulation by sulphate-reducing bacteria, *CORROSION 93*, New Orleans, Paper No. 303, NACE, Houston, TX, 1993.
62. J. A. Hardy and J. L. Bown, The corrosion of mild steel by biogenic sulphide films exposed to air, *Corrosion 40*:650–654 (1984).
63. R. A. King, J. D. A. Miller, and D. S. Wakerley, Corrosion of mild steel in cultures of sulphate reducing bacteria: Effect of changing the soluble iron concentration during growth, *Br. Corros. J. 8*:89–93 (1973).
64. J. A. Costello, Cathodic depolarization by sulphate reducing bacteria, *S. Afr. J. Sci. 70*:202–204 (1974).
65. E. Schaschl, Elemental sulphur as a corrodent in deaerated, neutral aqueous solutions, *Mater. Performance 7*:9–12 (1980).
66. W. P. Iverson, Anaerobic corrosion mechanisms, *CORROSION 83*, Anaheim, CA, Paper No. 243, NACE, 1983.
67. J. Telegdi, Z. Keresztes, G. Pálinkás, E. Kalman, and W. Sand, Microbially influenced corrosion visualized by atomic force microscopy, *Appl. Phys. A66*:S639–S642 (1998).
68. G. Cragnolino and O. H. Tuovinen, The role of sulphate reducing and sulphur oxidizing bacteria in the localized corrosion of iron-base alloys—A review, *Int. Biodeterior. 20*:9–26 (1984).
69. D. Thierry, Microbial corrosion in a closed cooling water system: The role of *Nitrobacter*, Swedish Corrosion Institute, Stockholm. Internal report, 1987, pp. 1–4.
70. D. Thierry, Field observations of microbiologically induced corrosion in cooling water systems, *Mater. Performance 26*:35–41 (1987).
71. J. W. Costerton and G. G. Geesey, The microbial ecology of surface colonization and of consequent corrosion, *Proceedings of Biologically Induced Corrosion* (S. C. Dexter, ed.), NACE-8, Houston, TX, 1986, pp. 223–232.
72. H. Lappin-Scott and J. W. Costerton, *Microbial Films*, Cambridge University Press, Cambridge, U.K., 1995.

73. I. W. Suntherland, The biofilm matrix—An immobilized but dynamic microbial environment, *Trends Microbiol. 9*:222 (2001).
74. C. C. Gaylarde and J. M. Johnston, The effect of *Vibrio anquillarum* on the anaerobic corrosion of mild steel by *Desulfovibrio vulgaris*, *Int. Biodeterior. 18*:111–116 (1982).
75. C. C. Gaylarde and J. M. Johnston, Anaerobic metal corrosion in cultures of bacteria from estuarine sediments, *Proceedings of Biologically Influenced Corrosion* (S. C. Dexter, ed.), NACE-8, Houston, TX, 1986, pp. 137–143.
76. C. C. Gaylarde and H. A. Videla, Localized corrosion induced by a marine vibrio, *Int. Biodeterior. 23*:91–104 (1987).
77. R. J. G. Edyvean, C. J. Thomas, and I. M. Austen, The use of biologically active environments for testing corrosion fatigue of offshore structural steel, *Proceedings of Biologically Induced Corrosion* (S. C. Dexter, ed.), NACE-8, Houston, TX, 1986, pp. 254–267.
78. M. Walsh and R. Mitchell, The role of microorganisms in the hydrogen embrittlement, *CORROSION 93*, New Orleans, LA, Paper No. 249, NACE, 1983.
79. M. Walsh and R. Mitchell, Microbial influence on hydrogen uptake by metals, *Proceedings of Biologically Induced Corrosion* (S. C. Dexter, ed.), NACE-8, Houston, TX, 1986, pp. 201–208.
80. C. M. Santegoeds, T. G. Ferdelman, G. Muyzer, and D. de Beer, Sructural and functional dynamics of sulfate-reducing populations in bacterial biofilms, *Appl. Environ. Microbiol. 64*:3731–3739 (1998).
81. F. P. van den Ende, J. Meier, and H. van Gemerden, Syntrophic growth of sulfate-reducing bacteria during oxygen limitation, *FEMS Microbiol. Ecol. 23*:65–80 (1997).
82. J. W. Santo Domingo, C. J. Berry, M. Summer, and C. B. Fliermans, Microbiology of spent nuclear fuel storage basins, *Curr. Microbiol. 37*:387–394 (1998).
83. U. P. Sinha, J. H. Wolfram, and R. D. Rogers, (1990). Microbially influenced corrosion of stainless steels in nuclear power plants, *Proceedings of Microbially Influenced Corrosion and Biodeterioration* (N. J. Dowling, M. W. Mittleman, and J. C. Danko, eds.), University of Tennessee, Knoxville, TN, 1967, pp. 4/51–4/60.
84. S. Stroes-Gascoyne and J. M. West, Microbial studies in the Canadian nuclear fuel waste management program, *FEMS Microbiol. Rev. 20*:573–590 (1997).
85. K. M. Wiencek and M. Fletcher, Effects of substratum wettability and molecular topography on the initial adhesion of bacteria to chemically defined substrata, *Biofouling 11*(4):293–311 (1997).
86. R. E. Tatnall, Case histories: Bacterial induced corrosion, *Mater. Performance 8*:41–48 (1981).
87. R. E. Tatnall, Fundamentals of bacterial induced corrosion, *Mater. Performance 9*:32–38 (1981).
88. R. Brown and G. S. Pabst, *Proceedings of 3rd International Symposium on Biodegradation* (J. M. Sharpley and A. M. Kaplan, eds.), Applied Science Pub., Barking, U.K., 1976, pp. 875–882.
89. G. Kobrin, Corrosion by microbiological organisms in natural waters, *Mater. Performance 7*:38–43 (1976).
90. S. W. Borenstein, Microbiologically influence corrosion of austenitic steel weldments, *Mater. Performance 1*:52–54 (1991).
91. D. Walsh, J. Seagoe, and L. Williams, Microbiologically influenced corrosion of stainless steel weldments: Attachment and film evolution, *CORROSION 92*, Nashville, TN, Paper No. 165, NACE, 1992.
92. W. H. Dickinson and Z. Lewandowski, Manganese biofouling and the corrosion behavior of stainless steel. *Biofouling 10*(1–3):79–93 (1996).
93. B. J. Webster, R. J. Kelly, and R. C. Newman, The electrochemistry of SRB corrosion in austenitic stainless steel, *Proceedings of Microbially Influenced Corrosion and Biodeterioration* (N. J. Dowling, M. W. Mittleman, and J. C. Danko, eds.), University of Tennessee, Knoxville, TN, 1990, pp. 2/9–2/18.
94. B. J. Webster, R. J. Kelly, and R. C. Newman, SRB-induced localized corrosion of stainless steels, *CORROSION 91*, Cincinnati, OH, Paper No. 106, NACE, 1991.
95. S. Da Silva, R. Basséguy, and A. Bergel, The role of hydrogenases in the anaerobic microbiologically influenced corrosion of steels, *Bioelectrochemistry 56*:77 (2002).

96. S. Da Silva, R. Basséguy, and A. Bergel, Electron transfer between hydrogenase and 316L stainless steel: Identification of a hydrogenase-catalyzed cathodic reaction in anaerobic MIC, *J. Electroanal. Chem. 93*:561 (2004).
97. S. J. Yuan and S. O. Pehkonen, AFM study of microbial colonization and its deleterious effect on 304 stainless steel by *Pseudomonas* NCIMB 2021 and *Desulfovibrio desulfuricans* in simulated seawater, *Corros. Sci. 51*:1372–1385 (2009).
98. P. J. B. Scott, J. Goldie, and M. Davies, Ranking alloys for susceptibility to MIC-a preliminary report on high-Mo alloys, *Mater. Performance 1*:55–57 (1991).
99. J. P. Audouard, C. Compere, N. J. E. Dowling, D. Feron, A. Mollica, T. Rogne, V. Scotto, U. Steinsmo, C. Taxen, and D. Thierry, Effect of Marine Biofilms on Stainless Steel: Results from a European Exposure Program. EFC Publication No. 19, The Institute of Materials, London, U.K., 1996.
100. S. C. Dexter and G. Y. Gao, Effect of seawater biofilms on corrosion potential and oxygen reduction of stainless steel, *Corrosion 44*:717–723 (1988).
101. W. H. Dickinson and Z. Lewandowski, Electrochemical concepts and techniques in the study of stainless steel ennoblement, *Biodegradation 9*:11–21 (1998).
102. R. Johnsen and E. Bardal, Cathodic properties of different stainless steels in natural sea water, *Corrosion 41*:296–301 (1985).
103. R. Johnsen and E. Bardal, The effect of microbiological slime layer on stainless steel in natural sea water, *CORROSION 86*, Paper No. 227, NACE, Houston, TX, 1986.
104. A. Mollica and A. Trevis, Correlation between the formation of a primary film and the modification of the cathodic surface of stainless steel in seawater, *Proceedings of 4th International Congress on Marine Corrosion and Fouling*, Antibe, France, 1976, pp. 351–355.
105. V. Scotto, R. Di Cento, and R. Marcenaro, The influence of marine aerobic microbial films on stainless steel corrosion behaviour, *Corros. Sci. 25*:185–194 (1985).
106. Le Bozec, M L' Her, C. Compère, A. Laouenan, D. Costa and P. Marcus, Influence of stainless steel surface treatment on the oxygen reduction reaction in seawater, *Corros. Sci. 43*:765–786 (2001).
107. C. Marconnet, Y. Wouters, F. Miserque, C. Dagbert, J. P. Petit, A. Galerie, and D. Feron, Chemical composition and electronic structure of the passive layer formed on stainless steels in a glucose-oxidase solution, *Electrochem. Acta 54*:123–132 (2008).
108. P. Chandrasekaran and S. C. Dexter, Mechanism of potential ennoblement on passive metals by seawater biofilms, *CORROSION 93*, New Orleans, LA, Paper No. 493, NACE, Houston, TX, (1993).
109. H. Amaya and H. Miyuki, Development of accelerated evaluation method for microbially influenced corrosion resistance of stainless steel, *Corros. Eng. 44*:123–133 (1995).
110. I. Dupont, D. Feron, and G. Novel, Effect of glucose oxidase activity on corrosion potential of stainless steels in seawater, *Int. Biodeterior. Biodegrad. 41*:13–18 (1998).
111. V. Scotto and M. E. Lai, The ennoblement of stainless steel in seawater: A likely explanation coming from the field, *Corros. Sci. 40*:1007–1018 (1998).
112. A. Mollica, E. Traverso, and D. Thierry, On oxygen reduction depolarisation induced by biofilm growth on stainless steels in sea water, *Aspects of Microbially Induced Corrosion* (D. Thierry, ed.), University Press, Cambridge, U.K., 1996.
113. N. Washizu, Y. Katada, and T. Kodama, The role of H_2O_2 in microbially influenced ennoblement of open circuit potentials for type 316L stainless steel in seawater, *Corros. Sci. 46*:1291 (2004)
114. A. Mollica et al., Crevice corrosion testing of stainless steel in natural and synthetic biochemical seawater, results from a round robin test involving 19 laboratories, Stainless steel world, Maastricht, the Netherlands, 2004.
115. N. Le Bozec, M. L' Her, C. Compère, P. Marcus, and D. Costa, Evidence for the effect of hydrogen peroxide produced by marine biofilms on the electrochemical behaviour evolution of stainless steel immersed in natural seawater, *Proceedings of Eurocorr 2000*, EFC, London, U.K., 2000.
116. H. Dhar and D. Howell, The use of in-situ electrochemical reduction of oxygen in the diminution of adsorbed bacteria on metals in seawater, *J. Electrochem Soc. 129*:2178 (1982).

117. J. Landoulsi, K. El Kirat, C. Richard, D. Feron, and S. Pulvin, Enzymatic approach of microbial influenced corrosion, a review based on stainless steel in natural seawater, *Environ. Sci. Technol., 42*:2233–2242 (2008).
118. P. Gümpel and R. Kreikenbohm, Betrachtungen über den Zusammenhang zwischen bakteriellem Wachstum und der mikrobiell induzierten Potential verschiebung bei nichtrostenden Stählen, *Mater. Corros. 50*:219–226 (1999).
119. M. Geiser, R. Avci, and Z. Lewandowski, Microbially initiated pitting on 316 L stainless steel, *Int. Biodeterior. Biodegrad. 49*:235 (2002).
120. X. Shi, R. Avci, M. Geiser, and Z. Lewandowski, Comparative study in chemistry of microbially and electrochemically induced pitting on 316 L, *Corros. Sci. 45*:2577 (2003).
121. W. H. Dickinson and Z. Lewandowski, Manganese biofouling and the corrosion behaviour of stainless steel, *Biofouling 10*:79 (1999)
122. W. H. Dickinson and Z. Lewandowski, Manganese biofouling of stainless steel: Deposition rates and influence on corrosion processes, *Proceedings of Corrosion 1996*, Paper No. 291, NACE, Houston, TX, 51996.
123. B. Little, M. McNeil, and F. Mansfeld, The impact of alloying elements on microbiologically influenced corrosion—A review, *Proceedings of 12th International Congress on Metal Corrosion*, NACE, Houston, TX, 1993, pp. 3680–3686.
124. A. V. Churchill, Microbial fuel tank corrosion, *Mater. Protect. 6*:19–23 (1963).
125. H. G. Hedrick, Microbiological corrosion of aluminum, *Mater. Protect. 1*:27–31 (1970).
126. M. B. McNeil, J. M. Jones, and B. J. Little, Production of sulphide minerals by sulphate reducing bacteria during microbiologically influence corrosion of copper, *Corrosion 9*:674–677 (1991).
127. D. Wagner, M. Tietz, O. von Franqué, and W. R. Fischer, Remedial measures versus microbiologically influenced corrosion in copper potable water installations, *Proceedings of 13th International Corrosion Congress*, Melbourne, Australia, Paper No. 143, 1996.
128. B. J. Webster, D. B. Wells, and P. J. Bremer, Potable water biofilms, copper corrosion and copper by-products release, *Proceedings of 13th International Corrosion Congress,* Melbourne, Australia, Paper No. 408, 1996.
129. A. Chamberlain and P. Angell, Influences of microorganisms on pitting of copper tube, *Proceedings of Microbially Influenced Corrosion and Biodeterioration* (N. J. Dowling, M. W. Mittleman, and J. C. Danko, eds.), University of Tennessee, Knoxville, TN, 1990, pp. 3/65–3/69.
130. W. E. Krumbein, Zur Frage der biologischen Verwitterung: Einfluß der Mikroflora auf die Bausteinverwittemng und ihre Abhángigkeit von edaphischen Faktoren, *Z. All. Mikrobiol. 8*:107–117 (1968).
131. F. Lewis and E. May, Isolation and enumeration of autotrophic and heterotrophic bacteria from decayed stone, *Proceedings of Fifth International Congress on Deterioration and Conservation of Stone*, Lausanne, Switzerland, 1985, pp. 633–642.
132. N. N. Microorganisms in nuclear waste disposal. I. Multi-author review, *Experientia 46*:777–851 (1990).
133. N. N. Spécial corrosion, *Mater. Tech. 78*:1–16 (1990).
134. K. Pedersen, Microbial life in deep granitic rock, *FEMS Microbiol. Rev. 20*:399–414 (1997).
135. J. Pochon, P. Tardieux, J. Kajudie, and M. Charpentier, Degradation des temples d'angkor et processus biologiques, *Ann. Inst. Pasteur. 98*:457–461 (1960).
136. W. Sand, E. Bock, and D. C. White, Biotest system for rapid evaluation of concrete resistance to sulfur-oxidizing bacteria, *Mater. Performance 26*:14–17 (1987).
137. W. Sand, B. Ahlers, T. Krause-Kupsch, M. Meincke, E. Kreig, M. Diercks, F. Same-luck, and E. Bock, Mikroorganismen und ihre Bedeutung für die Zerstörung von mineralischen Baustoffen, *UWSF Z. Umweltchem. Ökotox. 3*:36–40 (1989).
138. W. Sand and E. Bock, Mikrobielle Zerstörung kerami-scher Werkstoffe, *Werkst. Korros. 41*:64–68 (1990).
139. W. Sand and E. Bock, Biodeterioration of ceramic materials by biogenic acids, *Int. Biodeterior. 27*:175–183 (1991).
140. J. W. McCoy, *Microbiology of Cooling Waters*, Chemical Publishing Co., New York, 1980.
141. L. J. Seed, The significance of organisms in corrosion, *Corros. Rev. 9*:3–75 (1990).

142. D. Thierry, Methods for combating microorganisms in cooling water systems—A literature review and market inventory, Swedish Corrosion Institute, Stockholm, Sweden, report 1989:3, pp. 1–23.
143. K. Tiller, Biocorrosion in civil engineering, *Proceedings of Microbiology in Civil Engineering* (P. Howsam, ed.), Chapman & Hall, London, U.K., 1990, pp. 24–38.
144. T. Emmel, E. Brill, W., Sand, and E. Bock, Screening for biocides to inhibit biogenic sulphuric acid corrosion in sewage pipelines, *Proceedings of Biodeterioration 7* (D. R. Houghton, R. N. Smith, and H. O. W. Eggins, eds.), Elsevier Applied Science, London, U.K., 1988, pp. 118–122.
145. E. Heitz, A. D. Mercer, W. Sand, and A. K. Tiller, *Microbiological Degradation of Material and Methods of Protection*, European Federation of Corrosion Publications Number 9, The Institute of Materials, London, U.K., 1992, pp. 1–88.
146. J. L. Crolet, From biology and corrosion to biocorrosion, *Oceanol. Acta 15*(1):87–94 (1992).
147. O. Wanner, Modelling of biofilms. *Biofouling 10*(1–3):31–41 (1996).

18

Corrosion in Nuclear Systems: Environmentally Assisted Cracking in Light Water Reactors

F.P. Ford
General Electric Global Research Center

P.L. Andresen
General Electric Global Research Center

CONTENTS

18.1 Introduction 777
18.2 Problem and the Proposed Solution 778
18.2.1 Approaches to Life Prediction 782
18.3 Life Prediction Based on Empirical Correlations 784
18.4 Life Prediction Based on an Understanding of the Mechanism of Cracking 787
18.4.1 Slip-Oxidation Mechanism 789
18.4.2 Film-Induced Cleavage Mechanism 791
18.4.3 Hydrogen Embrittlement Mechanisms 792
18.5 Prediction Methodology for Ductile Structural Alloys in BWR Systems 794
18.5.1 Rationale for Choice of Slip-Oxidation Model as "Working Hypothesis" for Crack Propagation of Ductile Alloys in BWR Systems 795
18.5.2 Definition of Crack Tip Alloy/Environment System 795
18.5.3 Evaluation of Reaction Rates at Crack Tip 797
18.5.4 Definition of Crack Tip Strain Rate 798
18.5.5 Prediction of Stress Corrosion Crack Propagation of Unirradiated Stainless Steel in 288°C Water 801
18.5.6 Prediction of Cracking in Unirradiated Welded Plant Components 807
18.5.7 Irradiation Effects on Stress Corrosion Cracking of Stainless Steels 811
18.6 Conclusions 817
References 819

18.1 Introduction

The phenomena of crack initiation and subcritical propagation in structural materials due to the conjoint actions of stress, material microstructure, and environment have been recognized for many years, and the dependencies and mechanisms have been extensively researched. This is especially the case for alloys in relatively low-temperature (<150°C) environments containing high anionic concentrations of, for instance, chlorides, phosphates, hydroxides, etc.; such concentrations may be either in the bulk environment (e.g., paper mills, chemical plant, petrochemical, marine) or in localized environments where a high anionic activity may be created due to crevice or liquid-/gas-phase distribution effects on heat transfer surfaces. The mechanisms of such cracking in concentrated environments have been reviewed by, for example, Newman [1].

However, it has been recognized that environmentally assisted cracking under static or cyclic loading (e.g., stress corrosion and corrosion fatigue, respectively) can occur in ultrahigh-purity water at more elevated temperatures (>150°C) where the concentration of non-OH^- anions may be <10 ppb. Although the cracking susceptibility is generally lower than that in the concentrated environments, it is sufficient to give concerns in situations where extended lives or high levels of plant availability or safety are required.

Such is the case for the power generation industry, especially light water reactors (LWR) in which serious technical problems can arise as the worldwide fleet is increasingly being operated under power uprate and extended life conditions. Such conditions can increase the cracking susceptibility due to, for instance, increased time and neutron fluence. These problems also apply to new LWR designs that are under certification and construction, and are expected to operate with an extended fuel cycle (and, hence, decreased inspection frequency) for 60, or even 80, year design lives [2].

Unfortunately, the environmentally assisted material degradation issues in LWRs have been managed reactively in the past; that is, identifying the cause of cracking and developing mitigation actions have been conducted *after* the incidents have occurred. This reactive nature of the response has had two major consequences. First, the safety systems and barriers are at risk while the preventative or mitigation actions are being developed, which can take considerable time, especially since it also involves the concurrence of the regulator on the appropriate quantitative control criteria. Second, there are unplanned monetary and time commitments due to the unpredicted nature of the incidents.

As a result, there has been a drive both by the USNRC [3] and the U.S. utilities [4] to develop the capability to manage material degradation problems proactively, that is, to identify a problem well before costly and potentially safety-significant incidents occur and, thereby, to give a more extended time to develop the necessary understanding, mitigation, and inspection strategies. A life prediction capability for the various environmentally assisted degradation modes is key to all of these elements.

Attention, in this paper, is focused on the prediction of stress corrosion cracking in two systems: austenitic stainless steels in boiling water reactors (BWR) and nickel-base alloys in pressurized water reactor (PWR) primary systems. These two issues are chosen not only because the various assessment studies [3,4] have indicated that they pose significant challenges to reactor operation in the future, but also because they illustrate two different approaches to life prediction.

18.2 Problem and the Proposed Solution

It is now well recognized that the development of stress corrosion cracking in ductile alloys in LWR systems follows a sequence of four distinct periods (Figure 18.1), namely,

1. *A precursor period* is the stage during which specific metallurgical or environmental conditions may develop at the metal–solution interface that are conducive to subsequent crack initiation [5]. For example, the condition at the end of

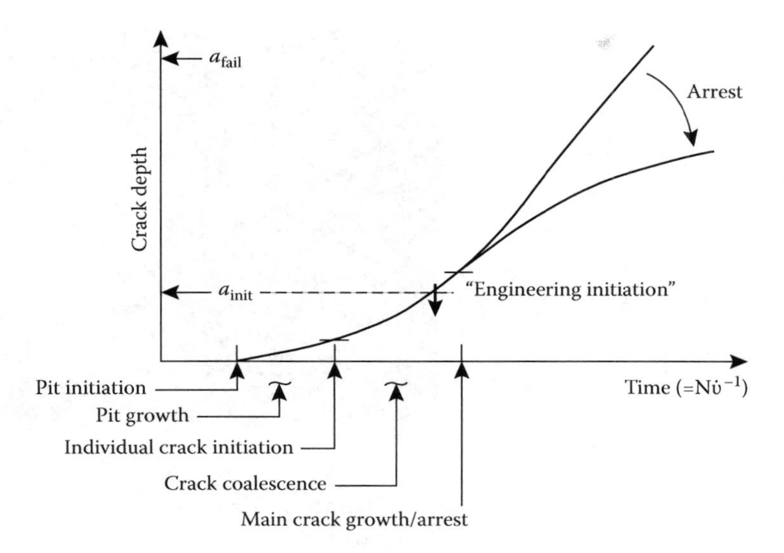

FIGURE 18.1
Sequence of crack initiation, coalescence, and growth during subcritical cracking in aqueous environments. Note the arbitrary definition of "engineering initiation," which generally coincides with the NDE resolution limit.

the precursor period may be slip offsets or pitting corrosion that causes breakdown of the normally protective surface oxide film leading (Figure 18.1); other precursor phenomena might be associated with the creation of localized environments via intergranular attack, crevices, etc., with diffusion or oxidation along grain boundaries. These "precursor" periods may be very short if stresses are high or severe water impurity transients during initial reactor operations occur. An example would be the cracking of weld-sensitized Type 304 stainless steel BWR piping whose surface had been abusively ground following welding. On the other hand, the precursor events may take years if they are associated with a change in metallurgical microstructure involving thermal aging, irradiation embrittlement, grain boundary diffusion or oxidation, or the creation of the necessary stress due to corrosion product formation and expansion. While the mechanism of degradation during the precursor period can be closely related to the processes that control cracking during subsequent stages, they can also be entirely different.

2. *The initiation of cracks* happens when the local environment, microstructure, stress, and crack geometry conditions have reached a critical state. An example is shown in Figure 18.2, where a transgranular corrosion fatigue crack in a mild steel, stressed in partially oxygenated 288°C water (symptomatic of BWR coolant), had initiated at a pit that was formed at lower temperatures (symptomatic of reactor shutdown or lay-up conditions). The size and spacing of the crack initiation sites will depend on the distribution of surface breaking MnS precipitates, machining marks, cold work, and stress raising sites such as sharply radiused bolt heads and the presence of reentrant angles at welds between misaligned pipes.

   The morphology of the cracks during this period and the subsequent short crack growth stage depends on the precursor event and the initiating site. For instance,

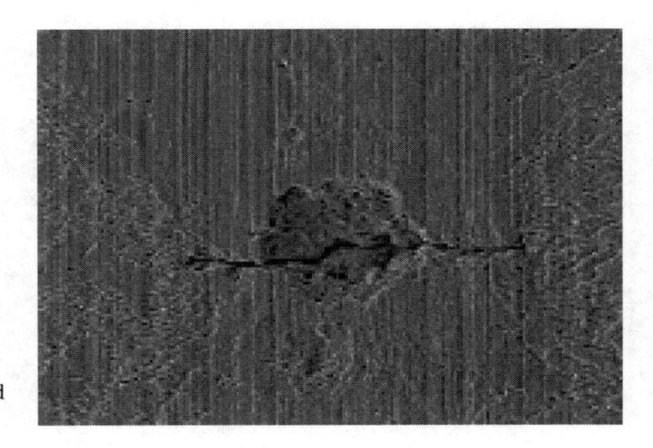

FIGURE 18.2
Initiation of a stress corrosion crack in mild steel strained in 200 ppb O_2 water at 288°C.

for cold-worked surfaces, or for cracks that are initiated at randomly spaced pits, the cracking morphology is usually transgranular, while those cracks that are initiated at sites of intergranular attack will often propagate intergranularly. However, it should be noted that, as the crack progresses, the cracking morphology may change depending on the specifics of the loading mode, the material microstructure, and the environmental condition. Thus, caution is required because it isn't always correct to associate stress corrosion cracking of sensitized stainless steels in high-temperature water with *exclusively* intergranular morphology, and this is more true of more resistant materials like Alloy 690 and stabilized or L-grade stainless steels.

The criteria for crack initiation (and the subsequent short crack growth) have been extensively reviewed elsewhere [6].

3. *Growth of the shallow individual cracks* is followed by their *coalescence* to form a "major" crack [6]. It is usually during this stage that the crack has reached dimensions that are resolvable by very sensitive nondestructive examination techniques or there has been an observable drop in stress in a strain-controlled test or component (such as a bolt). It is at this point that, *in engineering terms*, the crack is deemed to have "initiated" (see Figure 18.1). It is important to recognize, however, that the extent of damage at this stage is relatively large metallurgically (i.e., cracks may be much greater than grain size dimensions), and a lot has happened in terms of degradation phenomena during this "engineering initiation" time.
4. *Propagation of a single dominant crack* at a rate that, as will be discussed below, is very dependent on the material, stress, and environmental conditions. The growth rate is rarely constant, as is too often assumed, but may accelerate or decelerate over time depending on the stress and strain profiles (which can be complicated adjacent to welds) and material and irradiation conditions.

It is apparent that the development of a life prediction algorithm of the general form

$$\text{Time to failure} = f(\text{material, environment, stress}) \tag{18.1}$$

can conceptually take into account all of the factors listed above, yet poses considerable challenges. For instance, there are complicated interactions between the various system

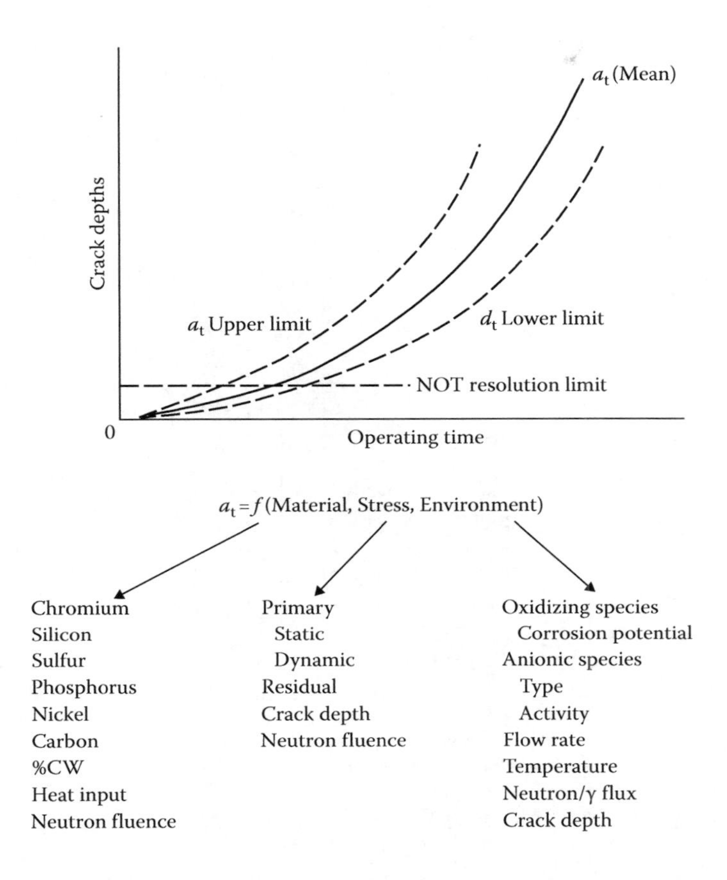

FIGURE 18.3
Material, stress, and environmental parameters relevant to environmentally assisted cracking of stainless steels in BWRs. (From Ford, F.P. et al., On-line BWR materials monitoring & plant component lifetime prediction, in *Proc. Nuclear Power Plant Life Extension*, Snowbird, UT, June 1988, Published by American Nuclear Society, Vol. 1, pp. 355–366.)

parameters (Figure 18.3) that control the various submodes (e.g., pitting, intergranular attack, crack coalescence, crack propagation, etc.). Moreover, it becomes apparent that small changes in these interacting factors can significantly affect the stress corrosion cracking susceptibility.

This immediately leads to the conclusion that there will be a measure of uncertainty (albeit quantifiable) in the life prediction capabilities and this uncertainty can be attributed to two sources. The first are *aleatory* uncertainties that arise out of random, stochastic events (such as pitting, intergranular attack, crack coalescence) that occur on a smooth surface, and these may dominate the onset of *engineering initiation* and the failure time of thin specimens or components. Such uncertainties are analyzable via, for instance, Weibull statistics, as illustrated in Figure 18.4 for the case of failure of smooth specimens of mill-annealed Alloy 600 in PWR primary environments. The second class of uncertainties are *epistemic* and arise out of incompleteness in the nature of the degradation model that is adopted, the dispersion in the model inputs, and the inadequacy of the system definition or control. This is illustrated in Figure 18.5 for crack propagation rates for nickel-base Alloys 82, 182, and 132 in PWR primary water. These epistemic uncertainties, unlike the aleatory uncertainties, may be reduced by improvements in the degradation model and the control/definition of the model inputs.

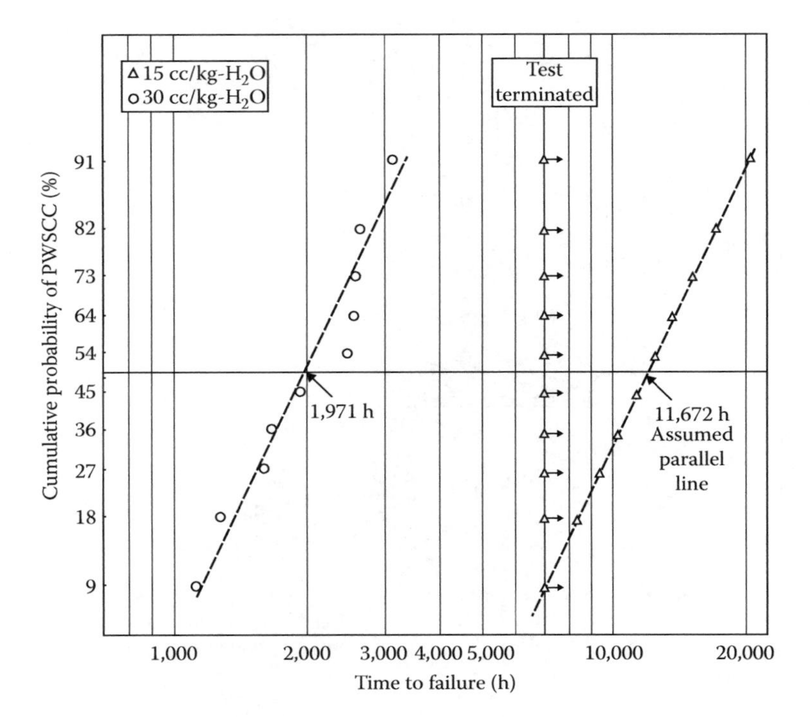

FIGURE 18.4
Distribution of times to failure for initially smooth specimens of Alloy 600MA in simulated PWR primary water at 360°C containing different concentrations of hydrogen. (From Sato, K. et al., Evaluation of PWRSCC susceptibility for alloy 600 under low hydrogen concentration and the benefits for various components in primary circuit, in *Proceedings of Conference on Optimization of Dissolved Hydrogen in PWR Primary Coolant*, Tohoku University, Japan, July 18–19, 2007.)

18.2.1 Approaches to Life Prediction

There are various approaches for predicting the locus of the degradation kinetics shown schematically in Figure 18.1 and these will be discussed in some detail when addressing the specific examples of stress corrosion cracking in various LWR alloy/environment systems. The approaches that have been used may be categorized as follows:

- Past-plant experience
- Empirical correlations based on plant and laboratory data
- A quantitative understanding of the mechanism of cracking

There is no question that past-plant experience is of crucial importance, since this acts as a "calibration point" that must be explained by any life prediction methodology. For instance, the methodology must be able to explain quantitatively the relative contributions of various system parameters (such as cold work, sensitization, weld repairs, etc.) to the incidents of stress corrosion cracking in piping in BWRs that have been observed in a given operating period (Figure 18.6).

However, plant experience is of limited value if it is the *sole* input to the algorithm for predicting the development of cracking over time. This is because the required definition of the system components and conditions is rarely available, and this leads to a very large dispersion in the data that are required for life prediction purposes. This is illustrated in

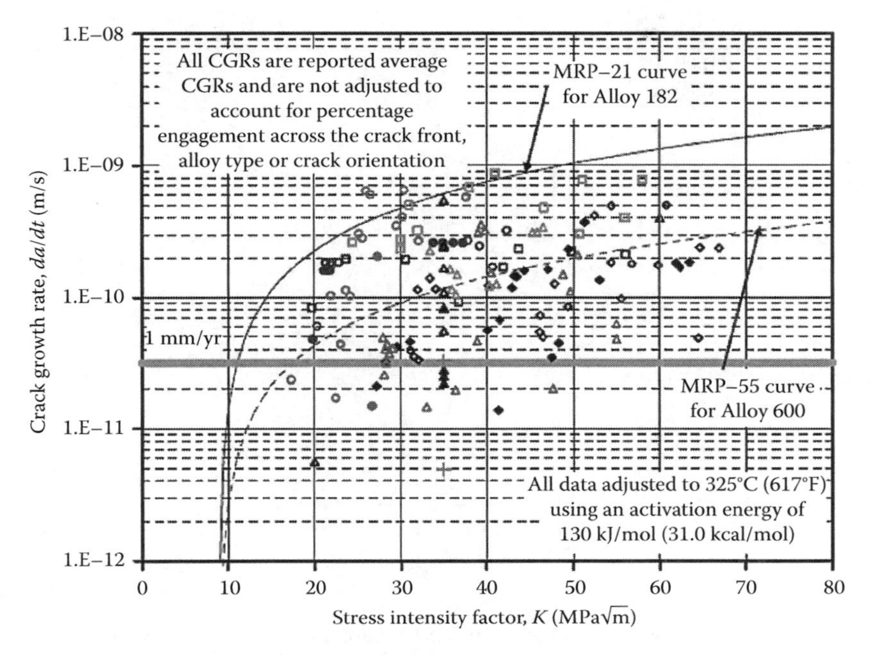

FIGURE 18.5
Crack propagation rate vs. stress intensity factor data for welding alloys 182/132/82 in PWR primary environment. The original data have been corrected for temperature to 325°C using an activation enthalpy of 130 kJ/mol. (From White, G.A. et al., Development of crack growth rate disposition curves for primary water stress corrosion cracking (PWSCC) of alloy 82, 182, 132 weldments, in *Proceedings of Twelfth International Conference on Environmental Degradation in Nuclear Power Systems–Water Reactors*, Skamania Lodge, Eds. L. Nelson and P. King, The Metallurgical Society, August 5–9, 2005, pp. 511–531.)

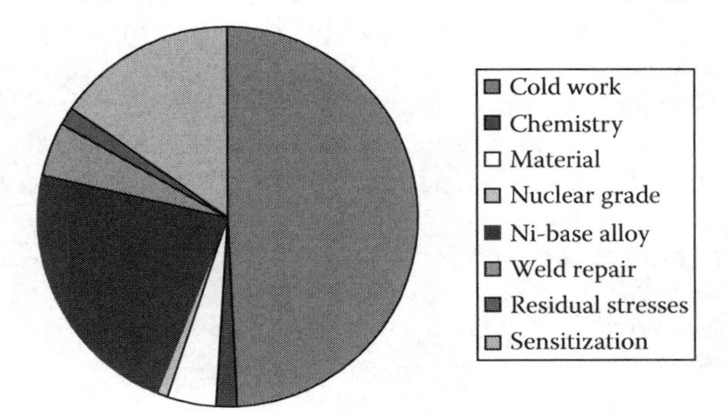

FIGURE 18.6
Root causes of failure of *BWR* piping in Swedish plants. (From Gott, K., Cracking data base as a basis for risk informed inspection, in *Proceedings of Tenth International Conference on Environmental Degradation in Nuclear Power Systems–Water Reactors*, Eds. F.P. Ford and G. Was, Published by National Association of Corrosion Engineers, Lake Tahoe, August 5–9, 2001.)

Figure 18.7 for a system defined as *stainless steel piping in BWRs*. It is seen that the dispersion of the plant data based on such a general definition of the system makes it impossible to predict accurately any future trends.

This situation can often be improved from an analytical viewpoint via the application of Weibull statistics to the plant data [12,13], which assumes that the life of the component is dominated by the time spent in the early crack initiation and "short" crack growth phases

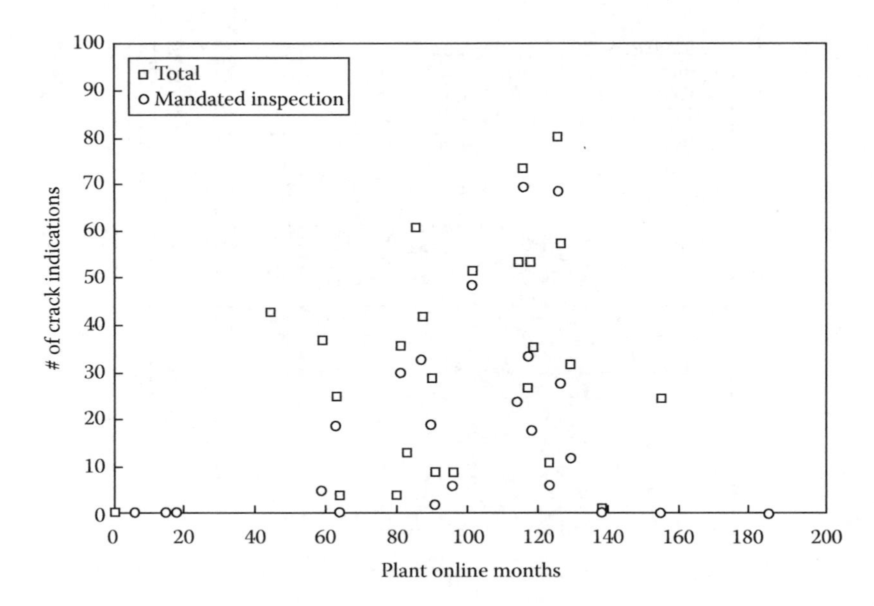

FIGURE 18.7
Relationship between incidences of cracking in BWR piping and operational time. (From Klepfer, H.H. et al., Investigation of cause of cracking in austenitic stainless steel piping, NEDO-21000, General Electric, July 1975.)

that are controlled by random, stochastic events. Such an analytical approach for the evolution of damage is reasonable for thin-walled components such as tubing. However, it may not be always justifiable for thick-section components where weld defects, high stresses, machining and grinding, and other factors conspire to accelerate or circumvent initiation, and where considerable time may be involved in the propagation of "deep" cracks.

In summary, it is apparent that developing a life prediction methodology from plant experience alone is unlikely to prove useful for large dimension components since that experience does not necessarily take into complete account the very specific effects of coolant purity, corrosion potential, material sensitization, stress intensity factor, etc. (Figure 18.3). Consequently, attention will now be focused on life prediction methodologies that are based on either empirical correlations between the component life and the governing stress, material, and environment parameters and/or an understanding of the mechanism of cracking.

18.3 Life Prediction Based on Empirical Correlations

The empirical correlation approach relies on an analysis of the effects of all (known) system variables that can impact the kinetics of the cracking process, and an experimental determination of the magnitude of these dependencies via "separate effect tests." Such tests can be conducted, at least in principle, under laboratory conditions where all the system parameters are adequately defined and controlled, and the crack development can be monitored with good resolution and accuracy.

In practice, this has proven extremely difficult to do [2] and, although improved testing techniques and material analytical procedures have *now* been developed, the fact is that many of the empirical life prediction approaches based on *historical* data may be questionable. This is illustrated in Figure 18.8, which compares the empirically derived prediction

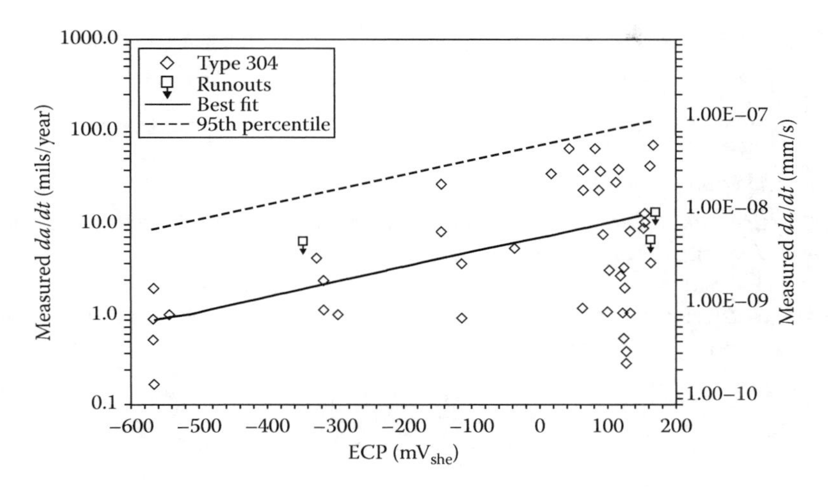

FIGURE 18.8
Comparison between prediction (according to Equation 18.2) and observation of the crack propagation rate vs. corrosion potential relationship for Type 304 stainless steel in BWR water. (From BWR water chemistry guidelines—2000 revision, EPRI Report TR-103515-R2, 2000.)

algorithm (Equation 18.2) with the data for crack propagation rates in stainless steel in high-temperature water as a function of the corrosion potential. The prediction algorithm in Figure 18.8 is a multiparameter least squares fit to the fracture mechanics data base for sensitized Type 304 stainless steels [12].

$$\ln(V) = \frac{2.181\ln(K) - 0.787\kappa - 0.586 + 0.00362\varphi_c + 6730}{T - 35.567} \quad (18.2)$$

where
V is the crack propagation rate, mm/s
K is the stress intensity factor, MPa$\sqrt{\text{m}}$
κ is the solution conductivity at 25°C
φ_c is the corrosion potential, mV_{she}
T is the absolute temperature, K

It is apparent that in this case the correlation factor between prediction and observation is very low ($R^2 < 0.1$), thereby raising questions regarding the quality of the data upon which the correlation was based and/or the completeness of the model (i.e., whether it takes into account the interactions between all the relevant system variables).

More advanced empirical approaches have been used in France [15,16] to predict cracking in nickel-base Alloy 600 in thick-section PWR primary side components such as the control rod drive mechanism (CRDM) nozzles. To be relevant the approach must be able to address all of the following interacting factors:

- Grain boundary carbide morphology as a function of various heat treatments
- Alloy composition, taking into account the differences in susceptibility between Alloys 600, 182, 82, 690, and 800, plus the fact that (for currently unknown reasons) cracking may be significantly greater in some heats compared with other nominally identical heats

- Corrosion potential and its proximity to the Ni/NiO equilibrium potential
- Hydrogen fugacity and the different effects that this parameter can have on crack initiation and crack propagation
- Yield strength and cold work
- Stress and stress intensity factor
- Temperature dependence, with possible variations in activation enthalpy for crack "initiation" and propagation

The initial life prediction algorithm for Alloy 600 was developed for steam generator tubing based on the following empirical algorithm:

$$t_f = C\left\{\frac{\sigma^{-4}}{I_m}\right\}\exp\left\{\frac{E}{RT}\right\} \tag{18.3}$$

where
t_f is the failure time, h
C is a constant
σ is the applied stress, MPa
I_m is a material susceptibility index
E is the activation energy, J/mol/K
R is the universal gas constant, 1.987 cal/mol or 8.314 J/mol/K
T is the absolute temperature, K

The choice of the fourth power dependency on stress was based on, for instance, separate laboratory time-to-failure data for steam generator tubing. The material parameter (I_m) was strictly an empirical derivation, based on fitting the prediction algorithm to past steam generator cracking. The values of I_m, defined by the annealing conditions and their effect on the carbide morphology, ranged from 0.2 for high-temperature mill-annealed and thermally treated condition to 1.0 for the low-temperature mill-annealed condition. This relatively coarse metric was subsequently refined as the vessel head penetration problems arose in the United States, with either more subdivisions of the microstructural effect due to the annealing temperature and carbon content [17] or to the observed grain boundary carbide morphology from nondestructive replication tests [18,19]. The activation enthalpy exhibits a variability depending on the stress/strain conditions and, possibly, the potential within the susceptible potential range around the Ni/NiO equilibrium value. However, values of 180 kJ/mol (44 kcal/mol) have been generally adopted for initiation and component life calculations and 130 kJ/mol (30 kcal/mol) for crack propagation estimations.

It should be noted, however, that the prediction algorithm (Equation 18.3) did not take into account the interrelated conditions of corrosion potential and hydrogen content listed above or other interdependencies.

It is apparent, therefore, that there is some uncertainty in this life prediction algorithm in terms of the completeness of the algorithm (e.g., the treatment of corrosion potential). On top of these epistemic uncertainties, there are the aleatory uncertainties associated with the random stochastic phenomena that dominate during the early precursor, crack initiation, and coalescence periods and which can lead to very long (engineering) crack initiation times.

Consequently, there has been a concerted effort to apply Weibull statistics to the laboratory and field data to assess [16,20] the future development of damage, its dispersion, and the benefit of various mitigation actions. An example of this development is shown via the Monte Carlo predictions in Figure 18.9 for the cumulative percentage of cracked vessel head

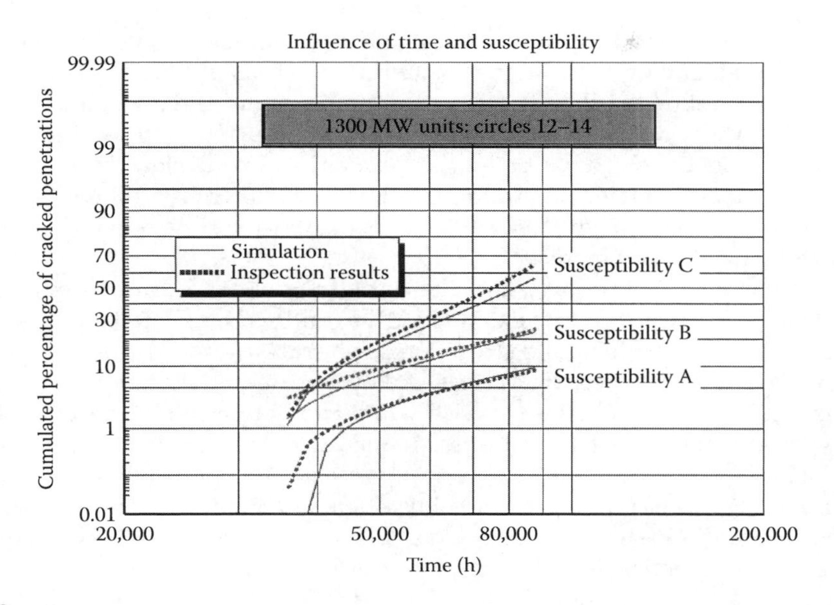

FIGURE 18.9
Results of Monte Carlo simulations of IGSCC in upper head penetrations of 1300 MWe French PWRs and comparison with inspection results for each class of Alloy 600. (From Scott, P.M. and Benhamou, C., An overview of recent observations and interpretations of IGSCC in nickel-base alloys in PWR primary water, in *Proceedings of Tenth International Conference on Environmental Degradation in Nuclear Power Systems–Water Reactors*, Lake Tahoe, Eds. F.P. Ford and G. Was, National Association of Corrosion Engineers, August 5–9, 2001.)

penetrations as a function of operating time. In this example, the inputs to Equation 18.3 have been the distributions of temperature, stress, activation enthalpy, and material susceptibility expected for the targeted penetrations (i.e., the reactor design, the alloy, and the position of the penetrations on the vessel head). The material condition in this example is defined in terms of the carbide morphology in increasingly susceptible condition as

- Susceptibility A—materials with primarily intergranular carbide precipitates
- Susceptibility B—materials with recrystallized grains with carbides mainly on a prior grain boundary
- Susceptibility C—materials with recrystallized grains with randomized intragranular carbides

It is seen that the calculated accumulation of damage as a function of time and material condition is in reasonable agreement with the observations, and this success has been the analytical basis for prioritizing vessel head replacements.

18.4 Life Prediction Based on an Understanding of the Mechanism of Cracking

It is recognized that no LWR vendor, utility, or regulator will accept a prediction of cracking based solely on a quantitative understanding of the mechanism of cracking, and that operational decisions must be based on high-quality cracking data obtained under well-controlled conditions. In spite of this fact, the development of a mechanism-informed life-prediction methodology for ductile structural materials in BWRs was undertaken 30 years ago in the belief that such a "fundamental" approach would give guidance in understanding of the complex system interactions that occur during the cracking process, and would give confidence to strictly data-based operational or design decisions. Indeed, a fully empirical model was deemed unfeasible, because with 20 variables evaluated at five levels—with interdependencies—over 10^{14} experiments would be required, not counting replicates and flawed experiments.

This belief has been borne out, since the initial model development for unirradiated stainless steel piping was logically extended [21] to address irradiated stainless steel core components, nickel-base alloys, and stress corrosion cracking and corrosion fatigue of carbon and steel piping and pressure vessels. A key feature is the ability to anticipate and hypothesize the primary effects that arise when irradiation is present or when nickel-base alloys or low-alloy steels are addressed. Such insights also lead to novel mitigation approaches. By comparison, early papers and proposals to evaluate irradiation effects on stress corrosion cracking (SCC) proposed hundreds of complex, interacting factors, with no sensible way to distinguish which to study first or how the interactions might operate.

The intent in this section is to outline the approach that has been taken to develop a mechanism-based life prediction capability for stainless steels in both unirradiated and irradiated conditions in BWRs.

In these mechanism-based approaches, attention has been focused on the crack propagation period in the chronology of phenomena discussed earlier. This was done for two reasons. First, there was an urgent need to develop justifiable propagation rate vs. stress intensity factor disposition relationships, which provide an estimate of the crack advance that might occur before the next inspection. Second, from the viewpoint of predicting overall life, there was experimental evidence (Figure 18.10) that indicated that crack propagation of a single dominant crack occurred relatively quickly after loading, with long crack

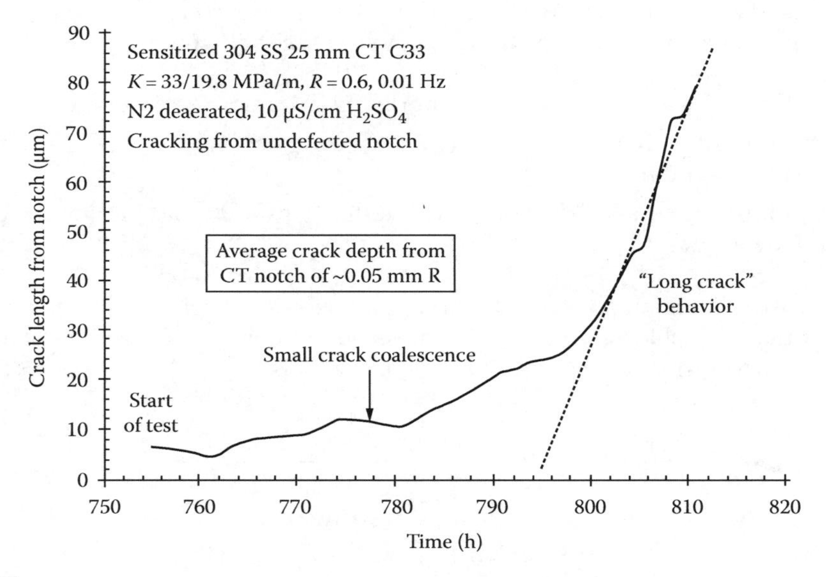

FIGURE 18.10
Crack depth–time relationship for intergranular cracks initiating, coalescing, and propagating in notched 1T CT specimen of sensitized stainless steel in 288°C water. (Adapted from Andresen, P.L. et al., Behavior of short cracks in stainless steel at 288°C, in Paper 495, *NACE Conference*, Las Vegas, NV, April 1990.)

response observed when the crack depth was about 50 μm. In other words, the initial precursor, crack initiation, and coalescence periods were less significant than the propagation period in these BWR systems.

There have been various hypotheses for the mechanism of crack propagation once the degradation has passed the early initiation and coalescence stages. Numerous hypotheses were proposed in a series of conferences during the period 1965–1988 where the mechanisms ranged from those that relied on preexisting or strain-assisted active paths to those depending on various adsorption/absorption phenomena (e.g., hydrogen embrittlement mechanisms). There was considerable debate concerning the dominant mechanism in a given system, but Parkins [23] pointed out that it was likely that there was a "stress corrosion spectrum" that logically graded the cracking systems between those that were mechanically dominated (for instance, hydrogen embrittlement of high-strength steels) and those that were environmentally dominated (for instance, preexisting active path attack in the severely sensitized grain boundary stainless steel/water system). Indeed, it was suggested that several mechanisms were feasible in one alloy/environment system with the dominant mode changing with relatively small alterations in the material, environment, or stressing conditions. This viewpoint was followed by the suggestion (e.g., [24–27]) that a similar spectrum of behavior occurs between constant load (stress corrosion), where creep is a significant factor in determining the crack tip strain rate, and situations where the strain rate is applied, as with dynamic load (strain-induced corrosion cracking) and cyclic load (corrosion fatigue) conditions.

With the advent of extremely sensitive analytical and crack monitoring capabilities, many of the earlier cracking hypotheses have been shown to be untenable, and the candidate mechanisms for environmentally assisted crack propagation for alloys in water-cooled reactors have narrowed down to slip-oxidation, film-induced cleavage, and hydrogen embrittlement.

18.4.1 Slip-Oxidation Mechanism

Various crack advancement theories have been proposed to relate crack propagation to oxidation and the stress/strain conditions at the crack tip, and these have been supported by a correlation between the average oxidation current density on a straining surface and the crack propagation rate for a number of systems [27,28]. There have been various hypotheses for the precise atom–atom rupture process at the crack tip, including (a) the effect that the environment has on the ductile-fracture process (i.e., the tensile ligament theory [29]), (b) the increase in the number of active sites for dissolution because of the strain concentration [30] and, (c) the preferential dissolution of mobile dislocations because of the inherent chemical activity of the dislocation core or the solute segregation that can occur there [31].

Experimentally validated elements of these earlier proposals have been incorporated into the current slip-oxidation model, which relates crack propagation to the oxidation that occurs when the protective film at the crack tip is ruptured [32–34]. Different types of protective film have been proposed varying from oxides, mixed oxides, salts, or noble metals left on the surface after selective dissolution of a more active component in the alloy.

Quantitative predictions of the crack propagation rate via the slip-oxidation mechanism are based on the Faradaic relationship between the oxidation charge density on a surface and the amount of metal transformed from the metallic state to the oxidized state (e.g., MO or M^{+}_{aq}). The change in oxidation charge density with time following the rupture of

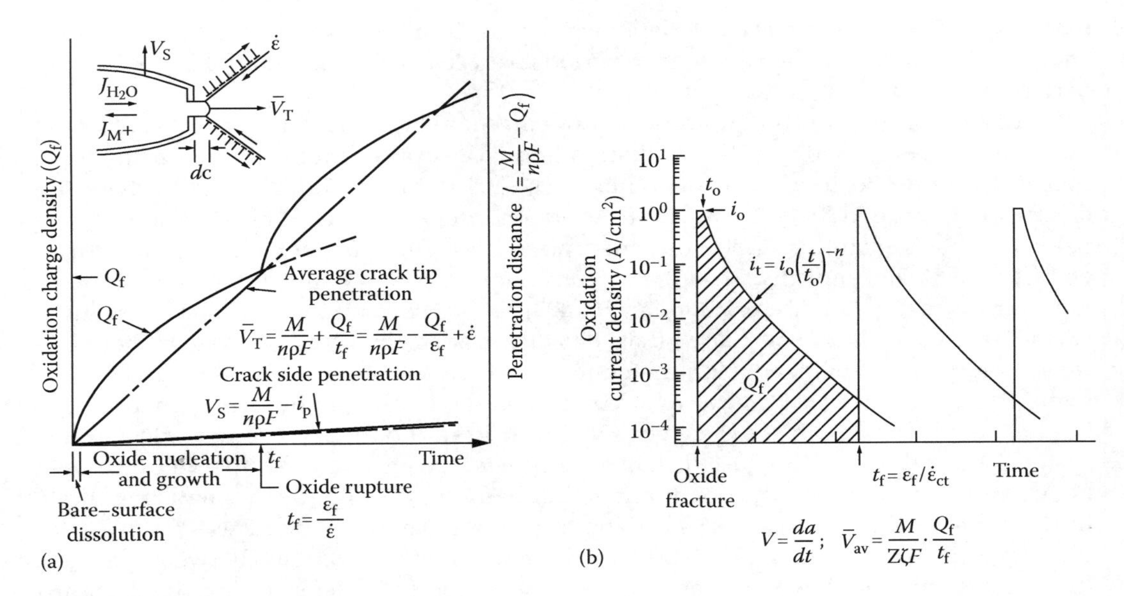

FIGURE 18.11
(a) Schematic oxidation charge density–time relationship for strained crack tip and unstrained crack sides. (From Ford, F.P., Mechanisms of environmental cracking peculiar to the power generation industry, Report NP2589, EPRI, Palo Alto, CA, September 1982.) (b) Changes in oxidation current density following the rupture of the oxide at the crack tip. The crack advance is related Faradaically to the oxidation charge density, Q_f, which will be a function of the bare surface dissolution rate, i_o, the passivation rate, n, the fracture strain of the oxide, ε_f, and the crack tip strain rate. (From Ford, F.P., The crack tip system & it's relevance to the prediction of cracking in aqueous environments, in *Proceedings of First International Conference on Environmentally Assisted Cracking of Metals*, Kohler, WI, October 2–7, 1988, Eds. R. Gangloff and B. Ives, Published by NACE, pp. 139–165.)

a protective film at the crack tip is shown schematically in Figure 18.11a and the associated oxidation current density vs. time relationship is shown in Figure 18.11b. Once the protective oxide is ruptured, the crack will advance rapidly into the metal by bare surface dissolution but will, within a matter of milliseconds, begin to slow down as the thermodynamically stable and protective oxide reforms at the crack tip. Sustained crack advance depends, therefore, on maintaining a strain rate in the vicinity of the crack tip that will promote repeated rupture of the oxide film.

Thus, [24] the crack propagation rate, V, is governed by the relationship

$$V = \left[\frac{MQ_f}{z\rho F}\right]\left(\frac{1}{\varepsilon_f}\right)\left(\frac{d\varepsilon}{dt}\right)_{ct} \tag{18.4}$$

and if the passivation relationship below is used

$$i = i_o\left[\frac{t}{t_o}\right]^{-n} \tag{18.5}$$

then Equation 18.4 becomes

$$V = \left(\frac{M}{zpF}\right)\left(\frac{i_o t_o^n}{(1-n)\,\varepsilon_f^n}\right)\left(\frac{d\varepsilon}{dt}\right)_{ct}^n \tag{18.6}$$

where

M, p are atomic weight and density of the crack tip metal
F is the Faraday's constant
Z is the number of electrons involved in the overall oxidation of an atom of metal
i_o, t_o are bare surface oxidation (dissolution) current density and its duration time
ε_f is the fracture strain of the oxide at the crack tip
$(d\varepsilon/dt)_{ct}$ is the crack tip strain rate

In general form Equation 18.6 collapses to

$$V = A\left(\frac{d\varepsilon}{dt}\right)_{ct}^{n} \tag{18.7}$$

where the parameters A and n are related to the dissolution and passivation kinetics at the strained crack tip and to the mechanical properties of the oxide at the crack tip. The crack tip strain rate, $(d\varepsilon/dt)_{ct}$, is a function of the stress/strain/time conditions, and this will be discussed later.

18.4.2 Film-Induced Cleavage Mechanism

It has been observed that in some incidences of transgranular cracking [37] the Faradaic equivalent of the oxidation change density at a strained crack tip is insufficient to explain the observed crack advance. Moreover, the cleavage-like crystallographic features on the fracture surfaces are hard to rationalize in terms of a dissolution/oxidation model by itself. Consequently, it has been proposed by several authors [37–41] that transgranular environmentally controlled crack propagation can occur by a combination of oxidation-related and brittle-fracture mechanisms. Specifically, it has been suggested that, initially, the crack front moves forward by an oxidation process that is controlled by the same rate-determining steps as those in the slip-oxidation model, but when the film rupture event occurs, the crack in the film may rapidly penetrate a small amount, a^*, into the underlying ductile metal matrix (Figure 18.12).

Thus, Equation 18.4 is modified as follows:

$$V = \left[\left[\left(\frac{MQ_f}{z\rho F}\right) + a^*\right]\right]\left(\frac{1}{\varepsilon_f}\right)\left(\frac{d\varepsilon}{dt}\right)_{ct} \tag{18.8}$$

The extent of the additional "film-induced cleavage" component of crack advance, a^*, may be governed [37] by the state of coherency between the surface film and matrix, the fracture toughness of the substrate, the film thickness, and the initial velocity of the cleavage crack emerging from the surface film. Although traditionally the surface film has been considered to be an oxide, more recent investigations [40,41] have refocused on the role played by dealloyed surface films, for example, copper-rich film in Cu-Zn or nickel-rich films in Fe-Cr-Ni alloys. The extent of the cleavage propagation into the matrix may be on the order of 1 µm, with the exact value being a function of a combination of the various plasticity and microstructural factors mentioned above. Although there is evidence for such a mechanism in copper-base alloys, austenitic nickel-base alloys, and stainless steels in low-temperature environments (i.e., <115°C), it has not been extensively evaluated for other alloy systems and especially those relevant to LWRs. It is attractive because

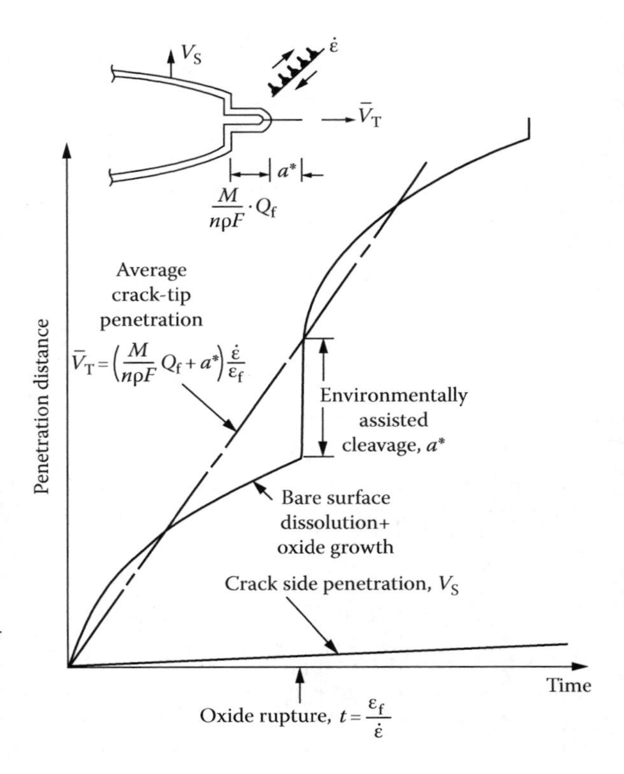

FIGURE 18.12
Schematic illustration of the elements of the film-induced cleavage mechanism of crack propagation. (From Ford, F.P., Mechanisms of environmental cracking peculiar to the power generation industry, Report NP2589, EPRI, Palo Alto, CA, September 1982; Ford, F.P., The crack tip system & it's relevance to the prediction of cracking in aqueous environments, in *Proceedings of First International Conference on Environmentally Assisted Cracking of Metals*, Kohler, WI, October 2–7, 1988, Eds. R. Gangloff and B. Ives, Published by NACE, pp. 139–165.)

it provides a rational basis for quantitatively explaining the interrelationships between the electrochemical parameters and the transgranular fractographic features, as seen in, for example, the carbon and low-alloy steel/high-temperature water systems used for nuclear reactor pressure vessels and steam generator shells. Elements of this approach have also been proposed for intergranular cracking of Alloy 600 in PWR primary environments, where the "cleavage event" is associated with internal oxidation occurring in the chromium-depleted zone in front of the advancing intergranular crack [42,43].

18.4.3 Hydrogen Embrittlement Mechanisms

The general concepts and concerns behind the various hydrogen models have been reviewed by Thompson and Bernstein [44], Hirth [45], Nelson [46] and Birnbaum [47]. In brief, the subcritical crack propagation rate due to hydrogen embrittlement in aqueous environments depends on a sequence of events in the following order (Figure 18.13) [48]:

- Diffusion of a reducible hydrogen-containing species (e.g., H_3O^+) to the crack tip region
- Reduction of the hydrogen-containing ions to give adsorbed hydrogen atoms
- Absorption of the hydrogen adatoms followed by interstitial diffusion of these hydrogen atoms to a "process" zone at a distance, X, in front of the crack tip
- Once the hydrogen concentration in a "process" zone has reached a critical level, C_{crit}, over a critical volume, d_{crit} [49],then localized crack initiation can occur within this zone followed by rapid propagation back to the main crack tip

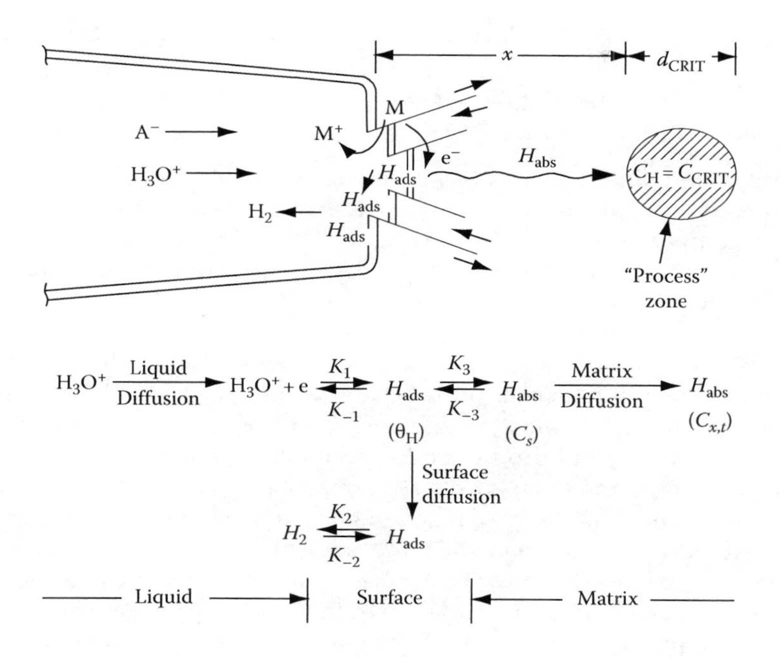

FIGURE 18.13
Schematic of various reactions at crack tip associated with hydrogen embrittlement mechanisms in aqueous environments. (From Ford, F.P., The crack tip system & it's relevance to the prediction of cracking in aqueous environments, in *Proceedings of First International Conference on Environmentally Assisted Cracking of Metals*, Kohler, WI, October 2–7, 1988, Eds. R. Gangloff and B. Ives, Published by NACE, pp. 139–165.)

Thus, disregarding the specifics of these localized fracture mechanisms in the "process" zone, it is apparent that hydrogen embrittlement models predict discontinuous crack propagation at an average rate:

$$\bar{V}_t = \frac{X}{t_c} \quad (18.9)$$

where

X is the distance from the main crack tip to the process zone, which, in turn, is defined by the values of C_{crit} and d_{crit}

t_c is the time for the concentration of absorbed hydrogen, $C_{x,t}$, to reach a critical value, C_{crit}, over the volume, d_{crit}

To evaluate the validity of Equation 18.9, quantitative data for X and t_c are needed. Unfortunately, these are difficult to define from first principles [50] and rely on the validity of various atom–atom rupture hypotheses that have been made (e.g., decohesion [51,52], gas rupture [53,54], enhanced plasticity [55,56], hydride formation [57], martensite formation [58], etc.).

Although traditionally such hydrogen embrittlement mechanisms have been applied qualitatively to high-strength alloys, Hanninen, Torronen, and coworkers [59,60] have suggested that a hydrogen embrittlement mechanism is operating in the relatively ductile pressure vessel steels in water at 288°C. The primary experimental evidence is the observation of "brittle" cracks that are associated with elongated MnS stringers ahead of the main crack tip. Moreover, the degree of environmental enhancement in fatigue crack growth rates may be directly correlated with the extent of these "brittle" fracture areas on the fracture surface.

18.5 Prediction Methodology for Ductile Structural Alloys in BWR Systems

Thermodynamic and kinetic criteria have been used to evaluate which of these candidate crack propagation mechanisms are valid for a given alloy/environment system [28,36]. However, it is comparatively rare that a candidate cracking mechanism can be categorically disallowed on such reasoning. Thus, *qualitative* predictions of cracking have centered around the observation that the rate-determining step in all of the cracking mechanisms is not necessarily the atom–atom rupture process itself, but is one or a combination of the following: mass transport of species to and from the crack tip and the oxidation or reduction reactions and the dynamic strain processes at the crack tip [28,36]. Thus, changes in cracking susceptibility for most ductile alloy/aqueous environment systems with, for instance, changes in temperature, electrode potential, stressing mode (dynamic or static stress, plane strain or plane stress, dislocation morphology, etc.), or environmental composition, can be explained [28,36] by a reaction rate surface (Figure 18.14), regardless of the specific atom–atom rupture mechanism at the crack tip.

Knowledge of the controlling mass transport within the crack enclave, the crack tip oxidation and reduction rates, and the crack tip strain rates has proven of great value in justifying the design and operational engineer the system changes that are likely to reduce the extent of a particular environmentally controlled cracking problem. For instance, increasing the passivation rate at the crack tip by control of the corrosion potential/anionic activity combinations and by material composition changes, reduction in crack tip strain rate

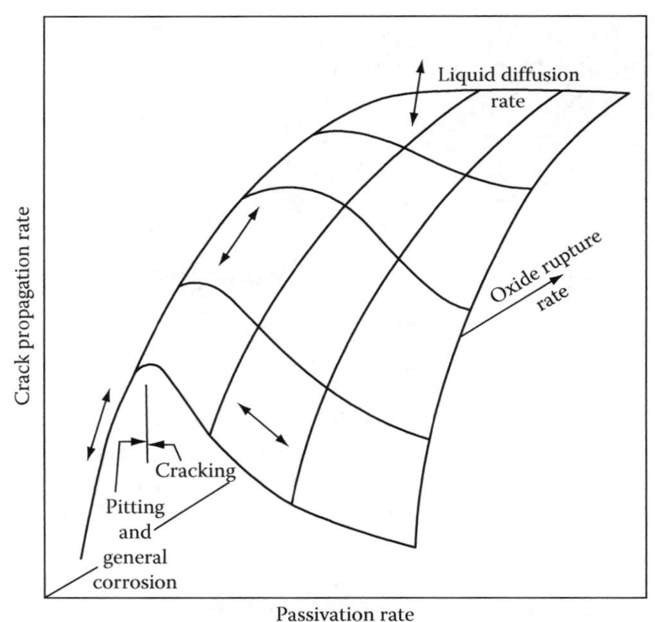

FIGURE 18.14
Schematic "reaction surface" relating the crack propagation rate to the fundamental phenomena of liquid diffusion rate, oxide rupture rate, and passivation rate. (Adapted from Ford, F.P., Stress corrosion cracking, in *Corrosion Processes*, Ed. R.N. Parkins, Applied Science, 1982; Ford, F.P., Mechanisms of environmental cracking peculiar to the power generation industry, Report NP2589, EPRI, Palo Alto, CA, September 1982.)

by attention to dislocation morphology, etc., and the resultant beneficial effects of increasing cracking resistance can be understood within a skeleton of mechanistic understanding [28,35]. Moreover, this knowledge can be used to predict limiting system conditions below which cracking will be minimal; for instance, the evaluation of "threshold stresses" for cracking via their influence on a minimal crack tip strain rate and, thereby, crack propagation rate [61]. In recent years, as economic and technical pressures dictate longer design lives, the emphasis has been on studying cracking in dilute environments where the relevant crack propagation rates are $<10^{-6}$ mm/s and on developing the prediction capability shown in Figure 18.1. Thus, in the following section, the advances in *quantifying* the crack tip atom–atom rupture mechanisms and their rate-determining reactions are discussed.

To limit discussion, attention is focused on the quantitative prediction of cracking in austenitic type 304/316 stainless steels in 288°C high-purity water (e.g., BWR environments) in which it is assumed [62] that the slip-dissolution model is a reasonable working hypothesis for the crack propagation mechanism.

18.5.1 Rationale for Choice of Slip-Oxidation Model as "Working Hypothesis" for Crack Propagation of Ductile Alloys in BWR Systems

It has long been recognized that, based on plant and laboratory observations, there are conjoint material, stress, and environment requirements that have to be met to sustain stress corrosion cracking in stainless steels (and other materials) in BWR systems (Figure 18.15a). The reason why the slip-oxidation model was adopted as the "working hypothesis" was that the basic controlling parameters at the crack tip in that model (creation of a localized environment, the periodic rupture of the protective oxide, and the oxidation kinetics on the bared crack tip surface) could be correlated with these empirical observations (Figure 18.15b).

To reduce the prediction algorithm (Equation 18.6) to practical usefulness, it is necessary to redefine that algorithm in terms of measurable engineering or operational parameters. This involves the following:

1. Defining the crack tip alloy/environment composition in terms of bulk alloy composition, anionic concentration or conductivity, dissolved oxygen content or corrosion potential, etc.
2. Measuring the reaction rates for the crack tip alloy/environment system that corresponds to the "engineering" system
3. Defining the crack tip strain rate in terms of continuum parameters such as stress, stress intensity factor, loading frequency, etc.

There has been extensive work conducted in these areas, and the progress will be reviewed only briefly in this article, prior to illustrating how these advances have been incorporated into verified, quantitative life prediction methodologies.

18.5.2 Definition of Crack Tip Alloy/Environment System

On the basis of direct measurements for Inconel, stainless steel and low-alloy steel/water systems at 288°C, it is known that the electrode potential and pH conditions at the tip of a crack can differ markedly from those at the exposed crack mouth [62–67]. These variations are understood [68–71] in terms of the thermodynamics of various metal oxidation and metal cationic hydrolysis reactions, and how these are influenced by the reduction processes of, for instance, dissolved oxygen at the crack mouth. From a practical viewpoint,

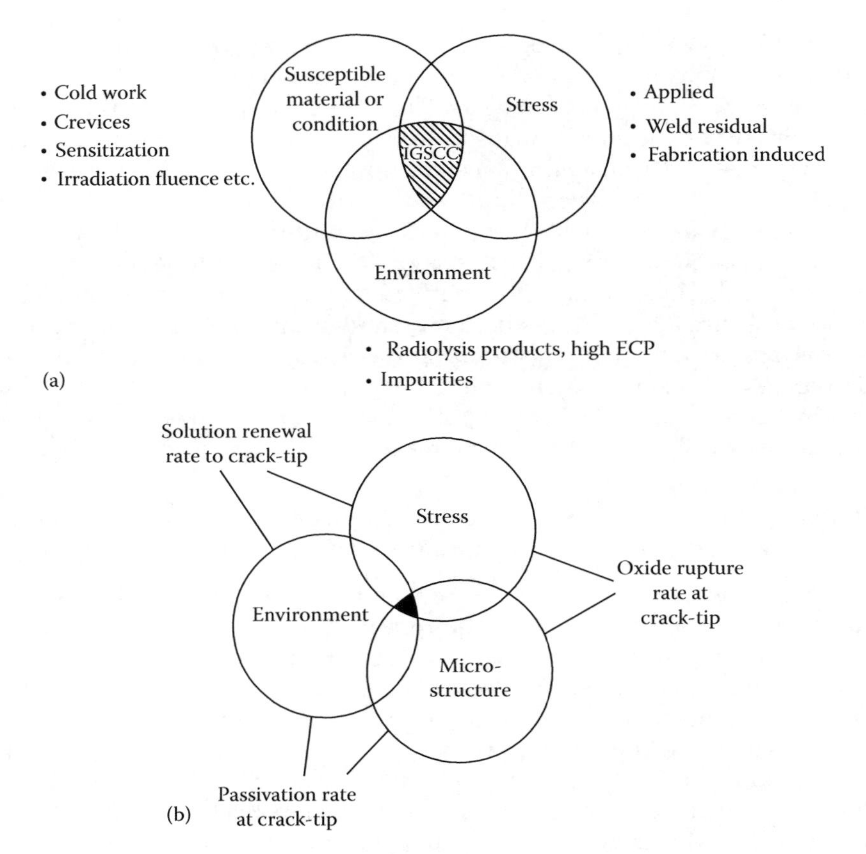

FIGURE 18.15
(a) The three conjoint factors necessary for producing IGSCC. Appropriate changes in the stress environment or material conditions make such a Venn diagram applicable to other modes of environmentally assisted cracking. (From Klepfer, H.H. et al, Investigation of cause of cracking in austenitic stainless steel piping, NEDO-21000, General Electric, July 1975.) (b) Relationship between the empirical conjoint factors necessary for EAC, stress, environment, and material conditions and the crack tip mechanistic factors, oxide rupture frequency, passivation rate, and solution renewal rate. (Adapted from Ford, F.P. et al., Corrosion-assisted cracking of stainless steel & low-alloy steels in LWR environments, Report NP5064S, February 1987, EPRI, Palo Alto, CA.)

the crack tip potential conditions can be defined [62] in terms of the measurable dissolved oxygen content in the high-purity external water environment (Figure 18.16) or, more preferably, by the measured corrosion potential of the external metal surface.

The transient and steady-state concentrations of non-OH^- anions in the crack have also been experimentally measured and analytically modeled [62,63,71]. The anion level at the crack tip is directly dependent on the external anionic activity, the dissolvable metallurgical impurities (e.g., MnS) level, the corrosion potential difference between the crack mouth and tip, and convective influences. For example, the steady-state sulfur anion concentration at the crack tip in low-alloy steels may be defined by the MnS content [72,73], aspect ratio and orientation, the solution flow rate, and degree of oxygenation in the water [62,71]; under these influences, the dissolved sulfur concentration may be on the order of 2 ppm compared with <10 ppb in the bulk solution. For stainless steel exposed under normal BWR oxygenated operating conditions, the potential drop down the crack length of approximately 600 mV will lead to a concentration of anions at the crack tip such that typical crack tip anion concentrations between 0.1 and 1.0 ppm are expected; under deaerated operating conditions

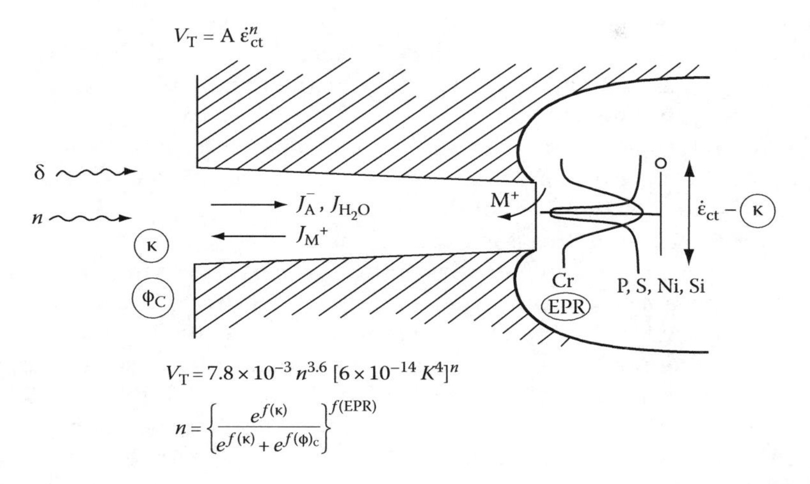

FIGURE 18.16
Schematic of the crack enclave and the relevant phenomena associated with the slip-oxidation mechanism of crack advance. (Adapted from Ford, F.P., The crack tip system & it's relevance to the prediction of cracking in aqueous environments, in *Proceedings of First International Conference on Environmentally Assisted Cracking of Metals*, Kohler, WI, October 2–7, 1988, Eds. R. Gangloff and B. Ives, Published by NACE, pp. 139–165; Ford, F.P. et al., Corrosion-assisted cracking of stainless steel & low-alloy steels in LWR environments, Report NP5064S, February 1987, EPRI, Palo Alto, CA.)

where no potential drop will exist down the crack length, the anion content at the crack tip will approximate that in the bulk environment (e.g., ~10 ppb). To maintain electroneutrality in the crack enclave, it is necessary that there be a corresponding concentration of cations at the crack tip. Because of the low solubilities of most of the metal cations, this electroneutrality criterion dictates a decrease in pH at the crack tip (for bulk oxygenated systems), but it is important to recognize that this acidification is only on the order of 1–2 pH units at most, that is, excessive acidification, as sometimes suggested, cannot occur [63] in these BWR systems where the coolant purity level is approaching theoretical purity water.

The alloy composition at the tip of transgranular cracks is generally assumed to be that of the bulk alloy. However, the alloy composition at the tip of an intergranular crack may be considerably different from the bulk composition if chromium depletion at the grain boundary exists; such compositional heterogeneity will occur at rates controlled by, for example, thermally induced diffusion and/or irradiation-assisted damage. Discussion of these latter metallurgical aspects is outside the range of this particular chapter, but certainly there is adequate knowledge, backed up by extensive analytical electron microscope studies, to allow definition of the grain boundary composition in terms of the thermal or irradiation history during fabrication or reactor operation and the bulk alloy composition (for instance, Refs. [74–79]).

Thus, on the basis of such investigations, the crack tip alloy/environment system may be defined in terms of measurable, or definable, bulk system parameters.

18.5.3 Evaluation of Reaction Rates at Crack Tip

Various techniques have been used for creating the macroscopic analogy to the crack tip bare surface upon which the oxidation and reduction rates can be measured. These have included mechanical methods to rupture the surface oxide such as slowly [80–84] or rapidly [85,86] straining the alloy, completely fracturing the specimen to create a bare fracture surface [32,87], cyclic straining [81,87,88], scratching the alloy surface [89–94], or grinding [95]. Alternatively, electrochemical methods have been used that involve cathodically reducing the oxide [62,96–98] and then rapidly pulsing to a potential associated with the predicted crack tip conditions.

Most of these techniques have been applied in the study of environmentally assisted cracking and the various experimental difficulties have been reviewed [36,99,100], as have the interpretations of the atomistics of the reaction rate [36] at a strained crack tip. These will not be covered further in this particular chapter.

The main conclusions vis-à-vis stainless and low-alloy steels in 288°C environments symptomatic of those at the crack tip are that both the oxidation and reduction reaction rates are increased when the protective oxide is removed. The bare surface oxidation rate, i_0, is a function of the electrode potential and also the dissolved anion content. Explanations for this behavior range from those by Ford and Andresen [62,63] who argue that the rate-controlling process for bare surface dissolution is the diffusion of solvating water molecules to the oxidizing surface (and how this is affected by the anion/pH combinations that affect the solubility of impeding oxide precipitates), to those of Combrade et al. [81], who argue that adsorbed sulfur on the surface impedes the incipient solid-state passivating oxide nucleation.

The subsequent oxide formation leads to a decrease in the overall oxidation rate, according to Equation 18.5. The value of n in this equation (which is the same as in the crack propagation rate, Equation 18.6) varies with the alloy chemistry (e.g., chromium content for a denuded grain boundary of Type 304 stainless steel), corrosion potential at the crack mouth, and the anionic activity in the bulk environment.

Thus, the passivation rate parameter n in Equation 18.5 is controlled by the crack tip environmental (e.g., pH, potential, anionic activity) and material (e.g., % chromium depletion in grain boundary) conditions. To make this practically useful, n has been reformulated in terms of measurable bulk system parameters such as water conductivity (κ), corrosion potential (φ_c)—which, in turn, is a function of the dissolved oxygen and hydrogen peroxide content—and the electrochemical potentiokinetic repassivation (EPR) parameter, which is related to the chromium denudation in the grain boundary [101].

The choice of using water conductivity as a measure of the anionic impurity concentration was again based on practicality since this remains the primary operational control metric for reactor coolant purity. Of course, this is directly relatable to, for instance, the sulfate and chloride activities, as will be mentioned later. An example of the variation of n with corrosion potential and water conductivity for a steel sensitized to an EPR value of 15 C/cm^2 (corresponding to a 0.05% C stainless steel after a 600°C for 4 h heat treatment)

$$n = \left\{ \frac{e^{f(\kappa)}}{e^{f(\kappa)} + e^{g(\varphi_c)}} \right\} \quad (18.10)$$

is shown in Figure 18.17.

18.5.4 Definition of Crack Tip Strain Rate

It has been recognized for many years that the oxide rupture prerequisite for crack advance leads to an expected interrelationship between cracking susceptibility and slip morphology, since coarse slip process will be more likely to rupture a brittle film of given thickness and expose bare metal [102] than those associated with fine slip. Such a relationship has been observed [28] for various alloys in both aqueous and gaseous environments, where

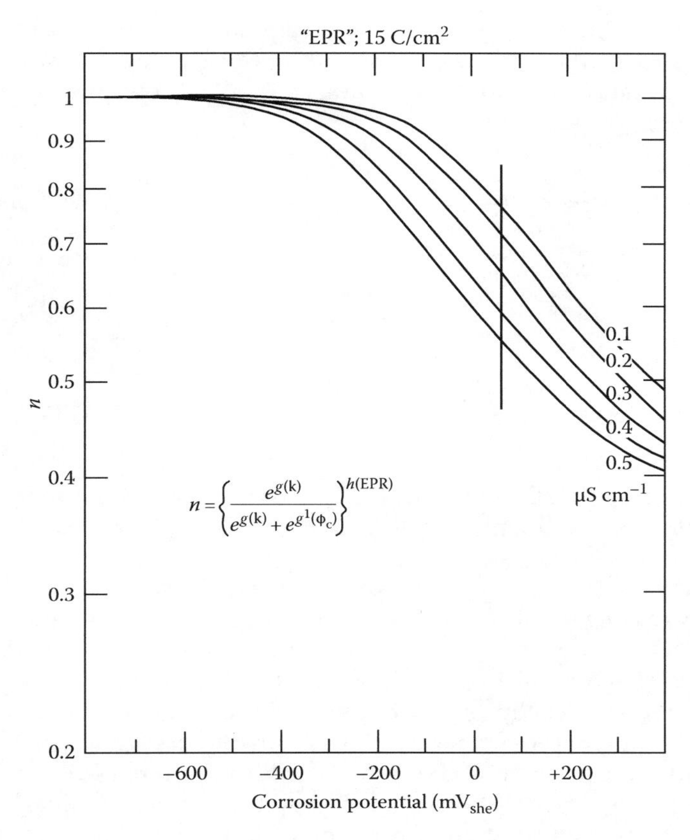

FIGURE 18.17
Relationships between n in Equations 18.5 and 18.6 and the corrosion potential and bulk solution conductivity for a sensitized (EPR = 15C/cm^2) Type 304 stainless steel in water at 288°C. (Adapted from Ford, F.P. et al., Corrosion-assisted cracking of stainless steel & low-alloy steels in LWR environments, Report NP5064S, February 1987, EPRI, Palo Alto, CA.)

the different dislocation morphologies are related to changes in stacking-fault energy, short-range order, precipitate/matrix coherency, and precipitate distribution.

Despite these known effects of microscopic heterogeneity of plastic flow at a crack tip on the cracking susceptibility, the main emphasis in formulating the periodicity of oxide or film rupture has been in terms of continuum parameters such as strain rate, stress intensity factor, and oxide fracture strain, with the success criteria being their ability to address such questions as

- Can the strain rate formulations account for the limiting stress conditions for cracking defined by σ_{th}? This aspect has been covered by, for example, Parkins and coworkers [61] in assessing the criteria for maintaining a critical creep rate and how this might be achieved by various stressing conditions.
- How will the crack tip strain rate vary with, for instance, dynamic stressing conditions and the degree of plastic constraint? A subsidiary aspect is the examination of the belief that σ_{th} and K_{ISCC} below which cracking will not occur are system constants.
- How does the fact that the crack is propagating affect the crack tip stain rate when dislocation movement is governed by an exhaustion theory of creep? Associated with this is the definition of the criteria that determine the onset of crack arrest.

Numerous formulation approaches have been suggested and reviewed [36,103–105] but currently without resolution, largely because the crack tip strain rate is ultimately controlled by localized dislocation movements that are inconsistent with continuum fracture mechanics analyses used in structural integrity assessments. Because of the urgency in developing a useable crack propagation algorithm, the following general relationships for the crack tip strain rate have been proposed:

$$\text{For constant load: } \dot{\varepsilon}_{ct} = A\,\dot{\varepsilon}_{creep} + B\left(\frac{\bar{V}_t}{x^*}\right) \tag{18.11}$$

$$\text{For slow strain rate: } \dot{\varepsilon}_{ct} = C\dot{\varepsilon}_{appl} + D\left(\frac{\bar{V}_t}{x^*}\right) \tag{18.12}$$

$$\text{For cycle load: } \dot{\varepsilon}_{ct} = \left(\frac{\delta\varepsilon}{\delta K}\right)\dot{K} + \left(\frac{\delta\varepsilon}{\delta a}\right)\bar{V}_t \tag{18.13}$$

In all of these equations, it is recognized that the crack tip strain rate is a function not only of the applied stress and strain rate or stress intensity factor, but also of the crack propagation rate. In other words, it is recognized that the movement of the crack tip stress field into the underlying metal matrix by an amount, x^*, is activating new dislocation sources in a given time period, thereby increasing the strain rate over and above that which would exist if the crack were stationary.

Despite the rationale for the above general formulations, it has proven remarkably difficult to quantify them in terms of crack tip plasticity and to independently verify their accuracy. For instance, uniaxial creep deformation laws at low homologous temperatures are not necessarily applicable to the multiaxial stress conditions in the surface region adjacent to the crack tip, and the use of linear-elastic fracture mechanics must have limitations for the near-crack-tip, high-plastic-deformation conditions.

Because of these fundamental difficulties, empirical formulations between the crack tip strain rate and "engineering" parameters evolved (Table 18.1), which have proven useful for a wide range of stressing conditions for structural alloys in 288°C water [62].

Although these formulations of the crack tip strain rate proved adequate in the early development and application of the model, they oversimplified the physical fact that as the crack tip advanced the associated stress field activated further dislocation sources. Such activation would increase the crack tip strain rate by an amount proportional to the crack propagation rate and the strain profile in front of the crack tip. The resultant crack tip strain rate would,

TABLE 18.1

Crack Tip Strain Rate (s^{-1}) Formulations for Unirradiated Stainless Steel in 288°C Water

	Type 304 Stainless Steel	A533B Low-Alloy Steel
Constant load	$4.1 \times 10^{14} K^4$	$3.3 \times 10^{-13} K^4$
Slow strain rate	$5\dot{\varepsilon}_{app}$	$10\dot{\varepsilon}_{app}$
Cyclic load	$68.3\dot{\nu}A_R\Delta K^4$	$547\dot{\nu}A_R\Delta K^4$

Notes: $K(\Delta K)$ = stress intensity (amplitude in MPa $\sqrt{m}$; $\dot{\varepsilon}_{app}$ = applied strain rate in s^{-1}; $\dot{\nu}$ = frequency of (symmetrical) load cycle in s; A_R = constant in "dry" Paris Law = 2.44×10^{-11} (for $R \leq 0.42$) = $-2.79 \times 10^{-11} + 1.115 \times 10^{-10} R + 5.5 \times 10^{-11} R^2$ (for $R > 0.42$); R = minimum load/maximum load.

therefore, be the result of a dynamic equilibrium between this crack propagation rate-dependent strain rate and the contribution due to exhaustion creep, which would be decreasing with time at the relatively low homologous temperatures relevant to BWR operations, that is

$$\left(\frac{d\varepsilon}{dt}\right)_{ct} = \left(\frac{d\varepsilon}{dt}\right)_{creep} + \left(\frac{d\varepsilon}{da}\right) \times \left(\frac{da}{dt}\right) \quad (18.14)$$

where a is the distance in front of the crack tip.

Although Equation 18.14 qualitatively took this dynamic equilibrium into account, Shoji and coworkers [106–108] reformulated the crack tip strain rate relationship to take into account these complex strain rate factors in front of an advancing crack tip and to include the expected contributions due to work-hardening coefficient, yield strength, degree of plastic constraint, and dynamic applied loads:

$$\left(\frac{d\varepsilon}{dt}\right)_{ct} = \beta\left(\frac{\sigma_y}{E}\right)\left(\frac{m}{(m-1)}\right)$$

$$\times\left(\left(2\left(\frac{dK}{dt}\right)\Big/K\right) + \left(\frac{da}{dt}\right)\Big/r\right)\left(\ln\left(\left(\frac{\lambda}{r}\right)\left(\frac{K}{\sigma_y}\right)^2\right)\right)^{\frac{1}{(m-1)}} \quad (18.15)$$

where

m is the strain hardening coefficient in Gao equation [109]

σ_y, E are the yield strength and elastic modulus respectively

K is the stress intensity factor

λ, β are constants

18.5.5 Prediction of Stress Corrosion Crack Propagation of Unirradiated Stainless Steel in 288°C Water

The relevant fundamental reactions for stress corrosion cracking propagation in austenitic stainless steels in 288°C water pertinent to crack tip systems have been discussed above. As a result the theoretical crack propagation rate is given by the following simplified version of Equation 18.6:

$$\frac{da}{dt} = 7.8\times10^{-3} n^{3.6}\left(\frac{d\varepsilon}{dt}\right)_{ct}^{n} \quad (18.16)$$

where da/dt is in units of cm s^{-1}, n varies between 0.33 and 1.0 depending on the material sensitization, anionic impurity concentration, and corrosion potential factors discussed above, and the crack tip strain rate is in units of s^{-1} and is quantified by the relationships in Table 18.1.

The general validity of the prediction methodology is indicated in Figure 18.18, which covers data obtained on sensitized Type 304 stainless steel in 8 ppm oxygenated water at 288°C stressed over a wide range of constant load, monotonically increasing load, and cyclic load conditions [36,62]. The solid line is the theoretical relationship from Equations 18.10 and 18.13 and the appropriate strain rate formula for the stressing mode (Table 18.1).

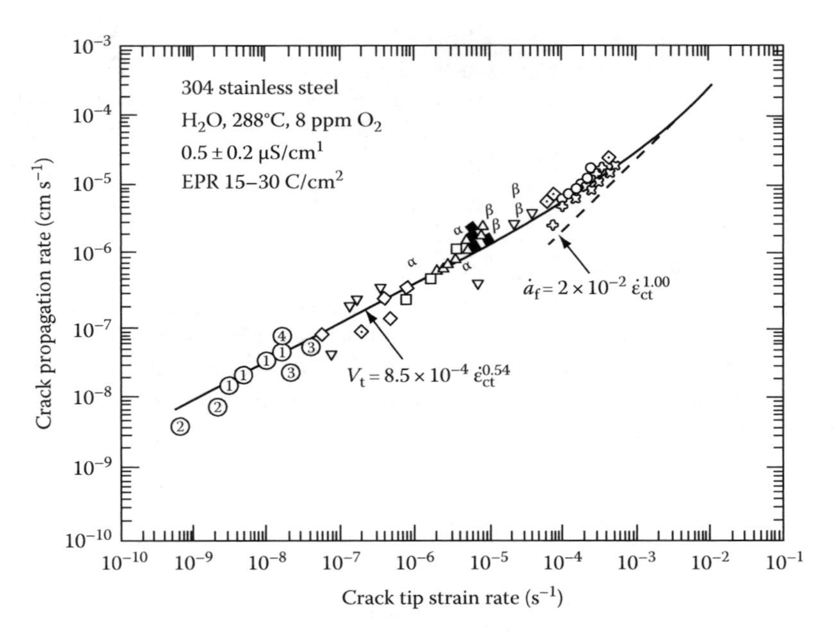

FIGURE 18.18
Observed and theoretical crack propagation rate vs. crack tip strain rate relationships for Type 304 sensitized 304 stainless steel in oxygenated water at 288°C. (Adapted from Ford, F.P., The crack tip system & it's relevance to the prediction of cracking in aqueous environments, in *Proceedings of First International Conference on Environmentally Assisted Cracking of Metals,* Kohler, WI, October 2–7, 1988, Eds. R. Gangloff and B. Ives, Published by NACE, pp. 139–165; Ford, F.P. et al., Corrosion-assisted cracking of stainless steel & low-alloy steels in LWR environments, Report NP5064S, February 1987, EPRI, Palo Alto, CA.) The data points that are numbered were obtained under constant load, the Greek letter points under constant applied strain rate and the geometric symbols under fatigue conditions with varying mean stress, stress amplitude, and frequency.

The agreement between prediction and observation illustrates the applicability of the methodology to the whole stress corrosion–corrosion fatigue spectrum.

The effect of coolant deaeration on the crack propagation rate–crack tip strain rate relationship is shown in Figure 18.19, and this is compared with the data produced under aerated conditions illustrated in Figure 18.18. The degree of the predicted and observed beneficial effect of deaeration is very dependent on the crack tip strain rate (i.e., stress level and/or degree of dynamic loading) and this has an impact on evaluating the benefit of deaeration and the choice of the test technique used to quantify the benefit. For instance, the factor of improvement (defined as the ratio of the crack propagation rates in the two environments at a given crack tip strain rate) decreases with increasing strain rate. Thus, factors of improvement determined by an accelerated "slow-strain-rate test" (with applied strain rates in the range 10^{-4}–10^{-6} s^{-1}) will significantly underestimate the improvement expected in a plant component subjected to weld residual stresses (with crack tip strain rates less than 10^{-8} s^{-1}).

The specific effects of changes in corrosion potential and solution conductivity on the propagation rate under constant load are shown in Figure 18.19a and b; the beneficial effects of improved water chemistry control (e.g., by controlling the coolant conductivity or anion content) and coolant deaeration are predicable and quantifiable and thereby lay a solid mechanistic basis for the definition (Figure 18.20) of water chemistry specifications [14].

It is seen in Figures 18.20a and b, and 18.21 that the sensitivities of the crack propagation rate to the corrosion potential and anionic impurity concentration are certainly not

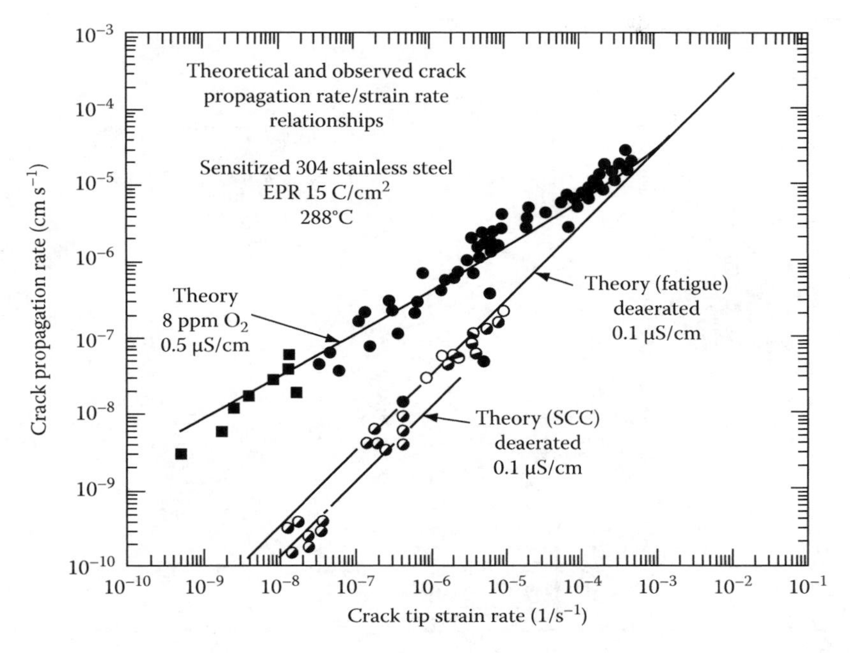

FIGURE 18.19
Observed and theoretical crack propagation rate vs. crack tip strain rate relationships for stainless steel in aerated and deaerated water. (Adapted from Ford, F.P., The crack tip system & it's relevance to the prediction of cracking in aqueous environments, in *Proceedings of First International Conference on Environmentally Assisted Cracking of Metals*, Kohler, WI, October 2–7, 1988, Eds. R. Gangloff and B. Ives, Published by NACE, pp. 139–165; Ford, F.P. et al., Corrosion-assisted cracking of stainless steel & low-alloy steels in LWR environments, Report NP5064S, February 1987, EPRI, Palo Alto, CA.)

linear, but are sigmoidal, reflecting the complex interactions between the various parameters that control the crack tip chemistry. From a practical viewpoint it is apparent that there is a high sensitivity to solution conductivity in certain corrosion potential ranges (associated with those in which hydrogen injection is not used). Thus, effective analysis of plant behavior in these cases is highly dependent on adequate control or definition of the water chemistry conditions. For instance, unmonitored changes in corrosion potential of ~100 mV in 200 ppb O_2 water (which is possible because of flow rate effects, etc.) would give a predictable change in crack propagation rate by a factor of four. It follows, therefore, that the practical use of such a prediction methodology for life prediction, codification, etc., hinges around an adequate definition of the actual system via the use of system monitors for corrosion potential, solution conductivity, etc. and prediction algorithms that relate these local measurements to conditions at remote unmonitored locations. On the other hand, the relatively smaller sensitivity at low potentials associated with "hydrogen water chemistry" indicates that larger impurity transients may be tolerated under these reducing environment conditions.

The comparisons between the theoretical and observed crack propagation rate–stress intensity factor relationships are shown for Type 304 sensitized stainless steel under constant load in a somewhat impure (0.3–0.5 μS/cm) but historically relevant BWR environment in Figure 18.22a. In this case, the predicted curves are for the "best" and "worst" combinations of corrosion potential and conductivity within the specified test conditions and bound the observed variance in the observed data.

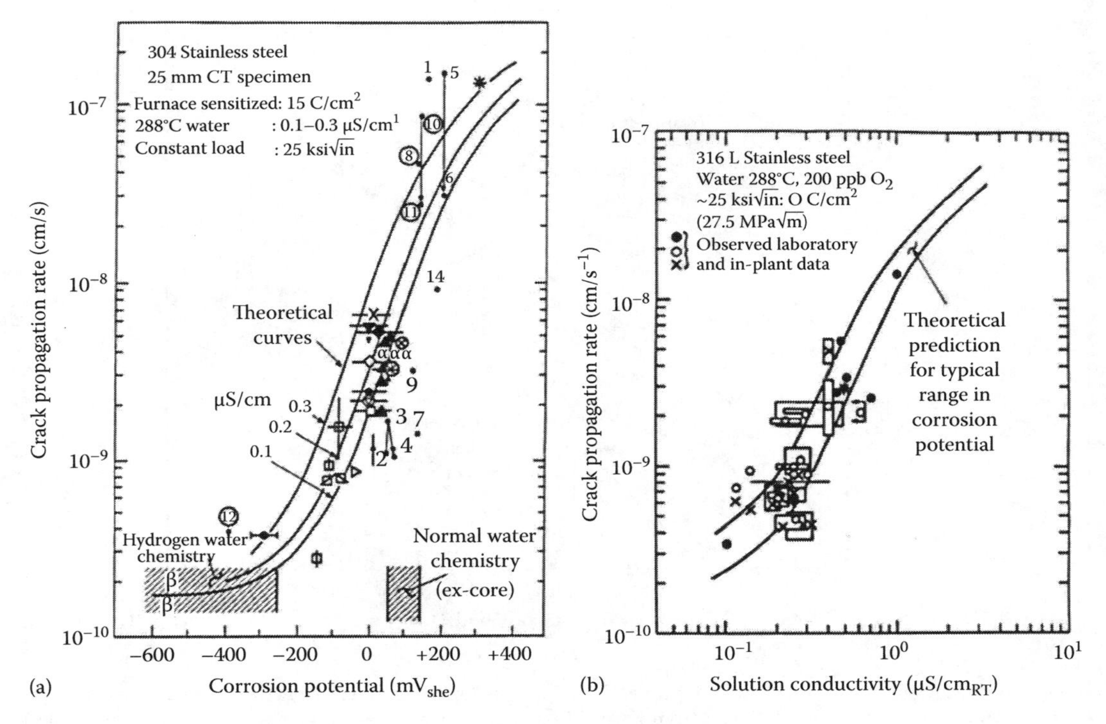

FIGURE 18.20
(a) Observed and predicted relationships between the crack propagation rate and corrosion potential for sensitized Type 304 stainless steel in 288°C water under constant load. Water conductivity is in range 0.1–0.3 µS/cm. (b) Observed and predicted relationships between the crack propagation rate and water conductivity for Type 316 L stainless steel under constant load (25 ksi√in.) in 288°C water containing 200 ppb oxygen. (Adapted from Ford, F.P. et al., Corrosion-assisted cracking of stainless steel & low-alloy steels in LWR environments, Report NP5064S, February 1987, EPRI, Palo Alto, CA.)

Similar observations vs. prediction comparisons are shown in Figure 18.22b for a more modern "hydrogen water chemistry" BWR environment where the corrosion potential is maintained below $-230\,mV_{she}$. It is notable in this case that although the propagation rates are significantly lower than those observed under the "normal water chemistry" conditions addressed by the NRC disposition curve, cracking is still possible at measurable rates at higher stress intensity factors.

It is apparent from these observations that the predicted and observed stress corrosion (and corrosion fatigue) crack propagation rates are very dependent on the specific combined system conditions. Thus, it is a mistake to base design or operating specifications that define only, for instance, mechanical engineering parameters (e.g., stress, strain, stress intensity factor, etc.) without taking into account the relevant material and environmental conditions.

The sources of tensile stress include pressurization, fit-up, and residual stresses associated with welding. Two further sources of stress and strain receiving renewed interest are those due to surface cold work and the realization that there can be considerable local strains immediately adjacent to weld fusion lines (Figure 18.23) in unsensitized stainless steels (Figure 18.24). Although surface cold work has long been recognized as an exacerbating factor for cracking in sensitized stainless steels, the extent of the phenomenon in unsensitized stainless steels is important since cracking in such alloy systems have been assumed to be mitigated.

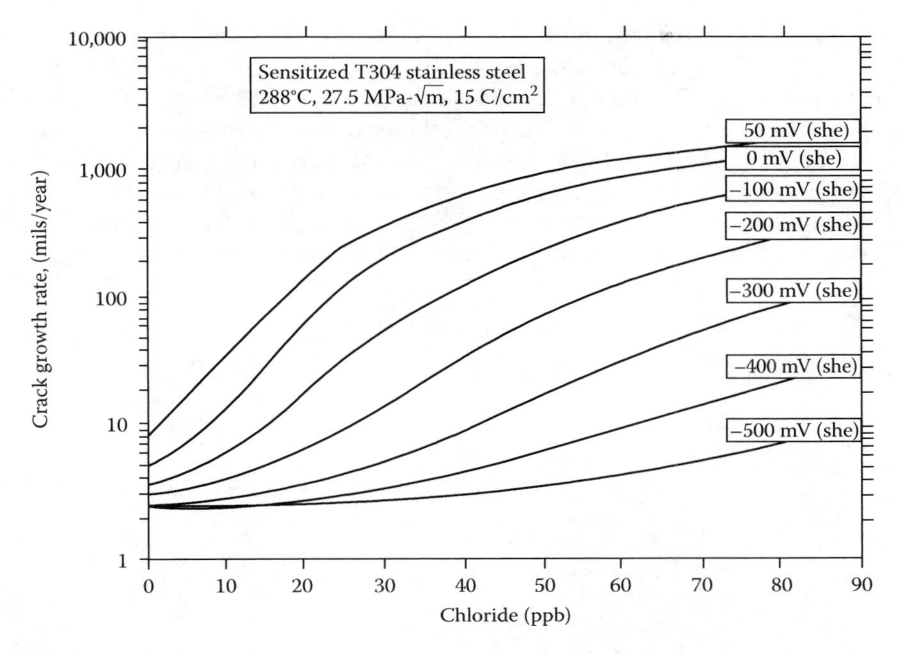

FIGURE 18.21
Crack growth rate predictions as a function of chloride concentration for sensitized stainless steel stressed at 27 MPa√m in 288°C water at various corrosion potentials. (From BWR water chemistry guidelines—2000 revision, EPRI Report TR-103515-R2, 2000.)

It is unfortunate that, from a prediction viewpoint, there is a wide variety of interacting phenomena of relevance (Figure 18.25) spanning from an understanding of the role that cold work plays on dislocation morphologies, athermal and thermal transformation kinetics (i.e., martensite formation, intergranular precipitation of $Cr_{23}C_6$), residual stress profiles, and the increase in yield stress due to work hardening.

The changes in the cracking susceptibility due to interactions between all of these factors cannot be quantified, although the effects are certainly recognized. For instance, investigations at the University of Michigan [113] have indicated that the cracking susceptibility of various Fe-Cr-Ni stainless steels in BWR and PWR environments is related to the dislocation stacking-fault energy, which controls the localization of deformation. Low-stacking-fault energies, found in low-chromium alloys and in silicon-containing steels lead to coplanar dislocation morphologies and increased cracking susceptibility. A further localized deformation phenomenon is strain-induced grain boundary creep [114] that has been advanced as a reason for the observation of *intergranular* stress corrosion cracking in cold-worked, *unsensitized* Type 304 and 316 stainless steel in both BWR and PWR primary side environments. Finally, dynamic strain aging, a phenomenon that leads to discontinuous yielding due to interactions between dislocation solute and interstitial atoms over specific temperature and strain rate ranges, has been advanced as an accelerant to stress corrosion crack propagation. Such a relationship has certainly been noted for aluminum alloys in room temperature saline environments, and for environmentally assisted crack propagation in low-alloy steels in BWR environments. It has been suggested more recently [115] that such localized deformation may play a role in the cracking susceptibility of cold-worked unsensitized stainless steels in BWR and PWR environments.

However, the role of the change in yield stress can be predicted as indicated in Figure 18.26 in an analysis by Shoji and coworkers [116] through the use of the crack tip strain rate algorithm, Equation 18.12, which accounts specifically for yield stress. Of interest in these predictions (and observations) is the fact that the solution-annealed stainless steel is more sensitive to increases in yield stress due to cold work than is the sensitized stainless steel, thereby confirming the reason why, although the surface cold work phenomenon was recognized on "sensitized" components as early as the 1970s, that observation has been marked more recently on replacement L-grade and non-sensitized steel components [117].

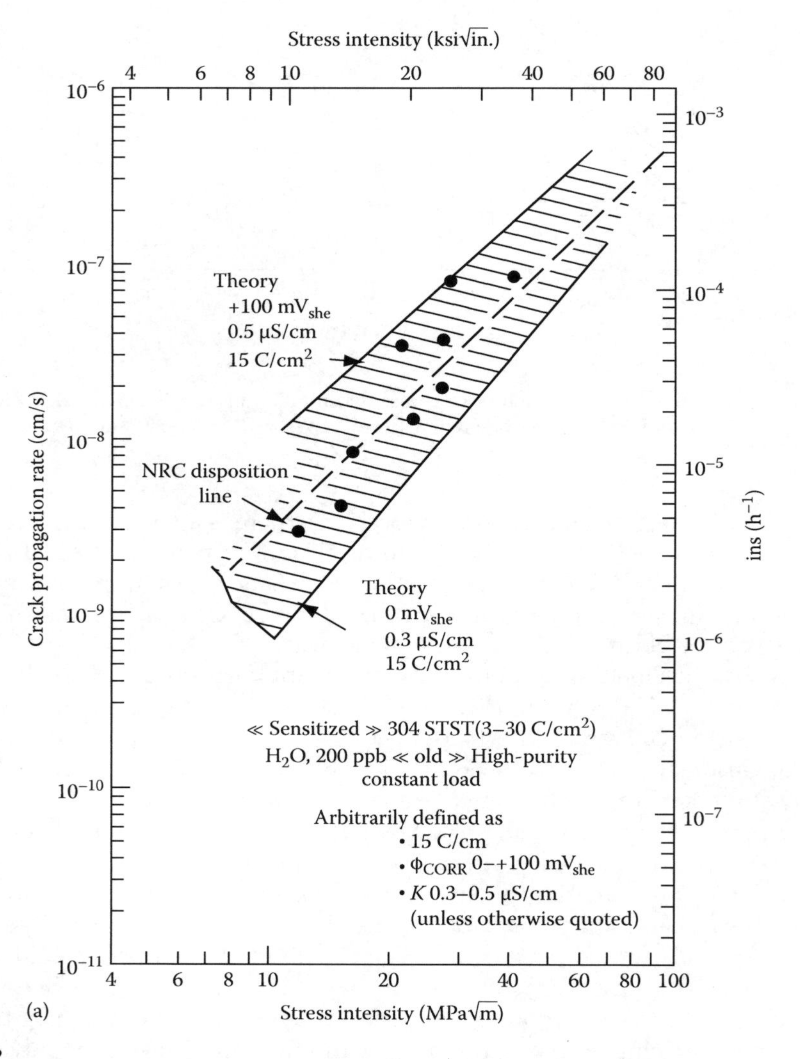

FIGURE 18.22
(a) Crack propagation rate/stress intensity factor dependence for sensitized stainless steel in water at 288°C water with 0.2 ppm oxygen and conductivity 0.3–0.5 μS/cm conductivity. (Adapted from Ford, F.P. et al., Corrosion-assisted cracking of stainless steel & low-alloy steels in LWR environments, Report NP5064S, February 1987, EPRI, Palo Alto, CA.) Note the insertion of theoretical relationships and the NRC regulatory relationship used for structural integrity evaluations. (From USNRC Generic Letter 88-01, NRC position on intergranular stress corrosion cracking in BWR austenitic stainless steel piping, January 25, 1988.)

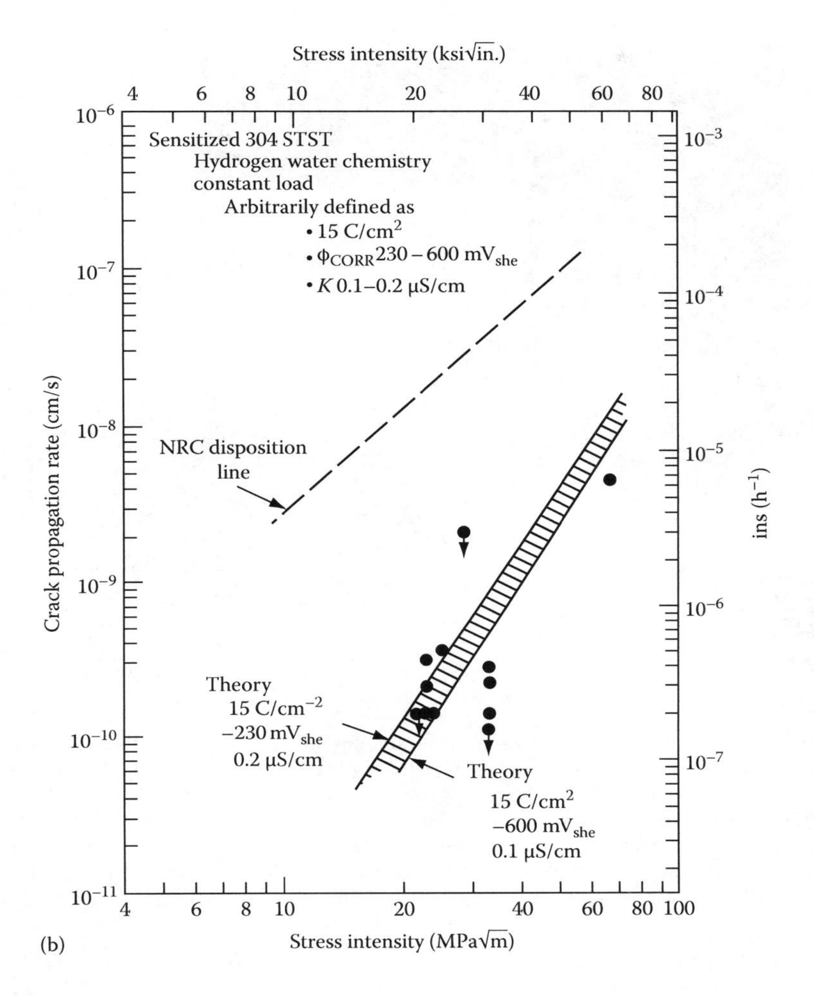

FIGURE 18.22 (continued)
(b) Observed and theoretical crack propagation rate vs. stress intensity factor relationships for sensitized type 304 stainless steel in deaerated water.

Such replacement steels are more resistant to cracking than the original sensitized Type 304 steels, but that advantage is more than counteracted if there are sources of strain localization and, importantly, the reason for this reversal is understood.

18.5.6 Prediction of Cracking in Unirradiated Welded Plant Components

Integration of the crack propagation rate algorithm, Equation 18.6 (with appropriate values of the parameter n, Equation 18.10, and crack tip strain rate, Table 18.1), with respect to time from an "intrinsic" defect of 50 μm depth (i.e., depths associated with the propagation of a single crack not complicated by the aleatory uncertainties associated with random surface phenomena such as pitting, etc.) leads to the appropriate crack depth–time relationship. When the strain rate component in the algorithm is driven by the residual stress, which may vary through the component dimensions (Figure 18.27), the corresponding

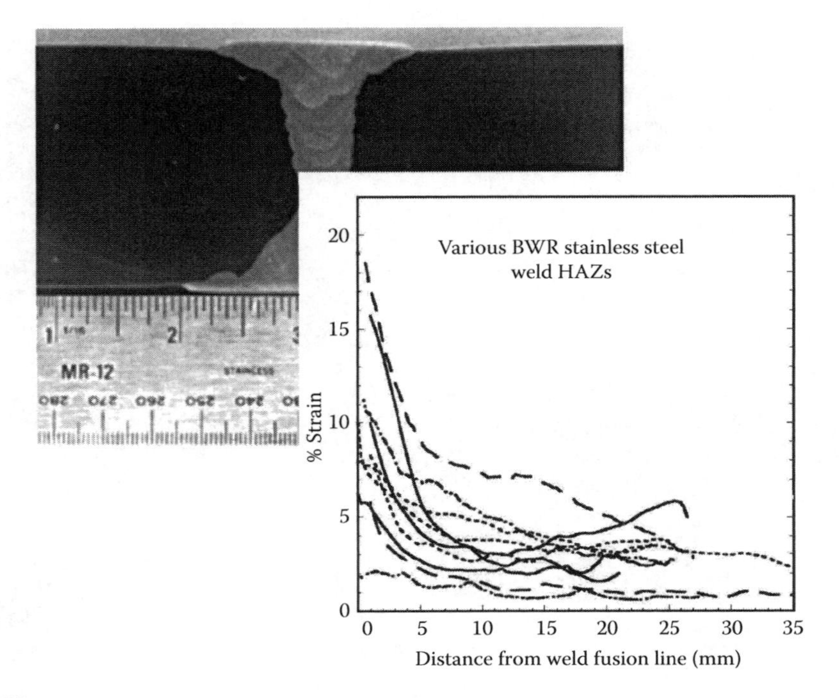

FIGURE 18.23
Weld residual strain vs. distance from the weld fusion line for stainless steel welds. If the number of welding passes is limited, the peak residual strain may be below 15% equivalent room temperature tensile strain. However, most pipe welds that have been analyzed show peak residual strains in the range of 15%–25%, with some slightly above 25%. The residual strain is also highest near the root of the weld. (From Angeliu, T.M. et al., Intergranular stress corrosion cracking of unsensitized stainless steels in BWR environments, in *Proceedings of Ninth International Symposium on Environmental Degradation in Nuclear Power Systems–Water Reactors*, Newport Beach, CA, Eds. S. Bruemmer and F.P. Ford, The Metallurgical Society, August 1–5, 1999, pp. 311–318.)

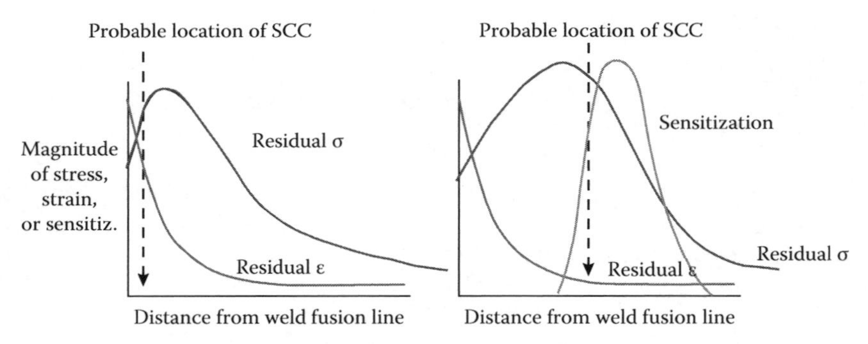

FIGURE 18.24
Schematic distribution of residual stress and strain, and sensitization that exist adjacent to the weld fusion line in stainless steels. Note that, for the unsensitized steel (the left-hand diagram), the expected crack path is associated with the maximization in the local stress and strain close to the weld fusion line, whereas in the sensitized case the crack path is more removed from the weld fusion line corresponding to the temperature–time combinations during welding that maximize the grain boundary sensitization. (From Andresen, P.L. et al., Effects of yield strength, corrosion potential, stress intensity factor, silicon and grain boundary character on SCC of stainless steels, in *Proceedings of Eleventh International Conference on Environmental Degradation in Nuclear Power Systems–Water Reactors*, Skamania Lodge, Eds. G. Was and L. Nelson, American Nuclear Society, August 5–9, 2003, pp. 816–832.)

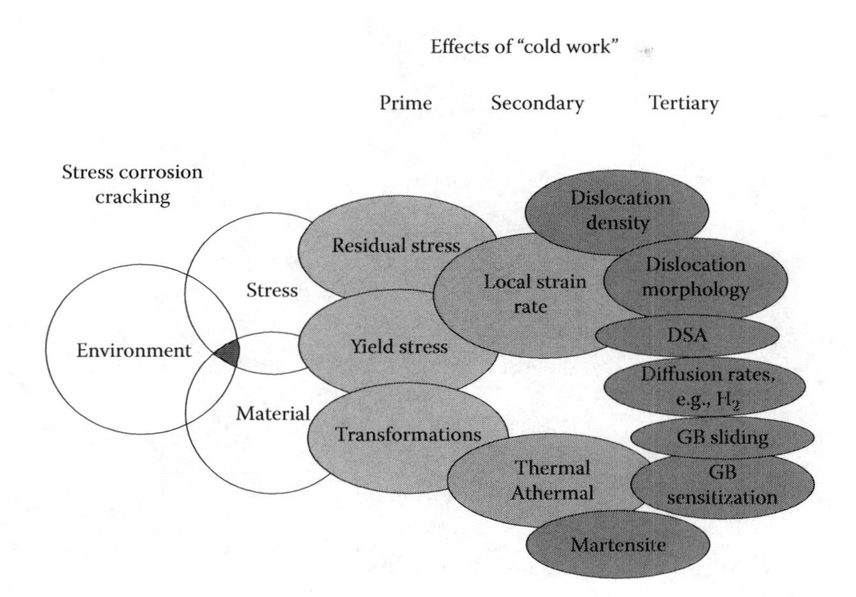

FIGURE 18.25
Interactions between the various parameters associated with "cold work" and their effect on the conjoint "material" and "stress" conditions for stress corrosion crack propagation.

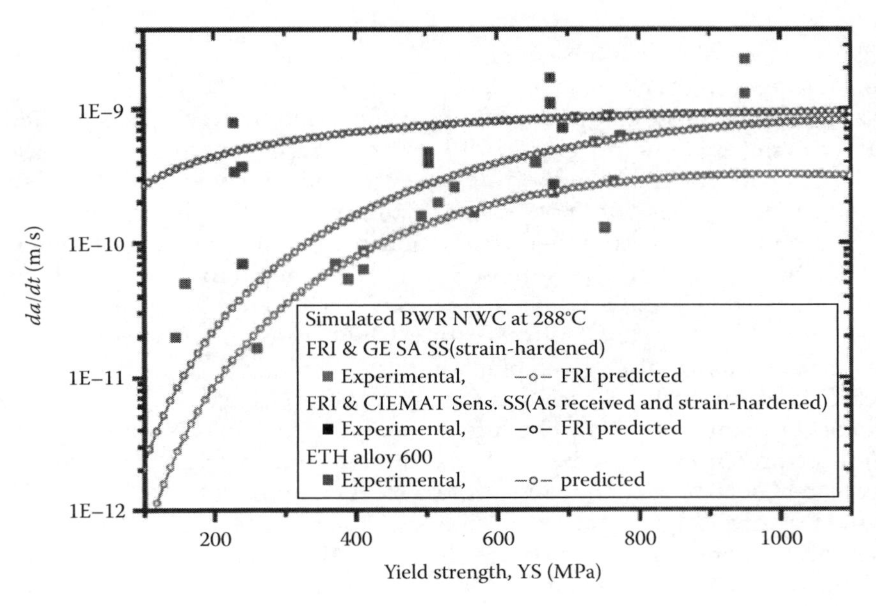

FIGURE 18.26
Observed and predicted crack propagation rate vs. yield strength relationships for cold-worked Alloy 600, solution-annealed 316 L SS, and sensitized 304 *SS* in BWR-simulated environment at 288°C. (From Shoji, T. et al., Experimental investigations and theoretical modeling of the effect of material hardening on SCC of austenitic alloys in LWR environments, in *Proceedings of Work Shop on Cold Work in Iron and Nickel-Base Alloys Exposed to High Temperature Water Environments*, Sponsored by CANDU Owners Group, AECL and EPRI', Toronto, Ontario, Canada, June 3–8, 2007.)

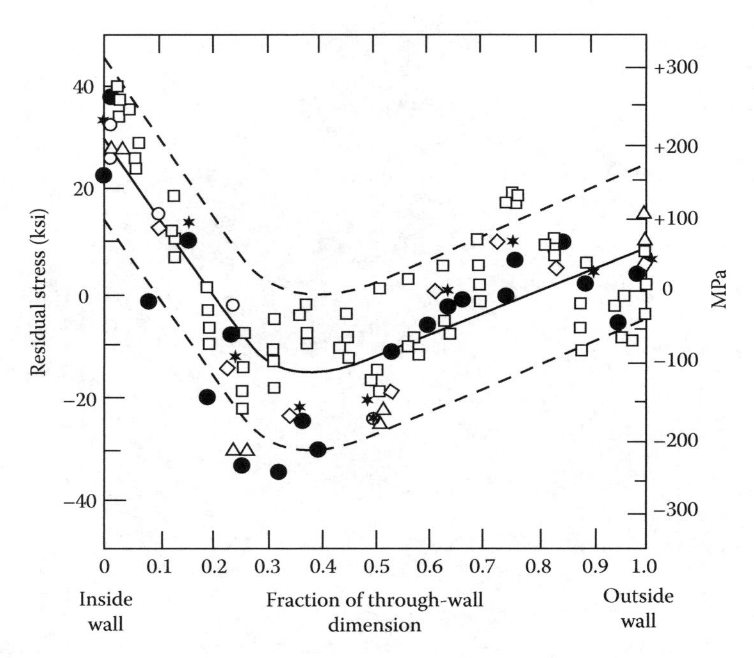

FIGURE 18.27
Variation in through-wall residual stress in 28 in. dia sch. 80 piping close to the fusion line. (From Shack, W.J., Measurement of through wall residual stresses in large diameter piping butt welds using strain gage techniques, in *Proceedings of Seminar on Countermeasures for Pipe Cracking in BWRs*, Vol. 2, EPRI Reports WS 79-174, May 1980.)

stress intensity factor input as a function of defect size (Figure 18.28) may be computed by procedures developed by, for instance Buchelet and Bamford [119].

An example is shown in Figure 18.29 of the comparison between the predicted and the observed cracking adjacent to welds in 28 in. schedule 80 recirculation lines at two BWR plants that operated with markedly different water purity control. In this instance, the residual stress profile was not known; as a result, the best that the prediction can do is evaluate the range in crack depth–time relationships for the range in residual stresses expected for this classification pipe (Figure 18.27). It is apparent that in spite of the lack of detailed knowledge of the residual stress it is possible to predict the time before cracks will be observed (i.e., the crack depth exceeds the NDE resolution limit) and to predict the range of crack depths within a given plant piping.

The specific effect of water purity control on the predicted and observed cracking at a number of plants for 28 in. schedule 80 recirculation lines is illustrated in Figure 18.30. Similar predictions may be made for the cracking propensity for different system parameters such as a change in stainless steel composition (304 vs. 316 NG vs. 321) or the effect of surface grinding and its effect on surface residual stresses. Obviously, there will be epistemic uncertainties in the predictions associated with uncertainties in the values of the model parameters (e.g., residual stress, coolant conductivity, corrosion potential, sensitization, etc.), but these can be treated via knowledge of the distributions of these parameters and their input, via a Monte Carlo assessment, into the deterministic algorithm, Equation 18.6. This was done early on for BWR piping systems [120] in the development of the PRAISE-B and PRAISE-CC codes using an empirically derived damage accumulation algorithm rather than the mechanism-informed algorithm described above.

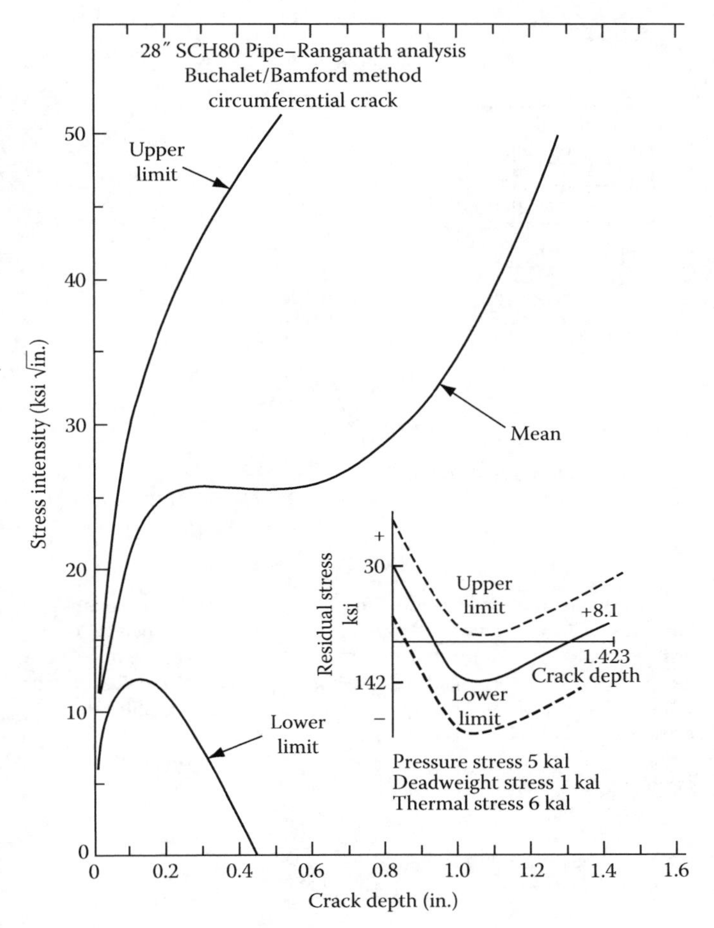

FIGURE 18.28
Variation of stress intensity factor as crack progresses through wall thickness for 28 in. sch. 80 piping associated with the scatter of residual stress shown in Figure 18.27.

Thus, mechanism-informed predictions, together with an assessment of the uncertainty in the prediction associated with epistemic uncertainties in the system definition, may be made about the cracking susceptibility of components in a specific plant for proposed material, stress, or environmental modifications (Figure 18.31), and the cost effectiveness of these various mitigation actions may be assessed. It should be noted that these predicted crack depth/time loci are exactly those needed for the proactive life management decisions discussed in Section 18.1.

18.5.7 Irradiation Effects on Stress Corrosion Cracking of Stainless Steels

The earliest incidents of stress corrosion cracking in BWRs occurred during the 1960s and were associated with cracking of type 304 stainless steel fuel cladding [121]; the driving forces for cracking were the tensile hoop stress in the cladding due to the swelling fuel and the highly oxidizing conditions in the water.

Since the mid-1980s, there have been an increasing number of observations of intergranular cracking in stainless steel core components, which have been exposed to neutron irradiation (>1 MeV) beyond a critical fluence level (>5×10^{20} N/cm^2) and which were highly stressed (Figure 18.32). Initially these incidents were confined to easily replaced components (e.g., in-core instrumentation tubes) but, more recently, IASCC has been observed at

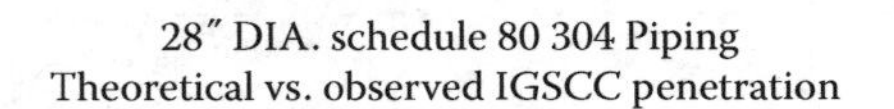

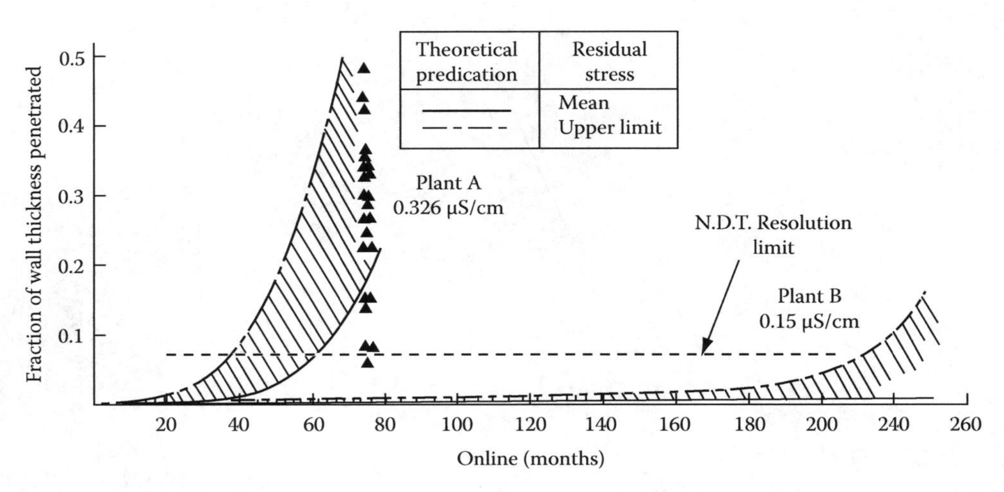

FIGURE 18.29
Theoretical and observed crack depth vs. operational time relationships for 28 in. diameter schedule 80 Type 304 stainless steel piping for two BWRs operating at different mean coolant conductivities. Note the bracketing of the maximum crack depth in the lower purity plant by the predicted curve that is based on the maximum residual stress profile and the predicted absence of observable cracking in the higher purity plant (in 240 operating months). (Adapted from Ford, F.P. et al., Corrosion-assisted cracking of stainless steel & low-alloy steels in LWR environments, Report NP5064S, February 1987, EPRI, Palo Alto, CA.)

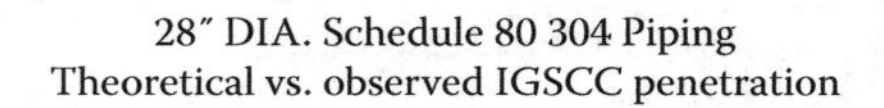

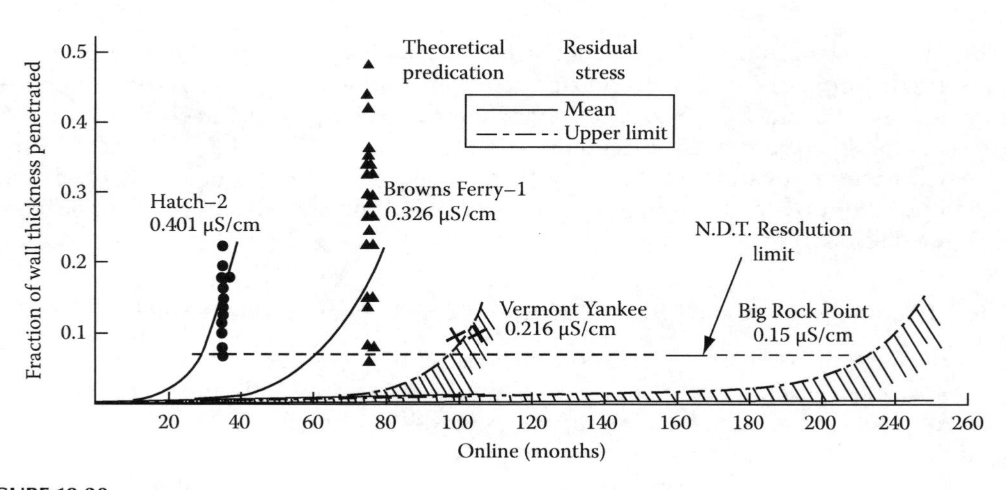

FIGURE 18.30
Theoretical and observed crack depth vs. operational time relationships for 28 in. diameter schedule 80 Type 304 stainless steel piping for various BWRs operating at different mean coolant conductivities.

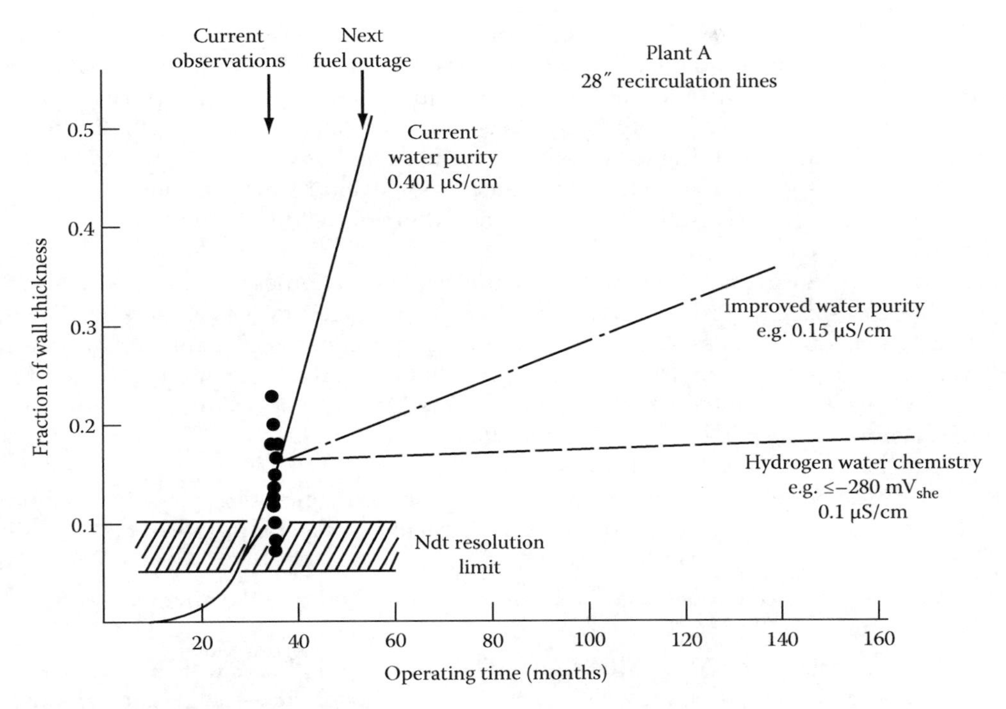

FIGURE 18.31
Predicted response of defected piping for defined changes in water chemistry in BWR plant.

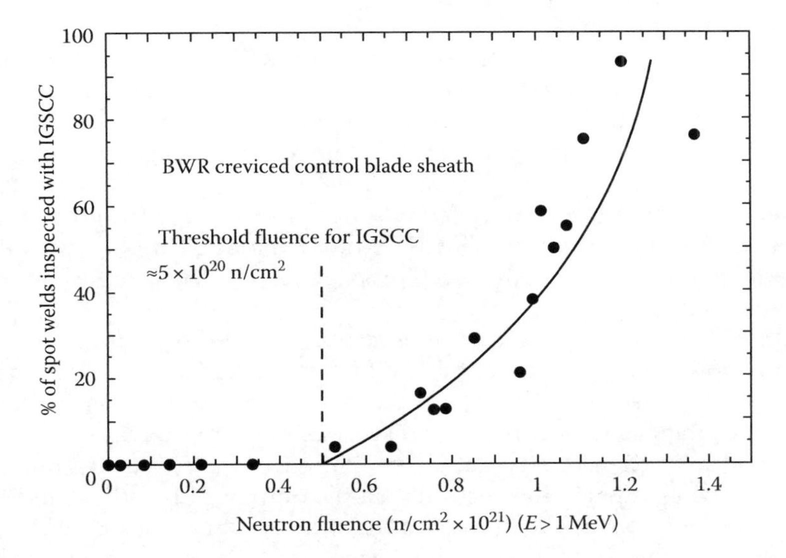

FIGURE 18.32
Dependence of IASCC on fast neutron fluence for creviced control blade sheath in high-conductivity BWRs where the tensile stress is high due to the spot welds and dynamic due to the swelling B_4C absorber tubes.

lower fluence levels in larger welded core structures, with the extent of cracking depending on the specific combinations of prior thermal sensitization due to the initial fabrication procedure, fast neutron flux and fluence, weld residual stress, and coolant purity [122–126]. Thus, it is apparent that the initial hypothesis that there is a critical "threshold fluence" required for IASCC depending on whether the component is at "high" or "low" stress is probably in error and there was, therefore, a need to develop a mechanism-based prediction methodology that accounted for irradiation on top of the other parameters discussed in the sections above.

Based on the arguments made above for unirradiated stainless steels, it was hypothesized that the slip-oxidation mechanism of crack propagation was a relevant cracking model for *irradiated* stainless steels and that the basic Equations 18.6 and 18.10 apply, but with modifications to account for the effect of irradiation on the inputs to that model. Thus, the relevant system parameters that might be affected by irradiation and which are important to the slip-oxidation mechanism of crack propagation are

- The corrosion potential and its dependency on neutron and gamma irradiation (since this controls the potential-driven diffusion of anions to the crack tip, and the associated change in crack tip pH)
- The yield stress due to irradiation hardening, and therefore the effective crack tip strain rate
- Grain boundary chemistry due to irradiation-induced chromium depletion and nickel enrichment and impurity (e.g., silicon, phosphorous) segregation, since these can affect the oxidation kinetics at the strained crack tip
- Radiation enhanced creep and stress relaxation for displacement-loaded components (e.g., welds)
- Irradiation effects on void swelling and fracture toughness

Each of these irradiation-induced effects can be evaluated in "separate effects" tests and their significance on crack propagation can be formulated. This extensive amount of work has been reviewed [122,123] and only the major conclusions are given. For instance, the effect of irradiation on corrosion potential and hence crack propagation is shown in Figure 18.33, which compares the observed and predicted crack depth vs. time relationships for thermally sensitized stainless steel specimens exposed in the unirradiated recirculation line and in an irradiated core instrumentation tube of a BWR. The observed and predicted effect of irradiation in increasing the crack propagation rate via its effect on the corrosion potential alone is apparent.

It is apparent that fast neutron irradiation can affect all of the controlling parameters for stress corrosion cracking and that these effects can take place over different time and geometry scales. For instance, the neutron flux will have an immediate effect on the corrosion potential, whereas radiation-induced segregation at grain boundaries, increases in yield stress, and relaxation of residual stress will all take effect over time (i.e., fluence). With regard to the geometrical aspect, the neutron flux will be attenuated through the thickness dimension of a thick component such as a core shroud; thus, not only is the fluence changing over time, but it is changing at the crack tip as it propagates into the component. The result is that the predicted crack depth vs. time (or fluence) is not necessarily monotonic where there can be competing effects of irradiation-assisted stress relaxation and hardening. This is illustrated in Figure 18.34 for crack propagation through a thick-section (2 in.)-welded core shroud.

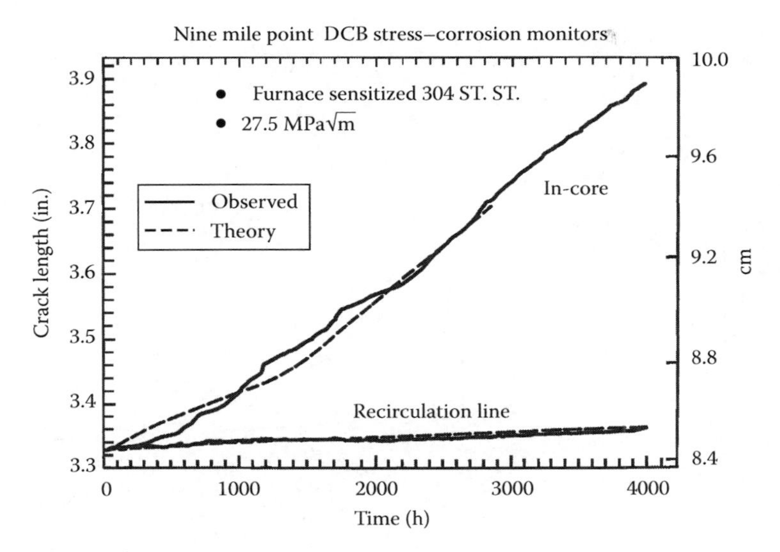

FIGURE 18.33
Crack length vs. time for DCB specimens of sensitized type 304 stainless steel exposed in core (high corrosion potential from radiolysis) and in the recirculation system.

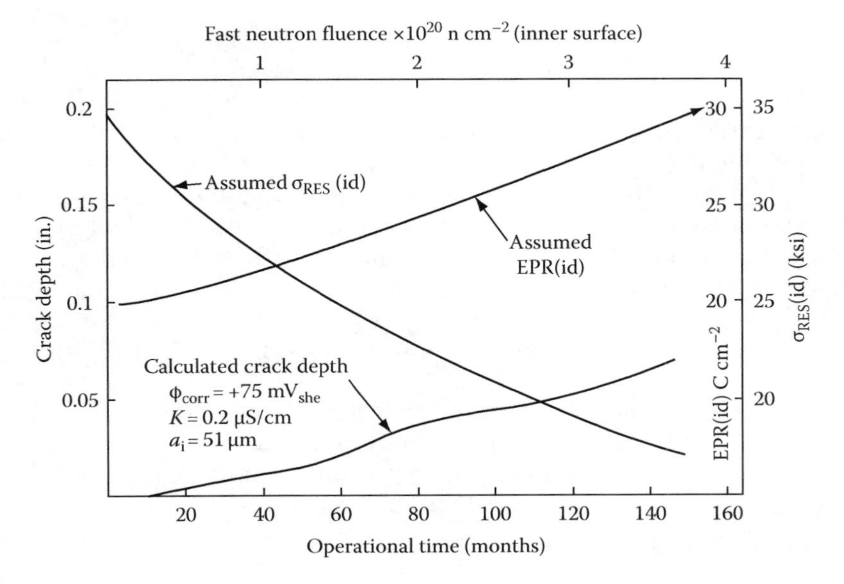

FIGURE 18.34
Predicted crack depth vs. time response for cracks propagating in the beltline weld of a core shroud, illustrating the non-monotonic variation in crack propagation due to the competing effects of fast neutron irradiation on grain boundary sensitization and on residual stress relaxation.

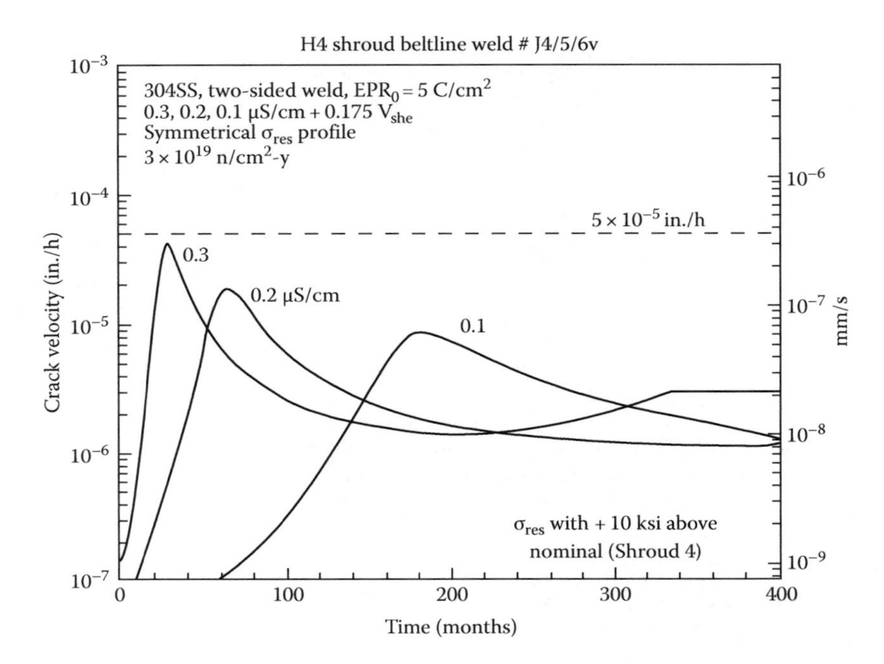

FIGURE 18.35
Predicted variation in crack propagation rate in an irradiated core shroud beltline weld as a function of time (fluence), indicating that the NRC disposition propagation rates cannot be sustained.

In this case, the sensitization is attributed (Equation 18.17) to the initial thermally induced chromium depletion (EPR_0) at the grain boundary due to the welding process (i.e., 20 C/cm^2), plus the additional depletion that occurs with fluence at the particular point in the shroud:

$$EPR = EPR_0 + 3.36 \times 10^{-24}(\text{fluence})^{1.17} \tag{18.17}$$

The variation in crack propagation *rate* with time can become complicated as illustrated in Figure 18.35 due to the interactions between, for instance, the competing effects of a decrease in susceptibility associated with stress relaxation and an increase in susceptibility associated with grain boundary sensitization. As indicated in Figure 18.35, these crack propagation rate responses will change with water conductivity (purity). They will also change with the initial degree of thermal sensitization (e.g., the use of L-grade steels), extent of surface cold work, position in the reactor core (e.g., flux), etc.

The comparison between prediction and observation is limited by two considerations:

1. The uncertainty in the definition of the system. Parameters such as the initial material conditions (e.g., thermally induced grain boundary sensitization, coolant purity, corrosion potential) are important, but the most significant uncertainty rests with the definition of the weld residual stress.
2. The uncertainty in defining the observed crack depths and, to a lesser extent, lengths.

The effect of these uncertainties is mirrored in the comparison (Figure 18.36a) between the observed and predicted crack depths adjacent to a horizontal H-4 beltline weld in a BWR core shroud, where the observations have been made during consecutive reactor outages.

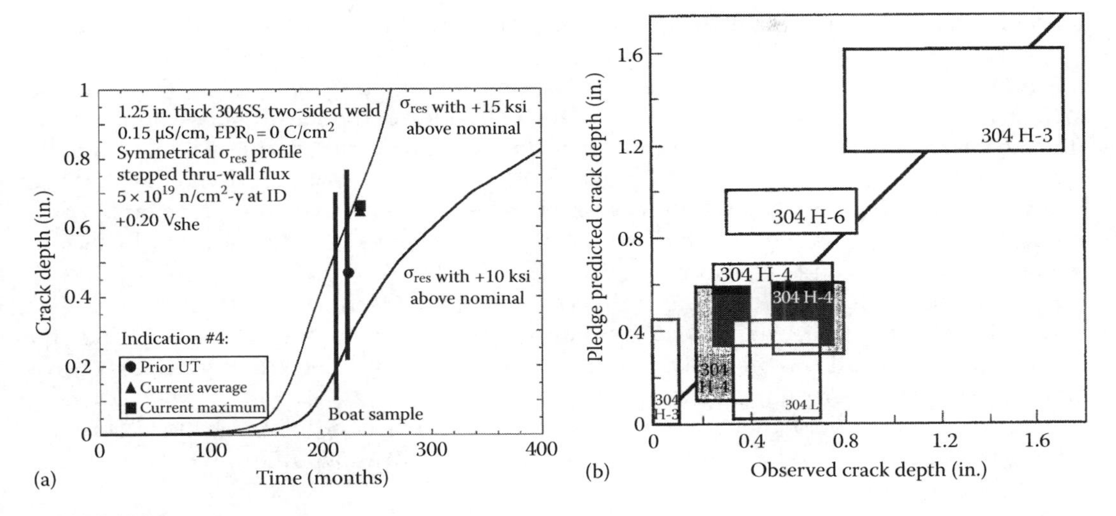

FIGURE 18.36
(a) Crack length vs. time predictions and observations for a Type 304 stainless steel BWR core shroud with multiple inspections and multiple cracks. (b) Comparison of observed and predicted crack depth for a number of welds in BWR core shrouds spanning various stainless steels (Type 304, 304 L), neutron fluxes, and water purity.

There is a range in *observed* crack depths adjacent to the horizontal welds due to the uncertainty in NDE measurements. It is also apparent that there is a range in *predicted* crack depths due to the uncertainty in defining the residual stress profile both around the circumference of the weld as well as through the 2 in. thickness of the shroud.

Given these uncertainties, the only rationale way of assessing the accuracy of the prediction methodology is to compare the *ranges* in observed and predicted crack depths for different welds and stainless steels. This comparison is made in Figure 18.36b for various BWR core shroud welds (H-3, H-4, H-6) manufactured with different steels (Type 304, 304 L). Over this population the agreement between observation and prediction is reasonable, given the uncertainties in observation and definition of the system.

18.6 Conclusions

The development of life prediction capabilities for LWR components subject to stress corrosion cracking are crucial to the success of proactive management of these problems and economic and safe operation of LWRs. Various approaches have been taken, including those based on (a) past-plant experience, (b) correlations based on the analysis of the effect of key stress, environment and material parameters on the cracking kinetics and, finally, (c) those that draw on an understanding of the mechanism of cracking.

All three approaches have uncertainties associated with, for example, the stochastic nature of many random processes that occur on the surface of the component (e.g., pitting, intergranular attack) and which may dominate the crack initiation stage. A second source of uncertainty is epistemic and is associated with both the completeness of the empirical and mechanistically based models and the definition of the inputs to these models. Examples of these uncertainties are discussed in this chapter for the specific situations associated with SCC of nickel-base alloys in PWR primary side environments and of austenitic stainless steel components in BWRs that are either unirradiated (e.g., piping) or irradiated (e.g., core internals).

Models based on the slip-oxidation mechanism of crack propagation for the stainless steel systems in BWRs give a sound basis for the prediction of the sensitivities of cracking to diverse parameters, such as stress (both static and dynamic) and the interactions between the corrosion potential and anionic impurities and, finally, material properties such as the degree of grain boundary sensitization, yield stress, etc. This mechanism-informed life prediction capability gives confidence to the adoption of various mitigation actions, including water purity control and efficient methods for assuring low corrosion potentials [127,128].

In addition, this life prediction capability of unirradiated stainless steels gives a basis for expanding that capability to irradiated systems of nickel-base alloys [63,129] in BWRs and to carbon and low-alloys steels [105,130–131] all of which are subject to both stress corrosion cracking and corrosion fatigue.

Although such developments are of practical use, there is no question that there are some generic unanswered questions that apply to a range of alloy/environment systems in LWRs. These concerns include

- The fact that the crack tip dimensions appear to be on the 5–10 nm scale, which raises the question of the effect on mass transport rates of reactants in this confined space. Such considerations have not been taken into account in the development of the current mechanism-based life prediction methodologies.
- The formulation of the crack tip strain rate, which controls the oxide fracture periodicity. This strain rate is currently treated via continuum fracture mechanics approaches, since these are the metrics upon which structural integrity calculations are based. However, the physical processes at the crack tip are dependent on localized deformations that are controlled by, for example, stacking-fault energies, grain boundary creep, and sliding and are quite inhomogeneous from grain to grain. Currently, there are no formulations of the oxide rupture frequency that take these physical realities into account.
- Although there is an adequate *qualitative* understanding of the physical processes associated with the development of a crack to resolvable dimensions (i.e., "engineering crack initiation"), there is no quantitative capability for predicting the preceding events associated with "precursors," "microscopic crack initiation," and "short crack growth and coalescence" in LWR systems.
- There is increasing evidence that the fracture resistance of a component that contains a crack that is propagating subcritically can be reduced, as has been observed in fracture resistance measurements made in the operating environment. The mechanism of this effect is unknown, and yet the phenomenon has a significant impact on structural integrity evaluations.
- There also remain outstanding issues regarding the current understanding of stress corrosion cracking, which are system specific [132]. These include, for example, (a) the deleterious effect in BWR environments of radiation-induced segregation of silicon to the grain boundaries in stainless steels, (b) the fundamental reason for the beneficial effect of increased chromium in nickel-base alloys (i.e., Alloy 690 as a replacement for Alloy 600) in PWR primary environments and, (c) the different effects of the dissolved hydrogen content on crack initiation and propagation for nickel-base alloys in PWR primary environments.

References

1. R.C. Newman, Stress-corrosion cracking mechanisms, Chapter 11 in this book.
2. F.P. Ford, Technical and management challenges associated with structural and materials degradation in nuclear reactors in the future, in *Proceedings of the Thirteenth Environmental Degradation Conference*, Whistler, British Columbia, Canada, August 2007.
3. J. Muscara, Expert panel report on proactive materials degradation assessment, USNRC NUREG Report, CR-6923, February 2007.
4. R. Pathania, EPRI materials degradation matrix revision 1, EPRI Technical Report 1016486 Final Report, May 2008.
5. R.W. Staehle, Predicting failures which have not yet been observed, in Ref. [3].
6. J. Hickling, Status review of initiation of environmentally assisted cracking and short crack growth, EPRI Report # 1011788, December 2005.
7. F.P. Ford, P.L. Andresen, M.G. Benz, and D. Weinstein, On-line BWR materials monitoring and plant component lifetime prediction, in *Proc. Nuclear Power Plant Life Extension*, Snowbird, UT, June 1988, published by American Nuclear Society, Vol. 1, pp. 355–366.
8. K. Sato, T. Tsuruta, T. Kobayashi, and Umehara, Evaluation of PWRSCC susceptibility for alloy 600 under low hydrogen concentration and the benefits for various components in primary circuit, in *Proceedings of Conference on Optimization of Dissolved Hydrogen in PWR Primary Coolant*, Tohoku University, Japan, July 18–19, 2007.
9. G.A. White, N.S. Nordmann, J. Hickling, and C.D. Harrington, Development of crack growth rate disposition curves for primary water stress corrosion cracking (PWSCC) of alloy 82,182,132 weldments, in *Proceedings of Twelfth International Conference on Environmental Degradation in Nuclear Power Systems–Water Reactors*, Warrendale, PA, Skamania Lodge, Eds. L. Nelson and P. King, The Metallurgical Society, August 5–9, 2005, pp. 511–531.
10. K. Gott, Cracking data base as a basis for risk informed inspection, in *Proceedings of Tenth International Conference on Environmental Degradation in Nuclear Power Systems–Water Reactors*, Eds. F.P. Ford and G. Was, Published by National Association of Corrosion Engineers, Lake Tahoe, August 5–9, 2001.
11. H.H. Klepfer et al., Investigation of cause of cracking in austenitic stainless steel piping, NEDO-21000, General Electric, July 1975.
12. R.W. Staehle, J.A. Gorman, K.D. Stavropoulos, and C.S. Welty, Application of statistical distributions to characterizing and predicting corrosion of tubing in steam generators of pressurized water reactors, in *Proceedings of Life Prediction of Corrodible Structures*, Ed. R. N. Parkins, NACE International, Houston, TX, 1994, pp. 1374–1439.
13. K.D. Stavropoulos, J.A. Gorman, R.W. Staehle, and C.S. Welty, Selection of statistical distributions for prediction of steam generator tube degradation, in *Proceedings of Fifth International Symposium on Environmental Degradation in Nuclear Power Systems–Water Reactors*, Monterey, CA, Eds. D. Cubicciotti and E. Simonen, American Nuclear Society, August 25–29, 1991, pp. 731–744.
14. BWR water chemistry guidelines—2000 revision, EPRI Report TR-103515-R2, 2000.
15. P.M. Scott and P. Combrade, *Corrosion in Pressurized Water Reactors*, 2nd edn. of ASM Handbook on Corrosion, ASM International, Vol. 13C, pp. 362–385.
16. P.M. Scott, Y. Meyzaud, and C. Benhamou, Prediction of stress corrosion cracking of alloy 600 components exposed to PWR primary water, in *Proceedings of International Symposium on Plant Aging and Life Prediction of Corrodible Structures*, Sapporo, Japan, Eds. T. Shoji and T. Shibata, Published by JSCE and NACE, May 15–18, 1995, pp. 285–293.
17. C.A. Campbell and S. Fyfitch, PWSCC ranking model for alloy 600 components, in *Proceedings of Sixth International Symposium on Environmental Degradation in Nuclear Power Systems–Water Reactors*, San Diego, CA, Eds. E Simonen and R. Gold, National Association of Corrosion Engineers, August 1–5, 1993, pp. 657, 863–870.

18. G.V. Rao, Methodologies to assess PWSCC susceptibility of primary component alloy 600 locations in pressurized water reactors, in *Proceedings of Sixth International Symposium on Environmental Degradation in Nuclear Power Systems–Water Reactors*, San Diego, CA, Eds. E. Simonen and R. Gold, National Association of Corrosion Engineers, August 1–5, 1993, pp. 657, 871–882.
19. G.V. Rao, Development of surface replication technology for the field assessment of alloy 600 microstructures for primary loop penetrations, Westinghouse report, WCAP-13746, June 1993.
20. P.M. Scott and C. Benhamou, An overview of recent observations and interpretations of IGSCC in nickel-base alloys in PWR primary water, in *Proceedings of Tenth International Conference on Environmental Degradation in Nuclear Power Systems–Water Reactors*, Lake Tahoe, Eds. F.P. Ford and G. Was, National Association of Corrosion Engineers, August 5–9, 2001.
21. F.P. Ford, B.M. Gordon, and R.M. Horn, Corrosion in boiling water reactors, in ASM Metals Handbook Vol. 13C, *Corrosion-Environments and Industries*, 2006, pp. 341–361.
22. P.L. Andresen, I.P. Vasatis, and F.P. Ford, Behavior of short cracks in stainless steel at 288°C, in Paper 495, *NACE Conference*, Las Vegas, NV, April 1990.
23. R.N. Parkins, *Brit. Corros. J.* 7, 15 (1972).
24. F.P. Ford, *Corrosion* 52, 375–395 (1996).
25. R.N. Parkins and B.S. Greenwell, *Met. Sci.* 12, 405 (August 1977).
26. C. Laird and D.J. Duquette, in *Surface Effects on Crystal Plasticity*, Hohegeiss, Germany, September 1975, Eds. R.M. Latanision and J.T. Fourie, Noordhof-Leyden, the Netherlands, 1977, p. 88.
27. R.N. Parkins, Environment-sensitive fracture—Controlling parameters, in *Proceedings of Third International Conference on Mechanical Behavior of Materials*, Eds. K.J. Miller and R.F. Smith, Cambridge, August 20–24, 1979, Pergamon, Vol. 1, pp. 139–164.
28. F.P Ford, Stress corrosion cracking, in *Corrosion Processes*, Ed. R.N. Parkins, Applied Science, London, U.K., 1982.
29. J.M. Kraft and J.H. Mulherin, *Trans. ASM* 62, 64 (1969).
30. T.P. Hoar, *Corrosion* 19, 331 (1963).
31. P.R. Swann, *Corrosion* 19, 102 (1963).
32. T.R. Beck, *Corrosion* 30, 408 (1974).
33. H.L. Logan, *J. Natl. Bur. Stand* 48, 99 (1952).
34. D.A. Vermilyea, *J. Electrochem. Soc.* 119, 405 (1972).
35. F.P. Ford, Mechanisms of environmental cracking peculiar to the power generation industry, Report NP2589, EPRI, Palo Alto, CA, September 1982.
36. F.P. Ford, The crack tip system & it's relevance to the prediction of cracking in aqueous environments, in *Proceedings of First International Conference on Environmentally Assisted Cracking of Metals*, Eds. R. Gangloff and B. Ives, Kohler, WI, October 2–7, 1988, NACE, pp. 139–165.
37. K. Sieradzki and R.C. Newman, *Phil. Mag.* A51, 95–132 (1985).
38. C. Edeleanu and A.J. Forty, *Philos. Mag.* 5, 1029 (1960).
39. J.A. Beavers and E.N. Pugh, *Met. Trans.* 11A, 809 (1980).
40. K. Sieradzki, Atomistic & micromechanical aspects of environment-induced cracking of metals, in *Proceedings of First International Conference on Environment-Induced Cracking of Metals*, Kohler, WI, October 1988, Published by NACE, pp. 125–138.
41. R.C. Newman, Stress corrosion of austenitic steels, in *Proceedings of First International Conference on Environment-Induced Cracking of Metals*, Kohler, WI, October 1988, Published by NACE, pp. 489–510.
42. P.M. Scott and P. Combrade, On the mechanism of stress corrosion crack initiation and growth in alloy 600 exposed to PWR primary water, in *Proceedings of Eleventh International Conference on Environmental Degradation in Nuclear Power Systems–Water Reactors*, Skamania Lodge, Eds. G. Was and L. Nelson, American Nuclear Society, August 5–9, 2003, pp. 29–36.
43. P. Combrade et al., Oxidation of nickel base alloys in PWR water; oxide layers and associated damage in *Proceedings of Twelfth International Conference on Environmental Degradation in Nuclear Power Systems–Water Reactors*, Snowbird, Eds. L. Nelson and P.J. King, The Metallurgical Society, August 5–9, 2005, pp. 883–890.

44. A.W. Thompson and I.M. Bernstein, The role of metallurgical variables in hydrogen-assisted environmental fracture, Rockwell Science Center Report SCPP-75-63.
45. J.P. Hirth, *Met. Trans.* 11A, 861–890 (1980).
46. H.G. Nelson, Hydrogen embrittlement, in *Treatise on Material Science and Technology*, Eds. C.L. Briant and S.K. Banerji, Academic Press, New York, 1983, Vol. 25, pp. 275–359.
47. H.K. Birnbaum, Hydrogen embrittlement, in *Proceedings of First International Conference on Environment-Induced Cracking of Metals*, Kohler, WI, October 1988, Published by NACE, pp. 21–29.
48. F.P. Ford, *Met. Sci.* 13, 326 (July 1978).
49. R. Raj and V.K. Varadan, The kinetics of hydrogen assisted crack growth, in *Mechanisms of Environment Sensitive Cracking of Materials*, Eds. P.R. Swann, F.P. Ford, and A.R.C. Westwood, University of Surrey, Surrey, U.K., April 1977, The Metals Society, p. 426.
50. R.A. Oriani, Hydrogen effects in high strength steels, in *Proceedings of First International Conference on Environment-Induced Cracking of Metals*, Kohler, WI, October 1988, Published by NACE, pp. 439–448.
51. R.A. Oriani, Bertichte der Bunsenges for *Phys. Chem.* 76, 848 (1972).
52. W.W. Gerberich, Y.T. Chen, and C. St. John, *Met. Trans.* 6, 1485 (1975).
53. F. de Kazinski, *J. Iron Steel Inst.* 177, 85 (1954).
54. C.A. Zapffe and C.E. Sims, *Trans. AIME* 145, 225 (1941).
55. C.D. Beachem, *Met. Trans.* 3, 437 (1972).
56. S.P. Lynch, *Met. Forum* 2, 189 (1979).
57. R.A. Gilman, The role of surface hydrides in stress corrosion cracking, in *Stress-Corrosion Cracking and Hydrogen Embrittlement of Iron-Base Alloys*, Firminy, France, June 1973, Eds. R.W. Staehle, J. Hochmann, R.D. McCright, and J.E. Salter, NACE, Houston, TX, 1977, p. 326.
58. C.L. Briant, *Met. Trans.* 10A, 181 (1979).
59. H. Hanninen, K. Torronen, M. Kempainen, and S. Salonen, *Corros. Sci.* 23, 663 (1983).
60. K. Torronen, M. Kemppainen, and H. Hanninen, Fractrographic evaluation of specimens of A533B pressure vessel steel, Final Report of EPRI Contract RP 1325-7, Report NP 3483, May 1984.
61. W.R. Wearmouth, G.P. Dean, and R.N. Parkins, *Corrosion* 29, 251 (1973).
62. F.P. Ford, D.F. Taylor, P.L. Andresen, and R.G. Ballinger, Corrosion-assisted cracking of stainless steel and low-alloy steels in LWR environments, Report NP5064S, February 1987, EPRI, Palo Alto, CA.
63. P.L. Andresen, Modeling of water and material chemistry effects on crack tip chemistry and resulting crack growth kinetics, in *Proc. Third Int. Conf. Environmental Degradation of Materials in Nuclear Power Systems–Water Reactors*, Traverse City, AIME, 1987, pp. 301–314.
64. D.F. Taylor and C.A. Caramihas, High temperature aqueous crevice corrosion in alloy 600 and 304 L stainless steel, in *Proceedings of Embrittlement by Localized Crack Environment*, Ed. R. Gangloff, AIME/ASM Symposium, Philadelphia, PA, October 1983, pp. 105–114.
65. D.F. Taylor and C.C. Foust, *Corrosion* 44, 204–208 (1988).
66. G. Gabetta and G. Buzzanaca, Measurement of corrosion potential inside and outside growing crack during environmental fatigue tests, in *Proceedings of Second International IAEA Specialists Meeting on Subcritical Crack Growth*, Sendai, Japan, May 1985, NUREG CP0067, Vol. 2, pp. 201–218.
67. P. Combrade, M. Foucault, and G. Slama, About the crack tip environment chemistry in pressure vessel steel exposed to primary PWR coolant, as Ref. [66], Vol. 2, pp. 201–218.
68. A. Turnbull, *Rev. Coat. Corros.* 5, 43–171 (1982).
69. A. Turnbull, *Corros. Sci.* 23, 833–870 (1983).
70. A. Turnbull, Progress in the understanding of the electrochemistry in cracks, as Ref. [64], pp. 3–31.
71. P.L. Andresen, Transition and delay time behavior of high temperature crack propagation rates resulting from water chemistry changes, in *Proceedings Second Int. Symposium on Environmental Degradation of Materials in Nuclear Power Systems–Water Reactors*, Joint Sponsorship of American Nuclear Soc., TMS-AIME, and National Association of Corrosion Engineers, Monterey, CA, September 9–12, 1985.

72. F.P. Ford and P.L. Andresen, Stress corrosion cracking of low-alloy steels in 288°C water, in Paper 498, *Corrosion-89*, New Orleans, LA, April 1989.
73. L. Young and P.L. Andresen, Crack tip microsampling and growth rate measurements in a 0.01%S low alloy steel, *Proceedings of Seventh International Symposium on Environmental Degradation in Nuclear Power Systems–Water Reactors*, Breckenridge, CO, Eds. R. Gold and A. McIlree, National Association of Corrosion Engineers, August 7–10, 1995, pp. 1193–1204.
74. S.M. Bruemmer, L.A. Charlot, and B.W. Arey, *Corrosion* 44, 328–333 (1988).
75. S.M. Bruemmer, L.A. Charlot, and D.G. Atteridge, *Corrosion* 44, 427–434 (1988).
76. H.D. Solomon, *Corrosion*, 40, 51–60 (1984).
77. G.S. Was and V.B. Rajan, *Corrosion* 43, 576 (1987).
78. T.M. Devine, *Acta Met.* 36, 1491 (1988).
79. D.I.R. Norris, Ed., *Proceedings of Symposium on Radiation-Induced Sensitization of Stainless Steels*, CEGB, Berkeley Nuclear Laboratories, Berkeley, CA, September 1986.
80. T.P. Hoar and F.P. Ford, *J. Electrochem. Soc.* 120, 1013 (1973).
81. P. Combrade, Prediction of environmental crack growth on reactor pressure vessels, EPRI Contract RP2006-8, UNIREC Report #1667, February 1985.
82. D. Engseth and J.C. Scully, *Corros. Sci.* 15, 505 (1975).
83. T. Zakroczmski and R.N. Parkins, *Corros. Sci.* 20, 723 (1980).
84. Y.S. Park, J.R. Galvele, A.K. Agrawal, and R.W. Staehle, *Corrosion* 34, 413 (1978).
85. R.B. Diegle and D.A. Vermilyea, *J. Electrochem. Soc.* 122, 180 (1975).
86. F.P. Ford and M. Silverman, *Corrosion* 36, 558–565 (1980).
87. R.P. Wei and A. Alavi, *Scr. Met.* 22, 969–974 (1988).
88. C. Patel, *Corros. Sci.* 21, 145 (1981).
89. F.P. Ford, G.T. Burstein, and T.P. Hoar, *J. Electrochem. Soc.* 127, 6 (1980).
90. T. Hagyard and W.B. Earl, *J. Electrochem. Soc.* 115, 623 (1968).
91. J. Pagetti, D. Lees, F.P. Ford, and T. Hoar, *Comp. Rend. Acad. Sci. Paris* 273, 1121 (1971).
92. D.J. Lees and T.P. Hoar, *Corros. Sci.* 20, 723 (1980).
93. R.C. Newman and G.T. Burstein, *Corros. Sci.* 21, 119 (1981).
94. R.B. Diegle and D.M. Lineman, *J. Electrochem. Soc.* 131, 106 (1984).
95. J.R. Ambrose and J. Kruger, *Corrosion* 28, 30 (1972).
96. Z. Szklarski-Smialowska and W. Kozlowski, *J. Electrochem. Soc.* 131, 234 (1984).
97. B. McDougall, *J. Electrochem. Soc.* 130, 114 (1983).
98. G.M. Bulman and A.C.C. Tseung, *Corros. Sci.* 12, 415 (1972).
99. A.T. Cole and R.C. Newman, *Corros. Sci.* 28, 109–118 (1988).
100. M. Cid, M. Puiggali, H. Fatmaoui and M. Petit, *Corros. Sci.* 28, 61–68 (1988).
101. W.L. Clarke, R.L. Cowan, and W.L. Walker, Comparative methods for measuring degree of sensitization in stainless steels, ASTM STP656, ASTM, 1978 [107].
102. R.W. Staehle, J.J. Royuela, T.L. Raredon, E. Serrate, C.R. Morin, and R.V. Ferrar, *Corrosion* 26, 451 (1970).
103. R.N. Parkins, G.P. Marsh, and J.T. Evans, Strain rate effects in environment sensitive fracture, in *Proceedings of EPRI Conference on Predictive Methods for Assessing Corrosion Damage to BWR Piping and PWR Steam Generators*, Mt. Fuji, Japan, May/June 1978, Eds. H. Okada and R. Staehle, Pub. NACE, 1982.
104. D.P.G. Lidbury, The estimation of crack tip strain rate parameters characterizing environment assisted crack data, as Ref. [64], pp. 149–172.
105. F.P. Ford, *J. Pressure Vessel Technol.* 1, 113 (1988).
106. T. Shoji, S. Suzuki, and R.G. Ballinger, Theoretical prediction of stress corrosion crack growth—Threshold and plateau growth rate, in *Proceedings of Seventh International Symposium on Environmental Degradation in Nuclear Power Systems–Water Reactors*, Breckenridge, Eds. R. Gold and A. McIlree, National Association of Corrosion Engineers, August 7–10, 1995, pp. 881–891.
107. T. Shoji, H. Takahashi, and M. Suzuki, Corrosion fatigue aspects in BWR pipe cracking, in *Predictive Methods for Assessing Corrosion Damage in BWR Piping and PWR Steam Generators*, Eds. H. Okada and R.W. Staehle Fuji, May 1978, p. 58.

108. T. Shoji, Progress in mechanistic understanding of BWR SCC and implication to predictions of SCC growth behaviour in plants, in *Proceedings of Eleventh International Conference on Environmental Degradation in Nuclear Power Systems–Water Reactors*, Skamania Lodge, Eds. G. Was and L. Nelson, Published by American Nuclear Society, August 5–9, 2003.
109. Y.C. Gao and K.C. Hwang, Elastic plastic fields in steady crack growth in a strain hardening material, in *Proceedings of Fifth Int. Conf. on Fracture*, Cannes, France, 1981, pp. 669–682.
110. USNRC Generic Letter 88-01, NRC position on intergranular stress corrosion cracking in BWR austenitic stainless steel piping, January 25, 1988.
111. T.M. Angeliu, P.L. Andresen, J.A. Sutliff, and R.M. Horn, Intergranular stress corrosion cracking of unsensitized stainless steels in BWR environments, in *Proceedings of Ninth International Symposium on Environmental Degradation in Nuclear Power Systems–Water Reactors, Newport Beach*, Eds. S. Bruemmer and F.P. Ford, The Metallurgical Society, August 1–5, 1999, pp. 311–318.
112. P.L. Andresen, P.W. Emigh, M.M. Morra, and R.M. Horn, Effects of yield strength, corrosion potential, stress intensity factor, silicon and grain boundary character on SCC of stainless steels, in *Proceedings of Eleventh International Conference on Environmental Degradation in Nuclear Power Systems–Water Reactors*, Skamania Lodge, Eds. G. Was and L. Nelson, American Nuclear Society, August 5–9, 2003, pp. 816–832.
113. Z. Jiao, G. Was, and J. Busby, The role of localized deformation in IASCC of proton irradiated austenitic stainless steels, in *Proceedings of Thirteenth International Conference on Environmental Degradation in Nuclear Power Systems–Water Reactors*, Whistler, British Columbia, Canada, August 2007.
114. K. Arioka, T. Yamada, T. Terachi, and G. Chiba, Cold work and temperature dependence on crack growth behaviours of austenitic stainless steels in hydrogenated and oxygenated high temperature water, in *Proceedings of the International Symposium Fontevraud VI*, SFEN, Paris, France, September 18–22, 2006, pp. 51–58.
115. H. Hanninen, U. Ehrnsten, and J. Talonen, Environmentally-assisted cracking of cold worked stainless steels—A mechanistic point of view, in *Proceedings of Work Shop on Cold Work in Iron and Nickel-Base Alloys Exposed to High Temperature Water Environments*, Sponsored by CANDU Owners Group, AECL and EPRI', Toronto, Ontario, Canada, June 3–8, 2007.
116. T. Shoji, Z. Lu, and Y. Takeda, Experimental investigations and theoretical modelling of the effect of material hardening on SCC of austenitic alloys in LWR environments, in *Proceedings of Work Shop on Cold Work in Iron and Nickel-Base Alloys Exposed to High Temperature Water Environments*, Sponsored by CANDU Owners Group, AECL and EPRI', Toronto, Ontario, Canada, June 3–8, 2007.
117. R.M. Horn, G.M. Gordon, F.P Ford, and R.L. Cowan, Experience and assessment of stress corrosion cracking in L-grade stainless steel BWR internals, *Nucl. Eng. Des.* 174, 313–325 (1997).
118. W.J. Shack, Measurement of through wall residual stresses in large diameter piping butt welds using strain gage techniques, in *Proceedings of Seminar on Countermeasures for Pipe Cracking in BWRs*, Vol. 2. EPRI Reports WS 79-174, May 1980.
119. C.B. Buchelet and W.H. Bamford, Stress intensity factor solutions for continuous surface flaws in reactor pressure vessels, in *Mechanics of Crack Growth ASTM STP S90*, American Society for Testing and Materials, 1976, pp. 385–402.
120. D.O. Harris, D.D. Dedia, E.D. Eason, and S.D. Patterson, Probability of failure in BWR reactor coolant piping; probabilistic treatment of stress corrosion cracking in 304 and 316 NG BWR piping weldments, NUREG/CR-4792, December, 1986.
121. R.N. Duncan et al., *Nucl. Appl.*, 1(5), 413–418 (October 1965).
122. P.L. Andresen, F.P. Ford, S.M. Murphy, and J.M Perks, State of knowledge of radiation effects on EAC in LWR core materials, in *Proceedings of Fourth International Symposium on Environmental Degradation in Nuclear Power Systems–Water Reactors*, Jekyll Island, August 6–10, 1989, Eds. D. Cubicciotti, E. Simonen, Published by National Association of Corrosion Engineers, ISBN 1-877914-04-5, pp. 1.83–1.121.
123. P.L. Andresen and F.P. Ford, Modeling and prediction of irradiation assisted cracking, in *Proceedings of Seventh International Symposium on Environmental Degradation in Nuclear Power Systems–Water Reactors*, Breckenridge, Eds. R. Gold and A. McIlree, Published by National Association of Corrosion Engineers, ISBN 0-877914-95-9, August 7–10, 1995, pp. 893–908.

124. J.L. Nelson and P.L. Andresen, Review of current research and understanding of irradiation assisted stress corrosion cracking, in *Proceedings of Fifth International Symposium on Environmental Degradation in Nuclear Power Systems–Water Reactors*, Monterey, Eds. D. Cubicciotti and E. Simonen, Published by American Nuclear Society, ISBN 0-89448-173-8, August 25–29, 1991, pp. 10–26.
125. S.M. Bruemmer et al., Critical issues reviews for the understanding and evaluation of irradiation assisted stress corrosion cracking, EPRI TR-107159, November 1996.
126. J. Medoff, Status report; intergranular stress corrosion cracking of BWR core shrouds and other internals components, USNRC NUREG-1544, March 1996.
127. S. Hettiarachchi, BWR SCC mitigation experiences with hydrogen water chemistry, in *Proceedings of Twelfth International Conference on Environmental Degradation in Nuclear Power Systems–Water Reactors*, Snowbird, Eds. L. Nelson and P.J. King, The Metallurgical Society, August 5–9, 2005, pp. 685–702.
128. P.L. Andresen, Y.J. Kim, T.P. Diaz, and S. Hettiarachchi, On-line Noblechem mitigation of SCC, in *Proceedings of Twelfth International Conference on Environmental Degradation in Nuclear Power Systems–Water Reactors*, Snowbird, Eds. L. Nelson and P.J. King, The Metallurgical Society, August 5–9, 2005, pp. 727–741.
129. P.L. Andresen, *Corrosion* 44, 376 (1988).
130. F.P. Ford, R.M. Horn, J. Hickling, R. Pathania, and G. Bruemmer, Stress corrosion cracking of low alloy steels under BWR conditions; assessment of crack growth rate algorithms, in *Proceedings of Ninth International Symposium on Environmental Degradation in Nuclear Power Systems–Water Reactors*, Newport Beach, CA, Eds. S. Bruemmer and F.P. Ford, The Metallurgical Society, August 1–5, 1999.
131. F.P. Ford, P.L. Andresen, D. Weinstein, S. Ranganath, and R. Pathania, Stress corrosion cracking of low alloy steels in high temperature water, in *Proceedings of Fifth International Symposium on Environmental Degradation in Nuclear Power Systems–Water Reactors*, Monterey, Eds. D. Cubicciotti and E. Simonen, American Nuclear Society, August 25–29, 1991, pp. 561–570.
132. P.L. Andresen and M. Morra, Emerging issues in environmental cracking in hot water, in *Proceedings of Thirteenth International Conference on Environmental Degradation in Nuclear Power Systems–Water Reactors*, Whistler, British Columbia, Canada, August 2007.

19

Corrosion of Microelectronic and Magnetic Data-Storage Devices

Gerald S. Frankel
The Ohio State University

Jeffrey W. Braithwaite
Sandia National Laboratories

CONTENTS

19.1 Introduction ... 826
19.2 Overview of Hardware Technologies ... 827
    19.2.1 Microelectronic Devices ... 827
        19.2.1.1 Integrated Circuits ... 827
        19.2.1.2 IC Packaging ... 828
        19.2.1.3 Macro Interconnects ... 829
    19.2.2 Magnetic-Storage Components ... 830
19.3 Microelectronic Corrosion ... 832
    19.3.1 Environmental Factors ... 832
        19.3.1.1 Water Adsorption ... 832
        19.3.1.2 Contamination ... 833
    19.3.2 Mechanisms ... 834
        19.3.2.1 Under Electrical Bias ... 834
        19.3.2.2 Open Circuit (without Electrical Bias) ... 836
    19.3.3 Device-Specific Corrosion Behavior and Concerns ... 836
        19.3.3.1 Integrated Circuits ... 836
        19.3.3.2 Macro Interconnects ... 840
    19.3.4 Product Qualification/Reliability Testing and Analysis ... 841
        19.3.4.1 Techniques ... 842
        19.3.4.2 Mathematical Relationships for PEM Aging ... 844
        19.3.4.3 Concerns and Limitations ... 847
19.4 Corrosion of Magnetic Data-Storage Components ... 850
    19.4.1 Thin-Film Magnetic Disks ... 850
    19.4.2 Thin-Film Inductive Head Materials ... 853
    19.4.3 Magnetoresistive (MR) Head Materials ... 855
    19.4.4 Magneto-Optic (MO) Alloys ... 855
    19.4.5 Corrosion Mitigation of Magnetic-Storage Components ... 856
19.5 Concluding Remarks ... 856
References ... 857

19.1 Introduction

Technological advances in microelectronic and magnetic-storage devices continue at an incredible rate. For example, over the past 30 years, microprocessor power has doubled every 1.5–2 years. Associated with this performance improvement is the miniaturization of integrated circuit (IC) packaging. The width and separation of conductor lines have been cut in half approximately every 5 years and are now less than 0.5 μm on some devices. Relative to data-storage media, the aerial bit density of thin magnetic films has doubled every 2 years and the separation distance (fly height) between heads and disks has been reduced to ~250 Å. These accomplishments have been coupled with increasingly long service life and more stringent reliability requirements.

Intuitively, because of their general use in relatively benign external environments, the vulnerability of electronic components to corrosion-induced failure could be considered to be less of an issue than that associated with other equipment and structures that are commonly exposed to severe environments (e.g., automobiles, bridges, airplanes, chemical processing, and oil production). However, there are several reasons why corrosion in electronics remains a major reliability concern. First and foremost, corrosion itself can directly cause failure of the device. This first-order result is in contrast to many other corrosion effects in which a material property is degraded and another subsequent process actually causes the failure (e.g., metal thinning that leads to mechanical overload or pitting that breaches housing hermeticity, thus permitting contamination of internal parts). Second, the two main factors needed for significant corrosion vulnerability can certainly exist: susceptible metallic materials and aggressive environments. Metals are selected primarily for their electrical or magnetic properties. Standard approaches to improving corrosion resistance, such as alloying with chromium, are often not possible because of deleterious effects on either these properties or indirectly on the properties of other nearby materials. Moisture and corrosion-enhancing contaminants (e.g., particulate, halogen compounds, oxidants) are almost always present around electronics. The electrical functioning of many devices, such as ICs, can also promote corrosion because metallic lines are often biased electrically relative to nearby lines. Given the dimensions and voltages involved, electric fields can reach in excess of 100,000 V/cm. Finally, the dimensions of the metal features are so small that the allowable loss of material to corrosion can be minuscule. Whereas a car can function with even pounds of metal lost to corrosion, an electronic device may fail after the loss of less than 1 pg, a factor of 10^{15} smaller!

The ability to predict and control the effects of corrosion on the service life or reliability of electronic devices must be based on a physical understanding of the relevant corrosion processes and controlling mechanisms. The state of knowledge of this key area for microelectronic and magnetic-storage devices is reviewed in this chapter. The microelectronic discussions focus on the primary components of present concern: packaged ICs are component interconnect technologies (printed circuit boards, connectors, and contacts). Corrosion of the other major components of microelectronics—discrete devices (e.g., resistors, diodes, and capacitors) and semiconductor materials—has been observed under certain conditions but is currently not viewed as important. Similarly, the section on magnetic-storage media addresses the most significant reliability concern: high-density data storage.

The body of this chapter is divided into three main sections. Initially, to enable a better understanding of the application of corrosion science to this subject, a very brief description of the hardware technologies and physical structures of these devices is presented. In the subsequent section on microelectronic corrosion, important environmental factors and relevant corrosion mechanisms are discussed. This is followed by an account of the observed impact of these mechanisms on three specific types of microelectronic hardware; finally, the application of accelerated-aging techniques and their effectiveness in qualifying product acceptance and determining reliability are considered. The concluding section addresses corrosion of magnetic data-storage devices.

19.2 Overview of Hardware Technologies

19.2.1 Microelectronic Devices

19.2.1.1 Integrated Circuits

Metals are used for several functions in integrated circuits. On-chip metal lines provide local interconnection to join a cluster of elements, connect circuit elements globally to far regions on a chip, and provide power and input/output (I/O) signals to the elements [1,2]. Because many of the prime conductor materials form deep levels in the band gap of Si, other metals are sometimes used as diffusion barriers to prevent unwanted doping of the Si. To form dense circuitry on a chip, also called a die, multiple layers of metal are commonly separated by insulating dielectrics layers (referred to as IDLs). Metals form the vias to connect the different layers and also act as thin adhesion layers. Connector lines can be as small as 0.5 μm wide (but typically are >1 μm) and from 1 to 1.5 μm thick. The top metallic layer on the die is protected by a passivation layer that is patterned to expose pads for wirebonding (~100 μm^2). A schematic view of these general features and structures is shown in Figure 19.1.

Al based alloys containing small amounts of Si to reduce interdiffusion with the substrate and Cu to strengthen the metal and limit electromigration were used in the past as

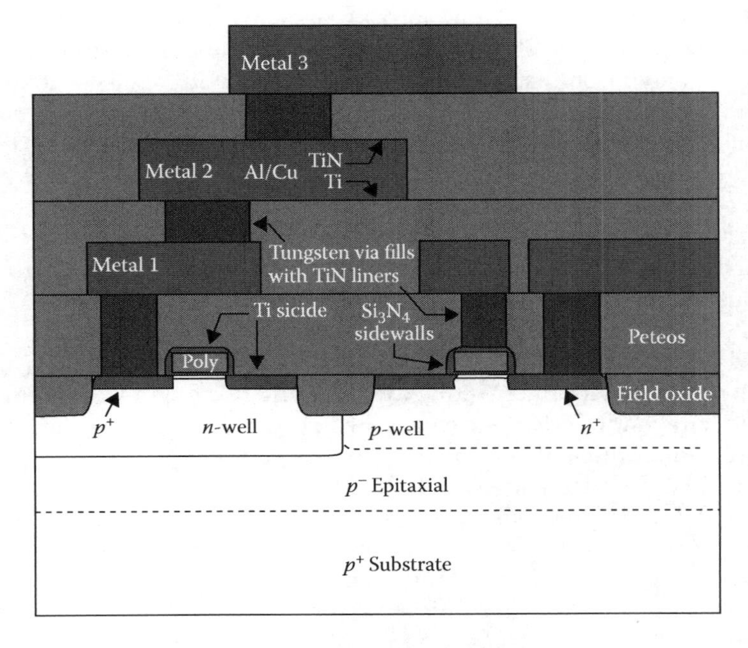

FIGURE 19.1
Interconnection structure of typical submicrometer integrated circuit technology that implements chemical-mechanical polished (CMP) planarized dielectric layers.

the conductor. Passivation layers are often sputter-deposited SiO_2, chemical-vapor-deposited (CVD) phosphosilicate glasses (that provide better coverage), and plasma-deposited silicon nitride. For corrosion engineers, this passivation layer should not be confused with the intrinsic oxide layer that protects many engineering alloys. Patterning of the Al lines is often achieved by subtractive techniques such as wet chemical etching or reactive ion etching (RIE). Metal selection is sometimes dictated by processing requirements. For example, tungsten metallization has been used because it can be deposited by a selective CVD process in which deposition occurs selectively on metal surfaces. Thus, high-aspect-ratio holes can be filled, an important consideration as the line widths continue to shrink. Refractory metals, such as tungsten, are also used as an underlying diffusion barrier. With the decreasing dimensions of very large scale integration (VLSI) devices, the on-chip interconnections account for an increasingly large fraction of the total circuit delay time. As a result, there has been a trend to more conductive metals and lower dielectric insulators. Cu is now commonly used as the conductor in IC devices because of its high conductivity [3].

19.2.1.2 IC Packaging

Electronic components (e.g., computer mainboard or motor controller) typically use a multiple-level packaging hierarchy to ensure proper power and signal distribution among the various discrete devices [1,4,5]. The core, or first level of packaging, is within the IC itself [6]. Besides permitting electrical communication to the outside, secondary packaging roles include providing robust mechanical support and environmental protection to the chip. Two general types of IC packaging technology exist: ceramic-hermetic package (CHP) and plastic-encapsulated microelectronics (PEM). As the terminology implies, CHP devices have ceramic bodies and the chip enclosure is hermetically sealed, normally by welding, brazing, or glassing on a metal or ceramic lid. Hence, the chip environment can be tightly controlled. This technology was dominant 25 years ago and is still preferred by many for high-reliability applications. With PEM technology, the chip is attached to a metal support/lead frame fixture and the entire assembly is encapsulated with a plastic molding compound (Figure 19.2) [7]. The molding material is usually a water-permeable polymer such as epoxy or silicone that is filled with inorganic particles. The encapsulant material is normally injection molded but can also be cast as a liquid (glob). Initially, PEM devices were considered unreliable because of corrosion and encapsulant cracking/delamination issues [8]. However, substantial technological improvements have been made over the past 20+ years to the point that their reduced cost, size, and weight; improved reliability; and now high level of availability have pushed the PEM worldwide market share to over 97% [9]. An excellent detailed description of the PEM technology and its application is available [9].

The exposed pads on the chip top-level passivation windows are connected to fingers on the metallic lead frame with fine 25-μm Au or Al wire (Cu or Ag is sometimes also used) by a solid-state wirebonding process using thermocompression and/or ultrasonic welding [9,10]. The lead frame, which also forms the external, board-level interconnects, is usually made from an Ni-Fe alloy, Cu alloy, or Cu-clad stainless steel. As with discrete connectors, the bare frame is usually given an Ni or Ni-Co strike and then plated with Au, Ag, or Pd for improved corrosion resistance and reduced contact resistance [9].

Tape automated bonding (TAB) and controlled collapse chip connection (C4) are two other common first-level packaging technologies that allow increased I/O density compared with wirebonding in molded plastic packages. In TAB, chips are thermocompression bonded to the inside ends of radially patterned metal leads on polymer tape. The outside ends of the leads are then soldered to a second-level package and the chip is covered with an encapsulant,

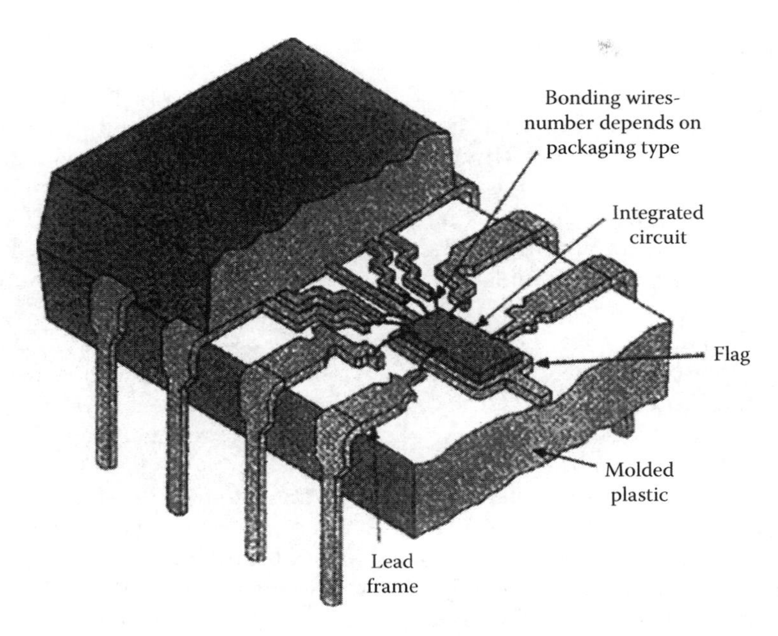

FIGURE 19.2
Sectioned schematic view of a typical plastic-encapsulated integrated circuit. (From Gallace, L. and Rosenfield, M., *RCA Rev.*, 45, 249, 1984.)

so the reliability concerns are similar to those of PEM devices. In C4 bonding, an array of solder bumps deposited on a chip is joined to a matching array of pads on the substrate. This technology, which allows maximum I/O density, has been used primarily in hermetic high-end applications but is now becoming popular in low-end, nonhermetic applications.

19.2.1.3 Macro Interconnects

Functional electrical circuits are formed by interconnecting packaged ICs with other devices using printed circuit boards (PCB). These boards can be viewed as second-level packaging. Individual devices are attached to the PCB using some form of automated or manual soldering process [9]. PbSn alloys are still the predominant choice for solder, although more environmentally compatible alternatives are emerging. The conducting lines on PCBs are normally made from copper and are protected with an Sn or SnPb plating, an organic coating, or an inorganic passivation layer. Often, a protective, organic-based conformal coating (e.g., acrylic, polyurethane, epoxy, or silicone based) is applied to the entire board to ensure cleanliness and reduce direct exposure to the external environment [9].

Low-force, low-voltage separable connectors and contacts also have wide use in electronic systems [11,12]. A variety of different configurations exist, but the substrate of most is made from copper, brass, bronze, or copper-beryllium. The use of copper alloys ensures some susceptibility to environmental degradation. To reduce interfacial resistance and corrosion, the substrate is plated with a nickel diffusion barrier and then with a precious metal (e.g., gold, palladium) [11,13]. However, a trade-off exists between plating cost (thickness) and reliability [14]. Hence, physical defects (pores and cracks) are usually present. In the past, connectors and contacts could not be coated because of interference with their function, although some new coating inhibition formulations have been developed for this purpose [11,12].

19.2.2 Magnetic-Storage Components

All computers, from mainframes to portables, use some form of technology for long-term data storage and retrieval. This chapter focuses on corrosion in high-performance magnetic disk storage because this technology is extensively used and has significant vulnerability to corrosion. The primary material components of a disk drive are the disk, which has a hard magnetic layer that stores the information, and a head, which uses a soft magnetic material to write and read the information to and from the disk. In obsolete hard drive technology and current low-end devices such as floppy drives, both heads and disks are made from ceramic materials such as ferrites and iron oxide, so corrosion is not a concern. In the past 25 years, however, technological advances in the field of magnetic recording that have driven the improvements in performance have also introduced metals to both disks and heads. As a result, the components in disk drives are now susceptible to failure by corrosion. An excellent review of magnetic-storage technology is contained in the book by Mee and Daniel [15].

Disk drives now use thin-film metal disks as the storage medium. Figure 19.3 is a schematic representation of the thin-film disk structure as of about 1999, with current dimensions of the layers given in the caption. Glass substrates are now widely used instead of Al alloy disks for small format mobile devices. The magnetic layer is a Co-based alloy that is covered by a thin C overcoat. Disks used to be intentionally roughened or "textured" to reduce sticking of the head to the disk when the drive was turned on. That roughness prevented complete coverage of the magnetic layer by the C overcoat, leading to the possibility of corrosion. Corrosion product formed on the disk surface could lead to collisions with the magnetic recording heads that are located very close to the spinning disk surface for highest magnetic recording density. Such head "crashes" can cause head misalignment and disk failure and they could, long before the corrosion process, cause a significant information error rate. At the present, the data zone on disks is almost atomically flat. In practice, the COC and thus the disk is also covered with a thin layer of lubricant that can further reduce corrosion by repelling and/or displacing water. However, Figure 19.3a

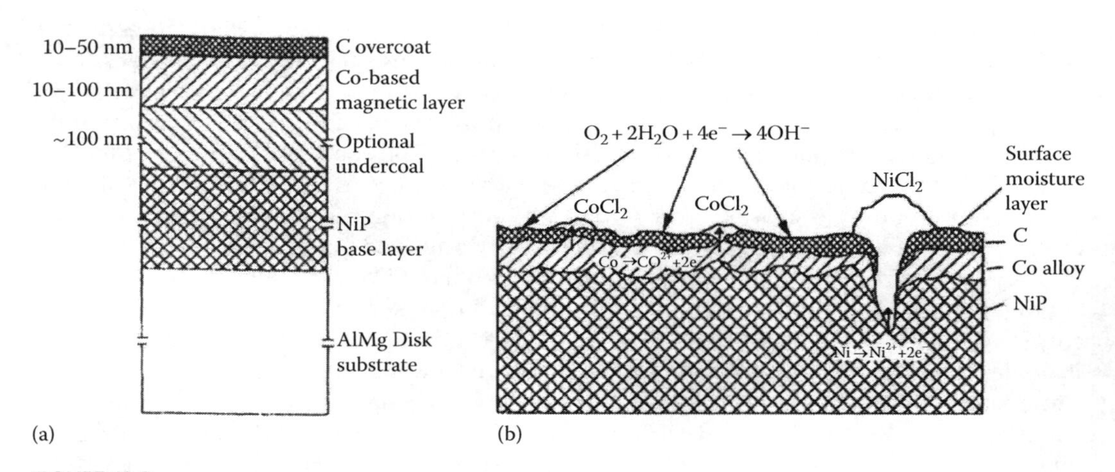

FIGURE 19.3
Schematic representation of thin-film disk structure, (a) layers and thickness as of about 1999. Dimensions as of 2010 are C overcoat of 3nm, magnetic layer 20nm, under layers 60nm. (b) Schematic view of corrosion of roughened disk structure in a chloride-contaminated environment.

is idealistic in that the layers are not smooth and the carbon overcoat does not completely cover the underlying metallic layers. A more realistic view is given in Figure 19.3b. The NiP base layer is intentionally roughened or "textured" to reduce sticking of the head to the disk when the drive is turned on. The magnitude of this texturing or roughness can be comparable to the thickness of the COC. As a result, the thin COC is often unable to cover completely and seal the underlying layers, thereby exposing susceptible metals to the environment. The corrosion that occurs can result in the formation of relatively large corrosion product deposits on the disk surface. The collision of heads with these corrosion deposits on the surfaces of spinning disks can cause head misalignment and disk failure. Such head "crashes" typically occur in practice long before the corrosion process causes a significant information error rate.

Several configurations of recording heads are used in disk drives. Thin-film inductive heads are standard components consisting of miniaturized horseshoe electromagnets manufactured by wafer-level fabrication techniques. These devices had been used for both reading and writing. Now they are used primarily as write elements that are combined with more sensitive sensors for reading data. The device is imbedded in an insulator, such as sputtered alumina. It is thus protected from the environment, except for the two ends of the pole piece (pole tips) and gold bond-pads connected to the coil. The pole piece is often made from electroplated permalloy (Ni-20%Fe) and the exposed permalloy pole tips are a few μm^2 in area. As in microelectronics, loss of material from corrosion can have a large influence on such small structures.

The read elements in advanced magnetic heads make use of the magnetoresistance (MR) phenomenon in which the resistance of a material changes as a function of magnetic field. The various types of magnetoresistance that are currently used in magnetic storage devices or under development include, in order of increasing sensitivity, anisotropic magnetoresistance (AMR), giant magnetoresistance (GMR), and tunneling magnetoresistance (TMR). At the time of this writing, TMR-based recording heads dominate the market. This read head is used for perpendicular magnetic recording, which allows higher density. There are several designs for the various MR-type sensors that all contain a stack of numerous thin-film layers, such as Co, Cu, NiFe, and MnFe, each with thickness on the order of several nm [16]. Most MR sensors use antiferromagnetic layers to pin the domains of adjacent soft magnetic layers by exchange coupling, although many AMR devices use a hard magnetic layer instead. MnFe, containing about 50% of each element, was a commonly used antiferromagnetic layer and is extremely corrodible. To be used in a disk drive, a cross section of the thin films in the sensor must be exposed at the device surface. This is achieved by abrasive lapping in an aqueous environment. Lapping is potentially very aggressive because fresh metal area is continually exposed to the aqueous lapping environment. Like disks, MR heads are often covered with a thin carbon overcoat after lapping, and this COC provides corrosion protection once it is in place.

Magneto-optic (MO) recording is another type of data-storage technology that is not as widely used as magnetic storage but poses special corrosion concerns. MO disks are erasable and have very high bit density. Because the disks are removable and thus handled by users, they are exposed to an uncontrolled range of conditions. The active layer in some MO drives is composed of a rare earth-transition metal alloy such as Fe-25Tb and is several hundred angstroms thick. The high Tb content makes these alloys extremely reactive. The MO film is deposited onto glass substrates and then covered with a thin oxide layer and a thin metallic layer as such as Al. Two such substrates sealed back to back with epoxy constitute an MO disk. Moisture can penetrate through the permeable epoxy and cause corrosion of the Al layer or of the MO layer where it is exposed.

19.3 Microelectronic Corrosion

Because corrosion has been and continues to be one of the prime degradation modes observed in microelectronic devices, a significant number of in-depth corrosion studies have been performed during the past two decades and several noteworthy review articles have been written [7,17–23]. For reference and to be consistent with the terminology used in the electronics community, corrosion is defined as the degradation of metallic features resulting from interactions with environments containing water or moisture. An important consideration to remember when assessing some of the historical information that follows is that the microelectronic technology has improved substantially over the years. Many of the early concerns and problems are no longer relevant to modern microelectronics. Also, because microelectronic devices are exposed to atmospheric conditions only during use, much of our existing knowledge of atmospheric corrosion is applicable. This subject is addressed by Leygraf in Chapter 15. Leygraf and Graedel [24] have written a textbook on atmospheric corrosion. An additional reference is the comprehensive review of this subject was that was compiled by Ailor [25] and contains information on fundamental aspects, behavior of specific metal systems, and testing.

19.3.1 Environmental Factors

Regardless of the type of device or the offending corrosion mechanism, corrosion-induced degradation involves interactions with the environment. The critical factors that must always exist are a susceptible metallization, the presence of moisture, and some type of contaminating ionic species. This subsection describes some of the aspects of the latter two factors common to corrosion of all microelectronic devices.

19.3.1.1 Water Adsorption

Under noncondensing atmospheric conditions, water adsorbs on metal and metal-oxide surfaces and forms a thin layer. The thickness of the layer ranges from <1 to tens of monolayers and depends primarily on the relative humidity (RH) and the substrate material [26,27]. At 20% RH, approximately one monolayer exists on average, whereas at 75% RH, the thickness increases to about five monolayers. Often the water layers cluster and so local thicknesses are probably greater. The corrosion rate for most metals is a function of the thickness of the adsorbed water layer and therefore the RH. Usually, the corrosion rate as a function of RH has a sigmoid-shaped response. The critical humidity level above which the corrosion rate significantly accelerates (lower inflection point in sigmoidal curve) ranges from 15% to 90% RH but is typically >50% and is thought to be the RH at which at least three monolayers of water exist. In general, if the humidity is lower than the critical level, the corrosion rate is minimal and the reaction process is essentially dry (requiring a direct gas-metal ambient-temperature interaction). Above this critical level, the adsorbed water phase becomes "quasi-aqueous" and is thus capable of supporting faster electrochemical charge transfer reactions. The final important water-adsorption aspect is associated with the extremely small physical features found in microelectronics where capillary condensation can occur. Examples include under the lip of a wirebond, between connector surfaces, and in defects in the plastic encapsulant. This condensation phenomenon can produce very bulk-like localized aqueous conditions at humidity levels 100%. A more in-depth description of water adsorption is presented in Chapter 15.

The temperature of the device can have a dramatic effect on water adsorption. For a given absolute humidity level, higher surface temperatures result in desorption of water (localized decrease in relative humidity according to the psychrometric law). As noted in the previous paragraph, an associated decrease in the corrosion rate could be produced. For many ICs, these higher temperatures occur when their power is on because of internal ohmic heating. Besides surface desorption, water is also driven out of the package. Ajiki et al. [28] did a benchmark study in 1979 that is still relevant today. They showed that device reliability increases with increasing power dissipation and with an increasing ratio of time on to time off. This correlation is simply due to moisture effects and a related worst-case condition could then be long-term dormant storage [29].

19.3.1.2 Contamination

Introduction of corrosive contaminants that may result in long-term reliability concerns can occur during manufacturing and/or use. Many of the fabrication process steps involve exposure to aggressive species and can leave corrosive residues. For example, reactive-ion etching (RIE) of Al metallization is performed in plasmas containing combinations of gases such as Cl_2, CCl_4, $CHCl_3$, and BCl_3 [30–36]. If removed untreated from the etcher, patterned structures are covered with high levels of $AlCl_3$, which is hygroscopic and can later form HCl in the presence of moisture. The sidewalls of the patterned photoresist that is used to mask areas that are not to be etched can also retain large amounts of chloride contamination. Other typically benign process solutions, such as deionized (DI) water and organic solvent rinses, can become contaminated over time from process chemical residues [37,38].

In addition, airborne particles or particles that are too small to be filtered out by the best clean room facilities can be deposited on susceptible surfaces. Studies performed at AT&T Bell Laboratories have highlighted the potential role of airborne particles in the corrosion of electronic devices that are relevant to both manufacturing and actual use conditions [20,39,40]. Sinclair [20] has shown that there exist in the environment a large number of fine particles (with aerodynamic diameters of 0.05–2 μm) that are rich in ammonium acid sulfate, derived from anthropogenic sources, and difficult to remove by filtration. Ionic compounds, such as these, often deliquesce above a certain RH level [41,42]. Another significant contaminant that can be introduced during manufacturing is solder flux/residue that may not be completely removed or reacted during PCB soldering operations [43]. Some materials in the microelectronics themselves can be a source of undesired chemical species. For example, plastic epoxy encapsulants can contain a fire-retarding agent that includes an organic bromine compound. Phosphorus is a component in many of the passivation glasses. With time, interaction and leaching with adsorbed moisture can produce corrosive species such as bromide salts or phosphoric acid [10,17].

Leygraf (see Chapter 15) discusses the environmental differences between indoor and outdoor exposure conditions and why these differences are important to corrosion of electronics. Although the service environments for electronic devices are typically less severe than for many other types of equipment, industrial atmospheres do contain harmful oxidizing (e.g., Cl_2, SO_2, NO_x) and halide-containing (e.g., NaCl) species that can diffuse or migrate to microelectronic metallization, especially over an extended service life. Because of the importance of contamination control, techniques to quantify microscopic levels have been developed [30,44].

19.3.2 Mechanisms

Because of the wide range of possible environments and the large number of metallic materials used in microelectronic devices, corrosion occurs by a number of different physical mechanisms that can be grouped into two general categories: those driven by the application of an applied potential (under electrical bias) and those occurring under static (open-circuit) conditions.

19.3.2.1 Under Electrical Bias

Although the potential that is applied to microelectronic devices is normally relatively limited (<5 V), very large electric fields can result because of the very small separation distances between conductors. When condensed water and ionic contamination are present between the lines (on the surface of separating insulators), these fields produce undesirable results by means of the three separate, but related mechanisms described next and shown schematically in Figure 19.4. For reference, these processes are often referred to as electrolytic. An extensive review of this subject was compiled by Steppan et al. [45]. The voltage driver is normally externally applied, but another source is the semiconductor

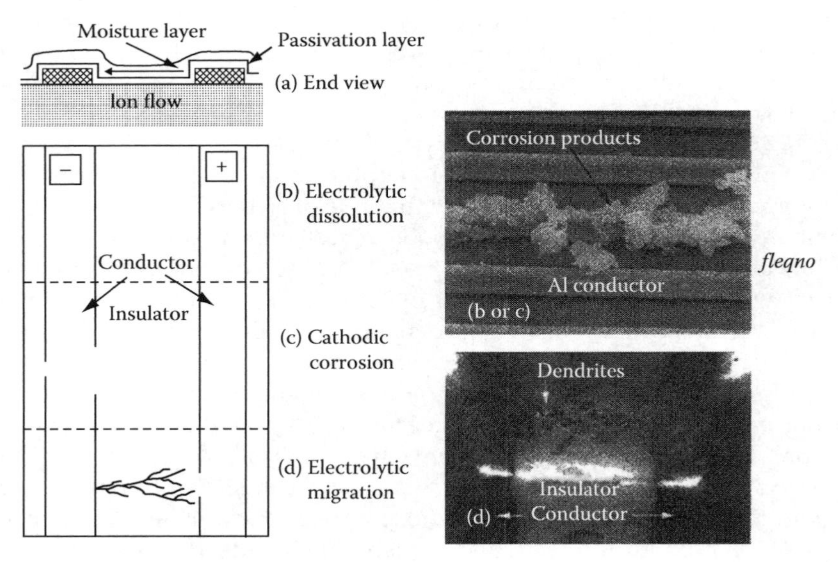

FIGURE 19.4
Schematic diagram of the effect of the three primary corrosion mechanisms that occur in integrated circuits due to the presence of moisture and an applied electrical bias. The photograph in the upper right is a typical result of either electrolytic dissolution or cathodic corrosion (depends on location of defects in the passivation layer). The lower right photograph shows a short-producing electrolytic migration of solder that occurred in a ceramic capacitor in the presence of high humidity and a chloride-containing flux residue.

junctions that exist within an IC [17]. Most environmentally induced failures in ICs that are observed in practice, and especially during accelerated aging, are caused by the applied electrical bias.

If an ionic path is present between two oppositely biased metal lines that are otherwise isolated and the path resistance is adequately low, then sufficient voltage will exist to enable an electrical current to flow between the lines (Figure 19.4a). A portion of the applied voltage difference will exist at each metal-electrolyte interface, permitting electrochemical oxidation/reduction reactions to occur. The extent of the resulting corrosion depends on many factors, but the resistance of the ionic pathway is the most important. The influence of increasing moisture and contamination on decreasing the ohmic resistance of the ionic path is further explained by Osenbach [17] and has been phenomenologically modeled by Comizzoli [46]. Contamination has a further role because of its effect on the breakdown of the passive oxide on many metals.

19.3.2.1.1 Electrolytic Dissolution

In this mechanism, the current leak results in electrochemical oxidation of the positive, anodic conductor. Eventually, the device can fail from an open circuit in the anodic line (Figure 19.4b). Given the small dimensions and volume of the adsorbed electrolyte, severe concentration gradients exist and voluminous corrosion products can precipitate near the surface (e.g., $Al(OH)_3$). Most of the metals used in microelectronics are susceptible to electrolytic dissolution.

19.3.2.1.2 Cathodic Alkalization (Cathodic Corrosion)

This degradation mechanism is specific to active/passive, amphoteric metals such as aluminum. In the absence of chloride or other aggressive contamination, the leakage current through the conductive path causes anodization (oxide buildup) at the anode instead of localized dissolution. The potential drop across the oxide will consume part of the applied potential and tend to reduce the leakage current. However, while current is flowing, proton and water reduction reactions will take place at the cathodically biased line, causing an increase in local pH by the following reactions: $2H^+ + 2e^- \rightarrow H_2$ or $2H_2O + 2e^- \rightarrow 2OH^- + H_2$. Because the metal is amphoteric, the increase in pH results in dissolution of the protective oxide on the surface of the cathodic line. The metal can then dissolve rapidly, even with a net negative current at the surface. Analogous to the case of electrolytic dissolution at anodic lines, devices fail by an open circuit, except now on the cathodic line (Figure 19.4c). Similarly, a corrosion product such as $Al(OH)_3$ will precipitate near the dissolution region.

19.3.2.1.3 Electrolytic Migration

In this degradation mechanism, metallic ions dissolved from the anode (electrolytic dissolution) are transported through the surface electrolyte film and are electrolytically reduced at the cathode [45]. Metal dendrites grow from the cathode toward the anode and lead to failure via a short of the adjacent conductors (Figure 19.4d). Important factors that affect dendritic growth include (1) properties of the metal (some metals such as Ag and solder readily electroplate with a dendritic morphology), (2) local current densities that, if high enough, deplete cations in the surface electrolyte film in the vicinity of the cathodic lines, and (3) mass transport–limited conditions that promote the stability and propagation of dendrites. Notably, metals such as aluminum that have a deposition potential that is outside the stability limit of water are not vulnerable to this mechanism. Steppan et al. [45] listed the order of susceptibility to electrolytic migration for many commonly used metals:

Ag > Mo > Pb > solder > Cu > Zn > bronze. A small volume of metal dissolved from the anode can result in comparatively large dendrites, so the lines may short long before an open would have formed in the anodic line [45]. As dimensions of microelectronics shrink, this failure mechanism becomes more important. Warren, Wynblatt, and coworkers performed a series of studies on electrolytic migration of Cu [47–51]. Their work, which developed a good understanding of the phenomenon, showed that contamination greatly accelerates electrolytic migration. The phenomenon of electrolytic migration should be distinguished from another failure mechanism with a similar name, electromigration, which is not electrochemical in nature [45].

19.3.2.2 Open Circuit (without Electrical Bias)

Although deleterious environmental interactions normally occur in microelectronics under electrical bias, more traditional atmospheric corrosion mechanisms certainly do take place in unbiased regions during the powered-off state and/or during storage. As noted previously in this section, the power-off condition for ICs may actually be more aggressive in some circumstances because a greater amount of water may exist at the metal surface. Relevant mechanisms, with examples given in parentheses, include uniform attack (Cu contacts), pitting (Al lines and bondpads), crevice corrosion (Al under wirebonds and under passivation layers), and intergranular corrosion (Al bondpads). Under these unbiased conditions, chemical incompatibility is the primary driver. Given the wide variety of metals that are present in intimate contact with each other (e.g., Au/Al wirebonds), all of these mechanisms can be enhanced in specific configurations by galvanic interactions. To help characterize the potential galvanic effects, Griffin and coworkers [52] developed a galvanic series specifically for the common thin-film metallization and barrier layers used in microelectronic devices. The corrosion potential in 2000 ppm NH_4Cl solution decreased in the following order: Au-40%Pd, Au, Ag, Cu, Al-0.5%Cu-l%Si, CVD W, W-10%Ti, Al-2%Cu-l%Si, Al-2%Cu, Al-0.4%Pd-l%Si, Si, Al-l%Si on CVD W, Al on CVD W, Al on W-10%Ti, Al, Mg.

These traditional corrosion mechanisms are described in detail in standard corrosion textbooks [53] and elsewhere in this book. However, care must be exercised during the application of this conventional understanding to microelectronics because of their very small dimensions. This caution can be demonstrated by considering the atmospheric pitting of aluminum. A standard technique for characterizing the atmospheric corrosion susceptibility of Al involves a salt fog test (ASTM B117-90). An associated military specifications (mil-spec) (Mil-C-5541E) states that no more than five pits can exist on a panel and normal practice is not to count a pit that is less than 125 μm in diameter. As noted previously, many IC feature are <1 μm. Clearly, even small metastable pits in the wrong place can be disastrous.

19.3.3 Device-Specific Corrosion Behavior and Concerns

19.3.3.1 Integrated Circuits

19.3.3.1.1 Aluminum Metallization

Because of the predominant use of aluminum alloys in integrated circuits and their vulnerability to atmospheric corrosion, the majority of the environmentally induced problems that have historically occurred have involved aluminum. As introduced in Section 19.3.1.2, corrosion has been observed even during manufacture due to exposure to aggressive processing chemicals. During the early phases of the microelectronic industry, yield loss from in-process corrosion was a costly problem. Chemical and RIE processes have been the most damaging. The preferred use of Al-Cu alloys exacerbates the problem because Cu is enriched in the subsurface region during RIE due to the lower volatility of Cu chlorides compared with Al chlorides [36]. At the high etching temperatures, Al_2Cu θ-phase particles precipitate and accelerate the subsequent corrosion of the Al matrix. Several postetch process steps have been developed to reduce corrosion, including CF_4 plasma cleaning, O_2 plasma cleaning, DI water rinsing, wet etching, and heat treatments [30–36]. Brusic et al. [30] evaluated the efficacy of several of these cleaning steps. By simply rinsing in water before stripping the photoresist, the impurity level and associated corrosion rate were reduced by more than two orders of magnitude compared with parts cleaned only with an O_2 plasma. Elimination of $CHCl_3$ from the RIE gas in the last step resulted in still lower corrosion rates and, unlike those with etching in $CHCl_3$, the properties did not degrade during subsequent storage in air. Any in-process corrosion or staining that does take place on bondpads during manufacture can adversely affect long-term device reliability because the integrity of the subsequent wirebond can be reduced [54].

During service, biased Al metallization features most commonly fail at defects or breaks in the passivation layer (upper right photograph in Figure 19.4). If chloride contamination is present, electrolytic dissolution occurs in the positive (or anodically) biased lines [55–57]. As implied in this photograph, the production of corrosion products causes more passivation layer defects and, in turn, more corrosion. As with pitting corrosion of aluminum, the presence of a halide is very important because it induces breakdown of the Al oxide film and permits rapid dissolution. Biased Al metallization that is not exposed to chloride-contaminated environments may still exhibit extensive corrosion and fail by cathodic corrosion [58–64]. This phenomenon is promoted for structures having a top passivation layer of phosphosilicate glass (PSG) with a high P content. The addition of P to SiO_2 improves coverage by lowering the stress in the oxide and reducing cracks. However, cracks along the sidewall and lack of coverage at steps and pinholes can still occur. When PSG containing more than about 5% P is exposed to moisture, phosphate dissolves to form a highly conductive surface layer. Cathodic corrosion is promoted because of the high surface conductivity and the absence of chloride ions. The anodic aluminum anodizes rather than dissolves. The activation energy for cathodic corrosion has been found to be similar to that for dissolution of Al in an NaOH solution [62]. Other ions besides phosphate can cause this form of corrosion, even those that do not form a strong base [62]. The effect of P in PSG was identified in the early 1980s and has been addressed by keeping the P content low or by using silicon nitride passivation, which has been found to be more protective [65–67].

Even in the absence of an applied bias, galvanic-induced electrochemical driving forces for corrosion can exist because unpassivated wirebonds often contain couples of Al and Au [57,68]. In the presence of moisture and chloride contamination, the galvanic driving force due to the Au can be sufficient to cause pitting and intergranular attack of Al and eventually produce device failure. Several researchers have studied the corrosion of Au/Al wirebonds [69,70]. Recent work at Sandia National Laboratories has focused on galvanic corrosion under dormant, un-powered storage conditions and has specifically been exploring the role of Al-Au intermetallic compounds that form during encapsulant curing operations. This study has shown that, under mildly accelerating conditions, unbiased Al bondpad corrosion occurs only in the presence of an Au wire and that it initiates under the wirebond and propagates along an Al/intermetallic interface (Figure 19.5). Also, considerable variability exists in the distribution and structure of the intermetallic phases

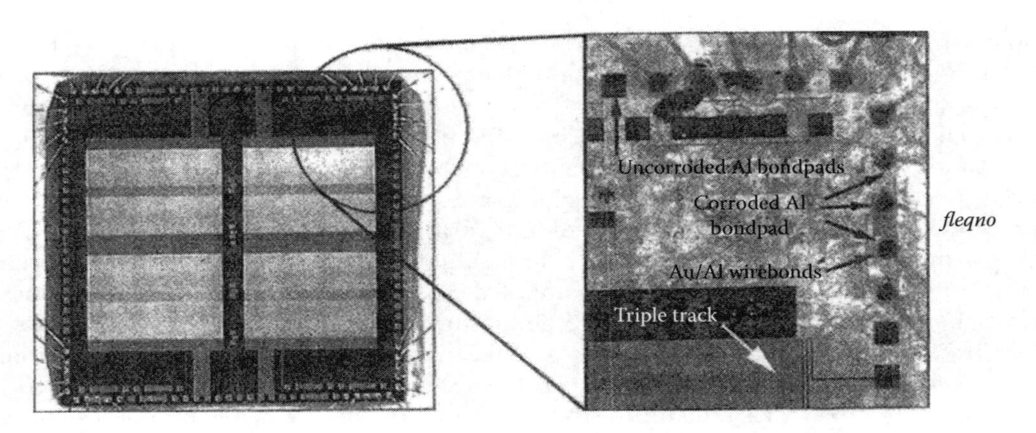

FIGURE 19.5
Corroded aluminum-bondpad metallization layer taken under back-lit conditions. A photograph of the entire test device prior to any corrosion is shown on the right. The photograph of the corroded structure shows the importance of galvanic interactions and the existence of crevice corrosion (under the passivation layer) and was taken after underlying silicon and glass insulation layer were etched away.

(Figure 19.5). This latter finding suggests that the often-observed stochastic nature of corrosion may actually be the result of manufacturing process variability. If the galvanic influence of Au is not present, θ-phase particles in Al-Cu alloys can still initiate pitting or intergranular corrosion on an unbiased surface via local galvanic action [38,71,72]. Pits associated with θ phase may be limited in size but are large enough to cause failure of micrometer-sized features.

19.3.3.1.2 Gold Metallization

Anodically biased gold metallization can fail in the absence of a contaminating salt by electrolytic dissolution if the ionic pathway contains an effective gold complexing agent. The dissolution is accompanied by the formation of a voluminous $Au(OH)_3$. However, examples of this type of failure have not been widely reported [73]. In the presence of a halide contaminant, failure can also result from electrolytic migration because gold metallization can form dendrites [73]. Marderosian [74] found a humidity threshold below which no migration of Au occurred. The threshold value was a function of the type and amount of purpose applied halide salt contamination but ranged from about 30% to 50% RH.

19.3.3.1.3 Plastic Encapsulated Microelectronics

Because the plastic encapsulant materials are relatively permeable to water, chip metallization in PEM devices is vulnerable to corrosion when exposed to or operated in humid environments. Nguyen and colleagues [75–77] used an in situ capacitance monitor to study moisture permeation in plastic packages [5]. The uptake and transport of moisture in epoxy were Fickian in nature and exhibited a diffusivity of about 2×10^{9} cm^2/s, meaning that water can penetrate a polymeric package and reach the metals on the die relatively quickly. Moisture can also migrate along defects in the lead frame–polymer interface and condense in undesirable voids near metallization surfaces. Because the bulk permeability of the plastic to typical external contaminants is very low, these lead frame defects appear to be a primary contaminant transport path. Historically, the plastic encapsulants themselves have been a source of significant contamination, with liquid/glob materials having higher corrosion-enhancing levels than the injection molded compounds.

Once sufficient moisture and contamination reach the IC metallization and a void for adsorption/condensation exists, corrosion can occur and degradation may ensue by any of the mechanisms described previously, including electrolytic, pitting, galvanic, and crevice corrosion. It should be noted again that top-level conductor lines are normally under passivation layers that, if impermeable, would prevent corrosion [78,79]. Historically, defects such as cracks, pinholes, and inadequate edge coverage were commonly present [55]. During the past 10–15 years, processes and materials have improved to the point that passivation defects are no longer a primary concern. Now, the unpassivated bondpads themselves are the most susceptible.

In practice, the key factors that influence PEM corrosion vulnerability are probably the defects in the plastic encapsulant, the level of leachable contaminant in the encapsulant, and the degree of drying during power-on cycles. The first two factors have been adequately addressed in modern, best commercial practice devices to the point that passivation detects and residual contamination are essentially inconsequential. Pecht and coworkers [9,29] have now concluded that modern, properly fabricated PEM devices are very reliable under a variety of service conditions.

19.3.3.1.4 Ceramic Hermetic-Packaged Microelectronics

The meaningful differences between CHP and PEM devices in the context of corrosion are the following: (a) Al wires are used instead of Au so that the prime corrosion vulnerability is the ultrafine 25-μm wire instead of the bondpad, (b) the internal environment can theoretically be controlled such that a benign environment will always exist, and (c) the cavity containing the die is unfilled so encapsulation defects are not needed as sites for water condensation. To achieve the environmental control advantage, three factors must be satisfied: (a) a hermetic seal, (b) a clean and dry assealed internal environment, and (c) control of outgassing from internal materials. If they are properly designed and manufactured, the reliability of CHP devices is not a concern. However, in practice, all three of these aspects have produced problems as evidenced by field-returned "hermetic" packages in which corrosion has occurred [80–82].

The specifications on hermetic seals are not always adequate to ensure total isolation from the environment for extended periods of time. For example, as noted by Pecht and Ko [83], hermeticity is defined by a maximum leak rate (e.g., 10^{-7} atm cc/s). If the device leaks at this particular maximum rate, the critical moisture content (three monolayers of water) can diffuse through the seal in as little as 2500 h. Supporting this finding is a study in which 20% of field failures of equipment sited in a tropical environment were due to corrosion of interconnects in packages that met the hermeticity specification [17]. Also, the seals and/or ceramic bodies can crack and create larger leaks during handling, soldering, or qualification thermal cycling, which permits both water and contamination to enter easily. If care is not exercised during fabrication, moisture adsorbed on the inside walls of the package can desorb during subsequent processing steps or during use. To address this possibility, mil-spec CHP parts are normally sealed in an atmosphere containing no more than 5000 ppm of water, ensuring that no liquid phase will exist at temperatures down to 0°C, where ice forms. The sealing glass used in some packages or a popular organic die-attach material can actually be sources of moisture. Devitrifying glass is one type of sealant with a high moisture content that is evolved during closure, whereas vitreous glass contains little moisture and can result in a ceramic package with less than 500 ppm moisture in the cavity [84]. Moisture trapped inside a sealed cavity can leach ions from the sealing glass or other sources to form a conductive electrolyte. Once an ionic path exists between conductors, corrosion and failure by any of the mechanisms described previously can ensue. Finally, one researcher found that residual stresses in Al-containing wirebonds could be an important factor in failure (e.g., by stress-induced corrosion) [85].

19.3.3.2 Macro Interconnects

Solder, copper conductors, and plated-copper connectors constitute the major metallic components of second-level packaging for functional electronic circuits. These metals are all susceptible to corrosion. Typically, interconnects are less protected and more exposed to ambient environments. However, their dimensions are much larger than those of chip metallization and more corrosion can therefore be tolerated before failure is produced.

19.3.3.2.1 Solder

PbSn alloys, ranging from Pb-rich to Sn-rich compositions, are the most solder materials used in electronic applications. Pure Sn forms a protective oxide film, but Pb forms an oxide that is not stable and can easily react with chlorides, borates, and sulfates [86]. Frankenthal and Siconolfi [87,88] found that both Sn- and Pb-rich PbSn alloys form an Sn oxide, most likely SnO, during the initial exposure of oxide-free surfaces to oxygen. Lead is oxidized on the surface of PbSn and a mixed oxide results only after all the metallic Sn is totally depleted from the surface. Similarly, the corrosion product formed on Pb-50 In solder during exposure to an aggressive gaseous environment was found to be rich in In [89]. Not surprisingly, the corrosion resistance of PbSn in various aqueous solutions and gaseous environments depends strongly on the alloy composition, improving greatly as the Sn content increases above 2 wt% [86].

In the presence of moisture and contamination, the most common degradation mechanism of solder is electrolytic migration (shown previously in lower right photograph in Figure 19.4). Thus, the cleanliness of fabrication steps and the effective removal of flux residues are critical factors. Some IC package types (e.g., surface mount) are very hard to clean effectively. Manko [43] summarized the corrosion behavior of PbSn solders in various chlorides that are typically used as activators in fluxes and noted their deleterious effect in destroying the native passivating oxide layers. He also noted that corrosion related to flux use might result from flux or flux residue that is trapped in inaccessible locations or from fumes liberated during soldering and subsequent condensation in uncleaned regions. Specifically, water-white rosin flux has been found to leach Sn from the solder and therefore decrease its corrosion resistance [90]. Soldering traditionally uses nonaqueous fluxes that require cleaning with Freon-based solvents. In an effort to reduce chlorofluorocarbon usage, water-soluble fluxes have been substituted in some applications. Cleaning of water-soluble fluxes can normally be accomplished in deionized water. Sn-Pb-In solders corrode in water-soluble fluxes containing chloride-based activators but are not attacked in fluxes containing phosphate-based activators [91]. Finally, in-process corrosion can result in poor quality solder joints or decreased thermal fatigue life. This latter result is especially significant for C4 solder connections [90].

19.3.3.2.2 Copper Conductors

Although precautions are taken to protect the copper conducting lines in printed wiring boards from environmental exposure, defects exist and corrosion does occur under many atmospheric conditions. For most environments that are encountered, copper is not thermodynamically stable. However, its native cuprous oxide surface film does offer some limited protection. Atmospheric corrosion of copper is briefly described in Chapter 15.

Examples of typical industrial atmospheric pollutants that are harmful to copper include SO_2, H_2S, COS, NO_x, Cl_2, and CO_2.

Contamination of copper surfaces with atmospheric particulate matter can accelerate the corrosion process. For example, the corrosion of copper in 100°C air in the presence of submicrometer ammonium sulfate particles has been found to depend strongly on RH [40]. Below the critical RH, Cu_2O formed uniformly on the Cu surface. At the critical RH of 75%, $Cu_4(SO_4)(OH)_6$ formed in the region where the $(NH_4)_2SO_4$ particles had been deposited. At 85% RH, a thick corrosion product covered the entire surface. Particulate contamination such as this is typically believed to be an issue during manufacturing. However, it can also cause field failures in electronic installations, such as telecommunications centers [92].

The formation of conductive anodic filaments (CAFs) has been observed during accelerated aging testing to cause failure in epoxy-glass printed circuit boards containing copper conductors, a process clearly related to electrolytic dissolution [93]. Here, a conductive ionic path consisting of precipitated corrosion products presumably forms between the oppositely biased conductors. These fibers grow from the corrosion-producing anodic line to the cathodic line. The degradation was found to be most severe between two nearby through-holes. Because CAF has not been observed in field-returned parts, it may be only an artifact of the test conditions.

19.3.3.2.3 Au-Plated Connectors and Contacts

As noted earlier, the substrates of connectors and contacts are normally fabricated from copper or copper alloys and then plated with a noble metal. The presence of the thin plating that almost always contains some level of physical defects (e.g., cracks and pores) enhances the corrosion susceptibility of the substrate and enables other processes to take place. Slade [22] and Abbott [23] recently published a comprehensive review of connector/contact corrosion. One form of observed degradation occurs when corrosion products creep across a noble metal surface. When Au-plated Cu is exposed to a sulfide-containing atmosphere, a tarnish film composed of copper oxide and copper sulfide forms in the plating pores and cracks where the Cu layer is exposed [94,95]. With time, a predominately copper sulfide tarnish film migrates over the Au surface, resulting in increased contact resistance. This phenomenon, termed creep corrosion by Tierney [94], is becoming increasingly important as the thickness of the gold plating is reduced for economic reasons and as the edges of exposed Cu become closer to contact points as a result of miniaturization. A second relevant degradation mode is often referred to as pore corrosion and simply consists of accelerated corrosion of locally exposed regions of the substrate. Two important processes are probably responsible: (a) capillary condensation of water in cracks, pores, and between mated surfaces and (b) galvanic interactions with a potentially large cathode-to-anode area. Ming et al. [14] showed that contact force and electrical loading can also affect corrosion. For example, contact force can improve performance in stationary contacts and applied voltage can accelerate growth of surface films and decrease contact service life. Finally, in recent years, a few special contact lubricants have been developed that can also inhibit corrosion [23].

19.3.4 Product Qualification/Reliability Testing and Analysis

A significant effort has been made over the past quarter century to develop effective techniques to test and assess the effect of corrosion on the performance and life of electronic devices. Such techniques are desired for two specific uses: (a) as a means of rapidly characterizing product quality during manufacturing (acceptance testing) and (b) to provide customers and users with an accurate estimate of expected service life. In this context, corrosion degradation is primarily characterized in terms of reliability, which is the probability that a device will not perform its intended function (i.e., electrically fail). Such device-level studies are required because the direct measurement of corrosion is very difficult or even impossible in actual operating environments using standard corrosion characterization techniques (e.g., electrochemical and gravimetric). A device manufacturer wants to be certain that a product will have at least a certain functional lifetime in service with a minimum number of failures. A typical historical requirement is 100 failures in 10^9 device hours (100 FIT) or 0.01% failure in 1000 h [96]. In modern practice, most electronic devices are expected to function reliably for at least 10 years [17], although obsolescence can certainly set in much sooner. Therefore, reliability engineers are forced to perform accelerated aging exposures on large numbers of parts. For general reference, Nelson [97] has published a comprehensive overview of accelerated aging, and Osenbach [17] describes factors and issues associated with accelerated aging of microelectronics. A clearly recognized deficiency of the reliability approach is that such tests do not permit a fundamental understanding of the corrosion mechanisms and processes to be easily identified. Truly predictive reliability assessments must have this type of physical basis. Nevertheless, accelerated reliability testing is of tremendous practical importance because susceptible materials, design flaws, and processing problems can be identified in a timely manner and a qualitative understanding of the degradation processes can be generated.

The underlying principle behind the use of accelerated aging is that the functional behavior of real devices or the response of test structures with similar materials and dimensions can be characterized using various combinations of environmental stress factors that include temperature, humidity, bias, and contamination. Each of these variables is typically used at levels more severe than actual operating conditions. Osenbach also notes that all of the accelerated aging strategies include two significant assumptions: (a) the parameters that characterize failure under high-stress conditions (where failure occurs in short time periods of days to months) can be extrapolated to actual use conditions where failure occurs after years of exposure, and (b) the test population is representative of the entire population. This section contains brief descriptions of (a) the aging techniques commonly being used, (b) resultant models of acceleration factors, and (c) the concerns and potential deficiencies associated with this subject.

19.3.4.1 Techniques

19.3.4.1.1 Microelectronics

Peck and Zierdt [98] led the way for adopting elevated humidity and temperature as the primary stress factors. The most common accelerated test used over the past 25 years is referred to as THB (temperature, humidity, and bias). In this test, parts are exposed for extended periods (>1000 h) at 85°C, 85% RH, with the conductor lines biased at, or sometimes above, the operating voltage [68,99]. As PEM device reliability improved during the 1980s due primarily to a reduction in residual chloride contamination and molding compound cleanliness, exposure times under THB conditions required to generate failures started approaching 10,000 h. In order to formulate predictive relationships, failure during accelerated testing must be observed. The extremely long times needed in conventional THB tests led to the development of tests with more severe conditions to reduce needed test time. A pressure-cooker test at 100% RH and temperatures above 100°C has been used for this purpose [99–102]. Its efficiency can be demonstrated by the observation that for one specific device, results equivalent to those obtained in 1000 h at 85°C, 85% RH could be produced in 20 h at 140°C, 100% RH [99]. However, artifacts associated with condensation and difficulties in applying voltage in saturated atmospheres complicated widespread adoption of this procedure. Because of attractive equipment cost and availability, many manufacturers still use this type of testing and the standard conditions are 121°C, 100% RH and no bias. More recently, the highly accelerated stress test (HAST) has become the standard. HAST tests typically use temperatures up to 150°C and humidity levels less than 100% [63,103,104]. Several other variations of the standard THB aging techniques have been developed that involve additions of internal and external contamination and/or cyclic application of the environmental parameters to better simulate actual use [28,105–109]. Failure analysis techniques to supplement the reliability information have also been reported [110].

Although most accelerated aging studies are performed on actual functional devices and the time to failure is directly measured, some studies have used test structures designed specifically to better address particular degradation modes. For corrosion under electrical bias, common configurations include two interdigitated combs, a triple track with three parallel but meandering lines, and combinations of tracks and bondpads/wirebonds [8,65,66,103,111,112]. Example of triple-track structures are shown in Figure 19.6 [8]. The two outer track conductors are normally biased positively and the center one is negative or grounded. The portion of the triple-track structure in Figure 19.6b is contained within the test device previously shown in Figure 19.5. To assess wirebond degradation, the design permits conventional four-point resistance measurements to be made. Such test devices can be tested bare or as packaged.

19.3.4.1.2 Macro Interconnects

Because connectors and contacts are usually exposed without protection to the environment during operation, testing typically has been performed in the context of atmospheric corrosion [113]. Various forms of accelerated tests for indoor office environments have been developed for this purpose [26,114–117]. For example, procedures have been developed at IBM (the Gl(T) test) [26,114,117], Battelle Institute (Flowing Mixed Gas Class II or FMG II) [23,115,116], and in Europe by the International Electrotechnical Commission (IEC Test 68-2-60) [118] for simulating an indoor office environment. These techniques vary slightly but involve exposure to an atmosphere containing a combination of dilute pollutants such

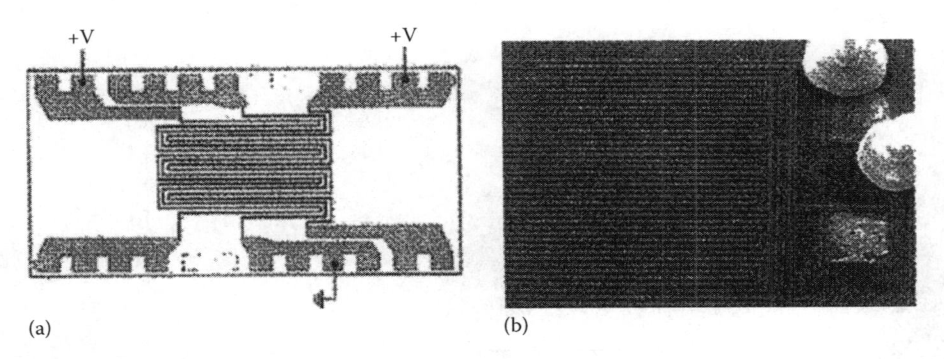

FIGURE 19.6
Schematic diagram (a) and an SEM photograph (b) of two triple-track test structures that are used to study electrolytic corrosion mechanisms along with the effectiveness of passivation layers. The structure shown on the right was encapsulated and then exposed to HAST conditions until failure. This particular structure is part of the integrated test device shown previously in Figure 19.5. These test devices contain eight triple-track sections (the left set with windows in the passivation layer) and exposed wirebond pads.

TABLE 19.1

Specifications for Three Standardized Atmospheric Corrosion Test Environments

ID	Cl_2 (ppb)	SO_2 (ppb)	NO_2 (ppb)	H_2S (ppb)	RH (%)	*T* CQ
GI(T)	3	350	610	40	70	30
FMG II	10	—	200	10	70	30
IEC	10	200	20	10	75	25

as H_2S, SO_2, NO_2, and Cl_2 and humidity (Table 19.1) and are usually performed at near-ambient temperatures because of complications arising from reactions between the various components at higher temperatures. Acceleration factors from 10 to 1000 are possible depending on the chosen conditions [115]. These aging environments are now used to test and qualify many electronic parts other than connectors and contacts, including printed circuit boards. The reliability of PCBs is also routinely tested using THB testing as discussed in the previous paragraph.

19.3.4.2 Mathematical Relationships for PEM Aging

When plastic-encapsulated microelectronic devices are exposed to aggressive environments, there is often a bimodal distribution in the time to fail resulting from latent manufacturing defects and long-term wearout [119]. These two causative mechanisms are very different directly correlate to the two primary reasons for performing accelerated aging noted in the first paragraph of this section. The elapsed time to fail and cumulative failure percentile are often plotted using a lognormal distribution (Figure 19.7). Two statistical measures result: the mean time to failure (MTTF, the time for 50% of the population to fail) and the population standard deviation (calculated from the slope) [8,119]. Because the goal of accelerated aging is to establish a relationship between the environmental parameters

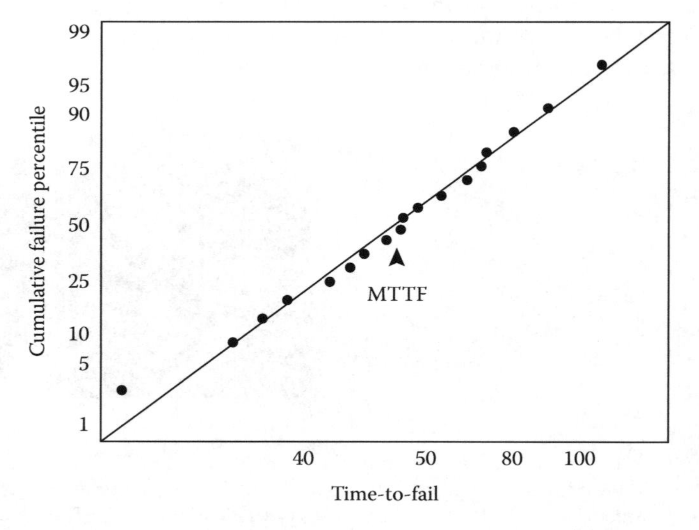

FIGURE 19.7
Example of a lognormal distribution of failure times and the definition of the mean-time-to-failure (MTTF) metric.

and service life under operating conditions, the MTTF metric can be used to calculate an important parameter known as the acceleration factor (AF). In terms of MTTF,

$$\mathrm{AF} = \frac{\mathrm{MTTF}_{\mathrm{field}}}{\mathrm{MTTF}_{\mathrm{test}}} \tag{19.1}$$

The acceleration factors are experimentally measured as a function of the environmental stresses using the techniques described in the previous subsection.

Using triple-track test devices, the surface conductance, G, can be directly measured. This parameter is a more fundamental indicator of the susceptibility to electrolytic degradation mechanisms compared with MTTF because, as noted earlier, electrolytic corrosion will occur only if the ohmic resistance between lines does not consume the applied voltage difference and leakage current flows [111,120]. G is calculated by multiplying the leakage current by the area of insulator and dividing by the applied voltage. The resistance change of the lines can also be used as a measure of degradation (either electrolytic or open circuit). The important assumption inherent in these types of tests is that the specific metric being measured (e.g., surface conductance or wirebond resistance) is directly related (proportional or inversely proportional) to failure rate and thus to MTTF [66,111,120].

The cumulative values for MTTF determined by any of these techniques for a range of conditions permit the formulation of general relationships that describe the dependence of failure rate on the stressing parameters. The effect of temperature, T, on reaction or failure rate, R, can usually be described by an Arrhenius-type relationship:

$$R = A\exp\left(-\frac{E_{\mathrm{a}}}{kT}\right) \tag{19.2}$$

where
- E_{a} is activation energy
- k is Boltzmann's constant
- A is an empirical constant

A range of activation energies for failure of ICs has been reported, from 0.4 to 1.2 eV. The acceleration factor associated with temperature becomes

$$\mathrm{AF}_{\mathrm{T}} = \exp\left[\frac{E_{\mathrm{a}}}{k\left(1/T_{\mathrm{field}} - 1/T_{\mathrm{test}}\right)}\right] \tag{19.3}$$

Because of the concern that higher voltage levels will increase the internal temperature (ohmic heating) and dry the device, most investigators use the normally applied operating voltage during testing. However, for the corrosion mechanisms that occur due to electrical bias, the corrosion rate is proportional to the current between the conductors and therefore scales with applied voltage. As such, the MTTF decreases as bias increases and an inverse relationship may exist [7,60,121,122]. For the limited cases in which a higher voltage level is used, Shirley [123] has proposed a general linear acceleration model where a and b are constants:

$$\mathrm{AF}_{\mathrm{v}} = \frac{\left(a + bV_{\mathrm{test}}\right)}{\left(a + bV_{\mathrm{field}}\right)} \tag{19.4}$$

TABLE 19.2

Empirically Derived Models Describing the Influence of Temperature (T) and Relative Humidity (RH) on Mean Time to Failure (MTTF)

Expression for MTTF	Constants	Reference
$A \exp [B(T + \mathrm{RH})]$	$B = -0.06$ to -0.09	[126]
$A(\mathrm{RH})^{n} \exp(E_a/kT)$	$E_a = 0.77$–$0.81\,\mathrm{eV}$	[99]
	$N = -2.5$ to -3	
	$E_a = 0.9\,\mathrm{eV}$	[127]
	$n = -3$	
	$E_a = 0.8\,\mathrm{eV}$	[123]
	$n = -4.64$	
$A \exp [E_a/kT + B(\mathrm{RH})^2]$	$E_a = -0.7$–$0.95\,\mathrm{eV}$	[112]
	$B = -0.0004$	
$A \exp (E/kT + B/\mathrm{RH})$	$E_a = 0.65\,\mathrm{eV}$	[68]
	$B = 304$	
$A \exp [B/kT + C(\mathrm{RH})/kT + D(\mathrm{RH})]$	$B = 1.0$–$1.1\,\mathrm{eV}$	[66]
	$C = -0.00444$ to -0.0077	
	$D = 0.076$–0.13	

Unfortunately, the influences of RH and impurities are most poorly established than the effects of temperature and bias. Many investigators have measured MTTF or G for actual devices or test structures in high-temperature and -humidity environments and developed empirical relationships by fitting the data to various equations. Several commonly referenced variants are presented in Table 19.2. Each expression in this table has a parameter A that represents a proportionality constant. These equations can be used to determine an acceleration factor by considering the ratio of field to test MTTF as was done in Equation 19.3. Using this technique, the proportionality constants cancel. Then by using Equation 19.1, the service life under operational conditions may be predicted after measuring lifetime under stress conditions. The first four expressions in Table 19.2 are "Eyring" models because the influences of T and RH are considered separately and then multiplied together to get a combined equation. The last expression does not assume that the effects of temperature and humidity are independent and has an activation energy that is dependent on RH [124]. These models have been used in a number of investigations to fit experimental data, and different values for the constants have been determined for various systems and degradation modes [96,98,99,102–105,119,121,125], An example of how such models are applied is shown in Figure 19.8, which is based on surface conductivity measurements that were made with an Al-Cu triple-track structure in a HAST chamber [103]. The temperature and humidity combinations were fit to various equations and those shown in Figure 19.8 are for the last expression in Table 19.2, which contains an RH-dependent activation energy. The correlation coefficient for this expression was found to be higher than those for other models [103].

As will be discussed in the next subsection, to be truly effective, predictive capabilities must be based on fundamental physical understanding. The models presented in Table 19.2 are simply empirical correlations. Limited progress has been made to date to improve this situation. Comizzoli [46] has provided a mechanistic explanation and associated mathematical expression for the exponential dependence of surface conductance (and MTTF) on relative humidity that is contained in some of the models listed above. Pecht and Ko [83] developed a comprehensive model for predicting absolute time to failure when microelectronic die metallization is corroded by an electrolytic process. Their phenomenological

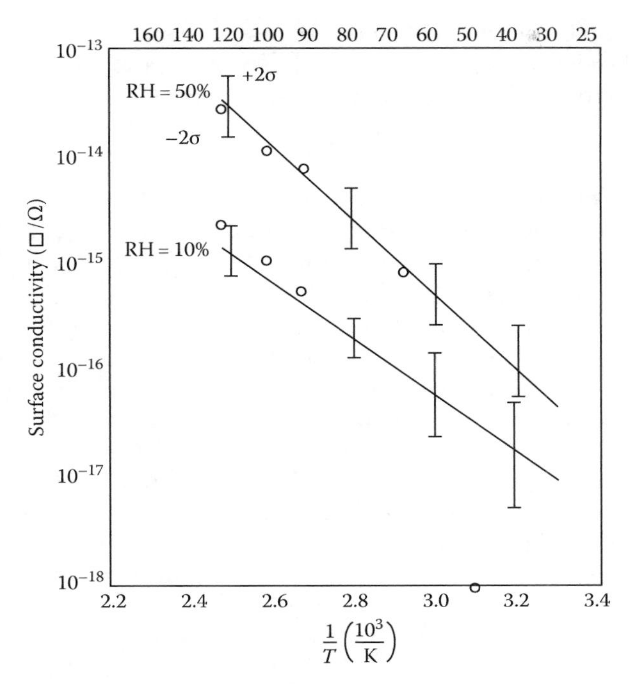

FIGURE 19.8
Surface conductivity as a function of temperature. RH is the relative humidity and the lines are fits of the last equation shown in Table 19.2. (Reproduced from Weick, W.W., *IEEE Trans. Reliab.*, R-29, 109, 1980. (© [1980, IEEE]).)

underpinning involves ion transfer based on Ohm's and Faraday's laws. The other key feature is their treatment of the critical process involving moisture ingress that permits this model to be useful for both PEM and CHP devices. A final reference that is relevant to this subject is the work being performed by Graedel and coworkers [128]. This team is developing and exercising a physically based atmospheric corrosion model referred to as GILDES that conceptually can include all microelectronic corrosion processes (e.g., gas transport, adsorbed water layer, corrosion product layer, electrochemistry).

19.3.4.3 Concerns and Limitations

In general, the reliability engineer cannot ignore failures observed during testing. The expense of accelerated testing and the often costly effort required to address the problems highlighted by such tests are rationalized by the assumption that failures have a finite probability of occurring during service if they are observed at all during testing. In other words, the accepted belief is that improved test yields correlate with longer service life. A device may actually be redesigned without any knowledge of whether the failure mechanism occurs under operating conditions at a rate that can be extrapolated from stress conditions or if it is just an artifact of the test. The prediction of device lifetime in the field by extrapolation of data obtained under high-stress conditions is extremely sensitive to the model chosen. The basic issue here goes back to the use of empirical equations—one cannot reliably extrapolate outside the range of testing conditions. Thus, the goal has to be to obtain data on the behavior in the mildest conditions possible [97]. The authors of this chapter do not know of a sole expert in this field who believes at this time that actual service life can be accurately predicted and, as such, in practice, accelerated aging is presently useful only for product qualification and identification of latent manufacturing defects and design deficiencies. The remainder of this subsection provides support for these statements.

19.3.4.3.1 Changing Failure Mechanisms as a Function of Environmental Stress

This classical issue is relevant to all applications of accelerated aging [97]. The complexity inherent in microelectronic devices results in a multitude of coupled processes that must be properly identified and accounted for in accelerated aging models. Every process that affects a failure mechanism could have a different sensitivity to the environmental parameters. Lall [129] has documented some of the problems associated with the common use of temperature as a stress agent for microelectronics. The wide variability in activation energy mentioned before is indicative of different or changing mechanisms. Three relevant examples further illustrate the issue of changing mechanisms. The first involves the formation of conductive anodic filaments (CAFs) on PCBs described earlier. Work performed subsequent to the discovery of CAF formation found that a threshold in the temperature/humidity phase space may exist below which CAF growth does not take place [130]. This threshold is apparently not encountered in typical use environments and thus the phenomenon of CAF may be only an artifact of accelerated testing. The second example is from an investigation of conduction in printed circuit boards by Takahashi [130]. He studied the various conduction paths in boards using an AC impedance technique during humidified exposures without bias and identified both ohmic and diffusion-controlled processes. He then made the point that because multiple conduction paths and mechanisms exist at a single temperature and humidity condition, it is possible for conduction processes determined in stress tests to have no bearing on failure mechanisms under field conditions. The final example concerns the observation that the glass-transition temperature of the encapsulating plastic in PEM devices cannot be exceeded in testing because a large difference in water permeability results. Nevertheless, sometimes this requirement is satisfied. A few studies have shown that data obtained from THB and HAST/pressure cooker testing can be fit to a single mathematical relationship, indicating that the failure mechanisms may be the same [68,99,104].

19.3.4.3.2 Evolving and Improving Technology

The continuing rapid evolution of microelectronic technology leads to several related difficulties. The first is that the dominant failure mechanisms change and thus long-term field failure information is not necessarily relevant to state-of-the art devices. Historically, during accelerated aging, two types of moisture-related phenomena have been observed: distributed Al track corrosion and Au wire/A1 bondpad interfacial degradation. The root cause of track corrosion was probably moisture penetration through defects in the protective passivation layer and the presence of contamination. Modern best commercial practice has effectively eliminated passivation defects and thus track corrosion as a significant failure mechanism. Now, the exposed wirebonds are the prime susceptibility. Lall [129] documents another important factor: field failure information is becoming more limited because the reliability of state-of-the art devices has now improved to the point that it no longer limits useful system lifetime. The final factor is that posttest analyses of aged devices have often been limited to simply confirming a failure, not determining root cause. Thus, the ability to correlate the results with true service life or identify actual failure mechanisms has not been developed to any significant extent.

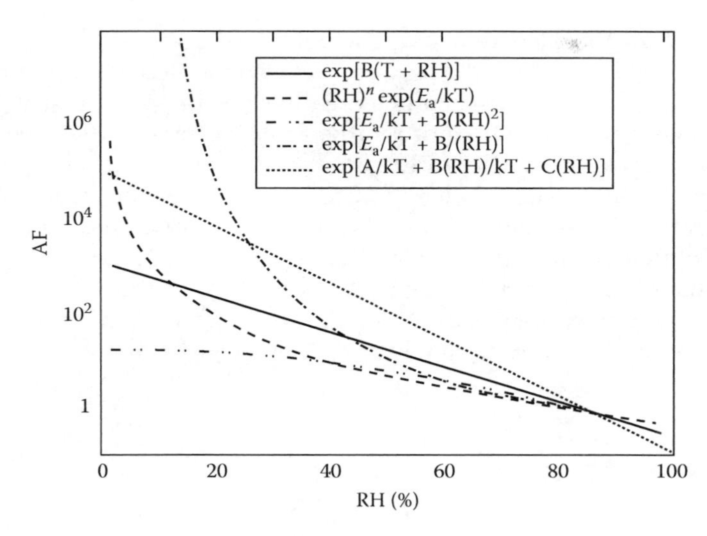

FIGURE 19.9
Humidity acceleration factor as a function of RH relative to 85% for the various models shown in Table 19.2.

19.3.4.3.3 Effect of Humidity and Device Type

A specific disconcerting aspect of the existing T/H models that is easy to identify is their wild divergence at low RH values. Figure 19.9 shows humidity acceleration factors relative to 85% RH calculated for each of the expressions in Table 19.2. In these calculations, temperature was normalized out by setting its value to 25°C. There is relatively good correspondence of the models in the high-RH range where the experimental data exist. However, at low RH, the models differ by many orders of magnitude. One possible explanation is that none of the mathematical relationships proposed have a phenomenological basis. For example, water adsorption on the metallization may be the key process and, if so, the rate response should be sigmoidal (which none of these equations simulate). Osenbach [17] describes another physical basis that involves the observation that surface leakage current is minimal below about 50% RH and has an exponential dependence above this level. Few experimenters examine the RH range less than 50% because of the long times needed to obtain failure. Unfortunately, this RH range is where most operating conditions exist, especially considering the local heating associated with power dissipation. Furthermore, there is no reason to expect different systems to exhibit similar temperature or, for this topic, different humidity relationships. For instance, silicone-encapsulated devices fail at much lower rates in a given environment than devices encapsulated in epoxy [60,96,112]. This occurs despite the faster transport of moisture in silicones and could result from low impurity content and improved adhesion to the die surface [75,112]. Silicone-encapsulated parts may therefore display a different humidity acceleration factor than unencapsulated or epoxy-encapsulated parts [66].

19.3.4.3.4 Effect of Contamination

As discussed previously, contamination at metallization surfaces is a critical corrosion factor. For example, the level of contamination affects the ionic strength in the adsorbed water layer that, in turn, influences surface conductance. Contamination also affects the critical relative humidity, which, for some salts, can be quite low. Finally, the type of contamination often determines the corrosion mechanism (e.g., whether electrolytic anodic or cathodic corrosion of A1 predominates). In many accelerated aging approaches, this dominant factor is uncontrolled and often uncharacterized. In fact, T/H testing may really be a measure of the cleanliness of the part prior to testing. This observation is an example of why Osenbach's second assumption noted earlier (testing of representative devices) is probably not, in general, valid. To address this problem, many investigators have either purposely contaminated the part prior to exposure or added controlled contamination to the stressing environment [42,65,74,106,107,122,131–133]. These approaches are useful for making comparative assessments of various structures. However, it is a difficult task to generate a model that predicts behavior under conditions of lower contamination levels. Pitting potentials of various metals have been found to decrease linearly with a logarithmic increase in chloride concentration [134]. Predicting part lifetime for a given surface contamination concentration from such information, however, remains quite challenging.

19.4 Corrosion of Magnetic Data-Storage Components

The critical metallic components of advanced magnetic and magneto-optic (MO) storage devices—thin-film metal disks, inductive or magnetoresistive heads, and MO layers—are all susceptible to corrosion and each has been a subject of considerable study. Several review articles covering corrosion of magnetic-storage media may be found in the literature [135,136].

19.4.1 Thin-Film Magnetic Disks

As described in the technology overview section, the carbon overcoat layer on thin-film disks typically did not in the past fully cover the underlying layers as a result of intentional roughening of the disk. The lack of coverage has two implications for the corrosion behavior. First, the Co-based magnetic layer and perhaps even the NiP substrate are exposed at small regions and can corrode. Furthermore, the overcoat layer, which is often sputter-deposited carbon, can be somewhat conductive and quite noble in comparison with the exposed areas of magnetic alloy. The unfavorable anode-to-cathode area ratio can therefore result in aggressive galvanic corrosion.

Figure 19.10 shows potentiodynamic polarization curves measured in DI water [135,137]. The plated Co-8%P material was considered for use as the magnetic alloy in thin-film disks when they were first developed. Like pure Co, it is not very corrosion resistant and does not readily passivate. The corrosion potential of two different sputter-deposited carbon thin films is seen to be about 60OmV higher than that of plated CoP. Although the nature of C thin films can change drastically as a function of deposition conditions, the two C films sustain reasonably large cathodic currents. In fact, galvanically induced corrosion of the small active areas of the magnetic alloy by the C overcoat is the primary mechanism of corrosion of thin-film disks. When an uncoated CoP film is exposed to an aggressive gaseous environment at 25°C containing 70% RH and 10 ppb Cl_2 gas, a 30-Å-thick uniform corrosion product forms [135,137]. However, a C-coated CoP sample forms a high density of localized corrosion product particles that are several μm in diameter. This dimension is 100 times larger than the head-disk separation in advanced disk files.

Corrosion can thus be minimized by decreasing the galvanic mismatch between the magnetic layer and the overcoat. As shown in Figure 19.10, a sputter-deposited Co-17%Cr film has a corrosion potential much closer to C and is spontaneously passive in the DI water droplet. In deaerated Na_2SO_4, the corrosion current was found to decrease and the

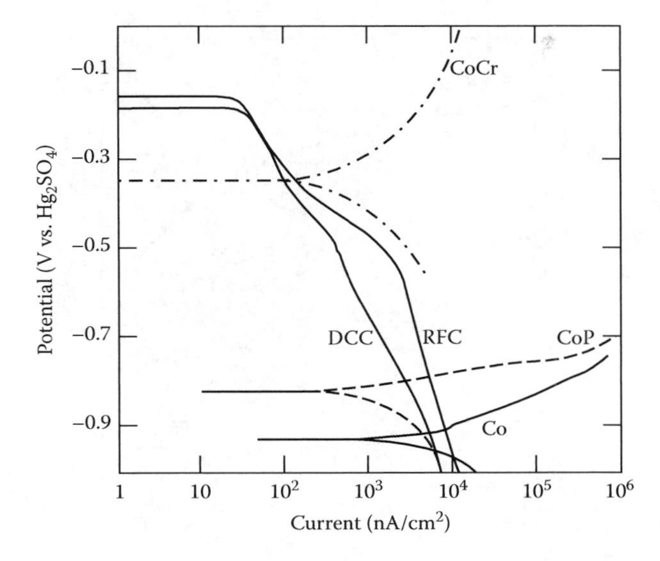

FIGURE 19.10
Potentiodynamic polarization curves of thin-film disk materials in a droplet of DI water. Parameter definitions: DCC DC sputtered C; RFC RF sputtered C; CoCr, Co + 17%Cr; CoP, CO + 8%P; Co, pure Co. (Reprinted from Brusic, V. et al., *Electrochemistry in Transition*, Murphy, O.J. et al. (eds.), Plenum, New York, 1992, p. 547. With permission.)

corrosion potential increased as the Cr content in CoCr alloys increased from 0% to 20% [136,138]. Nonconducting overcoats can also reduce corrosion but do not eliminate it if they are not totally covering, a requirement that is quite difficult given the thickness limitations of the overcoat and the roughness of most disks.

The trend in magnetic recording is, of course, to continually higher densities. As with microelectronic devices, higher density is achieved by shrinking the dimensions, including the separation of the magnetic medium and the sensor in the head, which has implications relative to corrosion. In order to bring the head and disk closer, the head must fly closer to the disk, and the carbon overcoat thickness must decrease. The decrease in fly height means that less corrosion product accumulated on the disk surface will cause detrimental interactions with the head. A decrease in carbon thickness is also potentially deleterious to corrosion resistance because it is harder to cover the magnetic layer with a very thin carbon overcoat. On the other hand, in order to fly closer, the disk roughness must also decrease. The trend toward smoother disks is beneficial to corrosion resistance because the magnetic layer can then be more easily covered by the carbon overcoat. Figure 19.11 shows the time trends for carbon overcoat thickness and disk roughness average (RA), as well as the ratio of the two. The average roughness of magnetic disks has decreased faster than the carbon overcoat thickness. Therefore, the ratio of COC thickness to disk roughness increased significantly during the 1990s, indicating that the COC should now provide a better covering despite being much thinner. Eventhough the COC is a few nm thick, the disk is almost atomically smooth, and the coverage is quite good. In essence, the disk structure now tends to be more like Figure 19.3a than b. As a result, disk corrosion should be less of a problem than in the past, and this trend will apparently continue.

Figure 19.12 presents polarization behavior in a droplet of water for more recent disk materials and structures than those shown in Figure 19.10 [139]. The disk had low roughness and a thin carbon overcoat. First note the two curves from full disk structures. One is from a lubricated disk, the other from an unlubed disk. The lube decreases the current

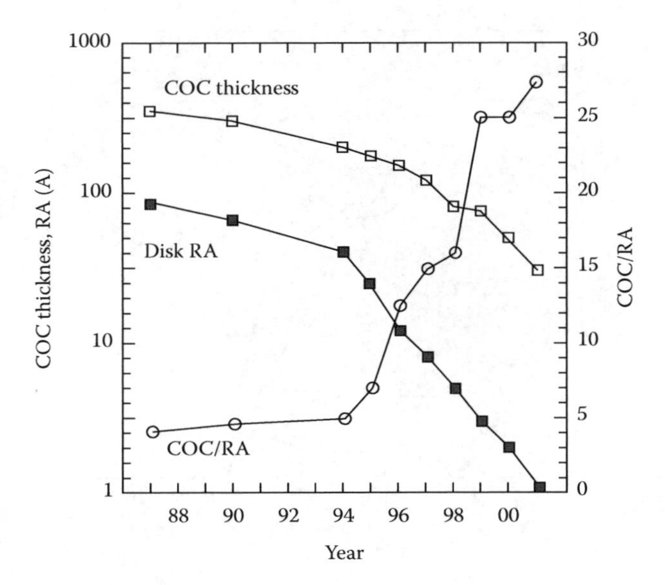

FIGURE 19.11
Trend of carbon overcoat (COC) thickness and disk texture or average roughness (RA) with time. Also shown, with right axis, is the ratio COC/RA. (Data from HMT Technology, Inc., Fremont, CA.)

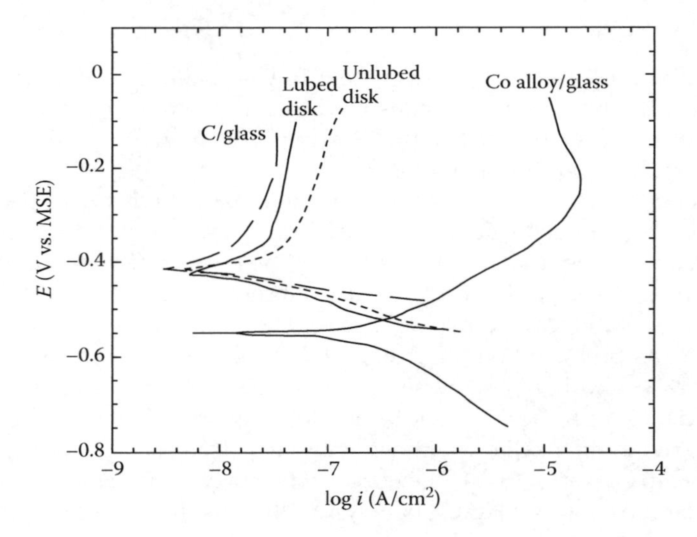

FIGURE 19.12
Polarization curves in a water droplet for a lubed and unlubed disk as well as a blanket magnetic layer and carbon layer on glass. (From Anoikin, E.V. et al., , *IEEE Trans. Magn.*, 54, 1717, 1988.)

only slightly, perhaps as a result of water displacement. The anodic current flowing at a given potential from the unlubed disk is about 100 times less than that of the blanket magnetic layer on glass. It is clear that the magnetic layer is rather well covered by carbon. In the past, it was assumed that the anodic current measured from a disk originated largely from spots not covered by COC. However, for the disks studied here, the anodic signal is not coming only from the small amount of uncovered magnetic layer because a blanket layer of C (unlubed) on glass exhibits an electrochemical signal that is a large fraction (about half as large) of the signal measured on the unlubed disk. Carbon is not really an extremely noble material. More accurately, the oxidation of C is kinetically hindered. Carbon can, in fact, oxidize to form various species, such as carbonyl, carboxyl, phenol, quinone, or lactone [140]. The polarization curve for the carbon film is similar to that of a spontaneously passive metal. It supports a reasonable amount of cathodic current, which is the cause of galvanic corrosion. At anodic potentials, there is some steady-state oxidation current, which is probably associated with a low rate of C oxidation. From the polarization curves, it could be suggested that the difference in current between the unlubed disk and the C blanket is the corrosion current of the magnetic layer. The finding that carbon oxidation is now a significant fraction of the total current flowing to a thin-film disk is reasonable given the trend of the increasing ratio of carbon thickness to disk roughness. However, despite the good coverage, very small areas of magnetic layer are still exposed in the disk structure. Therefore, corrosion will still occur if the disks are exposed to aggressive environments (such as chloride) or if problems in disk texture control occur.

An electrochemical study of the protectiveness of carbon overcoats as thin as 5 nm has been reported [141]. Polarization resistance of disk structures in 0.5 M sulfuric acid was measured. Disks coated with hydrogen-containing ion beam deposited carbon were found to have much lower corrosion rates than those coated with standard sputtered C of equivalent thickness. The composition of the sputtered carbons was varied by changing the sputtering gas. The corrosion rate depended on the type of carbon: ion beam C-H < sputtered C-H, < sputtered C-N< sputtered C. The different corrosion rates may be a result of differences in both coverage and catalytic activity of the carbon layers.

Novotny and colleagues [136,142] have described methods for estimating lifetimes of disks in drives using accelerated T/H tests. In the first approach, the extent of corrosion of a disk following T/H exposure was determined by a measurement of the surface Co concentration on the disk surface using Auger electron spectroscopy [142]. The surface Co concentration is a measure of the extent of corrosion. An unexposed disk surface generates no measurable Co signal despite small pores in the C overcoat that expose the Co-based alloy to the environment. Following T/H exposure, a Co-containing corrosion product forms on top of the overcoat as a result of attack of the exposed areas (Figure 19.3b). Using the data of Novotny et al., the Co surface concentration is found to vary with exp(RH/21) and $t^{1/2}$, with a thermal activation energy of 0.25 eV. In order to estimate disk lifetime with these data, one must assume that lifetime is inversely proportional to surface Co concentration. A later work describes an approach to lifetime estimation that uses a functional test, a head flying over a disk after T/H exposure [136]. If the head contains a magnetoresistive sensor, the number of asperities on the disk can be determined. This is a very relevant measurement because, as mentioned earlier, disk drives fail by head crashes long before they develop measurable increases in error rate from corrosion of magnetic material. In that work, the disk lifetime was found to vary with exp(−RH/13), with an activation energy of 0.27 eV. Another study determined activation energy by electrical resistance changes for sputter-deposited CoCr films exposed to T/H conditions [143]. Depending on composition and deposition conditions, values ranging from 0.08 to 0.3 eV were determined. The very low thermal activation energies reported in this study call into question the usefulness of performing stress tests at elevated temperatures for this system.

19.4.2 Thin-Film Inductive Head Materials

The structure of thin-film inductive recording heads results in the exposure to the environment of small areas of metal, often permalloy (Ni-20Fe), which is a relatively corrosion-resistant alloy. In fact, in-service corrosion of permalloy is not a problem and thin-film inductive heads have no history of corrosion-induced field failure. However, corrosion of permalloy can occur during manufacturing, leading to yield losses.

Exposure of freshly deposited permalloy films to the atmosphere or oxidation in pure O_2 up to 250°C results in surface segregation and preferential oxidation of Fe to Fe_2O_3 [144–146]. However, anodically formed films are enriched in Ni and consist of an inner layer of NiO and an outer layer of mixed nickel and iron hydroxides [147,148]. The more reactive element Fe oxidizes first and is enriched at the surface during atmospheric oxidation but dissolves into solution during anodic oxidation.

The method of deposition has been reported to have a large influence on the corrosion properties of permalloy [149–151]. In particular, plated permalloy films have been found to be more susceptible to corrosion than either bulk material or vacuum-deposited films of nominally the same composition. For instance, sputter-deposited permalloy was found to passivate in pH 2 solutions, but plated films did not [150]. Another study in neutral chloride solutions showed that plated permalloy had a lower pitting potential than bulk permalloy [151]. These reports suggested that plated films either have regions locally enriched in Fe or have a crystallographic orientation that is more susceptible to attack. However, little supporting evidence was provided.

An explanation for the enhanced corrosion susceptibility of plated permalloy films was offered in a study that examined the walls of pits in thin-film samples using Auger electron spectroscopy [149]. The pit walls showed a large enhancement of S compared with top surfaces at regions away from pits. C and N were also enriched at the pit walls. Analogous experiments with sputter-deposited permalloy films indicated that S was present in negligible amounts at both the pit wall and top surface, N was absent at both locations, and less C was present on the newer pit wall surface than on the top surface. The coincidental appearance of S, C, and N at pit walls in plated permalloy suggested that saccharin ($C_7H_5NO_3S$), a stress-relieving agent added to the plating bath, was responsible for the poor corrosion resistance. Apparently, a small amount of the surface-active molecule is incorporated in the alloy during plating, reduces corrosion resistance, and enriches at the pit surface during pit growth. Similar behavior has been described for the case of small S additions to Ni-25Fe single crystals [148].

As described before, current magnetic recording technology uses only inductive heads for writing. However, modern high-density disks with high magnetic coercivity require a write head fabricated from a material with a higher magnetic moment than permalloy. Many current production devices make use of a NiFe alloy having higher Fe than permalloy (45Ni-55Fe). The corrosion characteristics of high-Fe NiFe have not been reported, but it is certainly more susceptible to corrosion than permalloy given the higher Fe content. Co alloys have also been suggested for use as a high-moment soft magnetic material in write heads [152]. The corrosion behavior of electrodeposited CoNiFe alloys has been reported [153,154]. As with permalloy, sulfur incorporation into CoNiFe was shown to be detrimental to the corrosion behavior in 2.5% NaCl. The presence of 0.3 at % S resulted in a 40OmV lower pitting potential than for a similar alloy with <0.1 at % S [153]. An alloy deposited with thiourea as an additive instead of saccharin and an extremely high S content (3.8 at %) exhibited only active dissolution in 2.5% NaCl with no evidence of passivity [154].

A corrosion protection scheme developed for Co and Co-based alloys may be useful for reducing corrosion during processing [155]. The concept takes advantage of the remarkable corrosion resistance imparted to Cu by benzotriazole (BTA). A Cu-BTA film is generated on the surface of a Co sample when it is immersed in a solution containing cupric ions and BTA. Due to its relative nobility, Cu tends to exchange with Co and plate on the surface.

However, the two-step sequential reduction of cupric is stopped before metallic Cu is produced because cuprous ions react with the BTA to form a Cu(I)-BTA protective film. This film is robust in the sense that it provides protection even after the part is removed from the BTA-containing solution and is subsequently exposed to environments devoid of BTA. Such treatments have resulted in a 100-fold reduction of corrosion rate for Co samples in a DI water droplet.

19.4.3 Magnetoresistive (MR) Head Materials

A cross section of the nm-thick layers in MR heads is exposed during the lapping process. If all of the metals used in the device do not corrode slowly or spontaneously passivate in the potentially aggressive lapping environment, severe corrosion can result. In addition, given the variety of the metals used, their susceptibility to corrosion, and their intimate contact, galvanic corrosion can be a severe problem. Considerable work has been performed to address this liability. In particular, the corrosion behavior of the antiferromagnetic layer has been extensively characterized because it is usually the most susceptible layer in the structure. MnFe, a commonly used antiferromagnetic layer, has been shown to be extremely corrodible; for instance, it exhibits only active dissolution 0.1 N Na_2SO_4 [139]. In a buffered borate solution of higher pH, it forms a protective passive film at high potentials after a rather large active peak, and the corrosion rate, peak current density, and passive current density all decrease with increasing pH. These results are interesting because the possibility exists that environment control may be an effective corrosion mitigant in certain process steps (e.g., lapping) that might otherwise be aggressive.

Given the corrosion susceptibility of MnFe, the suitability of using several alternative antiferromagnetic layers has been investigated, including MnIr [139,156–158], MnRh [158], MnNi [157–160], MnNiCr [159], MnPd [160], MnPdPt [160], and NiO [161]. The noble metal substitutions for Fe make the potential of these new alloys much more positive, which decreases galvanic interactions with other metals in the structure such as Cu. The corrosion rates of MnIr and MnRh are apparently only slightly lower than that of MnFe in many environments [139,157,158]. The pitting potentials in chloride for MnNi, MnPd, and MnPdPt have been reported to be significantly higher than that of MnFe [160]. Replacement of the most susceptible material in the structure with an oxide certainly is desirable from the corrosion standpoint. For example, spin valves made with NiO antiferromagnetic layers were shown to exhibit remarkable corrosion resistance [161].

19.4.4 Magneto-Optic (MO) Alloys

The alloys used as MO recording media are extremely reactive because of their high rare-earth (e.g., Tb) content. Given the free energy of formation of terbium oxide (–446 kcal/mol), there exist few metal oxides that are thermodynamically stable in the presence of Tb metal, even at room temperature [162]. Therefore, attempts to protect Tb by using an oxide overlayer result in oxidation of the Tb and reduction of the overlayer. The oxides that form on Fe-26Tb in air at 200°C have been studied by AES [163,164]. Upon exposure to air at room temperature, a 20-Å outer layer of iron oxide covered a 60-Å layer of terbium oxide with approximate compositions of Fe_2O_3 and Tb_2O_3, respectively. Oxidation at 200°C resulted in the formation of an internally oxidized zone consisting of terbium oxide and metallic FeTb that was enriched in Fe. This zone grew initially with parabolic kinetics, and then the growth rate accelerated. At that point, the two outer oxide layers started to thicken.

Exposure of MO films to aqueous solutions or high-humidity conditions results in localized attack [162,165]. Pits can form even during brief immersion in DI water or long-term storage in office ambient conditions. Analysis of pits in Fe-23Tb-12Co by scanning Auger microscopy found depletion of Fe and Co from the pitted region [162]. This apparently reflects the low solubility of Tb_2O_3, which nonetheless does not provide corrosion protection. Such pits formed in aqueous solutions are circular in shape. In contrast, the localized corrosion formed under atmospheric conditions is wormlike, similar in nature to filiform corrosion.

Many different alloying elements, when added to FeTbCo in the amount of about 5%, have been found to improve the corrosion resistance [135,165]. Both improved passivity and reduced pitting susceptibility have been achieved in this fashion. The primary protection of MO disks, however, derives from protective overlayers. The MO layer is typically covered by a dielectric layer for optical and thermal reasons. Some designs employ a metallic layer such as Al as a reflector, which covers the dielectric layer. Sputter-deposited Al binary alloys, which display for superior pitting resistance compared with pure Al, have also been used. The metallic reflective layer in MO disk is the first application of these remarkably corrosion-resistant alloys that have generated significant attention [166–171].

19.4.5 Corrosion Mitigation of Magnetic-Storage Components

The preceding subsections describe the corrosion concerns and liabilities of magnetic-storage components and devices, as well as efforts taken to reduce corrosion. As a summary, care must be taken in materials selection, in particular to minimize galvanic corrosion. Heads and disks are covered with carbon layers in service, so it is important to consider issues such as electrochemical reactivity of the carbon layers (which will vary with composition and details of the deposition method) and coverage (which depends on the thickness of the carbon layer and the roughness of the head or disk). Certain processes, such as lapping and rinsing, are amenable to environmental control, and it is critical to understand and control the response of all exposed metals to process fluids. On the disk drive level, sealing is a critical issue. It is possible to seal drives using gaskets in an attempt to exclude the external environment, although rubber gaskets are permeable to water to a certain extent. Also, breather holes are often incorporated into drives to allow pressure equilibration, which is important, for instance, in laptops used in airplanes. One can envision how a portable device might be transported between a cool air-conditioned environment and a hot and humid one. The relative rates of heat and humidity transport will control whether deleterious water condensation occurs under such conditions.

19.5 Concluding Remarks

Significant progress has been made during the past quarter century in reducing environmentally induced damage in microelectronic and magnetic data-storage devices. These improvements have been made even though metallization features continue to become more susceptible to corrosion. Important advancements responsible for this progress have involved such aspects as improved manufacturing control, elimination of contamination sources, use of improved materials, and development of less vulnerable designs. One remaining significant shortcoming is the lack of a capability to predict reliability and service life using accelerated-aging data. This situation primarily exists because we still have not developed an adequate physical understanding of the relevant corrosion processes. Several factors have contributed to this "corrosion science" inadequacy including (a) the existence of numerous coupled and complex phenomena, (b) a rapidly evolving technology (moving target), and (c) the success with eliminating actual corrosion issues. Challenges to satisfy this need along with new emerging engineering-level problems will certainly continue well into the future as dimensions continue to shrink, more susceptible materials such as copper are used, and higher service life and reliability are demanded.

References

1. H. B. Bakoglu, *Circuits, Interconnections, and Packaging for VLSI*, Addison-Wesley, Reading, MA, 1990.
2. M. B. Small and D. J. Pearson, *IBM J. Res. Dev. 34*:858 (1990).
3. P. C. Andricacos, *Interface 8*(l):32(1999).
4. D. P. Seraphim, R. Lasky, and C. Li, *Principles of Electronic Packaging*, McGraw-Hill, New York, 1989.
5. R. R. Tummala and E. J. Rymaszewski, *Microelectronics Packaging Handbook*, Van Nostrand Reinhold, New York, 1989.
6. K. R. Kinsman, *Electronic Packaging and Corrosion in Microelectronics* (M. E. Nicholson, ed.), ASM International, Metals Park, OH, 1987, p. 1.
7. L. Gallace and M. Rosenfield, *RCA Rev. 45*:249 (1984).
8. D. R. Johnson, D. W. Palmer, D. W. Peterson, D. S. Shen, J. N. Sweet, J. T. Hanlon, and K. A. Peterson, *Microelectronics Plastic Molded Packaging*, Sandia National Laboratories, Albuquerque, NM, 1997.
9. M. G. Pecht, L. T. Nguyen, and E. B. Hakim, *Plastic-Encapsulated Microelectronics*, John Wiley & Sons, New York, 1995.
10. G. G. Harman, *Wire Bonding in Microelectronics: Materials, Processes, Reliability, and Yield*, McGraw-Hill, New York, 1997.
11. P. G. Slade, ed. *Electrical Contacts: Principles and Applications*, Marcel Dekker, New York, 1999.
12. R. Holm, *Electric Contacts*, Springer-Verlag, New York, 1967.
13. F. Chen and A. J. Osteraas, *Electronic Packaging and Corrosion in Microelectronics* (M. E. Nicholson, ed.), ASM International, Metals Park, OH, 1987, p. 175.
14. S. Ming, M. Pecht, and M. Natishan, *Microelectron. J. 30*:217 (1999).
15. C. D. Mee and E. D. Daniel, *Magnetic Recording*, Vol. I (Technology), McGraw-Hill, New York, 1987.
16. J. A. Brug, T. C. Anthony, and J. H. Nickel, *MRS Bull. 21*:23 (1996).
17. J. W. Osenbach, *Semiconductor Sci. Tech. 11*:155 (1996).
18. R. B. Comizzili, R. P. Frankenthal, and K. J. Hanson, *Mater. Sci. Eng. A A198*:153 (1995).
19. R. P. Frankenthal, *Solid State Electron. 33*:69 (1990).
20. J. D. Sinclair, *J. Electrochem. Soc. 135*:89C (1988).
21. S. C. Kolesar, *Ann. Proc. Reliab. Phys. 12*:155 (1974).
22. P. G. Slade, *Electrical Contacts: Principles and Applications* (P. G. Slade, ed.), Marcel Dekker, New York, 1999, p. 89.
23. W. H. Abbott, *Electrical Contacts: Principles and Applications* (P. G. Slade, ed.), Marcel Dekker, New York, 1999, p. 113.
24. C. Leygraf and T. Graedel, *Atmospheric Corrosion*, John Wiley & Sons, New York, 1999.
25. W. H. Ailor, ed. *Atmospheric Corrosion*, Wiley Interscience, New York, 1982.

26. P. B. P. Phipps and D. W. Rice, *Corrosion Chemistry* (J. R. Brubaker and P. B. P. Phipps, eds.), American Chemical Society, Washington, DC, 1979, p. 235.
27. S. P. Sharma, *J. Vac. Sci. Technol. 15*:1557 (1979).
28. T. Ajiki, M. Sugimoto, H. Higuchi, and S. Kumada, *IEEE Ann. Proc. Reliab. Phys. 17*:118(1979).
29. E. B. Hakim, J. Fink, S. M. Tarn, P. McCluskey, and M. Pecht, *Circuit World 23*:26 (1997).
30. V. Brusic, G. S. Frankel, C.-K. Hu, M. M. Plechaty, and B. M. Rush, *Corrosion 47*:35 (1991).
31. C. Hoge, *IEEE Trans. Comp. Hybrids Manuf. Tech. CHMT-13*:1098 (1990).
32. J. Maa, H. Gossenberger, and R. H. Paff, *J. Vac. Sci. Technol. B 8*:1052 (1990).
33. N. Parekh and J. Price, *J. Electrochem. Soc. 137*:2199 (1990).
34. S. Mayumi, Y. Hata, K. Hujiwara, and S. Ueda, *J. Electrochem. Soc. 137*:2534 (1990).
35. J. M. Eldridge, *Third Conference on Electronic Packaging, Materials, and Processing in Microelectronics* (M. E. Nicholson, ed.), ASM International, Metals Park, OH, 1987, p. 283.
36. W.-Y. Lee, J. M. Eldridge, and G. C. Schwartz, *J. Appl. Phys. 52*:2994 (1981).
37. P. A. Totta, *J. Vac. Sci. Technol. 13*:26 (1976).
38. S. Thomas and H. M. Berg, *IEEE Trans. Comp. Hybrids Manuf. Tech. CHMT-10*:252 (1987).
39. J. D. Sinclair, L. A. Psota-Kelty, C. J. Weschler, and H. C. Shields, *J. Electrochem. Soc. 137*:1200 (1990).
40. R. P. Frankenthal, R. Lobnig, D. J. Siconolfi, and J. D. Sinclair, *J. Electrochem. Soc. 140*:1902(1993).
41. W. H. J. Vernon, *Trans. Faraday Soc. 31*:1668 (1935).
42. J. E. Anderson, V. Markovac, and P. R. Troyk, *IEEE Trans. Comp. Hybrids Manuf. Tech. CHMT-11*:152 (1988).
43. H. Manko, *Solders and Soldering*, McGraw-Hill, New York, 1979.
44. J. Brous, *Electronic Packaging and Corrosion in Microelectronics* (M. E. Nicholson, ed.), ASM International, Metals Park, OH, 1987, p. 161.
45. J. J. Steppan, J. A. Roth, L. C. Hall, D. A. Jeannotte, and S. P. Carbone, *J. Electrochem. Soc. 134*:175 (1987).
46. R. B. Comizzoli, *Materials Developments in Microelectronic Packaging Conference Proceedings* (P. J. Singh, ed.), ASM International, Materials Park, OH, 1991, p. 311.
47. B. D. Yan, G. W. Warren, and P. Wynblatt, *Corrosion 43*:118 (1987).
48. S. L. Meilink, M. Zamanzadeh, G. W. Warren, and P. Wynblatt, *Corrosion 44*:644 (1988).
49. M. Zamanzadeh, Y. S. Liu, P. Wynblatt, and G. W. Warren, *Corrosion 45*:643 (1989).
50. M. Zamanzadeh, S. L. Meilink, G. W. Warren, P. Wynblatt, and B. D. Yan, *Corrosion 46*:665 (1990).
51. G. W. Warren, P. Wynblatt, and M. Zamanzadeh, *J. Electron. Mater. 18*:339 (1989).
52. A. J. Griffin, S. E. Hernandez, F. R. Brotzen, and C. F. Dunn, *J. Electrochem. Soc. 141*:807 (1994).
53. D. A. Jones, *Principles and Prevention of Corrosion*, Prentice-Hall, Englewood Cliffs, NJ, 1992.
54. R. Brownson, K. Butler, S. Cadena, and M. Detar, *Microelectronic Manufacturing Yield, Reliability, and Failure Analysis II*, International Society for Optical Engineering, Bellingham, WA, 1996, p. 95.
55. G. L. Schnable, R. B. Comizzoli, W. Kern, and L. K. White, *RCA Rev. 40*:416 (1979).
56. R. Padmanaghan, *IEEE Trans. Comp. Hybrids Manuf. Tech. CHMT-8*:435 (1985).
57. P. R. Engel, T. Corbett, and W. Baerg, *Elec. Comp. Conf. 33*:245 (1983).
58. R. B. Comizzoli, *RCA Rev. 37*:483 (1976).
59. W. M. Paulson and R. W. Kirk, *Ann. Proc. Reliab. Phys. 12*:172 (1974).
60. R. C. Olberg and J. L. Bozarth, *Microelectron. Reliab. 15*:601 (1976).
61. R. B. Comizzoli, *J. Electrochem. Soc. 1235*:386 (1976).
62. E. P. G. T. v. d. Ven and H. Koelmans, *J. Electrochem. Soc. 123*:143 (1976).
63. T. Wada, H. Higuchi, and T. Ajiki, *Ann. Proc. Reliab. Phys. 23*:159 (1985).
64. V. Bhide and J. M. Eldridge, *Ann. Proc. Reliab. Phys. 21*:44 (1983).
65. M. Iannuzzi, *IEEE Trans. Comp. Hybrids Manuf. Tech. 6*:191 (1983).
66. N. L. Sbar and R. P. Kozackiewicz, *IEEE Trans. Electron. Dev. ED-26*:56 (1979).
67. R. K. Ulrich, D. Yi, W. D. Brown, and S. S. Ang, *Corros. Sci. 33*:403 (1992).
68. N. Lycoudes, *Solid State Tech.* October:53 (1978).

69. C. G. Shirley and M. S. DeGuzman, *IEEE/Ann. Proc. Reliab. Phys. 31*:217 (1993).
70. V. Koeninger, H. H. Uchida, and E. Fromm, *IEEE Trans. Components Packaging Manufac. Tech. Part A 18*:835 (1995).
71. J. R. Scully, R. P. Frankenthal, K. J. Hanson, D. J. Siconolfi, and J. D. Sinclair, *J. Electrochem. Soc. 137*:1365 (1990).
72. J. R. Scully, D. E. Peebles, J. A. D. Romig, D. R. Frear, and C. R. Hills, *Metall. Trans. 23A*:2641 (1992).
73. R. P. Frankenthal and W. H. Becker, *J. Electrochem. Soc. 126*:1718 (1979).
74. A. D. Marderosian, *IEEE 15th Ann. Proc. Reliab. Phys. 15*:92 (1997).
75. L. T. Nguyen and C. A. Kovac, *SAMPLE Electronics Materials and Processes Conference*, Los Angeles, CA, 1987.
76. L. T. Nguyen, *SPE 46th ANTEC*, 1988.
77. J. J. Lietkus, L. T. Nguyen, and S. L. Buchwalter, *SPE 46th ANTEC*, 1988, p. 462.
78. T. Wada, M. Sugimoto, and T. Ajiki, *J. Electrochem. Soc. 136*:732 (1989).
79. J. S. Osenbach and J. L. Zell, *IEEE Trans. Comp. Hybrids Man Tech. 16*:350 (1993).
80. R. W. Thomas, *Proc. Elec. Comp. Conf. 26*:272 (1976).
81. W. E. Swartz, J. H. Linn, J. M. Ammons, M. Kovac, and K. Wilson, *Ann. Proc. Reliab. Phys. 21*:52 (1983).
82. M. Pecht, *IEEE Trans. Comp. Hybrids Manuf. Tech. 13*:383 (1990).
83. M. Pecht and W. C. Ko, *Int. J. Hybrid Microelectron. 13*:41 (1990).
84. R. K. Lowry, C. J. V. Leeuwen, B. L. Kennimer, and L. A. Miller, *Ann. Proc. Reliab. Symp. 16*:207 (1978).
85. A. H. Rawics, *Microelectron. Reliab. 34*:875 (1994).
86. V. Brusic, D. D. DiMilia, and R. MacInnes, *Corrosion 47*:509 (1991).
87. R. P. Frankenthal and D. J. Siconolfi, *J. Vac. Sci. Technol. 17*:1315 (1980).
88. R. P. Frankenthal and D. J. Siconolfi, *Corros. Sci. 21*:479 (1981).
89. K. J. Puttlitz, *IEEE Trans. Comp. Hybrids Manuf. Tech. CHMT-13*:188 (1990).
90. H. L. Yeh and H. Dalai, *Reliability of Semiconductor Devices and Interconnection and Multilevel, Metallization, Interconnection and Contact Technologies* (H. S. Rathore, G. C. Schwartz, and R. Susko, eds.), The Electrochemical Society, Pennington, NJ, 1989, p. 81.
91. J. R. White, D. D. Coolbaugh, and M. A. Chopra, *Corrosion of Electronic Materials and Devices* (J. D. Sinclair, ed.), The Electrochemical Society, Pennington, NJ, 1991, p. 319.
92. R. B. Comizzoli, J. P. Franey, G. W. Kammlott, A. E. Miller, A. J. Muller, G. A. Peins, L. A. Psota-Kelty, J. D. Sinclair, and R. C. Wetzel, *J. Electrochem. Soc. 139*:2058 (1992).
93. D. J. Lando, J. P. Mitchell, and T. L. Welsher, *Ann. Proc. Reliab. Phys. 17*:51 (1979).
94. V. Tierney, *J. Electrochem. Soc. 128*:1321 (1981).
95. D. W. Noon, *Third Conference on Electronic Packaging, Materials, and Processing in Microelectronics* (M. E. Nicholson, ed.), ASM International, Metals Park, OH, 1987, p. 49.
96. K. M. Striny and A. W. Schelling, *Elec. Comp Conf. 313*:238 (1981).
97. W. Nelson, *Accelerated Testing—Statistical Models, Test Plans, and Data Analysis*, John Wiley & Sons, New York, 1990.
98. D. S. Peck and J. C. H. Zierdt, *Ann. Proc. Reliab. Phys. 11*:146 (1973).
99. D. S. Peck, *IEEE/Ann. Proc. Reliab. Phys. 24*:44 (1986).
100. P. W. Peterson, *IEEE Trans. Comp. Hybrids Manuf. Tech. CHMT-2*:422 (1979).
101. R. P. Merrett, J. P. Bryant, and R. Studd, *Ann. Proc. Reliab. Phys. 21*:73 (1983).
102. C. F. Dunn and J. W. McPherson, *J. Electrochem. Soc. 135*:661 (1988).
103. W. W. Weick, *IEEE Trans. Reliab. R-29*:109 (1980).
104. J. E. Gunn, R. E. Camenga, and S. K. Malik, *Ann. Proc. Reliab. Phys. 21*:66 (1983).
105. J. W. Osenbach and J. L. Zell, *Reliability of Semiconductor Devices and Interconnection and Multilevel, Metallization, Interconnection and Contact Technologies* (H. S. Rathore, G. C. Schwartz, and R. Susko, eds.), The Electrochemical Society, Pennington, NJ, 1989, p. 53.
106. D. R. Sparks, *Thin Solid Films 235*:108 (1993).
107. T. Yoshida and T. Takahashi, *IEEE Proc. IRPS* 268 (1982).
108. C. G. Shirley and C. E. C. Hong, *IEEE/Ann. Proc. Reliab. Phys. 29*:12 (1991).

109. K. D. Cluff and D. B. Barket, *J. Inst. Environ. Sci. Tech. Jul/Aug*:36 (1998).
110. L. C. Wagner, *Electronic Packaging and Corrosion in Microelectronics* (M. E. Nicholson, ed.), ASM International, Metals Park, OH; 1987, p. 275.
111. H. Koelmans, *Ann. Proc. Reliab. Phys. 12*:168 (1974).
112. S. P. Sim and R. W. Lawson, *IEEE 17th Ann. Proc. Reliab. Phys. 17*:103 (1979).
113. R. P. Frankenthal, *Properties of Electrodeposits: Their Measurement and Significance* (R. Sard, J. H. Leidheiser, and F. Ogburn, eds.), The Electrochemical Society, Pennington, NJ, 1975, p. 142.
114. D. W. Rice, P. B. P. Phipps, and R. Tremoureux, *J. Electrochem. Soc. 126*:1459 (1979).
115. W. H. Abbott, *IEEE Trans. Comp. Hybrids Manuf. Tech. CHMT-11*:22 (1988).
116. W. H. Abbott, *IEEE Trans. Comp. Hybrids Manuf. Tech. CHMT-13*:40 (1990).
117. R. R. Gore, R. Witska, J. R. Kirby, and J. L. Chao, *IEEE Trans. Comp. Hybrids Manuf. Tech. CHMT-13*:27 (1990).
118. T. F. Reichert and K.-F. Ziegahn, *Corrosion and Reliability of Electronic Materials and Devices* (R. B. Comizzoli and J. D. Sinclair, eds.), The Electrochemical Society, Pennington, NJ, 1993, p. 364.
119. R. T. Howard, *IEEE Trans. Comp. Hybrids Manuf. Tech. CHMT-5*:454 (1982).
120. R. T. Howard, *IEEE Trans. Comp. Hybrids Manuf. Tech. CHMT-4*:520 (1981).
121. I. Lerner and J. M. Eldridge, *J. Electrochem. Soc. 129*:2270 (1982).
122. H. M. Berg and W. M. Paulson, *Microelectron. Reliab. 20*:247 (1990).
123. C. G. Shirley, *IEEE/Ann. Proc. Reliab. Phys. 32*:72 (1994).
124. J. W. McPherson, *Ann. Proc. Reliab. Phys. 24*:12 (1986).
125. D. Stroehle, *IEEE Trans. Comp. Hybrids Manuf. Tech. CHMT-6*:537 (1983).
126. B. Reich and E. B. Hakim, *Solid State Tech.* September:65 (1972).
127. O. Hallberg and D. S. Peck, *Qual. Reliab. Eng. Int. 7*:169 (1991).
128. T. E. Graedel, *Corros. Sci. 38*:2153 (1996).
129. P. Lall, *IEEE Trans. Reliab. 45*:3 (1996).
130. K. M. Takahashi, *J. Electrochem. Soc. 138*:1587 (1991).
131. N. L. Sbar, *Proc. Elec. Comp. Conf. 26*:277 (1976).
132. W. M. Paulson and R. P. Lorigan, *IEEE Ann. Proc. Reliab. Phys. 14*:42 (1976).
133. S. K. Fan and J. W. McPherson, *Ann. Proc. Reliab. Phys. 26*:50 (1988).
134. Z. Szklarska-Smialowska, *Pitting Corrosion of Metals*, NACE, Houston, TX, 1986.
135. V. Brusic, J. Horkans, and D. J. Barclay, *Electrochemistry in Transition* (O. J. Murphy, S. Srinivasan, and B. E. Conway, eds.), Plenum, New York, 1992, p. 547.
136. V. J. Novotny, *Adv. Info. Storage Syst. 4*:255 (1992).
137. V. Brusic, M. Russak, R. Schad, G. Frankel, A. Selius, D. DiMilia, and D. Edmonson, *J. Electrochem. Soc. 136*:42 (1989).
138. V. J. Novotny and N. Staud, *J. Electrochem. Soc.135*:2931 (1988).
139. G. S. Frankel, *Electrochemical Synthesis and Modification of Materials* (S. G. Corcoran et al., eds.), Materials Research Society, Warrendale, PA, 1997, p. 541.
140. K. Kinoshita, *Carbon, Electrochemical and Physiochemical Properties*, John Wiley & Sons, New York, 1988.
141. E. V. Anoikin, G. S. Ng, M. M. Yang, J. L. Chao, and J. R. Elings, *IEEE Trans. Magn. 54*:1717(1988).
142. V. Novotny, G. Itnyre, A. Homola, and L. Franco, *IEEE Trans. Magn. MAG-23:3645* (1987).
143. K. Tagami and H. Hayashida, *IEEE Trans. Magn. MAG-23*:3648 (1987).
144. R. A. Pollak and C. H. Bajorek, *J. Appl. Phys. 46*:1382 (1975).
145. W.-Y. Lee and J. Eldridge, *J. Electrochem. Soc. 124*:1747 (1977).
146. W.-Y. Lee, G. Scherer, and C. R. Guarnieri, *J. Electrochem. Soc. 1262*:1533 (1979).
147. G. Dagan, W. -M. Shen, and M. Tomkiewicz, *J. Electrochem. Soc. 139*:1855 (1992).
148. P. Marcus, A. Tessier, and J. Oudar, *Corros. Sci. 4*:259 (1984).
149. G. S. Frankel, V. Brusic, R. G. Schad, and J. -W. Chang, *Corros. Sci. 35*:63 (1993).
150. J. G. Bornstein, C. H. Lee, L. A. Capuano, and D. A. Stevenson, *J. Appl. Phys. 65*:2090 (1989).
151. C. H. Lee, D. A. Stevenson, L. C. Lee, R. D. Bunch, R. G. Walmsley, M. D. Juanitas, E. Murdock, and J. E. Opfer, *Corrosion of Electronic and Magnetic Materials* (P. Peterson, ed.), AIME, Philadelphia, PA, 1991, p. 102.

152. J.-W. Chang, P. C. Andricacos, B. Petek, and L. T. Romankiw, *Magnetic Materials, Processes, and Devices* (L. T. Romankiw and D. A. Herman eds.), The Electrochemical Society, Pennington, NJ, 1992, p. 275.
153. T. Osaka, M. Takai, K. Hayashi, K. Ohashi, M. Saito, and K. Yamada, *Nature 392*:796 (1998).
154. T. Osaka, M. Takai, Y. Sogawa, T. Momma, K. Ohashi, M. Saito, and K. Yamada, *J. Electrochem. Soc. 146*:2092 (1999).
155. V. Brusic, G. S. Frankel, A. G. Schrott, T. A. Petersen, and B. M. Rush, *J. Electrochem. Soc. 140*:2507 (1993).
156. H. N. Fuke, K. Saito, Y. Kamiguchi, H. Iwasaki, and M. Sahashi, *J. Appl. Phys. 81*:4004 (1997).
157. A. J. Devasahayam, P. J. Sides, and M. H. Kryder, *J. Appl. Phys. 83*:7261 (1998).
158. A. Veloso, P. P. Freitas, N. J. Oliveira, J. Fernandes, and M. Ferreira, *IEEE Trans. Magn. 34*:2343 (1998).
159. T. Lin, D. Mauri, N. Staud, C. Hwang, and J. K. Howard, *Appl. Phys. Lett. 65*:1183 (1994).
160. H. Kishi, Y. Kitade, Y. Miyake, A. Tanaka, and K. Kobayashi, *IEEE Trans. Magn. 32*:3380 (1996).
161. S. L. Burkett, S. Kora, J. L. Bresowar, J. C. Lusth, B. H. Pirkle, and M. R. Parker, *J. Appl. Phys. 81*:4912 (1997).
162. M. M. Farrow and E. E. Marinero, *J. Electrochem. Soc. 137*:808 (1990).
163. R. B. v. Dover, E. M. Gyorgy, R. P. Frankenthal, M. Hong, and D. J. Siconolfi, *J. Appl. Phys.59*:1291 (1986).
164. R. P. Frankenthal, D. J. Siconolfi, R. B. v. Dover, and S. Nakahara, *J. Electrochem. Soc. 134*:235 (1987).
165. G. Kirino, N. Ogihara, and N. Ohta, *J. Electrochem. Soc. 138*:2259 (1991).
166. G. S. Frankel, M. A. Russak, C. V. Jahnes, M. Mirzamaani, and V. A. Brusic, *J. Electrochem. Soc.136*:1243 (1989).
167. W. C. Moshier, G. D. Davis, J. S. Ahearn, and H. F. Hough, *J. Electrochem. Soc. 133*:1063 (1986).
168. W. C. Moshier, G. D. Davis, J. S. Ahearn, and H. F. Hough, *J. Electrochem. Soc. 134*:2677 (1987).
169. W. C. Moshier, G. D. Davis, and G. O. Cote, *J. Electrochem. Soc. 136*:356 (1989).
170. G. D. Davis, W. C. Moshier, T. L. Fritz, and G. O. Cote, *J. Electrochem. Soc. 137*:422 (1990).
171. B. A. Shaw, T. L. Fritz, G. D. Davis, and W. C. Moshier, *J. Electrochem. Soc. 137*:1317 (1990).

20

Organic Coatings

J.H.W. de Wit
Delft University of Technology

D.H. van der Weijde
Corus Research and Development

G. Ferrari
TNO Industries

CONTENTS

20.1 General Introduction 864
20.2 The Composition of Coating Systems 865
 20.2.1 Binder 866
 20.2.2 Pigments 866
 20.2.3 Fillers 867
 20.2.4 Additives 867
 20.2.5 Solvents 867
20.3 Complexcoating Systems 868
 20.3.1 Pretreatment: Conversion Layers 868
 20.3.2 Metallic Layers, Conversion Layers, and Organic Topcoats 870
20.4 Application Techniques 873
 20.4.1 Brushing 873
 20.4.2 Rolling 873
 20.4.3 Dip Coating 873
 20.4.4 Conventional (or Air) Spraying 873
 20.4.5 Airless Spraying 874
 20.4.6 Electrostatic Spraying 874
 20.4.7 Powder Coating 875
 20.4.8 Electrodeposition 875
20.5 Protective Mechanisms 875
 20.5.1 Water Permeation 876
20.6 Corrosion Underneath Organic Coatings 877
 20.6.1 Clustering of Water 878
 20.6.2 Cathodic Delamination and Anodic Undermining 878
  20.6.2.1 The Model as Presented in the Literature 880
  20.6.2.2 Extended Model for Cathodic Delamination 882
  20.6.2.3 Anodic Undermining 884
 20.6.3 Filiform Corrosion 885

20.6.4 Flash and Early Rust....885
20.6.4.1 Early Rusting....885
20.6.4.2 Flash Rusting....885
20.7 Specialties....886
20.7.1 Coatings for Marine Applications....886
20.7.2 Coatings for Fouling Prevention....886
20.7.2.1 Organotin-Based Antifoulings....887
20.7.2.2 Nonstick Coatings....887
20.8 Measuring and Monitoring Methods....888
20.8.1 Introduction....888
20.8.2 Electrochemical Impedance Spectroscopy....889
20.8.2.1 Atmospheric Water Uptake....891
20.8.2.2 Impedance Measurements and Delamination....893
20.8.2.3 Impedance Measurements and Artificial Defects, Laser Ablation....896
20.8.3 Local Electrochemical Test Methods....898
20.8.4 Other Electrochemical Methods....899
20.8.5 The Value of Enhanced Weathering Tests....900
20.8.6 Adhesion Tests....901
References....902

20.1 General Introduction

Construction metals are expected to have excellent mechanical properties and machinability at a low price, while at the same time they should be corrosion resistant. These properties can seldom be met in one and the same material. Separating the base metal with good mechanical properties from the corrosive surrounding by applying a surface layer can solve this problem. These layers can be:

1. Metallic layers
2. Nonmetallic inorganic layers such as conversion layers, anodized layers, some ceramic chemical vapor deposition (CVD) and physical vapor deposition (PVD) layers, and enamel layers
3. Organic layers such as paints, lacquers, and polymer sheets

In this chapter we will discuss organic layers. The use of organic coatings for the protection of metals against corrosion is widespread in building, construction, food packaging, automotive, and marine applications. For these applications, typical and mostly successful types are developed. The choice of coating also depends on the type of substrate—steel, galvanized steel or aluminum—combined with specific surface treatments. Evaluation of coating quality on site on a ship, a building, a car, or structural steelworks normally takes place when visible deterioration or defects occur. Moreover, apart from cleaning schedules, coating maintenance is usually a matter of planned repainting or unexpected repair when the coating fails. The development of coatings is mainly based on years of experience and accelerated testing in salt fog and humidity chambers, Q-UV, Weather-O-meters, Kesternich, etc. The conditions in these tests are based on the physical or chemical load in practice that is assumed to be the most important, and moreover the load is increased in order to speed up coating aging or corrosion. In these tests that are also used for screening of existing coatings, a reference sample with well-known properties is included to allow estimation of the quality. Outdoor exposures have to be used for a final verification of the quality, often in parallel with the actual use of the newly developed coating product. Cyclic weathering tests have been developed. Examples are the Hoogovens Cyclic Test (HCT) [1], the Hoogovens Atmospheric Corrosion Test (ABC) and the Volvo Indoor Corrosion test. The conditions in these tests are much more representative of the actual situation of practical use of the coatings. Still results are to be validated with actual exposition under practical conditions. When evaluating the coating systems, the total system should be taken into account.

First of all, in order to achieve good protective action of the applied layers, it is important to obtain excellent adhesion of the coating to the base metal. For this reason the substrate surface must be cleaned very well before further surface treatment and application of the surface layer.

Cleaning is often performed in two stages. First, the organic impurities such as oil, grease, and paint are removed from the surface. Then solid inorganic material such as rust, mill scales and other corrosion products can be removed. The organic impurities can be removed in various ways:

- With organic solvents
- With strongly alkaline solution
- With emulsion baths
- By steam cleaning

The solid inorganic material can be removed by:

- Mechanical treatment including brushing, grinding, polishing, and sandblasting and shot peening with various kinds of shot, such as metallic particles, corundum, and glass beads
- Heat treatment with flames or induction heating followed by fast cooling to obtain scaling
- Chemical picking with strong acids

Pickling in acid will probably be used less intensively in the coming years because of environmental problems. In many cases, optimized mechanical treatment (e.g., micropeening, an optimized form of shot peening with glass beads) will then be the substitute [2].

In most cases, after cleaning a further surface pretreatment, which may involve metallic and/or nonmetallic layers, is necessary in order to obtain good adhesion and thus protective properties. Also, the quality of the application procedures determines the final protective properties [3–18]. Of course, the structure of the polymeric network, the chemical composition, flaws, and the ease with which the coating is damaged on mechanical impact are also relevant for the final effectiveness of the protection. In fact, we have to take into account the properties of the whole system.

20.2 The Composition of Coating Systems

Organic coatings consist of four basic constituents: binder, pigments and fillers, additives, and solvents. Therefore the properties depend strongly on the actual recipe and the procedures and precautions taken during application of the layer.

Organic coatings may protect metal structures against a specific otherwise corrosive environment in a relatively economical way. The efficiency with which protection is provided is determined by a number of properties of the total coated system, which consists of the paint film(s), the metal substrate, and its pretreatment.

The composition of the paint film itself has a major influence on the corrosion protection provided by the coatings. Therefore the major constituents of organic coating systems are given here in a very concise form. More elaborate information can be found in the literature [12,13].

20.2.1 Binder

The binder forms the matrix of the coating, the continuous polymeric phase in which all other components may be incorporated. Its density and composition largely determine the permeability, chemical resistance, and ultraviolet (UV) resistance of the coating.

A continuous film is formed out of a recently applied liquid coating through physical curing, chemical curing, or a combination of these. An example of the physical curing process is the sintering of thermoplastic powder coatings. Before application, this type of paint consists of a large number of small binder particles. These particles are deposited on a metal surface using special application techniques. Subsequently, the paint is baked in an oven to form a continuous film by sintering.

Chemical curing involves film formation through chemical reactions. This may be oxidative curing, where oxygen from the atmosphere reacts with the binder monomers, thus causing polymerization. Another example is reactive curing, where a polymer network is formed through polycondensation or polyaddition reactions. This may be the case with multicomponent coatings, where the binder reacts with cross-linkers.

Most often, both physical and chemical curing take place, as is the case with the film formation of thermosetting powders. At elevated temperatures, physical sintering of the particles takes place, followed by chemical reactions between different components in the powder. Another example is film formation of solvent-based reactive coatings, such as the common house paints. In this case, the solvent (physically) evaporates from the curing film, causing the binder molecules to coalesce and start chemical polymerization reactions.

20.2.2 Pigments

Pigments may be added to a coating for two reasons: (a) They can provide color to the coating system, which is an important issue for aesthetic reasons. (b) Pigments may be used to improve the corrosion protection properties of a coating. This improvement may be obtained, for example, by incorporation of flake-shaped pigments parallel to the substrate surface. If a sufficiently large pigment volume concentration (PVC) is used, the flakes will hinder the permeation of corrosive species into the coating by elongating their diffusion pathways, resistance inhibition.

Furthermore, so-called anticorrosion pigments may provide active protection against corrosion phenomena. These pigments may dissolve slowly in the coating and may provide protection by covering corrosion-sensitive sites under the coating; by sacrificially corroding themselves, thus protecting the substrate metal; or by passivating the surface.

Blocking pigments may adsorb at the active metal surface. They reduce the active area for corrosion and form a transport barrier for ionic species to and from the substrate. An example of this type is a group of alkaline pigments (lead carbonate, lead sulfate, zinc oxide) that may form soaps through interaction with organic oils.

Galvanic pigments are metal particles that are non-noble relative to the substrata. On exposure, these particles (zinc dust on steel) corrode preferentially, while at the original metal surface only the cathodic reaction occurs.

Passivating pigments reconstruct and stabilize the oxide film on the exposed metal substrate. Commonly, chromates with limited water solubility are used as passivating pigments (zinc chromate, strontium chromate). In aqueous solutions they may cause anodic passivation of a metal surface with a very stable chromium- and oxygen-containing passive layer.

20.2.3 Fillers

The main function of fillers in organic coatings is to increase the volume of the coating through the incorporation of low-cost materials (e.g., chalk, wood dust). They may also be used to improve coating properties such as impact and abrasion resistance and water permeability.

20.2.4 Additives

The term additives refers to a large group of components with very specific properties, which typically are added to a paint in very small quantities. The functions of the different additives may be quite diverse. Examples are thickeners, antifungal agents, dispersing agents, antifoam agents, anticoalescence agents, UV absorbers, and fire-retarding agents.

20.2.5 Solvents

The solvent has the function of reducing the viscosity of the binder and other components and enabling their homogeneous mixing. Furthermore, the reduced viscosity makes it possible to apply the coating in a thin, smooth, continuous film on a specific surface. The roles of the solvent in a coating prior to and after application are quite contradictory. In the liquid state, prior to application, paint should form a solution or a stable dispersion or emulsion of binder, pigments, and additives in the solvent. In other words, all solid components should remain more or less homogeneously distributed in the liquid phase. This requires high compatibility between solvent and components and the presence of repulsive forces between components to avoid clustering. In contrast, after the paint has been applied, a major attractive force between the components is necessary for the formation of a continuous film. The interaction with the solvent should decrease to enable the solvent to evaporate from the curing film. In order to achieve optimum storage and application properties, a correct choice of additives is vital. Correct material selection for coating formulation is often a complicated operation, where elaborate practical experience is needed.

In some specific cases, organic paint may be mixed and applied without the presence of solvents. These paint systems are referred to as solvent free. Examples of this are low-viscosity two-component epoxies and powder coatings. The application and curing of the powder coating were mentioned earlier. The epoxy coatings may be mixed and applied without the use of a solvent, as the two components typically have low viscosity. Mixing and application of these coatings are often done at elevated temperatures to reduce the viscosity as much as possible.

For environmental reasons, waterborne coatings are being used increasingly. The binder may be either dispersed or dissolved in the water phase. Dispersion coatings show lower gloss and less good adherence than the dissolved types such as alkyds, ureum and melamine, phenol, and acrylic-type coatings.

Also, high-solids coatings are used to minimize the loss of organic solvent to the atmosphere. These coatings consist of a larger amount of solid particles than normal organic solvent-based coatings. In order to reach an acceptable viscosity despite the high content of solid particles, the molecular weight of the binder molecules is lowered. Of course, this will also influence other properties of the coating such as sagging.

20.3 Complexcoating Systems

20.3.1 Pretreatment: Conversion Layers

The corrosion protection of metallic substrates by simple organic layers is often not good enough because of, e.g., poor adhesion. Conversion layers are then applied. For these layers the substrate metal provides ions that become part of the protective coating after (electro-) chemical reaction of the substrate with a reactive medium. Well-known conversion layers are phosphate layers on steel and zinc, chromate layers on zinc and aluminum, and anodized layers on aluminum, the latter mostly without an organic topcoat. Anodizing is electrochemical treatment of a metal (mainly aluminum) while the metal itself is the anode. In this way a reasonably thick oxide layer (passive layer) is formed, with the thickness ranging from 1 to 30 μm, depending on the process conditions. Anodizing is performed in sulfuric acid, chromic acid, boric acid, oxalic acid, and other organic acids. Most important are anodizing processes in sulfuric acid and in chromic acid.
Pretreatment layers are used for a variety of reasons:

- To improve the adherence of the organic layer
- To obtain electrically insulating barrier layers
- To provide a uniform grease-free surface
- To provide active corrosion inhibition by passivating the metallic substrate or by reducing the rate of the oxygen reduction reaction

For an overview of this extensive field we refer to the literature [19–21]. Two important examples will be treated here very briefly, and some recent developments will also be mentioned.

Chromate coatings can be produced on aluminum and its alloys, magnesium, cadmium, and zinc. They offer good protection and improve the adhesion of the organic layer [22–26]. The corrosion resistance provided by chromate is based on the following three properties [27]:

1. Cr(III) oxide, which is formed by the reduction of Cr(VI) oxide, has poor solubility in aqueous media and thus provides a barrier layer.
2. Cr(VI) will be included in the conversion coating and will be reduced to Cr(III) to repassivate the surface when it is damaged, thereby preventing hydrogen gas from developing.
3. The rate of the cathodic oxygen reaction is strongly reduced.

Chromate baths always contain a source of hexavalent chromium ion (e.g., chromate, dichromate, or chromic acid) and an acid to produce a low pH. Typical pH levels range from zero to around 3. Most chromating baths also contain a source of fluoride ions, which effectively attack the original (natural) aluminum oxide film and thus expose the bare metal substrate to the bath solution. Fluoride also prevents the aluminum ions (which are released by the dissolution of the oxide layer) from precipitating by forming complex ions. The fluoride concentration is critical. A low concentration does not result in a conversion layer at all because of failure to attack the natural oxide layer, and a high concentration results in poor adherence of the coating due to reaction of the fluoride with the aluminum metal substrate.

During the reaction hexavalent chromium is partially reduced to trivalent chromium, forming a complex mixture consisting largely of hydrated (hydr) oxides of both chromium and aluminum:

$$6H^+ + H_2Cr_2O_7 + 6e \rightarrow 2Cr(OH)_3 + H_2O$$

The formation of the chromate layer according to this reaction preferentially occurs along grain boundaries and other heterogeneities at the aluminum substrate.

Chromate coatings are generally amorphous, nonporous, and gel-like when initially formed. If necessary, they can be readily dissolved in nitric acid at this stage. As the coating dries, it slowly hardens and becomes hydrophobic, less soluble, and more abrasion resistant. Up to 20% (Alodine 1200S) of unreacted hexavalent chromium is also included in the coating, which provides good corrosion resistance as described earlier.

Chromate conversion coatings are based on two types of processes: chromic acid processes and chromic acid–phosphoric acid processes. The reaction for the formation of a chromic acid–based conversion coating is given by the following overall equation:

$$6H_2Cr_2O_7 + 30HF + 12Al + 18HNO_3 \rightarrow 3Cr_2O_3 + Al_2O_3 + 10AlF_3 + 6Cr(NO_3)_3 + 30H_2O$$

The formed oxide, given as Cr_2O_3, is probably better described as amorphous chromium hydroxide, $Cr(OH)_3$. The color of the formed conversion coating is yellow to brown and typical coating weights range from 0.3 to 2.0 g/m^2. Those layers can be used as a primer for a paint layer or powder coating but can also be applied un-painted.

A possible overall reaction for the formation of a phosphoric acid-processed coating is

$$2H_2Cr_2O_7 + 10H_3PO_4 + 12HF + 4Al \rightarrow CrPO_4 + 4AlF_3 + 3Cr(H_2PO_4)_3 + 14H_2O$$

The resulting greenish layer consists mainly of hydrated chromium phosphate with hydrated chromium oxide concentrated toward the metal. At the conversion coating–aluminum substrate interface, aluminum oxides and other aluminum salts are present. The thinner chromic acid-phosphoric acid conversion coatings (Lg/m^2) are an excellent base for paint layers. Thicker coatings are often applied unpainted.

In most countries today extreme care is taken when applying chromate coatings due to toxicological and environmental hazards. It is possible that legislation will prevent the widespread use of these excellent layers in the future. A review of key literature that describes the toxicological and health effects of chromium, chromium (II), and chromium (VI) can be found elsewhere [28]. As a result of these constraints, new developments of nonchromate conversion layers have been taking place. However, it will be difficult to find a good substitute for the chromate layers. A more extensive review of new developments in corrosion prevention by nonchromate conversion coatings has been given by Hinton [29].

Treating the metal surface in a weak phosphoric acid solution of iron, zinc, or manganese phosphate produces phosphate layers. Phosphate layers offer only limited protection of their own but greatly improve the adhesion of the organic layers. A good example of a phosphate coating is the treatment generally used in the car industry. After degreasing the (galvanized) steel substrate in mild alkaline phosphate baths, the surface is phosphated by a mixed spray-dip process. A crystalline phosphate layer is formed with a total weight of about 2–5 g/m^2. The composition and structure of the layer depend on the substrate and the bath. Two common phosphate compounds are $Zn_3(PO_4)_2 \cdot 4H_2O$ and $Zn_2Fe(PO_4)_2 \cdot 4H_2O$. After phosphating, the material is given a standard chromate rinsing procedure.

During the cathodic electrodeposition of the primer layer the phosphated layer is attacked by the alkalis (pH of about 12 or higher) at the interface [30]. The phosphate is then treated at 200°C for 30 min when the painted metal is baked. During this treatment the phosphate compounds lose two water molecules. When the total coating system is in service later, water from the environment permeates the paint film and reacts with the partially dehydrated phosphate, and it may dissolve partially or completely in the water phase depending on the pH of the penetrating water [31]. When anodic and cathodic reaction sites emerge during propagation of the undercoating corrosion process, the phosphate layer can even completely dissolve either in the acid anodic medium or in the basic cathodic medium, especially when NaCl is present [32].

Modification of the actual chromium-based pretreatments has been undertaken by using trivalent chromium, especially on aluminum substrates. By forming trivalent chromium compounds on aluminum it has been shown possible to oxidize part of the film to hexavalent chromium, attaining a corrosion-resistant film comparable to the normal Cr(VI)-based layer [33].

Some no-rinse chromate-free formulations based on solutions of chromium trifluoride, phosphoric acid, and small quantities of polyacrylic acid give good coating adhesion [34].

Also, molybdates have been used as alternative passivity- and adhesion-enhancing oxidants, but with limited success (e.g., Ref. [35]).

Cerium compounds have been found effective in reducing the corrosion rate of aluminum alloys by inhibition of the cathodic reaction. The cerium can be incorporated in the passive oxide film to provide a protective conversion coating [36,37].

Zirconium, titanium, and hafnium compounds have been used in combination with fluoride as alternatives to chromates with reasonable success but are never as good as the Cr(VI)-treated surfaces [38,39].

Recent developments to be mentioned are the application of lithium salts, permanganates, self-assembling monolayers, and finally silanes to promote adhesion of the organic layers [40–43].

20.3.2 Metallic Layers, Conversion Layers, and Organic Topcoats

The complete coating system in automotive materials is quite complicated. A schematic overview is given in Figure 20.1. From this picture it is clear that any description of corrosion mechanisms in such complicated systems should encompass the possibility of a variety of defects caused by impact or degradation. A selection has been made here to describe the initial processes, damage exposing the primer, damage exposing the zinc coating, and damage exposing the steel.

The environmental factors determining the degradation of car body panels are salt (from deicing salt and from the sea in marine environments), humidity, temperature, impact by pebbles, sand, and acid rain. Of course, the most direct cause of corrosion initiation is

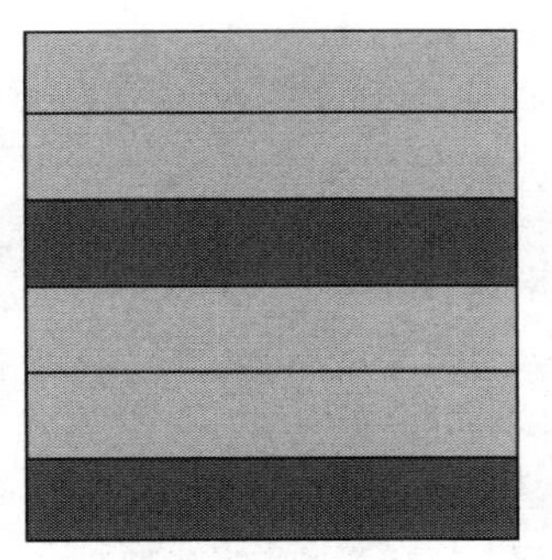

FIGURE 20.1
Schematic representation of the general buildup of the coated specimen.

physical damage due to impact by any kind of material. The propagation of any corrosion process depends on the depth of the damage.

During exposure, water permeates an intact paint film and reaches the interface, as we have seen before. On its way in, the water dissolves organic and inorganic compounds, which are later deposited on the metal surface, because the whole system goes through wet and dry cycles and also through mostly anticyclic temperature cycles. The water displaces the polymer from the metal surface during the wet period, resulting in poor wet adhesion behavior. Adhesion is sometimes partially restored during dry periods. After many cycles, the area of metal surface where the coating has delaminated, while the water phase at the metal surface incorporates more impurities, will be considerable. The increasing concentration of dissolved species at the interface also stems from the huge variety of inorganic water-soluble species present in the galvanized layer. Oxygen will also diffuse to the zinc layer, which is a rather good catalyst for the oxygen reduction reaction. However, no real corrosion damage normally results because of the active corrosion inhibition of the Zn-containing layer.

After damage by impact, this picture changes considerably. In Figure 20.2, a schematic picture is given of a coated system after impact resulting only in damage in the topcoats. Aggressive species can directly reach the primer surface. When water and oxygen have reached the phosphate layer by diffusion, the cathodic delamination process can start. Different authors found that the diffusion rate of water in the typical primers used for car body coatings is high and above the rate for epoxy-poly-amide resins [44–46]. The diffusion rate of oxygen being of similar magnitude, it must be concluded that the permeation

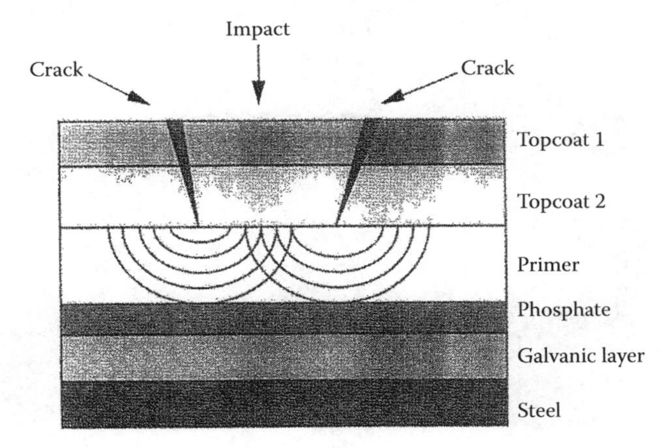

FIGURE 20.2
Schematic picture of the partially damaged coating system on a car body panel. Cracks after impact reach the primer. Diffusion of water and oxygen through the primer. (Picture drawn free after Granata, G.D., *Corrosion/91*, NACE, Houston, TX, paper 382, 1991.)

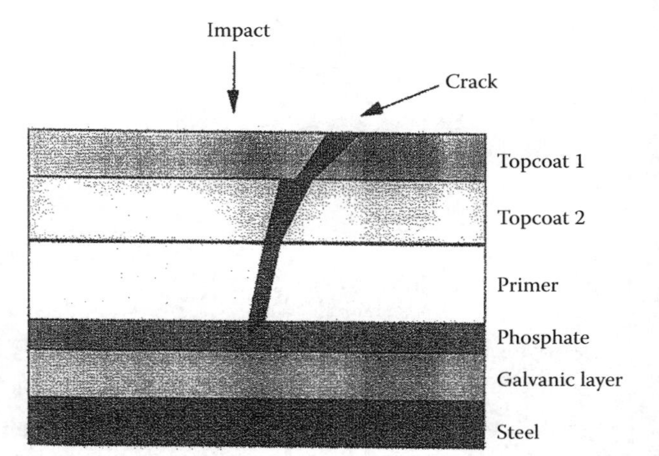

FIGURE 20.3
Schematic picture of the partially damaged coating system on a car body panel. Cracks after impact reach the phosphate and through porosity the zinc layer. (Picture drawn free after Granata, G.D., *Corrosion/91*, NACE, Houston, TX, paper 382, 1991.)

rate through the primer is not rate determining for the corrosion process. Corrosion proceeds as discussed for the cathodic delamination process. The corrosion spots spread radially. The pH increases at the cathodic sites, and the phosphate layer dissolves. Eventually a blister will form and further damage to the system will result in the mechanism that takes place at an earlier stage when the coating is damaged to the galvanic layer, as given in Figure 20.3.

In that case the phosphate layer provides hardly any protection. The zinc layer is consumed locally by the anodic reaction, and the steel substrate is exposed. The corrosion products are not taken up in the paint film. The cathodic delamination takes place along the zinc–paint interface. The corrosion process continues where the damaged site reaches the steel substrate as shown in Figure 20.4. All contaminants reach the paint–phosphate–metal interface. Zinc dissolves anodically at the tip while oxygen reduction takes place at the exposed steel surface. At cathodic sites the pH increases, leading to loss of adhesion. Also, chloride ions can now easily reach the corroding surface. The formation of zinc hydroxychloride further damages the interface by a wedging action. The anodic reaction of zinc lowers the pH. The primer is not very stable at low pH and is hydrolyzed. Phosphate also partially dissolves, thus further weakening the paint–phosphate–metal interfaces. It is clear from this short description that the overall reaction mechanism is very complicated and not yet unambiguously resolved. Opinions differ considerably and research is going on [47,48].

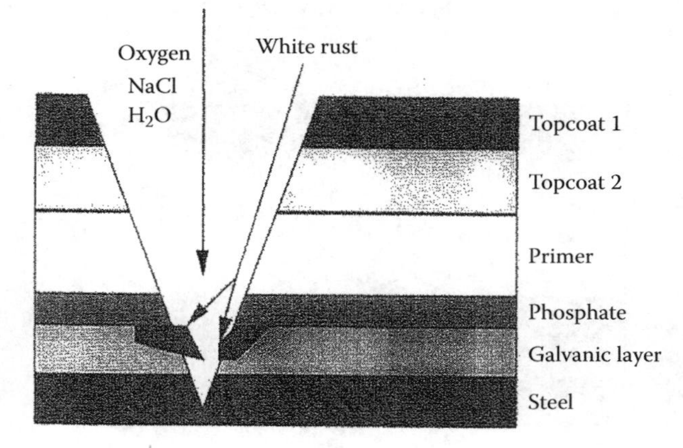

FIGURE 20.4
Schematic picture of the partially damaged coating system on a car body panel. Cracks after impact reach the steel substratum. (Picture drawn free after Granata, G.D., *Corrosion/91*, NACE, Houston, TX, paper 382, 1991.)

20.4 Application Techniques

Organic coatings can be applied to surface by different methods. The application depends mainly on the

- Purpose of application: protection or decorative
- Type of coating, target layer thickness, number of layers
- Size and geometry of the object to be coated
- Site of application: in a conditioned room, confined space, or in the field

Moreover, environmental and health aspects have a large influence on the application techniques. A short overview of the different techniques with (dis-) advantages is given next.

20.4.1 Brushing

Application by brushing is labor intensive and consequently time consuming and expensive. The layers obtained by brushing depend on the painter and can vary greatly. Brushing is still indispensable for small areas, repair purposes, and for less accessible places for parts that are difficult to protect, such as edges, angular parts, and welds. An advantage of this technique is the very low wastage. Also, it is generally believed that brushing gives better penetration of paints into pores, cracks, and crevices [49]. In the case of a solvent-based paints, health risks have to be taken into account, in particular in confined spaces. However, the technique is suitable for waterborne and high-solid systems.

Brushing is less suitable for very fast drying products; brush marks remain visible and the paint film is not homogeneous. It is also difficult to apply more layers of some types of physically drying paints: the solvent of the last coat layer may soften the previous layer.

20.4.2 Rolling

Painting by roller is much faster than brushing. It is suitable for large flat surfaces. On the other hand, it is not suitable in corners, edges, and welds. Additional brushing is necessary in these cases. Further, the same restrictions apply for the type of paint as in the case of brushing.

20.4.3 Dip Coating

The object to be coated is dipped in the coating. This method is very simple, cheap, and can be quick. The quality depends on a number of factors: time of withdrawal, viscosity, the rate of solvent loss, and the mixing of the coating. The method can be automated to deal with a large quantity of objects to be coated.

20.4.4 Conventional (or Air) Spraying

The paint is forced from a container to the spray gun by pressurized air or by a pump. As the paint passes through the nozzle, it is mixed with air and atomized at pressures of about 2–4 bar. Spray guns can be equipped with fluid air supplied by a compressor with sufficient extra capacity to compensate for the drop in the long hoses. The air is freed from oil, moisture, and dust by means of an oil-water separator.

Air spraying is a much faster application technique than brushing or rolling; moreover, the layers obtained are more uniform in thickness. The disadvantage of this technique is the larger waste. Part of the paint is lost in the so-called overspray: missing the object to be coated. Another cause of paint loss is the rebound of the air containing paint spray from the surface of the object to be painted. Paint losses can be excessively high when spraying is done outdoors.

The directions for spraying distance, pressure, and nozzle type are rather critical and should always be followed. Distance from surface should normally be 45–60 cm. If the spray gun is held too far from the substrate, the paint droplets are already dry when they reach the surface and will not aggregate. This phenomenon, called dry spray, gives a discontinuous, porous film.

Dry spray may also be expected in the case of highly pigmented products (for example, zinc silicate) or paints containing very volatile solvents (for example, vinyl paints). This phenomenon may also be expected when the paint material is insufficiently thinned or spraying is done during hot or windy weather.

20.4.5 Airless Spraying

The paint is brought to a high pressure (75–300 bar) by means of a pump and forced though a narrow nozzle, which atomizes the paint. In contrast to air spraying, in this case the paint is not atomized by air and consequently the amounts of spray mist, overspray, and rebound are much less than in air spraying. Because the paint is not mixed with air, there is almost no risk of contamination with impurities such as oil or water. Further advantages are:

- Significantly higher production rates: faster spraying.
- Paint can be applied at a higher viscosity, resulting in higher dry film thickness.
- Loss of solvents and thinner by evaporation is much lower than in air spraying.

The technique is also suitable for more volatile solvents.

Airless spraying is in particular appropriate for painting surfaces, such as buildings, ships, and offshore structures. It has the following disadvantages:

- The worker has to work in a very fast and concentrated way because of the high rate of deposition.
- It is unsuitable for spraying small objects.
- To obtain thin layers, very narrow nozzles are requited. Consequently, they become warm (particularly with paints that contain hard pigments and extenders) and malfunction.
- The paint jet has such a high velocity that it can injure workers.

20.4.6 Electrostatic Spraying

In this process atomized paint and the object to be painted are respectively negatively and positively charged. This results in a good adhesion and a uniform layer.

Further, it is an economical process because of the low waste, and application of water-based paints is possible. Disadvantages are the restriction to coat objects only on the outside and the occurrence of defects due to an incorrect viscosity/solvent balance.

20.4.7 Powder Coating

Powder coating can be considered as a variation of the previous technique. A constant flow of powder, fluidized in an air stream, passes through a highly charged (40–150 V) region of ionizing discharge. The charged paint particles are directed to the (grounded) substrate. Together with the electrostatic attraction, an enforced airflow increases the velocity of the particles to the substrate. After deposition, the layer is fused in an oven at 180° for 10–15 min, depending on the characteristics of the powder. Coated thicknesses of up to 500 μm are possible with these coatings without loss of flexibility; thickness between 300 and 450 μm are more common.

20.4.8 Electrodeposition

Charged paint particles in a solution of water-soluble resins are directed by a potential field between cathode and anode (respectively the object to be painted and bath tank or vice versa) to the surface of the object.

Advantages are that a uniform coating is obtained, penetration in narrow places (for example, seams) is possible, the process is fast, and there is no fire risk. The main disadvantage is the limitation of the thickness (due to the electrical resistance of the formed coating). The limit is 25 μm and only one coat is possible.

20.5 Protective Mechanisms

The corrosion protection provided by organic coatings results either from the barrier action of the layer, which may be supported by inert inorganic particles especially if these are arranged like tiles (e.g., aluminum flakes or iron phosphates), or from active corrosion inhibition provided by pigments in the coating. Ideally, a coating provides protection by forming a physical barrier between the metal substrate and the aqueous corrosive environment. In practice, however, these physical barrier properties are limited, as all organic coatings are permeable to water and oxygen to a certain extent. This, however, mostly does not influence the protective action of the coating as long as the coating adheres well. In fact, adhesion is the primary protection criterion. Permeation of the coating with water is not damaging as long as no condensation of water at the metal/coating interface takes place.

Corrosion under a coating can take place only after an electrochemical double layer has been established at the metal surface. For this to occur, the adhesion between the coating and the substrate must be broken, after which a separate thin water layer at the interface can be formed when the water permeates the coating.

Under normal, outdoors conditions an organic coating is saturated with water at least half its service life. For the remainder of the time it contains a quantity of water comparable in its behavior to an atmosphere of high humidity [50]. Furthermore, the average transmission rate of water through a coating is about 10–100 times larger than the water consumption rate of a freely corroding surface [13]. Because it has also been established that in most cases the diffusion of oxygen through the coating is large enough to facilitate unlimited corrosion, it is clear that the physical barrier properties alone may not account for the protective action of imperfectly adherent coatings.

Resistance inhibition, which is also part of the barrier mechanism, may supply additional protection. Inhibiting the charge transport between cathodic and anodic sites retards the corrosion reaction. An increase in the electronic resistance and/or the ionic resistance in the corrosion cycle may reduce the reaction rate. The electronic resistance may, for example, be increased by the formation of an oxide film on the metal. This is the case for aluminum substrates. The application of organic coatings on a metal surface results in an increase of the ionic resistance.

One of the weak points of organic coatings in corrosion prevention is the fact that these coatings are relatively easily damaged under mechanical and thermal load. This may cause corrosion under the paint film at and near the damaged site. The otherwise adequate barrier properties of the coating will no longer give sufficient protection. Active pigments are then often incorporated in the polymer matrix of the first coating layer near the substrate: the primer. These pigments (passivating, blocking, or galvanic) provide protection through an active inhibitive mechanism immediately when water and some corrosive agent reach the metal surface.

Again, protection can result only if the adhesion of the coating is good. Also, the mechanical properties (e.g., the glass transition temperature) of the polymer reflect to some extent the quality of the coating, as these determine the formability of coated substrates (e.g., for coil-coated products) and their sensitivity to external damage.

20.5.1 Water Permeation

All organic coatings are to some extent permeable to water. The effective permeability is closely related to the polymer structure and composition. The permeability of a coating is often given in terms of the permeation coefficient *P*. This is defined as the product of the solubility of water in the coating (S, kg/m^3), the diffusion coefficient of water in the coating (D, m^2/s), and the specific mass of water ρ (kg/m^3). Different coatings can have the same permeation coefficient even though the solubility and the diffusion coefficient, both being material constants, are very different. Therefore the usefulness of the permeation coefficient is limited.

Water permeation occurs under the influence of various driving forces:

- A concentration gradient, e.g., during immersion or during exposure to a humid atmosphere, resulting in true diffusion through the polymer
- Osmosis due to impurities or corrosion products at the interface between the metal and the coating
- Capillary forces in the coating due to poor curing, improper solvent evaporation, bad interaction between binder and additives, or entrapment of air during application

When a coated system is exposed to an aqueous solution or a humid atmosphere, water molecules eventually reach the coating substrate interface. Normally, a coating under immersion will be saturated after a relatively short time (of the order of 1 h), depending on the values for D and S and the thickness of the layer. Typically values for D and S are 10^{13} m^2/s and 3% [7–9]. For atmospheric exposure the actual cyclic behavior of the temperature and the humidity determines largely the periods of saturation. In any case, situations will result in which water molecules reach the coating–metal interface, where they can interfere with the bonding between the two phases, eventually resulting in loss of adhesion and corrosion

TABLE 20.1

Permeability of Oxygen and Water Vapor in Several Resin and Coating Films: Results Obtained for Free Films That Were Cast on Glass

	Permeability[a]	
Polymer	**Oxygen [cm^3 100 μm (m^2 day atmr)^1] at 23°C and 85% RH**	**Water vapor [g 10 μm (m^2 day)$^{-1}$] at 38°C and 95% RH**
Resin films		
VC-VDC copolymer latex	22 ± 9	28 ± 10
Chloropolymer, solvent borne[b]	82 ± 19	100 ± 20
Epoxy/polyamide	130 ± 33	155 ± 20
Chlorinated rubber, plasticized	183 ± 7	95 ± 5
Styrene acrylic latex	1464 ± 54	2300
Coating films		
VC-VDC copolymer latex	12 ± 5	68 ± 12
Chlorinated rubber unmodified	30 ± 7	50 ± 8
Chloropolymer, solvent borne[b]	33 ± 2	65 ± 12
Aluminum epoxy mastic	110 ± 37	105 ± 15
Coal tar epoxy	213 ± 28	75 ± 3
Acrylic water-borne primer	500	1800 ± 92
TiO_2 pigmented alkyd	595 ± 49	645 ± 15
Red lead oil-based primer	734 ± 42	535 ± 8

Source: Thomas, N.L., *J. Prot. Coat. Linings* 6(12), 63, 1989.

[a] The permeability of oxygen is given as the number of cm^3 gas of 1 atm permeating through a coating of 100 μm thickness per m^2 per day.

[b] Product under development.

initiation if a cathodic reaction can take place. For a corrosion reaction to proceed, a constant supply of a cathodic species such as water or oxygen is required. Some authors suggest that the transport rate of oxygen through the coating is higher using the interconnected water phases in a saturated polymer [51,52], whereas other experiments lead to the conclusion that the oxygen permeation is independent of the presence of water [53].

Dickie [12] determined the quantities of water and oxygen consumed during corrosion of a freely corroding unpainted steel substrate. Both were of the same order of magnitude and also comparable to the average permeability of coatings for oxygen. An overview of some data is given in Table 20.1. This leads to the conclusion that for some coatings with somewhat lower permeability for oxygen the transport of oxygen may be rate determining. Water permeation may also result in the buildup of high osmotic pressures, which are responsible for blistering and delamination.

20.6 Corrosion Underneath Organic Coatings

The causes of proceeding delamination underneath organic coatings can be roughly categorized in two types: delamination due to clustering of water at the interface and delamination caused by specific corrosion processes that produce either low or high pH values at the interface. In the case of blisters underneath an intact paint layer, the first mechanism will often be the starting point for the more "chemical" delamination caused by processes of the second category. Within the first category wet adhesion and osmotic blistering are the main mechanisms. The second category is mainly referred to as cathodic or anodic delamination.

20.6.1 Clustering of Water

Clustering of water at the metal–polymer interface and the subsequent formation of an electrochemical double layer can occur only if the adhesion between metal and coating is weaker than the bond between metal and water or polymer and water. For a proper evaluation, however, a distinction must be made between wet adhesion and osmotic blistering.

Wet adhesion is the complete deadherence of the coating without actual blistering. From a thermodynamic point of view this implies that the whole interface must have very weak binding energy. Therefore the bonds will be easily broken by water that can be transported through the coating at a sufficient rate. Fast temperature cycling of saturated coatings may, however, also lead to this type of delamination although the adherence under static conditions is sufficient.

Typical of wet adhesion is the absence of real blisters in the first stages. This is due to the fact that the water at the interface is relatively pure and as long as corrosion processes have not generated enough ions, osmotic mechanisms will not cause a further growth of the water layer. Therefore the adhesion may also be partly restored when coatings are completely dried after immersion. Only when ions are present due to contamination or are formed by corrosion processes may further blistering occur.

Osmotic blistering is caused by the presence of soluble species underneath the coating. Water that normally penetrates the coatings will dissolve these species at the interface and form a highly concentrated solution. Also, some acid fumes containing acetic on hydrochloric acid may penetrate through polymers and condense together with water in small voids or places with minor adhesion. Due to osmotic forces, more water will then be attracted to these solutions. This causes an internal pressure within the blisters, which has been calculated to be as high as 3 Pa [5,6]. These high internal pressures may even cause further delamination. Because corrosion processes within the blister tend to produce their own ions, the osmotic pressures may remain high for long times. In many cases these blisters will eventually burst or gradually develop into the electrochemical delamination mechanisms described in the following.

20.6.2 Cathodic Delamination and Anodic Undermining

In the literature two mechanisms are proposed to describe the propagation of underfilm corrosion in case of intact and defective coatings. These mechanisms are:

- Cathodic delamination
- Anodic undermining

In cathodic delamination the adhesion of the coating fails and causes lateral blister growth as a result of a high pH at the delamination front [6,55,56]. The loss of adhesion in anodic undermining is caused by the dissolution of the metal or the metal oxide at the interface with the coating [6].

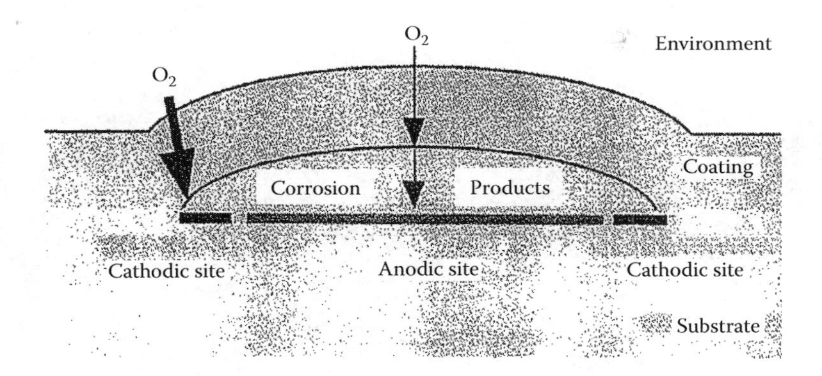

FIGURE 20.5
Oxygen transport paths in a blister in case of an intact coating. The resulting anodic and cathodic sites are shown. The transport paths for oxygen are shown.

Both cathodic delamination and anodic undermining are the result of a specific type of electrochemical cell: the differential aeration cell. In this cell separation of anodic and cathodic reaction sites takes place, but for both mechanisms this happens in different ways depending on the path for oxygen transport to the metal interface. This transport is, of course, also determined by whether the coating is intact of defective. With intact coatings, oxygen is mainly supplied through the coating or through both coating and formed solid corrosion products as schematically shown in Figure 20.5. The length of the oxygen transport path to the interface at the edge of the blister is shorter than that at other places in the blister where oxygen also has to diffuse through the corrosion products. According to the differential aeration cell theory, the cathodic reaction will occur at the edge of the blister and the anodic reaction in the center of the blister as shown in Figure 20.5.

The separation of the anodic and cathodic reaction sites in Figure 20.5 can be promoted by the nature of the corrosion products. When corrosion products consist of species that can be further oxidized, oxygen may be reduced during the transport through the corrosion products. The amount of oxygen that will reach the metal will therefore collapse and in the center of the blister less oxygen will reach the metal surface. An example is the formed corrosion product of iron, which initially consists mainly of Fe(II). In the presence of oxygen, Fe(II) will be oxidized to Fe(III) and will therefore consume the oxygen [7,12]. Because the solubility product of Fe(III) (hydr)oxides is very low, a solid film is formed. The proceeding corrosion process leads to film growth at the center of the blister, which hinders further oxygen transport at the center effectively.

From the preceding discussions it is concluded that blisters under intact coatings grow due to cathodic delamination.

When defective coatings are considered, the situation is much more complex. Part of the substrate is now directly exposed to the corrosive solution. Corrosion will initiate at the defect and subsequently the formed corrosion products will block the pore. Corrosion propagation depends on the nature of the corrosion products in the pore. These stages are shown in Figure 20.6a and b.

Whether anodic undermining or cathodic delamination will develop at defective coatings depends on the ratio of the transport rates of oxygen through the coating and that through the formed corrosion products. When the rate of oxygen transport through the coating exceeds that through the corrosion products, the mechanism is cathodic delamination; the opposite leads to anodic undermining.

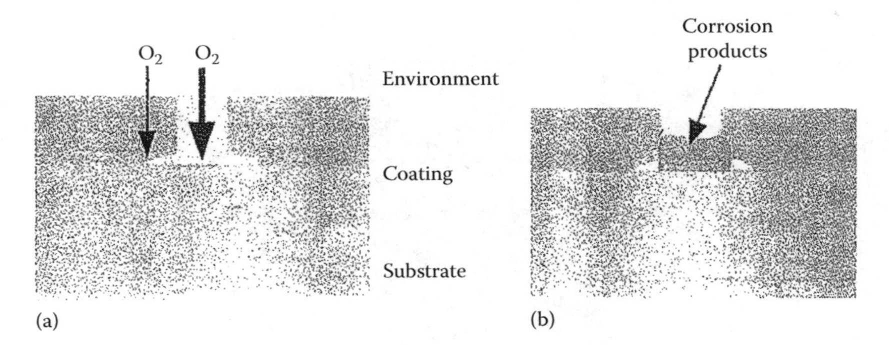

FIGURE 20.6
(a) Corrosion initiation at defective coatings; (b) blocking of the pore. Initial loss of coating adhesion at the edges of the pinhole is shown.

20.6.2.1 The Model as Presented in the Literature

In more detail, the model of cathodic delamination can be described and understood as follows [3,57]: the delamination starts with randomly distributed anodes and cathodes in a defect or at a dalaminated area. When iron is taken as an example, the dissolved Fe^{2+} at the anode is further oxidized to Fe^{3+} by oxygen and forms insoluble corrosion products in the defect that often adhere to the polymer (in the case of a blister) and at the edges of the defect (where oxygen enters the defect). In what way the defect is blocked and a cap of corrosion products is formed on the top of the blister; see Figure 20.7. In this cap oxygen

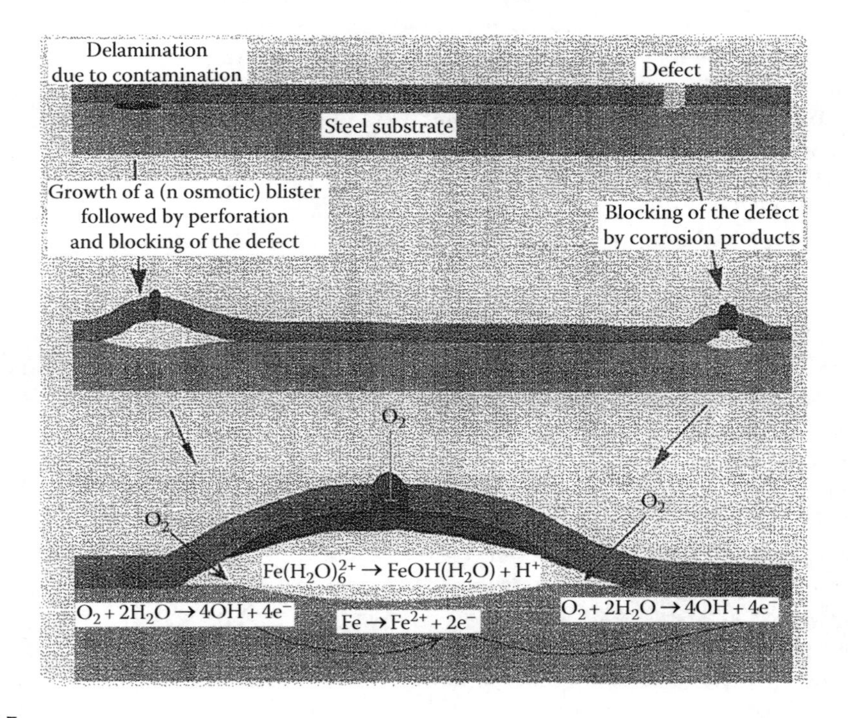

FIGURE 20.7
Schematic representation of the mechanism of cathodic delamination starting from a defect and from an osmotic blister.

is consumed through oxidation of Fe^2 (the cap is sometimes said to be impermeable to oxygen [7]). Because of this, the area underneath the cap will have a lower oxygen concentration than the regions at the edges where no cap is present and where oxygen "easily" penetrates the coating. This stimulates the separation of anodic and cathodic regions, which finally leads to large pH differences due to the different reactions that take place at anode and cathodes ($\frac{1}{2}O_2 + H_2O + 2e \rightarrow 2OH^-$ at the cathode and $Fe \rightarrow Fe^{2+} + 2e^-$ followed by subsequent hydrolysis at the anode). This process is shown schematically in Figure 20.7 for a delaminated and a defective coating. In both cases the same end situation results.

Taking a closer look at this model, some questions arise that are not directly answered by the model as presented normally:

What is the mechanism of growth of these blisters?

How can these large pH differences that are reported in literature exist for prolonged periods of time in such a confined space (differences of 10 pH units over less than a few mm)?

Despite these remaining questions, the model could explain recent model experiments on test panels that were deliberately contaminated with sodium chloride. On most panels a large (set of) blister(s) was present at the contaminated site, but also other, small blister were present at other places on the panel. In between these blisters, the polished metal underneath the transparent coating still has a mirror-like surface. Closer examination showed, however, that even on these apparently undamaged sites there was a thin layer of liquid present between coating and metal. When the completely deadhered coating was lifted off, the pH of this thin layer of liquid was "measured" with a pH indicator liquid. It was approximately 12.

Underneath the larger blisters the metal was etched as can be seen in Figure 20.8. This can only be the result of the attack by a quite acid solution. Below pH 3, however, hydrogen evolution may occur. Sometimes the presence of hydrogen bubbles on steel is reported in the literature [58], but in all cases the cathodic dis-bondment is caused by cathodic protection. As very little solutions be available within the center of the blister, it is extremely difficult to measure the pH in this area. Judging from the metal attack is should, however, be around pH 3.

Figure 20.8 also shows the undamaged cathodic areas around the blisters. These observations lead to the conclusion that the large pH differences do indeed exit.

Observations on smaller blisters give the answer to the question how these large pH differences can exist in such small areas. It appeared that in the case of very small blisters a closed sphere of corrosion products remained when the coating was removed. So there is not only a "cap" as presented in Figure 20.7 but also "walls" between anodic and cathodic sites.

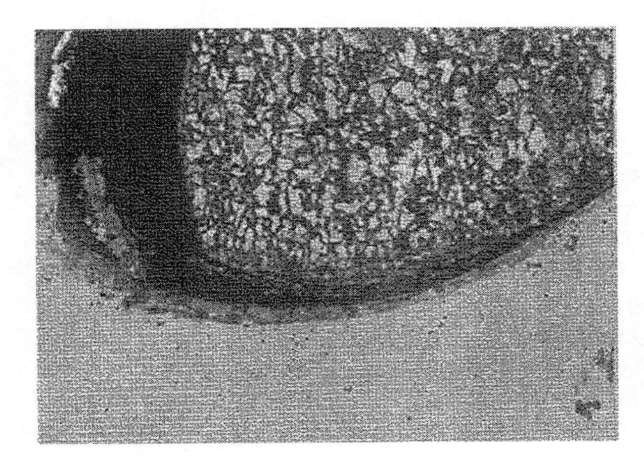

FIGURE 20.8
Photograph of a large blister after removal of the coating and the corrosion product (enlargement 20 times).

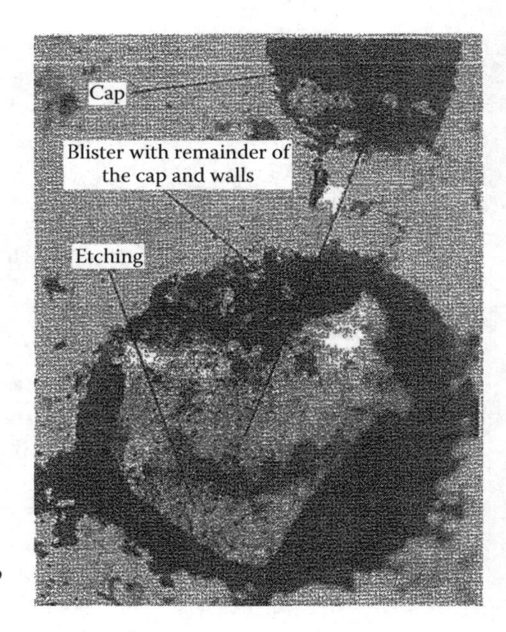

FIGURE 20.9
Photograph of a small blister. The top cap is visible at the top (enlargement 50 times).

Together with the cap, the walls form a closed sphere of iron (hydr)oxides. This result can be observed only on very small blisters because larger ones are damaged when the coating is removed. A photograph of such a small blister is shown in Figure 20.9. In this picture one can also see the beginning of the etched surface that is also present in Figure 20.8. To look inside the blister, the top cap was in this case removed with a needle (it is still visible at the top of the picture). Around the blister again a fully reflective surface is visible; only some debris from the opening of the sphere of corrosion products is present.

The formation of the walls of the sphere can be understood considering the ions that are present in the system: at the cathode OH^- will be present in rather high concentrations. In the center, the anodic region, Fe ions are present in rather high concentrations. Low soluble iron (hydr)oxides will precipitate after interdiffusion of OH^- and Fe^{2+} at the boundary region [59]. These precipitates form a wall that is connected with the cap, thus forming a closed sphere.

20.6.2.2 Extended Model for Cathodic Delamination

From the preceding observations the model for cathodic delamination can be extended in the following way:

1. Initially, the normal model as presented by Funke and Haagen [3,57] describes the delamination process.
2. After some time the anodic and cathodic areas become separated as a barrier of corrosion products is formed in between them. This wall stabilizes the pH differences that occur due to the separation of the anodic and cathodic reactions. Wall and cap together form the observed sphere
3. The wall slowly dissolves from the inside in the increasingly acid environment within the sphere. Fe ions diffuse through the wall, resulting in the formation of new $Fe(OH)_2$-like products on the outside. In this way the anodic area of the blister is able to grow. The high pH on the cathodic edges causes the delamination of the organic coating to proceed (as was already assumed in the existing model)

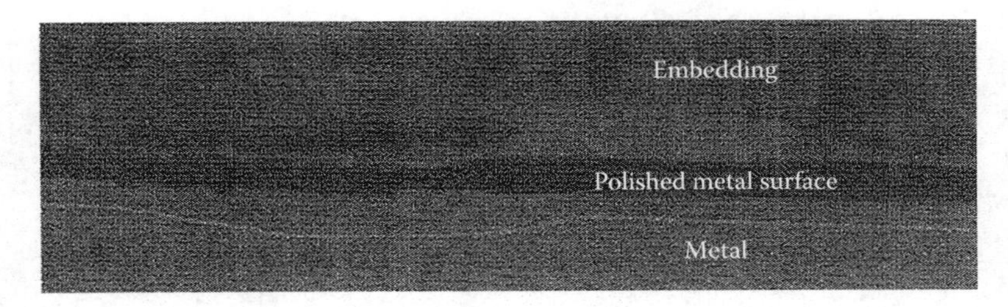

FIGURE 20.10
Cross section of a metal with a shallow "crater" due to cathodic delamination of a coating. The coating was pulled off before the sample was embedded and the cross section was made. The black areas are formed by a crevice between the metal and the embedding. The white line parallel to the metal surface indicates the (exaggerated) formation of a crater. The cross section was embedded as a metallographic specimen and photographed with an optical microscope.

The mechanism of growth implies that the center of the anodic area has been exposed for a longer period to the acid corrosive environment than the edges of the anodic area. As a consequence of this a shallow "crater" must be formed under every blister. Figure 20.10 shows that this is indeed the case.

Another importance consequence of the introduction of the wall is that contaminating ions such as Na^+ and Cl^- are necessary in both the anodic and the cathodic region for cathodic delamination to occur: the Fe ions cannot compensate the charge of the OH^-; they would directly produce a solid product in the cathodic region. On the other hand, in the anodic areas the H^+ and Fe^{n+} should be compensated by a negative ion such as Cl^- (OH^- would, of course, react directly with H^+). So the positive and negative ions of the contaminating salt will be separated. This further implies that the final pH values are also determined by the amount of contaminating ions. The importance of the contaminating ions for the delamination process is also known from literature in which the influence of cations of different sizes is described. It proved that the size (and therefore the diffusion coefficient) of the cations has a relation to the delamination rate [15,60–62].

Further proof of the existence of a membrane-like structure can be obtained through simple model experiments: it proved to be possible to produce a similar film in a test tube [63]. When such a tube is filled halfway with an NaOH solution of pH 12, an HCl solution of pH 3 containing Fe^{2+} ions can be injected underneath (the density of the acid is higher). When this is done, a film of $Fe(OH)_2$-like products is formed instantly and the geometry shown in Figure 20.11 is created.

This situation is stable for several weeks. Depending on the disturbance during the injection of the acid, the film can be thick (several millimeters) and colorful or very thin (hardly visible) and pale white. If the iron ions are omitted from the acid solution, the situation can also be created, but after 10 min the two solutions have completely mixed. So this Fe-containing film indeed has a stabilizing effect as was predicted from the observations on small blisters.

Further proof of the existence of the stabilizing film was also found in other tests: it was also possible to create it in a neutral NaCl solution by electrochemical dissolution of the central electrode in Figure 20.12. In this experiment a disk of carbon steel was placed in a sheet of carbon steel. The whole panel was embedded in epoxy and on the surface a container was glued. A 3% sodium chloride solution was put in the container and the central electrode was connected to a potentiostat and was made the anode, the rest of the panel being the cathode. One volt was applied, causing rapid dissolution of the disk (Fe to Fe^{2+}).

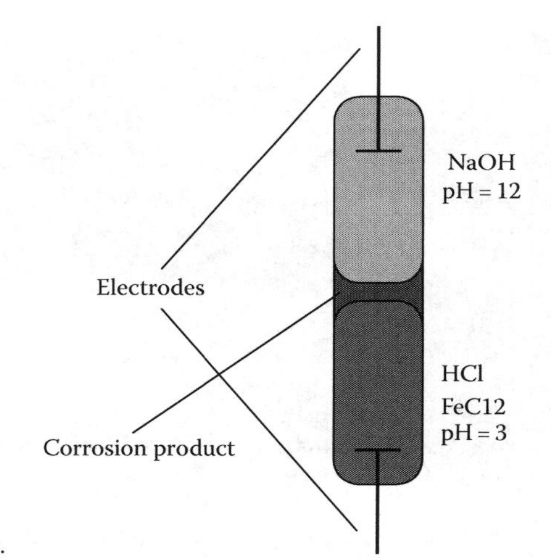

FIGURE 20.11
Schematic picture of the situation in a test tube.

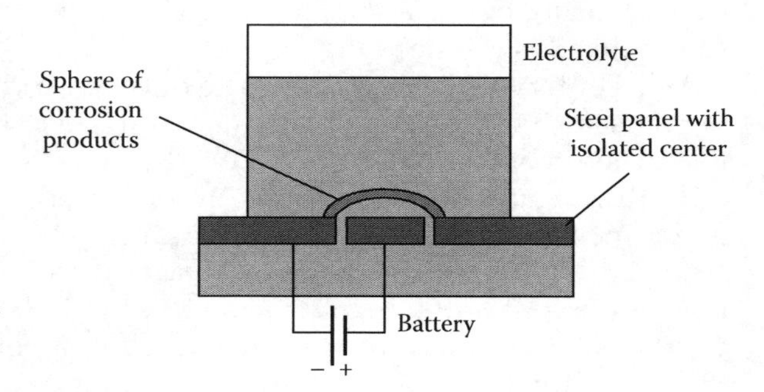

FIGURE 20.12
Production of the artificial sphere of corrosion products; 3% sodium chloride is poured into a container that is glued on the surface.

On the cathode hydrogen gas was formed (and maybe also oxygen consumed). In that case the pH is increased at the outer ring, resulting in the formation of the sphere of corrosion products.

20.6.2.3 Anodic Undermining

The model for anodic undermining shows large similarities to cathodic delamination. The main difference is the fact that oxygen transport through an open defect or porous corrosion products is faster than through the coating. As a result of this, the location of cathodic and anodic sites underneath the deadhering coating is reversed. Delamination now proceeds through dissolution of the interface between coating and metal in the acidic anolyte. In some cases also the osmotic pressure that is caused by the formation of voluminous solid corrosion products may help the delamination process by "lifting" the coating.

Anodic undermining is mainly reported on aluminum, which forms only Al^{3+} ions and porous corrosion products. Especially in the presence of chlorides, the natural oxide is easily attacked. Similarly to cathodic delamination, in this case also the anions and cations are separated leading to the formation of an HCl solution at the delamination front.

20.6.3 Filiform Corrosion

A specific type of delamination that is also driven by a differential aeration cell is filiform corrosion [5,6,62,63]. In most cases it is related to anodic undermining, especially because it occurs mainly on aluminum alloys. Filiform corrosion is a specific type of delamination that occurs only under atmospheric condition with relative humidity between approximately 50% and 90%. Small threadlike delamination tracks are formed that in general follow irregularities on the surface (either rolling or grinding marks). New interest in this special form of corrosion was stimulated by many practical cases along the coastal areas in Europe. It may be expected that new mechanistic information will become available within the next 2 years.

20.6.4 Flash and Early Rust

Two special types of corrosion failure that occur exclusively with water-based coatings are flash and early rust. Both look similar and are often treated similarly in the literature. They are, however, caused by different mechanisms and although they occur under similar circumstances they should be treated differently.

20.6.4.1 Early Rusting

A measles-like rusty appearance may occur when a cold steel substrate is pained under high-moisture conditions with a latex paint [6,50,62] immediately when the coating is touch dry. This is called early rusting. This corrosion form occurs only when the following conditions are met:

A thin (up to 40 μm) latex coating has been applied.

The substrate temperature is low.

The air humidity is high.

Early rusting is a result of the special curing mechanism of latex coatings. Water evaporates from the coating, while dispersed latex particles coalesce. Capillary forces and surface tension deform the particles and diffusion of the polymer chains among latex particles finally results in film hardening. It occurs under conditions in which water-soluble iron salts diffuse through the coating to the outer surface before the coating is completely hardened, which explains the conditions given before. After final evaporation of the moisture at the outer surface, the typical measles-like corrosion spots become visible.

To prevent early rust, inhibitors are often added to the paint formulation.

20.6.4.2 Flash Rusting

The appearance of brown rust stains on a blasted steel surface immediately after applying a water-based primer is called flash rusting [62]. The cause of this type of corrosion is the remaining contaminants after the blast cleaning. The grit on the surface provides crevices or local galvanic cells that activate the corrosion process as soon as the water-based primer wets the surface. This type of corrosion generally also occurs when coatings have remained wet for longer times due to lower temperature, high relative humidity, or extreme thickness of the layer.

Although inhibitors might provide a certain degree of protection, the most logical choice to prevent this type of corrosion is to prevent the presence of contaminants on the surface.

20.7 Specialties

20.7.1 Coatings for Marine Applications

Seawater and the atmosphere above seawater are extremely corrosive environments. Organic coatings are widely used to protect steel constructions in these cases, so the choice and buildup of the coating system are often illustrative of the possibilities of protection by organic coatings [49].

Protection of the underwater part of a ship's hull or offshore construction requires corrosion inhibition, water resistance, antifouling properties, and compatibility with cathodic protection. Cathodic protection is used to protect the steel in case the coating is damaged.

The corrosion inhibition and water resistance are obtained by paint systems formed by several anticorrosive coats based on epoxy, vinyl, chlorinated rubber, or bituminous coatings.

Coal-tar epoxy systems and vinyl tar coating have been used rather successfully from a technical point of view, but their use is now forbidden in some industrialized countries because of their effect on human health. In their place, high solid tar-free epoxy and glass flake-reinforced epoxies are used.

Aluminum flakes are usually added to bituminous coatings, but these flakes have also been added to epoxy systems to act as active anticorrosive agents [9].

Normally, the buildup of a protective system is as follows:

- Primer (indication 50 μm)
- One or two layers of anticorrosive coating (indication 250–400 μm)
- Tiecoat/sealer (indication 75 μm) to serve as intermediate between anti-corrosive system and antifouling (for example, to improve adherence or to prevent bleeding of the tar)
- Antifouling (indication 200 μm)

20.7.2 Coatings for Fouling Prevention

Some attention is given to a particular group of organic coatings, intended for the protection of ships and marine structures against plant and animal growth: the antifoulings.

In the sea and in estuaries, fouling chiefly consists of microorganisms, algae, barnacles, shellfish, tube worms, etc. Fouling develops most rapidly on stationary and slow-moving vessels (speed < 4 knots).

The disadvantages of fouling for vessels are evident in higher fuel consumption, reduction of maximum speed and maneuverability, corrosion, mechanical damage, and high maintenance costs [64]. Even with a limited amount of fouling (~1% of the underwater surface of the vessel), the friction is increased to such an extent that significantly more fuel must be consumed in order to maintain the same sailing speed. This can lead to substantial financial loss, depending upon fuel prices; e.g., during the 1970s oil crisis, a loss of

$300,000 per ship, per year, was not uncommon [65,66]. Fouling on ships between 1,850 and 35,000 tons causes an increase of 35%–50% in fuel consumption after 6 months in temperate waters [67]. Even a thin biofilm results in higher resistance and thereby a fuel loss of up to 20% [68–70].

Fouling on seagoing vessels and pleasure craft is mostly treated by paints from which toxic components with an antifouling action are encouraged to leach [66–74]. The active ingredients are freed at the coating–water interface and any organisms that try to stick to the hull are killed.

20.7.2.1 Organotin-Based Antifoulings

Organotin-based antifouling have been technically the most successful product for years. They owe this success to a unique mechanism to prevent fouling. They combine the toxic properties of TBT (tributyltin) with a controlled-release mechanism, which guarantee constant activity up to 5 years (coating thickness is the limitation).

A controlled and slow chemical reaction with the seawater at the paint surface occurs and guarantees a constant but very low TBT release: The TBT binder, a copolymer of tributyltin methacrylate and methyl methacrylate, hydrolyzes in sea-water at a constant linear rate. The binder becomes water soluble as soon as enough TBT has been hydrolyzed. Seawater flowing against the ship's hull polishes off the hydrolyzed remainder of the binder.

The major disadvantage of TBT and other biocide-containing antifouling coatings is the environmental problems that they cause [71], which have resulted in legal restrictions on their use. For vessels under 25 m in length TBT has already been prohibited in many countries, and a complete ban starting from 2003 is in preparation by the International Maritime Organization.

Consequently, a great deal of research is going on all over the world. However, so far alternatives technically and economically comparable to the TBT-based systems have not been developed. The best new systems, based on copper acrylate with some erodable properties, have a lifetime of only 3 years. Moreover, they contain a large amount of copper with addition of "antifouling" compounds from agriculture. In any case, it is clear that the environmental problem is not solved with these new systems.

However, some useful methods for certain classes of ships have now become available. These methods can serve in the future as an alternative to toxic paints for all ships, "can serve" because a gulf exists between scientific developments and practical use for shipping. Ship owners are understandably wary of completely new and untried products and will make the change only when the risk is minimal. This is, for example, currently the case with nonstick coatings.

20.7.2.2 Nonstick Coatings

Nonstick coatings, mostly based on silicon polymers, have, where fouling organisms are concerned, an unattractively low surface energy [65,75–79]. The free surface energy of the material determines whether an organism attaches or not. Only limited prevention of fouling is provided, but the adhesion between the fouling and the coating is weak; therefore, fouling can easily be removed or will fall off while sailing if the speed is high enough (5–10 knots). In theory, the lifetime is unlimited except when substances are used that have an additional antifouling effect through leaching, e.g., "sweating" silicon oil or paraffin.

Systems were introduced on the market and presented as wonder products with promised effective long-lasting protection, enduring for decades. However, reports about the effectiveness of nonstick coatings became very contradictory. The materials often showed insufficient antifouling properties, but fouling was easily removed. Weakness of the coatings and difficulties with their application are often reported. The best results were achieved with polysiloxanes (silicon elastomers). In recent years, with the exception of small improvements in mechanical strength and adhesion to subsurfaces, no decisive developments have occurred and attention is now focused on practical efficiency. Even though the antifouling effect is obviously improved, some major obstacles are difficult to overcome: price (estimates vary: 5–100 times higher than the classic antifouling paints), difficult application (special apparatus is needed; normal appliances are not suitable), mechanical frailty (with, as a result, damaged areas where fouling quickly occurs), bad adhesion, and the necessity to remove fouling if sailing is insufficient.

Pollution when using nonstick coatings is low in comparison with that of systems based on biocide leaching, but emissions as a result of damage, pieces of the coating chipping off, and possible leaching of coating ingredients must be taken into account. Rapidly biodegradable silicon products could possibly minimize pollution.

There are no products freely available on the market as yet. Practical application has been effected in other areas (especially cooling water systems). Recent or current sea trials with ships—e.g., in the United States, Australia, and France—confirm the potential suitability, but because of the previously mentioned obstacles, the step to commercial use will take several years at least, especially for larger vessels.

20.8 Measuring and Monitoring Methods

20.8.1 Introduction

An abundance of measuring techniques for characterizing or monitoring the performance or organic coatings is available. In the last few years, several "new" techniques were introduced to study the behavior of organic coatings at an almost microscopic level. The standard techniques include electrical or electrochemical techniques such as electrochemical impedance spectroscopy (EIS), dielectric measurements, corrosion potential measurements, or polarization resistance measurement. The new, local techniques are in general called "scanning probes" and include the Kelvin probe, scanning vibrating electrode (SVET), scanning reference electrode (SRET), and local EIS. Various techniques for developing a detailed understanding of the polymer properties, including FTIR, DSC, DMTA, TG, hardness measurements, and AFM, are also frequently used, and naturally also test methods for dry and wet adhesion are important tools [7,9]. It is important to combine the results of these various techniques in order to compose suitable models for the different chemical and physical phenomena in these very complex systems. In most cases, a specific method measures only one specific parameter of the coated system that has no direct relation to the subject of interest. The water diffusion coefficient (D), for instance, can easily be estimated with EIS [80]. The direct relation of D to underfilm corrosion, however, is questionable. Therefore the selection of a set of measuring techniques depends very much on the specific aim of the investigator, who may be interested in a variety of phenomena such as water uptake, curing, and underfilm corrosion that have a relation in a complete, complex systems. These techniques, with their specific possibilities and limitations for research and monitoring of organic coatings, could well be the subject of a complete monograph. Within the framework of this chapter, a limited choice had to be made.

Because impedance measurements in the form of dielectric sensors, also in situ, can in principle give a rather extensive package of information on all charge- and dielectric-related phenomena including charge transfer reactions, diffusion processes, and delamination, these will be described first. A second reason for this choice is the possibility offered by this technique to arrive at lifetime predictions based on short-term measurements [80–85]. After relatively short exposure times, an indication of long-term behavior may be obtained. A third and not unimportant reason is the fact that in the literature a rather extensive discussion is going on the possibilities and impossibilities of this technique.

Second, we will focus briefly on electrochemical techniques in general. This will be used as a starting point for a brief description of the more recently developed local techniques. In the part also the information on corrosion mechanisms as presented earlier in this chapter will be used to evaluate the different methods.

Besides these measuring techniques, various weathering tests are, of course, used. These may vary from the natural weathering, via the well-known and often discussed salt spray test, to more sophisticated cyclic weathering tests that are designed specially, for instance for atmospheric corrosion of buildings. The combination of the previously mentioned measuring techniques with these weathering tests is a subject that is discussed more and more. Finally, adhesion measurements are briefly addressed.

20.8.2 Electrochemical Impedance Spectroscopy

Impedance spectroscopy offers the possibility to study various phenomena. Here we first focus on delamination. In the next section we discuss the study of corrosion initiation. The study of the curing of coatings and special monitoring techniques has been left out of the discussion [7,9,86–90].

Electrochemical impedance spectroscopy (EIS) has been described by several authors as a technique for measuring all kinds of properties of organic coatings. But corrosion processes underneath coatings also have their own response in an impedance measurement. If corrosion occurs underneath coatings, electrons are transferred between molecules and metals in the corroding system. This charge transfer can be studied with normal electronic entities such as the impedance. The impedance Z as the ratio between a small sinusoidal potential perturbation [$V = V_0 \sin(\omega t)$] and the current resulting from this perturbation [normally $I = I_0 \sin(\omega t + \varphi)$]:

$$Z = \frac{V_0 \sin(\omega t)}{I_0 \sin(\omega t + \varphi)} \tag{20.1a}$$

Often this equation is represented using complex numbers:

$$Z = \frac{V_0 \exp(j\omega \mathrm{t})}{I_0 \exp(j\omega t + \varphi)} \tag{20.1b}$$

This impedance Z is frequency dependent and is characteristic of the system that is measured. In case of measurements of electrochemical reactions (e.g., corrosion), Z is called the electrochemical impedance. Normally Z is measured for many frequencies to create an impedance spectrum.

If a system that contains an electrochemical reaction is characterized by electrochemical impedance spectroscopy, there are often also nonelectrochemical parameters contributing to the total impedance. These might be the resistance of the electrolyte but also the dielectric behavior of an intact coating. In fact, part of the literature describes only the dielectric behavior of coatings during water absorption in the absence of corrosion.

The evaluation of the measured impedance is normally performed using complex numbers. In that case Z is written as complex number [9]:

$$Z = \mathrm{Re}(Z) + j\,\mathrm{Im}(Z) \tag{20.2}$$

with Re(Z) as the real part and Im(Z) as the imaginary part of the impedance. Re(Z) and Im(Z) are normally used to represent the impedance results.

In the analysis and interpretation of EIS, two fundamentally different approaches are possible, one using a complete transfer function and another simplified approach using equivalent circuits based on an assumed simplified physical model.

If the reaction mechanism is known, a transfer function may be calculated using the kinetics of the reactions, the diffusion equations, and so on. This transfer function couples the input from a system to the output. It includes all possible variables and is directly fitted to the measurements. In this way direct information on reaction parameters is obtained from the impedance. If the reaction mechanism is known and the system is not too complex, this method is favorable [91].

It is obvious that a drawback of this method is that the reaction should be almost fully understood in order to produce a reliable transfer function. Another drawback is the fact that for more complex reactions the number of variables is very high. For a reliable fit the less important processes have to be ignored or variables have to be fixed; in other words, the important parameters of the system have to be estimated prior to the fitting procedure. In many cases the interpretation of measurements on coatings will be extremely complex due to the unknown geometries and local variation in electrolyte concentrations.

In this approach, one has to start with at least some idea of a physical model of what is happening in the measured system. This physical model is then transformed into an equivalent electrical circuit using only electrical components such as resistors and capacitors and some special components representing transport phenomena. An objection to this method is the fact that often more than one circuit is possible and circuits might also be rewritten in a different configuration without changing the response. This makes the interpretation in terms of chemical and physical parameters difficult. Figure 20.13, for example, shows two different circuits that give exactly the same response if the right values for resistors and capacitors are used.

Despite these objections, it is often possible to configure reasonable circuits that can be fitted to the measured data. Especially for intact coatings, the dielectric behavior and deviations from ideal behavior may give valuable information. Due to the precision of the measurements, it is in many cases also possible to measure small changes in coating properties that result from changes due to weathering, aging, etc.

Three examples of the use of EIS will be given next: atmospheric water uptake, measurements during delamination of intact and perforated coatings, and measurement of reproducible laser defects.

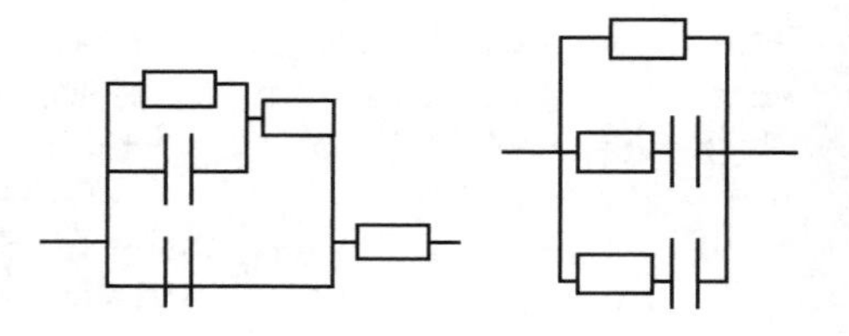

FIGURE 20.13
Two different equivalent circuits with the same impedance response. Rectangles represent resistors, parallel segments capacitors.

20.8.2.1 Atmospheric Water Uptake

The water uptake of organic coating can easily be monitored using the strong dipole of the water molecule. Because of this dipole character, the dielectric constant of water is high compared with that of most polymers [92]. Therefore a small amount of water absorbed in an organic coating will give a strong change in the dielectric constant of the system coating plus water. Dielectric constants can be studied using impedance spectroscopy. In this situation the term "electrochemical impedance spectroscopy" is not quite correct because no electrochemical but only physical parameters are measured. It is, however, widely accepted to use the term EIS instead of IS and for simplicity this terminology will be followed here too.

It is also interesting to follow water uptake under atmospheric conditions. Because no bulk electrolyte is present in this situation, special arrangements have to be made in order to make impedance measurements possible. This can, however, easily be done with the setups presented next.

If grids of a conduction metal are applied by, for instance, PVD on top of the coatings, an EIS measurement can be performed using either the two combs of the grid as electrodes or the grid and the metal substrate as the electrodes. The grids may also be placed underneath a coating if the substrate is not conductive. In the case of multilayer coating systems they may be placed in between the different layers to ensure the properties of the individual layers. A possible configuration for a grid is given in Figure 20.14.

The geometry of the grids may of course differ from the one presented here. Normally, measurements are performed between the metal substrate and both arms of the grid. It is, however, also possible to measure between the two arms of the grid.

A disadvantage of this type of setup is the fact that an impermeable metal coating covers part of the surface. This changes the transport route for water. Grids arms should therefore be as thin as possible.

Grids can easily be produced on top of coatings by PVD. Often gold is used for this purpose. If coatings are applied to inert and nonconductive substrates (e.g., glass) the grid can be placed underneath the coating and the higher precision of lithographic techniques can be used produce finer grid lines.

Using these grids, a series of impedance measurements was performed on panels coated with a clear two-components epoxy. The panels were put for at least 3 days in a furnace at 40°C to dry completely and after this period they were exposed to an atmosphere with an

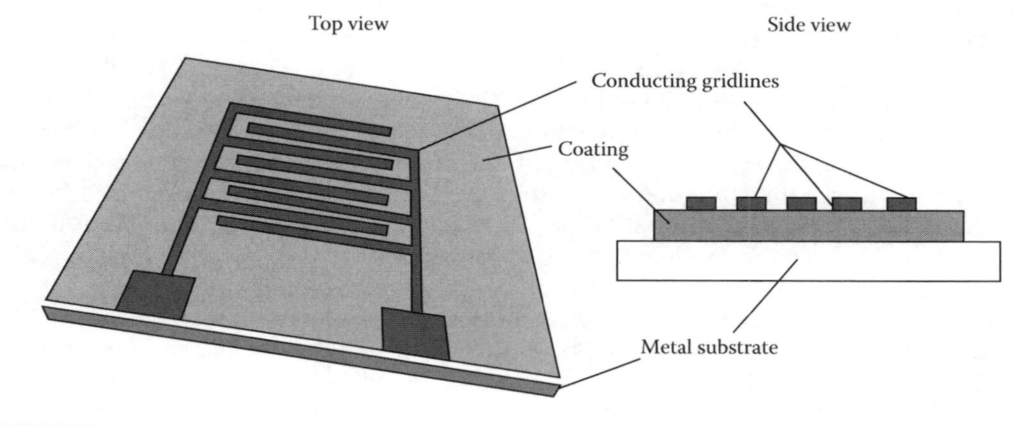

FIGURE 20.14
A possible configuration for a grid used to study atmospheric water uptake. (From van der Weijde, D.H., Impedance spectroscopy and organic barrier coatings, Thesis, TU-Delft, Delft, the Netherlands, 1996.)

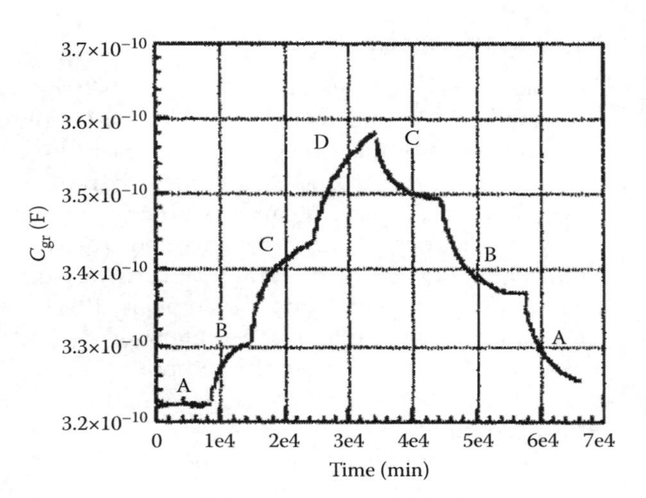

FIGURE 20.15
Two examples of the capacitance (C_{gr}) as determined from impedance measurements performed on the grids as a function of time. A, B, C, and D refer to RH values of 20%, 60%, 80%, and 90%, respectively. Thickness of the coatings was 92 μm (left) and 98 μm (right). The cycle in Figure 20.15 took about 7 weeks, and in Figure 20.16 about 6 weeks.

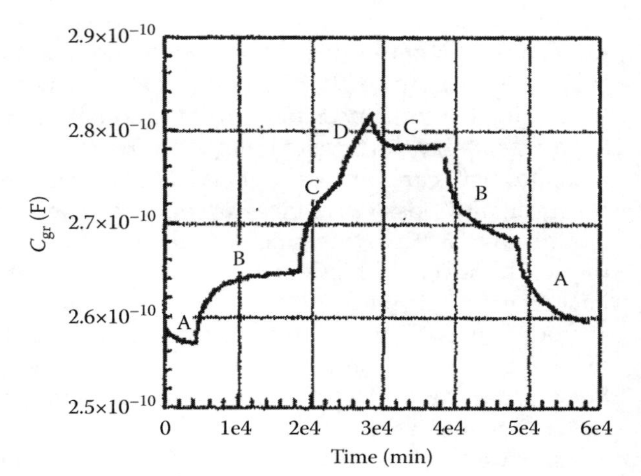

FIGURE 20.16
Two examples of the capacitance (C_{gr}) as determined from impedance measurements performed on the grids as a function of time. A, B, C, and D refer to RH values of 20%, 60%, 80%, and 90%, respectively. Thickness of the coatings was 92 μm (left) and 98 μm (right). The cycle in Figure 20.15 took about 7 weeks, and in Figure 20.16 about 6 weeks.

RH of 20%. Starting from 20% RH, the humidity was increased stepwise after some days of exposure. This stepwise increase was repeated until in three steps an RH of 90% was obtained. Then the atmosphere was brought back again in the same steps to 20% RH. During these periods the impedance of the grids was measured continuously. The results of these measurements are shown in Figures 20.15 and 20.16. In these figures the capacity values as determined from impedance data are presented.

Both series of measurements in Figures 20.15 and 20.16 seem to give approximately the same picture. The differences that are visible between the two cycles are caused by differences in exposure times in the different atmospheres. One big difference between the two series is, however, the absolute value of the capacitance. This is caused by the different surface area of the two grids.

For the transport of water the diffusion coefficient is also an important factor. The cyclic experiments can now be used to test whether D is independent of the amount of water absorbed in a coating. This is normally assumed in studies of water uptake by organic coatings. To describe the diffusion in a one-dimensional system, Fick's law is used:

$$J = -D\frac{\delta c}{\delta x} \tag{20.3}$$

Because all coatings in the cyclic experiments have a thickness (δx) of around 90 μm and within the cycles the thickness is assumed to be constant (no swelling occurs), a comparison of the diffusion coefficient of water in the cycles can be made. Because of the constant thickness δx can be set to unity. Furthermore, the flux J is equal to the increase in water content in time, which is approximately proportional to $\delta C/\delta t$ (C = capacitance), which is the derivative of the curves in Figures 20.15 and 20.16. Finally, δC is the change in relative humidity going from one atmosphere to another. Combining these assumptions, this leads to a comparative diffusion coefficient of water D^*, which can be calculated using:

$$D^* = \frac{\delta(\text{capacitance})}{\delta(\text{time})} \times (\text{RH2}(\%) - \text{RH1}(\%))^{-1} \tag{20.4}$$

In which RHl and RH2 are the relative humidities of the atmospheres (of course, this D^* is not a real diffusion coefficient, but it can be used for comparison of the results). When this equation is used to calculate the D^* value of each step in the cycles of Figures 20.15 and 20.16, it gives the results presented in Table 20.2.

From these data, despite the rough method, it is clear that there are quite large variations in D^* as a function of the RH. The higher the concentration of water in the coating, the higher the D^*. This can be expected because of the plasticizing effect of water on epoxies [93]. As expected, both grids do have comparable values for D^*.

20.8.2.2 Impedance Measurements and Delamination

EIS can also be used to measure and characterize defects in coating. This because a defect will also have a specific impedance response. Not very defect is, however, visible in measurements and care must therefore be taken in the interpretation of the data. To show this, two essentially identical panels were tested and will be referred to as cases A and B. On both panels a 100-μm-thick industrial coating was applied over a salt film that was present on the metal surface. As a result of this, both panels showed delamination and osmotic blistering as soon as they were exposed to an electrolyte.

Directly after immersion, case A showed a Nyquist and Bode plot that deviated from the expected plots for intact coatings. Representative measurements after 1 h and 10 days are given by Figure 20.17a and b.

Analysis of the data with equivalent circuits showed that at least four overlapping time constants were present. This implies that a rather complex equivalent circuit should be

TABLE 20.2

Comparative Diffusion Coefficient for Two Cycles and Different Changes in Atmospheres as Calculated from the Results of Figures 20.15 and 20.16[a]

From RH to RH	δC	D^* Cycle 1 (m^2/s)	D^* Cycle 2 (m^2/s)
20% to 60%	40	2.76e−15	3.00e−15
60% to 80%	20	8.98e−15	7.7e−15
80% to 90%	10	1.64e−14	9.67e−15
90% to 80%	10	9.41e−15	6.56e−15
80% to 60%	20	7.95e−15	6.9e−15
60% to 20%	40	3.75e−15	2.99e−15

[a] Values can be used only for comparison within this table.

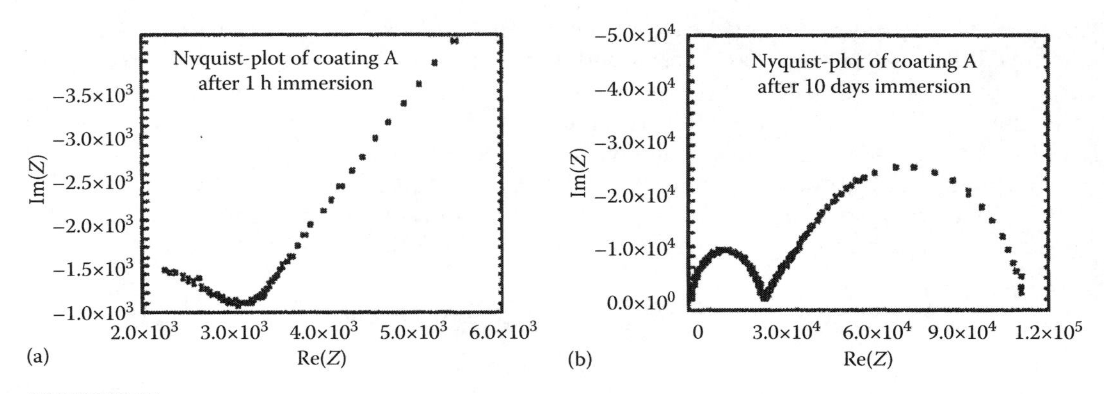

FIGURE 20.17
Two Nyquist plots of measurement on coating A: (a) after 1 h and (b) after 10 days of exposure to water (plots are not isotropic).

used to explain the data and that it is not sufficient for a fundamental study to monitor only one frequency or phase angle [94]. A complete analysis of these data is beyond the scope of this chapter, but an important conclusion is that the corrosion component in the equivalent circuit is in parallel with the coating component. This implies a real defect that penetrates through the coating. Prolonged exposure indeed showed corrosion products coming out of a small pore in the coating.

The almost identical panel B has a completely different start upon exposure. Here we see a normal, capacitive behavior after 1 h of immersion as can be seen in Figure 20.18a. After 2 days this still has not changed, as can be seen in Nyquist plot of Figure 20.18b. This behavior is observed for several days. However, after 2 days delamination is already clearly visible as a small blister above the contaminated area. So in this case it is clear that, although a defect is present, it is not (easily) visible in the EIS data.

The corrosion potential of panel B also appeared to be unstable. Several times the open-circuit potential drifted to −3000 mV versus SCE. Every time this happened it was polarized at a normal potential for steel in the electrolyte. This unstable behavior is due to the very high resistance of the coating, which in this case exceeds the input impedance of equipment used.

These observations support the conclusion that the coating is completely intact although partly delaminated due to the contamination underneath.

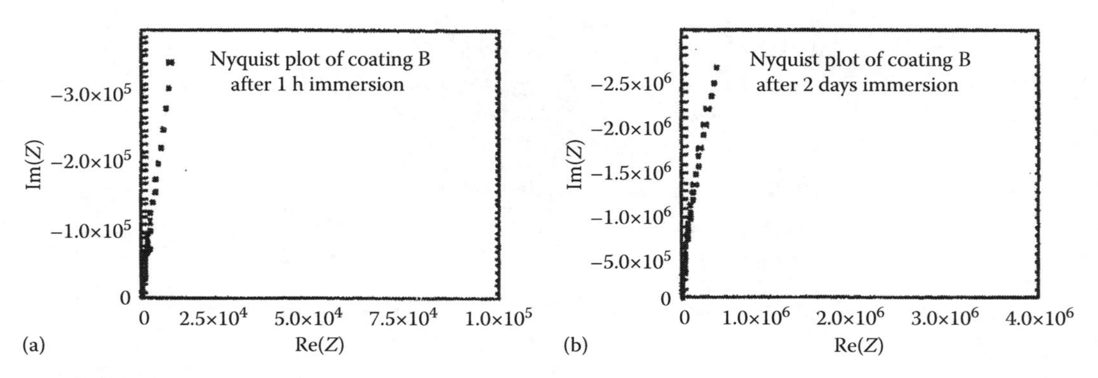

FIGURE 20.18
(a and b) Two Nyquist plots of measurement on coating B (plots are not isotropic).

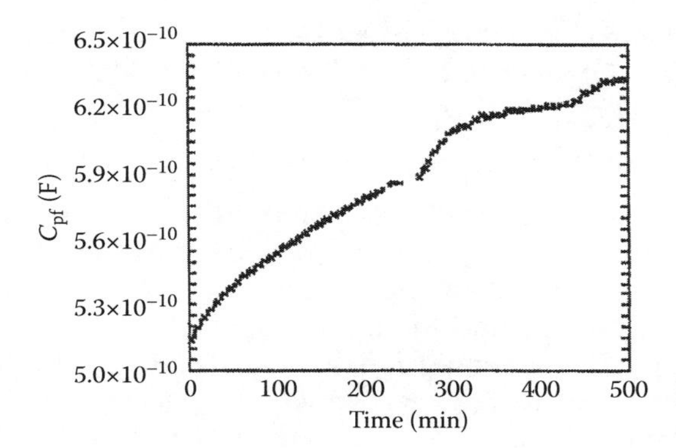

FIGURE 20.19
Coating capacitance as a function of time during water uptake of coating B.

From the measurement of the water uptake in the first hours of immersion, some qualitative information on this delamination can be obtained. From the coating capacitance (C_{pf}) against time plot of Figure 20.19 it is clear that the coating does not show stable Fickian water uptake but shows deviations that are probably caused by the delamination. Because the delamination is forced and is already starting when the coating is still taking up water, the water uptake and the deviation from normal saturation behavior are not visible separately.

To further prove the similarity between panels A and B, the blister on panel B was perforated deliberately with a needle. Directly after this the impedance spectrum of Figure 20.20a and b was measured, which is rather similar to the data measured on panel A. Because it is known that defect and delamination are in this case combined, an equivalent circuit with a corrosion component in parallel to the coating impedance must be used if further analysis is to be performed.

From the results of the measurements just described it is clear that quantitative measurements on delamination under an intact coating are not possible with impedance measurements as long as the impedance of the systems remains high, which is often the case for thicker heavy-duty systems. In terms of equivalent circuits that are used to explain impedance measurements, this implies that the corrosion component of the circuit has to be in series with the intact coating components. In the literature this corrosion component

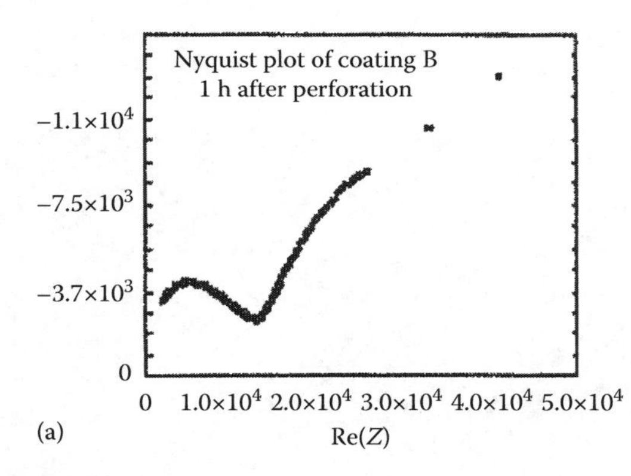

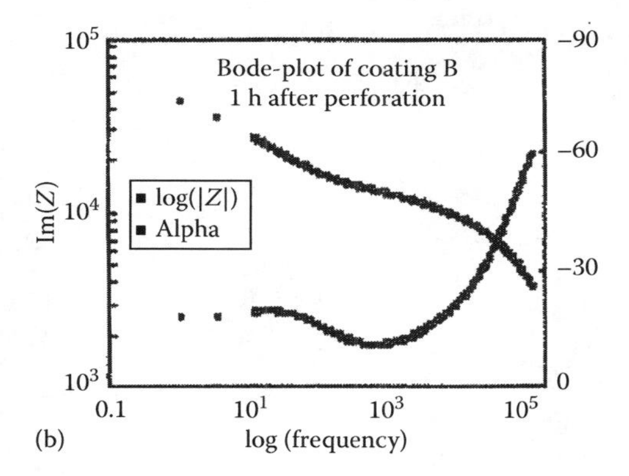

FIGURE 20.20
Nyquist (a) and Bode (b) plots measured on coating B directly after perforation, in (b) log Z is given on the left axis and alpha on the right axis.

is often placed in series with a relatively low resistor, meaning that the properties of the coating have already changed considerably (e.g., porosity and other flaws have developed).

20.8.2.3 Impedance Measurements and Artificial Defects, Laser Ablation

Often organic coatings do not fail early even under severe laboratory conditions. Therefore it is necessary to make artificial defects in a coating system to study the corrosion initiation under a coating. Both standard testing by exposure of standard panels with or without mechanically induced scribes and laboratory research into the mechanisms of the reactions suffer from the rather uncontrolled application method for these artificial defects. The width and the height of the scribe largely determine the mechanisms. Sometimes, for complicated systems with more layers as in car body panels, one wants to make sure that a certain layer is still intact, and for mechanistic studies in the laboratory one wants to study the very early stages of the delamination. In both cases, artificial defects made with the laser ablation technique may provide the answer [95–97].

Defects in surface coatings are used to study the protective properties of these coatings. Defects made by ablation with excimer lasers have important advantages with respect to traditional techniques. A resolution of the geometry down to the micrometer scale with high precision and good edge definition is feasible, yielding little, if any, distortion of the remaining coating layer: no coloring, foaming, or melting. Moreover, the results show no ruptures and no delamination of the coating layer from the substrate material. Normally, subsequent laser pulses perform the etching of a defect until the underlying metal substrate is reached. The etching is automatically stopped at the metal surface, as the threshold for ablation of the metal higher than the threshold for organic materials. The laser beam will not change the remaining metal surface. The fine control over the ablation process makes another feature possible. If the etching is interrupted just before the metal is reached, a defect with a thin (several μm) residual layer on the bottom can be produced, as illustrated schematically in Figure 20.21. It is this defect that proved to have ideal characteristics for broad characterization of organic coatings, including electrochemical impedance spectroscopy.

After impedance measurements have been performed, meaningful physical information on the system may be obtained only from a correct analysis and interpretation of the data as discussed before. Impedance analysis may be performed at different levels of complexity. The easiest form is a visual inspection of the measured impedance diagrams or changes in these diagrams with time. This may give first insight into the performance of a specific system. More detailed information on mechanisms and corrosion phenomena

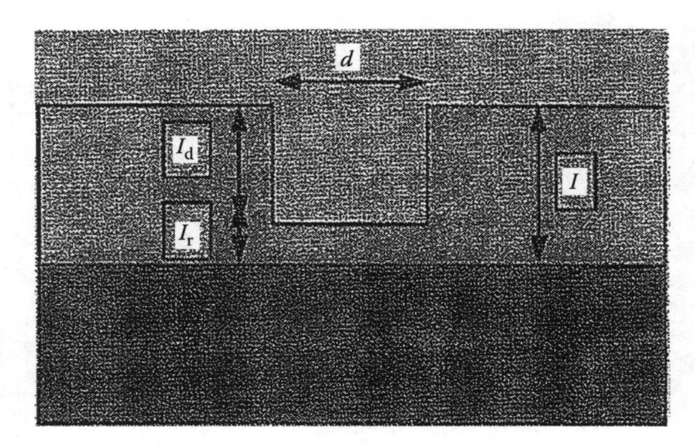

FIGURE 20.21
Typical geometry of an artificial laser defect in an organic coating, d = defect diameter, I_d = defect depth, and I_r = thickness of the residual coating layer in the defect.

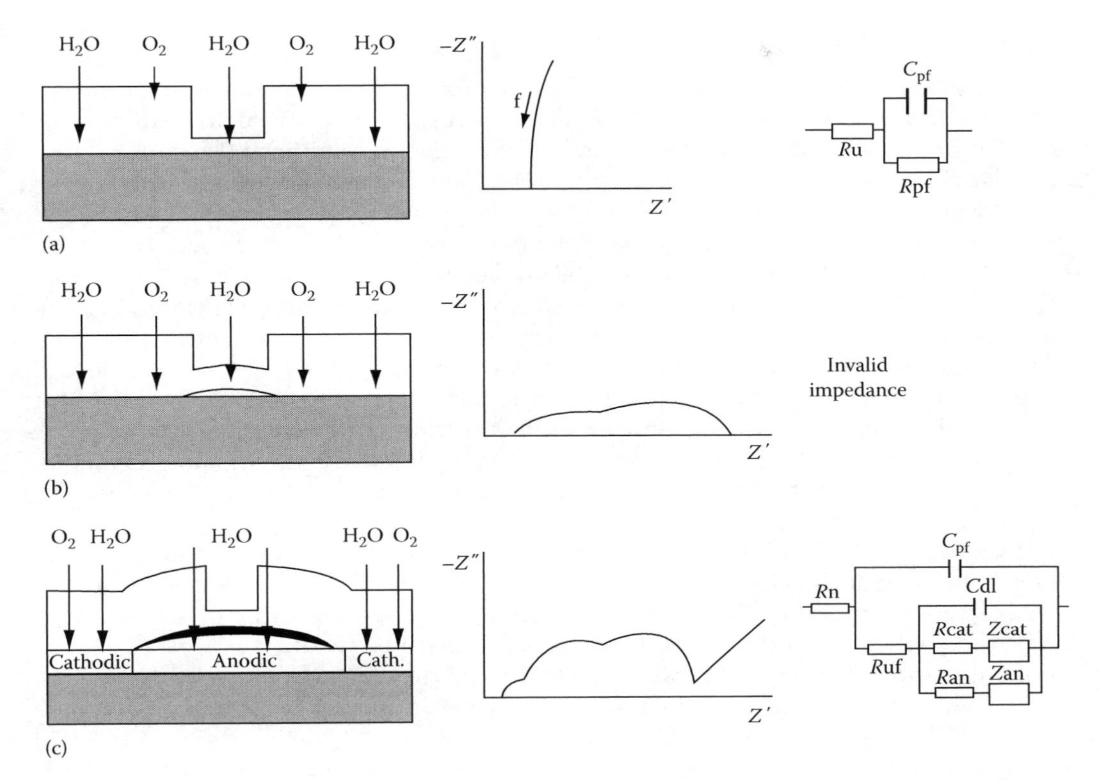

FIGURE 20.22
Different stages in behavior of a coated system with progressing exposure time to water. (Left to right) System, impedance response, and equivalent circuit Stages: (a) Water permeation; (b) corrosion initiation; (c) quasi-stationary corrosion. (R = resistance, C = capacitance, Z = diffusion impedance, pf = paint film, cat = cathodic, an = anodic, dl = double layer, u = electrolyte.)

may be obtained from an elaborate analysis procedure. In Figure 20.22 the different stages that may be distinguished in the behavior of a coated system as a function of exposure time to water are illustrated schematically. At different exposure times the coated system, with laser defect, is shown with the typical shape of the impedance response and the corresponding general equivalent circuits. The following stages are distinguished:

A. Water permeation
B. Corrosion initiation
C. Quasi-stationary corrosion

It should be stressed that the very start of the corrosion process is visible in the changing appearance of the Nyquist plot as given in Figure 20.22, but no quantitative interpretation of the impedance data is possible because of the relatively fast changes of the system. When the quasi-stationary stage, C, has been reached, detailed quantitative analysis is again possible, leading to a more detailed equivalent circuit supported by a physical model. It must be stressed that data from different techniques are always needed when searching for a detailed physical model. Also, the changes in time during exposure of the various components of the equivalent circuit can be used to check the practical value of a proposed model.

20.8.3 Local Electrochemical Test Methods

The local electrochemical techniques that are used for coatings research are mainly used in the more fundamental studies of corrosion mechanisms. They can be divided into techniques in immersion (SVET end SRET) and in atmospheric conditions (Kelvin). All techniques are based on the observation that corrosion mechanisms underneath coatings are generally mechanisms in which anodic and cathodic processes take place on separated sites with a slightly different local "corrosion potential." The techniques are used to visualize these differences. SRET and SVET can also be used on polarized samples to detect pinholes.

SRET is a technique in which a reference electrode is scanned over a surface. A second reference electrode is used to monitor the overall corrosion potential. In this way, small differences in the potential can be visualized. An alternative type of SRET has a tip with two platinum needles that are both placed close to the surface. If local potential differences exist over the surface, a corrosion current will run through the solution. In that case a small potential drop will occur in the solution between the two platinum wires. Assuming an ohmic behavior for the solution resistance, this potential difference is proportional to the potential differences on the surface.

A drawback of the SRET is the fact that it has to measure very small differences in a direct current mode. It is therefore not very sensitive and in most cases the substrate has to be polarized to increase the current that is flowing through the solution. In this mode it can be used only to show the location and size of defects in a coating.

A more sensitive technique is measuring the same *IR* drop in the solution by the SVET. In the SVET the potential of a small vibrating tip is measured. If this tip is vibrating in a potential gradient, the resulting potential vibration in the tip is easily detected (AC mode) with sensitive electronics. Although more accurate, this method is much slower than the SRET.

A drawback of both methods is that complex corrosion mechanisms underneath coatings including total delamination cannot be determined. Calculations with current distribution models [98] can easily show that in the case of underfilm corrosion almost no corrosion currents will be present in the outside electrolyte. Polarization of the substrate is therefore necessary to show the presence of defects. This will, however, at the same time influence the corrosion mechanisms.

The Kelvin probe is, unlike the other techniques, able to measure through an insulating coating layer. Lord Kelvin used the principle of the Kelvin probe to show that a Volta potential difference exists between metals. Apart from the obviously modern electronics, the principle of the Kelvin probe is still the same: if two metal plates are coupled through a wire and are brought close to each other, a potential difference that is equal to the Volta potential difference between the two metals exists over the narrow gap between the plates. As a result of this, the metal plates will have a capacitive charge. If one of the plates is now vibrated, the capacity between the plates is varied and as a result of this an AC current will flow through the external circuit with a magnitude of

$$I_{ac} = \frac{\Delta V \delta C}{\delta t}$$

in which ΔV is the Volta potential difference between the two metals and $\delta C/\delta t$ the variation of the capacitance in time due to the vibration. If in this circuit an external DC voltage source U is introduced, this equation changes to

$$I_{ac} = \frac{(\Delta V - U)\delta C}{\delta t}$$

From this equation it becomes clear that for the case in which the external DC voltage U is equal to ΔV, the AC current I_{ac} goes to zero, independent of $\delta C/\delta t$. This implies that with this measurement it is even possible to measure through vacuum, air, or an organic coating as this will influence only the term $\delta C/\delta t$.

Stratmann showed in several articles published around 1987 that for metals with a thin moisture layer the Volta potential difference was proportional to the corrosion potential of that system [99]. Using a calibration, it is therefore possible to measure the corrosion potential of a corroding system through an air gap and without making contact.

Although powerful, this technique is still under further development. Especially the interpretation of the measured results is still a matter of discussion. Without doubt it will in the future bring more and more results, especially on specific corrosion mechanisms underneath coatings arid on atmospheric corrosion. Recent developments have shown that it is even possible to combine the extremely high resolution of the AFM with the measuring principle of the Kelvin probe. For special systems (uncoated), this may reveal even more details of corrosion mechanisms.

20.8.4 Other Electrochemical Methods

Among the other electrochemical methods are the measurement of the corrosion potential and the determination of the polarization resistance (R_p). Both methods are relatively simple but may give relevant information.

The polarization resistance is determined as the slope of a polarization curve at the corrosion potential at the point where the external current is zero. In this way it is a linearization of the more complex polarization curve. The value of R_p/cm^2 is directly related to the reactivity of the exposed metal substrate through the Stern and Geary relation. For a coated system the total value of R_p is also determined by the amount of metal that is exposed. Both a higher reactivity and a larger exposed area lead to a lower R_p. If for a given system the R_p is rather constant in time, this technique can be used to predict the corrosion rate. In closed metal containers as used for beer and soft drinks only a minimum amount of oxygen in present in a normally rather acidic electrolyte. In that situation, no solid corrosion products are formed and the R_p is constant over time. In these drinks the amount of iron after a certain shelf life is an important quality check. Figure 20.23 shows

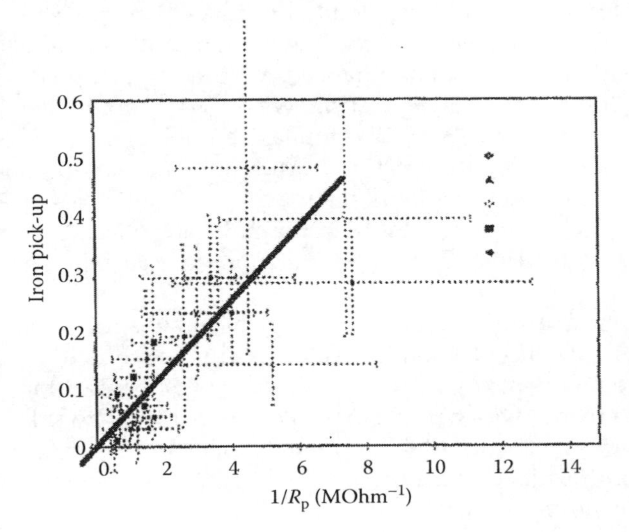

FIGURE 20.23
Iron content of the electrolyte after 6 months of exposure versus polarization resistance. Iron pickup in mg/L. Different points represent different series of filled cans corresponding to various closing conditions.

the correlation between $1/R_p$ as measured after 2 weeks and the iron pickup of the drinks after 6 months. The R_p was measured over a 20-mV range around the corrosion potential of the system in the beverage itself (in this case cola, which has a pH of around 2.5).

In the case that no defect is present in the coating, an R_p measurement is measuring only the normally very high resistance of the polymer layer. In that case, the obtained value should be similar (if measurable) to the values obtained from fitting an impedance measurement.

20.8.5 The Value of Enhanced Weathering Tests

The recent developments of electrochemical testing procedures, of which we have discussed two, may result in better system lifetime prediction methods and monitoring techniques for coating systems than are now available. Many new coating systems are being developed as a result of the various environmental demands and legislation together with demands for high quality for improved lifetime.

The industrial need for fast testing methods is therefore enormous because testing by, e.g., atmospheric exposure of the systems takes too much time. Many fast testing procedures exist. One of the first enhanced weathering tests was the exposure of coated panels in Florida, where 3 years of exposure would be more or less similar to 10 years of exposure in Europe because of the high temperature and humidity in Florida. However, many new tests have been developed, and the exposure in Florida is now not regarded as a fast test. Shortening the test period means that various parameters determining the corrosion initiation and propagation processes are considerably changed compared with natural conditions. These include:

- The temperature
- The composition of the corrosive including NaCl, pH, SO_2, and additions such as copper
- The relative humidity UV radiation

For modern cyclic testing some of these parameters are changed in time at regular intervals, simulating more natural exposure conditions.

The aim of any testing procedure is to find a simple, preferably linear relation between natural exposure and enhanced testing results. However, according to many authors there is no such simple relation [100]. Some authors state that enhanced weathering just gives some indication of the relative quality of coating during testing, without any predictive value for the lifetime. Even standard tests (e.g., ASTM B117) emphasize the lacking relation between, e.g., the salt spray test and corrosion protection in different surroundings due to the enormous differences in the reaction mechanisms. It has been stated that coated galvanized steel always performs worse in a salt spray test than normal cold-rolled steel, whereas in reality it always performs better [101]. Cyclic testing improves the relation between test and practice [102].

It has been reported [103] that standard salt spray tests give reasonable results for coatings in use in surroundings with a high salt concentration (seawater, deicing salt) and high RH but not for other applications. The high salt concentration would be responsible for decreased action of the anticorrosive pigments, which perform well under natural low-salt conditions. Also, salt spray tests often give blistering, which is often not relevant except for marine applications. Cyclic testing again results in less blistering, which at least seems to indicate a better simulation of natural conditions.

A comparison of a cyclic test newly developed by Hoogovens for automotive panels with the salt spray test, the SCAB (simulated corrosive atmospheric breakdown) test, and the VDA (Verein der Deutschen Automobilindustrie) test showed that some differences can be observed in the test results for the ranking as far as undermining is concerned. A negative relation exists between the results of the new cyclic test and the salt spray test. The new cyclic test shows reasonable agreement in many cases with the SCAB test. Detailed research showed that the different test procedures indeed result in different corrosion mechanisms [I].

Therefore is must be concluded that enhanced weathering tests normally should be cyclic, simulating many practical parameters with their changes in time. Interpretation should be performed with extreme care. Improvements of fast testing procedures, e.g., by incorporating electrochemical and dielectric sensors, are of extreme importance in order to arrive at laboratory-based lifetime prediction.

20.8.6 Adhesion Tests

As mentioned before, adhesion is one of the major properties of a protective coating. Testing of adhesion seems therefore of prime importance for the characterization of organic coatings. In practice it is, however, rather difficult to measures the adhesion of organic coatings.

One group of simple and straightforward techniques can be categorized as "deformation techniques." Among these are T-bend, direct impact, and Erichsen cup tests. In all these tests the specimen of interest is subjected to a nonuniform deformation (mainly stretching). The amount of deformation at which the coating deadheres is in these cases a measure of the adhesion. Although important for many mainly practical applications, these methods cannot be used to gain more fundamental knowledge of adhesion. This is because the measured value is a combination of the mechanical properties of the coating and the adhesion. Besides that, these methods mainly look at shear stresses.

Another method that is often used is a "Crosshatch cut." In this method a cross pattern is scratched into the coating. This operation also causes shear stresses on the coating–metal interface. As a result of this the smallest squares will completely deadhere. The minimum size of the squares that remains attached is in this case a measure of the adhesion.

Both these techniques (deformation and Crosshatch) are generally used in combination with a sort of "tape test" to determine the exact degree of delamination. Any part of the coating that has higher adhesion to the tape that is glued on top is removed in this test. This means that it is a binary method and is not able to quantify the exact adhesive properties.

One method that is designed for these more quantitative adhesion measurements is a pull test. For this test a dolly is glued unto the coating surface. The coating around the dolly is scratched down to the bare metal to isolate the area underneath the dolly from the other parts. After complete curing of the glue, the dolly is pulled from the coating in normal tension tests. In the case that the coating–metal interface fails, an exact value is obtained for the adhesion in tensile mode. In the case that either the coating polymer itself or the glue fails, of course, only a minimum value is obtained. A disadvantage of this method is the fact that the dolly has to be glued on the surface. Apart from the chemical effects this may cause, it effectively screens the interface from, for instance, water. It is therefore difficult (although maybe not impossible) to measure under wet conditions.

A last technique that will be mentioned here is also a quantitative method but it can be used under wet conditions. It is testing under a relatively realistic load, which is a combination of tensile and shear stress. In this method a strip of the coating is scratched and at one side freed from the substrate. This lip is then pulled from the surface at a constant speed and a constant angle to the substrate. The force necessary to do this is a measure of the adhesion of the coating. This method also gives the possibility to determine the adhesion along a part of the substrate. For the region just in front of underfilm corrosion this was recently illustrated by Fuerbeth and Stratmann and also discussed in Ref. [62]. The only disadvantage of this method is the fact that the tensile force is applied on the coating. This makes it suitable only for thicker coatings. Thin coatings (such as used in the canning industry) will in many cases break before delamination starts.

As a conclusion to this section on adhesion measurements, we can say that the despite the importance of this topic only a limited number of relevant techniques is available. When selecting a method it is important to use a method with a relevant force mode (shear, tensile, or a combination) and relevant conditions.

References

1. F. Blekkenhorst and E. Nagel Soepenberg, *Proceedings Corrosion '88*, St. Louis, paper 354, 1988.
2. J. H. W. de Wit and C. B. M. Nagel, Micropeenbehandlung mit guten Resultaten, *Chem. Anlagen Verfahren* June: 5 (1992); J. H. W. de Wit, Refinement of stainless steel surface through micropeening, *Proceedings of ACHEMA*, Frankfurt, Germany, 1991, p. 16.
3. W. Funke, *Prog. Org. Coatings 9*:29 (1981).
4. K. R. Gowers and J. D. Scantlebury, *Corros. Sci. 23*:935 (1983).
5. W. Funke, *Ind. Eng. Chem. Prod. Res. Dev. 24*:343 (1985).
6. J. R. H. Leidheiser, Corrosion of painted metals—A review, *Corrosion 38*:374–383 (1982).
7. F. M. Geenen, Characterization of organic coatings with impedance measurements, thesis, Delft University of Technology, Delft, the Netherlands, 1991.
8. F. M. Geenen, Corrosion protection by organic coatings, Lecture notes (Dutch), Delft University of Technology, Delft, the Netherlands, 1989.
9. E. van Westing, Determination of coating performance with impedance measurements, thesis, Delft University of Technology, Delft, the Netherlands, 1992, and a series of four papers in *Corros. Sci.*: Part 1, *34*:1511 (1993); Part 2, *36*:957 (1994); Part 3, *36*:979 (1994); Part 4, *36*:1323 (1994).
10. F. Geenen, E. van Westing, and J. de Wit, *Prog. Org. Coatings 18*:295 (1990).
11. F. Geenen and J. de Wit, *Proceedings, 30th ACA Conference*, Auckland, NZ, paper 59, 1990.
12. R. A. Dickie, *ACS Symp. Ser. 285*:773 (1985).
13. A. D. Wilson, J. W. Nicholson, and H. Y. Prosser, eds., *Surface Coatings 1 and 2*, Vol. 1, 1987, Vol. 2, 1988, Elsevier Applied Science, Amsterdam, the Netherlands.
14. H. Leidheiser, *Polymeric Materials for Corrosion Control* (R. A. Dickie and F. Louis Floyd, eds.), *ACS Symp. Ser.*, 1986, p. 124.
15. H. Leidheiser, W. Chang, and L. Ingetoft, *Prog. Org. Coatings 11*:19 (1983).
16. J. S. Thornton, J. F. Cartier, and R. W. Thomas, *Polymeric Materials for Corrosion Control* (R. A. Dickie and F. Louis Floyd, eds.), *ACS Symp. Ser.*, 1986, p. 169.
17. E. L. Koehler, *Localised Corrosion* (R. W. Staehle, B. F. Brown, J. Kruger, and A. Agarwal, eds.), NACE, Houston, TX, 1974, p. 117.
18. G. M. Kogh, *Localised Corrosion* (R. W. Staehle, B. F. Brown, J. Kruger, and A. Agarwal, eds.), NACE, Houston, TX, 1974, p. 134.
19. W. Loven, Pretreatments for aluminium, Literature Review (Dutch), Division of Corrosion Technology, Delft University of Technology, Delft, the Netherlands, 1992.
20. T. Van der Klis, *Vademécum—Surface Techniques for Metals*, 5th ed., Bilthoven, the Netherlands, 1989.

21. D. B. Freeman, *Phosphating and Metal Pretreatment*, Woodhead, Cambridge, U.K., 1986.
22. *Tool and Manufacturing Engineers Handbook*, Vol. 3, *Materials, Finishing and Coating*, 3rd edn., Society of Manufacturing Engineers, Dearborn, MI, 1985.
23. G. M. Brown, K. Shimizu, G. E. Kobayashi et al., The growth of chromate conversion coatings on high purity aluminum, *Corros. Set. 34*:1045–1054 (1993).
24. M. W. Kendig, A. J. Davenport, and H. S. Isaacs. The mechanism of corrosion inhibition by chromate conversion coatings from x-ray absorption near edge spectroscopy (XANES), *Corros. Sci. 34*:41–49 (1993).
25. A. H. Reed, For performance and economy: Chromate coatings, *Prod. Finish. 51*:60–64 (1986).
26. M. Koudelkova, J. Augustynski, and H. Berthou, On the composition of the passivating films formed on aluminum in chromate solutions, *J. Electrochem. Soc. 124*:1165–1168 (1977).
27. H. Leidheiser, Mechanism of corrosion inhibition with special attention to inhibitors in organic coatings, *J. Coat. Technol. 53*:29 (1981).
28. Toxicological profile for chromium (Public Health Service Report ATSDR/TP-88/10), Agency for Toxic Substances, Atlanta, 1989.
29. B. R. W. Hinton, Corrosion prevention and chromates: The end of an era? *Proceedings Asia Interfinish '90*, paper 30, Singapore, 1990.
30. D. G. Anderson, E. J. Murphy, and J. Tuccir, *J. Coat. Technol. 50*:38 (1978).
31. W. J. van Ooij, *Polym. Mater. Sci. Eng. 53*:698 (1985).
32. E. N. Soepenberg, H. G. Vreijberg, J. A. F. M. S. van Westrum, W. J. van Ooij, and O. T. de Vries, *Corrosion/85*, NACE, Houston, TX, paper 388, 1991.
33. F. Delaunois, V. Paulaini, and J. P. Petjtjean, *Mater. Sci. Forum 142*:213 (1997).
34. R. Mady, C. Reis, and R. Morlock, *Eur. Pat. Appl.* 831126107, 1993.
35. B. A. Shaw, G. D. Davis, T. L. Friz, and K. A. Olver, *J. Electrochem. Soc. 137*:359 (1990).
36. B. Hinton, A. Hughes, R. Taylor, M. Anderson, K. Nelson, and L. Wilson, *International Symposium on Aluminium Surface Science and Technology*, Antwerp, Belgium, May 12–15, 1997, p. 165.
37. F. Mansfeld, Y. Wang, and H. Shih, *Electrochim. Acta 37*:2211 (1992).
38. L. Fedrizzi, F. Florian, and S. Rossi, *International Symposium on Aluminium Surface Science and Technology*, Antwerp, Belgium, May 12–15, 1997, p. 243.
39. M. J. Rijkhoff, R. Bleeker, and J. Bottema, *International Symposium on Aluminium Surface Science and Technology*, Antwerp, Belgium, May 12–15, 1997, p. 125.
40. C. M. Rangel and M. A. Travassos, *Corros. Sci. 33*:327 (1992).
41. J. W. Bibber, *Corros. Rev. 15*:303 (1997).
42. R. Feser and Th. Schmidt-hansberg, *Eurocorr '97*, Trondheim, Norway, September 22–25, 1997, p. 291.
43. W. van Ooij, *Corrosion 54*:204–215 (1998).
44. W. J. van Ooij and T. J. Ossic, *SAE International Congress and Exposition*, Atlanta, GA, paper 870647, 1987.
45. C. Barreau, D. Massinon, and D. Thierry, *Proceedings, 5th Automotive Corrosion and Prevention Conference*, SAE, Bahrain, paper 327, 1991.
46. D. Massinon and D. Thierry, *Corrosion/91*, NACE, Houston, TX, paper 574, 1991.
47. R. D. Granata, *Corrosion/91*, NACE, Houston, TX, paper 382, 1991.
48. J. H. W. de Wit, E. P. M. van Westing, and D. H. van der Weijde, *Mater. Sci. Forum 247*:69 (1997).
49. *Marine Paint Manual*, Graham and Trotman Ltd., London, U.K., 1989.
50. J. E. O. Mayne, *Corrosion*, 2nd edn. (L. Shreier, edn.), Vols. 2,15, Newnes-Butterworth, London, U.K., 1976, p. 24.
51. F. Belluci, L. Nicodemo, and R. M. Latanision, *J. Mater. Sci. 25*:1097 (1990).
52. U. R. Evans, *The Corrosion and the Oxidation of Metals*, St. Martins Press, New York, 1960.
53. C. Haberer, C. M. Wolf, Y. P. Collin, Y. L. Leibenguth, and P. Schwing, *Electrochim. Acta 6*:755 (1981).
54. N. L. Thomas, *J. Prot. Coat. Linings 6*(12):63 (1989).
55. J. S. Hammond, J. W. Hobluka, and R. A. Dickie, *J. Coat. Technol. 51*:45 (1979).
56. J. E. Castle and J. F. Watts, *Proceedings Corrosion Control by Organic Coatings*, NACE, Houston, TX, 1981, p. 78.

57. W. Funke and H. Haagen, Empirical or scientific approach to evaluate the corrosion protective performance of organic coatings, *Ind. Eng. Chem. 17*: (1978).
58. R. N. Parkins, A. J. Markworth, J. H. Holbrook, and R. R. Fessler, *Corrosion '84*, New Orleans, LA, 1984.
59. CRC *Handbook of Chemistry and Physics*, 56th edn., CRC Press, Cleveland, OH, 1976.
60. W. S. Tait and K. A. Handrich, *Corrosion 50*: (1994).
61. T. N. Nguyen and J. B. Hubbard, A mathematical model for the cathodic blistering of organic coatings on steel immersed in electrolytes, *J. Coat. Technol. 63*(794):43–52 (1991).
62. W. Fuerbeth and M. Stratmann, *Corros. Sci. 3* (1999).
63. D. H. van der Weijde, Impedance spectroscopy and organic barrier coatings, Thesis, TU-Delft, Delft, the Netherlands, 1996.
64. P. R. Willemsen and G. M. Ferrari, A review of antifouling methods. TNO report CA/95.1026 (study carried out for the Netherlands Ministry of Housing, Physical Planning and Environment), 1995.
65. P. R. Willemsen and G. M. Ferrari, Possibilities and impossibilities of alternative antifouling techniques, *An International One-Day Symposium on Antifouling Paints for Ocean-Going Vessels*, The Hague, 21 February 21, 1996, pp. 60–67.
66. P. R. Willemsen and G. M. Ferrari, The removal of biofouling from ships' hulls: State of the art, TNO report CA/96.9906 (in Dutch) (study carried out for the Netherlands Ministry of Housing, Physical Planning and Environment), 1996.
67. E. C. Haderlie, A brief overview of the effects of macrofouling, Marine biodeterioration: An Interdisciplinary Study, *Proceedings of Symposium on Marine Biodeterioration* (J. D. Costlow and R. C. Tipper, eds.), Univ. Health Sciences, Berlin, Germany, April 20–23, 1981, pp. 163–166.
68. F. H. de Ia Court, Aangroeiwerende verven. *Chemische feitelijkheden—Aktuele chemische encyclopedic*, 1990, pp. 2–6.
69. G. I. Loeb, D. Laster, T. Gracik, and D. W. Taylor, The influence of microbial fouling films on hydrodynamic drag of rotating discs, Marine Biodeterioration: An Interdisciplinary Study, *Proceedings of Symposium on Marine Biodeterioration* (J. D. Costlow and R. C. Tipper, eds.), Univ. Health Sciences, Berlin, Germany, April 20–23, 1981, pp. 88–94.
70. E. C. Fischer, V. J. Castelli, S. D. Rodgers, and H. R. Bleile, Technology for control of marine biofouling—A review, Marine Biodeterioration: An Interdisciplinary Study. *Proceedings of Symposium on Marine Biodeterioration*, Univ. Health Sciences, Berlin, Germany, April 20–23, 1981, pp. 261–299.
71. J. D. Pidgeon, Critical review of current and future marine antifouling coatings, Lloyd's Register, Engineering Service. Propulsion and Environmental Engineering Department, report 93/TIPEE/478719, 1993.
72. Anonymous, TBT copolymer antifouling paints: The facts, CEFIC Brochure (based on paper presented to MEPC in November 1990), 1992.
73. C. C. ten Hallers-Tjabbes, TBT in the open sea: The case for a total ban on the use of TBT antifouling paint, *North Sea Monitor*, September 12–14, 1994.
74. R. De Nys, P. D. Steinberg, P. R. Willemsen, S. A. Dworjanyn, C. L. Gabelish, and R. J. King, Broad spectrum effects of secondary metabolites from the red algae Delisea pulchra in antifouling assays, *Biofouling, 8*:259–271 (1995).
75. A. S. Clare, D. J. Gerhart, and D. Rittschof, Development of nontoxic antifoulants based on natural product chemistry, Poster presented at the *International Symposium on Marine Biofouling and Corrosion*, Portsmouth, U.K., June 13–16, 1993.
76. D. J. Gerhart, D. Rittschof, and S. W. Mayo, Chemical ecology and the search for marine antifoulants, *J. Chem. Ecol. 14*:1905–1917 (1988).
77. M. Fletcher and J. H. Pringle, The effect of surface free energy and medium surface tension on bacterial attachment to solid surfaces, *J. Colloid Interface Sci. 104*:5–14 (1985).
78. R. C. Wallis and K. R. Strudwick, Nontoxic marine fouling release coatings, *Surf. Coat. Aust.* March: 14–17 (1989).

79. R. F. Brady and J. R. Griffith, Nontoxic alternatives to antifouling paints, *J. Coat. Technol. 59*:113–119 (1987).
80. J. H. W. de Wit and E. P. M. van Westing, Paper at UK Corrosion, London, U.K., October 1993.
81. J. H. W. de Wit, *Proceedings of 12th International Corrosion Congress*, Houston, TX, paper 324, September 19–24, 1993.
82. E. P. M. van Westing, G. M. Ferrari, F. M. Geenen, and J. H. W. de Wit, *Prog. Org. Coatings 23*:85 (1993).
83. J. H. W. de Wit, *Proceedings of 10th European Corrosion Congress*, Barcelona, July 1993.
84. D. H. van der Weijde, E. P. M. van Westing, and J. H. W. de Wit, *Corros. Sci. 36*:6935 (1994).
85. F. Deflorian, L. Fedrizzi, and P. L. Bowona, *Electrochim. Acta 38*:1609 (1993).
86. F. Geenen, E. P. M. van Westing, and J. H. W. de Wit, Study of the degradation mechanism of epoxy coatings on steel using impedance spectroscopy, *Proceedings of the IHh International Corrosion Conference*, Vol. 2, Florence, Italy, April 2–6, 1990, p. 231.
87. E. P. M. van Westing, F. M. Geenen, G. M. Ferrari, and J. H. W. de Wit, The influence of solvents on the water uptake and dielectrical and mechanical properties of epoxy coatings investigated using electrochemical impedance spectroscopy, *Proceedings of 16th International Conference in Organic Coating Science and Technology*, Athens, Greece, July 1990.
88. F. M. Geenen, H. J. W. Lenderink, E. P. M. van Westing, and J. H. W. de Wit, A detailed impedance spectroscopy investigation into the performance of different coated systems in weathering tests, *Proceedings of XXth FATIPEC Congress*, Nice, 1990.
89. F. M. Geenen, H. J. W. Lenderink, and J. H. W. de Wit, The structure of epoxy coatings studied with impedance measurements, *Proceedings of 5th International Conference on Electrochemical Methods for Corrosion Research*, Helsinki, Finland, July 1991.
90. F. M. Geenen and J. H. W. de Wit, Revêtements sur l'aluminium, *Eurocoat* 3:101 (1992).
91. J. A. L. Dobbelaar, PhD thesis, TU Delft, Delft, the Netherlands, 1990.
92. J. B. Hasted, *Aqueous Dielectrics*, C&H, London, U.K., 1973.
93. W. Bosch and W. Funke, Peasticization of organic coatings by water, *Farbe Lack 98*(8):589–592 (1992).
94. R. Hirayama and S. Haruyama, *Corrosion 42*:952 (1991).
95. M. Dings, R. de Jonge, S. M. Peters, E. J. H. Koot, F. M. Geenen, E. P. M. van Westing, and J. H. W. de Wit, Un nouveau test pour characteriser les produits de revétements á l'aide des défusts artificiéis apportés au laser et de la spectroscopie d'impédance électrochimique, *Double Liaison, Chimie des Peintures*, no. 425–426, 1991, 61/13.
96. M. M. M. Dings, R. de Jonge, S. M. Peters, E. J. H. Koot, F. M. Geenen, E. P. M. van Westing, and J. H. W. de Wit, An novel characterisation test for organic coatings based on artificial laser defects and electrochemical impedance spectroscopy, *Proceedings, XX FATIPEC Congress*, Nice, 1990.
97. F. M. Geenen, M. M. M. Dings, E. J. H. Koot, R. de Jonge, S. Peters, and J. H. W. de Wit, Characterisation of organic coatings using artificial laser defects and impedance measurements, *Proceedings, 5th International Conference on Electrochemical Methods for Corrosion Research*, Helsinki, Finland, July 1990.
98. B. van den Bossche, L. Bortels, J. Deconinck, S. Vandeputte, and A. Hubin, *J. Electroanal. Chem. 411*:129 (1996).
99. W. Fuerbeth and M. Stratmann, *Corros. Sci.*, *43*:207,229,243 (2001).
100. J. Mazia, *Metal Fin. 75*:49 (1977).
101. G. D. Kent, K. J. Hacias, and N. A. Fotinos, *Proceedings, 2nd Automotive Corrosion Prevention Conference*, Ann Arbor, MI, 1983, p. 23.
102. J. B. Harrison and T. C. K. Tickle, *JOCCA 45*:571 (1962).
103. W. Funke, *Farbe Lack 84*:389 (1978).

Index

A

Adsorption site blocker (ASB)
  blocking effect, 141
  coverage, 140–141
  on H entry
    local *vs.* overall coverage, 141
    quasiequilibrium, 142
    theoretical variations, 143–144
    Volmer and Heyrovsky steps, 142
    Volmer and Tafel steps, 143
Alloys, anodic behavior
  Fe-Cr
    analytical and mass balance approaches, 197–200
    current-potential curves, 194
    electronic interaction, 194
    reaction mechanism, dissolution, 195–197
  selective dissolution, 194
  simultaneous dissolution and formalism
    electrovalences, 193
    linear relationship, 193
  thermodynamics and rate constant
    activated state theory, 192
    free energy, 192
    single-phase alloys, dissolution, 193
Aluminum alloys corrosion
  description, 706
  EAC
    CF cracking, 727–728
    electrochemical process, 715
    environmental influence, 728–729
    hydrogen embrittlement (HE), 728
    LME, 726–727
    SCC, 725–726
  forms and causes
    bimetallic/galvanic corrosion, 718–719
    crevice, 721
    dealloying, 719–721
    description, 717
    filiform, 721–722
    pitting and localized corrosion, 718–719
  hydride formation, 716
  intergranular forms
    boundary structure, 724
    exfoliation, 725
    sensitization level, 724
  metallurgical aspects
    physical metallurgy, 707–712
    physical properties, 712
    processing, 713–715
    production, 707
  microstructure effects
    chloride-containing solutions, 722
    corrosion activity, 723
    dominant features, 722
    intermetallic particles, 723
    pitmorphology, 723
    precise morphology, 724
  preventions strategies
    inhibitors, 731
    inorganic coatings, 729–730
    organic coatings, 730
    specialized coatings, 731–732
  protective oxide film, 715
Aluminum metallization, 836–838
Anodic dissolution
  alloys
    Fe-Cr, 194–200
    selective, 194, 201–206
    simultaneous dissolution and formalism, 193
    thermodynamics and rate constant, 192–193
  corrosion
    passivity domain, 150–151
    potential dependence, 150
    transpassive potentials, 151
  definition, 150
  electrochemical techniques
    classification, 154–155
    metals and inductive behaviors, 161–164
    nonsteady-state *vs.* steady-state techniques, 155–159
    time/frequency-resolved measurements, 159–160
  intermediate surface species
    activated state, 151–152
    atomic crystallographic sites, 154
    FeOH, 153
    gas-phase and surface catalysis, 152
    hydroxo-ligand, 153
    mechanisms, 152–153
    pH and anion dependence, 152

kinetics, 151
pure metals
film relaxation and dissolution, passive state, 178–188
heterogeneous reaction mechanism, active state, 165–178
mass transport control, 188–191
Anodic undermining
adhesion, loss of, 878
aluminum, 884
filiform corrosion, 885
transport rates, oxygen, 879
Aqueous layer
acidifying pollutants
acid-dependent dissolution rate, 676
Henry's law, 677–678
infrared reflection-absorption spectroscopy, 676
molar concentration, 678
electrochemical reactions
anodic reaction rate, 675–676
atmospheric corrosion conditions, 675
conductive medium, 675
formation
bond strength, 675
metal–hydroxyl bonds, 674
water–metal interaction, 674
water monolayers, 674–675
pollutants deposition
air flow velocity, 680
corrosion-stimulating gases, 680
dry deposition velocity, 679
emission sources, 678–679
mass transport limitations, 680
ASB, *see* adsorption site blocker (ASB)
Atmospheric corrosion
aqueous layer
acidifying pollutants, 676–678
electrochemical reactions, 675–676
formation, 674–675
pollutants deposition, 678–680
chlorine-containing compounds, 673
description, 669–670
environmental effect, 670
indoors
exposure conditions, 693
ISO classification system, 693
particles deposition, 694
metals dispersion
bioavailable fraction, 696
copper corrosion rate, 695
long-term corrosion rates, 694
rain characteristics, 695
REACH, 696–697
realistic risk assessment, 694
nitrogen-containing compounds, 673
outdoors
corrosion rates, 687–689
dose–response functions, 690–691
field exposures, 689–690
gases and aerosol particles, 692–693
region, gases, 670–671
solid phase
acid-dependent dissolution, 680–681
aluminum, 685
copper, 685
corrosion products formation, 682–683
electrolyte, 686
HSAB principle, 684
instant formation, 684
iron, 685
Lewis acid–base concept, 682
nickel, 686
sensitive analytical techniques, 683
silver, 686
zinc, mechanisms, 683–684
sulfur-containing compounds, 672
Austenitic stainless steels passivity
alloy surface layers
alloying elements, 339–341
apparent metal content *vs.* potentials, 343
composition, metal phase, 342
Engel–Brewer model, 344
high-alloyed, 330, 341–342
intermetallic surface phase, 344
metal phase composition, 341–343
Mo role, 345
pits growth and repassivation, 344–345
pitting resistance, 346
polarization diagram, 341
XPS spectra, 341–342
austenite stabilizers, 327
barrier and deposit layers
argon ion sputter, 330–331
AVESTA composition, 329
ion content *vs.* etch depth, 330
polarization behavior, alloy, 329
surface analysis, 328–329
XPS spectra, 329–330
boundwater, passive films
Okamoto model structure, 331
vs. potential, 331–332
molybdenum (Mo)
bipolar model, 333–335
description, 332

insoluble salt and surface alloy models, 335
soluble stable Mo-oxo-chloro complexes, 333
nitrogen
Auger depth profile, 337
chemical composition, steel elements, 336–337
measured intensity ratios, 336
Mo and, 335
nitride spectra, 339, 340
potentiodynamic polarization, 338–339
oxidation states, 328
passivation, 327–328

B

Bernhardsson model, 485, 486
Binary alloys passivity
AlCu
anodic oxidation, 285–286
film Galvanostatic oxidation, 285
RBS, 286, 287
CuNi
film structure, 283–284
polarization curve, 283
CuSn
acidic electrolyte, 284–285
layer chemical structure, 284
polarization curves, 284
FeAl, 281, 282
FeCr
dissolution rat, 276
film structure, alkaline solutions, 277–278
hydrogen evolution, 275–276
ISS depth profiles, layers, 278
layer composition, 0.5 M H_2SO_4, 276–277
polarization curve, 275, 276
FeNi
Fe-53Ni, 1 M NaOH, 278–280
polarization curves, 243, 278
XPS, 280
FeSi
corrosion, 281, 283
XPS and ISS depth, 283
NiCr, 280–281
passive layer model, 275
Biodeterioration mechanisms
acid excretion, 747–748
biofilm, 749
chelatization, 748
description, 747
exoenzymes and emulsifying agents, 750
metabolic compounds, 748–749
organic solvents, 748
salt stress, 749–750
Boiling water reactors (BWR) systems
atom–atom rupture process, 789
crack tip alloy/environment system
austenitic stainless steel piping, 795–796
electrode potential and pH conditions, 795
transgranular cracks, 797
transient and steady-state concentrations, 796
liquid diffusion rate, 794
mass transport controlling, 794
reaction rates, crack tip
anionic impurity concentration, 798
bulk solution conductivity, 798, 799
macroscopic analogy, 797
subsequent oxide formation, 798
vis-à-vis stainless, 798
resultant beneficial effect, 794–795
slip-oxidation model, 795
stainless steels, stress corrosion cracking
core components, 811
crack length *vs.* time predictions, 816–817
grain boundary sensitization, 816
IASCC, 811, 813
irradiation-induced effects, 814
observed crack depths, 817
predicted crack depth *vs.* time response, 814–815
slip-oxidation mechanism, 814
strain rate, crack tip
aqueous and gaseous environments, 798
control parameters, 803
crack propagation rate, 800
growth rate predictions, 804–805
L-grade and non-sensitized steel components, 806
microscopic heterogeneity, 799
plastic constraint, 801
predicted and observed beneficial effect, 802–803
PWR environments, 805
stress intensity factor relationships, 806–807
structural integrity assessments, 800
tensile stress include pressurization, 804
theoretical crack propagation rate, 801–802
theoretical *vs.* observed crack propagation rate-stress, 803

stress corrosion crack propagation
intensity factor, 804, 806
L-grade and non-sensitized steel components, 806
observed and predicted relationships, 802, 804
predicted and observed beneficial effect, 802
PWR environments, 805
theoretical and observed crack propagation, 803
theoretical crack propagation rate, 801
unirradiated welded plant component
crack depth *vs.* operational time relationships, 810, 812
damage accumulation algorithm, 810
residual stress, 807, 810
water chemistry, 811, 813
Brown–Hobson model, 558
BWR systems, *see* Boiling water reactors systems

C

Cabrera–Mott expression, 219, 223, 225
Cathodic delamination
adhesion, coating, 878
corrosion products, 879
model
anodes and cathodes, randomly distributed, 880
blisters, 881, 882
extended, 882–884
Fe^2 oxidation, 881
low soluble iron (hydr)oxides, 882
mechanism, 880
CF, *see* Corrosion fatigue
CF cracking, 727–728
CF propagation mechanisms
anodic dissolution effects
hydrogen absorption, 560
restricted slip reversibility model, 559
hydrogen effects
cracking mechanism, 560–561
dilatation strains, 563
dissociated screw dislocation, nickel, 561, 562
partíais recombination, nickel, 563–564
plasticity model, 561, 562
screening, pair interactions, 563
phenomenology
Brown–Hobson model, 558
carbon steels, 557
characteristics, crack growth, 557
crack growth rate and tip stress intensity, 556–557
electrochemical reactions, 558
environment impact, crack growth rate, 555–556
Chemical equilibria thermodynamics
electrochemical equilibrium
hydrogen electrode, 36
Nernst equation, 36–37
standard potential/Gibbs free energy, 37
energy and enthalpy
Fe metal, 33
HCl formation, 32
internal, 30
KCl, 32–33
standard, 30–32
entropy, Helmholtz and Gibbs free energy, 33–36
potential–pH diagrams
Fe_3O_4 reduction to Fe, 37, 39
metals, 40
passivation, 40–41
Pourbaix, 37, 39
standard potentials, 38
reference electrodes (REs)
Ag/AgCl, 41–42
calomel, 41
Complex coating systems
conversion layers
chromate coatings, 868–869
chromic acid and chromic acid–phosphoric acid processes, 869
molybdates and cerium compounds, 870
nonchromate, 869
phosphate layers, 870
metallic, conversion layers and organic topcoats, 870–872
Conducting polymers, ultrathin coatings
anions expulsion, 657
chloride-doped PPy, 656
electrochemical conditions, 656
immersion tests, 655
inherent property, 658
intelligent corrosion protection
anodic corrosion, 660
cation incorporation, 659
extended percolation networks, 660
ICP, 658
macroscopic percolation networks, 658
PPy nanoparticle, 659
SKP delamination profiles, 659

intrinsically conducting polymers (ICPs), 654
polymerization conditions, 655
Conductive anodic filaments (CAFs) formation, 841, 848
Contact electric resistance (CER), 187
Control rod drive mechanism (CRDM) nozzles, 785
Corrosion
acid, 4–5
chemical and physical conditions, 3
conductivity, 5–6
description, 2
electrochemical double layer
charged metal surface, 23–24
composition, XPS, 25–27
Debye–Hückel theory, 24–25
differential capacity, 29
Gibbs' adsorption isotherm, 29–30
Helmholtz layer, 24
liquid metal, 27
mercury, electrocapillary curves, 28
model, 24
electrochemical methods
cells, 55–57
galvanostatic measurements, 58–59
impedance spectroscopy, 62–65
potentiostatic measurements, 57–58
RDEs and RRDEs, 59–60
transients, 61–62
electrochemistry
electrochemical cell electrodes, 21–22
equilibrium, 23
metal/metal ion electrodes, 22
electrode kinetics
charge transfer overvoltage, 44–49
charge transfer process, 43–44
diffusion overvoltage, 49–52
Ohmic drops and microelectrodes, 52–55
Tafel reaction, 44
electrolyte
diluted strong, 12–21
structure, 6–12
water, 3–4
heterogeneous metal surfaces and local elements
cathodic processes, 83
crystallites, 83
polarization, 83–84
potential profiles, 84–85
inhibition
adsorption, 88–91
compound precipitation, 91–92
inhibitor efficiency, 87–88
passivation, 92–93
liquid junction/diffusion potentials
electrical work, 42
HCl concentration, 42–43
KCl, 43
mobility, 42
osmotic work, 42–43
metal dissolution
alloys, selective, 76–79
and complex formation, 79–80
and deposition, 73
description, 71
free energy, 71–72
high rates and salt precipitation, 75–76
ionic product, 74
iron, OH^- ions, 73–74
model, 71
reduction reactions, 80–83
STM, 72–73
Tafel slope, 74–75
nonaqueous solvents, 5
photoelectrochemistry
electron-hole pairs, 100–101
light absorption, 100
photocurrent efficiency, 101–102
photocurrent mechanism, 99–100
protection
anodic, 87
cathodic, 85–87
stray currents, 87
redox system reduction
hydrogen evolution (HER), 67–70
oxygen, 70–71
research, 2–3
semiconductor electrochemistry
electrodes, electron transfer, 95–99
material degradation, 93
properties, 94–95
thermodynamics, chemical equilibria
electrochemical equilibrium and Nernst equation, 36–37
energy and enthalpy, 30–33
entropy, Helmholtz and Gibbs free energy, 33–36
potential–pH diagrams, 37–41
reference electrodes, 41–42
transfer coefficient, 6
Corrosion fatigue (CF) mechanisms, metallic materials
crack initiation
classical approaches, damage, 546–551
cyclic plasticity impact, 551–553

mechanical and electrochemical coupling effects, 555
PSB configurations, 554
softening effect, 553–554
electrochemical potential, 545
interaction, SCC
behavior comparison, 569–570
crack growth rates *vs.* frequency, 570
Fe-Cr-Mn alloy, 569
Monte Carlo-type simulations, 564–568
propagation mechanisms
anodic dissolution effects, 559–560
hydrogen effects, 560–564
phenomenology, 555–558
threshold stress intensity factors, 545–546
Countermeasures, MIC
biocides, 767
biogenic attack simulation, 768
cathodic protection, 767
changing/modifying material, 766
microbiological methods, 767
organic coatings, 766–767
physical methods, 767
process parameters, 766
CPT, *see* Critical pitting temperature
Crack initiation, CF
cyclic plasticity impact
applied strain rate, 552, 553
dissolution current density transients, 551
PSB formation, 552
electrochemical approaches, damage
calculation, cycle number, 548
critical pit depth, 547–548
electrochemical corrosion, 546
experimental *vs.* theoretical fatigue lives, 548
Faraday's law, 549
Fe-Mn-Cr alloy, 548, 549
film formation mechanisms, 546–547
limitations, 551
passive conditions, 549, 550
polarization curve, austenitic stainless steel, 547
mechanical and electrochemical coupling effects, 555
PSB configurations, 554
softening effect
free potential *vs.* air behavior, peak stress evolution, 553–554
macroscopic cycling, 554
Crevice corrosion, metallic materials
differential aeration, 450
vs. experimental characterization, alloys
crevice former devices, 488
electrochemical tests, 489–493
exposure tests, representative environments, 488–489
ferric chloride test, 489
tests and criteria, 487–488
mechanisms and parametric effects
environment modification, 451–452
external surfaces, 454
geometry, 453–454
initiation and propagation, 452
IR drop, 452–455
local change, anodic behavior, 452
zone specificity, 451
modeling and results
corrosion damage prediction, 485, 487
electrical charge balance, 482–483
geometry effect, initiation time, 485, 486
H^+ activity correction, 485
initiation criteria, 483
limitations, 486
mass balance, 482
pH drop *vs.* time, 486, 487
predictions, Sharland model, 483
propagation model, 487
severity factor effect, pH drop, 485, 486
solution chemistry, 479–481
steady state model, 478–479
surface reactions, 479
transient model, 479
transport process, 481–482
passive alloys
phenomenology, 455–461
process, 461–478
prevention
design, 493
fabrication procedure and surface finish, 493
maintenance procedures, 493–494
material selection and cathodic protection, 493
zones, 450–451
Crevice environment evolution, passivated alloys
deaeration
cathodic reaction, 461–462
IR drop, 462
dissolution, sulfide inclusions, 468
Fe-Ni-Cr alloys
analyses, passive films, 468
concentrated solutions, 467
formation, chloro-hydroxy complexes, 465, 467

pH evolution and concentration
anodic dissolution, 462
artificial crevice and synthetic solution, 465, 467
cathodic polarization effect, 465, 466
chromium content effect, 463, 464
C-steel crevice, 463, 465
dissolution current, 465, 466
hydrolysis reaction, 462–463
measurements, 463, 464
relationship, chromium content, 465, 467
stainless steel, 463, 465
time, 465, 468
Crevice solution chemistry
hydrolysis and complexation
aluminum alloys, 481
chromium cations, 480
equilibria, Gartland model, 481
models
activity coefficient, H^+ ions, 480
equilibrium constants, 479–480
problems, 479
Critical pitting temperature (CPT), 427
Critical potential, initiation and protection
determination
potentiokinetic techniques, 489–490
potentiostatic techniques, 490, 491
effect
polarization curves, alloy, 456, 457
protection potential, 458
stainless steel, 456, 458
time, 456, 457

D

Debye–Hückel–Onsager theory
activity coefficient
acids bases and salts, 21
charging, electrical work, 19–20
chemical potential, 19
neutral solutes, 21
cation, ionic cloud, 13
central ion charge and ionic cloud, 15
Debye length determination, 14–15
equivalent conductivity, strong dissociated electrolytes
coefficient effect, 16–17
concentration, 16
electrophoretic effect, 16
ionic cloud, 15–16
platinum, 17
values, 17–18
ionic strength, 14
weakly dissociated electrolytes, equivalent conductivity
acid HA, dissociation constant, 18
function, 18–19
Debye–Hückel theory, 5, 6, 15, 18–21, 24, 25, 96

E

Early rusting, 885
EIS, *see* Electrochemical impedance spectroscopy
Electrochemical cell, 55–57
Electrochemical impedance spectroscopy (EIS)
artificial defects, laser ablation
etching, 896
geometry, 896
stages, coated system behavior, 897
atmospheric water uptake
capacitance, 892
dielectric constants, 891
diffusion coefficient, 892–893
grids, 891
description, 889
equivalent circuits, same impedance response, 890
measurements and delamination
coating capacitance, 895
equivalent circuits, 895–896
industrial coating, salt film, 893
Nyquist plots, 894
nonelectrochemical parameters, 889–890
transfer function, 890
Electrochemical methods
cells, 55–57
impedance spectroscopy
Bode plot, 65
charge transfer and polarization resistance, 62–63
Faraday, Warburg and Nernst, 65–66
impedance, 63–64
Nyquist diagram, 64–65
Randles circle, 66
resistances, 63
voltage and current, 63
measurements
galvanostatic, 58–59
potentiostatic, 57–58
RDEs
electrolyte flow, 59
split, 60
RRDEs
electrolyte flow, 59–60
time resolution, 60
transfer efficiency, 60

transients
potentiostatic and galvanostatic, 61
redox reaction, 61–62
Electrochemical quartz crystal microbalance (EQCM), 155, 158
Electrochemical techniques, anodic dissolution
current–voltage techniques
RDE and RRDE, 155
steady-state/voltage sweep current-voltage curves, 154
metals and inductive behaviors
admittance and impedance plane, 161
flat (2D) and roughened (3D) interfaces, 163–164
frequency, inductive impedance, 162
impedance, 162
inductive/capacitive reaction impedance, 161
interfacial potential drop, distribution, 163
ion-transfer steps, 163
low frequency capacitive loop, rough case, 164
MICTF data, 163
Nyquist diagram, electrochemical impedances, 164
self-inductance, 161
time and frequency domain, 161
β values, 162
nonsteady-state *vs.* steady-state techniques
dissolution–passivation processes, 158
double-layer capacitance, 158
electrogravimetric transmittance, 158–159
faradic impedance, 157, 158
fundamentals, 155
instantaneous intensity, faradic current, 156
linear system theory, 156–157
resolution, Kramers method, 157
SVET, LEIS and SECM, 158
time dependence, system, 156
voltage sweep, 155
time/frequency-resolved measurements
collection efficiencies, 159–160
electrical charge balance, 159
emission efficiency and true capacitance, 160
Electrochemical tests
critical potential determination
measured parameters, 489
potentiokinetic techniques, 489–490
potentiostatic techniques, 490–491
material behavior determination
activation pH, 490–491
drawbacks, 493
peak current evolution, 491–492
spontaneous film breakdown pH, 490–492
Electrode kinetics
charge transfer overvoltage
band model, 49, 50
Butler–Volmer equation, 45
carriers, electrical work, 45–46
density, 44–45, 47
Fe^{2+}/Fe^{3+}, 48
free enthalpy, 46
metal, 45
oxygen reduction, 49
redox reaction, 44
Tafel plot, 47–48
transfer resistance, 48
charge transfer process, 43–44
diffusion overvoltage
anodic metal dissolution, 52
cathodic reaction, 52
Nernst, 51–52
reactant transport, 49, 51
Ohmic drops and microelectrodes
current density, 53–55
resistance, electrolyte, 52–53
Tafel reaction, 44
Electrode kinetics, passive state
ion transfer
activation energy, 250–251
cations inward and anions outward migration, 249–250
free activation enthalpy, 252
jump distance, 250, 251–252
layer growth, 251
nanocrystals, 252–253
SIMS/BS, 249
layer-electrolyte interface
charge transfer, 246
corrosion current density and layer formation, 247–248
Fe dissolution, 248–249
overvoltage determination, 248
pH dependence, 247
metal dissolution, 244–245
potential drops, 245–246
Electro-hydro-dynamic (EHD) impedance, 189–191

Electrolytes
and solvents
acid, 4
acidity/alkalinity, 4–5
dissociation constant, 3–4
gases dissolution, water, 4
nonaqueous solvents, 5
structure
H_3O^+ and OH^- ions, 7, 8
ice, tetrahedral coordination and short-range order, 7, 8
short-range order, dissolved species, 11–12
water distribution function, 7
XAS and EXAFS, 9–11
Electronic properties, passive layers
anodic oxide layers, 294, 295
bulk oxides band gap, 289
description, 288–289
electron transfer, metallic surfaces
capacity, 291
charge transfer, 293
oxide layer, 290
potential drop energy, 291
redox systems, 292–293
semiconductors, 289–290
space charge layer depth, 290–291
oxides band gap, 289
photoeffects, Cu
band gap, 297–298
band structure model, 299–300
cathodic photocurrents, 296–297
charge transfer reaction, Cu_2O layers, 300–301
Cu_2O layer, electron transfer, 297
duplex passive layer, 301
HeI-UP spectra, 298–299
work functions and threshold energies, 299
photoelectrochemical measurements, 294
redox currents, 296
Tafel plots, iron redox currents, 294, 296
Engel–Brewer model, 344
Environmentally assisted cracking (EAC)
CF cracking, 727–728
environmental influence, 728–729
hydrogen embrittlement (HE), 728
liquid metal embrittlement (LME), 726–727
stress corrosion cracking (SCC), 725–726
Extended x-ray absorption fine structure (EXAFS)
Ag^+ ions, 12
FEFF/data analysis packages, 10–11
Ni metal, 9–10
XA spectrum data evaluation, 10

F

Faraday's law, 549, 559
Fe-Cr alloys
analytical and mass balance approaches
channel flow arrangements and data processing, 197
dissolution, passive state, 199–200
impedance and frequency-resolved RRDE, 197–199
current-potential curves, 194
electronic interaction, 194
reaction mechanism, dissolution
current-potential curves and impedance diagram, 196
interaction types, 195
nonlinear interaction, 197
passivating feature, 195
Ferric chloride test, 489
Film-induced cleavage model, 501
Flade potential, 177, 178, 179, 181
Flash rusting, 885–886
Fouling prevention, organic coatings
disadvantages, vessels, 886–887
nonstick coatings
pollution, 888
silicon polymers, 887
organotin-based antifoulings, 887
sea and estuaries, 886
seagoing vessels and pleasure craft, 887

G

Galvele's surface mobility model, 535–536
Gartland model
corrosion damage prediction, 485, 487
H^+ activity correction effect, 485
solubility and hydrolysis equilibria, 481
Geochemical cycles
carbon, 743
description, 742
iron, 746–747
nitrogen, 743–745
sulfur, 745–746
Grain refined surface layers (GRSLs)
formation, 713
mechanical processing, 714
oxide composition and environment, 714
GRSLs, *see* Grain refined surface layers

H

H absorption promoters
adsorption blocking effect, 139
ASB modeling effects, HER and HAR, 140–144
conditions
elements, 136
hydride paramount role, 136
permeation rate, 136, 137
steels embrittlement, 136–137
proposed mechanism
AsH_3, 139
IPZ model, 138
Vomer–Tafel path, 137–138
suggestions, promoting effects, 144–145
Hardware technologies
magnetic-storage components
COC, 830–831
disk drives, 830
magneto-optic (MO) recording, 831–832
magnetoresistance (MR) phenomenon, 831
thin-film disk structure, 830
microelectronic devices
IC packaging, 828–829
integrated circuits, 827–828
macro interconnects, 829
HER and HAR
absolute rates, elementary surface steps
chemical combination and dissociative adsorption, 127–128
electroadsorption and electrodesorption, 127
electrocombination and electrodissociation, 128
Gibbs free energy, 126
surface-bulk transfer, 128–129
dependence rates, 131–134
steady-state equations, 129–131
Heterogeneous reaction mechanism, anodic dissolution
anions, iron
acetate, 173
Cl^- and OH^-, 173–174
mechanistic changes, pH, 174
chromium, 174–175
cobalt and nickel, 174
iron, acidic solutions
$(FeOH)_{ads}$, 165–166
atomistic interpretation, catalytic mechanism, 167–168
hydroxyl ions, 165
iron dissolution, impedance techniques, 169–173
steady-state and transient polarization data, 166–167
passivation, downstream collector electrode techniques
active–passive transition, 175–176
freshly generated iron surfaces, 176–177
impedance and AC RRDE study, 177–178
titanium and copper, 175
High-field migration model
chromium enrichment, 200
current, 179
features, 179
iron dissolution and transient overvoltage, 181
nonsteady-state passive iron, 179
positional changes, potential, 180
High-temperature corrosion
description, 573
liquid phases
chloride corrosion, 599–602
general aspects, 594–596
sulfate corrosion, 596–599
materials behavior, 574
solid corrosion products
alloy composition role, 584–588
formation and growth kinetics, 578–584
internal corrosion, 588–591
metal dusting, 591–594
thermodynamics, formation
alloying elements, 576
boundary-separating condition, 576
oxidation reaction, 574
solid and liquid phases, 578
stability diagrams, 576, 577
types, 574
volatile phases
gas-phase chlorine, 602–610
general aspects, 602
water vapor, 610–612
Hopping mechanism, *see* High-field migration model
H surface-bulk transfer mechanisms
adsorption sites, 121
aqueous electrolyte
metal–aqueous solution interface, 124–126
potential *vs.* RHE1, 124, 126
gas phase
potential energy *vs.* distance curves, 124, 125
TDS, 123–124
interstitial sites, 121–122
surface structure role, absorption, 122–123

Hydrogen electroabsorption reaction (HEAR)
equilibrium equation, 118
HER, 111
Hydrogen embrittlement (HE), 502, 728
Hydrogen-induced plasticity model, 561, 562
Hydrogen surface reactions
in aqueous electrolyte
elementary surface reactions, 109–110
overall electrode reactions, 111–113
Pd(111) surface model, 109
in gas phase, 107–108

I

Inhibition, corrosion
adsorption
anodic dissolution, 89, 90
capacity, 91
current densities, 91
efficiency, 89
Freundlich isotherm, 90–91
inhibitors, 88
Langmuir isotherm, 90, 91
steel efficiency and surface coverage, 92
Tafel line, 88–89
compound precipitation
benzotriazole (BTA), 91–92
polyphosphates and organophosphates, 92
inhibitor efficiency, 87–88
passivation
pH, 92–93
polarization curve, 93
types, 93
Inner Helmholtz plane (IHP), 24
Insulating dielectrics layers (IDLs), 827
Iron dissolution, impedance techniques
current-potential curves and complex impedance diagrams, 172
inductive character, faradic impedances, 169
numerical simulations, 171
partial current densities, 169
pH dependence, rate constants, 171
potential dependence, collection efficiency, 173
steady-state current-voltage characteristics, 169
transfer function analyzer (TFA), 170
IR potential drop
crack electrolyte resistance, 521
crevice corrosion
initiation, 474
parametric effects, 454–455
unalloyed steel, 452, 453
growing crack, 520

J

Johnson–Mehl–Avrami–Kolmogorov theory, 231

L

Langmuir adsorption model, 410
Langmuir–Blodgett (LB) technique, 638–639
Liquid metal embrittlement (LME), 726–727
Liquid phases, high-temperature corrosion
chloride corrosion
molten chlorides, 601, 602
reaction iron, 601
types, 601
general aspects
description, 594
high-molybdenum-containing steels, 595
high-nickel–containing materials, 595
vanadate-induced corrosion, 595
sulfate corrosion
combustion processes, 597
Cr-and Al-containing phases, 598–599
"dry" oxidation, 599–600
fluxing, defined, 596
oxidation reaction, 599
salt melts basic fluxing, 598–599
types, 596
Local electrochemical impedance spectroscopy (LEIS), 158
Low-energy electron diffraction (LEED), 224
Lynch's environmental cracking model, 534–535

M

Magnetic data-storage components corrosion
magneto-optic (MO) alloys
improved passivity and reduced pitting, 856
terbium oxide formation, 855
magnetoresistive (MR) head materials, 855
mitigation, 856
thin-film inductive head materials
Cu-BTA film, 854–855
magnetic recording technology, 854
permalloy, 853–854
sputter-deposited permalloy, 854
thin-film magnetic disks
anodic current, 852
carbon overcoat layer, 850
corrosion potential, 850–851
C oxidation, 853

higher density, 851
polarization curves, lubed and unlubed disk, 852
potentiodynamic polarization curves, 850, 851
T/H exposure, 853
trend, COC thickness, 852
Magnin's SCC model, 534, 535
Mass balance, 482
Mass transport control, anodic dissolution
active dissolution, iron
concentration field calculation, 190
current-potential curves, 189
electro-hydro-dynamic (EHD) impedance, 189
corrosion
H^+ depletion and anion overconcentration, 188
self-stabilizing process, 189
solid layers
$CuCl_{ads}$, 191
steady-state RDE and RRDE, 190
Tafel kinetics, 191
Metal dissolution
and complex formation, 79–80
and deposition, 73
description, 71
free energy, 71–72
high rates and salt precipitation
Fe and Ni galvanostatic transients, 75–76
ohmic drop, 76
transition time, 76
ionic product, 74
iron, OH^- ions, 73–74
model, 71
reduction reactions
corrosion current density, 81
current, 82
hydrogen evolution and oxygen, 80–81
iron Tafel plot, 81
oxygen reduction, 82–83
polarization, 81
selective, alloys
AuCu, 77–78
brasses, 78
oxidation, 78
single-phase *vs.* multiphase, 76–77
STM, 72–73
Tafel slope, 74–75
Metal–hydrogen systems
in aqueous electrolyte
electroadsorption/electrodesorption, 118
electrodissociation/electrocombination, 118
HEAR, 118
H–H combination, 117–118
surface-bulk transfer, 119
energetic and structural relationship, 120
Gibbs free energy, UPD H and OPD H, 119
isotherm
dissociative absorption, 114
dissociative adsorption, 113–114
surface-bulk transfer, 114
pressure/potential correspondence, 119
thermo chemical data
H absorption, 116–117
H adsorption, 114–115
underpotential electroadsorption, Pt, 120
UPD adsorption/desorption, 120
Metal oxidation, low temperatures
chromium
alloys, 227
in situ STM and Auger studies, 229–230
oxygen uptake, 229
polycrystalline, 228
Torr oxygen, 229
copper
electrical applications, 230
oxygen partial pressure adjustment, 231
"passive" oxide film, 230
iron, 226
nickel, 227, 228
silicon
Cabrera–Mott expression, 226–227
metal oxide semiconductor (MOS) transistors, 225
tantalum, 231
Metal oxide semiconductor (MOS), 225
MFMs, *see* Microstructural fracture mechanisms
MIC, *see* Microbially influenced corrosion
Microbially influenced corrosion (MIC)
biodeterioration mechanisms
acid excretion, 747–748
biofilm, 749
chelatization, 748
description, 747
exoenzymes and emulsifying agents, 750
metabolic compounds, 748–749
organic solvents, 748
salt stress, 749–750
countermeasures
biocides, 767
biogenic attack simulation, 768
cathodic protection, 767
changing/modifying material, 766
microbiological methods, 767

organic coatings, 766–767
physical methods, 767
process parameters, 766
description, 738
diagnosis
microorganisms, 768
problems, 769
requirements, 768–769
geochemical cycles
carbon, 743
description, 742
iron, 746–747
nitrogen, 743–745
sulfur, 745–746
metallic materials
aerobic conditions, corrosion, 754–756
aluminum and aluminum alloys, 763
anodic reaction, 750–751
biodeterioration, 751
cathodic reaction, 751
copper and copper alloys, 765
corrosion, anaerobic conditions, 752–754
microbial biofilm, 756–757
oxygen concentration cells, 752
stainless steels and titanium, 757–763
water-containing solutions, 750
microorganisms
characteristics, 738–739
chemoorganoheterotrophs, 739
exopolymers, 742
hydrogen ion concentration, 741
metabolism, 739
microbial omnipotence, 740
PCR technique, 739
phototrophs, 739
psychrophiles, 741
redoxpotential, 741
trace elements, 740
mineral materials, 764
natural and synthetic origin
construction materials, 764
guaiacylpropane, 765
lateral chains and ramification, 764–765
microbial growth, 766
natural compounds, 765
polyadducts, 765
Microelectronic corrosion
contamination
airborne particles, 833
fabrication process, 833
microelectronic metallization, 834
electrical bias mechanisms
cathodic alkalization, 835
electrolytic dissolution, 835
electrolytic migration, 835–836
electrolytic processes, 834
ionic path, 835
integrated circuits
aluminum metallization, 836–838
ceramic hermetic-packaged microelectronics, 839–840
gold metallization, 838
plastic encapsulated microelectronics, 838–839
macro interconnects
connectors and contacts, Au-plated, 841
copper conductors, 840–841
solder, 840
open circuit mechanisms, 836
product qualification/reliability testing and analysis
accelerated aging, 842
accelerated testing expenses, 847
changing failure mechanisms, environmental stress function, 848
contamination, 849–850
highly accelerated stress test (HAST), 843
humidity and device type, 849
macro interconnects, 843–844
PEM aging, mathematical relationships, 844–847
technology, evolution and improvement, 848
THB accelerated test, 842
triple-track structures, 843
uses, 841
water adsorption
relative humidity (RH), 832
temperature, device, 833
Microstructural fracture mechanisms (MFMs), 557
Modulation of interfacial capacitance transfer function (MICTF), 159, 163
Monte Carlo-type simulations, CF
damage process investigation, 564–565
high and low cycle fatigue regime, 564
numerical modeling
crack density evolution, 566
crack spatial distribution, 567, 568
fatal crack evolution, 566, 567
prediction *vs.* experimental lifetime, 566, 567
short crack densities *vs.* cycle number, 567, 568

physical description, damage
Coffin–Manson relationship, 566
surface short crack types, 565

N

Notched hollow tube experiment, 523
Nuclear system corrosion
cracking mechanism
BWRs, 788
film-induced cleavage, 791–792
hydrogen embrittlement, 792–793
intergranular cracks, 788
monitoring capabilities, 789
SCC, 788
slip-oxidation, 789–791
stress corrosion spectrum, 789
ductile structural alloys, BWR systems
atom-atom rupture process, 789
crack tip alloy/environment system, 795–797
liquid diffusion rate, 794
mass transport controlling, 794
reaction rates, crack tip, 797–798
resultant beneficial effect, 794–795
slip-oxidation model, 795
stainless steels, stress corrosion cracking, 811–817
strain rate, crack tip, 798–801
stress corrosion crack propagation, 801–807
unirradiated welded plant component, 807–811
empirical correlation, life prediction
Alloy 600, 786
carbide morphology, 787
crack propagation rates, 784–785
CRDM nozzles, 785
Monte Carlo simulations, 786–787
past steam generator cracking, 786
system parameters, 784
Weibull statistics, 786

O

Oldfield model, 486, 487
OPD, *see* Overpotential deposition
Organic coatings
application techniques
airless spraying, 874
brushing, 873
conventional spraying, 873–874
electrodeposition, 875
electrostatic spraying, 874
powder coating, 875
rolling and dip coating, 873
complexcoating systems
metallic, conversion layers and organic topcoats, 870–872
pretreatment, conversion layers, 868–870
composition
binder, 866
constituents, 865–866
fillers and additives, 867
pigments, 866–867
solvents, 867–868
corrosion underneath
cathodic delamination and anodic undermining, 878–884
clustering, water, 878
early rusting, 885
filiform corrosion, 885
flash rusting, 885–886
types, 877–878
fouling prevention
disadvantages, vessels, 886–887
nonstick coatings, 887–888
organotin-based antifoulings, 887
sea and estuaries, 886
seagoing vessels and pleasure craft, 887
marine applications, 886
measuring and monitoring methods
adhesion tests, 901–902
EIS, 889–897
enhanced weathering tests, 900–901
impedance measurements, 889
iron content, electrolyte, 899
Kelvin probe, 898
polarization resistance, 899–900
potential measurement, 899
SVET and SRET, 898
techniques, 888
protective mechanisms
barrier action, layer, 875
electrochemical double layer, 875
resistance inhibition, 876
water permeation, 876–877
Organic monolayers corrosion prevention
defective monolayer films
atomic force microscopy (AFM), 638
corrosion protection and adhesion promotion, 637
delamination, industrial paints, 640–641
inhibition efficiency, 637–638
Langmuir–Blodgett–Kuhn method, 639
LB technique, 638–639

saturation coverage, 637
self-assembled monolayer, 638
substrate surface, 640
oxide-free metal surfaces, thiol
adhesion promotion, phosphonate films, 634–636
adsorption, zinc, 626–627
iron, 627–629
phosphonate self-assembly, 629–634
self-assembly, gold, 623–625
self-assembled films
adsorption and self-organized formation, 622
polymer/metal interface, 621
Osmotic blistering, 878
Overall electrode reactions
hydrogen electro absorption reaction, 113
hydrogen evolution reaction, 111
underpotential and overpotential H electroadsorption, 111–113
Overpotential deposition (OPD)
characterization, 141
HER process, 112–113

P

Passive alloys, crevice corrosion
environment evolution, gap
deaeration, 461–462
dissolution, sulfide inclusions, 468
Fe-Ni-Cr alloys, 465–468
pH evolution and concentration, 462–465
passivity breakdown
high chloride concentration, 473–474
low pH and high chloride solutions, 469–471
mechanisms, 468
microcrevices, 471
pitting, crevice gap, 471–473
role, *IR* drop, 474
sulfide inclusion role, 473
phenomenology
alloy composition, 460–461
environment effect, 458, 459
geometric factors, 458, 460
incubation period, 455
initiation and propagation period, 455–456
potential effect, 456–458
propagation, 461
stainless steel bolts, 456
propagation
anodic dissolution, 475, 476
crevice mouth, 474
dissolution current, 475
evolution rate, environment, 474–475
pits, 475
repassivation
potential, 476–478
spontaneous crevice arrest, 476
Passive nickel and point defect model
dissolution current, 183
vs. high-field migration model, 184
Passivity
adsorption mechanism
$CrCl^{2+}$ complex, 365
current-time dependence, 362
electrochemical equilibrium, oxide/electrolyte interface, 363
Fe–Cr alloys, 365
oxide thickness, 364, 365
pit nucleation, 361–362
stability constants, 363, 364
XPS measurements, 364
binary alloys
AlCu, 285–286
CuNi, 283–284
CuSn, 284–285
FeAl, 281
FeCr, 275–278
FeNi, 278–280
FeSi, 281–283
NiCr, 280–281
chromium
current density, 272
oxide thickness, anodic charge and inverse capacity, 273
cobalt
passive layer composition and thickness, 269–270
polarization curve, 269
XPS, 270–271
copper
anodic oxidation, alkaline solutions, 265
electrochemical reduction charges, 264
ISS depth profiling, 264–265
polarization curve, 244, 264
reduction, alkaline solutions, 265–266
x-ray-induced AES L_3MM spectrum, 264
current density potential curve as, 236, 237
definitions, 237
electrochemical studies, 238
electrode kinetics
ion transfer, 249–253
layer–electrolyte interface, 246–249
metal dissolution, 244–245
potential drops, 245–246

electronic properties, layer
anodic oxide layers, 294, 295
description, 288–289
electron transfer, metallic surfaces, 289–293
oxides band gap, 289
photoeffects, 296–301
photoelectrochemical measurements, 294
redox currents, 296
Tafel plots, iron redox currents, 294, 296
ex situ surface analysis methods, 238–239
film-breaking mechanism
chemical changes/electrostriction, 359
disk and ring currents, 359
electrochemical current noise measurement, 360
Fe^{2+} ions, 359–360
passivation times, 360–361
films structure
Ag, 311–314
2D passive, 320
3D passive, 320–321
electron diffraction methods, 302
Fe, Ni, Cr, and alloys, 314–317
passive monolayer Co(0001) and reduction, 317–320
STM, 302–303
x-ray analytical methods, 302
iron
composition change, passive layer, 262
disk electrode polarization curve, 257–258
Fe(II) and Fe(III) cationic fraction, 260, 261
Fe_3O_4/Fe_2O_3 bilayer structure, 256
film composition, passive layer, 262–263
inverse capacity and oxide thicknesses, 260–262
layer thickness chemical structure, 259–260
ring currents, 258
thermodynamic values, 255–256
XPS, 258–259, 263
kinetic effects, polarization curves
Cu, 243–244
Fe, Cr and Ni, 242–243
layer chemical composition
and structure, 286, 288
valent species, 288
XPS and ISS depth profiles, 287–288
metal ions transfer, 351–352
nickel
acid solution, 272
NiO film growth, 274
passive layer thickness, 272–274
valence change, 273–274
nucleation, 366
oxides and hydroxides, 350–351
passive metal, 353–354
penetration mechanism
anion transfer, 354
bilayer/multilayer structure, 356–357
cation vacancies migration, 355
Cl^- presence, 355, 356
current density, increase, 357, 358
double logarithmic plot, 358
interdependence, cation and anion vacancies, 356
voltage drop, 359
silver
AgO formation, potentiostatic transients, 267–268
oxide formation, 266–267
polarization curve, 266
XPS, 268
strongly acidic electrolytes, 351
thermodynamics, pH
copper, 241–242
equilibriums, 239–240
negative potentials, iron, 240
neutral and alkaline solution, Fe, 241
tin, 268–269
UHV methods (*see* Ultra-high vaccum (UHV) methods)
valve metals, 350
vanadium, 271–272
Passivity breakdown mechanism
control factors, 468
high chloride concentration, 473–474
low pH and high chloride solutions
anodic characteristics, 469
applied potential, 470–471
depassivation pH measurements, 470
microcrevices
formation, 471
localized corrosion, 471, 472
pitting, crevice gap
initiation, 473
micropitting, 472
moiré fringe technique, 472
stainless steel, 471–472
role, *IR* drop, 474
sulfide inclusion role, 473
Passivity films structure and growth
Ag
STM, 311–312
XAS, 312–314
chromium, 317

Cu(111) and Cu(001)
Cu_2O formation, 307–309
duplex layer, 310
OH electrosorption, 303–307
2D passive, 320
3D passive, 320–321
electron diffraction methods, 302
Fe
Debye–Waller factor, 315
in situ STM and GIXD, 314–315
FeCr and Fe-Cr-Ni alloys, 317
monolayer, Co(0001) and reduction
$Co(OH)_2$ reduction process, 319–320
in situ STM images, 317–319
Ni
AFM image, 316–317
lattice, 315–316
NaOH solutions, 316
in situ STM image, 315
STM, 302–303
x-ray analytical methods, 302
Persistent slip band (PSB)
corrosion impact, 554
formation, dissolution current, 552
grain boundary interactions, 546
Pit growth
chloride concentration, 377
composition, electrolyte
aggressive anions, 385, 386
concentration, metal ions, 384
electroneutrality equation, 383
ionic concentration, metal surface, 383
pH shifts, 384, 385
local current density, 376–377
localized acidification, 378
metastable, stop-and-go sequence, 375
potential drops, 382
precipitation, salt films
concave hemispherical pit, 380
convection, 382
electropolishing effect, 381
galvanostatic transients, 378, 380
$NiCl_2$ concentration, 379
supporting electrolyte absence, 379
repassivation potential, 377
stabilizing factors, 387–390
Pit nucleation
mechanistic consequences, nanometer resolution
crystallinity, 366
depressions, 368, 369
enhanced dissolution, passive layer defect, 368
increased cation formation, 367–368
OH replacement, Cl, 369, 370
potential drop, 367
vacancies, accumulation, 369
nickel, 357
transition, pit growth
corrosion pits, development stages, 372–373
metal dissolution, 372
nonaggressive anions, 375
polycrystalline Fe, 374, 375
polycrystalline Ni electrodes, 374
polygonal shape, 374
triangular shape, 373
tunnels, corrosion, 376
Pitting corrosion
aggressive anions, 350
inhibition potentials
aggressive properties, perchlorate, 353
chloride concentration, 352
localized corrosion, conditions
Cu addition, 371
disordered hydroxide and ordered oxide layer, 371
silver, 370
passivity
breakdown, 353–366
metal ions transfer, 351–352
oxides and hydroxides, 350–351
strongly acidic electrolytes, 351
valve metals, 350
pit growth
chloride concentration, 377
composition, electrolyte, 383–386
local current density, 376–377
localized acidification, 378
potential drops, 382
precipitation, salt films, 378–382
repassivation potential, 377
stabilizing factors, 387–390
pit nucleation
mechanistic consequences, nanometer resolution, 366–370
transition, pit growth, 372–376
repassivation, corrosion pits
Einstein–Smoluchowski relation, 386
rate-determining step, 387
time, 386, 387
Pitting corrosion, stainless steels
depassivation pH, 419–420
electrochemical aspects
CPT, 427
current fluctuations, 427–428

pitting potential, 423–425
probabilistic behavior, 425–426
repassivation potential, 426–427
rest potential fluctuations, 429
inclusion effects
non-soluble inclusions, 444–445
sulfide solubility, 443–444
titanium and manganese sulfides, 434–443
metallurgical aspects
alloying element effect, 429–430
processing, 430–431
sulfide inclusions, 432–434
metastable pitting, 420
nucleation to stable pit
multistep mechanism, 421
passive film effect, 422–423
process stages, 421
stochasticity, 423
resistance, 421
semi-developed pit, 420
Pitting potential
described, 423
measurement, 424–425
pit stabilization, 424
Plastic-encapsulated microelectronics (PEM)
aging, mathematical relationship
acceleration factor (AF), 845
empirically derived models, 846
GILDES model, 847
lognormal distribution, 844
MTTF metric, 844–845
surface conductance, 845
surface conductivity, temperature function, 847
temperature and humidity, 846
chip metallization, 838
vs. CHP, 839
corrosion vulnerability, 839
reliability, 828, 842
Potentiokinetic techniques
pitting and crevice corrosion, 457
time-dependent evolution, 490
Potentiostatic techniques
critical and repassivation potential, 490
stepwise potential backscan, 490, 491
Pourbaix diagram, 151
Power spectral density (PSD)
birth and death frequencies, 441
potential fluctuations, 429
pre-pitting noise, 428
Pressurized water reactor (PWR)
primary systems, 778
side components, 785
Printed circuit boards (PCB), 829, 833, 844
Processing, aluminum alloys
castings, 713
continuous casting, 715
direct chill (DC) casting, 713–714
extrusion, 714–715
hot and cold rolling, 714
PSB, *see* Persistent slip band

R

Reactive ion etching (RIE)
Al
lines patterning, 828
metallization, 833
$CHCl_3$ elimination, 837
Redox system reduction
Fe^{3+}, 66
hydrogen evolution (HER)
current density, 67–68
neutral and alkaline solutions, 69–70
polarization, 70
Tafel reaction, 68–69
Volmer–Heyrovsky mechanism, 67
Volmer–Tafel mechanism, 67
oxygen
cathodic overvoltages, 71
mechanism, 70–71
Reflection high-energy electron diffraction (RHEED), 224, 229
Repassivation potential
anodic current, 426–427
"arrest" potential, 477
chloride content effect, 477–478
corrosion damage, 477–478
deactivation potential, 476–477
geometry, 478
Rotating disk electrode (RDE)
convection, 189
$CuCl_{ads}$, 191
steady-state, 190
Rotating ring-disk electrode (RRDE)
AC impedance, 190
active-passive range, iron, 177–178
branching reaction pattern existence, 171
frequency-resolved, 197–199
steady-state, 187
Rutherford backscattering (RBS), 224

S

"Scanning probes", 888
Scanning reference electrode technique (SRET), 84
Scanning tunneling microscopy (STM)
and AFM, 224, 239
dealloying thresholds, 204
dissolution, sulfur-modified Ni, 398
intermediate potential range, 624
and SFM, 302
Scanning vibrating electrode technique (SVET) technology, 85, 158
SCC mechanisms, *see* Stress-corrosion cracking (SCC) mechanisms
Secondary ion mass spectrometry (SIMS), 224
Selective dissolution
atomistic modeling and passivation, percolation theory
computer simulations, 203–204
critical potential, 205–206
curvature effects and roughening transition, 206
dealloyed structures morphology, 2D lattices, 204–205
dealloying kinetics, 205
model, binary alloy, 206
thresholds, dealloying, 204
binary alloys
alloy–electrolyte (LiCl) combinations, 201
anodic polarization curves, 203
atomic arrangements, 203
dependence, 202
dynamic polarization curves and current-time decays, 202
dissolution–redeposition mechanism, α-brass dealloying, 194
electrochemical measurements, 201
iron, transient passivation stages, 199
single-phase alloys, 193
Semiconductors
electrodes, electron transfer
capacity, 97
diffuse double layer, 96
electrochemical redox reactions, 97–98
Fermi level, 95–96
n-type, 95, 99
p-type and n-type, 96–97
properties
intrinsic, 94
n-type and p-type, 94–95
Sharland model, 483
Slip-dissolution model, 500–501, 523–524
Slip-oxidation mechanism, 814
Solid corrosion products
alloy composition role
chromium level, 587–588
Cr depletion kinetics, 587
homogenous continuous oxide, 584
nonmetal partial pressure, 585
oxide roses, 587–588
protective scale-forming element, 585
scale formation and growth, 586
scale-forming element, 586
Si and Cr concentration, 587
formation and growth kinetics
chemical defects, 578
diffusion-controlled surface scale growth, 581–582
high-temperature corrosion, 582
inert markers position, 583–584
nickel vacancies, 579
parabolic rate constant, 583
product layer, 580
protective capability, 583–584
internal corrosion
continuous surface layer, 588
diffusion-controlled process, 589
metal grain boundaries, 590
protective effect, 588–589
reaction equilibrium, 590–591
metal dusting
actual driving force, 593
carbon-containing environments, 591
carbon-rich environment, 594
crystallographic orientation, 592
initial stages, 591–592
morphology, 592–593
Steady state model, 479
STM, *see* Scanning tunneling microscopy
Stress-corrosion cracking (SCC) mechanisms
alloys, 706, 725–726
appearance
arrest marks and striation formation, 503, 505
fracture surfaces, 503, 504
CF
behavior comparison, 569–570
crack growth rates *vs.* frequency, 570
Fe-Cr-Mn alloy, 569
corrosion properties, 714
crack velocity, 537
definition, 499
electrochemistry
acidification, 528–529
anodic current density, 517

atomistic action, hydrogen, 528
austenitic/duplex stainless steels, 525
behavioral conditions, 515
cathodic protection, 529–530
chloride, 523–525
cleavage failure, α-brass, 525, 526
crevice potential and pH, 529
cyclical process, 518–519
dealloying, gold and brass, 525–526, 527
double-layer capacitance measurements, 525, 528
electrolyte resistance, 521
ferrite and austenite, 524
film-induced cleavage model, 527
half-angle, 520–521
hydrogen effects, 530
IR potential drop, 520
localized corrosion, 522, 523
monolayer adsorption, 527–528
notched hollow tube experiment, 523
plastic strain distribution, 517, 518
potential drop and metal ion concentration, 519, 520
potential effect, 529, 530
potentiodynamic pitting scans, 524
potentiostatic techniques, 515–516
repassivation rate, 515, 516
resistive system, 521
scratch method, 516
silver alloys, 526
strain rate effect, 519, 520
sulfur adsorption, 524–525
tip opening displacement, 519
tip strain rate, 518
type B occurrence, 522
type D, 526–527
Vermilyea's concept, 519
Galvele's surface mobility model, 535–536
immunity, 711
inorganic glasses and ceramics, 499–500
intensity factor effects
classical phenomena, 502
duplex components, 503
propagation velocity curves, 502–503
irradiation effects, 788
Lynch's model, 534–535
Magnin's model, 534, 535
mechanical aspects, 726
metallurgy
austenitic stainless steels, 509, 510
carbon and low-alloy Cr-Mo steel, 509, 510
cold work, 514
dealloying role, 508–509
electrochemistry and pattern, chloride, 513, 514
evaporating seawater, 513–514
grain boundary segregation, 510
intergranular corrosion and precipitation, 511
martensite and bainite, 512
martensitic steels, 512
minor alloying elements, 509
phase transformations, 509–510
resistance, duplex stainless steels, 512–513
sensitized grain boundaries, 511
solid solution composition, 508
transverse resistance, 512, 513
variables, 507
models
film-induced cleavage, 501
hydrogen embrittlement, 502
slip-dissolution, 500–501
plasticity and metal-induced fracture, 500
propagation
intensity factor, 804, 806
L-grade and non-sensitized steel components, 806
observed and predicted relationships, 802, 804
predicted and observed beneficial effect, 802
PWR environments, 805
theoretical and observed crack propagation, 803
theoretical crack propagation rate, 801
role, 726
Sieradzki and Friedersdorf analysis, 537
stainless steels, 808
susceptibility, 725
testing
C-Mn steel fractures, 530, 531
CO_2-induced, 531–532
effect, tip strain rate, 530, 532
elastic slow strain rate, 530–531
fractography, cathodic hydrogen embrittlement, 531, 533
microcrack coalescence, 533–534
precracked specimens, 532–533
role, crevices, 534
stress–strain curves, 531, 532
threshold stresses and localized corrosion
hydrogen embrittlement, 505, 507
low-alloy steels, 505, 507
low-amplitude cyclic loading, 504–505, 506

nucleation, 505, 506
smooth surfaces, 504
surface films, 507
universal models, 535
unsolved systems, 537–538
Sulfide inclusions
free machining steels, 432
MnS dissolution, 432–433
modeling, 433–434
pitting initiation role, 432
steelmaking process, 433
Sulfur adsorption thermodynamics, metal surfaces
E-pH diagrams and relations
chemical potential, adsorbed water, 410–411
dissolved and adsorbed species, 411–413
electrochemical experiments and surface analyses, 411
Langmuir model, 410
potential-pH diagrams
bidimensional layers, 409–410
chromium and copper, 413, 415
iron and nickel, 413, 414
stability domains and thiosulfate effects, 413, 416
Sulfur-assisted corrosion mechanisms and alloyed elements
anodic dissolution activation
acceleration factor, 397, 398
catalytic effect and potential shift, 396–397
in situ STM, 398
metal–metal bond weakening, 397–398
polarization, nickel-iron single-crystal alloy, 396, 397
sulfur-covered Ni electrodes, 398, 399
anodic segregation
Ni sample, 401–402
observation and experiment, 400–401
surface enrichment, 401
blocking/retarding effect, adsorbed sulfur
passivation kinetics, 399–400
passive film growth, 399
structure and properties, passive film, 400
chromium role
antagonistic effects, 407, 408
Ni-based alloys, 406–407
joint action, chloride ions
passivity breakdown and localized corrosion, 403, 404
pitting corrosion, 403–404
sulfide inclusions, 404
molybdenum role
anodic segregation, 406
binary and ternary single-crystal alloys, 405
corrosion resistance, 404
measurement, desorption kinetics, 405–406
sulfur coverage variation, 406
sulfur removal, single-crystal alloy, 406, 407
passivity breakdown
chloride ion effects, 395–396
enrichment rate and critical concentration, 402
metal–passive film interface, 402–403
practical importance
areas, 407–408
categories, sulfur species, 408–409
sulfide inclusions, 409
surface science techniques, 396
thermodynamics, sulfur adsorption, 409–416
Surface effects, hydrogen entry into metals
H absorption promoters action mechanisms, 135–145
HAR, 106
H_2 dissociative adsorption
in aqueous electrolyte, 108–113
in gas phase, 107–108
HER and HAR kinetics
absolute rates, 126–129
dependence rates, 131–134
steady-state equations, 129–131
H surface bulk transfer mechanisms, 121–126
metal–hydrogen systems, 113–121
modifiers impact on H entry
electronegative species effects, 135
metal oxide film effects, 135

T

Tafel reaction, 44
TDS, *see* Thermal desorption spectroscopy
Thermal desorption spectroscopy (TDS), 123–124
Thin organosilane films adsorption
alkyl chains, 647–648
bond formation, 648
dip coating, 649
functionalized organic chain, 648
silanol formation, 649
silicon–oxygen–metal bonds, 649
SKP line scan measurements, 650
sol-gel chemistry, 651
upper polymer layers, 650

Thin oxide film formation, metal
  controlling factors
    impurities, 223
    ion entry, 220
    metal structure, 221–222
    oxide structure, 222–223
    single oxide bond strength values, 221
  low temperatures
    chromium, 227–230
    copper, 230–231
    iron, 226
    nickel, 227
    silicon, 225–226
    tantalum, 231
  measurement techniques
    kinetics, oxide growth, 224
    oxide structure, chemical and physical, 224
  mechanism
    Cabrera–Mott expression, 219
    cation and anion self-diffusion coefficients, 220
    direct logarithmic kinetics, 220
    inverse logarithmic law, 219
    low-temperature oxidation, 218, 219
    quantum mechanical concept, electron tunneling, 218
  protective layer, 217
Titanium and manganese sulfides
  aging effects
    16 h potentiostatic, 441–442
    prepolarization treatment, 441, 442
    rest potential evolution, 442–443
  inhibitive effect, sulfate ions, 439
  melting process, 434–435
  non metallic inclusions
    scanning electron microscopy, 435
    STEM, thin foils, 435, 436
  pH effects and pitting sites
    crevice corrosion, 438–439
    potential measurement, 436–437
    scanning electron microscopy, pitted samples, 437–438
  prepitting events
    anodic current variations, 439–440
    noise, 439, 440
    polarization time and solution pH effect, 440
    PSD calculation, 440–441
  steel composition, 434
Transient model, 479

U

Ultra-high vaccum (UHV) methods
  characteristics, 253
  ISS, 254–255
  RBS, 255
  XPS and AES, 254
Ultrathin plasma polymer films corrosion prevention
  aqueous solutions
    classical surface treatments, 642
    conducting polymers, ultrathin coatings, 654–651
    conversion process, 642
    film formation process, 642–643
    inorganic amorphous conversion films, 644–645
    inorganic/organic hybrid films, 645–647
    rare earth metal salts (REMS), 642
    thin organosilane films adsorption, 647–651
    ultrathin protective films, 651–654
    Zr-oxyhydroxide film formation, 643–644
  conversion treatments, 641
  functional materials properties, 641
  traditional surface treatments, 642
Ultrathin protective films
  passive films plasma modification
    chemisorption process, 651
    hydrogen plasma, 652
    oxide layer covering, 652
    zinc surfaces, 651
  plasma polymer deposition
    chemical and physical properties, 652–653
    direct and remote, 652
    electrochemical impedance measurements, 654
    microwave reactor, 653
    polymerization process, 652
Underpotential deposition (UPD)
  adsorption, 120
  detection, 112
UPD, *see* Underpotential deposition

V

Volatile phases formation
  gas-phase chlorine
    active oxidation mechanism, 607–608
    alloying elements, 609
    chloridizing-oxidizing environment, 604–606

conventional stability diagram, 603, 605
description, 602
dynamic quasi-stability diagram, 608–609
Hertz–Langmuir equation, 603
laboratory tests, 603
metal halide evaporation, 603
oxygen-containing Cl-environments, 606–607
permanent protection, 609–610
quasi-stability diagrams, 603, 606
surface geometry, 609
general aspects, 602
water vapor
metal oxides and hydroxyls, 611–612
potential reaction products, 610
pressurized hydrogen, 612

W

Water permeation, organic coatings
driving forces, 876
permeability, oxygen and water vapor, 877
Watson model, 485, 486

X

XPS, *see* X-ray photoelectron spectroscopy
X-ray absorption spectroscopy (XAS), 6, 9–10, 239
X-ray photoelectron spectroscopy (XPS)
analysis, 334, 335
passive film, 330, 333
use, 329